TABLE OF ATOMIC WEIGHTS AND NUMBERS

Based on the 1977 Report of the Commission on Atomic Weights of the International Union of Pure and Applied Chemistry. Scaled to the relative atomic mass of carbon-12.

W9-DBA-860

Element	Symbol	Atomic Number	Atomic Weight	
Actinium	Ac	89	227.0278	(e)
Aluminum	Al	13	26.98154	
Americium	Am	95	(243)	(f)
Antimony	Sb	51	121.75	(a)
Argon	Ar	18	39.948	(a, b, c)
Arsenic	As	33	74.9216	
Astatine	At	85	(210)	(f)
Barium	Ba	56	137.33	(c)
Berkelium	Bk	97	(247)	(f)
Beryllium	Be	4	9.01218	
Bismuth	Bi	83	208.9804	
Boron	B	5	10.81	(b, d)
Bromine	Br	35	79.904	
Cadmium	Cd	48	112.41	(c)
Calcium	Ca	20	40.08	(c)
Californium	Cf	98	(251)	(f)
Carbon	C	6	12.011	(b)
Cerium	Ce	58	140.12	(c)
Cesium	Cs	55	132.9054	
Chlorine	Cl	17	35.453	
Chromium	Cr	24	51.996	
Cobalt	Co	27	58.9332	
Copper	Cu	29	63.546	(a, b)
Curium	Cm	96	(247)	(f)
Dysprosium	Dy	66	162.50	(a)
Einsteinium	Es	99	(252)	(f)
Erbium	Er	68	167.26	(a)
Europium	Eu	63	151.96	(c)
Fermium	Fm	100	(257)	(f)
Fluorine	F	9	18.998403	
Francium	Fr	87	(223)	(f)
Gadolinium	Gd	64	157.25	(a, c)
Gallium	Ga	31	69.72	
Germanium	Ge	32	72.59	(a)
Gold	Au	79	196.9665	
Hafnium	Hf	72	178.49	(a)
Helium	He	2	4.00260	(c)
Holmium	Ho	67	164.9304	
Hydrogen	H	1	1.0079	(b)
Indium	In	49	114.82	(c)
Iodine	I	53	126.9045	
Iridium	Ir	77	192.22	(a)
Iron	Fe	26	55.847	(a)
Krypton	Kr	36	83.80	(c, d)
Lanthanum	La	57	138.9055	(a, c)
Lawrencium	Lr	103	(260)	(f)
Lead	Pb	82	207.2	(b, c)
Lithium	Li	3	6.941	(a, b, c, d)
Lutetium	Lu	71	174.967	(a)
Magnesium	Mg	12	24.305	(c)
Manganese	Mn	25	54.9380	
Mendelevium	Md	101	(258)	(f)
Mercury	Hg	80	200.59	(a)
Molybdenum	Mo	42	95.94	
Neodymium	Nd	60	144.24	(a, c)
Neon	Ne	10	20.179	(a, d)
Neptunium	Np	93	237.0482	(e)
Nickel	Ni	28	58.70	
Niobium	Nb	41	92.9064	
Nitrogen	N	7	14.0067	
Nobelium	No	102	(259)	(f)
Osmium	Os	76	190.2	(c)
Oxygen	O	8	15.9994	(a, b)
Palladium	Pd	46	106.4	(c)
Phosphorus	P	15	30.97376	
Platinum	Pt	78	195.09	(a)
Plutonium	Pu	94	(244)	(f)
Polonium	Po	84	(209)	(f)
Potassium	K	19	39.0983	(a)
Praseodymium	Pr	59	140.9077	
Promethium	Pm	61	(145)	(f)
Protactinium	Pa	91	231.0359	(e)
Radium	Ra	88	226.0254	(c, e)
Radon	Rn	86	(222)	(f)
Rhenium	Re	75	186.207	
Rhodium	Rh	45	102.9055	
Rubidium	Rb	37	85.4678	(a, c)
Ruthenium	Ru	44	101.07	(a, c)
Samarium	Sm	62	150.4	(c)
Scandium	Sc	21	44.9559	
Selenium	Se	34	78.96	(a)
Silicon	Si	14	28.0855	(a)
Silver	Ag	47	107.868	(c)
Sodium	Na	11	22.98977	
Strontium	Sr	38	87.62	(c)
Sulfur	S	16	32.06	(b)
Tantalum	Ta	73	180.9479	(a)
Technetium	Tc	43	(98)	(f)
Tellurium	Te	52	127.60	(a, c)
Terbium	Tb	65	158.9254	
Thallium	Tl	81	204.37	(a)
Thorium	Th	90	232.0381	(c, e)
Thulium	Tm	69	168.9342	
Tin	Sn	50	118.69	(a)
Titanium	Ti	22	47.90	(a)
Tungsten	W	74	183.85	(a)
(Unnilhexium)	(Unh)	106	(263)	(f, g)
(Unnilpentium)	(Unp)	105	(262)	(f, g)
(Unnilquadium)	(Unq)	104	(261)	(f, g)
Uranium	U	92	238.029	(c, d)
Vanadium	V	23	50.9415	(a)
Xenon	Xe	54	131.30	(c, d)
Ytterbium	Yb	70	173.04	(a)
Yttrium	Y	39	88.9059	
Zinc	Zn	30	65.38	
Zirconium	Zr	40	91.22	(c)

Except as noted in the footnotes that follow, the atomic weight values are good to ±1 unit in the last place.

(a) Precise to ±3 units in the last place.

(b) Atomic weight cannot be expressed more precisely because among normal terrestrial materials there are known variations in isotopic compositions.

(c) Geological samples of this element have been found with anomalous isotopic composition and atomic weights different from this value.

(d) Considerable variations from this atomic weight value can occur in commercial samples because of changes in isotopic composition.

(e) The atomic weight of the radioisotope of longest half-life.

(f) The mass number of the radioisotope of longest half-life.

(g) The official name and symbol has not been agreed to. Element 105 is unofficially called hahnium. Element 104 is called rutherfordium by American scientists and kurchatovium by Russian scientists.

GENERAL CHEMISTRY

USED
PIRR 1
124B
$23.96
U. OF D. BOOKSTORE
5ACDC3G10103 1

JAMES E.
BRADY
St. John's University
Jamaica, New York

GERARD E.
HUMISTON
Widener University
Chester, Pennsylvania

THIRD EDITION

GENERAL CHEMISTRY

PRINCIPLES AND STRUCTURE

175 YEARS OF PUBLISHING
1807 1982

JOHN WILEY & SONS

NEW YORK • CHICHESTER • BRISBANE • TORONTO • SINGAPORE

Cover and Text Design: Judith Fletcher Getman

Cover Photograph: Reginald Wickham

Cover: When seen through polarizing filters, brilliant colors are produced when melted crystals of sodium thiosulfate solidify between layers of glass. Sodium thiosulfate is the chemical that photographers call "hypo".

Photo Research: Ilene Cherna

Photo Editor: Stella Kupferberg

Copy Editor: Rosemary Wellner

Production Supervisor: Linda Indig

Copyright © 1975, 1978, 1982 by John Wiley & Sons, Inc.

All rights reserved. Published simultaneously in Canada.

Reproduction or translation of any part of this work beyond that permitted by Sections 107 and 108 of the 1976 United States Copyright Act without the permission of the copyright owner is unlawful. Requests for permission or further information should be addressed to the Permissions Department, John Wiley & Sons.

Library of Congress Cataloging in Publication Data:

Brady, James E.
  General chemistry, principles and structure.

  Includes index.
  1. Chemistry. I. Humiston, Gerard E.
II. Title.
QD31.2.B7 1982   540   81-16162
ISBN 0-471-07806-9   AACR2
Printed in the United States of America
10 9 8 7 6 5 4 3 2

PREFACE

The continued success of this book through both its first and second editions has been gratifying, for it tells us that teachers have found it effective and that students have found it useful and informative. In preparing this Third Edition, we have been guided by our desire to retain the qualities that users of the text have found worthwhile in the past, yet have responded to suggestions for changes and further improvements. In this spirit, the general theme and level remain unchanged. The book continues to be written for the first year chemistry course for science majors.

In revising the text, we have had two principal goals: to make the text more useful, readable, and interesting for students and to ensure that the text continues to cover topics that teachers wish to present to their classes. In satisfying the first goal, we examined the second edition line by line with an eye toward improving the readability and clarity of presentations. At the same time, more examples of common chemicals and applications of chemistry have been woven into discussions. The visual appearance of the book has also been enhanced and a large number of photographs have been added to make chemistry seem more alive to students. In addition, the way in which topics discussed in individual chapters relate to the world around us is emphasized by chapter-opening photographs. We have also used marginal comments to add interest, to highlight salient points, and to further clarify discussions.

In meeting our second goal, we have responded to feedback from users of the Second Edition. Those who are familiar with the previous editions will notice at once that the stereo illustrations have been dropped. An extensive survey of users revealed that most students rarely worked with them, and there was general agreement that a large majority of students would benefit more from a useful full-color insert and a more effective range of photographs, both of which have now been incorporated.

In response to user suggestions, we have also made some changes in topic sequence. Many teachers prefer to teach all of bonding theory together. We have therefore moved the second bonding chapter forward to follow the first, but we have made it sufficiently independent so that those who prefer to teach the more sophisticated aspects of bonding during the second semester can easily continue to do so. We have also reordered topics within the second bonding chapter. It now begins with a general discussion of molecular structure, followed by VSEPR theory, and then finally the VB and MO theories.

Nomenclature of inorganic compounds, formerly presented in an appendix, now is included within the text in Chapter 4.

Former chapters dealing with solids and with liquids and changes of state have been combined in a single new chapter titled States of Matter and Intermolecular Forces. The extent of coverage of solids has been reduced from the previous edition.

We have also taken note of the nationwide trend toward increasing the amount of descriptive inorganic chemistry presented in the first-year course. We recognize, however, that considerable controversy exists over how much should be added and what should be dropped to make room for the additional topics. In preparing the revision, it was our desire to satisfy those wishing to teach more descriptive chemistry, yet to provide flexibility by permitting instructors considerable latitude in their depth of coverage. Consequently, we have completely rewritten and expanded the chapters dealing with descriptive inorganic chemistry. Now, in addition to the descriptive chemistry that is part of examples in the chapters dealing with principles, there are more than five chapters devoted entirely to this subject. To begin, much of the "nuts and bolts" chemistry so important in the laboratory is collected in Chapter 6, where metathesis and redox reactions in aqueous solutions are discussed. Three chapters later, after students have had backgrounds in bonding and the properties of the states of matter, the periodic table is revisited with a focus on trends in the chemical and physical properties of the elements and some of their compounds. This overview of the properties of the elements serves as background for the chapters on chemical principles that follow. Later, after a chapter on electrochemistry, there are four more chapters on descriptive inorganic chemistry that deal with specific properties of the elements and their compounds.

Although we have made some significant changes in this edition, features that have appealed to users in the past have been retained. As in previous editions, we assume no prior student background in chemistry. New terms are set in bold type when first encountered and are carefully defined before being used in subsequent discussions. A mathematical background sufficient to handle only simple algebra is assumed, and a review of some basic mathematical concepts is included in an appendix. The already large number of worked examples has been increased, and end-of-chapter exercises continue to be divided into Review Questions and Review Problems. The set of Review Problems has been enlarged by adding more simple drill-type problems as well as some more difficult ones, marked by an asterisk. As in the Second Edition, we have included in each chapter an Index to Questions and Problems to assist teachers in assigning homework and students in planning their study.

The question of how many SI units to use in the first-year course continues. We have carefully described SI units, but for practical laboratory reasons have retained the use of the atmosphere and torr as pressure units and the liter and milliliter as volume units. For energy, we have again taken a dual approach. Tables include both joules (or kJ) and calories (or kcal), but the emphasis in calculations is on joules—only occasionally are calories used. The reason for this is twofold. First, except in the most recent literature, energies are expressed in calories, so students should be familiar with both units. Second, some disciplines whose students we serve have not moved as far in the direction of SI units as most chemists have.

Despite some changes that have been made in topic sequences, concepts continue to be developed in a logical order that permits an early introduction of quantitative experiments in the laboratory. In Chapters 1 and 2, the concepts of atoms, molecules, atomic weights, and the mole are introduced. We have also included the concept of molar concentration in Chapter 2 because the second bonding chapter has been moved forward, and therefore the discussion of details of solution stoichiometry has been postponed slightly.

Following a discussion of electronic structure and the periodic table in Chapter 3, there are two chapters on chemical bonding. The first deals with an elementary treatment of ionic and covalent bonding; the second deals with molecular structure and modern theories of covalent bonding.

Next follows a chapter that focuses on aqueous solutions as a medium for carrying out chemical reactions. These topics, at a relatively early stage, prepare students for qualitative and quantitative experiments in the laboratory. In fact, if instructors wish, parts of this chapter dealing with solution stoichiometry could easily be covered immediately after Chapter 2.

Chapters 7 and 8 discuss the properties of the states of matter and interconversions among them. Chapter 7 covers the properties of gases; in Chapter 8 the effects of intermolecular attractions on the properties of liquids and solids is emphasized. This is followed by a descriptive chapter that, as mentioned earlier, focuses on trends in properties within the periodic table.

Chapter 10 once again turns to a discussion of solutions, but this time the emphasis is on the effects that the solute has on the physical properties of solutions.

Thermodynamics, kinetics, and equilibrium are discussed sequentially because they address the questions: "Is a reaction possible, how fast does it occur, and what is the system like when it reaches equilibrium?" In the thermodynamics chapter, the section on the First Law has been made less mathematical and its relation to chemical systems more apparent. Practical applications have also been integrated into the thermodynamic discussions. In the kinetics chapter, there is a new section dealing with half-lives and integrated rate laws. The chapter on equilibrium concentrates on gaseous and heterogeneous systems, the way that thermodynamics relates to equilibrium, and a thorough discussion of Le Châtelier's principle.

Chapter 14, Acids and Bases, presents the various ways that these substances can be defined, as a prelude to the chapter on acid-base equilibria in aqueous solutions. This is followed by a chapter on solubility and complex ion equilibria.

Chapter 17, Electrochemistry, includes practical examples of electrolysis reactions and galvanic cells. The next four chapters are concerned with descriptive inorganic chemistry introduced earlier, and then there are chapters on organic and biochemistry. The text concludes with a discussion of nuclear chemistry that includes illustrations of how nuclear phenomena can be used to study chemical systems, as well as discussions of nuclear fission and fusion and their relationship to the production of usable energy.

The order of topics naturally reflects our own bias, tempered by the comments of users and those who have assisted us as reviewers. In writing the text we have sought, as much as possible, to make chapters sufficiently independent so that their order of presentation can be easily modified. For example, the second bonding chapter can be taught in either the first or second semester. Similarly, Chapter 9, which deals with periodic trends, can also be postponed if desired, and both solution chapters can be taught in sequence.

In the first paragraph of this preface we mentioned how pleased we were that teachers and students have found the text useful. We hope that this edition will also please you, and we invite your comments or suggestions.

JAMES E. BRADY

GERARD E. HUMISTON

ACKNOWLEDGMENTS

It is our pleasure to thank all those who have in one way or another contributed to this edition. First, to our wives and children we express our heartfelt appreciation for their constant inspiration, encouragement, and patience. We thank June Brady, our typist, for her careful work and cheerful spirit in the face of many deadlines. We express special appreciation to the staff at Wiley for their diligence, attention to detail, and good humor, especially our Editor, Cliff Mills; his assistant, Francine Fielding; our Production Supervisor, Linda Indig; our Copy Editor, Rosemary Wellner; our Picture Editor, Stella Kupferberg; our Illustrator, John Balbalis; and our Designer, Judy Getman. We are grateful to our colleagues and students for their many constructive suggestions and discussions, especially Drs. Ernest Birnbam, Neil Jespersen, Eugene Holleran, William Pasfield, John Skarulis, and Siao Sun. Finally, our special thanks go to the following colleagues who have helped shape this book by their thoughtful reviews and criticisms of the manuscript and their many valuable suggestions: Professor Russell Trimble, Southern Illinois University; Professor Melvin Hanna, University of Colorado; Professor Richard Palmer, Manhattan College; Professor Frank Gomba, U.S. Naval Academy; Professor Ron Ragsdale, University of Utah; Dr. Ruth L. Sime, Sacramento City College; Professor Jo Beran, Texas A & I University; Professor G. G. Long, North Carolina State University; Professor Delwin Johnson, St. Louis Community College; Professor John Weyh, Western Washington University; Professor Leo J. Malone, St. Louis University; Professor Don Roach, Miami–Dade Community College.

J.E.B.
G.E.H.

SUPPLEMENTS

A complete package of supplements to accompany this text is available to assist both the teacher and the student.

Study Guide to Accompany General Chemistry, Principles and Structure, Third Edition, by James E. Brady. This softcover book has been carefully structured to assist students in mastering concepts developed in the text. It is keyed section by section to the text, and for each section there is a set of objectives, a brief review (sometimes with additional worked examples), a self-test with answers, and a list of new terms introduced in that section. The *Study Guide* also contans a complete glossary.

Laboratory Manual for General Chemistry, Principles and Structure, Second Edition, by Jo Beran and James E. Brady. This manual features a thorough techniques section with photographs illustrating important apparatus and manipulations, and 45 experiments sequenced to follow the topical development of the text. For the teacher, an instructor's manual accompanies the laboratory manual.

Solutions Manual for General Chemistry, Principles and Structure, by Theodore W. Sottery. This softcover supplement provides detailed solutions to all the numerical problems in the textbook.

Problem Exercises for General Chemistry, Second Edition, by G. Gilbert Long and Forrest C. Hentz, Jr.. Intended to bridge the gap between textbook-style exercises and those that students encounter on examinations, this softcover book features over 1000 problems and questions in multiple-choice format. Throughout the book students are taught to use the basic "tools of the chemical trade."

Teacher's Manual for General Chemistry, Principles and Structure, by James E. Brady. This manual, available to teachers only, provides all the usual services.

Transparency Masters. Instructors who adopt this book may obtain from Wiley, without charge, a set of $8\frac{1}{2} \times 11$-inch black and white line drawings that duplicate key figures in the text. They can be used to prepare transparencies.

Test File for General Chemistry, Principles and Structure, by David Becker. This card file of multiple-choice test questions is available from Wiley at no charge for teachers who adopt this book.

Programs in Computer-Aided Instruction, by William Butler. There are 20 self-contained programs on chemistry principles, paralleling the text. Most programs have 5 or 6 parts, and each part is usually divided into 10 to 12 entries consistng of problems, questions, and so on. Available for PET, Apple II, and TRS-80 in diskette form.

CONTENTS

1

INTRODUCTION

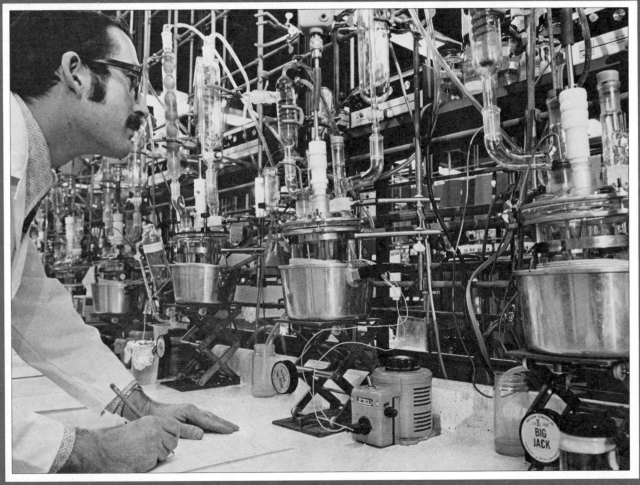

Every science relies on observations of nature. The laboratory is a place where these observations are made under controlled conditions so that the results can be reproduced. In this chapter we will see how science operates and begin to learn about the tools of measurements and how they apply to chemical systems.

Webster's defines science as "a branch of study concerned with observation and classification of facts."

Never before in history have people found themselves so able to influence their physical environment, for good or bad, as today. This has come about, in large measure, as the result of a form of human endeavor called science—an interest in the workings of nature and a study of its laws. Over a period of time, as the volume of facts about nature has grown, science has gradually evolved into a number of closely related specialties, such as biology, chemistry, and physics. **Chemistry,** the subject of this book, concerns itself with the composition of substances, the ways in which their properties are related to their composition, and the interaction of these substances with one another to produce new materials.

The degree to which chemistry has changed civilization is evident everywhere. A good part of the clothing we wear, the automobiles we drive, and other products we encounter daily are composed of materials that simply did not exist at the turn of the century. Medicines created in the laboratory have made us healthier and have prolonged our lives. In recent years the realization that a living organism is a complex chemical "factory" has generated a strong interest in the life sciences, especially biology and medicine. As a result, the study of biochemistry has brought great advances in our knowledge of the nature of life. Today, biochemists are tinkering with the most elementary processes of heredity by chemically altering genes and creating new life forms. It has been only recently, however, that we have also become painfully aware of a host of problems arising from this growth of technology. An example is the much publicized problem of the disposal of hazardous waste chemicals. Solving such problems poses much of the challenge for chemistry in the future.

In this chapter we consider how science operates, see the materials and concepts with which chemists and chemistry students work, and learn how the concept of the atom became firmly established. We will also introduce you to some of the jargon used by chemists. It is important to become familiar with chemical terminology (which will undoubtedly require some memorization), because many of the difficulties that students encounter in studying chemistry can be traced to an inability to "speak the language."

1.1 THE SCIENTIFIC METHOD

Many of the most important advances in science, such as the discoveries of radioactivity by Henri Becquerel and penicillin by Alexander Fleming, have come about by accident. These discoveries were really only partly accidental, however, because the people involved had learned to think "scientifically" and were aware that they had observed something new and exciting.

Progress in chemistry, as well as in other sciences, is usually much less spectacular than the discoveries by Becquerel or Fleming. It is accomplished by many hours of careful work that follows a more or less systematic approach toward answering scientific questions. This approach is called the **scientific method.**

The scientific method is really nothing more than a formal statement of the steps that we follow as we logically approach any problem. Consider, for example, how a TV technician handles the repair of a disabled TV set. First, the defective component is located by observing the results of a series of tests. Then the bad component is replaced, and finally the TV set is turned on to check that the proper repair had been made.

When we approach a problem in the sciences, we proceed in much the same way. The first step in the scientific method can be called **observation.** That is the purpose of the experiments that you, or any other scientist, perform in the laboratory. There nature is observed under controlled conditions so that the results of the experiments are reproducible. The bits of information you obtain are called **data** and can be classified as either qualitative or quantitative. **Qualitative** observations do not have numbers associated with them. An example is the observation that when sodium bicarbonate (baking soda) is added to acetic acid (vinegar), the mixture bubbles vigorously as the two chemicals react. However, if we measured the *amount* of sodium bicarbonate needed to react with a given quantity of acetic acid, we would be making a **quantitative** observation because it would result in numerical data. We will see that quantitative measurements are generally more useful to a scientist than qualitative observations because they provide more information.

After a large amount of data has been collected, it is desirable to find a way to summarize the information in a concise way. Statements that accomplish this goal are called **laws** and, in a sense, simply serve as a convenient means of storage for vast quantities of experimental facts. They also provide a means of predicting the results of some as yet untried experiments. For instance, it is always found that when hydrogen gas and oxygen gas at the same temperature and pressure combine to produce water, two volumes of hydrogen are needed to completely use up one volume of oxygen. This simple statement is a law dealing with the reaction of hydrogen with oxygen. If we had five cubic feet of oxygen gas, we would predict that 10 cubic feet of hydrogen gas would be needed to react with it completely.

A law may be expressed in a simple verbal statement, such as the law we just discussed regarding the reaction of hydrogen with oxygen. However, it is often more useful to have a law stated in the form of an equation. For example, it is observed that the force of attraction between oppositely charged particles decreases as their distance of separation increases. This is more accurately stated by Coulomb's equation, or law,

$$F = \frac{q_1 q_2}{r^2}$$

in which F is the force of attraction between the two oppositely charged particles, q_1 and q_2 are the charges on the particles, and r is their distance of separation. Laws quite commonly are expressed in equation form.

As we have noted, a law simply correlates large quantities of information. Laws in themselves do not explain why nature behaves as it does. Scientists, being human (despite what you may have heard to the contrary), are not satisfied with simple statements of fact and seek to explain their observations. Thus the second step in the scientific method is to propose tentative explanations, or **hypotheses,** that may be tested by experiment. If they are not disproven by repeated experiment, they develop into **theories.** Theories themselves always serve as guides to new experiments and are constantly being tested. When a theory is proven incorrect by experiment, it must either be discarded in favor of a new one or, as is often the case, modified so that all the experimental observations may be accounted for. Science develops, then, through a constant interplay between theory and experiment.

We should note that theories can seldom be *proven* to be correct. Usually the best we can do is to fail to find an experiment that disproves the theory. A scientist must always be careful not to confuse theory and experimental fact. In the past too many times incorrect theory has been accepted as fact, and the progress of science been slowed because of it.

Physicians apply the scientific method when they use the results of lab tests and a patient's symptoms to diagnose a disease and follow its treatment by various medications.

1.2 MEASUREMENT

No science can proceed very far without resorting to quantitative observations. This means that we must make measurements. The process of measurement

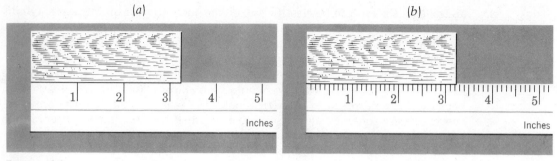

Figure 1.1

Measuring the length of a piece of wood with two different rulers. (a) Length = 3.2 in. (b) Length = 3.24 in.

usually involves reading numbers from some device; because of this, there is nearly always some limitation on the number of meaningful digits that may be obtained in an experimentally determined quantity. For example, let's consider measuring the length of a piece of wood with two different rulers, as shown in Figure 1.1.

Using the ruler in Figure 1.1a, we might estimate the length of the piece of wood to be 3.2 inches (abbreviated 3.2 in.). Notice that to arrive at this number we are forced to estimate the second digit; that is, we must decide whether the length lies closer either to 3.2 or to 3.3 in. Because we are making an estimate, some uncertainty exists in the second digit (the 2), and the third digit, for all practical purposes, is completely unknown. Therefore, for measurements made with the ruler in Figure 1.1a, we are not justified in reporting numbers containing more than two figures.

Digits that are obtained as the result of a measurement are called **significant figures.** When a number is written to represent the result of a measurement, it is always assumed that, unless stated otherwise, only the rightmost digit is uncertain. It is also assumed that any digits further to the right are unknown. Thus the measurement illustrated in Figure 1.1a yields a number with two significant figures.

In Figure 1.1b, we see the same piece of wood measured with a ruler subdivided by additional graduations. Now we can observe that both the 3 and the 2 are known for sure, and we can try to estimate the third digit. One estimate of the length might be 3.24 in., although some people might judge it to be either 3.23 in. or 3.25 in. The measurement 3.24 in. contains three significant figures because the 3 and the 2 are known positively and only the 4 has some uncertainty. Digits to the right of the 4 can't be estimated at all using the ruler in Figure 1.1b, so none are written.

The importance of significant figures is that they indicate to us the reliability of our measurements. In the determination of the length of the piece of wood previously described, we saw two different values, obtained with two different measuring devices; our intuition tells us that we can place more confidence in the value with the greater number of significant figures. Laws and theories are derived from measured quantities, and our confidence in them is directly related to the quality of the data on which they are based.

In discussing measured quantities, the words precision and accuracy are often encountered. The term **precision** refers to how closely two measurements of the same quantity come to each other. For example, repeating a measurement using the ruler in Figure 1.1a could be expected to give values that differ by about 0.1 in.; so we can consider lengths measured with this ruler to be uncertain by ±0.1 in. Repeated measurements with the ruler in Figure 1.1b will give values that differ by about 0.01 in.; the uncertainty in measurements made with it is about ±0.01 in. Values obtained with the second ruler have a smaller un-

The uncertainty in a measurement is sometimes reported along with the value of the measured quantity — for example, 3.2 ± 0.1 in. or 3.24 ± 0.01 in.

certainty and are considered more precise. In general, the more significant figures that there are in a measured quantity, the greater the precision of measurement. The value 3.24 in. suggests that if the measurement is repeated, the result will lie within a few hundredths of an inch of 3.24 in. On the other hand, a reported value of 3.2 in. implies that if the measurement is made again, the result may differ from 3.2 in. by as much as a few tenths of an inch. The reported value of 3.24 in., which has three significant figures, thus implies a greater precision of measurement than 3.2 in.

EXAMPLE 1.1 The length of a room was measured to be 26.0 ft using a ruler graduated in tenths of a foot. (a) How many significant figures are in this measurement? (b) What would be wrong with reporting the length simply as 26 ft?

SOLUTION (a) There are three significant figures. Both digits to the left of the decimal point can be assumed to be known for sure, and only the zero to the right of the decimal point is uncertain. In other words, the length could lie between 25.9 ft and 26.1 ft.

(b) If the length is reported as 26 ft, it is implied that the measurement is uncertain by at least ±1 ft (i.e., the length lies between 25 ft and 27 ft). But we know the measurement is more reliable than that. If you've done the work to get that third significant figure, don't waste it by reporting only two.

The term **accuracy** refers to how close an experimental observation lies to the true value. Generally, a more precise measurement will also be a more accurate measurement. In our example above, the value 3.24 in. has greater precision than 3.2 in. and probably also lies closer to the true length (whatever that may be).

There are instances where a number may be precise but not particularly accurate. The ruler in Figure 1.2, for instance, is improperly calibrated. Failure to notice this error in calibration (and, consequently, failure to adjust appropriately the measurements made with this ruler) would lead to results that would all be wrong by 1 in., even though three significant figures can be read from the scale. In any science, of course, it is important that measuring instruments be carefully calibrated to ensure their accuracy.

Significant figures and calculations

In almost all cases, the numbers we measure are used to calculate other quantities, and we must exercise care to report the proper number of significant figures in the computed result. This is particularly important when an electronic calculator is used to perform arithmetic, since it usually gives answers with eight or ten digits. Generally, most of these digits are meaningless in the sense that most of them shouldn't be counted as significant figures. To see how problems might arise, suppose that we wished to calculate the area of a rectangle whose sides have been measured, with two different rulers, to be 6.2 and 7.00 in. long. We know that the area is the product of these two numbers; that

Figure 1.2

An improperly calibrated ruler. All measurements made with this ruler will be in error by 1 in.

is, 6.2 in. × 7.00 in. = 43.4 in.² (in.² means "square inches"). How many significant figures are justified in the answer? To determine this, let's consider the number of significant figures in each of the measured lengths.

The number 6.2 has two significant figures, which implies that there is some uncertainty in the 2 (i.e., the length of the side could actually be either 6.1 or 6.3 in.). We are confident in the length of this side to, at best, one part in 62, or about 1.6%.

$$\tfrac{1}{62} \times 100 = 1.6\%$$

As for the other side of the rectangle, the minimum uncertainty implied by the number 7.00 is ±0.01 in.; therefore the *percent uncertainty* is (0.01/7.00) × 100 = 0.14%. Our degree of confidence in the value computed for the area of the rectangle depends on how reliable the measured lengths of the sides are. If one side could be in error by at least 1.6%, it is reasonable to expect that the area could also be in error by the same amount. Now, if we report the area to be 43.4 in.², we imply an uncertainty of one part in 434, which is 0.23%. This considerably overstates our confidence in the area, so we *must* round off[1] the answer to 43 in.² (which implies an uncertainty of 2.3%).

The analysis that we just performed to arrive at the proper way to report the computed area is clearly tedious and time consuming. Fortunately, there is a simple rule that we can follow to avoid such an analysis each time we perform a computation. In general, *for multiplication or division, the product or quotient should not possess any more significant figures than the least precisely known factor in the calculation.* In the example above, the factor 6.2 in. has two significant figures and the factor 7.00 in. has three. The rule says that the answer can't have more than two, because that's the number of significant figures in the least precise factor. Therefore, we must round the calculated answer of 43.4 in.² to 43 in.²

For addition and subtraction, the procedure used to determine the number of significant figures in an answer is slightly different. Here the number we write as the result of a calculation is determined by the figure with the fewest number of decimal places. Thus, in the sum of these measured quantities,

$$\begin{array}{r} 4.371 \\ \underline{302.5} \\ 306.871 \end{array}$$

we must round off the answer to 306.9. The reason for this is that the two digits that follow the 5 in the number 302.5 are completely unknown; that is, they could conceivably have any value from 0 to 9. As a result, the last two digits in the answer 306.871 must also be completely uncertain. If we follow our rules for writing numbers such that *only* those digits with real significance are included, we are not justified in writing these last two digits and must round off the answer to 306.9. For addition and subtraction the rule is that *the absolute uncertainty in a sum or difference cannot be smaller than the largest absolute uncertainty in any of the terms in the calculation.* In the example above, the absolute uncertainty implied by the number 4.371 is ±0.001, whereas the uncertainty in 302.5 is implied to be ±0.1. According to our rule, the sum of these two cannot have an uncertainty less than ±0.1; therefore the sum 306.871 must be rounded to 306.9 in order to suggest an uncertainty of ±0.1.

---

[1] When we wish to round off a number at a certain point, we simply drop the digits that follow if the first of them is less than five. Thus 6.2317 rounds to 6.23, if we wish only two decimal places. If the first digit following the point of round off is greater than 5, or if it is 5 and other nonzero digits follow after that, we add 1 to the preceding digit. Thus 6.236 and 6.2351 both round off to 6.24. Finally, when the digit following the point of round off is 5, and no other nonzero digits follow the 5, then we drop the 5 if the preceding digit is even and we add 1 if it is odd. Thus 8.165 rounds to 8.16 and 8.175 rounds to 8.18.

In some calculations we employ numbers that come from definitions (such as 3 feet = 1 yard) or that are the result of a direct count (such as the number of people in a room). These numbers are called **exact numbers** and contain no uncertainty (e.g., there are exactly 3 feet in 1 yard, or the number of people in a room must be a whole number). When such quantities are used in computation, they may be considered to possess an infinite number of significant figures. Thus the conversion of a measured length of 4.27 yards into feet would be accomplished as

$$4.27 \text{ yd} \times \left(\frac{3 \text{ ft}}{1 \text{ yd}} \right) = 12.8 \text{ ft}$$

3 significant figures

Units undergo the same kinds of mathematical operations that numbers do. If you are unfamiliar with this concept, be sure to read Appendix A.1 at the back of the book.

Notice that the number of significant figures in the product is determined by the number of significant figures in the measured length. Also notice that in performing this calculation we have cancelled the units, yards. We will see more of this **factor-label method** of setting up arithmetic in future problems. The factor-label method is explained in detail in Appendix A.1 at the back of the book.

EXAMPLE 1.2

Perform the following computations and report the results to the proper number of significant figures. Assume all the numbers are the result of measurement

(a) 3.142 ÷ 8.05 (b) 29.3 + 213.87 (c) 144.3 + (2.54 × 8.3)

SOLUTION

(a) Using a calculator,

$$\frac{3.142}{8.05} = 0.390310559$$

Sometimes a calculator gives too few significant figures. If you multiply 0.500 by 6.00, a calculator gives 3 as the answer instead of 3.00.

The number 3.142 contains four significant figures; the number 8.05 contains three significant figures. The result must be rounded off to 0.390 because for division the answer should not contain more significant figures than are found in the factor with the fewest number of significant figures.

(b) In performing this computation, note that the 7 must be added to a question mark.

$$\begin{array}{r} 29.3? \\ +\,213.87 \\ \hline 243.1? \end{array}$$

The only thing that can be said about the question mark in the answer is that it must be at least 7, and therefore we should round off the answer to

243.2

In this kind of situation you may be tempted to make the error of writing a zero after the 3 in 29.3 (that is, 29.30). The digit following the 3 is unknown. If it were known to be a zero, a zero would have been written there!

(c) In mixed computations such as this, we perform multiplications and divisions before additions and subtractions.

$$2.54 \times 8.3 = 21.082 \quad \text{(using a calculator)}$$

Since 8.3 contains only two significant figures, the product must be rounded to 21. Now the appropriate addition is performed.

$$\begin{array}{r} 144.3 \\ +\ 21.? \\ \hline 165.? = 165 \end{array}$$

1.3 UNITS OF MEASUREMENT

Units form an integral part of any measurement. For instance, to say that the distance between two points is "three" is meaningless unless a specific unit or units (inches, feet, miles, etc.) is associated with the number. In the United States we have traditionally used the English system of units, with length measured in inches, feet, or miles; volume in ounces, quarts, or gallons; and weight in ounces, pounds, or tons. All the other industrialized nations of the world have used a metric based system for many years, and this is also the system used in the sciences, including chemistry. The United States is in a gradual transition to these units and you have probably noticed some metric units appearing on consumer goods.

In 1960, the General Conference of Weights and Measures, an international body, adopted and recommended a modified version of the older metric system called the **International System of Units** (abbreviated **SI** from the French, *Le Système International d'Unités*) for worldwide use. As a foundation, SI metrics defines seven **base units** that are given in Table 1.1. All other units needed for measurement are derived from these base units by appropriate combinations, which depend on the dimensions of the measured quantity. For instance, if you wished to calculate the area of a rectangular carpet, you would multiply its length by its width. The unit for area is likewise the product of two length units. Since the SI base unit for length is the meter (m), the SI derived unit for area is m² (meter squared or square meter).

$$m \times m = m^2$$

Similarly, speed is a ratio of distance to time. It is computed as the distance traveled divided by the elapsed time. The SI unit for speed is therefore meter/second, or m/s.

Often, either the base units or the derived units are of a size that makes them inconvenient for ordinary measurements. The SI unit for volume, for instance, is the cubic meter, m³, which is somewhat larger than a cubic yard. For expressing volumes measured in a laboratory, however, the cubic meter is awkward. An ordinary glass of water, for example, is about 0.00025 m³. The SI solves this problem by modifying units with decimal factors and prefixes to obtain multiples or submultiples of them. A complete list of the SI prefixes is given

$0.00025 \text{ m}^3 = 250 \text{ cm}^3$
(That's about 8 ounces.)

Table 1.1
The seven SI base units

Physical Quantity	Name of Unit	Symbol
Mass	Kilogram	kg
Length	Meter[a]	m
Time	Second	s
Electric current	Ampere	A
Temperature	Kelvin	K
Luminous intensity	Candela	cd
Quantity of substance	Mole	mol

[a] All English-speaking nations except the United States have adopted the -re spelling for metre. Both -re and -er forms have support within the United States, although the National Bureau of Standards currently favors the -er spelling.

Table 1.2
The sixteen SI prefixes

Factor	Prefix	Symbol	Factor	Prefix	Symbol
10^{18}	exa	E	10^{-1}	deci	d
10^{15}	peta	P	10^{-2}	centi	c
10^{12}	tera	T	10^{-3}	milli	m
10^{9}	giga	G	10^{-6}	micro	μ
10^{6}	mega	M	10^{-9}	nano	n
10^{3}	kilo	k	10^{-12}	pico	p
10^{2}	hecto	h	10^{-15}	femto	f
10^{1}	deka	da	10^{-18}	atto	a

in Table 1.2. Be sure to learn those that are printed in color because they are the ones you will encounter most frequently in this course. Table 1.3 illustrates how these multipliers work and how the prefixes associated with them are used to name modified units. Example 1.3 shows how conversions from one unit to another are easily accomplished.

Table 1.3
Modifying the size of SI units with prefixes

Prefix	Multiplication Factor	Examples	Symbol
kilo-	1000 (10^3)	1 kilometer = 1000 meter (10^3 m) 1 kilogram = 1000 gram (10^3 g)	km kg
deci	1/10 (10^{-1})	1 decimeter = 0.1 meter (10^{-1} m)	dm
centi-	1/100 (10^{-2})	1 centimeter = 0.01 meter (10^{-2} m)	cm
milli-	1/1,000 (10^{-3})	1 millimeter = 0.001 meter (10^{-3} m) 1 millisecond = 0.001 second (10^{-3} s) 1 milligram = 0.001 gram (10^{-3} g)	mm ms mg
micro-	1/1,000,000 (10^{-6})	1 micrometer = 0.000 001 meter (10^{-6} m) 1 microgram = 0.000 001 gram (10^{-6} g)	μm μg
nano-	1/1,000,000,000 (10^{-9})	1 nanometer = 0.000 000 001 meter (10^{-9} m) 1 nanogram = 0.000 000 001 gram (10^{-9} g)	nm ng

EXAMPLE 1.3   A certain person is 172 cm tall. Express this height in decimeters.

SOLUTION   This is really a simple problem, but it illustrates how unit conversions of this type can be easily accomplished using the factor-label method. The first step is to analyze the problem, which can be restated

$$172 \text{ cm} = ? \text{ dm}$$

We need a way of relating the units cm to dm. Reviewing Table 1.2, we can't see a way of doing this directly, but we can write the relationships

$$1 \text{ cm} = 10^{-2} \text{ m} \quad \text{or} \quad 100 \text{ cm} = 1 \text{ m}$$

$$1 \text{ dm} = 10^{-1} \text{ m} \quad \text{or} \quad 10 \text{ dm} = 1 \text{ m}$$

Notice that these provide a path from cm to dm. Centimeters can be converted to meters using the first relationship, and then meters can be changed to decimeters using the second. When we use them, we keep our eye on the units to be cancelled, because that controls how the factors are set up. Thus

$$172 \ \cancel{cm} \times \left(\frac{1 \ m}{100 \ \cancel{cm}}\right) = 1.72 \ m$$

$$1.72 \ \cancel{m} \times \left(\frac{10 \ dm}{1 \ \cancel{m}}\right) = 17.2 \ dm$$

We can also "string together" the conversion factors and obtain the same net result.

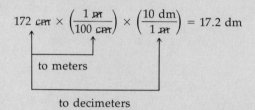

$$172 \ \cancel{cm} \times \left(\frac{1 \ \cancel{m}}{100 \ \cancel{cm}}\right) \times \left(\frac{10 \ dm}{1 \ \cancel{m}}\right) = 17.2 \ dm$$

to meters

to decimeters

EXAMPLE 1.4   Calculate the number of cubic centimeters in 0.255 cubic decimeters.

SOLUTION   This time we need a relationship between cubic units, because our problem can be restated as

$$0.225 \ dm^3 = ? \ cm^3$$

Table 1.2 doesn't give us a way of converting dm^3 to cm^3 directly, but we saw that it does provide a means of relating dm to cm.

$$1 \ dm = 10^{-1} \ m$$

$$1 \ cm = 10^{-2} \ m$$

To obtain the relationships between the corresponding cubic units, we simply cube each side of the equation.

$$(1 \ dm)^3 = (10^{-1} \ m)^3$$

$$1 \ dm^3 = 10^{-3} \ m^3$$

A review of arithmetic operations involving numbers expressed in powers of 10 is located in Appendix A.

and

$$(1 \ cm)^3 = (10^{-2} \ m)^3$$

$$1 \ cm^3 = 10^{-6} \ m^3$$

These can now be used to construct conversion factors to solve the problem.

$$0.225 \ \cancel{dm^3} \times \left(\frac{10^{-3} \ \cancel{m^3}}{1 \ \cancel{dm^3}}\right) \times \left(\frac{1 \ cm^3}{10^{-6} \ \cancel{m^3}}\right) = 255 \ cm^3$$

    The SI units are slowly being accepted; however, the older metric system is slow to fade away and is still used by many practicing scientists. Furthermore, the existence of the older units in the scientific literature demands that we be aware of both the old and the new. Some SI units are used in this book, but in a number of cases the more familiar metric units are retained. In some places we present both the old and the new units.

    In chemistry it is necessary to measure, on a routine basis, mass, length, and volume. The units that we ordinarily use to express these quantities are based on the **gram** (abbreviated **g**), the **meter** (**m**), and the **liter** (**l**, **L**), respectively. The gram itself is a conveniently sized unit for most laboratory measurements of mass (about which we will say more in Section 1.4). The meter, however, is slightly longer than a yard, so for most laboratory purposes we measure lengths in smaller units of centimeters or millimeters.

$$1 \text{ m} = 100 \text{ cm} = 1000 \text{ mm}$$

It is useful to remember that 1 cm = 10 mm. The liter is defined by the SI as exactly 1000 cubic centimeters (cm³), which is the same as one cubic decimeter.

$$1 \text{ liter} = 1000 \text{ cm}^3$$

Most laboratory glassware is graduated in milliliters.

Since there are 1000 milliliters in 1 liter, the size of the milliliter and cubic centimeter are the same—they are identical.

$$1 \text{ ml} = 1 \text{ cm}^3$$

Sometimes you may see the abbreviation cc used for cubic centimeter instead of cm³.

Table 1.4 gives the approximate English equivalents of some of these units. Try to develop a feel for the magnitude of these quantities—it will help you with your calculations.

Table 1.4
Comparison of the English and metric systems

Length	1 meter = 39.37 inches (in.)
	2.540 centimeters = 1 inch
Mass	1 kilogram = 2.204 pounds (lb)
	453.6 grams = 1 pound
Volume	1 liter = 1.057 quarts (qt)
	29.57 milliliters = 1 fluid ounce (oz)
	28.32 liters = 1 cubic foot (ft³)

EXAMPLE 1.5

Calculate the number of meters in 0.200 miles.

SOLUTION

Let's use the factor-label method again. First we list the relationships that we know among the units.

$$1 \text{ mile} = 5280 \text{ ft}$$
$$1 \text{ ft} = 12 \text{ in.}$$
$$1 \text{ m} = 39.37 \text{ in. (from Table 1.4)}$$

Now we use these relationships to construct conversion factors that enable us to cancel unwanted units. Our problem involves converting the units "miles" into "meters."

$$0.200 \text{ miles} = ? \text{ m}$$

$$0.200 \text{ miles} \times \left(\frac{5280 \text{ ft}}{1 \text{ mile}}\right) \times \left(\frac{12 \text{ in.}}{1 \text{ ft}}\right) \times \left(\frac{1 \text{ m}}{39.37 \text{ in.}}\right) = 322 \text{ m}$$

If you set up the problem so that you obtain the correct units for the answer, you can be confident your answer is also correct.

Note that the answer has been rounded to give three significant figures.

Scientific notation

When we express measurements in some units, either very large or very small numbers are often encountered. To avoid having to write down a large number of zeros, it is convenient to express these quantities as the product of a number lying between 1 and 10 multiplied by 10 raised to some power. This type of representation is called **exponential notation** or **scientific notation** and is discussed in detail in Appendix A. For example, using this notation we saw that we can write $1 \text{ kg} = 1 \times 10^3 \text{ g}$ rather than $1 \text{ kg} = 1000 \text{ g}$. If we write numbers in this fashion, the real usefulness of the metric system is apparent because converting from one unit to another merely involves changing the exponent on the 10. For example,

$$2.5 \text{ km} = 2.5 \times 10^3 \text{ m} = 2.5 \times 10^5 \text{ cm} = 2.5 \times 10^6 \text{ mm}$$

You might compare this to the conversions among miles, yards, feet, and inches in the English system.

There are occasions when the presence of zeros leads to difficulties in determining the number of significant figures in a number. The use of exponential notation allows us to eliminate any confusion that might arise. For example, suppose that we had measured the length of an object with a ruler and had found the object 1.2 m long. This number possesses two significant figures and implies an uncertainty of about one part in 12 ($0.1/1.2 = 1/12$). We could also express this length as 1200 mm and, because it still represents the same measurement, it must still possess only two significant figures, the 1 and the 2. The two zeros in 1200 are used only to locate the decimal point. Unfortunately, someone unfamiliar with our experiment might think that all four digits are significant, implying an uncertainty of only one part in 1200, which is not at all what we intended to convey.

By writing numbers using exponential notation we can eliminate this ambiguity. Thus our 1200 mm could be written as 1.2×10^3 mm; here the first portion of the number expresses only two significant figures. Had we wanted to specify four significant figures, we could have done so by writing the number as 1.200×10^3 mm.

To summarize, the only time that zeros are considered as significant figures is when they are not present for the sole purpose of locating the decimal point. Thus, the quantity 0.0072 has only two significant figures while 0.007020 has four, since these are written as 7.2×10^{-3} and 7.020×10^{-3}, respectively. In this example, the zero between the 7 and the 2, as well as the rightmost zero, are significant because they are not needed to position the decimal point.

1.4 MATTER

All the things that we can see or touch, whether books, pencils, telephones, hamburgers, or people, have something in common. They are all composed of **matter,** which is defined as anything that takes up space and has mass. In setting down this definition, we are very careful to specify the term *mass* rather than *weight,* even though we often use the terms as if they were interchangeable. Mass and weight are not really the same. The **mass** of a body is a measure of its resistance to a change in velocity. A Ping-Pong ball moving at 20 miles/hr, for example, is easily deflected by a soft breeze, but a cement truck moving at the same speed is not. Quite clearly, the mass of the cement truck is considerably greater than that of the Ping-Pong ball. The term **weight** refers to the force with which an object of a certain mass is attracted by gravity to the earth or to some other body that it may be near, such as the moon. Force and mass are related to each other by Newton's equation (Newton's law),

$$F = ma$$

where F = force, m = mass, and a = acceleration. In order to accelerate a body, a force must be applied to it. When an object is dropped, it accelerates because

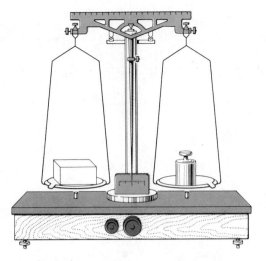

Figure 1.3
A traditional two-pan balance. Known masses are added to the right pan until the pointer is centered. The contents of each pan then have the same weight and therefore also possess the same mass.

of the gravitational attraction of the earth. An object resting on the earth or moon exerts a force (its weight, W) that is equal to its mass, *m*, multiplied by the acceleration due to gravity, *g*, that is,

$$W = mg$$

On the moon the gravitational acceleration is only about one-sixth of that on earth, so the same object weighs only one-sixth as much on the moon as on the earth, even though its mass is the same in both locations. Even on the earth the value of *g* has been found to vary slightly from place to place. This means that if very precise measurements are made, an object's weight also varies slightly according to its location on the earth. Because of this, we specify an object's mass instead of its weight when we wish to report the amount of matter in the object.

No one ever talks about "massing" an object.

The determination of mass (a process, oddly enough, called *weighing*) is actually performed by comparing the weights of two objects, one of known mass, the other of unknown mass. The apparatus used for this is called a **balance.** Figure 1.3 is a drawing of a traditional two-pan balance. The object to be weighed is placed on the left pan of the balance and objects of known mass are added to the other until the pointer comes to the center of the scale. At this point the contents of both pans weigh the same and, since they both experience the same gravitational acceleration, both pans contain equal masses. In chemistry, as mentioned previously, we generally measure mass in grams.

A modern single-pan analytical balance is shown in Figure 1.4. To use this balance, the object to be weighed is placed on the pan and the weights are adjusted internally by turning the knobs on the face until balance is achieved. Although this modern apparatus looks quite different from its older version, they both operate on exactly the same prinicple—that is, a balance beam adjusted to have the same weight on either side of the knife-edge pivot.

1.5 PROPERTIES OF MATTER

Substances are recognized by their characteristics, or **properties.** For example, you recognize your chemistry book by the way it looks and by the printing on its cover. But how could you tell what material was used to construct the objects in Figure 1.5? Their shiny surfaces suggest that they are metallic, and if you brought a magnet near them, you would feel the magnet pull at the objects. This tells you that they might be made of iron. If you left them overnight in the rain,

Figure 1.4

A modern analytical balance capable of measurements to the nearest 0.0001 g. Within the upper part of the balance case the weights rest on the same arm of the balance beam that also supports the pan. Both the weights and the pan are balanced by a counterweight attached to the arm that extends in the opposite direction from the knife edge pivot. When an object is placed on the pan, weights have to be removed by turning the dials until balance is reestablished. This method, called substitutional weighing, maintains a constant load on the balance beam and produces very accurate mass measurements.

A sample of a liquid can be identified by its density — an intensive property — but not by either its mass or its volume alone.

Figure 1.5

What properties could you use to determine what these objects are composed of?

they would begin to rust, so now you would be even more confident in saying that the objects are composed of iron.

Luster, color, magnetism, and the tendency to corrode are just some of the many properties that we use to recognize and classify different samples of matter. These properties themselves can be divided into two broad categories: **extensive properties,** which depend on the size of a sample of matter; and **intensive properties,** which are independent of the sample size. Of the two, intensive properties are the more useful, because a substance[2] will exhibit the same intensive property regardless of how much of it we examine.

Examples of extensive properties are mass and volume—as the amount of a substance increases, its mass and volume also increase. Some examples of intensive properties are melting point and boiling point. Another intensive property is **density,** which is defined as the ratio of an object's mass to its volume. Water, for instance, has a density of 1 g/ml. This means that if we had 1 g of water, it would occupy a volume of 1 ml. If instead we had 20 g of water, we would find that it occupied a volume of 20 ml, but the ratio of the water's mass to volume, 20 g/20 ml, is still the same as 1 g/1 ml. Therefore we see that density is a property that does not depend on sample size.

Normally, when a substance is heated or cooled its volume expands or contracts. This means that its density also changes, so for accurate work, the temperature corresponding to a reported density must be specified. For example, at 25°C (room temperature) the density of water is actually 0.9970 g/ml, while at 35°C its density is 0.9956 g/ml. For most purposes, however, we can take the density of water to be 1.00 g/ml.

Notice that the intensive property, density, is computed as the ratio of two extensive properties, mass and volume. Later in our discussion of chemistry we will encounter quite a few intensive properties defined in a similar fashion. The following examples show how density is calculated and how it can be used.

---

[2] *Substance* is a frequently used term in chemistry. It is normally taken to mean the material of which an object is composed. For example, an ice cube is composed of the substance water. Complicated objects, of course, can be composed of many substances.

EXAMPLE 1.6   An aluminum bar was weighed and found to have a mass of 14.2 g. Its volume was measured to be 5.26 ml. What is the density of aluminum?

SOLUTION   To calculate the density, d, we simply take the ratio of the object's mass to its volume.

$$d = \frac{14.2 \text{ g}}{5.26 \text{ ml}}$$

$$= 2.70 \text{ g/ml}$$

We could also express the density of aluminum as 2.70 g/cm³. (Why?)

EXAMPLE 1.7   A certain copper penny has a mass of 3.14 g. The density of copper is 8.96 g/ml. What is the volume of the penny?

SOLUTION   Density provides a relationship between an object's mass and its volume. It tells us, in this instance, that if we have 1.00 ml of copper, its mass is 8.96 g. It also tells us that if we have 8.96 g of copper, its volume is 1.00 ml. We can use the density as a conversion factor in two ways.

$$\frac{8.96 \text{ g copper}}{1.00 \text{ ml}} \quad \text{or} \quad \frac{1.00 \text{ ml}}{8.96 \text{ g copper}}$$

Since we are given the mass of copper, we must multiply by the second factor to eliminate the units *grams*.

$$3.14 \text{ g copper} \times \left(\frac{1.00 \text{ ml}}{8.96 \text{ g copper}} \right) = 0.350 \text{ ml}$$

The volume of the penny is 0.350 ml (or 0.350 cm³).

A property closely related to density is **specific gravity** (often abbreviated *sp. gr.*), which is defined as the ratio of the density of a substance to the density of water.

Specific gravity has no units.

$$\text{sp. gr.} = \frac{d_{\text{substance}}}{d_{\text{water}}}$$

Its usefulness is that it allows us to compute the density of the substance in various units simply by multiplying its specific gravity by the density of water expressed in those units. In this way two tables—one containing the specific gravities of substances and the other containing the density of water in a variety of units—take the place of the many tables that would otherwise be needed to express the densities of substances in all those different units.

EXAMPLE 1.8   Hexane, a solvent used for rubber cement, has a specific gravity of 0.668. What is the density of hexane in g/ml and in pounds/gallon? Water has a density of 1.00 g/ml or 8.34 lb/gal.

SOLUTION   By definition

$$\text{sp. gr. hexane} = \frac{d_{\text{hexane}}}{d_{\text{water}}}$$

Therefore

$$(\text{sp. gr. hexane}) \times d_{\text{water}} = d_{\text{hexane}}$$

In units of grams per milliliter

$$d_{\text{hexane}} = (0.668) \times (1.00 \text{ g/ml}) = 0.668 \text{ g/ml}$$

Notice that in units of g/ml the numerical values of density and specific gravity are the same. In units of pounds per gallon,

$$d_{hexane} = (0.668) \times (8.34 \text{ lb/gal}) = 5.57 \text{ lb/gal}$$

In speaking of the properties of substances, we also distinguish between physical properties and chemical properties. A **physical property** can be specified without reference to any other substance. Density, color, magnetism, mass, and volume are all examples of physical properties. A **chemical property,** on the other hand, states some interaction between chemical substances. When iron is exposed to oxygen and water, it corrodes and produces a new substance called iron oxide—rust. This is a chemical property of iron. We also say, for instance, that sodium is very reactive toward water. Reactivity is a chemical property that refers to the tendency of a substance to undergo a particular chemical reaction. However, to say simply that a substance is very reactive, without specifying "with what" or under what conditions, is not particularly helpful. Sodium, for example, is very reactive with water but quite unreactive toward the gas helium.

1.6 ELEMENTS, COMPOUNDS, AND MIXTURES

The three words that form the title to this section lie very close to the heart of chemistry, because we work with elements, compounds, and mixtures in the laboratory. We must therefore understand what they are and how to distinguish among them.

Elements are the simplest forms of matter that can exist under conditions that we encounter in a chemical laboratory; they thus are the simplest forms of matter with which the chemist deals directly. Elements serve as the building blocks for all of the more complex substances that we encounter, from common table salt to extremely complex proteins. All are composed of a limited set of elements. At present there are 106 known elements, but only a much smaller number will be of real interest to us.

Elements combine with each other to form compounds. A **compound** is characterized by having its constituent elements always present in the same proportions. For example, you probably know that water is composed of two elements: hydrogen and oxygen. All samples of pure water contain these two elements in the proportion of one part by weight hydrogen to eight parts by weight oxygen (for example, 1.0 g of hydrogen to 8.0 g of oxygen). Also, when hydrogen is allowed to react with oxygen to produce water, the relative amounts of hydrogen and oxygen that combine are always the same. Thus whenever 1.0 g of hydrogen reacts, it is always observed that only 8.0 g of oxygen are consumed, even if more than that amount of oxygen is available.

Mixtures differ from elements and compounds in that they may be of variable composition. A solution of sodium chloride (table salt) in water is a mixture of two substances. We know that by dissolving varying quantities of salt in water (or a bowl of soup) we can obtain solutions with a wide range of compositions. Most materials found in nature, or prepared in the laboratory, are not pure but, instead, are mixtures.

In nature, there are very few materials that even approach being pure compounds — nearly everything is a mixture.

A difficult problem often faced by the chemist is the separation of mixtures into their components. This can usually be accomplished by some **physical process**—a process that doesn't alter the chemical characteristics of the components. Our solution of sodium chloride, for instance, if left to evaporate will leave the salt behind as a solid. If we wished to recover the water as well, we could boil the solution in an apparatus similar to the one in Figure 1.6 and col-

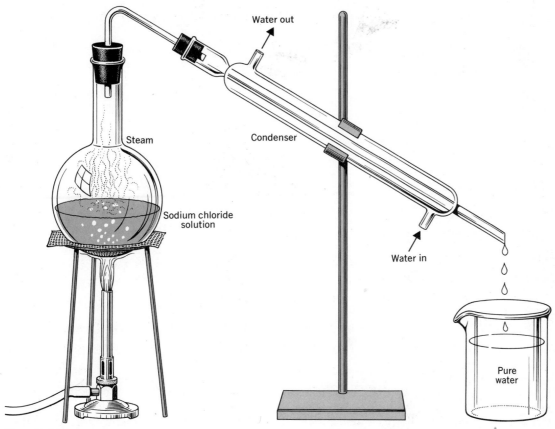

Water out

Steam

Condenser

Sodium chloride solution

Water in

Pure water

Figure 1.6

Simple distillation apparatus. The sodium chloride solution is boiled in the flask and the steam is converted to the liquid in the condenser.

Adsorption means adherence to a surface.
Absorption means swallowing up like a sponge.

lect the water after it has condensed from the steam. This process is called distillation. It is one method, incidentally, that is used to desalinate sea water.

Another method of separating mixtures, called **chromatography,** makes use of the different tendencies that substances have for being adsorbed onto the surface of certain solids. For example, in thin-layer chromatography (Figure 1.7), a sample of a mixture is spotted onto a material, such as silica gel, that thinly coats a glass plate. A solvent is then allowed to creep up the coating from a reservoir. As the solvent flows past the spot, the different components tend to be lifted from the silica gel surface with different degrees of ease. This causes the components of the mixture to move through the silica gel at different rates, with the more strongly adsorbed components moving more slowly. The result is a separation of the original spot into a set of spots each containing (we hope) one component. This technique is widely used today by chemists who synthesize new compounds.

Mixtures can be described as being either **homogeneous** or **heterogeneous.** *A homogeneous mixture is called a* **solution** *and has uniform properties throughout.* If we were to sample any portion of a sodium chloride solution, we would find that it has the same properties (e.g., composition) as any other portion of the solution; we say that it consists of a single **phase.** Thus, we define a phase as any part of a system that has uniform properties and composition.

A heterogeneous mixture, such as oil and water, is not uniform (Figure 1.8). If we were to sample one portion of the mixture, it would have the properties of water, while some other part of the mixture would have the properties of oil. This mixture consists of two phases: the oil and the water. If we shook the mixture so that the oil was dispersed throughout the water as small droplets (as

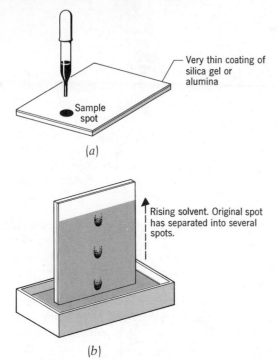

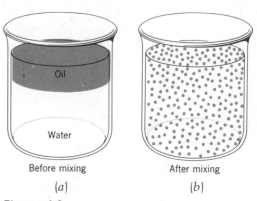

Before mixing (a)   After mixing (b)

Figure 1.8
Oil and water—a heterogeneous mixture. (a) Before mixing. (b) After mixing.

Figure 1.7
Thin-layer chromatography (TLC). (a) A solution containing a mixture is spotted near one end of a glass plate coated with silica gel or alumina. (b) The plate is placed with the end nearest the spot in a tray of solvent that rises through the coating by capillary action. The mixture is separated, with substances least strongly adsorbed by the coating moving farthest.

in a salad dressing), all the oil droplets taken together would still constitute only a single phase, since the oil in one droplet has the same properties as the oil in another. If we added an ice cube to this "brew," we would then have three phases: the ice (a solid), the water (a liquid), and the oil (another liquid). In all these examples, we can detect the presence of two or more phases because a boundary exists between them.

A useful feature of pure compounds is that they undergo **phase changes** (e.g., solid to liquid or liquid to gas) at constant temperature. Ice, for instance, melts at a temperature of 32°F, a temperature that remains constant while the water undergoes the change from solid to liquid. When mixtures undergo phase changes, they generally do so over a range of temperatures. This phenomenon provides us with one experimental test to determine when we have obtained a pure compound. The relationship among elements, compounds, and mixtures is summarized in Figure 1.9.

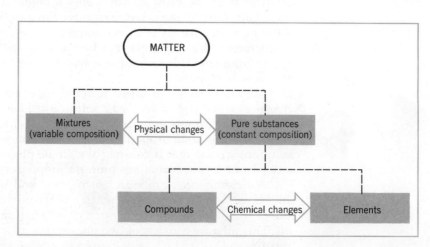

Figure 1.9
Classification of matter.

1.7 THE LAWS OF CONSERVATION OF MASS AND DEFINITE PROPORTIONS

The early history of chemistry was marked, not surprisingly, by incorrect theories about what occurred during chemical reactions. It had long been observed, for example, that when the combustion of wood took place, the resulting ash was very light and fluffy. Metals, too, when heated in air changed their appearance. The resulting material was less dense than the original metal and thus appeared lighter. These observations led to the conclusion that "something," which the German chemists Becher and Stahl called **phlogiston,** was lost by substances when they burned. Even when it was pointed out that metals gained weight when they were heated in air, the theory was salvaged by concluding that phlogiston simply had negative mass!

The reluctance to abandon the crumbling phlogiston theory demonstrates a general human phenomenon. New theories are difficult to come by, and old ones become so thoroughly entrenched that it is often tempting to try to shore up a sagging theory rather than to dream up a new one that will do a better job of explaining all the observed facts.

It was Antoine Lavoisier (1743–1794),[3] a French chemist, who finally laid the phlogiston theory to rest and set chemistry on the proper course again. He demonstrated by his experiments that the combustion process actually occurred by the reaction of substances with oxygen. He also showed, *through careful measurements,* that if a reaction is carried out in a closed container, so that none of the products of reaction escape, the total mass of all substances present after the reaction has occurred is the same as before the reaction began. These observations form the basis of the **law of conservation of mass,** which states that *mass is neither created nor destroyed in a chemical reaction.*[4] This is one of the most important principles in chemistry, even today.

The work of Lavoisier clearly demonstrated the importance of careful measurement. After his book, *Traité Elémentaire de Chimie,* appeared in 1789, many chemists were inspired to investigate the quantitative aspects of chemical reactions. These investigations led to another important chemical law, the law of definite proportions.

The **law of definite proportions** (also called the **law of definite composition**) states that, *in a pure chemical substance, the elements are always present in definite proportions by mass.* In the substance water, for instance, the ratio of the mass of hydrogen to the mass of oxygen is always 1/8, regardless of the source of the water. Thus, if 9.0 g of water are decomposed, 1.0 g of hydrogen and 8.0 g of oxygen are always obtained. If 18.0 g of water are broken down, 2.0 g of hydrogen and 16.0 g of oxygen are produced. Furthermore, if 2.0 g of hydrogen are mixed with 8.0 g of oxygen and the mixture ignited, 9.0 g of water are formed and 1.0 g of hydrogen remains unreacted. In the water that is formed, the hydrogen-to-oxygen mass ratio once again is 1/8. Thus, varying the amounts of hydrogen and oxygen that are present during the reaction does not alter the composition of the water produced.

> The definition of <u>compound</u> is a statement of the law of definite proportions.

---

[3] Lavoisier was a victim of the French Revolution. He went to the guillotine on May 8, 1794.

[4] Einstein showed that there is a relationship between mass and energy, $E = mc^2$, where c is the speed of light. Energy changes that take place during chemical reactions therefore are also accompanied by mass changes; however, the changes in mass are far too small to be detected experimentally. For example, the energy change associated with the reaction of 2 g of hydrogen with 16 g of oxygen happens to be equivalent to a mass change of approximately 10^{-9} g. Sensitive analytical balances can only detect mass differences of 10^{-6} to 10^{-7} g. Consequently, as far as the average chemist is concerned, there are no observable mass increases or decreases accompanying chemical reactions.

1.8 THE ATOMIC THEORY OF DALTON

The real father of modern chemistry could well be considered to be the Eng-lishman, John Dalton (1766–1844), who proposed his atomic theory of matter around 1803. The concept of the atom (from the Greek *atomos*, meaning indivis-ible) did not originate with Dalton. The Greek philosophers Leucippus and De-mocritus suggested, as early as 400 to 500 B.C., that matter cannot be forever di-vided into smaller and smaller parts and that ultimately particles would be en-countered that would be indivisible. These early proposals, however, were not based on the results of experiments and were little more than exercises in thought. Dalton's theory was different because it was based on the laws of con-servation of mass and definite proportions, laws that had been derived from many direct observations.

The theory Dalton proposed can be expressed by the following postulates:

1. Matter is composed of indivisible particles called atoms.
2. All atoms of a given element have the same properties (e.g., size, shape, and mass), which differ from the properties of all other elements.
3. A chemical reaction merely consists of a reshuffling of atoms from one set of combinations to another. The individual atoms themselves, how-ever, remain intact.

The test of any theory, of course, is how well it explains already existing fact and whether it can predict as yet undiscovered laws. Dalton's theory proved successful on both counts.

First, it accounts for the law of conservation of mass. If a chemical reaction does nothing more than redistribute atoms and no atoms are lost from the system, it follows that the total mass must remain constant when the reaction occurs.

Second, it explains the law of definite proportions. To see this, let's imag-ine a substance formed from two elements, say, A and B, in which each molecule of the substance is composed of one atom of A and one atom of B. We define a **molecule** as *a group of atoms bound tightly enough together that they behave as, and can be recognized as, a single particle* (just as a car is composed of many parts held together tightly enough so that we identify them as *a car*). Let's also suppose that the mass of an atom of A is twice the mass of an atom of B. Then, in one molecule of this substance, twice as much mass is contributed by A as by B, and the ratio of the mass of A to the mass of B in this molecule is 2/1. If we take a large collection of these molecules, we will always have equal numbers of A and B atoms; therefore, regardless of the size of the sample, we still always have a mass ratio (A to B) of 2/1. Also, if we were to react A and B together to form this compound, each atom of A would combine with only one atom of B. If we were to mix 100 atoms of A with 110 atoms of B, after the reaction was over, we would be left with 100 molecules of AB and 10 atoms of unreacted B. The substance AB would still have a mass ratio, A to B, of 2 to 1.

Third, Dalton's theory predicted the **law of multiple proportions.** This law can be stated as follows: *Suppose that we have samples of two different compounds formed by the same two elements. If the mass of one of the elements is the same in the two samples, the masses of the other element are in a ratio of small whole numbers.* Actually this may seem more complicated than it really is. Let's consider the two compounds formed by carbon and oxygen. In one of them (carbon mon-oxide) we find 1.33 g of oxygen combined with 1.00 g of carbon, while in the second (carbon dioxide) there are 2.66 g of oxygen combined with 1.00 g of carbon. If we examine the ratio of the masses of oxygen (1.33 g/2.66 g) that com-

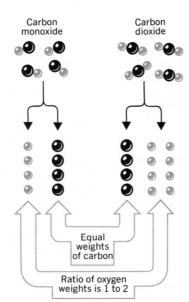

Figure 1.10

The law of multiple proportions. Here are four molecules each of carbon monoxide and carbon dioxide. Each contains the same number of carbon atoms, ◉, and thus the same mass of carbon. The masses of oxygen that are combined with the carbon are in a whole number (1 to 2) ratio.

bines with a fixed mass of carbon (1.00 g), we observe a ratio of small whole numbers

$$\frac{1.33}{2.66} = \frac{1.33/1.33}{2.66/1.33} = \frac{1}{2}$$

This is consistent with the atomic theory if we consider that carbon monoxide contains *one* atom of carbon and *one* of oxygen whereas carbon dioxide contains *one* atom of carbon and *two* atoms of oxygen. Because carbon dioxide has twice as many oxygen atoms bound to a carbon atom as does carbon monoxide, the weight of oxygen in a molecule of carbon dioxide *must* be twice the weight of oxygen in a molecule of carbon monoxide. (See Figure 1.10.)

EXAMPLE 1.9   Nitrogen forms several different compounds with oxygen. In one (called laughing gas), it is observed that 2.62 g of nitrogen are combined with 1.50 g of oxygen. In another (a major air pollutant), 0.655 g of nitrogen is combined with 1.50 g of oxygen. Show that these data demonstrate the law of multiple proportions.

SOLUTION   In both cases we are dealing with a weight of nitrogen that combines with 1.50 g of oxygen. If these data do fit the law of multiple proportions, the ratio of the masses of nitrogen in the two compounds should be a ratio of small whole numbers. Let's take the ratio

$$\frac{2.62}{0.655}$$

Dividing the numerator and denominator by 0.655, we get

$$\frac{4.00}{1.00}$$

which is indeed a ratio of small whole numbers.

EXAMPLE 1.10

Sulfur forms two compounds with fluorine. In one of them it is observed that 0.447 g of sulfur is combined with 1.06 g of fluorine and in the other, 0.438 g of sulfur is combined with 1.56 g of fluorine. Show that these data illustrate the law of multiple proportions.

SOLUTION

We can't examine the ratio of sulfur weights immediately because the weights of fluorine with which they are combined are different. We therefore must calculate how much sulfur will combine with the *same weight* of fluorine in the two compounds.

The data given in the problem provide us with relationships between the weights of sulfur and fluorine that combine with each other to form each compound. In the first compound, we could say that 1.06 g of fluorine are *equivalent* to 0.447 g of sulfur—equivalent in the sense that whenever we find 1.06 g of fluorine in a sample of this compound, we know that we will also find 0.447 g of sulfur. We will find it convenient to express this kind of an equivalence—a *chemical equivalence*—in the following way:

Some people prefer to write this as
1.06 g fluorine = 0.447 g sulfur.
The equals sign in this case should be read "is equivalent to."

$$1.06 \text{ g fluorine} \sim 0.447 \text{ g sulfur}$$

where the symbol ~ means "is equivalent to."

For the second compound we can write the following equivalence.

$$1.56 \text{ g fluorine} \sim 0.438 \text{ g sulfur}$$

For the purpose of calculation, the equivalence implied by the ~ behaves the same as an equality, so this sort of equivalence relationship can be used to construct conversion factors useful in mathematical computations. For the second compound, let us calculate the weight of sulfur that would have been combined with 1.06 g of fluorine.

$$1.06 \text{ g fluorine} \times \left(\frac{0.438 \text{ g sulfur}}{1.56 \text{ g fluorine}} \right) \sim 0.298 \text{ g sulfur}$$

We now know the weights of sulfur that combine with the *same weight* of fluorine (1.06 g) in these two compounds. The next step is to look at the ratio of these weights, that is,

$$\frac{0.447}{0.298} = \frac{0.447/0.298}{0.298/0.298} = \frac{1.50}{1.00}$$

which is the same as 3/2, a ratio of small whole numbers.

The use of chemical equivalencies in computations will be found extensively in later chapters. You should make a special effort now to understand the concept of "chemical equivalence" by working the problems on the laws of definite proportions and multiple proportions at the end of the chapter.

1.9 ATOMIC WEIGHTS

Dalton's atomic theory proved so successful at explaining the laws of chemistry that it was accepted almost immediately. Since the key to the success of the theory had been the concept that each element had a characteristic atomic mass, chemists at once set out to measure them. It was at this point that a serious problem arose.

Because of their small size, there certainly was no way to determine the masses of individual atoms. All that scientists could hope to do was to arrive at a set of relative atomic weights. (Note that the terms atomic mass and atomic weight are used interchangeably.) We might determine, for instance, that in one compound formed between carbon and oxygen, 3.0 g of carbon were combined with 4.0 g of oxygen. In other words, oxygen contributed $\frac{4}{3}$ (or $1\frac{1}{3}$) times as much mass toward the compound as the carbon did. If this substance is made up of molecules, each containing one atom of carbon and one atom of oxygen, then it follows that each oxygen atom must weigh $1\frac{1}{3}$ times as much as one carbon atom; thus it appears that we have established the relative masses of these two elements.

The arguments just presented rest on a very critical assumption, that is, that the compound we were discussing is composed of one atom of carbon and one atom of oxygen. If this assumption about the number of carbon and oxygen atoms in a molecule is false, we have arrived at the wrong relative weights. Because of this kind of difficulty a self-consistent table of atomic weights could not be developed until a method was devised for determining chemical formulas. This didn't happen until about 60 years after Dalton had introduced his theory.

A complete table of atomic weights appears on the inside front cover of this book. This is a table of relative atomic weights with the masses of atoms expressed in units called **atomic mass units** (amu). The size of this unit is chosen in a rather arbitrary way. To see this, let's return to our discussion of the compound between carbon and oxygen.

Let us assume, for the moment, that molecules of the substance are in fact composed of one atom of carbon and one of oxygen; then, as mentioned, one oxygen atom weighs 1.33 times as much as one carbon atom. If we were to define the atomic mass unit as equal to the mass of one carbon atom, we would then assign carbon an atomic weight of 1.00 amu and oxygen an atomic weight of 1.33 amu. Since we might find it more convenient to have these numbers as whole numbers (or as near to whole numbers as possible), we might define the amu as equal to one-third of the mass of a carbon atom. If we did this, we would assign carbon an atomic weight of 3.00 amu. Oxygen, which is $\frac{4}{3}$ times heavier, would have an atomic weight equal to 4.00 amu. Thus the size of the atomic mass unit, and hence the values that appear in a table of atomic weights, are really quite arbitrary.

As originally defined by chemists, the amu was taken to be $\frac{1}{16}$ of the mass of naturally occurring oxygen. In this way the masses of nearly all the elements come out to be approximately whole numbers. Oxygen was chosen as a standard because it forms compounds with nearly all the elements. However, as we will discuss in Chapter 3, not all atoms of an element have precisely the same mass, and naturally occurring oxygen was later found to be composed of a mixture of atoms of slightly different masses. Since the relative proportion of these different atoms (called **isotopes**) could conceivably change over a period of time, or be different at different locations, the entire atomic weight table could also change because the size of the amu was based on this mixture. To avoid this problem, the atomic mass unit is currently defined as $\frac{1}{12}$ of the mass of one particular isotope of carbon.

Because atoms are so very tiny, the actual mass of the amu is extremely small. For example, one atom of the reference isotope of carbon (called carbon-12) has a mass of 1.992×10^{-23} g. The mass of the amu is one-twelfth of this, or 1.660×10^{-24} g. You might write this out in decimal form to get a feel for how really small the masses of atoms are.

Scientists who work with substances having very large molecules, such as proteins, often refer to an atomic mass unit as a **dalton** (1 amu ≡ 1 dalton).

1.10 SYMBOLS, FORMULAS, AND EQUATIONS

In a certain sense, learning chemistry is like learning a strange new language such as Greek (or Russian, if you happen to be Greek!). We might compare the chemical symbols for the elements with an alphabet, the chemical formulas that we construct from the symbols with words, and the chemical equations that we write with sentences. In learning any new language, we must start at the beginning with the alphabet.

At the present time a total of 106 different elements are known. Each element is identified by its name and can also be represented by its **chemical symbol.** Usually, the symbol bears a resemblance to the English name for the element. For instance, carbon = C, chlorine = Cl, nitrogen = N, and zinc = Zn. Some elements, however, have symbols that do not seem to correspond at all with their names. In nearly all these cases, the elements have been known since the early history of chemistry when Latin was used as the universal language among scientists. For this reason symbols for these elements are derived from their Latin names—for example, potassium (L. *kalium*) = K, sodium (*natrium*) = Na, silver (*argentum*) = Ag, mercury (*hydrargyrum*) = Hg, and copper (*cuprum*) = Cu. Regardless of the origin of the symbol, the first letter is always capitalized. A complete list of the elements with their chemical symbols appears on the inside front cover of the book.

A chemical compound is represented symbolically by its **chemical formula.** Water is represented by H_2O, carbon dioxide by CO_2, methane (natural gas) by CH_4, and aspirin by $C_9H_8O_4$. Chemical formulas also show the composition of substances. The subscripts in a formula give the relative numbers of atoms of each element that appear in the compound (when no subscript is written, a subscript of 1 is implied). The formula H_2O, for example, describes a substance containing two hydrogen atoms for every one oxygen atom. Similarly, the compound CH_4 contains one atom of carbon for every four atoms of hydrogen. Some compounds are more complex, and their formulas are written containing parentheses. For example, ammonium sulfate (a fertilizer) has the formula $(NH_4)_2SO_4$, which specifies the presence of two NH_4 units—for a total of two nitrogen atoms and eight hydrogen atoms—plus one sulfur atom and four oxygen atoms. As we will see later, there are good reasons for writing this formula as $(NH_4)_2SO_4$ instead of $N_2H_8SO_4$, although both represent the same number of atoms.

There are certain substances that form crystals that contain water molecules when their aqueous solutions are evaporated. These crystals are called **hydrates.** For example, copper sulfate—an agricultural fungicide—forms blue crystals having five molecules of water for each copper sulfate ($CuSO_4$). Its formula is written $CuSO_4 \cdot 5H_2O$. If the blue crystals are heated, water can be driven off to leave pure $CuSO_4$, which appears almost white. (See Color Plate 1.)

Chemical equations are written to show the chemical changes that occur during chemical reactions. For example, the equation

$$Zn + S \longrightarrow ZnS$$

describes a reaction in which zinc (Zn) reacts with sulfur (S) to produce zinc sulfide (ZnS),[5] a substance used on the inner surfaces of TV screens. The substances on the left side of the arrow are present before the reaction begins and are known as the **reactants.** The substances on the right of the arrow are formed by the reaction and are called the **products.** (In the example above there is only one product.) The arrow is read as "react to yield" or simply "yield." This equation is read as "zinc plus sulfur react to yield zinc sulfide" or "zinc plus sulfur yield zinc sulfide." Sometimes it is also desirable (or necessary) to indicate whether the reactants and products in a chemical reaction are solids, liquids, gases, or are dissolved in a solvent such as water. This is accomplished by

A chemical equation gives a type of "before-and-after" view of a chemical reaction.

---

[5] The naming of chemical compounds is discussed in Chapter 4. For now, we use these names simply as labels.

Figure 1.11

The fierce reaction of hydrogen with oxygen (from the air) led to the fiery destruction of the Hindenburg, a German airship filled with hydrogen, in Lakehurst, New Jersey, in 1937. Thirty people died in the disaster.

placing the letters s = solid, l = liquid, g = gas, aq = aqueous (water) solution in parentheses following the formulas of the substances in the equation. For instance, the equation

$$CaCO_3 \ (s) + H_2O \ (l) + CO_2 \ (g) \longrightarrow Ca(HCO_3)_2 \ (aq)$$

describes a reaction between solid calcium carbonate (limestone), liquid water, and gaseous carbon dioxide to give an aqueous solution of calcium bicarbonate. This is the reaction responsible for the dissolving of limestone by ground water containing carbon dioxide. It is one of the causes of "hard" water and the formation of limestone caves.

Many of the equations that we write will contain **coefficients** preceding the chemical formulas. An example is the reaction of hydrogen (H_2) and oxygen (O_2) shown in Figure 1.11.

$$2H_2 + O_2 \longrightarrow 2H_2O$$

This equation is interpreted to mean that two hydrogen molecules plus one oxygen molecule (a coefficient of 1 is assumed when none is written) react to yield two molecules of H_2O. Such an equation is said to be **balanced** because it contains the same number of atoms of each element on both sides of the arrow. The techniques we employ to help us balance equations are discussed later.

1.11 ENERGY

When a chemical change occurs, it almost always is accompanied by either an absorption or release of energy. In fact, in Chapter 11 we will see that whether a particular chemical reaction is even possible is determined, in part, by the energy change that accompanies it. Energy changes also tell us a great deal about the nature of the substances that react and, equally important to us, chemical reactions serve as a way of providing the energy that our bodies and our society need to function.

Energy is usually defined as the capacity to do work. When an object possesses energy, it can affect other objects by doing work on them. A moving car possesses energy because it can do work on another car by moving it some distance in a collision. Coal and oil possess energy because they can be burned and the heat that is liberated can be harnessed to do work. Because energy can be transferred from one object to another as work, the *units* of energy and work are the same.

An object can possess energy in two ways: as kinetic energy and as potential energy. **Kinetic energy (K.E.)** is associated with motion and is equal to one-half of an object's mass, m, multiplied by its velocity, v, squared.

$$\text{K.E.} = \tfrac{1}{2}mv^2$$

Thus we see that the amount of work a moving body can do depends on both its mass and its velocity. For example, a truck moving at 20 miles/hr can do more "work" on the rear end of a car than a bicycle moving at the same speed. We also know that a truck moving at 80 miles/hr can do more "work" on a car than one traveling at only 5 miles/hr.

Potential energy (P.E.) represents energy a body possesses because of the attractive or repulsive forces it experiences with other objects. If there are no attractive or repulsive forces, the body does not possess potential energy.

To see how potential energy is affected by these forces, consider two balls connected by a coiled spring as shown in Figure 1.12. When the balls are pulled apart, the spring is stretched and the energy spent in separating the balls is stored in the stretched spring. We say, therefore, that the potential energy of the two balls has increased. This energy is released and converted to kinetic energy when the balls are allowed to move toward one another. In this case there was an attractive force (the stretched spring) between the two objects.

Using this same apparatus, energy can also be stored by compressing the spring, thereby producing a repulsive force between the two balls. This stored energy is released as kinetic energy as the balls are permitted to move away from one another.

The kinds of changes in potential energy that accompany changes in the relative position of objects that either attract or repel one another are very important to remember. They will help us analyze many physical and chemical changes in the chapters ahead.

As we have indicated, energy contained in chemical substances can be released through chemical reaction. Wood, for example, can react with oxygen, present in the air, in the process known as combustion. As the products of the reaction are formed, rather sizable quantities of energy are released. This chemical energy is initially present in the wood and oxygen and is released as the reaction proceeds. A change that results in the release of energy, such as the

Doubling the speed of an auto increases its kinetic energy by a factor of four.

An object has potential energy only if it experiences attractive or repulsive forces.

Figure 1.12

Potential energy exists between objects that either attract or repel each other. When the spring is either stretched or compressed, the P.E. of the two balls increases.

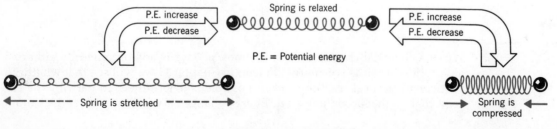

combustion of wood, is said to be **exothermic.** Changes that absorb energy, on the other hand, are termed **endothermic.** Exothermic changes (like that in Figure 1.11) heat up their surroundings and endothermic changes cool their surroundings.

The amount of energy released or absorbed in a chemical reaction depends on the amounts of materials that react. The burning of a match, for example, releases only a very small amount of energy while a large bonfire produces much more. Energy, therefore, is an extensive quantity.

The total amount of energy that a body possesses is equal to the sum of its kinetic energy and its potential energy. In an isolated system, such as our universe, the total energy is constant and *energy is neither created nor destroyed but, instead, can only be transformed from one kind of energy to another.* This statement is called the **law of conservation of energy** and is another example of the many conservation laws that appear to govern our physcial world.

Energy is transmitted from one body to another in a variety of ways, for example, as light, sound, electricity, and/or heat. These various forms of energy can be converted from one to another and therefore are ultimately equivalent. The SI unit for energy is the **joule (J),** and is defined as 1 kg m^2/s^2 in terms of the SI base units. It represents the kinetic energy possessed by an object with a mass of 2.0 kg traveling at 1 m/s (in English units, an object weighing 4.4 lb traveling at a speed of 197 ft/min, or about 2.2 miles/hr).[6] A smaller unit, the erg, is equal to 1×10^{-7} J (or 0.1 μJ). In referring to energies involved in chemical reactions it is useful to use the term **kilojoule (kJ).** One kilojoule is equal to one thousand joules (1 kJ = 1000 J).

All forms of energy ultimately end up as heat, and when chemists measure energy it is usually in the form of heat. For example, we might measure the heat liberated during the combustion of a gas such as methane. The measurement of heat energy involves the concept of **temperature.** It is important to remember that heat is not the same as temperature. Heat *is* energy, while temperature is a measure of the intensity of heat or hotness. Another useful definition of temperature is that it is an intensive quantity that defines the *direction* and *rate* of heat flow. We know that heat always flows by itself from a region of high temperature to one of low temperature, as when a hot cup of coffee becomes cool. We also know that the rate of heat transfer depends on the difference in temperature between two objects. If we want to warm something slowly, we place it in contact with an object just slightly warmer; if we wish to heat it quickly, we place it in contact with something very hot.

The measurement of temperature is usually accomplished with a thermometer (Figure 1.13), which consists of a narrow capillary tube connected to a thin-walled reservoir filled with some liquid (usually mercury). As the temperature is raised, the liquid expands. The volume increase causes the liquid to rise in the capillary, and the height of liquid in the capillary is proportional to the temperature.

A number of temperature scales have been devised. The Fahrenheit scale, in common use in the United States, is defined by the freezing point and boiling point of water.

At its freezing point (which is also its melting point), a pure substance such as water can exist as both solid (ice) and liquid in contact with each other *at the same temperature.* If heat is added to this mixture, some solid melts and more liquid is formed, but the temperature remains constant while the solid is melting. If heat is removed from the mixture, liquid freezes to produce more solid, once again without a temperature change. Similarly, at its boiling point a

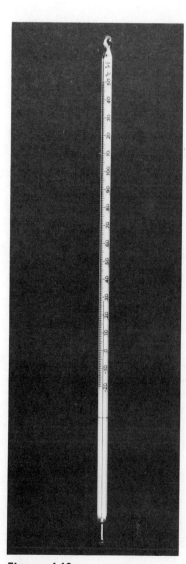

Figure 1.13

A typical laboratory thermometer.

[6] Remember, K.E. = $\frac{1}{2}mv^2$. In this case K.E. = $\frac{1}{2}$(2 kg)(1 m/s)^2, or K.E. = 1 kg m^2/s^2 = 1 J.

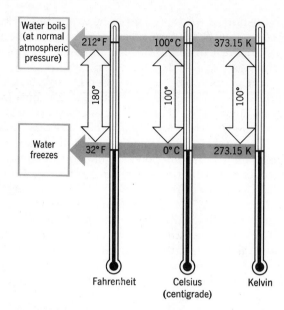

Figure 1.14

Comparison of temperature scales.

pure liquid can exist in contact with its vapor at the same temperature. In this case, addition of heat causes more liquid to evaporate, all at a constant temperature. The constancy of temperature during these phase changes makes them ideal temperatures to serve as calibration points on a temperature scale.

On the **Fahrenheit scale** the freezing point of water is assigned a temperature of 32°F and the boiling point a temperature of 212°F. The difference between these two reference points, 180 Fahrenheit degrees, thus defines the size of the degree unit. In the sciences the temperature scale that is employed is the Celsius scale (formerly called the centigrade scale). You have probably noticed the growing use of Celsius temperatures in weather reports on radio and TV. The **Celsius scale** defines 0°C as the freezing point of water and 100°C as the boiling point of water. This means that 100 Celsius degrees are equal to 180 Fahrenheit degrees, so the Celsius degree is nearly twice as large as a degree on the Fahrenheit scale. Temperatures in °C are related to temperatures in °F by the equations

$$°C = \tfrac{5}{9}(°F - 32)$$

$$°F = \tfrac{9}{5}°C + 32$$

The SI temperature unit is the **kelvin (K).** The size of the degree unit on the Kelvin scale is the same as that on the Celsius scale; the difference between the two scales is the location of the zero point. On the Kelvin scale water freezes at 273.15 K (note that the degree symbol, °, is not used in the SI unit), so the relationship between the Celsius and Kelvin scales is

$$°C = K - 273.15$$

Figure 1.14 illustrates the differences between these three temperature scales.

The original definition of the unit of heat energy called the **calorie** (abbreviated **cal**) relied on the concept of temperature. The calorie was defined as the amount of heat needed to raise the temperature of 1 gram of water, initially at 15°C, by 1°C.[7] The **kilocalorie (kcal),** like the kilojoule, is a more appropriately

Zero kelvin is known as absolute zero — the lowest temperature possible. We will say more about it later in the book.

---

[7] It is necessary to specify the temperature of the water because the amount of heat required to raise the temperature of 1 g of water by 1°C varies slightly with the temperature of the water. For example, at 25°C (room temperature) it takes 0.998 cal to raise the temperature of 1 g of water by 1°C.

sized unit for dealing with energy changes in chemical reactions. It is also the unit used in nutrition for reporting the energy content of foods—the Calorie (written with a capital C) is the same as a kilocalorie. Thus when we read that a serving of mashed potatoes contains 230 Cal, we are being told that 230 kcal of energy are liberated when the body metabolizes this food.

Until recently, nearly all the chemical scientific literature has used the calorie (or kilocalorie) in reporting energy changes. With the introduction of SI units, the joule (or kilojoule) is now preferred. Unfortunately, at this time in the sciences we must be able to handle both sets of units. The conversion between calories and joules, or kilocalories and kilojoules, is given by

$$1 \text{ cal} = 4.1840 \text{ J}$$

$$1 \text{ kcal} = 4.1840 \text{ kJ}$$

An intensive property of matter associated with energy is **specific heat,** the amount of heat required to raise the temperature of 1 g of a substance by 1°C. Because of the definition of the calorie, the specific heat of water is 1.00 cal/g °C. Most other substances have smaller specific heats. Iron, for example, has a specific heat of only 0.108 cal/g °C. This means that it takes less heat to raise the temperature of 1 g of iron by 1°C than it does to cause the same temperature change for a gram of water. It also means that a given amount of heat will raise the temperature of 1 g of iron more than it will raise the temperature of 1 g of water.

The large specific heat of water is responsible for the moderating effect the oceans have on weather, since they cool very slowly in winter and warm up slowly in the summer. Air moving over the oceans in winter never gets very cold, and in the summer the air never gets very hot.

EXAMPLE 1.11 How many calories are required to raise the temperature of a 7.5-cm (3.0-inch) iron nail weighing 7.05 g from room temperature (25°C) to 100°C? How many joules does this correspond to? The specific heat of iron is 0.108 cal/g °C.

SOLUTION To solve this problem we must multiply the specific heat by mass (g) and by the temperature change (°C) to eliminate these units and obtain calories as the units of the answer.

$$\text{specific heat} \times \text{mass} \times \text{temperature change} = \text{heat energy}$$

$$\left(\frac{\text{cal}}{g\,°C}\right) \times (g) \times (°C) = \text{cal}$$

Using our data (the temperature change is 75°C),

$$\left(\frac{0.108 \text{ cal}}{g\,°C}\right) \times (7.05\,g) \times (75\,°C) = 57 \text{ cal}$$

Notice that the answer has been rounded to two significant figures. Do you know why? If not, review the significant figure rules for multiplication.

To answer the final part of this question we must convert calories to joules.

$$57 \text{ cal} \times \left(\frac{4.184 \text{ J}}{1 \text{ cal}}\right) = 240 \text{ J}$$

(Again, the answer is rounded to two significant figures.)

INDEX TO QUESTIONS AND PROBLEMS (problem numbers in **bold type**)

REVIEW QUESTIONS

1.1 What is the difference between a theory and a law?

1.2 Why is it important for scientists to be concerned about reporting the correct number of significant figures in their measurements?

1.3 Define, in your own words, the terms *precision* and *accuracy*.

1.4 The SI units for mass and volume are the kilogram and cubic meter. However, in the laboratory we generally use the units gram and milliliter. Why?

1.5 What is the difference between an extensive and an intensive property? Can you think of any examples not mentioned in the text?

1.6 What is the difference between density and specific gravity? What are their common metric units?

1.7 There are many examples of homogeneous and heterogeneous mixtures in the world around us. How would you classify sea water, air (unpolluted), smog, smoke, club soda, black coffee, a penny, a ham sandwich?

1.8 Identify the phases that exist in a copper pan containing two iron nails, a quart of water, and four glass marbles.

1.9 Why is mass (instead of weight) used to specify the amount of matter in an object?

1.10 Distinguish among the terms element, compound, and mixture.

1.11 What is the most important atomic property in Dalton's atomic theory?

1.12 Distinguish between the terms atom and molecule.

1.13 Write chemical symbols for the following:
(a) iron
(b) sodium
(c) potassium
(d) phosphorus
(e) bromine
(f) calcium
(g) nitrogen
(h) neon
(i) manganese
(j) magnesium

1.14 Give the name of each of these elements:
(a) Ag
(b) Cu
(c) S
(d) Cl
(e) Al
(f) Au
(g) Cr
(h) W
(i) Ni
(j) Hg

1.15 How many atoms of each kind are represented in the following formulas:
(a) K_2S
(b) Na_2CO_3
(c) $K_4Fe(CN)_6$
(d) $(NH_4)_3PO_4$
(e) $Na_3Ag(S_2O_3)_2$

1.16 Plaster is composed of calcium sulfate ($CaSO_4$), which exists as a hydrate containing two water molecules for each $CaSO_4$. Write the formula for the hydrate.

1.17 State, in words, the following laws:
(a) the law of conservation of mass
(b) the law of definite proportions
(c) the law of multiple proportions

1.18 How is the currently accepted atomic mass unit defined?

1.19 Which of the following equations are not balanced?
(a) $ZnCl_2 + NaOH \rightarrow Zn(OH)_2 + NaCl$
(b) $CuCO_3 + 2HCl \rightarrow CuCl_2 + CO_2 + H_2O$
(c) $Fe_2O_3 + 2CO \rightarrow 2Fe + 2CO_2$
(d) $NH_4NO_3 \rightarrow N_2O + 2H_2O$

1.20 How does potential energy differ from kinetic energy?

1.21 What is meant by endothermic and exothermic?

1.22 If the potential energy of an object decreases as it is moved away from another object, what kind of force (attractive or repulsive) must exist between the two?

1.23 What is the difference between heat and temperature?

1.24 What are the units of specific heat? What is the value of the specific heat of water? Why do oceans and large lakes affect the temperatures of surrounding land masses?

REVIEW PROBLEMS (More difficult problems are marked by an asterisk.)

1.25 How many significant figures are there in the following numbers: 1.0370, 0.000417, 0.00309, 100.1, 9.0010?

1.26 Perform the following computations, rounding answers to the proper number of significant figures. All values are measured.
(a) 2.41×3.2
(b) 4.025×18.2
(c) $81.8 \div 104.2$
(d) $3.476 + 0.002$
(e) $81.4 - 0.002$

1.27 Express each of the following in scientific notation. Assume that any digits to the right of the last nonzero digit are *not* significant figures.
(a) 1,250
(b) 13,000,000
(c) 60,230,000,000,000,000,000,000
(d) 214,570
(e) 31.47

1.28 Express each of the following in scientific notation.
(a) 0.00040
(b) 0.0000000003
(c) 0.002146
(d) 0.0000328
(e) 0.00000000000091

1.29 Write the following numbers in decimal notation.
(a) 3×10^{10}
(b) 2.54×10^{-5}
(c) 122×10^{-2}
(d) 3.4×10^{-7}
(e) 0.0325×10^{6}

1.30 The length of a piece of land was measured to be 3000 m. Using scientific notation, express the measurement
(a) to two significant figures
(b) to three significant figures
(c) in cm, to two significant figures

1.31 Perform the following computations, expressing the answers in scientific notation rounded to the proper number of significant figures. Assume all values are from measurements.
(a) $(3.42 \times 10^{8}) \times (2.14 \times 10^{6})$
(b) $(1.025 \times 10^{6}) \times (14.8 \times 10^{-3})$
(c) $(143.7) \times (84.7 \times 10^{16})$
(d) $(5274) \times (0.33 \times 10^{-7})$
(e) $(8.42 \times 10^{-7}) \times (3.211 \times 10^{-19})$

1.32 Perform the following computations, expressing the answers in scientific notation rounded to the proper number of significant figures. All values are measured.
(a) $(12.45 \times 10^{6}) \div (2.24 \times 10^{3})$
(b) $822 \div 0.028$
(c) $(635.4 \times 10^{-5}) \div (42.7 \times 10^{-14})$
(d) $(31.3 \times 10^{-12}) \div (8.3 \times 10^{-6})$
(e) $(0.74 \times 10^{-9}) \div (825.3 \times 10^{18})$

1.33 Perform the following computations, expressing the answers in scientific notation rounded to the proper number of significant figures. All values are from measurements.
(a) $(2.047 \times 10^{8}) + (14.33 \times 10^{8})$
(b) $(12.4 \times 10^{8}) + (92.3 \times 10^{7})$
(c) $(42.003 \times 10^{5}) - (3.25 \times 10^{3})$
(d) $118.45 - (0.033 \times 10^{3})$
(e) $1.00 + (3.75 \times 10^{-8})$

1.34 Perform the following computations, rounding the answers to the correct number of significant figures. All values are measured.
(a) $(341.7 - 22) + (0.00224 \times 814{,}005)$
(b) $(82.7 \times 143) + (274 - 0.00653)$
(c) $(3.53 \div 0.084) - (14.8 \times 0.046)$
(d) $(324 \times 0.0033) + (214.2 \times 0.0225)$
(e) $(4.15 + 82.3) \times (0.024 + 3.000)$

1.35 Perform the following computations, expressing the answers in scientific notation rounded to the proper number of significant figures. All values are from measurement.
(a) $(8.3 \times 10^{-6}) \times (4.13 \times 10^{-7}) \div (5.411 \times 10^{-12})$
(b) $[(3.125 \times 10^{-6}) + (5.127 \times 10^{-5})] \times (6.72 \times 10^{8})$
(c) $[14.39 + (2.43 \times 10^{1})] \div 1275$
(d) $[(1.583 \times 10^{-4}) - (0.00255)] \times [(142.3) + (0.257 \times 10^{2})]$
(e) $(0.0000425) \div [0.0008137 + (2.65 \times 10^{-5})]$

1.36 Perform the following conversions.
(a) 1.40 m to cm
(b) 2800 mm to m
(c) 185 ml to liters
(d) 18 g to kg
(e) 10 sq yards (10 yd^2) to m^2
(f) 100 miles to inches
(g) 20 ft/sec to mi/hr
(h) 1 mi^3 to m^3
(i) 40 mi/hr to cm/sec
(j) 25 liters to dm^3

1.37 Fill in the blank with the correct unit.
(a) 3.2 _____ $= 3.2 \times 10^{-9}$ m
(b) 42 _____ $= 4.2 \times 10^{-2}$ m
(c) 7.3 _____ $= 0.0073$ g
(d) 12.5 _____ $= 125$ mm
(e) 3.5 _____ $= 3.5 \times 10^{-3}$ ml
(f) 0.84 _____ $= 840$ cm^3

1.38 The newton (N) is the SI derived unit for force. We saw that force is mass times acceleration. Acceleration has dimensions of (distance) $\div$ (time)^2. What would the units for the newton be in terms of the SI base units? (A large lemon has a weight—its force caused by gravity—of about one newton.)

1.39 After shopping for a sports car, a certain wealthy freshman decided that she would choose between a new two-passenger, six-cylinder Smokebelcher, which gives 21 miles/gal, and an old 1974, eight-cylinder, 10-passenger Pferdburper (with automatic transmission), which the owner guaranteed would deliver 10 km/liter. On the basis of gas mileage, which car will be more economical to operate?

1.40 A student has just returned from Germany with a car that he purchased while on vacation. The speedometer is calibrated in kilometers per hour (km/hr). As he drives away from the pier, he notices a sign that posts the speed limit at 35 mph. What is the maximum speed that he can reach, in km/hr, without having to worry about receiving a speeding ticket?

1.41 After receiving a speeding ticket, the student in the preceding question is informed by the police officer that the courthouse is located "4.3 miles straight down the road. You can't miss it." The odometer in the student's car measures kilometers (km). How far must he travel, in km, before arriving at the courthouse?

*1.42 In the country of Ferdovia, the Ferds thrive on potatoes. The average Ferd earns 142 thrubs (the local currency of Ferdovia) per week and spends approximately $\frac{1}{14}$ of this yearly income on potatoes. If potatoes cost 2 thrubs per pound, how many pounds of potatoes does the average Ferd consume each year?

1.43 A block of magnesium had a mass of 14.3 g and a volume of 8.46 cm³. What is the density of magnesium?

1.44 Water was placed into a graduated cylinder until the volume read 25.0 ml. An irregularly shaped piece of metal weighing 50.8 g was placed in the cylinder and completely submerged. The water level rose to the 36.2-ml mark. What is the density of the metal?

*1.45 Titanium is an important structural metal used in aircraft because of its strength and "light weight." A solid cylinder of titanium 2.48 cm in diameter and 4.75 cm long was weighed and found to have a mass of 104.2 g. Calculate the density of titanium.

1.46 Lead is a well-known "heavy" metal. It has a density of 11.35 g/cm³.
(a) What is the mass of 12.0 cm³ of lead?
(b) What is the volume occupied by 155 g of lead?

1.47 Chloroform, $CHCl_3$, a liquid once used as an anesthetic, has a density of 1.492 g/ml.
(a) What is the volume of 10.00 g of $CHCl_3$?
(b) What is the mass of 10.00 ml of $CHCl_3$?

*1.48 A glass vessel that can be repeatedly filled with precisely the same volume of liquid is called a *pycnometer*. A certain pycnometer, when empty and dry, weighed 25.296 g. When filled with water at 25°C the pycnometer and water weighed 34.914 g. When filled with a liquid of unknown composition the pycnometer and its contents weighed 33.485 g. At 25°C the density of water is 0.9970 g/ml.
(a) What is the volume of the pycnometer?
(b) What is the density of the unknown liquid?

1.49 Suppose that you were going to prepare a table of densities for 100 substances in which the density of each substance is to be expressed in five sets of units. How many numerical values must appear in this table? How many numerical values would be needed to express this same information using specific gravities and the density of water in different units?

1.50 The density of water is 8.34 lb/gal. Isopropyl alcohol, sold in drug stores as rubbing alcohol, has a density of 6.56 lb/gal.
(a) What is the specific gravity of isopropyl alcohol?
(b) What is the density of isopropyl alcohol in the units g/ml?

1.51 Propylene glycol, a substance used in nontoxic antifreeze for freshwater systems of campers and boats, has a specific gravity of 1.04. Water has a density of 8.34 lb/gal. What would be the weight of the contents of a 10,000-gallon tank car filled with propylene glycol?

1.52 Cyclopropane, a very effective anesthetic, contains the elements carbon and hydrogen combined in a ratio of 1.0 g of hydrogen to 6.0 g of carbon. If a given sample of cyclopropane was found to contain 24 g of hydrogen, how many grams of carbon would it contain?

1.53 Copper forms two oxides. In one of them there are 1.26 g of oxygen combined with 10.0 g of copper. In the other there are 2.52 g of oxygen combined with 10.0 g of copper. Show that these data illustrate the law of multiple proportions.

1.54 If the atomic mass unit were defined in such a manner that a single fluorine atom weighed 1 amu, what would be the atomic weights of carbon and hydrogen?

1.55 A truck having a mass of 4500 kg is moving at a speed of 1.79 m/s. Calculate its kinetic energy in joules. How many calories does this correspond to?

*1.56 Calculate the kinetic energy, in joules, of a 145-lb athlete running at a speed of 15.3 miles/hr.

1.57 Gallium metal has one of the largest liquid ranges of any element. It melts at 30°C and boils at 1983°C. What are its melting and boiling points in degrees Fahrenheit?

*1.58 Two samples of Freon (a coolant used in refrigerators and air conditioners) were analyzed. In one sample, 1.00 g of carbon was found to be combined with 6.33 g of fluorine and 11.67 g of chlorine. In the second sample, 2.00 g of carbon were found to be combined with 12.66 g of fluorine and 23.34 g of chlorine. What are the ratios of the weights of carbon to fluorine, carbon to chlorine, and fluorine to chlorine in the two samples? Show that these data support the law of definite composition.

*1.59 Three samples of a solid substance composed of elements X and Y were prepared. The first was found to contain 4.31 g X and 7.69 g Y; the second was composed of 35.9% X and 64.1% Y; it was observed that 0.718 g X reacted with Y to form 2.00 g of the third sample. Show how these data demonstrate the law of definite composition.

*1.60 In Example 1.9, the first compound (2.62 g of N, 1.50 g O) has the formula N_2O (the substance is nitrous oxide). Suggest a possible formula for the second compound.

*1.61 Two compounds are formed between phosphorus and oxygen. 1.50 g of one compound was found to contain 0.845 g of phosphorus while a 2.50-g sample of the other contained 1.09 g of phosphorus. Show that these data are consistent with the law of multiple proportions.

*1.62 In a certain compound, 6.92 g of X were found combined with 0.584 g of carbon. If the atomic weight of carbon = 12.0 amu and if four atoms of X are combined with one atom of carbon, calculate the atomic weight of X.

1.63 Tungsten, used as filaments in electric light bulbs, has a melting point of 6152°F. What is its melting point expressed in degrees Celsius and in kelvins?

1.64 Solid carbon dioxide (Dry Ice) has a temperature of −78°C. What is its temperature on the Fahrenheit scale? What is its temperature on the Kelvin scale?

1.65 Normal human body temperature is 98.6°F. In Europe, and in many hospitals in the United States, the Celsius temperature scale is used almost universally so that clinical thermometers are calibrated in °C. If you had one of these thermometers, what temperature would you expect, in °C, for a normal healthy individual? If your temperature registered 39°C, what would your temperature be in °F?

1.66 How many calories are required to raise the temperature of 500 g of liquid water by 24°C? How many joules are required?

1.67 How much will the temperature of 150 g of liquid water change if it loses 35.0 cal? How much will its temperature change if it gains 40.0 J?

1.68 Calculate the kinetic energy, in joules and in kilojoules, of a 255-kg motorcycle traveling at 80.0 km/hr.

*1.69 An automobile weighing 2.4 tons is traveling at 35 miles/hr. What is its kinetic energy in joules? If all this energy were converted to heat, how much water could have its temperature raised by 10°C?

1.70 A car, whose mass is 1500 kg, skids to a halt from a speed of 60 m/s. How many calories of heat are generated by friction between the car and the pavement?

*1.71 A packing crate full of machinery weighs 1400 kg. It is moved a distance of 40 m by a person exerting a force of 510 N (1 N = 1 kg m/s²). How many joules of work have been accomplished?

*1.72 Naphthalene (used in mothballs) has a melting point of 80°C and a boiling point of 218°C. Suppose that this substance were used to define a new temperature scale on which the melting point of naphthalene was 0°N and its boiling point was 100°N. What would be the freezing point and boiling point of water in °N? What general equation could we use to relate temperatures in °C and °N?

*1.73 A copper penny weighing 3.14 g is heated to 100°C in boiling water. It is quickly dried and placed into a Styrofoam cup containing 10.00 g of water at 25.0°C. Copper has a specific heat of 0.0920 cal/g °C. What will be the final temperature of the penny and water? (Assume that a negligible amount of heat is absorbed by the cup.) *Hint:* The law of conservation of energy requires that the number of calories lost by the metal is equal to the number of calories gained by the water.

2

STOICHIOMETRY: CHEMICAL ARITHMETIC

The proper fuel-to-air mixture for a race car engine can be as important as the driver's skill in winning a race. Chemicals, like the fuel and oxygen of the air, always combine in definite fixed ratios. In this chapter we will learn how these ratios are determined and used.

In Chapter 1 you were introduced to many important basic concepts in chemistry: the ideas of atoms and atomic weight, elements and compounds, and several major chemical laws. Chemists, physicists, and biologists have developed these ideas so that today they routinely think of chemical, physical, and biological processes taking place between atoms and molecules on a submicroscopic, atomic scale. However, we cannot see individual atoms or molecules, and in the laboratory it is necessary to work with huge numbers of these small particles. In this chapter we will see that there are ways of carrying out chemical reactions so that observations on a large (macroscopic) scale can be readily translated into the language of the atomic world. That this can be done with relative ease is really quite amazing.

In order to study chemical compounds effectively in the laboratory, it is necessary to have a knowledge of the quantitative relationships that exist among the amounts of the substances that enter into chemical reactions. **Stoichiometry** (derived from the Greek *stoicheion* = element and *metron* = measure) is the term that we use to refer to all the quantitative aspects of chemical composition and reaction. We will now see how chemical formulas are determined and how chemical equations prove useful for predicting the proper amounts of reactants that must be mixed to get a complete reaction—one in which there is not an excess amount of any reactant.

2.1 THE MOLE

We know that atoms react to form molecules in simple whole-number ratios. Hydrogen and oxygen atoms, for instance, combine in a 2-to-1 ratio to form water, H_2O; carbon and oxygen atoms combine in a 1-to-1 ratio to form carbon monoxide, CO. However, it is impossible to work with individual atoms because they are so tiny. Therefore, in any real-life laboratory situation we must increase the size of these quantities to the point where we can see them and weigh them.

One way of enlarging the reaction is to work with dozens, instead of individual atoms.

$$1 \text{ atom C} + 1 \text{ atom O} \longrightarrow 1 \text{ molecule CO}$$

$$1 \text{ dozen C} + 1 \text{ dozen O} \longrightarrow 1 \text{ dozen CO}$$

$$(12 \text{ atoms C}) \quad (12 \text{ atoms O}) \quad (12 \text{ molecules CO})$$

Notice that the 1-to-1 ratio of dozens is exactly the same as the required 1-to-1 ratio of atoms. This means that if we had some way of counting atoms by the dozen, we could take dozens of them in the same ratio as the ratio by which the individual atoms react and be sure that there were enough atoms of each kind to react completely, with nothing left over. For instance, since 2 hydrogen atoms combine with 1 oxygen atom to form a water molecule,

$$2 \text{ atoms of H} + 1 \text{ atom of O} \longrightarrow 1 \text{ molecule } H_2O$$

we can scale up the reaction by taking 2 dozen atoms of H and 1 dozen atoms of O, or 6 dozen atoms of H and 3 dozen atoms of O, or *any* combination of dozens as long as their ratio (H to O) is 2 to 1. This will always give exactly the right numbers of individual H and O atoms to react completely, with no excess atoms of either H or O remaining. Unfortunately, a dozen atoms or molecules is still much too small to work with, so we must find a still larger unit. The "chemist's dozen" is called the **mole** (abbreviated **mol**). It is composed of 6.022×10^{23} objects (we will say more about the origin of this number, called **Avogadro's number,** later).

$$1 \text{ dozen} = 12 \text{ objects}$$

$$1 \text{ mol} = 6.022 \times 10^{23} \text{ objects}$$

The same reasoning that we can use with the dozen applies equally to the mole. The mole is simply a larger "box."

$$1 \text{ mol C} + 1 \text{ mol O} \longrightarrow 1 \text{ mol CO}$$
(6.022 × 10²³   (6.022 × 10²³   (6.022 × 10²³
atoms C)   atoms O)   molecules CO)

We see that when we take 1 mole of carbon and 1 mole of oxygen we have equal numbers of carbon and oxygen atoms and can construct enough CO molecules to fill our mole "box" exactly.

A very important thing to notice here is that the same whole-number ratios that apply to the individual atoms and molecules also apply exactly to the numbers of *moles* of atoms and molecules. Everything is simply increased by the same factor. For example, to form carbon tetrachloride, CCl_4, we know that

$$1 \text{ atom C} + 4 \text{ atoms Cl} \longrightarrow 1 \text{ molecule } CCl_4$$

We can immediately enlarge this to moles.

$$1 \text{ mol C} + 4 \text{ mol Cl} \longrightarrow 1 \text{ mol } CCl_4$$

To repeat this very important idea, the *ratio* by which moles of substances react is the same as the *ratio* by which their atoms and molecules react. This simple idea forms the basis for *all* quantitative chemical reasoning.

Students who have difficulty learning chemistry often have not learned to think problems through in terms of moles.

EXAMPLE 2.1   What mole ratio of carbon to chlorine must be chosen to prepare the substance C_2Cl_6 (hexachloroethane), which is a solvent used to manufacture explosives.

SOLUTION   The atom ratio in C_2Cl_6 is

$$\frac{2 \text{ atoms C}}{6 \text{ atoms Cl}} = \frac{1}{3}$$

To prepare C_2Cl_6 the carbon to chlorine *atom* ratio must be maintained at $1:3$. *The mole ratio must also be $1:3$.* Carbon and chlorine must be combined in a ratio of 1 mol of C to 3 mol of Cl.

The atom ratio in a chemical formula such as C_2Cl_6 establishes a number of mole ratios that are useful for constructing conversion factors that can be employed in solving problems. We've already seen one of these, which can be written as either

Of course, we could also use $\frac{2 \text{ mol C}}{6 \text{ mol Cl}}$ and $\frac{6 \text{ mol Cl}}{2 \text{ mol C}}$, without simplification.

$$\frac{1 \text{ mol C}}{3 \text{ mol Cl}} \quad \text{or} \quad \frac{3 \text{ mol Cl}}{1 \text{ mol C}} \qquad [2.1]$$

There are others as well. The formula tells us that 1 molecule of C_2Cl_6 contains 2 atoms of C and that it also contains 6 atoms of Cl. We can scale this up immediately to moles: 1 mol C_2Cl_6 contains 2 mol C and 6 mol Cl. This gives two equivalencies.

$$1 \text{ mol } C_2Cl_6 \sim 2 \text{ mol C}$$

$$1 \text{ mol } C_2Cl_6 \sim 6 \text{ mol Cl}$$

In other words, any time we have a mole of C_2Cl_6 there will be two moles of carbon in it, and any time we have a mole of C_2Cl_6 there will be six moles of chlorine in it. That's what we mean by an equivalence ($\sim$). As in Chapter 1, we can use these equivalencies to form conversion factors,

$$\frac{1 \text{ mol } C_2Cl_6}{2 \text{ mol } C} \quad \text{or} \quad \frac{2 \text{ mol } C}{1 \text{ mol } C_2Cl_6} \qquad [2.2]$$

and

$$\frac{1 \text{ mol } C_2Cl_6}{6 \text{ mol } Cl} \quad \text{or} \quad \frac{6 \text{ mol } Cl}{1 \text{ mol } C_2Cl_6} \qquad [2.3]$$

The following examples show how we use them.

EXAMPLE 2.2

How many moles of carbon are needed to combine with 4.87 mol chlorine to form the substance C_2Cl_6?

SOLUTION

We use one of the conversion factors relating moles of carbon to moles of chlorine in this compound, and we choose the factor to make mol Cl cancel.

$$4.87 \text{ mol } Cl \times \left(\frac{1 \text{ mol } C}{3 \text{ mol } Cl}\right) = 1.62 \text{ mol } C$$

We need 1.62 mol C. Notice that we are entitled to report three significant figures. The numbers in the mole ratio are exact numbers because the atom ratio is exact—atoms always combine in exact whole number ratios when they form compounds.

EXAMPLE 2.3

How many moles of carbon are in 2.65 mol C_2Cl_6?

SOLUTION

This time we use the equivalence,

$$1 \text{ mol } C_2Cl_6 \sim 2 \text{ mol } C$$

which can be used to construct a conversion factor to set up the arithmetic.

$$2.65 \text{ mol } C_2Cl_6 \times \left(\frac{2 \text{ mol } C}{1 \text{ mol } C_2Cl_6}\right) \sim 5.30 \text{ mol } C$$

Thus, 2.65 mol C_2Cl_6 contains 5.30 mol C.

A mole of atoms or molecules is a sufficiently large sample to work with experimentally. The mole, however, is a *chemical unit,* because moles react in the same ratios as atoms and molecules. We still must have a way to translate this to *laboratory units,* something that can be measured directly in the laboratory.

Earlier it was stated that a mole consists of 6.022×10^{23} objects. This rather odd number is chosen because this number of atoms of any element has a weight in grams that is precisely numerically equal to its atomic weight.[1] For example, the atomic weights of C and O are 12.011 amu and 15.9994 amu respectively. Therefore,

$$1 \text{ mol } C = 12.011 \text{ g } C$$

$$1 \text{ mol } O = 15.9994 \text{ g } O$$

[1] Historically, the mole was first defined as the amount of an element having a mass in grams numerically equal to its atomic weight. Later it was discovered experimentally that there are 6.022×10^{23} atoms in 1 mole of an element. We will see how Avogadro's number can be measured in later chapters.

Since there are equal numbers of atoms when the weights of elements are taken in proportion to their atomic weights, we can also speak of a *pound-mole* or a *ton-mole.* Thus, 12 lb of C (1 lb-mol C) contain the same number of atoms as 16 lb of O (1 lb-mol O) because the weights of C and O are taken in the same ratio as their atomic weights. Similarly, 12 tons of C (1 ton-mol C) contain the same number of atoms as 16 tons of O (1 ton-mol O). These are useful ideas when dealing with very large quantities of reactants in industrial processes.

1 mol copper (63.5g)

1 mol mercury (201g)

1 mol lead (207g)

1 mol iron (55.8g)

One mole of four different elements:—copper, iron, lead, and mercury. Each sample contains the same number of atoms.

A table of atomic weights gives us the number of grams in 1 mol of any element.

The balance serves as our tool for measuring moles. If we take 12.011 g C and 15.9994 g O, we know that we have taken 1 mol of C and 1 mol of O which can produce exactly 1 mol of CO. With this as background, let's look at some sample problems dealing with the mole.

EXAMPLE 2.4	How many moles of silicon, Si, are in 30.5 grams of Si? Silicon is an element used to make transistors.
SOLUTION	Our problem is one of converting the units of grams of Si to moles of Si, that is, 30.5 g Si = (?) mol Si. We know from the table of atomic weights that

$$1 \text{ mol Si} = 28.1 \text{ g Si}$$

To convert from g Si to moles we must multiply 30.5 g Si by a factor that contains the units g Si in the denominator, that is,

$$\frac{1 \text{ mol Si}}{28.1 \text{ g Si}}$$

When we set up this problem,

$$30.5 \text{ g Si} \times \left(\frac{1 \text{ mol Si}}{28.1 \text{ g Si}}\right) = 1.09 \text{ mol Si}$$

Thus 30.5 g Si = 1.09 mol Si.

EXAMPLE 2.5	How many grams of Cu are there in 2.55 mol of Cu?
SOLUTION	From the table of atomic weights,

$$1 \text{ mol Cu} = 63.5 \text{ g Cu}$$

Our conversion factor must have the units, mol Cu in the denominator, that is,

$$\frac{63.5 \text{ g Cu}}{1 \text{ mol Cu}}$$

Setting up the problem, we have

$$2.55 \text{ mol Cu} \times \left(\frac{63.5 \text{ g Cu}}{1 \text{ mol Cu}}\right) = 162 \text{ g Cu}$$

EXAMPLE 2.6

How many moles of Ca are required to react with 2.50 mol of Cl to produce the compound $CaCl_2$ (calcium chloride), used to melt ice on roads in the winter.

SOLUTION

Our problem: 2.50 mol Cl ~ (?) mol Ca.

We know from the formula that 1 atom of Ca combines with 2 atoms of Cl. Thus,

$$1 \text{ atom Ca} \sim 2 \text{ atoms Cl}$$

We also know that 1 mol Ca and 1 mol Cl contain the same number of atoms; therefore, to maintain a ratio of 1 atom of Ca to 2 atoms of Cl, we state that 1 mol of Ca combines with 2 mol of Cl.

$$1 \text{ mol Ca} \sim 2 \text{ mol Cl}$$

We obtain our answer as follows:

$$2.50 \text{ mol Cl} \times \left(\frac{1 \text{ mol Ca}}{2 \text{ mol Cl}}\right) \sim 1.25 \text{ mol Ca}$$

EXAMPLE 2.7

How many grams of Ca must react with 41.5 g Cl to produce $CaCl_2$?

SOLUTION

Our problem: 41.5 g Cl ~ (?) g Ca.
What quantities do we need and what do we know?
We know that

$$1 \text{ mol Ca} \sim 2 \text{ mol Cl} \qquad \text{(Why?)}$$

Also,

$$\left.\begin{array}{l} 1 \text{ mol Cl} = 35.5 \text{ g Cl} \\ 1 \text{ mol Ca} = 40.1 \text{ g Ca} \end{array}\right\} \text{ (atomic weight table)}$$

Here is the solution to the problem:

$$41.5 \text{ g Cl} \times \left(\frac{1 \text{ mol Cl}}{35.5 \text{ g Cl}}\right) = 1.17 \text{ mol Cl}$$

$$1.17 \text{ mol Cl} \times \left(\frac{1 \text{ mol Ca}}{2 \text{ mol Cl}}\right) \sim 0.585 \text{ mol Ca}$$

$$0.585 \text{ mol Ca} \times \left(\frac{40.1 \text{ g Ca}}{1 \text{ mol Ca}}\right) = 23.5 \text{ g Ca}$$

And here are all these steps written together.

$$41.5 \text{ g Cl} \times \left(\frac{1 \text{ mol Cl}}{35.5 \text{ g Cl}}\right) \times \left(\frac{1 \text{ mol Ca}}{2 \text{ mol Cl}}\right) \times \left(\frac{40.1 \text{ g Ca}}{1 \text{ mol Ca}}\right) \sim 23.5 \text{ g Ca}$$

to moles of Cl

to moles of Ca

to grams of Ca

Stringing together conversion factors makes the arithmetic simple to perform using a calculator.

In this last example we have strung together a series of conversion factors. When we set up the problem, we link these together by considering what units must be eliminated by cancellation. Thus, the first factor had to have g Cl in the denominator and took us as far as moles of Cl. The second factor had to have moles of Cl in the denominator and would give us moles of Ca if we stopped at that point. The third factor was required to have moles Ca in the denominator. At this point we stop since our units are now those of the answer, and we have only to perform the arithmetic to obtain the correct result.

EXAMPLE 2.8

What is the mass of 1 atom of calcium?

SOLUTION

Avogadro's number provides us with a means of relating the submicroscopic world of atoms and molecules to the large world of the laboratory. Our problem can be stated as

$$1 \text{ atom Ca} \sim (?) \text{ g Ca}$$

We have the relationships,

$$1 \text{ mol Ca} = 40.1 \text{ g Ca}$$

$$1 \text{ mol Ca} = 6.02 \times 10^{23} \text{ atoms Ca} \qquad \text{(using 3 significant figures)}$$

The solution to the problem can then be set up as

$$1 \text{ atom Ca} \times \left(\frac{1 \text{ mol Ca}}{6.02 \times 10^{23} \text{ atoms Ca}}\right) \times \left(\frac{40.1 \text{ g Ca}}{1 \text{ mol Ca}}\right) \sim 6.66 \times 10^{-23} \text{ g Ca}$$

2.2 MOLECULAR WEIGHTS AND FORMULA WEIGHTS

As with elements, the balance can also be used to measure moles of compounds. The simplest way of obtaining the weight of one mole of a substance is merely to add up the atomic weights of all the elements present in the com-

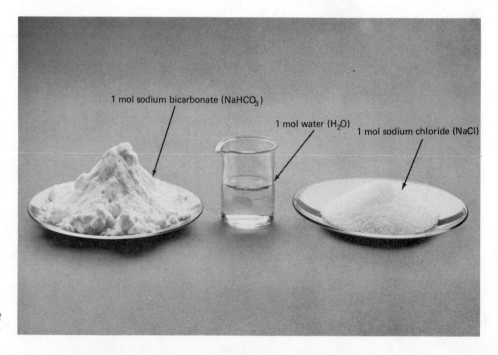

1 mol sodium bicarbonate (NaHCO$_3$)

1 mol water (H$_2$O)

1 mol sodium chloride (NaCl)

One mole of three different compounds. Each contains the same number of formula units, although the number of atoms is different from sample to sample. This is because the number of atoms per formula unit is different for each compound.

pound. If the substance is composed of molecules (for example, CO_2, H_2O, or NH_3), the sum of the atomic weights is called the **molecular weight.** Thus the molecular weight of CO_2 is obtained as

$$
\begin{array}{llll}
C & 1 \times 12.0 \text{ amu} & = & 12.0 \text{ amu} \\
2O & 2 \times 16.0 \text{ amu} & = & \underline{32.0 \text{ amu}} \\
CO_2 & \text{total} & & 44.0 \text{ amu}
\end{array}
$$

Similarly, the molecular weight of $H_2O = 18.0$ amu and that of $NH_3 = 17.0$ amu. The weight of 1 mol of a substance is obtained simply by writing its molecular weight followed by the units, grams. Thus,

$$1 \text{ mol } H_2O = 18.0 \text{ g}$$

$$1 \text{ mol } NH_3 = 17.0 \text{ g}$$

In later chapters we will encounter many compounds that do not contain separate, distinct molecules. We will find that when certain atoms react, they often gain or lose negatively charged particles called **electrons.** Sodium and chlorine happen to react this way. When sodium chloride, NaCl, is formed from the elements, each Na atom loses one electron and each Cl atom gains one. Since Na and Cl are electrically neutral to start, these atoms acquire a charge when NaCl is formed. These are written as Na^+ (positive because Na has lost a negatively charged electron) and Cl^- (negative because Cl has gained an electron). *Atoms or groups of atoms that have acquired an electrical charge are called* **ions.** Since solid NaCl is composed of Na^+ and Cl^- ions, it is said to be an **ionic compound.**

This entire topic is explored further in Chapters 3 and 4. For now, it is only necessary for you to know that compounds that are ionic do not contain molecules. Their formulas simply state the ratio of the different atoms in the substance. In NaCl the atoms are in a one-to-one ratio. In the ionic compound $CaCl_2$ the ratio of Ca to Cl atoms is 1 to 2 (relax—at this point you were not expected to know that $CaCl_2$ is ionic). Rather than refer to molecules of NaCl or $CaCl_2$, we use the term **formula unit** to specify the pair of ions (Na^+ and Cl^-) in NaCl, or the set of three ions in $CaCl_2$.

For ionic compounds the sum of the atomic weights of the elements present in one formula unit is known as the **formula weight.** For NaCl this is $22.99 + 35.45 = 58.44$. One mole of NaCl would contain 58.44 g NaCl (6.022×10^{23} formula units of NaCl). It is also true that 6.022×10^{23} formula units (1 mol) of NaCl contains 6.022×10^{23} Na^+ ions and 6.022×10^{23} Cl^- ions. Use of the term formula weight, of course, is not restricted to ionic compounds. It can also be applied to molecular substances, in which case the terms formula weight and molecular weight mean the same thing.

EXAMPLE 2.9   Sodium carbonate, Na_2CO_3, is a very important industrial chemical used in making glass.

(a) How many grams does 0.250 mol Na_2CO_3 weigh?
(b) How many moles of Na_2CO_3 are in 132 g Na_2CO_3?

SOLUTION   To answer these questions we need the formula weight of Na_2CO_3. We calculate this from the atomic weights of its elements:

$$
\begin{array}{lll}
2 \text{ Na} & 2 \times 23.0 = & 46.0 \\
1 \text{ C} & 1 \times 12.0 = & 12.0 \\
3 \text{ O} & 3 \times 16.0 = & \underline{48.0} \\
& \text{Total} & 106.0
\end{array}
$$

The formula weight is 106.0; therefore,

$$1 \text{ mol } Na_2CO_3 = 106.0 \text{ g } Na_2CO_3$$

This can be used to make conversion factors relating grams and moles of Na_2CO_3, which we need to answer the questions.

(a) To convert 0.250 mol Na_2CO_3 to grams we set up the units to cancel.

$$0.250 \cancel{\text{ mol } Na_2CO_3} \times \left(\frac{106.0 \text{ g } Na_2CO_3}{1 \cancel{\text{ mol } Na_2CO_3}}\right) = 26.5 \text{ g } Na_2CO_3$$

(b) Again, we set the units to cancel.

$$132 \cancel{\text{ g } Na_2CO_3} \times \left(\frac{1 \text{ mol } Na_2CO_3}{106.0 \cancel{\text{ g } Na_2CO_3}}\right) = 1.25 \text{ mol } Na_2CO_3$$

2.3 PERCENTAGE COMPOSITION

A very simple and also often very useful computation is the calculation of the **percentage composition** of a compound—that is, the percentage of the total mass (also called the percent by weight) contributed by each element. The procedure to determine the percent composition is illustrated in Example 2.10.

EXAMPLE 2.10   What is the percentage composition of chloroform, $CHCl_3$, a substance once used as an anesthetic?

SOLUTION   The total mass of 1 mol of $CHCl_3$ is obtained from the molecular weight,

$$(12.01 + 1.008 + 3 \times 35.45) \text{ amu} = 119.37 \text{ amu}$$

or,

$$(12.01 \text{ g} + 1.008 \text{ g} + 3 \times 35.45 \text{ g}) = 119.37 \text{ g}$$

In general:

$$\% \text{ by weight} = \frac{\text{weight of part}}{\text{weight of whole}} \times 100$$

$$\% \text{ C} = \frac{\text{weight carbon}}{\text{weight } CHCl_3} \times 100$$

$$\% \text{ C} = \frac{12.01 \text{ g C}}{119.37 \text{ g } CHCl_3} \times 100 = 10.06\% \text{ C}$$

$$\% \text{ H} = \frac{1.008 \text{ g H}}{119.37 \text{ g } CHCl_3} \times 100 = 0.844\% \text{ H}$$

$$\% \text{ Cl} = \frac{106.35 \text{ g Cl}}{119.37 \text{ g } CHCl_3} \times 100 = 89.09\% \text{ Cl}$$

$$\text{total } \% = 100.00$$

A similar calculation can be used to determine the mass of an element in a given sample of a compound. This is illustrated in Example 2.11.

EXAMPLE 2.11   Calculate the weight of iron (Fe) in a 10.0-g sample of rust, Fe_2O_3.

SOLUTION   To solve this problem we can first calculate the fraction of Fe_2O_3 that is Fe. From the formula we see that one formula unit of Fe_2O_3 contains 2 atoms of Fe. Therefore we know that 1 mol of Fe_2O_3 contains 2 mol of Fe. Also,

$$1 \text{ mol Fe} = 55.85 \text{ g Fe}$$

$$1 \text{ mol Fe}_2O_3 = 159.7 \text{ g Fe}_2O_3$$

(Remember, we can always calculate the weight of 1 mol of a substance if we know its chemical formula!) Therefore 1 mol Fe_2O_3 (159.7 g Fe_2O_3) contains 2 mol Fe (111.7 g Fe). The fraction of Fe_2O_3 that is Fe is thus

$$\frac{\text{weight of Fe}}{\text{weight of Fe}_2O_3} = \frac{111.7 \text{ g Fe}}{159.7 \text{ g Fe}_2O_3}$$

note × by 2

Finally, the weight of Fe in the 10.0-g sample Fe_2O_3 is equal to the sample weight multiplied by the fraction of the sample that is Fe.

$$10.0 \text{ g Fe}_2O_3 \times \left(\frac{111.7 \text{ g Fe}}{159.7 \text{ g Fe}_2O_3}\right) \sim 6.99 \text{ g Fe}$$

2.4 CHEMICAL FORMULAS

A formula conveys to us certain kinds of information, including elemental composition, relative numbers of each kind of atom present, the actual numbers of each kind of atom in a molecule of the substance, or the structure of the compound. We can classify formulas according to the amount of information they provide.

A formula that simply gives the relative number of atoms of each element present in a formula unit is called a **simplest formula**. It is also called an **empirical formula** because it is normally derived from the results of some experimental analysis. The formulas NaCl, H_2O, and CH_2 are empirical formulas.

A formula that states the actual number of each kind of atom found in a molecule is called a **molecular formula**. H_2O is a molecular formula (as well as an empirical formula) since a molecule of water contains 2 atoms of H and 1 atom of O. The formula C_2H_4 is a molecular formula for a substance (ethylene) containing 2 atoms of carbon and 4 atoms of hydrogen. Note that the simplest formula of this compound is CH_2 because the carbon-to-hydrogen ratio is 1:2. A substance whose empirical formula is CH_2 could have as a molecular formula CH_2, C_2H_4, C_3H_6, and so on. Molecular formulas for ionic substances do not exist, of course, because such compounds do not contain molecules.

A third type of formula is a **structural formula,** for example,

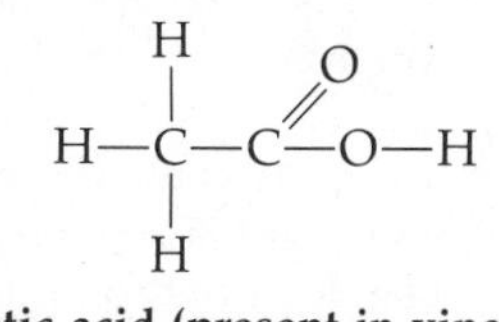

$$\begin{array}{ccccc} & & H & & O \\ & & | & & \parallel \\ H & - & C & - & C & - O - H \\ & & | & & \\ & & H & & \end{array}$$

acetic acid (present in vinegar)

In a structural formula the dashes between the different atomic symbols represent the "chemical bonds" that hold the atoms together in the molecule. We will see more of them in Chapter 4. A structural formula gives us information about the way in which the atoms in a molecule are linked together, and provides information that allows us to write the molecular and empirical formulas. Thus, for acetic acid shown above we can also write its molecular formula ($C_2H_4O_2$) and its empirical formula (CH_2O).

The most desirable kind of formula to have, of course, is the structural formula since it also contains all the information provided by the other two types. However, in chemistry, as in the rest of life, we never get something for nothing. The more information a formula conveys, the more difficult it is to arrive at experimentally. We will see how empirical and molecular formulas are derived; however, most of the procedures involved for the determination of structural formulas are beyond the scope of this book.

2.5 EMPIRICAL FORMULAS

Since the simplest formula gives the relative numbers of atoms present in a compound, it must also give the relative number of moles of each element. Therefore, to obtain the empirical formula for a compound we must determine the number of moles of each of its elements that are present in a particular sample. Then we can calculate the simplest whole-number ratio of the moles to find the subscripts. The following examples illustrate how this is done.

EXAMPLE 2.12

A sample of a brown-colored gas that is a major air pollutant is found to contain 2.34 g of N and 5.34 g of O. What is the simplest formula of the compound?

SOLUTION

We proceed by calculating the number of moles of each element present. We know that

$$1 \text{ mol N} = 14.0 \text{ g N} \qquad \text{(Why?)}$$

$$1 \text{ mol O} = 16.0 \text{ g O} \qquad \text{(Why?)}$$

Therefore,

$$2.34 \text{ g N} \times \left(\frac{1 \text{ mol N}}{14.0 \text{ g N}}\right) = 0.167 \text{ mol N}$$

$$5.34 \text{ g O} \times \left(\frac{1 \text{ mol O}}{16.0 \text{ g O}}\right) = 0.334 \text{ mol O}$$

We might write our formula $N_{0.167}O_{0.334}$. It does indeed tell us the relative number of moles of N and O; however, since the formula should have meaning on a molecular level where whole numbers of atoms are combined, the subscripts must be integers. If we divide each subscript by the smallest one, we obtain

$$N_{\frac{0.167}{0.167}}O_{\frac{0.334}{0.167}} = NO_2$$

EXAMPLE 2.13

What is the empirical formula of a compound composed of 43.7% P and 56.3% O by weight?

SOLUTION

It is quite common to have a chemical analysis in the form of percentage composition by weight. The simplest way to proceed in such a case is to imagine having a 100-g sample of the compound. From the analysis, this sample would contain 43.7 g P and 56.3 g O (notice that the percents become grams of compound). Now that we know the weights of phosphorus and oxygen in the same sample, we convert the weights to moles and proceed as before.

$$43.7 \text{ g P} \times \left(\frac{1 \text{ mol P}}{31.0 \text{ g P}}\right) = 1.41 \text{ mol P}$$

$$56.3 \text{ g O} \times \left(\frac{1 \text{ mol O}}{16.0 \text{ g O}}\right) = 3.52 \text{ mol O}$$

Our formula is

$$P_{1.41}O_{3.52} = P_{\frac{1.41}{1.41}}O_{\frac{3.52}{1.41}} = PO_{2.50}$$

Whole numbers may be obtained by doubling each of these values. Thus the empirical formula is P_2O_5.

EXAMPLE 2.14

A 0.1000-g sample of ethyl alcohol (grain alcohol), known to contain only carbon, hydrogen, and oxygen, was allowed to completely react with oxygen to produce the products CO_2 and H_2O. These products were trapped separately and weighed. 0.1910 g of CO_2 and 0.1172 g of H_2O were found. What is the empirical formula of the compound?

SOLUTION

To many students this problem at first appears impossible; however, let us consider what we know and what we can calculate.

Following the same procedure used in Example 2.11, we can compute the weight of carbon and the weight of hydrogen in the CO_2 and H_2O that were formed from the compound when it reacted. Since the only source of C and H was the original compound, the difference between the weight of the compound taken (0.1000 g) and the total weight of carbon and hydrogen must be the weight of oxygen in the original 0.1000 g. (We must obtain the weight of oxygen in this way, rather than from the total weight of oxygen in the H_2O and CO_2, since only a portion of the oxygen in the products came from the original compound.) Once we know the weights of carbon, hydrogen, and oxygen in the 0.1000 g-sample, we can calculate how many moles there are of each in the 0.1000 g and, hence, the empirical formula of the unknown compound. Thus, having planned our course of action, we now proceed with the computations. The formula weights of CO_2 and H_2O are 44.0 and 18.0, respectively. The fraction of the mass of CO_2 that is carbon is

In order to calculate the mole ratios, we must know the amounts of C, H, and O in the same sized sample.

$$\frac{12.0 \text{ g C}}{44.0 \text{ g } CO_2}$$

Likewise, the fraction of H_2O that is hydrogen is equal to

$$\frac{2.01 \text{ g H}}{18.0 \text{ g } H_2O}$$

The weight of carbon in the original compound is equal to the weight of CO_2 multiplied by the fraction of the weight that is due to carbon.

$$0.1910 \text{ g } CO_2 \times \left(\frac{12.0 \text{ g C}}{44.0 \text{ g } CO_2}\right) \sim 0.0521 \text{ g C}$$

Similarly,

$$0.1172 \text{ g } H_2O \times \left(\frac{2.01 \text{ g H}}{18.0 \text{ g } H_2O}\right) \sim 0.0131 \text{ g H}$$

The total weight contributed by the carbon and hydrogen in the sample is

$$0.0521 \text{ g C} + 0.0131 \text{ g H} = 0.0652 \text{ g}$$

The weight of oxygen = 0.1000 g − 0.0652 g = 0.0348 g O. Next, we calculate the number of moles of C, H, and O.

$$0.0521 \text{ g C} \times \left(\frac{1 \text{ mol C}}{12.0 \text{ g C}}\right) = 4.34 \times 10^{-3} \text{ mol C}$$

Similar calculations for hydrogen and oxygen give 1.31×10^{-2} mol H and 2.17×10^{-3} mol O. The empirical formula is therefore

$$C_{0.00434}H_{0.0131}O_{0.00217} = C_{\frac{0.00434}{0.00217}}H_{\frac{0.0131}{0.00217}}O_{\frac{0.00217}{0.00217}}$$

or

$$C_2H_6O$$

2.6 MOLECULAR FORMULAS

Ionic compounds do not have molecular formulas because they do not contain molecules.

Not only does the molecular formula provide the information contained in the empirical formula, but it also tells us how many atoms of each element are present in a molecule of a substance. Remember that an empirical formula of CH_2 is found for any molecule that possesses twice as many hydrogen atoms as carbon atoms. To distinguish among all the possible choices, we require the molecular weight of the compound. This is because *the molecular weight is always an integral multiple of the empirical formula weight* (see Table 2.1). To find the number of times that the empirical formula is repeated in the molecular formula, we simply divide the experimentally determined molecular weight[2] by the empirical formula weight.

EXAMPLE 2.15

A colorless liquid used in rocket engines, whose empirical formula is NO_2, has a molecular weight of 92.0. What is its molecular formula?

SOLUTION

The formula weight of NO_2 is 46.0. The number of times the empirical formula, NO_2, occurs in the compound is

$$\frac{92.0}{46.0} = 2$$

The molecular formula is then $(NO_2)_2 = N_2O_4$ (dinitrogen tetroxide).

N_2O_4 is the preferred answer because $(NO_2)_2$ implies a knowledge of the structure of the molecule (i.e., that two NO_2 units are somehow joined together).

2.7 BALANCING CHEMICAL EQUATIONS

Recall that a chemical equation is a shorthand description of the changes that occur during a chemical reaction—a sort of before-and-after picture of what happens. One of the most useful properties of a chemical equation is that it allows us to determine the quantitative relationships that exist among reactants and products. To be helpful in this way, however, the equation must be balanced; that is, it must obey the law of conservation of mass by having the same number of atoms of each kind on both sides of the arrow.

In order to minimize errors, writing a balanced chemical equation should always be considered a two-step process.

1. First write an unbalanced equation with correct formulas for all the reactants and products. At this time you're not expected to be able to write formulas for compounds, so they will be given to you. We will discuss how to write formulas later in the book.
2. Balance the equation by adjusting the coefficients that precede the formulas. *During this step you are not permitted to change the subscripts in any of the formulas!* To do so would change the nature of the substances.

There is never any excuse for having an improperly balanced equation, since it is always possible, by counting atoms on each side of the equation, to determine whether the equation is, in fact, balanced.

Most simple chemical equations can be easily balanced by inspection. This involves examining the equation and adjusting the coefficients until equal numbers of each element are present among both the reactants and products.

Table 2.1
Molecular weights as multiples of the empirical formula weight

Formula	Molecular Weight
CH_2	$14.0 = 1 \times 14.0$
C_2H_4	$28.0 = 2 \times 14.0$
C_3H_6	$42.0 = 3 \times 14.0$
C_4H_8	$56.0 = 4 \times 14.0$
C_nH_{2n}	$n \times 14.0$

---

[2] We will see how molecular weights may be determined in Chapters 7 and 10.

Hydrochloric acid is added to a solution of sodium carbonate.

For example, consider the reaction of sodium carbonate (Na_2CO_3) with hydrochloric acid (HCl) to produce sodium chloride (NaCl), carbon dioxide (CO_2), and water.

To obtain a properly balanced chemical equation we proceed as follows:

1. We write the unbalanced equation,

$$Na_2CO_3 + HCl \longrightarrow NaCl + H_2O + CO_2$$

2. Coefficients are introduced to balance the equation. This is where you will probably need some practice so that you can learn to balance an equation quickly. Although there are no set rules to tell you where to start, it is often best to seek out the most complex formula in the equation and begin there. In this case, we start with the Na_2CO_3. Since there are two Na atoms in this formula on the left, there must also be two Na atoms on the right, so place a 2 in front of the formula NaCl. This gives us

$$Na_2CO_3 + HCl \longrightarrow 2NaCl + H_2O + CO_2$$

Now there are two Cl atoms on the right but only one on the left, so we place a 2 in front of the HCl.

$$Na_2CO_3 + 2HCl \longrightarrow 2NaCl + H_2O + CO_2$$

A fast inspection reveals that this equation is now balanced.

Balancing an equation by inspection proceeds partly by trial and error, and becomes easier with practice.

It was stated above that a balanced equation obeys the law of conservation of mass. For the equation that we have just balanced, and in fact for any equation, there are an infinite number of sets of coefficients that fulfill this requirement. Thus, the equations

$$\tfrac{1}{2}Na_2CO_3 + HCl \longrightarrow NaCl + \tfrac{1}{2}CO_2 + \tfrac{1}{2}H_2O$$

$$5Na_2CO_3 + 10HCl \longrightarrow 10NaCl + 5CO_2 + 5H_2O$$

are also balanced. The usual practice, however, is to use the smallest possible set of whole-number coefficients, although there are occasions, as we will see, where other choices are advantageous.

EXAMPLE 2.16   Balance the following equation for the combustion of octane, C_8H_{18}, a component of gasoline:

$$C_8H_{18} + O_2 \longrightarrow CO_2 + H_2O$$

SOLUTION   Inspection of the equation quickly suggests that we must adjust the coefficients preceding CO_2 and H_2O to balance C and H. Carbon can be balanced by placing an 8 before the CO_2; hydrogen can be balanced by placing a 9 before the H_2O ($9H_2O$ contains 18 H atoms because each H_2O contains 2 H atoms). This gives

$$C_8H_{18} + O_2 \longrightarrow 8CO_2 + 9H_2O$$

Now we can work on the oxygen. On the right there are 25 O atoms ($2 \times 8 + 9 = 25$). On the left the O atoms come in pairs. This means that we must have $12\tfrac{1}{2}$ pairs (O_2 molecules) to have 25 O atoms on the left. This gives us

$$C_8H_{18} + 12\tfrac{1}{2}O_2 \longrightarrow 8CO_2 + 9H_2O$$

Finally, we can eliminate the fractional coefficient by doubling each coefficient.

$$2C_8H_{18} + 25O_2 \longrightarrow 16CO_2 + 18H_2O$$

2.8 CALCULATIONS BASED ON CHEMICAL EQUATIONS

A chemical equation can be interpreted in several ways. Consider, for example, the balanced equation for the combustion of ethanol, C_2H_5OH, the alcohol used in gasohol.

$$C_2H_5OH + 3O_2 \longrightarrow 2CO_2 + 3H_2O$$

On a molecular, submicroscopic level we can view this as the reaction between individual molecules.

1 molecule C_2H_5OH + 3 molecules $O_2 \longrightarrow$ 2 molecules CO_2 + 3 molecules H_2O

But we can just as easily scale this up to lab-sized amounts because moles react in the same ratios as the individual molecules.

1 mol C_2H_5OH + 3 mol $O_2 \longrightarrow$ 2 mol CO_2 + 3 mol H_2O

The key point is that *the coefficients in a chemical equation provide the ratios in which moles of one substance react with or form moles of another.* The coefficients provide us with a whole host of different chemical equivalencies that can be used to form conversion factors. In this example, we can write six such equivalencies:

$$1 \text{ mol } C_2H_5OH \sim 3 \text{ mol } O_2$$

$$1 \text{ mol } C_2H_5OH \sim 2 \text{ mol } CO_2$$

$$1 \text{ mol } C_2H_5OH \sim 3 \text{ mol } H_2O$$

$$3 \text{ mol } O_2 \sim 2 \text{ mol } CO_2$$

$$3 \text{ mol } O_2 \sim 3 \text{ mol } H_2O$$

$$2 \text{ mol } CO_2 \sim 3 \text{ mol } H_2O$$

Thus, the balanced equation provides quantitative relationships among all the reactants and products. Let's look at a few sample calculations based on the reaction for the combustion of ethanol.

EXAMPLE 2.17 How many moles of oxygen are needed to burn 1.80 mol C_2H_5OH according to the balanced equation,

$$C_2H_5OH + 3O_2 \longrightarrow 2CO_2 + 3H_2O$$

SOLUTION The coefficients of the equation give us the relationship

$$1 \text{ mol } C_2H_5OH \sim 3 \text{ mol } O_2$$

which we can use for a conversion factor. We set up the arithmetic so the units mol C_2H_5OH cancel.

$$1.80 \text{ mol } C_2H_5OH \times \left(\frac{3 \text{ mol } O_2}{1 \text{ mol } C_2H_5OH}\right) = 5.40 \text{ mol } O_2$$

We need 5.40 mol O_2.

EXAMPLE 2.18 How many moles of CO_2 will be formed when 0.274 mol C_2H_5OH burns?

SOLUTION Now we look at the coefficients for C_2H_5OH and CO_2, which give us

$$1 \text{ mol } C_2H_5OH \sim 2 \text{ mol } CO_2$$

Then we set up the arithmetic to get the proper units in the answer.

$$0.274 \ \text{mol } C_2H_5OH \times \left(\frac{2 \ \text{mol } CO_2}{1 \ \text{mol } C_2H_5OH}\right) = 0.548 \ \text{mol } CO_2$$

EXAMPLE 2.19   How many moles of water will form when 3.66 mol CO_2 are produced during the combustion of C_2H_5OH?

SOLUTION   The coefficients in the equation for the combustion of C_2H_5OH tell us that

$$2 \ \text{mol } CO_2 \sim 3 \ \text{mol } H_2O$$

Therefore,

$$3.66 \ \text{mol } CO_2 \times \left(\frac{3 \ \text{mol } H_2O}{2 \ \text{mol } CO_2}\right) = 5.49 \ \text{mol } H_2O$$

When we actually carry out a chemical reaction in the lab, we usually are interested in laboratory units—grams. But we can always translate from grams to moles or moles to grams if we have the chemical formulas of the substances involved. We learned to do that in Section 2.2.

EXAMPLE 2.20   Freshly exposed aluminum surfaces react with oxygen to form a tough oxide coating that protects the metal from further corrosion. The reaction is

$$4Al + 3O_2 \longrightarrow 2Al_2O_3$$

How many grams of O_2 are required to react with 0.300 mol of Al?

SOLUTION   Our problem can be stated simply as

$$0.300 \ \text{mol Al} \sim (?) \ \text{g } O_2$$

To solve it, we first convert moles of Al to moles of O_2. Then we can change moles of O_2 to grams of O_2. According to the balanced equation,

$$4 \ \text{mol Al} \sim 3 \ \text{mol } O_2$$

Therefore, the number of moles of O_2 required is

$$0.300 \ \text{mol Al} \times \left(\frac{3 \ \text{mol } O_2}{4 \ \text{mol Al}}\right) \sim 0.225 \ \text{mol } O_2$$

To convert moles of O_2 to grams of O_2 we use the relationship

$$1 \ \text{mol } O_2 = 32.0 \ \text{g } O_2$$

$$0.225 \ \text{mol } O_2 \times \left(\frac{32.0 \ \text{g } O_2}{1 \ \text{mol } O_2}\right) = 7.20 \ \text{g } O_2$$

We could have combined these two steps.

$$0.300 \ \text{mol Al} \times \left(\frac{3 \ \text{mol } O_2}{4 \ \text{mol Al}}\right) \times \left(\frac{32.0 \ \text{g } O_2}{1 \ \text{mol } O_2}\right) \sim 7.20 \ \text{g } O_2$$

EXAMPLE 2.21 From the chemical equation in the last example, calculate the number of grams of Al_2O_3 that could be produced if 12.5 g of O_2 completely react with aluminum.

SOLUTION Our problem:

$$12.5 \text{ g } O_2 \sim (?) \text{ g } Al_2O_3$$

The solution to this problem requires that we establish the chemical equivalence between O_2 and Al_2O_3. This kind of equivalence is *always* found from the coefficients in the balanced equation and is expressed in our chemical units, moles. From the equation

$$3 \text{ mol } O_2 \sim 2 \text{ mol } Al_2O_3$$

We now need the relationships permitting us to translate between laboratory units, grams, and chemical units, moles. These are

$$1 \text{ mol } O_2 = 32.0 \text{ g } O_2$$

$$1 \text{ mol } Al_2O_3 = 102 \text{ g } Al_2O_3$$

These three relationships can now be applied as conversion factors. We arrange them so that the proper units cancel.

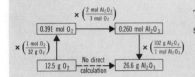

$$12.5 \text{ g } O_2 \times \left(\frac{1 \text{ mol } O_2}{32.0 \text{ g } O_2}\right) \times \left(\frac{2 \text{ mol } Al_2O_3}{3 \text{ mol } O_2}\right) \times \left(\frac{102 \text{ g } Al_2O_3}{1 \text{ mol } Al_2O_3}\right) \sim 26.6 \text{ g } Al_2O_3$$

2.9 LIMITING-REACTANT CALCULATIONS

If arbitrary quantities of reactants are chosen when a chemical reaction is carried out, it is quite likely that one of the reactants will be completely consumed before the others are used up. For example, if 5 mol of H_2 and 1 mol of O_2 are mixed and allowed to react following the equation

$$2H_2 + O_2 \longrightarrow 2H_2O$$

After the reaction is finished, there will be 3 mol of H_2 left over.

only 2 mol of H_2 will react, completely consuming the 1 mol of O_2. After the O_2 is gone, no further reaction can occur and no further product can be formed. The amount of product is therefore limited by the reactant that disappears first. The reactant that disappears first is called the **limiting reactant.** The following two examples illustrate how we can determine which is the limiting reactant and then calculate how much product is formed.

EXAMPLE 2.22 Zinc and sulfur react to form zinc sulfide, a substance used in phosphors that coat the inner surfaces of TV picture tubes. The equation for the reaction is

$$Zn + S \longrightarrow ZnS$$

How many grams of ZnS can be formed when 12.0 g of Zn are allowed to react with 6.50 g of S? Which is the limiting reactant? How much of which element will remain unreacted?

SOLUTION The equation gives us the mole relationship

$$1 \text{ mol } Zn \sim 1 \text{ mol } S$$

To determine the limiting reactant we first convert our reactant quantities to moles.

$$12.0 \; \text{g Zn} \times \left(\frac{1 \text{ mol Zn}}{65.4 \text{ g Zn}}\right) = 0.183 \text{ mol Zn}$$

$$6.50 \; \text{g S} \times \left(\frac{1 \text{ mol S}}{32.1 \text{ g S}}\right) = 0.202 \text{ mol S}$$

0.202 mol S requires 0.202 mol Zn. Since we only have 0.183 mol Zn, all the Zn will be used up, so Zn is the limiting reactant.

Because these elements combine in a 1-to-1 mole ratio, 0.183 mol Zn requires 0.183 mol S. We see that there is more S than is required and that all of the Zn is able to react. *Zn is therefore the limiting reactant.*

The amount of product formed depends only on the amount of limiting reactant. After the Zn has been used up, no more ZnS can form, so we use the amount of Zn (0.183 mol) to figure out the amount of ZnS that is produced. From the equation, we can write

$$1 \text{ mol Zn} \sim 1 \text{ mol ZnS}$$

Therefore, 0.183 mol Zn will form 0.183 mol ZnS. The weight of product is

$$0.183 \; \text{mol ZnS} \times \left(\frac{97.5 \text{ g ZnS}}{1 \text{ mol ZnS}}\right) = 17.8 \text{ g ZnS}$$

To calculate the weight of unreacted sulfur, we first subtract the number of moles of sulfur that reacted from the number of moles of sulfur initially available.

$$\text{moles S unreacted} = (0.202 - 0.183) = 0.019 \text{ mol S}$$

Now we convert to grams.

$$0.019 \; \text{mol S} \times \left(\frac{32.1 \text{ g S}}{1 \text{ mol S}}\right) = 0.61 \text{ g S} \quad \text{left over}$$

EXAMPLE 2.23

Ethylene, C_2H_4, burns in air to form CO_2 and H_2O according to the equation

$$C_2H_4 + 3O_2 \longrightarrow 2CO_2 + 2H_2O$$

How many grams of CO_2 will be formed when a mixture containing 1.93 g C_2H_4 and 5.92 g O_2 is ignited?

SOLUTION

The relationship between C_2H_4 and O_2, as specified by the chemical equation, is in units of moles.

$$1 \text{ mol } C_2H_4 \sim 3 \text{ mol } O_2$$

To solve our problem we first convert our given quantities to moles.

$$1.93 \; \text{g } C_2H_4 \times \left(\frac{1 \text{ mol } C_2H_4}{28.0 \text{ g } C_2H_4}\right) = 0.0689 \text{ mol } C_2H_4 \text{ available}$$

$$5.92 \; \text{g } O_2 \times \left(\frac{1 \text{ mol } O_2}{32.0 \text{ g } O_2}\right) = 0.185 \text{ mol } O_2 \text{ available}$$

Let's see now whether there is sufficient O_2 to react with all the C_2H_4 (alternatively, we could see if there is enough C_2H_4 to react with all the O_2).

0.185 mol O_2 requires 0.0617 mol C_2H_4. Since there is excess C_2H_4, the limiting reactant is O_2.

$$0.0689 \; \text{mol } C_2H_4 \times \left(\frac{3 \text{ mol } O_2}{1 \text{ mol } C_2H_4}\right) \sim 0.207 \text{ mol } O_2 \text{ needed to consume all the } C_2H_4$$

The calculation tells us that 0.207 mol of O_2 is required, but only 0.185 mol of O_2 is available. Therefore, O_2 is the limiting reactant.

We now use the limiting reactant to calculate the amount of product formed.

$$0.185 \; \text{mol } O_2 \times \left(\frac{2 \text{ mol } CO_2}{3 \text{ mol } O_2}\right) \times \left(\frac{44.0 \text{ g } CO_2}{1 \text{ mol } CO_2}\right) \sim 5.43 \text{ g } CO_2$$

There is an interesting postscript to the last problem. Although not stated explicitly, our assumption was that even without sufficient oxygen to consume all the C_2H_4, any C_2H_4 that did react was converted completely to CO_2 and H_2O. If this were the case, one aspect of automotive air pollution would be removed. What actually happens when the hydrocarbon (in this case C_2H_4) is present in excess is that some of it is converted to CO. In the internal-combustion engine, gasoline (which is composed of a mixture of hydrocarbons) is burned in a limited supply of oxygen and the incomplete combustion therefore produces a mixture of CO, CO_2, and H_2O.

2.10 THEORETICAL YIELD AND PERCENTAGE YIELD

Sometimes a given set of reactants are able to produce more than one set of products, depending on reaction conditions. In the last paragraph, for instance, it was pointed out that the combustion of hydrocarbons in a limited supply of oxygen produces a mixture of products. Usually the formation of side products (products other than those being sought) is undesirable, and two quantities that chemists are concerned with under these circumstances are the theoretical yield and the actual percentage yield.

The **theoretical yield** *of a given product is the maximum yield that could be obtained if the reactants gave only that product.* In Example 2.23 we calculated the theoretical yield of CO_2, assuming that all the C_2H_4 that burned was converted entirely to CO_2 and H_2O.

The **percentage yield** is a measure of the efficiency of the reaction. It is defined as

The theoretical yield is a computed quantity; the actual yield is obtained by measuring the amount of product actually formed in an experiment.

$$\text{percentage yield} = \frac{\text{actual yield}}{\text{theoretical yield}} \times 100$$

where the actual yield is the amount of product that is actually produced in a given experiment. For example, suppose that in the reaction in Example 2.23 we had obtained only 3.48 g CO_2, with the remainder of the carbon as either CO or elemental carbon. The actual yield, then, was 3.48 g CO_2. The theoretical yield was 5.43 g CO_2, so the percentage yield of CO_2 would then be

$$\text{percentage yield } CO_2 = \frac{3.48 \text{ g } CO_2}{5.43 \text{ g } CO_2} \times 100$$

$$\text{percentage yield } CO_2 = 64.1\%$$

2.11 MOLAR CONCENTRATION

In many chemical reactions, both in the laboratory and in the world around us, one or more of the reactants are present in a solution—that is, they are dissolved in some fluid such as water. In our bodies, for instance, the blood dissolves nutrients and transports them to our cells where they undergo the complex chain of reactions called metabolism.

In Chapter 1 we learned that solutions may be of variable composition. This means that there can be different ratios of the amounts of **solvent**—the substance present in the largest proportion—and **solute**—a substance present in a smaller proportion. The term **concentration** is used to describe the relative amounts of solute and solvent in a solution. A solution in which a large amount of solute is dissolved in the solvent is said to have a high concentration of the solute.

As we will see later in the book, there are a variety of ways of expressing the concentration of a solution quantitatively. The one that is the most useful for dispensing specific amounts of a dissolved solute is called **molar concentration** or **molarity.** This is defined as the ratio of the number of moles of solute in the solution divided by the volume of the solution expressed in liters.

Molarity is an intensive quantity formed as a ratio of two extensive quantities: moles and volume.

$$\text{molarity} = \frac{\text{moles of solute}}{\text{liters of solution}} \qquad [2.4]$$

A solution that contains 1.00 mol of NaCl in 1.00 liter of solution is a 1.00 **molar,** or 1.00 M solution. If it contained 2.00 mol NaCl per liter, it would be a 2.00 M solution. Let's look at a simple calculation showing how the molarity of a solution is computed.

EXAMPLE 2.24

A 2.00-g sample of the crystals of sodium hydroxide, NaOH, found in a container of Drano® was dissolved in water to produce a total volume of exactly 200 ml of solution. What is the molarity of this sodium hydroxide solution?

SOLUTION

To calculate the molarity we must take the ratio of moles of solute to liters of solution. Therefore, we first convert the 2.00 g NaOH (the solute) into moles. Since the formula weight of NaOH is 40.0,

$$2.00 \text{ g NaOH} \times \left(\frac{1 \text{ mol NaOH}}{40.0 \text{ g NaOH}}\right) = 0.0500 \text{ mol NaOH}$$

The volume of the solution is 200 ml, which is the same as 0.200 liter. Therefore, the molarity of the solution is

$$\text{molarity} = \frac{0.0500 \text{ mol NaOH}}{0.200 \text{ liter}}$$

$$= 0.250 \text{ mol/liter}$$

$$= 0.250 \ M$$

Molarity is such a useful concentration unit because if we know the molarity of a particular solution, we can dispense a desired number of moles of the solute just by measuring out the proper volume. For instance, suppose that we had a large container filled with a 0.250 M solution of NaOH. Let's also suppose that in a particular reaction we needed exactly 0.250 mol of NaOH. The label on the bottle tells us that each liter of the solution contains 0.250 mol NaOH, so all we have to do is measure 1.00 liter of the solution and we have the necessary 0.250 mol of NaOH. If we only needed 0.125 mol NaOH, we would only need 0.500 liter of the solution. On the other hand, if we needed 0.500 mol NaOH, we would need 2.00 liter of the solution. The following examples show how we use molarity in simple calculations.

EXAMPLE 2.25

How many milliliters of 0.250 M NaOH solution are needed to provide 0.0200 mol NaOH?

SOLUTION

In solving problems of this type we use molarity as a conversion factor. A label that reads "0.250 M NaOH" can be translated into either of two ratios

$$\frac{0.250 \text{ mol NaOH}}{1.00 \text{ liter solution}} \quad \text{or} \quad \frac{1.00 \text{ liter solution}}{0.250 \text{ mol NaOH}}$$

Conversions between liters and milliliters is something you should be able to do effortlessly.

If we want the volume in milliliters, these can be expressed as

$$\frac{0.250 \text{ mol NaOH}}{1000 \text{ ml solution}} \quad \text{or} \quad \frac{1000 \text{ ml solution}}{0.250 \text{ mol NaOH}}$$

Now let's restate our problem as

$$0.0200 \text{ mol NaOH} \sim (?) \text{ ml NaOH}$$

The units tell us which conversion factor to use.

$$0.0200 \text{ mol NaOH} \times \left(\frac{1000 \text{ ml solution}}{0.250 \text{ mol NaOH}}\right) = 80.0 \text{ ml solution}$$

Thus, if we measure out 80.0 ml of this solution, it will contain the desired 0.0200 mol NaOH.

EXAMPLE 2.26

How many grams of NaOH are in 50.0 ml of 0.400 M NaOH solution?

SOLUTION

This time we want to find the amount of solute in a portion of the solution. We begin by finding the number of moles of NaOH in the solution; then we convert that to grams. First we have to translate the label.

$$0.400 \ M \text{ NaOH means } \frac{0.400 \text{ mol NaOH}}{1000 \text{ ml solution}}$$

Then,

$$50.0 \text{ ml solution} \times \left(\frac{0.400 \text{ mol NaOH}}{1000 \text{ ml solution}}\right) = 0.0200 \text{ mol NaOH}$$

Finally,

$$0.0200 \text{ mol NaOH} \times \left(\frac{40.0 \text{ g NaOH}}{1 \text{ mol NaOH}}\right) = 0.800 \text{ g NaOH}$$

Thus, 50.0 ml of this solution contains 0.800 g NaOH

Sometimes it is necessary to prepare a solution having a specific concentration while you're working in the lab. This is really not very difficult, as the following example shows.

EXAMPLE 2.27

How many grams of silver nitrate, $AgNO_3$, are needed to prepare 500 ml of a 0.300 M $AgNO_3$ solution?

SOLUTION

This kind of problem really asks: "How many grams of $AgNO_3$ are in 500 ml of a 0.300 M $AgNO_3$ solution?" If you know the answer, you would know how much solute to use to make the solution. We therefore proceed as before. First we translate the label.

$$0.300 \ M \text{ AgNO}_3 \text{ means } \frac{0.300 \text{ mol AgNO}_3}{1000 \text{ ml solution}}$$

Then,

$$500 \text{ ml solution} \times \left(\frac{0.300 \text{ mol AgNO}_3}{1000 \text{ ml solution}}\right) = 0.150 \text{ mol AgNO}_3$$

The formula weight of $AgNO_3$ is 170. Therefore,

$$0.150 \text{ mol AgNO}_3 \times \left(\frac{170 \text{ g AgNO}_3}{1 \text{ mol AgNO}_3}\right) = 25.5 \text{ g AgNO}_3$$

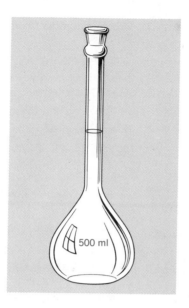

Figure 2.1

A 500-ml volumetric flask.

    To actually prepare the solution described in Example 2.27, we would have to dissolve the 25.5 g $AgNO_3$ in enough water to give a *final volume* of just 500 ml. Measuring volumes with this accuracy is accomplished using a volumetric flask (Figure 2.1). The flask contains the specified volume when it is filled to the mark etched around the neck. Figure 2.2 illustrates the sequence of steps involved in preparing the solution.

    As a final point, in preparing the solution in the preceding example, the volume of the solution is adjusted to a *final volume* of 500 ml. We didn't simply add 500 ml of water to the silver nitrate because this would give a final volume just a bit larger than 500 ml (both the solute and solvent take up space in the flask). If we had actually added 500 ml of water, the concentration would have been just a bit less than the desired 0.300 *M* because the solute would be spread over a slightly larger volume than expected.

(a)   (b)   (c)   (d)

Figure 2.2

Preparation of a solution having a particular molarity. (a) The solute is accurately weighed into a volumetric flask. (b) Distilled water is added. (c) The flask is stoppered and shaken to dissolve the solute. (d) Distilled water is added carefully to bring the volume up to the mark etched around the neck of the flask.

INDEX TO QUESTIONS AND PROBLEMS (Problem numbers in **bold type**)

REVIEW QUESTIONS

2.1 In what sense are the mole, the dozen, and the gross related to each other?

2.2 Why is the term formula weight preferred for some substances, rather than molecular weight?

2.3 In words, what is the difference between a structural formula, a molecular formula, and an empirical formula?

2.4 What is the empirical formula of each of the following?
(a) $(NH_4)_2S_2O_8$
(b) Fe_2O_3
(c) Al_2Cl_6
(d) C_6H_6
(e) $C_3H_8O_3$
(f) $C_6H_{12}O_6$
(g) Hg_2SO_4

2.5 Ethylene glycol, used as permanent antifreeze, has the structural formula

$$H-O-\overset{\overset{\displaystyle H}{|}}{C}-\overset{\overset{\displaystyle H}{|}}{C}-O-H$$
$$\underset{\underset{\displaystyle H}{|}}{}\quad\underset{\underset{\displaystyle H}{|}}{}$$

What is its molecular formula and empirical formula?

2.6 Why is the simplest formula for a substance called an empirical formula? What is the dictionary definition of empirical?

2.7 What important chemical law is obeyed by a balanced chemical equation?

2.8 Balance the following equations by inspection.
(a) $ZnS + HCl \rightarrow ZnCl_2 + H_2S$
(b) $HCl + Cr \rightarrow CrCl_3 + H_2$
(c) $Al + Fe_3O_4 \rightarrow Al_2O_3 + Fe$
(d) $H_2 + Br_2 \rightarrow HBr$
(e) $Na_2S_2O_3 + I_2 \rightarrow NaI + Na_2S_4O_6$

2.9 Balance the following equations.
(a) $FeCl_3 + Na_2CO_3 \rightarrow Fe_2(CO_3)_3 + NaCl$
(b) $NH_4Cl + Ba(OH)_2 \rightarrow BaCl_2 + NH_3 + H_2O$
(c) $Ca(OH)_2 + H_3PO_4 \rightarrow Ca_3(PO_4)_2 + H_2O$
(d) $Fe_2(CO_3)_3 + H_2SO_4 \rightarrow Fe_2(SO_4)_3 + H_2O + CO_2$
(e) $Na_2O + (NH_4)_2SO_4 \rightarrow Na_2SO_4 + H_2O + NH_3$

2.10 Balance the following equations.
(a) $C_4H_{10} + O_2 \rightarrow CO_2 + H_2O$
(b) $C_7H_6O_2 + O_2 \rightarrow CO_2 + H_2O$
(c) $P_4O_{10} + H_2O \rightarrow H_3PO_4$
(d) $FeS_2 + O_2 \rightarrow Fe_2O_3 + SO_2$
(e) $NH_3 + O_2 \rightarrow NO + H_2O$

2.11 What is meant by the term *limiting reactant*? In words, describe how the limiting reactant is identified.

2.12 What is meant by the *theoretical yield* for a given reaction mixture? What is the meaning of *percent yield*? When you carry out a reaction, what is the *actual yield*?

2.13 Define molar concentration.

2.14 To make 1.00 liter of a 1.00 M solution of the sugar glucose, $C_6H_{12}O_6$, requires 180 g $C_6H_{12}O_6$. Describe how you would actually go about preparing this solution.

2.15 Trisodium phosphate, Na_3PO_4, is a very powerful, but caustic, cleaning agent. If a bottle containing a solution of Na_3PO_4 were labeled 0.20 M Na_3PO_4, how could you use that information in the form of a conversion factor?

REVIEW PROBLEMS (More difficult problems are marked by an asterisk.)

2.16 Iron pyrite, FeS_2, forms beautiful golden crystals that are known as "fool's gold."
(a) How many moles of sulfur would be needed to combine with 1.00 mol Fe to form FeS_2?
(b) How many moles of iron are needed to combine with 1.44 mol S to form FeS_2?
(c) How many moles of sulfur are in 3.00 mol FeS_2?
(d) How many moles of FeS_2 are needed to give 3.00 mol Fe?

2.17 Ordinary sand is composed chiefly of silica, a compound in which there are two oxygen atoms for each silicon atom.

(a) What is the formula for silica?

(b) How many atoms of oxygen would be needed to combine with 25 atoms of silicon to form silica?

(c) How many moles of oxygen atoms would be needed to combine with 25 mol of silicon atoms to form silica?

(d) If you had 4.50 mol silica, how many moles of silicon and oxygen atoms would there be in it?

2.18 How many moles of S are in 1.00 mol of As_2S_3?

2.19 How many moles of O are in 1.50 mol of Cr_2O_3?

2.20 Based on the amount of carbon available, how many moles of CO_2 could be liberated from 1.00 mol of limestone, $CaCO_3$?

2.21 How many moles of $BaSO_4$ could be made from 1.25 mol of $Al_2(SO_4)_3$?

2.22 Give the weight of 1.00 mol of each of the following elements.

(a) magnesium   (d) chlorine
(b) carbon   (e) sulfur
(c) iron   (f) strontium

2.23 How many moles are in 50.0 g of each of the following elements?

(a) sodium   (d) aluminum
(b) arsenic   (e) potassium
(c) chromium   (f) silver

2.24 Give the formula weight of the following:

(a) MgO   (d) S_2Cl_2
(b) $CaCl_2$ (e) Na_3PO_4
(c) PCl_5

2.25 Give the formula weight of the following:

(a) SiO_2 (quartz)
(b) $Mg(OH)_2$ (milk of magnesia)
(c) $MgSO_4 \cdot 7H_2O$ (epsom salts)
(d) $Ca_2Mg_5(Si_4O_{11})_2(OH)_2$ (asbestos)
(e) $C_6H_8O_6$ (vitamin C)
(f) $C_{12}H_{22}O_{11}$ (sucrose—cane sugar)

2.26 What is the weight of 1.35 mol of caffeine, $C_8H_{10}N_4O_2$?

2.27 What is the weight of 2.33 mol of penicillin, $C_{16}H_{18}O_4N_2S$?

2.28 What is the weight of 6.30 mol of lead sulfate, $PbSO_4$?

2.29 What is the weight of 0.144 mol of TiO_2, a pigment used in white paint?

2.30 How many moles of sodium bicarbonate, $NaHCO_3$ (baking soda), are in a 242-g sample?

2.31 How many moles of butane, C_4H_{10}, the fluid in disposable lighters, are in 1.40×10^3 g of butane?

2.32 How many moles of sulfuric acid, H_2SO_4, are in 85.3 g of H_2SO_4?

2.33 A substance with the formula $PbHAsO_4$ has been used in insecticides, and on the label it appears under the name lead arsenate. How many moles of this poison are in 25.0 g?

2.34 How many moles of potassium are in 125 g of KCl?

2.35 How many moles of S are in 632 g of iron pyrite, FeS_2?

2.36 When coal containing iron pyrite, FeS_2, is burned, all the sulfur is converted to the air pollutant, sulfur dioxide, SO_2. How many moles of FeS_2 would have to react to produce 1.00 kg of SO_2?

2.37 Ordinary table sugar is sucrose, $C_{12}H_{22}O_{11}$. What is the weight of one molecule of sucrose? How much heavier is a molecule of sucrose than one atom of carbon? How many molecules of sucrose are in 25.0 g of sucrose? What is the total number of atoms in 25.0 g of sucrose?

2.38 How many atoms of carbon are in 4.00×10^{-8} g of propane, C_3H_8?

2.39 Carbon atoms have a diameter of approximately 1.5×10^{-8} cm. If carbon atoms were laid in a row 3.0 cm long, what would be the total mass of carbon?

2.40 Calculate the mass of Cu required to react with 5.00×10^{20} molecules of S_8 to form Cu_2S.

2.41 Calculate the percentage composition of each of the following:

(a) $FeCl_3$ (d) $(NH_4)_2HPO_4$
(b) Na_3PO_4 (e) Hg_2Cl_2
(c) $KHSO_4$

2.42 Calculate the percentage composition of each of the following:

(a) (benzene) C_6H_6
(b) (ethyl alcohol) C_2H_5OH
(c) (potassium dichromate) $K_2Cr_2O_7$
(d) (zenon tetrafluoride) XeF_4
(e) (calcium carbonate) $CaCO_3$

2.43 Calculate the weight of nitrogen in 30.0 g of the amino acid, glycine, CH_2NH_2COOH.

2.44 Calculate the weight of hydrogen in 12.0 g of NH_3.

2.45 Polystyrene, a common plastic, is composed of many styrene units linked together as shown below.

The basic styrene unit is shown within brackets. The subscript n means that this unit is repeated many times. A particular sample of this plastic was found to have an average molecular weight of 1 million. What is the average number of styrene units in a chain?

2.46 A sample of an air pollutant composed of sulfur and oxygen was found to contain 1.40 g sulfur and 2.10 g oxygen. What is the empirical formula of the compound?

2.47 The freon propellant from an aerosol can was analyzed. A sample of it contained 0.423 g C, 2.50 g Cl, and 1.34 g F. What is the empirical formula of this substance?

2.48 A dry-cleaning fluid composed of carbon and chlorine was found to have the composition: 14.5% C, 85.5% Cl (by weight). What is the empirical formula of this compound?

2.49 Arsenic reacts with oxygen to form a compound that is 75.7% arsenic and 24.3% oxygen, by weight. What is the empirical formula of this compound?

2.50 A 1.31-g sample of sulfur was allowed to react with an excess of chlorine to produce 4.22 g of a product that contains only sulfur and chlorine. What is the empirical formula of the compound?

2.51 A substance was found to be composed of 60.8% sodium, 28.5% boron, and 10.5% hydrogen. What is the empirical formula of the compound?

2.52 Vanillin is composed of carbon, 63.2%; hydrogen, 5.26%; and oxygen, 31.6%. What is the empirical formula of vanillin?

2.53 A 0.537-g sample of an organic compound containing only carbon, hydrogen, and oxygen was burned in air to produce 1.030 g of CO_2 and 0.632 g H_2O. What is the empirical formula of the compound?

*2.54 A 1.35-g sample of a substance containing carbon, hydrogen, nitrogen, and oxygen was burned to produce 0.810 g H_2O and 1.32 g CO_2. In a separate reaction all the nitrogen in 0.735 g of the substance was converted to ammonia. This gave 0.284 g of NH_3. Determine the empirical formula of the substance.

*2.55 An organic compound was synthesized and a sample of it was analyzed and found to contain C, H, N, O, and Cl. It was observed that when a 0.150-g sample of the compound was burned, it produced 0.138 g of CO_2 and 0.0566 g of H_2O. All the nitrogen in a different 0.200-g sample of the compound was converted to NH_3, which was found to weigh 0.0238 g. Finally, the chlorine in a 0.125-g sample of the compound was converted to Cl^- and by reacting it with $AgNO_3$ all the chlorine was recovered as AgCl. The AgCl, when dried, was found to weigh 0.251 g.

(a) Calculate the weight percent of each element in the compound.

(b) Determine the empirical formula for the compound.

2.56 The following are empirical formulas and molecular weights for five compounds. What are their molecular formulas?

(a) NaS_2O_3,   MW = 270 4

(b) C_3H_2Cl,   MW = 147.0

(c) C_2HCl,   MW = 181.4

(d) Na_2SiO_3,   MW = 732.6

(e) $NaPO_3$,   MW = 305.9

*2.57 Citric acid, the substance that makes lemon juice sour, is composed of only carbon, hydrogen, and oxygen. When a 0.5000-g sample of citric acid was burned, it produced 0.6871 g of CO_2 and 0.1874 g of H_2O. The molecular weight of the compound is 192. What is the empirical formula and the molecular formula for citric acid?

2.58 Acetylene, which is used as a fuel in welding torches, is produced in a reaction between calcium carbide and water,

$$CaC_2 + 2H_2O \longrightarrow Ca(OH)_2 + C_2H_2 \text{ (g)}$$
calcium   acetylene
carbide

(a) How many moles of C_2H_2 would be produced from 2.50 mol of CaC_2?
(b) How many grams of C_2H_2 would be formed from 0.500 mol of CaC_2?
(c) How many moles of water would be consumed when 3.20 mol of C_2H_2 are formed?
(d) How many grams of $Ca(OH)_2$ are produced when 28.0 g of C_2H_2 are formed?

2.59 Consider the following balanced equation:

$$6ClO_2 + 3H_2O \longrightarrow 5HClO_3 + HCl$$

(a) How many moles of $HClO_3$ are produced from 14.3 g of ClO_2?
(b) How many grams of H_2O are needed to produce 5.74 g of HCl?
(c) How many grams of $HClO_3$ are produced when 4.25 g of ClO_2 are added to 0.853 g of H_2O?

2.60 White phosphorus, composed of P_4 molecules, is used in military incendiary devices because it ignites spontaneously when exposed to air. The product of reaction with oxygen is P_4O_{10}. (Remember that oxygen in the air is present as O_2 molecules.)
(a) Write a balanced chemical equation for the reaction of P_4 with O_2.
(b) How many moles of P_4O_{10} can be produced using 0.500 mol O_2?
(c) How many grams of P_4 are needed to produce 50.0 g P_4O_{10}?
(d) How many grams of P_4 will react with 25.0 g O_2?

2.61 Hydrazine, N_2H_4, and hydrogen peroxide, H_2O_2, have been used as rocket propellants. They react according to the equation

$$7H_2O_2 + N_2H_4 \longrightarrow 2HNO_3 + 8H_2O$$

(a) How many moles of HNO_3 are formed from 0.0250 mol N_2H_4?
(b) How many moles of H_2O_2 are required if 1.35 mol H_2O are to be produced?
(c) How many moles of H_2O are formed if 1.87 mol HNO_3 are produced?
(d) How many moles of H_2O_2 are required to react with 22.0 g N_2H_4?
(e) How many grams of H_2O_2 are needed to produce 45.8 g HNO_3?

2.62 When iron is produced from its ore, Fe_2O_3, the reaction is

$$Fe_2O_3 + 3CO \longrightarrow 2Fe + 3CO_2$$

(a) How many moles of CO are needed to produce 35.0 mol Fe?

(b) How many moles of Fe_2O_3 react if 4.50 mol CO_2 are formed?

(c) How many grams of Fe_2O_3 must react to give 0.570 mol Fe?

(d) How many moles of CO are needed to react with 48.5 g Fe_2O_3?

(e) How many grams of Fe are formed when 18.6 g CO react?

2.63 During the naval battles of the South Pacific in World War II, the U.S. Navy produced smokescreens by spraying titanium tetrachloride into the moist air where it reacted according to the equation

$$TiCl_4 + 2H_2O \longrightarrow TiO_2 + 4HCl$$

The dense smoke was caused by the TiO_2.

(a) How many moles of H_2O are needed to react with 6.50 mol $TiCl_4$?

(b) How many moles of HCl are formed when 8.44 mol $TiCl_4$ react?

(c) How many grams of TiO_2 are formed from 14.4 mol $TiCl_4$?

(d) How many grams of HCl are formed when 85.0 g of $TiCl_4$ react?

2.64 The insecticide DDT (which ecologists now recognize as a serious environmental pollutant) is manufactured in a reaction between chlorobenzene and chloral,

$$\underset{\textbf{chlorobenzene}}{2C_6H_5Cl} + \underset{\textbf{chloral}}{C_2HCl_3O} \longrightarrow \underset{\textbf{DDT}}{C_{14}H_9Cl_5} + H_2O$$

How many kilograms of DDT can be produced from 1000 kg of chlorobenzene?

2.65 Aspirin (which many students take after working on chemistry problems) is prepared by the reaction of salicylic acid ($C_7H_6O_3$) with acetic anhydride ($C_4H_6O_3$) according to the reaction

$$C_7H_6O_3 + C_4H_6O_3 \longrightarrow \underset{\textbf{(aspirin)}}{C_9H_8O_4} + C_2H_4O_2$$

How many grams of salicylic acid must be used to prepare two 5 grain aspirin tablets (1 g = 15.4 grains)?

2.66 Dimethylhydrazine, $(CH_3)_2NNH_2$, has been used as a fuel in the Apollo lunar descent module, with liquid N_2O_4 as the oxidizer. The products of the reaction between these two in the rocket engine are H_2O, CO_2, and N_2.

(a) Write a balanced chemical equation for the reaction.

(b) Calculate the number of kilograms of N_2O_4 required to burn 50.0 kg of dimethylhydrazine.

2.67 The fermentation of sugar to produce ethyl alcohol follows the equation

$$C_6H_{12}O_6 \xrightarrow{\text{yeasts}} 2C_2H_5OH + 2CO_2$$

What is the maximum weight of alcohol that can be obtained from 500 g sugar?

2.68 Under appropriate conditions acetylene, C_2H_2, and HCl react to form vinyl chloride C_2H_3Cl. This substance is used to manufacture polyvinyl chloride (PVC) plastics and has recently been shown to be carcinogenic. The equation for the reaction is

$$C_2H_2 + HCl \longrightarrow C_2H_3Cl$$

In a given instance, 35.0 g of C_2H_2 are mixed with 51.0 g of HCl.

(a) Which is the limiting reactant?

(b) How many grams of C_2H_3Cl are formed?

(c) How many grams of the reactant in excess remain after the reaction is completed?

2.69 Phosgene, $COCl_2$, was once used as a war gas. It is poisonous because when it is inhaled, it reacts with water in the lungs to produce hydrochloric acid, HCl, which causes severe lung damage, leading ultimately to death. The chemical reaction is

$$COCl_2 + H_2O \longrightarrow CO_2 + 2HCl$$

(a) How many moles of HCl are produced by the reaction of 0.430 mol of $COCl_2$?

(b) How many grams of HCl are produced when 11.0 g of CO_2 are formed?

(c) How many moles of HCl will be formed if 0.200 mol of $COCl_2$ are mixed with 0.400 mol of H_2O?

2.70 In a typical experiment, a student reacted benzene, C_6H_6, with bromine, Br_2, in an attempt to prepare bromobenzene, C_6H_5Br. This reaction also produced, as a byproduct, dibromobenzene, $C_6H_4Br_2$. On the basis of the equation

$$C_6H_6 + Br_2 \longrightarrow C_6H_5Br + HBr$$

(a) What is the maximum amount of C_6H_5Br that the student could have hoped to obtain from 15.0 g of benzene? (This is the theoretical yield.)

(b) In this experiment the student obtained 2.50 g of $C_6H_4Br_2$. How much C_6H_6 was *not* converted to C_6H_5Br?

(c) What was the student's actual yield of C_6H_5Br?

(d) Calculate the percentage yield for this reaction.

*2.71 Freon-12, a gas used as a refrigerant, is prepared by the reaction

$$3CCl_4 + 2SbF_3 \longrightarrow \underset{\textbf{Freon-12}}{3CCl_2F_2} + 2SbCl_3$$

If 150 g of CCl_4 are mixed with 100 g of SbF_3,

(a) How many grams of CCl_2F_2 can be formed?

(b) How many grams of which reactant will remain after reaction has ceased?

*2.72 Acetylene, C_2H_2, can be reacted with two molecules of Br_2 to form $C_2H_2Br_4$ by the series of reactions

$$C_2H_2 + Br_2 \longrightarrow C_2H_2Br_2$$

$$C_2H_2Br_2 + Br_2 \longrightarrow C_2H_2Br_4$$

If 5.00 g of C_2H_2 are mixed with 40.0 g of Br_2, what weights of $C_2H_2Br_2$ and $C_2H_2Br_4$ will be formed? Assume that all the C_2H_2 has reacted.

*2.73 Silver tarnishes in the presence of hydrogen sulfide (rotten egg odor) because of the reaction

$$4Ag + 2H_2S + O_2 \longrightarrow 2Ag_2S + 2H_2O$$

hydrogen   (silver
sulfide   sulfide,
black)

How much Ag_2S could be obtained from a mixture of 0.950 g Ag, 0.140 g H_2S, and 0.0800 g O_2?

*2.74 White lead, a pigment used in lead-based paints, is manufactured by the reactions

$$2Pb + 2HC_2H_3O_2 + O_2 \longrightarrow 2Pb(OH)C_2H_3O_2$$

$$6Pb(OH)C_2H_3O_2 + 2CO_2 \longrightarrow$$
$$Pb_3(OH)_2(CO_3)_2 + 2H_2O + 3Pb(C_2H_3O_2)_2$$
white lead

(a) Starting with 20.0 g of Pb, how many grams of white lead can be prepared?
(b) How many grams of CO_2 will be required if 14.0 g of O_2 are consumed in the first reaction?
Assume that all the Pb in the first reaction is completely converted to the products of the second reaction.

*2.75 A chemist wishes to synthesize a certain compound that has a molecular weight of 100. The synthesis requires six consecutive steps, each giving a 50% yield (computed on a *mole* basis). If the chemist begins with 30.0 g of starting material having a molecular weight of 80.0, how many grams of final product will be obtained? How many grams of starting material will be required to produce 10.0 g of final product?

*2.76 In a reaction between methane, CH_4, and chlorine, Cl_2, four products can be formed: CH_3Cl, CH_2Cl_2, $CHCl_3$, and CCl_4. In a particular instance 20.8 g of CH_4 were allowed to react and gave 5.0 g CH_3Cl, 25.5 g CH_2Cl_2, and 59.0 g $CHCl_3$. All the CH_4 reacted.
(a) How many grams of CCl_4 were formed?
(b) On the basis of available CH_4, what is the theoretical yield of CCl_4?
(c) What is the percent yield of CCl_4?
(d) How many grams of Cl_2 reacted with the CH_4? (*Note:* The hydrogen that is displaced from the carbon also combines with Cl_2 to form HCl.)

2.77 Phosphate rock, $Ca_3(PO_4)_2$, is treated with sulfuric acid to produce phosphate fertilizer,

$$Ca_3(PO_4)_2 + 2H_2SO_4 + 4H_2O \longrightarrow$$
$$Ca(H_2PO_4)_2 + 2CaSO_4 \cdot 2H_2O$$
phosphate fertilizer

How many tons of sulfuric acid are required to react with 25.0 tons of phosphate rock? Work the problem using ton-moles.

2.78 How many pounds of ammonia can be produced from 650 lb of H_2 by the reaction $N_2 + 3H_2 \rightarrow 2NH_3$. Work out the problem using lb-moles.

2.79 Calculate the molar concentration (molarity) of the following solutions.
(a) 0.250 mol NaCl in 0.400 liter of solution
(b) 1.45 mol sucrose in 345 ml of solution
(c) 195 g H_2SO_4 in 875 ml of solution
(d) 80.0 g KOH in 0.200 liter of solution

2.80 Calculate the molar concentration (molarity) of the following solutions.
(a) 1.35 mol NH_4Cl in a total volume of 2.45 liters
(b) 0.422 mol $AgNO_3$ in a total volume of 742 ml
(c) 3.00×10^{-3} mol KCl in 10.0 ml of solution
(d) 4.80×10^{-2} g $NaHCO_3$ in 25.0 ml of solution

2.81 A millimole (mmol) is 10^{-3} mol. Suppose that each milliliter of a solution contained 0.250 mmol of NaCl—that is, its concentration is 0.250 mmol/ml.
(a) Calculate the concentration of this solution in units of mol/liter.
(b) What is the molarity of this solution?
(c) In what way are the numberical values of concentration related when the concentration is reported in molarity, mol/liter, and mmol/ml?

2.82 Suppose a solution of lithium carbonate, Li_2CO_3, a drug used to treat manic depression, is labeled 0.250 M.
(a) How many moles of Li_2CO_3 are present in 250 ml of this solution?
(b) How many grams of Li_2CO_3 are in 630 ml of the solution?
(c) How many milliliters of the solution would be needed to supply 0.0100 mol Li_2CO_3?
(d) How many milliliters of the solution would be needed to provide 0.0800 g of Li_2CO_3?

2.83 A solution was labeled 0.375 M KOH.
(a) How many milliliters of this solution are needed to give 0.100 mole KOH?
(b) How many moles of KOH are in 45.0 ml of this solution?
(c) How many milliliters of this solution are needed to give 10.0 g of KOH?
(d) How many grams of KOH are in each milliliter of the solution?

2.84 Calcium acetate, $Ca(C_2H_3O_2)_2$, is the substance used along with methyl alcohol to make "canned heat." How many grams of calcium acetate are needed to prepare 2.00 liter of 0.250 M solution?

2.85 How many grams of potassium nitrate, KNO_3, are needed to prepare 250.0 ml of a 3.00×10^{-2} M solution?

*2.86 The formula for epsom salts is $MgSO_4 \cdot 7H_2O$. How many grams of epsom salts are needed to prepare 500 ml of a solution that can be labeled "0.150 M $MgSO_4$"?

*2.87 The silver nitrate ($AgNO_3$) in 20.00 ml of solution was allowed to react with sodium chloride according to the equation $AgNO_3(aq) + NaCl(aq) \longrightarrow AgCl(s) + NaNO_3(aq)$. The AgCl was collected, dried, and weighed to give 0.2867 g AgCl. What was the molarity of the original $AgNO_3$ solution?

3

ATOMIC STRUCTURE AND THE PERIODIC TABLE

Much of our knowledge about the internal structures of atoms has, as its origin, the fact that atoms are only able to lose energy in certain specific amounts, and that this energy can appear as light of a particular color. In a **laser**, many atoms lose energy all at once, which produces a very intense light beam — intense enough to burn through metals. Here we see an industrial laser trimming a titanium panel that will become part of a Grumman F-14 Tomcat aircraft.

The atomic theory proposed by Dalton represented a major breakthrough in the development of chemistry. All the computations you learned to perform in the last chapter, in fact, are based on his idea that atoms of each element have a characteristic atomic mass. Although Dalton's theory accounted for the mass relationships observed in chemical reactions, it was not able to explain why substances react the way they do. It could be determined, for example, that one oxygen atom was able to react with a maximum of only two atoms of hydrogen, but no one understood why. Furthermore, as additional evidence came to light, it became increasingly clear that the simple picture of an indivisible atom was just not sufficient to account for all of the facts, and it was by a rather fascinating piecing together of bits of information that our current picture of the atom emerged. In this chapter we will see how our present knowledge of atomic structure was developed, and how an understanding of atomic structure can help to explain and correlate chemical and physical properties of the elements.

3.1 THE ELECTRICAL NATURE OF MATTER

In 1834, a British scientist named Michael Faraday reported the results of experiments in which he showed that chemical changes could be caused by the passage of electricity through water solutions of chemical compounds. These experiments demonstrated that matter was electrical in nature and led G. J. Stoney, 40 years later, to propose the existence of particles of electricity that he called **electrons.**

Toward the end of the nineteenth century, physicists began to investigate the flow of electric current in **gas discharge tubes.** These tubes were made of glass with a piece of metal (called an **electrode**) sealed into each end (Figure 3.1). It was found that when a high voltage was applied across the electrodes and air was partially removed from the tube, a flow of electricity—an *electrical discharge*—was observed and the residual air within the tube began to glow. (Neon signs, like those in Color Plate 2, are modern versions of gas discharge tubes in which neon or other gases are used in the tube instead of air.) If virtually all the gas was removed, the flow was no longer observed, but the electrical discharge continued. When a screen, coated with a zinc sulfide phosphor (similar to that found in a TV picture tube), was placed between the electrodes, it was found to glow on the side facing the negative electrode (the **cathode**). This demonstrated that the discharge originates at the cathode and flows toward the positive electrode (the **anode**). These "rays," as they were called, were thus named **cathode rays.**

Further investigations showed that cathode rays:

1. Normally travel in straight lines.
2. Cast shadows.

Ordinary fluorescent lamps are gas discharge tubes containing mercury vapor. Light given off by the mercury when the discharge occurs causes a phosphor on the inner surface of the lamp to give off white light.

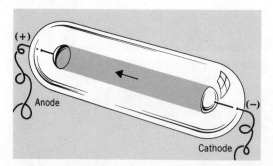

Figure 3.1
A gas discharge tube.

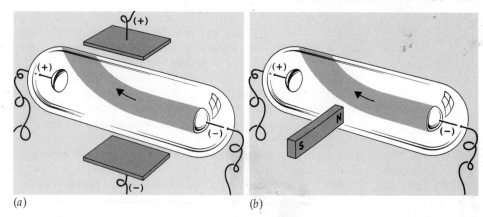

Figure 3.2

Some properties of cathode rays. (a) Cathode rays are deflected by an electric field and (b) by a magnetic field.

(a)　　　　　　　(b)

3. Can turn a pinwheel placed in their path, suggesting that cathode rays are composed of particles.
4. Heat a metal foil placed between the electrodes.
5. Can be bent by an electric or magnetic field in such a direction as to indicate that the particles are electrically charged and that the charge is negative (Figure 3.2).
6. Are always the same regardless of the nature of the material composing the electrodes or the kind of residual gas within the tube.

These findings suggested that cathode rays are composed of energetic, negatively charged particles that are building blocks of all the substances we know. Such particles are called **fundamental particles,** and the particles in cathode rays are, in fact, the electrons described by Stoney.

Quantitative information about the electron was discovered in 1897 when J. J. Thomson, another British scientist, used a cathode-ray tube quite similar to present-day TV picture tubes to measure the ratio of the charge to the mass of an electron. This device is shown schematically in Figure 3.3. Electrons generated at the cathode are accelerated toward the anode, which has a hole in it. Some electrons pass through the hole and continue on, striking the phosphor-coated face of the tube at B, producing a bright spot. If oppositely charged plates are placed above and below the tube, the beam is deflected toward the positive plate and strikes the face at A. The amount of deflection that the particle undergoes will be *directly proportional* to its charge—a particle with a large negative charge will be attracted to the positive plate more than one with

A modern television picture tube.

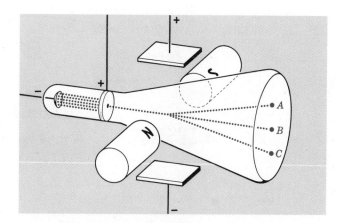

Figure 3.3

A cathode-ray tube used to measure the charge-to-mass ratio of the electron.

a small charge. The amount of deflection will also be *inversely proportional* to the mass of the particle, because a heavy particle will be less affected by the electrostatic attraction as it passes between the plates than a particle of smaller mass. This is similar to the differences between the way a breeze affects the path of a baseball and the way it affects the path of a much lighter Ping-Pong ball. The influences of charge and mass on the amount of deflection can be combined by saying that the observed deflection depends on the *ratio* of the particles charge, *e*, to its mass, *m*. This is the **charge-to-mass ratio,** symbolized as *e/m*.

> The path of a moving charged particle is affected by both electric and magnetic fields.

If a magnetic field is generated at right angles to the electric field, as shown in Figure 3.3, the electrons are deflected in a direction exactly opposite to that caused by the electrically charged plates. In the absences of the electric field, the electron beam is bent by the magnetic field so that it collides with the surface of the tube at C.

In practice, Thomson applied a magnetic field of known strength across the tube and noted the deflection of the electron beam. Charge was then applied to the plates until the beam was brought back to its original point of impact, B. From the magnitude of the electric and magnetic fields, Thomson calculated the charge-to-mass ratio, *e/m*, for the electron to be -1.76×10^8 coulombs/gram. The **coulomb (C)** is the SI unit of electric charge. It is equal to the amount of charge that moves past a given point in a wire when an electric current of 1 **ampere** (1 **A**) flows for 1 second. In more familiar terms, if you have an ordinary 100-watt light bulb burning, it is using a current of about 0.83 A. It takes 1.2 seconds for one coulomb of charge to pass through the bulb. A coulomb, therefore, represents a fairly large amount of electric charge, and the results of Thomson's experiment indicated that the electron has either a very large electric charge or that its mass is very small.

3.2 THE CHARGE ON THE ELECTRON

> X rays are a high-energy form of light and can eject electrons from particles that absorb them. That's what makes large doses of X rays dangerous.

The charge on the electron was determined by means of a rather clever experiment performed in 1908 by R. A. Millikan at the University of Chicago. In his apparatus, illustrated in Figure 3.4, a fine mist of oil droplets was sprayed above a pair of parallel metal plates. As droplets settled through a hole in the upper plate, the air between the plates was briefly irradiated with X rays. Electrons, knocked from gas atoms by the X rays, were picked up by the oil droplets, thereby giving them a negative charge. Millikan found that by placing an elec-

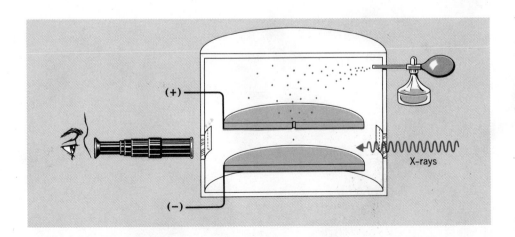

Figure 3.4

Millikan's oil drop experiment.

tric charge on the plates (upper plate positive, lower plate negative), the downward motion of negatively charged drops could be slowed or even stopped. A knowledge of the mass of a drop—measured by observing its rate of fall in the absence of the electric field—and the amount of charge on the plates required to keep that drop suspended permitted Millikan to calculate the amount of charge on the drop.

About 5.2 billion billion electrons pass through a 100-watt electric light bulb each second it is lit.

After he performed this experiment many times, Millikan observed that the amount of charge on the oil drops was always a multiple of -1.60×10^{-19} coulombs. He reasoned that since the oil drops could only pick up whole numbers of electrons, the total charge on any drop must be a multiple of the charge on a single electron. This suggested that the charge on the electron is -1.60×10^{-19} coulombs. Once the charge on the electron was known, its mass, 9.11×10^{-28} g, was obtained from the already known charge-to-mass ratio. If you write this number out in standard decimal form, you might be able to appreciate how incredibly tiny the electron is.

3.3 POSITIVE PARTICLES AND THE MASS SPECTROMETER

Ordinary things that we encounter every day are electrically neutral. Therefore, since negatively charged electrons are a part of everything, positively charged particles must also exist in all matter. The search for these positive particles began with experiments using specially designed gas discharge tubes with perforated cathodes. When an electric current flowed through the tube, streamers of light were observed coming from the holes at the rear of the negative electrode (Figure 3.5). These were called "canal rays."

During an electric discharge, electrons that are emitted from the cathode collide with neutral gas atoms, knocking electrons off them. When they lose electrons the atoms become positively charged ions (an **ion** is an electrically charged particle that is formed when electrons are added to or removed from a neutral atom or molecule). These positive ions are attracted toward the cathode. Although most of them collide with the cathode, some travel through the perforations and emerge at the rear where they are observed as the "canal rays." If the rear wall of the discharge tube is coated with a phosphor, flashes of light can also be seen where they hit the wall.

An instrument designed to measure the charge-to-mass ratio of positive ions is called a **mass spectrometer** (Figure 3.6). Gaseous material is introduced at A and ionized—converted to ions—by an electric discharge across electrodes B and C. The positive ions thus produced are accelerated through the wire grid, E. As they pass through the slits F and G a narrow beam is formed which is fed between the poles of a powerful magnet. The magnetic field acts to deflect the particles into a circular path, with the degree of curvature determined by the

Figure 3.5

Canal rays. Positive ions passing through holes in the cathode appear as canal rays at the rear of the electrode. Bursts of light are seen when they collide with a phosphor on the end of the tube.

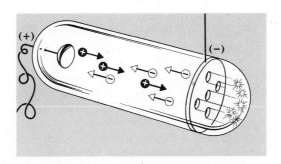

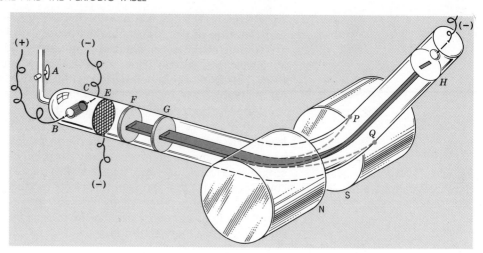

Figure 3.6

The mass spectrometer.

charge-to-mass ratio of the ions. For ions with the same charge the extent to which their paths are bent depends on their masses, with lighter particles being deflected more than heavier ones. For ions with the same mass, the degree to which their paths are curved is directly proportional to their charge. By adjusting the strength of the magnetic field, ions with any desired e/m ratio may be focused on the detector at H. Ions with higher e/m ratios are deflected more (for example, to P) while those of lower e/m ratios are deflected less (for example, to Q).

The measurement of e/m for positive particles reveals the following:

1. Positive ions always have e/m ratios that are *much smaller* than that of the electron. This means that they are either much more massive than the electron (that is, m is very large) or they carry very small positive charges (that is, e is small). Since they are formed from neutral atoms by the loss of electrons, the charge that they carry is either equal in magnitude to the electron's charge or is some whole-number multiple of it. Therefore, in order to have a much smaller e/m than the electron their masses must be much larger.

2. The value of the e/m ratio depends on the nature of the gas introduced into the mass spectrometer, which shows that not all positive ions have the same e/m.

When hydrogen, the lightest of all gases, is placed in the mass spectrometer the e/m ratio of the hydrogen ion is found to be $+9.63 \times 10^4$ coulombs/g. This is the largest e/m that is ever observed for any positive ion. Therefore, it is assumed that the hydrogen ion represents the fundamental particle of positive charge. The name of this particle is the **proton.** A neutral hydrogen atom, therefore, is composed of 1 electron and 1 proton. If we compare the charge-to-mass ratios of the proton and electron, we find the proton to be 1836 times heavier than the electron. Thus nearly all of the atom's mass is associated, somehow, with its positive charge.

Atoms that are heavier than hydrogen contain more than one proton, and each atom of a particular element has the same number of them. The number of protons in an atom of an element is called the element's **atomic number.** We will see later how atomic numbers were measured.

Because ions are formed from neutral particles by the gain or loss of electrons, each of which either adds or removes 1.60×10^{-19} coulombs of charge, it is convenient to express the charges on ions in units of this size. For example, the electron would possess 1 unit of negative charge. On this scale, this constitutes a charge of 1−. Two units of positive charge would be represented as 2+. We commonly indicate the charge on an ion formed from an atom by writing the number of units of positive or negative charge as a superscript on the right side of the chemical symbol. Thus the ion He^{2+} is an ion formed from the helium atom by the loss of two electrons. The ion, O^{2-}, is an ion formed from an oxygen atom by the addition of two electrons.

3.4 RADIOACTIVITY

The atoms of some elements are not stable. They spontaneously emit radiations of various types. This phenomenon, called **radioactivity,** was discovered by Henri Becquerel in 1896. Radioactive substances emit three important types of radiation.

We discuss radioactivity in more detail in Chapter 24.

1. Alpha radiation, composed of He^{2+} ions called alpha particles (α-particles).
2. Beta radiation, consisting of electrons, in this instance, called beta particles (β-particles).
3. Gamma radiation (γ rays), highly energetic, very penetrating light waves. (They are similar to X rays.)

The phenomenon of radioactivity was yet another bit of evidence that atoms were not indestructible particles and that they contained still simpler parts.

3.5 THE NUCLEAR ATOM

One of the most significant developments in our understanding of the structure of the atom was provided by Ernest Rutherford in 1911. Prior to this time it was thought that the atom had a nearly uniform density throughout, with the electrons being buried in a glob of positive charge, much like raisins in a pudding. With this picture of a rather mushy atom in mind, Rutherford assigned one of his students the task of measuring the scattering of alpha particles that were aimed at a thin gold foil. From his earlier experiments Rutherford expected the α-particles to pass through the foil virtually undisturbed, which would be consistent with a more or less uniform distribution of positive and negative charges. Nevertheless, he suggested that the student check to see if any α-particles were scattered at large angles, and he was astonished to learn that some were. In fact, it was found that some α-particles came almost straight back toward the source, which meant that they had encountered something positively charged and extremely massive (Figure 3.7).

Figure 3.7

(a) Scattering of α-particles by a metal foil. (b) Deflection of α-particles due to repulsions between positively charged α-particles and positively charged nuclei.

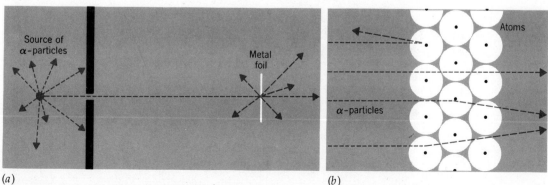

(a)
(b)

The only way Rutherford could explain why most of the α-particles passed easily through the foil, but a few were deflected at extremely large angles, was by concluding that the atom contained a very small, extremely dense positive **nucleus** that held all the protons and nearly all the atom's mass. Since the nucleus contains the positive charge in the atom, it follows that the electrons must be distributed somewhere in the remaining volume of the atom.

It's difficult to imagine how extremely small a nucleus is. Its diameter is approximately 10^{-13} cm, compared to the atom itself whose diameter is of the order of 10^{-8} cm. Since nearly all the mass of an atom is packed into this tiny nucleus, the density of nuclear material is enormous—about 10^{14} g/cm³. To give you some idea of how dense this is, if all of the nuclei in the crude oil cargo of one of the world's largest supertankers could be crammed together until they were touching, they would only occupy approximately 0.004 cm³—about the volume of a tenth of a drop of water—but they would weigh over 200,000 tons!

3.6 THE NEUTRON

Rutherford had observed that only about one-half of the nuclear mass could be accounted for by protons. He therefore suggested that particles of zero charge and of mass nearly the same as that of the proton are also present in the nucleus. The existence of these particles was confirmed in 1932 by an English scientist, J. Chadwick, who bombarded beryllium with α-particles and found that highly energetic, uncharged particles were emitted. These particles, called **neutrons,** have a mass only slightly larger than that of the proton.

The properties of the three major particles found in an atom are described in Table 3.1. In summary, an atom is composed of a dense nucleus that contains protons and neutrons. These particles provide nearly all the atom's mass. The nucleus is surrounded in some fashion by the atom's electrons, which are somehow distributed throughout the remaining volume of the atom.

3.7 ISOTOPES

In Chapter 1 we noted that not all atoms of the same element have identical masses, which is contrary to Dalton's original hypothesis. We referred to these different kinds of atoms as **isotopes.** The existence of isotopes is a common phenomenon, and most of the elements occur naturally as mixtures of isotopes.

As we will see, the properties of an element are determined almost entirely by the number and distribution of its electrons. *Therefore, it is the atomic number (or number of protons) that serves, indirectly, to distinguish an atom of one element from an atom of another, because the number of electrons must equal the number of protons in an electrically neutral atom.* In other words, an atom's atomic number

Each atom of a given element has the same atomic number, but the number of neutrons can vary.

Table 3.1
Some properties of subatomic particles

	Mass		Charge	
	Grams	Atomic Mass Units (amu)	Coulombs	Electronic Charge Units
Proton	1.67×10^{-24}	1.007276	$+1.602 \times 10^{-19}$	1+
Neutron	1.67×10^{-24}	1.008665	0	0
Electron	9.11×10^{-28}	0.0005486	-1.602×10^{-19}	1−

identifies which element it is. Any mass differences that exist between atoms of the same element arise from different numbers of neutrons.

A particular isotope of an element is identified by specifying its atomic number, Z, and its **mass number,** A. The mass number is the *sum* of the number of protons and neutrons in the atom. The number of neutrons present can therefore be obtained from the difference, $A - Z$.

We indicate an atom symbolically by writing its mass number as a superscript and its atomic number as a subscript. Both precede the atomic symbol,

$$_Z^A X$$

The atomic number is often omitted in specifying an isotope because the atomic symbol, in effect, supplies the same information. Thus $_6^{12}C$ and ^{12}C both represent carbon-12.

For example, a carbon atom (atomic number = 6) that has six neutrons would have a symbol of $_6^{12}C$. It is this isotope of carbon, incidentally, that serves as the basis of the current scale of atomic weights; that is, the mass of one atom of $_6^{12}C$ is defined as exactly 12 atomic mass units (amu).

The fact that atoms of a given element can differ in the number of neutrons they contain has had practical applications. One example is in archeological dating. Radioactive isotopes, such as ^{14}C, undergo nuclear changes that cause them to be transformed into other elements. Atoms of carbon-14, for example, are unstable, and over a period of time each of them emits a beta-particle and becomes a stable atom of $_7^{14}N$. This gradual transformation of the atoms of $_6^{14}C$ in a sample by the emission of beta-particles is called the **radioactive decay** of this isotope. From the known rate of this decay, and from the relative abundances of ^{14}C and ^{12}C (a nonradioactive isotope of carbon) in both living and fossil material, the age of the fossil can be estimated. The details of radioactive decay and how it is applied to archeological dating will be discussed in greater detail in Chapter 24.

Another application of radioactive isotopes is in chemotherapy. Treatment of thyroid cancer, for instance, is accomplished by administering carefully controlled doses of radioactive ^{131}I, which tends to concentrate in the thyroid gland where the radiation produced by the ^{131}I causes destruction of cancerous cells. In this case use is made of the body's natural tendency to concentrate iodine (either radioactive or nonradioactive) in the thyroid gland.

Very accurate measurements of the masses of isotopes and their relative abundances are made using the mass spectrometer.

As we noted above, nearly all elements as they are found in nature occur as mixtures of isotopes. For example, the element copper is found to contain the naturally occurring isotopes $_{29}^{63}Cu$ and $_{29}^{65}Cu$ whose masses have been accurately determined to be 62.9298 and 64.9278 amu, respectively. Their relative abundances are 69.09% and 30.91%. The observed atomic weight of copper, 63.55, is obtained as an average of the isotopic masses, weighted according to the relative abundances of each isotope, as shown in Example 3.1.

EXAMPLE 3.1   Using the data supplied in the paragraph above, calculate the average atomic mass of copper.

SOLUTION   The amount of mass that each isotope contributes toward the average atomic mass is equal to the product of its mass multiplied by its fractional abundance (percent abundance divided by 100).

^{63}Cu	62.9298 amu × 0.6909 =	43.48 amu
^{65}Cu	64.9278 amu × 0.3091 =	20.07 amu
	total	63.55 amu

In conclusion, observe the distinction between the *mass number* of an isotope and its *actual mass*. The mass number is simply the total count of protons plus neutrons and is not quite equal to the mass of the atom.

3.8 THE PERIODIC LAW AND THE PERIODIC TABLE

Scientists, even as early as 1800, had accumulated a significant amount of information concerning the physical and chemical properties of the known elements. This knowledge, however, existed for the most part as isolated and unrelated facts that needed to be correlated in some fashion before their total significance could be grasped. Early attempts at classification of the elements met with only limited success. It wasn't until 1869 that the forerunner of our modern periodic table was devised. This resulted from the efforts of two chemists, a Russian Dmitri Mendeleev, and a German, Julius Lothar Meyer, each of whom worked independently of the other and produced similar tables about the same time. Mendeleev presented the results of his work to the Russian Chemical Society in the early part of 1869. Meyer's table didn't appear until December of that same year, so Mendeleev is usually given credit for the periodic table.

Mendeleev was a chemistry teacher, and while he was preparing a textbook for his students, he discovered that if he arranged the elements in order of increasing atomic weight (atomic numbers weren't known yet), elements with similar properties occurred at periodic intervals. Mendeleev's genius was to divide this list of elements into a series of rows and stack them, with those elements having similar properties arranged in vertical columns called **groups** (Figure 3.8). Mendeleev insisted on having similar elements in columns even though it meant leaving some blanks in the table. He reasoned that the elements to fill them simply hadn't been discovered yet. It was also necessary for him to reverse the atomic-weight order of tellurium and iodine, whose atomic weights in 1869 were thought to be 128 and 127 amu. They were placed into the table in reverse order (according to atomic weight) because their properties dictated that tellurium be placed in Group VI and iodine in Group VII.

An advantage of Mendeleev's table was that it was possible to predict the properties of the missing elements because elements in any particular column

Figure 3.8

Mendeleev's Periodic Table (1871). The numbers appearing with the symbols are atomic weights.

	Group I	Group II	Group III	Group IV	Group V	Group VI	Group VII	Group VIII
1	H 1							
2	Li 7	Be 9.4	B 11	C 12	N 14	O 16	F 19	
3	Na 23	Mg 24	Al 27.3	Si 28	P 31	S 32	Cl 35.5	
4	K 39	Ca 40	— 44	Ti 48	V 51	Cr 52	Mn 55	Fe 56, Co 59 Ni 59, Cu 63
5	(Cu 63)	Zn 65	— 68	— 72	As 75	Se 78	Br 80	
6	Rb 85	Sr 87	?Yt 88	Zr 90	Nb 94	Mo 96	— 100	Ru 104, Rh 104 Pd 105, Ag 100
7	(Ag 108)	Cd 112	In 113	Sn 118	Sb 122	Te 128	I 127	
8	Cs 133	Ba 137	?Di 138	?Ce 140	—	—	—	— — — —
9	—	—	—	—	—	—	—	
10	—	—	?Er 178	?La 180	Ta 182	W 184	—	Os 195, Ir 517 Pt 198, Au 199
11	(Au 199)	Hg 200	Tl 204	Pb 207	Bi 208	—		
12	—	—	—	Th 231	—	U 240	—	— — — —

had to have similar properties. For example, germanium, which lies below silicon and above tin in Group IV, had not been discovered when Mendeleev constructed his table. Therefore, a blank space appears at this spot in the chart. On the basis of its position in the table, Mendeleev predicted that the properties of germanium, which he called "eka-silicon," should lie intermediate between those of silicon and tin. Table 3.2 shows how closely he predicted the properties that were found for germanium when it was discovered in 1886.

The necessity of reversing the order of atomic weights, which occurred when placing tellurium and iodine in the periodic table, was repeated after the noble gases were discovered. It was found that the atomic weight of argon (39.9 amu) was greater than that of potassium (39.1 amu); however, on the basis of physical and chemical properties, potassium clearly belonged in Group I (following argon) while argon had to be included in a separate group with the other noble gases. These reversals present no problems at all, and the proper arrangement is obtained when the elements are placed in the table in order of increasing atomic number instead of atomic weight. This leads us to the modern statement of the **periodic law:** *When the elements are arranged in order of increasing atomic number, there occurs a periodic repetition of physical and chemical properties.* Thus it is the charge on the nucleus, which we associate with the atomic number, and the number of electrons in the neutral atom that are important in determining the sequence in which the elements occur and that are responsible for their properties.

Table 3.2
**Predicted properties of eka-silicon
and observed properties of germanium**

Property	Silicon	Tin	Predicted for Eka-silicon (*Es*)	Currently Accepted for Germanium
Atomic weight (amu)	28	118	72	72.59
Density (g/ml)	2.33	7.28	5.5	5.3
Melting point (°C)	1410	232	High	947
Physical form at room temperature	Gray nonmetal	White metal	Gray metal	Gray metal
Reaction with acids and alkalies	Acid—no reaction Alkalies—slow reaction	Slow attack by acids and by alkalies	Very slow attack by acids and alkalies	Will react with concentrated acids and with concentrated alkalies
Number of chemical bonds usually formed	4	4	4	4
Formula of chloride	$SiCl_4$	$SnCl_4$	$EsCl_4$	$GeCl_4$
Density of chloride (g/ml)	1.50	2.23	1.9	1.88
Boiling point of chloride (°C)	57.6	114	100	84

Figure 3.9

Periodic table key: Atomic number (top), Symbol, Atomic mass (bottom). Example: H, 1, 1.0079.

Noble gases — Group 0.

IA	IIA	IIIB	IVB	VB	VIB	VIIB	VIII			IB	IIB	IIIA	IVA	VA	VIA	VIIA	0
1 H 1.0079																	2 He 4.00260
3 Li 6.941	4 Be 9.01218											5 B 10.81	6 C 12.011	7 N 14.0067	8 O 15.9994	9 F 18.998403	10 Ne 20.179
11 Na 22.98977	12 Mg 24.305											13 Al 26.98154	14 Si 28.0855	15 P 30.97376	16 S 32.06	17 Cl 35.453	18 Ar 39.948
19 K 39.0983	20 Ca 40.08	21 Sc 44.9559	22 Ti 47.90	23 V 50.9415	24 Cr 51.996	25 Mn 54.9380	26 Fe 55.847	27 Co 58.9332	28 Ni 58.70	29 Cu 63.546	30 Zn 65.38	31 Ga 69.72	32 Ge 72.59	33 As 74.9216	34 Se 78.96	35 Br 79.904	36 Kr 83.80
37 Rb 85.4678	38 Sr 87.62	39 Y 88.9059	40 Zr 91.22	41 Nb 92.9064	42 Mo 95.94	43 Tc (98)	44 Ru 101.07	45 Rh 102.9055	46 Pd 106.4	47 Ag 107.868	48 Cd 112.41	49 In 114.82	50 Sn 118.69	51 Sb 121.75	52 Te 127.60	53 I 126.9045	54 Xe 131.30
55 Cs 132.9054	56 Ba 137.33	57 *La 138.9055	72 Hf 178.49	73 Ta 180.9479	74 W 183.85	75 Re 186.207	76 Os 190.2	77 Ir 192.22	78 Pt 195.09	79 Au 196.9665	80 Hg 200.59	81 Tl 204.37	82 Pb 207.2	83 Bi 208.9804	84 Po (209)	85 At (210)	86 Rn (222)
87 Fr (223)	88 Ra 226.0254	89 †Ac 227.0278	104 Unq (261)	105 Unp (262)	106 Unh (263)												

* Lanthanides:

58 Ce 140.12	59 Pr 140.9077	60 Nd 144.24	61 Pm (145)	62 Sm 150.4	63 Eu 151.96	64 Gd 157.25	65 Tb 158.9254	66 Dy 162.50	67 Ho 164.9304	68 Er 167.26	69 Tm 168.9342	70 Yb 173.04	71 Lu 174.967

† Actinides:

90 Th 232.0381	91 Pa 231.0359	92 U 238.029	93 Np 237.0482	94 Pu (244)	95 Am (243)	96 Cm (247)	97 Bk (247)	98 Cf (251)	99 Es (254)	100 Fm (257)	101 Md (258)	102 No (259)	103 Lr (260)

The modern periodic table of the elements.

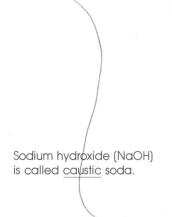

Members of a group in the periodic table bear resemblance to each other as do members of a family — hence the term, family of elements.

Sodium hydroxide (NaOH) is called caustic soda.

The periodic table in use today (sometimes called the "long" form of the periodic table) is shown in Figure 3.9. We see that, like Mendeleev's table, it is constructed of a number of vertical columns, called **groups,** each containing a *family of elements.* These groups are identified by a Roman numeral and a letter, either A or B. Groups IA through VIIA and Group 0 are referred to collectively as the **representative elements,** while Groups IB through VIIB and Group VIII (actually composed of the three short columns in the center of the table) constitute the **transition elements.** Similarities between properties of the A- and B-group elements exist, although the similarities are often very weak.

Finally, we see that there are two long rows of elements lying just below the main part of the table. These elements, called the **inner transition elements,** actually belong in the body of the table but are placed where they are simply to conserve space. The first of these rows, elements 58 through 71, fit into the chart following lanthanum and is collectively called the **lanthanides** or the **rare earths.** The second row, elements 90 through 103, belongs between actinium ($Z = 89$) and element 104. The elements in this series are termed the **actinides.**

The horizontal rows in the periodic table are called **periods** and are designated by means of Arabic numerals. The elements hydrogen and helium are members of the first period; lithium through neon are known as second-period elements; and so on.

Certain families of elements are characterized by names as well as by their group number. For example, the Group IA elements are frequently spoken of as the **alkali metals** because certain of their compounds are caustic or "alkaline." The Group IIA elements are called the **alkaline earth metals;** these elements are found in minerals and certain of their compounds are caustic, too. The Group VIIA elements are called the **halogens,** a name derived from the Greek meaning "salt former." Finally, the Group 0 elements are the **noble gases** (they are also sometimes called the *inert gases*) because of their extremely limited ability to react chemically.

The elements can also be broadly classified as **metals, nonmetals,** or **metalloids.** You are probably familiar with most of the physical properties that serve to identify metals: high electrical conductivity, luster, generally high melting points, ductility (ability to be drawn into wires), and malleability (ability to be hammered into thin sheets). Nonmetals, on the other hand, are uniformly very poor conductors of electricity, do not possess that characteristic luster found for metals, and, as solids, are brittle. Metalloids have properties that lie intermediate between those of metals and nonmetals. For example, they are useful as semiconductors of electricity. We will say more about these properties in Chapter 9.

In the periodic table, elements on the left are metals and those on the right are nonmetals. The heavy jagged line drawn from boron to astatine approximately represents the boundary between metallic and nonmetallic behavior, with elements lying immediately adjacent to the line generally having metalloid properties. Hydrogen, at the top of Group IA, is the only element that really fails to fit into this division, since it displays only nonmetallic behavior under normal conditions. It is interesting, however, that at extremely high pressures it has been found that hydrogen shows metallic properties similar to the other members of its group.

The periodic table is probably the most useful aid that chemists have at their disposal. We will see how it can be used to correlate much of the theoretical and factual information of chemistry that you should take with you when you leave this course. From the standpoint of the development of theory concerning the structure of the atom, the periodic table represents a compilation of experimental data that must be explained. A successful theory must somehow account for the way the table is structured. For example, it must explain *why* there are only two elements in period 1, *why* there are eight each in periods 2 and 3, and so on. It must also explain *why* it is that the elements in any given group exhibit similar properties.

Most of the elements are metals.

3.9 ELECTROMAGNETIC RADIATION AND ATOMIC SPECTRA

When atoms combine during chemical reactions, it is the electrons surrounding the nucleus that interact, because only the outer parts of the atoms come in contact with each other. Therefore, the chemical properties of an element are determined by the way in which the electrons in its atoms are arranged. We call this the atom's **electronic structure.** The nucleus serves mainly to determine the number of electrons that must be present to give a neutral atom. The key that has permitted the deduction of the electronic structure of the elements is an analysis of the light that atoms emit when they are energized by heating them in a flame or passing an electric discharge through them. Before we discuss this, however, let's learn what light is.

Light in all its forms—X rays, visible light, infrared and ultraviolet radiation, and radio and TV waves—is called **electromagnetic radiation.** It travels through space at a constant speed c called the speed of light, 3.00×10^8 m/s. These waves are characterized by their intensity or **amplitude,** by their **wavelength,** λ (Figure 3.10), which is the distance between consecutive peaks (or

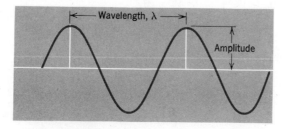

Figure 3.10

Properties of a wave.

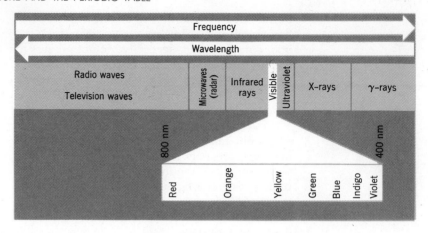

Figure 3.11
The electromagnetic spectrum.

troughs) in the wave, and by their frequency, ν, which is the number of peaks that pass by a given point per second. Wavelength and frequency are related to each other by the equation

$$\lambda \cdot \nu = c \qquad [3.1]$$

Our eyes are sensitive to only a very narrow band of wavelengths of light.

Wavelength is specified in length units that normally depend on the region of the spectrum in which the radiation occurs (see Figure 3.11). CB radios, for example, broadcast waves having wavelengths of about 11 meters. Electromagnetic radiation in the visible part of the spectrum has much shorter wavelengths, which are generally given in nanometers. The visible spectrum (Color Plate 3a) ranges from about 400 nm to about 800 nm. The SI unit of frequency is the **hertz (Hz)**, where 1 Hz = 1 (second)$^{-1}$, which we write as 1 s^{-1}.

EXAMPLE 3.2 A ham radio operator broadcasts at a frequency of 14.2 MHz (megahertz). What is the wavelength of the radio waves put out by the transmitter?

SOLUTION Since we wish to calculate the wavelength, let's solve Equation 3.1 for λ.

$$\lambda = \frac{c}{\nu}$$

The speed of light can be expressed as 3.00×10^8 m s^{-1}, and the frequency can be written $\nu = 14.2 \times 10^6$ Hz or 14.2×10^6 s^{-1} (remember, *mega* means $\times 10^6$). Substituting these values gives

$$\lambda = \frac{3.00 \times 10^8 \text{ m s}^{-1}}{14.2 \times 10^6 \text{ s}^{-1}} = 21.1 \text{ m}$$

EXAMPLE 3.3 What is the wavelength, in nanometers, of green light having a frequency of 6.67×10^{14} Hz?

SOLUTION Again we solve Equation 3.1 for λ ($\lambda = c/\nu$). Using $c = 3.00 \times 10^8$ m s^{-1} and $\nu = 6.67 \times 10^{14}$ s^{-1}, we get

$$\lambda = \frac{3.00 \times 10^8 \text{ m s}^{-1}}{6.67 \times 10^{14} \text{ s}^{-1}} = 4.50 \times 10^{-7} \text{ m}$$

The SI prefix nano means "$\times 10^{-9}$."

Since 1 nm = 10^{-9} m,

$$\lambda = 4.50 \times 10^{-7} \text{ m} \times \left(\frac{1 \text{ nm}}{10^{-9} \text{ m}}\right)$$

$$= 450 \text{ nm}$$

EXAMPLE 3.4 What is the frequency of infrared radiation that has a wavelength of 1.25×10^3 nm?

SOLUTION First we solve Equation 3.1 for ν ($\nu = c/\lambda$). If we use $c = 3.00 \times 10^8$ m s^{-1}, we must express the wavelength in meters.

$$\lambda = 1.25 \times 10^3 \text{ nm} \times \left(\frac{10^{-9} \text{ m}}{1 \text{ nm}}\right)$$

$$= 1.25 \times 10^{-6} \text{ m}$$

Now we can solve for the frequency.

$$\nu = \frac{3.00 \times 10^8 \text{ m s}^{-1}}{1.25 \times 10^{-6} \text{ m}}$$

$$= 2.40 \times 10^{14} \text{ s}^{-1}$$

The frequency is 2.40×10^{14} Hz.

If sunlight or the light from an ordinary incandescent light bulb is formed into a narrow beam by a slit and passed through a prism onto a screen, a rainbow of colors is observed (Figure 3.12). This spectrum is composed of visible light of all wavelengths and is called a **continuous spectrum** (see Color Plate 3a). However, if the source of light is a discharge tube containing a gas such as hydrogen, the observed spectrum consists of a number of *lines* projected on the screen as shown in Figure 3.13. These lines are the image of the slit, and the spectrum is called an **atomic emission spectrum** or a **line spectrum.** Typical line spectra are shown in Color Plate 3b. Obviously, the visible light emitted by hydrogen does not contain radiation of all wavelengths, as sunlight does, but rather only a relatively few wavelengths. Similar, yet distinctive line spectra are produced by all the elements when they are caused to emit light. The wavelengths of the lines are characteristic of a particular element, and can be used to identify new elements. Atomic spectra can also be used to identify the compositions of mixtures. For example, as you have probably seen and heard on TV crime shows, a sample of paint taken from the clothing of a hit-and-run victim can be analyzed for the elements it contains, as well as the relative amounts of each element; this is done by using atomic spectra. The results can be compared with a similar analysis of a paint scraping taken from a suspect's car. If they match, there is fairly strong evidence that the suspect is indeed the hit-and-run driver.

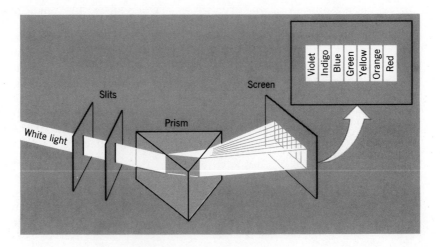

Figure 3.12

Production of a continuous spectrum.

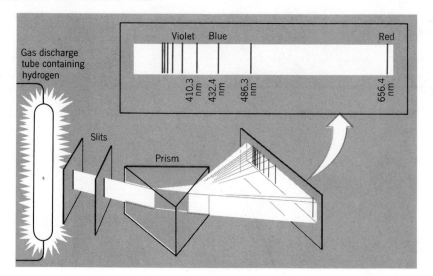

Figure 3.13

The line spectrum of hydrogen. Only the four lines whose wavelengths are given occur in the visible part of the spectrum.

The atomic spectra of certain elements are also used in modern high intensity street lighting. Sodium, for example, emits intense light at a wavelength of 589 nm, which is yellow. High- and low-pressure sodium vapor lamps are used in lighting because most of the energy supplied to the lamps appears as this visible light and relatively little of it is wasted as heat. You have probably seen examples of this golden-yellow lighting illustrated in Color Plate 4.

The occurrence of line spectra baffled physicists for many years. In 1885 Balmer found that there was a relatively simple equation that could be used to calculate the wavelengths of all the lines in the visible spectrum of hydrogen:

$$\frac{1}{\lambda} = 109,678 \text{ cm}^{-1} \left(\frac{1}{2^2} - \frac{1}{n^2}\right) \qquad [3.2]$$

where λ is the wavelength and n is an integer that can have values of 3, 4, 5, 6, . . . , ∞. By choosing a particular value of n, the wavelength of a line in the spectrum can be calculated. Thus, when $n = 3$,

$$\frac{1}{\lambda} = 109,678 \text{ cm}^{-1} \left(\frac{1}{4} - \frac{1}{9}\right)$$

$$\frac{1}{\lambda} = 15233 \text{ cm}^{-1}$$

or

$$\lambda = 6.565 \times 10^{-5} \text{ cm} = 656.5 \text{ nm}$$

Similarly, when $n = 4, 5,$ and 6, we compute λ to be 486.3, 432.4, and 410.3 nm. These values, as we can see from Figure 3.13, are equal to the wavelengths of the lines in the visible portion of the hydrogen spectrum. All the lines related by Equation 3.2 constitute what is called the *Balmer series*.

The hydrogen spectrum in Figure 3.13 lists only the lines that appear in the visible region of the spectrum. Light is also emitted by hydrogen in the infrared and ultraviolet regions. The wavelengths of these other series of lines can be fitted to the general equation (called the Rydberg equation)

$$\frac{1}{\lambda} = 109,678 \text{ cm}^{-1} \left(\frac{1}{n_1^2} - \frac{1}{n_2^2}\right) \qquad [3.3]$$

where n_1 and n_2 are integers that may assume values of 1, 2, 3, . . . , ∞ with the requirement that n_2 is always greater than n_1. Thus, where $n_1 = 1$, the values of

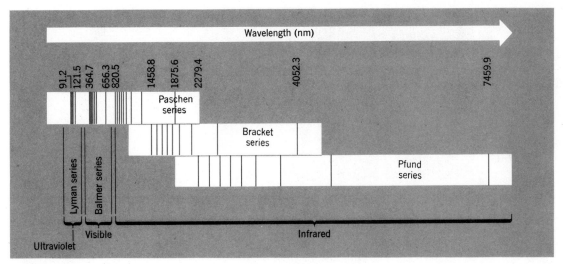

Figure 3.14
Series of lines in the hydrogen spectrum.

Table 3.3
Series of lines in the hydrogen spectrum

Series	n_1	n_2
Lyman	1	$2, 3, 4, \ldots, \infty$
Balmer	2	$3, 4, 5, \ldots, \infty$
Paschen	3	$4, 5, 6, \ldots, \infty$
Brackett	4	$5, 6, 7, \ldots, \infty$
Pfund	5	$6, 7, 8, \ldots, \infty$

n_2 can be $2, 3, 4, \ldots, \infty$ and the lines in the **Lyman series** are obtained. When $n_1 = 2$ and $n_2 = 3, 4, 5, \ldots, \infty$, we get the Balmer series. These and other series are summarized in Figure 3.14 and Table 3.3.

EXAMPLE 3.5   Calculate the wavelength of the third line in the Brackett series for hydrogen.

SOLUTION   The Rydberg equation gives the reciprocal of the wavelength

$$\frac{1}{\lambda} = 109{,}678 \text{ cm}^{-1} \left(\frac{1}{n_1^2} - \frac{1}{n_2^2} \right)$$

For the Brackett series (Table 3.3), $n_1 = 4$. The third line in the series would correspond to $n_2 = 7$. Substituting gives

$$\frac{1}{\lambda} = 109{,}678 \text{ cm}^{-1} \left(\frac{1}{4^2} - \frac{1}{7^2} \right)$$

$$= 109{,}678 \text{ cm}^{-1} (0.0420918)$$

$$= 4616.55 \text{ cm}^{-1}$$

Taking the reciprocal gives

$$\lambda = 2.16612 \times 10^{-4} \text{ cm}$$

Expressed in nanometers (1 nm = 10^{-9} m)

$$\lambda = 2166.12 \text{ nm}$$

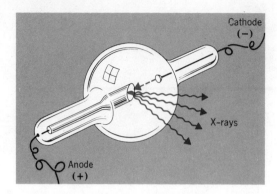

Figure 3.15

Production of X rays. A high-voltage electric discharge across the electrodes causes the anode to emit X rays.

For many years X rays were called roentgen rays.

Henry Moseley's death during the invasion of Gallipoli in WWI led Britain to assign non-combat duties to its scientists during WWII.

A postscript to this story of atomic spectra is the discovery of atomic numbers by Henry Moseley. In 1895, Wilhelm Roentgen (1845–1923) discovered that when high energy electrons in a discharge tube collide with the anode, a very penetrating kind of radiation is produced. Roentgen called them X rays. An X-ray tube is illustrated in Figure 3.15. Moseley discovered that the frequencies of the X rays produced by the tube depended on the material used for the anode. Thus, each element produces its own characteristic X-ray spectrum. In analyzing the frequencies of these X rays, Moseley found that they could be related to the location of the elements in the periodic table. He was able to assign an integer—the atomic number—which was the same as an element's position-number in the table. Experiments by Rutherford and his students enabled Moseley to conclude that this atomic number represented the number of protons in the nucleus.

3.10 THE BOHR THEORY OF THE HYDROGEN ATOM

Early attempts to account for the existence of line spectra on the basis of the motion of electrons in the atom met with complete failure. If an electron is moving around a nucleus, it must be following a curved path; otherwise it would simply leave the atom. However, a particle moving at a constant speed along a curved path undergoes an acceleration[1] and, according to the accepted laws of physics at that time, a charged particle (such as the electron) that undergoes an acceleration should continuously lose energy by emitting electromagnetic radiation. In fact, that's how radio and TV signals are broadcast. Electricity is pumped up and down an antenna at the proper frequency. The starting and

---

[1] For example, consider what happens to an object, such as a coin, when it is placed on the edge of a rapidly spinning phonograph record. Anyone who has attempted this has found that the object is thrown outward, away from the center of the record (Figure 3.16). The coin obviously experiences a force (called centrifugal force) that causes it to fly off the edge. Since the coin possesses mass, and since there is the relationship that *force = mass × acceleration,* the coin must also be experiencing an acceleration as it moves in its *circular* path at a constant speed at the edge of the record.

Figure 3.16

A coin is thrown outward from the center of a spinning record. The coin experiences an acceleration directed away from the center of the record.

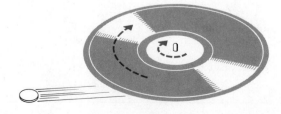

stopping (acceleration and deceleration) of the charge causes the antenna to radiate the electromagnetic waves. In terms of the atom, the known laws of physics implied that the electron should gradually lose energy and spiral in toward the nucleus, resulting in the collapse of the atom. Since atoms do not collapse, physicists were faced with a problem that challenged their most fundamental theories.

The way out of this problem found its origin in the work of Max Planck (1900) and Albert Einstein (1905). They had demonstrated that, in addition to possessing wave properties, light also has particle properties. Thus there are instances where light behaves as if it were composed of tiny packets, or **quanta,** of energy (later called **photons**). The energy, E, of the photon emitted or absorbed by a substance is proportional to the frequency of the light, ν. These two quantities are related by the equation

$$E_{\text{photon}} = h\nu \qquad\qquad [3.4]$$

where h is a proportionality constant called **Planck's constant,** which has a value of 6.63×10^{-34} joule seconds (the units are a product of energy × time).

In 1913 Niels Bohr developed a theory that incorporated the ideas of Planck and Einstein and successfully accounted for the spectrum of hydrogen. Unfortunately, the theory failed for atoms more complicated than hydrogen and it has since been replaced by a more successful one. Bohr's theory is instructive to look at briefly, though, because it illustrates how theories about the submicroscopic world of atoms develop and how they are tested.

Bohr's approach to the structure of the atom was simply to postulate that because atoms do not collapse and *because light is emitted by an atom only at certain frequencies* (which means that only certain specific energy changes occur), the electron in an atom can possess only certain, restricted amounts of energy. This is often phrased in a somewhat more esoteric way by saying that the energy of the electron is *quantized*. This means that the electron can have only certain discrete amounts of energy and none in between. We express this by saying that the electron is restricted to specific **energy levels** in the atom.

Bohr's theoretical model of the atom imagined that the electron travels around the nucleus in orbits of fixed size and energy (Figure 3.17a). From this

If the electron can't have energies between its quantized values, it cannot lose energy a little at a time and spiral into the nucleus.

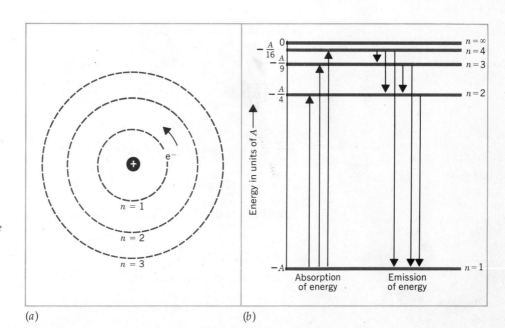

Figure 3.17

Bohr's view of the atom. (a) The electron is able to travel along certain specific orbits of fixed energy. (b) The energy of the electron changes by specific amounts when the electron goes from one orbit to another.

(a)   (b)

model he mathematically derived an equation for the energy of the electron that had the form

$$E = -A\frac{1}{n^2} \tag{3.5}$$

in which the constant A could be evaluated from a knowledge of the mass and charge of the electron and Planck's constant. The value of A is 2.18×10^{-18} joule. The quantity n is an integer, called a **quantum number,** that can have only whole-number values of 1, 2, 3, and so on, up to infinity. The quantum number serves to identify the orbit of the electron, and the energy of an electron in a particular orbit depends on the value of n (Figure 3.17b). The lowest energy level occurs when $n = 1$, since this yields the largest value for the fraction $1/n^2$ and thus the most negative (and therefore lowest) E. The idea of a negative energy seems rather odd at first glance. Actually, the minus sign occurs because of a rather arbitrary choice of the zero point on the energy scale. We will learn later that we can only measure differences in energy, so the choice of where the zero point is placed is really unimportant.

With his theory Bohr had created a model that describes how the electron behaves in an atom. His theory, just like any other theory, must be able to be checked experimentally; otherwise we cannot know if it's wrong. Of course, there is no way actually to observe the electron. Therefore, indirect evidence must be used to check the validity of the model. To do this, Bohr mathematically derived an equation for the wavelengths of the light emitted by hydrogen when it produces its atomic spectrum. According to Bohr, when energy is absorbed by an atom, for example in an electric discharge, the electron is raised in energy from one level to another, and when the electron returns to a lower energy level, a photon is emitted whose energy is equal to the difference between the two levels (see Figure 3.17). If we take n_2 to be the quantum number of the upper level and n_1 to be that of the lower (so that $n_2 > n_1$), the difference in energy, ΔE, between the two is

$$\Delta E = E_{n_2} - E_{n_1} \tag{3.6}$$

$$\Delta E = \left(-A\frac{1}{n_2{}^2}\right) - \left(-A\frac{1}{n_1{}^2}\right)$$

which can be written as

$$\Delta E = A\left(\frac{1}{n_1{}^2} - \frac{1}{n_2{}^2}\right) \tag{3.7}$$

If this energy difference appears as a photon, it would have a frequency, ν, that could be calculated from Equation 3.4,

$$\Delta E = h\nu$$

which, by incorporating Equation 3.1, can be expressed as

$$\Delta E = h\frac{c}{\lambda} = hc\frac{1}{\lambda}$$

Substituting this into Equation 3.7, we get

$$hc\frac{1}{\lambda} = A\left(\frac{1}{n_1{}^2} - \frac{1}{n_2{}^2}\right) \tag{3.8}$$

which, upon rearrangement, yields

$$\frac{1}{\lambda} = \frac{A}{hc}\left(\frac{1}{n_1{}^2} - \frac{1}{n_2{}^2}\right) \tag{3.9}$$

Using a theoretical model to derive an equation that allows a computed value to be compared with a measured value is an approach frequently followed by theoreticians.

The quantity A/hc has a value of 109,730 cm^{-1}, so our final equation is

$$\frac{1}{\lambda} = 109{,}730 \text{ cm}^{-1} \left(\frac{1}{n_1{}^2} - \frac{1}{n_2{}^2} \right)$$ [3.10]

Comparing Equations 3.3 and 3.10, we see that they are virtually identical. The Rydberg equation (3.3) is obtained from experimental observation while Equation 3.10 is derived from theory. This match between theory and experiment would suggest that Bohr was on the right track. Unfortunately, his approach was not at all successful with atoms more complex than hydrogen; however, his introduction of the notion of quantum numbers and quantized energy levels played a significant role in the development of our understanding of atomic structure.

EXAMPLE 3.6 Calculate the energy required to remove an electron from the lowest energy level of the hydrogen atom to produce the H$^+$ ion.

SOLUTION The lowest energy level has $n = 1$. The electron becomes free of the atom if it is raised to the level with $n = \infty$. The energy required to raise the electron from $n = 1$ to $n = \infty$ is given by Equation 3.6,

$$\Delta E = E_\infty - E_1$$

or, as shown in Equation 3.7,

$$\Delta E = A \left(\frac{1}{1^2} - \frac{1}{\infty^2} \right)$$

Substituting the value of A, 2.18×10^{-18} joule, gives

$$\Delta E = 2.18 \times 10^{-18} \text{ joule} \left(\frac{1}{1^2} - \frac{1}{\infty^2} \right)$$

The value of $1/\infty^2$ is zero and $1^2 = 1$. Therefore,

$$\Delta E = 2.18 \times 10^{-18} \text{ joule}$$

EXAMPLE 3.7 Calculate the energy liberated when an electron drops from the fifth to the second energy level in hydrogen.

SOLUTION We have $n_2 = 5$, $n_1 = 2$.

$$\Delta E = 2.18 \times 10^{-18} \text{ joule} \left(\frac{1}{2^2} - \frac{1}{5^2} \right)$$

$$\Delta E = 2.18 \times 10^{-18} \text{ joule} \left(\frac{1}{4} - \frac{1}{25} \right)$$

$$\Delta E = 4.58 \times 10^{-19} \text{ joule}$$

The energy liberated is 4.58×10^{-19} joule.

3.11 WAVE MECHANICS

The currently accepted theory that explains the behavior of electrons in atoms is called **wave mechanics,** which has its roots in a hypothesis put forward by Louis de Broglie in 1924. De Broglie suggested that if light can behave in some instances as if it were composed of particles, perhaps particles, at times, exhibit properties that we normally associate with waves.

De Broglie's argument proceeded as follows. Einstein had shown that the energy equivalent, E, of a particle of mass, m, is equal to

$$E = mc^2 \qquad [3.11]$$

where c is the speed of light. A photon whose energy is E could thus be said to have an effective mass equal to m. Max Planck had shown that the energy of a photon is given by Equation 3.4,

$$E = h\nu = \frac{hc}{\lambda}$$

Equating these two gives

$$\frac{hc}{\lambda} = mc^2$$

When we solve for λ, the wavelength, we obtain

$$\lambda = \frac{h}{mc}$$

If this equation also applies to particles, such as the electron, the equation can be written as

$$\lambda = \frac{h}{mv} \qquad [3.12]$$

where we have replaced c, the speed of light, by v, the speed of the particle.

Experimental evidence for this dual wave-particle nature of matter exists in the form of a phenomenon called diffraction, a property that can only be explained by wave motion. If light is allowed to pass through a very thin slit whose width is about the same as the wavelength of the light, the slit behaves as if it were a tiny light source, scattering light in all directions. This phenomenon is called **diffraction.** If two of these slits are placed alongside each other, each behaves as a separate source of light. When a screen is placed so that this light falls on it, we observe a pattern, called a **diffraction pattern,** that consists of light and dark areas as shown in Figure 3.18. In the bright areas, light waves that arrive from each hole are *in phase;* that is, the peaks and troughs of the two waves are lined up so that the amplitudes of the waves add together to produce a resultant wave of greater intensity. This is illustrated in Figure 3.19a. In the darkened areas, waves that arrive from the two pinholes are *out of phase* with each other, which means that the peaks of one wave coincide with the troughs of the other. When this happens, the amplitudes of the waves cancel (Figure 3.19b), so that zero intensity (darkness) is observed. Similar diffraction patterns can be produced with certain particles, including electrons, protons, and neutrons. Since diffraction can only be explained as a property of waves, this con-

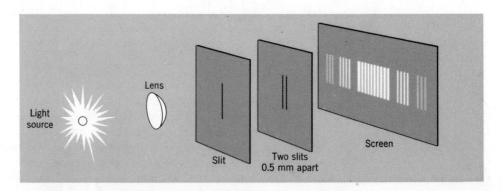

Figure 3.18

Production of a diffraction pattern.

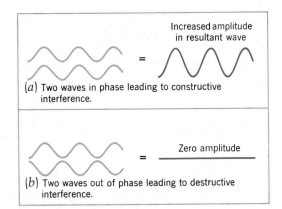

Figure 3.19

Constructive and destructive interference.

firms the wave nature of matter. The reason that the wave nature of matter was not discovered earlier is that objects large enough to see, either with the naked eye or with the aid of a microscope, possess so much mass that their wavelengths are much too short to be observed.

EXAMPLE 3.8 What is the wavelength of a grain of sand that weighs 0.000010 g and is moving at a speed of 0.010 m/s (approximately 0.02 miles/hr)?

SOLUTION We must use Equation 3.12,

$$\lambda = \frac{h}{mv}$$

Planck's constant has a value of

$$h = 6.63 \times 10^{-34} \text{ J s} \quad (\text{J s} = \text{joule} \times \text{second})$$

Since 1 joule = 1 kg m²/s², we can write h as

$$h = 6.63 \times 10^{-34} \text{ (kg m}^2/\text{s}^2) \text{ s} = 6.63 \times 10^{-34} \text{ kg m}^2/\text{s}$$

We are given that

$$m = 1.0 \times 10^{-5} \text{ g} = 1.0 \times 10^{-8} \text{ kg}$$

$$v = 0.010 \text{ m/s}$$

Substituting these quantities into our equation, we have

$$\lambda = \frac{6.63 \times 10^{-34} \text{ kg m}^2/\text{s}}{(1.0 \times 10^{-8} \text{ kg})(0.010 \text{ m/s})} = 6.6 \times 10^{-24} \text{ m}$$

This wavelength is far too small to be detected by any device existing at this time. Larger objects have even larger masses and therefore still smaller wavelengths.

We saw in Section 3.10 that one of the significant results of the Bohr theory of the atom was the introduction of integer quantum numbers. If we consider the electron to be a wave that surrounds the nucleus, we find that the appearance of integers occurs in a very natural way. To see how this might happen, let's first consider some simpler waves–those that occur on a guitar string. When a string is plucked it moves up and down in the center, but its ends remain motionless as shown in Figure 3.20a. At the ends, the height of the wave—its amplitude—is zero. These points of zero amplitude are called **nodes,** and any waves that have stationary nodes, whether on a guitar string or not, are called **standing waves.**

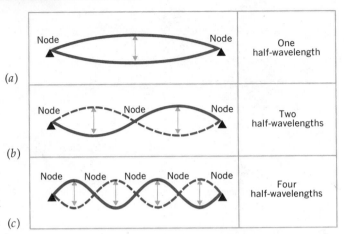

Figure 3.20

Vibrations on a guitar string. (a) An open string is plucked. (b) A harmonic produced by briefly touching the string at its midpoint as it is plucked. (c) A still higher harmonic.

If the length of the string is **L**, then the allowed wavelengths are given by

$$n\left(\frac{\lambda}{2}\right) = L$$

where **n** is a whole number.

Schrödinger shared the Nobel prize for physics in 1933 with Paul Dirac, another physicist, for their pioneering work in quantum mechanics.

If you've ever played the guitar, you know that even without using your finger to press the string at the frets along the neck of the instrument, it is possible to play a variety of notes called harmonics. For example, if the string is briefly touched at its midpoint as it is plucked, a note an octave higher is produced. Figure 3.20*b* shows that when this harmonic is played there are three nodes along the string. Still other harmonics produce even more nodes.

In examining the waves in Figure 3.20, we can see that there are certain restrictions on their allowed wavelengths along the string because each end of the string must always be a node. This means that there must be a *whole number* of *half-wavelengths* repeated along the length of string. Waves for which this is not true simply cannot exist on the string. What we see, therefore, is that a whole number—sort of a "musical quantum number"—arises automatically when we consider the standing waves that are possible on a guitar string.

A similar kind of situation exists in atoms because the electron's matter waves are standing waves, too. The shapes of the matter waves are much different than those on a guitar string, of course, because the atom is three-dimensional and the conditions that restrict the locations of the nodes are also much different. Nevertheless, the restrictions that determine what electron waves can exist give rise quite naturally to integer quantum numbers. In three dimensions, however, there are three quantum numbers instead of only one.

In 1926, Erwin Schrödinger (1887–1961) applied mathematics to the investigation of the standing waves in the hydrogen atom and began a field of study called wave mechanics or **quantum mechanics.** The mathematics here is quite advanced, so we will avoid it completely and only look at the results of the theory.

Schrödinger solved a mathematical equation called a wave equation.[2] He obtained a set of mathematical functions called **wave functions** (usually represented by the Greek letter psi, ψ) that described the shapes and energies of the electron waves. Each of these different possible waves is called an **orbital** (to distinguish it from Bohr's orbits). *Each orbital in an atom has a characteristic energy and is viewed as describing a region around the nucleus where the electron can be expected to be found.* The wave functions that describe the orbitals are themselves characterized by the values of three quantum numbers (as we hinted earlier).

---

[2] The wave equation can really be solved only for a very few species. Fortunately, one of them is the hydrogen atom, and the results obtained for hydrogen, as it turns out, can be extended quite successfully to the other elements in the periodic table.

According to wave mechanics, the various energy levels in an atom are composed of one or more orbitals and in atoms that contain more than one electron, the distribution of the electrons about the nucleus is determined by the number and kind of energy levels that are occupied. Therefore, in order to investigate the way the electrons are arranged in space, we must first examine the energy levels in the atom. This is best accomplished through a discussion of the quantum numbers.

1. The **principal quantum number,** n. The energy levels in an atom are arranged roughly into main levels, or **shells,** as determined by the principal quantum number, n. The larger the value of n, the greater the average energy of the levels belonging to the shell. We will also see that n determines the size of the orbitals. As in the Bohr theory, n may have values of 1, 2, 3, . . . , and so on up to infinity. Letters are also frequently associated with these shells as shown below.

The larger the value of **n**, the greater the average distance of the electron from the nucleus.

Principal quantum number	1	2	3	4 . . .
Letter designation	K	L	M	N . . .

For example, we would refer to the shell with $n = 1$ as the K shell.

2. The **azimuthal quantum number,** l. Wave mechanics predicts that each main shell is composed of one or more subshells, or sublevels, each of which is specified by a secondary quantum number, l, that is called the azimuthal quantum number. As we will see, this quantum number determines the shape of an orbital and, to a certain degree, its energy. For any given shell, l may have values of 0, 1, 2, and so on, up to a maximum of $n - 1$ for that shell. Thus, when $n = 1$, the largest (and only) value of l that is allowed is $l = 0$. Therefore, the K shell consists of only one subshell. When $n = 2$, two values of l occur, $l = 0$ and $l = 1$; hence the L shell is made up of two subshells. The values of l that occur for each value of n are summarized in the table at the left.

n	l
1	0
2	0, 1
3	0, 1, 2
4	0, 1, 2, 3
.	.
.	.
.	.
n	0, 1, 2, . . . , $n - 1$

Notice that *the number of subshells in any given shell is simply equal to its value of n.*

For the purposes of discussing the distribution of electrons in an atom, it is common practice to associate letters with the various values of l:

Value of l	0	1	2	3	4	5	6 . . .
Subshell designation	s	p	d	f	g	h	i . . .

The first four letters find their origin in the atomic spectra of the alkali metals (lithium through cesium). In these spectra four series of lines were observed and were termed the "sharp," "principal," "diffuse," and "fundamental" series, hence the letters s, p, d, and f. For $l = 4, 5, 6$, and so on, we just continue with the alphabet. For our purpose, however, we will be interested only in s, p, d, and f subshells, because they are the only ones populated by electrons in atoms in their **ground state** (state of lowest energy).

To specify a subshell within a given shell, we write the value of n for the shell followed by the letter designation of the subshell. For example, the s subshell of the second shell ($n = 2$, $l = 0$) would be called the $2s$ subshell. Similarly, the p subshell of the second shell ($n = 2$, $l = 1$) would be the $2p$ subshell.

3. The **magnetic quantum number,** m. Each subshell is composed of one or more orbitals. An orbital within a particular subshell is distinguished by its value of m, which serves to determine its orientation in space rela-

Table 3.4
Summary of quantum numbers

Principal Quantum Number, n (Shell)	Azimuthal Quantum Number, l (Subshell)	Subshell Designation	Magnetic Quantum Number, m (Orbital)	Number of Orbitals in Subshell
1	0	$1s$	0	1
2	0	$2s$	0	1
	1	$2p$	$-1\ 0\ +1$	3
3	0	$3s$	0	1
	1	$3p$	$-1\ 0\ +1$	3
	2	$3d$	$-2\ -1\ 0\ +1\ +2$	5
4	0	$4s$	0	1
	1	$4p$	$-1\ 0\ +1$	3
	2	$4d$	$-2\ -1\ 0\ +1\ +2$	5
	3	$4f$	$-3\ -2\ -1\ 0\ +1\ +2\ +3$	7

tive to the other orbitals. The magnetic quantum number derives its name from the fact that it can be used to explain the appearance of additional lines in atomic spectra produced when atoms are caused to emit light while in a magnetic field. It has integer values that range between $-l$ and $+l$. When $l = 0$, only one value of m is permitted, $m = 0$; therefore an s subshell consists of only one orbital (we call it an s orbital). A p subshell ($l = 1$) contains three orbitals corresponding to m equal to -1, 0, and $+1$. In a similar fashion, we find that a d subshell ($l = 2$) is composed of five orbitals and an f subshell ($l = 3$), seven. This is summarized in Table 3.4. Notice the simple progression in the number of orbitals per subshell: 1, 3, 5, 7, etc.

In describing the electron waves in an atom, we can assign each one a set of values for n, l, and m. For example, one wave will have $n = 1$, $l = 0$, and $m = 0$. This lets us identify it as a $1s$ orbital, and we think of the $1s$ orbital as being *occupied* by that electron. In a certain sense, it is as if we viewed the atom as a sort of "parking garage" for electrons, where each orbital is a possible "parking space" having a particular wave shape and energy. When an electron is in one of these parking spaces, it has the wave shape and energy of that particular orbital.

The energies of these shells, subshells, and orbitals in atoms having more than one electron are perhaps best illustrated by means of Figure 3.21. There are several points about this diagram worth noting. First, we see that the energy of the shells increases with increasing value of the principal quantum number, n. Thus the K shell, with $n = 1$, lies lowest in energy; above that there is the L shell with $n = 2$ (composed of the $2s$ and $2p$ subshells); higher still we find the M shell ($n = 3$), and so on.

Also note that as n becomes larger, the spacing between successive shells becomes less, as illustrated on the right side of Figure 3.21. Because of this narrowing energy separation, we begin to observe overlap among the subshells of the third and higher shells. The $4s$ subshell, for example, lies lower in energy than does the $3d$ subshell. This overlap is even more pronounced in higher shells, where the $5s$ subshell lies below the $4d$, the $6s$ and $4f$ below the $5d$, and the $7s$ and $5f$ below the $6d$.

Figure 3.21

Electronic energy-level diagram for atomic orbitals.

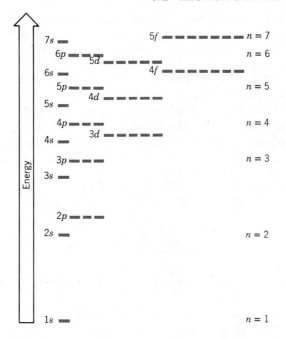

In Figure 3.21 we have indicated each orbital by means of a dash. Each s subshell is shown as a single dash to stress that it is composed of only one orbital. Likewise, p subshells are shown as three dashes, d subshells as five dashes, and f subshells as seven dashes. Observe that each orbital of a given subshell is shown to have the same energy. This applies for isolated atoms but not always for atoms in chemical compounds.

The sequence of energy levels described by Figure 3.21 turns out to be of critical importance in determining the arrangement of electrons in the atom. Before discussing this, however, we must look at yet another quantum number.

Hydrogen is the only element in which all orbitals having the same value of **n** have the same energy. That's why Bohr only needed one quantum number in his theory and why the Bohr theory didn't work for other atoms.

3.12 ELECTRON SPIN AND THE PAULI EXCLUSION PRINCIPLE

In addition to the three quantum numbers, n, l, and m, which come directly from the solution of the wave equation, there is yet another, called the **spin quantum number,** s. This quantum number arises because the electron behaves as if it were spinning (in much the same manner as the earth spins about its axis). The circular motion of electric charge that results causes the electron to act as a tiny electromagnet, just as passing an electric current through a wire wrapped about a nail causes the nail to become magnetic (see Figure 3.22). Since the electron can only spin in either of two directions, s may have only two values. These turn out to be $+\frac{1}{2}$ and $-\frac{1}{2}$, although the actual values are not really important to us.

We find, therefore, that each electron in an atom can be assigned a set of values for its four quantum numbers, n, l, m, and s, which determine the orbital in which the electron will be found and the direction in which the electron will be spinning. There is a restriction, however, on the values that these quantum numbers can have. This is expressed as the **Pauli exclusion principle,** which states that *no two electrons in any one atom may have all four quantum numbers the*

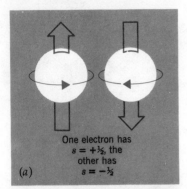

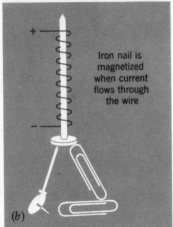

Figure 3.22

The spin of the electron. (a) The electron behaves as if it is spinning about an axis through its center. (b) A spinning charge produces a magnetic field just as the circulation of charge through a wire wrapped about a nail causes the nail to become magnetic.

same. This means that if we choose a particular set of values for n, l, and m corresponding to a particular orbital (for example, $n = 1$, $l = 0$, $m = 0$; the 1s orbital), we are able to have only two electrons with different values of the spin quantum number, s (that is, either $s = +\frac{1}{2}$ or $s = -\frac{1}{2}$). In effect, *this limits the number of electrons in any given orbital to two, and it also requires that the spins of these two electrons be in opposite directions.*

Because the Pauli exclusion principle leads to a restriction of a maximum of two electrons in any orbital, the maximum number of electrons that can be accommodated in s, p, d, and f subshells can be summarized as follows.

Subshell	Number of Orbitals	Maximum Number of Electrons
s	1	2
p	3	6
d	5	10
f	7	14

The maximum number of electrons permitted in any shell is equal to $2n^2$. For example, the K shell ($n = 1$) can hold up to two electrons and the L shell ($n = 2$) can hold a maximum of eight.

The spin of the electron is also responsible for most of the magnetic properties that we find associated with atoms and molecules. Materials that are **diamagnetic** experience no attraction for another magnet.[3] In these substances there are the same number of electrons of each spin, so their magnetic effects cancel. **Paramagnetic** substances, on the other hand, are weakly attracted to a magnetic field. In these materials there are more electrons of one spin than the other (as will always be true when an atom or molecule has an odd number of electrons) and total cancellation does not occur. The extra electrons of one spin cause the atom or molecule, as a whole, to behave as if it were itself a tiny magnet. **Ferromagnetic** substances, of which iron is the most common example, owe their very strong magnetic behavior to interactions between paramagnetic atoms in the solid state. Ferromagnetism is about 1 million times stronger than paramagnetism. This phenomenon is discussed in more detail in Chapter 21.

3.13 THE ELECTRON CONFIGURATIONS OF THE ELEMENTS

The way the electrons are distributed among the orbitals of an atom is its **electronic structure** or **electron configuration.** As we've suggested earlier, this is determined by the order in which the subshells occur on the scale of increasing energy. The reason is that in an atom in its ground state, the electrons will be found in the lowest energy levels available. In hydrogen, for instance, the single electron will be located in the 1s subshell because it is this level that has the lowest energy. To indicate that the 1s subshell is populated by one electron, we use a superscript (in this case, 1), on the subshell designation. Thus we would denote the electron configuration of hydrogen as 1s¹. As we proceed in this discussion, it will also be necessary to keep tabs on the electron spins. One method

[3] They are, in fact, repelled slightly by a magnetic field. This is a result of the motion of the electrons in the atom and is not associated with the electron's spin.

that is often employed is to symbolize an electron with its spin in one direction by an arrow pointing up, $\uparrow$, and an electron with opposite spin as an arrow pointing down, $\downarrow$. To indicate the distribution of electrons among the orbitals of the atom, we will place the arrows over bars that symbolize orbitals. Hydrogen, for example, is represented as

$$H \quad \frac{\uparrow}{1s}$$

This kind of representation of the electron configuration is usually called an **orbital diagram.**

To obtain the electron configurations of the other elements in the periodic table, let us imagine that we are able to proceed from one atom to the next by adding a proton and any necessary neutrons to the nucleus, followed by an electron that we place into the lowest available energy level. As you follow this discussion, refer both to the periodic table (Figure 3.9) and the energy-level diagram in Figure 3.21. A complete table of the electron configurations of the elements is contained in Table 3.5 on page 90.

Hydrogen is the simplest element, consisting of just a single proton and one electron. The next element, atomic number, 2, is helium. Here there are two electrons to consider and, because the $1s$ orbital can accommodate them both, the electronic structure of helium is $1s^2$ and its orbital diagram is

$$He \quad \frac{\uparrow\downarrow}{1s}$$

Note that in placing the electrons in the same orbital, we have indicated that their spins are in opposite directions as required by the Pauli exclusion principle. We refer to this by saying that their spins are *paired*, or simply that the electrons are paired.

The next two elements following He are Li and Be, which have three and four electrons, respectively. In each element, the first two electrons will enter the $1s$ subshell and, since no more than two electrons can occupy an s subshell, the remaining electron(s) must occupy the $2s$ subshell. The electron configurations of Li and Be, then, are Li, $1s^2 2s^1$ and Be, $1s^2 2s^2$. We could also show this as

$$Li \quad \frac{\uparrow\downarrow}{} \; \frac{\uparrow}{}$$

$$Be \quad \frac{\uparrow\downarrow}{1s} \; \frac{\uparrow\downarrow}{2s}$$

Since both Li and Be have a completed $1s$ subshell, which corresponds to the electron configuration of He, they can also be written as

We can also write these configurations as

Li   [He] 2**s**^{1}

Be   [He] 2**s**^{2}

$$Li \quad [He] \quad \frac{\uparrow}{}$$

$$Be \quad [He] \quad \frac{\uparrow\downarrow}{2s}$$

Here we focus our attention on the electronic structure of the outermost shell—the shell with highest n—which, in chemical reactions, is responsible for chemical changes. Electrons in shells below the outer shell are said to be **core electrons.** In this example, the inner filled $1s$ subshell is called the helium core. We will frequently find it useful to consider only those electrons that occur outside a core of electrons that corresponds to the electron configuration of one of the noble gases.

At beryllium, which has four electrons, the $2s$ subshell is completed. The fifth electron of boron ($Z = 5$), must then enter the next lowest available subshell, which is the $2p$. This gives boron the configuration $1s^2 2s^2 2p^1$. Similarly, the fifth and sixth electrons of carbon must enter the $2p$ subshell; thus, we rep-

Table 3.5
The electron configurations of the elements

Atomic Number				Atomic Number				Atomic Number			
1	H	$1s^1$		36	Kr	[Ar]	$4s^23d^{10}4p^6$	71	Lu	[Xe]	$6s^24f^{14}5d^1$
2	He	$1s^2$		37	Rb	[Kr]	$5s^1$	72	Hf	[Xe]	$6s^24f^{14}5d^2$
3	Li	[He]	$2s^1$	38	Sr	[Kr]	$5s^2$	73	Ta	[Xe]	$6s^24f^{14}5d^3$
4	Be	[He]	$2s^2$	39	Y	[Kr]	$5s^24d^1$	74	W	[Xe]	$6s^24f^{14}5d^4$
5	B	[He]	$2s^22p^1$	40	Zr	[Kr]	$5s^24d^2$	75	Re	[Xe]	$6s^24f^{14}5d^5$
6	C	[He]	$2s^22p^2$	41	Nb	[Kr]	$5s^14d^4$	76	Os	[Xe]	$6s^24f^{14}5d^6$
7	N	[He]	$2s^22p^3$	42	Mo	[Kr]	$5s^14d^5$	77	Ir	[Xe]	$6s^24f^{14}5d^7$
8	O	[He]	$2s^22p^4$	43	Tc	[Kr]	$5s^24d^5$	78	Pt	[Xe]	$6s^14f^{14}5d^9$
9	F	[He]	$2s^22p^5$	44	Ru	[Kr]	$5s^14d^7$	79	Au	[Xe]	$6s^14f^{14}5d^{10}$
10	Ne	[He]	$2s^22p^6$	45	Rh	[Kr]	$5s^14d^8$	80	Hg	[Xe]	$6s^24f^{14}5d^{10}$
11	Na	[Ne]	$3s^1$	46	Pd	[Kr]	$4d^{10}$	81	Tl	[Xe]	$6s^24f^{14}5d^{10}6p^1$
12	Mg	[Ne]	$3s^2$	47	Ag	[Kr]	$5s^14d^{10}$	82	Pb	[Xe]	$6s^24f^{14}5d^{10}6p^2$
13	Al	[Ne]	$3s^23p^1$	48	Cd	[Kr]	$5s^24d^{10}$	83	Bi	[Xe]	$6s^24f^{14}5d^{10}6p^3$
14	Si	[Ne]	$3s^23p^2$	49	In	[Kr]	$5s^24d^{10}5p^1$	84	Po	[Xe]	$6s^24f^{14}5d^{10}6p^4$
15	P	[Ne]	$3s^23p^3$	50	Sn	[Kr]	$5s^24d^{10}5p^2$	85	At	[Xe]	$6s^24f^{14}5d^{10}6p^5$
16	S	[Ne]	$3s^23p^4$	51	Sb	[Kr]	$5s^24d^{10}5p^3$	86	Rn	[Xe]	$6s^44f^{14}5d^{10}6p^6$
17	Cl	[Ne]	$3s^23p^5$	52	Te	[Kr]	$5s^24d^{10}5p^4$	87	Fr	[Rn]	$7s^1$
18	Ar	[Ne]	$3s^23p^6$	53	I	[Kr]	$5s^24d^{10}5p^5$	88	Ra	[Rn]	$7s^2$
19	K	[Ar]	$4s^1$	54	Xe	[Kr]	$5s^24d^{10}5p^6$	89	Ac	[Rn]	$7s^26d^1$
20	Ca	[Ar]	$4s^2$	55	Cs	[Xe]	$6s^1$	90	Th	[Rn]	$7s^26d^2$
21	Sc	[Ar]	$4s^23d^1$	56	Ba	[Xe]	$6s^2$	91	Pa	[Rn]	$7s^25f^26d^1$
22	Ti	[Ar]	$4s^23d^2$	57	La	[Xe]	$6s^25d^1$	92	U	[Rn]	$7s^25f^36d^1$
23	V	[Ar]	$4s^33d^3$	58	Ce	[Xe]	$6s^24f^15d^1$	93	Np	[Rn]	$7s^25f^46d^1$
24	Cr	[Ar]	$4s^13d^5$	59	Pr	[Xe]	$6s^24f^3$	94	Pu	[Rn]	$7s^25f^6$
25	Mn	[Ar]	$4s^23d^5$	60	Nd	[Xe]	$6s^24f^4$	95	Am	[Rn]	$7s^25f^7$
26	Fe	[Ar]	$4s^23d^6$	61	Pm	[Xe]	$6s^24f^5$	96	Cm	[Rn]	$7s^25f^76d^1$
27	Co	[Ar]	$4s^23d^7$	62	Sm	[Xe]	$6s^24f^6$	97	Bk	[Rn]	$7s^25f^9$
28	Ni	[Ar]	$4s^23d^8$	63	Eu	[Xe]	$6s^24f^7$	98	Cf	[Rn]	$7s^25f^{10}$
29	Cu	[Ar]	$4s^13d^{10}$	64	Gd	[Xe]	$6s^24f^75d^1$	99	Es	[Rn]	$7s^25f^{11}$
30	Zn	[Ar]	$4s^23d^{10}$	65	Tb	[Xe]	$6s^24f^9$	100	Fm	[Rn]	$7s^25f^{12}$
31	Ga	[Ar]	$4s^23d^{10}4p^1$	66	Dy	[Xe]	$6s^24f^{10}$	101	Md	[Rn]	$7s^25f^{13}$
32	Ge	[Ar]	$4s^23d^{10}4p^2$	67	Ho	[Xe]	$6s^24f^{11}$	102	No	[Rn]	$7s^25f^{14}$
33	As	[Ar]	$4s^23d^{10}4p^3$	68	Er	[Xe]	$6s^24f^{12}$	103	Lw	[Rn]	$7s^25f^{14}6d^1$
34	Se	[Ar]	$4s^23d^{10}4p^4$	69	Tm	[Xe]	$6s^24f^{13}$				
35	Br	[Ar]	$4s^23d^{10}4p^5$	70	Yb	[Xe]	$6s^24f^{14}$				

resent carbon as $1s^2 2s^2 2p^2$. However, if we examine the distribution of the electrons over the various orbitals, we face a choice: the electrons could be arranged in the following three ways:[4]

Notice that all the orbitals of the **p** subshell are shown, even though not all of them are occupied by electrons.

C   [He]   ⇅ ⇅ __ __ __

or

C   [He]   ⇅ ↑ ↓ __ __

or

C   [He]   ⇅ ↑ ↑ __
           2s   2p

The last two electrons can be paired in the same orbital, paired in different orbitals, or arranged so that their spins are in the same direction (unpaired).

As it turns out, experiments show that the last diagram gives the electron configuration that is lowest in energy. **Hund's rule** summarizes this experimental evidence: *electrons entering a subshell containing more than one orbital will be spread out over the available orbitals with their spins in the same direction.* For nitrogen $(Z = 7)$, therefore, the electron configuration would be written as $1s^2 2s^2 2p^3$, and its ground state would have the orbital diagram

N   [He]   ⇅ ↑ ↑ ↑
           2s   2p

Finally, the elements oxygen, fluorine, and neon $(Z = 8, 9,$ and 10, respectively) lead to the completion of the $2p$ subshell.

O   [He]   ⇅ ⇅ ↑ ↑

F   [He]   ⇅ ⇅ ⇅ ↑

Ne [He]   ⇅ ⇅ ⇅ ⇅
           2s   2p

After the $2p$ subshell is filled at Ne, the next lowest available energy level is the $3s$. This becomes populated with Na and Mg $(Z = 11$ and 12). After this the $3p$ subshell is gradually filled by the next six electrons as we complete the configurations of the atoms Al through Ar $(Z = 13$ through 18). Then, since the $4s$ subshell lies at lower energy than the $3d$, it is occupied next by the nineteenth and twentieth electrons of K and Ca $(Z = 19$ and 20).

Examination of Figure 3.21 reveals that after the $4s$ subshell is completed, additional electrons begin to populate the $3d$ subshell. Scandium, therefore, will have the electron configuration $1s^2 2s^2 2p^6 3s^2 3p^6 4s^2 3d^1$ or

Sc   [Ar]   ⇅   ↑ __ __ __ __ __
            4s   3d

As we proceed through Ti and V $(Z = 22$ and 23), two more electrons are added to the $3d$ subshell; however, when we get to Cr $(Z = 24)$, we find the structure

Cr   [Ar]   ↑   ↑ ↑ ↑ ↑ ↑
            4s   3d

instead of

Cr   [Ar]   ⇅   ↑ ↑ ↑ ↑ __
            4s   3d

---

[4] These are the only three possibilities that we have to consider because in an isolated atom, each of the p orbitals is equivalent in energy. Thus the arrangements,

⇅ __ __   __ ⇅ __   __ __ ⇅
   2p   2p   2p

are indistinguishable from one another experimentally.

This unexpected result occurs because a half-filled or completely filled sub-shell possesses an extra, added stability. The origin of this extra stability is very complex; so we cannot discuss it here. Nevertheless, the phenomenon is quite important and should be kept in mind. We see it again in period 4, for example, when we get to copper. On the basis of our energy-level diagram in Figure 3.21, we would predict copper to have the electron configuration

$$\text{Cu} \quad [\text{Ar}] \quad \underset{4s}{\uparrow\downarrow} \quad \underset{}{\uparrow\downarrow} \; \underset{3d}{\uparrow\downarrow} \; \underset{}{\uparrow\downarrow} \; \underset{}{\uparrow\downarrow} \; \underset{}{\uparrow}$$

The actual structure of the ground state is given by

$$\text{Cu} \quad [\text{Ar}] \quad \underset{4s}{\uparrow} \quad \underset{}{\uparrow\downarrow} \; \underset{3d}{\uparrow\downarrow} \; \underset{}{\uparrow\downarrow} \; \underset{}{\uparrow\downarrow} \; \underset{}{\uparrow\downarrow}$$

By transferring an electron from the $4s$ to the $3d$ subshell of copper, one filled and one half-filled subshell is produced, instead of the filled $4s$ and the neither filled nor half-filled $3d$ subshell that we initially would predict. Because the electron configurations of Cr and Cu are not predictable by our rules, they must be remembered as exceptions.

Similar exceptions occur else-where, too — for example, Ag and Au have filled **d** subshells just as copper does.

After the $3d$ subshell is completed at atomic number 30 (zinc), the $4p$ subshell is gradually filled as we proceed from Ga to Kr ($Z = 31$ to 36). This is followed by the completion of the $5s$ subshell from Rb to Sr ($Z = 37, 38$); the $4d$ subshell as we progress across the second row of transition elements ($Z = 39$ to 48); the $5p$ from In to Xe ($Z = 49$ to 54); and the $6s$ with Cs and Ba ($Z = 55, 56$).

Based on the energy-level sequence in Figure 3.21, we would expect that after the $6s$ subshell had been filled, we would begin to populate the $4f$ subshell next. Actually, at La ($Z = 57$) the last electron enters the $5d$ subshell instead. The $4f$ subshell is filled afterward, with a few minor irregularities. As we go to higher and higher shells, these irregularities become more frequent because the spacing between subshells becomes smaller and smaller. As we proceed from atom to atom, the energy of the various subshells shifts about somewhat as the nuclear charge increases. The result is that it is difficult to predict accurately the electron configuration of elements of very high atomic number. Nevertheless, we can account for the occurrence of the lanthanide elements by the filling of the $4f$ subshell (an f subshell can accommodate 14 electrons and the lanthanide series consists of 14 elements). Likewise, we can account for the actinide elements as the result of the filling of the $5f$ subshell.

3.14 THE PERIODIC TABLE AND ELECTRON CONFIGURATIONS

In the last section we saw that the results of wave mechanics could be used to predict the electron configurations of the elements. These electron configurations are based on theory and, to be considered useful and valid, they must somehow manifest themselves in obvious ways. One of the strongest supports for the assignment of electron configurations is the periodic table itself. Recall that, in constructing the current periodic table, elements were arranged under each other in groups because of their similar chemical properties. For example, all the elements in Group IA are metals that form ions with a charge of $1+$ when they react. If we examine the electron configurations of these elements, we see that the outer shell (shell of highest n) for each has only one electron in an s subshell.

Li	$[\text{He}] \, 2s^1$	Rb	$[\text{Kr}] \, 5s^1$
Na	$[\text{Ne}] \, 3s^1$	Cs	$[\text{Xe}] \, 6s^1$
K	$[\text{Ar}] \, 4s^1$		

Similarly, all the elements in Group IIA have an outer-shell electron configuration that we might generalize as ns^2. In fact, by examining any group within the periodic table, we see that all the elements in the group possess essentially identical outer-shell electronic structures—only their values of n differ. It is not really surprising that similar electronic structures lead to similar chemical and physical properties.

Because the properties of the elements depend on their electron configurations, it is important for you to develop the ability to write them down. There are a variety of ways to remember the sequence in which the various levels are filled; however, the best aid is the periodic table itself. As we have just seen, the order of filling the energy levels can be used to account for the structure of the periodic table. We can also work in the other direction and use the periodic table to deduce electronic structures.

If we look back over the procedure for determining electronic structures, we find that for any element in Groups IA and IIA the final electron is added to an s subshell, and that the principal quantum number of that subshell is the same as the period number. Sodium, for instance, is a period 3 element and has its outer electron in the $3s$ subshell. For elements in Groups IIIA to Group 0 the last electron is added to a p subshell whose value of n is also the same as the period number. For example, as we complete the electron configurations of the period 2 elements boron through neon, the final electron is placed in a $2p$ subshell. In the case of the transition elements, the final electron that we add is placed into a d subshell with n equal to *one less* than the period number. For example, with iron (a fourth-period element), the last electron enters a $3d$ subshell. Finally, notice that the electronic structure of an inner transition element (i.e., one from the lanthanide or actinide series) is completed by placing an electron in an f subshell whose principal quantum number is *two* less than the period number.

We can use these observations by making the periodic table tell us which subshells are filled as we build up the electron configuration of an atom. As before, we begin with hydrogen and we proceed through the elements in the periodic table in order of increasing atomic number until we arrive at the element in which we are interested. As we move across a given period, we add electrons to an s subshell when we pass through Groups IA and IIA, and to a p subshell when we pass through Groups IIIA through Group 0. The value of n for these subshells is the same as the period number. As we pass through a row of transition elements we fill a d subshell with n equal to the period number minus 1, and as we move across a row of inner transition elements we fill an f subshell with n equal to the period number minus two. This is summarized in Figure 3.23, and the following examples illustrate how this method works.

Figure 3.23

The use of the periodic table to predict electron configurations. Shaded areas depict the subshells that are filled to obtain the configuration of lead (see Example 3.10).

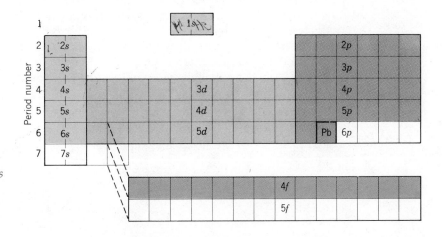

EXAMPLE 3.9

SOLUTION

What is the electron configuration of antimony (Sb)?

Antimony has atomic number 51. To reach this element we have to pass completely through periods 1, 2, 3, and 4, and part of the way across period 5. As we do this, here is what we get:

Period 1—fill the 1s subshell, which gives $1s^2$
Period 2—fill the 2s and 2p subshells $2s^2 2p^6$
Period 3—fill the 3s and 3p subshells $3s^2 3p^6$
Period 4—fill the 4s, 3d, and 4p subshells (in that order) $4s^2 3d^{10} 4p^6$
Period 5—to get as far as cadmium ($Z = 48$), we have to fill the
         5s and 4d subshells. Then we have to move three spaces
         into the "p region" to get to Sb. This gives: $5s^2 4d^{10} 5p^3$

Putting this all together, we have

$$1s^2 2s^2 2p^6 3s^2 3p^6 4s^2 3d^{10} 4p^6 5s^2 4d^{10} 5p^3$$

Some people like to collect all the subshells of a given shell together and prefer to write this as

$$1s^2 2s^2 2p^6 3s^2 3p^6 3d^{10} 4s^2 4p^6 4d^{10} 5s^2 5p^3$$

We can also write the configuration by showing the noble gas core plus the electrons outside of it

$$[Kr]4d^{10} 5s^2 5p^3$$

EXAMPLE 3.10

SOLUTION

What is the electron configuration of lead?

Lead has atomic number 82. This means that in building up the lead atom we cross through periods 1 to 5 and part of 6. Proceeding from left to right across one period after another, we fill, in order, the subshells 1s, 2s, 2p, 3s, 3p, 4s, 3d, 4p, 5s, 4d, 5p, 6s, 4f, 5d, and finally end by placing two electrons into the 6p subshell. Taking into account the maximum population of each subshell, we obtain as the electron configuration of lead,

$$1s^2 2s^2 2p^6 3s^2 3p^6 4s^2 3d^{10} 4p^6 5s^2 4d^{10} 5p^6 6s^2 4f^{14} 5d^{10} 6p^2$$

As before, we might prefer to write all subshells of a given shell together. Thus for lead we would have

$$1s^2 2s^2 2p^6 3s^2 3p^6 3d^{10} 4s^2 4p^6 4d^{10} 4f^{14} 5s^2 5p^6 5d^{10} 6s^2 6p^2$$

In Chapter 4 we will see that often we are only interested in the electron population of the outer shell of the atom. In that case we do not have to work through the entire electron configuration. Instead we can locate the element in the periodic table and immediately find the information we need. For instance, suppose that we wished to know only the outer-shell electron configuration for antimony. We find this element in Group VA and in period 5, so the outer shell is the fifth ($n = 5$). Crossing period 5 as far as Sb, we would place two electrons in the 5s subshell, ten in the 4d, and finally three in the 5p. If we are interested only in the subshells with highest n, we would conclude that the outer-shell configuration of antimony is

Sb $5s^2 5p^3$

EXAMPLE 3.11

SOLUTION

What is the outer-shell configuration of silicon?

Silicon (Si) is in period 3; therefore, the outer shell is the third shell. To get to Si in period 3 we fill the 3s subshell and place two electrons in the 3p subshell. This gives us

Si $3s^2 3p^2$

3.15 THE SPATIAL DISTRIBUTION OF ELECTRONS

In wave mechanics, our view of where the electron is likely to be found around the nucleus is far different from the idea of circular orbits imagined by Bohr. This is a consequence of the **uncertainty principle** of Heisenberg, which states that if we attempt to measure, at the same time, both the position and momentum of a particle, our measurements will be subject to errors that are related to one another by the equation

$$\Delta x \cdot \Delta(mv) \geq \frac{h}{4\pi} \qquad [3.13]$$

This equation states that the product of the uncertainty in the position of the particle, Δx, times the uncertainty in its momentum, $\Delta(mv)$, must be greater than or equal to Planck's constant divided by 4π. What this really means is that we are limited in our ability to know simultaneously where the electron is and where it is going. It leads us, instead, to refer to the probability of finding the electron in some small element of volume at various places around the nucleus. More specifically, it is the square of the wave function, ψ^2, that gives this probability.

On the basis of this concept let's look at the **probability distribution**—the way the probability varies throughout the volume of an atom—for the single electron in the 1s orbital of the hydrogen atom. A graph of ψ^2 as a function of the distance from the nucleus, r, is shown in Figure 3.24. Notice that those regions in which the probability of observing the electron is greatest lie close to the nucleus and that the probability decreases as we move away from the nucleus, gradually approaching zero as r approaches infinity. In other words, we might expect to find the electron in the hydrogen atom almost anywhere, but most of the time it stays fairly close to the nucleus, effectively surrounding it in a cloud of electronic charge (in fact, the terms *electron cloud* and *charge cloud* are frequently used when describing an electron distribution). The electron spends most of its time in regions where the probability of finding it is high, and there the concentration of charge, which we call **electron density,** is large. In other regions the charge is thinly spread and the electron density is small.

There are several ways to indicate the distribution of charge in an orbital. One way is to plot ψ^2 as we have already done. Another, in two dimensions, is to illustrate the charge cloud as shown in Figure 3.25, where the darker shaded areas represent regions of higher electron density.

Higher-energy s orbitals differ in some respects from the 1s orbital. Figure 3.26 compares the way the electron density varies for 1s, 2s, and 3s orbitals. For the 2s orbital, notice that as we move out from the nucleus the electron density drops to zero, then increases again before gradually decreasing once more. We learned earlier that those places where the amplitude or intensity of a wave drops to zero are called nodes. Thus the electron wave for an electron in a 2s orbital has a node just like waves on a guitar string. In Figure 3.26 we see that the 2s orbital has one node and the 3s orbital has two nodes.

Despite these differences, all s orbitals have an important property in common. If we draw a surface on which the probability of observing the electron is constant, the surface has the shape of a sphere. *All s orbitals have a spherical "shape."* The main difference is that as we go to higher values of n, the sphere within which we find most (for example, 90%) of the electron density becomes larger. In other words, the size of the charge cloud gets larger with increasing principal quantum number, not only for s orbitals, but also for p, d, and f orbitals. This means that electrons in orbitals of higher n will be at a greater average distance from the nucleus and that the atom gets larger as its higher energy subshells become populated.

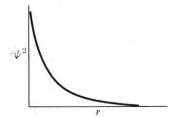

Figure 3.24

Probability of finding the electron (ψ^2) as a function of distance from the nucleus for the 1s orbital of hydrogen.

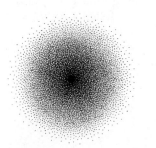

Figure 3.25

A representation of the charge cloud in hydrogen.

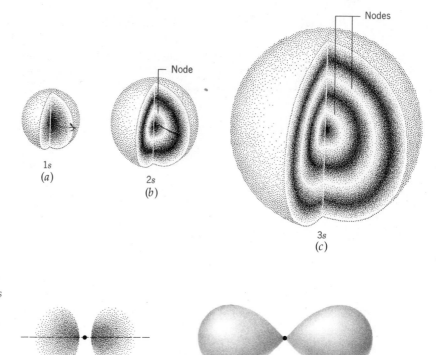

Figure 3.26

Electron density distribution in the 1s, 2s, and 3s orbitals of an atom.

Figure 3.27

The shape of a 2p orbital. (a) A "dot distribution" drawing shows that the electron density is concentrated in two regions that lie on opposite sides of the nucleus. (b) The kind of drawing that we will use in this book to represent a p orbital.

The "shape" of the electron cloud characteristic of a $2p$ orbital is illustrated in Figure 3.27. We see that for a $2p$ orbital the electron density is not distributed in a spherically symmetrical manner about the nucleus as it is in an s orbital. Instead, it is concentrated in particular regions along a straight line passing through the nucleus. Electron density occurs on both sides of the nucleus so that an electron in a $2p$ orbital spends part of its time on each side of the atom.

Higher-energy p orbitals contain nodes just as do the higher-energy s orbitals. A cross section of a $3p$ orbital is illustrated in Figure 3.28. Despite these differences there is a concentration of electron density along certain specific directions in all p orbitals. As a result, all p orbitals have definite directional properties. As we will see, these allow us to understand why molecules have the shapes they do. Because we are interested only in the directional properties

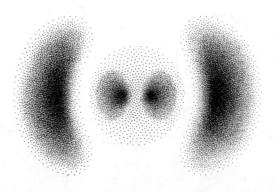

Figure 3.28

Cross section of a 3p orbital.

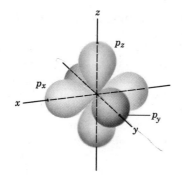

Figure 3.29

The three p orbitals of a given p subshell point in directions that are mutually perpendicular. By imagining that they point along a set of xyz axes, we can label them p_x, p_y, and p_z.

of the orbitals, we will represent all *s* orbitals as spheres and all *p* orbitals as a pair of dumbbell-shaped lobes pointing in opposite directions from the nucleus.

A *p* subshell is composed of three *p* orbitals, each having the same "shape." They differ from one another only in the directions in which their electron density is concentrated. These directions lie at right angles to one another as shown in Figure 3.29. Because the *p* orbitals can be drawn on a set of *xyz* axes, we identify the orbitals by the notation p_x, p_y, p_z.

We have seen that *s* and *p* orbitals differ in the "shapes" of their electron clouds. Orbitals in *d* and *f* subshells also have characteristic "shapes"; however, they are considerably more complicated than *p* orbitals. The *d* orbitals play an important role in the chemistry of the transition elements, and we will discuss them in some detail when we get to Chapter 21. The shapes of *f* orbitals are very complex and, since they are only required to discuss the chemistry of the inner transition elements, we will not attempt to describe them further.

3.16 THE VARIATION OF PROPERTIES WITH ATOMIC STRUCTURE

Many of the properties of the elements vary in a more or less regular fashion as we proceed from left to right within a period or from top to bottom within a group in the periodic table. Most variations can be accounted for directly in terms of variations in the electronic structures of the elements. In this section we will look briefly at three of these properties: atomic and ionic size, ionization energy, and electron affinity.

Atomic and ionic size

As we've mentioned previously, atoms are very tiny—they have diameters of about 10^{-10} m. A unit that has been used for many years to report the sizes of atoms, ions, or molecules is the **angstrom** (Å).

$$1 \text{ Å} = 10^{-8} \text{ cm} = 10^{-10} \text{ m}$$

This is not a recognized SI unit and in current scientific literature these dimensions are usually expressed either in nanometers (10^{-9} m) or picometers (10^{-12} m). It is useful to remember the conversions.

$$1 \text{ Å} = 0.1 \text{ nm}$$

$$1 \text{ Å} = 100 \text{ pm}$$

We will usually use angstroms for atomic dimensions because the numbers are a little easier to comprehend.

In any discussion of atomic size, we immediately face a basic problem of definition—exactly what do we mean by the size or radius of an atom? We have seen that the electron density in an atom does not end abruptly at some particular distance from the nucleus. Instead it trails off gradually, approaching zero at very large distances from the center of the atom. Because of this it is difficult to define precisely what we mean by the size of an atom. Since atoms never occur all by themselves in chemical systems, but are always in the neighborhood of other atoms, the radius of an atom could be taken to be half the distance between neighboring atoms when the element is present in its most dense form (i.e., most highly compacted form, which is usually the solid). Even this definition, however, is complicated because when atoms enter into a bond, as they do in molecules like H_2 or Cl_2, they approach each other more closely than nonbonded atoms do (e.g., the noble gases when they are frozen). Also, the atomic radius that we measure for atoms of a pure element will not necessarily be the same in compounds. For example, carbon atoms in diamond (pure carbon) are separated by a distance of 1.54 Å[5] and we would thus assign carbon a radius of 0.77 Å. In the ethane molecule, C_2H_6, the carbon–carbon distance is also 1.54 Å; however, in ethylene, C_2H_4, and acetylene, C_2H_2, we find carbon–carbon distances equal to 1.37 and 1.20 Å, respectively. These lead to atomic radii for carbon of 0.69 and 0.60 Å, both considerably smaller than 0.77 Å.

Despite this difficulty of definition, we can compare the atomic radii of the elements if they are measured under circumstances that lead to essentially similar kinds of bonds between their atoms. In Figure 3.30 we illustrate the variation of atomic radius with atomic number. We see that as we proceed down within a group, the size of atoms generally increases, and that as we proceed from left to right across a period, a gradual decrease in size is observed.

Figure 3.30

A graph of atomic radius versus atomic number.

[5] We will see how interatomic distances are obtained in Chapter 8.

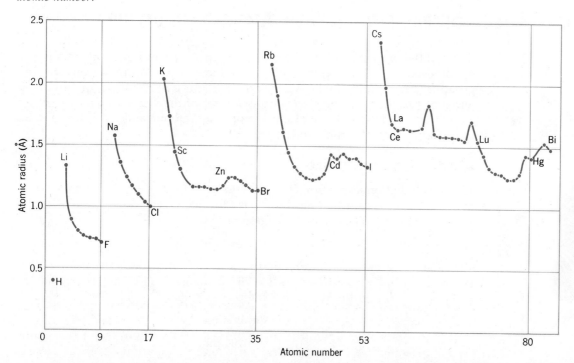

In order to interpret these trends within the periodic table in terms of electronic structure, we must look at the factors that determine the size of the outer shells of atoms, that is, the average distance at which electrons in the outer shell occur. As we have discussed in the preceding section, one factor is the principal quantum number of the outer shell (recall that the electron occurs at increasingly larger distances from the nucleus with increasing value of n).

The size of the outer shell also depends on the **effective nuclear charge** that an electron in that shell experiences. Electrons in inner shells tend to lie between the nucleus and those in the outer shell and thus shield the outer shell from the nuclear charge. In a sodium atom, for example, the 10 electrons of the neon core shield the outer $3s$ electron from the positive charge of the 11 protons in the nucleus. Therefore, the outer $3s$ electron feels an effective charge of only about $1+$. Electrons within the same shell also provide some shielding for one another; however, their ability to do so is not very great.

As we proceed from one atom to the next down within a group, each successive element has its outer electron in a shell with a larger value of n. The effective nuclear charge experienced by the outer electron(s) remains nearly the same so that the net effect is an increase in size with increase in atomic number within a group. For instance, among the alkali metals, Li through Cs, the principal quantum number of the outer electron increases from 2 for Li to 6 for Cs. The single outer electron in each of these elements, however, experiences a nearly constant effective nuclear charge of $1+$. Therefore the increase in size that occurs from Li to Cs is a result of the electron being in a shell with progressively higher n.

For the representative elements, as we move from left to right across a period, we add electrons to the same shell and simultaneously increase the nuclear charge. Since the outer-shell electrons do not shield each other from the nucleus very well, the effective nuclear charge experienced by any one electron in the outer shell increases. This increase in effective nuclear charge leads to a greater attraction for the outer-shell electrons moving from left to right in a period. As a result, the outer electrons are pulled in closer to the nucleus, and the sizes of the atoms decrease.

The variation in size as we pass through a row of transition or inner transition elements is much less than among the representative elements. This is so because electrons are being added to an inner shell as the nuclear charge gets larger. In the first row of the transition elements, for instance, the outer electrons occur in a $4s$ subshell, but each successive electron is added to the inner $3d$ subshell as we proceed across the table. The inner-shell electrons are nearly completely effective at shielding the outer shell from the nuclear charge, so the outer $4s$ electrons experience only a very gradual increase in effective nuclear charge across this region of the periodic table. Hence small changes in size occur.

The variation in atomic size across a period is not always a smooth one, as the irregularities seen for period 6 in Figure 3.30 demonstrate.

The gradual decrease in size that occurs with the filling of the $4f$ subshell in the lanthanides, which is termed the **lanthanide contraction,** has some very marked effects on the chemistry of the transition elements that follow the lanthanides in the sixth period. For example, because of the lanthanide contraction, the size of Hf is the same as the size of Zr. Since their outer-shell electron configurations are virtually identical, the chemical properties of these two elements are very similar, and it is extremely difficult to separate them from one another.

In the next chapter we will see that many elements react to produce compounds by the formation of ions. In general, positive ions are smaller than the neutral atoms from which they are formed, while negative ions are larger than neutral atoms (Table 3.6). The decrease in size that accompanies the creation of a positive ion is often a result of the removal of all the electrons from the outer shell of the atom. This gives the ion an electron configuration that is the same as

Table 3.6
Atomic and ionic radii (in angstroms)

		Positive Ions					Negative Ions		
		Atomic Radius	Ionic Radius	Charge			Atomic Radius	Ionic Radius	Charge
Group IA	Li	1.35	0.60	(+1)	Group VIIA	F	0.64	1.36	(−1)
	Na	1.54	0.95	(+1)		Cl	0.99	1.81	(−1)
	K	1.96	1.33	(+1)		Br	1.14	1.95	(−1)
	Rb	2.11	1.48	(+1)		I	1.33	2.16	(−1)
	Cs	2.25	1.69	(+1)	Group VIA	O	0.66	1.40	(−2)
Group IIA	Be	0.90	0.31	(+2)		S	1.04	1.84	(−2)
	Mg	1.30	0.65	(+2)		Se	1.17	1.98	(−2)
	Ca	1.74	0.99	(+2)		Te	1.37	2.21	(−2)
	Sr	1.92	1.13	(+2)	Group VA	N	0.70	1.71	(−3)
	Ba	1.98	1.35	(+2)		P	1.10	2.12	(−3)
Group IIIA	Al	1.43	0.50	(+3)					
	Ga	1.22	0.62	(+3)					
	In	1.62	0.81	(+3)					

		Elements That Form More Than One Ion			
Fe	1.26	Fe^{2+}	0.76	Fe^{3+}	0.64
Co	1.25	Co^{2+}	0.78	Co^{3+}	0.63
Cu	1.28	Cu^+	0.96	Cu^{2+}	0.69

a noble gas. For example, when sodium atoms react, each loses its single 3s electron to produce an Na^+ ion whose electronic structure consists of the neon core. The outer shell at this point has its principal quantum number equal to 2 and therefore the outer-shell electrons in an Na^+ ion are at a smaller average distance from the nucleus than the 3s electron in an Na atom.

When negative ions are produced from neutral atoms, electrons are added to the outer shell without any change in the nuclear charge. Each additional electron provides some degree of shielding for the other electrons, so the effective nuclear charge felt by any one electron in the outer shell decreases. At the same time, the presence of additional electrons in the outer shell increases the repulsions between electrons. Both of these factors tend to cause the outer shell to expand in size, causing the negative ion to be larger than the neutral atom.

Particles become smaller when electrons are removed from them, and larger when electrons are added.

Ionization energy

*The **ionization energy** is defined as the energy required to remove an electron from an isolated gaseous atom in its ground state.* We can represent this process by an equation such as

$$Na\ (g) \longrightarrow Na^+\ (g) + e^-$$

This is an endothermic process because the electron is attracted to the positive nucleus; therefore, energy must be supplied to remove it. Since all atoms other than hydrogen possess more than one electron, they also have more than one ionization energy. The amount of energy required to remove the first electron is called the first ionization energy, that required to remove the second electron is called the second ionization energy, and so forth. As we might expect, successive ionization energies increase in magnitude because the species from which the electron is removed becomes progressively more positively charged. For example, the first ionization energy involves the removal of an electron from a neutral atom, but the second ionization energy involves the removal of an electron from an ion whose charge is 1+.

Work must be done to pull the electron from the neutral atom.

Table 3.7
Ionization energies of the first 20 elements (kJ/mol)

	First	Second	Third	Fourth	Fifth	Sixth	Seventh	Eighth
H	1,312							
He	2,371	5,247						
Li	520	7,297	11,810					
Be	900	1,757	14,840	21,000				
B	800	2,430	3,659	25,020	32,810			
C	1,086	2,352	4,619	6,221	37,800	47,300		
N	1,402	2,857	4,577	7,473	9,443	53,250	64,340	
O	1,314	3,391	5,301	7,468	10,980	13,320	71,300	84,050
F	1,681	3,375	6,045	8,418	11,020	15,160	17,860	92,000
Ne	2,080	3,963	6,276	9,376	12,190	15,230	—	—
Na	495.8	4,565	6,912	9,540	13,360	16,610	20,110	25,490
Mg	737.6	1,450	7,732	10,550	13,620	18,000	21,700	25,660
Al	577.4	1,816	2,744	11,580	15,030	18,370	23,290	27,460
Si	786.2	1,577	3,229	4,356	16,080	19,790	23,780	29,250
P	1,012	1,896	2,910	4,954	6,272	21,270	25,410	29,840
S	999.6	2,260	3,380	4,565	6,996	8,490	28,080	31,720
Cl	1,255	2,297	3,850	5,146	6,544	9,330	11,020	33,600
Ar	1,520	2,665	3,947	5,770	7,240	8,810	11,970	13,840
K	418.8	3,069	4,600	5,879	7,971	9,619	11,380	14,950
Ca	589.5	1,146	4,941	6,485	8,142	10,520	12,350	13,830

Table 3.7 contains successive ionization energies for the first 20 elements in the periodic table. They are given in units of kilojoules per mole (kJ/mol) and represent energies needed to remove electrons from 1 mol of gaseous atoms. An examination of the data in this table points out the great stability associated with an electron core having a noble gas electron configuration. We see, for example, that for a Group IA element the first ionization energy is relatively low and that the second ionization energy is very much greater. For the Group IIA elements a large increase in ionization energy occurs after two electrons have been removed, while for Group IIIA elements the break occurs after the third electron has been lost. In fact, we see that in general a very large jump in ionization energy always occurs after an atom has lost a number of electrons that is numerically equal to its group number. Since a Group IA element contains one electron outside a noble gas electron configuration, a Group IIA element, two, and so on, these large increases in ionization energy must reflect the extreme difficulty that is encountered in trying to break into the noble gas structure that lies below the outer shell.

The variation of the first ionization energy across periods and down groups, illustrated in Figure 3.31, quite closely parallels the trends in atomic size. This really should not be too surprising, since the energy required to remove an electron from an atom completely should depend in part on how far away it is from the nucleus. In addition, the same factors that are responsible for causing an outer shell to contract in size as we proceed across a period will also lead to the electron being held more tightly. Thus, as we proceed down within a group (e.g., the alkali metals), the increase in size that occurs is accom-

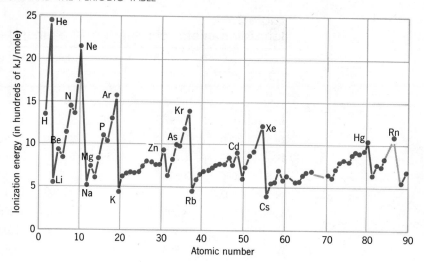

Figure 3.31

The variation of first ionization energy with atomic number.

panied by a decrease in ionization energy. As we move across a period, from left to right, the increased effective nuclear charge experienced by the outer-shell electrons causes the shell to shrink in size and also makes it more difficult to remove an electron.

If we examine more closely the trend in ionization energy across a period, we note some irregularities. In period 2, for example, we might expect a uniform increase in ionization energy as we go from Li to Ne. What we actually find, however, is that the ionization energy of beryllium is larger than that of boron and the ionization energy of nitrogen is larger than that of oxygen. These reversals can also be explained by the electronic structures of the elements.

In the case of beryllium, the first electron to be removed comes from the filled 2s subshell, whereas the electron that is removed first from boron is in the singly occupied 2p subshell. The 2p subshell is higher in energy than the 2s, so the 2p electron of boron is more easily removed than a 2s electron of beryllium.

When we get to nitrogen, we find that we have a half-filled 2p subshell (electronic structure of nitrogen is $1s^2 2s^2 2p^3$), while in oxygen the 2p subshell is occupied by four electrons. The fourth electron in this 2p subshell is in an orbital already occupied by another electron, so it experiences considerable electron-electron repulsion. As a result this electron is more easily removed than one of the electrons in a singly occupied orbital in the nitrogen atom. Note that the same inverted order of values for the ionization energy also occurs in periods 3 and 4, where the ionization energy of phosphorus is greater than for sulfur and that for arsenic is greater than for selenium.

Electron affinity

*The **electron affinity** is the energy that is released or absorbed when an electron is added to a neutral gaseous atom in its ground state.* Such a process occurs, for example, when a chlorine atom picks up an electron to become a negative ion

$$Cl\ (g) + e^- \longrightarrow Cl^-\ (g)$$

The electron affinity (like the ionization energy) applies to isolated atoms and usually represents an exothermic process. This is so because we are placing the electron into an environment where it experiences the attraction of the nucleus. We can see how the addition of an electron to an atom would release energy by considering the reverse process, pulling the electron away from the attractive force of the nucleus. If removing the electron requires work (i.e., is endothermic), the opposite process would release energy.

Table 3.8
Electron affinities[a] for the representative elements (kJ/mol)

IA	IIA	IIIA	IVA	VA	VIA	VIIA
H −73						
Li −60	Be ≈+100	B −27	C −122	N ≈+9	O −141	F −328
Na ≈−53	Mg ≈+30	Al −44	Si −134	P −72	S −200	Cl −348
K ≈−48	Ca —	Ga −30	Ge −120	As −77	Se −195	Br −325
Rb ≈−47	Sr —	In −30	Sn −121	Sb −101	Te −190	I −295
Cs −45	Ba —	Tl −30	Pb −110	Bi −110	Po −183	At −270

[a] Negative values mean that the process $M + e^- \rightarrow M^-$, is exothermic.

Adding more than one electron to an isolated atom is always endothermic overall.

There are instances where more than one electron is added to the outer shell of the atom. For example, oxygen reacts to form the ion, O^{2-}, which requires that an oxygen atom pick up two electrons. The first electron enters a neutral atom, but the second electron, which gives the ion a charge of $2-$, must be forced onto an already negative ion. This requires work; therefore we find that the second electron affinity of an atom is an endothermic quantity. Table 3.8 contains electron affinities for some representative elements. The table is not complete because electron affinities are difficult to measure and, for many elements, have not been accurately determined.

As with the ionization energy, the variations in electron affinity generally parallel the variations in atomic size. This is because we are considering the placement of an electron into the outer shell of the atom. The closer the electron can get to the nucleus, the greater the effect of the nuclear charge. Therefore, atoms that are very small and that have outer shells that experience a high effective nuclear charge (i.e., elements in the upper right of the periodic table) have very large electron affinities. On the other hand, atoms that are large and whose outer shells feel the effect of a small effective nuclear charge (such as the elements in Groups IA and IIA) have small electron affinities.

If we examine Table 3.8, we find that the nonmetals of the second period have electron affinities that are less exothermic than the elements of the same group just below them in period 3. This is a bit of a surprise because in a given group, size increases going down. Apparently, the crowding of electrons in the outer shell of a period 2 element makes the mutual repulsions of the electrons substantially greater than in the outer shell of a period 3 element. Therefore, even though the electron that is added to a period 2 element gets closer to the nucleus than one added to a period 3 element, the greater repulsions in the smaller outer shell lead to a lesser *net* amount of energy being evolved.

Finally, carbon has a rather substantial exothermic electron affinity, while that of nitrogen is actually endothermic. With carbon, the entering electron can occupy a vacant $2p$ orbital and therefore experiences only minimal electron-electron repulsions. With nitrogen, however, an additional electron must be placed into an orbital that is already occupied by an electron. The greater electron-electron repulsions that result, as well as the loss of the extra stability that a nitrogen atom has because of its half-filled p subshell, cause the electron affinity to be an endothermic quantity for nitrogen.

INDEX TO QUESTIONS AND PROBLEMS (Problem numbers in **bold type**)

REVIEW QUESTIONS

3.1 What properties are observed for cathode rays?

3.2 What is the definition of the coulomb?

3.3 Sketch and label a typical gas discharge tube.

3.4 What is the origin of the positive particles that exist in a gas discharge tube?

3.5 What are canal rays? What are they composed of?

3.6 Why was the hydrogen ion assumed to be a fundamental particle (a particle that cannot be broken into something simpler)?

3.7 Describe the three important kinds of radiation that are observed to be emitted by radioactive substances.

3.8 How are particles of different mass separated in the mass spectrometer?

3.9 Why did Rutherford conclude that the positive charge must be concentrated in a very dense nucleus within the atom?

3.10 What are the properties of the electron, the proton, and the neutron?

3.11 What are the numbers of protons, neutrons, and electrons in each of the following: $^{132}_{55}Cs$, $^{115}_{48}Cd^{2+}$, $^{194}_{81}Tl$, $^{105}_{47}Ag^+$, $^{78}_{34}Se^{2-}$?

3.12 What are the numbers of protons, neutrons, and electrons in each of the following: $^{131}_{56}Ba$, $^{109}_{48}Cd^{2+}$, $^{36}_{17}Cl^-$, $^{63}_{28}Ni$, $^{170}_{69}Tm$?

3.13 Write the appropriate symbol for each of the following isotopes:
(a) $Z = 26$, $A = 55$ (d) $Z = 71$, $A = 170$
(b) $Z = 37$, $A = 86$ (e) $Z = 70$, $A = 169$
(c) $Z = 81$, $A = 204$

3.14 How many neutrons are in each of the atoms in Question 3.13?

3.15 What isotope serves as the current standard for the atomic mass scale?

3.16 Explain why the atomic masses of some elements (such as Cl or Cu, for instance) are so far from whole numbers?

3.17 What is the difference between an atom's mass number and its actual atomic mass?

3.18 State the modern version of the periodic law.

3.19 Why were there blanks left in Mendeleev's periodic table?

3.20 What was the basis on which Mendeleev constructed his periodic table?

3.21 The element gallium was not known when Mendeleev constructed his table. Here are some properties of aluminum (just above gallium) and indium (just below gallium) along with those actually observed for gallium.

	Aluminum	Indium	Gallium
Atomic weight	27	115	70
Melting point	660°C	157°C	30°C
Boiling point	2467°C	2000°C	2403°C
Formula of chloride	$AlCl_3$	$InCl_3$	$GaCl_3$
Formula of oxide	Al_2O_3	In_2O_3	Ga_2O_3
Melting point of chloride	190°C	586°C	78°C

What would Mendeleev have predicted for these properties of gallium? How do they compare with those actually observed?

3.22 Make a sketch of the modern periodic table in which the lanthanides and actinides are placed in the body of the table in their proper locations.

3.23 Which of the following are representative elements: Mg, Ti, Fe, Se, Ni, Br?

3.24 Which of the following are transition elements: Sr, Ru, As, W, Ag, Al?

3.25 Which elements constitute the inner transition elements?

3.26 Which of these elements is a halogen: Na, Ca, Fe, F, As?

3.27 Which of these elements is an alkali metal: Br, K, O, S, N?

3.28 Which of these elements is an alkaline earth metal: Cu, B, Ba, Ne, Se?

3.29 Which of the following are metals: Ta, Nd, Se, F, Cs?

3.30 What properties are observed for nonmetals?

3.31 What significance does the periodic table have with regard to the development of theory about the electronic structure of the atom?

3.32 Sketch a wave and label its wavelength and its amplitude. What relationship exists between the wavelength and the frequency of a light wave?

3.33 What are the units of frequency in the SI? Why are the wavelengths of radio waves given in meters and those of visible light given in nanometers? What wavelength limits encompass the visible portion of the electromagnetic spectrum?

3.34 How do the wavelengths of infrared light and ultraviolet light compare to the wavelengths of visible light?

3.35 What is a line spectrum? How does it differ from a continuous spectrum? From the point of view of atomic structure, what is the significance of the occurrence of line spectra?

3.36 Describe Bohr's model of the atom. What initial evidence was there that Bohr's theory might be correct? Why was his theory eventually discarded?

3.37 What is a diffraction pattern? How is it produced?

3.38 Why don't we observe the wave properties of large objects such as baseballs or airplanes?

3.39 What direct evidence is there for the wave properties of the electron?

3.40 What is a standing wave?

3.41 What are the allowed values of the quantum number n? What restrictions are there on the allowed values of l? What values can the magnetic quantum number m have?

3.42 What does the term *ground state of an atom* mean?

3.43 How many electrons can be accommodated in each of the following types of subshells: s, p, d, f, g, h? What is the lowest value of n for a shell that has an h subshell? What are the allowed values of m for an h subshell?

3.44 What is the Pauli exclusion principle? What is Hund's rule?

3.45 Give the values of n, l, m, and s for each electron in a filled L shell.

3.46 How many electrons can be placed into the M shell of an atom?

3.47 Draw the orbital diagram for (a) phosphorus and (b) calcium.

3.48 Draw orbital diagrams for each element in the first row of transition elements ($Z = 21$ through 30). Indicate which are paramagnetic and which are diamagnetic.

3.49 Use the periodic table as a guide in writing the complete electron configurations of these elements: P, Ni, As, Ba, Rh, Ho, Sn.

3.50 Use the periodic table to arrive at the electronic structure of the outer shells of the atoms of Si, Se, Sr, Cl, O, S, As, and Ga.

3.51 Write the complete electron configuration for Rb, Sn, Br, Cr, and Cu.

3.52 Give the outer-shell electron configuration for K, Al, F, S, Tl, and Bi.

3.53 What is the major difference between a Bohr orbit and an orbital? In what way is the Heisenberg uncertainty principle involved in this comparison?

3.54 On a single set of Cartesian coordinate axes, sketch the "shapes" of the three p orbitals. Label them p_x, p_y, and p_z.

3.55 How do the $1s$ and $2s$ orbitals differ? How are they alike?

3.56 How does the shape of an s orbital differ from that of a p orbital? Sketch the shapes of these kinds of orbitals.

3.57 What is a node?

3.58 On the basis of the mutual repulsion of electrons and the spatial orientations of the p orbitals in a given subshell, the correct orbital diagram for the ground state of nitrogen seems quite reasonable. Why?

3.59 What difficulties are there in defining the size of an atom or an ion? What units are used for expressing atomic size?

3.60 Choose the largest atom: Ge, Sb, Sn, As.

3.61 Choose the larger species in each pair:

(a) S or Se
(b) C or N
(c) Fe^{2+} or Fe^{3+}
(d) O^+ or O^-
(e) S or S^{2-}

3.62 Explain the variation in ionic size observed for the series, N^{3-}, O^{2-}, and F^- (Table 3.6) in terms of the effective nuclear charge and electron-electron repulsions experienced by their outer-shell electrons.

3.63 What is the lanthanide contraction? How might this be used to explain why the elements in the sixth period following the lanthanides have higher ionization energies than the elements directly above them in the fifth period (e.g., the ionization energy of Pt = 870 kJ/mol, while that of Pd = 805 kJ/mol)?

3.64 In Table 3.6 we find some elements that form more than one positive ion. In each case the ion with the greater positive charge is smaller. Why is this so?

3.65 Define *ionization energy* and *electron affinity*.

3.66 How can we explain the variation in ionization energy across a period in the periodic table?

3.67 Choose the species with the larger ionization energy.
(a) Li or Be
(b) Be or B
(c) C or N
(d) N or O
(e) Ne or Na
(f) S or S^+
(g) Na^+ or Mg^+
Check your answers by referring to Table 3.7.

3.68 Draw a graph, on a set of axes like that below, of the ionization energy versus the number of electrons removed from the atom for each of the elements Li, C, O, S, and Ne.

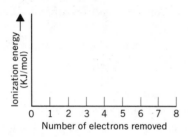

3.69 Choose the species with the more exothermic electron affinity:
(a) S or Cl   (b) S or S^- (c) P or As   (d) O or S

3.70 Why is the second electron affinity for an element always an endothermic quantity?

3.71 The Group VIIA elements have electron affinities that are considerably larger than those of the Group VIA elements. What does this suggest about the stability of the noble gas electron configuration?

3.72 Describe the contributions made by each of the following scientists toward the development of atomic theory: Faraday, Thomson, Rutherford, Millikan, Moseley, Schrödinger, de Broglie, and Planck.

REVIEW PROBLEMS (More difficult problems are marked by an asterisk.)

3.73 In Millikan's experiment, the charge on the oil droplets was always found to be a multiple of -1.60×10^{-19} coulomb. Suppose that this experiment were repeated and the following values were obtained:

-3.20×10^{-19} coulomb -2.40×10^{-19} coulomb

-5.60×10^{-19} coulomb -7.20×10^{-19} coulomb

-6.40×10^{-19} coulomb

On the basis of these data, what would be the charge on the electron?

3.74 The element Eu occurs naturally as a mixture of 47.82% $^{151}_{63}Eu$, whose mass is 150.9 amu, and 52.18% $^{153}_{63}Eu$, whose mass is 152.9 amu. Calculate the average atomic mass of Eu.

3.75 Naturally occurring boron consists of two isotopes, ^{10}B with a mass of 10.01294 and ^{11}B with a mass of 11.00931. The abundance of ^{10}B is 19.6% and that of ^{11}B is 80.4%. Calculate the average atomic mass of B.

3.76 Naturally occurring lead is composed of four isotopes. Their abundances and masses are given below. Calculate the average atomic mass of lead.

Isotope	Mass (amu)	Abundance (%)
^{204}Pb	203.973	1.48
^{206}Pb	205.9745	23.6
^{207}Pb	206.9759	22.6
^{208}Pb	207.9766	52.3

*3.77 Naturally occurring chlorine is composed of ^{35}Cl with atomic mass of 34.96885 and ^{37}Cl with atomic mass of 36.96590. The average atomic mass of Cl is 35.453. What are the percentages of each isotope in naturally occurring chlorine?

*3.78 Ordinary silver is a mixture of two isotopes, ^{107}Ag with a mass of 106.9041 amu and ^{109}Ag with a mass of 108.9047 amu. The average relative atomic weight of silver is 107.868 amu. What are the relative abundances, expressed as percents, of these two silver isotopes?

3.79 Calculate the wavelength of light whose frequency is 8.0×10^{15} Hz. Calculate the frequency of light whose wavelength is 200.0 nm.

3.80 Radio station WCBS in New York broadcasts their FM signal at a frequency of 101.1 megahertz (mHz). Their AM signal is broadcast at 880 kilohertz (kHz). What are the wavelengths of these signals expressed in meters?

3.81 Mercury vapor lamps, used for highway and street lights, produce light corresponding to the atomic spectrum of mercury. One of the lines in the spectrum is green and has a wavelength of 546 nm. What is the frequency of this line?

3.82 Verify the value for the mass of the electron using the experimentally determined charge-to-mass ratio of -1.76×10^8 coulombs/g and the measured charge, -1.60×10^{-19} coulomb.

*3.83 The charge-to-mass ratio (e/m) of the proton (a hydrogen nucleus) is 9.65×10^4 coulombs/g. The proton has a charge equal to 1.60×10^{-19} coulomb. Calculate the value of Avogadro's number.

3.84 Use the Rydberg equation (Equation 3.3) to calculate the wavelengths of the first two lines in the Pfund series of the hydrogen spectrum.

3.85 Use the Rydberg equation to calculate the wavelength of the spectral line of hydrogen that would result when an electron drops from the fourth Bohr orbit to the second, and from the sixth Bohr orbit to the third.

3.86 How much energy must be supplied to raise an electron from the first Bohr orbit to the third?

3.87 Use the Rydberg equation (Equation 3.3) to calculate the ionization energy of hydrogen and compare your answer to the results of Example 3.6.

3.88 Calculate the energy of a photon having a frequency of 3×10^{15} Hz. If a photon has an energy of 2×10^{-13} erg, what is its wavelength?

3.89 The proton (the nucleus of a hydrogen atom) has a mass of 1.67×10^{-24} g. Suppose that its diameter is 1.00×10^{-13} cm. Calculate the density of the nucleus, assuming it is spherical in shape.

*3.90 The earth has a mass of 6.59×10^{21} tons and a diameter of approximately 8000 miles. What would be the diameter of the earth (in miles) if it had the same mass but was composed entirely of nuclear material? (Use the density of nuclear material calculated in Question 3.89).

3.91 A potassium atom has a diameter of about 4.06 Å. Express this in meters, nanometers, and picometers.

*3.92 How many tons of water could be heated from 0°C to 100°C by converting 1.0 g of matter entirely into energy. Recall that it takes 1 cal of energy to raise the temperature of 1.0 g of water by 1°C. (See also Section 1.11.)

*3.93 Calculate the kinetic energy of an electron with a wavelength of 0.10 nm.

*3.94 How long would it take a 2.0 g bullet to travel the length of a 10-cm gun barrel if it had a wavelength of 0.10 nm?

*3.95 If all the energy required to remove the electrons from 1 mol of H atoms was used instead to heat water, how many grams of water could have their temperature increased by 25°C?

4

CHEMICAL BONDING: GENERAL CONCEPTS

A familiar scene. Yet how different are the properties of the substances shown here. These properties are related to the kinds of particles of which the substances are composed — the water in the soup consists of molecules while the salt in the shaker is composed of positive and negative ions. Why elements interact to form these different substances is the subject we turn to next.

In Chapter 3 we spent considerable time discussing the electronic structures of atoms and their relationships to some atomic properties. A property possessed by almost all atoms is the ability to combine with others to produce more complex species. We've seen that hydrogen atoms combine with oxygen atoms to form water molecules, and sodium atoms combine with chlorine atoms to form salt. In each of these substances, forces of attraction called **chemical bonds** hold the atoms together. The way atoms form these chemical bonds is related to their electronic structures.

The theories and language used to describe chemical bonds have evolved over the years. In this chapter we begin with the simplest concepts. Although these have been followed by more elaborate theories based on wave mechanics, they still serve many uses in discussing the structures and shapes of molecules. This is due directly to their simplicity.

4.1 LEWIS SYMBOLS

G. N. Lewis received his Ph.D. from Harvard and later served as professor of chemistry and dean at the University of California

When atoms interact to form a bond, only their outer portions come in contact; therefore, only their outer electron configurations are usually important. To keep tabs on the outer-shell (also called **valence-shell**[1]) electrons, we use a special type of notation called **Lewis symbols,** named after the American chemist, Gilbert N. Lewis (1875–1946).

To construct the Lewis symbol for an element, we write its atomic symbol surrounded by a number of dots (or X's or circles, etc.), each of which represents one electron in the atom's valence shell. For example, the element hydrogen, which has one electron in its valence shell, is given the Lewis symbol, H·. Any atom, in fact, with one electron in its outer shell has a similar Lewis symbol. This includes any element in Group IA of the periodic table, so that each of the elements Li, Na, K, Rb, Cs, and Fr has a Lewis symbol that we might generalize as X· (where X = Li, Na, etc.). Generalized Lewis symbols for the representative elements are given in Table 4.1.

Table 4.1
Lewis symbols for A-Group elements

Group	IA	IIA	IIIA	IVA	VA	VIA	VIIA
Symbol	$X\cdot$	$\cdot X\cdot$	$\cdot \overset{.}{X}\cdot$	$\cdot \overset{.}{\underset{.}{X}}\cdot$	$\cdot \overset{.}{\underset{.}{X}}:$	$\cdot \overset{..}{\underset{.}{X}}:$	$\cdot \overset{..}{\underset{..}{X}}:$

In general, the number of valence electrons that an atom of a representative element has is equal to its group number. Therefore, we see that the group number is also equal to the number of dots in the atom's Lewis symbol. This is useful to remember because it makes writing the Lewis symbol for an element very simple. Notice that in Table 4.1 the number of unpaired electrons for atoms in Groups IIA, IIIA, and IVA doesn't agree with the predictions you would make by writing their electron configurations. The Lewis symbols are written this way only because when the atoms form bonds, they *behave* as though they have the number of unpaired electrons shown by their Lewis symbols. We will see that Lewis symbols are useful in discussing bonds between atoms. The formulas we draw with them are called either *Lewis structures* or *electron-dot formulas.*

[1] *Valence* is a term sometimes associated with chemical bonding. It normally describes an atom's bond-forming capacity. Unfortunately, the term has been used with more than one meaning, so its definition has become ambiguous. For this reason, the term is rarely used today.

EXAMPLE 4.1   What is the Lewis symbol for germanium ($Z = 32$)?

SOLUTION   Germanium is in Group IVA and therefore has four valence electrons. Its Lewis symbol has four dots that we arrange symmetrically around the chemical symbol.

$$\cdot \overset{\displaystyle \cdot}{\underset{\displaystyle \cdot}{Ge}} \cdot$$

4.2   THE IONIC BOND

Chemical bonds can be divided into two general categories: **ionic** (or **electrovalent**) **bonds** and **covalent bonds.**

An ionic bond occurs when one or more electrons are transferred from the valence shell of one atom to the valence shell of another. The atom the loses electrons becomes a positive ion (a **cation**) while the atom that acquires electrons becomes negatively charged (an **anion**). The ionic bond results from the attraction between the oppositely charged ions.

An example of the formation of an ionic substance is the reaction between atoms of lithium and fluorine. The electronic structures of these are

$$\text{Li} \quad 1s^2 2s^1$$

and

$$\text{F} \quad 1s^2 2s^2 2p^5$$

When they react the lithium atom loses the electron from its $2s$ subshell and becomes Li^+. Notice that the lithium ion has an electron configuration that is the same as the noble gas helium.

> The configuration of helium is $1s^2$.

$$\text{Li} \ (1s^2 2s^1) \longrightarrow \text{Li}^+ \ (1s^2) + e^-$$

The electron lost by the lithium atom is picked up by the fluorine atom, which becomes a fluoride ion, F^-. This ion has an electron configuration identical to that of the noble gas neon.

> The configuration of neon is $1s^2 2s^2 2p^6$.

$$\text{F} \ (1s^2 2s^2 2p^5) + e^- \longrightarrow \text{F}^- \ (1s^2 2s^2 2p^6)$$

Once formed, the Li^+ and F^- ions attract one another because of their opposite charges. It is this attraction between the ions that constitutes the ionic bond.

When a reaction between lithium and fluorine actually takes place in real life, huge numbers of atoms are involved and equally huge numbers of ions are formed. These ions pack themselves together to form the ionic solid, LiF. It is important to remember that an ionic solid such as this does not contain discrete molecules, but instead contains ions packed so that the attractive forces between ions of opposite charge are maximized while repulsive forces between ions of the same charge are minimized. In LiF, for example, each cation (Li^+) is surrounded by, and attracted equally to six anions (F^-), as shown in Figure 4.1. Similarly, each anion is attracted equally to the six cations surrounding it.

The Lewis symbols introduced in the last section can be used to illustrate the transfer of electrons that occurs during the formation of an ionic compound. For instance, we can show the reaction of Li and F as

$$\text{Li} \overset{\frown}{\odot} + \cdot \overset{\displaystyle \cdot \cdot}{\underset{\displaystyle \cdot \cdot}{F}} : \longrightarrow \text{Li}^+ \left[: \overset{\displaystyle \cdot \cdot}{\underset{\displaystyle \cdot \cdot}{F}} : \right]^-$$

The brackets appearing around the fluorine on the right indicate that all eight electrons are the exclusive property of the fluoride ion, F^-.

Figure 4.1

In a crystal of lithium fluoride, LiF, each Li^+ ion is surrounded by six F^- ions and each F^- ion is surrounded by six Li^+ ions. The large spheres are F^- ions and the small spheres are Li^+ ions.

When Li and F react, electrons are lost or gained until a noble gas electron configuration is reached. Except for helium, this corresponds to an outer-shell configuration of ns^2np^6 (a total of eight electrons in the outer shell). In the last chapter we saw that the electronic structures of the noble gases possess a great deal of stability. The tendency for atoms to achieve this stable electron arrangement forms the basis of the so-called **octet rule,** which simply states that *an atom tends to gain or lose electrons until there are eight electrons in its valence shell.*

In the case of ionic bonding, the octet rule really works well only for the metals in Groups IA and IIA, and for the nonmetals. Here we find that the metals lose electrons and the nonmetals gain them. The charges on the ions that these elements form, which are given in Table 4.2, are determined by the numbers of electrons that the atoms must lose or gain to reach an octet.

When the transition elements, or metals such as tin and lead (which are to the right of the transition metals and are called **post-transition elements**), form ions, the octet rule doesn't usually apply. In general, when a positive ion is formed from an atom, the electrons are always lost first from the shell with the largest n. The transition elements therefore lose their outer s electrons before any underlying d electrons are removed. The Zn^{2+} ion, for instance, is formed when a zinc atom loses its outer $4s$ electrons. The outer-shell electron configuration of the Zn^{2+} ion is therefore $3s^23p^63d^{10}$. The $ns^2np^6nd^{10}$ outer-shell configuration is often called the **pseudonoble gas configuration.** Another outer-shell elec-

Table 4.2

Some common cations and anions

Ions with noble gas electron configurations

1+	2+	3+	3−	2−	1−
Li^+		Al^{3+}	N^{3-}	O^{2-}	F^-
Na^+	Mg^{2+}		P^{3-}	S^{2-}	Cl^-
K^+	Ca^{2+}			Se^{2-}	Br^-
Rb^+	Sr^{2+}			Te^{2-}	I^-
Cs^+	Ba^{2+}				

Table 4.3
Some cations of the transition and post-transition metals

Ions with pseudonoble gas electron configurations

1+	2+	3+
Cu^+	Zn^{2+}	Ga^{3+}
Ag^+	Cd^{2+}	In^{3+}
Au^+	Hg^{2+}	Tl^{3+}

Ions with other electron configurations

Cr^{2+}	$[Ne]3s^23p^63d^4$
Cr^{3+}	$[Ne]3s^23p^63d^3$
Mn^{2+}	$[Ne]3s^23p^63d^5$
Fe^{2+}	$[Ne]3s^23p^63d^6$
Fe^{3+}	$[Ne]3s^23p^63d^5$
Co^{2+}	$[Ne]3s^23p^63d^7$
Ni^{2+}	$[Ne]3s^23p^63d^8$
Cu^{2+}	$[Ne]3s^23p^63d^9$
Sn^{2+}	$[Kr]4d^{10}5s^2$
Pb^{2+}	$[Xe]4f^{14}5d^{10}6s^2$

tron distribution that is relatively stable is $ns^2np^6nd^5$, found in ions such as Fe^{3+} and Mn^{2+}.

Some ions occur with none of the electronic structures just discussed. Since it is very difficult to form a highly charged ion (ions with a charge greater than 3+ are rare), electron loss sometimes ceases before a noble gas electron configuration is achieved. Some examples are Ti^{2+} ($[Ar]3d^2$), V^{2+} ($[Ar]3d^3$), and Cr^{2+} ($[Ar]3d^4$). Ions of this type are common among the transition elements. A list of some common cations of the transition and post-transition elements is given in Table 4.3.

The ratio in which two elements react to form an ionic substance is usually determined by the numbers of electrons that must be lost or gained by the respective reactant atoms in order to attain a stable electron configuration. For instance, in the reaction between calcium (Group IIA) and chlorine (Group VIIA), each calcium atom must lose two electrons to achieve a noble gas structure, while each chlorine atom needs to acquire only one electron to obtain an octet. We can diagram this using Lewis symbols.

$$Ca: + \begin{matrix} \cdot \ddot{\underset{\cdot\cdot}{Cl}}: \\ \\ \cdot \ddot{\underset{\cdot\cdot}{Cl}}: \end{matrix} \longrightarrow Ca^{2+}, 2\left[:\ddot{\underset{\cdot\cdot}{Cl}}:\right]^-$$

or

$$CaCl_2$$

We see that two chlorine atoms must react with each calcium atom to produce one Ca^{2+} ion and two Cl^- ions. The neutral compound, calcium chloride, has the formula $CaCl_2$. Similar reasoning leads us to expect a compound between Li and O to have the formula Li_2O (the compound is called lithium oxide).

$$\begin{matrix} Li \cdot \\ \\ Li \cdot \end{matrix} + :\ddot{\underset{\cdot\cdot}{O}}: \longrightarrow 2Li^+, \left[:\ddot{\underset{\cdot\cdot}{O}}:\right]^{2-}$$

In writing the formulas for ionic compounds such as lithium fluoride, calcium chloride, or lithium oxide, two rules are followed.

1. The positive ion is always written first in the formula. This is simply a custom that chemists consistently follow.
2. The ratio of positive and negative ions is always chosen so that the total positive and negative charges are equal—the compound must be electrically neutral.

Applying the second rule sometimes requires a little juggling. Consider, for example, the compound formed when oxygen in the air reacts with an aluminum surface to form a thin film of aluminum oxide. Aluminum, in Group IIIA, loses three electrons to achieve a noble gas structure and produces the ion Al^{3+}. Oxygen, on the other hand, forms the ion O^{2-}. To produce a neutral compound two Al^{3+} ions must be combined with three O^{2-} ions. In that way the total positive charge $[2 \times (3+) = 6+]$ is the same as the total negative charge $[3 \times (2-) = 6-]$. Therefore, aluminum oxide has the formula Al_2O_3.

EXAMPLE 4.2   What is the formula of an ionic compound formed between Mg and P?

SOLUTION   Magnesium is in Group IIA and therefore will lose two electrons to achieve a noble gas configuration. The ion formed is Mg^{2+}.

Phosphorus is found in Group VA and must acquire three electrons to attain a noble gas structure. This produces the ion P^{3-}. To make the compound neutral, the total positive or negative charge must be a multiple of both the 2+ and 3−. The smallest number that is divisible by both 2 and 3 is 6, so

$$(2+) \times 3 = 6+$$
$$(3-) \times 2 = \underline{6-}$$
$$\text{net} \quad 0$$

The formula has to be Mg_3P_2.

There is a very simple shortcut that you can use to obtain the formula of an ionic compound. It involves just exchanging superscripts for subscripts.

The total positive charge is 6+; the total negative charge is 6−. If you use this method to obtain the formula, remember that ionic compounds are always represented by empirical formulas. Be sure that the subscripts are the simplest set of whole numbers. For example, using this method with Mg^{2+} and O^{2-} gives Mg_2O_2 initially. This must be reduced to MgO.

There are many substances that contain **polyatomic ions**—ions composed of more than one atom. The formulas of these compounds are also determined by the relative numbers of cations and anions that must be present in order to achieve a neutral solid. Table 4.4 lists some common polyatomic ions. You should be sure to learn their formulas, their charges, and their names.

The atoms within a polyatomic ion are held to each other by covalent bonds, which we will discuss later. A few of these ions are highly colored and

You must know the names of the polyatomic ions so that you can name compounds containing them, and you need to know their charges to write correct formulas for their compounds.

Table 4.4
Some common polyatomic ions

Cations	
NH_4^+	Ammonium
H_3O^+	Hydronium

Anions (alternate names in parentheses)			
CO_3^{2-}	Carbonate	ClO_2^-	Chlorite
HCO_3^-	Hydrogen carbonate (bicarbonate)	$ClO^-(OCl^-)$	Hypochlorite
$C_2O_4^{2-}$	Oxalate	PO_4^{3-}	Orthophosphate (phosphate)
NO_3^-	Nitrate	HPO_4^{2-}	Hydrogen orthophosphate (hydrogen phosphate)
NO_2^-	Nitrite		
OH^-	Hydroxide	$H_2PO_4^-$	Dihydrogen orthophosphate (dihydrogen phosphate)
SO_4^{2-}	Sulfate		
HSO_4^-	Hydrogen sulfate (bisulfate)	CrO_4^{2-}	Chromate
SO_3^{2-}	Sulfite	$Cr_2O_7^{2-}$	Dichromate
HSO_3^-	Hydrogen sulfite (bisulfite)	MnO_4^-	Permanganate
ClO_4^-	Perchlorate	$C_2H_3O_2^-$	Acetate
ClO_3^-	Chlorate		

impart their characteristic colors to compounds (and aqueous solutions) containing them—see Color Plate 5. Let us now look at some examples of how the formulas of ionic compounds containing this type of ion are obtained.

EXAMPLE 4.3   What is the formula for the ionic substance containing the ions Na^+ and CO_3^{2-}?

SOLUTION   In order for the compound to be neutral the number of positive charges must equal the number of negative charges. This requires two Na^+ ions per CO_3^{2-} ion. The compound (called sodium carbonate) therefore has the formula Na_2CO_3. One of its uses is in making glass.

EXAMPLE 4.4   Calcium phosphate, an ionic compound that is an important source of fertilizers, contains the ions Ca^{2+} and PO_4^{3-}. What is its formula?

SOLUTION   The total number of positive or negative charges represented in the formula must be divisible by both 2 and 3. The smallest number that meets this requirement is $2 \times 3 = 6$. Thus, there must be six positive and six negative charges in the formula. This is achieved by taking three Ca^{2+} ions and two PO_4^{3-} ions, so the formula of calcium phosphate is $Ca_3(PO_4)_2$.

We get this same answer exchanging superscripts for subscripts.

$$Ca^{②+} \underset{\displaystyle Ca_3(PO_4)_2}{\overset{\displaystyle \diagdown\diagup}{}} PO_4^{③-}$$

4.3 FACTORS INFLUENCING THE FORMATION OF IONIC COMPOUNDS

In general, low energy implies stability.

The driving force in the formation of an ionic bond is a lowering of the energy of the particles that come together to form the compound. We normally associate a lowering of the energy of a system with an increase in its stability. For example, a ruler standing on end will fall on its side. In the process, its potential energy decreases and it achieves a more stable position. In general, any system will seek its most stable configuration spontaneously, that is, without outside help. This applies to chemical reactions as well as to rulers.

The actual reaction between the elements lithium and fluorine is not as simple as we pictured it in the last section. Lithium does not exist as simple atoms, but rather as a solid; fluorine occurs as a gas composed of F_2 molecules. Nevertheless, an important feature of the reaction is the transfer of electrons from lithium to fluorine. Let's look briefly at the energy changes to see why it occurs.

Interpreting the coefficients as moles, the equation for the formation of 1 mol of lithium fluoride is

$$Li\ (s) + \tfrac{1}{2}F_2\ (g) \longrightarrow LiF\ (s)$$

The coefficient of $\tfrac{1}{2}$ is permitted here because we are dealing in mole-sized quantities, that is

$$1\ mol\ Li\ (s) + \tfrac{1}{2}\ mol\ F_2\ (g) \longrightarrow 1\ mol\ LiF\ (s)$$

To analyze the factors that contribute to the energy change in this reaction we will *imagine* that the change occurs in a sequence of steps, whose net result is simply the reaction we have written here. The law of conservation of energy allows us to do this. No matter how many steps it takes to get from one place to another along different paths, the net change in energy must be the same; otherwise we could create or destroy energy by running back and forth along different paths. This kind of application of the law of conservation of energy will appear frequently in the coming chapters.

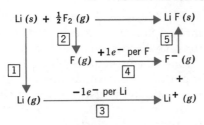

Figure 4.2

Born-Haber cycle for LiF.
Numbers correspond to processes
described in the text.

Let's suppose that instead of combining Li (s) with $\frac{1}{2}F_2$ (g) directly, we could follow the lower path in Figure 4.2. This path, called a Born-Haber cycle, consists of five different steps, each having an energy change associated with it. Let's look at each step.

Step 1. $$Li\ (s) \longrightarrow Li\ (g)$$

Vaporization of 1 mol of Li (s) to produce lithium atoms in a vapor where they are so far apart that they are effectively isolated atoms. The energy required for this endothermic process has been measured to be 37.1 kcal. (In this example we use kilocalories instead of kilojoules. Most tabulated data in the scientific literature have energies listed in kilocalories.)

Step 2. $$\tfrac{1}{2}F_2\ (g) \longrightarrow F\ (g)$$

Decomposition of $\frac{1}{2}$ mol of F_2 (g) to give 1 mol of fluorine atoms. The energy required for this endothermic process is 18.9 kcal.

Step 3. $$Li\ (g) \longrightarrow Li^+\ (g) + e^-$$

Removal of the outer electrons from the Li atoms. This is the first ionization energy of Li and has a value of 124.3 kcal. This produces 1 mol of Li^+ (g). This step is also endothermic.

Step 4. $$F\ (g) + e^- \longrightarrow F^-\ (g)$$

Addition of an electron to each F atom. This is the electron affinity of F and is equal to 78.4 kcal. This exothermic step produces 1 mol of F^- (g).

Step 5. $$Li^+\ (g) + F^-\ (g) \longrightarrow LiF\ (s)$$

Bringing together the mole of Li^+ and F^- to give 1 mol of LiF (s). Energy is also released in this step[2] and is called the **lattice energy.** It is equal to 242.8 kcal.

Just as the overall reaction is the net result of steps 1 through 5, the overall energy change is the net result of the energy changes in steps 1 through 5. Steps 1, 2, and 3 are endothermic—energy must be put into the system to vaporize Li (s), decompose F_2 (g), and remove electrons from Li (g). The total input is 180.3 kcal. The last two steps (4 and 5) are exothermic—energy is released from the system when electrons are added to F (g) and when Li^+ and F^- come together to form LiF (s). The total energy released is 321.2 kcal.

The net energy change is the difference between the energy put into the system and the energy released. Here we see that 140.9 kcal more are released than must be added. Therefore the net reaction is exothermic. This means that the system's energy is lowered and that LiF (s) is considerably more stable than Li (s) and F_2 (g). In fact, when we examine the numbers, we see that it is the large energy lowering effect of the lattice energy that is primarily responsible for the stability of this compound. This is the case with other ionic solids as well.

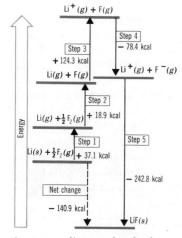

An energy diagram for the formation of LiF(s) from Li(s) and $\frac{1}{2}F_2(g)$. Going up corresponds to endothermic changes; going down corresponds to exothermic ones. Since more energy is released, overall, than is absorbed, the net reaction is exothermic (by 140.9 kcal):

---

[2] Since energy is required to separate positive and negative ions, energy must obviously be released when they are brought together. The reaction $Li^+ + F^- \rightarrow LiF$ is, therefore, exothermic.

By a similar analysis, we can also understand why electron transfer between lithium and fluorine ceases once Li^+ and F^- have been formed—that is, once these ions have acquired a noble gas configuration. Removing an additional electron from lithium requires breaking into the noble gas core, which demands the input of an enormous amount of energy. Placing an additional electron on the fluoride ion is also very expensive in terms of energy primarily because the electron must be placed into the next higher energy level (the $3s$ orbital), and also because the electron must be forced onto an already negative ion. These very endothermic processes—breaking into the noble gas core of Li^+ and exceeding the octet of F^-—require much more energy than can be gotten back in the form of lattice energy, so an ionic solid composed of Li^{2+} and F^{2-} is unstable and never even forms. Similar analyses for other ionic compounds of the representative elements show that the lattice energy is only able to offset the endothermic contributions to the net energy change up to the point where the ions have achieved a noble gas configuration.

At this point we can ask: What conditions most favor the formation of an ionic substance? From our analysis we see that the most stable ionic compounds will result when elements of low ionization energy combine with elements of high electron affinity, or when the lattice energy of the resulting compound is very large, or both. Under these conditions more energy is given off by the exothermic processes than is absorbed by the endothermic ones, with the net result that the total energy contained within the reacting species decreases.

Since metals generally have rather low ionization energies and electron affinities, they tend to lose electrons to form cations; nonmetals, on the other hand, have large ionization energies and electron affinities, so they usually acquire electrons to produce anions. For this reason, most compounds formed between metals and nonmetals are ionic, particularly the substances formed when an element from Group IA or IIA reacts with an element in the upper right corner of the periodic table (excluding Group 0).

> Ionic compounds tend to be formed between metals and nonmetals.

4.4 THE COVALENT BOND

In many instances the formation of an ionic substance is not energetically favorable. For example, the creation of a cation may require too large an energy input (ionization energy) to be recovered by the energy released when the anion is formed and the ionic solid is produced (electron affinity and lattice energy). In these situations a **covalent bond** is formed.

A covalent bond results from the *sharing* of a *pair* of electrons between atoms. The binding force results from the attraction between these shared electrons and the positive nuclei of the atoms entering into the bond. In this sense, the electrons serve as a sort of glue cementing the atoms together. Consider, for example, the formation of the H_2 molecule from two hydrogen atoms. As the atoms approach each other, the single $1s$ electron on each of them begins to feel the attraction of both nuclei. The electron density therefore begins to shift to the region between the nuclei, as shown in Figure 4.3.

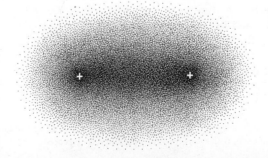

Figure 4.3

The electron distribution in the H_2 molecule.

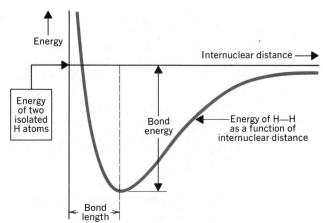

Energy

Internuclear distance ⟶

Energy of two isolated H atoms

Bond energy

Energy of H—H as a function of internuclear distance

Bond length

Figure 4.4

Energy diagram for the formation of H₂ from two hydrogen atoms. The bond energy is 435 kJ/mol of bonds and the bond length is 0.75 Å.

At small distances, the repulsions between nuclei outweigh the attractions that the nuclei have for the electron density between them.

A covalent bond is sometimes called an electron pair bond.

If we examine the energy changes that accompany the formation of the bond, we find that as the atoms come together the energy begins to decrease. This is because the electrons are coming closer to another positive nucleus to which they are also attracted (remember how potential energy changes between particles that attract each other). The energy curve for the molecule is shown in Figure 4.4. Notice that at small internuclear distances the energy rises steeply. This is caused by the repulsions between the nuclei. The most stable (lowest energy) distance of separation between the two nuclei occurs when the energy is a minimum. At this point the attractions and repulsions are balanced. The depth of this minimum is the amount of energy that must be supplied to separate the atoms and is called the **bond energy.** The distance between the nuclei when the energy is a minimum is called the **bond length** or **bond distance.**

When two atoms such as hydrogen share a pair of electrons, the spins of the electrons become paired. This is an important aspect of the creation of a covalent bond. Each H atom completes its valence shell by acquiring a share of an electron from another atom. We can indicate the formation of H_2, using Lewis symbols, as

$$\text{H} \cdot + \text{H} \cdot \longrightarrow \text{H} : \text{H}$$

in which the pair of electrons in the bond is shown as a pair of dots between the two H atoms. Often a dash is used instead of the pair of dots, so the H_2 molecule may be represented as H—H.

As with the ionic bond, the number of covalent bonds that an atom will form can frequently be predicted by counting the number of electrons required to achieve a stable electron configuration (usually that of a noble gas). For example, the carbon atom has four electrons in its valence shell; to attain a noble gas configuration it usually acquires, through sharing, four additional electrons. The carbon atom, therefore, is capable of forming four bonds with H atoms to form the molecule CH_4 (methane, the chief component of natural gas).

$$\cdot \overset{\displaystyle \cdot}{\underset{\displaystyle \cdot}{\text{C}}} \cdot + 4\text{H}\times \longrightarrow \text{H} \overset{\displaystyle \text{H}}{\underset{\displaystyle \text{H}}{\overset{\times}{\underset{\times}{\text{C}}}}} \text{H}$$

Nitrogen, which has five valence electrons, has to gain only three electrons through sharing to complete an octet; therefore, nitrogen forms three covalent bonds with hydrogen to form the ammonia molecule, NH_3. In a similar fashion it is easy to see why the formula for water is H_2O and that for hydrogen fluoride is HF.

$$\text{H} \overset{\times}{\underset{\displaystyle \overset{..}{\underset{\displaystyle \text{H}}{\text{N}}}}{\times}} \text{H} \qquad \text{H} \overset{\times}{\underset{\displaystyle \overset{..}{\underset{\displaystyle \text{H}}{\text{O}}}}{\times}} : \qquad \text{H} \overset{..}{\underset{\displaystyle ..}{\times}} \text{F} :$$

Unfortunately, it is not always possible to predict the formula of a covalent molecule on the basis of these simple rules. There are many examples of covalent compounds that fail to obey the octet rule. For instance, the molecule $BeCl_2$, which exists in the gas phase at high temperatures, is formed by the pairing of the two Be valence electrons with electrons on two chlorine atoms:[3]

$$:\ddot{C}l\cdot + {}^{\times}Be^{\times} + \cdot\ddot{C}l: \longrightarrow :\ddot{C}l\overset{\times}{\cdot}Be\overset{\times}{\cdot}\ddot{C}l:$$

In this molecule the Be atom has only four electrons in its valence shell. Another example is BCl_3:

$$\begin{array}{c} :\ddot{C}l: \\ :\ddot{C}l\overset{\times}{\underset{\times}{\cdot}}\overset{}{B}\overset{\times}{\underset{\cdot}{\cdot}}\ddot{C}l: \end{array}$$

Molecules having atoms that are surrounded by less than an octet of electrons are rare and must simply be learned as exceptions. The method we will learn later for drawing Lewis structures doesn't work for them.

There are many molecules in which the central atom has more than eight electrons in its valence shell. Two typical examples are PCl_5 and SF_6. To form covalent bonds between the central atom and each of the surrounding atoms, more than four pairs of electrons are needed. In PCl_5, for example, there are five covalent bonds; in SF_6 there are six. The central atom in each of these molecules uses all its valence electrons to form covalent bonds.

In these compounds, both phosphorus and sulfur have exceeded the number of electrons required for a noble gas electron configuration. This can occur with these elements because, in each case, the valence shell can accommodate more than eight electrons (both P and S are in the third period and the third shell can contain up to 18 electrons because of the availability of the relatively low-energy $3d$ subshell). Elements in the second period (Li to Ne) never form compounds with more than eight electrons in their valence shell because the second shell cannot have more than an octet.

In each of the molecules that we've discussed up to this point, the atoms have been joined by covalent bonds that each consist of a single pair of electrons. These are called **single bonds.** It is also possible for a pair of atoms to share two or even three pairs of electrons. For example, in carbon dioxide the carbon shares two pairs of electrons with each oxygen atom. These are **double bonds.**

$$\ddot{O}::C::\ddot{O}$$

The electrons in the bonds are counted as belonging to each atom, so we can count eight electrons around carbon and around each oxygen.

8 electrons

[3] Even though $BeCl_2$ is formed from elements in Groups IIA and VIIA, it is covalent instead of ionic.

Using dashes to represent a pair of electrons, we can also draw this as

$$\ddot{O}=C=\ddot{O}$$

Nitrogen is an example of a molecule with a **triple bond.**

$$:N⋮⋮N:\quad\text{or}\quad:N{\equiv}N:$$

If we count all six electrons between the atoms as belonging to both, each atom has an octet.

Drawing Lewis structures

Lewis structures for covalently bonded molecules and polyatomic ions are useful. One reason, as we will see in the next chapter, is that they allow us to predict the shapes of molecules or ions. Therefore, you should learn how to draw them.

The first step in drawing a Lewis structure is deciding what atoms are attached to each other. For example, in CO_2 we must know that there are two O atoms bound to the C atom and that the molecule does not have a structure such as O—O—C. In many instances the arrangement of atoms can be inferred from the formula, since it is common practice to write the central atom of a molecule first in the formula, followed by the atoms that surround the central atom. This is so with CO_2, for example. It is also true for species such as NH_3, NO_2, NO_3^-, SO_3, CO_3^{2-}, and SO_4^{2-}. Unfortunately, it is not true for H_2O and H_2S (in which H atoms are bound to O and S, respectively). Nor is it true for molecules such as HClO (in which the O is the central atom) or ions like SCN^- (in which C is central). The structure of the molecule is therefore not always obvious. If you must guess, the most symmetrical arrangement of atoms has the greatest chance of being correct. Once we know the arrangement of atoms in the molecule, however, we can then go about distributing the valence electrons. The procedure for doing this can be summarized in the following steps, which we will illustrate by some examples.

It's useful to remember that hydrogen can never be a central atom because it forms only one covalent bond.

1. *Count all the valence electrons of the atoms.* If the species is an ion, add an additional electron for each negative charge or subtract an electron for each positive charge.
2. *Place one pair of electrons in each bond.*
3. *Complete the octets of the atoms bonded to the central atom.* (Remember, however, that the valence shell of any hydrogen atom is complete with only two electrons.)
4. *Place any additional electrons on the central atom in pairs.*
5. *If the central atom still has less than an octet, you must form multiple bonds so that each atom has an octet.*

Now let's look at some examples that show how this procedure works.

EXAMPLE 4.5 What is the Lewis structure for CCl_4 (carbon tetrachloride, a substance once used as a dry cleaning solvent until it was found to be very toxic)?

SOLUTION First we need the arrangement of atoms. The formula suggests that it is

Cl
Cl C Cl
Cl

Now we count the valence electrons:

carbon (Group IVA) contributes $4e^-$ $4e^-$

chlorine (Group VIIA) contributes $7e^-$ each $\underline{28e^-}$

              Total $32e^-$

We begin distributing the electrons by placing a pair in each bond. This gives

$$\begin{array}{c} \text{Cl} \\ \ddot{} \\ \text{Cl} : \ddot{\text{C}} : \text{Cl} \\ \ddot{} \\ \text{Cl} \end{array}$$

This has used $8e^-$, so there are $32e^- - 8e^- = 24e^-$ left. We now complete the valence shells of the Cl atoms.

$$\begin{array}{ccc} : \ddot{\text{Cl}} : & & : \ddot{\text{Cl}} : \\ : \ddot{\text{Cl}} : \ddot{\text{C}} : \ddot{\text{Cl}} : & \text{or} & : \ddot{\text{Cl}} - \text{C} - \ddot{\text{Cl}} : \\ : \ddot{\text{Cl}} : & & : \ddot{\text{Cl}} : \end{array}$$

This has used all $24e^-$, so none are left. Each atom has an octet and, therefore, we can stop. This is the Lewis structure for CCl_4.

EXAMPLE 4.6   What is the Lewis structure for SF_4?

SOLUTION   We begin by choosing an arrangement of atoms. Once again, the formula provides a clue.

$$\begin{array}{c} \text{F} \\ \text{F S F} \\ \text{F} \end{array}$$

Next, we count valence electrons: 6 from sulfur and 7 from each fluorine gives a total of $6 + 28 = 34$ electrons. As before, we start by placing a pair in each bond.

$$\begin{array}{c} \text{F} \\ \text{F} : \ddot{\text{S}} : \text{F} \\ \ddot{\text{F}} \end{array}$$

This leaves us with $34 - 8 = 26$ electrons. Next we complete the valence shells of fluorine.

$$\begin{array}{c} : \ddot{\text{F}} : \\ : \ddot{\text{F}} : \ddot{\text{S}} : \ddot{\text{F}} : \\ : \ddot{\text{F}} : \end{array}$$

This has used $24e^-$, so there are still $2e^-$ left. Step 4 tells us to place them on the sulfur (the central atom). Rearranging the fluorines to make room for the extra dots on sulfur gives us the dot structure for SF_4.

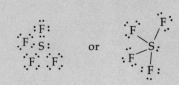

EXAMPLE 4.7   What is the Lewis structure for the carbonate ion, CO_3^{2-}?

SOLUTION   This time we have an ion whose formula suggests the atom arrangement

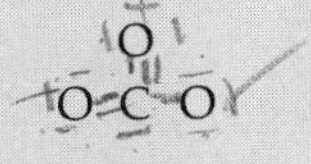

$$
\begin{array}{c}
O \\
\| \\
O\text{-}C\text{-}O
\end{array}
$$

Counting valence electrons; carbon supplies $4e^-$, each oxygen supplies $6e^-$, and the negative charge adds another $2e^-$. The total is $4 + 18 + 2 = 24$ electrons. Placing a pair in each bond gives

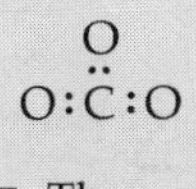

$$
\begin{array}{c}
O \\
.. \\
O:C:O
\end{array}
$$

This has used $6e^-$, leaving us with $18e^-$. These are used to complete the octets around oxygen.

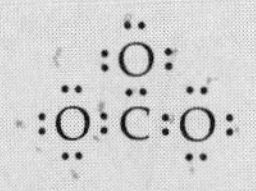

$$
\begin{array}{c}
:\ddot{O}: \\
.. \quad .. \quad .. \\
:\ddot{O}:\ddot{C}:\ddot{O}:
\end{array}
$$

This time we've run out of electrons, but carbon doesn't have an octet. What we must do now is move one of the unshared pairs on an oxygen into one of the bonds (it doesn't matter which one).[4]

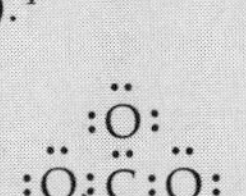

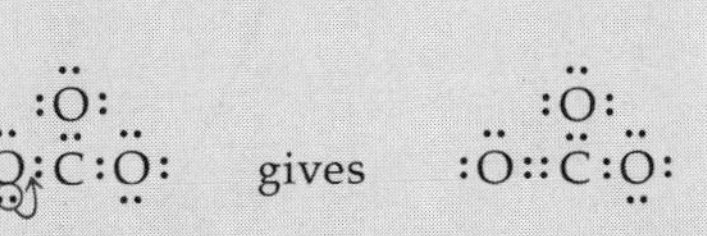

$$
\begin{array}{ccc}
:\ddot{O}: & & :\ddot{O}: \\
.. \quad .. \quad .. & \text{gives} & .. \quad \quad .. \\
:\ddot{O}:C:\ddot{O}: & & :\ddot{O}::C:\ddot{O}:
\end{array}
$$

Notice that the oxygen hasn't lost the pair, but carbon completes its octet by sharing it. Finally, we should be sure to indicate the charge on the ion by enclosing the Lewis structure in brackets with the charge outside.

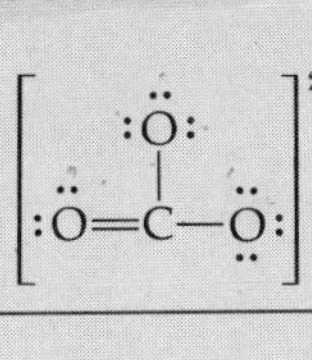

$$
\left[
\begin{array}{c}
:\ddot{O}: \\
| \\
:\ddot{O}=C-\ddot{O}:
\end{array}
\right]^{2-}
$$

EXAMPLE 4.8   Draw the electron-dot formula for the poisonous gas, hydrogen cyanide, HCN (C is the central atom).

SOLUTION   There are 10 valence electrons to distribute (1 from H, 4 from C, 5 from N). First we place a pair in each bond.

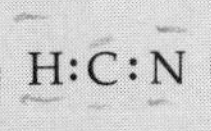

$$
H:C:N
$$

This accounts for 4 electrons. As we add the remaining $6e^-$, we must keep in mind that the valence shell of H can hold only $2e^-$. No more electrons can be placed around H because it already has two. Completing the octet of nitrogen gives

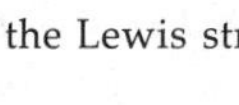

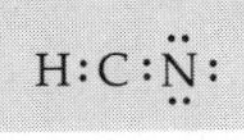

$$
H:C:\ddot{N}:
$$

[4] Applying this approach to drawing the Lewis structures of BCl_3 or $BeCl_2$ would give

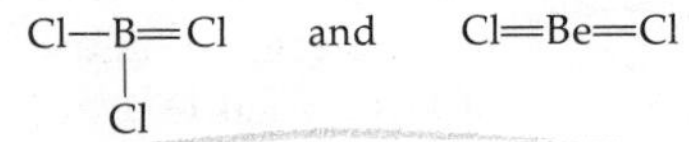

$$
\begin{array}{ccc}
Cl-B=Cl & \text{and} & Cl=Be=Cl \\
| \\
Cl
\end{array}
$$

But these molecules *behave* as though they only contain single bonds, so the structures with double bonds are not acceptable, even though they satisfy the octet rule. As mentioned earlier, BCl_3 and $BeCl_2$ are exceptions to the octet rule and must be learned as special cases.

Once again the central atom doesn't have an octet. This can be corrected as follows.

$$\text{H}:\text{C}:\ddot{\text{N}}: \longrightarrow \text{H}:\text{C}:::\text{N}:$$

The HCN molecule contains a triple bond.

4.5 BOND ORDER AND SOME BOND PROPERTIES

At the beginning of the last section we described two properties of covalent bonds: bond length and bond energy. Bond length, you recall, is the distance between the nuclei of the two atoms joined by the bond. The bond energy is the amount of energy necessary to break the bond to produce neutral fragments. For a diatomic molecule such as H_2 this represents the process

$$\text{H}:\text{H}\,(g) \longrightarrow \text{H}\cdot(g) + \text{H}\cdot(g)$$

while in a molecule such as C_2H_6 the carbon–carbon bond energy represents the energy needed to cause the reaction

$$\begin{matrix} \text{H} & \text{H} & & \text{H} & & \text{H} \\ | & | & & | & & | \\ \text{H}-\text{C}-\text{C}-\text{H} & \longrightarrow & \text{H}-\text{C}\cdot & + & \cdot\text{C}-\text{H} \\ | & | & & | & & | \\ \text{H} & \text{H}\,(g) & & \text{H}\,(g) & & \text{H}\,(g) \end{matrix}$$

It should not be surprising to learn that the magnitudes of the bond length and bond energy differ for bonds between different atoms. Some bonds are strong and some are weak; some have long bond lengths and some bonds are short.

There are ways of experimentally measuring bond lengths and bond energies.

One factor that affects the length of a bond and the bond energy is the amount of electron density between the nuclei. A convenient way of expressing this is by giving the **bond order**—*the number of covalent bonds that exist between a pair of atoms.* Consider, for example, the following molecules:

$$\begin{matrix} \text{H} & \text{H} & & \text{H} & \text{H} & & & & \\ | & | & & | & | & & & & \\ \text{H}-\text{C}-\text{C}-\text{H} & & \text{H}-\text{C}=\text{C}-\text{H} & & & \text{H}-\text{C}\equiv\text{C}-\text{H} \\ | & | & & & & & & & \\ \text{H} & \text{H} & & & & & & & \end{matrix}$$

      ethane   **ethylene**   **acetylene**

The carbon–carbon bond order in ethane is 1, in ethylene it is 2, and in acetylene it is 3.

As long as we are dealing with bonds between the same elements, we can relate bond length and bond energy to the bond order. As the bond order between a pair of atoms increases, additional electron density is placed between the two nuclei, which causes them to be pulled together. Therefore, the bond length decreases as the bond order increases. Increasing the bond order also makes it more difficult to pull the bonded atoms apart. Therefore, the bond energy increases as the bond order increases. Data that illustrate this are shown in Table 4.5.

Another bond property that is related to the bond order is the **vibrational frequency** of the atoms joined by the bond. The atoms within a molecule are not stationary; they are in constant motion. This motion can be resolved into two basic types: vibration in which a pair of atoms move toward and away from

Table 4.5
Variation of bond properties with bond order

Bond	Bond Order	Average Bond Length (Å)	Average Bond Energy		Average Vibrational Frequency (Hz)
			(kJ/mol)	(kcal/mol)	
C—C	1	1.54	370	88	3.0×10^{13}
C═C	2	1.37	699	167	4.9×10^{13}
C≡C	3	1.20	960	230	6.6×10^{13}
C—O	1	1.43	350	83	3.2×10^{13}
C═O	2	1.23	750	180	5.2×10^{13}
C—N	1	1.47	300	71	3.7×10^{13}
C≡N	3	1.16	730	175	6.8×10^{13}

each other along a line joining their centers, much as two balls connected by a spring (Figure 4.5a); and bending in which the angle between the three atoms alternately increases and decreases (Figure 4.5b). For simplicity we will restrict our discussion to vibrational motion.

There are two factors that affect the frequency of vibration (i.e., the number of vibrations per second). One is the masses of the atoms bonded together and the other is the bond order. *For a given pair of atoms, as the bond order increases, so does the vibrational frequency.* This is because increasing the bond order increases the attractive forces holding the nuclei together, in effect, stiffening the "spring" between the two atoms.

Today the measurement of the vibrational frequencies of bonds is really quite simple. It happens that these vibrational frequencies are about the same

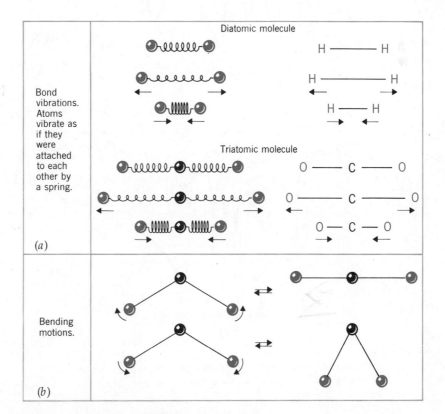

Figure 4.5

The motion of atoms within molecules.

Virtually every modern chemistry laboratory has one or more instruments to measure and record infrared absorption spectra.

as the frequency of infrared radiation, and when infrared light shines on a substance, radiation having the same frequencies as the vibrational frequencies of the bonds is absorbed. By observing which frequencies are selectively removed from the infrared spectrum, we can deduce these vibrational frequencies. The data recorded in the right column of Table 4.5 were obtained in this way.

In complex molecules there are many different vibrational modes available to the atoms, and many different frequencies are absorbed from the infrared "rainbow." The infrared absorption spectra of any but the most simple molecules are therefore quite complicated. Nevertheless, an experienced chemist often finds such absorption spectra extremely valuable as an aid in deducing molecular structure. In addition, each molecule, because of its unique structure, gives rise to its own characteristic absorption spectrum, which can be used to identify the compound and thus serves as a sort of "fingerprint." Examples of infrared absorption spectra of some drugs are shown in Figure 4.6.

4.6 RESONANCE

It often happens that a single satisfactory electron-dot formula for a molecule or ion cannot be drawn. For example, following our usual procedure, two

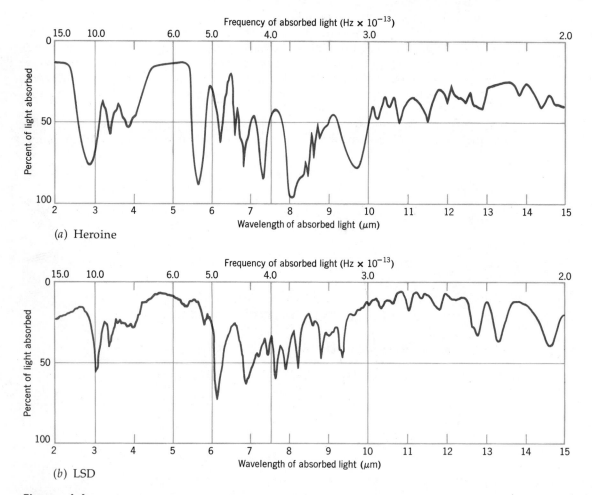

Figure 4.6

Infrared absorption spectra can be easily measured. The spectra are of two drugs: (a) heroine (b) LSD. Notice how distinctly different the spectra are.

electron-dot structures that obey the octet rule can be drawn for sulfur dioxide (a substance that is a major air pollutant). These are shown as structures **1** and **2**.

(1)   (2)

These structures have their nuclei in identical positions but differ in the arrangement of electrons. In both, one oxygen atom is bound through a single bond to the sulfur while the other oxygen atom is connected via a double bond.

Based on our discussion in the last section, we might expect the two sulfur-oxygen bonds in SO_2 to be different. The Lewis structures suggest that one should be shorter, stronger, and have a higher vibrational frequency than the other. However, all experimental evidence suggests that the two sulfur-oxygen bonds are identical! This means that neither of the electron-dot structures that we have drawn for SO_2 is satisfactory. In fact, it is impossible to draw a single electron-dot formula for SO_2 that obeys the octet rule and, at the same time, is consistent with all the experimental facts.

We get around this problem by the concept of **resonance.** We say that the actual electronic structure of SO_2 does not correspond to either **1** or **2** but, instead, to a structure somewhere in between that has properties of both. This true structure is said to be a **resonance hybrid** *of the contributing structures 1 and 2*. The bond properties of the true structure are also intermediate between the properties of the bonds of the contributing structures. In SO_2, for example, the actual bond lengths and bond energies lie between those expected for sulfur-oxygen single and double bonds, and suggest bond orders of about 1.5 for each bond.

It is really quite unfortunate that the word resonance was used to describe this phenomenon, because the impression is often received that the structure of SO_2 fluctuates between **1** and **2**. This is definitely not so. The structure of SO_2 is never **1** and it is never **2**. Sulfur dioxide has a single structure that is between **1** and **2**. Our difficulty is that we cannot draw it satisfactorily using Lewis symbols. The problem is somewhat like trying to describe the beast you would obtain if you were able to cross a cat and a dog. When you try to picture this hypothetical offspring, you visualize it with characteristics of both parents. However, you do not think of it as being a cat one instant and a dog the next.

Some species cannot be adequately explained with only two resonance structures. For example, even though SO_2 can be represented by two structures, the SO_3 molecule requires three.

Fractional bond orders are not only possible, but actually occur in many different compounds.

A double-headed arrow is used here to indicate resonance.

The electronic structure of some ions must also be represented by resonance. For example, the nitrate ion, NO_3^-, and the carbonate ion, CO_3^{2-}, have the same number of valence electrons as SO_3 and therefore have similar resonance structures.

Resonance is certainly not restricted to inorganic compounds. In proteins, for example, amino acids are linked together in long chains by "peptide bonds."

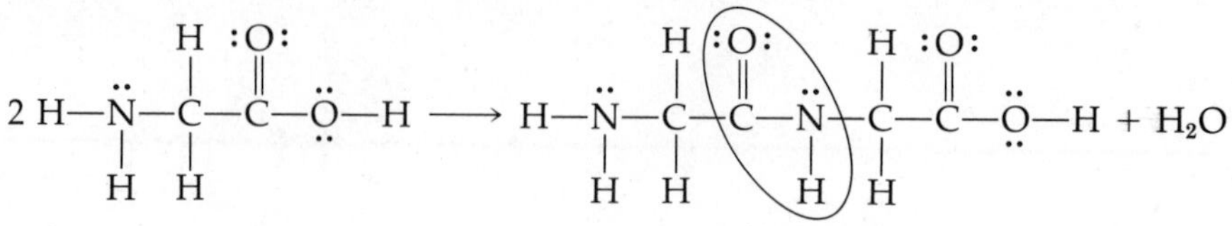

glycine
(an amino acid)

peptide bond
(a peptide)

There is evidence that the C—N bond in the peptide linkage actually lies somewhere between a single bond and a double bond. To explain this it is suggested that the peptide bond is a resonance hybrid of structures such as

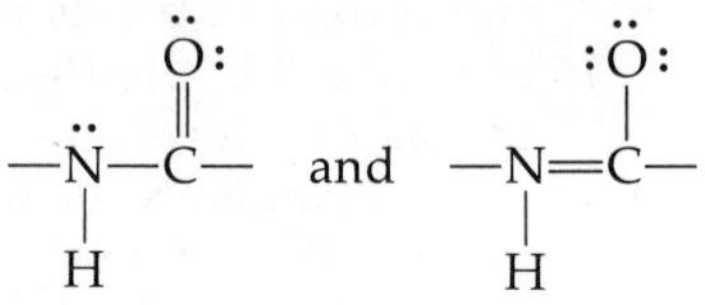

(1)   and   (2)

Structure **2** is obtained by rearranging the electrons in structure **1** in this way:

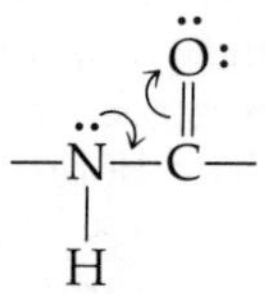

4.7 COORDINATE COVALENT BONDS

When a nitrogen atom combines with three hydrogen atoms to form the molecule NH_3, the nitrogen atom has completed its octet. We might expect, therefore, that the maximum number of covalent bonds that we would observe a nitrogen atom to form is three. There are instances, however, where nitrogen may have more than three covalent bonds. In the ammonium ion, NH_4^+, which is formed in the reaction

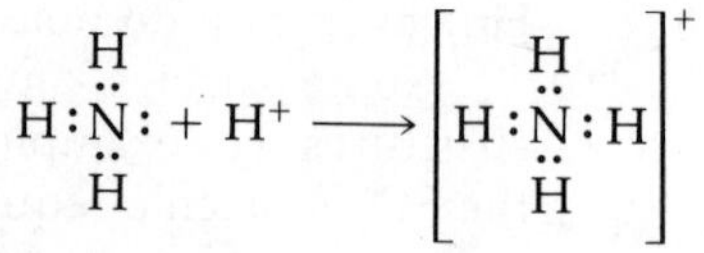

$$H:\ddot{N}: + H^+ \longrightarrow \left[H:\ddot{N}:H \right]^+$$

the nitrogen is covalently bound to four hydrogen atoms. When the additional bond between the H^+ and the N atom is created, both of the electrons in the bond come from the nitrogen. *This type of bond, where a pair of electrons from one atom is shared by two atoms, is called either a* **coordinate covalent bond,** *or a* **dative bond.** It is important that you remember that the coordinate covalent bond is really no different, once formed, than any other covalent bond and that our distinction is primarily aimed at keeping track of electrons; that is, it is "book-keeping."

When Lewis structures are written using dashes to represent electron pairs, the coordinate covalent bond is sometimes indicated by means of an arrow pointing away from the atom supplying the electron pair. For example, the product of the reaction of boron trichloride, BCl_3, and ammonia, NH_3, is a substance known as an **addition compound** because it is formed by simply adding one molecule to another.

$$H-\overset{\displaystyle H}{\underset{\displaystyle H}{N}}: + \overset{\displaystyle Cl}{\underset{\displaystyle Cl}{B}}-Cl \longrightarrow H-\overset{\displaystyle H}{\underset{\displaystyle H}{N}}:\overset{\displaystyle Cl}{\underset{\displaystyle Cl}{B}}-Cl$$

To show that the electron pair shared between the B and N originates on the nitrogen, the Lewis structure of this addition compound can be written

$$H-\overset{\displaystyle H}{\underset{\displaystyle H}{N}}\rightarrow\overset{\displaystyle Cl}{\underset{\displaystyle Cl}{B}}-Cl$$

Using this type of notation we are tempted to write the structure of the NH_4^+ ion as

$$\left[H-\overset{\displaystyle H}{\underset{\displaystyle H}{N}}\rightarrow H\right]^+$$

This gives the impression that one of the N—H bonds is different from the other three. It has been shown experimentally, however, that all four N—H bonds are identical. Therefore, to avoid conveying false impressions, the NH_4^+ ion is written simply as

$$\left[H-\overset{\displaystyle H}{\underset{\displaystyle H}{N}}-H\right]^+$$

Remember, once the bond is formed, it doesn't matter where the electrons in it came from.

4.8 POLAR MOLECULES AND ELECTRONEGATIVITY

When we realize that each element has a different nuclear charge and electron configuration, it is not unreasonable to expect that atoms of different elements have different abilities to attract electrons when they form chemical bonds. **Electronegativity** *is the term used to describe an atom's attraction for the electrons in a bond.* It is important not to confuse this term with electron affinity, which is an energy term and refers to an isolated atom.

When two identical atoms combine, as for instance in H_2, both have the same electronegativity. Since each atom is equally capable of attracting the electron pair in the bond, the pair will be shared equally and will spend, on the average, 50% of its time in the vicinity of each nucleus. Therefore, each H atom has around it two electrons 50% of the time. When averaged out, this is the same as one electron all of the time. The "averaged" one electron will completely neutralize the positive charge on each nucleus, and each atom in H_2 carries a net charge of zero.

If the electronegativities of the two atoms joined by a bond are different, the electron pair will spend more of its time around the more electronegative element. For example, consider the HCl molecule. It happens that chlorine is more electronegative than hydrogen, so the pair of electrons in the H—Cl bond spends more time around chlorine than around hydrogen. This means that the Cl atom acquires a slight negative charge and the H atom, a slight positive charge. We indicate this as

$$\overset{\delta+}{\text{H}}\overset{\delta-}{—\overset{..}{\underset{..}{\text{Cl}}}:}$$

where $\delta+$ and $\delta-$ are meant to indicate partial positive and negative charges.

Within a molecule, equal positive and negative charges separated by a distance constitute a **dipole.** Therefore, the HCl molecule, with its centers of positive and negative charge, is a dipole and is said to be **polar.** In fact, any diatomic molecule (a molecule formed from two atoms) formed from two elements of different electronegativity will be polar.

A dipole is defined quantitatively by its **dipole moment,** the product of the charge on either end of the dipole times the distance between the charges. A very polar molecule is one with a large dipole moment, while a nonpolar molecule will have no dipole moment at all.

When three or more atoms are bonded together, it is possible to have a nonpolar molecule even though there are polar bonds present. Carbon dioxide is an example. The CO_2 molecule is linear and may be represented as

$$\overset{..}{:\underset{\delta-}{\text{O}}}\overset{\delta+}{=\underset{\delta+}{\text{C}}}\overset{\delta-}{=\underset{..}{\overset{..}{\text{O}}}:}$$

showing that oxygen is more electronegative than carbon.

The overall dipole moment of a molecule arises as a sum of the individual bond dipoles within the molecule, which add together like vectors. In CO_2 these bond dipoles are oriented in opposite directions, and they exactly cancel each other.

$$:\overset{..}{\text{O}}=\text{C}=\overset{..}{\underset{..}{\text{O}}}:$$
$$\longleftarrow+\;+\longrightarrow$$

(An arrow with a plus sign on one end is used to represent the bond dipole.)

In the water molecule, which happens to have a bent or angular shape, the two bond dipoles do not cancel each other entirely, but rather are partially additive. As a result, the H_2O molecule does have a net dipole moment (indicated by the heavy arrow) and is polar.

In general, a polyatomic molecule will be nonpolar only if its bonds are nonpolar or if its structure is such that the effects of the bond dipoles cancel. Later in our studies we will see that many of the physical properties of substances are related to the polarity of their molecules. Therefore, these two examples—carbon dioxide and water—illustrate how very important molecular structures and the shapes of molecules are. In the next chapter we will learn how to predict molecular shapes.

We would like to have some quantitative measure of electronegativity so that we can make predictions concerning the polarity of bonds. One approach toward this, taken by R. S. Mulliken in 1934, uses the average of the ionization energy and electron affinity. A very electronegative element has a very high ionization energy, so it is difficult to remove its electrons. It also has a very high electron affinity, so a very stable species results when electrons are added. On the other hand, an element of low electronegativity will have a low ionization energy and low electron affinity so that it loses electrons readily and has

Elements with low electronegativity are often said to be **electropositive** because they tend to give their electrons away rather easily.

little tendency to pick them up. Unfortunately, it is very difficult to measure the electron affinity of an element. Therefore, this method of assigning electronegativities is not universally applicable.

The most widely used scale of electronegativities was developed by Linus Pauling (a winner of two Nobel prizes and an advocate of using vitamin C to ward off the common cold). He observed that when atoms of different electronegativities are combined, their bonds are stronger than expected. Presumably two factors contribute to the bond strength. One of them is the covalent bonding between the atoms. The other is an additional binding produced by an attraction between the oppositely charged ends of the bond dipole. The extra bond strength, then, was attributed to this additional binding and Pauling used this concept to develop his table of electronegativities (Table 4.6).

The actual numerical values in Table 4.6 are not very important. What is important is the *difference* in the electronegativities of the atoms joined by a bond. When the difference is small, the bond is relatively nonpolar; when the difference is large, a very polar bond is formed. And when the difference between the electronegativities of two combining atoms is very large, the electron pair will spend virtually 100% of its time about the more electronegative element. This is the same as saying that an electron is transferred from the atom of low electronegativity to that of high electronegativity. The result, of course, is an ionic bond. It is clear that the degree of ionic character in a bond can vary between zero (for example, H_2) to essentially 100% depending on the electronegativity difference between the bonded atoms. There is no sharp dividing line between ionic and covalent bonding. When the electronegativity difference is 1.7, the bond is about 50% ionic.

Finally, electronegativity trends within the periodic table are worth noting. We see that the most electronegative elements are located in the upper right portion of the table; the least electronegative are found in the lower left. This is consistent with the trends in ionization energy (IE) and electron affinity (EA) discussed in Chapter 3, where we saw that elements with the largest IE and EA are in the upper right region of the periodic table and those with the smallest IE and EA are in the lower left. It is also consistent with our observations that atoms from opposite ends of the periodic table—lithium and fluorine, for example—form bonds that are essentially 100% ionic, and that atoms such as carbon and oxygen form covalent bonds that are only somewhat polar.

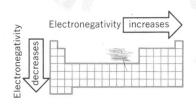

Table 4.6
The complete electronegativity scale[a]

Li 1.0	Be 1.5											H 2.1		B 2.0	C 2.5	N 3.0	O 3.5	F 4.0
Na 0.9	Mg 1.2													Al 1.5	Si 1.8	P 2.1	S 2.5	Cl 3.0
K 0.8	Ca 1.0	Sc 1.3	Ti 1.5	V 1.6	Cr 1.6	Mn 1.5	Fe 1.8	Co 1.8	Ni 1.8	Cu 1.9	Zn 1.6			Ga 1.6	Ge 1.8	As 2.0	Se 2.4	Br 2.8
Rb 0.8	Sr 1.0	Y 1.2	Zr 1.4	Nb 1.6	Mo 1.8	Tc 1.9	Ru 2.2	Rh 2.2	Pd 2.2	Ag 1.9	Cd 1.7			In 1.7	Sn 1.8	Sb 1.9	Te 2.1	I 2.5
Cs 0.7	Ba 0.9	La–Lu 1.1–1.2	Hf 1.3	Ta 1.5	W 1.7	Re 1.9	Os 2.2	Ir 2.2	Pt 2.2	Au 2.4	Hg 1.9			Tl 1.8	Pb 1.8	Bi 1.9	Po 2.0	At 2.2
Fr 0.7	Ra 0.9	Ac 1.1	Th 1.3	Pa 1.5	U 1.7	Np–No 1.3												

[a] Reprinted from Linus Pauling, *The Nature of the Chemical Bond.* Copyright 1939 and 1940 by Cornell University. Third edition © 1960 by Cornell University. Used by permission of Cornell University Press. British Commonwealth rights by permission of Oxford University Press.

4.9   OXIDATION AND REDUCTION, OXIDATION NUMBERS

In the formation of the ionic bond between Li and F, we saw that an electron was transferred from Li to F to produce Li^+ and F^-. With HCl, we saw that a polar covalent bond was formed in which an electron was partially transferred from the H to the Cl atom. Very many chemical reactions are of this type; that is, they involve some transfer of electronic charge from one atom to another. Because this is such a common and important process, we define terms that apply specifically to these changes. These are

> **Oxidation**—a loss of electrons
> **Reduction**—a gain of electrons

Thus in the formation of LiF, Li undergoes oxidation by losing an electron and F undergoes reduction by acquiring an electron. In a similar fashion, when the HCl molecule is formed by the reaction of hydrogen and chlorine, the H atom loses some electronic charge to the Cl atom and is oxidized while the Cl atom becomes reduced.

Lithium, in its reaction with fluorine, is said to be a **reducing agent** because it supplies the electron that the fluorine requires in order to be reduced—that is, it is the agent that has allowed reduction to occur. The fluorine, on the other hand, by accepting the electron from Li, permits oxidation to take place and thus is said to be an **oxidizing agent.** In a similar manner, we would consider hydrogen the reducing agent and chlorine the oxidizing agent when these two elements react to produce HCl. In general, *oxidizing agents acquire electrons and become reduced, while reducing agents lose electrons and become oxidized.*

Oxidation and reduction are always discussed together because, in any reaction, whenever one substance loses electrons, another substance picks them up. We know that this is true because we never observe electrons as a product of a chemical reaction, nor are electrons ever consumed when a chemical change occurs. Thus, oxidation is always accompanied by reduction. As a result, the simple abbreviated term, **redox,** is often used in discussing these reactions.

Chemists have devised a bookkeeping system using what are called **oxidation numbers** to keep track of electrons during chemical reactions. An oxidation number can be defined as *the charge that an atom would have if both of the electrons in each bond were assigned to the more electronegative element.* The term **oxidation state** is also used, interchangeably, with the term oxidation number.

In the substance LiF, because an electron has, in fact, been transferred to the F atom, the oxidation number assigned to Li^+ is 1+. The oxidation number of fluorine in the F^- ion is 1−.

In the HCl molecule, because Cl is more electronegative than hydrogen, we assign an oxidation number of 1+ to H and 1− to Cl, as if the electron pair were in the sole possession of the Cl atom.

In a nonpolar molecule such as H_2, where both atoms are the same and therefore have the same electronegativity, it is senseless to assign the electron pair to either atom since no electron transfer has occurred. In this case each H atom is assigned an oxidation number of zero.

The following set of rules has been developed to aid us in assigning oxidation numbers to the various atoms in a compound.

Rules for assigning oxidation numbers

1. The oxidation number of any element in its elemental form is zero, regardless of the complexity of the molecule in which it occurs. Thus the atoms in Ne, F_2, P_4, and S_8 all have oxidation numbers of zero.

Redox reactions are very common. For example, they occur in batteries and even within our bodies when we metabolize food.

2. The oxidation number of any simple ion (one atom) is equal to the charge on the ion. The ions Na^+, Al^{3+}, and S^{2-} have oxidation numbers of 1+, 3+, and 2−, respectively.

3. The sum of all the oxidation numbers of all the atoms in a neutral compound is zero. For a polyatomic ion, the algebraic sum of the oxidation numbers must be equal to the ion's charge.

In addition to these basic rules, we can use the fact that certain elements have the same oxidation numbers in all (or nearly all) their compounds.

4. Fluorine always has an oxidation number of 1− in its compounds.

5. The elements in Group IA (except hydrogen) always have an oxidation number of 1+ in compounds.

6. The elements in Group IIA always have an oxidation number of 2+ in compounds.

7. A Group VIIA element has an oxidation number of 1− in binary compounds (compounds that contain only two different elements) with metals. For example, Cl has an oxidation number of 1− in $FeCl_2$, $CrCl_3$, and NaCl.

8. Oxygen almost always has an oxidation number of 2− in its compounds.

9. Hydrogen almost always has an oxidation number of 1+ in its compounds.

10. For familiar polyatomic ions, such as SO_4^{2-} or NO_3^-, the charge on the ion can be taken as the ion's *net* oxidation number.

In rules 8 and 9 the words "almost always" were used because there are some exceptions. For example, in binary compounds with fluorine, oxygen must have a positive oxidation number because fluorine is *always* 1−. Somewhat more common are peroxides, such as the O_2^{2-} ion and H_2O_2 that contain an O—O bond, in which oxygen is assigned an oxidation number of 1−. Oxygen also forms compounds called superoxides that contain the O_2^- ion in which the oxidation number of oxygen is $\frac{1}{2}-$. Finally, hydrogen forms binary compounds with some metals—for example, NaH—in which hydrogen is given an oxidation number of 1−. Although you should know of these exceptions, they are rare, and we normally assign oxidation numbers of 2− to oxygen and 1+ to hydrogen when we see them in compounds.

EXAMPLE 4.9   What are the oxidation numbers of all of the atoms in KNO_3 (potassium nitrate, a substance used to make gun powder)?

SOLUTION   We know that the sum of the oxidation numbers of all of the atoms must be equal to zero (the charge on KNO_3—rule 3).

$$
\begin{array}{lll}
\text{K} & 1 \times (1+) = 1+ & \text{(rule 5)} \\
\text{N} & 1 \times (x) = x & \\
\text{O} & 3 \times (2-) = 6- & \text{(rule 8)} \\
\hline
\textit{sum of oxidation numbers} = 0 & \text{(rule 3)}
\end{array}
$$

x must equal 5+ in order for the sum to be zero.

EXAMPLE 4.10 What is the oxidation number of sulfur in $Na_2S_4O_6$ (sodium tetrathionate)?

SOLUTION Again, the sum of the oxidation numbers must be zero.

$$
\begin{array}{lrcl}
Na & 2 \times (1+) & = & 2+ \\
S & 4 \times (x) & = & 4x \\
O & 6 \times (2-) & = & \underline{12-} \\
& \text{sum} & = & 0
\end{array}
$$

$$4x = 10+ \quad \text{or} \quad x = \frac{10+}{4} = \frac{5}{2}+$$

Notice that the oxidation number of an atom need not be an integer.

EXAMPLE 4.11 What is the oxidation number of Cr in the $Cr_2O_7{}^{2-}$ ion?

SOLUTION This time the sum of the oxidation numbers must equal 2− (rule 3).

$$
\begin{array}{lrcl}
Cr & 2 \times (x) & = & 2x \\
O & 7 \times (2-) & = & \underline{14-} \\
& \text{sum} & = & 2-
\end{array}
$$

Therefore,

$$2x = 12+$$

$$x = 6+$$

It is very important to keep in mind that oxidation numbers were developed simply for bookkeeping. Except for simple ions like Na^+ or F^-, the charges are fictitious. For example, the Cl in HCl does not carry a 1− charge as its oxidation number would imply; it carries only a partial negative charge (the actual charge is only about 0.17−).

4.10 THE NAMING OF CHEMICAL COMPOUNDS

If you leaf through a chemical handbook, such as the *Handbook of Chemistry and Physics*, you will find an enormous number of compounds listed there. Yet these represent only a fraction of all the compounds that have been discovered, and each year the list grows longer. Naming these compounds represents a real challenge because it is important that each unique substance have its own unique name. In addition, the names can't be chosen in a haphazard way; otherwise there would be no possible way for anyone to remember them all. For this reason, a systematic procedure for naming chemical compounds has been developed.

In this section we will describe, in a somewhat abbreviated way, how **inorganic compounds** are named. These are compounds whose structures are not primarily determined by the linking together of carbon atoms. Such carbon compounds are called **organic compounds** and are discussed in Chapter 22.

You will discover during your study of chemistry that not every compound is named according to the system. Some very common substances, such as water (H_2O) and ammonia (NH_3), were known long before the systematic nomenclature was developed and are best recognized by their common (or *trivial*) names. Trivial names are also used for extremely complex compounds where the names derived on a systematic basis are very long, complex, and cumbersome.

Table 4.7
Names of anions derived from nonmetals

Group IVA	Group VA	Group VIA	Group VIIA
C^{4-}; *carbide*[a]	N^{3-} *nitride*	O^{2-} *oxide*	F^- *fluoride*
Si^{4-}; *silicide*	P^{3-} *phosphide*	S^{2-} *sulfide*	Cl^- *chloride*
	As^{3-} *arsenide*	Se^{2-} *selenide*	Br^- *bromide*
		Te^{2-} *telluride*	I^- *iodide*

[a] Carbon also forms a number of complex carbides, for example, C_2^{2-} in CaC_2.

Binary compounds

A **binary compound** is composed of atoms of only two different elements. In naming these substances the less electronegative (more metallic) element is specified first by giving its ordinary English name. The name of the second element (almost always a nonmetal) is obtained by adding the suffix *ide* to its stem, as shown in Table 4.7. Some typical examples are

NaCl	sodium chloride
SrO	strontium oxide
Al_2S_3	aluminum sulfide
Mg_3P_2	magnesium phosphide
HBr	hydrogen bromide

Many elements are commonly found to exist in more than one positive oxidation state. When the element is a metal, there are two methods that may be used to indicate its oxidation state. In the older of these methods the suffixes *-ic* and *-ous* are used to differentiate between high and low oxidation states. Thus the 3+ and 2+ oxidation states of chromium would be specified as

Cr^{3+}	chromic	$CrCl_3$	chromic chloride
Cr^{2+}	chromous	$CrCl_2$	chromous chloride

When the metal has a symbol derived from the Latin name for the element, its Latin stem is generally used. For example, with iron there are two common oxidation states, Fe^{3+} (ferric) and Fe^{2+} (ferrous). Other common examples are found in Table 4.8. Notice that this system *only* differentiates between high and low oxidation states; it does not specify what the oxidation state of the metal is.

The second and preferred method of indicating the oxidation state of the metal is called the *Stock system*, named after a German chemist, Alfred Stock (1876–1946). It involves placing a Roman numeral equal to the oxidation

Table 4.8
Metals commonly found in two oxidation states

Chromium	Manganese	Iron	Cobalt	Lead
Cr^{2+} chromous	Mn^{2+} manganous	Fe^{2+} ferrous	Co^{2+} cobaltous	Pb^{2+} plumbous
Cr^{3+} chromic	Mn^{3+} manganic	Fe^{3+} ferric	Co^{3+} cobaltic	Pb^{4+} plumbic

Copper	Tin	Mercury		
Cu^+ cuprous	Sn^{2+} stannous	Hg_2^{2+} mercurous (note that there are two Hg atoms)		
Cu^{2+} cupric	Sn^{4+} stannic	Hg^{2+} mercuric		

number of the metal in parentheses following the regular English name of the element. Thus Fe^{2+} and Fe^{3+} would be iron(II) and iron(III). The alternative names for the compounds $FeCl_2$ and $FeCl_3$ are therefore

$FeCl_2$	ferrous chloride	or iron(II) chloride
$FeCl_3$	ferric chloride	or iron(III) chloride

Even though the Stock system is preferred today, it is necessary to know the older system as well. For example, if an experiment calls for $FeCl_3$, iron(III) chloride, it is likely that you will only find it in a reagent bottle labeled "ferric chloride."

When naming binary covalent compounds formed between two nonmetals, a third system of nomenclature is preferred in which the numbers of each atom in a molecule is specified by a Greek prefix: *di-* (2), *tri-* (3), *tetra-* (4), *penta-* (5), *hexa-* (6), *hepta-* (7), *octa-* (8), *nona-* (9), *deca-* (10), and so on. The Stock system is usually *not* used for these compounds because it does not distinguish between molecular formulas such as the first two examples below.

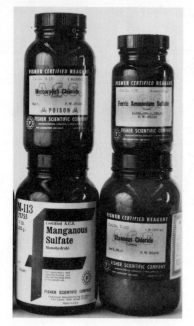

Bottles of chemicals labeled using the older system of nomenclature.

NO_2	nitrogen(IV) oxide	nitrogen dioxide
N_2O_4	nitrogen(IV) oxide	dinitrogen tetroxide
N_2O_5	nitrogen(V) oxide	dinitrogen pentoxide
PCl_3	phosphorus(III) chloride	phosphorus trichloride
PCl_5	phosphorus(V) chloride	phosphorus pentachloride

In some instances the prefix *mono-* (1) is also used to avoid ambiguity.

CO_2	carbon dioxide
CO	carbon monoxide

Compounds containing polyatomic ions

In Section 4.2 we saw that many ions contain more than one atom and are therefore referred to in general as polyatomic ions. These species enter into ionic compounds as discrete units and generally stay intact in most chemical reactions. A list of these is given in Table 4.4. As with binary compounds, substances that contain these ions are always named with the positive ion first. Some examples are:

Na_2CO_3	sodium carbonate	$Ba(OH)_2$	barium hydroxide
$Ca(C_2H_3O_2)_2$	calcium acetate	$(NH_4)_2SO_4$	ammonium sulfate

The Stock system is also preferred when the metal can exist in more than one oxidation state.

	Stock System	Old System
$MnSO_4$	manganese(II) sulfate	manganous sulfate
$Fe_2(C_2O_4)_3$	iron(III) oxalate	ferric oxalate

Binary acids

Among the important classes of compounds that we will discuss later in the book are substances called acids. Perhaps you have already encountered some common ones in the laboratory—hydrochloric acid, for example—and you surely have experienced the sour taste of citric acid in lemon juice and acetic acid in vinegar. As we will see, acids are substances that release H^+ ions when they are dissolved in water.

An important kind of acid is formed when a binary compound of hydrogen and a nonmetal (for example, hydrogen chloride, HCl, or hydrogen sulfide, H_2S) is dissolved in water. Water solutions of these compounds are called binary acids (or sometimes *hydro acids*). They are named as *hydro . . . ic acid*, where the stem of the name of the nonmetal is inserted in place of the dotted line. Examples are

HF	hydro*fluor*ic acid
HCl	hydro*chlor*ic acid
HBr	hydro*brom*ic acid
HI	hydr*iod*ic acid
H_2S	hydro*sulfur*ic acid

When these acids are allowed to react with hydroxide ion (a reaction called **neutralization**), an anion is formed; for example,

$$HCl + OH^- \longrightarrow H_2O + Cl^-$$

*hydro**chlor**ic acid*   **chlor**ide

Note that *hydro . . . ic acids* give *. . . ide* salts. (**Salt** is a general term that we will use for any ionic compound not containing oxide or hydroxide ion, for example, sodium chloride.)

Anions such as Cl^- can be formed directly from the element, as when Cl_2 reacts with Na to form NaCl, or they can be formed by neutralization of an acid with a base.

Oxoacids

Oxoacids are acids that contain hydrogen, oxygen, and at least one other element. Sulfuric acid, H_2SO_4 is an example. Notice that the hydrogen is specified first in the formula (we will say more about this in Chapter 6). When the third element (S in the case of H_2SO_4) can exist in more than one oxidation state, more than one oxoacid is possible. For example, two common oxoacids of sulfur are H_2SO_4 and H_2SO_3, containing sulfur in the 6+ and 4+ oxidation states, respectively. The acid containing the element in the higher oxidation state is given the suffix *-ic*, whereas the acid having the element in the lower oxidation state is given the ending *-ous*. Thus we have

H_2SO_4	sulfuric acid
H_2SO_3	sulfurous acid

Notice that the prefix *hydro* is not used in naming oxoacids.

Compounds produced by neutralization of these acids contain the polyatomic anions given in Table 4.4. The anion derived from the "ic" acid ends in *-ate*, whereas the anion from the "ous" acid ends in *-ite*:

Although polyatomic ions are formed by neutralization of oxoacids, they can also be made by other reactions.

H_2SO_4	sulfuric acid	SO_4^{2-}	sulfate
H_2SO_3	sulfurous acid	SO_3^{2-}	sulfite
HNO_3	nitric acid	NO_3^-	nitrate
HNO_2	nitrous acid	NO_2^-	nitrite
$HClO_3$	chloric acid	ClO_3^-	chlorate
$HClO_2$	chlorous acid	ClO_2^-	chlorite

Some elements form oxoacids in more than two oxidation states. In this case the prefixes *hypo-* and *per-* are used to designate a lower or higher oxidation state, respectively. A good example occurs among the oxoacids of the halogens.

hypochlorous	HClO	ClO^-	hypochlorite
chlorous	$HClO_2$	ClO_2^-	chlorite
chloric	$HClO_3$	ClO_3^-	chlorate
perchloric	$HClO_4$	ClO_4^-	perchlorate

Acid salts

Parent Acid	Typical Acid Salts
H_2SO_4	$NaHSO_4$
H_2CO_3	$NaHCO_3$
H_3PO_4	NaH_2PO_4
	Na_2HPO_4

Partial neutralization of an acid that is capable of furnishing more than one H^+ per acid molecule gives salts that are called acid salts. Some examples are shown in the table at the left. When only one acid salt is formed (as with H_2SO_4 or H_2CO_3), the salt can be named by adding the prefix *bi-* to the name of the anion of the acid.

$NaHSO_4$	sodium **bi**sulfate
$NaHCO_3$	sodium **bi**carbonate

The salt can also be named by specifying the presence of the H by writing "hydrogen."

$NaHSO_4$	sodium hydrogen sulfate
NaH_2PO_4	sodium dihydrogen phosphate
Na_2HPO_4	sodium hydrogen phosphate
	(disodium hydrogen phosphate)

Note the use of the prefix *di-* to indicate the number of hydrogen atoms (as well as to remove ambiguity as to the number of Na in the last formula).

INDEX TO QUESTIONS AND PROBLEMS (Problem numbers in **bold type**)

REVIEW QUESTIONS

4.1 Write Lewis symbols for Se, Br, Al, K, Ba, Ge, and P.

4.2 What is the purpose of using Lewis symbols?

4.3 Define the terms *cation* and *anion*.

4.4 Draw Lewis structures for the ionic compounds, BaO, Na_2O, KF, $MgBr_2$.

4.5 Why is KF (s) more stable than K (s) and F_2 (g)?

4.6 Indicate how the electron configuration changes for each atom when the following ionic compounds are formed from the elements: K_2O, Mg_3N_2, Na_2S, $BaBr_2$.

4.7 Give the electron configuration of each of the following ions: Ba^{2+}, Se^{2-}, Al^{3+}, Li^+, Br^-, Fe^{2+}, Cu^+, Ni^{2+}.

4.8 Why do elements from period 2 never exceed an octet in their valence shells?

4.9 What is a pseudonoble gas electron configuration? List some ions that possess this kind of electronic structure.

4.10 Write formulas for compounds composed of the following pairs of ions: (a) Na^+, CO_3^{2-}; (b) Ca^{2+}, ClO_3^-; (c) Sr^{2+}, S^{2-}; (d) Cr^{3+}, Cl^-; (e) Ti^{4+}, ClO_4^-.

4.11 Write formulas for compounds composed of the following pairs of ions:
(a) Fe^{3+}, HPO_4^{2-}
(b) K^+, N^{3-}
(c) Ni^{2+}, NO_3^-
(d) Cu^{2+}, $C_2H_3O_2^-$
(e) Ba^{2+}, SO_3^{2-}

4.12 On the basis of their electron configurations, why do many of the transition elements form ions with a charge of $2+$?

4.13 What is a post-transition element?

4.14 Construct a Born-Haber cycle for the formation of KBr (s) from K (s) and Br_2 (l). Indicate which steps are endothermic and which are exothermic.

4.15 What is the name of the energy term associated with the reaction, Na^+ (g) + Cl^- (g) → NaCl (s)? Is the process indicated in this chemical equation endothermic or exothermic?

4.16 Use Lewis symbols to diagram the reaction for the formation of the covalently bonded molecules NH_3, H_2O, and HF.

4.17 Draw Lewis structures for the molecules PCl_3, SiH_4, BCl_3, H_2S, C_3H_8, and CO.

4.18 Draw Lewis structures for the molecules Cl_2, SO_2, OF_2, SnH_4, C_2H_4, and SCl_2.

4.19 Draw Lewis structures for the ions Cl^-, S^{2-}, ClO^-, ClO_4^-, SO_3^{2-}, and PO_4^{3-}.

4.20 Draw Lewis structures for the ions NO_3^-, NO^+, NO_2^-, and CO_3^{2-}.

4.21 Draw Lewis structures for SeF_6, SeF_4, ICl_3, $AsCl_5$, ICl_2^-, ICl_4^-, and XeF_4.

4.22 Which of the following compounds do not obey the octet rule: ClF_3, OF_2, SF_4, SO_2, IF_7, NO_2, BCl_3?

4.23 The C—C bond length in a series of compounds was found to be as follows: compound 1, 1.54 Å; compound 2, 1.37 Å; compound 3, 1.46 Å; and compound 4, 1.40 Å. Arrange these in order of increasing C—C bond order. How would you expect the C—C bond energies to vary?

4.24 Compare the C—O bond properties—bond order, bond energy, bond length, and vibrational frequencies—in the following. (Hint: draw resonance structures when necessary.)

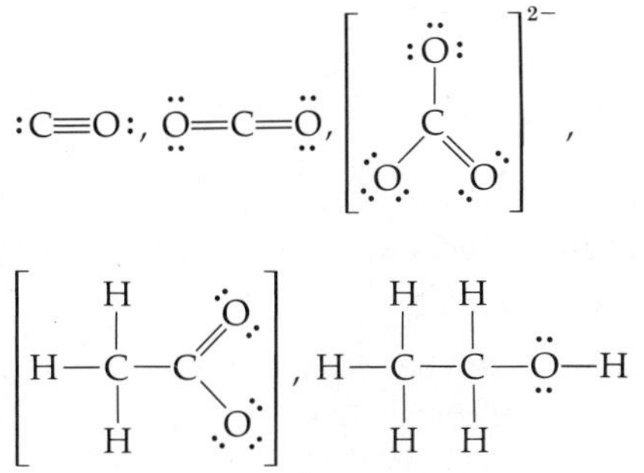

4.25 What kind of information is provided by an infrared absorption spectrum of a molecule?

4.26 What is a resonance hybrid? Why is the concept of resonance used?

4.27 Draw the resonance structures of SO_3, NO_3^- and CO_3^{2-}; SO_2 and NO_2^-.

4.28 Draw the resonance structures of HNO_3, SeO_2, SeO_3, and N_2O_4.

Structures:

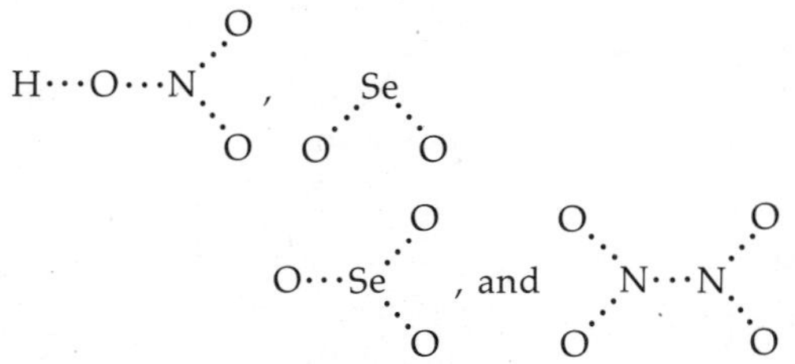

4.29 Draw resonance structures for $C_2O_4^{2-}$, CH_3COO^-, and N_3^-.

Structures:

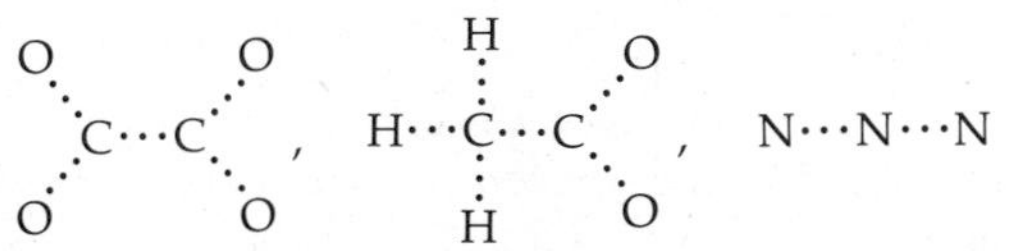

4.30 In Section 4.6 we noted that there was evidence that the C—N bond in the peptide linkage possessed some double-bond character. What kind of experimental evidence would be expected to confirm this?

4.31 For the molecules SO_2 and SO_3, compare the SO bond energies, bond lengths, and S—O vibrational frequencies.

4.32 How would you expect the N—O bond distance in NO_2^- to compare with that in NO_3^-?

4.33 What would you expect the value of the bond order in SO_3 to be?

4.34 What is a coordinate covalent bond? How does it differ from other covalent bonds?

4.35 Use Lewis symbols to show the formation of a coordinate covalent bond in the reaction

$$AlCl_3 + Cl^- \longrightarrow AlCl_4^-$$

4.36 What is another term used for a coordinate covalent bond?

4.37 Define electronegativity. What is the difference between electronegativity and electron affinity?

4.38 Define polar, dipole, and dipole moment.

4.39 What trends in electronegativity occur in the periodic table? What correlation, if any, exists between ionization energy and electronegativity?

4.40 Which of the following contain bonds that are predominantly ionic: $AlCl_3$, MgO, Al_2O_3, NF_3, CsF, $FeCl_2$, SO_2, Ca_3P_2, Mg_2Si?

4.41 Which of the following have bonds that are predominantly covalent: NH_3, MnF_2, BCl_3, $MgCl_2$, BeI_2, NaH?

4.42 Arrange the following compounds in order of increasing ionic character of their bonds: SO_2, H_2S, SF_2, OF_2, ClF_3, H_2Se, F_2.

4.43 Define oxidation, reduction, oxidation state, oxidizing agent, and reducing agent.

4.44 Assign oxidation numbers to each atom in $KClO_2$, $BaMnO_4$, Fe_3O_4, O_2F_2, IF_5, $HOCl$, $CaSO_4$, $Cr_2(SO_4)_3$, O_3, and Hg_2Cl_2.

4.45 Many biological processes involve oxidation and reduction. For example, ethyl alcohol (grain alcohol) is metabolized in a series of oxidation steps that involve the following carbon-containing compounds.

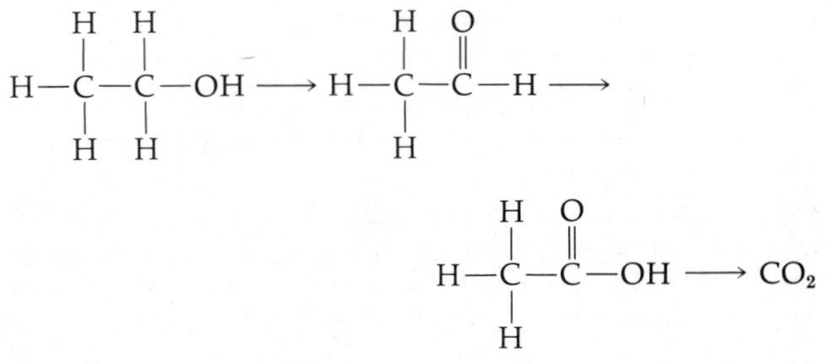

Assign an average oxidation number to carbon in each of these compounds.

4.46 Assign oxidation numbers to each atom in H_2SO_4, CBr_4, OF_2, H_2O_2 (hydrogen peroxide), $CrCl_3$, Mn_2O_7, $KMnO_4$, $H_2C_2O_4$, $KClO_3$, and $LiNO_3$.

4.47 Identify the following changes as either oxidation or reduction:
(a) MnO_2 to MnO_4^-
(b) BiO_3^- to Bi^{3+}
(c) SO_2 to SO_3
(d) OCl^- to ClO_3^-
(e) N_2O_4 to N_2O

4.48 Name the following:
(a) $NaBr$
(b) CaO
(c) $FeCl_3$
(d) $CuCO_3$
(e) CBr_4
(f) P_4O_6
(g) $AsCl_5$
(h) $Mn(HCO_3)_2$
(i) $NaMnO_4$
(j) O_2F_2

4.49 Write chemical formulas for the following compounds.
(a) Aluminum nitrate
(b) Iron(II) sulfate
(c) Ammonium dihydrogen phosphate
(d) Iodine pentafluoride
(e) Phosphorus(III) chloride
(f) Dinitrogen tetroxide

(g) Potassium permanganate
(h) Magnesium hydroxide
(i) Hydrogen selenide
(j) Sodium hydride

4.50 Name the following:
(a) Cr_2O_3
(b) $Mg(H_2PO_4)_2$
(c) $Cu(NO_3)_2$
(d) $CaSO_4$
(e) $Ba(OH)_2$
(f) $AlPO_4$
(g) Mg_3N_2
(h) PbC_2O_4
(i) $(NH_4)_2CO_3$
(j) $K_2Cr_2O_7$

4.51 Write formulas for the following:
(a) Titanium(IV) oxide
(b) Silicon tetrachloride
(c) Calcium selenide
(d) Potassium nitrate
(e) Aluminum sulfate
(f) Nickel(II) bicarbonate
(g) Sodium bisulfate
(h) Ammonium dichromate
(i) Calcium acetate
(j) Strontium hydroxide

REVIEW PROBLEMS (More difficult problems are marked by an asterisk.)

4.52 Use a Born-Haber cycle to show that the reaction

$$K \ (s) + \tfrac{1}{2}Cl_2 \ (g) \longrightarrow KCl \ (s)$$

is exothermic. The following energies are known (a negative sign indicates the process is exothermic): For K $(s) \rightarrow$ K (g), 21.5 kcal; for $\frac{1}{2}Cl_2$ $(g) \rightarrow$ Cl (g), 28.5 kcal; for K $(g) \rightarrow$ K^+ (g), 100.1 kcal; for Cl $(g) \rightarrow Cl^-$ (g), -83.2 kcal; for K^+ $(g) + Cl^-$ $(g) \rightarrow KCl$ (s), -168.3 kcal.

4.53 Below are calculated and experimental bond energies for the hydrogen halides. Use these data to show that the electronegativities of the halogens decrease from F to I.

	Calculated (kcal/mol)	Experimental (kcal/mol)
HF	70.5	135
HCl	80.5	103
HBr	75	87
HI	70	71

*4.54 Given the following data, calculate the lattice energy of $CaCl_2$ in kilojoules per mole. Energy needed to vaporize 1 mol of Ca (s) = 46.0 kcal; first ionization energy of Ca = 140.9 kcal/mol; second ionization energy of Ca = 273.8 kcal/mol; electron affinity of Cl = -83.2 kcal/mol; bond

energy of Cl_2 = 57.0 kcal/mol of Cl—Cl bonds; energy change for the reaction, Ca (s) + Cl_2 $(g) \rightarrow CaCl_2$ (s), -190.0 kcal/mol of $CaCl_2$ (s) formed.

*4.55 Given the following data, calculate the electron affinity of Br. The energy change for the reaction, Na (s) + $\frac{1}{2}Br_2$ $(l) \rightarrow NaBr$ (s), is -86.0 kcal. The energy needed to vaporize 1 mol of Na (s) is 26.0 kcal. The energy needed to vaporize 1 mol of Br_2 (l) is 7.3 kcal. The ionization energy of Na (g) is 118.5 kcal/mol. The bond energy of Br_2 is 46.0 kcal/mol of Br—Br bonds. The lattice energy of NaBr is -175.5 kcal/mol.

*4.56 Hydrogen is more electronegative than any of the Group IA elements. Based on this statement and the data below, show that the electronegativities of the alkali metals decrease from Li to Rb.

	Calculated Bond Energy (kcal/mol)	Experimental Bond Energy (kcal/mol)
Li—H	65.1	56.9
Na—H	61.1	48
K—H	58.4	44
Rb—H	57.8	39

COVALENT BONDING AND MOLECULAR STRUCTURE

Subtle differences in the shapes of molecules can sometimes have very profound effects. Humans, for example, can digest the starch in the wheat shown here, but not the cellulose that forms the body of the plants. Starch and cellulose are practically identical, except for a small difference in their molecular structures. In this chapter we examine molecular structures and the theories used to explain them.

In Chapter 4 we found that chemical bonds can be broadly classified into two main categories: ionic bonds and covalent bonds. The ionic bond arises as a purely electrostatic attraction between oppositely charged particles and is therefore nondirectional. This means that the arrangement of ions in a cluster is determined simply by the balancing of attractive and repulsive forces between the ions, not by their electronic structures. The covalent bond, on the other hand, has very definite directional properties. Covalently bound substances, such as molecules or polyatomic ions, have characteristic shapes that are usually retained when these substances undergo physical changes such as melting or vaporization.

The shapes of molecules—how their atoms are arranged in space—affect many of their physical and chemical properties. In Chapter 4 we learned that molecular shape can determine whether or not a molecule is polar. Later we will see that this has a strong effect on such physical properties as melting point and boiling point. Molecular shape can also affect chemical properties. In biological systems, such as our own bodies, the chemical reactions that keep us alive (and even allow us to study chemistry) depend on the very precise fitting together of molecules. If this fit is destroyed, which generally is what happens in cases of poisoning, the organism dies. Thus, an understanding of molecular geometry and the factors that affect it are critical to our understanding of chemistry.

So far, the simple picture of a covalent bond as a pair of dots shared between two atoms has given us no information about molecular structure. In this chapter we will first see how Lewis structures can be used to predict molecular shapes with a surprisingly high degree of accuracy. Then we will examine modern theories of bonding that attempt to answer the "why" and "how" questions—*why* do molecules have the shapes that they do, and *how* can atoms actually share electrons with each other. You should keep in mind throughout these discussions that each theory represents an attempt to describe the *same* physical phenomenon. None of the theories is perfect—otherwise we would only have to consider one of them. Each has its usefulness, and each has its weaknesses. The theory you apply in a particular circumstance depends largely on what aspect of the covalent bond you are attempting to explain and, to some extent, your own feelings about the validity of the various theories.

5.1 MOLECULAR SHAPES

Although there is an enormous number of different molecules, the number of different ways that atoms arrange themselves around each other is rather limited. This makes understanding the shapes of molecules much easier, because we are able to describe these shapes with a relatively small number of terms.

Virtually all molecules have shapes that can be considered to be derived from a basic set of five different geometries. Therefore, before we discuss theories that predict them or explain them, it is important for you to fully understand these structures and have a feel for what they represent in three dimensions. Let's examine them closely before we proceed further.

1. *Linear.* A linear arrangement of atoms occurs when they are all in a straight line. The angle formed between two bonds that go to the same central atom, which we call the **bond angle,** is 180°.

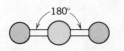

2. *Planar triangular.* A planar triangular arrangement of four atoms has them all in the same plane. The central atom is surrounded by three others located at the corners of a triangle. The bond angles are all 120°.

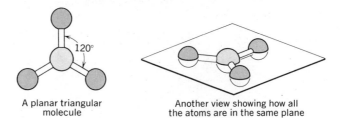

A planar triangular molecule

Another view showing how all the atoms are in the same plane

3. *Tetrahedral.* A tetrahedron is a four-sided pyramid having equilateral triangles as faces. In a tetrahedral molecule, the central atom is located in the center of this tetrahedron and four other atoms are located at the corners. The bond angles are all equal and have values of 109.5°.

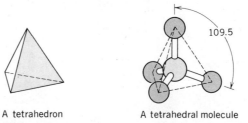

A tetrahedron   A tetrahedral molecule

4. *Trigonal bipyramidal.* A trigonal bipyramid consists of two triangular pyramids (similar to tetrahedrons) that share a common face.

A trigonal bipyramid

In a trigonal bipyramidal molecule, a central atom is surrounded by five others. The central atom is located in the center of the triangular plane that is shared by the upper and lower pyramids. The five atoms attached to it are located at the five corners. In this kind of molecule, the bond angles are not all the same. Between any two bonds that lie in the central triangular plane, the bond angle is 120°. The angle is only 90° between a bond in the central triangular plane and a bond that points to the top or bottom of the trigonal bipyramid.

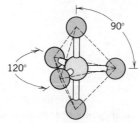

A trigonal bipyramidal molecule

When we draw a trigonal bipyramidal molecule we normally sketch a triangle as it would look tilted backward so that we would be looking at it from its edge. Then we draw lines to the top and bottom of the trigonal bipyramid.

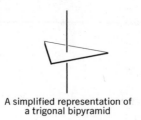

A simplified representation of
a trigonal bipyramid

5. *Octahedral.* An octahedron is a geometrical figure that has eight faces. We can think of it as two square pyramids that share a common square base. Notice that the figure has only six corners even though it has eight faces.

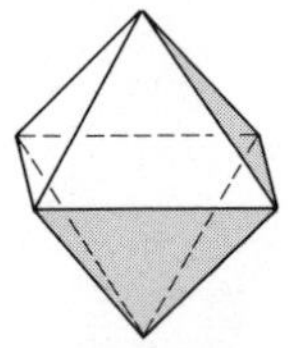

An octahedron

Remember, an octahedral molecule has only six atoms joined to a central atom.

In an octahedral molecule, the central atom is surrounded by six others. The central atom is located in the center of the square plane that passes through the middle of the octahedron. The six atoms bonded to it are at the six corners of the octahedron. The angle between any pair of adjacent bonds is the same, and has a value of 90°.

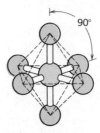

A simplified drawing of an octahedron generally shows the square plane in the center, as it would look tilted back, with lines running to the top and bottom of the octahedron.

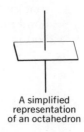

A simplified
representation
of an octahedron

Before going on the next section, <u>you should practice</u> drawing each of the five structures described here. If you understand them well, it will make the rest of this chapter easier to understand, too.

5.2 VALENCE SHELL ELECTRON-PAIR REPULSION THEORY

If necessary, review the procedures for writing Lewis structures on page 119.

One of the primary goals of chemical bonding theory is to explain and (we hope) to predict molecular structure. A theory that is exceedingly simple in its concept and remarkably successful in its ability to predict molecular geometry accurately is called the **valence shell electron-pair repulsion theory (VSEPR theory).** In applying it, it is not necessary to employ the notion of atomic orbitals at all. We will see, instead, that if an electron-dot structure can be drawn for a molecule, its general shape can be predicted.

The valence shell electron-pair repulsion theory proposes that the geometric arrangement of atoms, or groups of atoms (for which we use the general term *ligand*) about some central atom is determined *solely* by the repulsions between the electron pairs present in the valence shell of the central atom. To see how this works, let's begin by considering the molecule $BeCl_2$. Its electron-dot structure is given as

$$:\overset{..}{\underset{..}{Cl}} \! \overset{\times}{} \! Be \! \overset{\times}{} \! \overset{..}{\underset{..}{Cl}}:$$

where the crosses are beryllium electrons and the dots are chlorine electrons. This particular molecule, you recall, violates the octet rule, and there are only two pairs of electrons in the valence shell of Be. According to the theory, these electron pairs will arrange themselves to be as far apart as possible, so that the repulsions between them are at a minimum. When there are two electron pairs in the valence shell, this minimal repulsion occurs when the electron pairs are located on opposite sides of the nucleus, so that we have

$$\overset{\times}{} \! \! - \! Be \! - \! \overset{\times}{}$$

In the $BeCl_2$ molecule the ligands (i.e., the chlorine atoms) are attached to the Be by the sharing of these electron pairs. This means that the chlorines must be placed where the electron pairs are, and the molecule should therefore have the *linear* structure

$$\overset{180°}{Cl\overset{\frown}{-}Be\overset{\frown}{-}Cl}$$

This is, in fact, the shape of the $BeCl_2$ molecule in the gas phase.

We can also extend this reasoning to situations involving double or triple bonds. For instance, the CO_2 molecule has the dot structure

$$:\overset{..}{O}::C::\overset{..}{O}:$$

where we see that there are double bonds between C and O. Both pairs of electrons in a double bond must be confined to the same general region in the valence shell of an atom—otherwise, it wouldn't be a *double* bond. Therefore, in terms of their effect on determining molecular geometry, a group of four electrons in a double bond behaves much like a group of two electrons in a single bond. In the valence shell of carbon, therefore, we have *two* groups of four, and these groups will locate themselves on opposite sides of the carbon nucleus so that the repulsions between them are a minimum:

$$\overset{..}{} \! \! - \! C \! - \! \overset{..}{}$$

As before, the ligands (in this case, oxygen) are attached to the central atom through these electron pairs and we again have a *linear* structure.

$$:\overset{..}{O}=C=\overset{..}{O}:$$

Number of Electron Pairs	Geometric Arrangement of Electron Pairs	
2	Linear	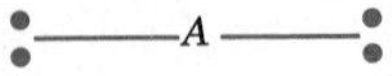
3	Planar triangular	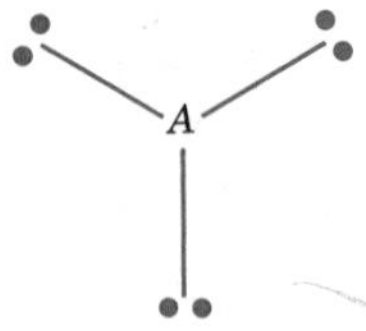
4	Tetrahedral	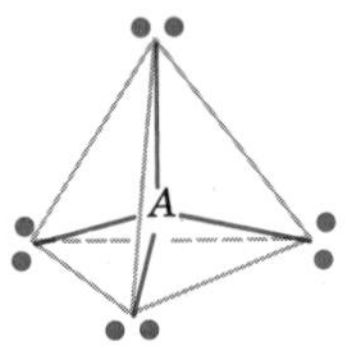
5	Trigonal bipyramidal	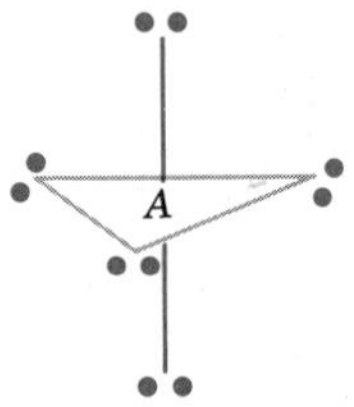
6	Octahedral	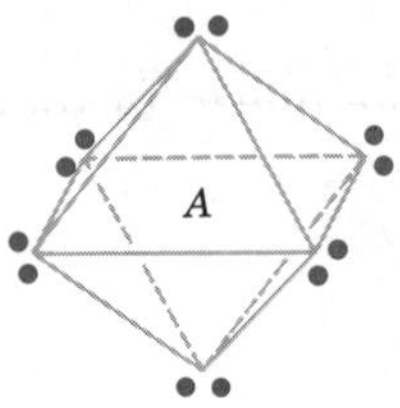 or 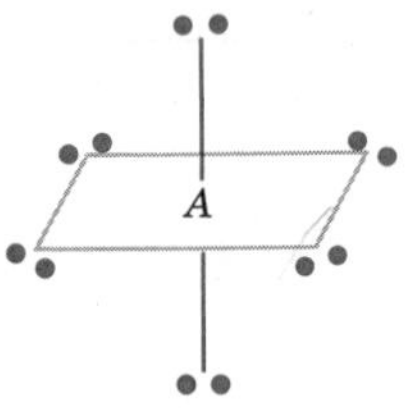

Figure 5.1

Arrangements of electron pairs that lead to minimum repulsions.

More than two pairs (or groups of pairs) of electrons

When there are more than two pairs (or groups of pairs) of electrons in the valence shell we find other geometric arrangements as shown in Figure 5.1. Electron pairs arranged in the valence shell in this manner lead to minimum repulsions. Let us see how we can use these electron-pair arrangements to predict molecular structure.

Three groups of electrons in the valence shell

In Chapter 4 we saw that the molecule BCl_3 has the dot structure

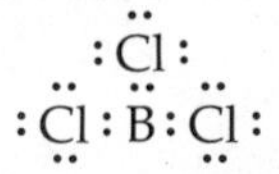

$$: \overset{\displaystyle \cdot \cdot}{\underset{\displaystyle \cdot \cdot}{Cl}} :$$
$$: \overset{\displaystyle \cdot \cdot}{\underset{\displaystyle \cdot \cdot}{Cl}} : B : \overset{\displaystyle \cdot \cdot}{\underset{\displaystyle \cdot \cdot}{Cl}} :$$

BCl_3 is said to be a planar triangular molecule.

Thus there are three electron pairs around boron. According to Figure 5.1, we therefore expect the three chlorine atoms to be arranged around the boron atom at the corners of a planar equilateral triangle. Experimentally, this is the structure that is found for BCl_3.

Now let's consider the molecule SO_2. The electron-dot structure for one of its two resonance structures is

$$:\ddot{O}:\ddot{S}::\ddot{O}:$$

Around the sulfur there are three groups of electrons—two groups each with one pair, and one group with two pairs (the double bond). To have minimum repulsions these groups of electrons are situated at the corners of a planar triangle with the sulfur in the center.

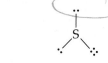

Attaching the oxygen atoms, one to a single pair and one to the double pair, we have

Thus the theory predicts that the two oxygen atoms and the sulfur do not lie in a straight line. How then, should we describe the structure?

When we give the shape of the SO_2 molecule, or of any other molecule, we describe how the atoms in the molecule are arranged relative to each other, *not* how the electrons are arranged around the central atom. Therefore, even though the electrons in the valence shell of sulfur are presumed to be in a triangle, we do not describe the SO_2 molecule as triangular. Instead, we say that it is nonlinear, or bent, or angular (If there are three atoms, either they are in a straight line and are linear, or they're not in a straight line and they are nonlinear.)

An important aspect of the structure of SO_2 is the presence of the unshared pair or **lone pair** of electrons in the valence shell of the sulfur. It is this lone pair that causes the bond pairs to be squeezed together and produces the nonlinear molecular shape. We will see that lone pairs have a strong influence on the structures of many molecules.

In summary, when there are three groups of electrons around an atom, they are arranged at the corners of a triangle. If they all are bonded to ligands, we have a molecule that we might generalize as AX_3, in which A stands for the central atom and X stands for a ligand. The structure of an AX_3 molecule is planar triangular. If only two groups are bonded, leaving one lone pair, we have a species AX_2E, where we use E to represent the lone pair. In an AX_2E molecule, the atomic nuclei are situated so as to give a nonlinear structure. Figure 5.2 illustrates the structures that we find for these kinds of molecules or polyatomic ions. In the figure, the lone pair on the AX_2E species is shown as an electron cloud. Notice that we have omitted from this figure, as well as from our discussion, molecules with the formula AXE_2. This would be a diatomic molecule, and when only two atoms are bonded to each other there's only one way for them to be connected. Only when there are three or more atoms in a molecule or ion do we have a choice of geometries.

Four groups of electrons in the valence shell

If an atom has four pairs of electrons in its valence shell, minimum repulsions occur if they are arranged tetrahedrally. We have just seen that when there are three electron pairs (or groups of pairs) in the valence shell of a central atom, two possible molecular shapes can occur depending on whether or not one of

We can determine the arrangement of atoms in a molecule experimentally, but not the geometrical arrangement of their electrons.

TYPE	EXAMPLE	STRUCTURE	DESCRIPTION
AX_3	BCl_3 :$\ddot{C}l$: :$\ddot{C}l$:B:$\ddot{C}l$:		Planer triangular
AX_2E	SO_2 :$\ddot{O}$: :$\ddot{S}$: :$\ddot{O}$:		Nonlinear angular bent

Figure 5.2

Geometries of molecules or ions in which the central atom has three electron pairs (or groups of pairs) in its valence shell.

the pairs is a lone pair. For molecules in which the central atom has four pairs in its valence shell there are three possible molecular shapes—all of which are derived from the tetrahedral arrangement of the electrons. Once again using A for the central atom, X for a ligand, and E for a lone pair, these can be represented as follows (see also Figure 5.3).

AX_4 These are tetrahedral molecules with ligands bonded by all four electron pairs. An example is methane, CH_4.

AX_3E When one lone pair is present a pyramidal (pyramid-shaped) molecule is formed. An example is NH_3. Notice that we describe the shape by the arrangement of atoms, not by the way the electrons are arranged.

AX_2E_2 Two lone pairs gives a nonlinear or angular structure—for example, water.

Five electron pairs

Five electron pairs will have minimum repulsions if they are arranged at the corners of a trigonal bipyramid, as shown in Figure 5.1. This gives four possible structures, depending on the number of lone pairs, as illustrated in Figure 5.4.

AX_5 All electron pairs are used in bonds and a trigonal bipyramidal molecule is formed.

AX_4E If one of the five pairs is a lone pair, one might suspect that there are two possible molecular structures, one with the lone pair in the central triangular plane and the other with the lone pair perpendicular to this plane.

It turns out that the repulsions are less in Structure I than in II. In fact, it is *always* found that the lone pairs prefer the triangular plane even when the structure has two or three of them.

     The shape of the AX_3E molecule is difficult to describe—the description we will use is *distorted tetrahedral*.

AX_3E_2 This structure has two lone pairs in the central triangular plane and the atoms are arranged in the form of the letter T drawn on its side (⊣). The molecule is described as being *T-shaped*.

AX_2E_3 Now there are three lone pairs in the triangular plane and the atoms are in a straight line—the structure is described as *linear*.

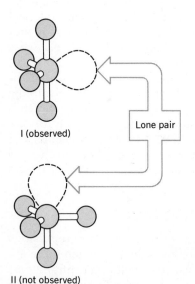

I (observed)

Lone pair

II (not observed)

TYPE	EXAMPLE	STRUCTURE	DESCRIPTION
AX_4	CH_4 H $H:\ddot{C}:H$ H	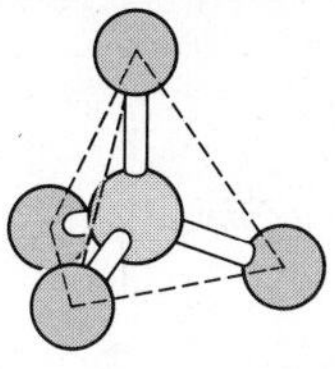	Tetrahedral
AX_3E	NH_3 $H:\ddot{N}:H$ H	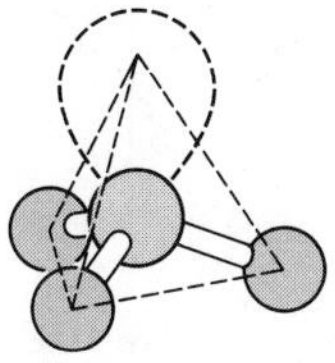	Pyramidal
AX_2E_2	H_2O $:\ddot{O}:H$ H	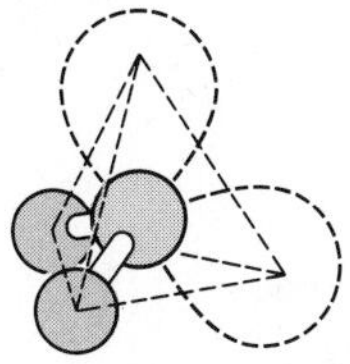	Nonlinear

Figure 5.3
Geometries of molecules in which the central atom has four pairs of electrons.

TYPE	EXAMPLE	STRUCTURE	DESCRIPTION
AX_5	PCl_5 Cl Cl P Cl Cl Cl	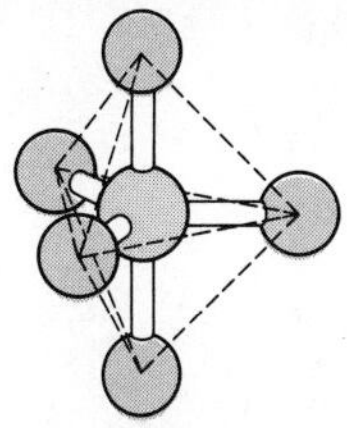	Trigonal bipyramidal
AX_4E	$TeCl_4$ Cl Cl Te Cl Cl	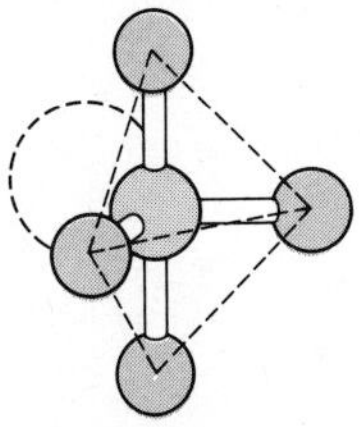	Distorted tetrahedral
AX_3E_2	ClF_3 $F\cdot\ddot{C}l:$ F F	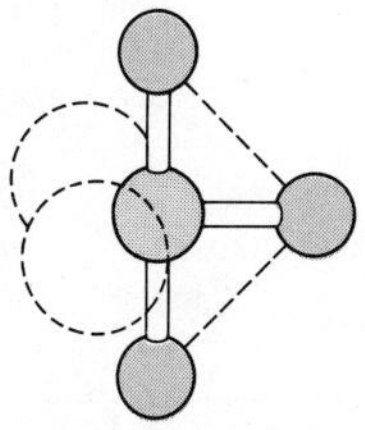	T-shaped
AX_2E_3	I_3^- $[I:\ddot{I}:I]^-$	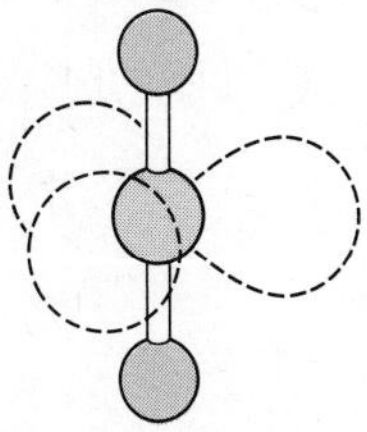	Linear

Figure 5.4
Molecular structures that result when the central atom has five electron-pair groups.

TYPE	EXAMPLE	STRUCTURE	DESCRIPTION
AX_6	SF_6  F F F:S:F F F	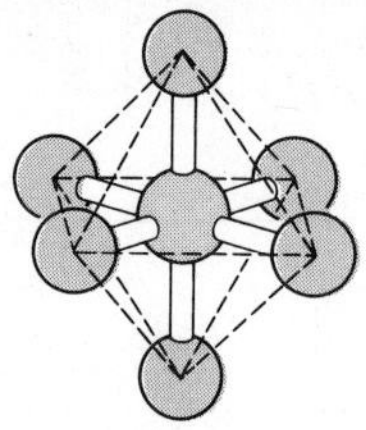	Octahedral
AX_5E	IF_5  F F :I:F F F	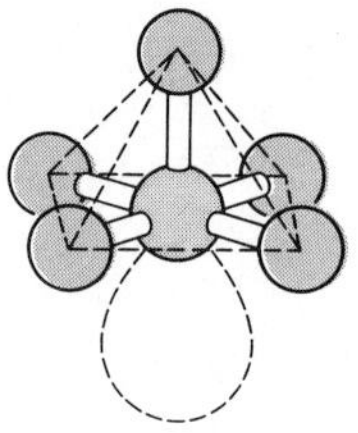	Square pyramidal
AX_4E_2	ICl_4^- 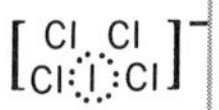 $\left[\begin{array}{cc} Cl & Cl \\ Cl:I:Cl \end{array}\right]^-$	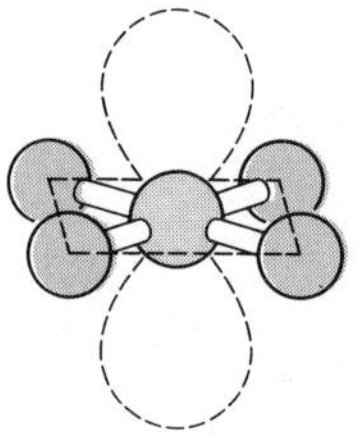	Square planar

Figure 5.5

Molecular structures that result when the central atom has six electron-pair groups.

Six electron pairs

These have minimum repulsions when arranged octahedrally (Figure 5.1). This gives *five* possibilities: AX_6, AX_5E, AX_4E_2, AX_3E_3, and AX_2E_4. Only the first three are actually observed to occur, however. These are shown in Figure 5.5.

AX_6 When all electron pairs are used for bonding, an *octahedral* structure is formed.

AX_5E The atoms in this structure are at the corners of a pyramid having a square base, so the structure is described as being *square pyramidal*.

AX_4E_2 With two lone pairs, minimum repulsions occur if they are as far apart as possible. This gives an arrangement of atoms that we describe as *square planar*.

Let's look now at some examples that illustrate how we use the VSEPR theory to predict the shapes of molecules and ions. In doing this, it is best if you are able to visualize the structure and identify it by its name. If this is an overwhelming problem for you, the structure can be found by identifying the number of ligands and lone pairs around the central atom, and then referring to Table 5.1.

Practice drawing the various structures. It will help you to visualize them.

EXAMPLE 5.1   What is the shape of the sulfate ion, SO_4^{2-}?

SOLUTION   We begin by drawing the Lewis structure for the ion following the procedure given in Chapter 4. This gives

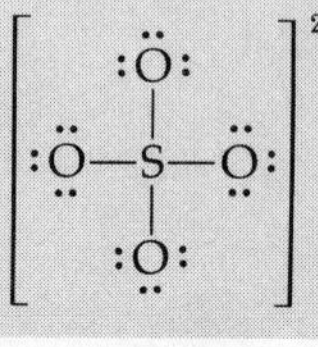

$$\left[\begin{array}{c} :\ddot{O}: \\ | \\ :\ddot{O}-S-\ddot{O}: \\ | \\ :\ddot{O}: \end{array}\right]^{2-}$$

Table 5.1
Summary of molecular shapes

Type of Molecule or Ion	Shape
AX_2	Linear
AX_3	Planar triangular
AX_2E	Nonlinear (angular, bent)
AX_4	Tetrahedral
AX_3E	Pyramidal
AX_2E_2	Nonlinear (angular, bent)
AX_5	Trigonal bipyramidal
AX_4E	Distorted tetrahedral
AX_3E_2	T-shaped
AX_2E_3	Linear
AX_6	Octahedral
AX_5E	Square pyramidal
AX_4E_2	Square planar

The sulfur has four electron pairs in its valence shell, which must be arranged tetrahedrally.

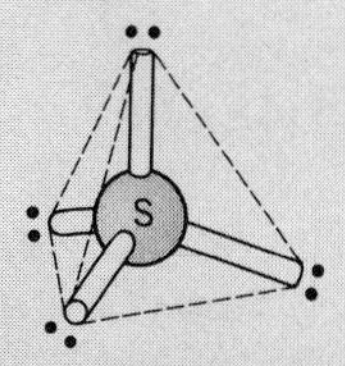

Attaching the oxygen atoms gives a tetrahedral ion.

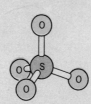

We can get the same answer, of course, by recognizing (from the Lewis structure) that SO_4^{2-} is an AX_4 species, which is tetrahedral.

EXAMPLE 5.2

Formate ion, HCO_2^-, comes from formic acid, the substance that produces the stinging sensation in bites from fire ants. The dot structure of formate ion is

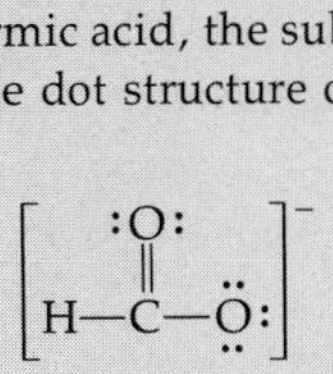

$$\left[\begin{array}{c} :\text{O}: \\ \| \\ \text{H}-\text{C}-\ddot{\text{O}}: \end{array} \right]^-$$

What is its shape?

SOLUTION

The double bond behaves just like a single bond for purposes of predicting molecular shape. This ion has three groups of electrons around the carbon and they are arranged in a planar triangular fashion. Since all groups are used in bonding, the ion (an AX_3 type) has a planar triangular shape.

EXAMPLE 5.3   Arsenic, a well-known poison, can be detected by converting its compounds to the very unstable substance AsH$_3$ (arsine), which decomposes easily on a clean hot glass surface where it deposits a mirrorlike coating of pure arsenic. What is the shape of an AsH$_3$ molecule?

SOLUTION   The first step is always to draw the Lewis structure. This is

$$H—\overset{..}{A}s—H$$
$$|$$
$$H$$

The four electron pairs are arranged tetrahedrally, but only three of them are used in bonds; one is a lone pair. The molecular structure that we get is

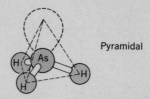

Pyramidal

We get the same answer recognizing AsH$_3$ as an AX_3E molecule.

5.3   VALENCE BOND THEORY

The electron-pair repulsion theory is a useful device for predicting molecular geometry, but it still doesn't answer the basic questions: "How do atoms share electrons between their valence shells and how are these electrons able to avoid each other?" To find the answers we must look to the results of quantum mechanics to see how the orbitals of atoms interact with each other when bonds are formed.

There are two important approaches to chemical bonding based on the results of quantum mechanics. One of these, called the **valence bond theory,** permits us to retain our picture of individual atoms coming together to form a covalent bond. The other, called **molecular orbital theory,** views a molecule as a set of positive nuclei with orbitals that extend over the entire molecule. The electrons that populate these molecular orbitals do not belong to any individual atoms but, instead, to the molecule as a whole. We will look at the molecular orbital theory in more detail in Section 5.7.

When the valence bond and molecular orbital theories are extended and refined, both give essentially the same results.

The basic postulate of the valence bond theory is that when two atoms come together to form a covalent bond, an atomic orbital of one atom overlaps with an atomic orbital of the other. By **overlap,** we mean that the two orbitals share some common region in space. The pair of electrons that we have come to associate with a covalent bond is shared between the two atoms in this region of overlap, and the strength of the covalent bond, as measured by the amount of energy needed to break it, is proportional to the *extent* of overlap of the atomic orbitals. As a consequence, the atoms in a molecule tend to position themselves so that there is a maximum amount of orbital overlap.

Let's see how this theory applies to some familiar compounds. The simplest of these is the hydrogen molecule, which is formed from two hydrogen atoms that each has a single electron in a 1s orbital. According to valence bond theory, we would view the H—H bond as resulting from the overlap of the two

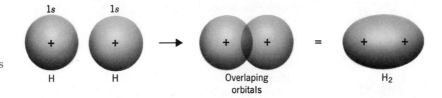

Figure 5.6

Formation of H₂ by overlap of 1s orbitals.

1s orbitals, as shown in Figure 5.6.[1] The electron density that this gives in the molecule is the same as they described in Chapter 4 (Figure 4.3).

In the HF molecule we have a somewhat different state of affairs. Fluorine has the valence-shell electron configuration

$$F \quad \underset{2s}{\uparrow\downarrow} \quad \underset{2p}{\uparrow\downarrow \ \uparrow\downarrow \ \uparrow}$$

where we find one of the 2p orbitals occupied by a single electron. It is with this partially occupied 2p orbital that the hydrogen 1s orbital overlaps, as illustrated in Figure 5.7. In this case the hydrogen electron and the fluorine electron can pair up and be shared between the two nuclei. Note that the 1s orbital of the hydrogen atom does not overlap with an already filled atomic orbital on fluorine because then there would be three electrons in the bond (two from the fluorine 2p orbital and one from the hydrogen 1s orbital). This situation is not permitted. *Only two electrons with their spins paired may be shared in one set of overlapping orbitals.*

Suppose that we now consider the molecule H₂O. Here we have two hydrogen atoms bound to a single oxygen atom. The outer-shell electron configuration of oxygen

$$O \quad \underset{2s}{\uparrow\downarrow} \quad \underset{2p}{\uparrow\downarrow \ \uparrow \ \uparrow}$$

shows that there are two unpaired electrons in *p* orbitals. This allows the two hydrogen atoms, with their electrons in 1s orbitals, to bond to the oxygen by means of overlap of their 1s orbitals with these partially filled oxygen *p* orbitals (Figure 5.8). We can represent this using the following orbital diagram:

$$O \text{ (in } H_2O) \quad \underset{2s}{\uparrow\downarrow} \quad \underset{2p}{\uparrow\downarrow \ \uparrow\downarrow \ \uparrow\downarrow}$$

where the colored arrows represent the electrons from the hydrogen atoms. Since the *p* orbitals are oriented at 90° to one another, we expect the H—O—H bond angle in water to also be 90°. Actually, this angle is 104.5°. One explanation for this discrepancy (we will see another one later) is that since the O—H bonds are highly polar, the H atoms carry a substantial positive charge and therefore repel one another. This factor tends to increase the H—O—H angle. However, since the best overlap between the hydrogen 1s orbitals and the oxygen 2p orbitals occurs at an angle of 90°, the H—O—H angle cannot increase too much without a considerable decrease in overlap, which would produce a substantial loss in bond strength. There are thus two factors working in opposition to each other, one tending to increase the bond angle and one tending to reduce it to 90°. It appears that a balance is obtained when the angle is 104.5°.

Qualitatively, the valence bond theory can account for the geometry of the water molecule. We can also apply the theory to the ammonia molecule with

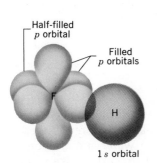

Figure 5.7

Formation of HF by the overlap of the partially filled fluorine 2p orbital with the 1s orbital of hydrogen.

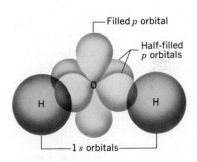

Figure 5.8

Bonding in H₂O. Overlap of two half-filled oxygen 2p orbitals with the hydrogen 1s orbitals.

---

[1] In this illustration, as well as in others throughout this chapter, it is important to keep in mind that we are looking at schematic representations of orbital wave functions.

reasonable success. Nitrogen, being in Group VA, has three unpaired electrons in it p subshell.

$$N \quad \underset{2s}{\underline{\uparrow\downarrow}} \quad \underset{2p}{\underline{\uparrow}\ \underline{\uparrow}\ \underline{\uparrow}}$$

Three hydrogen atoms can form bonds to nitrogen by overlapping their $1s$ orbitals with the half-filled p orbitals as shown in Figure 5.9a. The orbital diagram shows how the nitrogen completes its valence shell by this process.

$$N \text{ (in NH}_3) \quad \underset{2s}{\underline{\uparrow\downarrow}} \quad \underset{2p}{\underline{\uparrow\downarrow}\ \underline{\uparrow\downarrow}\ \underline{\uparrow\downarrow}} \quad \text{(Colored arrows are H electrons)}$$

As in the water molecule, the H—N—H angles are larger than the expected 90°, having values in this case of 107°; as with H_2O, we might attempt to explain this angle in terms of repulsions between the hydrogens. In any event, we obtain a picture of the NH_3 molecule that has a pyramidal shape, with the nitrogen atom at the apex of the pyramid and the three hydrogen atoms at the corners of the base (Figure 5.9b).

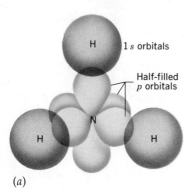

(a)

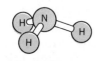

(b)

Figure 5.9

Bonding in NH₃ gives a pyramidal molecule. (a) Overlap of the 2p orbitals of nitrogen with 1s orbitals of hydrogen. (b) Pyramidal shape of the NH₃ molecule.

5.4 HYBRID ORBITALS

The very simple picture of the overlap of half-filled atomic orbitals that we have just developed cannot be used to account for all molecular structures. It works well for H_2 and HF, but is only marginally acceptable for water and ammonia. When we get to methane, it breaks down completely. With carbon, we would initially expect only two bonds to be formed with hydrogen since the valence shell of carbon contains only two unpaired electrons.

$$C \quad \underset{2s}{\underline{\uparrow\downarrow}} \quad \underset{2p}{\underline{\uparrow}\ \underline{\uparrow}\ \underline{\quad}}$$

The species CH_2, however, does not exist as a stable molecule. Instead, the simplest compound between carbon and hydrogen is methane, which has the formula CH_4. Attempting to explain the structure of this molecule by spreading the electrons out to give

$$\underset{2s}{\underline{\uparrow}} \quad \underset{2p}{\underline{\uparrow}\ \underline{\uparrow}\ \underline{\uparrow}}$$

Figure 5.10

The structure of methane, CH₄.

There are no simple atomic orbitals arranged in a trigonal bipyramidal fashion.

suggests that three of the C—H bonds will be formed by overlap of hydrogen $1s$ orbitals with carbon $2p$ orbitals, while the remaining bond would be the result of the overlap of the carbon $2s$ orbital with a hydrogen $1s$ orbital. This fourth C—H bond should certainly be different from the other three bonds, because it is formed from different kinds of orbitals. It has been found experimentally, however, that *all* four C—H bonds are identical and that the molecule has a structure in which the carbon atoms lies at the center of a tetrahedron with the hydrogen atoms located at the four corners (Figure 5.10). Apparently, the orbitals that carbon uses to form bonds in molecules like CH_4, and that other atoms use to form bonds in the more complex trigonal bipyramidal and octahedral structures, are not just simple atomic orbitals. The question is: "What kinds of orbitals are they?"

The solution to this apparent dilemma is found in the mathematics of quantum mechanics. In that theory, the solution of Schrödinger's wave equation provides us with a series of wave functions, ψ, each of which describes a different atomic orbital. It is the property of these mathematical functions that when they are squared, they enable us to calculate the probability of locating the electron at some point in space around the nucleus and, in fact, the spheres

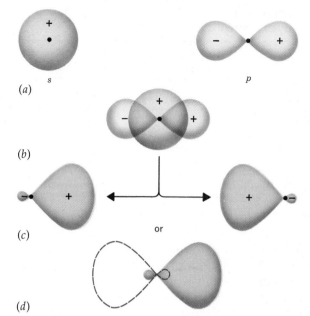

Figure 5.11

Formation of two sp *hybrid orbitals from an* s *and a* p *orbital. (a)* s *and* p *orbitals drawn separately. (b)* s *and* p *orbitals before hybridization. (c) Two* sp *hybrid orbitals are formed (drawn separately). (d) The two* sp *hybrid orbitals drawn together to show their directional properties. Note that one orbital points to the left, the other to the right.*

and figure-eights that we have been drawing roughly correspond to pictorial representations of the probability distributions predicted by the wave functions for *s* and *p* orbitals.

What is important to us here is that it is possible to combine these wave functions by appropriately adding or subtracting them to give new functions that are referred to as **hybrid orbitals.** In this way, two or more atomic orbitals are mixed together to produce a new set of orbitals and, invariably, these hybrid orbitals possess different directional properties than the pure atomic orbitals from which they are created. For example, Figure 5.11 illustrates the result of combining a 2*s* orbital with a 2*p* orbital to give a new set of two *sp* hybrid orbitals. In this drawing, notice that we have indicated that the wave function for a *p* orbital has positive numerical values in some regions about the nucleus and negative values in others.[2] The *s* orbital, on the other hand, has the same algebraic sign everywhere. Therefore, when these wave functions are alternatively added and subtracted, the new orbitals that result become larger in those regions where both functions have the same sign and smaller in regions where they are of opposite sign. In effect, by constructive and destructive interference of the electron waves corresponding to the *s* and *p* orbitals, two new orbitals are formed.

Hybrid orbitals such as these possess some very interesting properties. We see for each orbital that one lobe is much larger than the other, and because of this a hybrid orbital can overlap well in only one direction—the direction in which the orbital protrudes the most. A hybrid orbital is therefore very strongly directional in its ability to enter into covalent bond formation. Furthermore, because the large lobe of a hybrid orbital extends out farther from the nucleus than an unhybridized orbital does, it is able to overlap much more effectively with an orbital of another atom. Consequently, bonds formed from hybrid orbitals tend to be stronger than those formed from ordinary atomic orbitals.

It takes energy to form hybrid orbitals, but this energy is more than paid back by the much stronger bonds formed by them.

[2] Although difficult to visualize, the amplitude of the electron wave in certain orbitals can be positive in some regions and negative in others. This is similar to the displacement of the string on a guitar being positive (upward) in some places and negative (downward) in others for the different harmonics that we looked at in Figure 3.20 on Page 84.

Table 5.2
Hybrid orbitals

Hybrid Orbitals	Number of Orbitals	Orientation
sp	2	Linear
sp^2	3	Planar triangle
sp^3	4	Tetrahedral
sp^2d	4	Square planar
sp^3d	5	Trigonal bipyramidal
sp^3d^2	6	Octahedral

Thus far we have examined what occurs when one s and one p orbital are mixed together. Other combinations of orbitals are also possible, with the number of orbitals in a hybrid set, as well as their orientations, determined by which atomic orbitals are combined. Table 5.2 contains a listing of the sets of hybrid orbitals that can be used to explain most of the molecular structures that we encounter in this book. Their directional properties, which you no doubt recognize by now, are illustrated in Figure 5.12. Notice that the number of each kind of atomic orbital included in a combination is specified by an appropriate superscript on the atomic orbital type. Thus the sp^3d^2 hybrids are formed from one s orbital, three p orbitals, and two d orbitals.

Let us see now how we can use the information contained in Table 5.2 and Figure 5.12 to account for the structures of some typical molecules. We might begin with a molecule such as BeH_2. It would have the dot structure

$$H \overset{\times}{\cdot} Be \overset{\times}{\cdot} H$$

VSEPR theory would predict that BeH_2 should be linear.

where the dots are Be electrons and the crosses are H electrons. The electronic structure of beryllium's valence shell is

$$Be \quad \underset{2s}{\underline{\uparrow\downarrow}} \quad \underset{2p}{\underline{\quad}\,\underline{\quad}\,\underline{\quad}}$$

In order to form two covalent bonds with H atoms, the Be atom must provide two half-filled (i.e., singly occupied) orbitals. This can be accomplished by creating a pair of sp hybrids and placing one electron in each of them.

$$Be \quad \underbrace{\underline{\uparrow}\;\underline{\uparrow}}_{sp} \quad \underbrace{\underline{\quad}\;\underline{\quad}}_{\text{Unhybridized } 2p \text{ orbitals}}$$

The two H atoms can then bond to the beryllium atom by overlap of their respective singly occupied s orbitals with the singly occupied Be sp hybrids as shown in Figure 5.13. The orbital diagram for the molecule is

$$Be \text{ (in } BeH_2) \quad \underbrace{\underline{\uparrow\downarrow}\;\underline{\uparrow\downarrow}}_{sp} \quad \underbrace{\underline{\quad}\;\underline{\quad}}_{p} \qquad \text{(Colored arrows are H electrons)}$$

Because of the orientation of the sp hybrid orbitals, the H atoms are forced to lie on opposite sides of the Be, and a linear H—Be—H molecule results.

sp	Linear	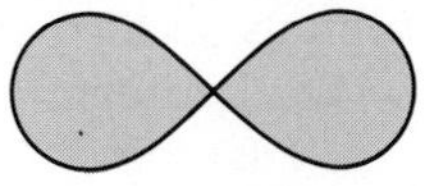
sp^2	Planar triangular	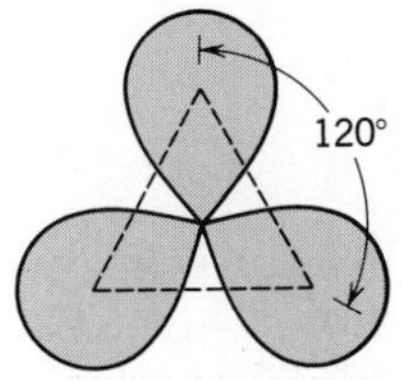 120°
sp^3	Tetrahedral	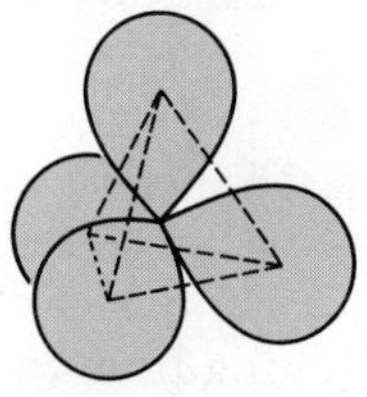
sp^2d	Square planar	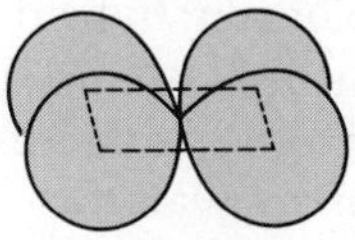
sp^3d	Trigonal bipyramidal	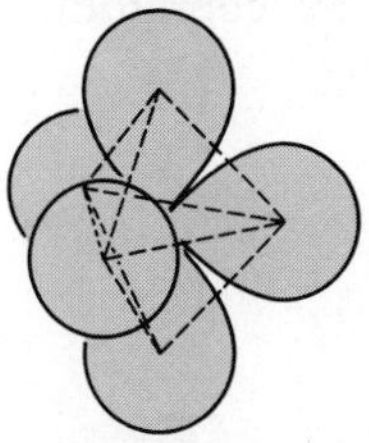
sp^3d^2	Octahedral	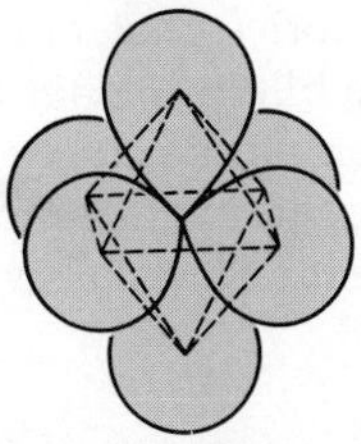

Figure 5.12

Directional properties of hybrid orbitals. The minor lobes have been omitted for the sake of clarity.

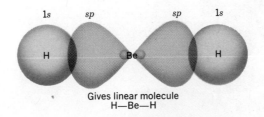

Gives linear molecule
H—Be—H

Figure 5.13

The bonding in BeH$_2$.

Let us now return to our problem of the structure of CH_4. If we use hybrid orbitals on the carbon atom, we find that in order to provide four orbitals with which hydrogen $1s$ orbitals can overlap, we must use a set of sp^3 hybrids.

C $\underset{2s}{\underline{\uparrow\downarrow}} \quad \underset{2p}{\underline{\uparrow}\ \underline{\uparrow}\ \underline{}}$ (Unhybridized)

$\underset{sp^3}{\underline{\uparrow}\ \underline{\uparrow}\ \underline{\uparrow}\ \underline{\uparrow}}$ (Hybridized)

In Figure 5.12 we see that these orbitals point toward the vertices of a tetrahedron. Therefore, when the four hydrogen atoms are attached to the carbon by orbital overlap with these sp^3 hybrids

C (in CH_4) $\underset{sp^3}{\underline{\uparrow\downarrow}\ \underline{\uparrow\downarrow}\ \underline{\uparrow\downarrow}\ \underline{\uparrow\downarrow}}$

a tetrahedral molecule results, as shown in Figure 5.14. This is in agreement with the structure that is found by experiment.

In our earlier discussion we viewed the structure of H_2O and NH_3 as resulting from the use of half-filled p atomic orbitals on the oxygen and nitrogen atoms, respectively. An alternative view of the bonding in these molecules employs sp^3 hybrid orbitals on the central atom. In the tetrahedral set of hybrids, the orbitals are oriented at angles of 109.5° to each other. The bond angles in water (104.5°) and ammonia (107°) are not too different from the tetrahedral angle and, using water as an example, we might consider the molecule to result from the overlap of hydrogen $1s$ orbitals with two partially occupied sp^3 orbitals on the oxygen atom.

O $\underset{2s}{\underline{\uparrow\downarrow}} \quad \underset{2p}{\underline{\uparrow\downarrow}\ \underline{\uparrow}\ \underline{\uparrow}}$ (Unhybridized)

$\underset{sp^3}{\underline{\uparrow\downarrow}\ \underline{\uparrow\downarrow}\ \underline{\uparrow}\ \underline{\uparrow}}$ (Hybridized)

O (in H_2O) $\underset{sp^3}{\underline{\uparrow\downarrow}\ \underline{\uparrow\downarrow}\ \underline{\uparrow\downarrow}\ \underline{\uparrow\downarrow}}$ (Colored arrows are H electrons)

Notice that only two of the hybrid orbitals are involved in bond formation while the other two contain nonbonded "lone pairs" of electrons. In the case of ammonia, three of the sp^3 orbitals are used in bonding while the fourth orbital contains a lone pair of electrons (Figure 5.15). There is, in fact, rather strong experimental evidence to indicate that this lone pair does indeed project out

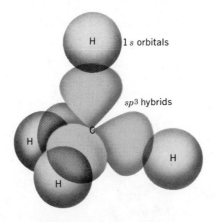

Figure 5.14

The formation of methane by overlap of hydrogen 1s orbitals with carbon sp³ hybrids.

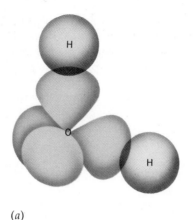

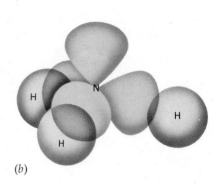

Figure 5.15
The use of sp³ *hybrids for bonding in (a) H₂O and (b) NH₃.*

(a)

(b)

from the nitrogen atom as implied in this picture of the NH_3 molecule. It is worth noting that in our previous description of NH_3 we found this lone pair of electrons in an *s* orbital that would have spread the electron pair symmetrically about the nucleus.

In the case of H_2O and NH_3, the H—X—H bond angles (104.5° and 107°, respectively) are less than the tetrahedral angle of 109° that is observed in the molecule, CH_4. One way to account for this is through the influence of the lone-pair electrons present in hybrid orbitals of the central atom. A pair of electrons in a bond is attracted to two nuclei and, therefore, might be expected to occupy a smaller effective volume than a pair of electrons in a nonbonded orbital, which experience the attraction of only one nucleus. The lone-pair electrons, then, because of their greater space requirement, tend to crowd together the electron pairs located in the bonds and hence reduce the bond angle to something less than 109°. On this basis we anticipate a greater reduction in bond angle for water than for ammonia, since water has two lone pairs while ammonia has only one.

As another example, consider the molecule SF_6. Sulfur, being in Group VIA, has six valence electrons distributed over the $3s$ and $3p$ subshells

$$S \quad \underset{3s}{\underline{\uparrow\downarrow}} \quad \underset{3p}{\underline{\uparrow\downarrow}\,\underline{\uparrow}\,\underline{\uparrow}} \quad \underset{3d}{\underline{\quad}\,\underline{\quad}\,\underline{\quad}\,\underline{\quad}\,\underline{\quad}}$$

Here we have shown the empty $3d$ subshell as well as the $3s$ and $3p$ subshells that contain electrons. In order for sulfur to form six covalent bonds to fluorine, six half-filled orbitals must be created. This can be accomplished by using two of the unoccupied $3d$ orbitals, and an sp^3d^2 hybrid set is formed.

$$S \quad \underset{sp^3d^2}{\underline{\uparrow}\,\underline{\uparrow}\,\underline{\uparrow}\,\underline{\uparrow}\,\underline{\uparrow}\,\underline{\uparrow}} \quad \underset{3d \text{ (Unhybridized)}}{\underline{\quad}\,\underline{\quad}\,\underline{\quad}}$$

$$S \text{ (in } SF_6) \quad \underset{sp^3d^2}{\underline{\uparrow\downarrow}\,\underline{\uparrow\downarrow}\,\underline{\uparrow\downarrow}\,\underline{\uparrow\downarrow}\,\underline{\uparrow\downarrow}\,\underline{\uparrow\downarrow}} \quad \underset{3d}{\underline{\quad}\,\underline{\quad}\,\underline{\quad}} \quad \begin{array}{l}\text{(Colored arrows = F} \\ \text{electrons)}\end{array}$$

The sp^3d^2 orbitals point toward the corners of an octahedron, which explains the octahedral geometry of SF_6.

By now you have probably noticed that the orientations of the hybrid orbitals in Figure 5.12 are the same as the orientations that give minimum repulsions between electron pairs described in the valence shell electron-pair repulsion theory, and that the two theories give identical results. For example, the

Until now we've omitted the **3d** subshell when writing the electron configuration of a period 3 element.

shapes of methane, water, and ammonia predicted by the VSEPR theory are the same as those accounted for by using sp^3 hybrids in valence bond theory. Both theories use a tetrahedral arrangement of electron pairs in these molecules. This useful correlation gives us a simple way of anticipating the kinds of hybrid orbitals that an atom will use in a particular molecule. For instance, let's look again at SF_6. If we draw a dot structure for the molecule, we get

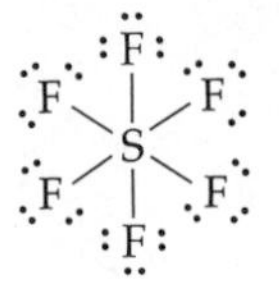

$$\begin{array}{c} :\ddot{F}: \\ :\ddot{F}\diagdown \mid \diagup\ddot{F}: \\ S \\ :\ddot{F}\diagup \mid \diagdown\ddot{F}: \\ :\ddot{F}: \end{array}$$

The VSEPR theory predicts that the six electron pairs around sulfur should be arranged octahedrally. Now we can ask ourselves, "What kinds of hybrid orbitals have an octahedral geometry?" The answer, of course, is sp^3d^2, and that is exactly what we used in our explanation of the structure of SF_6 by valence bond theory. We see, therefore, that the VSEPR theory can be used to help us choose the right kinds of hybrid orbitals to use in the valence bond theory. The two theories complement one another very nicely in explaining the bonding in molecules.

EXAMPLE 5.4   Determine the kind of hybrid orbitals used by sulfur in SF_4 and explain the bonding in this molecule according to valence bond theory.

SOLUTION   Let's use the VSEPR theory to help us choose hybrid orbitals. This means that we first need the Lewis structure for SF_4. Following our usual procedure, we get

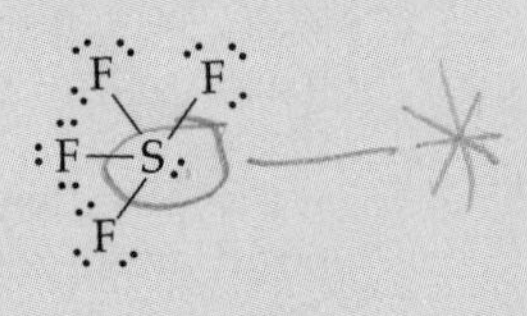

$$:\ddot{F}\underset{:\ddot{F}:}{\overset{:\ddot{F}\quad\ddot{F}:}{-S}} \longrightarrow *$$

Notice that there are five electron pairs around the sulfur. The VSEPR theory tells us that they should be located at the corners of a trigonal bipyramid, and the hybrid orbital set that is trigonal bipyramidal is sp^3d.

Now we examine the electronic structure of sulfur.

$$S \quad \underset{3s}{\underline{\uparrow\downarrow}} \quad \underset{3p}{\underline{\uparrow\downarrow}\ \underline{\uparrow}\ \underline{\uparrow}} \quad \underset{3d}{\underline{\quad}\ \underline{\quad}\ \underline{\quad}\ \underline{\quad}\ \underline{\quad}}$$

Forming sp^3d hybrid gives

$$S \quad \underset{sp^3d}{\underline{\uparrow\downarrow}\ \underline{\uparrow}\ \underline{\uparrow}\ \underline{\uparrow}\ \underline{\uparrow}} \quad \underset{3d}{\underline{\quad}\ \underline{\quad}\ \underline{\quad}\ \underline{\quad}}$$

Notice that we have enough half-filled orbitals to form the four bonds to fluorine.

$$S\ (\text{in}\ F_4) \quad \underset{sp^3d}{\underline{\uparrow\downarrow}\ \underline{\uparrow\downarrow}\ \underline{\uparrow\downarrow}\ \underline{\uparrow\downarrow}\ \underline{\uparrow\downarrow}} \quad \underset{3d}{\underline{\quad}\ \underline{\quad}\ \underline{\quad}\ \underline{\quad}} \qquad \text{(Colored arrows = F electrons)}$$

This gives us our bonding "picture" for SF_4 in which a lone pair occupies one of the hybrid orbitals. The structure of SF_4 is shown in Figure 5.16.

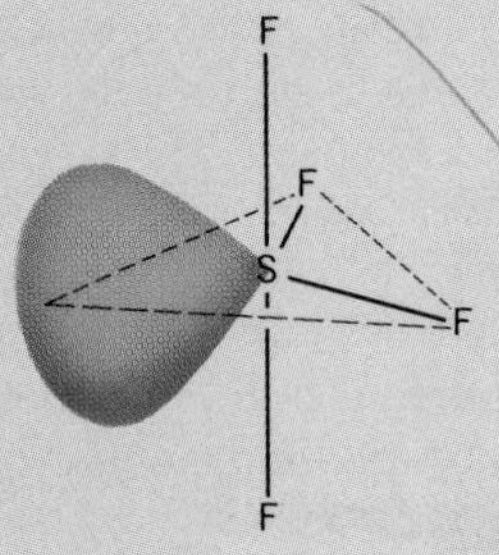

Figure 5.16

The structure of SF_4. Note the lone pair of electrons in the sp^3d hybrid orbital. The VSEPR theory, by predicting the structure, even tells us which of the hybrid orbitals houses the lone pair.

Before moving on, a word should be said about the coordinate covalent bond. An example of this, you remember, is provided by the ammonium ion.

$$\left[\begin{array}{c} H \\ \overset{\times}{\underset{\times}{H:N:H}} \\ H \end{array}\right]^{+} \quad \text{or} \quad \left[\begin{array}{c} H \\ | \\ H-N\rightarrow H \\ | \\ H \end{array}\right]^{+}$$

According to valence bond theory, two electrons are shared between two over-lapping orbitals. It doesn't matter, however, where the electrons come from. If one comes from each of the overlapping orbitals, an "ordinary" covalent bond is formed. If one orbital is empty and the other is filled, both electrons can come from the filled orbital and a coordinate covalent bond is formed. Thus we can imagine the coordinate covalent bond in the ammonium ion to be formed by the overlap of an *empty* 1s orbital centered on a proton (a hydrogen ion, H$^+$) with the *completely filled* lone-pair orbital on the nitrogen of an ammonia molecule. The electron pair is then shared in the region of orbital overlap. Once the bond is formed, of course, it is a full-fledged covalent bond whose properties do not depend on its origin. Consequently, the four N—H bonds in NH$_4^+$ are identical and the ion is usually represented as simply

$$\left[\begin{array}{c} H \\ | \\ H-N-H \\ | \\ H \end{array}\right]^{+}$$

This same argument can be extended to other coordinate covalent bonds as well.

5.5 MULTIPLE BONDS

Double and triple bonds occur when two and three pairs of electrons, respectively, are shared between two atoms. As examples, we have seen the molecules ethylene, C$_2$H$_4$, and acetylene, C$_2$H$_2$.

$$\underset{\text{ethylene}}{\overset{\displaystyle{\overset{H}{\diagup}\,C=C\,\overset{H}{\diagdown}}}{\underset{\displaystyle{H\diagdown\qquad\diagup H}}{}}} \qquad \underset{\text{acetylene}}{H-C\equiv C-H}$$

The bonding in ethylene is usually interpreted in the following way. In order to form bonds to three other atoms (two hydrogens and one carbon), each carbon atom employs a set of sp^2 hybrids.

$$C \quad \underset{2s}{\uparrow\downarrow} \quad \underset{2p}{\uparrow\ \uparrow} \quad \underline{\quad}$$

gives

$$C \quad \underset{sp^2}{\uparrow\ \uparrow\ \uparrow} \quad \underset{p}{\uparrow} \quad \underline{\quad} \text{ (Unhybridized)}$$

Two of these hybrid orbitals are used for overlap with hydrogen 1s orbitals while the third sp^2 orbital overlaps with a similar orbital on the other carbon atom, as shown in Figure 5.17a. This, then, accounts for all of the C—H bonds in C$_2$H$_4$ as well as *one* of the electron pairs shared between the two carbons.

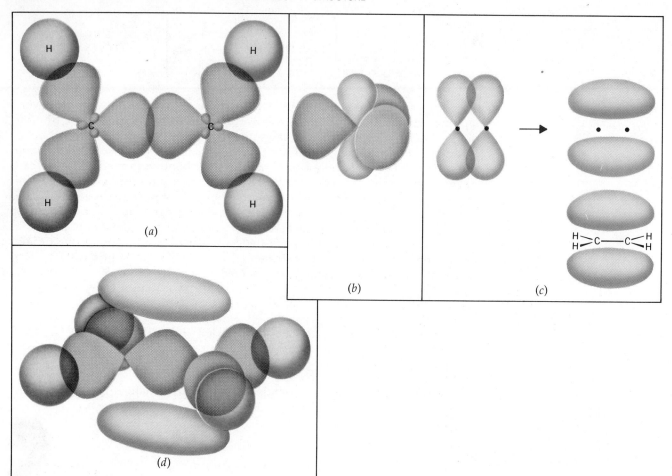

Figure 5.17

The bonding in ethylene, C_2H_4. (a) Overlap of hydrogen 1s orbitals with sp² hybrid orbitals on carbon. The carbon atoms are also bound to each other by overlap of sp² hybrid orbitals. (b) Unhybridized p orbital is perpendicular to the plane of the sp² hybrid orbitals. (c) Formation of π bond by sideways overlap of p orbitals. (d) A drawing showing the entire ethylene molecule.

Because of the way the sp^2 orbitals are created, each carbon atom also has an unhybridized p atomic orbital that is perpendicular to the plane of the sp^2 orbitals and that projects above and below the plane of these hybrids (Figure 5.17b). When the two carbon atoms are joined together, these p orbitals approach each other sideways and, in addition to the bond formed from the overlap of sp^2 orbitals, a second bond is formed in which the electron cloud is concentrated above and below the carbon–carbon axis. This we see illustrated in Figure 5.17c.

In terms of this interpretation, the double bond in ethylene consists of two distinctly different kinds of bonds, and to differentiate between them a specific notation is employed. A bond that concentrates electron density along the line joining the bound nuclei is called a **σ bond (sigma bond).** The overlap of the sp^2 orbitals of adjacent carbons therefore gives rise to a σ bond. The bond that is formed by the sideways overlap of two p orbitals, and that provides electron density above and below the line connecting the bound nuclei, is called a **π bond (pi bond).**[3] Thus in ethylene we find the double bond to consist of one σ

---

[3] In later chapters we refer to a π bond formed by the sideways overlap of two p orbitals as a $p\pi$-$p\pi$ bond.

Figure 5.18

*The triple bond in acetylene con-
sists of one σ bond and two π
bonds. (a) Two π bonds. (b) Cylindrical electron distribution
about the bond axis.*

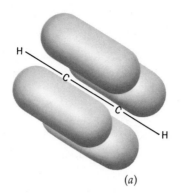

(a)

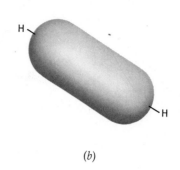

(b)

The σ bond is sandwiched
between the two halves of the
π bond like a frankfurter
between the two halves of a
bun.

bond and one π bond. Notice that in this double bond the two electron pairs
manage to avoid one another by occupying different regions in space.

Another point to note is that bonds formed by the overlap of the hydrogen
$1s$ orbitals with carbon sp^2 hybrid orbitals (Figure 5.17a) also concentrate electron density along a line joining bound atoms. Therefore, these C—H bonds
would also be termed σ bonds.

In acetylene each carbon is bound to only two other atoms, a hydrogen and
a carbon atom. Two orbitals are needed for this purpose, and a pair of sp hybrid
orbitals are used.

$$C \quad \underline{\uparrow\downarrow} \quad \underline{\uparrow} \; \underline{\uparrow} \quad \underline{}$$

$$\underbrace{\underline{\uparrow}\;\underline{\uparrow}}_{sp} \quad \underbrace{\underline{\uparrow}\;\underline{\uparrow}}_{p}$$

This leaves two singly occupied unhybridized p orbitals on each carbon that are
mutually perpendicular as well as being perpendicular to the sp hybrids. When
the carbon atoms join by way of σ bond formation between an sp hybrid orbital
on each carbon, the p orbitals can also overlap to yield two π bonds that surround the axis joining the carbon nuclei (Figure 5.18). A triple bond therefore
consists of one σ and two π bonds. The two π bonds in acetylene (or in any
triple bond) give a total electron distribution that is cylindrical about the bond
axis. This is shown in Figure 5.18b.

In arriving at the structure of a molecule, such as ethylene or acetylene, *the
shape of the molecular framework is determined by the σ bonds that arise from
overlap of the hybrid orbitals.* Double and triple bonds in a structure result from
additional π bonds. In summary, we find the following:

single bond—one σ bond
double bond —one σ + one π bond
triple bond—one σ + two π bonds

EXAMPLE 5.5 Identify the kinds of hybrid orbitals used by the atoms in acetic acid whose structure is

$$
\begin{array}{ccc}
 & \text{H} & :\!\overset{\displaystyle ..}{\text{O}}\!: \\
 & | & || \\
\text{H} - & \!\!\text{C} - \text{C} - & \!\!\overset{\displaystyle ..}{\underset{\displaystyle ..}{\text{O}}} - \text{H} \\
 & | & \\
 & \text{H} &
\end{array}
$$

What kinds of bonds (σ, π) exist between the atoms?

SOLUTION To identify the kind of hybrid orbitals that an atom uses, we simply count the number of groups of electrons around the atom. Then we choose a hybrid set that has that number of orbitals. For example, the carbon on the left has four bonds (four pairs) and uses sp^3 orbitals. The carbon next to it has three groups of electrons, which means it uses sp^2 orbitals. The doubly bonded oxygen has three groups of electrons, so its uses sp^2 orbitals, too. Finally, the singly bonded oxygen has four pairs around it and must use sp^3 hybrids. (No doubt you've noticed that we haven't said anything about hydrogen—hydrogen always uses only its $1s$ orbital for bonding.)

Hydrogen only forms σ bonds.

Now we can identify the kinds of bonds in the molecule.

5.6 RESONANCE

In Chapter 4 we saw that there are instances in which we cannot draw a single satisfactory Lewis dot formula for a molecule or ion. Some examples, you might remember, are SO_2, SO_3, and NO_2^-. Sulfur dioxide, for instance, was drawn as

and it was stated that the actual electronic structure of this molecule corresponds to a resonance hybrid of these two structures.

Electron dot formulas, as we have drawn them, closely correspond to the valence bond pictures that were developed in the preceding sections. Each pair of dots drawn between two atoms represents a pair of electrons shared in a region where atomic orbitals of the bonded atoms overlap. When we draw one of the resonance structures of SO_2, we are therefore referring to a bonding picture in which one S—O bond consists of a single σ bond while the other is composed on one σ and one π bond.

When the valence bond theory was developed, it was recognized that there were numerous instances where a single valence bond structure was inadequate in accounting for the molecular structure, so the concept of resonance evolved. The inability in these cases to draw a single picture to describe the electron density in the molecule is one drawback of valence bond theory. Nevertheless, the correspondence between the valence bond structures that are based on orbital overlap and the simple electron-dot formulas make the valence bond concept a very useful one.

5.7 MOLECULAR ORBITAL THEORY

In our discussion of atomic structure in Chapter 3 we saw that around an atomic nucleus there exists a set of atomic orbitals. The electronic structure of a particular atom was derived by feeding the appropriate number of electrons into this set of atomic orbitals such that (1) no more than two electrons populated a single orbital and (2) each electron was placed into the lowest energy orbital available, and (3) electrons were spread out as much as possible, with unpaired spins, over orbitals of the same energy.

Molecular orbital theory proceeds in much this same way. According to this theory, a molecule contains a certain arrangement of atomic nuclei, and spread out over these nuclei is a set of **molecular orbitals.** The electronic structure of the molecule is obtained by feeding the appropriate number of electrons into these molecular orbitals following the same rules that apply to the filling of atomic orbitals.

No one is quite sure what shapes molecular orbitals have in any particular molecule or ion. What *appears* to be an approximately correct picture is obtained by combining the atomic orbitals that reside on the nuclei making up the molecule. These combinations are achieved by considering the constructive interference and destructive interference of the electron waves of the atoms in the molecule. This is shown in Figure 5.19 for the $1s$ orbitals on two identical nuclei. Notice that when the amplitudes of the two waves are *added*, the resulting molecular orbital has a shape that concentrates electron density between the two nuclei. Electrons placed in such a molecular orbital tend to hold the nuclei together and stabilize a molecule. For that reason this orbital is called a **bonding molecular orbital.** Because the electron density in the orbital is centered along the line joining the atomic nuclei, it is also a σ type of orbital. Since it is derived in this case from two $1s$ atomic orbitals, we refer to it as the σ_{1s} **molecular orbital.**

You will also observe in Figure 5.19 that a second molecular orbital is obtained by destructive interference of the electron waves. In this instance a molecular orbital is produced that places the maximum electron density outside the region between the two nuclei. If the electrons of a molecule are placed into this molecular orbital, they do not help to cement the nuclei together and, in fact, the unshielded nuclei repel one another. Consequently, electrons placed into this molecular orbital lead to a destabilization of the molecule and, as a result, the orbital is said to be **antibonding.** This antibonding orbital also has its greatest electron density along the line that passes through the two nuclei and is thus a σ-type orbital. Its antibonding character is denoted by an asterisk superscript; thus it is called the σ_{1s}^{*} **molecular orbital.** As you might expect, we can also draw similar pictures for the combination of any pair of s orbitals; therefore, in a diatomic molecule we have $\sigma_{2s}, \sigma_{2s}^{*}, \sigma_{3s}, \sigma_{3s}^{*}, \ldots$, molecular orbitals.

Figure 5.19

The combination of atomic $1s$ orbitals to give bonding and antibonding molecular orbitals.

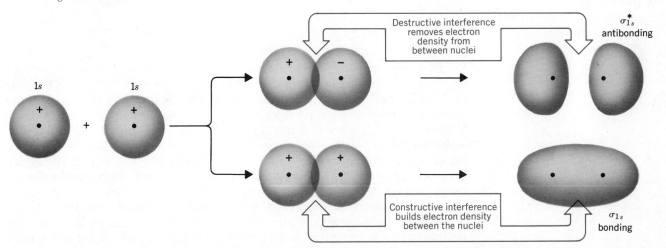

1s 1s

Destructive interference removes electron density from between nuclei

σ_{1s}^{*} antibonding

Constructive interference builds electron density between the nuclei

σ_{1s} bonding

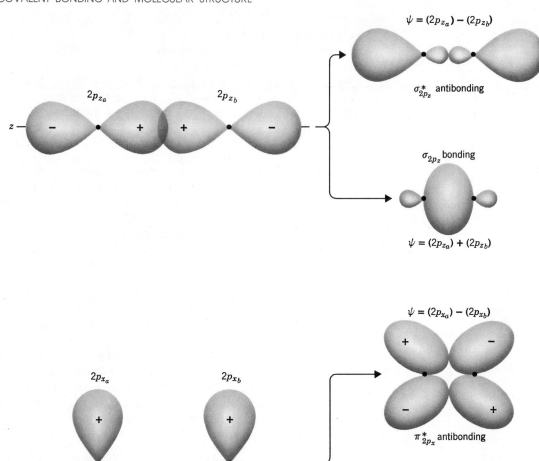

Figure 5.20

Formation of molecular orbitals from atomic p orbitals.

In a molecule the p orbitals are also capable of interacting to produce bonding and antibonding molecular orbitals, as illustrated in Figure 5.20. Here we have arbitrarily chosen to denote the internuclear axis as the z axis of our coordinate system, so the p orbitals that point toward one another correspond to p_z orbitals. Again we find that one combination of orbitals gives a bonding molecular orbital, with electron density placed between the two nuclei, while the second combination places most of the electron density outside the region between the nuclei. The p_z orbitals, like s orbitals, form σ-type molecular orbitals, and for $2p_z$ orbitals they would be labeled σ_{2p_z} and $\sigma_{2p_z}^*$.

Having chosen the z axis as the internuclear axis, we find that the p_x and p_y orbitals on the two nuclei of our molecule are forced to overlap in a sideways fashion to produce π and π^* molecular orbitals (Figure 5.20). Also keep in mind

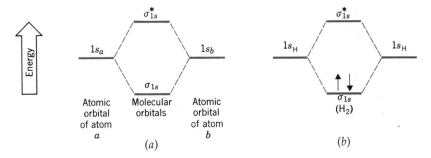

Figure 5.21

(a) *The energies of the bonding and antibonding σ_{1s} molecular orbitals. (b) Bonding in H_2.*

that the π_{p_x} and $\pi_{p_x}^*$ orbitals are the same as the π_{p_y} and $\pi_{p_y}^*$ orbitals, except that they are situated at 90° to each other when viewed down the molecular axis.

For a diatomic molecule, we have now examined the shapes of the molecular orbitals that can be considered to arise as a consequence of the overlap of atomic orbitals. To discuss the electronic structure of a diatomic molecule, however, we must know the relative energies of these orbitals. Once this has been established, we can then proceed with filling the orbitals with electrons, following the rules mentioned earlier.

Let us first consider the σ_{1s} and σ_{1s}^* orbitals. Electrons placed into the bonding orbital lead to stable bond formation and, therefore, to an energy lower than that of two separate atoms. On the other hand, electrons placed into the antibonding orbital lead to a destabilization of the molecule and thus to a state higher in energy than the atoms from which the molecule is formed. We can represent this schematically as shown in Figure 5.21a, where the energies of the atomic orbitals of the separate atoms appear on either side of the energy-level diagram while the energies of the molecular orbitals appear in the center.

Using this simple diagram we can examine the bonding in the H_2 molecule. There are two electrons in H_2 that we place in the lowest-energy molecular orbital, the σ_{1s}, (Figure 5.21b). The electron distribution in H_2 is therefore that described by the shape of the σ_{1s} orbital. Notice that this picture is the same as that developed in the valence bond view of H_2. This should not be too surprising since both theories are attempting to describe the same molecular species.

Before moving on, let us also see why the molecule He_2 does *not* exist. The species He_2 would have four electrons, two of which would be placed into the σ_{1s} orbital. The other two would be forced to occupy the σ_{1s}^* orbital. The pair of electrons in the antibonding orbital would cancel out the stabilizing influence of the bonding pair. As a result, the **net bond order,** which we can define as

$$\begin{pmatrix} \text{Net bond} \\ \text{order} \end{pmatrix} = \frac{(\text{no. of } e^- \text{ in bonding MOs}) - (\text{no. of } e^- \text{ in antibonding MOs})}{2}$$

has a value of zero for He_2. Since the bond order in He_2 is zero, He_2 is not a stable molecule and is not observed to exist under normal conditions.

For diatomic molecules of second period elements we really only need to consider molecular orbitals that are derived from the interaction of the $2s$ and $2p$ orbitals. The $1s$ orbitals are essentially buried beneath the valence-shell orbitals and are therefore not involved to any appreciable extent in the bonding in these species. The energy-level diagram for the molecular orbitals created from the $2s$ and $2p$ orbitals is shown in Figure 5.22a.[4] Let's see how this energy-level diagram can be used to account for the bonding in the molecules N_2, O_2, and F_2.

---

[4] The relative energies of the σ_{2p_z} and the π_{2p_x}, π_{2p_y} set actually shift about somewhat as we cross the second period. This diagram will give correct results for any species you will encounter here.

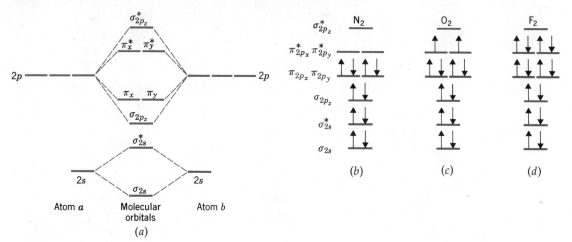

Figure 5.22

(a) Energies of molecular orbitals formed from atomic orbitals having n=2 in diatomic molecules. (b–d) Molecular orbital electron configurations of N_2, O_2, and F_2.

Nitrogen is in Group VA and, therefore, each nitrogen atom contributes five electrons to the N_2 molecule from its valance shell. This means that we must place ten electrons into our set of molecular orbitals. As shown in Figure 5.22b, two electrons enter the σ_{2s}, two go into the σ_{2s}^*, two more into the σ_{2p_z} and, finally, two into each of the bonding π orbitals, π_{2p_x} and π_{2p_y}. As before, the two σ_{2s}^* antibonding electrons cancel the effect of the σ_{2s} bonding electrons, leaving us with a net total of six bonding electrons (two each in the σ_{2p_z}, π_{2p_x}, and π_{2p_y} orbitals). If, as usual, we take two electrons to represent a "bond," we find that N_2 is held together by a triple bond that is composed of one σ and two π bonds. As with H_2, we arrive at the same resultant description of the bonding in N_2 with both the valence bond and molecular orbital theories.

The real mark of success for molecular orbital theory is seen in its description of the O_2 molecule. This species is found experimentally to be paramagnetic with two unpaired electrons. In addition, its bond length and bond energy suggest that there is a double bond between the two oxygen atoms. An attempt to derive a valence bond picture for O_2, however, gives us

$$\ddot{O}::\ddot{O}$$

where, to satisfy the octet rule and to give a double bond, all of the electrons appear in pairs.

The molecular orbital description of O_2 is seen in Figure 5.22c. The first 10 of the 12 valence electrons populate all the same molecular orbitals as in N_2. The final two electrons must then be placed in the $\pi_{2p_x}^*$ and $\pi_{2p_y}^*$ antibonding orbitals. However, because these two orbitals are of the same energy, the electrons spread themselves out with their spins in the same direction. These two antibonding π electrons cancel the effects of two of the π-bonding electrons, so in the final analysis we see that O_2 is held together by a *net* double bond (one σ and one net π bond). Also notice that the molecule is predicted to have two unpaired electrons, in precise agreement with experimental evidence.

Finally, with F_2 (which contains two more electrons than O_2), we find that the two π^* antibonding orbitals are filled (Figure 5.22d). This leaves one net single bond, and once again the valence bond and molecular orbital theories give the same result.

The success of molecular orbital theory is not restricted merely to diatomic molecules. In more complex molecules, however, the energy-level diagrams are more difficult to predict, and we will not attempt to extend the theory much further. One useful concept in molecular orbital theory that we can look at further,

A dot structure such as

$$\cdot\ddot{O}:\ddot{O}\cdot$$

has the correct number of unpaired electrons, but it is unsatisfactory because there is only an O—O single bond.

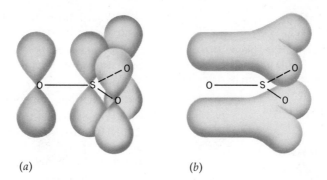

Figure 5.23
Simultaneous overlap of atomic p orbitals in the SO_3 molecule. (a) p orbitals on the sulfur and oxygen atoms. (b) Delocalized π molecular orbital.

(a)   (b)

however, is the idea that molecular orbitals may extend over more than two nuclei. It is this aspect of molecular orbital theory that allows one to avoid the concept of resonance.

Consider, for example, the molecule SO_3. From experiment we know this to be a planar molecule (all four atoms lie in the same plane) with all three S—O bonds the same. This structure can be explained if we assume that the sulfur employs a set of sp^2 hybrid orbitals to form σ bonds with the three oxygen atoms. This leaves one unhybridized p orbital on the sulfur that can overlap *simultaneously* with p orbitals on the three oxygen atoms, as shown in Figure 5.23. The result is the creation of a molecular orbital that extends over all four nuclei in such a way that the electron densities in the S—O bonds are all the same. Obviously, there is no need to draw more than one bonding picture for the molecule; molecular orbital theory thus is able to explain the bonding in SO_3 satisfactorily without resorting, as valence bond theory does, to the rather awkward concept of resonance.

Molecular orbitals that spread over more than two nuclei are said to be delocalized.

INDEX TO QUESTIONS

REVIEW QUESTIONS

5.1 Sketch the shapes of the five common molecular shapes. (If necessary, refer to the drawings in Section 5.1.)

5.2 What is the bond angle in a linear molecule?

5.3 What are the bond angles in planar triangular, tetrahedral, and octahedral molecules?

5.4 What are the bond angles in a trigonal bipyramidal molecule?

5.5 What is the basic postulate of the valence shell electron-pair repulsion theory?

5.6 Use the valence shell electron-pair repulsion theory to predict the geometry of each of the following (in each case, the central atom is written first).

(a) NF_3

(b) $PH_4{}^+$

(c) $CO_3{}^{2-}$

(d) NO_2

(e) $SeCl_4$

(f) $ICl_2{}^-$

(g) BrF_5

(h) CCl_4

(i) $AlCl_6{}^{3-}$

(j) $SbCl_5$

(k) $SnCl_2$

5.7 For each of the following, use the valence shell electron-pair repulsion theory to predict (1) the arrangement of electron-pair groups around the central atom (written first in the formula) and (2) the molecular shape.

(a) CH_2O

(b) $ClO_3{}^-$

(c) $SiF_6{}^{2-}$

(d) PCl_3

(e) $SOCl_2$

(f) $SnCl_4$

(g) SO_3

(h) BF_4^-

(i) OF_2

(j) XeF_4

5.8 Use the valence shell electron-pair repulsion theory to predict for each of the following: (1) the geometric arrangement of electron pairs around the central atom (written first in the formulas), (2) the molecular shape.

(a) ClO_2^-

(b) SCl_2

(c) $SbCl_6^-$

(d) PCl_4^+

(e) IF_4^-

(f) PO_4^{3-}

(g) CH_3^+

(h) ICl_3

(i) NO_3^-

(j) AsH_3

(k) $POCl_3$

5.9 Describe the changes in molecular geometry that take place during the following reactions:

(a) $BF_3 + F^- \rightarrow BF_4^-$

(b) $PCl_5 + Cl^- \rightarrow PCl_6^-$

(c) $ICl_3 + Cl^- \rightarrow ICl_4^-$

(d) $SF_2 + F_2 \rightarrow SF_4$

(e) $C_2H_2 + H_2 \rightarrow C_2H_4$

(f)
$$\text{Cl}-\overset{\overset{\textstyle O}{\|}}{\text{C}}-\text{Cl} \longrightarrow \text{COCl}^+ + \text{Cl}^-$$

5.10 What is the basic concept on which the valence bond theory is based?

5.11 Use valence bond theory to explain the bonding in the Cl_2 molecule.

5.12 From valence bond theory, predict the geometry of the $SnCl_2$ molecule.

5.13 What orbitals would you expect to be involved in bonding in the AsH_3 molecule? The H—As—H bond angles are very close to 90°.

5.14 How can you account for the fact that the H—S—H bond angle in H_2S is approximately 92°?

5.15 Diagram the outer-shell electronic structures of P and F. Indicate how bonding occurs between P and F to give PF_3. What molecular shape would you expect?

5.16 What is a hybrid orbital?

5.17 Why is it necessary to employ hybrid orbitals when attempting to account for the structure of methane, CH_4?

5.18 On the basis of the electronic structure of the central atom, suggest what kind of hybrid orbitals would be involved in the bonding in each of the following molecules or ions:

(a) BCl_3

(b) NH_4^+

(c) PCl_5

(d) $AlCl_6^{3-}$

(e) $BeCl_2$

(f) $SbCl_6^-$

(g) PCl_3

(h) TeF_4

(i) ClO_4^-

5.19 From a knowledge of the hybrid orbitals used for bonding, predict the structures of each of the species in Question 5.18.

5.20 Use your predictions from Question 5.6 to suggest the type of hybrid orbitals that would be used in valence bond theory to account for these geometries.

5.21 What kind of hybrid orbitals would be used by the central atom in each species in Question 5.7?

5.22 Diagram the bonding in $SbCl_5$. What kind of hybrid orbitals are involved in the bonding?

5.23 It is possible for SiF_4 to react with F^- to give SiF_6^{2-} but it is not possible for CF_4 to form CF_6^{2-}. Why?

5.24 We have discussed the reaction, $BCl_3 + NH_3 \rightarrow Cl_3BNH_3$ earlier (Chapter 4). What kind of hybrid orbitals are used by B and N before and after reaction? How does the geometry change about B and N as the reaction occurs?

5.25 Which of the species in Question 5.18 have one or more bonds that would be considered to have been formed by way of coordinate covalent bonding? How, if at all, does a coordinate covalent bond differ from a normal covalent bond once it has been formed?

5.26 What angles exist between the orbitals in

(a) sp^3 hybrids

(b) sp^2 hybrids

(c) sp hybrids

(d) sp^3d^2 hybrids

5.27 $SnCl_4$ is a volatile liquid composed of individual $SnCl_4$ molecules. Describe the bonding that is expected in this molecule.

5.28 Describe a σ bond; a π bond. What constitutes a double bond? A triple bond?

5.29 What kind of hybrid orbitals are used by each atom in the molecule below? What kinds of bonds (σ, π) occur between the atoms?

$$\text{H}-\text{C}\equiv\text{C}-\overset{\overset{\textstyle H}{|}}{\text{C}}=\overset{\overset{\textstyle H}{|}}{\text{C}}-\overset{\overset{\textstyle :O:}{\|}}{\text{C}}-\overset{..}{\underset{..}{\text{O}}}-\overset{\overset{\textstyle H}{|}}{\underset{\underset{\textstyle H}{|}}{\text{C}}}-\text{H}$$

5.30 Draw the resonance structures for (a) NO_3^- and (b) NO_2^-.

5.31 How does molecular orbital theory view the formation of a molecule? How does molecular orbital theory differ from valence bond theory?

5.32 Describe the bonding in the N_2 molecule

(a) according to valence bond theory.

(b) according to molecular orbital theory.

5.33 Predict the relative stabilities of the species N_2^+, N_2, N_2^-. From the discussion in Section 4.5 (p. 122), how would you expect the bond lengths in these species to compare?

5.34 Predict the relative stabilities of the species O_2^+, O_2, O_2^-. From the discussion in Section 4.5 (p. 122), how would you expect the bond lengths in these species to compare?

5.35 What is the difference between a bonding and an antibonding molecular orbital? How do their energies compare?

5.36 Use Figure 5.22*a* to draw molecular orbital energy-level diagrams for Li_2, Be_2, B_2, and C_2. Which of these should not exist, which should be paramagnetic?

5.37 What can you predict about the stabilities of the species in Question 5.36 when one electron is (a) removed from each, (b) added to each?

5.38 How does molecular orbital theory avoid the concept of resonance?

5.39 The species H_2^+ and He_2^+ have been observed. Use molecular orbital theory to account for their existence.

5.40 Give the valence bond and molecular orbital descriptions for the following species that can be drawn as two or more resonance structures: (a) SO_2 (b) NO_3^-

(c)

$$H-C\begin{array}{c} \nearrow O \\ \searrow O^- \end{array}$$

5.41 On p. 157 it was suggested that lone electron pairs tend to repel rather strongly electron pairs between bonded atoms; in other words, lone-pair/bond-pair repulsions are greater than bond-pair/bond-pair repulsions. This will lead to structural distortions of some of the idealized molecular geometries pictured in Figures 5.2, 5.3, 5.4, and 5.5. Predict the nature of these distortions and sketch the shapes of the resulting molecules.

5.42 Hybrid orbitals are not symmetrical about the nucleus. They concentrate electron density on the side of the nucleus where the orbital is "large." Lone electron pairs in hybrid orbitals therefore are expected to contribute to the dipole moment of the molecule. It is observed experimentally that NF_3 is nearly a nonpolar molecule; NH_3 is very polar. The electronegativity difference between N and F is nearly the same as that between N and H (see Table 4.6). How does this support the view that in both NF_3 and NH_3 the nitrogen uses sp^3 hybrid orbitals?

6

CHEMICAL REACTIONS IN AQUEOUS SOLUTION

Water serves as a medium for many chemical reactions. Producing a photographic image on film or paper, as shown here, involves a number of chemical reactions in which at least one reactant is present in an aqueous solution.

Figure 6.1

(a) Two solutions, one containing silver nitrate and the other containing sodium chloride. (b) As one solution is added to the other, a white solid composed of silver chloride, AgCl, is formed.

In the last three chapters we learned about the structures of atoms and how chemical bonds hold atoms together in ionic and molecular compounds. Now it is time to begin our study of the kinds of reactions that these substances undergo.

As we might expect, for a chemical reaction to occur between two substances, the ions or molecules constituting the reactants must come in contact with one another. For this reason the speed at which a reaction takes place depends on how freely the reacting species are able to intermingle. For instance, if crystals of NaCl and AgNO$_3$ are mixed together, no noticeable chemical changes are observed. However, if the NaCl and AgNO$_3$ are first dissolved in water and their solutions are then mixed, a white solid is produced that has the formula AgCl. This reaction is shown in Figure 6.1. Here, the formation of silver chloride requires that silver ions and chloride ions meet. When the two solids are mixed, this can only occur at the surfaces where the crystals touch one another. Because of the homogeneous nature of solutions, however, dissolved substances are intimately mixed at the molecular or ionic level, and chemical changes can occur rapidly. This is why we routinely use solutions to carry out chemical reactions.

Water is one of the most abundant chemicals in nature and serves as a good solvent for many substances, both ionic and molecular. Our preoccupation with reactions in aqueous systems stems from the general availability of water as a solvent and, particularly in recent times, the recognition of the importance of water as a medium in which biochemical reactions take place. In this chapter we will discuss various types of chemical reactions that occur in aqueous solution and learn how the quantitative principles developed in Chapter 2 can be applied to these reactions.

6.1 SOLUTION TERMINOLOGY

There are certain terms that apply to all kinds of solutions and that should be understood before we proceed further. The words **solvent** and **solute** are two of these. The general practice is to refer to the substance present in greatest proportion in a solution as the solvent, with all the other substances in the solution considered solutes. In solutions that contain water, however, the solvent is nearly always considered to be the water, even when it is present in relatively small amounts. For example, a mixture of 96% H_2SO_4 and 4% H_2O by weight is called "concentrated sulfuric acid," which implies that a large quantity of sulfuric acid is *dissolved* in a small amount of water; that is, H_2O is taken to be the solvent and H_2SO_4 the solute.

It is often necessary to express the proportions of solute and solvent in a solution. This is done by specifying the **concentration** of the solute in the mixture. Concentration can be stated quantitatively in a variety of ways, as we will see. The terms **concentrated** and **dilute** are used when we wish to speak, in qualitative terms, of the relative proportions of solvent and solute. In a concentrated solution there is a relatively large amount of solute present in the solvent; a dilute solution, on the other hand, possesses only a small quantity of solute. These two terms have meaning only in relationship to one another; they do not imply any specific quantities of solute in solvent. For example, concentrated sulfuric acid contains, as we said above, 96% H_2SO_4 and 4% H_2O. By comparison, a solution containing 20% H_2SO_4 would be dilute. This latter solution would be considered concentrated in comparison to a 5% H_2SO_4 solution.

In most cases there is a limit to the amount of solute that will dissolve in a fixed quantity of solvent at any particular temperature. For example, if we add sodium chloride to 100 ml of water at 0°C, only 35.7 g of the salt will dissolve, regardless of the total amount that we place into the water. A solution that contains as much dissolved solute as it can hold while in contact with excess solute is said to be **saturated.** If it contains less solute than required for saturation, it is said to be **unsaturated.** The **solubility** of the solute is taken to be the amount required to achieve a saturated solution in a given amount of the solvent. For example, the solubility of NaCl in water at 0°C is 35.7 g per 100 ml of H_2O. Usually, a solute's solubility changes with temperature. For example, at 100°C the solubility of NaCl is 39.1 g/100 ml of H_2O. This means that we must always specify the temperature when stating the solubility.

The terms saturated and unsaturated are in no way directly related to the terms concentrated and dilute. For example, a saturated solution of silver chloride at room temperature contains only 0.000089 g AgCl/100 ml of water and is certainly considered dilute. On the other hand, it would take about 500 g of lithium chlorate, $LiClO_3$, to form a saturated solution in 100 ml of water at the same temperature. A solution containing 400 g of $LiClO_3$ in 100 ml of water is unsaturated but, nevertheless, is quite concentrated. Thus a saturated solution can be dilute and an unsaturated solution can be concentrated.

Finally, there are some substances, such as sodium acetate, that frequently form **supersaturated** solutions—solutions that contain more solute than ordinarily required for saturation. Sodium acetate is soluble to the extent of 119 g per 100 ml of water at 0°C, and becomes more soluble at higher temperatures. If an unsaturated hot solution containing more than 119 g of sodium acetate per 100 ml is cooled slowly to 0°C, the excess solute remains dissolved and the solution becomes supersaturated. Such solutions are unstable; if a small crystal of the solute is added, additional solute crystallizes on this "seed" crystal until the concentration drops to the point of saturation (see Figure 6.2).

We write this as
35.7 g/100 ml H_2O.

6.2 ELECTROLYTES

At the beginning of this chapter we noted that water is generally a good solvent for ionic compounds. In the solid state, these substances are composed of posi-

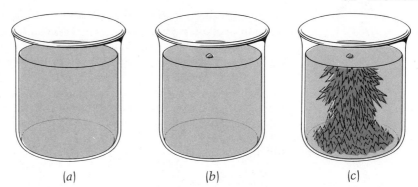

Figure 6.2

Supersaturation. (a) Supersaturated solution. (b) Introduction of a seed crystal. (c) Excess solute crystallizes on the seed.

(a)   (b)   (c)

tive and negative ions held together in a rigid framework by electrostatic forces. However, when they dissolve in water these solids break apart, or **dissociate** to yield ions that are more or less free to roam about in the solution. As the solid dissolves, the ions become surrounded by water molecules and are said to be **hydrated.** We examine the solution process in more detail in Chapter 10.

The presence of ions imparts to the water the ability to conduct electricity, which we can demonstrate by using an apparatus such as the one shown in Figure 6.3. When electrical contact is made across the two electrodes, an electric current can flow and the light bulb will light. If we immerse these electrodes in pure water, no conductivity is observed (i.e., the bulb will not light) because water is a very poor conductor of electricity. However, if a typical ionic solid such as NaCl is added to the water, the bulb will begin to burn brightly as soon as the NaCl begins to dissolve. Substances (such as NaCl) that give electrically conducting solutions that contain ions are called **electrolytes.**

The dissociation of NaCl that occurs when the solid is dissolved can be represented by the equation

$$NaCl \ (s) \longrightarrow Na^+ \ (aq) + Cl^- \ (aq)$$

where the symbols s and aq in parentheses denote that the solid (s) produces ions in aqueous (aq) solution. Often, for simplicity, we will leave off the labels (s) or (aq) when doing so creates no confusion.

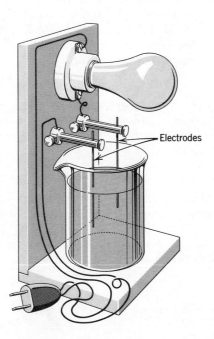

Electrodes

Figure 6.3

A conductivity apparatus.

The production of ions in solution is not limited to ionic compounds. There are many covalent substances that react with water to produce ions and therefore yield conducting solutions. Hydrogen chloride is a typical example. When HCl gas is dissolved in water the following reaction takes place:

$$HCl\ (aq) + H_2O \longrightarrow H_3O^+\ (aq) + Cl^-\ (aq)$$

This reaction occurs by the transfer of a proton or hydrogen ion (H^+) from the HCl molecule to the water molecule to produce a **hydronium ion,** H_3O^+, and a chloride ion.

The name proton is frequently used to mean hydrogen ion. Removal of hydrogen's single electron gives just a bare proton.

$$H:\overset{..}{\underset{..}{Cl}}: + :\overset{..}{O}:H \longrightarrow \left[H:\overset{..}{O}:H \right]^+ + \left[:\overset{..}{\underset{..}{Cl}}: \right]^-$$
$$ \overset{|}{H} \overset{|}{H}$$

Thus, even though hydrogen chloride exists by itself as discrete molecules of HCl (liquid HCl does not conduct electricity), when it dissolves in water it produces ions and becomes an electrolyte.

As we will see, the hydronium ion is one of the most important species to consider in discussions of chemical reactions in aqueous solutions. It is useful to think of it as a proton that has associated itself with a water molecule. We can do this because when the hydronium ion reacts it gives up the proton, and a water molecule is formed. In chemical reactions, therefore, the H_2O of the hydronium ion merely serves as a carrier for the H^+ ion. For that reason, the H_3O^+ ion is very often written simply as H^+, and we often speak of H_3O^+ as hydrogen ion. Thus the dissociation that occurs when HCl is dissolved in water is often represented simply as

H^+ is the active ingredient in H_3O^+.

$$HCl \longrightarrow H^+ + Cl^-$$

Even though we write H^+, *always* remember that there is at least one, and probably several more, H_2O molecules associated with the proton in aqueous solution.[1]

The two examples of electrolytes discussed above, NaCl and HCl, are essentially completely dissociated in aqueous solution; that is, 1 mol of NaCl gives 1 mol of Na^+ and 1 mol of Cl^-, and 1 mol of HCl gives 1 mol of H_3O^+ and 1 mol of Cl^-. Substances such as NaCl and HCl that, for all practical purposes, are completely dissociated in aqueous solution are said to be **strong electrolytes.**

There are many compounds, such as acetic acid ($HC_2H_3O_2$, found in vinegar), for example, that dissociate to only a limited extent in water. Only a small fraction of all the acetic acid molecules that are placed into a solution actually exist as ions formed by the reaction

$$HC_2H_3O_2 + H_2O \longrightarrow H_3O^+ + C_2H_3O_2^-$$

These substances are called **weak electrolytes** because their solutions contain relatively few ions and conduct electricity weakly. Additional examples are given in Table 6.1. In the table we see that water itself is a very weak electrolyte by virtue of the reaction

$$H_2O + H_2O \longrightarrow H_3O^+ + OH^-$$

This slight dissociation of water plays a very important role in many chemical reactions in which water is the solvent. Special attention is given to this topic in Chapter 15.

Finally, there are many molecular compounds that do not dissociate into ions at all when they are dissolved in water. Sugar and ethyl alcohol are two

---

[1] There is, in fact, evidence that suggests that the H^+ ion may exist as $H_9O_4^+$, that is, $H_3O(H_2O)_3^+$, in aqueous solution.

Table 6.1
Some weak electrolytes

Substance	Dissociation Reaction	Percent Dissociation of the Solute in a 1.00 M Solution
Water	$H_2O + H_2O \rightarrow H_3O^+ + OH^-$	1.8×10^{-7} (55.5 moles of H_2O per liter)
Acetic acid	$HC_2H_3O_2 + H_2O \rightarrow H_3O^+ + C_2H_3O_2^-$	0.42
Ammonia	$NH_3 + H_2O \rightarrow NH_4^+ + OH^-$	0.42
Hydrogen cyanide	$HCN + H_2O \rightarrow H_3O^+ + CN^-$	2.0×10^{-3}
Mercury(II) chloride	$HgCl_2 \rightarrow HgCl^+ + Cl^-$	1
Cadmium sulfate	$CdSO_4 \rightarrow Cd^{2+} + SO_4^{2-}$	7

common examples. These are called **nonelectrolytes.** Since solutions of nonelectrolytes contain no ions, they do not conduct an electric current.

6.3 CHEMICAL EQUILIBRIUM

The reason for the limited degree of dissociation of weak electrolytes is worth discussing at this point because it illustrates one of the most important concepts in chemistry, one to which we will devote three chapters at a later time (Chapters 13, 15, and 16).

In a solution of acetic acid, molecules of $HC_2H_3O_2$ are constantly colliding with molecules of water and, in each encounter, there is a certain probability that a proton will be transferred from an $HC_2H_3O_2$ molecule to a water molecule to yield H_3O^+ and $C_2H_3O_2^-$ ions. There are also encounters, in this solution, between acetate ions and hydronium ions. When these ions meet, there is a high probability that an H_3O^+ ion will lose a proton to a $C_2H_3O_2^-$ ion to reform $HC_2H_3O_2$ and H_2O molecules. Thus, in this solution we have two reactions occurring simultaneously:

$$HC_2H_3O_2 + H_2O \longrightarrow H_3O^+ + C_2H_3O_2^- \tag{I}$$

and

$$H_3O^+ + C_2H_3O_2^- \longrightarrow HC_2H_3O_2 + H_2O \tag{II}$$

When the rate at which the ions are formed by reaction I is equal to the rate at which they disappear by reaction II, their concentrations in the solution will no longer change with time. In fact, the concentrations of all the species will remain constant from this point on, even though if we followed any particular $C_2H_3O_2$ unit in the solution, it would sometimes exist as a $C_2H_3O_2^-$ ion and at other times as a $HC_2H_3O_2$ molecule. Such a state of affairs is called **equilibrium.** It is said to be a **dynamic equilibrium** because things are continually happening in the solution—two reactions are taking place: ions reacting to yield molecules and molecules reacting to produce ions.

To indicate chemical equilibrium in a reacting system, we use a set of double arrows, $\rightleftharpoons$, in the chemical equation. Thus the equilibrium that we have been discussing is expressed as

$$HC_2H_3O_2 + H_2O \rightleftharpoons H_3O^+ + C_2H_3O_2^-$$

The use of this notation implies that the forward reaction (the reaction going from left to right) is occurring at the same rate as the reverse reaction (the reaction from right to left). In a solution of acetic acid these rates become equal when only a small fraction of all the acetic acid exists as ions. We say that the extent of dissociation is small and that the **position of equilibrium**—the relative proportions of the reactants and products expressed in the equation—lies in the direction of the molecular form of the substance. In other words, almost everything is present in undissociated form.

For strong electrolytes the reaction of the ions to produce molecules has almost no tendency to occur. The position of equilibrium, therefore, lies almost completely toward the ions, and the strong electrolyte is essentially fully dissociated in the solution. When we write an equation to represent what takes place when a strong electrolyte is dissolved in water, we omit the arrow for the reverse reaction because, for all practical purposes, the reverse reaction does not occur. For the strong electrolyte NaCl we would write simply

$$NaCl \ (s) \longrightarrow Na^+ \ (aq) + Cl^- \ (aq)$$

The concept of a dynamic equilibrium is very important. All processes, both chemical and physical, tend to move toward a state of equilibrium. We will use this concept at numerous times in later chapters to analyze physical changes as well as chemical reactions.

6.4 IONIC REACTIONS

Most chemical reactions encountered in the laboratory portion of an introductory chemistry course involve reactions between ions in solution. In fact, anyone who uses water as a solvent eventually encounters such reactions. What we wish to examine in this section is *how* and *why* reactions between electrolytes take place in solution.

Let's begin by considering the reaction that takes place when a solution containing 1 mol of sodium chloride is added to a solution of 1 mol of silver nitrate. This is the reaction that was shown in Figure 6.1. When these two solutions are combined, 1 mol of the white solid, silver chloride, is formed, and the solution that remains contains 1 mol of sodium nitrate. If we wished, we could separate the silver chloride from the solution by filtering the mixture as shown in Figure 6.4. When the clear solution—the **filtrate**—is evaporated, sodium nitrate remains.

The chemical equation that represents the change that has occurred during this reaction is

$$NaCl \ (aq) + AgNO_3 \ (aq) \longrightarrow AgCl \ (s) + NaNO_3 \ (aq)$$

This kind of reaction, in which cations and anions have changed partners, is known as **metathesis,** or **double replacement** (Cl^- has replaced NO_3^- or Na^+ has replaced Ag^+). The equation we have written to describe the reaction is called a **molecular equation,** because all the reactants and products are written as if they were molecules. A more accurate representation of the reaction as it actually occurs in solution is given by the **ionic equation.** We have seen that a solution of NaCl does not contain molecules but, instead, consists of Na^+ and Cl^- ions dispersed throughout the solvent. Similarly, a silver nitrate solution contains Ag^+ and NO_3^- ions. When these two solutions are mixed, solid AgCl is formed by a combination of the Ag^+ and Cl^- ions. We refer to such a solid formed in a solution as the result of a chemical reaction as a **precipitate.** The solution of sodium nitrate that remains contains Na^+ and NO_3^- ions and we can write the ionic

NaCl exists as ions in both the solid and in its aqueous solutions.

Figure 6.4

Separating a precipitate from a solution by filtration. (a) The mixture is guided into the filter by pouring it down a glass rod. The beaker below catches the filtrate that passes through the filter. (b) The last traces of the precipitate are rinsed from the beaker by a stream of water from a wash bottle.

equation as

$$Na^+ \ (aq) + Cl^- \ (aq) + Ag^+ \ (aq) + NO_3^- \ (aq) \longrightarrow AgCl \ (s) + Na^+ \ (aq) + NO_3^- \ (aq)$$

In this equation we have shown all the soluble ionic substances as being dissociated in solution. The formula for silver chloride is written in molecular form because its ions are no longer separate—they are together in the solid.

If we examine the ionic equation we have just written, we see that Na^+ and NO_3^- do not actually undergo any change during the course of the reaction. The same Na^+ and NO_3^- ions are present after the chemical reaction as before and they have, in a sense, just "gone along for the ride." For this reason, ions that do not change during reaction are frequently called **spectator ions.** Since they do not take part in the reaction, we may eliminate them from the equation to arrive at the **net ionic equation,** that is, the equation for the net change that has taken place:

$$Ag^+ \ (aq) + Cl^- \ (aq) \longrightarrow AgCl \ (s)$$

This net ionic equation is useful in more than one way. First, it focuses our attention on the species that participate in the important changes that are occurring in the solution. Second, it tells us that any substance that produces Ag^+ ions in solution will react with any other substance that gives Cl^- ions in solution to yield a precipitate of AgCl. For example, we would predict that a precipitate of AgCl would also form if we mixed solutions of potassium chloride and silver fluoride. (AgF is soluble in water even though AgCl is not.[2]) This is, in

[2] The solubility of AgCl is very low, 0.000089 g/100 ml, and AgCl can be considered, for most purposes, to be insoluble; that is, the amount of AgCl in solution can generally be considered to be negligible. For comparison, the solubility of AgF in water is approximately 185 g/100 ml at room temperature.

fact, precisely what takes place. The molecular, ionic, and net ionic equations for this reaction are as follows:

$$AgF\ (aq)\ +\ KCl\ (aq)\ \longrightarrow\ AgCl\ (s)\ +\ KF\ (aq) \qquad \text{(Molecular)}$$

$$Ag^+\ (aq)\ +\ F^-\ (aq)\ +\ K^+\ (aq)\ +\ Cl^-\ (aq)\ \longrightarrow\ AgCl\ (s)\ +\ K^+\ (aq)\ +\ F^-\ (aq) \qquad \text{(Ionic)}$$

$$Ag^+\ (aq)\ +\ Cl^-\ (aq)\ \longrightarrow\ AgCl\ (s) \qquad \text{(Net Ionic)}$$

Each of these kinds of equations is useful in its own way; none is the "best" way of representing the reaction. The form that we use in any particular instance depends on what aspect of the reaction we wish to focus our attention. If we want to think about the net chemical change that occurs, the net ionic equation is best. However, if we want to work with these chemicals in carefully measured amounts, it is the stoichiometry of the molecular equation that is important.

In the two metathesis reactions that we have considered, the *driving force* was provided by the formation of an insoluble precipitate of AgCl. If silver chloride were soluble in water, no chemical reaction of this type would occur. For instance, if solutions of KCl and $NaNO_3$ are combined, we might be tempted to write the equation,

$$KCl\ +\ NaNO_3\ \longrightarrow\ KNO_3\ +\ NaCl$$

Both KNO_3 and NaCl are soluble, and in almost all instances ionic compounds (**salts**) undergo virtually complete dissociation when they dissolve. If we write this equation in ionic form, we have

$$K^+\ +\ Cl^-\ +\ Na^+\ +\ NO_3^-\ \longrightarrow\ K^+\ +\ NO_3^-\ +\ Na^+\ +\ Cl^-$$

When we compare the left and right sides of this equation, we see that they are identical except for the sequence in which we have written the ions. If we cross out all spectator ions, there is nothing left; that is, there is no net chemical change. Therefore, we would say that KCl does not react with $NaNO_3$ when their solutions are mixed. All we get is a mixture of the two salts.

We have now seen that when a precipitate is formed on mixing two solutions of electrolytes, a net chemical change takes place. In principle, by knowing the solubilities of all the compounds that can be formed between pairs of cations and anions, we could *predict*, based on the formation of a precipitate, when chemical reactions would occur. This is not as simple as it may seem at first, however, because there is no sharp distinction between soluble and insoluble compounds. Substances such as lithium chlorate or sodium chloride, both of which were mentioned in Section 6.1, are certainly considered soluble. Silver chloride, on the other hand, would undoubtedly be termed insoluble. Some salts, such as $PbCl_2$ and $AgC_2H_3O_2$, are of intermediate solubility and are referred to as *partially soluble*, or *slightly soluble* compounds. Whether or not a precipitate of a particular salt will form on mixing solutions of reactants depends on whether the concentration of the ions that constitute the salt exceeds that required to achieve a saturated solution of the salt. If dilute solutions of potential reactants are mixed, this might not occur and no precipitate would be formed in the reaction mixture. Generally, a compound is considered insoluble when the combination of even very dilute solutions of its constituent ions leads to the formation of a precipitate.

This entire discussion has been on a qualitative level. A quantitative treatment of the solubilities of ionic solids is reserved until Chapter 16. For now, we will use the following solubility rules as a rough guide in predicting the course of metathesis reactions. To help you learn them, they have been divided into two groups: *soluble compounds*, with exceptions, and *insoluble compounds*, with exceptions.

Solubility Rules

Soluble compounds

1. All salts of the alkali metals are *soluble*.
2. All ammonium salts are *soluble*.
3. All salts containing the anions, NO_3^-, ClO_3^-, ClO_4^-, and $C_2H_3O_2^-$ are *soluble* ($AgC_2H_3O_2$ and $KClO_4$, however, are slightly soluble).
4. All chlorides, bromides, and iodides are *soluble except* those of Ag^+, Pb^{2+}, and Hg_2^{2+} (note that mercury in the 1+ oxidation state exists as the ion Hg_2^{2+}). $PbCl_2$ is slightly soluble.
5. All sulfates are *soluble except* those of Pb^{2+}, Sr^{2+}, and Ba^{2+}. The sulfates of Ca^{2+} and Ag^+ are slightly soluble.

Insoluble compounds

6. All metal oxides *except* those of the alkali metals and Ca^{2+}, Sr^{2+}, and Ba^{2+} are *insoluble*. Metal oxides, when they dissolve, react with the solvent to form hydroxides; for example,

$$CaO + H_2O \longrightarrow Ca^{2+} + 2OH^-$$

7. All hydroxides are *insoluble except* those of the alkali metals, Ba^{2+} and Sr^{2+}. $Ca(OH)_2$ is slightly soluble.
8. All carbonates, phosphates, sulfides, and sulfites are *insoluble except* those of NH_4^+ and the alkali metals.

[handwritten margin note: $AgCl + PbCl, Hg(I, Br)$ $PbSO_4 \ SrSO_4 \ BaSO_4$]

EXAMPLE 6.1 Would a chemical reaction occur if solutions of $FeCl_3$ and KOH were mixed? If so, give the net ionic equation.

SOLUTION The first step is to write an equation for the metathesis reaction that might occur. In doing this, be very careful to write the correct formulas of the products. Some students, for example, might be tempted to write the products as $FeOH$ and KCl_3; *these are wrong!* You must take into account the charges on the ions to be sure that the formulas represent electrically neutral compounds. Since $FeCl_3$ is composed of Fe^{3+} and Cl^- ions and KOH is composed of K^+ and OH^- ions, the correct formulas of the products are $Fe(OH)_3$ and KCl. Assembling this into a balanced molecular equation gives

$$FeCl_3 + 3KOH \longrightarrow 3KCl + Fe(OH)_3$$

On the basis of our solubility rules, KCl is soluble but $Fe(OH)_3$ is not. Therefore, we would expect a precipitate of $Fe(OH)_3$ to form in the mixture and the KCl would remain dissociated in solution. The ionic equation for the reaction is

$$Fe^{3+} + 3Cl^- + 3K^+ + 3OH^- \longrightarrow Fe(OH)_3 \ (s) + 3K^+ + 3Cl^-$$

Cancelling spectator ions (K^+ and Cl^-), we have the net ionic equation

$$Fe^{3+} + 3OH^- \longrightarrow Fe(OH)_3 \ (s)$$

EXAMPLE 6.2 Would a chemical reaction be expected to occur if solutions containing NH_4NO_3 and $Pb(C_2H_3O_2)_2$ were mixed?

SOLUTION To answer the question we must attempt to write a net ionic equation. First we start with the molecular equation, exchanging NO_3^- for $C_2H_3O_2^-$ and vice versa when we write the products.

$$2NH_4NO_3 + Pb(C_2H_3O_2)_2 \longrightarrow 2NH_4C_2H_3O_2 + Pb(NO_3)_2$$

Rules 2 and 3 tell us that all the reactants _and products_ are soluble. The ionic equation can then be written

$$2NH_4^+ + 2NO_3^- + Pb^{2+} + 2C_2H_3O_2^- \longrightarrow 2NH_4^+ + 2C_2H_3O_2^- + Pb^{2+} + 2NO_3^-$$

If we cancel ions that are the same on both sides of the equation, everything disappears. Consequently, the answer to the question is that there is no net chemical reaction.

For ionic reactions that do not involve oxidation-reduction there are two other factors, in addition to the formation of a precipitate, that can lead to a net chemical change. One of these is _the formation of a weak electrolyte;_ the other is _the formation of a gaseous product._ Let's look at each.

Formation of a weak electrolyte

In any solution of acetic acid, we find mostly $HC_2H_3O_2$ and very little H^+ and $C_2H_3O_2^-$. This is because the reaction of the ions to form molecules occurs much more readily than the reaction of molecules to produce ions. Therefore, if H^+ and $C_2H_3O_2^-$ ions are mixed together in large numbers, an unstable situation exists, and the ions immediately combine to form molecules of $HC_2H_3O_2$. This is exactly what happens when solutions of the strong electrolytes HCl and $NaC_2H_3O_2$ are mixed. The ionic reaction is

$$H^+ + Cl^- + Na^+ + C_2H_3O_2^- \longrightarrow Na^+ + Cl^- + HC_2H_3O_2$$

and if we cancel spectator ions, the net reaction is

$$H^+ + C_2H_3O_2^- \longrightarrow HC_2H_3O_2$$

The "driving force" for this reaction—the reason we say a reaction occurs—is the decrease in the number of ions that takes place when the two fully dissociated reactants form the partially dissociated product. The _weak acids_ discussed in the next section form an important class of weak electrolytes.

Formation of a gas

Frequently, the molecular species formed in a reaction can escape as a gas, either directly or by decomposing to produce a gaseous product. For example, when HCl is added to Na_2S, the reaction that occurs is

$$2HCl + Na_2S \longrightarrow H_2S\ (g) + 2NaCl$$

for which the net ionic equation is

$$2H^+ + S^{2-} \longrightarrow H_2S\ (g)$$

Hydrogen sulfide gas has a limited solubility in water and when that solubility is exceeded, the H_2S bubbles out of the solution.

Another example is the reaction between HCl and Na_2CO_3,

$$2HCl + Na_2CO_3 \longrightarrow H_2CO_3 + 2NaCl$$

for which the net ionic equation is

$$2H^+ + CO_3^{2-} \longrightarrow H_2CO_3$$

Carbonic acid, H_2CO_3, is not stable in large concentrations and decomposes readily to produce CO_2 and H_2O,

$$H_2CO_3 \longrightarrow H_2O + CO_2\ (g)$$

Table 6.2
Gases that are only slightly soluble in water

Gas	Typical Reaction in Which It Is Produced
CO_2	$Na_2CO_3 + 2HCl \rightarrow H_2CO_3 + 2NaCl$ $H_2CO_3 \rightarrow H_2O + CO_2 \ (g)$ *Net equation:* $CO_3^{2-} + 2H^+ \rightarrow CO_2 \ (g) + H_2O$
SO_2	$Na_2SO_3 + 2HCl \rightarrow H_2SO_3 + 2NaCl$ $H_2SO_3 \rightarrow H_2O + SO_2 \ (g)$ *Net equation:* $SO_3^{2-} + 2H^+ \rightarrow H_2O + SO_2 \ (g)$
NH_3	$NH_4Cl + NaOH \rightarrow NH_3 \ (g) + H_2O + NaCl$ *Net equation:* $NH_4^+ + OH^- \rightarrow NH_3 \ (g) + H_2O$
H_2S	$Na_2S + 2HCl \rightarrow H_2S \ (g) + 2NaCl$ *Net equation:* $S^{2-} + 2H^+ \rightarrow H_2S \ (g)$
NO NO_2	$NaNO_2 + HCl \rightarrow HNO_2 + NaCl$ $2HNO_2 \rightarrow H_2O + NO_2 \ (g) + NO \ (g)$ *Net equation:* $2NO_2^- + 2H^+ \rightarrow H_2O + NO_2 \ (g) + NO \ (g)$

Both soluble and insoluble carbonates react with acids to liberate carbon dioxide.

Therefore, we can write the net ionic equation for the overall reaction between hydrogen ion and carbonate ion as

$$2H^+ + CO_3^{2-} \longrightarrow H_2O + CO_2 \ (g)$$

A list of some other substances that are gases having a limited solubility in water, or that decompose to produce gaseous products, is given in Table 6.2.

EXAMPLE 6.3

SOLUTION

Write a net ionic equation for the reaction between NH_4NO_3 and $Ba(OH)_2$

We begin as usual by writing a double replacement reaction.

$$2NH_4NO_3 + Ba(OH)_2 \longrightarrow 2NH_4OH + Ba(NO_3)_2$$

The molecular species NH_4OH (ammonium hydroxide) does not really exist; it is ficti-
tious. It actually is $NH_3 + H_2O$ (note that there are one N, five H, and one O in both NH_4OH and $NH_3 + H_2O$). We should therefore rewrite the equation as

$$2NH_4NO_3 + Ba(OH)_2 \longrightarrow 2NH_3 + 2H_2O + Ba(NO_3)_2$$

The ionic equation is obtained by applying the solubility rules, and also noting that NH_3 is a gas (Table 6.2).

$$2NH_4^+ + 2NO_3^- + Ba^{2+} + 2OH^- \longrightarrow 2NH_3 \ (g) + 2H_2O + Ba^{2+} + 2NO_3^-$$

Finally, cancelling spectator ions gives

$$2NH_4^+ + 2OH^- \longrightarrow 2NH_3 \ (g) + 2H_2O$$

which becomes

$$NH_4^+ + OH^- \longrightarrow NH_3 \ (g) + H_2O$$

when the coefficients are reduced to the simplest set of whole numbers.

In this last section we have developed methods of predicting the outcome of a large number of ionic reactions. *To apply these methods you must learn the solubility rules and the contents of Tables 6.1 and 6.2!* You probably should learn them now, because we will use these methods again later in this chapter.

6.5 ACIDS AND BASES IN AQUEOUS SOLUTION

Acids and bases are among our most common and important chemicals. The sour taste of vinegar and lemon juice results from acids. They contain acetic acid and citric acid, respectively. We all need ascorbic acid in our diets—it's also called vitamin C. Sulfuric acid, the substance used as the electrolyte in the fluid in automobile batteries, ranks well above any other industrial chemical in production volume each year.

Among important bases we find household ammonia. Sodium hydroxide, also called lye, is available on supermarket shelves and is found in such common drain cleaners as Drano. Even sodium bicarbonate and milk of magnesia, which we take for an upset (acid) stomach, are bases.

A property of the substances that we call acids and bases is that they always react with each other, so the acid-base concept is extremely useful in chemistry for classifying substances. In fact, nearly all chemical reactions can be broadly classified as either reactions between acids and bases, or as reactions involving oxidation and reduction. Because the acid-base concept is so important, an entire chapter (Chapter 14) is devoted to a detailed discussion of acid-base behavior. For now, we will limit ourselves to a simple definition of acids and bases, adequate for the treatment of reactions in aqueous solution.

We define an **acid** as *any substance that increases the concentration of hydronium ion in solution.* Thus, HCl is an acid because when it is dissolved in water, it reacts with the solvent to produce H_3O^+.

$$HCl + H_2O \longrightarrow H_3O^+ + Cl^-$$

Aqueous solutions of hydrogen chloride are called *hydrochloric acid.* Because it is a strong electrolyte, HCl is called a **strong acid.**

Carbon dioxide is also an acid because its aqueous solutions contain more H_3O^+ than pure water does. In the last section we described the decomposition of H_2CO_3 that occurs when it is produced in large quantities as the result of a chemical reaction. The reverse reaction occurs to a limited extent when CO_2 is dissolved in water (as in a carbonated beverage).

$$CO_2 + H_2O \longrightarrow H_2CO_3$$

Carbonic acid is able to dissociate slightly (actually by further reaction with water).

$$H_2CO_3 + H_2O \rightleftharpoons H_3O^+ + HCO_3^-$$

It is a weak electrolyte, and is also called a **weak acid.** This particular acid, incidentally, is responsible for the large limestone caves that exist in different parts of the world. Groundwater, made acidic by CO_2 dissolved in it from the atmosphere, trickles through the limestone ($CaCO_3$) with which it reacts.

$$CaCO_3 \ (s) + H_2CO_3 \ (aq) \longrightarrow Ca^{2+} \ (aq) + 2HCO_3^- \ (aq)$$

The formation of soluble calcium bicarbonate gradually dissolves the rock, leaving huge caverns (Figure 6.5).

We have seen that acids do not necessarily have to contain hydrogen; HCl does but CO_2 does not. The only requirement is that solutions of these substances contain more H_3O^+ than is found in pure H_2O. Table 6.3 contains a list of common acids and the reactions that they undergo to produce acidic solutions. You might note that nonmetal oxides, when dissolved in water, produce acids that dissociate by further reaction with the solvent. This provides us with one chemical distinction between metals and nonmetals (we will shortly see that soluble metal oxides react with water to yield bases).

Figure 6.5

Huge limestone caverns such as this are created as water—made acidic by dissolved carbon dioxide—trickles through a limestone deposit.

Table 6.3
Some acids and bases

Acids (Those followed by an asterisk are weak electrolytes and exist primarily as undissociated molecules in aqueous solution.)

Monoprotic acids: $HX \rightarrow H^+ + X^-$	HF	Hydrofluoric acid (*)	$HClO_3$	Chloric acid
	HCl	Hydrochloric acid	$HClO_4$	Perchloric acid
	HBr	Hydrobromic acid	HIO_4	Periodic acid
	HI	Hydriodic acid	HNO_3	Nitric acid
	HOCl	Hypochlorous acid (*)	HNO_2	Nitrous acid (*)
	$HClO_2$	Chlorous acid (*)	$HC_2H_3O_2$	Acetic acid (*)

Diprotic acids: $H_2X \rightarrow H^+ + HX^-$ $HX^- \rightarrow H^+ + X^{2-}$	H_2SO_4	Sulfuric acid^a	H_2S	Hydrosulfuric acid (*)
	H_2SO_3	Sulfurous acid (*)	H_3PO_3	Phosphorous acid (only
	H_2CO_3	Carbonic acid (*)		two hydrogens can be
	$H_2C_2O_4$	Oxalic acid (*)		removed as protons) (*)

Triprotic acids: $H_3X \rightarrow H^+ + H_2X^-$ $H_2X^- \rightarrow H^+ + HX^{2-}$ $HX^{2-} \rightarrow H^+ + X^{3-}$	H_3PO_4	Orthophosphoric acid (*)

Typical acidic oxides *(nonmetal oxides)*	SO_2	$SO_2 + H_2O \rightarrow H_2SO_3$
	SO_3	$SO_3 + H_2O \rightarrow H_2SO_4$
	N_2O_3	$N_2O_3 + H_2O \rightarrow 2HNO_2$
	N_2O_5	$N_2O_5 + H_2O \rightarrow 2HNO_3$
	P_4O_6	$P_4O_6 + 6H_2O \rightarrow 4H_3PO_3$
	P_4O_{10}	$P_4O_{10} + 6H_2O \rightarrow 4H_3PO_4$

Bases

Molecular bases	NH_3	Ammonia^b (*)	$(NH_3 + H_2O \rightleftarrows NH_4^+ + OH^-)$
	N_2H_4	Hydrazine (*)	$(N_2H_4 + H_2O \rightleftarrows N_2H_5^+ + OH^-)$
	NH_2OH	Hydroxylamine (*)	$(NH_2OH + H_2O \rightleftarrows NH_3OH^+ + OH^-)$

Ionic bases	Metal hydroxides (see solubility rules—Section 6.4)	$M(OH)_n \rightarrow M^{n+} + nOH^-$

Typical basic oxides *(metal oxides)*	Na_2O K_2O	$M_2O + H_2O \rightarrow 2MOH$
	CaO SrO BaO	$MO + H_2O \rightarrow M(OH)_2$

^{a} The second dissociation of H_2SO_4 is that of a weak acid.

^{b} Aqueous solutions of ammonia are sometimes referred to as ammonium hydroxide. Commercially prepared solutions of NH_3, for example, are labeled in this way.

A **base** will be defined as *any substance that increases the concentration of hydroxide ion in aqueous solutions.* Sodium hydroxide, an example of an ionic compound containing metal ions and hydroxide ions, is a base because when it dissolves in water it dissociates to yield Na^+ and OH^- ions.

$$NaOH \longrightarrow Na^+ + OH^-$$

The dissociation of NaOH is complete, and therefore NaOH is said to be a **strong base.** Ammonia, NH_3, is also a base because it reacts with water to produce hydroxide ions.

$$NH_3 + H_2O \rightleftharpoons NH_4^+ + OH^-$$

In this case, a proton is transferred from a water molecule to the ammonia molecule.

$$H:\overset{\overset{H}{\cdot\cdot}}{\underset{\cdot\cdot}{N}}: + \overset{H}{\underset{\overset{\cdot\cdot}{H}}{O}}: \rightleftharpoons \left[H:\overset{\overset{H}{\cdot\cdot}}{\underset{H}{N}}:H \right]^{+} + \left[:\overset{\cdot\cdot}{\underset{\cdot\cdot}{O}}:H \right]^{-}$$

This reaction is indicated as an equilibrium because only a small fraction of the NH_3 in solution is present as NH_4^+ and OH^- at any given instant. Ammonia is therefore called a **weak base** because its solutions contain only relatively small amounts of OH^-.

Metal oxides, when they are dissolved in water, produce basic solutions. For example, when CaO is treated with water, it undergoes reaction to yield calcium hydroxide.

$$CaO + H_2O \longrightarrow Ca(OH)_2$$

When $Ca(OH)_2$ dissolves, it dissociates fully to Ca^{2+} and OH^-.

$$Ca(OH)_2 \longrightarrow Ca^{2+} + 2OH^-$$

With NH_3 and CaO, we see that a base does not necessarily have to contain hydroxide ions to produce a basic solution. Furthermore, the presence of the OH unit in a compound does not guarantee that the substance will behave as a base. For example, the compounds, sodium hydroxide, sulfuric acid, and methyl alcohol can be represented by the following structures and formulas:

$$Na^+ \left[:\overset{\cdot\cdot}{\underset{\cdot\cdot}{O}}-H \right]^{-} \qquad H-\overset{\cdot\cdot}{\underset{\cdot\cdot}{O}}-\overset{:\overset{\cdot\cdot}{O}:}{\underset{:\overset{\cdot\cdot}{O}:}{S}}-\overset{\cdot\cdot}{\underset{\cdot\cdot}{O}}-H \qquad H-\overset{\overset{H}{|}}{\underset{\underset{H}{|}}{C}}-\overset{\cdot\cdot}{\underset{\cdot\cdot}{O}}-H$$

$$\textbf{NaOH} \qquad\qquad \textbf{SO}_2\textbf{(OH)}_2 \qquad\qquad \textbf{CH}_3\textbf{OH}$$
$$\text{or } \textbf{H}_2\textbf{SO}_4$$

Each of these contains an OH group; however, only the first compound is basic. Sulfuric acid dissociates by the breakage of O—H bonds, with the transfer of protons to water molecules, and methyl alcohol does not dissociate at all in water. Therefore, we must exercise care in drawing conclusions about the acid-base behavior of compounds based on their chemical formulas or structures.

Acids and bases, in general, exhibit certain properties that lead us to characterize them as acids or bases. The most important of these properties is that they react with one another in a process called **neutralization.** In aqueous solution the neutralization reaction takes the form of the net ionic equation,

$$H_3O^+ + OH^- \longrightarrow 2H_2O$$

As mentioned previously, the hydronium ion can be considered a hydrated proton; it is therefore common practice to leave out the water molecule that is attached to the proton and simply to write the equation as

$$H^+ + OH^- \longrightarrow H_2O$$

This reaction is the driving force behind the reaction of all acids and bases with each other in aqueous solution.

A typical reaction between an acid and a base occurs between hydrochloric acid and sodium hydroxide.

$$HCl + NaOH \longrightarrow NaCl + H_2O$$

The driving force behind this reaction is strong enough to cause insoluble bases to dissolve in acids and insoluble acids to dissolve in bases.

We see that *the products of the neutralization reaction are a salt and water.* The hydrochloric acid used in this particular reaction is said to be a **monoprotic** acid because 1 mol of HCl furnishes only 1 mol of H^+ ions. There are also many acids that are capable of furnishing more than 1 mol of H^+ per mole of acid; as a group, these are called **polyprotic acids.** Examples are H_2SO_4 (a **diprotic acid**) and H_3PO_4 (a **triprotic acid**). These substances dissociate in a series of steps, each providing one proton. Sulfuric acid, for example, dissociates as follows:

$$H_2SO_4 \longrightarrow H^+ + HSO_4^-$$

$$HSO_4^- \longrightarrow H^+ + SO_4^{2-}$$

By controlling the quantity of base available for the neutralization of a diprotic acid, such as H_2SO_4, it is possible to form either salts that are the product of complete neutralization or salts in which only one of the available protons of the acid has been neutralized. For example, when 2 mol of NaOH are reacted with 1 mol of H_2SO_4, 2 mol of H_2O and 1 mol of Na_2SO_4 are produced.

$$2NaOH + H_2SO_4 \longrightarrow 2H_2O + Na_2SO_4$$

However, if only 1 mol of NaOH is available to react with 1 mol of H_2SO_4, the products are 1 mol of water and 1 mol of $NaHSO_4$ (called sodium bisulfate or sodium hydrogen sulfate).

$$NaOH + H_2SO_4 \longrightarrow H_2O + NaHSO_4$$

Similar reactions can occur with a triprotic acid such as H_3PO_4, and salts such as the following can be obtained.

Na_3PO_4	trisodium phosphate (sodium phosphate)
Na_2HPO_4	disodium hydrogen phosphate
NaH_2PO_4	sodium dihydrogen phosphate

Salts that are the product of the partial neutralization of a polyprotic acid are called **acid salts** because they can furnish protons for the neutralization of additional base. Thus $NaHSO_4$ can react with NaOH in a neutralization reaction.

$$NaHSO_4 + NaOH \longrightarrow Na_2SO_4 + H_2O$$

EXAMPLE 6.4   Write the chemical equation for the reaction of 1 mol of H_3PO_4 with 2 mol of NaOH.

SOLUTION $$H_3PO_4 + 2NaOH \longrightarrow Na_2HPO_4 + 2H_2O$$

6.6   THE PREPARATION OF INORGANIC SALTS BY METATHESIS REACTIONS

In the past two sections we have described typical metathesis reactions that occur in aqueous solutions. These reactions can often be useful to us in the preparation of inorganic salts. For example, suppose that we were faced with the problem of preparing relatively pure NaBr using other inorganic compounds as starting materials. How would we proceed? We cannot achieve success if we mix together solutions of NaCl and KBr because, although the mixture would contain the constituents of NaBr, all the ions, Na^+, K^+, Cl^-, and Br^-, are in solution together, and it would be nearly impossible to isolate pure NaBr from the mixture. *We are looking for reactions in which one product is easily separated from the other, a condition that is met if one of the products is either a precipitate, a gas, or the solvent, water.*

Precipitation reactions

Precipitation reactions are particularly useful when the desired product is insoluble in water. For instance, suppose that we wished to prepare silver bromide, AgBr, the light sensitive substance used in most photographic film. This substance is insoluble and could be formed by the reaction

$$Ag^+ + Br^- \longrightarrow AgBr$$

Knowing this, the trick now is to transform this net ionic equation into a molecular equation so that we can choose the right chemicals with which to work. In doing this, let's look at the requirements we wish the chemicals to have.

1. Both reactants must be soluble. This means we must choose a soluble silver salt and a soluble bromide salt.
2. The other product in the metathesis reaction must be soluble so that the AgBr can be separated from the reaction mixture without being contaminated by another precipitate. This means that the salt formed from the anion of the silver compound and the cation of the bromide compound must also be soluble.

To be able to choose chemicals that fit these requirements we must know the solubility rules well. One likely choice of reactants is $AgNO_3$ (all nitrates are soluble, so both conditions 1 and 2 are sure to be fulfilled) and NaBr (all salts of Na^+ are soluble). The molecular equation would then be

$$AgNO_3\ (aq) + NaBr\ (aq) \longrightarrow AgBr\ (s) + NaNO_3\ (aq)$$

The last step in preparing the AgBr would be to separate it from the mixture by filtration. In this preparation it isn't necessary to worry too much about stoichiometry because if we want to prepare 1 mol of AgBr, we can add 1 mol of $AgNO_3$ to a solution containing *at least* 1 mol of NaBr. Any excess NaBr beyond that required to react completely with the $AgNO_3$ remains dissolved and does not contaminate the AgBr precipitate.

When the desired product remains in solution, greater attention must be paid to the stoichiometry of the reaction. For example, to prepare NaBr (which is water-soluble) by a precipitation reaction, we might mix solutions of $NaSO_4$ and $BaBr_2$, since $BaSO_4$ is insoluble. The net reaction is

$$Ba^{2+}\ (aq) + SO_4^{2-}\ (aq) \longrightarrow BaSO_4\ (s)$$

and the molecular equation is

$$Na_2SO_4 + BaBr_2 \longrightarrow BaSO_4 + 2NaBr$$

To prepare pure NaBr we must react equal numbers of moles of Na_2SO_4 and $BaBr_2$. The solid $BaSO_4$ could then be filtered from the mixture and the volume of the resulting solution could then be reduced by evaporation to the point where nearly all the desired product crystallizes.

There are some instances where we can use relative degrees of insolubility in synthetic procedures. For instance, the silver halides decrease in solubility in this order: AgCl, AgBr, and AgI. If a solution containing KBr is stirred with solid AgCl, the solid is converted to the less soluble AgBr, and Cl^- replaces Br^- in solution. The net effect is that AgCl is converted to AgBr.

$$AgCl\ (s) + Br^- \longrightarrow AgBr\ (s) + Cl^-$$

EXAMPLE 6.5   How could lead sulfate, $PbSO_4$ be prepared by a precipitation reaction?

SOLUTION *Answering questions of this kind requires a thorough knowledge of the solubility rules.* Lead sulfate is insoluble and can be formed by the net ionic reaction

$$Pb^{2+} + SO_4^{2-} \longrightarrow PbSO_4 \ (s)$$

To prepare it we need a soluble lead salt and a soluble sulfate. Furthermore, the other product in the metathesis must be soluble. There are undoubtedly several alternatives. One of them is

$$Na_2SO_4 + Pb(NO_3)_2 \longrightarrow PbSO_4 \ (s) + 2NaNO_3$$

$Pb(NO_3)_2$ was chosen because all nitrates are soluble. This ensures a soluble lead salt as a reactant *and* a soluble second product. After choosing $Pb(NO_3)_2$ we can pick any soluble sulfate as the other reactant.

Neutralization reactions

Another useful method of preparing salts is by the neutralization of an acid with a base. Suppose we wanted to prepare NaBr. In this case, we choose an acid that gives us the anion (HBr) and a base that gives us the cation (NaOH). When they are combined, the following reaction occurs.

$$NaOH + HBr \longrightarrow H_2O + NaBr$$

If we are careful to react precisely 1 mol of NaOH with 1 mol of HBr, the only product in the aqueous solution is NaBr.

The reaction, $O^{2-} + 2H^+ \rightarrow H_2O$, has such a strong tendency to occur that insoluble oxides dissolve in acids.

Metal oxides can be used in place of hydroxides in similar reactions. For example, the reaction

$$ZnO + 2HClO_4 \longrightarrow Zn(ClO_4)_2 + H_2O$$

could be used to prepare zinc perchlorate.

EXAMPLE 6.6 How could we prepare calcium acetate, $Ca(C_2H_3O_2)_2$, one of the substances used to make "canned heat," by a neutralization reaction?

SOLUTION Since metal oxides react with acids, we could accomplish our goal by reacting CaO with acetic acid, $HC_2H_3O_2$

$$CaO + 2HC_2H_3O_2 \longrightarrow Ca(C_2H_3O_2)_2 + H_2O$$

Reactions in which one product is a gas

A very convenient way to prepare inorganic salts is provided by the reaction between an acid and a metal carbonate. Recall that the carbonate ion reacts with hydrogen ion to produce carbon dioxide and water,

$$CO_3^{2-} + 2H^+ \longrightarrow H_2O + CO_2 \ (g)$$

This reaction proceeds readily regardless of whether the source of carbonate ion is a soluble or insoluble salt.

The fact that most metal carbonates are insoluble leads to a very simple method for the preparation of pure salts. The synthesis of copper(II) bromide, for example, can be accomplished by the addition of HBr to a suspension of the insoluble $CuCO_3$ until nearly all the starting material reacts according to the equation

$$CuCO_3 \ (s) + 2HBr \ (aq) \longrightarrow CuBr_2 \ (aq) + H_2O + CO_2 \ (g)$$

When the mixture is filtered to remove the undissolved $CuCO_3$, the solution that remains contains $CuBr_2$, which is virtually free of contaminants. The product can then be recovered by evaporation of the water.

The preparation of salts from soluble carbonates is also feasible; once again, however, attention must be paid to the stoichiometry of the reaction to obtain pure products. For example, NaBr can be made from Na_2CO_3 and HBr.

$$Na_2CO_3 \ (aq) + 2HBr \ (aq) \longrightarrow 2NaBr \ (aq) + H_2O + CO_2 \ (g)$$

The quantity of *pure* NaBr that can be crystallized from the reaction mixture depends on how closely a 1:2 mole ratio is maintained between the reactants Na_2CO_3 and HBr.

EXAMPLE 6.7	How could we prepare $Ca(ClO_4)_2$ by a reaction in which one product is a gas?
SOLUTION	We can react $CaCO_3$ with $HClO_4$.

$$CaCO_3 \ (s) + 2HClO_4 \ (aq) \longrightarrow Ca(ClO_4)_2 \ (aq) + CO_2 \ (g) + H_2O$$

To obtain a pure product we use an excess of $CaCO_3$, which is filtered from the mixture after reaction has ceased.

EXAMPLE 6.8	How could we prepare $Cu(ClO_4)_2$ from $CuSO_4$?
SOLUTION	Often we can combine a series of reactions to give a final desired product. In this case, we know that we can prepare $Cu(ClO_4)_2$ from $Cu(OH)_2$ and $HClO_4$. Also, $Cu(OH)_2$ is insoluble and can be prepared from $CuSO_4$ by a precipitation reaction. One set of reactions is

$$CuSO_4 \ (aq) + 2NaOH \ (aq) \longrightarrow Cu(OH)_2 \ (s) + Na_2SO_4 \ (aq)$$

$$Cu(OH)_2 \ (s) + 2HClO_4 \ (aq) \longrightarrow Cu(ClO_4)_2 \ (aq) + 2H_2O$$

In practice, we would add the NaOH to the $CuSO_4$ and filter the insoluble $Cu(OH)_2 \ (s)$. This solid would then be allowed to react with less than a stoichiometric amount of $HClO_4$ so that a little excess $Cu(OH)_2$ remains unreacted. This would be filtered to give a solution containing nearly pure $Cu(ClO_4)_2$.

6.7 OXIDATION-REDUCTION REACTIONS

Oxidation-reduction processes form a very important class of chemical reactions. They take place between many inorganic and organic compounds, and they are extremely important in biochemical systems where they provide the mechanism for energy transfer in living organisms.

As a rule, the stoichiometry of redox reactions tends to be more complicated than for reactions that do not involve electron transfer. As a result, chemical equations for oxidation-reduction reactions are often complex and are difficult to balance by inspection. Fortunately, there are methods that can be applied to aid in balancing these equations.

In Section 4.9 it was noted that during redox reactions electrons are never produced as a product, nor are they necessary as a reactant. *When a chemical reaction involves oxidation-reduction, the total number of electrons lost in the oxidation process must equal the total number gained during reduction.* We can use this fact to help us balance equations for this type of reaction. One procedure we can use is called the **oxidation number change method.** To illustrate the steps involved, let's balance the equation

If necessary, review the rules for assigning oxidation numbers given on page 130.

$$HCl + K_2Cr_2O_7 \longrightarrow KCl + CrCl_3 + Cl_2 + H_2O$$

Step 1 *Assign oxidation numbers to all the atoms in the equation.* We will write them below the chemical symbols to avoid confusing them with actual charges.

$$H \quad Cl \; + \; K_2 \; Cr_2 \; O_7 \longrightarrow K \quad Cl \; + \; Cr \; Cl_3 \; + \; Cl_2 \; + \; H_2 \quad O$$
$$1+ \quad 1- \quad 1+ \;\; 6+ \;\; 2- \quad\quad 1+ \;\; 1- \quad 3+ \;\; 1- \quad\; 0 \quad\quad 1+ \;\; 2-$$

Step 2 *Identify which atoms change oxidation number and insert* temporary coefficients *so that we have the same number of each on both sides.*

We see that chromium changes from 6+ to 3+. We also see that some chlorine changes from 1− to 0 and some doesn't change at all; for now, we will only look at the chlorine that does change. To make the number of chromium atoms and chlorine atoms that change the same on both sides, we put a 2 in front of HCl and a 2 in front of $CrCl_3$. This gives

These temporary coefficients help us figure the number of electrons transferred in step 3.

$$2H \quad Cl \; + \; K_2 \; Cr_2 \; O_7 \longrightarrow K \quad Cl \; + \; 2Cr \; Cl_3 \; + \; Cl_2 \; + \; H_2 \quad O$$
$$1+ \quad 1- \quad 1+ \;\; 6+ \;\; 2- \quad\quad 1+ \;\; 1- \quad 3+ \;\; 1- \quad\; 0 \quad\quad 1+ \;\; 2-$$

Step 3 *Compute the total change in oxidation number for both the oxidation and reduction.*

We have two chlorines changing from 1− to 0; the total change corresponds to a loss of $2e^-$. Next we see that two chromiums change from 6+ to 3+; this total change corresponds to a gain of $6e^-$ $(2 \times 3e^-)$.

$$2e^- \text{ lost (total)}$$

$$6e^- \text{ gained (total)}$$

$$2HCl \; + \; K_2Cr_2O_7 \longrightarrow KCl \; + \; 2CrCl_3 \; + \; Cl_2 \; + \; H_2O$$
$$1- \quad\quad 6+ \quad\quad\quad\quad\quad 3+ \quad\quad 0$$

Step 4 *Make the total loss and gain of electrons the same by multiplying coefficients by appropriate factors.*

If we multiply the number of electrons lost by 3, there will be a total of $6e^-$ lost as well as gained. We accomplish this by also multiplying the coefficients of HCl and Cl_2 by 3. This gives

$$3 \times 2e^- = 6e^- \text{ lost}$$

$$6e^- \text{ gained}$$

$$6HCl \; + \; K_2Cr_2O_7 \longrightarrow KCl \; + \; 2CrCl_3 \; + \; 3Cl_2 \; + \; H_2O$$

Step 5 *Finally, balance the rest of the equation by inspection.* There are 2 potassiums on the left, which require a coefficient of 2 for KCl. Then we add up all the chlorines on the right (there are fourteen), so we change the coefficient of HCl to 14. Finally, we place a 7 in front of H_2O to balance the H and O. The final equation is

$$14HCl \; + \; K_2Cr_2O_7 \longrightarrow 2KCl \; + \; 2CrCl_3 \; + \; 3Cl_2 \; + \; 7H_2O$$

The procedure outlined above will work on most oxidation-reduction equations, although simple equations (for example, $H_2 + Cl_2 \rightarrow 2HCl$) are more easily balanced by inspection.

6.8 BALANCING REDOX EQUATIONS BY THE ION-ELECTRON METHOD

In addition to the oxidation-number-change method discussed in the last section, there is still another method particularly well suited for balancing net ionic equations for oxidation-reduction reactions in solution. The procedure is called the **ion-electron method** and involves breaking the overall reaction into two **half-reactions,** one for the oxidation step and one for reduction. Each half-reaction is first balanced materially (that is, in terms of atoms), and then electrically by adding electrons to the side of the half-reaction deficient in negative charge. Finally, the balanced half-reactions are added together in such a way that the electrons cancel from both sides of the final equation.

As a simple example, let's consider the reaction of Sn^{2+} with Hg^{2+} in the presence of chloride ion to produce Hg_2Cl_2 and Sn^{4+} as products. The unbalanced equation is

$$Sn^{2+} + Hg^{2+} + Cl^- \longrightarrow Hg_2Cl_2 + Sn^{4+}$$

The first step is to divide the equation into the two half-reactions. When you do this, it is important to remember that the atoms on each side of each half-reaction should be of the same kind. The half-reactions here are

$$Sn^{2+} \longrightarrow Sn^{4+}$$

$$Hg^{2+} + Cl^- \longrightarrow Hg_2Cl_2$$

Next, we balance each half-reaction in terms of atoms. We don't have to do anything to the first one, but we must adjust the coefficients in the second.

$$Sn^{2+} \longrightarrow Sn^{4+}$$

$$2Hg^{2+} + 2Cl^- \longrightarrow Hg_2Cl_2$$

The third step is to balance the charge by adding electrons to the more positive (or less negative) side. In the first half-reaction the net charge on the left is 2+ and on the right it is 4+, so we add $2e^-$ to the right so that the net charge on both sides is the same. In the second half-reaction, the net charge is 2+ on the left $[2 \times (2+) + 2(1-) = 2+]$ and zero on the right. Therefore, we add $2e^-$ to the left. This gives

$$Sn^{2+} \longrightarrow Sn^{4+} + 2e^-$$
$$2e^- + 2Hg^{2+} + 2Cl^- \longrightarrow Hg_2Cl_2$$

Step four is to make the number of electrons gained equal the number lost. In this example, we don't have to actually do anything because the necessary condition is already fulfilled. Now the half-reactions can be added together.

$$Sn^{2+} \longrightarrow Sn^{4+} + 2e^-$$
$$\underline{2e^- + 2Hg^{2+} + 2Cl^- \longrightarrow Hg_2Cl_2}$$
$$2e^- + Sn^{2+} + 2Hg^{2+} + 2Cl^- \longrightarrow Sn^{4+} + Hg_2Cl_2 + 2e^-$$

Finally, we cancel anything that is the same on both sides. In this and every other reaction balanced by this method the electrons must cancel. The finished equation is

$$Sn^{2+} + 2Hg^{2+} + 2Cl^- \longrightarrow Sn^{4+} + Hg_2Cl_2$$

Notice that both atoms and charge are in balance.

In many oxidation-reduction reactions that take place in aqueous solution, water plays an active role. Any aqueous solution contains the species H_2O, H^+, and OH^-. In acidic solutions the predominant species are H_2O and H^+; in basic solutions they are H_2O and OH^-. When balancing half-reactions that occur in solution, we can use these species to achieve material balance. For example, let us consider the reaction between dichromate ion, $Cr_2O_7^{2-}$, and hydrogen sulfide in acidic solution to produce chromium(III) ion and elemental sulfur.

$$Cr_2O_7^{2-} + H_2S \longrightarrow Cr^{3+} + S$$

To balance the equation we write individual half-reactions for the changes that occur for chromium and sulfur. For the sulfur we write

$$H_2S \longrightarrow S$$

The half-reaction is balanced materially by adding two H^+ ions to the right.

$$H_2S \longrightarrow S + 2H^+$$

Electrical balance is achieved by adding two electrons to the right so that the net charge on both sides is zero.

$$H_2S \longrightarrow S + 2H^+ + 2e^- \qquad [6.1]$$

For chromium we begin by writing

$$Cr_2O_7^{2-} \longrightarrow Cr^{3+}$$

The chromium atoms are balanced first by placing the coefficient 2 before the Cr^{3+} on the right. The oxygens that appear on the left are balanced by adding H_2O (seven of them) to the right and the hydrogen imbalance that results is removed by placing $14H^+$ on the left.

$$14H^+ + Cr_2O_7^{2-} \longrightarrow 2Cr^{3+} + 7H_2O$$

Now we have to balance the half-reaction electrically. The net charge on the left is $14 \times (1+) + (2-) = 12+$. On the right the net charge is $6+$. The algebraic difference between these is the number of electrons we add to the more positive side. Thus we need $6e^-$ on the left.

$$6e^- + 14H^+ + Cr_2O_7^{2-} \longrightarrow 2Cr^{3+} + 7H_2O \qquad [6.2]$$

The next step is to add these half-reactions (Equations 6.1 and 6.2) so that all the electrons in the final equation cancel. This is accomplished by multiplying Equation 6.1 through by 3 before adding.

$$3(H_2S \longrightarrow S + 2H^+ + 2e^-)$$

$$\underline{6e^- + 14H^+ + Cr_2O_7^{2-} \longrightarrow 2Cr^{3+} + 7H_2O}$$

$$3H_2S + \cancel{14}H^+ + Cr_2O_7^{2-} + \cancel{6e^-} \longrightarrow 3S + \cancel{6H^+} + 2Cr^{3+} + 7H_2O + \cancel{6e^-}$$
$$8$$

The six electrons cancel, as do six of the H^+, so the final balanced equation reads

$$3H_2S + 8H^+ + Cr_2O_7^{2-} \longrightarrow 3S + 2Cr^{3+} + 7H_2O$$

In summary, when balancing half-reactions in acid solution:

(a) *To balance a hydrogen atom we add a hydrogen ion, H^+, to the other side of the equation.*

(b) *To balance an oxygen atom we add a water molecule to the side deficient in oxygen and then two H^+ ions to the opposite side to remove the hydrogen imbalance.*

Redox reactions in basic solutions

We have just seen that H_2O and H^+ are used to balance half-reactions that occur in acidic solution. In a basic solution the dominant species are H_2O and OH^-, so these are the species that should be used to achieve material balance.

Except for hydrogen and oxygen, each side of a given half-reaction must begin with the same elements; otherwise it can't be balanced.

Although you can use H_2O and OH^- directly (see footnote 3), the simplest technique is to first balance the reaction as if it occurred in acidic solution, and then perform the "conversion" described below to adjust it to conform to conditions in basic solution.

Suppose we wished to balance the following half-reaction taking place in a basic solution.

$$Pb \longrightarrow PbO$$

First we balance it as if it occurred in an acidic solution.

$$H_2O + Pb \longrightarrow PbO + 2H^+ + 2e^-$$

The conversion to basic solution follows these three steps:

Step 1 For each H^+ that must be eliminated from the equation, add an OH^- to both sides of the equation. In this example, we have to eliminate $2H^+$, so we add $2OH^-$ to each side.

$$H_2O + Pb + 2OH^- \longrightarrow PbO + 2H^+ + 2OH^- + 2e^-$$

Step 2 Combine H^+ and OH^- to form H_2O. We have $2H^+$ and $2OH^-$ on the right, which give $2H_2O$.

$$H_2O + Pb + 2OH^- \longrightarrow PbO + \underbrace{2H_2O}_{2H^+ + 2OH^-} + 2e^-$$

Step 3 Cancel any H_2O that are the same on both sides. We can cancel one H_2O from each side. The final balanced half-reaction in basic solution is

$$Pb + 2OH^- \longrightarrow PbO + H_2O + 2e^-$$

EXAMPLE 6.9   Balance the following reaction for the oxidation of plumbite ion, $Pb(OH)_3^-$, to lead dioxide by hypochlorite ion in basic solution.

$$Pb(OH)_3^- + OCl^- \longrightarrow PbO_2 + Cl^-$$

SOLUTION   First, the equation is balanced as though it occurred in acidic solution. We begin by dividing it into half-reactions.

$$Pb(OH)_3^- \longrightarrow PbO_2$$

$$OCl^- \longrightarrow Cl^-$$

Now we balance them according to atoms.

$$Pb(OH)_3^- \longrightarrow PbO_2 + H_2O + H^+$$

$$2H^+ + OCl^- \longrightarrow Cl^- + H_2O$$

Then we balance the charge by adding electrons.

$$Pb(OH)_3^- \longrightarrow PbO_2 + H_2O + H^+ + 2e^-$$
$$2e^- + 2H^+ + OCl^- \longrightarrow Cl^- + H_2O$$

Since the number of electrons lost in the first half-reaction is already equal to the number gained in the second, we can add them.

$$2e^- + 2H^+ + OCl^- + Pb(OH)_3^- \longrightarrow Cl^- + 2H_2O + PbO_2 + H^+ + 2e^-$$

Next, we cancel electrons and any other substances that are the same on both sides. The

---

[3] To balance half-reactions in basic solution, the following rules can be used:
(a) *To balance a hydrogen atom we add* **one** H_2O *molecule to the side of the half-reaction deficient in hydrogen, and to the other side we add* **one** *hydroxide ion.*
(b) *To balance one oxygen atom we add* **two** *hydroxide ions to the side deficient in oxygen and* **one** H_2O *molecule to the other side.*

equation is now balanced for acidic solution.

$$H^+ + OCl^- + Pb(OH)_3^- \longrightarrow Cl^- + 2H_2O + PbO_2$$

Next, we perform the three-step conversion to basic solution. First we add to each side the same number of OH^- as there are H^+ in the equation.

$$\underbrace{H^+ + OH^-}_{\text{forms } H_2O} + OCl^- + Pb(OH)_3^- \longrightarrow Cl^- + 2H_2O + PbO_2 + OH^-$$

On the left, H^+ and OH^- become H_2O.

$$H_2O + OCl^- + Pb(OH)_3^- \longrightarrow Cl^- + 2H_2O + PbO_2 + OH^-$$

Then we cancel one H_2O from each side to get the final balanced equation.

$$OCl^- + Pb(OH)_3^- \longrightarrow Cl^- + H_2O + PbO_2 + OH^-$$

6.9 QUANTITATIVE ASPECTS OF REACTIONS IN SOLUTION

In Section 6.6 we saw that the preparation of pure compounds often depends on the mixing together of reactants in the proper ratio, as determined by the stoichiometry of the reaction. For example, if we wish to prepare $NaHSO_4$ from solutions containing $NaOH$ and H_2SO_4, the two must be combined in such a way as to maintain a 1:1 mole ratio between the base and the acid. This means that we must have a means at our disposal of expressing the concentrations of reactants so that appropriate quantities of the various solutions may be mixed.

There are many ways of expressing the concentration of a solute in a solution, and each has its advantages for specific applications. One way to express concentration, for instance, is percent composition by weight, or *weight/weight percent*. Concentrated sulfuric acid, we have said, is composed of 96% H_2SO_4. We write this as 96% (w/w) H_2SO_4 to indicate weight/weight percent. This gives the composition of the solution in parts per hundred by weight. In other words, it tells us how many grams of solute there are in 100 g of the solution.

EXAMPLE 6.10	How would you prepare a solution that is 5.00% (by weight) NaCl in water?
SOLUTION	The label tells us that there should be 5.00 g NaCl in 100 g of the solution. To prepare the solution, we add 95.0 g of water to 5.00 g of NaCl. Since the density of water is very close to 1.00 g/ml, we can use 95.0 ml of water and save the trouble of weighing the water.

Analytical methods are becoming so sensitive that some substances can be detected at concentration levels of parts per billion (ppb).

A somewhat similar unit that is frequently used to express very small concentrations (e.g., the concentration of impurities or pollutants in air or water) is **parts per million, ppm,** by volume or by weight. For instance, a typical carbon monoxide level in heavy smog is approximately 40 ppm by volume, whereas a typical nitrogen oxide concentration is about 0.2 ppm by volume.

EXAMPLE 6.11	Atmospheric SO_2, produced by the combustion of high-sulfur fuels, is an important air pollution problem. The amount of SO_2 in the air may be determined by bubbling the air through an acidic solution of $KMnO_4$, which oxidizes the SO_2 to SO_4^{2-}. In a typical analysis, 500 liters of air with a density of 1.20 g/liter are passed through the $KMnO_4$ solution and 1.50×10^{-5} mol of $KMnO_4$ are reduced to Mn^{2+}. What is the concentration of SO_2 in parts per million (by weight)?

SOLUTION   We first must have a balanced equation. The ion-electron method gives us

$$2MnO_4^- + 5SO_2 + 2H_2O \longrightarrow 2Mn^{2+} + 5SO_4^{2-} + 4H^+$$

(If 1.50×10^{-5} mol of MnO_4^- is reduced, this requires (based on the stoichiometry of the reaction)

$$1.50 \times 10^{-5} \text{ mol } MnO_4^- \times \left(\frac{5 \text{ mol } SO_2}{2 \text{ mol } MnO_4^-}\right) \sim 3.75 \times 10^{-5} \text{ mol } SO_2$$

The weight of SO_2 in the air is

$$3.75 \times 10^{-5} \text{ mol } SO_2 \times \left(\frac{64.1 \text{ g } SO_2}{1 \text{ mol } SO_2}\right) \sim 2.40 \times 10^{-3} \text{ g } SO_2$$

The 500-liter sample of air contained 2.40×10^{-3} g SO_2.
The concentration in parts per million is obtained as

$$\frac{\text{weight of } SO_2}{\text{total weight of air sample}} \times 10^6$$

The weight of the air sample can be obtained from its density

$$500 \text{ liters} \times \left(\frac{1.20 \text{ g}}{\text{liter}}\right) \sim 600 \text{ g air}$$

Therefore, the concentration of the SO_2 pollutant is

$$\frac{2.40 \times 10^{-3} \text{ g } SO_2}{600 \text{ g air}} \times 10^6 = 4.00 \text{ ppm}$$

Expressing concentration in percent (as in concentrated H_2SO_4) or in parts per million (as in the solution of gases in Example 6.11) conveys information about the relative proportions of solute and solvent. However, for many applications to stoichiometry these concentration units are inconvenient. The most useful concentration unit for working problems in stoichiometry is molar concentration, or molarity, which was discussed in Section 2.11. Molar concentration, you recall, gives us the ratio of the number of moles of solute to the total volume of the solution (in liters). The following examples should be studied to refresh your memory.

Molarity is useful because it expresses concentration in chemical units: moles.

EXAMPLE 6.12   How many grams of NaOH are required to prepare 350 ml of 1.25 *M* NaOH solution?

SOLUTION   We must first determine how many moles of NaOH are in 350 ml of 1.25 *M* NaOH. This can be accomplished by multiplying molarity by volume.

$$\left(\frac{1.25 \text{ mol NaOH}}{1000 \text{ ml}}\right) \times 350 \text{ ml} \sim 0.438 \text{ mol NaOH}$$

It is helpful to think of molarity as a conversion factor.

Note that molarity functions as a conversion factor that allows us to convert *volume of solution to moles of NaOH.*

The weight of NaOH can then be obtained from the formula weight.

$$0.438 \text{ mol NaOH} \times \left(\frac{40.0 \text{ g NaOH}}{1 \text{ mol NaOH}}\right) \sim 17.5 \text{ g NaOH}$$

EXAMPLE 6.13   A solution of copper sulfate is labeled 0.100 *M* $CuSO_4$. How many milliliters of this solution contain 4.25 g $CuSO_4$?

SOLUTION First, let's calculate the number of moles of $CuSO_4$ in 4.25 g of this salt.

$$4.25 \text{ g CuSO}_4 \times \left(\frac{1 \text{ mol CuSO}_4}{159.6 \text{ g CuSO}_4} \right) = 2.66 \times 10^{-2} \text{ mol CuSO}_4$$

Now we use molarity as a conversion factor to change moles $CuSO_4$ to milliliters of solution.

$$2.66 \times 10^{-2} \text{ mol CuSO}_4 \times \left(\frac{1000 \text{ ml solution}}{0.100 \text{ mol CuSO}_4} \right) \sim 266 \text{ ml solution}$$

EXAMPLE 6.14 What is the molar concentration of concentrated sulfuric acid that contains 96% (w/w) H_2SO_4 and has a density of 1.84 g/ml?

SOLUTION We must determine the number of moles of H_2SO_4 in 1 liter of the concentrated acid. We know, from the density, that

$$1.00 \text{ ml} \sim 1.84 \text{ g solution}$$

or

$$1000 \text{ ml} = 1.00 \text{ liter} \sim 1840 \text{ g solution}$$

Only 96.0% of this weight is contributed by H_2SO_4. Thus

$$1840 \text{ g solution} \times \left(\frac{96.0 \text{ g H}_2\text{SO}_4}{100 \text{ g solution}} \right) \sim 1770 \text{ g H}_2\text{SO}_4$$

The formula weight of $H_2SO_4 = 98.08$; therefore,

$$1770 \text{ g H}_2\text{SO}_4 \times \left(\frac{1 \text{ mol H}_2\text{SO}_4}{98.08 \text{ g H}_2\text{SO}_4} \right) \sim 18.0 \text{ mol H}_2\text{SO}_4$$

The concentration is

$$\frac{18.0 \text{ mol H}_2\text{SO}_4}{1.00 \text{ liter}} = 18.0 \ M \ H_2SO_4$$

The usefulness of molarity, as pointed out in Chapter 2, is that it allows us to measure out moles of a reactant by simply measuring out the necessary volume of a solution. Let's see how this applies to some problems in stoichiometry.

EXAMPLE 6.15 How many milliliters of 2.00 M NaCl solution are required to react with exactly 5.37 g of $AgNO_3$ to form AgCl?

SOLUTION The first step in any problem involving the stoichiometry of a chemical reaction should be to write a chemical equation (balanced, of course).

$$NaCl + AgNO_3 \longrightarrow AgCl + NaNO_3$$

We see that 1 mol of NaCl reacts with 1 mol of $AgNO_3$. Therefore, 5.37 g $AgNO_3$ requires

$$5.37 \text{ g AgNO}_3 \times \left(\frac{1 \text{ mol AgNO}_3}{170 \text{ g AgNO}_3} \right) \times \left(\frac{1 \text{ mol NaCl}}{1 \text{ mol AgNO}_3} \right) \sim 3.16 \times 10^{-2} \text{ mol NaCl}$$

Now we use molarity as a conversion factor to change mol NaCl to milliliters of the solution.

$$3.16 \times 10^{-2} \text{ mol NaCl} \times \left(\frac{1000 \text{ ml solution}}{2.00 \text{ mol NaCl}} \right) = 15.8 \text{ ml solution}$$

EXAMPLE 6.16   How many grams of $Mg(OH)_2$ (found in milk of magnesia) are required to neutralize 50.0 ml of 0.0950 M HCl (stomach acid)?

SOLUTION   Again we begin with a balanced equation.

$$Mg(OH)_2 + 2HCl \longrightarrow MgCl_2 + 2H_2O$$

The equation tells us that

$$1 \text{ mol } Mg(OH)_2 \sim 2 \text{ mol HCl}$$

In 50.0 ml of 0.0950 M HCl there is

$$50.0 \text{ ml} \times \left(\frac{0.950 \text{ mol HCl}}{1000 \text{ ml}} \right) \sim 0.0475 \text{ mol HCl}$$

Now we know how much HCl has to be neutralized. Using the information from the equation,

$$0.0475 \text{ mol HCl} \times \left(\frac{1 \text{ mol } Mg(OH)_2}{2 \text{ mol HCl}} \right) \sim 0.0238 \text{ mol } Mg(OH)_2$$

Finally we calculate the weight of $Mg(OH)_2$.

$$0.0238 \text{ mol } Mg(OH)_2 \times \left(\frac{58.3 \text{ g } Mg(OH)_2}{1 \text{ mol } Mg(OH)_2} \right) \sim 1.39 \text{ g } Mg(OH)_2$$

EXAMPLE 6.17   How many milliliters of 0.250 M $Cr_2(SO_4)_3$ solution are needed to react completely with 300 ml of 0.400 M $BaCl_2$ according to the equation

$$Cr_2(SO_4)_3 \, (aq) + 3BaCl_2 \, (aq) \longrightarrow 3BaSO_4 \, (s) + 2CrCl_3 \, (aq)$$

SOLUTION   This is simply a problem in stoichiometry. We will first calculate how many moles of $BaCl_2$ are in the given solution. Then we will compute how many moles of $Cr_2(SO_4)_3$ are needed. Finally, we will calculate the volume of the $Cr_2(SO_4)_3$ solution that is required.

The number of moles of $BaCl_2$ given is

$$300 \text{ ml solution} \times \left(\frac{0.400 \text{ mol } BaCl_2}{1000 \text{ ml solution}} \right) \sim 0.120 \text{ mol } BaCl_2$$

Using the coefficients of the equation, the number of moles of $Cr_2(SO_4)$ needed in the reaction is

$$0.120 \text{ mol } BaCl_2 \times \left(\frac{1 \text{ mol } Cr_2(SO_4)_3}{3 \text{ mol } BaCl_2} \right) \sim 0.0400 \text{ mol } Cr_2(SO_4)_3$$

Next, we calculate the volume of $Cr_2(SO_4)_3$ solution required.

$$0.0400 \text{ mol } Cr_2(SO_4)_3 \times \left(\frac{1000 \text{ ml solution}}{0.250 \text{ mol } Cr_2(SO_4)_3} \right) \sim 160 \text{ ml solution}$$

Thus, we need 160 ml of the 0.250 M $Cr_2(SO_4)_3$ solution.

6.10   EQUIVALENT WEIGHTS AND NORMALITY

In certain types of reactions there is always the same clearly defined and predictable ratio in which the reactants combine. For example, in any acid-base neutralization, an amount of acid that furnishes 1 mol of H^+ *always* reacts completely with an amount of base that furnishes 1 mol of OH^-. This is assured by the stoichiometry of the neutralization reaction,

$$H^+ + OH^- \longrightarrow H_2O$$

Similarly, in a redox reaction the amount of the substance that loses 1 mol of electrons will *always* react exactly with the amount of another substance that gains 1 mol of electrons. This is because the numbers of electrons lost and gained *must* be exactly the same.

Recognition of such similarities in the stoichiometry of acid-base reactions and of redox reactions leads us to the concept of an **equivalent** (we will use the abbreviation **eq**). If we have a reaction between two substances—call them A and B—the equivalent is always defined so that 1 equivalent of A (1 eq A) reacts with exactly 1 equivalent of B (1 eq B). Equivalents *always* react in a 1-to-1 ratio.

$$1 \text{ eq } A \sim 1 \text{ eq } B$$

The equivalent is not an SI unit, and many practicing chemists have gotten away from the concept. Nevertheless, the idea of equivalents is still widely used in other sciences, so it is necessary for you to learn about it.

Remember, equivalents always react one for one.

Acid-base reactions

An equivalent of an acid is defined as the amount of acid that furnishes 1 mol of H^+. An equivalent of a base furnishes 1 mol of OH^-. Notice that this definition assures that 1 eq acid will react with exactly 1 eq base. Let's see how this applies to some typical acids and bases.

$\left.\begin{array}{l}\text{1 eq acid}\\(\text{1 mol } H^+)\end{array}\right\} \sim \left\{\begin{array}{l}\text{1 eq base}\\(\text{1 mol } OH^-)\end{array}\right.$

Consider the reaction of H_2SO_4 with NaOH.

$$H_2SO_4 + 2NaOH \longrightarrow Na_2SO_4 + 2H_2O$$

In this reaction, 1 mol H_2SO_4 furnishes 2 mol H^+, so 1 mol H_2SO_4 must be the same as 2 eq H_2SO_4. For the base, 1 mol NaOH furnishes 1 mol OH^-, so 1 mol NaOH = 1 eq NaOH. For acids and bases, the number of equivalents per mole is the same as the number of H^+ supplied per molecule of acid or the number of OH^- supplied per formula unit of base. Another way of stating this is

The number of equivalents is never less than the number of moles.

$$\text{number of eq} = \text{number of mol} \times n \qquad [6.3]$$

where n is a whole number; it is the number of H^+ provided by one molecule of acid or the number of OH^- provided by one formula unit of base in the particular reaction being studied.

EXAMPLE 6.18

How many equivalents are there in 0.400 mol H_3PO_4 if this acid is completely neutralized to give PO_4^{3-}? How many equivalents are there if it is converted to HPO_4^{2-}?

SOLUTION

One molecule of H_3PO_4 must supply three H^+ if it is to become PO_4^{3-}; therefore $n = 3$. Substituting into Equation 6.3,

$$\text{number of eq} = 0.400 \times 3$$

$$= 1.20$$

Thus, 0.400 mol H_3PO_4 is 1.20 eq when it is converted to PO_4^{3-}.

The number of equivalents per mole of reactants is undefined unless the reaction is specified.

If the original H_3PO_4 were only converted to HPO_4^{2-}, each molecule of acid would give 2H^+, so $n = 2$. Therefore, 0.400 mol H_3PO_4 is 0.800 eq when it is converted to HPO_4^{2-}.

In Example 16.18 we see that the number of equivalents per mole of acid depends on the extent to which the acid is neutralized in a particular reaction. Although this is a concern with acids, you can assume that the number of equivalents per mole for a base is the same as the number of OH^- provided by one formula unit of the base. Thus, 1 mol $Ba(OH)_2$ would be 2 eq $Ba(OH)_2$, and 1 mol $Al(OH)_3$ would be 3 eq $Al(OH)_3$.

A quantity used in many stoichiometric calculations involving equivalents is the **equivalent weight**—the mass of one equivalent. In terms of the integer n defined in Equation 6.3,

$$\text{equivalent weight (eq wt)} = \frac{\text{mass of 1 mol}}{n} \qquad [6.4]$$

For example, for the complete neutralization of H_2SO_4, $n = 2$. The equivalent weight of H_2SO_4 is its formula weight divided by 2.

$$\text{eq wt } H_2SO_4 = \frac{98.0 \text{ g}}{2} = 49.0 \text{ g}$$

In other words, 1 eq H_2SO_4 = 49.0 g H_2SO_4.

EXAMPLE 6.19

How many grams of H_3PO_4 will be completely neutralized to give PO_4^{3-} by 20.0 g NaOH?

SOLUTION

Let's calculate the weight of 1 equivalent for each reactant. These weights will exactly react with each other because 1 eq H_3PO_4 reacts with exactly 1 eq NaOH.

$$1 \text{ eq NaOH} \sim 1 \text{ eq } H_3PO_4$$

Sodium hydroxide has a formula weight of 40.0. Therefore,

$$1 \text{ eq NaOH} = \frac{40.0 \text{ g NaOH}}{1} = 40.0 \text{ g NaOH}$$

The formula weight of phosphoric acid is 98.0.

$$1 \text{ eq } H_3PO_4 = \frac{98.0 \text{ g } H_3PO_4}{3} = 32.7 \text{ g } H_3PO_4$$

Therefore, we can say that

$$40.0 \text{ g NaOH} \sim 32.7 \text{ g } H_3PO_4$$

Now we use this in a conversion factor to find our answer.

$$20.0 \text{ g NaOH} \times \left(\frac{32.7 \text{ g } H_3PO_4}{40.0 \text{ g NaOH}}\right) = 16.4 \text{ g } H_3PO_4$$

Thus, 20.0 g NaOH will neutralize 16.4 g H_3PO_4. Notice that we did the problem without a chemical equation. That's the only real usefulness of the concept of equivalents—for certain types of reactions we don't need to write an equation to establish the stoichiometry.

Oxidation-reduction reactions

An equivalent of an oxidizing or reducing agent in a redox reaction is the amount of substance that gains or loses 1 mol of electrons. This definition ensures that 1 equivalent of oxidizing agent reacts with exactly 1 equivalent of reducing agent.

Equations 6.3 and 6.4 apply to redox, too. The value of n for an oxidizing agent or reducing agent is just the total number of electrons gained or lost by one formula unit of the substance. For example, suppose that calcium iodate, $Ca(IO_3)_2$, were converted to iodide ion, I^-, in a particular reaction.

$$Ca(IO_3)_2 \longrightarrow I^-$$
$$5+ \qquad\qquad 1-$$

The oxidation number of iodine in $Ca(IO_3)_2$ is $5+$ and in I^- it is $1-$. Each iodine changes by $6e^-$ and a formula unit of $Ca(IO_3)_2$ therefore changes by $12e^-$ because it contains two iodine atoms. In this reaction, $n = 12$ for $Ca(IO_3)_2$, which means that 1 mol of $Ca(IO_3)_2$ gains 12 mol of e^-. Therefore, 1 mol of $Ca(IO_3)_2$ is 12 eq $Ca(IO_3)_2$.

> To calculate equivalent weights in redox reactions, we must know the <u>changes</u> in oxidation numbers that the reactants undergo.

$$\text{number eq } Ca(IO_3)_2 = \text{number mol } Ca(IO_3)_2 \times 12$$

The equivalent weight of $Ca(IO_3)_2$ is $\frac{1}{12}$ of the weight of a mole.

$$\text{eq wt } Ca(IO_3)_2 = \frac{\text{mass of 1 mol } Ca(IO_3)_2}{12}$$

EXAMPLE 6.20 $FeSO_4$ reacts with $KMnO_4$ in H_2SO_4 to produce $Fe_2(SO_4)_3$ and $MnSO_4$. How many grams of $FeSO_4$ react with 3.71 g $KMnO_4$?

SOLUTION In this reaction,

$$1 \text{ mol of } FeSO_4 \text{ loses 1 mol of electrons;} \qquad \underset{2+}{Fe} \xrightarrow{-1e^-} \underset{3+}{Fe}$$

$$1 \text{ mol of } KMnO_4 \text{ gains 5 mol of electrons;} \qquad \underset{7+}{Mn} \xrightarrow{+5e^-} \underset{2+}{Mn}$$

The equivalent weight of $FeSO_4$ in this reaction would be the same as the weight of 1 mol because each mole $FeSO_4$ loses 1 mol of electron.

$$1 \text{ eq } FeSO_4 = \frac{1 \text{ mol } FeSO_4}{1} = 152 \text{ g } FeSO_4$$

The equivalent weight of $KMnO_4$ in this reaction is one-fifth of the weight of 1 mol, since each mole of $KMnO_4$ gains 5 mol of electrons:

$$1 \text{ eq } KMnO_4 = \tfrac{1}{5} \text{ mol } KMnO_4 = \tfrac{1}{5}(158.0 \text{ g } KMnO_4)$$

$$1 \text{ eq } KMnO_4 = 31.6 \text{ g } KMnO_4$$

Since 1 eq of $FeSO_4$ must react with precisely 1 eq of $KMnO_4$,

$$152 \text{ g } FeSO_4 \sim 31.6 \text{ g } KMnO_4$$

The solution to the problem can then be obtained as

$$3.71 \, \cancel{\text{g } KMnO_4} \times \left(\frac{152 \text{ g } FeSO_4}{31.6 \, \cancel{\text{g } KMnO_4}} \right) \sim 17.8 \text{ g } FeSO_4$$

Normality

Another concentration unit that can be used in the treatment of problems concerning stoichiometry is **normality,** defined as *the number of equivalents of solute per liter of solution.* Again, we have a ratio—the quantity of solute, expressed in equivalents, divided by the total volume of the solution. Like molarity, normality serves as a conversion factor relating equivalents of solute and liters (or milliliters) of solution. A solution labeled $1.00 \, N$ (a 1.00 *normal* solution) can be translated to mean

$$\frac{1.00 \text{ eq solute}}{1000 \text{ ml solution}} \qquad \text{or} \qquad \frac{1000 \text{ ml solution}}{1.00 \text{ eq solute}}$$

In the preceding discussion, an equivalent of an oxidizing or reducing agent was defined as the quantity that either gains or loses 1 mol of electrons.

We saw that when the permanganate ion in $KMnO_4$ is reduced to Mn^{2+}, five electrons are gained by each formula unit of $KMnO_4$. One mole of $KMnO_4$, therefore, consumes 5 mol of electrons, and therefore 1 mol $KMnO_4$ is equal to 5 eq $KMnO_4$. This means that a 1.00 M $KMnO_4$ solution, which contains 1.00 mol $KMnO_4$ per liter, can also be considered to contain 5.00 eq $KMnO_4$ per liter. We would say that the concentration of this solution is 5.00 normal, or 5.00 N. Notice that there is a very simple relationship between normality and molarity—the normality is always an integral multiple of the molarity,

The normality is never less than the molarity.

$$N = n \times M$$

where n is the same integer as before. For substances undergoing oxidation or reduction the integer, n, corresponds to the number of electrons transferred per formula unit. When $KMnO_4$ is reduced to Mn^{2+}, $n = 5$. If the product of the reduction is MnO_2, only three electrons are acquired by each $KMnO_4$ formula unit, and n would equal 3. Similarly, when oxalic acid, $H_2C_2O_4$, is oxidized to produce CO_2, two electrons are lost by each $H_2C_2O_4$ formula unit and the normality of a solution of $H_2C_2O_4$ is twice its molarity.

For solutions of acids and bases, n is once again the number of H^+ provided by a formula unit of the acid or the number of OH^- made available by a formula unit of the base. For complete neutralization, a 1.00 M H_3PO_4 solution could also be labeled 3.00 N H_3PO_4; a 1.00 M $Ba(OH)_2$ solution is also 2.00 N $Ba(OH)_2$.

Normality is useful because mixing equal volumes of solutions having the same normality will lead to complete reaction between their solutes; 1 liter of 1 N acid will completely neutralize 1 liter of 1 N base because 1 equivalent of acid reacts with 1 equivalent of base. In the same manner, 1 liter of 1 N oxidizing agent will completely oxidize the contents of 1 liter of 1 N reducing agent.

EXAMPLE 6.21

How many milliliters of 0.150 N $K_2Cr_2O_7$ are required to react with 75.0 ml of 0.400 N $H_2C_2O_4$?

SOLUTION

In 75.0 ml of 0.400 N $H_2C_2O_4$ there are

$$75.0 \text{ ml} \times \left(\frac{0.400 \text{ eq } H_2C_2O_4}{1000 \text{ ml}}\right) \sim 3.00 \times 10^{-2} \text{ eq } H_2C_2O_4$$

This requires 3.00×10^{-2} eq of $K_2Cr_2O_7$ for complete oxidation. The volume of potassium dichromate solution needed for this is

$$3.00 \times 10^{-2} \text{ eq } K_2Cr_2O_7 \times \left(\frac{1000 \text{ ml}}{0.150 \text{ eq } K_2Cr_2O_7}\right) \sim 200 \text{ ml}$$

In Example 6.21 we used the idea that equal numbers of equivalents of oxidizing and reducing agent must react. Since the product of volume $(V) \times$ normality (N) gives equivalents, when any two substances in solution (let us call them A and B) undergo complete reaction

$$V_A N_A = V_B N_B$$

which means

(number of equivalents of A) = (number of equivalents of B)

It is very important to remember that this equation only applies if the concentration is expressed in normality! If you use molarity instead, you may get wrong answers.

6.11 CHEMICAL ANALYSIS

Whether we synthesize chemicals in the laboratory or isolate them from natural sources, we must know their composition before we can proceed very far in understanding their properties. When a substance appears as the product of a reaction, it does not stand up and declare its composition—it must be analyzed. One very important facet of chemistry is the investigation of new and better ways to achieve chemical analyses, particularly of compounds present in mixtures where difficulties in separating the components present a major obstacle to success.

Chemical reactions in solution can frequently be used to advantage in performing chemical analyses. Example 6.22 illustrates how a precipitation reaction might be put to use.

EXAMPLE 6.22 A mixture of NaCl and BaCl$_2$ weighing 0.200 g was dissolved in water, and sulfuric acid was added until no further precipitate was formed. The solid BaSO$_4$ was filtered, dried, and found to weigh 0.0643 g. What percent of the mixture was BaCl$_2$?

SOLUTION In this example the barium is precipitated as BaSO$_4$.

$$BaCl_2 + H_2SO_4 \longrightarrow 2HCl + BaSO_4$$

From the weight of BaSO$_4$ we can compute the weight of BaCl$_2$ in the mixture and hence the fraction that was BaCl$_2$.

$$0.0643 \text{ g BaSO}_4 \times \left(\frac{1 \text{ mol BaSO}_4}{233.4 \text{ g BaSO}_4}\right) \sim 2.75 \times 10^{-4} \text{ mol BaSO}_4$$

According to the chemical equation, 1 mol of BaSO$_4$ is produced from 1 mol of BaCl$_2$; therefore, the original mixture must have contained 2.75×10^{-4} mol of BaCl$_2$. The weight of BaCl$_2$ in the mixture was

$$2.75 \times 10^{-4} \text{ mol BaCl}_2 \times \left(\frac{208.4 \text{ g BaCl}_2}{1 \text{ mol BaCl}_2}\right) \sim 5.73 \times 10^{-2} \text{ g BaCl}_2$$

$$\text{weight of BaCl}_2 = 0.0573 \text{ g}$$

$$\text{percent BaCl}_2 = \frac{\text{weight BaCl}_2}{\text{total weight of mixture}} \times 100$$

$$= \frac{0.0573 \text{ g}}{0.200 \text{ g}} \times 100 = 28.7\%$$

Often, the quantitative relationships introduced in our discussion of concentration can be employed directly in what are termed **volumetric analyses.** For example, the quantity of base in some sample of an unknown can be determined by a procedure called a **titration,** in which a solution of an acid is dispensed into the solution of the base from a **buret** (Figure 6.6) until reaction is complete, as signaled by some form of **indicator.** There are a series of organic dyes that can serve as indicators. They have one color in acid solution and a different color in basic solution. Litmus is a common example—it is red in acid solution and blue in the presence of base. Another indicator you may use in the laboratory is phenolphthalein; it is colorless in acidic solution and pink in basic solution. In a typical titration, the base is dissolved in water and a small amount of the indicator is added. Acid is then gradually added from the

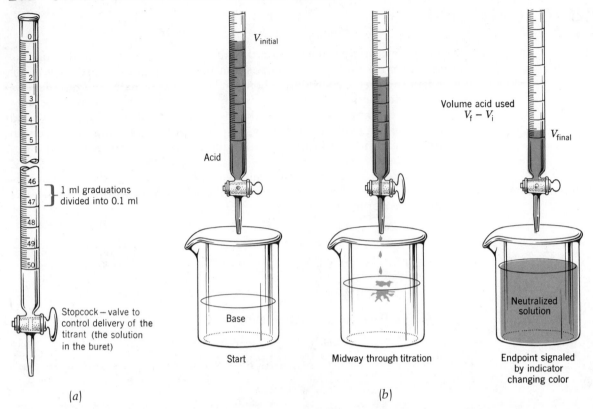

Figure 6.6

(a) *A buret.* (b) *Titration of a base with an acid.*

Ideally, the endpoint should occur at the equivalence point. How close they come depends on the choice of indicator, as we will see in Chapter 15.

buret until the indicator just changes color. This is called the **endpoint** of the titration because the addition of acid is stopped.

If the proper indicator has been chosen, at the endpoint we have combined equal numbers of equivalents of acid and base—we have reached the **equivalence point.** From the total volume of acid added and its concentration, the number of equivalents of acid, and, hence, the number of equivalents of base can be deduced.

EXAMPLE 6.23   A sample of an analgesic drug was analyzed for aspirin, a monoprotic acid, $HC_9H_7O_4$, by titration with base. A 0.500-g sample of the drug required 21.5 ml of 0.100 *N* NaOH. What percentage of the drug was aspirin?

SOLUTION   From the volume of base and its concentration we can calculate the number of equivalents of base consumed.

$$21.5 \text{ ml} \times \left(\frac{0.100 \text{ eq NaOH}}{1000 \text{ ml}}\right) \sim 0.00215 \text{ eq NaOH}$$

From the concept of equivalents we can write

$$0.00215 \text{ eq NaOH} \sim 0.00215 \text{ eq aspirin}$$

Since aspirin is a monoprotic acid

$$0.00215 \text{ eq aspirin} = 0.00215 \text{ mol aspirin}$$

The weight of aspirin in the sample is

$$0.00215 \; \text{mol aspirin} \times \left(\frac{180.2 \text{ g aspirin}}{1 \text{ mol aspirin}} \right) \sim 0.387 \text{ g aspirin}$$

The percent aspirin can be found as

$$\text{percent aspirin} = \frac{\text{weight aspirin}}{\text{weight sample}} \times 100$$

Substituting,

$$\text{percent aspirin} = \frac{0.387 \text{ g}}{0.500 \text{ g}} \times 100 = 77.4\%$$

Similar titrations are frequently performed in which the reaction involves oxidation-reduction. $KMnO_4$ is a useful **titrant** (the solution delivered from the buret) because it has a deep purple color, whereas the product of the reduction of the MnO_4^- ion in acid solution is the almost colorless Mn^{2+} ion. The end-point in the titration is signaled by the appearance of a pale purple color in the reaction mixture due to excess MnO_4^- ion.

EXAMPLE 6.24 A 1.000-gram sample of an iron ore containing Fe_2O_3 was dissolved in acid and all the iron converted to Fe^{2+}. The solution was titrated with 90.4 ml of 0.100 N $KMnO_4$ to give Mn^{2+} and Fe^{3+} as products. What percentage of the ore is Fe_2O_3?

SOLUTION The number of equivalents of $KMnO_4$ reduced was

$$90.4 \; \text{ml} \times \left(\frac{0.100 \text{ eq } KMnO_4}{1000 \text{ ml}} \right) \sim 9.04 \times 10^{-3} \text{ eq } KMnO_4$$

This amount of $KMnO_4$ has oxidized 9.04×10^{-3} eq Fe^{2+} to Fe^{3+}. Since 1 mol of Fe^{2+} loses 1 mol of electrons,

$$1 \text{ mol } Fe^{2+} = 1 \text{ eq } Fe^{2+}$$

Therefore, 9.04×10^{-3} eq $Fe^{2+} = 9.04 \times 10^{-3}$ mol Fe^{2+}. Each mol of Fe_2O_3 contains 2 mol of Fe.

$$1 \text{ mol } Fe_2O_3 \sim 2 \text{ mol Fe}$$

The number of mol of Fe_2O_3 in the sample was

$$9.04 \times 10^{-3} \; \text{mol Fe} \times \left(\frac{1 \text{ mol } Fe_2O_3}{2 \text{ mol Fe}} \right) \sim 4.52 \times 10^{-3} \text{ mol } Fe_2O_3$$

The number of grams of Fe_2O_3 was

$$4.52 \times 10^{-3} \; \text{mol } Fe_2O_3 \times \left(\frac{160 \text{ g } Fe_2O_3}{1 \text{ mol } Fe_2O_3} \right) \sim 0.723 \text{ g } Fe_2O_3$$

Since the sample weighed 1.000 g,

$$\text{percent } Fe_2O_3 = \frac{0.723 \text{ g}}{1.000 \text{ g}} \times 100 \doteq 72.3\%$$

Dilution

In the course of performing routine laboratory operations as well as volumetric analyses, it is not uncommon to have to dilute solutions—to make them less concentrated. For example, many common laboratory reagents are bottled in

Table 6.4
Concentrated laboratory reagents

Reagent	Density (g/ml)	Weight Percent	Molarity
Sulfuric acid	1.84	96	18
Hydrochloric acid	1.18	36.5	12.0
Phosphoric acid	1.7	85	15
Nitric acid	1.42	70.7	16.0
Acetic acid	1.05	100	17.5
Aqueous ammonia	0.90	28	14.8

concentrated form (Table 6.4) and usually must be diluted for use.[4] The process of dilution actually involves simply spreading a given quantity of solute throughout a larger volume of solution; when a concentrated solution is diluted by the addition of solvent, the number of moles of solute in the mixture remains constant. This means that the solution's molar concentration multiplied by its volume must remain constant as solvent is added, because

$$M \cdot V = \frac{\text{mol}}{\cancel{\text{liter}}} \times \cancel{\text{liter}} = \text{mol}$$

In other words, the product of the initial molarity and volume ($M_i V_i$) must equal the product of the final molarity and volume ($M_f V_f$).

$$M_i V_i = M_f V_f$$

This equation should only be used for dilution problems, not for problems relating volumes of solutions in a chemical reaction.

It is important to remember that this equation should *only* be used for dilution problems. For dilution problems in which the concentration unit is normality, the similar equation, $N_i V_i = N_f V_f$, applies.

EXAMPLE 6.25 How much water must be added to 25.0 ml of 0.500 M KOH solution to produce a solution whose concentration is 0.350 M?

SOLUTION Our equation is

$$M_i V_i = M_f V_f$$

$$M_i = 0.500 \ M \qquad M_f = 0.350 \ M$$

$$V_i = 25.0 \ \text{ml} \qquad V_f = ?$$

Solving for V_f and substituting, we get

$$V_f = \frac{(0.500 \ M)(25.0 \ \text{ml})}{0.350 \ M}$$

$$V_f = 35.7 \ \text{ml}$$

Since the initial volume was 25.0 ml, we must *add* 10.7 ml. (We are assuming that volumes are additive. When working with dilute solutions this assumption is generally quite valid.)

---

[4] **Caution!** When diluting concentrated reagents, the proper procedure is to **add the reagent to the solvent.** *Never* add water to a concentrated acid.

EXAMPLE 6.26 How many milliliters of concentrated H_2SO_4 (18.0 M) are required to prepare 750 ml of 3.00 M H_2SO_4 solution?

SOLUTION Again,

$$M_i V_i = M_f V_f$$

$$M_i = 18.0\ M \qquad M_f = 3.00\ M$$

$$V_i = ? \qquad\qquad V_f = 750\ ml$$

Solving for V_i gives

$$V_i = \frac{M_f V_f}{M_i}$$

$$V_i = \frac{(3.00\ M)(750\ ml)}{18.0\ M}$$

$$V_i = 125\ ml$$

To prepare the solution, 125 ml of the concentrated H_2SO_4 is diluted to a total final volume of 750 ml.

INDEX TO QUESTIONS AND PROBLEMS (Problem numbers in **bold type**)

REVIEW QUESTIONS

6.1 Why are solutions usually employed for carrying out chemical reactions?

6.2 Give the meaning of the terms: solvent, solute, concentrated, dilute, saturated, supersaturated, unsaturated.

6.3 What is the meaning of the term, solubility?

6.4 How can one distinguish between a strong and a weak electrolyte? What are present in solutions of electrolytes that are absent in solutions of nonelectrolytes?

6.5 Write chemical equations for the dissociation of the following strong electrolytes: KCl, $(NH_4)_2SO_4$, Na_3PO_4, NaOH, HCl.

6.6 Why is the hydronium ion often abbreviated H^+?

6.7 What is a dynamic equilibrium? Write an equation to represent the dissociation of the weak electrolyte, $CdSO_4$.

6.8 Write an equation to represent the equilibrium dissociation of water.

6.9 What is the meaning of the term *position of equilibrium?*

6.10 What can be said about the position of equilibrium in (a) the dissociation of water; (b) the dissociation of HCl?

6.11 What is the term used to describe a solid that is formed in a solution during a chemical reaction?

6.12 What is a metathesis reaction? What other name is often applied to this kind of reaction?

6.13 Write chemical equations for the equilibria that exist when the following weak electrolytes undergo dissociation: HCN, H_2S, NH_3, $HgCl_2$.

6.14 Write ionic and net ionic equations for the following:

(a) $CuCl_2$ (aq) + $Pb(NO_3)_2$ (aq) →
$$Cu(NO_3)_2 \ (aq) + PbCl_2 \ (s)$$

(b) $FeSO_4$ (aq) + $2NaOH$ (aq) →
$$Fe(OH)_2 \ (s) + Na_2SO_4 \ (aq)$$

(c) $ZnSO_4$ (aq) + $BaCl_2$ (aq) → $ZnCl_2$ (aq) + $BaSO_4$ (s)

(d) $2AgNO_3$ (aq) + K_2SO_4 (aq) →
$$Ag_2SO_4 \ (s) + 2KNO_3 \ (aq)$$

(e) $(NH_4)_2CO_3$ (aq) + $CaCl_2$ (aq) →
$$2NH_4Cl \ (aq) + CaCO_3 \ (s)$$

6.15 Write ionic and net ionic equations for the following:

(a) $Al(OH)_3$ (s) + $3HCl$ (aq) → $AlCl_3$ (aq) + $3H_2O$

(b) $CuCO_3$ (s) + H_2SO_4 (aq) →
$$CuSO_4 \ (aq) + H_2O + CO_2 \ (g)$$

(c) $Cr_2(CO_3)_3$ (s) + $6HNO_3$ (aq) →
$$2Cr(NO_3)_3 \ (aq) + 3H_2O + 3CO_2 \ (g)$$

(d) $3Cu$ (s) + $8HNO_3$ (aq) →
$$3Cu(NO_3)_2 \ (aq) + 2NO \ (g) + 4H_2O$$

(e) MnO_2 (s) + $4HCl$ (aq) →
$$MnCl_2 \ (aq) + Cl_2 \ (g) + 2H_2O$$

6.16 Without looking at the solubility rules, indicate whether the following are soluble or insoluble: KCl, $(NH_4)_2SO_4$, $AgNO_3$, $PbSO_4$, $Mn(OH)_2$, $FePO_4$, $CaCO_3$, $Zn(ClO_4)_2$, $Ba(C_2H_3O_2)_2$, NiO.

6.17 Write net ionic equations for each of the following?

(a) $NaBr + AgNO_3 → AgBr + NaNO_3$

(b) $CoCO_3 + 2HNO_3 → Co(NO_3)_2 + CO_2 + H_2O$

(c) $NaC_2H_3O_2 + HNO_3 → NaNO_3 + HC_2H_3O_2$

(d) $Pb(NO_3)_2 + (NH_4)_2SO_4 → PbSO_4 + 2NH_4NO_3$

(e) $H_2S + Cu(NO_3)_2 → 2HNO_3 + CuS$

(f) $NaOH + NH_4Cl → NH_3 + H_2O + NaCl$

6.18 Write molecular, ionic, and net ionic equations for the reaction (if any) between:

(a) Na_2SO_4 and $BaCl_2$ \qquad (d) $NaOH$ and $CuCl_2$

(b) $Ca(NO_3)_2$ and $(NH_4)_2CO_3$ \qquad (e) $(NH_4)_2CO_3$ and HNO_3

(c) $NaC_2H_3O_2$ and HNO_3

6.19 Write molecular, ionic, and net ionic equations for the reaction (if any) between:

(a) H_2SO_4 and $BaSO_4$ \qquad (d) $MgSO_4$ and $LiOH$

(b) NH_4Br and $MnSO_4$ \qquad (e) $AgC_2H_3O_2$ and KCl

(c) K_2S and $HC_2H_3O_2$

6.20 Write molecular, ionic, and net ionic equations for any reaction that would occur between:

(a) $AgBr$ and KI \qquad (d) K_2SO_3 and HCl

(b) $BaCl_2$, SO_2, and H_2O \qquad (e) $BaCO_3$ and H_2SO_4

(c) $Na_2C_2O_4$ and HCl

6.21 Define the terms *acid* and *base* as applied to aqueous solutions.

6.22 Write typical reactions for the dissociation of typical monoprotic, diprotic, and triprotic acids.

6.23 Write chemical equations for the following neutralization reactions:

(a) $HCl + KOH$

(b) $H_2SO_4 + NaOH$

(c) $H_3AsO_4 + Ba(OH)_2$

6.24 What is the net ionic equation characteristic of neutralization reactions in aqueous solution? Use Lewis symbols to diagram this reaction.

6.25 What is an acid salt? Give some examples.

6.26 A newly discovered element is found to react with oxygen to form an oxide that, when dissolved in water, causes litmus to turn blue. Would the element be classed as a metal or a nonmetal? Why?

6.27 Describe how you would prepare the following by a metathesis reaction in which one product is a precipitate: (a) $NH_4C_2H_3O_2$, (b) $Fe_3(PO_4)_2$, (c) $CuCO_3$, (d) Na_2SO_4, (e) $PbSO_4$.

6.28 Describe how you would prepare the following by a metathesis reaction in which one product is a precipitate: (a) $Mg(OH)_2$, (b) $BaCl_2$, (c) $Fe(C_2H_3O_2)_2$, (d) $Ni(ClO_4)_2$, (e) $BaSO_3$.

6.29 Describe how you would prepare the following by a metathesis reaction in which one product is a gas: (a) $CaCl_2$, (b) $Mn(ClO_4)_2$, (c) $BaSO_4$, (d) $NaNO_3$, (e) $NH_4C_2H_3O_2$.

6.30 Describe how you would prepare the following by a neutralization reaction: (a) $Ca(NO_3)_2$, (b) $Na_2C_2O_4$, (c) $Fe(HSO_4)_2$, (d) $Al(ClO_4)_3$, (e) $NiBr_2$.

6.31 How could you prepare

(a) $CuCl_2$ from $Cu(NO_3)_2$

(b) $BaCl_2$ from $BaBr_2$

(c) $NaClO_4$ from Na_2SO_4

(d) $Mg(C_2H_3O_2)_2$ from $MgCl_2$

(e) Na_2CO_3 from Na_2SO_3

6.32 Balance the following molecular equations by the oxidation-number-change method.

(a) $HNO_3 + Zn → Zn(NO_3)_2 + H_2O + NH_4NO_3$

(b) $K + KNO_3 → N_2 + K_2O$

(c) $C_3H_7OH + Na_2Cr_2O_7 + H_2SO_4 →$
$$Cr_2(SO_4)_3 + Na_2SO_4 + H_2O + HC_3H_5O_2$$

(d) $H_2S + HNO_3 → S + NO + H_2O$

(e) $Fe(OH)_2 + O_2 + H_2O → Fe(OH)_3$

6.33 Balance the following molecular equations by the oxidation-number-change method.

(a) $Cu + HNO_3 → Cu(NO_3)_2 + NO + H_2O$

(b) $MnO_2 + HBr → Br_2 + MnBr_2 + H_2O$

(c) $(CH_3)_2CHOH + Cr\,O_3 + H_2SO_4 →$
$$(CH_3)_2CO + Cr_2(SO_4)_3 + H_2O$$

(d) $PbO_2 + Sb + NaOH → PbO + NaSbO_2 + H_2O$

(e) $NO_2 + H_2O → HNO_3 + NO$ (NO_2 is both oxidized *and* reduced)

6.34 Identify the oxidizing agent in each reaction in Question 6.32.

6.35 Balance the following equations by the ion-electron method. All reactions occur in acidic solution.
(a) $Cu + NO_3^- \rightarrow Cu^{2+} + NO$
(b) $Zn + NO_3^- \rightarrow Zn^{2+} + NH_4^+$
(c) $Cr + H^+ \rightarrow Cr^{3+} + H_2$
(d) $Cr_2O_7^{2-} + H_3AsO_3 \rightarrow Cr^{3+} + H_3AsO_4$
(e) $I^- + SO_4^{2-} \rightarrow I_2 + H_2S$
(f) $Ag^+ + AsH_3 \rightarrow H_3AsO_4 + Ag$
(g) $S_2O_8^{2-} + HNO_2 \rightarrow NO_3^- + SO_4^{2-}$
(h) $MnO_2 + Br^- \rightarrow Br_2 + Mn^{2+}$
(i) $S_2O_3^{2-} + I_2 \rightarrow I^- + S_4O_6^{2-}$
(j) $IO_3^- + HSO_3^- \rightarrow I^- + SO_4^{2-}$

6.36 Balance the following equations by the ion-electron method. All the reactions take place in acidic solution.
(a) $Cr_2O_7^{2-} + CH_3CH_2OH \rightarrow Cr^{3+} + CH_3CHO$
(b) $PbO_2 + Cl^- \rightarrow Pb^{2+} + Cl_2$
(c) $Mn^{2+} + BiO_3^- \rightarrow MnO_4^- + Bi^{3+}$
(d) $ClO_3^- + HAsO_2 \rightarrow H_3AsO_4 + Cl^-$
(e) $PH_3 + I_2 \rightarrow H_3PO_2 + I^-$
(f) $MnO_4^- + S_2O_3^{2-} \rightarrow S_4O_6^{2-} + Mn^{2+}$
(g) $Mn^{2+} + PbO_2 \rightarrow MnO_4^- + Pb^{2+}$
(h) $As_2O_3 + NO_3^- \rightarrow H_3AsO_4 + N_2O_3$
(i) $P + Cu^{2+} \rightarrow Cu + H_2PO_4^-$
(j) $MnO_4^- + H_2S \rightarrow Mn^{2+} + S$

6.37 Balance the following equations by the ion-electron method. All the reactions take place in basic solution.
(a) $CN^- + AsO_4^{3-} \rightarrow AsO_2^- + CNO^-$
(b) $CrO_2^- + HO_2^- \rightarrow CrO_4^- + OH^-$
(c) $Zn + NO_3^- \rightarrow Zn(OH)_4^{2-} + NH_3$
(d) $Cu(NH_3)_4^{2+} + S_2O_4^{2-} \rightarrow SO_3^{2-} + Cu + NH_3$
(e) $N_2H_4 + Mn(OH)_3 \rightarrow Mn(OH)_2 + NH_2OH$
(f) $MnO_4^- + C_2O_4^{2-} \rightarrow MnO_2 + CO_3^{2-}$
(g) $ClO_3^- + N_2H_4 \rightarrow NO_3^- + Cl^-$

*6.38 Balance the following equations by the ion-electron method.
(a) $P_4 \rightarrow PH_3 + H_2PO_2^-$ (basic solution)
(b) $Cu + Cl^- + As_4O_6 \rightarrow CuCl + As$ (acidic solution)
(c) $IPO_4 \rightarrow I_2 + IO_3^- + H_2PO_4^-$ (acidic solution)

(d) $NO_2 \rightarrow NO_3^- + NO$ (acidic solution)
(e) $Br_2 \rightarrow Br^- + BrO_3^-$ (basic solution)
(f) $HSO_2NH_2 + NO_3^- \rightarrow N_2O + SO_4^{2-}$ (acidic solution)
(g) $ClO_3^- + Cl^- \rightarrow Cl_2 + ClO_2$ (acidic solution)
(h) $ClO_2 \rightarrow ClO_2^- + ClO_3^-$ (basic solution)
(i) $Se \rightarrow Se^{2-} + SeO_3^{2-}$ (basic solution)
(j) $ICl \rightarrow IO_3^- + I_2 + Cl^-$ (acidic solution)
(k) $FNO_3 \rightarrow O_2 + F^- + NO_3^-$ (basic solution)
(l) $Fe(OH)_2 + O_2 \rightarrow Fe(OH)_3$ (basic solution)

6.39 Balance the following equations that occur in basic solution
(a) $Zn \rightarrow Zn(OH)_4^{2-} + H_2$
(b) $CrO_2^- + HO_2^- \rightarrow CrO_4^{2-}$

6.40 Define molarity, normality, equivalence point.

6.41 Describe in detail how you would prepare a 1.50 M KCl solution.

6.42 What is the meaning of the concentration unit *percent by weight*?

6.43 What is the meaning of the concentration unit *parts per million*?

6.44 Modern analytical techniques have achieved such sensitivities that pollutants and potential cancer-causing agents can be detected at the levels of *parts per billion, ppb*. What does this concentration term mean?

6.45 How is the equivalent defined for acids and bases? How is it defined for oxidizing and reducing agents?

6.46 If two substances A and B react, what relationship exists between the number of equivalents of A and B that react?

6.47 Why is $KMnO_4$ a convenient titrant in redox reactions?

6.48 How does the number of moles of solute in a solution change when the solution is diluted?

6.49 What is the proper procedure for diluting a concentrated reagent?

REVIEW PROBLEMS (More difficult problems are marked by an asterisk)

6.50 What is the molarity of each of the following?
(a) 1.50 mol of NaCl in 2.00 liters of solution
(b) 0.248 mol of KCN in 250 ml of solution
(c) 0.750 mol of H_2SO_4 in 1.35 liters of solution
(d) 85.5 g of HNO_3 in 1.00 liter of solution
(e) 44.5 g of $NH_4C_2H_3O_2$ in 600 ml of solution

6.51 How many moles of solute are in: (a) 250 ml of 0.100 M KCl; (b) 1.65 liters of 1.40 M $HClO_4$; (c) 0.0250 liter of 0.0100 M $HC_2H_3O_2$.

6.52 How many grams of Na_2CO_3 are required to prepare 300 ml of a 0.150 M solution?

6.53 How many grams of $Ba(OH)_2$ are required to prepare 250 ml of a solution that has a hydroxide ion concentration of 0.300 M?

6.54 Pure nitric acid has a density of 1.513 g/ml. What is its molar concentration?

6.55 A solution of $MgSO_4$ contains 22.0% $MgSO_4$ by weight and contains 273.8 g of the salt per liter. What is the density of the solution? What is the molarity of the solution?

6.56 A particular pollutant was present in water at a concentration of 825 ppm. What was the concentration in percent by weight? If the pollutant was benzene, C_6H_6, what was its molar concentration (assume that 1000 ml of the solution weighs essentially 1000 g)?

6.57 How many milliliters of 0.100 M NaOH are needed to react with 5.00×10^{-3} mol of H_2SO_4 to give Na_2SO_4?

6.58 How many milliliters of 1.250 M HNO_3 contain

enough nitric acid to dissolve a copper penny weighing 3.22 g? The reaction is

$$3Cu + 8HNO_3 \rightarrow 3Cu(NO_3)_2 + 2NO + 4H_2O$$

6.59 Copper(II) carbonate dissolves in perchloric acid according to the equation: $CuCO_3 + 2HClO_4 \rightarrow Cu(ClO_4)_2 + CO_2 + H_2O$
(a) How many milliliters of 1.35 M $HClO_4$ are needed to prepare 5.25 g of $Cu(ClO_4)_2$?
(b) How many grams of $CuCO_3$ are needed to prepare 5.25 g $Cu(ClO_4)_2$?

6.60 How many milliliters of 0.300 M NaOH are required to react with 500 ml of 0.170 M H_3PO_4 to yield (a) Na_3PO_4, (b) Na_2HPO_4, (c) NaH_2PO_4?

6.61 How many milliliters of 0.100 M $BaCl_2$ are required to react with 25.0 ml of 0.200 M H_2SO_4?

6.62 How many milliliters of 0.1000 M $BaCl_2$ are required to react completely with 25.0 ml of 0.200 M $Fe_2(SO_4)_3$?

6.63 A 0.244-g sample of benzoic acid (a monoprotic acid) requires 20.0 ml of 0.100 M NaOH for complete neutralization. Calculate the molecular weight of the acid.

6.64 20.0 ml of 0.200 M $AgNO_3$ were added to 30.0 ml of 0.200 M NaCl.
(a) What chemical reaction occurs?
(b) How many moles of precipitate are formed?
(c) How much does the precipitate weigh?
(d) What are the concentrations of each of the remaining ions in the final solution?

6.65 What weight of AgCl will be formed if 25.0 ml of 0.050 M HCl is added to 100 ml of 0.050 M $AgNO_3$?

6.66 50.0 ml of 0.240 M $BaCl_2$ were added to 45.0 ml of 0.180 M $Fe_2(SO_4)_3$.
(a) What weight of $BaSO_4$ was formed?
(b) What are the concentrations of the remaining ions in the final solution?

6.67 A sample of $Cr_2(SO_4)_3$ weighing 0.500 g was dissolved in water and an excess of dilute NaOH was added, precipitating $Cr(OH)_3$. This precipitate was collected by filtration. How many milliliters of 0.400 M HNO_3 are needed to dissolve the precipitate?

6.68 Caproic acid, a foul-smelling substance found in certain excretions of male goats during breeding season, has an empirical formula of C_3H_6O. A 0.100-g sample of the acid required 17.2 ml of 0.0500 M NaOH for complete reaction. Assuming that the acid is monoprotic, calculate (a) its molecular weight; (b) its molecular formula.

6.69 If 380 ml of 0.273 M $Ba(OH)_2$ is added to 500 ml of 0.520 M HCl, will the mixture be acidic or basic? Calculate the concentration of H^+ (or OH^- if the solution is basic) in the final mixture. Assume that volumes are additive.

6.70 A 0.249-g sample of a compound containing titanium and chlorine was dissolved in water and treated with silver nitrate solution. The silver chloride that formed was found to weigh 0.694 g after being filtered, washed, and dried. What is the empirical formula of the original compound?

6.71 Mercury is an extremely toxic substance that deactivates enzyme molecules that promote biochemical reactions. A 25.0-g sample of tuna fish taken from a large shipment was analyzed for this substance and found to contain 2.1×10^{-5} mol of Hg. By law, foods having a mercury content above 0.50 ppm cannot be sold (they cannot even be given away!). Determine whether this shipment of tuna must be confiscated.

*6.72 An amino acid isolated from a piece of animal tissue was believed to be glycine, $NH_2CH_2CO_2H$. A 0.0500-g sample was treated in such a way that all the nitrogen in it was converted to ammonia. This NH_3 was added to 50.0 ml of 0.05000 M HCl, neutralizing part of the acid.

$$NH_3 + HCl \longrightarrow NH_4Cl$$

The acid remaining in the solution was titrated with 0.06000 M NaOH, which required 30.57 ml of the base for neutralization.
(a) How many moles of HCl were neutralized by the NH_3?
(b) How many grams of nitrogen were in the 0.0500-g sample?
(c) What was the percent nitrogen in the sample? How does it compare with the percentage nitrogen calculated for glycine?

*6.73 40.0 ml of 0.270 M $Ba(OH)_2$ are added to 25.0 ml of 0.330 M $Al_2(SO_4)_3$.
(a) Write the chemical equation for the reaction that occurs.
(b) What total weight of precipitate is formed?
(c) What is the concentration of each of the ions remaining in solution?

*6.74 A 1.850-g sample of a mixture of $CuCl_2$ and $CuBr_2$ was dissolved in water and mixed thoroughly with a 1.800-g portion of AgCl. After the reaction the solid, which now consisted of a mixture of AgCl and AgBr, was filtered, washed, and dried. Its mass was found to be 2.052 g. What percent of the original mixture was $CuBr_2$?

*6.75 Silver iodide is less soluble than silver bromide, which in turn is less soluble than silver chloride. A 0.2000-g sample that was known to be a mixture of NaCl, NaBr, and NaI was dissolved in water and an excess of $AgNO_3$ was added. The precipitate containing AgCl, AgBr, and AgI was filtered, dried, and found to weigh 0.4120 g. This solid was then placed in water and treated with NaBr solution, which converted any AgCl to AgBr (but it did not affect the AgI). The precipitate that now consisted of AgBr and AgI was filtered, dried, and found to weigh 0.4881 g. It was then placed in water and treated with NaI, which converted the AgBr to AgI. When the solid (now consisting entirely of AgI) was filtered and dried, it weighed 0.5868 g. What were the percentages of NaCl, NaBr, and NaI in the original sample?

6.76 How many equivalents are there in 0.200 mol $Ba(OH)_2$?

6.77 How many moles of H_3PO_4 must be used to give 5.00 eq H_3PO_4 if the acid is to be neutralized to give $PO_4{}^{3-}$?

6.78 How many equivalents are there in 0.140 mol H_3AsO_4 if it is neutralized to give $HAsO_4^{2-}$?

6.79 What is the equivalent weight of $MnSO_4$ when it is oxidized to produce (a) Mn_2O_3, (b) MnO_2, (c) K_2MnO_4, (d) $KMnO_4$?

6.80 How many grams of Na_2CrO_4 are needed to give 0.400 eq Na_2CrO_4 if it will be reduced to Cr^{3+}?

6.81 How many grams of $MnSO_4 \cdot 6H_2O$ are needed to prepare 300 ml of 0.100 N $MnSO_4$ if the Mn^{2+} will be oxidized to MnO_4^-?

6.82 How many grams of $NaBiO_3$ are required to react with 0.500 g of $Mn(NO_3)_2$ to produce $NaMnO_4$ and $Bi(NO_3)_3$?

6.83 What is the equivalent weight of

(a) H_3PO_4 when neutralized to HPO_4^{2-}
(b) $HClO_4$
(c) $NaIO_3$ when reduced to I^-
(d) $NaIO_3$ when reduced to I_2
(e) $Al(OH)_3$

6.84 What is the normality of each of the following solutions?

(a) 22.0 g of $Sr(OH)_2$ in 800 ml of solution
(b) 500 ml of 0.25 M H_2SO_4, for complete neutralization
(c) 0.150 M H_3PO_4, when neutralized to HPO_4^{2-}
(d) 41.7 g of $K_2Cr_2O_7$ in 600 ml of solution when used in a reaction where one product is Cr^{3+}
(e) 25.0 g of Na_2O dissolved in sufficient water to give 1.50 liters of solution
(f) 0.135 equivalents of H_2SO_4 in 400 ml of solution

6.85 A volume of 129 ml of 0.850 N $Ba(OH)_2$ was required to completely neutralize a 4.93-g sample of an acid. What is the equivalent weight of the acid?

*6.86 In acid solution, 45.0 ml of $KMnO_4$ solution is required to react with 50.0 ml of 0.250 N $H_2C_2O_4$ to give Mn^{2+} and CO_2 as products. How many milliliters of this same $KMnO_4$ solution are required to oxidize 25.0 ml of 0.250 N $K_2C_2O_4$ to yield, in basic solution, MnO_2 and CO_2 as products?

*6.87 A 10.0-ml portion of a solution of HCl was diluted to exactly 50.0 ml. If 5.0 ml of this solution required 41.0 ml of 0.255 N NaOH for complete neutralization, what was the concentration of the original HCl solution before dilution?

*6.88 Ascorbic acid (vitamin C) is a diprotic acid having the formula, $H_2C_6H_6O_6$. A sample of a vitamin supplement was analyzed by titrating a 0.1000-g sample dissolved in water with 0.0200 M NaOH. A volume of 15.2 ml of the base was required to completely neutralize the ascorbic acid. What was the percent ascorbic acid in the sample?

*6.89 A mixture of the monoprotic acids, latic acid, $HC_3H_5O_3$ (found in sour milk) and caproic acid, $HC_6H_{11}O_2$, (found in excretions from the goat) was titrated with 0.0500 M NaOH. A 0.1000-g sample of the mixture required 20.4 ml of the base. What is the weight of each acid in the sample?

*6.90 A mixture of $MgCO_3$ and $CaCO_3$ (a dolomitic limestone used in agriculture) was heated to produce MgO and CaO. A 2.000-g sample of this oxide mixture was reacted with 100 ml of 1.00 M HCl. The excess HCl required 19.6 ml of 1.00 M NaOH for complete neutralization. What were the percentages of $CaCO_3$ and $MgCO_3$ in the original limestone sample?

*6.91 A sample of rock containing limestone ($CaCO_3$) was heated, converting the $CaCO_3$ to CaO. This was treated with H_2O to give $Ca(OH)_2$, which was then titrated with HCl. In one analysis, a 0.2000-g sample, taken *after* converting the $CaCO_3$ to CaO as described above, required 30.3 ml of 0.1000 M HCl for complete neutralization. What was the weight percent $CaCO_3$ in the original rock?

*6.92 A 0.1000-g sample of a mixture of $FeSO_4$ and $Fe_2(SO_4)_3$ was dissolved in water and titrated with 0.00400 M $KMnO_4$. The titration required 15.8 ml of the $KMnO_4$ solution. What percent (by weight) of the mixture was $FeSO_4$?

6.93 What volume of 18.0 M H_2SO_4 must be added to 100 ml of H_2O to give a solution of 5.0 M H_2SO_4?

*6.94 How many milliliters of 1.00 M HCl must be added to 50.0 ml of 0.500 M HCl to give a solution whose concentration is 0.600 M?

6.95 What volume of concentrated NH_3 must be used to prepare 250 ml of 0.500 M NH_3?

6.96 What volume of concentrated H_2SO_4 must be used to produce 400 ml of 3.0 M H_2SO_4 solution?

6.97 To what volume must 100 ml of 0.500 M H_2SO_4 be diluted to give 0.200 M H_2SO_4?

6.98 How many milliliters of 0.500 N $K_2Cr_2O_7$ must be used to completely oxidize the contents of 120 ml of 0.850 N $H_2C_2O_4$?

6.99 How much water must be added to 85.0 ml of 1.00 N H_3PO_4 to produce 0.650 N H_3PO_4?

*6.100 To what volume must 250 ml of 1.40 M H_2SO_4 be diluted to give a solution, 25.0 ml of which is able to be completely neutralized by 15.0 ml of 0.750 M NaOH?

7

GASES

We are familiar with many properties of gases because we are surrounded constantly by a gas mixture called the atmosphere. Here we see a hang glider making use of one property of gases: they expand when heated. Rising currents of warm air keep this fellow aloft. In this chapter we will study this and other properties of gases.

Matter is capable of existing in three different physical forms or **states:** solid, liquid, and gas. In this chapter and the next we will examine the physical and chemical characteristics of these states and the transformations that occur among them. This chapter deals with the gaseous state, in which the **intermolecular forces of attraction**—the attractions one molecule experiences toward others around it—are weak. These weak forces allow rapid, independent movement of the molecules and cause the physical behavior of a gas to be nearly independent of its chemical composition. Instead, the behavior of a gas is controlled by its volume, pressure, temperature, and number of moles. Since these variables are of paramount importance, we will begin our discussion by taking a close look at them as they apply to the gaseous state.

7.1 VOLUME AND PRESSURE

When a gas is introduced into a container the molecules move freely within it and occupy the container's entire volume. Consequently, the volume of a gas is given simply by specifying the volume of the vessel in which it is held. Because gases mix freely with one another, when there are several gases in a mixture, the volume of each component is the same as the volume occupied by the entire mixture.

Pressure is defined as force per unit area; it is an intensive quantity formed as a ratio of two extensive quantities: force and area. For example, if a 100-lb force is exerted on a piston whose total area is 100 in.², the pressure acting on each square inch is 100 lb/100 in.² = 1 lb/in.² (1 pound per square inch or 1 psi). A force of only 1 lb pressing on an area of 1 in.² would exert exactly the same pressure because the ratio of force to area is the same.

The <u>unit area</u> in this discussion is 1 in.²

$$\frac{100 \text{ lb}}{100 \text{ in.}^2} = \frac{1 \text{ lb}}{1 \text{ in.}^2}$$

If the 100-lb force is exerted on a smaller area, for example, 1 in.², the pressure is greatly magnified (Figure 7.1). Now the pressure is 100 lb/in.² = 100 psi. The dependence of pressure on both force and the area over which it is spread has been experienced firsthand by anyone who has ever stepped on a nail. A 110-lb person stepping on even a dull nail having a point area of 0.01 in.² will experience a pressure of *11,000 psi!* This is more than enough to cause the nail to puncture the skin.

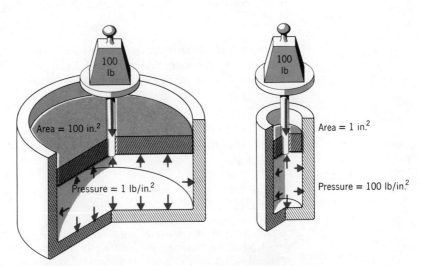

Figure 7.1

Pressure. The fluid in the cylinder exerts the same pressure on all the walls of the container.

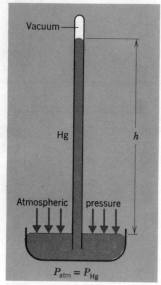

Figure 7.2

A barometer. The atmospheric pressure supports a column of mercury of height h.

If the pressure generated by a piston is applied to a fluid (a gas or liquid), as illustrated in Figure 7.1, it is transmitted uniformly in all directions so that all the walls of the container experience the same pressure. If the piston is supported by the fluid, then the fluid also exerts an equal pressure on the piston as well as the other walls of the container.

The ability of trapped gases to exert a pressure is demonstrated when you inflate an automobile tire. In most cases the four tires that support the car are inflated to a pressure of approximately 28 lb/in.² (psi). The reason a tire becomes flat when it "springs a leak" is because gases flow from a region of high pressure to a region of lower pressure; in our example, this flow is from inside the tire to the atmosphere.

The atmosphere of the earth is a mixture of gases that exert a pressure called the atmospheric pressure. We measure this pressure using a device called a **barometer.** Figure 7.2 shows a barometer constructed by filling a glass tube about 1 meter long with mercury and inverting it (without spilling any) into a dish of mercury so that the open end is submerged. In this figure we see that the mercury in the tube does not completely pour out when it is inverted; instead it maintains a particular height (h) above the reservoir. What keeps the mercury in the tube is the pressure of the atmosphere pushing on the surface of the mercury in the dish. The height of the mercury column is found to be independent of the diameter and length of the glass tube, as long as a space appears over the mercury. This space, for all practical purposes, is a vacuum ($P \approx 0$). The height of the column does change, however, when the atmospheric pressure changes. For example, when a storm approaches, the atmospheric pressure drops and the column becomes shorter. In fact, it is the height of such a mercury column that is reported as the "barometric pressure" in weather forecasts on radio and TV.

To determine atmospheric pressure we compare the various pressures acting along a reference level, which we choose as the surface of the reservoir. At this level outside the inverted tube, the pressure is caused by the downward force of the gases in the atmosphere (P_{atm}). Inside the tube the pressure at the reference level is caused by the downward pull of gravity on the mercury in the column (P_{Hg}). When these two opposing pressures are exactly equal ($P_{Hg} = P_{atm}$), the mercury in the column remains stationary. Atmospheric pressure is directly related to the length (h) of the column of mercury in a barometer and can therefore be expressed in units of centimeters (cm Hg) or, more commonly, millimeters of mercury (mm Hg). A standard unit of pressure, the **standard atmosphere (atm),** was originally defined as the pressure that will support a column of mercury 760 mm in length measured at 0°C and at sea level.[1]

$$1 \text{ atm} = 760 \text{ mm Hg}$$

In English units this corresponds to a pressure of 14.7 lb/in.² Another unit of pressure is the **torr,** named after Evangelista Torricelli, the inventor of the barometer. One torr is equal to the pressure exerted by a column of mercury 1 mm high.

$$1 \text{ torr} = 1 \text{ mm Hg}$$

The SI unit of pressure is the **pascal (Pa),** defined as 1 newton per square meter (1 N/m²). By definition, the standard atmosphere equals exactly 101,325 Pa.

$$1 \text{ atm} = 101,325 \text{ Pa}$$

$$= 101.325 \text{ kPa}$$

The newton (N) is the SI unit of force; the square meter (m²) is the SI unit of area.

[1] The length of the column of mercury, which is supported by atmospheric pressure, varies with both the density of the mercury and the pull of gravity on the mercury in the column. Since density varies with temperature and the pull of gravity varies with altitude, in the definition of the standard atmosphere, it is necessary to specify a reference temperature (0°C) as well as a reference altitude (sea level).

In the chemistry laboratory we find torr and atm to be more convenient units to work with than the pascal. Therefore, for simplicity we will not use this SI unit in our computations.

A barometer similar to that described above could also be constructed using water as a liquid. The length of the column of water would be considerably greater than that of a column of mercury, because the atmospheric pressure would be supporting a less dense liquid ($d_{water} = 1.00$ g/ml, $d_{Hg} = 13.6$ g/ml).

EXAMPLE 7.1	If water were the liquid in a barometer, what would be the length (h) of the water column at 1 atm of pressure?
SOLUTION	The density of Hg is 13.6 times that of H_2O. This means that to have equal masses of mercury and water, the volume of water must be 13.6 times as great as the volume of mercury. Since we are comparing columns of the same diameter, the water column would have to be 13.6 times as long as the Hg column to contain 13.6 times as much volume. Thus

$$1.00 \text{ mm Hg} \sim 13.6 \text{ mm H}_2\text{O}$$

Then

$$1 \text{ atm} \sim 760 \text{ mm Hg} \times \left(\frac{13.6 \text{ mm H}_2\text{O}}{1.00 \text{ mm Hg}} \right) \sim 1.03 \times 10^4 \text{ mm H}_2\text{O}$$

In general, as the density of the liquid being supported in a column by some external pressure increases, the length of the column of liquid decreases.

Often it is desirable to know the pressure of a gas present in a closed system (for example, the pressure of gases produced during a chemical reaction). The instrument normally used for these pressure measurements is called a **manometer.** An open-end manometer (Figure 7.3) is simply a U-shaped tube containing some liquid, such as mercury. One arm of the tube is connected to a system whose pressure is to be measured while the other arm remains open to the atmosphere. When the pressure of the gas inside the system (P_{gas}) is equal to P_{atm}, the level of the liquids in both arms will be the same as shown in Figure 7.3a. If the pressure of the gas is greater than P_{atm}, the mercury in the left arm will be forced downward, causing the mercury in the right arm to rise (Figure 7.3b). We obtain the pressure of the gas in this system by comparing the pressures exerted in both arms at a reference level, h_0, which is chosen to be the height of the shortest column. The pressure exerted on the left column when

Figure 7.3

An open-end manometer. (a) The pressure of the trapped gas is equal to the atmosphere pressure. (b) The gas pressure is greater than atmospheric pressure. (c) The gas pressure is less than atmospheric pressure.

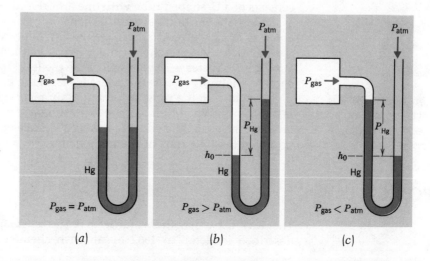

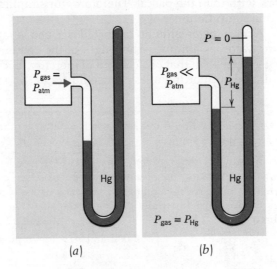

Figure 7.4

A closed-end manometer.

(a)   (b)

$P_{gas} > P_{atm}$ is simply P_{gas}, while at the same level in the right arm the pressure is P_{atm} plus the pressure exerted by the column of mercury that rises above the reference level, P_{Hg}. When the levels are stationary, the pressures at the reference level on both sides are equal and

$$P_{gas} = P_{atm} + P_{Hg}$$

The atmospheric pressure (P_{atm}) is found with a barometer, and P_{Hg} is simply the difference in the heights of the two mercury columns (Δh). Similarly, when $P_{gas} < P_{atm}$, shown in Figure 7.3c, the pressure in the left arm at the reference level is $P_{gas} + P_{Hg}$, while in the right column the pressure is P_{atm}. In this case, when the columns are stationary,

$$P_{gas} + P_{Hg} = P_{atm}$$

so that

$$P_{gas} = P_{atm} - P_{Hg}$$

Therefore, when $P_{gas} < P_{atm}$, the pressure of the gas in the system is found by subtracting the difference in the heights of the columns from atmospheric pressure.

A closed-end manometer is generally used for the measurement of low pressures (usually much smaller than atmospheric pressure). This manometer consists of a U-shaped tube with one arm closed and the other connected to the system as shown in Figure 7.4. When the pressure of the gas in the system is equal to P_{atm}, the right arm is completely filled while the left arm is only partially filled. If the pressure of the gas in the system is reduced, the level in the left arm will increase, which will cause the level in the right arm to decrease, as shown in Figure 7.4b. At the reference level the pressure exerted on the left arm is P_{gas}, while on the right arm the pressure is P_{Hg} (the space above the mercury is a vacuum). When the columns are stationary, $P_{gas} = P_{Hg}$, and the pressure exerted by the gas in the system is simply found as the difference in the heights of liquid in the two arms of the manometer.

A closed-end manometer is convenient because you don't have to measure the atmospheric pressure.

7.2 BOYLE'S LAW

The way in which the volume of a gas is related to its pressure was first described in detail in 1662 by an Irish chemist and physicist, Robert Boyle

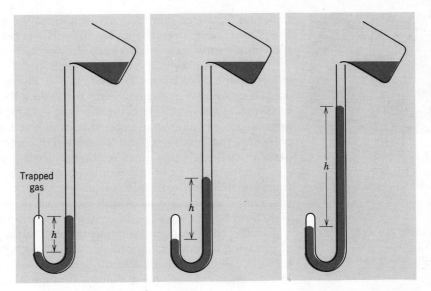

Figure 7.5

Boyle's law apparatus. The pressure on the trapped gas is increased by adding mercury to the U-tube. The volume of the trapped gas decreases as the pressure is raised.

Trapped gas

(1627–1691). Using an apparatus similar to that in Figure 7.5, Boyle found that at constant temperature the volume of a fixed quantity of trapped gas decreases as the pressure on the gas is increased. Anyone who has ever used a bicycle pump is aware of this inverse relationship between the pressure and volume of a gas. As the piston of the pump is forced downward, the gas is compressed into a smaller volume while its pressure is raised (Figure 7.6). If we allow the compressed gas to escape, we can, for instance, use it to inflate a tire.

Boyle repeated his experiments many times with several different gases and found his initial observation to be a universal property of all of them. The results of his experiments can be formulated into **Boyle's law,** which states that *at a constant temperature, the volume occupied by a fixed quantity of gas is inversely proportional to the applied pressure.* This can be expressed mathematically as

$$V \propto \frac{1}{P} \qquad [7.1]$$

The proportionality can be made into an equality by the introduction of a proportionality constant. Thus

$$V = \text{constant} \cdot \frac{1}{P}$$

or

$$PV = \text{constant} \qquad [7.2]$$

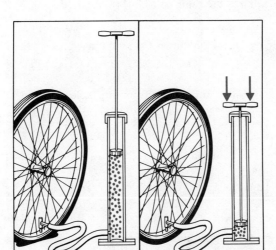

Figure 7.6

Compression of a gas by decreasing its volume using a bicycle pump is an example of Boyle's law.

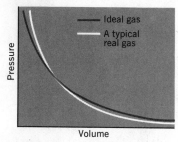

Figure 7.7

Boyle's law. The colored line illustrates how the pressure and volume of an ideal gas are related. The white line shows how a typical real gas behaves. At low pressure the behavior of the real gas approaches that of the ideal gas.

Equation 7.2 states that for a given quantity of a gas at constant temperature, the product of its pressure and volume is a constant. If the pressure increases, the volume must decrease to keep the product of $P \times V$ constant.

The inverse relationship between P and V predicted from Boyle's law is shown graphically by the colored line in Figure 7.7. Real gases, such as hydrogen or oxygen or nitrogen, do not follow this predicted behavior exactly, however, as shown by the white line. At very high pressures the measured volume is always somewhat larger than that calculated from Boyle's law. At low pressures the measured and calculated volumes approach each other—real gases obey Boyle's law quite well when their pressures are low.

A hypothetical gas that would obey Boyle's law perfectly under all conditions is called an **ideal gas.** Real gases approach ideal gas behavior at low pressures but they are said to exhibit nonideal behavior when their properties deviate from Boyle's law. The degree of nonideality differs for different real gases.

Under conditions usually encountered in the laboratory, and to the degree of precision of most of our calculations, gases generally behave ideally; that is, they obey Boyle's law. Equation 7.1, therefore, is useful in calculating the effect of a pressure change on the volume of a gas at constant temperature.

EXAMPLE 7.2

If 100 ml of a gas, originally at 760 torr, is compressed to a pressure of 800 torr at constant temperature, what will its final volume be?

SOLUTION

In problems of this type, which deal with two sets of conditions, it is best to begin by setting up a table containing all the data.

	Initial (i)	Final (f)
Pressure (P)	760 torr	800 torr
Volume (V)	100 ml	?

There are two ways to approach the solution of this problem. One method is to use Equation 7.2 and simply substitute numerical values. The other relies on an *understanding* of how P and V are related to each other to set up the arithmetic. Although they both give the same answers, you can improve your understanding of gas behavior by practicing the second method.

METHOD 1 According to Boyle's law, the PV products in both initial and final states are equal to the same constant; therefore they must be equal to each other.

$$P_i V_i = P_f V_f$$

Solving for the unknown final volume,

$$V_f = V_i \left(\frac{P_i}{P_f} \right) \tag{7.3}$$

We now substitute values from our table.

$$V_f = 100 \text{ ml} \times \left(\frac{760 \text{ torr}}{800 \text{ torr}} \right)$$

and

$$V_f = 95.0 \text{ ml}$$

METHOD 2 We have just seen (Equation 7.3) that V_f is related to V_i by a ratio of pressures,

$$V_f = V_i \times (\text{ratio of pressures})$$

We can use an understanding of Boyle's law to determine what this pressure ratio must be. In this problem, the pressure is increasing from 760 torr to 800 torr. We know that this will cause the volume to decrease, so V_f must be smaller than V_i. This means that we must multiply V_i by a factor smaller than one. Of the two possible pressures ratios, (800 torr/760 torr) and (760 torr/800 torr), only the second satisfies this condition. Therefore we multiply the initial volume, 100 ml, by 760 torr/800 torr.

$$V_f = 100 \text{ ml} \times \left(\frac{760 \text{ torr}}{800 \text{ torr}} \right)$$

$$V_f = 95.0 \text{ ml}$$

7.3 CHARLES' LAW

In 1787, a French chemist named Jacques Alexander Charles studied the effects of temperature changes on the volume of a given quantity of gas held at constant pressure. He found that, if a gas is heated in such a way that the pressure remains constant, the gas expands. If we plot data gathered in such experiments, we obtain a graph similar to that shown in Figure 7.8, where the volume is plotted against temperature in degrees Celsius. Each line represents the results of a series of measurements performed on a different quantity of gas. All real gases ultimately condense to a liquid if they are cooled to a sufficiently low temperature, and the solid portions of the lines in Figure 7.8 represent the temperature region above the condensation (liquefaction) point, where the substance is in the gaseous state. When these straight lines are extended back (extrapolated), they all intersect at the same point corresponding to zero volume and a temperature of −273.15°C. What is particularly significant is that this same behavior is observed for *any* gas! When the volume-temperature graph is extrapolated to $V = 0$, the temperature is always −273.15°C. This point represents a temperature at which all gases, if they did not condense, would have a volume of zero, and below which they would have a negative volume. Since negative volumes are impossible, this temperature must represent the lowest temperature possible and is called **absolute zero.**

Absolute zero represents the zero point on the kelvin temperature scale, as we saw in Chapter 1. The kelvin scale is also called the **absolute temperature scale,** and to obtain the kelvin temperature we add 273.15 to the Celsius temperature,

$$T \text{ (K)} = t(°C) + 273.15$$

For most purposes, we will use only three significant figures, so 0°C = 273 K

When you encounter a capitol *T* in an equation, it almost always stands for the absolute temperature.

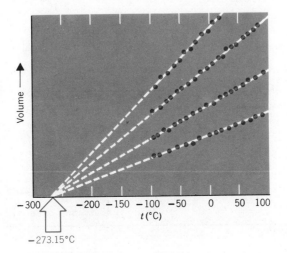

Figure 7.8

Charles' law plot of V *versus* t(°C).

and we simply use the approximate relationship,

$$T\,(K) = t(°C) + 273$$

The absolute temperature scale is *always* used when the temperature enters numerically into a computation involving pressures and volumes of gases.

The straight lines in Figure 7.8 suggest that at constant pressure the volume of a gas is directly proportional to its temperature, providing the temperature is expressed in the proper units. The relationship that exists between the volume of a gas and its temperature is summarized by **Charles' law:** *at constant pressure, the volume of a given quantity of a gas is directly proportional to its absolute temperature.* Writing Charles' law mathematically, we have

$$V \propto T \qquad\qquad [7.4]$$

Making the proportionality an equality and rearranging, we obtain

$$\frac{V}{T} = \text{constant} \qquad\qquad [7.5]$$

If Charles' law were strictly obeyed, gases would not condense when they are cooled. Therefore, condensation is considered nonideal behavior, and all real gases behave more and more nonideally as their condensation temperatures are approached. This means that *gases behave in an ideal fashion only at relatively high temperatures and low pressures.* An example of the application of Charles' law is shown below.

EXAMPLE 7.3

A sample of a gas occupies 250 ml at 27°C. What volume will it occupy at 35°C if there is no change in pressure?

SOLUTION

Once again we should set up our table of data.

	Initial (*i*)	Final (*f*)
Volume (*V*)	250 ml	?
Temperature (*T*)	27 + 273 = 300 K	35 + 273 = 308 K

Note that we have converted to absolute temperatures.

Equation 7.5 implies that

$$\frac{V_i}{T_i} = \frac{V_f}{T_f}$$

If we solve for V_f, we find that the initial and final volumes are related as

$$V_f = V_i \times (\text{ratio of absolute temperatures})$$

Without actually solving the equation, we can obtain the correct temperature ratio through reasoning. Gases expand when heated. Since the temperature is increasing from 300 K to 308 K, the final volume must be larger than the initial volume. The temperature ratio must be a fraction with a value greater than one. This requires the larger temperature in the numerator. Therefore,

$$V_f = 250 \text{ ml} \times \left(\frac{308\ \cancel{K}}{300\ \cancel{K}}\right)$$

$$V_f = 257 \text{ ml}$$

7.4 GAY-LUSSAC'S LAW

If you read the label on an aerosol can, you will find a warning such as "Do not incinerate" or "Do not store above 120°F." These warnings are necessary because the pressure of a confined gas rises when it is heated. If the pressure becomes too large, the container can burst and cause serious injury. Joseph Gay-Lussac, a contemporary of Jacques Charles, was the first to discover that the pressure of a fixed amount of gas is directly proportional to its absolute temperature if the volume of the gas is held constant,

$$P \propto T$$

or

$$\frac{P}{T} = \text{constant} \qquad [7.6]$$

This is a mathematical statement of the **law of Gay-Lussac.**

EXAMPLE 7.4   What would be the pressure of a gas, originally at 760 torr, if the temperature is lowered from 35°C to 25°C at a constant volume?

SOLUTION   Equation 7.6 implies that

$$\frac{P_i}{T_i} = \frac{P_f}{T_f}$$

We should now be able to write that

$$P_f = P_i \times (\text{ratio of temperatures})$$

Tabulating the data,

	(i)	(f)
P	760 torr	?
T	308 K	298 K

Since the temperature is decreasing, the pressure will decrease. This requires the temperature ratio to be smaller than one.

$$P_f = 760 \text{ torr} \times \left(\frac{298 \text{ K}}{308 \text{ K}}\right)$$

$$P_f = 735 \text{ torr}$$

7.5 THE COMBINED GAS LAW

The equations corresponding to Boyle's law, Charles' law, and Gay-Lussac's law can be incorporated into one single equation that is useful for many computations. This is

$$\frac{P_i V_i}{T_i} = \frac{P_f V_f}{T_f} \qquad [7.7]$$

Notice that if $T_i = T_f$, the temperature may be dropped and Equation 7.7 reduces to a statement of Boyle's law (that is, $P_i V_i = P_f V_f$ at constant temperature). Similarly, if $P_i = P_f$, the equation reduces to Charles' law and if $V_i = V_f$, the equation reduces to Gay-Lussac's law. As in each of the separate laws, the combined gas law holds only if the amount of gas is not changed.

When working with gases it is useful to define a reference set of conditions of temperature and pressure. These conditions, known as **standard temperature and pressure,** or simply **STP,** are 0°C (273 K) and 1 atm (760 torr).

EXAMPLE 7.5   What would be the volume of a gas at STP if it was found to occupy a volume of 255 ml at 25°C and 650 torr?

SOLUTION   In this problem both temperature and pressure are changing. To compute the final volume we must combine Boyle's law and Charles' law. Equation 7.7 does this,

$$\frac{P_i V_i}{T_i} = \frac{P_f V_f}{T_f}$$

If we solve this equation for V_f, we find

$$V_f = V_i \times (\text{pressure ratio}) \times (\text{temperature ratio})$$

Once again we can use reasoning to set up these ratios. First let's tabulate the data.

	Initial (*i*)	Final (*f*)	
V	255 ml	?	
P	650 torr	760 torr	STP
T	298 K	273 K	

There is a pressure increase, which should tend to cause the volume of the gas to decrease. The pressure ratio should therefore be smaller than one, which requires the larger pressure in the denominator.

$$V_f = 255 \text{ ml} \times \left(\frac{650 \text{ torr}}{760 \text{ torr}}\right) \times (\text{temperature ratio})$$

The temperature is decreasing. According to Charles' law this should cause a further decrease in the volume. The temperature ratio must also be smaller than one and the larger temperature must be in the denominator.

$$V_f = 255 \text{ ml} \times \left(\frac{650 \text{ torr}}{760 \text{ torr}}\right) \times \left(\frac{273 \text{ K}}{298 \text{ K}}\right)$$

$$V_f = 200 \text{ ml}$$

The volume at STP is 200 ml.

EXAMPLE 7.6   A sample of a gas exerts a pressure of 625 torr in a 300-ml container at 25°C. What pressure would the same gas sample exert in a 500-ml container at 50°C.

SOLUTION   As usual, begin by tabulating the data.

	Initial (*i*)	Final (*f*)
P	625 torr	?
V	300 ml	500 ml
T	298 K	323 K

This time the relationship is

$$P_f = P_i \times (\text{volume ratio}) \times (\text{temperature ratio})$$

The volume is increasing from 300 ml to 500 ml, which should favor a pressure decrease. Therefore the volume ratio must be smaller than one.

$$P_f = 625 \text{ torr} \times \left(\frac{300 \text{ ml}}{500 \text{ ml}}\right) \times (\text{temperature ratio})$$

The temperature is increasing from 298 K to 323 K. This favors a pressure increase, so the temperature ratio must be larger than one.

$$P_f = 625 \text{ torr} \times \left(\frac{300 \text{ ml}}{500 \text{ ml}}\right) \times \left(\frac{323 \text{ K}}{298 \text{ K}}\right)$$

$$P_f = 406 \text{ torr}$$

7.6 DALTON'S LAW OF PARTIAL PRESSURES

When two or more gases that do not react chemically are placed in the same container, the pressure exerted by each gas in the mixture is the same as it would be if it were the only gas in the container. The pressure exerted by each gas in a mixture is called its **partial pressure** and, as observed by John Dalton, the total pressure is equal to the sum of the partial pressures of each gas in the mixture. This statement, known as **Dalton's law of partial pressures,** can be expressed as

$$P_T = p_a + p_b + p_c + \cdots$$

We can't actually measure the individual partial pressures; we can only measure the total pressure.

where P_T is the total pressure of the mixture (which could be measured with a manometer) and p_a, p_b, and p_c are the partial pressures of gases a, b, and c, respectively. For example, if nitrogen, oxygen, and carbon dioxide were placed in the same vessel, the total pressure of the mixture would be

$$P_T = p_{N_2} + p_{O_2} + p_{CO_2}$$

Thus, if the partial pressure of nitrogen were 200 torr, that of oxygen 250 torr, and that of carbon dioxide 300 torr, the total pressure of the mixture would be

$$P_T = 200 \text{ torr} + 250 \text{ torr} + 300 \text{ torr}$$

$$P_T = 750 \text{ torr}$$

Dalton's law can be useful in determining the pressure resulting from the mixing of two gases that were originally in separate containers, as shown by the following example.

EXAMPLE 7.7   If 200 ml of N_2 at 25°C and a pressure of 250 torr are mixed with 350 ml of O_2 at 25°C and a pressure of 300 torr, so that the resulting volume is 300 ml, what would be the final pressure of the mixture at 25°C?

SOLUTION   From Dalton's law we know that we can treat each gas in the mixture as if it were the only gas present. Therefore, we can calculate *independently* the new pressures of N_2 and O_2 when they are placed in the 300-ml container. Since there is no temperature change, we have simply a Boyle's law calculation for each gas.

Following the methods used in Examples 6.2 through 6.4, we first set up our tables of data.

For N_2	(i)	(f)		For O_2	(i)	(f)
p	250 torr	?		p	300 torr	?
V	200 ml	300 ml		V	350 ml	300 ml

For each calculation we can write

$$p_f = p_i \times (\text{ratio of volumes})$$

Since the volume of the N_2 is increasing, its pressure must decrease; p_f must be less than p_i. This requires a volume ratio smaller than one, which means that the larger volume must be in the denominator. Thus

$$p_{N_2} = 250 \text{ torr} \times \left(\frac{200 \text{ ml}}{300 \text{ ml}}\right)$$

$$p_{N_2} = 167 \text{ torr}$$

For O_2 the volume is decreasing; p_f must be greater than p_i. This requires a volume ratio larger than one.

$$p_{O_2} = 300 \text{ torr} \times \left(\frac{350 \text{ ml}}{300 \text{ ml}}\right)$$

$$p_{O_2} = 350 \text{ torr}$$

The total pressure of the mixture is the sum of the partial pressures.

$$P_T = p_{N_2} + p_{O_2} = 167 \text{ torr} + 350 \text{ torr}$$

$$P_T = 517 \text{ torr}$$

Gases prepared in the laboratory are quite often collected by the displacement of water, as shown in Figure 7.9. A gas collected in this manner becomes "contaminated" with water molecules that evaporate into the gas. These water molecules also exert a pressure called the **vapor pressure.** For reasons that we will discuss in Chapter 8, the vapor pressure of water depends *only* on the temperature of the liquid water (Table 7.1). The pressure of the water vapor contributes to the total pressure of the "wet" gas, so we can write

$$P_T = p_{gas} + p_{H_2O}$$

If the level of the water is the same inside the collection flask as outside, as shown in Figure 7.9, then the pressure inside must also be the same as outside, namely, atmospheric pressure. The atmospheric pressure can be determined with a barometer, and the vapor pressure of water can be obtained from Table 7.1 if the temperature of the liquid is known. The partial pressure of the pure gas is therefore

$$p_{gas} = P_T - p_{H_2O}$$

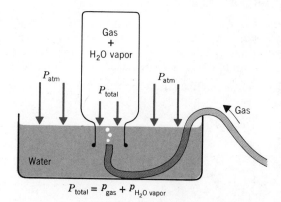

Figure 7.9

Collection of a gas by displacement of water.

Table 7.1
Vapor pressure of water
as a function of temperature

Temp. (°C)	Press. (torr)	Temp. (°C)	Press. (torr)	Temp. (°C)	Press. (torr)
0	4.6	18	15.5	40	55.3
1	4.9	19	16.5	45	71.9
2	5.3	20	17.5	50	92.5
3	5.7	21	18.7	55	118.0
4	6.1	22	19.8	60	149.4
5	6.5	23	21.1	65	187.5
6	7.0	24	22.4	70	233.7
7	7.5	25	23.8	75	289.1
8	8.0	26	25.2	80	355.1
9	8.6	27	26.7	85	433.6
10	9.2	28	28.3	90	525.8
11	9.8	29	30.0	95	634.1
12	10.5	30	31.8	96	657.6
13	11.2	31	33.7	97	682.1
14	12.0	32	35.7	98	707.3
15	12.8	33	37.7	99	733.2
16	13.6	34	39.9	100	760.0
17	14.5	35	42.2	101	787.6

EXAMPLE 7.8 A student generates oxygen gas in the laboratory and collects it in a manner similar to that shown in Figure 7.9. She collects the gas at 25°C until the levels of the water inside and outside the flask are equal. If the volume of the gas is 245 ml and the atmospheric pressure is 758 torr:

(a) What is the partial pressure of O_2 gas in the "wet" gas mixture at 25°C?
(b) What would be the volume of dry oxygen at STP?

SOLUTION (a) For gases collected in this manner we know that

$$P_T = p_{gas} + p_{H_2O}$$

Substituting p_{O_2} for p_{gas} and rearranging, we have

$$p_{O_2} = P_T - p_{H_2O}$$

According to Table 7.1, the partial pressure of water at 25°C is 23.8 torr. The atmospheric pressure was given as 758 torr. Therefore, the partial pressure of O_2 is

$$p_{O_2} = 758 \text{ torr} - 23.8 \text{ torr}$$

$$p_{O_2} = 734 \text{ torr} \qquad \text{(rounded to the correct number of significant figures)}$$

This is the pressure exerted by the oxygen alone.

(b) This part of the problem is a combined Boyle's law—Charles' law calculation.

First, tabulate the data as follows:

	(i)	(f)	
V	245 ml	?	
P	734 torr	760 torr	} STP
T	298 K	273 K	

Next, recall that

$$V_f = V_i \times \text{(ratio of pressures)} \times \text{(ratio of temperatures)}$$

The pressure change should tend to decrease the volume; therefore, the pressure ratio should be less than one. The temperature decrease should also decrease the volume; therefore, the temperature ratio should be less than one. Thus

$$V_f = 245 \text{ ml} \times \left(\frac{734 \text{ torr}}{760 \text{ torr}}\right) \times \left(\frac{273 \text{ K}}{298 \text{ K}}\right)$$

$$V_f = 217 \text{ ml at STP}$$

7.7 CHEMICAL REACTIONS BETWEEN GASES

Many gases are able to undergo chemical reactions with each other. Hydrogen and oxygen can combine to form water. Nitrogen and hydrogen, under appropriate conditions, combine to form ammonia. If these reactions are carried out so that the volumes of the gaseous reactants and products are measured at the same temperatures and pressures, the volumes are related in a simple fashion. For example, when hydrogen and oxygen are placed in a vessel and allowed to react with each other to form gaseous water, two volumes of hydrogen always react with one volume of oxygen to form two volumes of gaseous water. This can be expressed in the form of an equation as

2 volumes hydrogen + 1 volume oxygen $\longrightarrow$ 2 volumes gaseous water

Similarly, when one volume of hydrogen reacts with one volume of chlorine, two volumes of hydrogen chloride gas are produced; that is,

1 volume hydrogen + 1 volume chlorine $\longrightarrow$ 2 volumes hydrogen chloride

For the reaction between hydrogen and nitrogen to form ammonia, we find this relationship.

1 volume nitrogen + 3 volumes hydrogen $\longrightarrow$ 2 volumes ammonia

Experimental observations such as these form the basis of **Gay-Lussac's law of combining volumes,** which states that *the volumes of gaseous substances that are consumed and produced in a chemical reaction are in ratios of small whole numbers, provided the volumes are measured under the same conditions of temperature and pressure.*

The significance of Gay-Lussac's observation was later recognized by Amadeo Avogadro. He proposed what is now known as **Avogadro's principle:** *Under conditions of constant temperature and pressure, equal volumes of gas contain equal numbers of molecules.* Since equal numbers of molecules mean equal numbers of moles, the number of moles of any gas is related directly to its volume:

$$V \propto n \tag{7.8}$$

Table 7.2
Molar volumes of several real gases at STP

Substance	Molar Volume (liters)
Oxygen, O_2	22.397
Nitrogen, N_2	22.402
Hydrogen, H_2	22.433
Helium, He	22.434
Argon, Ar	22.397
Carbon dioxide, CO_2	22.260
Ammonia, NH_3	22.079

where n is the number of moles of gas. On this basis Gay-Lussac's law is easily understood because the volumes of gaseous reactants and products occur in the same ratios as the coefficients in the balanced equation.[2] For example,

$$2H_2\ (g) + O_2\ (g) \longrightarrow 2H_2O\ (g)$$

$$2 \text{ volumes} + 1 \text{ volume} \longrightarrow 2 \text{ volumes}$$

From Avogadro's principle we expect 1 mol of any gas to occupy the same volume at a given temperature and pressure. It has been found by experiment that the average volume occupied by 1 mol of a gas at STP is 22.4 liters. We will assume this is the **molar volume** of an ideal gas at STP. For real gases the molar volume actually fluctuates about this average, as shown in Table 7.2.

EXAMPLE 7.9 What volume of O_2, at STP, is required for the complete combustion of 4.50 liters of butane, C_4H_{10}, at STP? Butane is the fuel in disposable cigarette lighters.

SOLUTION As with any problem in stoichiometry, we should first write a balanced equation for the reaction. This is

$$2C_4H_{10} + 13O_2 \longrightarrow 8CO_2 + 10H_2O$$

To solve the problem we can compute the number of moles of C_4H_{10} using the molar volume at STP.

$$4.50 \text{ liters } C_4H_{10} \times \left(\frac{1 \text{ mol } C_4H_{10}}{22.4 \text{ liters } C_4H_{10}}\right) \sim 0.201 \text{ mol } C_4H_{10}$$

Next we calculate the number of moles of O_2 required, using the coefficients in the equation,

$$0.201 \text{ mol } C_4H_{10} \times \left(\frac{13 \text{ mol } O_2}{2 \text{ mol } C_4H_{10}}\right) \sim 1.31 \text{ mol } O_2$$

Finally we can calculate the volume of O_2, again using the molar volume of a gas at STP.

$$1.31 \text{ mol } O_2 \times \left(\frac{22.4 \text{ liters } O_2}{1 \text{ mol } O_2}\right) \sim 29.3 \text{ liters } O_2$$

[2] In fact, these observations were used to show that gases such as H_2, O_2, and Cl_2 must be *at least* diatomic and that their subscripts must be even. The only way *two* volumes of hydrogen chloride could be formed from *one* volume of hydrogen and *one* volume of chlorine is if each molecule of hydrogen and chlorine contained two atoms of H and two atoms of Cl, respectively, and hydrogen chloride was HCl. If hydrogen chloride was H_2Cl_2, then each molecule of hydrogen and chlorine would have to be composed of four atoms, and so forth.

The volume of O_2 required is 29.3 liters.

In problems involving gaseous reactants or products at the *same temperature and pressure*, we can take a shortcut to the answer. If we set up the calculation above with all the conversion factors strung together, we have

$$4.50 \text{ liters } C_4H_{10} \times \left(\frac{1 \text{ mol } C_4H_{10}}{22.4 \text{ liters } C_4H_{10}}\right) \times \left(\frac{13 \text{ mol } O_2}{2 \text{ mol } C_4H_{10}}\right) \times \left(\frac{22.4 \text{ liters } O_2}{1 \text{ mol } O_2}\right) \sim 29.3 \text{ liters } O_2$$

The number 22.4 appears in both numerator and denominator and therefore cancels. The volumes of reactants (or products) are simply related by the coefficients in the equation. Thus in this problem we could state that

$$2 \text{ liters } C_4H_{10} \sim 13 \text{ liters } O_2$$

This is a consequence of Gay-Lussac's law. Realizing this, the solution to the problem could have been obtained as

Remember, this only works if the volumes are compared at the same T and P.

$$4.50 \text{ liters } C_4H_{10} \times \left(\frac{13 \text{ liters } O_2}{2 \text{ liters } C_4H_{10}}\right) \sim 29.3 \text{ liters } O_2$$

EXAMPLE 7.10   The drain cleaner, Drano, contains small bits of aluminum, which react with NaOH (the main ingredient in this product) to produce bubbles of hydrogen. These bubbles presumably are designed to stir the mixture and hasten its action. How many milliliters of H_2, measured at STP, will be released when 0.150 g of Al are dissolved? The chemical equation is

$$2Al + 2OH^- + 2H_2O \longrightarrow 3H_2 + 2AlO_2^-$$

SOLUTION   First we calculate the number of moles of Al that react.

$$0.150 \text{ g Al} \times \left(\frac{1 \text{ mol Al}}{27.0 \text{ g Al}}\right) \sim 0.00556 \text{ mol Al}$$

Next we calculate the number of moles of H_2 produced.

$$0.00556 \text{ mol Al} \times \left(\frac{3 \text{ mol } H_2}{2 \text{ mol Al}}\right) \sim 0.00834 \text{ mol } H_2$$

Since 1 mol $H_2 \sim 22.4$ liters H_2 at STP,

$$0.00834 \text{ mol } H_2 \times \left(\frac{22.4 \text{ liters } H_2}{1 \text{ mol } H_2}\right) \sim 0.187 \text{ liter } H_2$$

Expressed in milliliters, the answer is 187 ml H_2.

7.8   THE IDEAL GAS LAW

We have thus far discussed three volume relationships that an ideal gas obeys. These are

$$\text{Boyle's law} \quad V \propto \frac{1}{P}$$

$$\text{Charles' law} \quad V \propto T$$

$$\text{Avogadro's law} \quad V \propto n$$

We can combine these to obtain

$$V \propto n \left(\frac{1}{P}\right) (T)$$

or

$$V \propto \frac{nT}{P} \qquad [7.9]$$

Equation 7.9 reduces to an expression of Boyle's law during a process where the volume changes as a result of a pressure change only. Since n and T remain constant, volume is proportional only to pressure; that is,

$$V \propto \frac{1}{P} \qquad \text{at constant } n \text{ and } T$$

Similarly, we can see that Equation 7.9 becomes Charles' law for a volume change at constant n and P.

$$V \propto T \qquad \text{at constant } n \text{ and } P$$

Avogadro's principle is seen as

$$V \propto n \qquad \text{at constant } T \text{ and } P$$

The proportionality in Equation 7.9 is made into an equality by the introduction of a proportionality constant, R, called the **universal gas constant.** Equation 7.9 then becomes

$$V = \frac{nRT}{P}$$

or, as it is usually written,

$$PV = nRT \qquad [7.10]$$

Equation 7.10 is obeyed exactly only by the hypothetical ideal gas and is a mathematical statement of the **ideal gas law.** It is also called the **equation of state for an ideal gas** because it relates those variables (P, V, n, T) that specify the physical properties of the gas. If any three of these are given, the fourth variable can have only one value as determined by Equation 7.10.

When a real gas comes very close to obeying the ideal gas law its behavior is said to be ideal. Fortunately, most real gases are nearly ideal under conditions of temperature and pressure normally encountered in the laboratory, so the ideal gas law can be used quite accurately to describe their behavior. The only time Equation 7.10 should not be used is when it is necessary to perform very accurate computations.

To use the ideal gas law we must have a value for the gas constant, R. This can be computed by inserting appropriate values for P, V, n, and T into Equation 7.10 and solving for R. For 1 mol of an ideal gas at STP, $P = 1$ atm, $V = 22.4$ liters, $n = 1$ mol, and $T = 273$ K. Solving for R,

$$R = \frac{PV}{nT}$$

$$R = \frac{(1 \text{ atm})(22.4 \text{ liters})}{(1 \text{ mol})(273 \text{ K})}$$

$$R = 0.0821 \frac{\text{liter atm}}{\text{mol K}}$$

or

$$R = 0.0821 \text{ liter atm mol}^{-1} \text{ K}^{-1}$$

The constant R may have other numerical values depending on the units used to express pressure and volume. Example 7.11 illustrates how we can convert R from one set of units to another. The most useful values of R, with their corresponding units, are included on the inside back cover of this book.

EXAMPLE 7.11   What is the value of R when pressure is expressed in torr and volume is expressed in milliliters?

SOLUTION   This is simply a unit-conversion problem making use of the relationships

$$1 \text{ liter} = 1000 \text{ ml}$$

$$1 \text{ atm} = 760 \text{ torr}$$

$$R = \left(\frac{0.0821 \text{ liter atm}}{\text{mol K}} \right) \times \left(\frac{1000 \text{ ml}}{1 \text{ liter}} \right) \times \left(\frac{760 \text{ ml}}{1 \text{ atm}} \right)$$

$$= \frac{6.24 \times 10^{-4} \text{ ml torr}}{\text{mol K}}$$

When you use **R** in a computation, be sure the units cancel correctly.

The choice of the value of R to be used in a given computation is governed by the units of P and V. Most people find it best, however, to learn one value of R and to convert P and V to units that can be used with that value of R.

The following are some examples of the application of the ideal gas law.

EXAMPLE 7.12   What volume will 25.0 g of O_2 occupy at 20°C and a pressure of 0.880 atm?

SOLUTION   From the ideal gas law,

$$V = \frac{nRT}{P}$$

We will use $R = 0.0821$ liter atm mol^{-1} K^{-1}. Tabulating our data,

P	0.880 atm
V	?
n	25.0 g of $O_2 \times \dfrac{1 \text{ mol of } O_2}{32.0 \text{ g of } O_2} = 0.781$ mol
T	20 + 273 = 293 K

Substituting,

$$V = \frac{(0.781 \text{ mol}) \times (0.0821 \text{ liter atm mol}^{-1} \text{ K}^{-1}) \times (293 \text{ K})}{(0.880 \text{ atm})}$$

$$V = 21.3 \text{ liters}$$

EXAMPLE 7.13   A student collected natural gas from a laboratory gas jet at 25°C in a 250-ml flask until the pressure of the gas was 500 torr. The gas sample weighed 0.118 g at a temperature of 25°C. From these data, calculate the molecular weight of the gas.

SOLUTION   Determining molecular weights is one of the most useful applications of the ideal gas law. To do this, we need to know the number of moles of gas in the sample. We can determine the number of moles of gas present using the ideal gas law,

$$n = \frac{PV}{RT}$$

Again we use $R = 0.0821$ liter atm mol^{-1} K^{-1}. Our data are

P	$550 \text{ torr} \times \dfrac{1 \text{ atm}}{760 \text{ torr}} = 0.724 \text{ atm}$
V	$250 \text{ ml} \times \dfrac{1 \text{ liter}}{1000 \text{ ml}} = 0.250 \text{ liter}$
n	?
T	$25 + 273 = 298 \text{ K}$

Substituting,

$$n = \frac{(0.724 \text{ atm}) \times (0.250 \text{ liter})}{(0.0821 \text{ liter atm mol}^{-1} \text{ K}^{-1}) \times (298 \text{ K})}$$

$$n = \frac{0.00740}{\text{mol}^{-1}} = 0.00740 \text{ mol}$$

To calculate the molecular weight we must determine the weight of 1 mol of the substance. We now know that

$$0.118 \text{ g} = 0.00740 \text{ mol}$$

The weight of 1 mol, therefore, is $\;1 \text{ mol} \times \left(\dfrac{0.118 \text{ g}}{0.00740 \text{ mol}} \right) = 15.9 \text{ g}$

Thus the molecular weight is 15.9. Natural gas is really methane, CH_4, having a molecular weight of 16.0.

EXAMPLE 7.14   A student measured the density of a gas to be 1.34 g/liter at 25°C and 760 torr, and was told that the gas was composed of 79.8% carbon and 20.2% hydrogen.

(a) What is the empirical formula of the compound?
(b) What is its molecular weight?
(c) What is the molecular formula of the compound?

SOLUTION   (a) Following the procedure outlined in Section 2.5, we find the empirical formula of the carbon–hydrogen compound.

$$79.8 \text{ g C} \times \left(\frac{1 \text{ mol C}}{12.0 \text{ g C}} \right) = 6.65 \text{ mol C}$$

$$20.2 \text{ g H} \times \left(\frac{\text{mol H}}{1.01 \text{ g H}} \right) = 20.0 \text{ mol H}$$

The empirical formula is $C_{\frac{6.65}{6.65}} H_{\frac{20.0}{6.65}}$ or CH_3 which would give an empirical formula weight of $CH_3 = 15.0$.

(b) The density gives the weight of 1 liter of the gas. $1.00 \text{ liter} \sim 1.34 \text{ g}$

To calculate molecular weight we need a relationship between mass and moles. Following the procedure in Example 7.13,

P	$760 \text{ torr} = 1 \text{ atm}$
V	1.00 liter
n	?
T	$25 + 273 = 298 \text{ K}$

$$n = \frac{PV}{RT} = \frac{(1 \text{ atm}) \times (1.00 \text{ liter})}{(0.0821 \text{ liter atm mol}^{-1} \text{ K}^{-1}) \times (298 \text{ K})}$$

$$n = 0.0409 \text{ mol}$$

Thus

$$0.0409 \text{ mol} \sim 1.34 \text{ g}$$

The weight of 1 mol is

$$1 \text{ mol} \times \left(\frac{1.34 \text{ g}}{0.0409 \text{ mol}} \right) = 32.8 \text{ g}$$

The molecular weight is 32.8

(c) We see that the molecular weight is approximately twice the empirical formula weight, which means that the molecular formula must be

$$(CH_3)_2 \quad \text{or} \quad C_2H_6$$

This is a substance called ethane.

An equation relating the molecular weight of a gas to its density directly can easily be derived from the ideal gas law and used to solve problems such as part (b) of this last example. We know, for example, that the number of moles of a substance is obtained by dividing its mass, in grams, by the molecular weight.

$$\text{number of moles } (n) = \frac{\text{number of grams } (g)}{\text{molecular weight } (M)}$$

or simply,

$$n = \frac{g}{M} \tag{7.11}$$

Substituting for n in the ideal gas law gives

$$PV = \frac{g}{M} RT \tag{7.12}$$

which can be rearranged to solve for M.

$$M = \frac{g}{V} \frac{RT}{P} \tag{7.13}$$

Density (d) is a ratio of grams to volume, g/V. This allows us to write Equation 7.13 as

$$M = d \frac{RT}{P} \tag{7.14}$$

To solve part (b) of Example 7.14 we can substitute the given values for d, R, T, and P into this equation.

$$M = \frac{1.34 \text{ g}}{\text{liter}} \frac{(0.0821 \text{ liter atm mol}^{-1} K^{-1})(298 \text{ K})}{(1 \text{ atm})}$$

$$M = 32.8 \text{ g/mol}$$

7.9 GRAHAM'S LAW OF EFFUSION

The ability of a gas to mix spontaneously with and spread throughout another gas, a process known as **diffusion,** is demonstrated every time we drive near a skunk that didn't quite make it across the road. It isn't long before the occupants of the car are in need of "a little fresh air." **Effusion,** on the other hand, is the process by which a gas, under pressure, escapes from one chamber of a vessel to another by passing through a very small opening. This process is illustrated in Figure 7.10.

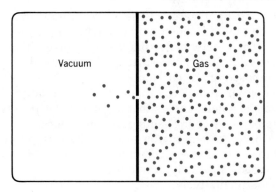

Figure 7.10

Effusion of a gas into a vacuum.

Thomas Graham demonstrated that when the rates of effusion of several gases are compared, the less dense gases (lighter gases) always effuse faster than the more dense ones. When the rates are compared under identical conditions of temperature and pressure, the best agreement between the rate of effusion and density is obtained when the rate is expressed as being inversely proportional to the square root of the density of the gas. This statement, known as **Graham's law,** can be expressed mathematically as

$$\text{rate of effusion} \propto \sqrt{\frac{1}{d}}$$

The rate of effusion of two gases (labeled simply A and B) can be compared by dividing the rate of one by the other; that is,

$$\frac{\text{rate of effusion }(A)}{\text{rate of effusion }(B)} = \sqrt{\frac{d_B}{d_A}}$$

Looking back at Equation 7.14, we see that the density of a gas is directly proportional to its molecular weight. This allows us to rewrite the ratio of effusion rates as

Graham's law works for diffusion, too.

$$\frac{\text{rate of effusion }(A)}{\text{rate of effusion }(B)} = \sqrt{\frac{d_B}{d_A}} = \sqrt{\frac{M_B}{M_A}} \qquad [7.15]$$

where M_A and M_B are the molecular weights of gases A and B, respectively.

EXAMPLE 7.15   Which gas will effuse faster, ammonia or carbon dioxide? What are their relative rates of effusion?

SOLUTION   The molecular weight of CO_2 is 44 and that of NH_3 is 17. Therefore NH_3 will effuse faster. We can calculate how much faster with Equation 7.15.

$$\frac{\text{rate of effusion }(NH_3)}{\text{rate of effusion }(CO_2)} = \sqrt{\frac{M_{CO_2}}{M_{NH_3}}} = \sqrt{\frac{44}{17}} = 1.6$$

Therefore the rate of effusion of NH_3 is 1.6 times faster than the rate of CO_2.

7.10   KINETIC MOLECULAR THEORY AND THE GAS LAWS

We have seen that the study of gases produced a variety of different gas laws. The people who discovered these laws could not help wondering what gases must be composed of to behave as they do. What is the origin of the pressure of a gas? Why are gases so compressible? Why do they expand when heated? And why do light gases effuse more rapidly than heavy ones? These questions, among others, illustrate the need for a theoretical model of a gas.

The **kinetic molecular theory** evolved in an attempt to explain the physical behavior of substances. According to this theory, any given sample of matter is composed of a huge number of small particles (molecules or individual atoms) that are in constant, random motion. In addition, the theory proposes that the average kinetic energy that these particles possess is proportional to the absolute temperature of the sample. As we will see, these two basic postulates apply to all three states of matter—solids, liquids, and gases. What we want to do now is to see how they explain the properties of a gas. Although the postulates can be used to actually derive the ideal gas law mathematically, we will only use them qualitatively.

The pressure-volume relationship: Boyle's law

The most striking quality of a gas is its compressibility. The molecules envisioned in the kinetic molecular theory must therefore be very tiny and very far apart in a gas so that there is plenty of empty space between them. Only in this way could they be so easily crowded together. As these tiny particles fly about, they collide with each other and with the walls of the container. Each impact with the wall exerts a tiny push, and the cumulative effects of enormous numbers of such impacts each second on each square centimeter of the wall gives rise to the pressure of the gas.

With this model of a gas in mind, we can now explain Boyle's law. If we halve the volume of a gas, we pack twice as many molecules into each cubic centimeter. Over each square centimeter of wall there are now twice as many molecules as before, and therefore there must be twice as many molecule-wall collisions each second. This means that the pressure has doubled, as illustrated in Figure 7.11. If halving the volume doubles the pressure, then pressure and volume are inversely proportional to each other, which is a statement of Boyle's law.

An ideal gas, you recall, would obey Boyle's law exactly under all conditions. This means that no matter how tightly packed the molecules were, it would always be possible to halve their volume by doubling the pressure. The only way this could happen over and over again, of course, is if the gas were composed of particles having no volume themselves, so that the entire volume would be empty space. Real molecules do have finite volumes, however, so no real gas could obey Boyle's law perfectly, especially at high pressure. We will discuss the consequences of this further in the next section.

At 0°C, the average speed of an oxygen molecule is about 1000 miles/hr.

Figure 7.11

When the volume of a gas is halved on going from (a) to (b), twice as many molecules are crowded into each unit of volume. This produces twice as many molecule-wall collisions each second, and that doubles the pressure.

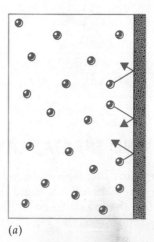
(a)

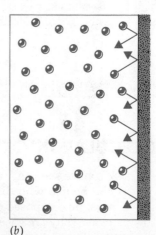
(b)

Distribution of molecular speeds

The second postulate of the kinetic molecular theory is that the average kinetic energy of a collection of molecules is proportional to their absolute temperature. Before we can see how this applies to the gas laws, we have to take a close look at the distribution of molecular energies implied by the use of the term *average kinetic energy.*

In a gas (or a liquid or a solid), molecules are in constant motion. Because they continually collide with each other, there is a wide range of different molecular speeds. For example, occasionally a collision might leave a particular molecule almost motionless, until it suffers another collision and is sent on its way again an instant later. Another molecule might receive several "rear-end" collisions in a row that gives it a very high speed. In this way the speeds of molecules are forever changing through collisions with others. At any instant, some molecules will be moving slowly and some will be going very fast, although most will have speeds in between.

Because each molecule has a kinetic energy equal to $\frac{1}{2}mv^2$, where m is its mass and v is its speed, there is also a distribution of kinetic energies associated with the distribution of molecular speeds. Figure 7.12 illustrates the shape of this kinetic energy distribution at three different temperatures. Plotted vertically are the fractions of all the molecules that have the particular kinetic energies given along the horizontal axis. For example, at zero K.E., which corresponds to molecules standing still, this fraction is essentially zero since very few, if any, molecules are motionless at any instant. The fraction having a particular K.E. increases as we move to higher energies (higher speeds) and eventually becomes a maximum. At still higher kinetic energies the fraction decreases and gradually approaches zero again at kinetic energies corresponding to very fast-moving molecules. The curve does not go all the way to zero, however, because there is virtually no upper limit to molecular speeds, other than the speed of light.

The maximum on this curve represents the kinetic energy possessed by the largest fraction of molecules. This kinetic energy would be found most frequently (that is, with the greatest probability) if we were able to examine mole-

This is called a Maxwell-Boltzmann distribution.

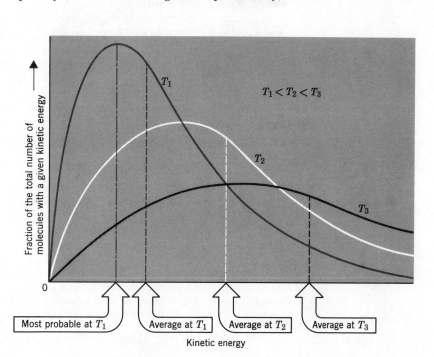

Figure 7.12

The distribution of kinetic energies in a collection of molecules at three different temperatures.

cules at random; hence it is called the *most probable kinetic energy*. The average kinetic energy occurs at a higher value than the most probable kinetic energy because the curve is not symmetrical. Just as a few "curve breakers" in a chemistry class tend to raise the class average on exams, the high-velocity molecules shift the average K.E. above the most probable K.E.

When the temperature of a substance is raised, the curve changes so that the average K.E. increases, as shown in Figure 7.12. More molecules have high speeds and fewer have low ones and, on the average, the molecules move faster. Thus when heat is added to a substance to raise its temperature, the energy goes into increasing the average kinetic energy and increases the speed of the particles.

The relationship between kinetic energy and temperature also leads quite naturally to the concept of an absolute zero. As kinetic energy is removed from a substance, its molecules move more and more slowly. If all the molecules were to cease moving, their average kinetic energy would be zero and, since negative kinetic energies are impossible (a molecule cannot be going slower than when it is standing still), the temperature of the substance would also be at its lowest point. It is this temperature that we refer to as absolute zero—the temperature when all molecular motion has ceased.[3] It should be understood, however, that electronic motion would still continue at absolute zero. Even though the molecules would be motionless, the electrons would still be "whizzing" about their respective nuclei.

Temperature and the gas laws

You've probably begun to wonder what all of this has to do with the gas laws. Let's look at Gay-Lussac's pressure-temperature law first. This law states that if we keep the volume constant, the pressure is directly proportional to the absolute temperature. In other words, when the temperature goes up, so does the pressure. How can we explain this? According to kinetic theory, raising the temperature increases the average kinetic energy of the molecules, so the molecules move faster. This means that they will strike the wall more frequently, and that when they hit the wall the average force of impact will be greater. These factors make the pressure increase.

We can also explain Charles' law, which says that the volume increases when we raise the temperature, provided we keep the pressure constant. We have just seen that raising the temperature causes more molecules to hit the wall each second and also causes the force of impact of the molecules with the walls to increase. The only way to keep the pressure constant is to simultaneously decrease the number of collisions per second with each square centimeter of wall. This can be accomplished by allowing the gas to expand so fewer molecules are over each square centimeter of wall. In other words, to keep the pressure of a gas constant as we raise its temperature, we have to allow it to expand and occupy a larger volume.

Now let's look at what happens if we cool a gas. Following the reasoning in the preceding paragraph, to keep the pressure constant as the gas cools, we must decrease the volume. The molecules move more and more slowly and the space between them becomes less and less. All real gases eventually condense to a liquid when they are cooled because attractive forces between the mole-

[3] In fact, even at absolute zero there is still a residual molecular motion that is required by the Heisenberg uncertainty principle (Chapter 3). This principle states that we cannot simultaneously know exactly the position and momentum of a particle. We will see in the next chapter that the average positions of particles in the solid state can be determined, and if the particles were motionless we would also know that their momentum was zero, thereby violating the uncertainty principle.

When the molecules stick together they settle to the bottom of the container as a liquid.

cules eventually cause "sticky" collisions. An ideal gas, however, would not condense regardless of how much we cooled it, so another property of the "molecules" of an ideal gas is that they have no intermolecular attractions. Consequently, *an ideal gas is a hypothetical substance whose molecules have no volume and no intermolecular attractive forces.*

How can Ar have no vol?

Graham's law

The postulate of the kinetic theory relating average kinetic energy to temperature can be used very simply to derive Graham's law. Suppose we have two different gases, A and B. If they are both at the same temperature, the average kinetic energies of their molecules must be the same. This means that

$$\overline{K.E._A} = \overline{K.E._B}$$

or

$$\tfrac{1}{2}m_A\overline{v_A^2} = \tfrac{1}{2}m_B\overline{v_B^2} \qquad [7.16]$$

where $\overline{v^2}$ is called the **mean square speed** of the molecules, and is the average of the speeds-squared of all the molecules; that is,

$$\overline{v^2} = \frac{v_1^2 + v_2^2 + v_3^2 + \cdots}{n_T}$$

where v_1, v_2, v_3, etc., represent the speeds of molecules 1, 2, 3, etc. and n_T is the total number of molecules present. Equation 7.16 can be rearranged to give

$$\frac{\overline{v_A^2}}{\overline{v_B^2}} = \frac{m_B}{m_A}$$

Taking the square root of both sides, we have that

m_A and m_B are the mass of molecules A and B.

$$\frac{\overline{v_A}}{\overline{v_B}} = \sqrt{\frac{m_B}{m_A}} \qquad [7.17]$$

where $\overline{v}$ is called the root-mean-square speed. We have seen earlier that, for a given number of moles of a gas, the weight of the gas present is directly related to the molecular weight by Equation 7.11, which means that

$$M \propto m$$

Therefore, we can substitute M_A and M_B for m_A and m_B into Equation 7.17 and arrive at

$$\frac{\overline{v_A}}{\overline{v_B}} = \sqrt{\frac{M_B}{M_A}}$$

The rate at which gases effuse should be directly proportional to the velocity of their molecules, with faster molecules effusing at a higher rate. Thus we are led to conclude that

$$\frac{\text{rate of effusion }(A)}{\text{rate of effusion }(B)} = \frac{\overline{v_A}}{\overline{v_B}} = \sqrt{\frac{M_B}{M_A}}$$

or simply

$$\frac{\text{rate of effusion }(A)}{\text{rate of effusion }(B)} = \sqrt{\frac{M_B}{M_A}}$$

which is Graham's law.

Avogadro's principle

Equal volumes of gas at the same temperature and pressure have equal numbers of molecules. That is how we previously stated Avogadro's principle. But we could have also stated this principle as follows: Equal numbers of gas molecules occupying the same volume at the same temperature exert the same pressure. We can explain this by noting that the average force of impact of the molecules colliding with a given area of the wall depends on their average kinetic energy, and therefore on their temperature. If the temperature of two gas samples is the same, then the average kinetic energy of their molecules must be equal, and if the number of molecules per unit volume is the same, then it follows that their pressures must also be the same.

Dalton's law of partial pressures

Molecules of an ideal gas are unaware of each other's existence, except when they collide, because they have no attractions for each other. In a mixture of gases, each behaves independently and exerts a pressure that is the same as it would exert if it were alone. The cumulative effect of the individual partial pressures is the total pressure.

Avogadro's principle can also be applied to gas mixtures. For example, consider the earth's atmosphere, where about 1 of every 5 molecules are O_2 and 4 of every 5 are N_2. Since one-fifth of the molecules are oxygen, only one-fifth of the pressure is contributed by O_2; the other four-fifths of the pressure is contributed by N_2.

The partial pressure of a gas is related quantitatively to the total pressure by its **mole fraction** (usually given by the symbol X)—the number of moles of the gas in question divided by the total number of moles of gas in the mixture. For some gas A,

$$X_A = \frac{\text{number of moles of } A}{\text{total number of moles of gas in the mixture}} \qquad [7.18]$$

and the partial pressure of A, p_A, is

$$p_A = X_A P_T \qquad [7.19]$$

In the atmosphere, for example, in each 5 mol of air there are 1 mol O_2 and 4 mol N_2. Therefore,

$$X_{O_2} = \frac{1 \text{ mol}}{5 \text{ mol}} = 0.2$$

$$X_{N_2} = \frac{4 \text{ mol}}{5 \text{ mol}} = 0.8$$

Simple but may need to know

If the total pressure of a sample of air were 500 torr, then

$$p_{O_2} = 0.2 \, (500 \text{ torr})$$

$$= 100 \text{ torr}$$

$$p_{N_2} = 0.8 \, (500 \text{ torr})$$

$$= 400 \text{ torr}$$

7.11 REAL GASES

The differences between a real gas and a hypothetical ideal gas became apparent in the last section. Molecules of an ideal gas are abstract points in space and have no volume, whereas a real gas is composed of actual molecules whose atoms occupy some space. The effects of this were seen in Figure 7.7, where we found that the volume occupied by a real gas at high pressure is larger than the volume an ideal gas would occupy under the same conditions.

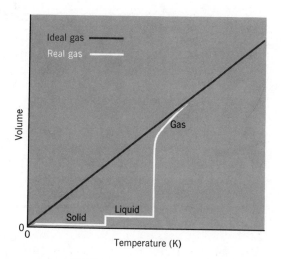

Figure 7.13

Variation of volume with temperature for real and ideal gases.

We also learned in the last section that molecules of an ideal gas would have no attractive forces between them and they could be cooled to absolute zero without condensing to a liquid. Molecules of a real gas do attract each other, however. Their behavior is similar to that shown in Figure 7.13. As the gas is cooled, its volume begins to fall below the Charles' law value. Then, suddenly, the substance condenses to a liquid with a much smaller volume. At still lower temperatures it freezes to a solid. Another manifestation of the attractive forces between gas molecules is the cooling that occurs when a compressed gas is allowed to expand freely into a vacuum. As the gas expands, the average distance of separation between the molecules increases. Since there are forces of attraction between them, moving the molecules apart requires work (energy). The source of this energy is the kinetic energy of the gas—in the process of expansion kinetic energy is converted to potential energy. This removal of kinetic energy leads, of course, to a decrease in the average kinetic energy of the gas and, since the average kinetic energy is directly related to temperature, the gas becomes cooler.

Because real gases deviate from ideal behavior, especially at high pressure and low temperature, the ideal gas law can't be used to make highly accurate calculations. One way to improve the accuracy is to modify the ideal gas law to take into account the factors that cause a real gas to differ from an ideal gas.

Suppose that the gas molecules in a container could be stopped and allowed to settle to the bottom. We would see that part of the volume of the container is occupied by the gas molecules. The remaining free space is somewhat less than the volume of the container. If, in this hypothetical case, another gas molecule were added, it could move in the free space, but not the entire volume of the container. This same situation exists when the molecules are moving.

In an ideal gas the molecules would have no volume themselves, so the ideal gas would be entirely empty space into which other molecules could be squeezed when it is compressed. If we associate the empty space available in a real gas with this *"ideal volume,"* V_{ideal}, then the measured volume of the real gas, V_{meas}, is actually slightly larger than V_{ideal} by an amount that is related to the size of the real molecules. According to J.D. van der Waals (1837–1923), a Dutch physicist, the measured volume is

$$V_{meas} = V_{ideal} + nb$$

The volume within which the molecules cannot move is called the excluded volume.

where b is the correction due to the excluded volume per mole and n is the number of moles of gas. Solving for the ideal gas volume we have

$$V_{ideal} = V_{meas} - nb$$

[7.20]

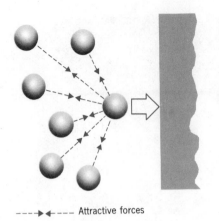

Figure 7.14

In a real gas the molecules collide with the walls with slightly less force than they would if there were no attractive forces tending to hold them back.

---►◄--- Attractive forces

Van der Waals also included a correction to the pressure that takes into account the attractive forces between the molecules of a real gas. A molecule in a gas that is just about to make a collision with the wall feels the attraction of all the molecules surrounding it. Since there are no molecules in front of it (or relatively very few), the greatest concentration of these forces is in a direction away from the wall (Figure 7.14). When the collision does take place, it is less energetic than it would be if there were no attractive forces present. The overall effect of these forces is a lowering of the pressure. The extent to which the pressure will be reduced is directly proportional to (1) the number of impacts per second with the wall, which in turn is directly proportional to the concentration of the molecules (n/V), and (2) the decrease in the force of each impact, which is also proportional to the concentration of the molecules. The decrease in the pressure, therefore, is directly proportional to the square of the concentration, or n^2/V^2. The ideal pressure—that is, the pressure the gas could exert in the absence of intermolecular attractive forces—is higher than the actual measured pressure by an amount that is directly proportional to n^2/V^2 or

$$P_{ideal} = P_{meas} + \frac{n^2 a}{V^2}$$ [7.21]

where a is a proportionality constant that depends on the strength of the intermolecular attractions.

Substituting these corrected pressures and volumes (from Equations 7.21 and 7.20, respectively) into the ideal gas equation gives us

$$\left(P + \frac{n^2 a}{V^2}\right)(V - nb) = nRT$$ [7.22]

Van der Waals received the 1910 Nobel prize in physics for this equation.

in which all symbols stand for measured quantities. This equation is called the **van der Waals equation of state for a real gas.** It is more complex than the ideal equation, but it does work well for many gases over fairly wide ranges of temperature and pressure.

The values of the constants a and b depend on the nature of the gas because the molecular volumes and the molecular attractions vary from gas to gas. Some typical values of a and b are found in Table 7.3. We see that molecules containing many atoms, such as C_2H_5OH, have large values of b. This is not surprising, since such molecules would be expected to be larger than molecules containing only a few atoms.

The variation among the values of a reflect variations in the strengths of the intermolecular attractions. It is easy to understand why polar molecules such as NH_3, H_2O, CH_3OH, and C_2H_5OH attract each other. These molecules are di-

Table 7.3
van der Waals constants for real gases

	a (liters² atm/mol²)	b (liters/mol)
He	0.034	0.0237
O_2	1.36	0.0318
NH_3	4.17	0.0371
H_2O	5.46	0.0305
CH_4	2.25	0.0428
C_2H_6	5.489	0.06380
CH_3OH	9.523	0.06702
C_2H_5OH	12.02	0.08407

poles that tend to align themselves so that the partial positive charge on one attracts the partial negative charge on another (Figure 7.15). The attractions between nonpolar molecules, such as O_2, CH_4, and C_2H_6, or between isolated atoms such as helium and the other noble gases, are more difficult to explain. We will study all these intermolecular attractions in detail in the next chapter.

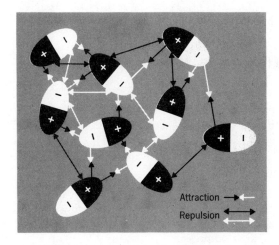

Figure 7.15

Electrostatic interactions between dipoles. The attractions outweigh the repulsions, so the molecules feel a net attraction toward each other.

From the discussion above, we see that the values of a and b enable us to expand our knowledge about the molecules of which a real gas is composed. The van der Waals constants for a gas are obtained by making careful measurements of P, V, and T and then choosing values of a and b that make the van der Waals equation produce the best match with the experimental data. In this sense, a and b are experimentally determined quantities that enable us to check our theories on molecular size and attractions.

INDEX TO QUESTIONS AND PROBLEMS (Problem numbers in **bold type**)

REVIEW QUESTIONS

7.1 Define pressure. Explain why the height of mercury in a barometer is independent of the cross-sectional area of the tube.

7.2 What is the SI unit of pressure and how is it related to the standard atmosphere?

7.3 An open-end manometer containing mercury was connected to a vessel containing a gas at a pressure of 740 torr. The atmospheric pressure was 765 torr. Sketch a diagram showing the relative heights of the mercury in each arm of the manometer.

7.4 What is the advantage of using a closed-end manometer for pressure measurements?

7.5 Why is mercury a useful substance for use in barometers and manometers?

7.6 State Boyle's law in words. Is the law obeyed exactly by all gases? What is a gas that exactly obeys Boyle's law called?

7.7 State Charles' law in words. In terms of Charles' law, why is −273°C the lowest possible temperature?

7.8 If you want to be sure that the tires of your car are properly inflated, you should check them before beginning a trip rather than after having driven on them for a long period. Why?

7.9 Why is it dangerous to incinerate an aerosol can?

7.10 Turbocharging is a way of increasing the power output from a gasoline or diesel engine. The exhaust gases are used to turn a turbine that is connected to a compressor. The compressor forces more air into the cylinders so that more fuel can be burned and more power can be produced. Compressing the air, however, heats it, and even more power can be produced if the air from the compressor is cooled before it is let into the cylinders. Why?

7.11 What makes a hot-air balloon rise?

7.12 What is STP?

7.13 What is Dalton's law of partial pressures? Can partial pressures actually be measured directly?

7.14 What is Gay-Lussac's law of combining volumes? What is Avogadro's principle?

7.15 Suppose that the ideal gas law were $PV^2 = nR/T^2$. What would be the units of R if P is in atm and V in liters?

7.16 What determines the choice of the value of R that you should use in an ideal gas law calculation?

7.17 What is Graham's law of effusion?

7.18 What are the postulates of the kinetic molecular theory?

7.19 In terms of kinetic theory, what is the origin of the pressure of a gas?

7.20 In qualitative terms, why do molecules of gases with low molecular weights diffuse faster than gases with high molecular weights, provided they are at the same temperature?

7.21 In terms of the kinetic theory, how should the rate of diffusion of a gas be affected by an increase in temperature? Explain.

7.22 Why does the pressure of a gas increase when the volume is decreased (at constant temperature)?

7.23 Why does the pressure of a gas increase when the temperature is increased, provided the volume is constant?

7.24 If a warm object is placed in contact with a cool object, heart transfer occurs until they both come to the same temperature. How can the kinetic molecular theory account for this heat transfer and these temperature changes that occur?

7.25 Sketch a graph showing the distribution of kinetic energies for a gas at two different temperatures. For each temperature indicate the most probable K.E. and the average K.E. Why is the curve not symmetrical?

7.26 What is meant by nonideal behavior of a gas? Under what conditions is this behavior most evident?

7.27 Explain why most gases cool upon expansion into a vacuum.

7.28 What physical significance do the constants a and b have in the van der Waals equation of state for real gases?

REVIEW PROBLEMS   (More difficult problems are marked by an asterisk.)

7.29 Perform the following conversions.
(a) 1.50 atm to torr
(b) 785 torr to atm
(c) 3.45 atm to Pa
(d) 3.45 atm to kPa
(e) 165 torr to Pa
(f) 342 kPa to atm
(g) 11.5 kPa to torr

7.30 Suppose that in the closed-end manometer shown in Figure 7.4 the mercury in the closed arm was 15.8 cm higher than in the arm connected to the container full of gas. What is the pressure of the gas in the container expressed in torr?

7.31 An open-end manometer was connected to a flask containing a gas at an unknown pressure. The mercury in the arm open to the atmosphere was 65 mm higher than in the closed end. The atmospheric pressure was 733 torr. What was the pressure of the gas in the flask expressed in torr?

7.32 An open-end manometer was connected to a vessel containing a gas at a pressure of 535 torr. The atmospheric pressure was 774 torr. What was the difference in height (in millimeters) between the mercury levels in the manometer?

7.33 A flask containing a gas was connected to both a closed-end and an open-end manometer. In the closed-end manometer the mercury in the sealed arm was 755 mm above the level in the arm connected to the flask. In the open-end manometer, the arm connected to the gas was 17 mm higher than the side open to the air. What was the atmospheric pressure in torr?

7.34 A manometer connecting two flasks (labeled A and B) contains an oil having a density of 0.847 g/ml. The oil in the arm connected to flask A is 74 cm higher than the oil in the arm connected to flask B. The gas in flask A has a pressure of 836 torr. What is the pressure of the gas in Flask B in torr?

7.35 A man digs a well and finds water 35 ft below the ground. If the average atmospheric pressure at the well is 1 atm, will the man be able to draw water from the well using a pump, mounted at ground level, that works by suction? Explain your answer.

7.36 A gas has a volume of 350 ml at 740 torr. How many milliliters will the gas occupy at 900 torr if the temperature remains constant?

7.37 A sample of SO_2 occupies 1.45 liters at 2.75 atm. Assuming no temperature change, how many liters will this gas occupy at 800 torr?

7.38 A gas is compressed at constant temperature from a volume of 540 ml to 320 ml. If the initial pressure was 475 torr, what is the final pressure in torr?

7.39 A gas exerts a pressure of 20.0 lb/in.^{2} in a container having a volume of 35.0 ft^3. What will its pressure be in lb/in.^{2} if the gas is transferred to a 40.0 ft^3 container at the same temperature?

7.40 A bicycle pump has a barrel that is 75.0 cm long (about 30 in.). Assuming that on the upstroke air is drawn in at a pressure of 1.00 atm, how long must the downstroke be (in centimeters) to raise the pressure of the air to 5.50 atm, if the temperature of the air remains constant? (That's about the pressure in the tire of a 10-speed bike.)

7.41 At 25°C and 1 atm a gas occupies a volume of 1.50 liters. How many liters will it occupy at 100°C and 1 atm?

7.42 A balloon has a volume of 2.0 liters indoors at a temperature of 25°C (77°F). If it is taken outdoors on a very cold winter day when the temperature is −28.9°C (−20°F), what will its volume be in liters? Assume constant air pressure within the balloon.

7.43 What would be the final volume of a 2.00-liter sample of a gas that is heated from 26 to 100°C at constant pressure?

7.44 A gas exerts a pressure of 350 torr at 20°C. How many torr will it exert if its temperature is raised to 40°C without a change in volume?

7.45 A gas has a pressure of 655 torr at 25°C. To what Celsius temperature must it be heated to raise its pressure to 825 torr?

*7.46 An automobile tire is inflated to a *gauge pressure* of 29 lb/in.^{2} at 65°F. (The actual pressure is 29 lb/in.^{2} *above* atmospheric pressure.) After a trip the temperature of the tire has risen to 130°F. What will be the gauge pressure of the air in the tire (in lb/in.^{2}), assuming that no air has leaked out and that the volume of the tire hasn't changed? Assume the atmospheric pressure is 14.7 lb/in.^{2}

7.47 A sample of O_2 occupies 285 ml at 25.0°C. At what Celsius temperature will it occupy 350 ml if the pressure remains constant?

7.48 At what Celsius temperature will a gas sample occupy 0.850 liter at 1 atm pressure if it occupies 400 ml at 32°C and 1 atm?

7.49 If a gas, originally in a 50-ml container at a pressure of 645 torr, is transferred to another container whose volume is 65 ml, what would be its new pressure in torr if
(a) There were no temperature change?
(b) The temperature of the first container was 25°C and that of the second was 35°C?

7.50 A 300-ml sample of a gas exerts a pressure of 450 torr at 27°C. What pressure (in torr) would it exert in a 200-ml container at 20°C?

7.51 A 2.00-liter sample of a gas originally at 25°C, and a pressure of 700 torr, is allowed to expand to a volume of 5.00 liters. If the final pressure of the gas is 585 torr, what is its final temperature in degrees Celsius?

7.52 The density of CO_2 is 1.96 g/liter at 0°C and 1 atm. Determine its density in grams per liter at 650 torr and 25°C.

7.53 If a 50.0-ml sample of gas exerts a pressure of 450 torr at 35°C, how many milliliters will it occupy at STP?

7.54 A 1.00-liter mixture of gases is produced from 1.00 liter of N_2 at 200 torr, 1.00 liter of O_2 at 500 torr, and 1.00 liter of Ar at 150 torr. What is the pressure of the mixture in torr?

7.55 A 1.00-liter flask is filled by placing in it the contents of a 2.00-liter flask of N_2 at 300 torr and a 2.00-liter flask of H_2 at 80 torr. What is the pressure (expressed in atm) of the mixture in the 1.00-liter flask?

7.56 What would be the total pressure in torr of a mixture prepared by adding 20.0 ml of N_2 at 0°C and 740 torr plus

30 ml of O_2 at 0°C and 640 torr to a 50-ml container at 0°C?

7.57 A mixture of N_2 and O_2 has a volume of 100 ml at a temperature of 50°C and a pressure of 800 torr. It was prepared by adding 50 ml of O_2 at 60°C and 400 torr with X ml of N_2 at 40°C and 400 torr. What is the value of X?

7.58 A gas is collected by the displacement of water until the total pressure inside a 100-ml flask is 700 torr at 25°C. How many milliliters would the dry gas occupy at STP?

7.59 A mixture of N_2 and O_2 in a 200-ml vessel exerts a pressure of 720 torr at 35°C. If there are 0.0020 mole of N_2 present:
(a) What is the mole fraction of N_2? (See Equation 7.19.)
(b) What is the partial pressure of N_2 in torr?
(c) What is the partial pressure of O_2 in torr?
(d) How many moles of O_2 are present?

7.60 The air exhaled by an average human being might have the following typical composition, expressed in terms of partial pressures: N_2, 569 torr; O_2, 116 torr; CO_2, 28 torr; water vapor, 47 torr. What are the mole fractions of each gas?

7.61 How many milliliters of CO_2 at 30°C and 700 torr must be added to a 500-ml container of N_2 at 20°C and 800 torr to give a mixture having a pressure of 900 torr at 20°C?

7.62 Calculate the number of liters occupied, *at STP*, by
(a) 0.200 mol O_2
(b) 12.4 g Cl_2
(c) A mixture of 0.100 mol N_2 and 0.050 mol O_2

7.63 Calculate the mass, in grams, of 245 ml of SO_2 at STP.

7.64 What is the density of butane, C_4H_{10}, at STP expressed in grams per liter?

7.65 The density of a gas was found to be 1.96 g/liter at STP. What is its molecular weight?

7.66 What is the value of the gas constant in the units Pa m³/mol K?

7.67 In the laboratory a student filled a 250-ml container with an unknown gas until a pressure of 760 torr was obtained. He then found that the sample of gas weighed 0.164 g. Calculate the molecular weight of the gas if the temperature in the laboratory was 25°C.

7.68 Calculate the pressure, in torr and in atmospheres, that would be exerted by 25 kg of steam (H_2O) in a 1000-liter boiler at 200°C assuming ideal gas behavior.

7.69 The density of a gas was found to be 1.81 g/liter at 30°C and 760 torr. What is its molecular weight?

7.70 How many milliliters would be occupied by 0.234 g of NH_3 at 30°C and a pressure of 0.847 atm.

7.71 A chemist observed a gas being evolved in a chemical reaction and collected some of it for analysis. It was found to contain 80.0% carbon and 20.0% hydrogen. It was also observed that 500 ml of the gas at 760 torr and 0°C weighed 0.6695 g.
(a) What is the empirical formula of the gaseous compound?

(b) What is its molecular weight?
(c) What is its molecular formula?

*7.72 A 0.2000-g sample of a fishy smelling liquid known to contain only carbon, hydrogen, and nitrogen was burned and produced 0.482 g CO_2 and 0.271 g H_2O. A second sample weighing 0.2500 g was treated in such a way that all the nitrogen in the substance was converted to N_2. This gas was collected and found to occupy 42.3 ml at 26.5°C and 755 torr.
(a) What are the percentages of carbon, hydrogen, and nitrogen in the compound?
(b) What is the empirical formula of the compound?

*7.73 120 ml of NH_3 at 25°C and 750 torr were mixed with 165 ml of O_2 at 50°C and 635 torr and transferred to a 300-ml reaction vessel where they were allowed to react according to the equation

$$4NH_3\,(g) + 5O_2\,(g) \longrightarrow 4NO\,(g) + 6H_2O\,(g)$$

What will be the total pressure (in torr) in the reaction vessel at 150°C after the reaction is over?

7.74 In the reaction, $N_2\,(g) + 3H_2\,(g) \rightarrow 2NH_3\,(g)$, how many milliliters of N_2, measured at STP, are required to produce 400 ml of NH_3, measured at STP? How many milliliters of H_2 at STP are required?

7.75 In the reaction,

$$2NO\,(g) + 2H_2\,(g) \longrightarrow 2H_2O\,(g) + N_2\,(g)$$

how many milliliters of N_2, measured at STP, would be produced from (a) 0.00140 mol NO, (b) 1.3×10^{-3} g H_2?

7.76 Oxygen gas, generated in the reaction, $KClO_3 \rightarrow KCl + O_2$ (unbalanced), was collected over water at 30°C in a 150-ml vessel until the total pressure was 600 torr.
(a) How many grams of dry O_2 were produced?
(b) How many grams of $KClO_3$ were consumed in the reaction?

7.77 Nitric acid is produced by dissolving NO_2 in water according to the equation,

$$3NO_2\,(g) + H_2O\,(l) \longrightarrow 2HNO_3\,(l) + NO\,(g)$$

How many milliliters of NO_2 at 25°C and 770 torr are required to produce 10.0 g of HNO_3?

7.78 Compare the rates of effusion of He and Ne. Which gas effuses faster, and how much faster?

7.79 If, at a particular temperature, the average speed of CH_4 molecules is 1000 miles/hr, what would be the average speed of CO_2 molecules at the same temperature expressed in miles per hour?

7.80 The rate of effusion of an unknown gas was determined to be 2.92 times faster than that of NH_3. What is the approximate molecular weight of the unknown gas?

7.81 Use the van der Waals equation to calculate the pressure, in atm, exerted by 1.000 mol of He at 0°C in a volume of 22.400 liters. Compare this to the pressure an ideal gas would exert under these same conditions.

7.82 Use the van der Waals equation to calculate the pressure, in atm, exerted by 1.000 mol of C_2H_6 at 0°C in a volume of 22.400 liters. Compare this to the pressure of an ideal gas under these same conditions.

*7.83 Mercury has a density of 13.6 g/ml. Calculate the value of the standard atmosphere in the units, $lb/in.^2$.

7.84 How many milliliters would be occupied by 0.0244 g of O_2 if it were collected over water at 23°C and at a total pressure of 740 torr.

*7.85 Three gases were added to the same 10-liter container to give a total pressure of 800 torr at 30°C. If the mixture contained 8.0 g of CO_2, 6.0 g of O_2, and an unknown amount of N_2, calculate the following.
(a) The total number of moles of gas in the container
(b) The mole fraction of each gas
(c) The partial pressure of each gas, in torr
(d) The number of grams of N_2 in the container

*7.86 A gas, at a total pressure of 800 torr and a volume of 500 ml over water at 35°C, is compressed to a volume of 250 ml, also over water at 35°C. Calculate the final pressure of the wet gas in torr.

*7.87 During a rainstorm in July in New York City the humidity was found to be 100%, which means that the air was saturated with water vapor. The atmospheric pressure was 740 torr and the temperature was 31°C. Dry air has an average molecular weight of 28.8. Calculate the weight of water in 1.00 liter of the air during the storm.

*7.88 280 ml of gas are collected over water at 20°C. The water level inside the collection bottle is 28.4 mm higher than the water level outside. The atmospheric pressure is 763 torr. How many milliliters would the dry gas occupy at STP?

*7.89 The product PV has the dimensions of energy. Given the data, 1 J = 1 N m, 1 atm = 101,325 Pa, and 1 Pa = 1 N/m^2, calculate the number of joules equal to 1 liter atm. What is the value of the gas constant, R, in $J\ mol^{-1}\ K^{-1}$ and $cal\ mol^{-1}\ K^{-1}$?

*7.90 Calculate the maximum number of milliliters of CO_2, at 750 torr and 28°C, that could be produced by reacting 500 ml of CO, at 760 torr and 15°C, with 500 ml of O_2 at 770 torr and 0°C.

*7.91 Ozone, O_3, is an important species in the chain of reactions that lead to the production of smog. In an ozone

analysis, 2.0×10^4 liters of air at STP were drawn through a solution of NaI where the O_3 undergoes the reaction,

$$O_3 + 2I^- + H_2O \longrightarrow O_2 + I_2 + 2OH^-$$

The I_2 formed was titrated with 0.0100 M $Na_2S_2O_3$ with which it reacts.

$$I_2 + 2S_2O_3{}^{2-} \longrightarrow 2I^- + S_4O_6{}^{2-}$$

In the analysis, 0.042 ml of the $Na_2S_2O_3$ solution was required to completely react with all of the I_2.
(a) Calculate the number of moles of I_2 that were reacted with the $S_2O_3{}^{2-}$ solution.
(b) How many moles of I_2 were produced in the first reaction?
(c) How many moles of O_3 were contained in the 20,000 liters of air?
(d) How many milliliters would the O_3 occupy at STP?
(e) What is the concentration of O_3, in parts per million by volume, in the air sample?

*7.92 An important reaction in the production of nitrogen fertilizers is the oxidation of ammonia,

$$4NH_3\ (g) + 5O_2\ (g) \xrightarrow{500°C} 4NO\ (g) + 6H_2O\ (g)$$

How many liters of O_2, measured at 25°C and 0.895 atm, must be used to produce 100 liters of NO at 500°C and 750 torr?

*7.93 A student collected 35.0 ml of O_2 over water at 25°C and a total pressure of 745 torr from the decomposition of a 0.2500-g sample known to contain a mixture of KCl and $KClO_3$. The reaction that produced the oxygen was

$$2KClO_3 \longrightarrow 2KCl + 3O_2$$

(a) How many moles of O_2 were collected?
(b) How many grams of $KClO_3$ were decomposed?
(c) What percentage by mass of the sample was $KClO_3$?

*7.94 Calculate the molar volume (in liters) of O_2 at STP from the van der Waals equation and compare it to the value in Table 7.2. (*Hint:* The volume can be obtained by successive approximations if you solve for the V in the term, $V - nb$.)

8

STATES OF MATTER AND INTERMOLECULAR FORCES

Gigantic icebergs such as this float because solid water is less dense than the liquid. This is caused by the way water molecules interact with each other in the liquid and solid states. The effect that intermolecular attractions have on the physical properties of substances is examined in this chapter.

The properties of gases, which we studied in the last chapter, are vastly different than those of the other two states of matter, liquids and solids. For one thing, we can't even see a gas, except if it's colored. We would be totally unaware that gases such as those in our atmosphere exist if we couldn't feel a breeze and see it rustle the leaves on a tree, or if we hadn't discovered that gases exert a pressure. Yet who would fail to recognize a lake full of water or a gigantic icy glacier?

The water vapor in the air (which we recognize as humidity), the water of a lake, and the water frozen in a glacier are all forms of the same chemical substance. They are each composed of molecules of water and have the same set of chemical properties. It is their physical properties that make them seem so different. In this chapter we will study why gases, liquids, and solids are so different, and what important properties liquids and solids have. We will also study how the three states of matter change from one to another and what this tells us about them.

8.1 COMPARING THE PROPERTIES OF GASES, LIQUIDS, AND SOLIDS

According to the kinetic molecular theory, all forms of matter are composed of small, rapidly moving particles. In Chapter 7 we saw how this postulate could be used to explain certain properties of gases. These same particles also exist in liquids and solids, and there are two main reasons why gases, liquids, and solids differ so much from each other. One is the tightness of the packing of the particles, and the other is the strengths of the attractive forces between them. Although both factors are interrelated and affect each of the physical properties, certain properties are influenced more by one than by the other. Two properties particularly affected by how tightly the particles are packed are the compressibility and the rates of diffusion. Properties particularly influenced by the strengths of the intermolecular attractive forces are shapes, volumes, and the ability to flow; surface tension; and rate of evaporation.

Compressibility

In a gas, the molecules are widely separated so there is a lot of empty space into which they may be crowded. As a result, gases are very compressible. The molecules in a liquid or a solid, however, are tightly packed and there is very little empty space between them. For this reason, increasing the pressure has hardly any effect on their volume, and liquids and solids are virtually incompressible.

The incompressibility of liquids is an important and useful property. Many types of hydraulic machinery depend on it to transmit enormous forces that lift

The incompressibility of liquids is used in hydraulic machinery to exert enormous forces. Here we see one of the largest hydraulic forging presses in the United States, located at the Alcoa plant in Cleveland, Ohio. It can exert forces up to 50,000 tons!

and move heavy things. You depend on it yourself when you step on the brakes of a car. The force exerted by your foot is first magnified and then transmitted by an oil through the brake lines to the wheels where it causes the brake shoes to rub against a surface and stop the car. If air gets into the brake lines, the force you exert with your foot simply compresses the air, and the car won't stop very quickly at all. (Disaster!)

Compressed gases are used in many practical applications, too.

Diffusion

Molecules diffuse rapidly in a gas, compared to a liquid or a solid, because they move relatively long distances between collisions. They get where they're going quickly because there are relatively few interruptions along their paths, as shown in Figure 8.1. When two liquids mix, however, the molecules of one diffuse throughout the molecules of the other at a much slower rate than when two gases are mixed. We can observe the diffusion of two liquids by dropping a small quantity of ink into some water. As shown in Figures 8.2a and 8.2b, when the ink drop strikes the water we see it as a concentrated "dot," which slowly spreads throughout the liquid. Diffusion takes place because the molecules in both liquids are able to move throughout the container. However, since the molecules in a liquid are so close together, the average distance that they travel between collisions—their *mean free path*—is very short. The molecules undergo billions of collisions before going very far and these constant interruptions in their paths keep them from spreading rapidly throughout the liquid.

Diffusion within solids is even much slower than in liquids. Not only are the molecules very tightly packed, but they are also held quite rigidly in place. They are not free to roam about, even though each may vibrate and bounce around inside the small cavity that it occupies within a crystal. Only molecules (or ions, if the solid is ionic) that have very large kinetic energies are able to muscle their way past their neighbors, so diffusion is extremely slow at ordinary temperatures.

High temperature solid-state diffusion is important in manufacturing electronic components such as transistors.

Volume and shape

The most obvious property of gases, liquids, and solids is the way they behave when transferred from one container to another. A gas, as we learned, expands to completely fill whatever container it is placed in. A liquid, however, retains a constant volume, but conforms to the shape of its container. Gases and liquids are both fluids; they flow and can be pumped from place to place. A solid, however, is not a fluid, and maintains both its shape and volume.

In a gas, the intermolecular attractive forces are so weak that the rapidly moving molecules can easily overcome them and expand to fill a container. The

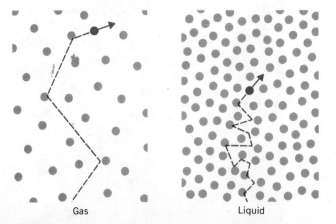

Figure 8.1

Diffusion can occur more rapidly in a gas than in a liquid because a molecule moves further between collisions in a gas.

Gas   Liquid

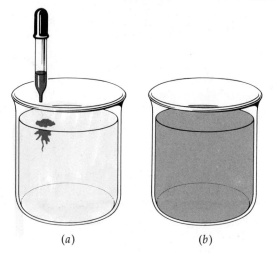

Figure 8.2

Diffusion in liquids (a) An Ink drop is placed into water. (b) Ink has spread throughout the liquid.

(a)   (b)

strengths of these forces are much larger in a liquid, however, and are responsible for holding the molecules close together. In a solid the attractive forces are still larger and hold the molecules more or less firmly in place so that they cannot move over and around each other. This prevents solids from flowing like gases and liquids.

Surface tension

Have you ever noticed how raindrops form beads of water on a freshly waxed car? Have you ever wondered why moist grains of sand stick together, but fall apart if either dry or completely submerged in water? These are phenomena that are caused by a property of liquids called surface tension.

In a liquid, each molecule moves about always under the influence of its neighbors. A molecule within the liquid is completely surrounded by others to which it is attracted (Figure 8.3a). However, a molecule at the surface is not completely surrounded, and feels attractions only to molecules below and beside it (Figure 8.3b). Molecules at the surface thus feel a net attraction in a direction toward the interior of the liquid. For a molecule to come to the surface, it must overcome this attraction. In other words, its potential energy must increase—in a sense, work must be done to pull it to the surface. Making the surface of a liquid larger, therefore, requires an input of energy, and the amount of energy needed is proportional to the liquid's **surface tension.**

The lowest energy (most stable) state for a given volume of liquid is when its surface area is a minimum. This gives the fewest high energy surface molecules. The shape that satisfies this condition is a sphere, which is why raindrops are nearly spherical. All liquids strive to minimize their surface areas and

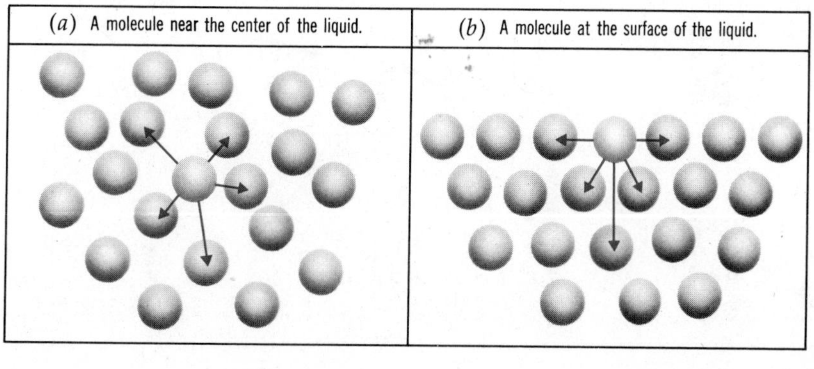

Manufacturing perfectly spherical ball bearings in the weightlessness of space is a potential dividend of the space program that relies on the surface tension of liquid metals.

(a) A molecule near the center of the liquid.	(b) A molecule at the surface of the liquid.

Figure 8.3

Intermolecular attractive forces in liquids. (a) A molecule near the center of the liquid. (b) A molecule at the surface of the liquid.

Figure 8.4

Grains of sand such as these are drawn together when wet because the film of water between them tends to reduce its surface area, thereby achieving a lower energy.

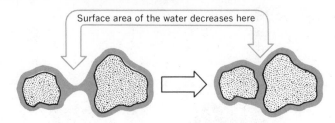

Surface area of the water decreases here

tend toward spherical shapes as much as possible. This natural tendency is what allows you to fire polish glass tubing in the lab. As the glass softens, the sharp edges become rounded because the attractive forces within the glass tend to reduce the surface area. In Figure 8.4 we see why moist grains of sand are drawn together. In doing so, the film of water between them achieves a lower energy by reducing its surface area. Pulling the grains of sand apart requires that the surface tension be overcome, so the particles tend to stay together.

The magnitude of a liquid's surface tension depends on the strengths of the attractive forces between its molecules. When the attractive forces are large, the surface tension is large. Surface tension is also a function of the temperature of the liquid. Increasing the temperature (which increases the kinetic energy of the individual molecules) decreases the effectiveness of the intermolecular attractive forces, so surface tension decreases as the temperature is raised.

A property that we often think of when a liquid is mentioned is its ability to **wet** things—to spread across their surfaces as a thin film. Water, for example, spreads smoothly across a clean glass surface. This ability reflects a similarity among the strengths of the attractive forces between individual water molecules and between water molecules and the glass surface. Because these attractive forces are similar, little effort must be expended to spread the water molecules across the surface of the glass, so the surface tension is easily overcome and the water wets the glass. However, if the glass has a thin film of grease or oil on it to which the water molecules are only weakly attracted, the weak attractions can't overcome the water's surface tension and beads of water form. The water can't wet the greasy surface—a fact that is especially annoying with an oily automobile windshield and greasy lab glassware.

Detergents contain substances called surfactants that lower the surface tension of water and allow the detergent solution to spread more easily over greasy surfaces.

Evaporation

In a liquid or a solid, just as in a gas, the molecules are constantly undergoing collisions, giving rise to a distribution of individual molecular velocities and, of course, kinetic energies. Even at room temperature, a small percentage of the molecules are moving with relatively high kinetic energies. If some of these faster-moving molecules possess enough kinetic energy to overcome the attrac-

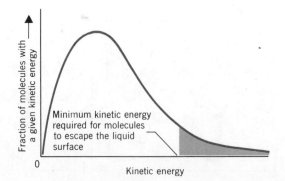

Minimum kinetic energy required for molecules to escape the liquid surface

Kinetic energy

Figure 8.5

Kinetic energy distribution in a liquid.

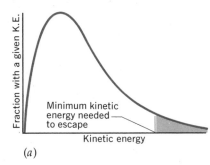

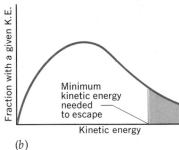

Figure 8.6

Kinetic energy distribution at low temperature (a) and high temperature (b). The same minimum kinetic energy is needed to escape at both temperatures, but the total fraction of molecules having at least this much energy (the shaded area) is larger at the higher temperature.

tive forces within the liquid or solid, they can escape through the surface into the gaseous state—they **evaporate.**

The evaporation of a liquid is something everyone has seen. A small amount of spilled gasoline or nail polish remover quickly disappears, and rain puddles evaporate after a summer shower. Solids can also evaporate, although most of them that we encounter under normal conditions don't seem to disappear rapidly, if at all. But have you ever seen Dry Ice? It is composed of solid carbon dioxide. It is called *Dry* Ice because the solid doesn't melt; it simply evaporates. Have you ever wondered what happens to the crystals of naphthalene (moth flakes) that you sprinkle in a drawer or garment bag? They evaporate, too. Direct conversion of a solid to a gas, without melting, is called **sublimation.** Solid carbon dioxide and naphthalene are two substances that readily sublime at atmospheric pressure.

Figure 8.5 represents a typical distribution of the kinetic energies of the molecules in a liquid. The shaded area corresponds to the fraction of the total number of molecules that possess sufficient kinetic energy to evaporate. Just as removing the smart students from a chemistry class will lower the class average on exams, the loss of the higher-energy fraction because of evaporation leads to a lowering of the average kinetic energy of the remaining molecules. Since the temperature is directly proportional to the average kinetic energy, this results in a decrease in the temperature of a liquid as it evaporates. For example, we have all felt cool after a bath, because the evaporation of water from the body has drawn heat from us. In fact, evaporation of perspiration provides the body with a mechanism for controlling body temperature.

Several things affect the rates at which liquids evaporate. One, of course, is the temperature. We know that warm water evaporates faster than cold water. The reason can be seen in Figure 8.6. Notice that the shaded area under the curve, which is the *total* fraction of the molecules that have enough kinetic energy to escape from the liquid, is larger at the higher temperature. In a given period of time, more molecules evaporate simply because more have enough energy to get away.

Another factor affecting the rate of evaporation is surface area. Increasing the surface area brings more high-speed molecules close to the surface so they can escape before losing kinetic energy through collisions.

A third factor is the strengths of the intermolecular attractions. Figure 8.7 compares two liquids at the same temperature. In one (liquid *A*) the attractive

Steam issues from cooling towers at the Three Mile Island nuclear reactor prior to the accident there. The temperature lowering effect produced by the evaporation of water is often used commercially for cooling purposes.

Figure 8.7

Two liquids, A and B, at the same temperature. The total fraction of molecules (shaded area) having enough energy to escape from the liquid is greater for liquid B than for A.

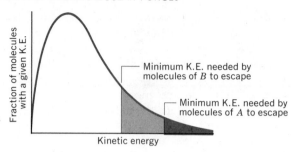

Minimum K.E. needed by molecules of *B* to escape

Minimum K.E. needed by molecules of *A* to escape

forces are very strong, so only very high-speed molecules have enough kinetic energy to overcome the attractions and escape. There are relatively few of these molecules, so the rate of evaporation is slow. In the other liquid the attractive forces are weak, and a much larger fraction have the energy needed to escape. Therefore, this liquid evaporates quickly. The attractive forces in most solids are much greater than in liquids, which explains why solids usually do not evaporate easily.

8.2 INTERMOLECULAR ATTRACTIVE FORCES

The nonideal behavior of real gases described in Chapter 7, and the properties of liquids and solids, such as those discussed in the last section, illustrate the importance of understanding the origins and relative strengths of intermolecular attractions. These ultimately determine *all* the physical properties of a substance. Knowing how the strengths of intermolecular attractions relate to chemical composition and structure therefore allows us to understand the behavior of substances and to even anticipate the kinds of materials that have particular desirable properties.

Polar molecules have ends that are oppositely charged. In a collection of these molecules, the individual dipoles tend to orient themselves so that the partial positive charge on one is near the partial negative charge on others. Because the molecules are constantly moving and colliding with each other, this alignment is far from perfect, particularly in liquids and gases. Nevertheless, the attractions between the oppositely charged ends of the dipoles outweigh the repulsions between like-charged ends, and a net overall attraction exists between them (Figure 8.8).

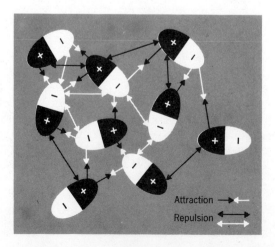

Attraction

Repulsion

Figure 8.8

Electrostatic interactions between dipoles.

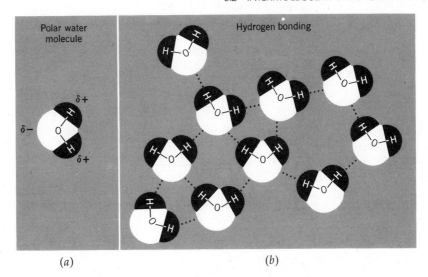

(a) (b)

Figure 8.9

Hydrogen bonding in water. (a)
The water molecule is very polar.
(b) Hydrogen bonding produces
strong attractions between the
water molecules.

The energy required to separate a pair of dipoles is proportional to $\frac{1}{d^3}$, where **d** is the
distance between dipoles.

Dipole-dipole attractions are normally considerably weaker than ionic or covalent bonds—they are only about 1% as strong. Their strength also decreases very rapidly as the distance between the dipoles increases, so their effects between the widely spaced molecules in a gas are very much less than between tightly packed molecules in a liquid or a solid. This is why the molecules of a gas behave almost as though there were no attractive forces at all.

Hydrogen bonding

A particularly strong dipole-dipole attraction occurs when hydrogen is covalently bonded to a very small highly electronegative element such as fluorine, oxygen, or nitrogen. In these instances extremely polar molecules result in which the small hydrogen atom carries a substantial positive charge. Because the positive end of this dipole can closely approach the negative end of a neighboring dipole, the force of attraction between the two is quite large. This special kind of dipole interaction is called a **hydrogen bond,** and is about 5 to 10% as strong as an ordinary covalent bond.

Hydrogen bonding is a very important type of weak attractive force. In water, for example, the molecules interact strongly with each other by hydrogen bonding (Figure 8.9). This produces attractive forces that are much stronger than those between other molecules of similar size and mass, and is what makes water a liquid at room temperature. Hydrogen bonding is also responsible for controlling the orientation of water molecules in ice (Figure 8.10), where we see that each water molecule is surrounded tetrahedrally by four others to which it is held by hydrogen bonds. This causes ice to have a very "open" structure, and makes ice less dense than liquid water. That's why ice cubes and icebergs float (much to the distress of the captain of the *Titanic*).

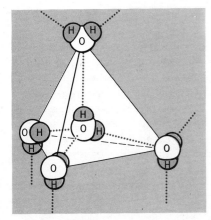

Figure 8.10

Hydrogen bonding (dotted lines)
between water molecules in ice.

London forces

Even uncombined atoms and nonpolar molecules experience weak attractions. We know this because substances like helium and hydrogen can be condensed to a liquid if they are cooled to a sufficiently low temperature. Attractive forces must exist to hold them together in the liquid state. These attractions are called **London forces,** after the German physicist Fritz London who provided an explanation of them.

When electrons move about in an atom or molecule, their motion is somewhat random, so that at any given instant there is a chance that more electrons will be on one side of the particle than on the other. At that particular instant

The strength of London forces,
as measured by the energy
needed to separate the par-
ticles, is proportional to $\frac{1}{d^6}$.

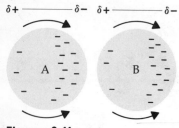

Figure 8.11

London forces. As the instantaneous dipole in atom A forms, it induces a dipole in atom B.

the particle will be a dipole. We call it an *instantaneous dipole* because its existence is only momentary. In a collection of atoms or molecules the electron motions in neighboring particles are not entirely independent. As the negative end of an instantaneous dipole begins to form in one of them, it pushes electrons away in the particle alongside, as shown in Figure 8.11. We say that the instantaneous dipole induces a dipole in its neighbor. As you can see, because of the way the dipoles are formed, they attract each other, and that attraction produces a momentary tug that helps hold them together.

Normally, London forces are rather weak because of their fleeting existence. They are present between all particles—ions as well as polar and nonpolar molecules—but play a very minor role in the attractions between ions. This is because the forces of attraction between ions are so strong. London forces do play a significant role in the attractions between all kinds of molecules, especially nonpolar ones.

8.3 HEAT OF VAPORIZATION

The **molar heat of vaporization,** which we will refer to as $\Delta H_{\text{vaporization}}$ (or simply ΔH_{vap}), *represents the amount of energy that must be supplied to 1 mol of liquid to convert it to a vapor at the same temperature.* The Greek letter, Δ, is usually used to symbolize a change. In this case it is a change in the heat content (the total amount of heat energy) of a substance as it undergoes a change from liquid to vapor. This change in heat content is equal to the energy contained in the substance in its final state (vapor) minus the energy that the substance possessed in its initial state (liquid). Thus

$$\Delta H_{\text{vaporization}} = H_{\text{vapor}} - H_{\text{liquid}}$$

In actual practice, neither H_{vapor} nor H_{liquid} can be measured; however, their difference (ΔH_{vap}) can be.

The molar heat of vaporization is an important physical property of a substance. Engineers have to know about it when they design chemical plants because they must know how much energy will be needed to vaporize the solvents used in various processes. The heat of vaporization of water is important to meteorologists because much of the solar energy absorbed by the earth goes to evaporating water from the oceans. In a sense, this energy becomes stored in the water vapor and when the water condenses it is released. This is the origin of the enormous amounts of energy contained in violent hurricanes and other rain storms.

The term heat of vaporization usually means molar heat of vaporization.

EXAMPLE 8.1   The heat of vaporization of water is 40.6 kJ/mol. How much heat energy is required to convert 1.00 liter of water to steam?

SOLUTION   Water has a density of 1.00 g/ml. Therefore, 1.00 liter of water weights 1.00×10^3 g. The problem, as is generally the case, is one of unit conversion.

$$1.00 \times 10^3 \text{ g } H_2O \sim (?) \text{ kJ}$$

This can be set up as

$$1.00 \times 10^3 \text{ g } H_2O \times \left(\frac{1 \text{ mol } H_2O}{18.0 \text{ g } H_2O}\right) \times \left(\frac{40.6 \text{ kJ}}{1 \text{ mol } H_2O}\right) \sim 2260 \text{ kJ (to three significant figures)}$$

Thus 2260 kJ of heat are required.

Table 8.1
**Heats of vaporization and boiling points
of various substances**

Compound	ΔH_{vap} kJ/mol	kcal/mol	Boiling Point (°C)
CH_4	9.20	2.20	−161
C_2H_6	14	3.3	−89
C_3H_8	18.1	4.32	−30
C_4H_{10}	22.3	5.32	0
C_6H_{14}	28.6	6.83	68
C_8H_{18}	33.9	8.10	125
$C_{10}H_{22}$	35.8	8.56	160
F_2	6.52	1.56	−188
Cl_2	20.4	4.88	−34.6
Br_2	30.7	7.34	59
HF	30.2	7.21	17
HCl	15.1	3.60	−84
HBr	16.3	3.90	−70
HI	18.2	4.34	−37
H_2O	40.6	9.71	100
H_2S	18.8	4.49	−61
NH_3	23.6	5.63	−33
PH_3	14.6	3.49	−88
SiH_4	12.3	2.95	−112

A useful feature of the heat of vaporization is that its magnitude provides a good measure of the strengths of the attractive forces in a liquid. This is easy to understand. When the intermolecular attractions are strong, a lot of energy must be supplied to pull the molecules apart as the liquid is changed to a gas.

Table 8.1 lists some values of ΔH_{vap} for various substances. We can use these data to gain some insight into the way chemical composition and structure influence intermolecular attractions. For example, if we look at the series of hydrocarbons, CH_4 through $C_{10}H_{22}$, we observe a steady increase in ΔH_{vap} with an increase in molecular weight. These compounds are nonpolar; therefore the only attractive forces that exist between their molecules are London forces, so these must also increase from CH_4 to $C_{10}H_{22}$. The reason can be seen by examining the structure of the molecules. These hydrocarbons are chainlike molecules, as shown in Figure 8.12, and as we proceed from CH_4 to $C_{10}H_{22}$ the length of the carbon chain increases. This has the effect of increasing the number of locations along the molecule where London forces may occur with other mole-

Figure 8.12

Attractive forces increase with increasing chain length. There are more points along the molecule that can be attracted to other molecules nearby.

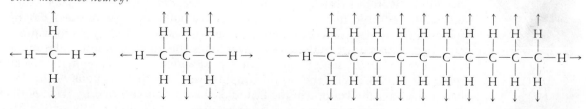

cules. A long chainlike molecule is therefore held in more places than a short molecule, and more energy must be supplied to remove such long-chain molecules from the liquid. The result is that as the chain length increases, ΔH_{vap} increases.

Another factor that influences the strengths of London forces is molecular size. If we examine molecules of the same general formula, such as the halogens (F_2, Cl_2, Br_2), we find that large molecules have a greater ΔH_{vap} than small molecules. As we proceed from F_2 to Br_2, the atoms that make up the molecules become larger, so the molecules also become larger. As the size increases, the outer electrons become further from the nuclei and are not held as tightly. Because of this the electron cloud of a large molecule is more easily distorted, or polarized, and it is easier to create the instantaneous dipoles that are responsible for the London forces. (The ease of distortion of the electron cloud is referred to as **polarizability**.) The result is that the London forces are stronger between molecules composed of large, easily polarized atoms such as bromine than between molecules composed of small atoms such as fluorine. That is why ΔH_{vap} increases from F_2 to Br_2.

When we look at the hydrogen halides, HF through HI, we find that the expected variation of ΔH_{vap} with molecular size is reversed between HF and HCl. In fact, HF has a considerably higher heat of vaporization than any of the other HX compounds. This anomalous behavior is attributed to the presence of hydrogen bonding. We see the same inverted order of ΔH_{vap} for H_2O and H_2S and for NH_3 and PH_3, where, again, hydrogen bonding is significant for H_2O and NH_3 but not for H_2S and PH_3. Oxygen, fluorine, and nitrogen are all very small and are the most electronegative elements in the periodic table, while the elements below them are much larger and much less electronegative. It is not surprising, therefore, that hydrogen bonding is important only for H_2O, HF, and NH_3. "Normal" behavior is reached in Group IVA hydrides, where ΔH_{vap} for CH_4 is less than ΔH_{vap} for SiH_4. Here, neither CH_4 nor SiH_4 have any tendency to hydrogen bond because they are nonpolar.

X = halogen atom in HX.

8.4 VAPOR PRESSURES OF LIQUIDS

When a liquid evaporates from an open container, all the liquid will eventually disappear because the molecules that have escaped simply diffuse into the atmosphere. What will happen, however, if the same quantity of liquid at the same temperature is placed in a closed container? In this case the volume of the liquid will initially decrease and then eventually become constant. If we monitor the pressure of the gas above the liquid, we find that it initially increases and then it too levels off at a constant value. These observations can be explained in the following way. The molecules with higher kinetic energies begin to leave the liquid, evaporating into the vapor phase where they become trapped. The loss of molecules from the liquid must, of course, be accompanied by a volume decrease. As time passes the space above the liquid becomes filled with more and more gaseous molecules, and the pressure of the vapor increases. As the number of gas molecules increases, the number of collisions with the walls in this restricted volume also increases. One of these walls is the surface of the liquid itself, which will trap any bombarding molecules having low kinetic energies. Thus, condensation as well as evaporation (vaporization) take place at the surface of the liquid. Eventually the number of molecules in the vapor becomes large enough so that the rate at which the gas condenses exactly equals the rate at which the liquid evaporates. Once this happens no further change in either the volume of the liquid or the pressure exerted by its vapor is observed. To emphasize this point again, vaporization and condensation are

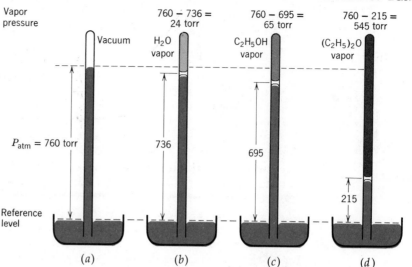

Figure 8.13

Measurement of vapor pressure. When a small amount of liquid is introduced above the mercury in a barometer, the vapor pressure of the liquid forces the mercury down. (a) No liquid above the mercury. (b) H_2O. (c) Ethyl alcohol, C_2H_5OH. (d) Diethyl ether, $(C_2H_5)_2O$.

A similar dynamic equilibrium was described in Chapter 6 for the dissociation of weak acids.

still taking place, but with no change in the liquid volume or the vapor pressure. At this point the liquid is said to be in *dynamic equilibrium* with its vapor. The pressure exerted by the vapor above the liquid when equilibrium is established is called the **equilibrium vapor pressure** of the liquid.

The vapor pressure of a liquid (usually, the term *vapor pressure* is used to mean *equilibrium vapor pressure*—the word "equilibrium" is understood) depends on the ease with which its molecules can leave the liquid. In liquids where the intermolecular attractive forces are strong, the vapor pressure will be low, and vice versa. Since increasing the temperature of a liquid increases the number of molecules possessing sufficient energy to overcome the attractive forces, the vapor pressure must increase with increasing temperature. Therefore, whenever the vapor pressure is given, the temperature at which it was measured must also be specified.

One way that we can determine the vapor pressure of a liquid is by the use of a barometer, as shown in Figure 8.13. First, the height of the mercury in the barometer is measured accurately. Next, the liquid whose vapor pressure is to be determined is carefully added to the barometer by means of an eye dropper, and allowed to rise to the top of the mercury in the column, as shown in Figures 8.13b, 8.13c, and 8.13d (most liquids are less dense than mercury and will therefore float on the mercury surface). The space above the mercury column in Figure 8.13a is, for all practical purposes a vacuum[1], so there is no pressure exerted on the top of the mercury column. The space above the mercury in Figures 8.13b, 8.13c, and 8.13d, however, contains a very small amount of liquid plus its vapor. As the liquid evaporates, the pressure of the trapped vapor pushes mercury out of the barometer and causes the level of the mercury in the column to fall, and when the liquid and vapor are finally in equilibrium, the height of the mercury column becomes stationary.

The total pressure exerted at the reference level outside each barometer will be the atmospheric pressure, P_{atm}. The total pressure exerted within the barometer is P_{Hg}, the pressure resulting from the pull of gravity on the mercury in the column, plus P_{vapor}, the pressure exerted by the vapor in equilibrium with its liquid. The additional pressure exerted by the weight of the small amount of liquid on top of the column is negligibly small. Therefore, at equilibrium in each barometer

$$P_{atm} = P_{Hg} + P_{vapor}$$

[1] Mercury itself does have a finite vapor pressure (about 10^{-3} torr at room temperature) and, therefore, should never be left in an open container because of its high toxicity.

In Figure 8.13a, $P_{vapor} = 0$; therefore, $P_{atm} = P_{Hg} = 760$ torr. In Figure 8.13b, 8.13c, and 8.13d, $P_{Hg} = 736$ mm, 695 mm, and 215 mm, respectively. Therefore, at 25°C the vapor pressure of water is 24 torr, that of ethyl alcohol is 65 torr, and that of diethyl ether is 545 torr. Water has the lowest vapor pressure of the three liquids in our example; therefore, it must have the strongest intermolecular attractive forces. Diethyl ether, on the other hand, has the highest vapor pressure of the three liquids, which means that relatively weak attractive forces exist in it.

We have seen that the vapor pressure of a liquid depends on the nature of the liquid and its temperature. What happens to the vapor pressure if the volume of the vapor is changed? Suppose that a liquid-vapor equilibrium has been established in the apparatus in Figure 8.14a. The arrows indicate that evaporation and condensation are occurring at equal rates. Now suppose that the piston is pulled up, as shown in Figure 8.14b. This expansion will cause a sudden drop in the pressure of the vapor, which means that fewer collisions will be occurring each second with each square centimeter of the walls. Each square centimeter of the liquid's surface will now be receiving fewer returning vapor molecules, but we've done nothing to affect the rate of evaporation. As a result, evaporation will be occurring faster than condensation. More liquid will therefore evaporate until the concentration of molecules in the vapor is sufficient to make the rate of condensation equal to the rate of evaporation once again. At that point, equilibrium will be reestablished and, provided the temperature hasn't changed, the vapor pressure will have returned to its previous value. At this newly established equilibrium the larger volume of gas is now occupied by more molecules. The vapor pressure will be the same as before the volume change occurred, but the volume of the liquid will be slightly smaller.

Decreasing the volume of the vapor by lowering the piston (Figure 8.14c) will also disturb the equilibrium. Increasing the pressure of the vapor will cause an increase in the number of collisions each second with each square centimeter of the walls. This, in turn, will lead to an increase in the rate of condensation, but will have essentially no effect on the rate of evaporation. The rate at which the molecules leave the vapor will consequently be greater than the rate at which molecules leave the liquid. This imbalance in rates causes the pressure exerted by the vapor to decrease and the volume of the liquid to increase. Eventually, the rate of condensation will decrease to a point where it exactly equals the rate of evaporation, reestablishing equilibrium. At this new equilibrium the smaller vapor volume, caused by the movement of the piston, will be occupied by fewer gaseous molecules. The vapor pressure will have returned to its initial value, and the volume of the liquid will have increased slightly.

The net result of this discussion is that the *vapor pressure of a liquid is independent of the volume of the container, provided that there is some liquid present so that an equilibrium can be established.*

Recall that in Chapter 7 we saw that the vapor pressure of water depends only on its temperature.

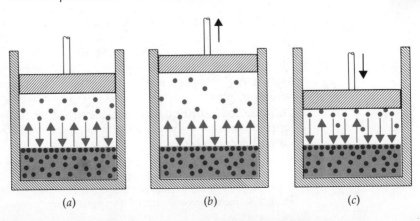

Figure 8.14

Effect of volume changes on vapor pressure. (a) Equilibrium between liquid and vapor. (b) No equilibrium. Rate of evaporation is greater than rate of condensation. (c) No equilibrium. Rate of condensation is greater than rate of evaporation.

(a)   (b)   (c)

Le Châtelier's principle

The dynamic equilibrium between a liquid and its vapor can be represented by the equation

$$\text{liquid} \rightleftharpoons \text{vapor} \qquad [8.1]$$

Here the double half-arrows mean that the rates of evaporation and condensation are equal. If we in any way disturb this system so that it is no longer at equilibrium (we say that a "stress" is applied to the system), a change occurs that will, if possible, bring the system back to equilibrium. In our example above, an increase in the volume of the vapor caused the system to no longer be at equilibrium. We saw that more liquid evaporated until equilibrium was reestablished. In Equation 8.1, this corresponds to the process as read from left to right—that is, liquid → vapor—and results in a new "position of equilibrium" in which there is less liquid and more vapor. In this sense the position of equilibrium has shifted to the right when we applied the stress. The action taken by any system at equilibrium when a stress is applied can be described by **Le Châtelier's principle,** which states that *when a system in a state of dynamic equilibrium is acted on by some outside stress, the system will, if possible, shift to a new position of equilibrium to minimize the effect of the stress.*

To see how this works, let's apply Le Châtelier's principle to the effect that a volume change has on a liquid-vapor equilibrium. When the piston in the apparatus in Figure 8.14 is pulled upward, the volume of the vapor is increased and the pressure of the vapor drops, so the system is no longer at equilibrium. This pressure decrease is the stress that we've placed on the system. How can the system respond to counteract the stress—that is, how can it bring the pressure up again? The answer, of course, is that the pressure of the vapor will rise if more liquid evaporates. Le Châtelier's principle tells us, therefore, that decreasing the pressure (by increasing the volume) causes more liquid to evaporate. After equilibrium has been reestablished, there will be less liquid and more vapor in the container, and we say that the position of equilibrium represented by Equation 8.1 has shifted to the right.[2]

The effect of temperature changes on an equilibrium can also be described using Le Châtelier's principle. Increasing the temperature of a system at equilibrium favors the absorption of energy (an endothermic change). In a liquid-vapor equilibrium, this means that a temperature increase causes more liquid to evaporate, because evaporation absorbs heat. Decreasing the temperature (removing heat), on the other hand, favors the release of energy—an exothermic change. As the temperature is decreased in a liquid-vapor equilibrium, more molecules condense into the liquid phase, releasing heat and thereby minimizing the effect of the applied stress.

In summary, Le Châtelier's principle predicts that a temperature increase will shift the position of equilibrium in the direction of the endothermic process. Similarly, a decrease in temperature will favor the exothermic change.

Vapor pressure curves for liquids

The vapor pressures of liquids as a function of temperature can be determined by using the same apparatus described in Figure 8.13 and varying the surrounding temperature. Data accumulated in such experiments performed on water, ethyl alcohol, and diethyl ether are illustrated in Figure 8.15. Points along a curve in Figure 8.15 represent combinations of pressures and temperatures that must be satisfied for the liquid to be in equilibrium with its vapor.

---

[2] If the volume of the vapor is increased sufficiently, all the liquid will evaporate and equilibrium cannot be reestablished. This will occur, for example, if the piston in Figure 8.14 is removed entirely so that the liquid is open to the atmosphere.

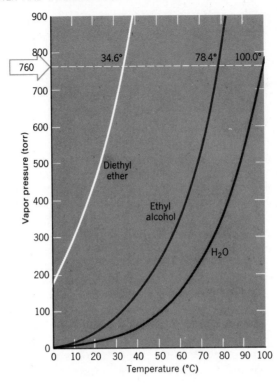

Figure 8.15

Vapor pressure curves.

Above its critical temperature, a substance behaves as a gas.

Vapor pressure-temperature curves like those in Figure 8.15 do not continue indefinitely. Consider, for example, what happens as a sealed container half full of liquid is heated. At first there is a distinct boundary or surface between the more dense liquid and the less dense vapor. As the temperature is raised, more liquid evaporates and more molecules accumulate in the vapor. Since the number of molecules per cubic centimeter is increasing, the density of the vapor is also increasing. At the same time, the liquid becomes less dense—it expands just as the liquid mercury in a thermometer does. As the temperature rises, therefore, the increasing density of the vapor gradually approaches the decreasing density of the liquid. Eventually a temperature is reached at which they become identical: both have the same number of molecules per cubic centimeter and there is really no difference between them. If we were watching the liquid-vapor boundary in the sealed container, we would see that it suddenly disappears when the two phases become the same.

The highest temperature at which a distinct liquid phase exists is called the **critical temperature,** symbolized T_c. The vapor pressure at the critical temperature is called the **critical pressure,** P_c. Vapor pressure-temperature curves terminate at T_c and P_c and the point at the end of the curve is the **critical point.** Critical temperatures and pressures of some substances are given in Table 8.2.

When the temperature of a substance is below its critical temperature its vapor can be liquefied by raising the pressure. Above the critical temperature, however, it can't be liquefied no matter what the pressure is because only one phase can exist. Helium, for example, can't be liquefied until it is first cooled to at least $-267.8°C$.

Each substance has its own characteristic values for T_c and P_c that are controlled by the strengths of the intermolecular attractions. When these attractions are very weak, as with helium, the molecules must be slowed down a great deal before they will stick together when they collide. This means that the gas must be cooled to a low temperature.

Table 8.2
Some critical temperatures and pressures

Compound	T_c (°C)	P_c (atm)
Methane (CH_4)	−82.1	45.8
Ethane (C_2H_6)	32.2	48.2
Benzene (C_6H_6)	288.9	48.6
Ammonia	132.5	112.5
Carbon dioxide	31	72.9
Water	374.1	217.7
Helium	−267.8	2.3

8.5 BOILING POINT

A glance at the photograph in Figure 8.16 tells you that the liquid in the beaker is boiling. Large bubbles are forming within the liquid and rising to the surface. When a bubble is formed, the liquid that originally occupied this space is pushed aside and the level of the liquid in the container is forced to rise against the downward pressure exerted by the atmosphere. In other words, it is the pressure exerted by the vapor inside the bubble that pushes the surface of the liquid up against the atmospheric pressure. This can occur only when the vapor pressure of the liquid becomes equal to the prevailing atmospheric pressure. If it were less, the atmospheric pressure would cause the bubble to collapse. The temperature at which the liquid boils—its **boiling point**—is therefore the temperature at which its vapor pressure equals the atmospheric pressure.

As long as bubbles are forming within the liquid—that is, as long as the liquid is boiling—the vapor pressure of the liquid is equal to the atmospheric pressure. Since the vapor pressure remains constant, the temperature of the boiling liquid also stays the same. An increase in the rate at which heat is supplied to the boiling liquid simply causes bubbles to form more rapidly. The liquid boils away more quickly, but the temperature does not increase.

Figure 8.16

Just a casual glance at the water in this beaker would tell you that it is boiling. Why?

The normal boiling points of water, ethyl alcohol, and diethyl ether can be read from the graph in Figure 8.15.

It is obvious, from this discussion, that the boiling point of a liquid depends on the prevailing atmospheric pressure. The boiling point of a liquid at 1 atm (760 torr) is referred to as its **normal boiling point.** For water, the normal boiling point is 100°C. At higher pressures its boiling point is greater; at lower pressures (for example, on a mountaintop) its boiling point is less. Boiling points given in reference tables are always normal boiling points, unless otherwise stated.

The constant temperature maintained by a boiling liquid is relied on when we use water to cook foods. Once water boils, its temperature remains at 100°C, which is ideal for cooking foods evenly and rapidly. As long as water surrounds the food we know that the food won't burn. A pressure cooker also takes advantage of the fact that the boiling point changes with pressure. These cookers are time-savers because they allow foods to be prepared at a much faster rate than possible in an open pot. The lid on a pressure cooker forms a tight seal on the pot and is equipped with a pressure-relief valve to prevent the pot from exploding. The heat supplied by the stove causes more and more liquid water to evaporate; as a result, the pressure inside the kettle increases until steam begins to exit from the relief valve. Since the pressure inside the cooker at this point is higher than 760 torr, the water boils at a higher temperature and foods cook faster.

The temperature at which a liquid boils is another example of a property that gives a good estimation of the strength of the attractive forces operating within a liquid. Liquids whose attractive forces are relatively high have correspondingly high boiling points, while liquids with weak attractive forces boil at relatively low temperatures. In Table 8.1, for example, we see that the trends in boiling points follow trends in ΔH_{vap}.

The dependence of boiling point on intermolecular attractive forces is also seen in Figure 8.17, in which the boiling points of some hydrogen compounds of the elements in Groups IVA, VA, VIA, and VIIA are compared. Let's look at the compounds of Group IVA first because they form a nearly ideal pattern. We see from the figure that as the atomic weights of the elements in Group IVA increase, so do the boiling points of their hydrogen compounds. Using reasoning similar to that in Section 8.3, we know that as the size of the molecules increase from CH_4 to SnH_4, the London forces also increase. We expect, therefore, that the boiling points of these compounds should increase, as they actually do, in the direction of increasing molecular weights.

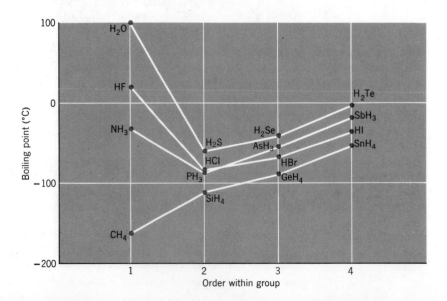

Figure 8.17

Boiling points of hydrogen compounds of Group IVA, VA, VIA, and VIIA elements.

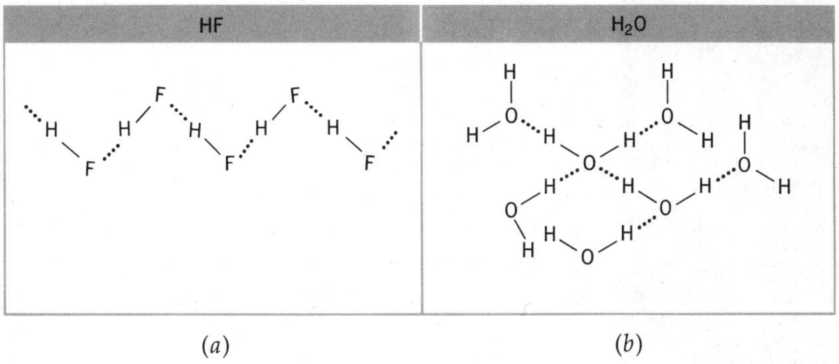

Figure 8.18

Hydrogen bonding in HF and H_2O. (a) Each HF molecule has only one hydrogen atom that can hydrogen bond to something else. (b) Each H_2O molecule has two hydrogen atoms that can hydrogen bond to other H_2O molecules.

Except for the first members of the hydrogen compounds of Groups VA, VIA, and VIIA, the same trend is also observed—that is, increasing boiling point with increasing atomic weight of the element in the group. The first member of each of these groups, however, has a relatively high boiling point, much higher than expected from molecular weights alone. They, therefore, must possess attractive forces in addition to those of the London type. The position of each of the first members in relation to the rest of the group can be attributed to the presence of hydrogen bonding. We saw in Section 8.3 that the strongest hydrogen bonds form when hydrogen is bonded to a very small electronegative element. Therefore, the contribution of hydrogen bonding is expected to be the strongest for the first member of Groups VA, VIA, and VIIA, and becomes relatively unimportant for the remaining members. Methane, CH_4, which is nonpolar and cannot hydrogen bond (the electronegativity of carbon is too low and there are no lone electron pairs on the carbon to which hydrogen bonds can be formed) follows the normal, nearly straight-line pattern, within its group.

You may have noticed that water has a higher boiling point than hydrogen fluoride, even though HF is more polar than H_2O. The reason seems to be the number of hydrogen bonds that they can form. In HF, each molecule is hydrogen bonded to two others, whereas in water, each molecule can be hydrogen bonded to four others (Figure 8.18). Therefore, even though HF forms somewhat stronger hydrogen bonds than water, the total strength of four hydrogen bonds to a water molecule exceeds the total strength of two hydrogen bonds to an HF molecule.

The hydrogen bonding in NH_3 is much weaker than in H_2O or HF because of the considerably lower electronegativity of nitrogen. In addition, the nitrogen in NH_3 has only one lone pair to which a hydrogen bond can be formed, so each molecule, on the average, can be held by only two hydrogen bonds: one *from* another NH_3 molecule and one *to* another NH_3 molecule. Altogether, the total strength of the hydrogen bonding in NH_3 is so small that NH_3 has a lower boiling point than either HF or H_2O.

8.6 FREEZING POINT

Anyone who has ever made a tray of ice cubes in a refrigerator realizes that liquids freeze if heat is removed from them. You also know that ice cubes melt when they absorb heat. For any substance at a given pressure there is a characteristic temperature at which the liquid and solid can coexist in equilibrium. This is called either the **freezing point** or **melting point,** depending on whether you imagine approaching it from high or low temperature. At the freezing point (melting point) the rate at which particles leave the solid and enter the liquid is the same as the rate at which particles leave the liquid and join the solid. If heat

Fusion means melting. The thin metal band in an electrical fuse protects a circuit by melting if too much current is passed through it. On the right is a fuse that has done its job.

is added, some solid melts and more liquid is formed, but the temperature stays the same as long as both phases are present. Similarly, if some heat is removed, some liquid freezes and more solid forms—again without a temperature change.

As with evaporation and condensation, there are energy changes associated with freezing and melting. The **molar heat of crystallization,** ΔH_{cryst} *is the amount of energy that must be removed from 1 mol of a liquid to convert it to a solid at the same temperature.* For melting, which is also called *fusion*, there is the **molar heat of fusion,** ΔH_{fus}—*the energy needed to melt 1 mol of a solid.* Following the definition given for the molar heat of vaporization, we can express these as

$$\Delta H_{cryst} = H_{solid} - H_{liquid}$$

$$\Delta H_{fus} = H_{liquid} - H_{solid}$$

We see that the numerical values of ΔH_{cryst} and ΔH_{fus} must be identical; one is simply the negative of the other.

The size of the molar heat of fusion (or crystallization) is a measure of the difference in the strengths of the attractive forces between the liquid and solid, and it is always much smaller than the molar heat of vaporization, as shown in Table 8.3. When a solid melts there are relatively small changes in the distances between the molecules. Therefore, only small changes in potential energy are involved. When a liquid is converted to a gas, however, the intermolecular distances increase tremendously and large energy changes occur. This means that the amount of energy (ΔH_{fus}) required to cause the molecules of a solid to overcome their attractive forces and form a liquid is small compared to the energy (ΔH_{vap}) required for liquid molecules to move apart, forming a gas.

Table 8.3
Heats of fusion and vaporization

Substance	ΔH_{fus}		ΔH_{vap}	
	kJ/mol	kcal/mol	kJ/mol	kcal/mol
Water	5.98	1.43	40.6	9.71
Benzene	9.92	2.37	30.7	7.35
Chloroform	12.4	2.97	31.9	7.62
Diethyl ether	6.86	1.64	26.0	6.21
Ethanol	7.61	1.82	38.6	9.22

EXAMPLE 8.2

Calculate the energy, in kilojoules, necessary to melt 1.00 g of ice.

SOLUTION

As before, we have a unit conversion requiring ΔH_{fus}. The problem reduces to

$$1.00 \text{ g } H_2O \sim (?) \text{ kJ}$$

From Table 8.3, $\Delta H_{fus} = 5.98$ kJ/mol for H_2O. Therefore,

$$1.00 \text{ g } H_2O \times \left(\frac{1 \text{ mol } H_2O}{18.0 \text{ g } H_2O}\right) \times \left(\frac{5.98 \text{ kJ}}{1 \text{ mol } H_2O}\right) \sim 0.332 \text{ kJ}$$

The energy needed is 0.332 kJ.

Figure 8.19
Snowflakes are crystals of ice. Like other crystals, they have regular features that reflect the ordered arrangement of the particles within them.

8.7 CRYSTALLINE SOLIDS

When most substances freeze, or when they are created in a precipitation reaction, they form crystals that have highly regular, symmetrical shapes. You have probably seen photographs of snowflakes similar to those in Figure 8.19. Notice how symmetrical each ice crystal is and how each has a characteristic hexagonal form, although the individual fine details are different.

The highly regular surface features of a crystal are a reflection of an orderly repeating pattern of atoms, molecules, or ions that exist within it. This order has made possible detailed analyses of the structures of solids and has led to much of our knowledge of the shapes of molecules and the sizes of atoms and ions.

X-ray diffraction

In 1912, a German physicist named Max von Laue pointed out that a crystal could serve as a three-dimensional diffraction grating if the wavelength of the incident radiation were of the same order of magnitude as the distance between particles in the solid. This condition is fulfilled by X rays, which have wavelengths of approximately 1 Å (0.1 nm or 10 pm).

When a crystal is bathed in X rays, each atom of the crystal within the path of an X ray absorbs some of its energy and then reemits it in all directions. Thus each atom is a source of secondary wavelets, and the X rays are said to be scattered by the atoms. These secondary wavelets from the different sources interfere with each other, either reinforcing or cancelling each other. In certain directions the waves emanating from nearly all the atoms in any orderly array are in phase—that is, the peaks and troughs of the waves coincide as shown in Figure 8.20*a*—and intense beams of X rays are observed in these directions. In all other directions the waves from various atoms are out of phase (Figure 8.20*b*) and cancel each other; thus no intensity is detected.

In the older scientific literature, X-ray wavelengths and atomic dimensions are given in Ångstroms. Today, the SI units nanometers or picometers are preferred.

1 Å = 0.1 nm

1 Å = 10 pm

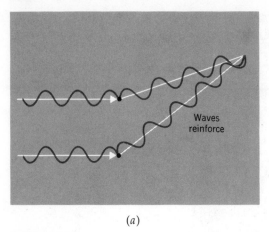

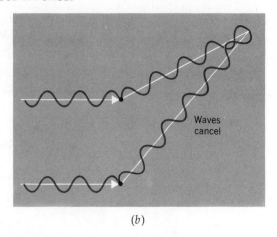

(a)   (b)

Figure 8.20

Diffraction. Scattered waves re-inforce each other only at certain angles. (a) In phase. (b) Out of phase.

The Braggs received the 1915 Nobel Prize for physics for their work.

Two English scientists, William Bragg and his son Lawrence, treated the diffraction of X rays as if the process were reflection. In Bragg's treatment the X rays that penetrate a crystal are thought of as being reflected by successive layers of particles within the substance (Figure 8.21). We can see from this diagram that beams reflected from deeper layers must travel further to reach the detector. For there to be any intensity at the detector these waves have to be in phase with those reflected from the upper layers, which must mean that the extra distance traveled by the more penetrating beam has to be some integral multiple of the wavelength of the X rays.

Bragg showed that in order to observe any intensity in the emerging X rays, a relatively simple relationship had to be fulfilled. This relationship, known as the **Bragg equation,** is

$$2d \sin \theta = n\lambda \qquad\qquad [8.2]$$

where d is the spacing between the successive layers that are reflecting the X rays, θ is the angle at which the X rays enter and leave the particular set of layers, λ is the wavelength of the X rays, and n is an integer (that is, $n = 1$, or 2, or 3, etc.). The Bragg equation serves as the basis for the study of crystalline structure by X-ray diffraction.

In practice, X rays of known wavelength are directed at a crystal and the angles at which they are reflected are recorded—for example, on a piece of photographic film (Figure 8.22). By measuring the angles at which the X rays are re-

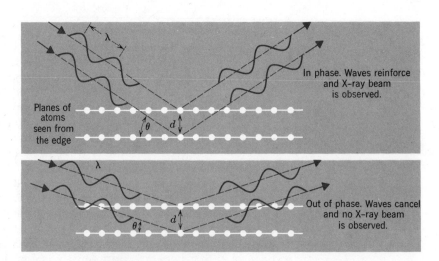

Figure 8.21

Bragg's law. Bragg showed that the waves from different planes of atoms are in phase only when 2d sin θ = nλ.

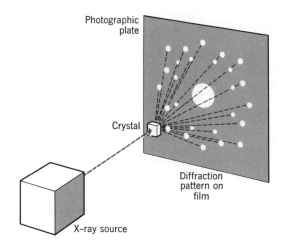

Figure 8.22

Production of an X-ray diffraction pattern.

flected, it is a simple matter to calculate the distances between planes of atoms within a crystal, as illustrated in Example 8.3. If, in addition, the intensities of the reflected X rays are measured, a crystallographer may be able to deduce, through a rather complex procedure, the actual positions of atoms within the solid. In this way the molecular structures of many substances have been found. In recent years, X-ray diffraction has become a powerful tool in biochemistry by which the structures of even very complex molecules have been investigated. For example, Rosalind Franklin's X-ray data led James Watson, Francis Crick, and Maurice Wilkins to their deduction of the double-helix structure of DNA—a feat that won Watson, Crick, and Wilkins the Nobel Prize in 1962.

Dorothy Hodgkin won the Nobel Prize in 1964 for her X-ray structure determination of vitamin B_{12}.

EXAMPLE 8.3   X rays of wavelength 1.54 Å strike a crystal and are observed to be reflected at an angle of 22.5°. Assuming that $n = 1$, calculate the spacing between the planes of atoms that are responsible for this reflection.

SOLUTION   We wish to calculate d. Solving Equation 8.2 for d, we have

$$d = \frac{n\lambda}{2 \sin \theta}$$

From the data, $n = 1$, $\lambda = 1.54$ Å, $\theta = 22.5°$. Substituting,

$$d = \frac{(1)(1.54 \text{ Å})}{2 \sin(22.5)}$$

$$d = \frac{1.54 \text{ Å}}{2(0.383)}$$

$$= 2.01 \text{ Å}$$

8.8   LATTICES

Any repetitive pattern has a symmetrical aspect about it, whether it be the stacking of bricks in a wall, a wallpaper design, or the orderly packing of particles in a crystal. For example, certain repeating distances between the elements of the pattern can be easily recognized, and the lines along which the elements of the pattern are repeated are at certain angles to each other.

Symmetry in some
repetitive patterns

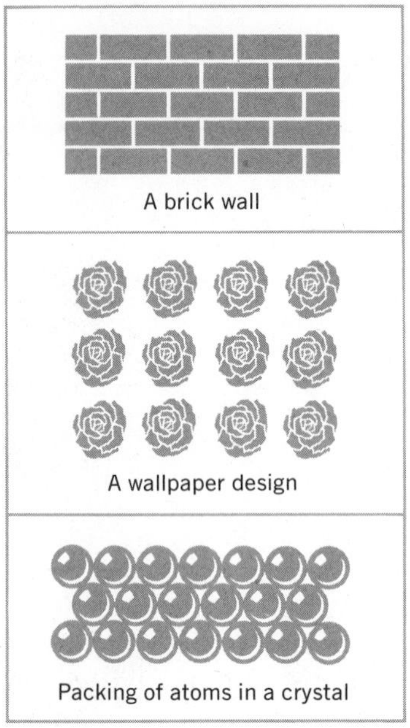

A brick wall

A wallpaper design

Packing of atoms in a crystal

To avoid having to deal with the details of a repeating structure, and to concentrate on its symmetrical features, it is convenient to describe the structure simply in terms of a set of points that have the same repeat distances as the structure, arranged along lines oriented at the same angles. This kind of pattern of points is called a **lattice,** and when applied to the description of the packing of particles in a solid, we often use the term **crystal lattice.**

In a crystal the number of particles is enormous. If you could imagine being in the center of even the tiniest crystal, you would find that the particles go on as far as you could see in every direction. Describing the positions of all these particles or their lattice points is impossible and, fortunately, unnecessary. All we need to do is to describe the basic repeating unit of the lattice. To see this, let's begin in two dimensions.

Figure 8.23a illustrates a two-dimensional lattice. It is a *square lattice* because the lattice points lie at the corners of squares. The repeating unit of this lattice—its **unit cell**—is the square drawn in white. If we began with this unit cell, we could produce the entire lattice simply by moving it repeatedly left and right, and up and down by distances equal to its edge length. In this sense, all the properties of a lattice are contained in the properties of its unit cell.

An important fact about lattices is that the same *kind* of lattice can be used to describe many different designs. For example, if we decided to place a diamond at each lattice point, we could create a wallpaper design like that shown in Figure 8.23b. A different wallpaper pattern could be created by placing a rose at each lattice point, or by changing the lengths of the edges of the unit cell.

The extension of the lattice concept to crystals in three dimensions is quite straightforward. In Figure 8.24 we see an example of a simple cubic lattice in which a unit cell is shaded in color. By associating a particular chemical environment with each lattice point we can arrive at a chemical structure, and by varying the chemical environment about each point we can create an infinite number of chemical structures all based on the same lattice.

The unit cells of all three-dimensional lattices are similar in that they all have eight corners, as shown in Figure 8.25. Unit cells differ by the lengths of their edges (a, b, and c) and by the angles opposite them (α, β, and γ). In 1848, Auguste Bravais showed that *only* 14 different kinds of lattices are possible. These can be divided among seven basic crystal systems, whose properties are described in Table 8.4. What this means is that the crystals of all the millions of different chemical compounds ever discovered can be described by just this small set of lattices. It is this fact that has made a detailed understanding of the solid state possible.

Three common kinds of lattices are characterized by the cubic unit cells shown in Figure 8.26. The simplest is called a **simple cubic unit cell** or a **primitive cubic unit cell.** It has lattice points only at the corners. When oxygen is frozen it has a primitive cubic lattice with this kind of unit cell.

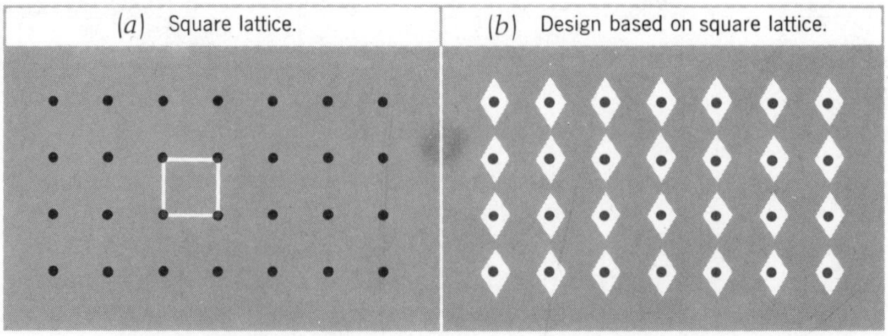

Figure 8.23

(a) A portion of a two-dimensional square lattice with its unit cell drawn in white. (b) A design based on a square lattice.

| (a) Square lattice. | (b) Design based on square lattice. |

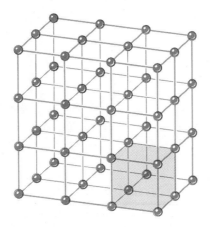

Figure 8.24
A simple cubic lattice.

Table 8.4
Properties of the unit cells
of the seven crystal systems

System	Edge Lengths	Angles
Cubic	$a = b = c$	$\alpha = \beta = \gamma = 90°$
Tetragonal	$a = b \neq c$	$\alpha = \beta = \gamma = 90°$
Orthorhombic	$a \neq b \neq c$	$\alpha = \beta = \gamma = 90°$
Monoclinic	$a \neq b \neq c$	$\alpha = \beta = 90° \neq \gamma$
Triclinic	$a \neq b \neq c$	$\alpha \neq \beta \neq \gamma$
Rhombohedral	$a = b = c$	$\alpha = \beta = \gamma \neq 90°$
Hexagonal	$a = b \neq c$	$\alpha = \beta = 90°; \gamma = 120°$

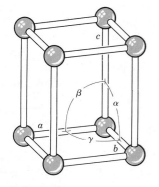

Figure 8.25
A three-dimensional unit cell.
The edges a, b, and c intersect at
angles α, β, and γ.

A **body-centered cubic unit cell** has lattice points at the corners, plus a lattice point in the center of the cell. Some common metals—chromium, iron, and tungsten, for example—form crystals with a body-centered cubic lattice.

A **face-centered cubic unit cell** has lattice points at the eight corners and one in the center of each of its six faces. This produces a very common kind of lattice that is found in crystals of such metals as nickel, copper, silver, gold, and aluminum.

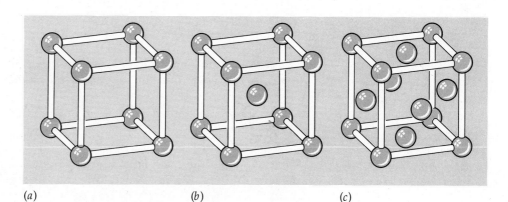

Figure 8.26
The three unit cells belonging to
the cubic system. (a) Simple
cubic. (b) Body-centered cubic.
(c) Face-centered cubic.

(a) (b) (c)

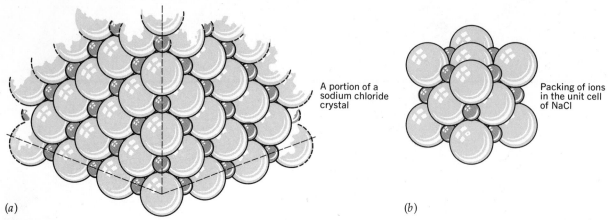

A portion of a sodium chloride crystal

Packing of ions in the unit cell of NaCl

(a)

(b)

Figure 8.27

The crystal structure of sodium chloride. (a) A portion of a NaCl crystal showing the packing of chloride ions (large spheres) and sodium ions (small spheres). (b) The arrangement of ions in the face-centered cubic unit cell of NaCl.

The mineral rock salt is NaCl.

The unit cell could also be drawn with Na^+ at the lattice points and Cl^- in between.

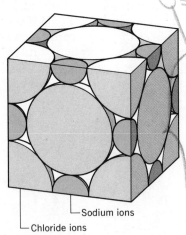

Sodium ions

Chloride ions

Figure 8.28

Counting ions in the face-centered cubic unit cell of sodium chloride.

Cubic lattices are not only characteristic of many elements. Some important and familiar compounds also have cubic lattices. For example, one of our most common compounds, sodium chloride, forms crystals having a face-centered cubic lattice. A portion of a NaCl crystal is shown in Figure 8.27 along with its unit cell. Notice how the chloride ions are located at positions corresponding to the lattice points, with the sodium ions squeezed in between. This kind of packing of cations and anions is called the rock salt structure. Similar structures are found for many of the other alkali halides as well—for instance, KCl and LiCl.

Among the factors that determine the kind of lattice and structure an ionic compound can form are the relative sizes of the ions and the ratio of the numbers of anions to cations in the crystal. The question of size is rather complex, so we won't discuss it further, but the importance of the anion/cation ratio is not very difficult to see.

Since a crystal is composed of a large number of unit cells, whatever the anion/cation ratio is in the crystal as a whole, it must also be the same in the unit cell. Let's count the number of chloride and sodium ions in the unit cell of NaCl to show that their ratio is one-to-one. In doing this, however, we have to be careful, because ions at the corners, along the edges, and in the face-centers are shared with one or more other unit cells.

An ion at the corner of the unit cell is shared with seven others. In Figure 8.28 we see that only one-eighth of such an ion is in a given unit cell. We also see that an ion along an edge, which is shared among four unit cells, has only one-fourth of itself in a given unit cell. An ion in the center of a face contributes half to a given unit cell, because it is shared between two of them. In addition to these, there is one Na^+ ion that can't be seen in Figure 8.28. It is located in the center of the unit cell, as shown in Figure 8.29.

Now we can count the ions. For the chloride ions, we see that they are at the eight corners and in the centers of the six faces.

$$8 \text{ corners} \times \tfrac{1}{8}Cl^- \text{ per corner} = 1Cl^-$$
$$6 \text{ faces} \times \tfrac{1}{2}Cl^- \text{ per face} = 3Cl^-$$
$$\text{Total} \quad \overline{4Cl^-}$$

Thus altogether there are four chloride ions contained within the unit cell. For the sodium ions, we have one along each of the 12 edges of the cube, which

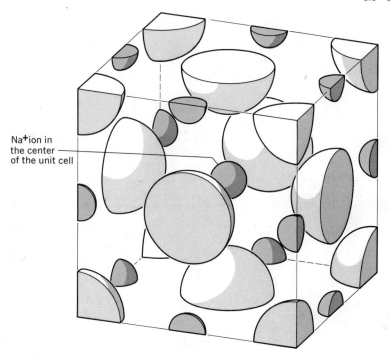

Na⁺ion in the center of the unit cell

Figure 8.29

Exploded view of the NaCl unit cell showing the Na⁺ ion in the center.

each contribute $\frac{1}{4}$ to the unit cell, plus one in the center that is entirely within the unit cell.

$$12 \text{ edges} \times \tfrac{1}{4}Na^+ \text{ per edge} = 3Na^+$$
$$1Na^+ \text{ in center} = \underline{1Na^+}$$
$$\text{Total} \quad 4Na^+$$

The number of sodium ions in the unit cell is also four. This means, therefore, that the Na^+ and Cl^- ions are in a one-to-one ratio, which is necessary for the crystal to be electrically neutral.

Any substance that crystallizes with the rock salt structure *must* have a one-to-one anion/cation ratio. Sodium chloride and the other alkali halides have formulas that satisfy this condition, and many of them form crystals having this structure. Calcium oxide, CaO, also has the rock salt structure, which is allowed because of its formula. However, $CaCl_2$ or Al_2O_3 could not possibly form crystals with the rock salt structure because their anion/cation ratios forbid it. Thus, we see that the formula of a compound places certain restrictions on the kinds of crystal structures it can and cannot have.

This does not mean that all substances with a 1-to-1 anion/cation ratio must crystallize with the rock salt structure.

EXAMPLE 8.4   Sodium forms crystals having a body-centered cubic lattice. How many sodium atoms are contained within one unit cell?

SOLUTION   The unit cell has sodium atoms at the eight corners, each of which is shared among a total of eight cells, plus one sodium atom in the center.

$$8 \text{ corners} \times \tfrac{1}{8}Na \text{ per corner} = 1Na$$
$$1 \text{ Na in the center} = \underline{1Na}$$
$$\text{Total} = 2Na$$

Each unit cell has two sodium atoms within it.

Atomic and ionic radii

The study of the structures of crystals has yielded much useful information. The applications to the determination of molecular structures—even those as complex as DNA—have already been mentioned, although they are far too complex to discuss in any detail here. We can see an illustration of the usefulness of studying crystals, however, by examining how information about unit cells can be used to calculate the radii of atoms and ions.

It was noted in the last section that copper is a metal that crystallizes with a face-centered cubic lattice. X-ray diffraction measurements show that the unit cell has an edge length of 362 pm (3.62 Å) (this is length AC in Figure 8.30). Copper atoms are in contact along the line joining points A and B (the face diagonal). This distance corresponds to four times the radius, r, or a copper atom. From geometry we know that

$$\overline{AB} = \sqrt{2}\,(\overline{AC}) = \sqrt{2}\,(362 \text{ pm})$$

$$\overline{AB} = 512 \text{ pm}$$

Therefore,

$$4r = 512 \text{ pm}$$

$$r = 128 \text{ pm}$$

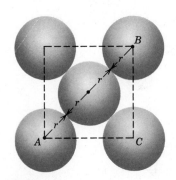

Figure 8.30

Copper atoms in the face of a unit cell.

The radius of a copper atom in metallic copper is therefore 128 pm (1.28 Å or 0.128 nm). In a similar fashion we can also use the results of X-ray diffraction to determine the radii of ions in ionic crystals.

8.9 TYPES OF CRYSTALS

We have seen that there are only a limited number of ways of arranging particles in a crystalline solid. The particular arrangements, as well as the physical properties of the solid, are determined by the types of particles present at the lattice points and the nature of the attractive forces between them. We can divide crystals into types: molecular, ionic, covalent, and metallic.

Molecular crystals

In molecular crystals, either molecules or individual atoms occupy lattice sites. The attractive forces between them are of the type described in Section 8.2 and are much weaker than the covalent bonds that exist within individual molecules. London forces are present in crystals of nonpolar substances such as Ar, O_2, naphthalene (moth crystals), and CO_2 (Dry Ice). In crystals of polar molecules such as SO_2, the dominant forces are a result of dipole-dipole attractions, and in solids such as ice (H_2O), NH_3, and HF, the molecules are held in place by hydrogen bonding. Since these are relatively weak forces (compared to covalent or ionic attractions), molecular crystals tend to have small lattice energies and are easily deformed; we say that they are soft. Also, relatively little thermal energy is required to overcome these attractions, and molecular solids generally tend to have low melting points.

Molecular crystals are poor conductors of electricity because the electrons are bound to individual molecules and are not able to move freely through the solid.

Ionic crystals

In an ionic crystal, such as NaCl, there are ions located at lattice sites and the binding between them is mainly electrostatic (which is essentially nondirec-

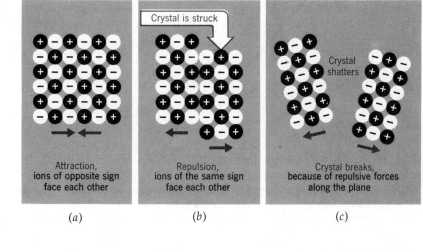

Figure 8.31

An ionic crystal breaks when struck. (a) Attraction between ions opposite each other. (b) When struck, part of the crystal slips past the rest. Ions of the same sign face each other. (c) The repulsive forces push the crystal apart.

Crystal is struck

Attraction, ions of opposite sign face each other

Repulsion, ions of the same sign face each other

Crystal shatters

Crystal breaks, because of repulsive forces along the plane

(a)   (b)   (c)

tional). As a result, the kind of lattice that is formed is determined mostly by the relative sizes of the ions and their charges. When the crystal forms, the ions arrange themselves to maximize attractions and minimize repulsions.

Because electrostatic forces are strong, ionic crystals have large lattice energies. They are hard and are characterized by high melting points. They are also very brittle. When struck they tend to shatter, because as planes of ions slip by one another they pass from a condition of mutual attraction to one of mutual repulsion. This is illustrated in Figure 8.31.

In the solid state, ionic compounds are poor conductors of electricity because the ions are held rigidly in place. When melted, however, the ions are free to move about and ionic substances become good conductors.

Covalent crystals

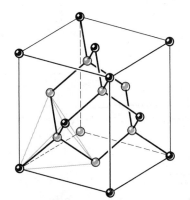

Figure 8.32

The unit cell of diamond. Note how the carbon atoms inside the unit cell are each covalently bonded to four others. The same applies to each of the carbon atoms in the crystal.

In a covalent crystal there is a network of covalent bonds between the atoms that extends throughout the entire solid. An example of such a substance is diamond, whose structure is shown in Figure 8.32. Diamond is a form of elemental carbon in which each atom is covalently bonded to four nearest neighbors. Other common examples are carborundum (silicon carbide, SiC) and quartz (silicon dioxide, SiO_2, commonly recognized as sand).

Because of the interlocking framework of covalent bonds, covalent crystals have very high melting points and are usually extremely hard. Diamond, of course, is the hardest substance known and is used in grinding and cutting tools. Silicon carbide is like diamond, except that half of the carbon atoms in the structure have been replaced by silicon atoms. It too is very hard and is used as an abrasive in sandpaper, as well as in other grinding and cutting applications.

Covalent crystals are poor conductors of electricity because the electrons in the solid are localized in the covalent bonds and are not free to move through the crystal.

Metallic crystals

The simplest picture of a metallic crystal has positive ions (nuclei plus core electrons) situated at lattice points, with the valence electrons belonging to the crystal as a whole instead of to any single atom. The solid is held together by the electrostatic attraction between the lattice of positive ions and this sort of "sea of electrons." These electrons may move freely, so we find metals to be good conductors of electricity. Since the melting points and hardness of metals vary

Table 8.5
Types of solids

	Molecular	Ionic	Covalent	Metallic
Chemical units at lattice sites	Molecules or atoms	Positive and negative ions	Atoms	Positive ions
Forces holding the solid together	London forces, dipole–dipole, hydrogen bonds	Electrostatic attraction between + and – ions	Covalent bonds	Electrostatic attraction between + ions and electron "sea"
Some properties	Soft, generally low melting, nonconductors	Hard, brittle, high melting, non-conductors (but conduct when melted)	Very hard, high melting, non-conductors	Hard to soft, low to high melting, high luster, good conductors
Some examples	CO_2 (Dry Ice) H_2O (ice) $C_{12}H_{22}O_{11}$ (sugar) I_2	NaCl (salt) $CaCO_3$ (limestone; chalk) $MgSO_4$ (in Epsom salt)	SiC (carborundum) C (diamond) WC (tungsten carbide, used in cutting tools)	Na, Fe, Cu, Hg

over wide ranges, there must also be, at least in some cases, some degree of covalent bonding between atoms in the solid.

Table 8.5 summarizes the important properties of these different kinds of solids.

8.10 LIQUID CRYSTALS

In a typical liquid, the molecules have no orderly pattern. They are able to move past each other easily, so liquids are able to flow. In a typical solid, the molecules are highly ordered, but they are also held rigidly in place, so solids do not flow. Certain substances, in a range of temperatures just above their melting points, exhibit properties characteristic of both of these states, and are therefore called **liquid crystals.** They are fluid, but their molecules are arranged in a highly ordered way. At temperatures above this range this order is lost and they become like any other liquid.

There are three kinds of liquid crystals, called nematic, smectic, and cholesteric. They are all composed of rodlike molecules, but they differ in the kind of order that exists within them. In a **nematic liquid crystal** the rodlike molecules are arranged like loosely packed soda straws (Figure 8.33a). The order within a

Figure 8.33
Liquid crystals. (a) Packing in a nematic liquid crystal. (b) Packing in a smectic liquid crystal. (c) Packing in a cholesteric liquid crystal.

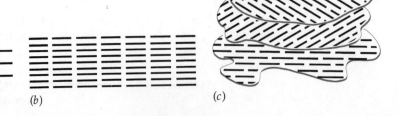

(a) (b) (c)

Liquid crystal displays have become common on pocket calculators and on digital wristwatches such as the one shown here.

smectic liquid crystal is even greater. Here there are parallel rodlike molecules arranged in layers, as shown in Figure 8.33b. In a **cholesteric liquid crystal** the molecules are in layers too, but they are aligned parallel to the layers in a nematic fashion, with slightly different orientations from one layer to the next, as illustrated in Figure 8.33c.

What makes liquid crystals so interesting and potentially useful is the way they affect light. If you've ever seen an oil slick on water, you've probably noticed a rainbow of colors that seems to be reflected from it (Color Plate 6). This is actually caused by diffraction of light from the surfaces of the oil and water. When the oil layer is very thin, all but a narrow band of wavelengths is lost as reflected waves by destructive interference. We see only those waves that are reflected in phase from the two layers—the top of the oil layer and the top of the water layer beneath the oil. This same phenomenon occurs when light reflects from the layers in a cholesteric liquid crystal, but what is especially interesting is that the reflected color changes with temperature. The reason is that the distance between the layers changes with temperature. This allows these substances to be used for a sort of "color mapping" of temperature regions on an object. One medical application is locating a vein by the warmer skin temperature that it produces (Color Plate 7).

The substances used in the liquid crystal displays found in pocket calculators and wristwatches are nematic liquid crystals. Their optical characteristics are affected by an electric field. In this case a thin film of the liquid crystal is sandwiched between transparent electrodes that are arranged on glass in special patterns. When a particular electrode segment is energized, the orientations of the molecules in the liquid crystal are changed and the substance becomes opaque. By activating appropriate segments in this way, various numbers or letters can be formed. An important advantage of these displays is that they use very little energy, so the batteries used to power them last a long while.

8.11 HEATING AND COOLING CURVES; CHANGES OF STATE

When heat is added to a solid, initially at some temperature below its melting point, the temperature begins to rise. Once the melting point is reached, the temperature levels off—it stays constant until all the solid has melted. When more heat is added, the liquid becomes warmer until it reaches its boiling point. As heat is supplied to the boiling liquid, the temperature stays constant again until the liquid has all been converted to a gas. Then the temperature can rise further as still more heat is added. Figure 8.34, which illustrates this graphically for 1 mol of a typical substance, is called a **heating curve.**

In those portions of the heating curve where the temperature is rising, heat that is added is increasing the average kinetic energy of the molecules. But what is happening during melting and boiling—those portions of the curve where the temperature stays the same? Because the temperature stays constant, the average kinetic energy also remains the same. This means that the heat energy being added must be increasing the potential energies of the molecules.

If the K.E. doesn't change, then the P.E. must change if energy is added.

When most solids melt, there is a slight expansion in volume. This means that the particles must be getting slightly further apart. Since there are attractive forces between the particles, energy must be supplied to separate them.[3] This energy is the heat of fusion. Similarly, when a liquid vaporizes, the molecules

---

[3] A most important exception to this, of course, is ice, in which the solid is less dense (more expanded) than the liquid. Here energy supplied to the solid disrupts some of the hydrogen bonding that exists in the solid ice and the "open" structure of ice collapses to give a more dense liquid.

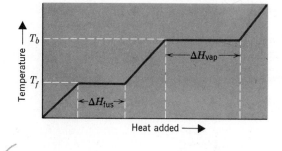

Figure 8.34

A typical heating curve for 1 mol of a substance. T_f is the freezing point and T_b is the boiling point.

You haven't forgotten what you learned in Chapter 1, have you?

go from the closely packed liquid state to their widely spaced distribution in a vapor. This too requires an energy input to overcome the attractive forces—the heat of vaporization. In both melting and vaporization it is the distance between molecules that attract each other that is changing, and, as we learned in Chapter 1, this involves changes in potential energy. Thus, ΔH_{fus} and ΔH_{vap} are energy changes that affect the potential energies of particles.

A cooling curve, like that in Figure 8.35, is essentially the opposite of a heating curve. During condensation and freezing, the temperature stays constant while the potential energies of the particles decrease as they come together.

Some liquids do not follow a smooth transition into the solid state, but instead give rise to a cooling curve such as that shown in Figure 8.36. As the temperature of the liquid drops, it eventually reaches point A, the expected freezing point of the substance. The molecules, however, may not be oriented properly to fit into the crystalline lattice, and random motion continues as heat is further withdrawn from the liquid. Consequently, the temperature of the liquid drops below its expected freezing point and the liquid is said to be **supercooled.** Once a small number of molecules have achieved the correct pattern, a tiny crystal is formed that serves as a seed on which additional molecules may rapidly accumulate. Potential energy is suddenly released as this crystal quickly grows, and the energy that is evolved increases the average kinetic energy of the molecules in the liquid and solid. As a result, the temperature of the system rises again until it returns to the freezing point, after which the substance behaves normally. Further removal of heat eventually leads to complete conversion of the liquid to a solid.

Some substances, such as glass, rubber, and many plastics, never do achieve a crystalline state when their liquids solidify on cooling. These compounds consist of long chainlike molecules that intertwine in the liquid. As they are cooled their molecules move so slowly that they never do find the proper orientation to form a crystalline solid, and an amorphous solid results instead. Thus amorphous solids are actually supercooled liquids and in fact

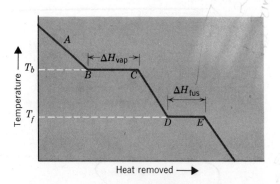

Figure 8.35

A typical cooling curve for 1 mol of a substance.

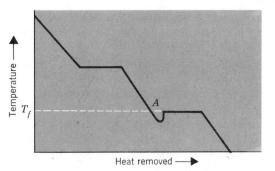

Figure 8.36

Supercooling. As the liquid is cooled, its temperature drops below the freezing point. After a short time, freezing begins and the temperature rises to the freezing point.

continue to flow, although very slowly, even at room temperature. For example, very old glass exhibits greater crystallinity, when examined by X-ray diffraction, than freshly formed glass does, showing that molecules are slowly finding their way into a crystalline lattice. A supercooled liquid familiar to many children is Silly Putty. In many ways it behaves like a solid, particularly when forced to flow rapidly (for example, it breaks when pulled apart suddenly). In other ways it flows like a liquid. Supercooled liquids such as glass, Silly Putty, and plastics in general, do not have sharp, well-defined melting points, but instead gradually soften when heated.

(a) Silly Putty flows slowly like a very viscous liquid. (b) If it is stretched quickly it fractures like a solid.

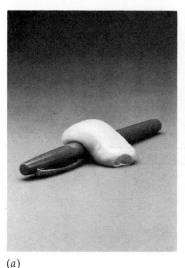

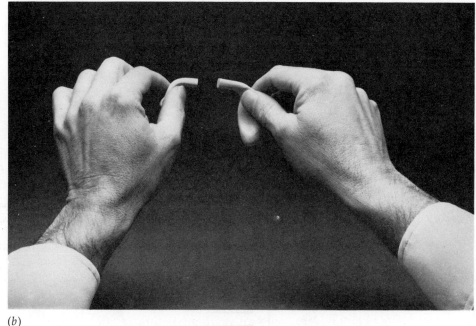

(a) *(b)*

8.12 VAPOR PRESSURE OF SOLIDS

In Section 8.1 you learned that solids are able to evaporate just like liquids do—the process is called sublimation. In a crystalline solid the molecules vibrate about their lattice positions and continually undergo collisions with their nearest neighbors. This gives rise to a distribution of kinetic energies in the solid, just as it does in a liquid or a gas. A small fraction of the molecules at the surface of a solid possess large kinetic energies, large enough to be able to overcome the attractive forces within the solid and break away from the surface, entering the gaseous phase above. That's how sublimation occurs.

When sublimation takes place in a closed container, more and more molecules enter the gaseous state and the pressure exerted by the vapor increases. Meanwhile, the slower-moving gaseous molecules that are colliding with the surface of the solid become trapped and return to the solid state. In time, the rate at which the molecules leave will exactly equal the rate at which they return, and a dynamic equilibrium will be established. The pressure exerted by a vapor in equilibrium with its solid is known as the **equilibrium vapor pressure of the solid.** Just as in liquids, the vapor pressure of a solid depends on the ease with which its molecules are able to enter the gaseous state. For example, the attractive forces are stronger in ionic solids than in molecular solids and, as expected, we find that the vapor pressures of ionic solids are generally very much lower than those of molecular solids.

A practical application of sublimation that you've probably heard of is freeze drying. Freeze-dried instant coffee, for example, is manufactured by first freezing a batch of brewed coffee and then removing the ice component by vacuum. The vacuum creates an atmosphere of diminished pressure in which ice readily sublimes. Removing the water in this way preserves the delicate heat-sensitive molecules that give coffee its taste, so the quality of the product is improved. Solid foods—even entire meals—are also freeze dried, both to protect their flavor as well as to prevent spoilage, since bacteria that might cause harm cannot grow and multiply in the absence of moisture. This allows freeze-dried foods to be stored without refrigeration, and they are easily reconstituted by adding water to them. Campers often carry these products because of their convenience.

8.13 PHASE DIAGRAMS

The vapor pressure of a solid, like that of a liquid, is a function of its temperature. Increasing the temperature of a solid-vapor equilibrium, according to Le Châtelier's principle, leads to a shift in the position of the equilibrium that will occur with the absorption of heat. The production of vapor from the solid is an endothermic process; therefore as the temperature rises, more of the solid will evaporate and more of the vapor will be produced until equilibrium is once again attained. As a result, the equilibrium vapor pressure of a solid increases with increasing temperature until eventually a temperature is reached at which the solid melts. Further increases in temperature, beyond this point, will then give rise to a liquid-vapor equilibrium curve that terminates at the critical temperature of the substance. If, using water as an example, we plot the vapor pressure versus temperature for the solid-vapor equilibrium and for the liquid-vapor equilibrium on the same graph, we produce Figure 8.37. Each point along the "solid" curve represents specific combinations of temperatures and pressures that must be achieved for the solid to be in equilibrium with its vapor. Likewise, points along the "liquid" curve represent combinations of temperatures and pressures required for the liquid to be in equilibrium with its vapor. The point of intersection of these two curves, called the **triple point,** corresponds to a unique temperature and pressure where all *three* states of matter (solid, liquid, and gas) coexist in equilibrium with each other simultaneously. The triple point occurs at a temperature and pressure that depend on the nature

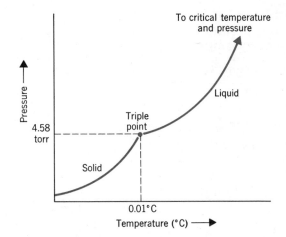

Figure 8.37

Solid and liquid vapor pressure curves for water.

The temperature at the triple point of water is the SI reference temperature 273.16 K.

of the substance in question. For example, the triple point of water occurs at a temperature of 0.01°C and a pressure of 4.58 torr, while the triple-point temperature of carbon dioxide is −57°C and the triple-point pressure is 5.2 atm.

There is still another equilibrium that can be represented on the same graph. This line corresponds to the combinations of temperatures and pressures that must be maintained to achieve a solid-liquid equilibrium. At a pressure of 1 atm, the melting point of water is 0°C; therefore, the solid-liquid equilibrium line passes through both the triple point and the normal melting point as shown in Figure 8.38. This graph is called a **phase diagram** because it allows us to pinpoint temperatures and pressures at which the various phases exist, as well as those conditions under which equilibrium can occur. For instance, at a pressure of 1 atm, water exists as a solid at all temperatures below 0°C, and in fact the region bounded by the solid-liquid equilibrium line and the solid-vapor equilibrium line corresponds to all the temperatures and pressures at which water exists as a solid. Similarly, in the region bounded by the solid-liquid and liquid-vapor equilibrium lines, the substance can exist only as a liquid, while to the right of both the solid-vapor and liquid-vapor lines the substance must be a gas.

Figure 8.38

Phase diagram for water (somewhat distorted). T_f *is the normal freezing point and* T_b *is the normal boiling point. The slope of the solid-liquid line is exaggerated. The actual slope is much less to the left (33 atm is required to lower the melting point of ice by only 1°C).*

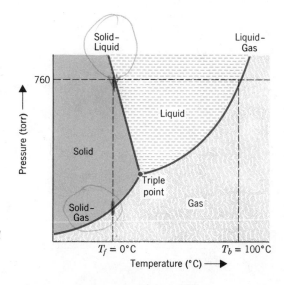

Table 8.6
Physical state of water at
random temperatures and pressures

Temperature (°C)	Pressure (atm)	State
25	1.0	Liquid
0	2.0	Liquid
0	0.5	Solid
100	0.5	Gas

Table 8.6 lists some randomly chosen temperatures and pressures and the physical states of water that we can predict from its phase diagram. You might verify these predictions to illustrate how to use a phase diagram.

To gain a further insight into the meaning of a phase diagram, let us follow the changes that take place as we move along a line of constant pressure, say 1 atm, by varying the temperature. In Figure 8.39, point A lies in the region of the diagram where a sample of the substance would exist entirely as a solid, as shown in Figure 8.40a. When the temperature rises to point B in Figure 8.39, the solid begins to melt and an equilibrium between the solid and liquid can occur (Figure 8.40b). At a still higher temperature, point C, all the solid will have been converted to a liquid (Figure 8.40c); and, when the liquid-vapor line is encountered at point D in Figure 8.39, vapor may at last begin to form and an equilibrium can exist (Figure 8.40d). Finally at a sufficiently high temperature, such as point E, all the water will exist in the vapor state (Figure 8.40e).

We could also proceed with a similar analysis in which the temperature is held constant and the pressure is permitted to change. For example, at point F in Figure 8.39, the water would exist entirely as a gas (Figure 8.41a). At a higher pressure, point G in Figure 8.39, a solid-vapor equilibrium would exist (Figure 8.41b), and above that pressure, at point H, all the water would be converted to a solid (Figure 8.41c). As the pressure is increased further, we encounter the solid-liquid line at point B in Figure 8.39, where we again have an equilibrium as represented by Figure 8.41d. At still higher pressures, the water will melt so that at point I all the water is present in the liquid state (Figure 8.41e).

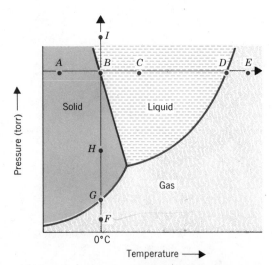

Figure 8.39

Phase diagram for water (not drawn to scale).

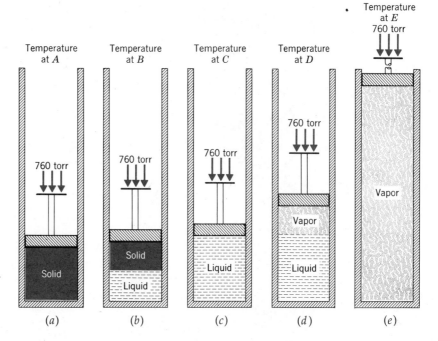

Figure 8.40

Raising the temperature at a constant pressure of 760 torr. Temperatures correspond to points A to E in Figure 8.39.

In the phase diagram for water we see that the solid-liquid equilibrium line slants to the left. This is a direct consequence of the fact that liquid water at 0°C has a higher density than the solid does. Le Châtelier's principle requires that an increase in pressure on a system at equilibrium will lead to the production of the more dense phase; that is, a rise in pressure favors the packing together of molecules—quite a reasonable expectation. This means that if we have solid and liquid water at equilibrium and we increase the pressure while holding the

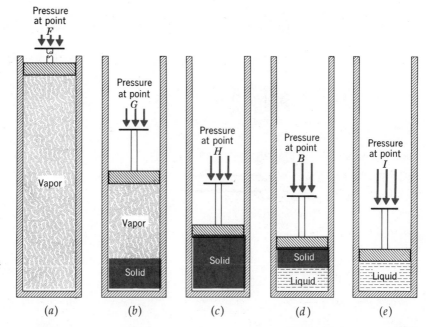

Figure 8.41

Raising the pressure at a constant temperature of 0°C. Pressures correspond to points on Figure 8.39.

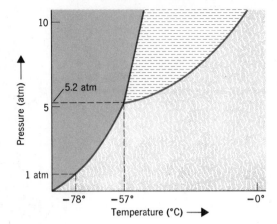

Figure 8.42

Phase diagram for carbon dioxide.

At room temperature, CO_2 exists as a liquid at the high pressure inside a CO_2 fire extinguisher.

temperature at 0°C, we should produce the higher-density, liquid phase.[4] On the phase diagram, a rise in pressure at constant temperature amounts to moving upward along a vertical line. We can move from the solid-liquid equilibrium line upward into a region of all liquid only if the solid-liquid line leans to the left.

Water is quite an unusual substance. For nearly all other compounds the solid phase is more dense than the liquid and for these substances the solid-liquid line slants to the right, as is shown in the phase diagram for CO_2 that appears in Figure 8.42. An interesting feature of this phase diagram is that the entire liquid range lies above a pressure of 1 atm; therefore, it is impossible to form liquid CO_2 at atmospheric pressure. Instead, as the gas is cooled, the solid-vapor equilibrium line is encountered at −78°C and the vapor is converted directly to the solid. This also explains why Dry Ice sublimes rather than melts at ordinary pressures.

[4] In fact, it had long been thought that this melting of water that occurs at high pressure is responsible for our ability to skate on ice. It was believed that the high pressure produced by the skater's weight concentrated on the sharp edge of a blade causes the ice just beneath the blade to melt, producing a thin film of liquid water that serves as a lubricant and allows the skate to slide smoothly on the ice. Current feeling, however, is that this film of water is most probably the result of melting due to friction between the moving skate blade and the ice.

INDEX TO QUESTIONS AND PROBLEMS (Problem numbers in **bold type**)

REVIEW QUESTIONS

8.1 State how the following physical properties differ for the three states of matter—solid, liquid, and gas: (a) density, (b) rate of diffusion, (c) compressibility, (d) ability to flow.

8.2 Which of the properties in Question 8.1 are determined primarily by how tightly the molecules are packed?

8.3 Explain why the rate of diffusion in a liquid is less than in a gas.

8.4 What is surface tension? Why do liquids tend to form spherical droplets?

8.5 As a child, you probably had the experience of filling a glass above its rim with water. Why doesn't the water simply overflow?

8.6 Sketch the kinetic energy distribution for a liquid at two temperatures. Indicate the minimum K.E. required for molecules to escape the liquid. On the basis of this diagram, explain why liquids evaporate faster at higher temperatures.

8.7 Why does evaporation lead to a lowering of temperature?

8.8 Clothes dry more rapidly on a dry day than on a humid day. They also dry more rapidly if there is a breeze blowing than if the air is still. Why?

8.9 At high altitudes, ice and snow gradually disappear without melting. Why?

8.10 Why does warm water evaporate more quickly than cold water?

8.11 Solid iodine vaporizes without melting if warmed gently. What is this process called?

8.12 Salt, NaCl, is harvested from seawater by evaporation of water from large shallow ponds. What is the purpose of spreading the water over such large areas?

8.13 At a given temperature, methyl alcohol—the fuel in canned heat—evaporates much more rapidly than propylene glycol—a food additive. Which substance has the weaker intermolecular attractions?

8.14 What are dipole-dipole attractions? Why are they weaker in a gas than in a liquid or solid?

8.15 What are hydrogen bonds? Why are they most important in compounds containing N—H, O—H, and F—H bonds?

8.16 What are London forces?

8.17 What type (or types) of intermolecular attractive forces are found in the following?

(a) HCl
(b) Ar
(c) CH_4
(d) HF
(e) NO
(f) CO_2
(g) H_2S
(h) SO_2

8.18 How is the structure of ice affected by hydrogen bonding?

8.19 If you were asked to compare the strengths of the intermolecular attractive forces in liquid A with those in liquid B, what types of data would you collect?

8.20 Suppose that two substances, X and Y, have heats of vaporization equal to 9.0 and 6.5 kcal/mol, respectively. Which compound would you expect to have the higher boiling point? Which compound would be less likely to exhibit hydrogen bonding?

8.21 Among the hydrocarbons, CH_4 through $C_{10}H_{22}$, both ΔH_{vap} and boiling points increase with increasing molecular weight, even though the molecules are composed of the same kinds of atoms. How can this be explained?

8.22 How would you expect the heats of vaporization to vary among the compounds, PH_3, AsH_3, SbH_3?

8.23 How would you expect the heats of vaporization to vary among the compounds, H_2S, H_2Se, H_2Te?

8.24 What is a reasonable explanation for the fact that ΔH_{vap} is larger for H_2O than for HF?

8.25 The molar heat of vaporization of water is greater than that of any of the components of gasoline. Yet, if gasoline is spilled on your hand, it gives a greater cooling effect than if water is spilled. Why?

8.26 How should the strengths of London forces vary from helium to argon in Group 0?

8.27 What is polarizability? How is it related to the strengths of intermolecular attractions?

8.28 What is meant by the term vapor pressure?

8.29 A glass of cola, or other beverage, with ice in it often becomes wet on the outside. Explain why.

8.30 At 0°F there is less water in 1 liter of air at 100% humidity than at 75°F when the humidity is 100%. Why is this so?

8.31 When warm moist air is forced to rise over mountains, clouds form and rain frequently falls. On the basis of the concepts in this chapter, explain why this happens.

8.32 What is the source of energy in a thunderstorm?

8.33 Steam at 100°C produces a much more severe burn than an equivalent amount of liquid water at 100°C. Why?

8.34 Why is the vapor pressure a measure of the strengths of intermolecular attractive forces in a liquid?

8.35 Explain why decreasing the volume of a container does not alter the vapor pressure of a liquid.

8.36 Using Le Châtelier's principle, predict the effect of a change in temperature on the equilibria:
(a) solid + heat $\rightleftharpoons$ liquid
(b) liquid + heat $\rightleftharpoons$ vapor

8.37 How will an increase in pressure effect the equilibrium: solid $\rightleftharpoons$ vapor

8.38 Define the terms: critical temperature, critical pressure.

8.39 What happens to the boundary between liquid and vapor at temperatures above the critical temperature?

8.40 Methane (natural gas) can be shipped across the oceans as a liquid in huge specially made ships. To what temperature must methane be cooled before it can be liquefied by being compressed?

8.41 If you shake a CO_2 fire extinguisher on a cool day, you can feel a liquid sloshing around inside. On a hot summer day, when the temperature is around 90°F, there is no sloshing when the same extinguisher is shaken. Why?

8.42 Why does blowing gently across the surface of a hot cup of coffee help to cool it?

8.43 Define: boiling point, normal boiling point.

8.44 From Figure 8.16, estimate the boiling point of water at a pressure of 500 torr.

8.45 At the top of Mount Everest in the Himalayas, which is 29,000 ft above sea level, the atmospheric pressure is approximately 270 torr. At what Celsius temperature would water boil at that altitude?

8.46 Explain why compounds with strong intermolecular attractive forces have higher boiling points than compounds with weak intermolecular attractive forces.

8.47 What evidence is there for hydrogen bonding in H_2O, HF, and NH_3?

8.48 Explain why, for any given substance, ΔH_{fus} is smaller than ΔH_{vap}.

8.49 What external features do crystals exhibit?

8.50 What is an amorphous solid? How does it differ from a crystalline solid?

8.51 What is a lattice? What is a unit cell? Why can one kind of lattice be used to describe many different chemical structures?

8.52 What is the Bragg equation? What do the symbols in the equation stand for?

8.53 What quantities determine the kind of lattice to which a particular unit cell belongs?

8.54 Sketch and name the three types of cubic lattices

8.55 Describe the rock salt structure. What kind of lattice does it belong to? How many formula units are there per unit cell? Could a salt like K_2S crystallize in the rock salt structure? Explain your answer.

8.56 Identify the kinds of chemical units associated with each of the following kinds of crystals: molecular, metallic, covalent, ionic. Describe the properties of each of them. What kinds of attractive forces exist between the chemical units in these crystals?

8.57 Indicate which type of crystal (ionic, covalent, etc.) each of the following would form upon solidification: (a) O_2 (b) H_2S (c) Pt (d) KCl (e) Ge (f) $Al_2(SO_4)_3$ (g) Ne

8.58 Indicate which type of crystal (ionic, covalent, etc.) each of the following would form upon solidification: (a) Br_2 (b) LiF (c) MgO (d) Cr (e) SiO_2 (f) PH_3 (g) NaOH.

8.59 $SnCl_4$ is a colorless liquid having a boiling point of 114°C and a melting point of −33°C. $SnCl_2$, on the other hand, is a white solid that melts at 246°C. What type of solid (ionic, covalent, etc.) is most likely formed when $SnCl_4$ solidifies?

8.60 Elemental boron is extremely hard (nearly as hard as diamond) and has a melting point of 2300°C. It is a poor conductor of electricity at room temperature. What kind of solid would you expect for boron based on these properties?

8.61 Parafin (wax) is low-melting, soft, and is a nonconductor in both the solid and liquid states. What kind of solid is expected for parafin?

8.62 OsO_4 has a melting point of 39.5°C and is a nonconductor of electricity in the molten state. It boils at 130°C. What kind of solid is expected for OsO_4?

8.63 $CaCO_3$ (calcite) is hard and brittle. It decomposes, before it melts, at a temperature of about 900°C. What kind of solid is likely for calcite?

8.64 What are the three kinds of liquid crystals? How do they differ?

8.65 What is the origin of the color of cholesteric liquid crystals? Why does the color change with temperature?

8.66 Aluminum has a melting point of 660°C and a boiling point of 1800°C. Its $\Delta H_{fus} = 2.55$ kcal/mol and $\Delta H_{vap} = 53.8$ kcal/mol. Sketch and label a heating curve for aluminum.

8.67 Iodine, I_2, sublimes without melting when heated in an open container at atmospheric pressure. What can be said about the triple point of I_2?

8.68 Can you think of any common household products that sublime?

8.69 At a pressure of 760 torr a new compound was found to melt at 25°C and boil at 95°C. The triple point of the substance was determined to occur at a pressure of 150 torr and at a temperature of 20°C. Sketch the phase diagram for this substance. Label, on your drawing, the solid, liquid, and vapor regions as well as the solid–liquid, liquid–vapor, and solid–vapor equilibrium lines.

8.70 On the basis of the phase diagram in Question 8.69, describe the changes you would observe if, at a constant temperature of 22°C, the pressure on a sample of the compound is gradually increased from 10 to 1000 torr. What would be observed if the same process were to occur at a constant temperature of 10°C?

8.71 Sketch the heating curve you would expect to find when 1 mol of the compound described in Question 8.69 is heated at a constant rate under a constant pressure of 1.00 atm. On your drawing, indicate the melting point and boiling point of the substance. Also, label the intervals that correspond to ΔH_{fus} and ΔH_{vap}.

8.72 What can we conclude about the relative densities of the liquid and solid phases of the compound in Question 8.69?

8.73 When a supercooled liquid begins to freeze, its temperature rises. Why doesn't the temperature ever rise above the melting point of the substance?

8.74 Why doesn't glass have a sharp melting point?

8.75 Is it possible to have only *liquid* water in a container at 32°F (0°C)?

8.76 Explain, on a molecular level, why the temperature remains constant as heat is added to vaporize a liquid at its boiling point.

8.77 Use Le Châtelier's principle to predict how variations in pressure will affect the melting point of: (a) water, (b) carbon dioxide.

8.78 With the aid of the phase diagram in Figure 8.42, predict the physical state of carbon dioxide under the following conditions of temperature and pressure:

Temperature (°C)	Pressure (atm)
−80	1.0
−60	1.0
−56	10.0
−56	2.0
−65	5.0
−40	10.0

REVIEW PROBLEMS (More difficult problems are marked by an asterisk.)

8.79 Trouton's rule states that the ratio of the heat of vaporization to the boiling point (in kelvins) is approximately a constant. Verify this for the hydrocarbons CH_4 through $C_{10}H_{22}$ in Table 8.1. What conclusions can you draw concerning the relationship between ΔH_{vap} and boiling point?

8.80 Calculate the heat necessary to convert 55.0 g of ethanol (ethyl alcohol, C_2H_5OH) from liquid to vapor. $\Delta H_{vap} = 9.22$ kcal/mol.

8.81 How many kilojoules of energy are necessary to melt 35.0 g of benzene (C_6H_6)? $\Delta H_{fus} = 2.37$ kcal/mol.

8.82 A 14.5-g sample of liquid mercury required 4.29 kJ to completely convert it to vapor at the same temperature. What is ΔH_{vap} of Hg in kJ/mol? What is it in kcal/mol?

*8.83 A 150-lb (68.2-kg) skater skids to a halt from a speed of 10 miles/hr. Assuming that all of his energy appears as frictional heat transferred to ice at 0°C, how many grams of ice will be melted?

*8.84 A student (with very slow reflexes) holds his hand in a stream of steam at 100°C until exactly 1.00 g of water has condensed. If this water then cools to 40°C, how many calories have been absorbed by the student's hand?

*8.85 A cube of solid benzene (C_6H_6) at its melting point weighing 10.0 g is introduced into 50.0 g of H_2O at 30° C. Given that ΔH_{fus} for C_6H_6 is 2.37 kcal/mole, to what temperature will the water have cooled by the time all of the benzene has melted?

*8.86 A 50.0-g ice cube at 0°C is added to 10.0 g of steam at 100°C. What will be the final temperature of the 60.0 g of water?

8.87 Calculate the angles at which X rays of wavelength 2.29 Å will be observed to be reflected from crystal planes spaced (a) 10 Å apart, (b) 2 Å apart. Assume that $n = 1$.

8.88 Calculate the interplanar spacings (in picometers) that correspond to reflections at $\theta = 20°$, 27.4°, and 35.8° by X rays of wavelength 0.141 nm. Assume that $n = 1$.

8.89 From the following list of angles, determine the angles at which X rays of wavelength 1.41 Å, diffracted from planes of atoms 2.0 Å apart, are in phase: $\theta = 17.3°$, 20.5°, 44.4°, and 55.3°. Assume that $n = 1$.

8.90 Chromium, used to protect and beautify other metals, crystallizes in a body-centered cubic structure in which the Cr atoms are in contact along the body diagonal of the unit cell. The edge of the unit cell is 288.4 pm. Calculate the atomic radius of a Cr atom in picometers.

*8.91 Chromium crystallizes with a body-centered lattice. Its density is 7.19 g/cm³ and the unit cell edge is 288.4 pm. Use these data to compute Avogadro's number.

*8.92 Gold crystallizes with a face-centered cubic lattice. The length of the unit cell edge is 407.86 pm. What is the atomic radius of a gold atom in picometers?

8.93 Aluminum crystallizes in a face-centered cubic structure. If the Al atom has an atomic radius of 1.43 Å, what is the length of the unit cell edge in Al in Ångstroms?

8.94 CsCl forms a simple cubic lattice in which there are Cs^+ ions at the corners of the unit cell and a Cl^- ion in the center of the cell. The cation/anion contact occurs along the body diagonal of the unit cell. The length of the unit cell edge is 412.3 pm. The Cl^- ion has a radius of 181 pm. What is the radius of the Cs^+ ion in picometers?

8.95 RbCl has the rock salt structure shown in Figure 8.27. The unit-cell edge length is 6.58 Å. Cations and anions are in contact along the edges. The ionic radius of the chloride ion is 1.81 Å. Calculate the ionic radius of the Rb^+ ion in Ångstroms. What is its radius in picometers?

8.96 Silver has an atomic radius of 144 pm. What would be the density of Ag in g/cm³ if it were to crystallize in the following structures: (a) simple cubic, (b) body-centered cubic, (c) face-centered cubic? The actual density of Ag is 10.6 g/cm³. Which of these corresponds to the correct structure for Ag?

*8.97 Refer to the diagram on page 284 to derive the Bragg equation. Remember that the extra distance traveled by the more penetrating beam must be an integral multiple of the wavelength in order to have constructive interference.

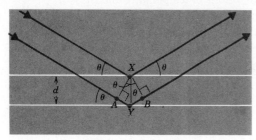

$$\angle XAY = \angle XBY = 90°$$

*8.98 Calculate the amount of vacant (unoccupied) space (in Å³) in a primitive cubic, a body-centered cubic, and a face-centered cubic packing of identical spheres of diameter 1.0 Å.

*8.99 LiBr has the rock salt structure in which Br⁻ ions, centered at lattice points, are in contact. Calculate the ionic radii of Br⁻ and Li⁺ in picometers if the unit-cell edge is 550 pm. Why is the accepted value for the ionic radius of Li⁺ (60 pm) smaller than the value that you just computed?

*8.100 CsCl crystallizes with a cubic unit cell of edge length 4.123 Å. The density of CsCl is 3.99 g/cm³. Show that the unit cell cannot be face-centered or body-centered.

*8.101 Metallic sodium crystallizes with a body-centered lattice. The element has a density of 0.97 g/cm³. What is the length of the edge of the unit cell in Na expressed in nanometers?

*8.102 Calcium fluoride crystallizes with a cubic lattice. The unit cell has an edge length of 5.4626 Å. The density of CaF₂ is 3.180 g/cm³. How many formula units of CaF₂ must there be per unit cell?

*8.103 NaCl (which has the rock salt structure) has a density of 2.165 g/cm³. The ionic radius of Cl⁻ is 1.81 Å. What is the ionic radius of Na⁺ in Ångstroms?

9

THE PERIODIC TABLE REVISITED

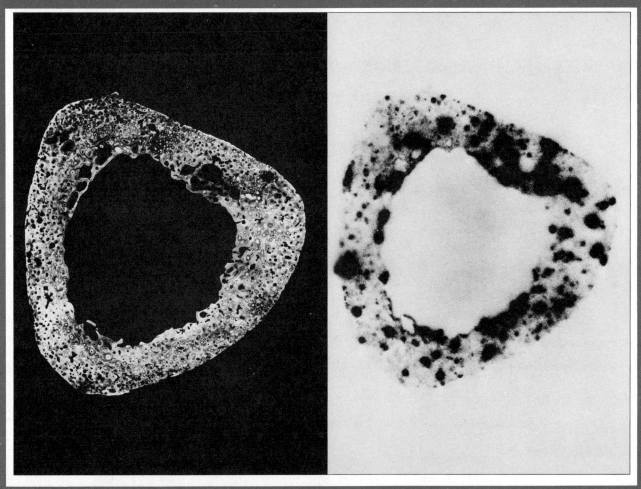

Elements in the same group of the periodic table have similar chemical properties, sometimes with frightening consequences. On the left is a photograph of a bone slice from the body of a former radium watch-dial painter. On the right is the image it produced on photographic film by radiation from radium that has replaced another Group IIA element, calcium, in the worker's bone. In this chapter we return for a closer look at the periodic table and some trends in the properties of the elements.

We learned in Chapter 3 that the structure of the periodic table is based on chemical facts—observed similarities and differences among the elements. We also saw how the periodic table could be explained in terms of the electronic configurations of the elements, and that the structure of the periodic table adds support to quantum theory. The periodic table is useful in many more ways than that, however. It is the single most useful device that we have to help us correlate chemical and physical properties, not only of the elements themselves but of many of their compounds as well. In serving this function, the periodic table aids us in remembering large quantities of chemical facts that would otherwise be isolated bits of information.

In this chapter we will study a number of *trends* in the physical and chemical properties of the elements and their compounds. We will study how these properties vary across a period or down a group. It is not our purpose in this chapter to dwell too long on the specific chemical properties of individual elements—we will get to that later in the book. Our goal here is to understand how the chemistries of the elements relate to each other and to their locations in the periodic table. With this "big picture" in mind, you will find the study of the individual elements much more rewarding and interesting later on.

9.1 METALS, METALLOIDS, AND NONMETALS

Figure 9.1 illustrates how the elements are divided among the three common classifications. On the left of the periodic table we find the metals and in the upper right we find the nonmetals. The metalloids—elements with properties in between those of metals and nonmetals—are found in a narrow band running from boron to astatine.

Most elements are metals, and most metals are transition elements in which there is a partially filled *d* and *f* subshell just below the outer shell. If, for the moment, we focus our attention on the representative elements and bring Groups

Transition elements are in the B-groups and Group VIII.

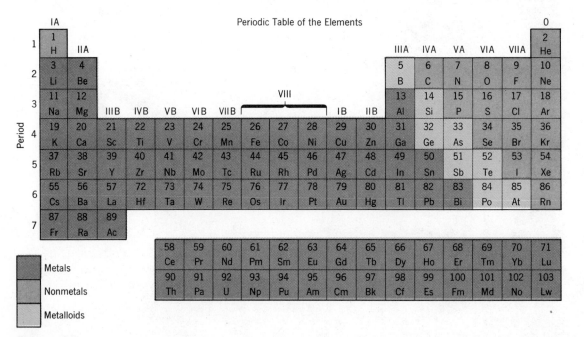

Figure 9.1

Distribution of metals, metalloids, and nonmetals in the periodic table.

						VIIA	O
IA	IIA	IIIA	IVA	VA	VIA	H	He
Li	Be	B	C	N	O	F	Ne
Na	Mg	Al	Si	P	S	Cl	Ar
K	Ca	Ga	Ge	As	Se	Br	Kr
Rb	Sr	In	Sn	Sb	Te	I	Xe
Cs	Ba	Tl	Pb	Bi	Po	At	Rn
Fr	Ra						

Figure 9.2

The representative elements. Metalloids are indicated by gray squares.

IA and IIA over against Group IIIA, we find the division between metals and nonmetals a rather even one, as shown in Figure 9.2.

The representative elements are characterized by the filling of the *s* and *p* orbitals of their outer shells. None of their inner subshells are partially occupied, so their outer-shell electrons are the only ones involved in bonding. For this reason, we will sometimes find it convenient to leave out the transition elements when we examine certain periodic trends.

Representative elements are in the A-groups.

9.2 PHYSICAL PROPERTIES OF METALS

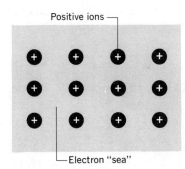

Figure 9.3

A metallic lattice. Positive ions that are formed when metal atoms lose their valence electrons are surrounded by a freely moving electron sea made up of these electrons.

There are certain properties that immediately come to mind when we think of metals, and they are usually physical properties. One is the ability of metals to conduct electricity, a property they owe to mobile electrons in a metallic lattice. This kind of lattice, you recall, was discussed in the last chapter. It consists of positive ions, formed from metal atoms, that are surrounded by a loosely held "sea" of electrons (Figure 9.3).

The metallic lattice is a result of the electronic structure of metals. In general, metal atoms have few electrons in their valence shells, so they would have to acquire many electrons to achieve an octet. Since their neighbors in a metallic crystal also have few electrons, the metal atoms simply cannot achieve a noble gas configuration by electron sharing. In addition, metals generally have low ionization energies (IE), so they are able to lose electrons without much difficulty. When they become part of a metallic crystal, therefore, the metal atoms lose their valence electrons to the lattice rather than form bonds to each other. In a sense, they all have a share of an enormous pool of electrons and thereby achieve a certain degree of stability.

If you look around the room, you will see objects that you immediately recognize as metals based on their appearance—we say they have a *metallic luster*. These objects could be shiny new copper pennies, iron nails, a bit of chrome decoration, or gold or silver jewelry. Lead and aluminum also shine brightly if buffed to remove the layer of corrosion on their surfaces. Even sodium, an extremely reactive metal that quickly combines with oxygen and moisture in the air, exhibits a metallic luster when a fresh surface is exposed (see Color Plate 8).

The characteristic sheen that all metals have is no mere coincidence. It too is a consequence of the metallic lattice. When light shines on a metal's surface,

The alkali metals are so reactive that hardly anyone but chemists ever sees them.

Silver is used for the reflective coating on mirrors because it reflects light so well.

the loosely held electrons are set into vibration by the light wave. When the electrons vibrate they re-emit the light at the same frequency, so the metal reflects whatever color light it is bathed in.

The mobility of the electrons also accounts for the **thermal conductivity** of metals—their ability to transport heat. Have you ever noticed, in cool weather, how much colder a metal object feels than nonmetallic objects alongside? This is because the metal's surface is able to conduct heat away from your hand much more rapidly than the surface of a nonmetal. For this same reason, a metal will feel burning hot if left in the sun but nonmetallic objects nearby will only feel warm. The high thermal conductivity of metals results because kinetic energy is quickly transported through the metal by the loosely held electrons.

Other important properties of metals include their malleability and ductility. **Malleability** is the ability of a metal to be hammered or rolled into thin sheets. A blacksmith uses this property when he fashions a horseshoe from a metal bar (Figure 9.4). Modern steel mills use it when they roll huge bars of steel into thin sheets that are used to manufacture household appliances and automobiles. Gold can be hammered into extremely thin sheets used for gold leaf—so thin that light can be seen through it.

Ductility is the ability of a metal to be drawn into wire. Figure 9.5 shows how copper wire is made. The wire is pulled through a series of dies of ever decreasing size so that the wire becomes thinner and thinner.

Another property we usually associate with a metal is a high melting point. This is because most metals that are useful also melt at high temperatures. Actually, the melting points of metals cover a broad range. The lowest melting metal of all is mercury—it melts at $-38.9°C$ and is commonly used as the fluid in thermometers. Tungsten has the highest melting point of any metal, $3400°C$ ($6150°F$), which accounts for its use as the filament in electric light bulbs. Even when heated to the point where it is white hot, the tungsten still remains a solid.

As we learned in the last chapter, melting points are controlled by the attractive forces within the solid. Here we can see some trends within the peri-

Figure 9.4

A blacksmith makes use of the malleability of iron as he pounds an iron bar into the shape of a horseshoe.

Figure 9.5

Copper wire is pulled through smaller and smaller dies and becomes thinner. The ductility of copper makes this possible.

read chapter all closely

The filament in an electric light bulb is made of tungsten.

Table 9.1
Melting points of group IA and IIA metals

Group IA		Group IIA	
Metal	Mp (°C)	Metal	Mp (°C)
lithium	180.5	beryllium	1278
sodium	97.8	magnesium	650
potassium	63.7	calcium	843
rubidium	38.98	strontium	769
cesium	28.59	barium	725

odic table. The metals of Group IA have generally low melting points. Sodium for instance, melts at about 98°C, which is below the boiling point of water. As we can see in Table 9.1, the alkaline earth metals of Group IIA have somewhat higher melting points than their neighbors in Group IA. This is because the cations with a 2+ charge produced by the Group IIA metals in their metallic lattices hold onto the electron "sea" more tightly than do the 1+ cations of the Group IA metals in their lattices.

An interesting variation in melting points is found among the transition elements, as shown in Figure 9.6 for those in Period 4. Notice that the melting points rise toward the center of the row and then drop off again at the end of the row. This variation parallels, approximately, the number of unpaired *d* electrons in the atoms of these elements and suggests that some degree of covalent bonding may exist between them.

The metals with the highest melting points are found in Period 6 following lanthanum. See Figure 9.7. This is a consequence of the lanthanide contraction discussed in Chapter 3 (page 99). Recall that the elements that follow lanthanum in Period 6 are unexpectedly small. Hafnium, for example, is the same size as the element just above it (zirconium), even though it has a nucleus containing 32 more protons, and therefore its atoms have 32 more electrons. Because of their small size and large effective nuclear charges, these Period 6 metals hold onto the electron "sea" very tightly and have exceptionally high melting points.

Figure 9.6

Melting points of the transition metals of Period 4.

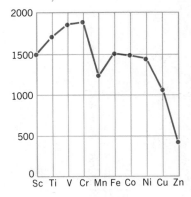

Figure 9.7

Variation among the melting points of the metallic elements.

A property of metals that <u>parallels</u> their <u>melting points</u> is *hardness*. Metals that have low melting points tend to be soft. Sodium, as can be seen in Color

Properties of some typical metals

Metal	Melting Point (°C)	Relative Hardness (room temperature)
Mercury	−38.4	(Liquid)
Sodium	97.8	0.07
Lead	327	4.2
Magnesium	650	30
Aluminum	660	16
Nickel	1453	90
Iron	1536	77
Platinum	1769	64
Molybdenum	2610	150
Tungsten	3410	350

The hardness of metals can sometimes be improved by mixing them with other metals to make alloys.

Plate 8, is easily cut with a knife. <u>High-melting metals</u> <u>tend</u> to <u>be hard</u>, whereas those with intermediate melting points are of intermediate hardness. Thus the transition metals in the center of the periodic table, such as chromium and iron, are often used commercially because they are hard and strong. The metals

Because silver is a soft metal this skilled craftsman can work intricate designs into a silver tray.

on either side of them are much softer. Intricate designs, for example, can be easily worked into gold and silver—elements that occur toward the end of rows of transition elements.

9.3 CHEMICAL PROPERTIES OF METALS

With very few exceptions, when metals react with other substances they are oxidized—the metal atoms lose electrons and take on positive oxidation numbers. In many cases, actual cations are formed, as we saw in Chapter 4. When the metal is in a very high oxidation state, however, it is usually found in a polyatomic anion. Examples that you have seen before are permanganate ion, MnO_4^-, chromate ion, CrO_4^{2-}, and dichromate ion, $Cr_2O_7^{2-}$.

Ease of oxidation

The ease of oxidation is an important property of metals. Many of their practical applications depend on it (or rather the *lack* of it). This is because oxidation of the metals, which we often call *corrosion,* produces compounds that lack metallic properties. Corrosion therefore wipes out the desirable properties for which metals are often chosen. For this reason, the extremely reactive elements of Group IA have very few practical uses, and none that would require exposure of the metals to air or moisture.

We are able to tolerate a moderate ease of oxidation in some metals—iron, for example—because they have such desirable physical properties, but if severely corrosive conditions are to be encountered, iron must somehow be protected. Huge sums of money are spent annually, for example, to paint steel structures such as bridges.

The tendency of steel to react with air and moisture to form rust is a problem often solved by applying a protective coat of paint. Here we see the Golden Gate Bridge receiving its periodic paint job.

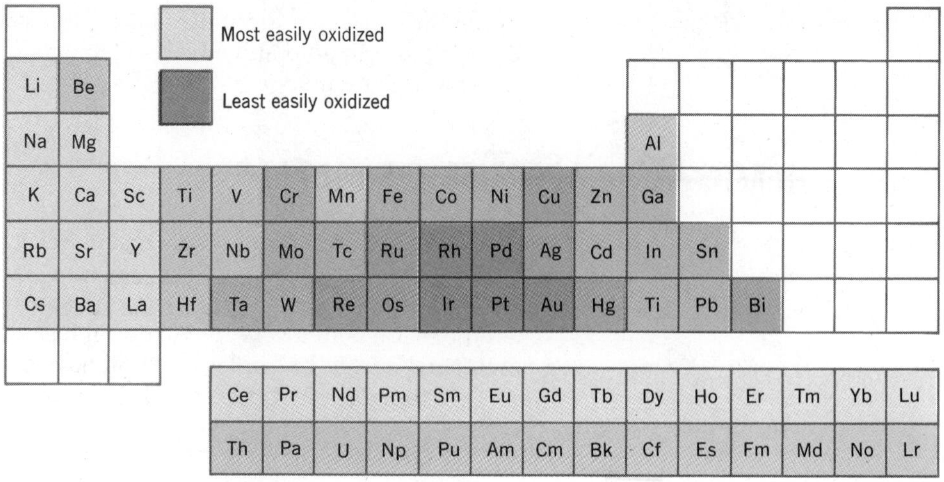

Li	Be																
Na	Mg											Al					
K	Ca	Sc	Ti	V	Cr	Mn	Fe	Co	Ni	Cu	Zn	Ga					
Rb	Sr	Y	Zr	Nb	Mo	Tc	Ru	Rh	Pd	Ag	Cd	In	Sn				
Cs	Ba	La	Hf	Ta	W	Re	Os	Ir	Pt	Au	Hg	Ti	Pb	Bi			

☐ Most easily oxidized

☐ Least easily oxidized

Ce	Pr	Nd	Pm	Sm	Eu	Gd	Tb	Dy	Ho	Er	Tm	Yb	Lu
Th	Pa	U	Np	Pu	Am	Cm	Bk	Cf	Es	Fm	Md	No	Lr

Figure 9.8

General trends in the ease with which metals are oxidized. Elements with generally low ionization energies are easily oxidized.

Electrical contacts in low-voltage circuits are gold plated because even a small amount of corrosion can stop the flow of electricity.

how can you tell?

Some metals, such as gold and platinum, are very resistant to oxidation, and their uses in jewelry as well as in more practical commercial applications depend on this. They are called **noble metals** because of their indifference to oxidizing agents that affect most other metals.

The general trends in the ease of oxidation of the metals are illustrated in Figure 9.8. You should be aware of the locations of the very reactive metals as well as the very unreactive ones. If we study Figure 9.8, we see that trends in the ease of oxidation roughly parallel those of ionization energy, although not perfectly. One reason is because ionization energy refers to isolated atoms and in chemical reactions involving the free elements, metal atoms normally occur in solids.

One way of comparing the ease of oxidation of metals is to study their reactions with acids. Many metals, such as the iron nail in Figure 9.9, react with acids and liberate hydrogen. The ionic equation for the reaction is

$$Fe\ (s) + 2H^+\ (aq) + 2Cl^-\ (aq) \longrightarrow Fe^{2+}\ (aq) + 2Cl^-\ (aq) + H_2\ (g)$$

In this case, hydrogen ion is the oxidizing agent.

It would be extremely dangerous to place an alkali metal such as sodium into hydrochloric acid because the Group IA metals, due to their low IE, are so easily oxidized. In fact, they *all* react vigorously with water to produce H_2, as

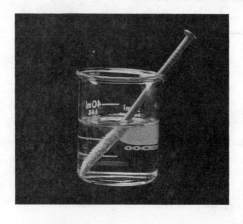

Figure 9.9

An ordinary iron nail dissolving in hydrochloric acid.

shown in Figure 9.10. For sodium, the reaction is

$$2Na\ (s) + 2H_2O\ (l) \longrightarrow 2NaOH\ (aq) + H_2\ (g)$$

The violence of this reaction increases going down the group. Potassium, for example, reacts explosively with water.

The Group IIA elements, which have slightly larger ionization energies than the elements in Group IA, react less vigorously. Calcium and the elements below it react with water, although not as quickly as the alkali metals.

$$Ca\ (s) + 2H_2O\ (l) \longrightarrow Ca(OH)_2\ (aq) + H_2\ (g)$$

Magnesium doesn't react with water because of an oxide coating on its surface, and beryllium is less reactive than magnesium.

Most of the transition metals in Period 4 react with dilute acids such as HCl or H_2SO_4 in much the same way as iron.

$$Mn\ (s) + 2HCl\ (aq) \longrightarrow MnCl_2\ (aq) + H_2\ (g)$$

$$Zn\ (s) + 2HCl\ (aq) \longrightarrow ZnCl_2\ (aq) + H_2\ (g)$$

An exception is copper.

Three common metals that are resistant to attack by acids such as HCl are copper, silver, and mercury. Hydrogen ion is simply not a strong enough oxidizing agent to react with them. They require, in addition to H^+, an oxidizing anion such as nitrate ion—these metals will dissolve in HNO_3. The products depend, to a degree, on how concentrated the acid is.

$$3Cu + 2NO_3^- + 8H^+ \longrightarrow 3Cu^{2+} + 2NO + 4H_2O \qquad \text{(Dilute } HNO_3\text{)}$$

$$Cu + 2NO_3^- + 4H^+ \longrightarrow Cu^{2+} + 2NO_2 + 2H_2O \qquad \text{(Concentrated } HNO_3\text{)}$$

Figure 9.10

Sodium reacts violently with water. The heat of reaction ignites the sodium metal and the hydrogen gas produced by the reaction.

Relative rates of reaction of HCl with (a) iron (b) zinc, and (c) magnesium. Zinc reacts more rapidly than iron, and magnesium reacts more rapidly than zinc.

(a)

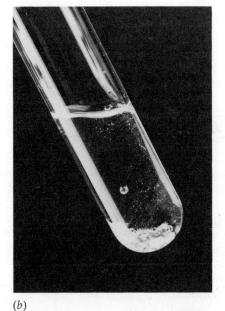

(b)

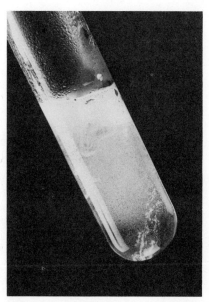

(c)

Aqua regia is 3 parts conc. HCl to 1 part conc. HNO_3, by volume.

The least reactive metals are those in the lower center portion of the block of transition elements. Here we find platinum, iridium, and gold—metals so unreactive that they are not even attacked by concentrated nitric acid. They do dissolve slowly, however, in a mixture of concentrated HCl and HNO_3 called **aqua regia.** (Alchemists, using Latin, named this mixture *royal water* because it alone attacked gold—a *noble* metal.)

Oxidation states of metals

It is not difficult to correlate the oxidation states of metals with their locations in the periodic table. This is because the oxidation states found for individual metals depend on their electron configurations. Consider, for example, the representative metals: those in the A-groups. Table 4.2 on page 111 lists their observed oxidation states. Notice that the Group IA and IIA metals form ions whose charges correspond to the loss of the outer s electrons from the atoms. The metals of each of these groups show only one oxidation state.

In Groups IIIA, IVA, and VA we find two oxidation states, especially for the members at the bottom of the groups. The lower oxidation state in each case corresponds to the loss of only the p electrons from the valence shell; the higher state results from the loss of both the s and p electrons. *As we go down a given group of representative metals, the lower oxidation states become progressively more stable.* In Group IIIA, for example, only the 3+ state is found for aluminum. At the bottom of the group, however, the 1+ state of thallium is much more stable than the 3+ state. Apparently, as we go down a group, the energy needed to remove the additional two s electrons is less well compensated by the energy released on forming two additional "bonds."

The transition metals characteristically have multiple oxidation states. One of the most common is 2+. Recall that these elements have a partially filled d subshell just below the next higher *filled s* subshell (although we saw that there are some exceptions). When electrons are removed from an atom, the first to go always come from the shell with highest principal quantum number. This means that these outer s electrons are lost first. Iron, for example, has the configuration

$$Fe \quad [Ar]\, 3d^6 4s^2$$

Removal of the $4s$ electrons gives the Fe^{2+} ion with the configuration

$$Fe^{2+} \quad [Ar]\, 3d^6$$

We will discuss the oxidation states of the transition elements and their relative stabilities in greater detail in Chapter 21.

9.4 TRENDS IN METALLIC BEHAVIOR

So far we have looked at a number of physical and chemical properties normally associated with metals. But not all metals are equally "metallic," especially in their chemical properties. For example, we commonly think of compounds formed from a metal and a halogen, such as chlorine or bromine, as being ionic. Beryllium chloride, however, does not fit this mold. When melted, $BeCl_2$ is a poor conductor of electricity, which suggests that molecules are present instead of ions. Similarly, aluminum forms molecular species such as Al_2Cl_6 and Al_2Br_6 in which the atoms are held together by covalent rather than ionic bonds.

Within the periodic table there are clear trends in the tendency of atoms to form covalent bonds. In Chapter 4 we learned that ionic bonds are preferred when a metal from the far left of the table combines with a nonmetal from the upper right corner. Covalent bonds are preferred, however, if two nonmetals combine. The deciding factor in each case is the difference in electronegativity between the bonded atoms. Metals have low electronegativities, which reflect their low ionization energies and electron affinities. Nonmetals, on the other hand, have high electronegativities. An atom's electronegativity, therefore, is a measure of how metallic it is chemically.

We learned in Chapter 4 that electronegativity increases from left to right across a period. If we compare the bonds formed by a metal with a given nonmetal—chlorine, for example—they should become less ionic moving from left to right across the table. This is because the difference in electronegativity between the metal and nonmetal becomes progressively smaller. In Period 2 we see this by comparing LiCl, $BeCl_2$, and BCl_3. Lithium chloride is distinctly ionic—it conducts electricity well in the molten state. As we've noted, $BeCl_2$ is a poor conductor when molten, which suggests that Be—Cl bonds are covalent. Boron trichloride is also covalent and exists as distinct molecules of BCl_3.

We also learned that electronegativity decreases going down a group. Metal–nonmetal compounds should therefore become more ionic as we descend a group, which they do. In Group IIA, for example, $BeCl_2$ is covalent, but $MgCl_2$ is ionic. In Group IIIA, BCl_3 is molecular and boron shows no evidence of forming B^{3+} ions. Aluminum chloride, on the other hand, appears to be ionic in the solid, although it exists in the vapor as molecules of Al_2Cl_6.

The elements of Group IVA, perhaps more than any others, illustrate the transition toward increasing metallic character as we descend a group. Carbon, at the head of the group, is nonmetallic in virtually every way.[1] Below carbon are the metalloids, silicon and germanium, and below these the metals, tin and lead.

Tin is unusual. At high temperatures it forms crystals that are metallic in appearance. At low temperatures (below 13.2°C) these very gradually change to a nonmetallic, powdery form in which tin has a diamond-type lattice. Tin articles left for long periods in the cold therefore gradually crumble. At one time it was thought that some pest attacked the tin, and this disintegration of tin articles was called "tin disease."

Lead, at the bottom of the group, demonstrates only metallic properties in the elemental state.

The trends in the metallic character of the elements that we have just seen—decreasing from left to right across a period and increasing from top to bottom in a group—can also be demonstrated by considering acid-base properties. Recall that metal oxides such as Na_2O are typically basic. They react with water to form hydroxides.

$$Na_2O + H_2O \longrightarrow 2NaOH$$

Nonmetal oxides are acidic. They form acids if they are able to react with water.

$$SO_3 + H_2O \longrightarrow H_2SO_4$$

The acid-base properties of the oxide of an element can therefore be used as an indicator of its metallic character.

If we look again at Period 2, the trend in metallic character is clearly seen. Lithium oxide reacts with water to form LiOH. It also reacts with acids to neutralize them.

$$Li_2O \ (s) + 2HCl \ (aq) \longrightarrow 2LiCl \ (aq) + H_2O$$

---

[1] One form of carbon, graphite, does conduct electricity. This, however, is more a result of the bonding in graphite than evidence for metallic behavior. The structure of graphite and the mechanism of its electrical conductivity are discussed in Section 9.9.

Metals with low electronegativities are said to be **electropositive.**

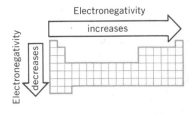

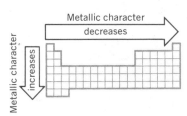

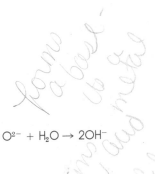

$O^{2-} + H_2O \rightarrow 2OH^-$

Lithium hydroxide reacts with acids, but it does not react with bases. This purely basic behavior of Li_2O and LiOH identifies lithium as a distinctly metallic element.

When we examine beryllium oxide, we find quite a different state of affairs. Although BeO is rather inert (resistant to chemical attack), under appropriate conditions it will react with *either acids or bases*. The hydroxide, $Be(OH)_2$, is also able to react with either acids or bases. Substances that behave this way are said to be **amphoteric**—they show both acidic and basic characteristics. Because beryllium compounds show some acidic characteristics, while the corresponding lithium compounds are purely basic, beryllium is less "metallic" than lithium. Moving further to the right, boron is even less metallic than beryllium. Its oxide shows only acidic properties.

B_2O_3 reacts with water to give $B(OH)_3$, boric acid.

The same trend toward decreasing metallic character is also seen in moving from left to right in Period 3. The hydroxides of sodium and magnesium are only basic—they react with acids but not with other bases. Aluminum, on the other hand, is amphoteric, indicating that it is less metallic than either sodium or magnesium. The free element as well as its oxide and hydroxide reacts with both acids and bases.

$$2Al\ (s) + 6H^+\ (aq) \longrightarrow 2Al^{3+}\ (aq) + 3H_2\ (g)$$

$$2Al\ (s) + 2OH^-\ (aq) + 2H_2O \longrightarrow 2AlO_2^-\ (aq) + 3H_2\ (g)$$

The ion AlO_2^- is called aluminate ion.

The reaction of metallic aluminum with a base explains why you will find warnings on many oven cleaners not to use them on aluminum pots and pans. These oven cleaners contain lye (NaOH), which would quickly attack aluminum cookware. The popular drain cleaner Drano® also makes use of this reaction, as mentioned in Example 7.10. It consists of crystals of lye mixed with bits of metallic aluminum. When the Drano® is placed into a plugged drain, the reaction of the NaOH with the aluminum releases bubbles of H_2 that stir the mixture and help the cleaner dissolve any hair and grease that might be causing the stoppage.

9.5 IONIC–COVALENT CHARACTER OF METAL–NONMETAL BONDS

In the last section we saw that there is a gradual transition in the properties of the elements from metallic to nonmetallic as we move from left to right across a period. We also saw that the elements become more metallic going down a group. Although the changes in the covalence of metal–nonmetal compounds can be related to the variations in electronegativity, there is another way of looking at this that is useful in accounting for many properties.

Metal cations are almost always smaller than nonmetal anions. The positive charge on the cation is therefore concentrated in a rather small volume, while the negative charge on the anion is spread over a much larger volume. When a cation is placed near an anion, it can distort the electron cloud of the anion, pulling part of the electron density into the region between the two nuclei as shown in Figure 9.11. This causes the anion to become something of a dipole (besides being an anion) and we say that the electron cloud of the anion has been *polarized* by the cation. Since a covalent bond consists of electrons shared *between* nuclei, polarization of the anion results in the partial formation of a covalent bond. Furthermore, the greater the degree of polarization, the greater the amount of covalent character of the bond.

There are two major factors that contribute to a cation's ability to polarize a given anion. One of these is the number of positive charges on the cation. All

Figure 9.11

Polarization of an anion by the positive charge on the cation. (a) Unpolarized anion and cation. (b) Electrical charge on the anion is distorted by the cation.

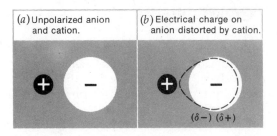

other things being equal, a cation with a 2+ charge will distort the electron cloud of an anion more than one with a 1+ charge. The second factor is the size of the cation. In a small cation, the positive charge is highly concentrated and has a strong effect on the anion. The same positive charge on a large cation will be more spread out and diffuse, so it won't distort the anion's electron cloud as well. A cation's effect on an anion is therefore directly proportional to its positive charge and inversely proportional to its size. We could also say that the effect of the cation is proportional to the *ratio* of its charge to its size. This ratio is called the **ionic potential,** ϕ,

$$\phi = \frac{q}{r}$$

where q is the cation's charge and r is its radius.

To see how this relates to the covalent character of metal–nonmetal bonds, let's look once more at the chlorides of the Period 2 elements: $LiCl$, $BeCl_2$, and BCl_3. Suppose, for the moment, that these compounds could all be made ionic. The cations would then be Li^+, Be^{2+}, and B^{3+}, and each would be surrounded by one or more chloride ions. Suppose we now ask the question, "How would the ionic potentials of these cations vary?" Studying their electronic structures, we find that each has just a filled $1s$ subshell. All the valence electrons have been stripped away. As the nuclear charge increases from Li ($Z = 3$) to boron ($Z = 5$), this $1s$ subshell is drawn more and more closely to the nucleus, so the size of the cations should decrease from Li^+ to B^{3+}. At the same time, the charge increases. The ionic potentials, therefore, increase very rapidly from Li^+ to B^{3+}, which means that their polarizing effects on the Cl^- ions surrounding them should also increase dramatically. This would cause a rapid increase in the amount of covalent bonding going from Li^+—Cl^- to B^{3+}—Cl^-, which, of course, is exactly what we've found. In LiCl the bonding is ionic, in $BeCl_2$ it is largely covalent, and in BCl_3 the bonds are even more covalent.

Going down a group, the charge on the cations stays the same (for example, Be^{2+}, Mg^{2+}, Ca^{2+}, etc). The sizes of the cations increases, however, so their ionic potentials decrease. This means that if we compare $BeCl_2$ with $MgCl_2$, we would expect Mg^{2+} to have less of a polarizing effect on Cl^- than Be^{2+}. The result is that $MgCl_2$ should be more ionic than $BeCl_2$, which is also what we've seen before.

The usefulness of this approach toward discussing the covalent character of metal–nonmetal bonds is that it lets us easily understand why certain metal compounds are so clearly covalent. Consider, for instance, tin(IV) chloride, $SnCl_4$. This substance is a clear colorless liquid at room temperature. It boils at 114°C, and its colorless cubic crystals are soft and melt at −33°C. These are properties we learned to identify with molecular substances, so the Sn—Cl bonds must be covalent. This is really not surprising in light of our previous discussions. If $SnCl_4$ were intially ionic, the cation would be Sn^{4+}. The large charge on this cation would give it a very large ionic potential and we would expect very extensive polarization of the chloride ions to occur. This would lead to quite covalent Sn—Cl bonds.

True ions with large positive charges are rare; their ionic potentials are too large.

Of course, we can also explain the bonding in $SnCl_4$ by valence bond or molecular orbital theory.

9.6 COLORS OF METAL COMPOUNDS

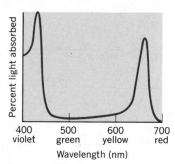

Figure 9.12

The absorption spectrum of chlorophyll. The light that is not absorbed is in the green and yellow parts of the spectrum—which is why the leaves of plants appear green.

Recall that the energy of a photon is **E = h**ν.

Many inorganic compounds, such as common table salt or sodium bicarbonate, appear white, which means that when they are bathed in visible white light, their crystals reflect all the wavelengths of visible light equally well. You've only to look around, of course, to also appreciate that there is a huge number of very brightly colored compounds. Although many of these are organic substances, many also are inorganic compounds that are formed from a metal and a non-metal.

A substance appears colored if it absorbs some wavelengths (or a band of wavelengths) from white light. The light that is reflected has these wavelengths missing and the color we perceive results from the wavelengths that remain.[2] Color Plate 9 illustrates a color wheel that we can use to anticipate the colors that result when selected wavelength regions are absorbed. Across from each other are complementary colors. Green is the complementary color to red; yellow is the complementary color to violet. If a substance absorbs a particular color when bathed in white light, the reflected light is perceived as its complement. A substance that absorbs yellow light, for example, will appear violet. Chlorophyll, the substance that gives plants their green color, absorbs both red and blue light as shown by its absorption spectrum, Figure 9.12. The light that is reflected is yellow-green, and that's the color we see.

When a substance absorbs light of a particular color, the energy of the absorbed photon raises an electron from one energy level to another. Thus the energy levels in substances affect the wavelengths (colors) of light absorbed just as we saw in Chapter 3 that they affect the wavelengths of light emitted. In molecules or polyatomic ions, these are molecular orbital energy levels somewhat like those discussed in Section 5.7.

Typical ionic compounds such as NaCl or Na_2S don't absorb visible light. They have no electronic transitions with energies corresponding to visible wavelengths, so they appear white or colorless. The absorption that does take place occurs in the shorter-wavelength (higher-frequency) ultraviolet region of the spectrum. Here the energy that is absorbed is used to shift an electron from the anion to the cation.

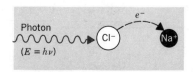

This is a **charge transfer process,** and the band of wavelengths absorbed from the ultraviolet "rainbow" is called a **charge transfer absorption band.**

As the bond between a metal and a nonmetal becomes somewhat covalent (that is, as electron density shifts away from the anion *toward* the cation), less energy is required to produce the charge transfer. This means that light of lower frequency (longer wavelength) is required, and the absorption band shifts from the high frequency ultraviolet region toward the violet and blue end of the visible spectrum. As a portion of the absorption band moves into the visible region, some violet light is absorbed and the compound takes on a pale yellow color. As the bond becomes even more covalent, the band shifts further into the

---

[2] The perception of color is actually somewhat more complex than this because of the varying sensitivity of the human eye to various wavelengths. For example, the eye is much more sensitive to green than to red. If a compound reflects both of these colors with equal intensity, it will appear greenish simply because the eye sees green better than it sees red.

Table 9.2
Colors of the silver halides

Compound	Color	Anion Radius (Å)
AgF	White	1.36
AgCl	White	1.81
AgBr	Cream	1.95
AgI	Yellow	2.16

The cream color of unexposed photographic film results from AgBr that is used as the light sensitive agent.

visible region and even more violet light is absorbed, so the compound appears even more intensely yellow. The depth or intensity of the color, therefore, is related to the degree of covalent character of the metal–nonmetal bonds.

Using color as a way to gauge the covalent character of bonds, we can now look at two other factors related to the ease with which a cation is able to polarize an anion. One is the size of the anion. For anions of the same charge the ease of polarization increases with size. This occurs because as the size gets larger, the outer electrons are further from the nucleus and are not held as tightly, so the cloud becomes more easily deformed. We can see this effect by looking at the colors of the silver halides (Table 9.2). Comparing AgCl, AgBr, and AgI, we see that as the anion becomes larger, the colors of the compounds become progressively deeper. This indicates a progressive increase in the covalent character of the Ag—X bonds as the halide ions become easier to polarize.[3]

We also find a similar relationship between covalent character and anion size if we compare metal oxides ($r_{O^{2-}} = 1.40$ Å) and sulfides ($r_{S^{2-}} = 1.84$ Å), as shown in Table 9.3. Both of these anions are colorless as evidenced by the fact that both Na_2O and Na_2S are colorless. However, with aluminum we find Al_2O_3 to be white while Al_2S_3 is yellow, suggesting a greater degree of ionic bonding in the oxide.

A second factor related to the ease of anion polarization is the charge on the anion. For anions of the same size, the ease of polarization increases as the negative charge becomes greater. For example, we can compare the colors of some compounds containing chloride ($r = 1.81$ Å) and sulfide ($r = 1.84$ Å) as shown in Table 9.4. In each case the sulfide has a deeper color than the corresponding chloride salt, indicating that the compounds containing the more highly charged sulfide ion are more covalent. We also see that, in general, most metal sulfides, except those of the alkali and alkaline earth metals, are deeply colored and possess quite a substantial degree of covalent bonding.

The presence of these colored compounds can often be seen. For instance, when silver tarnishes, it reacts with traces of H_2S in the air to give a dull film of

[3] The salts NaF, NaCl, NaBr, and NaI, which are predominantly ionic, are colorless. This indicates that the halide ions themselves are colorless. The colors of the AgX compounds are therefore a reflection of covalent bonding.

Table 9.3
Colors of some metal oxides and sulfides

Oxides	Color	Sulfides	Color
Al_2O_3	White	Al_2S_3	Yellow
Ga_2O_3	White	Ga_2S_3	Yellow
Sb_2O_3	White	Sb_2S_3	Yellow-red
Bi_2O_3	Yellow	Bi_2S_3	Brown-black
SnO_2	White	SnS_2	Yellow

Table 9.4
Colors of metal chlorides and sulfides

Chloride ($r = 1.81$ Å)		Sulfide ($r = 1.84$ Å)	
AgCl	White	Ag_2S	Black
CuCl	White	Cu_2S	Black
AuCl	Yellow	Au_2S	Brown-black
$CdCl_2$	White	CdS	Yellow
$HgCl_2$	White	HgS	Black or red, depending on crystal structure
$PbCl_2$	White	PbS	Black
$SnCl_2$	White	SnS	Black
$AlCl_3$	White	Al_2S_3	Yellow
$GaCl_3$	White	Ga_2S_3	Yellow
$BiCl_3$	White	Bi_2S_3	Brown-black

black Ag_2S. This hydrogen sulfide, produced generally from decomposing organic matter, also darkens lead-based paint. The pigment, $Pb_3(OH)_2(CO_3)_2$, called *white lead,* reacts with H_2S to form black PbS.

The fact that many of these compounds possess rather striking colors has also been put to use throughout history. For example, the brilliant yellow of natural CdS and the "vermilion" red of HgS have led them to be employed as pigments for the oil paints used by artists. Not long ago, several sticks of black PbS that had been used as a type of mascara were recovered from an ancient Egyptian burial ground.

9.7 SOME PHYSICAL PROPERTIES OF NONMETALS AND METALLOIDS

When we look at the world around us we see many practical applications of metals in their elemental forms. These applications exploit the physical properties of metals we discussed in Section 9.2. The elemental forms of the nonmetals and metalloids are less widely used, however, because their properties are so different from those of metals. At room temperature, a number of them are gases, one is a liquid, and the rest are solids. Photographs of some nonmetals appear in Color Plate 10. They are all poor conductors of electricity and heat, and their physical appearance is such that most would certainly not be mistaken for a metal. In addition, their solids are brittle.

Except for the noble gases (Group 0), which exist as single atoms, all the nonmetals and metalloids consist of at least two atoms covalently bonded together. Hydrogen, the simplest of all molecules, exists as diatomic H_2. Nitrogen and oxygen, you have learned, exist as the diatomic molecules, N_2 and O_2, and are the chief constituents of the atmosphere. The halogens, although not found free in nature, also form diatomic molecules: F_2, Cl_2, Br_2, and I_2. Fluorine and chlorine are gases at room temperature, bromine is a volatile red liquid, and iodine is a solid. Table 9.5 lists their melting and boiling points. The increase from fluorine to iodine is consistent with the rise in the strengths of the London forces that is expected as the molecules become larger.

The remainder of the nonmetals and metalloids exist as solids at room temperature in which various numbers of atoms are linked covalently. You have undoubtedly encountered several of them. Carbon exists in two forms: diamond and graphite. Graphite is the form of carbon we see most frequently. It is the "lead" in a pencil and the black substance in charcoal. Sometimes it is difficult to appreciate that a charcoal briquette and the glittering diamonds in Color Plate 10 are composed of the same substance. Phosphorus also occurs in more than one form. The one you have seen is called red phosphorus—it is mixed with fine sand on the striking surfaces of matchbooks. Silicon also has commercial applications, although you probably haven't actually seen the element. It is the semiconductor material used to make solid-state electronic devices, including the ones that power pocket calculators, digital watches, or the electronic exposure meters in cameras.

The increase in melting points from F_2 to I_2 is caused by an increase in the strengths of the London forces that accompanies the increase in molecular size.

Table 9.5
Melting and boiling points of the halogens

Element	Melting Point (°C)	Boiling Point (°C)
Fluorine, F_2	−219.6	−188.1
Chlorine, Cl_2	−101.0	−34.6
Bromine, Br_2	−7.2	58.8
Iodine, I_2	113.5	184.4

9.8 CONDUCTORS, NONCONDUCTORS, AND SEMICONDUCTORS

The most striking difference between the physical properties of metals and those elements that we classify as nonmetals and metalloids is their electrical conductivity. Metals conduct electricity well, nonmetals essentially not at all, and metalloids are semiconductors—their conductivity is between metals and nonmetals.

In a very elementary way, we can see why metals and nonmetals are so different. In a metal, the valence electrons belong to the lattice as a whole and are not localized on any particular atom, so they can move more or less freely through the solid. In nonmetals, however, the valence electrons are used to form covalent bonds to neighboring atoms. These bonds cause the electrons to be localized between the atoms that they help hold together, so they are unable to move through the solid and the substance is an insulator.

To understand semiconductor behavior we must take a slightly more sophisticated look at the energies of electrons in a solid. As you know, the atoms in a solid are very close together and evenly spaced. According to the **band theory of solids,** in a crystal, atomic orbitals of similar energy combine to produce *bands* of very closely spaced energy levels. For example, in sodium the 1s atomic orbitals, one from each atom, combine to form a single 1s band that extends in three dimensions throughout the entire solid. The same thing occurs with the 2s, 2p, etc., orbitals so that we also have 2s, 2p, etc., bands within the lattice.

Sodium atoms have filled 1s, 2s, and 2p orbitals; therefore the corresponding bands in the solid are also filled. The 3s orbital of sodium, however, is only half-filled, which leads to a half-filled 3s band. Following the same logic, in the ground state the 3p and higher energy bands are completely empty.

When a voltage is applied across sodium, electrons in filled bands are not able to move through the solid because orbitals in the same band on neighboring atoms are already filled and thus cannot acquire an additional electron. In the 3s band, which is half-filled, an electron may hop from atom to atom with ease. In Figure 9.13 the localized nature of electrons in filled bands is illustrated.

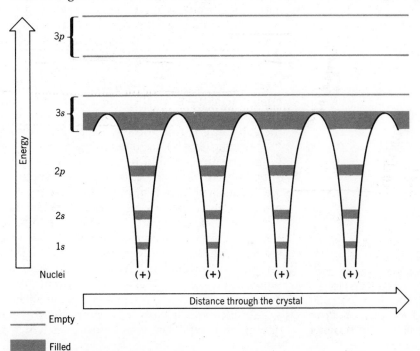

Figure 9.13

Bands in metallic sodium.

Figure 9.14

Energies of the valence band and conduction band in three kinds of solids. (a) Magnesium—a conductor. (b) An insulator. (c) A semiconductor.

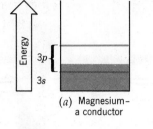
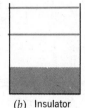
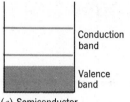

(a) Magnesium—a conductor   (b) Insulator   (c) Semiconductor

We can also see that the 3s band extends continuously through the solid, which accounts for the high conductivity of sodium.

The band containing the outer-shell (valence-shell) electrons is called the **valence band.** Any band that is either vacant or partially filled and is uninterrupted throughout the lattice is called a **conduction band.**

In metallic sodium, the valence band and conduction band are the same. In magnesium, the 3s valence band is filled and therefore cannot be used to transport electrons. However, the vacant 3p conduction band actually overlaps the valence band and can easily be populated by electrons when a voltage is applied (Figure 9.14a). This causes magnesium to be a conductor.

In an insulator (for example, glass, diamond, or rubber) the energy separation between the filled valence band and the conduction band is very large, as illustrated in Figure 9.14b. This large separation prevents electrons from populating the conduction band under an applied voltage.

A semiconductor, such as Si and Ge, has a relatively small gap between the filled valence band and the empty conduction band (Figure 9.14c). Thermal energy is sufficient to promote some electrons to the conduction band where they are then able to move through the solid. The conductivity of these substances increases with temperature because at the higher temperature a greater fraction of electrons populate the conduction band.

One of the most significant scientific discoveries made in this century is that the conductivity of semiconductors such as silicon and germanium can be altered in a controllable way by adding very small amounts of certain impurities to them. For example, Ga or As atoms may be introduced into very pure germanium to yield substances with enhanced semiconductor properties.

Germanium has the diamond structure (Figure 8.32) with four covalent bonds to each Ge atom. When a Ga atom ($4s^2 4p^1$) replaces a Ge atom ($4s^2 4p^2$), one of the Ga—Ge bonds is deficient by one electron. Under an applied voltage an electron from a neighboring atom may move to fill this deficiency and thus leave a positive "hole" behind. When this "hole" is filled, another is created elsewhere and electrical conduction results from the migration of this positive "hole" through the crystal. Because of the positive nature of the charge carrier, the substance is said to be a **p-type semiconductor.**

Arsenic ($4s^2 4p^3$) has one more electron in its valence shell than germanium. When arsenic is added to germanium as an impurity (we say that the Ge is "doped" with As), the extra electrons supplied are able to move through the solid when a voltage is applied. Because the conduction is the result of negative electrons, the crystal is called an **n-type semiconductor.**

Since their discovery, n- and p-type semiconductors have served as the nucleus for the explosive growth of solid-state electronics. The transistor, used in so many electronic gadgets that we now take for granted, is made of n- and p-type semiconductors. Your pocket calculator, for instance, contains thousands of transistors. Of even more current interest perhaps is the ability of these materials to harness the sun's energy in solar batteries (solar cells).

A silicon solar battery is composed of a silicon wafer doped with arsenic (giving an n-type semiconductor) over which is placed a thin layer of silicon

William Shockley, a controversial figure because of his opinions on intelligence and race, shared the 1956 Nobel Prize in physics with John Bardeen and William Brattain for their invention of the transistor.

A

B

Plate 1 (a) Blue crystals of copper sulfate, $CuSO_4 \cdot 5H_2O$. (b) When they are heated, the crystals lose water. The nearly white powder that remains is pure $CuSO_4$.

Plate 2 The familiar neon sign is a discharge tube containing a gas at a low pressure through which an electric current is passed. Different colors are produced by using different gases in the tube.

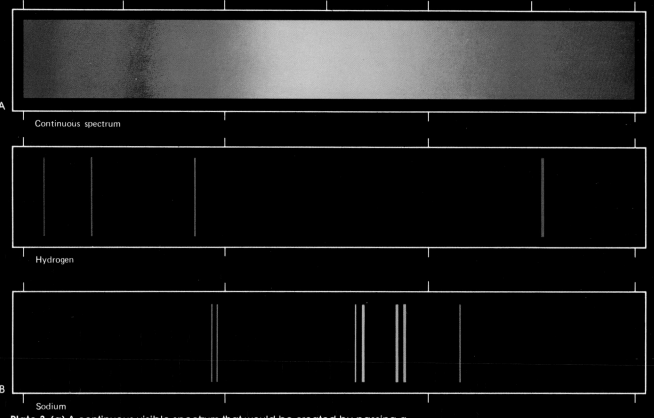

A

Continuous spectrum

Hydrogen

B

Sodium

Plate 3 (a) A continuous visible spectrum that would be created by passing a beam of white light from an incandescent light bulb through a prism. (b) Atomic emission spectra of hydrogen and sodium. Each element has its own characteristic atomic spectrum.

Plate 4 Energy-efficient high pressure sodium lamps light Park Avenue in New York City. The golden-yellow tone of the lighting is caused by the intense emission of yellow light by sodium at 589 nm.

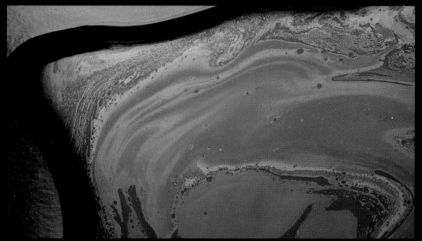

Plate 5 Some polyatomic ions have characteristic colors that they impart to their compounds. On the left is a solution of potassium chromate containing the yellow CrO_4^{2-} ion. In the center is a solution of potassium dichromate containing the red-orange $Cr_2O_7^{2-}$ ion. On the right is a solution of potassium permanganate containing the violet MnO_4^- ion.

Plate 7 Temperature affects the colors of light reflected from cholesteric liquid crystals. This phenomenon is used here to locate veins in a patient's arm. Dark reddish brown areas are cool, but the warm skin just above a vein shows up as a dark blue line, marking the vein's position.

Plate 6 A rainbow of colors is reflected from a thin film of oil on water. The colors arise from constructive interference of the light waves reflected from the upper and lower surfaces of the oil film. Those colors not reflected pass through the film into the water below.

Plate 8 The typical metallic luster of this freshly exposed surface of sodium will quickly fade as the metal tarnishes as a result of the reaction of the sodium with oxygen and moisture in the air.

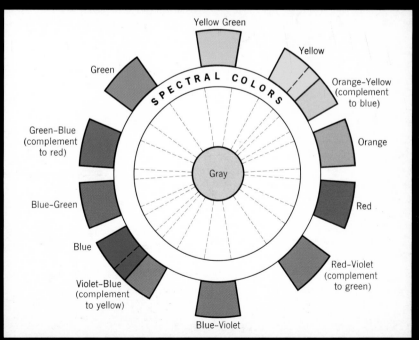

Plate 9 A color wheel. Colors that are across from each other are called complementary colors. When a substance absorbs a particular color, light that is reflected or transmitted has the color of its complement. Thus, something that absorbs red light appears green-blue, and vice versa.

Plate 10 Some of the nonmetallic elements as they occur at room temperature: (a) carbon in its diamond form, (b) carbon in its graphite form, (c) red phosphorus, (d) sulfur crystals, (e) gaseous chlorine, (f) liquid bromine, (g) solid iodine.

A

B

C

D

E

F

Plate 12 Electrolysis of a solution of sodium sulfate, Na_2SO_4, in the presence of an indicator. (a) Initially, the solution is neutral and the indicator gives it a yellow color. (b) During electrolysis, H^+ formed along with O_2 at the anode causes the solution there to become pink. Around the cathode the solution is made bluish violet by OH^- that is formed along with H_2. (c) After electrolysis, the solution is stirred. The H^+ and OH^- neutralize each other, which restores the original yellow color.

A

B

Plate 11 The reaction of zinc with copper ion. (a) A piece of shiny zinc next to a beaker containing a copper sulfate solution. (b) When the zinc is placed in the solution, copper is reduced to the free metal while the zinc dissolves. (c) After a while the zinc becomes coated with a red-brown layer of copper. Notice that the solution is a lighter blue than before.

Plate 13 Emerald (top) and aquamarine (bottom) are both forms of the mineral beryl, $Be_3Al_2(SiO_3)_6$.

Plate 14 Bicentennial fireworks display in Washington, D.C. The brilliant colors are produced by metal salts mixed with the burning chemicals. Red is produced by strontium salts, green by barium salts, blue by copper salts, and yellow by sodium salts.

A

B

C

D

Plate 15 Gems composed of aluminum oxide; (a) ruby, (b) sapphire (c) topaz, (d) amethyst. The different colors result from different trace impurities in nearly pure Al_2O_3.

Plate 16 In this illustration of the Ostwald process, ammonia escaping from a concentrated aqueous solution reacts with oxygen on a catalytic platinum surface, which glows from the heat of reaction. The colorless NO produced in this reaction is further oxidized to give the reddish-brown gas, NO_2.

Plate 17 Smog hangs over New York City. The reddish-brown color is caused by nitrogen dioxide, NO_2, one of the principal constituents of smog.

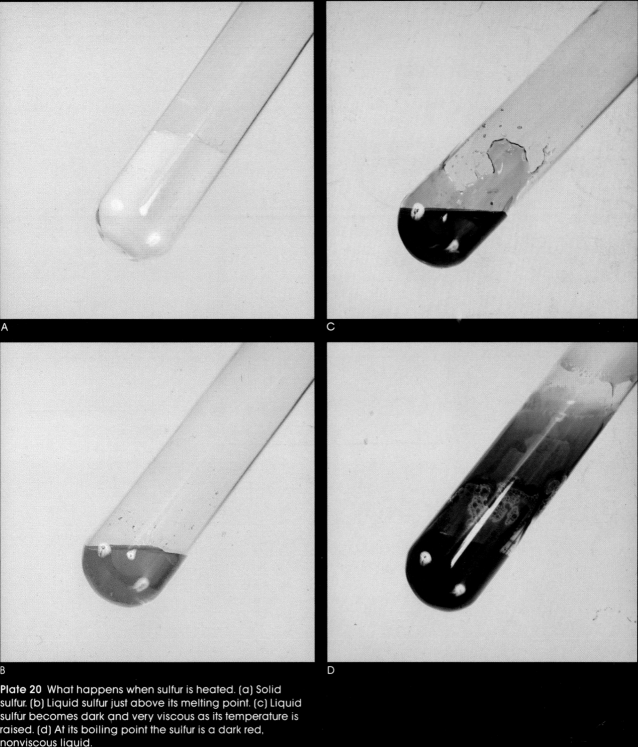

A

C

B

D

Plate 20 What happens when sulfur is heated. (a) Solid sulfur. (b) Liquid sulfur just above its melting point. (c) Liquid sulfur becomes dark and very viscous as its temperature is raised. (d) At its boiling point the sulfur is a dark red, nonviscous liquid.

Plate 21 Zircon, $ZrSiO_4$, is found in a wide variety of colors. When cut and polished, zircon crystals find uses as semiprecious gems.

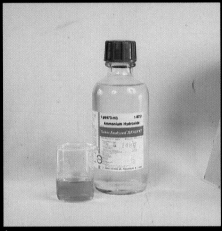

A

B

Plate 22 Complex ions of copper. (a) The pale blue $Cu(H_2O)_4^{2+}$ ion in a solution of copper sulfate. (b) Addition of aqueous ammonia gives the deep blue $Cu(NH_3)_4^{2+}$ ion.

Plate 23 Each of these brightly colored compounds contains a complex ion of Co^{3+}. The variety of colors arises because of the different ligands (molecules or anions) that are bonded to the cobalt ion in the complexes.

Figure 9.15
A solar battery.

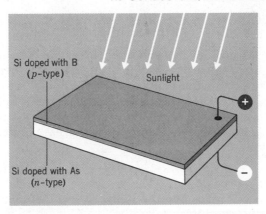

doped with boron (a *p*-type semiconductor). This is illustrated in Figure 9.15. In the absence of light there is an equilibrium between electrons and holes at the interface between the two layers (called a *p-n junction*). Some electrons from the *n*-type layer diffuse into the holes in the *p*-layer and are trapped. This leaves positive holes behind in the *n*-layer. Equilibrium is achieved when the positive holes in the *n*-layer prevent further movement of electrons to the *p*-layer. When light is allowed to fall on the surface of the cell, the equilibrium is upset. Energy is absorbed that permits electrons trapped in the *p*-layer to return to the *n*-layer. As these released electrons move across the *p-n* junction, other electrons leave the *n*-layer through the wire, pass through the electrical circuit, and enter the *p*-layer. Thus an electrical current flows when light falls on the cell and the external circuit is completed.

Solar batteries have been successfully used to provide electrical power to spacecraft. Today, much research is underway in an attempt to improve their efficiency and reduce their cost. Commercial units are already available that can be used to charge an automobile battery, but they are expensive. Someday, some scientists believe, we may receive electrical energy beamed to earth by microwaves from enormous satellites that collect solar energy with solar batteries (Figure 9.16).

This solar panel on the deck of a sailboat is used to charge the vessel's batteries when the engine isn't running.

Figure 9.16
This is an artist's concept of one of several configurations of Space-Based Power Conversion Systems now under study by the NASA-Marshall Space Flight Center (MSFC). Such systems would collect pollution-free energy from the Sun and transmit it in the form of microwaves to receiving stations on earth for conversion to electricity. These systems could provide electric power in the order of 5,000 to 10,000 megawatts on Earth.

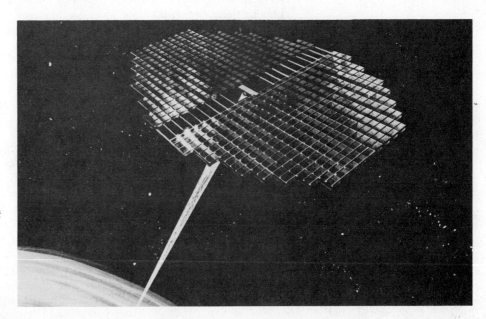

9.9 MOLECULAR STRUCTURES OF THE NONMETALS AND METALLOIDS

We learned in Chapter 4 that when atoms react they have a strong tendency to acquire a noble gas electron configuration, either by mutual sharing of electrons or by electron transfer. Except for the noble gases themselves, the nonmetals and metalloids have incomplete valence shells and, therefore, their individual atoms tend to combine until a noble gas structure is achieved. For example, hydrogen atoms unite to form H_2 so that each H atom acquires the helium configuration

$$H\cdot + H\cdot \longrightarrow H:H$$

In a similar fashion, the atoms of the other nonmetals and metalloids combine to give structures containing two or more atoms.

One of the controlling factors in determining the complexity of the molecular structures of the nonmetals and metalloids is their ability to form double and triple bonds by the sideways overlap of p orbitals on adjacent atoms ($p\pi$-$p\pi$ double and triple bonds). The period 2 elements (C, N, and O) are able to form these kinds of bonds rather easily, but as it turns out, *there is very little tendency for elements in the third and succeeding periods to form multiple bonds of this type.* Presumably this is because their larger size prevents them from approaching each other too closely. Consequently, for these heavier elements, the sideways overlap of p orbitals required for $p\pi$-$p\pi$ multiple bonding is not very effective, and the formation of two (or three) separate single bonds tends to be preferred energetically over the formation of one double (or triple) bond. The net result, regardless of how we justify it, is that *elements of the second period are able to form multiple bonds fairly readily, while the elements below them in the following periods have a tendency to prefer single bonds.* This phenomenon is particularly striking when we examine the structures of the elements in their free states.

We have seen that some elements in the second period form stable diatomic molecules, for example, N_2, O_2, and F_2. In these three cases the valence shells of the atoms are completed by sharing three, two, and one electron, respectively, with nitrogen and oxygen participating in $p\pi$-$p\pi$ bonding. It is because these elements are capable of achieving a stable electron configuration by sharing electrons with a *single* neighbor that they are able to form simple diatomic molecules in their elemental state.

Oxygen, in addition to forming the stable species O_2, also can exist in another exceedingly reactive molecular form, O_3, called ozone. The structure of ozone can be represented as a resonance hybrid

This unstable molecule can be generated by the passage of an electric discharge through ordinary O_2, and its pungent odor can often be detected in the vicinity of electrical equipment. It is also formed in limited quantities in the upper atmosphere by the action of ultraviolet radiation from the sun on O_2. Its presence there shields the earth and its creatures from exposure to intense, and harmful, ultraviolet light.

The existence of an element in more than one form, either as the result of differences in molecular structure, as with O_2 and O_3, or as a consequence of differences in the packing of atoms or molecules in the solid, is a phenomenon called **allotropism.** Oxygen is only one of several nonmetals that exist in different allotropic forms, although the phenomenon is not limited to the non-

π bonding was discussed in Chapter 5, page 160.

notes

The properties of ozone, a chief constituent of smog, are discussed further in Chapter 19.

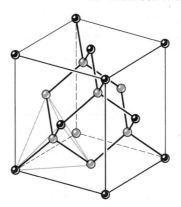

Figure 9.17

The structure of diamond.

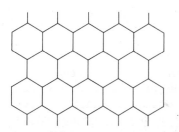

Figure 9.18

Sigma-bond framework of a graphite sheet.

C $\uparrow\downarrow$ $\uparrow$ $\uparrow$
   $\overline{2s}$ $\overline{2p}$

hybridization

$\uparrow$ $\uparrow$ $\uparrow$ $\uparrow$
   $\overline{sp^2}$ $\overline{p}$

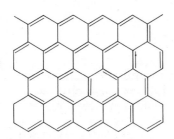

Figure 9.19

One of the resonance structures for graphite.

metals. For example, we saw earlier that tin, an element with mostly metallic properties, exhibits a metallic lattice at high temperatures and a nonmetallic lattice at low temperatures.

If we next turn our attention to carbon, also a second-period element, we see that it must share four electrons to complete its octet. There is no way for carbon to form a "quadruple bond," so a simple C_2 species is not stable under ordinary conditions. Instead, carbon tends to complete its octet in either of two ways so that two allotropic forms of elemental carbon are found. One of these is diamond. In diamond each carbon atom is covalently bonded to four others located at the corners of a tetrahedron. Each of those atoms, in turn, is bonded to three more, and so on, as illustrated in Figure 9.17. In this way a three-dimensional network is created. A diamond crystal (for example, a gem-quality diamond like those in Color Plate 10) consists of a huge number of carbon atoms covalently bonded together in one gigantic molecule. Since breaking a diamond crystal involves rupturing a very large number of covalent bonds, diamond is very hard.

In the second allotrope of carbon, graphite, the atoms are arranged in the form of hexagonal rings fused together in large planar sheets, perhaps somewhat reminiscent of chicken wire (Figure 9.18). Each carbon atom is surrounded by three nearest neighbors located at angles of 120° from one another, and the molecular framework is therefore based on σ bonds produced by overlap of sp^2 hybrid orbitals on the carbon atoms. On each of the carbon atoms throughout the entire structure there remains an unhybridized p orbital, each containing one electron, and these p orbitals are situated ideally for $p\pi\text{-}p\pi$ overlap. The result is a huge delocalized π electron cloud extending across the graphite sheet above and below the plane of the carbon atoms.[4] The free movement of electrons in this π cloud accounts for the electrical conductivity of graphite. An electron can be pumped into the cloud at one end of the sheet and another removed from the other end to give a net transfer of electrons through the solid.

In the total graphite structure the planes of carbon atoms are stacked in layers so that each carbon atoms lies above another in every second layer, as shown in Figure 9.20. Within any given layer, adjacent carbon atoms are fairly close together (1.41 Å) while the spacing between successive planes is much greater (3.35 Å). The different planes of carbon atoms are not held together by covalent bonds, but, instead, by very much weaker London forces. As a result, the layers are able to slide over one another with relative ease and, as you may

[4] In the valence bond approach we imagine that adjacent carbon p orbitals overlap to form simple π bonds and that the total bonding picture can be viewed as a composite of resonance structures such as that shown in Figure 9.19.

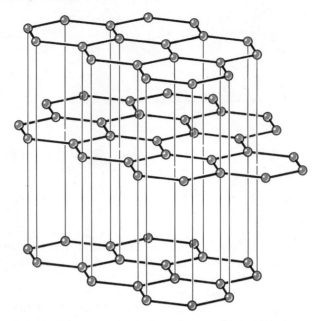

Figure 9.20
Total graphite structure.

know, graphite has many applications as a dry lubricant. As every school child
is taught, the "lead" in a "lead pencil" is actually graphite, and when one
writes with a pencil, the layers of graphite slide off one another onto the paper.

In graphite, carbon exhibits multiple bonding, as do nitrogen and oxygen
in their molecular forms. When we move now to the third and successive
periods, a different state of affairs exists. Here we have atoms that prefer to
form only single bonds with other atoms. They have relatively little tendency to
participate in $p\pi$-$p\pi$ multiple bonds. The molecular structures of the free elements reflect this.

Chlorine, because it only needs to form a single covalent bond to complete
its octet, exists as diatomic molecules. Bromine and iodine form diatomic Br_2
and I_2 for the same reason. The structures of the remaining nonmetals, however, are considerably more complex. With sulfur, for instance, we find that
each atom must share two electrons to fill its valence shell, and it does so by
forming two single bonds to *two different* sulfur atoms. These, in turn, must also
be bonded to two separate S atoms and a —S—S—S—S— sequence is produced. Actually, in its most stable form the sulfur atoms are arranged in puckered, eight-membered S_8 rings with a crownlike structure, illustrated in Figure
9.21. Selenium also forms Se_8 rings in one of its allotropic forms. And selenium
and tellurium also exist in a gray form in which there are long Se_x and Te_x
chains, respectively (the subscript x is a large number).

Now let's examine the element phosphorus. Being in Group VA, an atom of
phosphorus has three unpaired electrons

Therefore, to achieve a noble gas structure, it must acquire three more. Since
there is little tendency for this element to form multiple bonds, as nitrogen does
when it forms N_2, the octet is completed by the formation of three single covalent bonds to three different phosphorus atoms.

The simplest elemental form of phosphorus is a waxy solid called **white
phosphorus.** It consists of P_4 molecules in which each phosphorus atom lies at a
corner of a tetrahedron, as illustrated in Figure 9.22. In this structure we see that
each phosphorus atom is nicely bound to three others. This particular allotrope
of phosphorus is very reactive because of the highly strained P—P—P bond

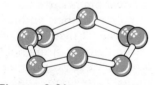

Figure 9.21
The structure of the S_8 ring.

Figure 9.22
*The structure of white phos-
phorus, P_4.*

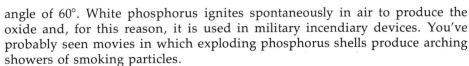

Streamers of white smoke arc through the air as a phosphorus artillery shell explodes.

angle of 60°. White phosphorus ignites spontaneously in air to produce the oxide and, for this reason, it is used in military incendiary devices. You've probably seen movies in which exploding phosphorus shells produce arching showers of smoking particles.

A second allotrope that is much less reactive is **red phosphorus** (Color Plate 10). At the present time its structure is unknown, although it has been suggested that it contains P_4 tetrahedra linked at the corners. Red phosphorus is used in explosives and, as noted earlier, it is mixed with fine sand on the striking surface used to light matches. In this case, friction caused by the match being drawn across the surface ignites the phosphorus, which in turn ignites the ingedients in the match head.

The third allotropic form is black phosphorus, formed by heating the white variety at very high pressures. It has a layer structure in which each phosphorus atom in a layer is singly bonded to three others. The layers are stacked on each other with only weak London forces between them. Consequently, black phosphorus looks and behaves physically much like graphite. Like red phosphorus, it is quite unreactive.

The elements below phosphorus—arsenic and antimony—are also able to form somewhat unstable yellow allotropic forms containing As_4 and Sb_4 molecules. The most stable forms have a metallic appearance with structures similar to black phosphorus.

Finally, we look at the heavier elements in Group IVA: silicon and germanium. To complete their octets these elements must each acquire four electrons through sharing, and they do so by the formation of single bonds. As a result, Si and Ge (as well as the low-temperature nonmetallic form of tin) have the diamond structure shown in Figure 9.17. Since silicon and germanium do not appear able to enter into $p\pi$-$p\pi$ bonding, graphitelike structures do not exist.

One element whose structure we have not considered is boron. This period 2 element, found in Group IIIA, is quite unlike any of the others, since there is no simple way for it to complete its valence shell. Sharing electrons in such a way as to give three ordinary single bonds still would leave each boron atom with only six electrons around it. As a result, understanding its structure is not easy.

Boron exists in several different crystalline forms, each of which is characterized by clusters of 12 boron atoms located at the vertices of an icosahedron (a 20-sided geometric figure) as shown in Figure 9.23*a*. Each boron atom within a

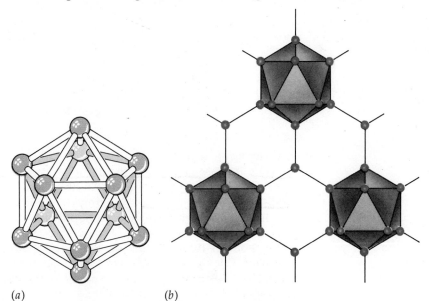

Figure 9.23

(*a*) *The icosahedral* B_{12} *unit.* (*b*) B_{12} *units connected together by other boron atoms.*

(*a*)   (*b*)

given cluster is equidistant from five others and, in the solid, each of these is also joined to yet another boron atom outside the cluster (Figure 9.23*b*). The electrons available for bonding are therefore delocalized to a large extent over many boron atoms.

The linking together of B_{12} units produces a large three-dimensional network solid that is very difficult to break down. Boron, therefore, is very hard (it is the second hardest element) and has a very high melting point (about 2200°C).

9.10 CHEMICAL PROPERTIES OF THE NONMETALS AND METALLOIDS

Although relatively few in number, the nonmetals and metalloids form many more compounds than the metals. The reason is that these elements combine not only with metals but with each other as well. With metals they often form anions, such as Cl^- or O^{2-}, and participate in ionic bonding. With each other, however, they form covalent bonds of varying degrees of polarity. This produces molecules that range from such simple species as carbon monoxide, CO, to the gigantic molecules of DNA that control heredity and the synthesis of proteins in our bodies.

The chemical properties of nonmetals and metalloids are conveniently discussed together because the elements show many similarities. Chemically, metalloids behave like nonmetals, with hardly any metallic characteristics. In the remainder of this discussion, therefore, we will not make a formal distinction between these two classes of elements—we will refer to them all as nonmetals.

In Section 9.4 we discussed trends in metallic character by examining the basicity of metal oxides. Nonmetal oxides are acidic, and there are trends in their acidity, too. This is most easily studied by looking at the acidity of the oxoacids formed when nonmetal oxides are dissolved in and react with water, because trends in the acidity of the oxides parallel trends in the acidity of the oxoacids. Let's begin by investigating the factors that affect how acidic these substances are.

An oxoacid such as H_2SO_4 is acidic in water because a hydrogen from one of its O—H bonds is released as a hydrogen ion, H^+. This becomes attached to a water molecule to produce a hydronium ion.

In a molecule of H_2SO_4, or in any oxoacid, the O—H bonds are polar because oxygen is more electronegative than hydrogen. This means that the hydrogen carries a partial positive charge. Anything that can affect how polar this bond is will also affect the acidity. Making the O—H bond more polar, for instance, gives the hydrogen a greater partial positive charge and makes it easier for it to come off as H^+.

There are two major factors that affect the polarity of the O—H bonds in an oxoacid. One is the number of lone oxygen atoms bonded to the central nonmetal atom. For example, we can compare the two oxoacids of sulfur, H_2SO_4 and

H_2SO_3, whose structures are

Each undergoes ionization in water by transferring its protons, in two successive steps, to water molecules. Thus, for H_2SO_4 we have in the first step,

H_2SO_4 is a strong acid.

With H_2SO_4 this step proceeds essentially to completion, while for H_2SO_3, on the other hand, the first step,

H_2SO_3 is a weak acid.

takes place only to a very limited degree (about 11% for a 1 M solution). In each of these substances the sulfur is bonded to two O—H groups. In H_2SO_4, however, the sulfur is also attached to two other lone oxygen atoms, while the sulfur in H_2SO_3 is bonded to only one such O atom. We know that oxygen is a very electronegative element, and we expect these S—O bonds to be polar, with the negative charge concentrated about the oxygen end of the dipole. In other words, the oxygen atoms attached to the sulfur draw electron density away so that the sulfur acquires a partial positive charge, the magnitude of which will be greater in H_2SO_4 than in H_2SO_3,

This positive charge on the sulfur tends to draw electron density from the S—OH bonds. Since the electronegative oxygen does not wish to lose electrons, the overall net effect is that some electronic charge will be withdrawn from the O—H bonds. This electron-withdrawing effect will be greater in H_2SO_4 than in H_2SO_3 because in H_2SO_4 the sulfur bears a greater positive charge and is therefore better able to draw electrons to itself. This means that in H_2SO_4 the hydrogen atoms carry a higher positive charge than they do in H_2SO_3 and are thus more easily removed as H^+. As a result, H_2SO_4 is a stronger acid than H_2SO_3.

This phenomenon is illustrated even more graphically if we consider the oxoacids of chlorine:

Hypochlorous acid   HOCl   H—O—Cl:

Chlorous acid $HClO_2$ H—O—Cl—O:

Chloric acid $HClO_3$ $H-\overset{..}{\underset{..}{O}}-\overset{..}{\underset{|}{Cl}}-\overset{..}{\underset{..}{O}}:$

$:\overset{..}{\underset{..}{O}}:$

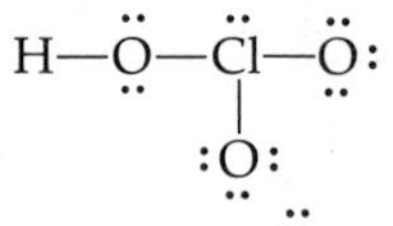

Perchloric acid $HClO_4$ $H-\overset{..}{\underset{..}{O}}-\overset{:\overset{..}{\underset{..}{O}}:}{\underset{:\underset{..}{O}:}{Cl}}-\overset{..}{\underset{..}{O}}:$

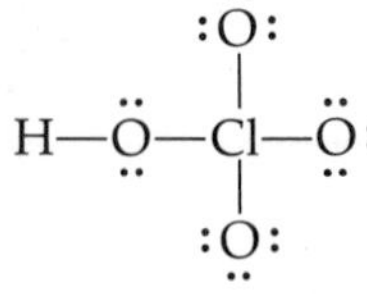

As the number of oxygen atoms surrounding the chlorine increases, so does the acidity. Thus HOCl is a relatively weak acid, $HClO_2$ is stronger, $HClO_3$ is essentially fully dissociated in water, and $HClO_4$ is just about the strongest acid there is.[5]

EXAMPLE 9.1

Nitrous acid, HNO_2, is formed in the stomach from nitrites that are used as preservatives in meats. There is considerable concern that it may be able to react with certain biological molecules to produce carcinogenic (cancer-causing) agents. How does the acid strength of HNO_2 compare with that of nitric acid, HNO_3?

SOLUTION

First let's draw the Lewis structures for HNO_2 and HNO_3. Each would have one O—H bond (the hydrogen is bonded to an oxygen in an oxoacid) and any other oxygens are "lone oxygens."

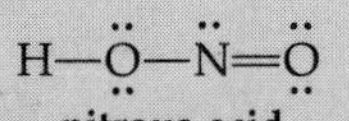

$:\overset{..}{O}:$

$H-\overset{..}{\underset{..}{O}}-\overset{..}{N}=\overset{..}{O}$ $H-\overset{..}{\underset{..}{O}}-\overset{|}{\underset{..}{N}}=\overset{..}{O}$

nitrous acid 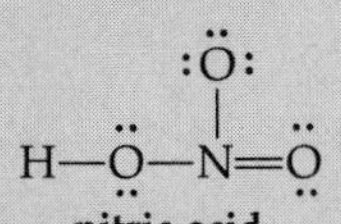   **nitric acid**

Since HNO_3 has more lone oxygens than HNO_2, the O—H bond in HNO_3 should be more polar than the one in HNO_2. Therefore HNO_3 should be the stronger acid (which it is).

[5] This reasoning can also be extended to other compounds. For example, acetic acid has the formula

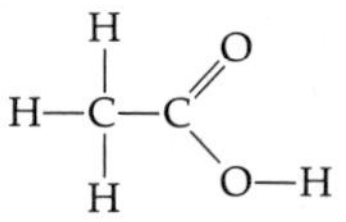

$H-\overset{\overset{H}{|}}{\underset{\underset{H}{|}}{C}}-C\overset{\diagup O}{\diagdown O-H}$

acetic acid

and is a weak acid. At any instant in a 1 M solution, only about 0.4% of the acetic acid molecules have reacted to form H_3O^+ and $CH_3CO_2^-$ ions. However, if a hydrogen in the CH_3 group is replaced with a very electronegative element such as chlorine,

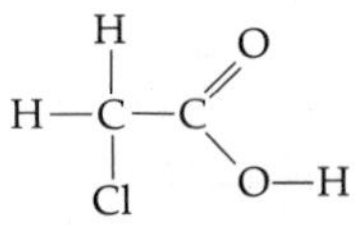

$H-\overset{\overset{H}{|}}{\underset{\underset{Cl}{|}}{C}}-C\overset{\diagup O}{\diagdown O-H}$

chloroacetic acid

the electron-attracting power of the chlorine increases the polarity of the O—H bond and makes the acid stronger. In a 1 M solution of chloroacetic acid, about 3% of the molecules have reacted to form H_3O^+ and $CH_2ClCO_2^-$ ions.

The second major factor that affects the acidity of an oxoacid is the electronegativity of the central atom, and here we can find some trends in the periodic table. As the central atom, X, in a molecule such as

$$
\begin{array}{c}
\ddot{\text{O}}\text{:} \\
| \\
\text{:}\ddot{\text{O}}\text{—X—}\ddot{\text{O}}\text{—H} \\
| \\
\text{:}\ddot{\text{O}}\text{:}
\end{array}
$$

becomes more electronegative, electron density is funneled toward X and away from the O—H bond. Therefore, as the electronegativity of X increases, so does the acidity of the molecules.

Within a group, we know that the electronegativity increases from bottom to top. This means that the acidity of oxoacids having the same general formula should also increase from bottom to top in a group, which they do. For instance, phosphoric acid (H_3PO_4) is a stronger acid than arsenic acid (H_3AsO_4). Similarly, sulfuric acid (H_2SO_4) is stronger than selenic acid (H_2SeO_4).

We also learned that the electronegativity increases from left to right in a period, so the strengths of the oxoacids should increase in that direction, too. In fact, there is a very rapid increase in acid strength from left to right, not only because of electronegativity changes, but also because the formulas of the acids change. In Period 3, for example, we find the following oxoacids in which the central atoms are each surrounded by four oxygen atoms.

$$
\begin{array}{ccc}
H_3PO_4 & H_2SO_4 & HClO_4 \\
\end{array}
$$

$$
\begin{array}{ccc}
\ddot{\text{O}}\text{:} & \ddot{\text{O}}\text{:} & \ddot{\text{O}}\text{:} \\
| & | & | \\
\text{H—}\ddot{\text{O}}\text{—P—}\ddot{\text{O}}\text{—H} & \text{H—}\ddot{\text{O}}\text{—S—}\ddot{\text{O}}\text{—H} & \text{H—}\ddot{\text{O}}\text{—Cl—}\ddot{\text{O}}\text{:} \\
| & | & | \\
\text{:}\ddot{\text{O}}\text{:} & \text{:}\ddot{\text{O}}\text{:} & \text{:}\ddot{\text{O}}\text{:} \\
| \\
\text{H}
\end{array}
$$

The electronegativity increases from phosphorus to chlorine and the number of lone oxygens also increase—both are factors favoring an increase in acid strength. Thus, sulfuric acid is stronger than phosphoric acid, and perchloric acid is stronger than sulfuric acid. Figure 9.24 summarizes the trends in the strengths of the oxoacids.

We can also find periodic trends in the strengths of the binary acids of the nonmetals. Recall that these acids are water solutions of nonmetal hydrides—

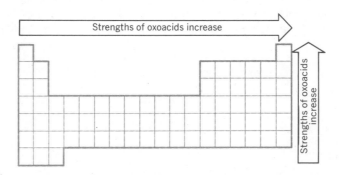

Figure 9.24
Variations within the periodic table of the strengths of oxoacids having similar formulas.

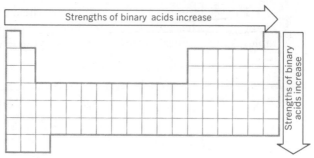

Figure 9.25

Variations within the periodic table of the strengths of the binary acids of the nonmetals.

for example, HF, HCl, and H_2S. They produce acidic solutions by reacting with water. For instance,

$$H-\overset{\cdot\cdot}{\underset{H}{O}}: + H:\overset{\cdot\cdot}{\underset{\cdot\cdot}{Cl}}: \longrightarrow \left[H-\overset{\cdot\cdot}{\underset{H}{O}}-H \right]^{+} + \left[:\overset{\cdot\cdot}{\underset{\cdot\cdot}{Cl}}: \right]^{-}$$

When we consider binary acids derived from elements belonging to the same group, an increase in acidity going down the group is observed experimentally. Thus the hydrohalides increase in acidity in the order HF < HCl < HBr < HI.

At first glance this order of acidity seems opposite to what we might predict. We know, for instance, that fluorine is more electronegative than chlorine; therefore the HF bond is more polar than the HCl bond. Consequently, the H in HF is more positively charged than the H in HCl. Thus we are tempted to predict that HF should lose a proton more readily than HCl. This, however, is precisely the reverse of their acid strengths in water, where HF is weak and HCl is essentially 100% ionizied.

The answer to this dilemma can be understood by realizing that there are actually two opposing factors that contribute to the acidity of these compounds. One is, indeed, the ionic character of the H—X bond; the other is the H—X bond strength. As we proceed down within a group, the nonmetal becomes progressively larger and there is an accompanying rapid decrease in the strength of the H—X bond.[6] This weakening of the H—X bond turns out to be more than suffcient to compensate for the decrease in the polarity of the bonds, and a net increase in acid strength is observed.

When we look at the acidity of the binary hydrogen compounds of elements in the same period, for example, NH_3, H_2O, and HF, the dominant factor becomes the polarity of the H—X bond. As we go from left to right within the period there is little change in the size of the nonmetal and relatively little change in the H—X bond energy. There is, however, a very dramatic increase in the ionic character of the H—X bond, which is reflected in a rapid increase in acidity from ammonia to hydrogen fluoride. This is summarized in Figure 9.25.

In general, bond strengths tend to decrease as the bonded atoms become larger.

---

[6] A decrease in bond energy going down a group was mentioned earlier as a reason for the increasing stability of the lower oxidation states of the metals in Groups IIIA and IVA.

INDEX TO QUESTIONS

REVIEW QUESTIONS

9.1 Why is the periodic table so useful to chemists and chemistry students?

9.2 Make a rough sketch of the periodic table and, without peeking at Figure 9.1, mark off the regions where you would expect to find the metals, metalloids, and nonmetals.

9.3 Among the representative elements, how do the number of metals and nonmetals compare.

9.4 Write the chemical symbol for

(a) a representative element
(b) an alkali metal
(c) an alkaline earth metal
(d) an inner transition element
(e) a halogen
(f) a noble gas

9.5 Name and describe four physical properties of metals.

9.6 For each of the properties named in Question 9.5, give a specific example of a practical application.

9.7 Describe a metallic lattice.

9.8 Why, under ordinary conditions, don't metals form simple molecular species with each other by sharing electrons in covalent bonds?

9.9 Why do metals have a characteristic luster?

9.10 Why are metals better conductors of heat than nonmetals?

9.11 What property of metals is used in manufacturing wire?

9.12 Have you ever seen a penny that's been run over by a railroad train? What property of metals is responsible for its new shape?

9.13 Which metal has the highest melting point? Which has the lowest?

9.14 How do the melting points of the transition elements in Period 4 vary?

9.15 Where are the metals with the highest melting points found in the periodic table?

9.16 Why does sodium have a lower melting point than magnesium?

9.17 Which element would have the greater hardness, potassium or calcium?

9.18 What kinds of oxidation states do metals have in virtually all their compounds?

9.19 Based on trends in ionization energy, which element in each of the following pairs should be more easily oxidized?
(a) Rb or Sr   (c) Na or Al
(b) Na or Rb   (d) Al or Ca

9.20 Why do the metals below magnesium in Group IIA have very few practical applications?

9.21 Why are gold and platinum known as noble metals?

9.22 Write a balanced equation for the reaction of hydrochloric acid with (a) magnesium and (b) aluminum.

9.23 Write balanced equations for these reactions:
(a) chromium + hydrochloric acid → chromium(III) chloride + hydrogen
(b) nickel + sulfuric acid → nickel(II) sulfate + hydrogen

9.24 Write net ionic equations for the reactions in Questions 9.22 and 9.23. What is the oxidizing agent in each of these reactions?

9.25 Which three common metals are not attacked by HCl but do dissolve in nitric acid?

9.26 How do the stabilities of the 4+ oxidation states compare for tin and lead? Which should have the greater tendency to acquire electrons? Which should be the stronger oxidizing agent, SnO_2 or PbO_2?

9.27 Among the representative elements, why are the lower oxidation states preferred by elements at the bottom of a given group in the periodic table?

9.28 In what portion of the periodic table are the least reactive metals found?

9.29 What is *aqua regia*?

9.30 Write electron configurations for the following ions:
(a) Cu^+ (f) Pt^{2+}
(b) Fe^{2+} (g) Cd^{2+}
(c) Ti^{4+} (h) Co^{3+}
(d) V^{3+} (i) Fe^{3+}
(e) Ag^+ (j) Sc^{3+}

9.31 What accounts for the occurrence of a 2+ oxidation state for a large number of transition elements?

9.32 Would you expect calcium to react more or less rapidly than mangesium with HCl? What is the basis for your answer?

9.33 How does the metallic character of the elements depend on electronegativity? What vertical and horizontal trends in metallic character exist in the periodic table? Illustrate these trends for the elements in the second period and in Group IVA.

9.34 In each pair below, choose the element expected to have the more metallic character.
(a) Li or Be
(b) B or Al
(c) Al or Cs
(d) Sn or P
(e) Ga or I

9.35 Write an equation for the reaction of water with (a) calcium and (b) potassium.

9.36 Which oxide should be more basic, Al_2O_3 or Ga_2O_3?

9.37 Which should be more acidic, MgO or Al_2O_3?

9.38 What is meant by amphoteric? Write chemical equations to illustrate the amphoteric behavior of beryllium and aluminum.

9.39 When aluminum oxide dissolves in base, the aluminate ion, which can be written as AlO_2^-, is formed. Write an equation for this reaction.

9.40 Define ionic potential. How is it related to the degree of covalent character in a metal–nonmetal bond?

9.41 Which should have the greater degree of covalent bonding?
(a) $GaCl_3$ or $GeCl_4$
(b) Bi_2O_3 or Bi_2O_5
(c) PbO or PbS
(d) $LiCl$ or Li_2S
(e) Na_2S or MgS

9.42 Which should be more ionic?
(a) SnO or SnS
(b) $AlCl_3$ or $AlBr_3$
(c) $BeCl_2$ or BeF_2
(d) SnS or PbS
(e) SnS or SnS_2

9.43 What is responsible for the color in compounds such as SnS_2 or PbS?

9.44 Which compound would you expect to be more deeply colored?
(a) Ag_2O or Ag_2S
(b) $CuCl$ or $CuBr$
(c) SnS or SnS_2
(d) MgS or Al_2S_3

9.45 If a particular compound absorbs green light, what color will it appear to be?

9.46 Which nonmetals occur as gases at room temperature?

9.47 Which two elements are liquids at room temperature?

9.48 In qualitative terms, why are crystals of the nonmetallic elements nonconductors of electricity?

9.49 How does the band theory of solids differentiate among conductors, semiconductors, and insulators?

9.50 What is an n-type semiconductor? What is a p-type semiconductor?

9.51 (a) Give two elements that would make Si a p-type semiconductor. (b) Give two elements that would make Si an n-type semiconductor.

9.52 Describe what happens when light falls on a silicon solar cell.

9.53 Why does the electrical conductivity of a semiconductor increase with increasing temperature?

9.54 What appears to be the dominant factor in determining the complexity of the molecular structures of the elemental nonmetals?

9.55 What explanation can be offered for the fact that the nonmetals in the third and succeeding periods do not appear to form stable $p\pi$-$p\pi$ bonds?

9.56 What are the two allotropic forms of carbon? How do they differ from one another in terms of structure and physical properties?

9.57 Compare the structures of N_2 and P_4. What is the P—P—P bond angle in P_4? How does this compare to the normal bond angles that would occur with overlap of p orbitals? Can you suggest how this might account for the high reactivity of P_4?

9.58 What are the structural differences between white and black phosphorus?

9.59 How do we account for the electrical conductivity of graphite?

9.60 What is the structural unit that occurs in elemental boron? Practice sketching its shape.

9.61 What is "tin disease"?

9.62 Which is the stronger acid?
(a) H_2SO_3 or $HClO_3$
(b) HNO_3 or H_2CO_3
(c) H_3AsO_4 or H_3PO_4

(g) $HClO_3$ or $HBrO_3$
(h) $HOBr$ or $HBrO_3$

9.63 Which is the stronger acid?
(a) H_2S or H_2Se
(b) H_2Se or HBr
(c) PH_3 or NH_3

9.64 Why is HNO_3 a stronger acid than HNO_2?

9.65 Why is HCl a stronger acid than HF?

9.66 Why is HCl a stronger acid than H_2S?

9.67 Which compound would you expect to be more acidic, CH_3OH or CH_3SH?

10

PROPERTIES OF SOLUTIONS

A common sight in northern climates, as snow plows spread salt behind them to melt ice and snow. This is just one of many practical applications of the effects that a solute has on the physical properties of a solution.

In Chapters 7 and 8 we discussed the properties associated with the three states of matter. For the most part, however, these discussions applied to pure substances, and it is only rarely that we encounter pure materials, either in our daily activities or in the laboratory. Usually, the chemicals with which we work occur in mixtures, and very often these are solutions. The presence of a solute in a solution has very marked effects on the properties of the substance in which it is dissolved, and many times these effects can provide us with useful information about the way substances interact with one another.

In this chapter we will take a close look at how solutions are formed to explore the changes that occur when one substance dissolves in another. We will also focus our attention on the way the solute affects the physical properties of the solution. Many of these phenomena have very useful applications, such as the determination of molecular weights. They are also applied to many practical problems, such as refining crude oil and desalinating seawater.

10.1 TYPES OF SOLUTIONS

The most common type of solution we encounter consists of a solute dissolved in a liquid, so most of our attention will be directed toward solutions of this kind. Liquid solutions can be prepared by dissolving a solid in a liquid (for example, NaCl in water), a liquid in a liquid (for example, ethylene glycol in water—antifreeze solution), or a gas in a liquid (for example, carbonated beverages, which contain dissolved carbon dioxide).

In addition to liquid solutions it is possible to have solutions of gases, such as the atmosphere that surrounds the earth, as well as solid solutions that are formed when a substance is dissolved in a solid. The properties of gaseous solutions were discussed in Section 7.6 under the heading "Dalton's Law of Partial Pressures," and nothing more need be said about them here. Solid solutions, of which many **alloys** (mixtures of metals) are examples, are of two types. **Substitutional solid solutions** exist in which atoms, molecules, or ions of one substance take the place of particles of another substance in a crystalline lattice, as shown in Figure 10.1a. Zinc sulfide and cadmium sulfide form such mixtures in which cadmium ions randomly replace zinc ions in the ZnS lattice. Another example is brass—a substitutional solid solution of copper and zinc.

n- and p-type semiconductors are solid solutions formed by high-temperature solid-state diffusion.

Interstitial solid solutions constitute the other type and are formed by placing atoms of one kind into voids, or interstices, that exist between atoms in the host lattice. This is illustrated in Figure 10.1b. Tungsten carbide, WC, an extremely hard substance that has found many uses in cutting tools designed for machining steels, is an example of an interstitial solid solution in which the

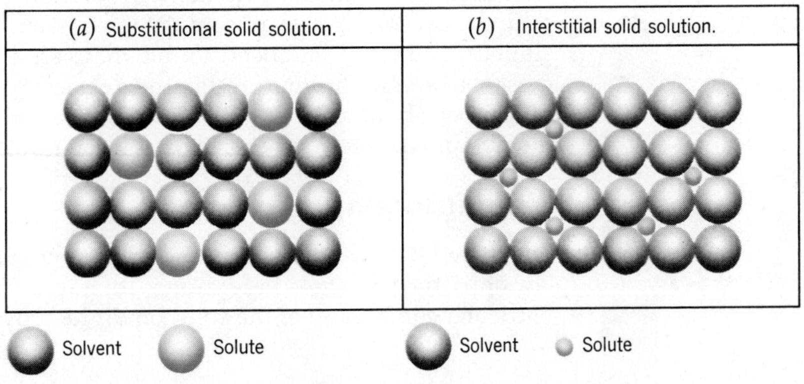

| (a) Substitutional solid solution. | (b) Interstitial solid solution. |

Solvent   Solute   Solvent   Solute

Figure 10.1

Two types of solid solutions.
(a) Substitutional solid solution.
(b) Interstitial solid solution.

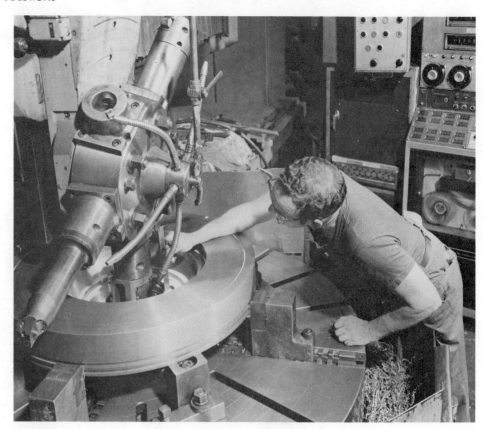

Cutting tools tipped with tungsten carbide are often used by machinists like the one shown here operating a metal lathe.

tungsten atoms are arranged in a face centered cubic pattern with carbon atoms in *octahedral holes*—spaces within the crystal where the carbon atoms are surrounded by six tungsten atoms at the vertices of an octahedron.

10.2 CONCENTRATION UNITS

The physical properties of solutions are determined by the relative proportions of the various components of which they are composed. We have already seen that there are a variety of ways of expressing the concentration of one substance in another. For example, in Chapter 6 we discussed the concentration units, molarity and normality, and we saw that they are particularly useful for the purpose of dealing with stoichiometry in solution. Molarity and normality, defined again below for the sake of completeness, were created to satisfy this need. In a similar fashion, it has been found that certain other concentration units are most convenient for interpreting the physical properties of solutions. A point to remember about all concentration units is that they are *ratios*. The way to remember them is to learn the units associated with numerator and denominator.

Mole fraction

The mole fraction unit of concentration appeared in our discussion of Dalton's law of partial pressures in Section 7.10. It is defined as the number of moles of a particular component of the solution divided by the total number of moles of all the substances present in the mixture,

Recall that the symbol **X** stands for mole fraction.

$$X_A = \frac{n_A}{n_A + n_B + n_C + \cdots}$$

For example, a solution composed of 2.0 mol of water and 3.0 mol of ethanol (C_2H_5OH) has a mole fraction of water given by

$$X_{H_2O} = \frac{2.0 \text{ mol } H_2O}{2.0 \text{ mol } H_2O + 3.0 \text{ mol } C_2H_5OH} = \frac{2.0 \text{ mol}}{5.0 \text{ mol}}$$

$$X_{H_2O} = 0.40$$

Similarly, the mole fraction of ethanol in the mixture is

$$X_{C_2H_5OH} = \frac{3.0 \text{ mol}}{5.0 \text{ mol}} = 0.60$$

We see that the sum of all the mole fractions is equal to one, as of course it must be.

Another frequently used term is **mole percent** (abbreviated **mol %**), which is simply equal to 100 × mole fraction. Thus, the mixture above is composed of 40 mol % water and 60 mol % ethanol.

Weight fraction

The weight fraction specifies the fraction of the total weight of a solution that is contributed by a particular component. A mixture composed of 12.5 g of water and 37.5 g of ethanol has a weight fraction of water, w_{H_2O}, given by

$$w_{H_2O} = \frac{12.5 \text{ g } H_2O}{12.5 \text{ g } H_2O + 37.5 \text{ g } C_2H_5OH} = \frac{12.5 \text{ g}}{50.0 \text{ g}}$$

$$w_{H_2O} = 0.250$$

In a similar fashion, we find that the weight fraction of ethanol in the mixture is 0.750.

Parts per million equals weight fraction × 10^6

Weight percent, which is perhaps best thought of as *parts per hundred by weight*, is equal to the weight fraction multiplied by 100. It is more frequently used than weight fraction. The solution above is described as composed of 25.0% water and 75.0% ethanol, by weight. This means that in 100.0 g of a solution with this concentration there are 25.0 g of water and 75.0 g of ethanol. If there were 100.0 kg of this solution, it would contain 25.0 kg of water and 75.0 kg of ethanol.

This weight percent can be indicated as follows:
25% w/w water
75% w/w ethanol

Molarity

Molarity is the ratio of the number of moles of solute to the *total volume* of the solution. It is expressed in moles per liter. If 0.500 mol of HCl is dissolved in 250 ml of solution, the molarity is

$$\frac{0.500 \text{ mol solute}}{0.250 \text{ liter solution}} = 2.00 \text{ mol/liter} = 2.00 \text{ } M$$

Normality

Normality is expressed as the ratio of the number of equivalents of solute to the volume of solution, in liters.

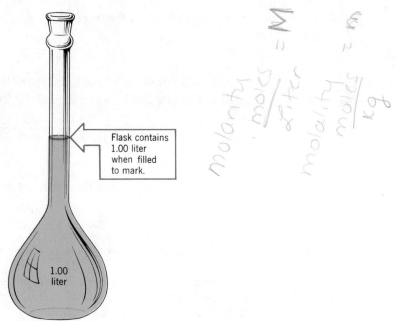

Figure 10.2
A volumetric flask.

Flask contains 1.00 liter when filled to mark.

1.00 liter

Molality

Molality is defined as the number of moles of solute per 1000 g (1.00 kg) of the solvent—that is, it is a ratio of *moles of solute to mass of solvent*. A 1.00 molal (written 1.00 *m*) solution would therefore contain 1.00 mol of solute for every 1.00 kg of the solvent. It is very important not to confuse molality with molarity—they are quite different. To see this difference, let's consider how we would prepare typical 1.00 *M* and 1.00 *m* solutions using, for example, sucrose ($C_{12}H_{22}O_{11}$) as the solute and water as the solvent.

> In dilute aqueous solutions, molarity and molality are nearly equal. Can you explain why?

To prepare the 1.00 *M* solution we place exactly 1.00 mol of sucrose (342 g) into a volumetric flask that is calibrated to contain precisely 1.00 liter when filled to the line etched around its neck (Figure 10.2). Water is added while the mixture is stirred to dissolve the solute, until the flask is filled to the mark. At this point we have exactly 1.00 mol of sugar in a total volume of 1.00 liter of solution, and the concentration is 1.00 mol/liter, or 1.00 molar (1.00 *M*).

> At 20°C (approximately room temperature), $d_{H_2O} = 0.9982$ g/ml

To prepare the 1.00 *m* solution, we place 1.00 mol of sucrose into a flask or beaker and *add to it* 1000 g of water. Since the density of water is nearly 1 g/ml, we are adding very nearly 1 liter of water to the 342 g (1 mol) of solute. However, the total final volume of this 1.00 *m* solution is somewhat larger than 1 liter—it is actually 1110 ml—because part of the volume of the final solution is taken up by molecules of the sucrose. The molarity of this 1 *m* solution is 1.00 mol/1.110 liter = 0.901 *M*. Because the mole of solute is contained in a larger volume in this 1 *m* solution, a 1-ml portion will contain a smaller amount of solute than a 1-ml portion of the 1 *M* solution of sucrose.

> This sucrose solution could be labeled as both 0.901 M and 1.00 m.

The difference between molarity and molality becomes even greater if we choose a solvent whose density is far from 1 g/ml. For instance, in Figure 10.3 we see a 1 *M* solution of a solute in carbon tetrachloride.[1] It contains 1 mol of solute in a total volume of 1 liter; however, because CCl_4 has a density of 1.59 g/ml (considerably greater than water), a 1 *m* solution would contain the 1 mol of solute in a volume of only about 630 ml. In other words, it takes much less than 1 liter of CCl_4 to weigh 1000 g. The molarity of this 1 *m* solution is actually about 1.6 *M*.

[1] Carbon tetrachloride is a very toxic solvent that should always be handled with care. It is absorbed through the skin and is a cumulative poison.

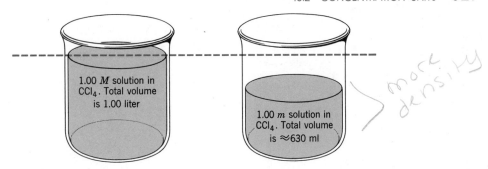

Figure 10.3
Molar and molal solutions in
CCl₄.

Conversions among concentration units

The three concentration units—mole fraction, weight fraction (or weight percent), and molality—can be easily converted from one to another. All that we
require is the formula weights of the solvent and solute. In performing these
conversions, we begin by imagining a certain quantity of the solution and then,
from the way the given concentration unit is defined, we "divide" the solution
into its components: solute and solvent. Once the amounts of solute and solvent
are known, we can use them to calculate the concentration in the new units desired.

EXAMPLE 10.1   An aqueous solution of Epsom salts is composed of 20.0% magnesium sulfate by
weight. What is (a) the molality of the solution and (b) the mole fractions of $MgSO_4$ and
H_2O?

SOLUTION   When given either weight percent or weight fraction, we begin by imagining having a
certain total *weight* of solution. This allows us to use the weight percent or weight fraction to divide the solution into the weights of solute and solvent. To make the arithmetic easy, let's imagine that we have 100.0 g of the magnesium sulfate solution. The
concentration—20.0% $MgSO_4$—tells us immediately that this 100.0 g of solution must
contain 20.0 g $MgSO_4$ and 80.0 g H_2O. Now that we have amounts of $MgSO_4$ and H_2O,
we can compute the other concentration units.

(a) Molality has units

$$\frac{\text{moles solute}}{\text{kg solvent}}$$

We therefore convert 20.0 g $MgSO_4$ to moles and 80.0 g H_2O to kilograms and take their
ratio to get molality.

$$20.0 \text{ g MgSO}_4 \times \left(\frac{1 \text{ mol MgSO}_4}{120.4 \text{ g MgSO}_4}\right) = 0.166 \text{ mol MgSO}_4$$

$$80.0 \text{ g H}_2\text{O} \times \left(\frac{1 \text{ kg}}{1000 \text{ g}}\right) = 0.0800 \text{ kg H}_2\text{O}$$

The molality is

$$\frac{0.166 \text{ mol MgSO}_4}{0.0800 \text{ kg H}_2\text{O}} = 2.08 \ m \text{ MgSO}_4$$

(b) to calculate mole fraction we must know the number of moles of each component in the solution. We have already found that 100 g of the solution contains
0.166 mol of $MgSO_4$. The number of moles of water in the solution is

$$80.0 \text{ g H}_2\text{O} \times \left(\frac{1 \text{ mol H}_2\text{O}}{18.0 \text{ g H}_2\text{O}}\right) = 4.44 \text{ mol H}_2\text{O}$$

Find total wts when working w/ fractions

The mole fraction of $MgSO_4$ is found by dividing the number of moles of $MgSO_4$ by the total number of moles of both the solute and solvent.

$$X_{MgSO_4} = \frac{0.166 \text{ mol}}{4.44 \text{ mol} + 0.166 \text{ mol}} = \frac{0.166 \text{ mol}}{4.61 \text{ mol}}$$

$$X_{MgSO_4} = 0.0360$$

It follows that the mole fraction of water must be

$$X_{H_2O} = 1.0000 - 0.0360 = 0.9640$$

EXAMPLE 10.2

Benzene (C_6H_6) and chloroform ($CHCl_3$) are solvents that have proven to be highly toxic. They are mutually soluble in each other. In a certain solution of benzene and chloroform, the mole fraction of C_6H_6 is 0.450. What is the weight percent of C_6H_6 in this mixture?

SOLUTION

When we are given the mole fraction, we begin by imagining that we have enough of the mixture so that there is exactly 1 mol of particles. In this 1 mol of mixture, therefore, we must have 0.450 mol of C_6H_6.

re: total must = 1 mol

$$1.000 \text{ mol mixture} \times \left(\frac{0.450 \text{ mol } C_6H_6}{1 \text{ mol mixture}}\right) \sim 0.450 \text{ mol } C_6H_6$$

The amount of $CHCl_3$ must be the difference between the total number of moles and the number of moles that are C_6H_6.

$$\text{moles } CHCl_3 = \text{total moles} - \text{moles } C_6H_6$$

$$= 1.000 \text{ mol} - 0.450 \text{ mol}$$

$$= 0.550 \text{ mol}$$

We have now divided the solution into its components. To compute weight percent C_6H_6, we need the weight of C_6H_6 and the total weight of the mixture.

wts you work w) grams

$$0.450 \text{ mol } C_6H_6 \times \left(\frac{78.1 \text{ g } C_6H_6}{1 \text{ mol } C_6H_6}\right) = 35.1 \text{ g } C_6H_6$$

$$0.550 \text{ mol } CHCl_3 \times \left(\frac{119.4 \text{ g } CHCl_3}{1 \text{ mol } CHCl_3}\right) = 65.7 \text{ g } CHCl_3$$

The total weight of the mixture is

$$35.1 \text{ g} + 65.7 \text{ g} = 100.8 \text{ g}$$

The percent of C_6H_6 can be found as

$$\%C_6H_6 = \frac{\text{weight } C_6H_6}{\text{weight of mixture}} \times 100$$

$$\%C_6H_6 = \frac{35.1 \text{ g}}{100.8 \text{ g}} \times 100 = 34.8\%$$

EXAMPLE 10.3

Calcium chloride has a strong tendency to absorb moisture from the air and dissolve in it to produce an aqueous solution. It is sold in hardware stores under a variety of trade names as a dehumidifying agent for use in drying out damp basements and other places with high humidity. Suppose that a certain solution of $CaCl_2$ has a concentration of 4.57 m. What are the mole fractions of $CaCl_2$ and water in the solution?

SOLUTION

This time we imagine that we have enough solution so that there is exactly 1.00 kg of solvent in it. Then we can use the units of molal concentration to divide the solution.

Always label

wts you work w) grams)

$$4.57 \ m \ CaCl_2 = \frac{4.57 \ mol \ CaCl_2}{1.00 \ kg \ H_2O}$$

Our sample of solution, therefore, has 4.57 mol $CaCl_2$ and 1000 g H_2O. To obtain mole fraction we have to first convert 1000 g H_2O to moles.

$$1000 \ g \ H_2O \times \left(\frac{1 \ mol \ H_2O}{18.0 \ g \ H_2O}\right) = 55.6 \ mol \ H_2O$$

Now we can compute mole fraction.

$$X_{CaCl_2} = \frac{4.57 \ mol \ CaCl_2}{4.57 \ mol \ CaCl_2 + 55.6 \ mol}$$

$$= \frac{4.57 \ mol}{60.2 \ mol} = 0.0759$$

The simplest way to calculate the mole fraction of water is

$$X_{H_2O} = 1.0000 - 0.0759$$

$$= 0.9241$$

To perform conversions among mole fraction, weight percent, and molality, molecular weights are the only data required. In order to convert any of these concentration units to molarity, we need to know the density of the solution.

EXAMPLE 10.4   The painful sting of ant bites is caused by formic acid injected under the skin by the ant. Calculate the weight percent of formic acid (HCO_2H) in a solution that is 1.099 M HCO_2H. The density of the solution is 1.0115 g/ml.

SOLUTION   To compute weight percent we need the weight of HCO_2H and the weight of the solution.

If we take 1.0000 liter of solution, its weight is

$$1.0000 \ liter \times \left(\frac{1000 \ ml}{1 \ liter}\right) \times \left(\frac{1.0115 \ g}{1 \ ml}\right) = 1011.5 \ g \ solution$$

From the molarity we know that this solution contains 1.099 mol of HCO_2H. The weight of solute is

$$1.099 \ mol \ HCO_2H \times \left(\frac{46.03 \ g \ HCO_2H}{1 \ mol \ HCO_2H}\right) = 50.59 \ g \ HCO_2H$$

The percent solute in the solution is

$$\%HCO_2H = \frac{50.59 \ g}{1011.5 \ g} \times 100$$

$$\%HCO_2H = 5.001\%$$

You've probably come to realize that in order to solve problems like those in the last three examples it is absolutely essential that you know how the various concentration units are defined. If you've gotten this knowledge and begin the problem correctly, these conversions are not difficult.

10.3   THE SOLUTION PROCESS IN LIQUID SOLUTIONS

Experience has taught us that substances differ widely in their solubilities in various solvents. For instance, we all know that oil and water don't mix, and that to remove an oil stain from clothing a solvent such as naptha must be used. It is also generally known that sodium chloride (table salt) will dissolve in water but not in gasoline. What accounts for these differences in behavior? The answer lies in a close examination of the solution process.

Solutions of liquids in liquids

When one substance dissolves in another, particles of the solute—either molecules or ions, depending on the nature of the solute—must be distributed throughout the solvent and, in a sense, the solute particles in the solution occupy positions that are normally taken by solvent molecules. In a liquid, molecules are packed together very closely and interact strongly with their neighbors. The ease with which a solute particle may replace a solvent molecule depends on the relative forces of attraction of solvent molecules for each other, solute particles for each other, and the strength of the solute-solvent interactions. For example, in a solution formed between benzene (C_6H_6) and carbon tetrachloride (CCl_4), both species are nonpolar and experience only relatively weak London forces. As it happens, the strengths of the attractive forces between pairs of benzene molecules and between pairs of carbon tetrachloride molecules are of nearly the same magnitude as between molecules of benzene and carbon tetrachloride. For this reason, molecules of benzene can replace CCl_4 molecules in solution with ease. As a result, these two substances are completely soluble in all proportions. We say they are **miscible.**

Now let's look at what happens when we attempt to dissolve water in CCl_4. Water is a very polar substance that interacts with other water molecules through the formation of hydrogen bonds. By comparison, the strength of the attractive forces between water and the nonpolar CCl_4 molecules is much weaker. If we attempt to disperse water molecules throughout CCl_4, we find

This explains why oil and grease are soluble in CCl_4 and other dry-cleaning agents.

Water and carbon tetrachloride form a two-phase mixture. The layer on the bottom is the more dense CCl_4.

Table 10.1
Solubilities of some alcohols in water

Substance	Formula	Solubility (mol of solute/100 g H_2O)
Methanol	CH_3OH	∞
Ethanol	C_2H_5OH	∞
Propanol	C_3H_7OH	∞
Butanol	C_4H_9OH	0.12
Pentanol	$C_5H_{11}OH$	0.031
Hexanol	$C_6H_{13}OH$	0.0059
Heptanol	$C_7H_{15}OH$	0.0015

less soluble (in notes!!)

that when the water molecules encounter one another they tend to stick together simply because they attract each other much more strongly than they do molecules of the solvent. This "clumping together" continues until the two substances have formed two distinct phases: one consisting of water with a very small amount of CCl_4 in it, and the other, CCl_4 containing a small quantity of H_2O.

When two polar substances, such as ethanol and water, are mixed, we again have a situation in which the solute-solute forces of interaction are of comparable strength to the solvent-solvent attractive forces, and where the solute and solvent molecules interact strongly with each other. Once again a condition exists where the solute particles can readily replace those of the solvent and, hence, water and ethanol are miscible.

Between the two extremes of complete miscibility (benzene-carbon tetrachloride; water-ethanol) and virtually total immiscibility (CCl_4-H_2O), we have many substances that are only partially soluble in one another. For instance, in Table 10.1 the solubilities of a series of different alcohols are listed in moles of solute (alcohol) per 100 g of water. As we proceed to higher molecular weights, the polar OH group represents an ever smaller portion of the molecules; as a result, alcohol molecules become less like water as they become larger. Paralleling the increasing size of the nonpolar hydrocarbon portions of these alcohols, we observe a corresponding decrease in their solubilities in water.

Water and ethanol, C_2H_5OH, interact with each other by hydrogen bonding, too.

Solutions of solids in liquids

When a solid dissolves in a liquid, somewhat different factors must be considered. In a solid, the molecules or ions are arranged in a very regular pattern and the attractive forces are at a maximum. In order for the solute particles to enter into a solution, the solute-solvent forces of attraction must be sufficient to overcome the attractive forces that hold the solid together. In molecular crystals these attractive forces are relatively weak, being of dipole-dipole or London type, and are rather easily overcome. Substances whose crystals are held together by London forces will, therefore, dissolve to appreciable extents in nonpolar solvents. They are not, however, soluble to any great degree in polar solvents for the same reasons that nonpolar liquids are not soluble in polar solvents—that is, the polar solvent molecules attract each other too strongly to be replaced by the molecules to which they are only weakly attracted. For example, solid iodine, which is composed of nonpolar I_2 molecules, is appreciably soluble in CCl_4 (giving rise to a beautiful violet solution) but only very slightly soluble in water (where it produces a pale yellow-brown solution). Very polar solutes and ionic solids are not soluble in nonpolar solvents. The weak

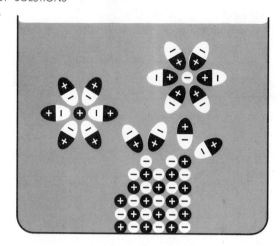

Figure 10.4
Hydration of ions in solution.

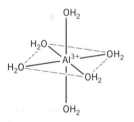

An octahedral hydrated
aluminum ion, $Al(H_2O)_6^{3+}$

This orientation of water molecules may actually extend
through several layers.

solute-solvent interactions, compared to the attractions between solute particles, are not sufficient to tear apart the lattice. Ionic solids, in particular, are
held together by the very strong electrostatic forces between ions; therefore a
very polar solvent, such as water, is required to rip apart an ionic lattice. This is
why NaCl is soluble in water but not in gasoline, which is a mixture of nonpolar
hydrocarbons (compounds containing only C and H).

When an ionic substance dissolves in water, the ions that are adjacent to
one another in the solid become separated and are surrounded by water molecules. In Chapter 6 we represented this dissociation of the solute by an equation
such as

$$NaCl\ (s) \longrightarrow Na^+\ (aq) + Cl^-\ (aq)$$

In Figure 10.4 we take a closer look at what takes place during this process. In
the immediate vicinity of a positive ion, the surrounding water molecules are
oriented so that the negative ends of their dipoles point in the direction of the
positive charge. The water molecules surrounding a negative ion have their
positive ends directed at the ion. An ion enclosed within this ''cage'' of water
molecules is said to be **hydrated** and, in general, when a solute particle becomes
surrounded by molecules of a solvent we say that it is **solvated;** hydration is a
special case of the more general phenomenon of solvation.

The layer of oriented water molecules that surrounds an ion helps to neutralize the ion's charge and serves to keep ions of opposite charge from attracting each other strongly over large distances within the solution. In a sense,
the solvent insulates the ions from each other. Nonpolar solvents do not dissolve ionic compounds because they can neither tear an ionic lattice apart nor
do they offer any shielding for the ions. In a nonpolar solvent, ions quickly con-
gregate and separate from the solution as the solid.

In summary, substances that exhibit *similar* intermolecular attractive forces
tend to be soluble in one another. This observation is often stated very simply
as ''like dissolves like.'' Nonpolar substances are soluble in nonpolar solvents,
while polar or ionic compounds dissolve in polar solvents.

10.4 HEATS OF SOLUTION

The solution process nearly always occurs with either an absorption or a release
of energy. For example, when potassium iodide is dissolved in water the mixture becomes cool, indicating that for potassium iodide the solution process is
endothermic. On the other hand, when lithium chloride is added to water the
mixture becomes warm, signifying that the solution process in this case evolves
heat and is therefore exothermic. *The amount of energy that is absorbed or released*

Table 10.2
Heats of solution

Substance	Heat of Solution[a] (kJ/mol of solute)
KCl	17.2
KBr	19.9
KI	20.3
LiCl	−37.0
LiI	−59.0
LiNO$_3$	−1.3
AlCl$_3$	−321
Al$_2$(SO$_4$)$_3$·6H$_2$O	−230
NH$_4$Cl	−16
NH$_4$NO$_3$	26

[a] At "infinite" dilution. The heat of solution depends, to an extent, on the concentration of the solution produced. A negative sign signifies an exothermic process.

when a substance enters solution is called the heat of solution and is given the symbol, ΔH_{soln}. As in our definitions of the heat of vaporization and heat of fusion, ΔH_{soln} represents a difference. It is the difference between the energy possessed by the solution after it has been formed and the energy that the components of the solution had before they were mixed; that is,

$$\Delta H_{soln} = H_{soln} - H_{components}$$

We will learn why in Chapter 11.

Neither H_{soln} nor $H_{components}$ can actually be measured, but their difference, ΔH_{soln}, can be. When energy is evolved during the solution process, the resulting solution possesses less energy than the components from which it was prepared, so the difference represented by ΔH_{soln} is a negative number. Conversely, an endothermic solution process would have a positive ΔH_{soln}. Heats of solution for some typical ionic solids in water are shown in Table 10.2.

The magnitude of the heat of solution can provide us with information about the relative forces of attraction between the various particles that make up a solution. To analyze the factors that contribute to the absorption or evolution of energy, let us imagine that we could create the solution in a stepwise fashion.

Solutions of liquids in liquids

When one liquid dissolves in another, we can imagine that the molecules of the solvent are caused to move apart so as to allow room for the solute molecules. Similarly, for the solute to enter solution, its molecules must also become separated so that they can take their places in the mixture. Since there are attractive forces between molecules in both the solvent and solute, the process of separating their molecules requires an input of energy—that is, work must be done on both the solute and solvent to separate their molecules from one another. Finally, as the solute and solvent, in their "expanded" states, are brought together, energy is released because of the attractions that exist between the solute and solvent molecules.[2] This sequence of steps we have just described is illustrated in Figure 10.5.

liquid as one-solv
one-solute ?

Although this three-step process is purely hypothetical, it allows us to analyze the factors that contribute to ΔH_{soln}.

[2] Recall from Chapter 1 that when particles that attract one another are pulled apart, work must be supplied to increase the potential energy. When these particles are brought together again, the same amount of energy is released.

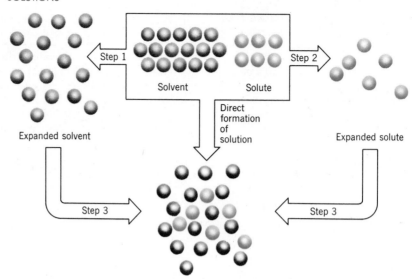

Figure 10.5

Formation of a liquid-liquid solution.

In some substances, such as benzene and carbon tetrachloride, the inter-molecular attractive forces are of very nearly the same magnitude; therefore, these compounds form solutions with virtually no evolution or absorption of heat. Solutions in which the solute-solute, solute-solvent, and solvent-solvent interactions are all the same are called **ideal solutions.** The energy changes that occur along the series of steps that we have devised to arrive at the solution are shown graphically in Figure 10.6. We see that for an ideal solution the energy released in the final step is the same as that absorbed in the first two; thus the net change is zero.

$\Delta H_{\text{soln}} = 0$ for an ideal solution.

When the molecules of the solute and solvent attract each other more strongly than they do molecules of their own kind, more energy can be released as the expanded solute and solvent are brought together than was required to separate them in the first place (Figure 10.7). Under these circumstances the overall solution process can result in the evolution of heat and be exothermic. This occurs when acetone, an important solvent (nail polish remover), and water are mixed.

When the solute-solvent attractive forces are weaker than those between pure solute or pure solvent, the formation of a solution requires a net input of

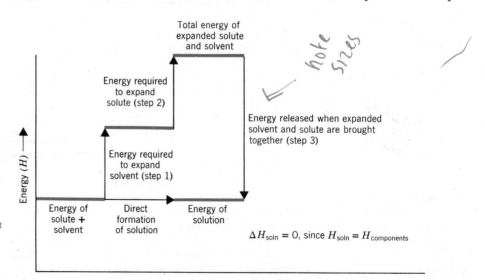

Figure 10.6

Energy changes to produce an ideal solution. The actual solution process follows the direct path, although the net result is the same both ways.

$\Delta H_{\text{soln}} = 0$, since $H_{\text{soln}} = H_{\text{components}}$

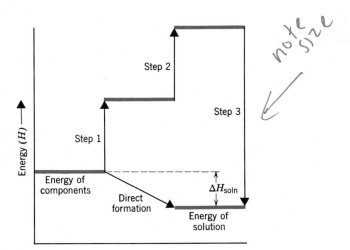

Figure 10.7

An exothermic solution process.

Figure 10.8

An endothermic solution process.

energy. This is because more energy is absorbed in separating the molecules in the first two steps than is recovered when the solute and solvent are brought together in the third (Figure 10.8). In this case the solution becomes cool as it is formed, signifying that an endothermic change has occurred. This happens, for example, when ethanol and a hydrocarbon solvent such as hexane are mixed. The nonpolar hexane molecules come between the hydrogen bonded ethanol molecules, effectively destroying the hydrogen bonds. This absorbs energy and the solution becomes cool.

Solutions of solids in liquids

We can approach the energetics of the solution of a solid in a liquid in much the same way as we did for a liquid in a liquid. We recognize that the act of dissolving a compound such as KI in water involves removing the K^+ and I^- ions from the solid and placing them in an environment where they are surrounded by water molecules. Suppose, now, that we could separate these two steps. The first step involves separating the ions so that they are infinitely far apart, and the second amounts to taking these now isolated ions and placing them into water where they become surrounded by molecules of the solvent. To accomplish the first step, it is necessary to break apart the lattice by pulling the ions away from one another, so that the process

$$KI\ (s) \longrightarrow K^+\ (g) + I^-\ (g)$$

requires an input of energy; that is, it is endothermic. Recall that the amount of energy required to tear apart the solid to obtain isolated particles is called the lattice energy. Ionic substances, because of the very strong electrostatic attractions between oppositely charged ions, have quite large lattice energies. Molecular solids, however, have small lattice energies because of their relatively weak intermolecular attractive forces.

In the second step of our solution process we imagine that the K^+ and I^- ions are brought into water where they become hydrated. As described in the last section, an ion in water is surrounded by a "cage" of oriented water molecules; therefore, a hydrated ion experiences a net attraction for the solvent dipoles. Because of this attraction, a quantity of energy, known as the **hydration energy,** is released when an ion is placed into water.[3] We can indicate the

This instant cold-pack contains capsules of ammonium nitrate, NH_4NO_3, and water. When they are broken and their contents mixed, the salt dissolves and produces an amazing cooling effect.

---

[3] In general, for a solvent other than water, this energy is called the solvation energy.

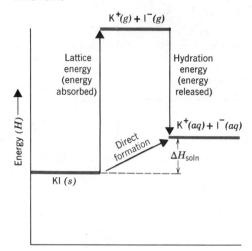

Figure 10.9

Energy changes that occur when KI dissolves in H_2O.

change that occurs in this hydration step as

$$K^+ (g) + I^- (g) + xH_2O \longrightarrow K^+ (aq) + I^- (aq)$$

which tells us that some number, x, of water molecules becomes "bound" to the potassium and iodide ions as they become hydrated in aqueous solution.

The overall change that takes place when KI dissolves in water can be represented as the sum of the two steps we have just considered.

$$KI (s) \longrightarrow K^+ (g) + I^- (g)$$
$$\underline{K^+ (g) + I^- (g) + xH_2O \longrightarrow K^+ (aq) + I^- (aq)}$$
$$KI (s) + K^+ (g) + I^- (g) + xH_2O \longrightarrow K^+ (g) + I^- (g) + K^+ (aq) + I^- (aq)$$

or, after canceling species that appear on both sides,

$$KI (s) + xH_2O \longrightarrow K^+ (aq) + I^- (aq)$$

The heat of solution, ΔH_{soln}, corresponds to the net energy change that occurs and is equal to the difference between the amount of energy supplied during the first step and the amount of energy evolved during the second (Figure 10.9). If the lattice energy is greater than the hydration energy, a net input of energy is required when the substance dissolves and the process is endothermic. This occurs with KI. Conversely, when the hydration energy exceeds the lattice energy, a situation that occurs with LiCl, more energy is released when the ions become hydrated than is required to break up the ionic lattice, and an exothermic change is observed when the solid dissolves.

The relationship between the lattice energy and hydration energy can be seen in Table 10.3. Note that the agreement between the calculated and experimentally determined heats of solution is far from perfect. This is a result of inaccuracies in the theoretical models used in computing these energy quantities. Nevertheless, we still can see that when theory predicts a large exothermic change the experimental quantity corresponds to a large release of energy. Similarly, when the theory predicts a trend in ΔH_{soln}, as with KCl, KBr, and KI, the measured values follow the same trend. The agreement is therefore sufficient to support the arguments made above.

The preceding explanations allow us to understand the energy changes that take place during the solution process. Unfortunately, however, it is very difficult to predict ahead of time, in any particular case, whether the formation of a solution will be exothermic or endothermic. This is because the same factors

Remember that the solution is actually formed by the direct path in Figure 10.9.

Table 10.3
Lattice energies, hydration energies, and heats of solution for some alkali halides

Compound	Lattice Energy (kJ/mol)	Hydration Energy (kJ/mol)	Calculated ΔH_{soln} (kJ/mol)	Measured ΔH_{soln} (kJ/mol)
LiCl	+833	−883	−50	−37.0
LiBr	+787	−854	−67	−49.0
NaCl	+766	−770	−4	+3.9
NaBr	+728	−741	−13	−0.602
KCl	+690	−686	+4	+17.2
KBr	+665	−657	+8	+19.9
KI	+632	−619	+13	+20.3

that lead to a high hydration energy also tend to produce a high lattice energy. The extent to which an ion is attracted to a solvent dipole increases as the ion becomes smaller, because a smaller ion can get closer to a solvent molecule than a larger one can. The interaction of the solvent with an ion also increases as the charge on the ion becomes greater. However, the degree to which ions attract each other in the solid also grows with decreasing size and increasing charge, so that as the hydration energy becomes larger so does the lattice energy. Thus we have two factors that are affected in the same way by changes in size and charge, and it is virtually impossible to predict in advance which effect will predominate.

10.5 SOLUBILITY AND TEMPERATURE

In Chapter 6 we defined the solubility of a substance as the amount of solute required to produce a saturated solution in some particular quantity of solvent. At a given temperature a saturated solution in contact with undissolved solute represents another example of a state of dynamic equilibrium. As illustrated in Figure 10.10, particles of the solute are constantly passing into the solution and, at the same time, the solute particles already in the solution are continually colliding with and sticking to the undissolved solute. Although we show this equilibrium here for a solid dissolved in a liquid, the same concept applies to any type of solution (except gases—all gases are completely miscible).

Since a saturated solution in contact with excess solute constitutes a state of dynamic equilibrium, when the system is disturbed the effect of the disturbance can be predicted using Le Châtelier's principle. A change in temperature corresponds to such a disturbance, and in Chapter 8 we saw that a rise in temperature favors a shift in the position of equilibrium in a direction that will absorb heat. Therefore, if the dissolving of additional solute into an already saturated solution absorbs energy, the solubility of that substance increases as the temperature is raised. Conversely, if placing additional solute into the saturated solution is an exothermic process, the solute will become less soluble as the temperature is increased.

In general, the solubility of most solid and liquid substances in a liquid solvent increases with increasing temperature. For gases in liquids, the opposite behavior is observed. The solution process for a gas in a liquid is nearly always

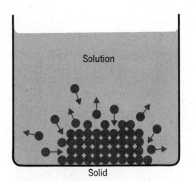

Figure 10.10
Solubility as an equilibrium state. The solute dissolves at the same rate as it crystallizes.

Solution

Solid

So you can't predict solution of liquid in liquid but not liq in solid?

exothermic, because the solute particles are already separated from each other and the dominant heat effect arises from the solvation that occurs when the gas dissolves. Le Châtelier's principle predicts that a rise in temperature will favor an endothermic change which, for a gas, takes place when it leaves solution. Therefore, we expect gases to become less soluble as the temperature of the liquid in which they are dissolved becomes higher. For example, in bringing water to a boil, tiny bubbles appear on the surface of the pot before boiling begins. These bubbles contain air that is driven out of solution as the water becomes hot. We also use this general solubility behavior of gases when we store opened bottles of carbonated beverages in the refrigerator. These liquids retain their dissolved CO_2 longer when they are kept cold because CO_2 is more soluble in them at low temperatures. Analysis of the quantity of dissolved gases in streams, lakes, and rivers reveals still another example of this phenomenon. The concentration of dissolved oxygen, which is imperative to marine life, decreases in the summer months, compared to when similar analyses are performed during the winter months—all other conditions being equal, of course.

Certain species of fish die if the water becomes too warm because the amount of dissolved oxygen becomes too small.

Fractional Crystallization

Figure 10.11 illustrates graphically the way in which solubility changes with temperature for a variety of typical solids in water. From these solubility curves it is evident that the variation of solubility with temperature is quite different for different substances. For some substances, such as KNO_3, the solubility changes very rapidly with temperature, while for others the change is more gradual. These differences in behavior provide the basis for a useful laboratory technique called **fractional crystallization,** which is often used to separate impurities from the products of a chemical reaction.

In this technique the impure product is first dissolved in a small amount of hot solvent—generally one in which the desired product is less soluble than the impurities. As the hot solution is allowed to cool, the pure product separates from the mixture, leaving the impurities behind. Finally, the crystals of the product are filtered from the cool solution and dried. The quantity of pure product that can be recovered in this fashion depends on the concentration of the impurities and their solubility relative to that of the desired material.

Rock candy is formed by the crystallization of sugar from a saturated solution that is slowly cooled.

10.6 THE EFFECT OF PRESSURE ON SOLUBILITY

In general, pressure has very little effect on the solubility of liquids or solids in liquid solvents. The solubility of gases, however, always increases with increasing pressure. Carbonated beverages, for example, are bottled under pressure to ensure a high concentration of CO_2 and once the bottle has been opened, the beverage quickly loses its carbonation unless it is recapped. The same phenomenon is responsible for decompression sickness, also known as "the bends." When a deep-sea diver or a tunnel worker comes to the surface too quickly, nitrogen and oxygen that has dissolved in the blood at high pressure is suddenly released in the form of bubbles in the blood vessels. This is very painful and in extreme cases can even cause death.

Astronauts have to worry about their space suits tearing for the same reason.

The effect of pressure on the solubility of a gas is not difficult to understand. Let's imagine that a liquid is saturated with a gaseous solute, and that this solution is in contact with the gas at some particular pressure. Once again we have a dynamic equilibrium where molecules of the solute are entering the vapor phase at the same rate at which molecules from the gas are entering the solution, as shown in Figure 10.12a. As we might expect, the rate at which molecules go into solution depends on the number of collisions per second that

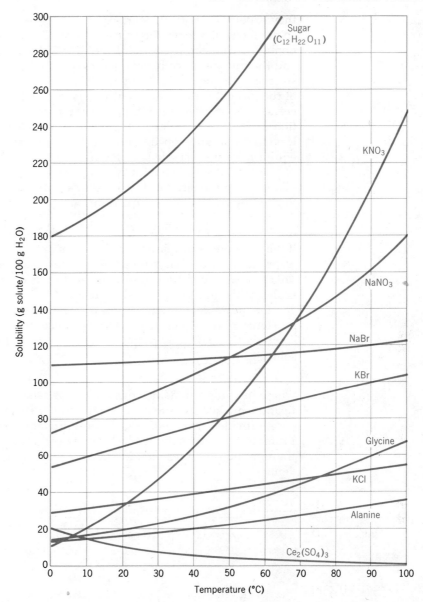

Figure 10.11
Solubility curves for typical solids in H$_2$O.

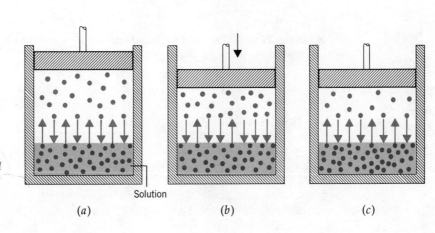

Figure 10.12
The effect of pressure on the solubility of a gas. (a) An equilibrium exists between a gas and its solution. (b) The equilibrium is upset. (c) Equilibrium is restored when more gas dissolves.

they experience with the surface of the liquid. Similarly, the rate at which the solute molecules leave the solution depends on their concentration. If we suddenly increase the pressure of the gas, we pack the molecules closer together and the number of collisions per second that the gas molecules make with the surface of the liquid gets larger. When this occurs, the rate at which molecules of the gas enter the solution also gets larger, but without a corresponding increase in the rate at which they leave (Figure 10.12b). As a result, the concentration of solute molecules in the solution rises until the rate at which they are leaving the solution once again equals the rate at which they enter. At this point equilibrium is reestablished (Figure 10.12c).

The solubility behavior of gases with respect to pressure also can be explained easily in terms of Le Châtelier's principle. We might represent the equilibrium by the following equation:

$$\text{solute } (g) + \text{solvent } (l) \rightleftharpoons \text{solution } (l)$$

If the pressure is increased, Le Châtelier's principle tells us that the system will respond in a way that brings the pressure back down toward its initial value. This can happen if more gas dissolves, because then there would be fewer molecules in the gas phase to exert pressure.

Quantitatively, the influence of pressure on the solubility of a gas is given by **Henry's law,** which states that the concentration of the gaseous solute in the solution, C_g, is directly proportional to the partial pressure of the gas above the solution; that is,

$$C_g = k_g p_g \qquad [10.1]$$

where the proportionality constant, k_g, is called the **Henry's law constant.** This relationship allows us to compute the solubility of a gas at some particular pressure, provided that we know its solubility at some other pressure, as shown in Example 10.5. Actually, Henry's law is accurate only for relatively low concentrations and pressures, and for gases that do not react significantly with the solvent.

EXAMPLE 10.5

At 25°C, oxygen gas collected over water at a *total* pressure of 1.00 atm is soluble to the extent of 0.0393 g/liter. What would its solubility be if its partial pressure over water were 800 torr?

SOLUTION

The data given us permits the calculation of the Henry's law constant if we know the partial pressure of oxygen above the solution. The total pressure is the sum of the partial pressures of the H_2O vapor and the oxygen,

$$P_{total} = p_{H_2O} + p_{O_2}$$

From Table 7.1 we find the vapor pressure of water to be 23.8 torr at 25°C; therefore, the partial pressure of oxygen is

$$p_{O_2} = P_{total} - p_{H_2O}$$

$$p_{O_2} = 760 \text{ torr} - 24 \text{ torr} = 736 \text{ torr}$$

The Henry's law constant is obtained as the ratio

$$k_{O_2} = \frac{C_{O_2}}{p_{O_2}}$$

$$k_{O_2} = \frac{0.0393 \text{ g/liter}}{736 \text{ torr}} = 5.34 \times 10^{-5} \frac{\text{g}}{\text{liter torr}}$$

Now we can use Henry's law to determine that at a partial pressure of 800 torr the solubility of oxygen is

$$C_{O_2} = \left(5.34 \times 10^{-5} \; \frac{g}{\text{liter} \cancel{\text{torr}}}\right) (800 \; \cancel{\text{torr}})$$

$$C_{O_2} = 0.0427 \; \frac{g}{\text{liter}}$$

10.7 VAPOR PRESSURES OF SOLUTIONS

The formation of a solution has very little effect on the *chemical* properties of its components. Sodium, for instance, reacts with the water in an aqueous solution to give exactly the same products as when it reacts with distilled water. The physical properties of substances, however, are often dramatically altered when they become part of a solution. The same water that can freeze and crack the block of an automobile engine at 0°F will remain liquid if it is mixed with ethylene glycol—antifreeze.

Among the many physical properties of liquids that are affected by the formation of a solution is the vapor pressure. When a nonvolatile solute—that is, one that has no tendency to escape from a solution—is dissolved in a liquid solvent, the solvent's vapor pressure is lowered. If, in addition, the solute is a nonelectrolyte and doesn't dissociate, the extent to which the vapor pressure is lowered depends on the mole fraction of the solute. The vapor pressure of the solvent above the solution, which in this case we can call the vapor pressure of the solution, P_{solution}, is given by **Raoult's law**

Raoult's law is easy to use if you remember the definitions of the various terms.

$$P_{\text{solution}} = X_{\text{solvent}} P^0_{\text{solvent}} \qquad [10.2]$$

where X_{solvent} is the mole fraction of the solvent in the solution and P^0_{solvent} is the vapor pressure of the pure solvent. For example, a solution that contains 95 mol % water and 5 mol % of a nonvolatile solute such as sugar has $X_{H_2O} = 0.95$. At a temperature where the vapor pressure of pure water is 100 torr, the vapor pressure of the solution will be

$$P_{\text{solution}} = 0.95 \, (100 \; \text{torr})$$

$$= 95 \; \text{torr}$$

It is not difficult to see why Raoult's law holds. Figure 10.13*a* illustrates the condition in which a pure solvent is in equilibrium with its vapor. This vapor, as described in Section 8.4, exerts a pressure that is ultimately determined by the fraction of the total number of molecules at the surface of the liquid that have enough kinetic energy to escape. If we now look at a solution containing a

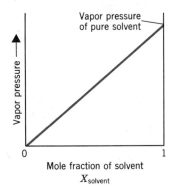

Figure 10.13

Molecular view of Raoult's law.
(a) Pure solvent (b) Solution.

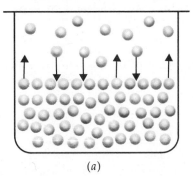

(*a*)

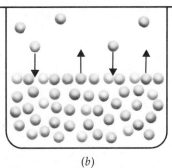

(*b*)

Solute

nonvolatile solute (Figure 10.13b), we find that a portion of the solvent molecules at the surface has been replaced by molecules of the solute. Since the entire system—solvent plus solute—is at a single temperature, all the molecules in the solution belong to a single distribution of kinetic energies. In both the solution and the pure solvent, the same fraction of surface molecules has more than the minimum kinetic energy that *solvent* molecules need to break away from the liquid, but in the solution only a *portion* of that fraction is actually composed of molecules of the solvent. The others are solute molecules. The result is that there are fewer molecules at the surface of the solution capable of leaving than at the surface of the pure solvent. Therefore, the rate of evaporation of solvent molecules from a solution is less than from the pure solvent. This means that for the solution, fewer molecules are needed in the vapor to give an equal rate of condensation, so the vapor pressure is lower.

Because only the solvent can evaporate, the fraction of molecules at the surface of the solution that can escape into the vapor depends on that *fraction* of all the molecules at the surface that are solvent molecules—that is, the ratio of the number of moles of solvent particles to the total number of moles of particles that compose the surface. This ratio, of course, is the mole fraction of the solvent. If the solution were composed of 95 mol % solvent, then we expect to find only 95% of the molecules at the surface to belong to the solvent. This means that the rate of evaporation from the solution is expected to be only 95% of that for the solvent alone. The equilibrium vapor pressure should therefore be reduced to 95% of that for the pure solvent, which is the same result we obtain by the application of Raoult's law.

Solutions containing more than one volatile component

In many solutions, such as benzene and carbon tetrachloride, for example, both solute and solvent have appreciable tendencies to undergo evaporation. In this case, the vapor will contain both solute and solvent molecules, and the vapor pressure of the solution will be the sum of the partial pressures exerted by each component. If we follow the same line of reasoning as above, we conclude that the partial pressure of any component above such a mixture is also given by Raoult's law. Thus the partial pressure of component A, p_A, is given by

$$p_A = X_A P_A^0 \qquad [10.3]$$

where P_A^0 is the vapor pressure of pure A and X_A is its mole fraction in the solution. Similarly, the partial pressure of a second component, p_B is given as

$$p_B = X_B P_B^0 \qquad [10.4]$$

Finally, the total vapor pressure of a mixture of A and B is given by Dalton's law,

$$P_T = p_A + p_B \qquad [10.5]$$

Substituting Equations 10.3 and 10.4 into Equation 10.5 gives

$$P_T = X_A P_A^0 + X_B P_B^0$$

Figure 10.14a is a plot of the partial pressures of A and B, and the total vapor pressure as a function of solution composition for such a two-component mixture.

Actually, very few mixtures really obey Raoult's law very closely over wide ranges of composition. Benzene and carbon tetrachloride, a pair of substances that do form such mixtures, are said to yield **ideal solutions**. Mixtures that deviate from Raoult's law are called **nonideal**. When the vapor pressure of a mixture is greater than that predicted, it is said to exhibit a **positive deviation** from Raoult's law (Figure 10.14b); conversely, when a solution gives a lower vapor

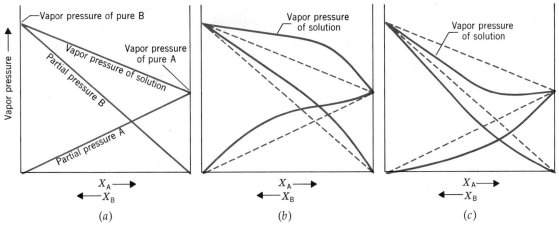

Figure 10.14

Vapor pressure of a two-component system. (a) Ideal solution. (b) Positive deviations from Raoult's law. (c) Negative deviations from Raoult's law.

pressure than we would expect from Raoult's law, it is said to show a **negative deviation** (Figure 10.14c).

As we saw in our discussion of heats of solution, the origin of nonideal behavior lies in the relative strengths of the interactions between molecules of the solute and solvent. When the attractive forces between the solute and solvent molecules are weaker than those between solute molecules or between solvent molecules, neither the solute nor solvent particles are held as tightly in the solution as they are in the pure substances. The escaping tendency of each is therefore greater in the solution than in the solute or solvent alone. As a result, the partial pressures of both of them over the solution are greater than predicted by Raoult's law, and the solution exhibits a larger vapor pressure than expected.

Just the opposite effect is produced when the solute-solvent interactions are stronger than the solute-solute or solvent-solvent interactions. Each substance, in the presence of the other, is held more tightly than in the pure materials, and their partial pressures over a solution are therefore less than Raoult's law would predict. The result is that such a solution exhibits a negative deviation from ideality.

Since, in a solution that shows positive deviations from ideal behavior, the forces of attraction between solute and solvent are weaker than those between both solute molecules and solvent molecules, the formation of these solutions occurs with the absorption of energy (Section 10.4). Conversely, of course, mixtures that exhibit negative deviations from Raoult's law are formed with the evolution of heat. This is summarized in Table 10.4.

Table 10.4
Summary of solution properties

Relative Attractive Forces	ΔH_{soln}	Temperature Change When Solution Is Formed	Deviations from Raoult's Law	Example
A−A, B−B = A−B	Zero	None	None (ideal solution)	Benzene−chloroform
A−A, B−B < A−B	Negative (exothermic)	Increase	Negative	Acetone−water
A−A, B−B > A−B	Positive (endothermic)	Decrease	Positive	Ethanol−hexane

10.8 FRACTIONAL DISTILLATION

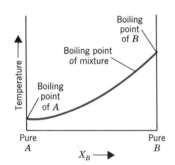

Figure 10.15

Boiling-point curve for a mixture of A and B.

In a simple distillation process—one that could be used to separate sodium chloride and water, for example—a volatile solvent is vaporized from a solution and subsequently condensed to provide a pure liquid (see Figure 1.6). If the process is continued, eventually all the solvent will be removed and only the solid solute will remain.

The separation of mixtures of volatile liquids into their components presents more of a problem. A technique that can frequently be used successfully to accomplish this task is called **fractional distillation.**

Let's suppose that we had a mixture of two volatile liquids, A and B, that form an ideal solution. This mixture will boil when the sum of the partial pressures of A and B equals the prevailing atmospheric pressure; that is, when

$$P_{atm} = p_A + p_B$$

The boiling points of various mixtures of A and B will increase gradually from that of the more volatile component (let us say, A) to that of the less volatile one, B, as shown in Figure 10.15.

Suppose, now, that when 1.00 mol of A is mixed with 2.00 mol of B, the resulting mixture boils (at 1 atm) at a temperature at which the vapor pressure of pure A is 1140 torr and that of pure B is 570 torr. Under these conditions the partial pressure of A is

$$p_A = X_A P_A^0$$

$$p_A = \left(\frac{1.00 \text{ mol } A}{1.00 \text{ mol } A + 2.00 \text{ mol } B}\right) \times (1140 \text{ torr})$$

$$p_A = \left(\frac{1.00 \text{ mol}}{3.00 \text{ mol}}\right) \times (1140 \text{ torr})$$

$$p_A = (0.333)(1140 \text{ torr}) = 380 \text{ torr}$$

Similarly, the partial pressure of B would be

$$p_B = \left(\frac{2.00 \text{ mol}}{3.00 \text{ mol}}\right) \times (570 \text{ torr})$$

$$p_B = 380 \text{ torr}$$

The sum of p_A and p_B is 760 torr as, of course, it must be if the solution is to boil.

What can we say about the composition of the vapor? In Section 7.10, under our discussion of Dalton's law of partial pressures, it was stated that the partial pressure of a gas in a mixture is equal to its mole fraction multiplied by the total pressure *exerted by the gas*. We can write this as

$$p_A = X_{A(\text{vapor})} P_{T(\text{vapor})}$$

where $X_{A(\text{vapor})}$ is the mole fraction of A in the vapor and $P_{T(\text{vapor})}$ is the total vapor pressure. In the vapor over our solution, the partial pressure of each gas is 380 torr and the total pressure is 760 torr. This means that the mole fraction of both A and B *in the vapor* must be 0.500. In the liquid the mole fraction of A was only 0.333, so the vapor contains a greater proportion of the more volatile component (A) than does the solution. In fact, any time we boil a mixture of these two substances, the vapor will be richer than the solution in the more volatile compound.

On our boiling-point diagram we can indicate the composition of the vapor by the upper curve drawn in Figure 10.16. Here, points corresponding to the compositions of liquid and vapor in equilibrium can be obtained by drawing a

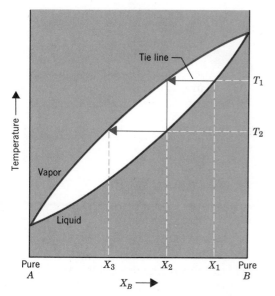

Figure 10.16

Boiling-point diagram for a two-component mixture.

Figure 10.17

Boiling-point diagram for water—ethanol mixtures (not drawn to scale).

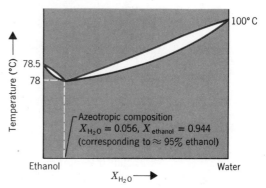

horizontal line, called a **tie line,** between the curve for the liquid and that for the vapor. When the composition of the mixture is X_1, it boils at a temperature T_1 to provide a vapor that has a composition X_2. If this vapor is condensed and then reheated, it will boil at a temperature T_2 and give a vapor whose composition is X_3. Repetition of this process will produce fractions ever richer in A. This procedure is called fractional distillation, and is useful not only in the laboratory, where it is employed for purifying the products of chemical reactions, but also industrially. For instance, the petroleum industry uses fractional distillation to separate crude oil into its various components, which include gasoline, kerosene, oils, and paraffin.

There are some solutions that exhibit very large deviations from ideality; as a result they cannot be totally separated into their components even by fractional distillation. Ethanol (grain alcohol) and water form such a mixture. Solutions of these two substances have such large positive deviations from Raoult's law that there is a maximum in the vapor pressure curve and hence a minimum in the boiling-point diagram as shown in Figure 10.17. A solution with such a minimum boiling point is called a **minimum-boiling azeotrope.** Fractional distillation of solutions lying on either side of this azeotropic composition is capable of separating them into, at best, one pure component plus a solution having the minimum boiling point. As any moonshiner will agree, ethyl alcohol-water mixtures (obtained by fermentation of sugars, for example) are

The alcohol obtained by distilling a water-alcohol mixture cannot be used for blending with gasoline to make gasohol because the alcohol must be water-free. Other methods must be used to make dry "absolute alcohol."

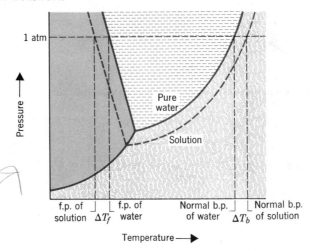

Figure 10.18

Effect of a nonvolatile solute on the phase diagram of H$_2$O.

rich in water. Fractional distillation is able to concentrate the alcohol to, at best, the azeotropic composition of approximately 95% by volume of ethyl alcohol.[4] Once this composition has been achieved, the liquid and vapor have the same composition, and no additional fractionation takes place.

There are also solutions that show large negative deviations from ideality and therefore have a minimum in their vapor pressure curves. This leads to a maximum on the boiling point diagram and hence to a **maximum-boiling azeotrope.** Hydrochloric acid, for instance, forms a maximum-boiling azeotrope having the approximate composition, 20% HCl and 80% H$_2$O by weight, with a boiling point of 109°C.

10.9 COLLIGATIVE PROPERTIES OF SOLUTIONS

Properties that depend on the number of particles of solute in a solution, instead of on their specific chemical nature, are called **colligative properties.**[5] Vapor pressure is one of these. We have seen that, according to Raoult's law, the addition of a nonvolatile solute to a substance causes its vapor pressure to be lowered. In our explanation of how this happened, nothing was said about the specific nature of the solute other than that it was incapable of escaping from the solution and that it was undissociated.

What effect does this vapor pressure lowering have on the phase diagram of a solvent such as water? In Figure 10.18, we see that the vapor pressure of the solution lies below that of the pure solvent at every temperature, as indicated by the dashed line. Because of this the solution must attain a higher temperature in order to have the same vapor pressure as water alone. Therefore the normal boiling point of the solution is higher, by an amount ΔT_b, than that of water by itself.

We also find, in Figure 10.18, that the vapor pressure curve of the solution intercepts the solid-vapor line of water at a temperature below the triple point of the solvent.[6] The solid-liquid equilibrium line of the solution must pass

---

[4] This is too strong to consume without dilution. A 95% solution of ethyl alcohol is 190 proof. Aged whiskey that is 86 proof is only 43% alcohol, by volume.

[5] Derived from the latin, *colligare*, to collect. These properties are determined by the number of particles in the entire "collection," not by what the particles are composed of.

[6] There is almost never a solid-vapor curve for the solution, because when it freezes the solid that is formed nearly always is composed of the pure solvent. Solute particles can only rarely be accommodated in the lattice of the solid solvent.

Table 10.5
K$_b$ and K$_f$ for some solvents

Solvent	Boiling Point (°C)	K_b (°C/m)	Melting Point (°C)	K_f (°C/m)
Water	100.0	0.51	0.0	1.86
Benzene	80.1	2.53	5.5	5.12
Camphor	—	—	179	39.7
Acetic acid	118.2	2.93	17	3.90

through this new triple point, so it lies to the left of the solid-liquid line for pure water. As a result, the freezing point of the solution, at 1 atm, lies below that of the solvent by an amount ΔT_f.

In summary, a nonvolatile solute increases the liquid range of a solution and results in a boiling-point elevation and freezing-point depression. In effect the solute reduces the escaping tendency of the solvent molecules both in the direction of the vapor and the solid. A common example of this is the effect that automotive antifreeze has on the liquid range of an antifreeze solution in the radiator of an automobile. The solute (usually ethylene glycol) lowers the freezing point so that the radiator doesn't freeze in cold weather, and it also raises the boiling point so that the radiator doesn't "boil over" as readily in very hot weather.

For dilute solutions it has been found that the extent to which the boiling point is raised, and the freezing point is lowered, depends on the *molality* of the solute in the solution,

$$\Delta T_b = K_b m \qquad [10.6]$$

and

$$\Delta T_f = K_f m \qquad [10.7]$$

where K_b and K_f are referred to as the **molal boiling-point elevation constant** and the **molal freezing-point depression constant,** respectively. The magnitudes of K_b and K_f are characteristic of each solvent. Table 10.5 contains a list of some typical solvents and their values of K_b and K_f.

If we know the concentration of the solution, in moles of solute per kilogram of solvent, the relationships expressed in Equations 10.6 and 10.7 permit us to calculate the extent to which the boiling point and freezing point are changed. For example, a solution containing 1 mol of sugar in 1000 g of water will have its freezing point lowered by 1.86°C and its boiling point raised by 0.51°C. The solution will therefore freeze at −1.86°C and boil at 100.51°C when the atmospheric pressure is 1 atm. Example 10.6 provides another sample calculation.

EXAMPLE 10.6   What would be the freezing point and boiling point of a solution containing 6.50 g of ethylene glycol ($C_2H_6O_2$), commonly used as an automotive antifreeze, in 200 g of water?

SOLUTION   To determine ΔT_f and ΔT_b we must know the molality of the solution—the ratio of moles of solute to kilograms of solvent. The number of moles of $C_2H_6O_2$ is

$$6.50 \text{ g } C_2H_6O_2 \times \left(\frac{1 \text{ mol } C_2H_6O_2}{62.1 \text{ g } C_2H_6O_2}\right) = 0.105 \text{ mol } C_2H_6O_2$$

The number of kilograms of solvent is

$$200 \text{ g H}_2\text{O} \times \left(\frac{1 \text{ kg}}{1000 \text{ g}}\right) = 0.200 \text{ kg H}_2\text{O}$$

The molality is therefore

$$\frac{0.105 \text{ mol C}_2\text{H}_6\text{O}_2}{0.200 \text{ kg H}_2\text{O}} = 0.525 \, m \text{ C}_2\text{H}_6\text{O}_2$$

For H_2O, $K_f = 1.86°\text{C}/m$ and $K_b = 0.51°\text{C}/m$. Therefore, the changes in the freezing point and boiling point are

$$\Delta T_f = \left(1.86 \frac{°\text{C}}{m}\right) \times (0.525 \, m) = 0.976°\text{C}$$

$$\Delta T_b = \left(0.51 \frac{°\text{C}}{m}\right) \times (0.525 \, m) = 0.27°\text{C}$$

The freezing and boiling points of the solution are then $-0.976°\text{C}$ and $100.27°\text{C}$. We see that solutions considerably more concentrated than this (approximately 3%) are necessary to protect an automobile's cooling system in frigid weather. See Problem 10.73.

If a knowledge of the molal concentration permits us to determine the extent to which the boiling point and freezing point differ from those of the pure solvent, then it should also be possible to calculate the molal concentration of a solution from measured values of ΔT_b and ΔT_f. This aspect of these colligative properties proves particularly useful, because with it we can measure molecular weights experimentally. This, combined with analytical data on percentage composition, enables us to find molecular formulas. Example 10.7 illustrates how to apply this concept.

EXAMPLE 10.7   A 5.50-g sample of a compound whose empirical formula is $\text{C}_3\text{H}_3\text{O}$ was dissolved in exactly 250.0 g of benzene. It was found that the freezing point of the solution was 1.02°C below that of pure benzene. Determine (a) the molecular weight and (b) the molecular formula of this compound.

SOLUTION   (a) From Table 10.5, K_f for benzene is 5.12°C/m. If we solve Equation 10.7 for the concentration, we obtain

$$m = \frac{\Delta T_f}{K_f}$$

Substituting the values for the freezing-point depression and K_f, we have

$$\text{molality} = \frac{1.02°\text{C}}{5.12°\text{C}/m}$$

$$\text{molality} = 0.199 \, m = \frac{0.199 \text{ mol solute}}{1.00 \text{ kg benzene}}$$

From the data given, we know that there are 5.50 g of solute per 250 g of benzene. From this we can obtain the number of grams of solute per kilogram of solvent.

$$\left(\frac{5.50 \text{ g solute}}{250.0 \text{ g benzene}}\right) \times \left(\frac{1000 \text{ g benzene}}{1.00 \text{ kg benzene}}\right) = \frac{22.0 \text{ g solute}}{1.00 \text{ kg benzene}}$$

We now have two expressions giving the quantity of solute per 1.00 kg of solvent. They must be equivalent because the solution cannot have two different concentrations at the same time. Therefore,

$$0.199 \text{ mol solute} = 22.0 \text{ g solute}$$

Now we can calculate the weight of 1 mol, which is numerically equal to the molecular weight

$$1 \text{ mol solute} \times \left(\frac{22.0 \text{ g solute}}{0.199 \text{ mol solute}} \right) = 111 \text{ g solute}$$

(b) Now that we know the molecular weight, we can determine the molecular formula as in Chapter 2. The molecular formula must contain the empirical formula repeated an integral number of times. The molecular weight is therefore an integral multiple of the empirical formula weight, which for C_3H_3O is 55.0 amu. Since the molecular weight that we have found is twice this value, the molecular formula must be $(C_3H_3O)_2$ or $C_6H_6O_2$.

In practice, molecular weights cannot be determined by this method as accurately as we have implied. However, an error even as large as 10% (that is, measured molecular weights ranging, in this case, from about 100 to 120 amu) certainly still permits us to choose among the possibilities, C_3H_3O, $C_6H_6O_2$, and $C_9H_9O_3$, with their corresponding molecular weights, 55.0, 110, and 165 amu. For this reason, this technique has proven to be a valuable tool in chemistry.

10.10 OSMOTIC PRESSURE

Osmosis is a process whereby a solvent passes from a dilute solution into a more concentrated one by moving through a thin film that selectively permits the passage of the solvent, but restricts the passage of the solute. Such films are called **semipermeable membranes,** and typical examples include certain types of parchment paper and some gelatinlike inorganic substances. A similar phenomenon called **dialysis,** which occurs at cell walls in plants and animals,

A patient uses a portable artificial kidney machine. It cleanses the blood of impurities by dialysis.

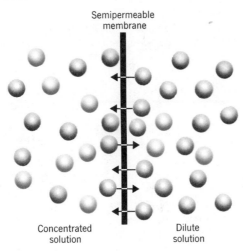

Figure 10.19

Osmosis.

allows the passage of water, small ions, and small molecules but restricts passage of large molecules such as proteins. Osmosis is the limiting case of dialysis.

In the process of osmosis there is a drive toward equalization of concentrations between the two solutions in contact with one another across the membrane. The rate of passage of solvent molecules through the membrane into the more concentrated solution is greater than their rate of passage in the opposite direction, presumably because at the surface of the membrane the solvent concentration is greatest in the more dilute solution (Figure 10.19). We observe a similar effect if two solutions, with unequal concentrations of a nonvolatile solute, are placed in a sealed enclosure, as shown in Figure 10.20. The rate of evaporation from the dilute solution is greater than that from the concentrated solution, but the rate of return to each is the same (both solutions are in contact with the same gas phase). As a result, neither solution is in equilibrium with the vapor. In the dilute solution molecules are evaporating faster than they are condensing, while in the concentrated solution the reverse occurs. Consequently, there is a gradual net transfer of solvent from the dilute solution into the more concentrated one until they both achieve the same concentration.

If we perform an osmosis experiment using the apparatus in Figure 10.21, in which we have a solution in compartment *A* and pure water in *B*, the passage

Figure 10.20

Unequal vapor pressures lead to a net transfer of solvent from the less concentrated solution to the more concentrated one.

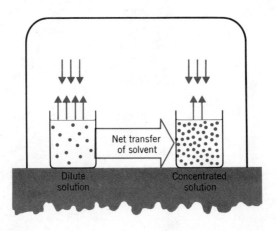

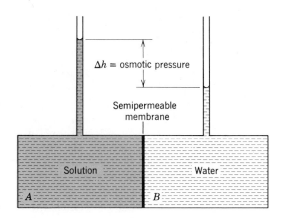

Δh = osmotic pressure

Semipermeable
membrane

Solution

Water

A

B

Figure 10.21
Measurement of osmotic pressure.

of solvent from B to A will slowly increase the volume of A and decrease the volume of B. As this occurs, the height of the liquid in the capillary of compartment A will rise while the height of the liquid in the other capillary will drop, and there will be a pressure difference between the two solutions that depends on the difference in these heights, Δh. Now, the ease with which a solvent molecule can be transferred from B to A, and from A to B, also depends on this pressure difference. As the pressure on side A increases, it becomes increasingly more difficult to squeeze another solvent molecule into this solution. It also becomes easier for a water molecule to "pop out" of the solution into the compartment with pure water. Therefore, as Δh becomes larger, the net rate of transfer of water from B to A decreases, and at some point a value of Δh is reached at which water molecules are moving across the membrane at equal rates—equilibrium is achieved. The pressure difference between the two compartments at equilibrium is called the **osmotic pressure** of the *solution* and is symbolized by the Greek letter, Π.

Π is the capital Greek letter pi.

For solutions that are dilute, it can be shown that the osmotic pressure is proportional to the molarity (M) of the solute, and that the proportionality constant is RT, where R is the gas constant and T is the absolute temperature.

$R = 0.0821$ liter atm/mol K.

$$\Pi = MRT \qquad [10.8]$$

Another form of this equation can be obtained if we realize that the molarity of the solute is obtained as the ratio of the number of moles of solute (n) to the volume of the solution (V).

$$M = \frac{n}{V} \qquad [10.9]$$

Substituting this into Equation 10.8 gives

$$\Pi = \frac{n}{V} RT$$

or

$$\Pi V = nRT \qquad [10.10]$$

Jacobus Henricus van't Hoff, a Dutch chemist, received the first Nobel Prize for chemistry in 1901.

It is interesting to note the similarity between Equation 10.10 (called the van't Hoff equation) and the ideal gas law,

$$PV = nRT$$

The magnitude of the osmotic pressure, even in very dilute solutions, is quite large. For instance, with a concentration of 0.010 mol of solute particles per liter (0.010 M) at room temperature (298 K), the osmotic pressure would be

$$\Pi = MRT$$

$$\Pi = \left(\frac{0.010 \text{ mol}}{1 \text{ liter}}\right) \times \left(\frac{0.0821 \text{ liter atm}}{\text{mol K}}\right) \times 298 \text{ K}$$

$$\Pi = 0.24 \text{ atm}$$

This pressure is sufficient to support a column of water 8.1 ft high!

Because the osmotic pressure that can be developed between solutions of only slightly different concentrations is so great, it is very important that fluids added to the body intravenously not alter significantly the osmotic pressure of the blood. If the blood fluids become too dilute, the osmotic pressure that develops within the blood cells can cause them to rupture. On the other hand, if the fluids are too concentrated, water will diffuse out of the cells and they will no longer function properly. For this reason care is taken to use solutions with the same osmotic pressure as the solution within the cells. Solutions that have the same osmotic pressure are called **isotonic** solutions.

The large differences in pressure developed between solutions of very similar concentrations provide us with a method of measuring the very large molecular weights of polymers (both of synthetic and biological origin). Freezing-point lowering and boiling-point elevation just won't work in these cases. For instance, a solution containing even as much as 150 g of a solute whose molecular weight is 30,000 in 1000 g of water produces a solution with a concentration of only 0.005 m. The freezing-point depression of this solution is approximately 0.009 degrees. Moreover, in most cases it is not even possible to dissolve this much solute, so that in practice the freezing-point changes are virtually undetectable.

$$\frac{150 \text{ g solute}}{1.00 \text{ kg H}_2\text{O}} \times \frac{1 \text{ mol}}{3 \times 10^4 \text{ g}} = 0.005 \text{ m}$$

A solution containing only 15 g of this solute per 1000 g of water at 25°C, however, would have an easily measured osmotic pressure. The concentration of the solution is 5.0×10^{-4} m. Because the solution is so dilute, the molality and molarity will be nearly the same, so we can express the concentration as 5.0×10^{-4} M with very little error. Now we can calculate the osmotic pressure.

$$\Pi = \left(\frac{5.0 \times 10^{-4} \text{ mol}}{\text{liter}}\right) \times \left(\frac{0.0821 \text{ liter atm}}{\text{mol K}}\right) \times 298 \text{ K}$$

$$\Pi = 1.2 \times 10^{-2} \text{ atm}$$

This corresponds to a pressure of 9.1 torr, or 9.1 mm Hg. If we assume that the solution has a density of 1 g/ml, this pressure will support a column of the liquid 12.3 cm high (about 4.8 in.). A height such as 12.3 cm is very easily measured with accuracy, so osmotic pressure measurements are a useful tool for the determination of molecular weights of large molecules.

The osmosis process can be stopped and even reversed by the application of pressure equal to or greater than the osmotic pressure of the solution. This reversal of osmosis is used to desalinate seawater. An apparatus for this process is shown schematically in Figure 10.22. If pressure were not applied to this system, osmosis would occur from left to right—that is, molecules would be transferred from freshwater to saltwater (Figure 10.22a). However, when a pressure exceeding Π is applied, the osmosis is driven in the reverse direction (Figure 10.22b), which forces water molecules out of the saline solution, leaving the impurities behind. One type of membrane that is strong enough to withstand these pressures is a film of cellulose acetate placed over a suitable support. Cellulose acetate is permeable to water but impermeable to the ions and impurities

Many large ships use reverse osmosis to produce freshwater from the sea.

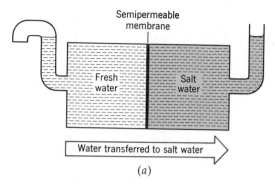

Semipermeable
membrane

Fresh
water

Salt
water

Water transferred to salt water

(a)

Figure 10.22

Desalination by reverse osmosis.
(a) Osmosis with no pressure
applied to the saline solution. (b)
Reverse osmosis occurs when a
pressure larger than Π *is applied*
to the solution.

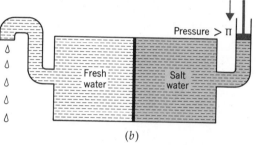

Pressure $> \Pi$

Fresh
water

Salt
water

(b)

in seawater. Desalination plants have been constructed thus far that can produce as much as 45,000 gallons of fresh water daily.

10.11 SOLUTIONS OF ELECTROLYTES

For simplicity, we have limited our discussion thus far to solutions that do not contain an electrolyte. The reason for this is that the vapor pressure lowering, freezing-point depression, boiling-point elevation, and osmotic pressure depend on the *number* of particles present in the solution. One mole of a nonelectrolyte, such as sugar, when placed in water yields 1 mol of particles, and a solution labeled "1 *m* sucrose" would, therefore, have a freezing point 1.86°C lower than pure water. However, a solution containing 1 mol of an electrolyte such as NaCl contains 2 mol of particles—1 mol of Na^+ ions and 1 mol of Cl^- ions. As a result, a solution labeled "1 *m* NaCl" actually contains 2 mol of particles per 1000 g of water and theoretically should have a freezing-point depression of $2 \times 1.86°C = 3.72°C$. Similarly, a 1 *m* solution of $CaCl_2$, which contains 3 mol of ions per 1000 g of water, should have a freezing-point depression three times as great as a 1 *m* solution of sucrose. (Actually, these predictions for NaCl and $CaCl_2$ are not entirely accurate, as we will see later, but they are close.) Similar predictions also hold fairly well for the other colligative properties.

EXAMPLE 10.8	A solution of calcium chloride was prepared by dissolving 25.0 g of $CaCl_2$ in exactly 500 g of H_2O. What is the expected vapor pressure of this solution at 80°C? At 80°C, water has a vapor pressure of 355 torr. What would the vapor pressure of the solution be if $CaCl_2$ were not an electrolyte?
SOLUTION	To solve the problem we must use Raoult's law.

$$P_{\text{solution}} = X_{\text{solvent}} P^0_{\text{solvent}}$$

so we need to calculate the mole fraction of water. For the $CaCl_2$,

$$25.0 \; \cancel{\text{g } CaCl_2} \times \left(\frac{1 \text{ mol } CaCl_2}{111 \; \cancel{\text{g } CaCl_2}} \right) = 0.225 \text{ mol } CaCl_2$$

and for water

$$500 \; \cancel{\text{g } H_2O} \times \left(\frac{1 \text{ mol } H_2O}{18.0 \; \cancel{\text{g } H_2O}} \right) = 27.8 \text{ mol } H_2O$$

Before we compute X_{H_2O}, we have to realize that the 0.225 mol $CaCl_2$ will produce three times as many moles of ions, that is, 0.675 mol ions. Therefore,

$$X_{H_2O} = \frac{27.8 \text{ mol } H_2O}{27.8 \text{ mol } H_2O + 0.675 \text{ mol ions}} = 0.975$$

The vapor pressure of the solution should therefore be

$$P_{solution} = 0.975 \times 355 \text{ torr}$$

$$= 346 \text{ torr}$$

If $CaCl_2$ were not an electrolyte, the mole fraction of water would have been

$$X_{H_2O} = \frac{27.8 \text{ mol } H_2O}{27.8 \text{ mol } H_2O + 0.225 \text{ mol } CaCl_2} = 0.993$$

and

$$P_{solution} = 0.993 \times 355 \text{ torr}$$

$$= 352 \text{ torr}$$

As you might expect, weak electrolytes produce effects on the colligative properties that lie between those of nonelectrolytes and those of strong electrolytes. For example, acetic acid undergoes reaction with water to establish the equilibrium

$$HC_2H_3O_2 + H_2O \rightleftharpoons H_3O^+ + C_2H_3O_2^-$$

If we ignore the H_2O molecule that serves as the carrier for the H^+, this can be written in a simpler way as

$$HC_2H_3O_2 \rightleftharpoons H^+ + C_2H_3O_2^-$$

Because there is this equilibrium, only a portion of the acetic acid is present as its ions while the rest exists as molecules, so at equilibrium, 1 mol of this solute exists as more than 1 mol of particles but less than 2. We can use the boiling-point elevation and freezing-point depression to determine the extent to which such a weak electrolyte is dissociated.

EXAMPLE 10.9

Very careful measurement reveals that a 1.00 m solution of HF has a freezing point of −1.91°C. What percent of the HF is dissociated into H^+ and F^- ions in this solution?

SOLUTION

The dissociation of hydrogen fluoride can be represented by

$$HF \rightleftharpoons H^+ + F^-$$

If HF were a nonelectrolyte, a 1.00 m solution would have a freezing point of $-1.86°C$; if it were a strong electrolyte, the solution should freeze at $-3.72°C$. Because the measured freezing point lies between these two extremes, only a part of the 1 mol of HF has dissociated. We want to know how much. Since this is presently unknown, let's give it a name—let's call the amount of HF that has undergone dissociation in 1.00 kg of water by the name, x.

$$\text{number of moles of HF dissociated} = x$$

The number of moles of HF remaining must be the difference between the amount of HF that we put into the solution and the amount dissociated; that is, $1.00 - x$.

$$\text{number of moles of HF remaining at equilibrium} = 1.00 - x$$

We see from the chemical equation above that for every mole of HF that dissociates, we produce 1 mol of H^+ and 1 mol of F^-. When x moles of HF dissociate, we must therefore form x mol of H^+ and x mol of F^-, so that at equilibrium we also have

$$\text{number of moles of } H^+ = x$$

$$\text{number of moles of } F^- = x$$

The *total* number of moles of particles in 1 kg of solvent is the sum of the moles contributed by the HF, H^+, and F^-. This total is

$$\text{total number of moles of particles} = (1.00 - x) + x + x$$

$$= 1.00 + x$$

Now, from the freezing-point depression of this solution, $1.91°C$, we can calculate the molal concentration of *particles*.

$$m = \frac{\Delta T_f}{K_f} = \frac{1.91°C}{1.86°C/m} = 1.03 \ m$$

This number *also* represents the total number of moles of particles per 1.00 kg of water. Therefore,

$$1.03 \text{ mol of particles} = (1.00 + x) \text{ mol of particles}$$

and thus

$$x = 0.03 \text{ mol}$$

The fraction of HF dissociated is equal to the number of moles that have broken apart (0.03 mol) divided by the total number of moles of HF placed in the solution (1.00 mol).

$$\text{fraction dissociated} = \frac{0.03}{1.00} = 0.03$$

The percentage of the HF dissociated is 3%.

Solutions of electrolytes have larger values of ΔT_f and ΔT_b than we might initially have expected because of dissociation. There are also instances where the freezing-point depression and boiling-point elevation are smaller than we would at first predict. This occurs when **association** (the opposite of dissociation) takes place between solute particles in the solution. For instance, a solution of 1 mol (122 g) of benzoic acid in 1.00 kg of benzene produces a freezing-point depression only slightly more than half the expected depression, implying that there are only about half as many particles in the solution as we anticipated. Since this approximately $\frac{1}{2}$ mol of particles weighs 122 g, the apparent molecular weight is about 240. Therefore, when association takes place, the measured molecular weights are actually higher than we would predict.

In this particular example, the anomalous behavior of benzoic acid in benzene is attributed to hydrogen bonding between benzoic acid molecules to form a **dimer** (a particle created from two identical simpler units).

benzoic acid   benzoic acid dimer—hydrogen bonds shown as dotted lines

Interionic attractions

van't Hoff discovered the effects of interionic attractions.

Earlier in this section we mentioned that the actual freezing points of solutions of electrolytes such as NaCl and CaCl$_2$ are not exactly the same as those calculated assuming complete dissociation. This happens because the ions in a solution are not really completely independent particles. Although the solvent shields them from each other's charge, this shielding is not perfect, and it becomes worse as the solution becomes more concentrated—that is, as the average distance between the ions becomes smaller. As a result, a concentrated solution behaves as if it has fewer ions than a dilute solution, and the ions' effectiveness at altering the properties of the solution (boiling point, freezing point, osmotic pressure) diminishes as the concentration of the solute gets larger. Consequently, ionic compounds behave as if they are less fully dissociated in concentrated solutions than when they are dilute.

Quantitatively, the degree to which an electrolyte behaves as if it were dissociated can be expressed by the van't Hoff factor, i. This quantity can be defined as the ratio of the observed freezing-point depression produced by a solution to the freezing point that the solution would exhibit if the solute were a nonelectrolyte.

$$i = \frac{(\Delta T_f) \text{ measured}}{(\Delta T_f) \text{ calculated as nonelectrolyte}}$$

Table 10.6 has values of the i factor for several "strong" electrolytes. For NaCl, KCl, and MgSO$_4$, i approaches 2 as the solution becomes more dilute. For K$_2$SO$_4$, i approaches 3, as we expect.

Table 10.6
Values of the van't Hoff factor at various concentrations

Salt	Concentration (mol/kg of H$_2$O)			i Factor if Completely Dissociated
	0.1	0.01	0.001	
NaCl	1.87	1.94	1.97	2.00
KCl	1.85	1.94	1.98	2.00
K$_2$SO$_4$	2.32	2.70	2.84	3.00
MgSO$_4$	1.21	1.53	1.82	2.00

It is interesting to compare the effects of the charges on the ions on the interionic attractions. For NaCl, the value of i changes by about 5% going from 0.1 m to 0.001 m. For K_2SO_4, which has a doubly charged SO_4^{2-} ion, i changes by about 22% for the same dilution. When there are two doubly charged ions in $MgSO_4$, the i factor changes by about 50% for the same dilution. These observations are not surprising, because as the charges on the cations and anions become larger, so should their attractions for each other. This leads to a lower degree of "independence" of the ions as their charges increase.

INDEX TO QUESTIONS AND PROBLEMS (Problem numbers in **bold type**)

REVIEW QUESTIONS

10.1 Describe substitutional and interstitial solid solutions.

10.2 State the definitions of the following concentration units: mole fraction, mole percent, weight fraction, weight percent, molarity, molality.

10.3 What feature do all concentration units have in common?

10.4 As applied to solubility, what is the significance, on a molecular level, of the phrase "like dissolves like"?

10.5 Frequently a liquid that is soluble in water can be made less soluble (and can therefore be made to separate as a distinct phase) by addition of salt to the solution. How can this be explained?

10.6 Small amounts of water in the fuel tank of an automobile can cause severe difficulties in engine performance. The problem can be overcome by adding "dry gas" to the fuel. The "dry gas" consists mostly of methyl alcohol, CH_3OH, which allows the water to dissolve in the gasoline. Can you explain how the dry gas accomplishes this?

10.7 What is meant when we say an ion is hydrated? What is the meaning of the term solvation?

10.8 Compare the definitions of an ideal gas and an ideal solution.

10.9 Discuss the relationship between lattice energy and hydration energy in determining the magnitude and sign of the heat of solution.

10.10 On the basis of size and charge, choose the ion in each of the following pairs with the larger hydration energy:
(a) Na^+ or K^+
(b) F^- or Cl^-
(c) K^+ or Ca^{2+}
(d) Fe^{2+} or Fe^{3+}
(e) S^{2-} or Cl^-

10.11 The solution process for KI in water is endothermic (ΔH_{soln} is positive). Would you expect KI to become more or less soluble as the temperature is increased? Explain.

10.12 Why is ΔH_{soln} for gases nearly always negative?

10.13 Describe, qualitatively, the procedure called fractional crystallization.

10.14 On the basis of the information provided in Figure 10.11, predict which solid will separate first from a solution containing equal weights of KNO_3 and KBr when the solution is gradually evaporated at a temperature of 70°C. What will occur if the solution is gradually evaporated at 20°C?

10.15 Why do gases become more soluble in liquids as the pressure is increased?

10.16 Why do pressure changes have very little effect on the solubility of a solid in a liquid?

10.17 Explain, on a molecular level, why the vapor pressure of the solvent is expected to be directly proportional to its mole fraction in the solution (Raoult's law).

10.18 What are meant by positive and negative deviations from Raoult's law?

10.19 How is the sign of ΔH_{soln} related to positive and negative deviations from Raoult's law?

10.20 Describe, qualitatively, the procedure called fractional distillation.

10.21 Referring to Figure 10.16, approximately how many times must boiling, followed by condensation of the resulting vapor, be repeated in order to obtain a portion of liquid having a mole fraction of A of at least 0.80 if the original mole fraction of A was 0.20?

10.22 Benzene has a boiling point of 80.1°C; carbon tetrachloride boils at 76.8°C. Sketch a boiling-point diagram for benzene-carbon tetrachloride mixtures. Assume ideal solution behavior.

10.23 Water and butyl alcohol form an azeotrope that boils at 92.4°C at 760 torr. At this same pressure butyl alcohol boils at 117.8°C. The composition of the azeotrope is 28.4 mol % butyl alcohol, 71.6 mol % water. Sketch the boiling-point diagram for butyl alcohol-water mixtures. Do these substances show positive or negative deviations from ideality?

10.24 We found that the addition of a nonvolatile solute to a solvent reduces the escaping tendency of the solvent from the solution, and in Section 10.9 we saw that this leads to a boiling-point elevation. On a molecular level, account for the fact that the presence of a solute also reduces the tendency of the solvent to escape from the liquid into the solid. Explain why a lower temperature must be achieved to establish equilibrium between the solid solvent and the solution than between the pure solid and liquid solvent.

10.25 Based on the data in Table 10.6, which 1:1 electrolyte (one positive ion to one negative ion) appears to be *least* fully dissociated in concentrated solutions? How does this agree (or disagree) with what might be predicted based on the charges on the ions involved?

10.26 On the basis of what you have learned in this chapter, how would you interpret an i factor having a value less than 1.00?

10.27 Assuming equal concentrations in water, which of the following salts would you expect to appear to be least fully dissociated: KCl, $NiCl_2$, or $Al_2(SO_4)_3$?

10.28 Assuming complete dissociation, what i factor would you expect for each of the salts in Question 10.27?

REVIEW PROBLEMS (More difficult problems are marked by an asterisk)

10.29 Calculate the mole fraction, weight fraction, weight percent, and molality of glycerin in a solution prepared by dissolving 45.0 g of glycerin, $C_3H_5(OH)_3$, in 100.0 g of H_2O.

10.30 A mixture is prepared from 45.0 g of benzene (C_6H_6) and 80.0 g of toluene (C_7H_8). Calculate (a) the weight percent of each component, (b) the mole fraction of each component, (c) the molality of the solution if toluene is taken to be the solvent.

10.31 A solution containing 121.8 g of $Zn(NO_3)_2$ per liter has a density of 1.107 g/ml. Calculate (a) the weight percent of $Zn(NO_3)_2$ in the solution, (b) the molality of the solution, (c) the mole fraction of $Zn(NO_3)_2$, (d) the molarity of the solution.

10.32 What are the mole fraction, molality, and weight percent of a solution prepared by dissolving 0.30 mol of $CuCl_2$ in 40.0 mol of H_2O?

10.33 A sample of drinking water was found to be severely contaminated with chloroform, $CHCl_3$, a known carcinogen. The level of contamination was 12.4 ppm.
(a) Express this in percent by weight.
(b) What is the molarity of the $CHCl_3$ in the water?

10.34 A solution of isopropyl alcohol (rubbing alcohol), C_3H_7OH, in water has a mole fraction of alcohol equal to 0.250. What is the weight percent alcohol and the molality of the alcohol in the solution?

10.35 The solubility of baking soda, $NaHCO_3$, in water at 20°C is 9.6 g/100 g of H_2O. What is the mole fraction of NaHCO₃ in a saturated solution? What is the molality of the solution?

10.36 A saturated solution of NaCl at 30°C has a molality of 6.25 m. What is the mole fraction and weight fraction of NaCl in the solution?

10.37 A solution of sodium carbonate was prepared containing 14.0% Na_2CO_3 by weight. What is the mole fraction and molality of Na_2CO_3 in this solution?

10.38 The World Health Organization's drinking water standards specify that the maximum permissible concentration of magnesium ion in drinking water is 150 mg per liter. What does this correspond to in terms of (a) molarity, (b) molality (assume the density of the water is 1.00 g/ml)?

10.39 Concentrated sulfuric acid is 96.0% H_2SO_4 by weight. What is the molality of the H_2SO_4? What are the mole fractions of H_2SO_4 and water?

10.40 The molality of an aqueous solution of ammonium nitrate was 2.25 m. What was the weight percent of ammonium nitrate in the solution and what were the mole fractions of ammonium nitrate and water?

10.41 A solution of benzene, C_6H_6, dissolved in chloroform, $CHCl_3$, had a mole fraction of C_6H_6 equal to 0.240.
(a) What was the mole percent CHCl₃ in the solution?
(b) What was the molality of the C_6H_6 in chloroform?
(c) Taking benzene to be the solvent, what was the molality of the CHCl₃ in the C_6H_6?
(d) What were the weight percents of CHCl₃ and C_6H_6 in the solution?

10.42 An antifreeze solution is prepared from 222.6 g of ethylene glycol, $C_2H_4(OH)_2$, and 200 g of water. Its density is 1.072 g/ml. Calculate the molality and molarity of the solution.

10.43 A 4.03 M solution of ethylene glycol, $C_2H_4(OH)_2$, has a density of 1.045 g/ml. Calculate the weight percent $C_2H_4(OH)_2$, mole fraction of $C_2H_4(OH)_2$ and the molality of the solution.

10.44 Use the data in Table 10.2 to calculate the amount of heat liberated by dissolving 10.0 g of $AlCl_3$ in 1.00 liter water.

10.45 How many calories are absorbed by dissolving 115 g of NH_4NO_3 in 100 ml of H_2O?

10.46 How many grams of $NaNO_3$ will precipitate if a saturated solution of $NaNO_3$ in 200 g of H_2O at 70°C is cooled to 25°C?

10.47 How many grams of water at 80°C are required to dissolve 35.0 g of NaBr?

10.48 The partial pressure of ethane over a saturated solution containing 6.56×10^{-2} g of ethane is 751 torr. What is its partial pressure when the saturated solution contains 5.00×10^{-2} g of ethane?

10.49 The Henry's law constant for a gas dissolved in water was found to be 6.50×10^{-5} g/liter torr at 25°C. In an experiment the gas was collected over water and its concentration was found to be 0.0478 g/liter. What was the *total* pressure of gas above the solution?

10.50 Methane (natural gas) has a solubility in water at 25°C of 2.09×10^{-4} g/liter at a pressure of 0.968 atm. Assuming Henry's law is valid, what would be the concentration of methane (in grams per liter) in ground water deep below the earth's surface at a pressure of 1000 atm and at a temperature of 25°C?

10.51 The vapor pressure of benzene (C_6H_6) at 25°C is 93.4 torr. What will be the vapor pressure in torr at 25°C, of a solution prepared by dissolving 56.4 g of the nonvolatile solute, $C_{20}H_{42}$, in 1000 g of benzene?

10.52 The vapor pressure of pure methyl alcohol at 30°C is 160 torr. What mole fraction of glycerol (a nonvolatile nondissociating solute) would be required to lower the vapor pressure to 130 torr?

*__10.53__ To what celsius temperature would a solution containing 150 g of glycerol, $C_3H_5(OH)_3$, in 100 g of H_2O have to be heated to have a vapor pressure of 91.1 torr? Assume ideal solution behavior, $C_3H_5(OH)_3$ to be a nonvolatile solute, and refer to Table 7.1.

10.54 Heptane (C_7H_{16}) has a vapor pressure of 791 torr at 100°C. At this same temperature, octane (C_8H_{18}) has a vapor pressure of 352 torr. What will be the vapor pressure of a mixture of 25.0 g of heptane and 35.0 g of octane? Assume ideal solution behavior.

10.55 What will be the freezing point and boiling point of an aqueous solution containing 55.0 g of glycerol, $C_3H_5(OH)_3$, dissolved in 250 g of water? Glycerol is a nonvolatile, undissociated solute.

10.56 What is the molecular weight and molecular formula of a nondissociating compound whose empirical formula is C_4H_2N if 3.84 g of the compound in 500 g of benzene gives a freezing-point depression of 0.307°C?

10.57 A solution containing 16.9 g of a nondissociating substance in 250 g of water has a freezing point of −0.744°C. The substance is composed of 57.2% C, 4.77% H, and 38.1% O. What is the molecular formula of the compound?

10.58 How many grams of glucose, $C_6H_{12}O_6$ (a nondissociating solute), are required to lower the freezing point of 150 g of H_2O by 0.750°C? What will be the boiling point of this solution?

10.59 An aqueous solution freezes at −2.47°C. What is its boiling point?

10.60 Calculate the freezing point in °C of a 0.100 *m* **aqueous solution of a weak electrolyte, HX, that is 7.5% dissociated.**

10.61 Calculate the osmotic pressure, in torr, of an aqueous solution containing 5.0 g of sucrose, $C_{12}H_{22}O_{11}$, per liter at 25°C.

10.62 A solution of 0.40 g of a polypeptide in 1.00 liter of an aqueous solution has an osmotic pressure at 27°C of 3.74 torr. What is the approximate molecular weight of this polymer?

10.63 What would be the osmotic pressure, in torr, of a 0.010 M aqueous solution of the electrolyte, NaCl, at 25°C? (Assume 100% dissociation of NaCl in water).

10.64 Calculate the *i* factor for the weak electrolyte, HF, in Example 10.9. What conclusions would you draw about the *i* factors for weak electrolytes?

10.65 Below is a list of the most abundant ions in seawater.

Ion	Molality
Chloride	0.566
Sodium	0.486
Magnesium	0.055
Sulfate	0.029
Calcium	0.011
Potassium	0.011
Bicarbonate	0.002

Calculate the weight, in grams, of each component contained in 3.78 liters (1.00 gallon) of seawater having a density of 1.024 g/ml. What is the total weight of ions in this sample?

10.66 What is the osmotic pressure of sea water at 25°C? What is the minimum pressure, in atm, needed to desalinate seawater by reverse osmosis? (See the data in Problem 10.65.)

10.67 Assuming complete dissociation, what would the expected freezing point be of a 0.10 m solution of $MgSO_4$? What would the actual freezing point be, taking into account interionic attractions?

10.68 Suppose that you wish to prepare a solution containing 10.0% Na_2CO_3 by weight. The bottle of chemical that you have lists the contents as $Na_2CO_3 \cdot 10H_2O$. How many grams of the hydrate would be needed to prepare 50.0 g of the 10.0% Na_2CO_3 solution?

*10.69 Air contains approximately 20% O_2 by volume. The Henry's law constant for O_2 at 25°C is 5.34×10^{-5} g/liter torr. Calculate the weight of O_2 per liter of water in a stream that has a temperature of 25°C if the atmospheric pressure is 760 torr. (Assume equilibrium with the atmosphere.)

*10.70 At 25°C the vapor pressures of benzene (C_6H_6) and toluene (C_7H_8) are 93.4 and 26.9 torr, respectively. At what applied pressure, in torr, will a solution prepared from 60.0 g of benzene and 40.0 g of toluene boil at 25°C?

*10.71 The vapor pressure of a mixture containing 400 g of carbon tetrachloride and 43.3 g of an unknown substance is 137 torr at 30°C. The vapor pressure of pure carbon tetrachloride at 30°C is 143 torr, while that of the pure unknown is 85 torr. What is the approximate molecular weight of the unknown?

*10.72 A solution containing 8.3 g of a nonvolatile nondissociating substance dissolved in 1 mol of chloroform, $CHCl_3$, has a vapor pressure of 511 torr. The vapor pressure of pure $CHCl_3$ at the same temperature is 526 torr. Calculate (a) the mole fraction of the solute, (b) the number of moles of solute, (c) the molecular weight of the solute.

*10.73 The cooling system of an automobile usually contains a solution of antifreeze prepared by mixing equal volumes of ethylene glycol, $C_2H_4(OH)_2$, and water. The density of ethylene glycol is 1.113 g/ml. Calculate the freezing point of this mixture. On the label of the antifreeze container it is said that this mixture will protect your engine to a temperature of $-34°F$. How does your computed freezing point compare with this?

*10.74 What is the percent dissociation of a weak electrolyte, HX, in water if a 0.250 m solution of it has a freezing point of $-0.500°C$?

11

CHEMICAL THERMODYNAMICS

The Kenilworth Hotel, a Miami Beach hotel famous in the 1950s, is demolished in 20 seconds by explosives placed strategically on its ground floor. Once begun, this spontaneous collapse was sure to happen because it was accompanied by a decrease in energy and an increase in disorder. In this chapter we will see that these two factors control the fate of all physical and chemical changes.

In a study of chemistry it's natural to question why certain chemical reactions take place and why others do not. Certainly, it would be nice if we could predict ahead of time what will happen when several chemicals are mixed. Then we could sit at home and do our chemistry, instead of going to the lab. Unfortunately (or fortunately, depending on your love for lab work), chemistry hasn't evolved to that point yet, but we do know what controls the outcome of a reaction.

There are two factors that determine whether or not we will observe a particular reaction, either in the laboratory or elsewhere. Thermodynamics, studied in this chapter, tells us whether a change is possible—it answers the question, "Will the reaction occur by itself, without outside help?" It also tells us what the position of equilibrium will be when the composition of the reaction mixture ceases to change. Chemical kinetics, the subject of Chapter 12, is concerned with the speeds at which chemical changes take place. Both of these factors, spontaneity and speed, must be in our favor if we hope to observe the formation of the products of a chemical change. For example, thermodynamics predicts that at room temperature hydrogen gas and oxygen gas should react to produce water. However, a mixture of H_2 and O_2 is stable virtually indefinitely (provided no one strikes a match). This is so because, at room temperature, hydrogen and oxygen react at such an extremely slow rate that even though their reaction to produce water is spontaneous, it takes nearly forever for the reaction to proceed to completion.

Thermodynamics is basically concerned with the energy changes that accompany chemical and physical processes. Historically, it evolved without a detailed knowledge of the structure of matter; in fact, this is one of its strongest points. In this chapter we will take a rather informal approach to the subject in an effort to avoid mathematical formalism, and we will develop many of the concepts of thermodynamics by considering changes that take place on a molecular level.

Thermo implies heat; dynamics implies movement or change.

11.1 SOME COMMONLY USED TERMS

Before we proceed, we must first establish the meaning of some frequently used terms. A word that has been used rather loosely in previous sections is **system.** By system we mean *that particular portion of the universe on which we wish to focus our attention.* Everything else we call the **surroundings.** For example, if we wished to consider the changes taking place in a solution of sodium chloride and silver nitrate, our system is the solution, while the beaker and everything else around the solution are considered the surroundings.

If a change occurs so that heat cannot be transferred across the interface, or boundary, between the system and its surroundings, we speak of it as an **adiabatic** process. An example is a reaction carried out in an insulated container, such as a Thermos bottle. An explosive reaction is another example of an adiabatic process. Such a reaction occurs so rapidly that the heat energy produced cannot be readily dissipated. The heat build-up raises the products to very high temperatures, and these products fly apart rapidly, pushing walls, ceilings, and so on (the surroundings) before them.

When thermal contact is maintained between a system and its surroundings, heat can flow between them and it is frequently possible to keep the system at a constant temperature while a change takes place. In this case the change is said to be **isothermal.** The human body possesses an elaborate temperature-control system that maintains a constant body temperature, so biochemical reactions taking place within us are therefore essentially isothermal.

To discuss the changes that occur in a system it is necessary to define its properties very precisely before and after the change occurs. We do this by specifying the **state** of the system—that is, some particular set of conditions of pressure, temperature, number of moles of each component, and their physical form (for example, gas, liquid, solid, or crystalline form). Once these variables are specified, all the properties of the system are fixed. Thus a knowledge of these quantities permits us to define unambiguously the properties of a system. For instance, if we have two samples of pure liquid water, each consisting of 1 mol and each at the same temperature and pressure, we know that all the properties of each sample will be identical (volume, density, surface tension, vapor pressure, etc.).

The quantities P, V, and T are called **state functions** or **state variables.** This is because they serve to determine the physical state of a given system, and in a particular state their values do not depend on the prior history of the sample. For example, the volume of 1 mol of water at 25°C and 1 atm does not depend on what its temperature or pressure might have been at some time in the past. Furthermore, on going from one state to another the changes in P, V, and T do not depend on how the sample is treated. If the temperature of a sample of water is changed from 25°C to 35°C, it doesn't matter if the sample is first cooled to 0°C and then warmed to 35°C, or whether the temperature is increased directly from 25 to 35°C. In the final state the temperature is the same regardless of the path taken between the initial and final conditions, and the change in temperature, ΔT, therefore depends *only* on the temperatures of those initial and final states.

There are some instances where the interrelationships between the state functions can be expressed in equation form to give an **equation of state.** The equation of state for an ideal gas, $PV = nRT$, is an example. We have also seen the van der Waals equation of state, which can be applied with reasonable success to real gases.

Another quantity we will use is called the **heat capacity**—*the amount of heat energy required to raise the temperature of a given quantity of a substance one degree Celsius.* The units of heat capacity are cal/°C or J/°C. The **specific heat** represents the heat capacity per gram; that is, it is *the amount of heat necessary to raise the temperature of 1 g of a substance by 1.0°C.* The specific heat of water is 1.00 cal/g°C or 4.18 J/g°C. We also speak of the **molar heat capacity:** *the heat necessary to raise the temperature of 1 mol of a substance by 1 degree.* Since the molecular weight of water is 18.0 g/mol, the molar heat capacity of water is 18.0 cal/mol°C or 75.3 J/mol°C.

EXAMPLE 11.1	What would be the heat capacity expressed in kJ/°C of a water bath containing 4.00 liters of water? The specific heat of water is 4.18 J/g°C.

SOLUTION

For simplicity, we've ignored the heat capacity of the container that holds the water.

We wish to calculate the number of joules needed to raise the temperature of the water in the bath by 1.00°C. Assuming the density of water to be 1.00 g/ml, the mass of water is 4000 g. Therefore,

$$\text{heat capacity of bath} = 4000 \text{ g} \times \frac{4.18 \text{ J}}{\text{g°C}} \times \frac{1 \text{ kJ}}{1000 \text{ J}}$$

$$= 16.7 \text{ kJ/°C}$$

11.2 THE FIRST LAW OF THERMODYNAMICS

In thermodynamics we study the energy changes that occur when systems pass from one state to another. Repeated observations by many scientists over many years have led to the conclusion that, in any process, energy is neither created nor destroyed. This conclusion forms the basis of the *law of conservation of energy* discussed in Chapter 1, and we used this notion in our discussions of the energy changes involved in the formation of ionic compounds as well as the energy changes accompanying the formation of a solution in Chapter 10.

The **first law of thermodynamics** is simply a statement of the law of conservation of energy in the form of a simple equation,

$$\Delta E = q - w \qquad [11.1]$$

Let's look at what the symbols mean. The quantity E is called the **internal energy.** It is the total energy of the system—the total of all the energies possessed by the system as a consequence of the kinetic energy of its atoms, ions, or molecules, plus the potential energy that arises from the binding forces between the particles that make up the system. ΔE is the difference between the energy contained in a system in some final state and the energy it possessed in an initial state. It corresponds to the change in the internal energy of a system that occurs when the system goes from an initial to a final state.

$$\Delta E = E_{\text{final}} - E_{\text{initial}}$$

Note that we have used the same convention here as in our previous discussions of energy changes (heats of vaporization, heats of solution). Here, too, we cannot actually determine E. The reason is that we don't know how fast a system or its particles are moving, and we have no way of knowing the effects of all the attractive forces on the system. For instance, any system that we study moves as the earth revolves around its axis. The earth, in turn, revolves around the sun, which moves through a galaxy, which moves through space. How fast, nobody knows, so we can't calculate the kinetic energy. Nor can we figure the total potential energy caused by all the attractive forces between the system and the rest of the universe. Even though we can't measure E, we can measure how E changes—that is, ΔE—and that is all that matters, because we are only really concerned with changes in energy.

The quantity q in Equation 11.1 represents the amount of heat that is *added* to the system as it passes from the initial to the final state, and w denotes the work done *by* the system on its surroundings. Thus Equation 11.1 simply states that the change in the internal energy is equal to the difference between the energy *supplied to* the system as heat and the energy *removed from* the system as work performed on the surroundings.[1]

Since the first law deals with the transfer of quantities of energy, it is necessary to establish sign conventions to avoid confusion in our bookkeeping. Heat *added to* a system and work *done by* a system are considered positive quantities. Thus, if a certain change is accompanied by the absorption of 50 joules of heat and the expenditure of 30 joules of work, $q = +50$ J and $w = +30$ J. The change in internal energy of the system is

$$\Delta E_{\text{system}} = (+50 \text{ J}) - (+30 \text{ J})$$

or

$$\Delta E_{\text{system}} = +20 \text{ J}$$

Thus the system has undergone a net increase in energy amounting to $+20$ J. How about the surroundings?

When the system gains 50 J, the surroundings lose 50 J; therefore $q = -50$ J for the surroundings. When the system performs work, it does so on the

ΔE is equal to the energy input minus the energy output.

---

[1] Energy, you recall, is the capacity to do work. When the system performs work, its capacity to do additional work diminishes, which means that its energy has diminished. Energy, equal to the work performed, has been lost by the system and, in the process, gained by the surroundings.

surroundings. We say that the surroundings have done negative work, and $w = -30$ J for the surroundings. The change in the internal energy of the surroundings is thus

$$\Delta E_{\text{surroundings}} = (-50 \text{ J}) - (-30 \text{ J})$$

$$\Delta E_{\text{surroundings}} = -20 \text{ J}$$

The change in internal energy of the system is thus equal, but opposite in sign, to ΔE for the surroundings. This has to be so in order to satisfy the law of conservation of energy.

In summary,

q positive ($q > 0$); heat is added to the system
q negative ($q < 0$); heat is evolved by (removed from) the system
w positive ($w > 0$); the system performs work—energy is removed
w negative ($w < 0$); work is done on the system—energy is added

The internal energy happens to be a state function, and the magnitude of ΔE therefore depends only on the initial and final states of the system and not on the path taken between them. This is very much the same as the change in your bank balance that occurs between the beginning and the end of a month. During any given month the change in the balance is brought about as the combined results of some number of deposits and withdrawals. If the total number of dollars provided by the deposits exceeds those removed by the withdrawals, your balance increases. However, the net change in your balance at the end of the month depends only on the initial and final amounts of money in the bank, not on the individual transactions during the month. There is an infinite number of combinations of deposits and withdrawals that could lead to the same change in your balance. The same sort of relationship exists among ΔE, q, and w. The sign and magnitude of ΔE is controlled only by the values of E in the initial and final states. For any given change, ΔE, there are many different paths that can be followed with their own characteristic values of q and w. However, for the same initial and final states, the difference between q and w is always the same. Therefore, even though ΔE is a state function, q and w are not.

q and w are like the deposits and withdrawals. E is like the bank balance.

Before we can study the chemical or physical implications of the first law of thermodynamics, we have to examine how a system can do work. One kind of work a system can perform is electrical work. That's what happens when energy is withdrawn from a battery to power a wristwatch or start a car. Chemical reactions in the battery produce energy that pushes electrons through a wire, and this electrical energy can do work for us.

Another kind of work is accomplished when a system expands against an opposing pressure. Work is always accomplished when an opposing force is pushed through some distance. Expressed mathematically,

$$\text{work} = \text{force} \times \text{distance}$$

Pressure is defined as force per unit area, and in Figure 11.1 we see an external pressure P exerted by a force F spread over the area A of a piston.

$$P = \frac{F}{A}$$

The volume of the gas in the cylinder is equal to its cross-sectional area, A, multiplied by the height of the column of gas, h.

$$V = Ah$$

When the gas expands and pushes back the piston, A remains the same but h changes. The volume change is therefore

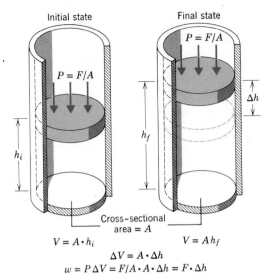

Figure 11.1
Pressure-volume work.

$$\Delta V = A \cdot \Delta h$$
$$w = P \Delta V = F/A \cdot A \cdot \Delta h = F \cdot \Delta h$$

$$\Delta V = V_f - V_i$$
$$\Delta V = Ah_f - Ah_i$$
$$\Delta V = A(h_f - h_i) = A(\Delta h)$$

The product of pressure times volume change is

$$P \, \Delta V = \frac{F}{A} A(\Delta h) = F \, \Delta h$$

We see that $P \, \Delta V$ is equivalent to force (F) times distance (Δh) and, therefore, equals work.

$$w = P \, \Delta V \qquad\qquad [11.2]$$

This expansion work is performed by any system—it doesn't have to be a gas—when it expands against an external pressure imposed by the surroundings. Conversely, when a system contracts under the influence of an external pressure, work is performed on the system. If the volume change is measured in liters and the pressure in atmospheres, $P \, \Delta V$ has the units, *liter atmosphere* (liter atm). We can, if we wish, convert this to the more familiar energy units, calories or joules, by the relationships

$$1 \text{ liter atm} = 24.2 \text{ cal}$$
$$1 \text{ liter atm} = 101.3 \text{ J}$$

In the SI, pressure is in pascals (newtons per square meter)

$$1 \text{ Pa} = 1 \text{ N/m}^2$$

and volume is in cubic meters, m^3. Therefore, the units of **P** Δ**V** are

$$\text{Pa m}^3 = \frac{N}{m^2} \cdot m^3 = N \cdot m$$

but $1 \text{ N} \cdot \text{m} = 1 \text{ J}$
Therefore,

$$1 \text{ Pa m}^3 = 1 \text{ J}$$

Let's look now at a couple of examples that illustrate what the first law of thermodynamics tells us. First, any particular change, whether it be physical or chemical, has a certain net energy change associated with it that is determined solely by the energies of the initial and final states. Once we have picked what these states are, the value of ΔE is fixed. But there are many different sets of q and w that produce the same ΔE. In other words, the amount of work that we can get from a given change depends on how we carry it out.

A physical example that illustrates this is the expansion of a gas. For simplicity, let's imagine an ideal gas expanding in such a way that its temperature remains constant. This can be accomplished, for example, by placing a cylinder of the gas in a large vat of water held at a constant temperature by a thermostat.

If the temperature is constant during the expansion, then the average kinetic energy of the molecules stays the same. Furthermore, since there are no attractive forces between the particles of an ideal gas, there is no potential en-

Since the change in K.E. and P.E. are both zero, $\Delta E = 0$

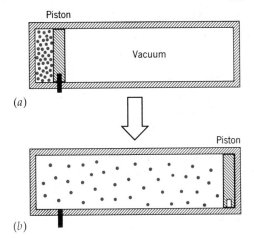

Figure 11.2

Expansion of a gas against an opposing pressure equal to zero. (a) Before expansion. P = 10 atm; V = 1 liter. (b) After expansion, final state. P = 1 atm; V = 10 liter.

ergy. This means that when an ideal gas expands or contracts at constant temperature there is no net change in its total energy, so $\Delta E = 0$. Therefore

This really only holds for an ideal gas. For a real gas ΔE is small, but not equal to zero.

$$\Delta E = 0 = q - w$$

$$q = w$$

This tells us that if the gas does work by expanding, it absorbs an equivalent amount of heat energy from the surroundings.

If you've ever lifted a heavy box, you know that the amount of work you have to do depends on how heavy the box is—that is, on how hard the box "pushes back." So it is with the expansion of a gas. In Figure 11.2*a* we see a compressed gas pushing against a piston held by a pin through the cylinder wall. On the other side of the piston is a vacuum. When the pin is pulled out of the piston, you know what will happen. The compressed gas will push the piston to the opposite end of the cylinder (Figure 11.2*b*). But the gas won't do any work, because the piston doesn't push back—there is no opposing pressure during the expansion. Thus $w = 0$ and $q = 0$.

Now let's look at the expansion pictured in Figure 11.3. Here we begin with the same gas, at the same temperature as before, and at the same initial pressure and volume (10.0 atm and 1.00 liter). We also finish at the same set of

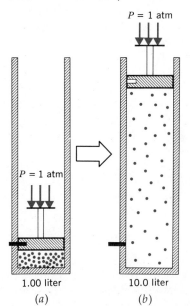

Figure 11.3

Expansion of an ideal gas against a constant opposing pressure of one atmosphere. (a) Initial state: $P_{gas} = 10.0$ atm, $V_{gas} = 1.00$ liter. (b) Final state: $P_{gas} = 1.00$ atm, $V_{gas} = 10.0$ liter.

$P_iV_i = P_fV_f$
(10 atm)(1 liter) = (1 atm)(10 liter)

conditions as before, so ΔE must again equal zero and $q = w$. This time, however, we have a constant opposing pressure of 1.00 atm. When the pin is pulled back and the gas expands, it has to push against this opposing pressure and it therefore does work. The amount of work is equal to $P \Delta V$. Since the final volume is 10.0 liters, ΔV equals 9.0 liters. Therefore,

$$w = P \Delta V = (1.00 \text{ atm})(9.0 \text{ liters}) \qquad w = 9.0 \text{ liter atm}$$

opposing pressure

(This is 912 joules or 218 calories.)

This 9.0 liter atm of work done by the gas is equal to 9.0 liter atm of heat energy that the gas absorbed from the surroundings.

The two expansions described by Figures 11.2 and 11.3 have the same ΔE and show that the values of q and w depend on how a change is carried out. But what has this got to do with chemistry? How does it apply to a chemical system? To answer this question, let's consider an ordinary automobile battery in which a redox reaction provides the source of electricity. What we discover here is that the energy that we get in the form of work depends on how we discharge the battery. If we simply take a steel bar and place it across the terminals, sparks fly and the battery gets very hot—all the energy produced by the chemical reaction inside the battery appears as heat and we've done no work. On the other hand, if we connect the battery to an electric motor, like the starter of a car, we can use some of the energy of the reaction in the battery to crank the engine. This rotates parts and moves pistons—work is accomplished because the motor offers some resistance to the flow of electricity. It is as if the motor supplied an opposing "pressure" against which the electrical "pressure" could work.

What is particularly interesting is that if very careful measurements are made, we find that the amount of work that can be done by a battery depends on how fast it is discharged. If we discharge it quickly by offering little resistance to the flow of electricity, heat is generated that can't be used as work. As we increase the resistance, we slow down the rate at which energy is removed, but more of it can appear as work. The limit of this is where the energy is withdrawn infinitely slowly by having the battery just barely able to overcome the opposing resistance—in this case the work done would be a maximum. Any change such as this where the driving force is virtually balanced by an opposing force is called a **reversible process.** This is because increasing the opposing force just slightly would reverse the direction of the change. An example in a physical system is the piston-cylinder apparatus in Figure 11.4. As the water evaporates one molecule at a time, the pressure opposing the expansion is very gradually decreased to match the internal pressure of the gas. The internal pressure, which is the driving force for the expansion, is continually balanced by the opposing pressure, and if a molecule of water vapor were to condense, the pressure would increase and a slight compression of the gas would occur.

A very important concept, which we will return to later, is that *the maximum work available from any change is obtained if the change occurs by a reversible process.* Unfortunately, reversible processes like the infinitely slow discharge of a battery take forever to occur. All real, spontaneous changes do not take place by a reversible process and the work that can be derived from them is always less than the theoretical maximum.

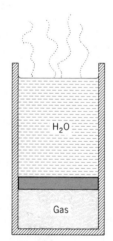

Figure 11.4

A reversible expansion of a gas. As water molecules gradually evaporate, the pressure decreases and the gas expands. The process will be reversed if water molecules begin to condense into the liquid instead of evaporate.

11.3 HEATS OF REACTION: THERMOCHEMISTRY

Virtually every chemical reaction occurs with either the absorption or evolution of energy. These changes reflect the differences between the potential energies

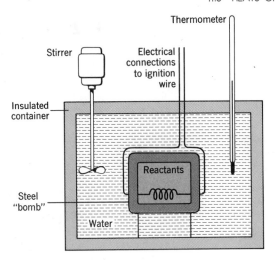

Figure 11.5
Bomb calorimeter.

associated with the bonds in the reactants and the products. For instance, when two hydrogen atoms come together to form an H_2 molecule, energy is evolved because the total potential energy of the nuclei and electrons in the H_2 molecule is lower than the total potential energy of these particles in two isolated H atoms. This same energy, when added to an H_2 molecule, can break it apart, and thus represents the bond energy discussed in Chapter 4. We see, therefore, that the measurement of the amount of energy evolved or absorbed when a chemical reaction occurs has the potential of providing us with very fundamental information concerning the stability of molecules and the strengths of chemical bonds.

If we carry out a chemical reaction in a closed container of fixed volume, the system undergoing reaction cannot perform pressure-volume work on the surroundings because $\Delta V = 0$; hence $P \Delta V = 0$. Any heat absorbed or evolved under these circumstances (let's call it q_v) is precisely equal to the change in the internal energy of the system,

$$\Delta E = q_v \qquad [11.3]$$

Stated another way, ΔE is equal to the heat absorbed or evolved by the system under conditions of constant volume (i.e., the *heat of reaction at constant volume*). When the reaction is endothermic, both q and ΔE are positive; for an exothermic process q and ΔE are negative.

Experimentally, ΔE can be measured by using a device called a **bomb calorimeter** (Figure 11.5). The apparatus consists of a strong steel "bomb" into which the reactants (for example, H_2 and O_2) are placed. The bomb is then immersed in an insulated bath containing a precisely known quantity of water. There the reaction is set off by a small heater wire within the bomb, and heat is evolved. The entire system is permitted to come to thermal equilibrium, at which point the calorimeter (bomb and water) will be at a higher temperature than before reaction. By carefully measuring the temperature of the water before and after reaction, and by knowing the heat capacity of the calorimeter (including the bomb and the water), the quantity of heat evolved by the chemical reaction can be computed.

A bomb calorimeter is normally used to measure ΔE for exothermic reactions.

EXAMPLE 11.2   Hydrogen (0.100 g) and oxygen (0.800 g) are compressed into a 1.00-liter bomb, which is then placed into the water in a calorimeter. Before the reaction is set off, the temperature of the water is 25.000°C; after reaction the temperature rises to 25.155°C. The heat capacity of the calorimeter (bomb, water, etc.) is 21,700 cal/°C. What is ΔE, in kilocalories, for this reaction?

SOLUTION   The change in temperature that occurs is 0.155°C. From the heat capacity we can find the number of calories evolved.

$$\text{Heat evolved, } q = \left(21,700 \, \frac{\text{cal}}{°C}\right)(0.155°C)$$

Therefore

$$q_v = -3360 \text{ cal}$$

Hence,

$$\Delta E = -3.36 \text{ kcal}$$

Notice that q_v and ΔE are both negative because the system—the H_2 and O_2—evolved energy when they reacted.

In Example 11.2, the magnitude of ΔE depends on the quantity of H_2 and O_2 reacted; that is, ΔE is an extensive quantity. We can convert this to an intensive property—one that is characteristic of the reaction between any amounts of H_2 and O_2—by calculating the heat evolved *per mole* of product formed. In Example 11.2 we produced 0.0500 mol of H_2O. Therefore, we say that $\Delta E = -3.36$ kcal/0.0500 mol of H_2O, or $\Delta E = -67.2$ kcal/mol.

In the past it was common practice always to express heats of reaction in calories or kilocalories. Since the acceptance of the SI system, joules and kilojoules are preferred. At the present time science is in the midst of a transition between these units, and therefore to provide familiarity with both we will work out some examples in joules, others in calories. Regardless of the units, the methods employed in solving the problems are the same. By now you should be able to convert readily between joules and calories.

The heat evolved at constant volume permits us to compute ΔE. However, most changes that are of practical interest to us take place in open containers at essentially constant atmospheric pressure. Under these conditions rather sizable volume changes can occur. For example, when 2 mol of gaseous H_2 react with 1 mol of gaseous O_2 to produce 2 mol of liquid water, at a constant pressure of 1 atm, the volume changes from about 67 liters to 0.036 liter. If we imagine this change to take place in a cylinder with a piston exerting a constant pressure of 1 atm, then as a reaction proceeds to completion, the surroundings perform work on the system, the magnitude of which is the $P \, \Delta V$ product of very nearly 67 liter atm (6.8 kJ or 1.6 kcal).

To avoid the necessity of considering PV work when heats of reaction are measured at constant pressure, we define a new thermodynamic function called the **heat content,** or **enthalpy,** H. This is given as

$$H = E + PV \qquad \qquad [11.4]$$

For a change at constant pressure,

$$\Delta H = \Delta E + P \, \Delta V \qquad \qquad [11.5]$$

If only PV work is involved in the change, we know that

$$\Delta E = q - P \, \Delta V$$

Substituting this into Equation 11.5, we have

$$\Delta H = (q - P \, \Delta V) + P \, \Delta V$$

$$\Delta H = q_p \qquad \qquad [11.6]$$

Remember: 1 cal = 4.184 J.

Enthalpy from the German word enthalen, meaning to contain.

Thus we see that ΔH is equal to the heat, q_p, absorbed or evolved at constant pressure.

The enthalpy of a system is a state function, just like the internal energy, so the magnitude of ΔH depends only on the enthalpies of the initial and final states. Thus we can write

$$\Delta H = H_{\text{final}} - H_{\text{initial}}$$

Here we use the same symbolism as in our earlier discussions of ΔH_{vap}, ΔH_{fus}, and so on. Those quantities, in fact, correspond to the enthalpy changes associated with vaporization, fusion, and so forth.

In many instances the differences between ΔH and ΔE are small, particularly for chemical reactions. When a reaction occurs in which all the reactants and products are liquids or solids, only very small changes in volume take place. As a result, $P \Delta V$ is very small and ΔH has very nearly the same magnitude as ΔE. When chemical reactions occur in which gases are either consumed or produced, much larger volume changes occur, and the $P \Delta V$ product is also much greater. Even in these cases, however, ΔE is usually so large compared to the $P \Delta V$ term that ΔE and ΔH are still nearly the same. We see this in Example 11.3.

EXAMPLE 11.3

When 2.00 mol of H_2 and 1.00 mol of O_2 at 100°C and 1 atm, react to produce 2.00 mol of gaseous water at 100°C and 1 atm, a total of 484.5 kJ are evolved. What are (a) ΔH, and (b) ΔE for the production of a single mole of H_2O (g)?

SOLUTION

(a) Since the reaction

$$2H_2\ (g) + O_2\ (g) \longrightarrow 2H_2O\ (g)$$

is occurring at constant pressure,

$$q_p = \Delta H = \frac{-484.5 \text{ kJ}}{2 \text{ mol } H_2O}$$

The minus sign, remember, signifies that the reaction is exothermic. For the production of 1 mol of water,

$$\Delta H = -242.2 \text{ kJ/mol}$$

(b) To calculate ΔE from ΔH, we have to use Equation 11.5. Solving for ΔE we have

$$\Delta E = \Delta H - P \Delta V$$

Thus we have to subtract the $P \Delta V$ product from ΔH. For simplicity, we will assume the gases to be ideal (this causes little error). Then, taking V_i and n_i to be the initial volume and initial number of moles of reactants at a given P and T,

$$PV_i = n_iRT$$

Similarly, for the final state (the products)

$$PV_f = n_fRT$$

The pressure-volume work in the process is given by

$$PV_f - PV_i = P(V_f - V_i) = P \Delta V$$

This is equal to

$$P \Delta V = n_fRT - n_iRT$$

$$P \Delta V = (n_f - n_i)RT = (\Delta n)RT$$

We are interpreting the equation on a mole basis.

The quantity Δn = (number of moles of gaseous products) − (number of moles of gaseous reactants). Using the coefficients in the chemical equation in part (a), there are two moles of gaseous products and three moles of gaseous reactants. Therefore,

$$\Delta n = 2.00 \text{ mol} - 3.00 \text{ mol} = -1.00 \text{ mol}$$

Next, taking $R = 8.31$ J/mol K to obtain the desired energy units,

$$P \, \Delta V = (-1.00 \text{ mol}) \, (8.31 \text{ J/mol K}) \, (373 \text{ K})$$

$$P \, \Delta V = -3100 \text{ J} = -3.10 \text{ kJ}$$

What we have calculated, based on the chemical equation, is the $P \, \Delta V$ work for the formation of 2 mol of $H_2O(g)$. For one mole, $P \, \Delta V = -1.55$ kJ. Now we can calculate ΔE in kJ/mol.

$$\Delta E = \Delta H - P \, \Delta V$$

$$\Delta E = -242.2 \text{ kJ/mol} - (-1.55 \text{ kJ/mol})$$

$$\Delta E = -240.6 \text{ kJ/mol}$$

Notice that ΔH and ΔE are really not very different, even in this reaction involving a change in the number of moles of gas. They differ by only about 0.6%.

In this last example we see that the PV work involved in a chemical reaction in which gases are either consumed or produced can be calculated by the simple expression

$$\text{pressure-volume work} = (\Delta n)RT \qquad [11.7]$$

Remember that Δn *is the change in the number of moles of* **gas** *on going from reactants to products.* Equation 11.7 does not apply to reactions in which only liquids or solids are involved. For such reactions the volume changes are extremely small and the $P \, \Delta V$ work is usually negligible compared to other energy changes that take place. For such reactions ΔE and ΔH are therefore essentially identical.

11.4 HESS'S LAW OF HEAT SUMMATION

Since enthalpy is a state function, the magnitude of ΔH for a chemical reaction does not depend on the path taken by the reactants as they proceed to form the products. Let's consider, for example, the conversion of 1 mol of liquid water at 100°C and 1 atm to 1 mol of vapor at 100°C and 1 atm. This process absorbs 9.7 kcal of heat for each mole of H_2O vaporized and, hence, $\Delta H = +9.7$ kcal. We can represent this "reaction" as

$$H_2O \, (l) \longrightarrow H_2O \, (g) \qquad \Delta H = +9.7 \text{ kcal}$$

An equation written in this manner, in which the energy change is also shown, is called a **thermochemical equation** and is nearly always interpreted on a mole basis. Here, for instance, we see that 1 mol of $H_2O \, (l)$ is converted to 1 mol of $H_2O \, (g)$ by the absorption of 9.7 kcal.

The value of ΔH for this process will always be $+9.7$ kcal, provided that we refer to the same pair of initial and final states. We could even go so far as to first decompose the 1 mol of liquid into gaseous hydrogen and oxygen and then recombine the elements to produce $H_2O \, (g)$ at 100°C and 1 atm. The net change in enthalpy would still be the same, $+9.7$ kcal. Consequently, it is possible to look at some overall change as the net result of a sequence of chemical reactions. The net value of ΔH for the overall process is merely the sum of all the enthalpy changes that take place along the way. These last statements constitute **Hess's law of heat summation.**

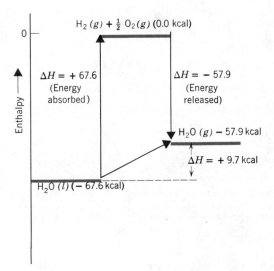

Figure 11.6

Enthalpy diagram for the reaction H_2O (l) $\rightarrow$ H_2O (g).

Thermochemical equations serve as a useful tool for applying Hess's law. For example, the thermochemical equations that correspond to the indirect path just described for the vaporization of water are[2]

$$H_2O \ (l) \longrightarrow H_2 \ (g) + \tfrac{1}{2}O_2 \ (g) \quad \Delta H = +67.6 \text{ kcal}$$

$$H_2 \ (g) + \tfrac{1}{2}O_2 \ (g) \longrightarrow H_2O \ (g) \qquad \quad \Delta H = -57.9 \text{ kcal}$$

These equations tell us that 67.6 kcal are required to decompose 1 mol of H_2O (l) into its elements and that 57.9 kcal are evolved when they recombine to produce 1 mol of H_2O (g). The sum of the two equations, after cancelling quantities that appear on both sides of the arrow, gives us the equation for the vaporization of 1 mol of water,

$$H_2O \ (l) + \cancel{H_2 \ (g)} + \cancel{\tfrac{1}{2}O_2 \ (g)} \longrightarrow H_2O \ (g) + \cancel{H_2 \ (g)} + \cancel{\tfrac{1}{2}O_2 \ (g)}$$

or

$$H_2O \ (l) \longrightarrow H_2O \ (g)$$

We also find that the heat of the overall reaction is equal to the algebraic sum of the heats of reaction for the two steps.

$$\Delta H = +67.6 \text{ kcal} + (-57.9 \text{ kcal})$$

$$\Delta H = +9.7 \text{ kcal}$$

Thus, *when we add thermochemical equations to obtain some net change, we also add their corresponding heats of reaction.*

To describe the nature of these thermochemical changes, we can also illustrate them graphically (Figure 11.6). This type of figure is frequently called an **enthalpy diagram.** Notice that we have chosen the enthalpy of the free elements as the zero point on the energy scale. This choice is entirely arbitrary because we are interested only in determining differences in *H*. In fact, we have no way at all of knowing absolute enthalpies, just as we have no way of knowing absolute internal energies. We can only measure ΔH. We used diagrams similar to Figure 11.6 in the last chapter in our discussion of heats of solution, so the concept of Hess's law should already be familiar.

[2] Note that fractional coefficients are permitted in thermochemical equations. This is because a coefficient such as $\tfrac{1}{2}$ is taken to mean $\tfrac{1}{2}$ mol. In ordinary equations fractional coefficients are avoided because they are meaningless on a molecular level. One cannot have half an atom or molecule and still retain the chemical identity of the species.

EXAMPLE 11.4

The thermochemical equation for the combustion of acetylene, a fuel used in torches, is given by Equation (1).

(1) $2C_2H_2 (g) + 5O_2 (g) \longrightarrow 4CO_2 (g) + 2H_2O (l)$ $\Delta H_1 = -2602$ kJ

Ethane, another hydrocarbon fuel, reacts as follows:

(2) $2C_2H_6 + 7O_2 (g) \longrightarrow 4CO_2 (g) + 6H_2O (l)$ $\Delta H_2 = -3123$ kJ

and hydrogen and oxygen combine following this equation

(3) $H_2 (g) + \frac{1}{2}O_2 (g) \longrightarrow H_2O (l)$ $\Delta H_3 = -286$ kJ

We've numbered the ΔH's simply for ease of identification in the solution of the problem.

All these data correspond to the same temperature and pressure: 25°C and 1 atm. Use these thermochemical equations to calculate the heat of hydrogenation of acetylene,

(4) $C_2H_2 (g) + 2H_2 (g) \longrightarrow C_2H_6 (g)$ $\Delta H_4 = ?$

at 25°C and 1 atm.

SOLUTION

To solve this problem we must combine the given equations (1), (2), and (3) in such a way that when they are added, everything cancels except the formulas in the desired equation (4). This involves keeping an eye on the final equation as we rearrange the given equations so that they add up correctly. For example, we have $1C_2H_2 (g)$ on the left of equation (4), so we want to use equation (1) with its coefficients divided by 2 (this gives equation 5). We must also divide its ΔH by 2 because there are now only half as many moles involved. We also have $2H_2 (g)$ on the left, so we have to take equation (3) multiplied by 2, and we have to multiply its ΔH by 2 (this gives equation 6). Finally, we have $1C_2H_6 (g)$ on the right. To get it there, we must reverse equation (2) and divide its coefficients by 2 (equation 7). When we reverse an equation, we change the sign of its ΔH because it changes an exothermic reaction into an endothermic one. In this case, we also must divide it by 2. This gives us, therefore,

(5) $C_2H_2 (g) + \frac{5}{2}O_2 (g) \longrightarrow 2CO_2 (g) + H_2O (l)$ $\Delta H_5 = -\dfrac{2602 \text{ kJ}}{2} = -1301$ kJ

(6) $2H_2 (g) + O_2 (g) \longrightarrow 2H_2O (l)$ $\Delta H_6 = 2(-286 \text{ kJ}) = -572$ kJ

(7) $2CO_2 (g) + 3H_2O (l) \longrightarrow C_2H_6 (g) + \frac{7}{2}O_2 (g)$ $\Delta H_7 = +\dfrac{3123 \text{ kJ}}{2} = +1561$ kJ

Adding equations (5), (6), and (7) gives

On the left, $\frac{5}{2}O_2 + O_2$ gives $\frac{7}{2}O_2$.

$C_2H_2 (g) + 2H_2 (g) + \frac{7}{2}O_2 (g) + 2CO_2 (g) + 3H_2O (l) \longrightarrow$
$\qquad\qquad\qquad\qquad 2CO_2 (g) + 3H_2O (l) + C_2H_6 (g) + \frac{7}{2}O_2 (g)$

Canceling elements that are the same on both sides, we get

$$C_2H_2 (g) + 2H_2 (g) \longrightarrow C_2H_6 (g)$$

which is the equation we want. Since it is obtained by adding equations (5), (6) and (7), its ΔH (ΔH_4 in the statement of the problem) is obtained by adding the ΔH's of (5), (6) and (7).

$$\Delta H_4 = \Delta H_5 + \Delta H_6 + \Delta H_7$$

$$\Delta H_4 = (-1301 \text{ kJ}) + (-572 \text{ kJ}) + (1561 \text{ kJ})$$

$$= -312 \text{ kJ}$$

The heat of hydrogenation of acetylene is therefore -312 kJ.

Heats of Formation

A particularly useful type of thermochemical equation corresponds to the formation of one mole of a substance from its elements. The enthalpy changes associated with these reactions are called **heats of formation** or **enthalpies of formation** and are denoted as ΔH_f. For example, thermochemical equations for the formation of liquid and gaseous water at 100°C and 1 atm are, respectively,

$$H_2 \ (g) + \tfrac{1}{2}O_2 \ (g) \longrightarrow H_2O \ (l) \qquad \Delta H_f = -67.6 \ \text{kcal/mol}$$

$$H_2 \ (g) + \tfrac{1}{2}O_2 \ (g) \longrightarrow H_2O \ (g) \qquad \Delta H_f = -57.9 \ \text{kcal/mol}$$

How can we use these equations to obtain the heat of vaporization of water? Clearly, we must reverse the first equation and then add it to the second. When we reverse this equation, we must also remember to change the sign of ΔH. If the formation of H_2O (l) is exothermic, as indicated by a negative ΔH_f, the reverse process must be endothermic.

(Exothermic) $\qquad H_2 \ (g) + \tfrac{1}{2}O_2 \ (g) \longrightarrow H_2O \ (l) \qquad \Delta H = \Delta H_f = -67.6 \ \text{kcal}$

(Endothermic) $\qquad H_2O \ (l) \longrightarrow H_2 \ (g) + \tfrac{1}{2}O_2 \ (g) \qquad \Delta H = -\Delta H_f = +67.6 \ \text{kcal}$

When this last equation is added to that for the formation of H_2O (g), we obtain

$$H_2O \ (l) \longrightarrow H_2O \ (g)$$

and the heat of reaction is

$$\Delta H = \Delta H_{f\,H_2O\,(g)} - \Delta H_{f\,H_2O\,(l)}$$

$$\Delta H = -57.9 \ \text{kcal} - (-67.6 \ \text{kcal}) = +9.7 \ \text{kcal}$$

We will see that Equation 11.8 is a particularly useful form of Hess's law.

Notice that the heat of reaction for the overall change is equal to the heat of formation of the product *minus* the heat of formation of the reactant. In general, we can write that for any overall reaction

$$\Delta H_{\text{reaction}} = (\text{sum of } \Delta H_f \text{ of products}) - (\text{sum of } \Delta H_f \text{ of reactants}) \quad [11.8]$$

11.5 STANDARD STATES

The magnitude of ΔH_f depends on the conditions of temperature, pressure, and the physical state (gas, liquid, solid, crystalline form) of the reactants and products. For instance, at 100°C and 1 atm, the heat of formation of liquid water is −67.6 kcal/mol, while at 25°C and 1 atm, ΔH_f for H_2O (l) is −68.3 kcal/mol. To avoid the necessity of always having to specify the conditions for which ΔH_f is recorded, and to permit comparisons between ΔH_f for various compounds, a standard set of conditions is chosen, usually 25°C and a pressure of 1 atm.[3] Under these conditions a substance is said to be in its **standard state.** Heats of formation of substances in their standard states are indicated as ΔH_f^0. For example, the standard heat of formation of liquid water, $\Delta H_{f\,H_2O\,(l)}^0 = -68.3$ kcal/mol, and represents the heat liberated when H_2 and O_2, each in their natural form at 25°C and 1 atm, react to produce H_2O (l) at 25°C and 1 atm.

We know energy is liberated because ΔH^0 is negative.

Table 11.1 contains standard heats of formation for a variety of different substances. Such a table is very useful because it permits us to use Equation 11.8 to calculate the standard heats of reaction, ΔH^0, for a very large number of different chemical changes. In performing these calculations, we arbitrarily take the ΔH_f^0 for an element in its natural, most stable form at 25°C and 1 atm to be equal to zero. Thus, in our computations, the zero point on the energy scale is again chosen to be that of the free elements. As mentioned before, because we speak only of *changes* in energy, the actual location of this zero point is unimportant. The following examples illustrate how the principles developed in the preceding two sections can be applied.

[3] Note that the choice of the standard temperature here differs from the standard temperature of 0°C used in calculations involving gases in Chapter 7.

Table 11.1
Standard heats of formation of some substances at 25°C and 1 atm

Substance	ΔH_f^0 kJ/mol	ΔH_f^0 kcal/mol	Substance	ΔH_f^0 kJ/mol	ΔH_f^0 kcal/mol
Al_2O_3 (s)	−1676	−400.5	HBr (g)	−36	−8.7
Br_2 (l)	0.00	0.00	HI (g)	+26	+6.3
Br_2 (g)	+30.9	+7.39	KCl (s)	−436.0	−104.2
C (s, diamond)	+1.88	+0.45	LiCl (s)	−408.8	−97.7
CO (g)	−110	−26.4	$MgCl_2$ (s)	−641.8	−153.4
CO_2 (g)	−394	−94.1	$MgCl_2 \cdot 2H_2O$ (s)	−1280	−306.0
CH_4 (g)	−74.9	−17.9	$Mg(OH)_2$ (s)	−924.7	−221.0
C_2H_6 (g)	−84.5	−20.2	NH_3 (g)	−46.0	−11.0
C_2H_4 (g)	+51.9	+12.4	N_2O (g)	+81.5	+19.5
C_2H_2 (g)	+227	+54.2	NO (g)	+90.4	+21.6
C_3H_8 (g)	−104	−24.8	NO_2 (g)	+34	+8.1
C_6H_6 (l)	+49.0	−11.7	NaF (s)	−571	−136.5
CH_3OH (l)	−238	−57.0	NaCl (s)	−413	−98.6
HCOOH (g)	−363	−86.7	NaBr (s)	−360	−86.0
CS_2 (l)	+89.5	+21.4	NaI (s)	−288	−68.8
CS_2 (g)	+117	+28.0	Na_2O_2 (s)	−504.6	−120.6
CCl_4 (l)	−134	−32.1	NaOH (s)	−426.8	−102.0
C_2H_5OH (l)	−278	−66.4	$NaHCO_3$ (s)	−947.7	−226.5
CH_3CHO (g)	−167	−39.8	Na_2CO_3 (s)	−1131	−270.3
CH_3COOH (l)	−487.0	−116.4	O_3 (g)	+143	+34.1
CaO (s)	−635.5	−151.9	PbO_2 (s)	−277	−66.3
$Ca(OH)_2$ (s)	−986.6	−235.8	$PbSO_4$ (s)	−920.1	−219.9
$CaSO_4$ (s)	−1433	−342.4	SO_2 (g)	−297	−70.9
CuO (s)	−155	−37.1	SO_3 (g)	−396	−94.6
Fe_2O_3 (s)	−822.2	−196.5	H_2SO_4 (l)	−813.8	−194.5
H_2O (l)	−286	−68.3	SiO_2 (s)	−910.9	−217.7
H_2O (g)	−242	−57.8	SiH_4 (g)	+34	+8.2
HF (g)	−271	−64.8	ZnO (s)	−348	−83.2
HCl (g)	−92.5	−22.1	Zn(OH) (s)	−642.2	−153.5

EXAMPLE 11.5

Many careful cooks keep sodium bicarbonate (baking soda) handy because it is a good extinguisher of oil or grease fires. Its decomposition products help smother the flames. The reaction is

$$2NaHCO_3 \ (s) \longrightarrow Na_2CO_3 \ (s) + H_2O \ (g) + CO_2 \ (g)$$

Calculate ΔH^0 for the reaction.

SOLUTION

Equation 11.8 implies that

$$\Delta H^0 = (\text{sum } \Delta H_f^0 \text{ products}) - (\text{sum } \Delta H_f^0 \text{ reactants})$$

This means that we must add up all the heat evolved during the formation of the products from their elements and then subtract the heat evolved by the formation of the reactants from their elements.

Using data from Table 11.1, the total enthalpy of formation of the products is

$$1 \text{ mol } Na_2CO_3 \text{ (s)} \times \left(\frac{-270.3 \text{ kcal}}{1 \text{ mol } Na_2CO_3 \text{ (s)}} \right) = -270.3 \text{ kcal}$$

$$1 \text{ mol } H_2O \text{ (g)} \times \left(\frac{-57.8 \text{ kcal}}{1 \text{ mol } H_2O \text{ (g)}} \right) = -57.8 \text{ kcal}$$

$$1 \text{ mol } CO_2 \text{ (g)} \times \left(\frac{-94.1 \text{ kcal}}{1 \text{ mol } CO_2 \text{ (g)}} \right) = -94.1 \text{ kcal}$$

$$\text{total } \Delta H_f^0 \text{ of products} = -422.2 \text{ kcal}$$

For the reactant,

$$2 \text{ mol } NaHCO_3 \text{ (s)} \times \left(\frac{-226.5 \text{ kcal}}{1 \text{ mol } NaHCO_3 \text{ (s)}} \right) = -453.0 \text{ kcal}$$

We have said that

$$\Delta H^0 = (\text{sum } \Delta H_f^0 \text{ products}) - (\text{sum } \Delta H_f^0 \text{ reactants})$$

Therefore

$$\Delta H^0 = -422.2 \text{ kcal} - (-453.0 \text{ kcal})$$

or

$$\Delta H^0 = +30.8 \text{ kcal}$$

Notice that in computing ΔH^0 for the overall reaction we have multiplied each ΔH_f^0 by the appropriate coefficient from the equation. This gives the total heat of reaction for the numbers of moles specified by the chemical equation.

EXAMPLE 11.6   Calculate ΔH^0 for the reaction

$$2Na_2O_2 \text{ (s)} + 2H_2O \text{ (l)} \longrightarrow 4NaOH \text{ (s)} + O_2 \text{ (g)}$$

How many kilojoules are liberated when 25.0 g Na_2O_2 react according to this equation?

SOLUTION   We use Equation 11.8.

$$\Delta H^0 = [4\Delta H_{f\,NaOH\,(s)}^0 + \Delta H_{f\,O_2\,(g)}^0] - [2\Delta H_{f\,Na_2O_2\,(s)}^0 + 2\Delta H_{f\,H_2O\,(l)}^0]$$

All the data are available in Table 11.1 except $\Delta H_{f\,O_2\,(g)}^0$, but we know (or *should* know) that the heat of formation of a pure element is zero. Therefore,

$$\Delta H^0 = \left[4 \text{ mol} \times \left(\frac{-426.8 \text{ kJ}}{1 \text{ mol } NaOH} \right) + 0.00 \right]$$

$$- \left[2 \text{ mol} \times \left(\frac{-504.6 \text{ kJ}}{1 \text{ mol } Na_2O_2} \right) + 2 \text{ mol} \times \left(\frac{-286 \text{ kJ}}{1 \text{ mol } H_2O \text{ (l)}} \right) \right]$$

$$= [-1707 \text{ kJ}] - [-1581 \text{ kJ}]$$

$$= -126 \text{ kJ}$$

To calculate the number of kilojoules liberated by 25.0 g Na_2O_2, we have to realize that the ΔH^0 that we calculated is the energy given off when 2 mol Na_2O_2 react. Therefore,

$$2 \text{ mol } Na_2O_2 \sim -126 \text{ kJ}$$

The formula weight of Na_2O_2 is 78.0, so

$$25.0 \text{ g } Na_2O_2 \times \left(\frac{1 \text{ mol } Na_2O_2}{78.0 \text{ g } Na_2O_2}\right) \times \left(\frac{-126 \text{ kJ}}{2 \text{ mol } Na_2O_2}\right) \sim -20.2 \text{ kJ}$$

The reaction of 25.0 g Na_2O_2 liberates 20.2 kJ.

It is frequently impossible to directly measure the heat of formation of a compound. For example, we cannot get hydrogen, oxygen, and graphite (the most stable crystalline form of carbon) to react directly together to produce ethyl alcohol, C_2H_5OH. If we could, we would convert coal to alcohol and alleviate the energy crisis! Consequently, in order to determine ΔH_f^0 for substances like C_2H_5OH, an indirect method must be used. One technique, which can be applied to most organic materials, is to burn the substance in a calorimeter to produce products whose heats of formation are known, as shown in Example 11.7.

EXAMPLE 11.7   The combustion of 1 mol of benzene, C_6H_6 (l), to produce CO_2 (g) and H_2O (l) liberates 3271 kJ when the products are returned to 25°C and 1 atm. What is the standard heat of formation of C_6H_6 (l) expressed in kilojoules per mole?

SOLUTION   The equation for the combustion of 1 mol of C_6H_6 is

$$C_6H_6 \text{ } (l) + 7\tfrac{1}{2}O_2 \text{ } (g) \longrightarrow 6CO_2 \text{ } (g) + 3H_2O \text{ } (l)$$

The standard heat of reaction, $\Delta H^0 = 3271$ kJ. From Equation 11.8, we know that

$$\Delta H^0 = [6\Delta H^0_{f \text{ } CO_2 (g)} + 3\Delta H^0_{f \text{ } H_2O (l)}] - [\Delta H^0_{f \text{ } C_6H_6 (l)}]$$

Solving for the heat of formation of benzene,

$$\Delta H^0_{f \text{ } C_6H_6 (l)} = 6\Delta H^0_{f \text{ } CO_2 (g)} + 3\Delta H^0_{f \text{ } H_2O (l)} - \Delta H^0$$

From Table 11.1 we can obtain the heats of formation of CO_2 and H_2O. Therefore,

$$\Delta H^0_{f \text{ } C_6H_6 (l)} = 6(-394) \text{ kJ} + 3(-286) \text{ kJ} - (-3271 \text{ kJ})$$

$$\Delta H^0_{f \text{ } C_6H_6 (l)} = +49 \text{ kJ}$$

Since 1 mol of C_6H_6 is involved,

$$\Delta H_f^0 = +49 \text{ kJ/mol}$$

11.6   BOND ENERGIES

We stated earlier that it should be possible to relate heats of reaction to changes in the potential energy associated with breaking and forming chemical bonds. Strictly speaking, we should use ΔE for this purpose; however, since $P \Delta V$ contributions to ΔH are relatively small for chemical reactions, we can use ΔH in place of ΔE and still expect to obtain quite reasonable results. Therefore, we will use the terms bond energy and bond enthalpy interchangeably.

In Chapter 4 the bond energy was defined as the energy required to break a bond to produce neutral fragments. For a complex molecule, the energy needed to reduce the gaseous molecule to neutral gaseous atoms is called the **atomization energy.** It is the sum of all the bond energies in the molecule. Simple diatomic molecules, such as H_2, O_2, Cl_2, or HCl, possess only one bond; therefore the atomization energy is the same as the bond energy. For these simple cases

Table 11.2
Heats of formation of
gaseous atoms from the
elements in their standard
states

Atom	ΔH_f^0 per mole of atoms	
	kJ/mol	kcal/mol
H	218	52.1
Li	161	38.4
Be	327	78.2
B	555	132.6
C	715	170.9
N	473	113.0
O	249	59.6
F	79.1	18.9
Na	108	25.8
Si	454	108.4
Cl	121	28.9
Br	112	26.7
I	107	25.5

the atomization energy can be obtained by studying the spectra produced when these molecules absorb or emit light. For more complex molecules, however, we employ an indirect method that makes use of measured heats of formation.

As an example, let's consider the molecule CH_4. Using the same technique that was followed in Example 11.7, the standard heat of formation of CH_4 (g) can be determined experimentally to be −17.9 kcal/mol. This corresponds to the enthalpy change, ΔH_f^0, for the reaction,

$$C \ (s, \text{ graphite}) + 2H_2 \ (g) \longrightarrow CH_4 \ (g)$$

We can envision an alternative path to take us from the free elements to the compound, methane, that follows a series of successive reactions,

$$(1) \qquad\qquad C \ (s, \text{ graphite}) \longrightarrow C \ (g) \qquad \Delta H_1$$

$$(2) \qquad\qquad 2H_2 \ (g) \longrightarrow 4H \ (g) \qquad \Delta H_2$$

$$(3) \qquad\qquad C \ (g) + 4H \ (g) \longrightarrow CH_4 \ (g) \qquad \Delta H_3$$

The sum of these three will give us our desired overall reaction.

Steps 1 and 2 each involve the heat of formation of gaseous atoms from an element in its standard state. For hydrogen the heat of formation of each mole of gaseous hydrogen atoms is half the atomization energy of H_2 (g). For carbon it amounts to the sublimation energy of graphite. Table 11.2 contains some heats of formation of gaseous atoms from typical elements in their standard states.

Applying Hess's law, we know that we can obtain ΔH for the overall reaction—that is, ΔH_f^0 for CH_4 (g)—by adding up the enthalpy changes for each step.

$$\Delta H_f^0 = \Delta H_1 + \Delta H_2 + \Delta H_3 \qquad\qquad [11.9]$$

From the data in Table 11.2, we see that ΔH_1 = +170.9 kcal—the heat of formation of gaseous carbon atoms. Similarly, ΔH_2 = 4(+52.1 kcal), that is, four times the heat of formation of 1 mol of H (g). The quantity ΔH_3 is the negative of the atomization energy of CH_4 (g).

$$\Delta H_3 = -\Delta H_{\text{atom}} \text{ for } CH_4 \ (g)$$

Solving for the atomization energy of methane in Equation 11.9 gives

$$\Delta H_{\text{atom}} = \Delta H_1 + \Delta H_2 - \Delta H_f^0$$

Substituting numerical values,

$$\Delta H_{\text{atom}} = (+170.9 + 208.4 + 17.9) \text{ kcal}$$

$$\Delta H_{\text{atom}} = +397.2 \text{ kcal}$$

This quantity is the total amount of energy that must be absorbed to break all 4 mol of C—H bonds in 1 mol of CH_4. Division by 4 provides us with an average bond energy of 99.3 kcal/mol of C—H bonds. This value, along with some other bond energies, appears in Table 11.3.

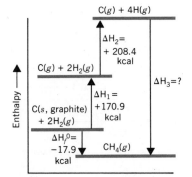

EXAMPLE 11.8   The heat of formation of ethylene, C_2H_4, is +51.9 kJ/mol. The structure of the molecule is

$$\begin{array}{ccc} H & & H \\ \diagdown & & \diagup \\ & C = C & \\ \diagup & & \diagdown \\ H & & H \end{array}$$

Assuming the C—H bond energy to be 415 kJ/mol, calculate the C=C bond energy.

SOLUTION

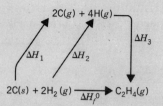

$$2C(g) + 4H(g)$$

ΔH_1 ΔH_2 ΔH_3

$$2C(s) + 2H_2(g) \xrightarrow{\Delta H_f^0} C_2H_4(g)$$

Figure 11.7
Alternative paths for the formation of ethylene from its elements in their standard states.

Let's examine the two alternative paths shown in Figure 11.7 that take us from the reactants [C (s) and H$_2$ (g)], to the product [C$_2$H$_4$ (g)]. The sum of the energy changes in steps 1, 2, and 3 must be the same as that for the direct path, ΔH_f^0.

$$\Delta H_1 + \Delta H_2 + \Delta H_3 = \Delta H_f^0$$

STEP 1. ΔH_1 is twice the heat of formation of a mole of carbon atoms.

$$\Delta H_1 = 2(715 \text{ kJ}) = 1430 \text{ kJ}$$

STEP 2. ΔH_2 is four times the heat of formation of a mole of hydrogen atoms.

$$\Delta H_2 = 4(218 \text{ kJ}) = 872 \text{ kJ}$$

STEP 3. ΔH_3 is the negative of the atomization energy.

$$\Delta H_3 = -\Delta H_{atom}$$

Therefore,

$$1430 \text{ kJ} + 872 \text{ kJ} + (-\Delta H_{atom}) = +51.9 \text{ kJ}$$

Solving for ΔH_{atom}

$$\Delta H_{atom} = 1430 \text{ kJ} + 872 \text{ kJ} - 51.9 \text{ kJ}$$
$$= 2250 \text{ kJ}$$

This is the energy required to break four moles of C—H bonds plus one mole of C=C bonds.

$$\Delta H_{atom} = 4\Delta H_{C-H} + \Delta H_{C=C}$$

Substituting and solving for $\Delta H_{C=C}$,

$$\Delta H_{C=C} = \Delta H_{atom} - 4\Delta H_{C-H}$$
$$= 2250 \text{ kJ} - 4(415 \text{ kJ})$$
$$= 590 \text{ kJ per mol of C=C bonds}$$

It's worth noting that this value is within 3% of the accepted average C=C bond energy of 607 kJ/mol.

Table 11.3
Average bond energies

Bond	Bond Energy		Bond	Bond Energy	
	kJ/mol	kcal/mol		kJ/mol	kcal/mol
H—C	415	99.3	C=O	724	173
H—O	463	110.6	C—N	292	69.7
H—N	391	93.4	C=N	619	148
H—F	563	134.6	C≡N	879	210
H—Cl	432	103.2	C—C	348	83.1
H—Br	366	87.5	C=C	607	145
H—I	299	71.4	C≡C	833	199
C—O	356	85.0			

A very important fact is that the average bond energies found in Table 11.3 can be used, in many cases, to compute heats of formation with a fair degree of accuracy, as illustrated in Example 11.9. It is very significant that a bond between two atoms has very nearly the same strength in one molecule as it does in another. This implies, for example, that nearly all C—H bonds are pretty much alike, whether in a small molecule such as CH_4 or in a large complex one such as $C_{42}H_{86}$. The same applies to many other bonds as well. This phenomenon has greatly simplified the development of the modern theories about chemical bonding that we discussed in Chapter 5.

EXAMPLE 11.9   Use the data in Tables 11.2 and 11.3 to compute the heat of formation of liquid ethyl alcohol in kilocalories per mole. This compound has a heat of vaporization, $\Delta H^0_{vap} = 9.4$ kcal/mol and the structural formula

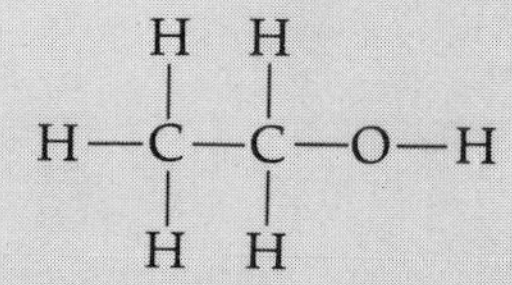

$$
\begin{array}{c}
\overset{\displaystyle H}{|}\ \ \overset{\displaystyle H}{|} \\
H-C-C-O-H \\
\underset{\displaystyle H}{|}\ \ \underset{\displaystyle H}{|}
\end{array}
$$

SOLUTION   We wish to determine ΔH^0 for the reaction,

$$2C\ (s,\ graphite) + 3H_2\ (g) + \tfrac{1}{2}O_2\ (g) \longrightarrow C_2H_5OH\ (l)$$

To compute ΔH_f, we follow the alternative path from reactants to products illustrated in Figure 11.8. Using the data in Table 11.2, we can compute the energy required to convert the reactants to gaseous atoms, that is, ΔH_A in Figure 11.8.

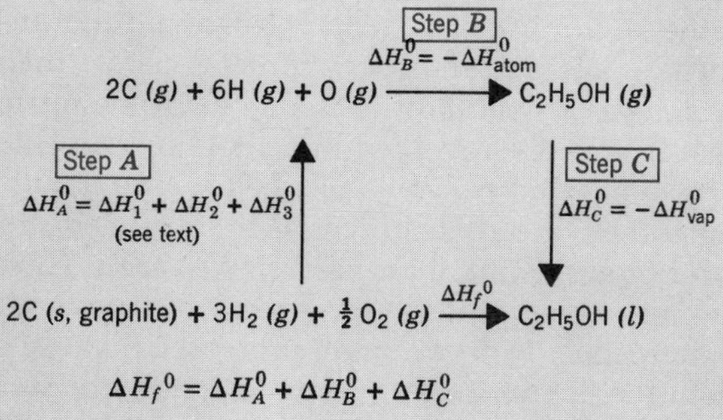

Figure 11.8

Alternative paths for the formation of liquid ethyl alcohol, C_2H_5OH (l).

$$
\begin{array}{ll}
2C\ (s,\ graphite) \longrightarrow 2C\ (g) & \Delta H_1{}^0 = 2\Delta H^0_{f\ C(g)} \\
 & \qquad\quad = 2(+170.9\ kcal) = +341.8\ kcal \\[6pt]
3H_2\ (g) \longrightarrow 6H\ (g) & \Delta H_2{}^0 = 6\Delta H^0_{f\ H(g)} \\
 & \qquad\quad = 6(+52.1\ kcal) = +312.6\ kcal \\[6pt]
\tfrac{1}{2}O_2\ (g) \longrightarrow O\ (g) & \Delta H_3{}^0 = \Delta H^0_{f\ O(g)} = +59.6\ kcal
\end{array}
$$

The total energy needed to give gaseous atoms, $\Delta H_A{}^0 = \Delta H_1{}^0 + \Delta H_2{}^0 + \Delta H_3{}^0 = +714.0$ kcal.

Next, we compute the energy liberated when these atoms combine to form 1 mol of gaseous C_2H_5OH. This is the negative of the atomization energy of C_2H_5OH, which involves five C—H bonds, one C—C bond, one C—O bond, and one O—H bond. The energies are obtained from Table 11.3.

5(C—H)	5(99.3 kcal)
1(C—C)	83.1 kcal
1(C—O)	85.0 kcal
1(O—H)	110.6 kcal
ΔH^0_{atom} for C_2H_5OH (g) =	775.2 kcal

Therefore, for step B, $\Delta H_B^0 = -775.2$ kcal

Finally, energy is liberated when C_2H_5OH (g) is condensed to a liquid. The ΔH^0 for this process is the negative of ΔH_{vap}^0. Therefore $\Delta H_C^0 = -9.4$ kcal.

Now we can compute ΔH_f^0 for C_2H_5OH (l) by adding the ΔH^0 values of each step in the alternative path.

$$\Delta H_f^0 = \Delta H_A^0 + \Delta H_B^0 + \Delta H_C^0$$

$$= +714.0 \text{ kcal} + (-775.2 \text{ kcal}) + (-9.4 \text{ kcal})$$

$$= -70.6 \text{ kcal}$$

Since the computation was performed for 1 mol, we can write

$$\Delta H_f^0 = -70.6 \text{ kcal/mol}$$

Comparing this to the value reported in Table 11.1,

$$\Delta H_f^0 = -66.4 \text{ kcal/mol}$$

we see that the agreement (within about 6%) is not really too bad, considering that we have assumed that each kind of bond has the same energy in *all* compounds.

11.7 SPONTANEITY OF CHEMICAL REACTIONS

At the beginning of this chapter we said that thermodynamics is able to tell us when a reaction can occur spontaneously—that is, without continued outside assistance. To see how this can be accomplished, let's begin by finding out what factors are involved in a spontaneous change. One spontaneous process that we have all observed is a ball rolling down a hill. When it finally comes to rest at the bottom, its potential energy has decreased and it is in a more stable, lower energy state than before. We might conclude, therefore, that a process leading to a decrease in the energy of a system (here, a ball) should tend to be spontaneous. Indeed, many processes that are spontaneous do occur with the evolution of energy. For example, a mixture of hydrogen and oxygen, when ignited, reacts very rapidly to produce water. This chemical change is accompanied by the release of a large quantity of heat, so much, in fact, that the hot water vapor produced expands explosively.

Exothermic changes like those occurring in this forest fire tend to be spontaneous.

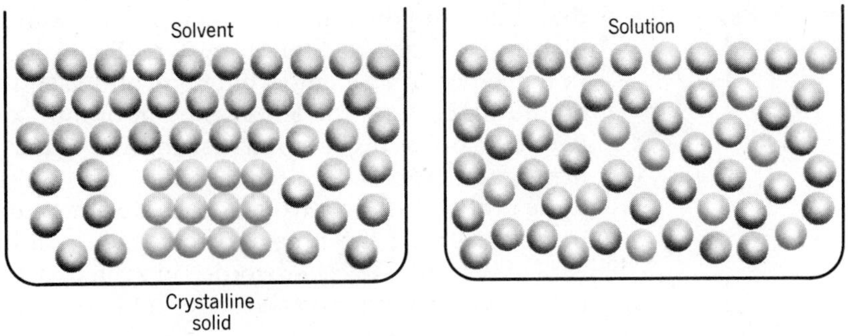

Solvent

Solution

Crystalline
solid

Figure 11.9

*An increase in disorder occurs
when a crystalline solid dissolves
to form a solution.*

Evolution of energy, however, is not the only criterion to be considered. There are many examples of changes that take place with the absorption of energy and yet are spontaneous. In the last chapter, we discussed heats of solution and saw that in many instances energy is absorbed when a salt dissolves in water. An example, you recall, is ammonium nitrate, the salt used in instant cold-packs. The formation of the solution, although endothermic, is nevertheless spontaneous. What, then, is the driving force for this process that is capable of outweighing the endothermic energy effect that occurs?

When a solid such as NH_4NO_3 dissolves in water, the particles of the solute leave the well-ordered crystalline state and gradually diffuse throughout the liquid to produce a solution. In this final state the particles of the solute are in a more disordered, random condition than they were before they dissolved, as shown in Figure 11.9. Similarly, the solvent is in a more random state in the solution because the solvent molecules are, in a sense, dispersed throughout those of the solute as well.

In any process there is a natural tendency or drive toward increased randomness, because a highly random distribution of particles represents a condition of higher statistical probability than an ordered one. To see the effects of statistical probability on spontaneity, let's suppose that we had 1 mol of a gas in the left compartment of the apparatus in Figure 11.10, and a vacuum in the other compartment. If the stopcock between them is opened, we know intuitively that the gas will rush into the right compartment, and that after a time each side will contain equal numbers of molecules, provided their volumes are the same. But why does this happen spontaneously?

To understand this, first imagine that there are only two molecules in the compartment on the left. When the stopcock is opened, both molecules are free

Figure 11.10

*(a) A gas expands spontaneously
to fill a vacuum. (b) Four ways
that two molecules can distribute
themselves in two compartments.*

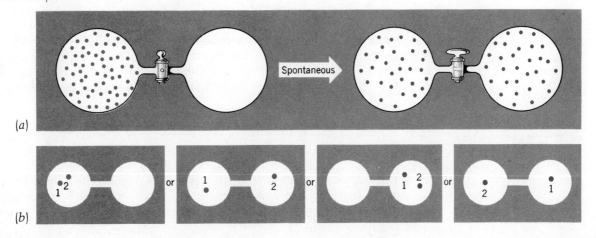

(a)

Spontaneous

(b)

to wander into the other bulb and, at this point, there are four different particle distributions possible for the system (Figure 11.10b). Since both particles reside in the left compartment in only one of these four distributions, the probability of finding both on the left is one out of four, or 1/4. Similarly, since there are two ways of having an even distribution of particles between the two compartments, the probability of an even distribution is two out of four, or 1/2. From these probabilities we would conclude that there will be an even distribution of particles 50% of the time and that there will be two particles on the left only 25% of the time. In other words, an even distribution is a more likely state than one in which both molecules are to be found in the compartment on the left.

When there are more particles in the system, the difference between the probabilities of these two types of molecular distributions is even more dramatic. With six particles, for instance, the probability of finding them all in one compartment is only 1/64; whereas, an even three-to-three distribution has a probability of 20/64.

In general, the probability of finding all the molecules in a gas in one compartment of this apparatus is $1/2^n$, where n is the number of particles. For 1 mol of gas the probability that all of the gas will be found in one bulb (that is, the probability that the gas will *not* spontaneously expand) is

$$ p = \frac{1}{2^{(6.02 \times 10^{23})}} $$

This can be shown to be the same as

$$ p = \frac{1}{10^{(1.81 \times 10^{23})}} = 10^{(-1.81 \times 10^{23})} $$

It is nearly impossible to appreciate how small this probability is. If we were to write this as a decimal fraction, it would be

$$ p = 0.\underbrace{0000 \ldots \ldots 00001}_{1.81 \times 10^{23} \text{ zeros}} $$

How fast can you write? At three zeros per second it would take about 19,000 billion centuries to write this number!

and if each zero were 0.10 inches in diameter, we would need a piece of paper 2,800,000,000,000,000,000 *miles* long just to write the number! This very small probability tells us that a state having all the molecules in one compartment cannot be maintained. The reason that the gas expands, therefore, is because it goes from a state of low probability (all the molecules on one side) to a state of higher probability corresponding to a nearly equal distribution. The final state is one of greater disorder, or randomness, because the particles of the gas are more widely distributed and have a greater degree of freedom of movement.

By way of summary, we see that there are two factors that influence the spontaneity of a physical or chemical change. One is the change in energy; the other is the change in randomness or disorder of the system. Any change that is accompanied by both a decrease in energy and an increase in randomness, like the collapse of the building shown in the photograph at the beginning of this chapter, is bound to occur spontaneously once it has begun.

11.8 ENTROPY

The degree of randomness in a system is represented by a thermodynamic quantity called the **entropy,** denoted by the symbol S; the greater the randomness, the greater the entropy. Entropy, like E and H, is also a state function,

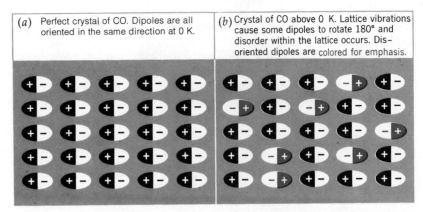

(a) Perfect crystal of CO. Dipoles are all oriented in the same direction at 0 K.	(b) Crystal of CO above 0 K. Lattice vibrations cause some dipoles to rotate 180° and disorder within the lattice occurs. Disoriented dipoles are colored for emphasis.

Figure 11.11
(a) *Perfect crystal of CO at 0 K.*
(b) *Crystal of CO above 0 K.*

which means that the magnitude of ΔS depends only on the entropies of the system in its initial and final states.

A change in entropy, given by ΔS, can be brought about by the addition of heat to a system. For example, consider a perfect crystal of carbon monoxide at absolute zero in which all the C—O dipoles are aligned in the same direction (Figure 11.11). Because of the perfect alignment of the dipoles there is essentially perfect order in the crystal, and the entropy of the system is at a minimum. When heat is added to this crystal, the temperature rises above 0 K and thermal motion (vibrations) within the lattice cause some of the dipoles to become oriented in the opposite direction (Figure 11.11*b*). As a result, there is less order (more disorder), and the entropy of the crystal has obviously increased. Logically, the more heat added to the system, the greater the extent of disorder afterwards. It should not be surprising, therefore, to find that the entropy change, ΔS, is directly proportional to the amount of heat added to the system. This heat is specified as q_{rev}—the amount of heat that would be added to the system if the change follows a reversible path.[4]

$$\Delta S \propto q_{rev}$$

The magnitude of ΔS is also *inversely* proportional to the temperature at which the heat is added. At low temperatures a given quantity of heat makes a large change in the relative degree of order. Near absolute zero, the addition of even a small amount of heat causes the system to go from essentially perfect order to some degree of randomness—a very substantial change, so ΔS is large. If the same amount of heat is added to the system at a high temperature, however, the system goes from an already highly random state to one just slightly more random. This constitutes only a very small change in the *relative* degree of disorder and, hence, to only a small entropy change. Thus, it can be shown that a change in entropy is finally given as

$$\Delta S = \frac{q_{rev}}{T} \qquad [11.10]$$

where T is the absolute temperature at which q_{rev} is transferred to the system. Note that entropy has units of *energy/temperature,* for example, calories per kelvin (cal/K) or joules per kelvin (J/K).

---

[4] In Section 11.2 we saw that, in general, q is not a state function. For a reversible change, however, q_{rev} does depend only on the initial and final states because the path (a reversible one) is clearly defined. ΔS can be a state function only if the value of q used to compute it is itself a quantity that depends only on the initial and final states, and therefore q_{rev} must be used.

11.9 THE SECOND LAW OF THERMODYNAMICS

The second law of thermodynamics provides us with a way of comparing the effects of the two driving forces involved in a spontaneous process—changes in energy and changes in entropy. One statement of the second law is that *in any spontaneous process there is always an increase in the entropy of the universe* (ΔS_{total} > 0). This increase takes into account entropy changes in both the system and its surroundings,

$$\Delta S_{total} = \Delta S_{system} + \Delta S_{surroundings}$$

The entropy change that occurs in the surroundings is brought about by the heat added to the surroundings divided by the temperature at which it is transferred. For a process at constant P and T, the heat added to the surroundings is equal to the *negative* of the heat added to the system, which is given by ΔH_{system}. Thus

$$q_{surroundings} = -\Delta H_{system}$$

The entropy change for the surroundings is therefore

$$\Delta S_{surroundings} = \frac{-\Delta H_{system}}{T}$$

and the total entropy change for the universe is

$$\Delta S_{total} = \Delta S_{system} - \frac{\Delta H_{system}}{T}$$

or

$$\Delta S_{total} = \frac{T \, \Delta S_{system} - \Delta H_{system}}{T}$$

This can be rearranged to give

$$T \, \Delta S_{total} = -(\Delta H_{system} - T \, \Delta S_{system})$$

T is positive and $\Delta\mathbf{S}_{TOTAL}$ is positive, so **T** $\Delta\mathbf{S}_{TOTAL}$ is positive.

Since ΔS_{total} must be a positive number for a spontaneous change, the product $T \, \Delta S_{total}$ must also be positive. This means that the quantity in parentheses on the right, $(\Delta H_{system} - T \, \Delta S_{system})$, must be negative so that $-(\Delta H_{system} - T \, \Delta S_{system})$ may be positive. Thus, in order for a spontaneous change to take place, the expression $(\Delta H_{system} - T \, \Delta S_{system})$ must be negative.

At this point it is convenient to introduce another thermodynamic state function, G, called the **Gibbs free energy.** This is defined as

$$G = H - TS$$

For a change at constant T and P, we write

$$\Delta G = \Delta H - T \, \Delta S \qquad [11.11]$$

From the argument presented in the preceding paragraph, we see that ΔG must be less than zero for a spontaneous process; that is, ΔG must have a negative value at constant T and P.

The Gibbs free energy change, ΔG, represents a composite of the two factors contributing to spontaneity, ΔH and ΔS. For systems in which ΔH is negative (exothermic) and ΔS is positive (increased disorder accompanying the change), both factors favor spontaneity and the process will occur spontaneously at all temperatures. Conversely, if ΔH is positive (endothermic) and ΔS is negative (increase in order), ΔG will always be positive and the change cannot occur spontaneously at any temperature.

It is thermodynamically impossible for a building to form all by itself from a pile of rubble. Why?

In situations where ΔH and ΔS are both positive, or both negative, Equation 11.11 shows that temperature plays the determining role in controlling whether or not a reaction will take place. In the first case (ΔH, $\Delta S > 0$), ΔG will be negative only at high temperatures, where $T\,\Delta S$ is greater in magnitude than ΔH; as a consequence, the reaction will be spontaneous only at elevated temperatures. An example is the melting of ice, which is nonspontaneous at low temperature (below 0°C) and spontaneous at high temperature (above 0°C). When ΔH and ΔS are both negative, ΔG will be negative only at low temperatures. An example of this is the freezing of water. We know that heat must be removed from the liquid to produce ice; hence the process is exothermic with a negative ΔH. Freezing is also accompanied by an ordering of the water molecules as they leave the random liquid state and become part of the crystal. As a result, ΔS is also negative. The sign of ΔG is determined both by ΔH, which in this case is negative, and by $T\,\Delta S$, which is also negative. To compute ΔG we must subtract a negative $T\,\Delta S$ from a negative ΔH. The result will be negative only at low temperature. Consequently, at 1 atm we observe H_2O to freeze spontaneously only below 0°C. Above 0°C the magnitude of $T\,\Delta S$ is greater than ΔH, and ΔG becomes positive. As a result, freezing is no longer spontaneous. Instead, the reverse process (melting) occurs. The effects of the signs of ΔH and ΔS and the effect of temperature on spontaneity, can be summarized as follows.

ΔH	ΔS	Outcome
(−)	(+)	Spontaneous at all temperatures
(+)	(−)	Nonspontaneous regardless of temperature
(+)	(+)	Spontaneous only at high temperature
(−)	(−)	Spontaneous only at low temperature

The significance of the Gibbs free energy change in determining spontaneity can be seen in the practical world around us. A classic example is in the production of synthetic diamonds. People had been fascinated by this problem ever since 1797, when it was found that diamond was simply a form of carbon. Over the years many experiments were devised in an attempt to convert graphite, the common form of carbon, into its much more valuable counterpart. However, as of 1938 no one had yet been able to accomplish this feat. At that time a careful thermodynamic analysis of the problem was performed, the results of which are summarized in Figure 11.12.

In this figure, $\Delta G/T$ is plotted along the vertical axis and temperature, in kelvins, is plotted along the horizontal axis. Since ΔG for the reaction,

$$\text{C (s, graphite)} \longrightarrow \text{C (s, diamond)}$$

must be negative in order for the process to be spontaneous, diamond can be produced only at temperatures and pressures that lie *below* the zero on the $\Delta G/T$

Figure 11.12

Thermodynamics of graphite to diamond conversion. From Chemical and Engineering News, *April 5, 1971. p. 51. Used by permission.*

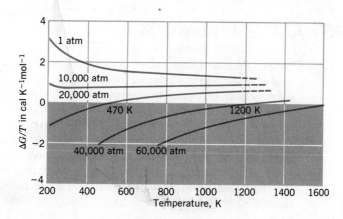

scale. For example, at 470 K the conversion of graphite into diamond can only take place at pressures greater than or equal to 20,000 atm. We can also conclude from the figure that at a constant pressure of 20,000 atm the reaction is not spontaneous above 470 K.

This analysis served to define the limits of temperature and pressure that would permit the conversion to take place. That was not the end of the problem, however, because suitable materials had to be found that would allow the reaction to proceed at a measurable rate. In fact, it was not until 1955 that success was finally achieved, and today synthetic diamonds are an important industrial abrasive used in grinding and cutting tools.

The conversion of graphite to diamond represents only one example of how thermodynamics can serve as a guide to answering practical questions. The principles of thermodynamics have been applied over the years to such diverse problems as the design of steam engines and the development of fuel cells that are now used as a source of electric power in spacecraft. Dr. Frederick Rossini, a noted thermodynamicist, has pointed out that the balance between the simultaneous opposing drives toward security (low energy) and freedom (high entropy) that control the fate of chemical systems also seem to determine equilibrium in a stable society, illustrating, perhaps, the truly wide scope of thermodynamics.

Another illustration of the impact of thermodynamics on our lives is the problem of thermal pollution, particularly in the neighborhood of power plants. It is a consequence of the second law of thermodynamics that any heat engine designed to convert heat into useful work must absorb heat from a high-temperature source. Some of this heat can be converted to work, but some of it must also be deposited in a low-temperature reservoir. Therefore, any device that tries to convert heat into useful work also discharges some heat into the environment. An automobile engine burns gasoline or diesel fuel, converts part of the resultant heat energy into work that propels the car, while the remainder is discharged to the radiator. Large power stations burn fuels (either fossil fuels such as coal or oil, or nuclear fuels such as uranium) to generate electrical power. In the process they too deposit relatively large quantities of heat into the environment. Cooling water from such power plants is heated to such an extent that it has been known to raise the temperature of surrounding rivers or lakes by as much as 25°F or more. These higher temperatures decrease the solubility of oxygen and kill some species of fish. It has also been found, however, that clams and oysters thrive in these warmer waters.

Several years ago a nuclear power station in New Jersey was forced to shut down for repairs during the winter. In this case, the halting of thermal pollution caused a large fish kill because many fish that normally live in warm water were suddenly "put out in the cold."

11.10 FREE ENERGY AND USEFUL WORK

One of the most important applications of chemical reactions is in the production of energy in the form of useful work. This can, for example, take the form of combustion, in which the heat generated is used to create steam for the production of mechanical work, or perhaps electrical work drawn from a dry cell or storage battery. The quantity G is called the *free energy* because ΔG represents the *maximum* amount of energy released in a process occuring at constant temperature and pressure that is free—or available—to perform useful work. We have already associated ΔG with the factors that lead to a drive for spontaneity. What we see now is that this driving force in a chemical change can be harnessed to perform work for us.

The interior of a modern steam powered electric generating plant. Although efficient by engineering standards, the thermodynamic efficiency is only about 35 to 40%.

The actual amount of work obtained from any real spontaneous process is always less than the maximum predicted by ΔG. This is because real processes are always irreversible, and we saw earlier that the maximum work can be extracted only from a truly reversible change. The free-energy change gives us a goal at which to aim. The closer a given process is to reversibility, the greater the amount of available work that can be used. However, even relatively efficient systems are able to harness only a small fraction of the available free energy. Living systems, for example, are able to convert only about 40% of the free energy available in the oxidation of glucose to other forms of stored chemical energy (for example, ATP).

Mechanical systems, such as electrical generators and gasoline engines, are even less efficient than living systems at converting energy into work.

11.11 STANDARD ENTROPIES AND FREE ENERGIES

The **third law of thermodynamics** states that the entropy of any pure crystalline substance at absolute zero is equal to zero. This makes sense because in a perfect crystal at absolute zero there is perfect order. Because of this, it is possible to actually measure the absolute amount of entropy that a substance has in its standard state by summing q_{rev}/T increments from 0 K to 298 K (25°C). Table 11.4 contains a number of these standard entropies. Standard entropies can be used to calculate standard entropy changes by a Hess's law-type of calculation.

$$\Delta S^0_{reaction} = (\text{sum of } S^0 \text{ of products}) - (\text{sum of } S^0 \text{ of reactants})$$

EXAMPLE 11.10   Calculate the standard entropy change (in units of joules/kelvin) for the reaction

$$2NaHCO_3 \, (s) \longrightarrow Na_2CO_3 \, (s) + CO_2 \, (g) + H_2O \, (g)$$

SOLUTION   We use Equation 11.12 and the data in Table 11.4.

$$\Delta S^0_{reaction} = [S^0_{Na_2CO_3 \, (s)} + S^0_{CO_2 \, (g)} + S^0_{H_2O \, (g)}] - [2S^0_{NaHCO_3 \, (s)}]$$

$$= [1 \text{ mol}(136 \text{ J/mol K}) + 1 \text{ mol}(213.6 \text{ J/mol K})$$
$$+ 1 \text{ mol}(188.7 \text{ J/mol K})] - [2 \text{ mol}(155 \text{ J/mol K})]$$

$$= (538 \text{ J/K}) - (310 \text{ J/K})$$

$$= +228 \text{ J/K}$$

Table 11.4
Absolute entropies at 25°C and 1 atm

Substance	S^0 J/mol K	S^0 cal/mol K	Substance	S^0 J/mol K	S^0 cal/mol K
Al (s)	28.3	6.77	Mg(OH)₂ (s)	63.1	15.09
Al₂O₃ (s)	51.0	12.19	N₂ (g)	191.5	45.77
Br₂ (l)	152.2	36.38	NH₃ (g)	192.5	46.01
Br₂ (g)	245.4	58.65	N₂O (g)	220.0	52.58
C (s, graphite)	5.69	1.36	NO (g)	210.6	50.34
C (s, diamond)	2.4	0.58	NO₂ (g)	240.5	57.47
CO (g)	197.9	47.30	Na (s)	51.0	12.2
CO₂ (g)	213.6	51.06	NaF (s)	51.5	12.3
CH₄ (g)	186.2	44.50	NaCl (s)	72.8	17.4
C₂H₆ (g)	230	54.9	NaBr (s)	83.7	20.0
C₂H₄ (g)	220	52.5	NaI (s)	91.2	21.8
C₂H₂ (g)	201	48.0	NaHCO₃ (s)	155	37.1
C₃H₈ (g)	269.9	64.51	Na₂CO₃ (s)	136	32.5
CCl₄ (l)	214.4	51.25	O₂ (g)	205.0	49.00
Cl₂ (g)	223.0	53.29	Pb (s)	64.9	15.5
F₂ (g)	202.7	48.44	PbO₂ (s)	68.6	16.4
H₂ (g)	130.6	31.21	PbSO₄ (s)	149	35.5
H₂O (l)	70.0	16.72	S (s, rhombic)	31.8	7.60
H₂O (g)	188.7	45.11	SO₂ (g)	248	59.3
HF (g)	173.5	41.47	SO₃ (g)	256	61.3
HCl (g)	186.7	44.62	H₂SO₄ (l)	157	37.5
HBr (g)	198.5	47.44	Si (s)	19	4.5
HI (g)	206	49.3	SiO₂ (s)	41.8	10.0
I₂ (s)	116.1	27.76	Zn (s)	41.8	10.0
Mg (s)	32.5	7.77	ZnO (s)	43.5	10.4

It's interesting to note that this reaction also has a positive ΔH^0 (Example 11.5), so it falls into that category in which both ΔH and ΔS are positive. The decomposition of $NaHCO_3$ is nonspontaneous at low temperature but becomes spontaneous at high temperature, and that is why it can be used as a fire extinguisher.

From standard heats of formation and standard entropies we can also calculate **standard free energies of formation,** ΔG_f^0. For example, consider the formation of CO_2 from the elements, with all reactants and products in their standard states,

$$C \text{ (s, graphite)} + O_2 \text{ (g)} \longrightarrow CO_2(g)$$

Table 11.1 gives us the standard enthalpy of formation of CO_2 (g), ΔH_f^0, as -94.1 kcal/mol. From the data in Table 11.4 we can calculate ΔS_f^0.

Notice that we have to calculate ΔS_f^0; they're not tabulated.

$$\Delta S_f^0 = S_{CO_2}^0 - (S_C^0 + S_{O_2}^0)$$

$$\Delta S_f^0 = 51.1 - (1.4 + 49.0) \text{ cal/mol K} = +0.7 \text{ cal/mol K}$$

Table 11.5
Standard free energies of formation at 25°C and 1 atm

Substance	ΔG_f^0 kJ/mol	kcal/mol	Substance	ΔG_f^0 kJ/mol	kcal/mol
Al_2O_3 (s)	−1577	−376.8	HBr (g)	−53.1	−12.7
$AgNO_3$ (s)	−32	−7.7	HI (g)	+1.30	+0.31
C (s, diamond)	+2.9	+0.69	H_2O (l)	−237	−56.7
CO (g)	−137	−32.8	H_2O (g)	−228	−54.6
CO_2 (g)	−395	−94.3	$MgCl_2$ (s)	−592.5	−141.6
CH_4 (g)	−50.6	−12.1	$Mg(OH)_2$ (s)	−833.9	−199.3
C_2H_6 (g)	−33	−7.9	NH_3 (g)	−17	−4.0
C_2H_4 (g)	+68.2	+16.3	N_2O (g)	+104	+24.8
C_2H_2 (g)	+209	+50.0	NO (g)	+86.8	+20.7
C_3H_8 (g)	−23	−5.6	NO_2 (g)	+51.9	+12.4
CCl_4 (l)	−65.3	−15.6	HNO_3 (l)	−79.9	−19.1
C_2H_5OH (l)	−175	−41.8	PbO_2 (s)	−219	−52.3
CH_3COOH (l)	−392	−93.8	$PbSO_4$ (s)	−811.3	−193.9
CaO (s)	−604.2	−144.4	SO_2 (g)	−300	−71.8
$Ca(OH)_2$ (s)	−896.6	−214.3	SO_3 (g)	−370	−88.5
$CaSO_4$ (s)	−1320	−315 6	H_2SO_4 (l)	−689.9	−164.9
CuO (s)	−127	−30.4	SiO_2 (s)	−805	−192.4
Fe_2O_3 (s)	−741.0	−177.1	SiH_4 (g)	−39	−9.4
HF (g)	−271	−64.7	ZnO (s)	−318	−76.1
HCl (g)	−95.4	−22.8			

We can then obtain ΔG_f^0 as

$$\Delta G_f^0 = \Delta H_f^0 - T\,\Delta S_f^0$$

At 25°C (298 K), then,

$$\Delta G_f^0 = -94.1 \text{ kcal/mol} - (298 \text{ K})(0.7 \text{ cal/mol K})$$

$$\Delta G_f^0 = -94.1 \text{ kcal/mol} - 200 \text{ cal/mol}$$

Converting entirely to kilocalories per mole gives

$$\Delta G_f^0 = (-94.1 - 0.2) \text{ kcal/mol}$$

$$\Delta G_f^0 = -94.3 \text{ kcal/mol}$$

This and other standard free energies of formation are given at Table 11.5.

Earlier in this chapter we saw that ΔH^0 for a reaction can be computed from standard heats of formation. The same rules also apply for the calculation of ΔG^0 using standard free energies of formation; that is,

$$\Delta G^0 = (\text{sum of } \Delta G_f^0 \text{ of products}) - (\text{sum of } \Delta G_f^0 \text{ of reactants}) \quad [11.13]$$

EXAMPLE 11.11 Silane, SiH_4, is the silicon analog of the main constituent in natural gas—methane, CH_4. Like methane, silane burns in air. The product is silica—a solid quite unlike carbon dioxide.

Silica is found in sand.

$$SiH_4 \ (g) + 2O_2 \ (g) \longrightarrow SiO_2 \ (s) + 2H_2O \ (g)$$

silicon dioxide
(silica)

Calculate ΔG^0 for this reaction in kilojoules.

SOLUTION   We have to apply Equation 11.13, using the data in Table 11.5.

$$\Delta G^0 = [\Delta G^0_{f \ SiO_2 \ (s)} + 2\Delta G^0_{f \ H_2O \ (g)}] - [\Delta G^0_{f \ SiH_4 \ (g)} + 2\Delta G^0_{fO_2(g)}]$$

(As with ΔH_f^0, we take the standard free energy of formation of a free element to be zero.) Therefore,

$$\Delta G^0 = \left[1 \ mol\left(\frac{-805 \ kJ}{mol}\right) + 2 \ mol\left(\frac{-228 \ kJ}{mol}\right)\right] - \left[1 \ mol\left(\frac{-39 \ kJ}{mol}\right) + 0\right]$$

$$= -1222 \ kJ$$

11.12   FREE ENERGY AND EQUILIBRIUM

Earlier we said that ΔG determines the maximum amount of energy that is available to perform useful work as a system passes from one state to another. As a reaction proceeds, its capacity to perform work, as measured by G, diminishes until finally, at equilibrium, the system is no longer able to supply additional work. This means that both reactants and products possess the same free energy, and therefore $\Delta G = 0$. We see, therefore, that the value of ΔG for a particular change determines where the system stands with respect to equilibrium. When ΔG is negative—meaning that the free energy of the system is decreasing—the reaction is spontaneous and proceeds in the forward direction toward a state of equilibrium. When ΔG is zero, the system is in a state of dynamic equilibrium, and when ΔG is positive, the reaction is really spontaneous in the reverse direction.

At this point, it should be reemphasized that although ΔG may predict that a particular process is spontaneous, nothing is implied about how rapid the change will be.

Figure 11.13 illustrates graphically the free energy changes in a chemical

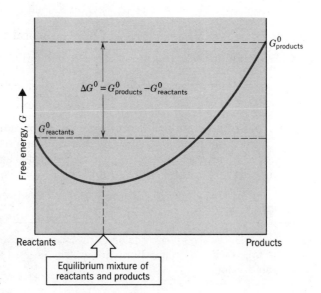

Figure 11.13

The variation of free energy in a homogeneous chemical system as the reaction proceeds from pure reactants on the left to pure products on the right. The minimum in the curve represents the extent of reaction required for the system to achieve equilibrium.

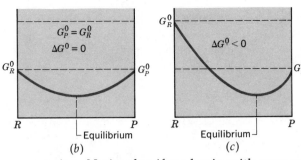

Figure 11.14

The position of equilibrium changes as the value of ΔG^0 changes. (a) Position of equilibrium in favor of the reactants. (b) Position of equilibrium intermediate between the reactants and products. (c) Position of equilibrium in favor of the products.

It is the change in **G** with a change in composition that corresponds to the ΔG that controls the spontaneity of a chemical reaction in one direction or the other.

reaction. Notice that if we begin with pure reactants, we proceed in the direction of the products because this gives a lowering of the free energy—ΔG is negative. Also notice that if we begin with pure products, we proceed in the direction of the reactants—this lowers the free energy, too. Therefore, whether we begin with reactants or products, we *always* head in the direction of the minimum on the free energy curve. When the reaction gets there, equilibrium is reached ($\Delta G = 0$). No change in the composition can occur because going toward either the reactants or products involves going "uphill" on the free energy curve, and that is nonspontaneous.

In Figure 11.13 we also see ΔG^0 for the reaction. It is simply the difference between the free energies of the pure products and the pure reactants, but it doesn't tell us (directly) which way the reaction will proceed when the reaction mixture has some given composition. For instance, in this particular reaction ΔG^0 is positive because $G^0_{products}$ is larger than $G^0_{reactants}$. The sign and numerical value of ΔG^0 for the reaction is fixed—it is a constant for the reaction. It is the way that G changes with composition—that is, it's the slope of the free energy curve—that determines the direction the reaction will go.

At this point, you may ask, "Why bother with ΔG^0 at all?" The answer can be seen in Figure 11.14. In Figure 11.14*a* we see a reaction for which ΔG^0 is positive. In Figure 11.14*b*, the reaction has $\Delta G^0 = 0$, and in Figure 11.14*c*, the reaction has a negative ΔG^0. Notice how the location of the minimum on the free energy curve varies with ΔG^0. When ΔG^0 is positive, the minimum is near the reactants—hardly any reaction has occurred by the time equilibrium is reached. When ΔG^0 equals zero, equilibrium is reached about halfway between reactants and products, and when ΔG^0 is negative the reaction goes almost to completion by the time equilibrium is reached.

Now we can see the usefulness of ΔG^0 calculations. The value of ΔG^0 allows us to predict the ultimate outcome of a reaction if it is allowed to proceed to equilibrium. When ΔG^0 is positive by about 5 kcal or so, so little product will be present at equilibrium that we might conclude that nothing has happened in the reaction mixture. On the other hand, when ΔG^0 is negative by about 5 kcal, the reaction will appear to have gone to completion. It is in this sense that ΔG^0 predicts the "spontaneity" of a reaction!

EXAMPLE 11.12   Is the reaction, $2SO_2\ (g) + O_2\ (g) \rightarrow 2SO_3\ (g)$ expected to be spontaneous at 25°C and 1 atm (i.e., will it proceed far toward completion by the time equilibrium is reached)?

SOLUTION   Let's calculate ΔG^0 for the reaction (we can use either kcal or kJ; it doesn't matter).

$$\Delta G^0 = 2\Delta G^0_{f\,SO_3\,(g)} - [2\Delta G^0_{f\,SO_2\,(g)} + \Delta G^0_{f\,O_2\,(g)}]$$

$$= 2\ \text{mol}\left(\frac{-88.5\ \text{kcal}}{\text{mol}}\right) - \left[2\ \text{mol}\left(\frac{-71.8\ \text{kcal}}{\text{mol}}\right) + 0.0\right] = -177\ \text{kcal} - (-144\ \text{kcal})$$

$$= -33\ \text{kcal}$$

This very negative ΔG^0 indicates that the position of equilibrium is very largely in favor of SO_3, so the reaction should appear to be quite spontaneous. Actually, this is true but the reaction is very slow in the absence of agents called catalysts that we will discuss in the next chapter.

INDEX TO QUESTIONS AND PROBLEMS (Problem numbers in **bold type**)

REVIEW QUESTIONS

11.1 What two factors ultimately determine whether we will be able to observe the formation of products in a chemical reaction?

11.2 What does the word *thermodynamics* imply?

11.3 Define the following: system, surroundings, isothermal change, adiabatic change, state function, heat capacity, molar heat capacity, specific heat.

11.4 State the first law of thermodynamics. On a molecular basis, why is E a state function?

11.5 Why is it not possible to measure or calculate E for a system?

11.6 A real gas usually cools slightly if it is allowed to expand adiabatically into an evacuated container.
(a) What should happen to the temperature of an ideal gas when it is treated in the same way?
(b) What conclusions can you draw about ΔE, q, and w for the *isothermal* expansion of a real gas?

11.7 Why is ΔE equal to zero for an isothermal expansion of an ideal gas?

11.8 What is an "equation of state?"

11.9 What are some different kinds of work that a system can perform on its surroundings?

11.10 What is meant by the term spontaneous change?

11.11 What is a reversible process?

11.12 In terms of converting energy into useful work, what advantage and disadvantage does a reversible process offer?

11.13 Suppose that a steel spring is compressed, tied so it cannot expand, and then dissolved in hydrochloric acid. What happens to the potential energy stored in the spring?

11.14 What advantages do the SI units of pressure and volume offer in calculations of pressure-volume work?

11.15 From an experimental point of view, how can a change be carried out isothermally? How can a change be carried out adiabatically?

11.16 Why is ΔE called the heat of reaction at constant volume? Why is ΔH called the heat of reaction at constant pressure?

11.17 Why are we usually concerned more about changes in enthalpy than changes in internal energy?

11.18 What is a bomb calorimeter?

11.19 What is Hess's law of heat summation? What is meant by the term, standard state?

11.20 The following are two reactions showing the formation of 1 mol of SO_3.

$$SO_2\ (g) + \tfrac{1}{2}O_2\ (g) \longrightarrow SO_3\ (g)$$

$$S\ (s) + \tfrac{3}{2}O_2\ (g) \longrightarrow SO_3\ (g)$$

Should their enthalpy changes both be labeled ΔH_f^0 if they occur at 25°C and 1 atm? Explain.

11.21 Which of the following should have a more negative ΔH^0?

$$2C\ (s) + 3H_2\ (g) + \tfrac{1}{2}O_2\ (g) \longrightarrow C_2H_5OH\ (g)$$

$$2C\ (s) + 3H_2\ (g) + \tfrac{1}{2}O_2\ (g) \longrightarrow C_2H_5OH\ (l)$$

11.22 For what kinds of reactions is it safe to ignore differences between ΔE and ΔH? In general, do ΔE and ΔH differ by much?

11.23 What is meant by atomization energy? Write an equation representing the process for which ΔH is the atomization energy of H_2O. Indicate the physical state (gas, liquid, or solid) for each substance in the equation.

11.24 Why don't values of ΔH_f^0 computed from tabulated bond energies agree precisely with ΔH_f^0 measured experimentally?

11.25 In what way is statistical probability related to spontaneity? How is entropy related to statistical probability?

11.26 In computing ΔS for a system, why must q_{rev} be used?

11.27 Explain why the entropy of a pure substance is zero at 0 K. Would the entropy of an alloy such as brass be zero at 0 K? Explain your answer.

11.28 What two criteria must be met in order for a process to be spontantous, regardless of the temperature?

11.29 State the second law of thermodynamics.

11.30 What is the sign of the entropy change for each of the following processes?
(a) A solute crystallizes from a solution.
(b) Water evaporates.
(c) A deck of playing cards is shuffled.
(d) A card player is dealt 13 spades.
(e) Solid $AgCl$ precipitates from a solution of $AgNO_3$ and $NaCl$.

(f) $^{235}_{92}U$ is extracted from a mixture of $^{235}_{92}U$ and $^{238}_{92}U$.
(g) $Na_2CO_3\ (aq) + HCl\ (aq) \rightarrow$
$$2NaCl\ (aq) + H_2O\ (l) + CO_2\ (g)$$

11.31 Why is it possible for a chemical reaction to occur spontaneously even though ΔG^0 for the reaction is positive?

11.32 Why can ΔG^0 be used to predict whether or not a reaction can be observed?

11.33 What relationship is there between ΔG^0 and the speed with which the products of a reaction are formed?

11.34 Describe the relationship between ΔG^0 and the position of equilibrium in a chemical reaction.

11.35 Referring to Figure 11.12, what is the minimum pressure necessary for the conversion of graphite to diamond at a temperature of 200 K? Is it theoretically possible to change graphite to diamond at 1 atm?

11.36 Explain why $\Delta G = 0$ for a system that is in a state of equilibrium.

11.37 ΔH and ΔS are nearly independent of temperature. Why is this not true for ΔG?

REVIEW PROBLEMS (More difficult problems are marked by an asterisk)

11.38 A gas, initially under a pressure of 15.0 atm and having a volume of 10.0 liters, is permitted to expand isothermally in two steps. In the first step the external pressure is held constant at 7.50 atm, and in the second step the external pressure is maintained at 1.00 atm. What are q and w for each step? What is the net change in the internal energy, expressed in liter atm, for both the system and its surroundings? Assume ideal gas behavior.

11.39 A gas having an initial volume of 50.0 m³ at an initial pressure of 200 kPa is allowed to expand against a constant opposing pressure of 100 kPa. Calculate the work done by the gas in kJ. If the gas is ideal and the expansion is isothermal, what is q for the gas in kilojoules?

11.40 If 500 ml of a gas is compressed to 250 ml under a constant external pressure of 3.00 atm, and if the gas also absorbs 3.00 kcal, what are the values of q, w, and ΔE for the gas expressed in kcal? What is the value of ΔE for the surroundings?

11.41 A certain reaction in a bomb calorimeter liberated 3.41 kcal. If the initial temperature of the calorimeter was 25.000°C and the heat capacity of the calorimeter and its contents was 4250 cal/°C, what was the final temperature of the calorimeter?

11.42 A 0.100-mol sample of propane, a gas used for cooking in many rural areas, was placed into a bomb calorimeter with excess oxygen and ignited. The reaction was

$$C_3H_8\ (g) + 5O_2\ (g) \longrightarrow 3CO_2\ (g) + 4H_2O\ (l)$$

The initial temperature of the calorimeter was 25.000°C and its total heat capacity was 23.2 kcal/°C. The reaction raised the temperature of the calorimeter to 27.282°C.
(a) How many calories were liberated by the combustion of the propane?
(b) What is ΔE for the reaction expressed in kcal/mol C_3H_8?

11.43 Calculate ΔH for the reaction in Problem 11.42 in the units kJ/mol C_3H_8.

11.44 At 25°C and a constant pressure of 1.00 atm, the reaction of 0.500 mol of OF_2 with water vapor, according to the equation

$$OF_2\ (g) + H_2O\ (g) \longrightarrow O_2\ (g) + 2HF\ (g)$$

liberates 38.6 kcal. Calculate ΔH and ΔE in units of kilocalories *per mole* of OF_2.

11.45 At 25°C and 1.00 atm, the reaction of 1.00 mol of CaO with water (shown below) evolves 15.6 kcal.

$$CaO\ (s) + H_2O\ (l) \longrightarrow Ca(OH)_2\ (s)$$

What are ΔH and ΔE, in kilocalories per mole of CaO, for this process given that the densities of $CaO\ (s)$, $H_2O\ (l)$, and $Ca(OH)_2\ (s)$ at 25°C are 3.25 g/ml, 0.997 g/ml, and 2.24 g/ml, respectively? What does this tell you about the relative values of ΔH and ΔE when all substances are either liquids or solids?

*11.46 At 25°C, burning 0.200 mol of H_2 with 0.100 mol of O_2 to produce $H_2O\ (l)$ in a bomb calorimeter raises the temperature of the apparatus 0.880°C. When 0.0100 mol of toluene, C_7H_8, is burned in this calorimeter, the temperature is raised by 0.615°C. The equation for the combustion reaction is

$$C_7H_8 (l) + 9O_2 (g) \longrightarrow 7CO_2 (g) + 4H_2O (l)$$

Calculate ΔE for this reaction in kilojoules. Use ΔH_f^0 for H_2O (l) found in Table 11.1 to compute ΔE_f^0 for H_2O (l) in kJ/mol.

11.47 The combustion of ethanol, C_2H_5OH (l), to give CO_2 (g) and H_2O (l) evolves 1.37×10^3 kJ per mol of C_2H_5OH (l) when the products are returned to 25°C and 1 atm. Use this information and the data in Table 11.1 to calculate ΔH_f^0, in kJ/mol, for C_2H_5OH (l).

11.48 Calculate ΔH_f^0 (in kJ/mol) for C_3H_8 (g) from the data in Problem 11.42.

***11.49** Based on the results of Problem 11.46, compute the standard heat of formation, ΔH_f^0 (in kJ/mol), of toluene. (Assume that the reaction was carried out at 25° C.)

11.50 The reaction,

$$Ca (s) + O_2 (g) + H_2 (g) \longrightarrow Ca(OH)_2 (s),$$

has $\Delta H^0 = -214.3$ kcal. What is ΔE for this reaction in kilocalories?

11.51 The heat of vaporization ΔH_{vap} of H_2O at 25°C is 10.5 kcal/mol. Calculate q, w, and ΔE for the process in kilocalories.

11.52 Use the data in Table 11.1 and Hess's law to calculate ΔH^0, in kilojoules, for each of the following reactions:
(a) $2Al (s) + Fe_2O_3 (s) \rightarrow Al_2O_3 (s) + 2Fe (s)$
(b) $SiH_4 (g) + 2O_2 (g) \rightarrow SiO_2 (s) + 2H_2O (g)$
(c) $CaO (s) + SO_3 (g) \rightarrow CaSO_4 (s)$
(d) $CuO (s) + H_2 (g) \rightarrow Cu (s) + H_2O (g)$
(e) $C_2H_4 (g) + H_2 (g) \rightarrow C_2H_6 (g)$

11.53 Use the data in Table 11.1 to calculate ΔH^0, in kilocalories, for each of the following reactions:
(a) $C_2H_2 (g) + H_2 (g) \rightarrow C_2H_4 (g)$
(b) $SO_3 (g) + H_2O (l) \rightarrow H_2SO_4 (l)$
(c) $Mg(OH)_2 (s) + 2HCl (g) \rightarrow MgCl_2 \cdot 2H_2O (s)$
(d) $CO_2 (g) + H_2 (g) \rightarrow CO (g) + H_2O (g)$
(e) $10N_2O (g) + C_3H_8 (g) \rightarrow$
$$10N_2 (g) + 3CO_2 (g) + 4H_2O (g)$$

11.54 Aerosol propellants are often chlorofluoromethanes (CFMs) such as Freon-11 ($CFCl_3$) and Freon-12 (CF_2Cl_2). It has been suggested that continued use of these may ultimately deplete the ozone shield in the stratosphere, with catastrophic results to the inhabitants of our planet. In the stratosphere CFMs absorb high-energy radiation and produce Cl atoms that have a catalytic effect on removing ozone, O_3.

$$
\begin{array}{lll}
O_3 + Cl \longrightarrow O_2 + ClO & \Delta H^0 = -30 \text{ kcal} \\
ClO + O \longrightarrow Cl + O_2 & \Delta H^0 = -64 \text{ kcal} \\
\hline
\text{net} \quad O_3 + O \longrightarrow 2 O_2
\end{array}
$$

The O atoms are present due to dissociation of O_2 molecules by high-energy radiation. Calculate ΔH^0 in kilocalories for the net reaction for the removal of the ozone.

11.55 Given the following thermochemical equations,

$$2H_2 (g) + O_2 (g) \longrightarrow 2H_2O (l) \qquad \Delta H^0 = -136.6 \text{ kcal}$$

$$N_2O_5 (g) + H_2O (l) \longrightarrow 2HNO_3 (l) \qquad \Delta H^0 = -18.3 \text{ kcal}$$

$$\tfrac{1}{2}N_2 (g) + \tfrac{3}{2}O_2 (g) + \tfrac{1}{2}H_2 (g) \longrightarrow HNO_3 (l)$$
$$\Delta H^0 = -41.6 \text{ kcal}$$

Calculate ΔH^0 for the reaction

$$2N_2 (g) + 5O_2 (g) \longrightarrow 2N_2O_5 (g)$$

11.56 Given the following thermochemical equations,

$$Fe_2O_3 (s) + 3CO (g) \longrightarrow 2Fe (s) + 3CO_2 (g) \quad \Delta H = -28 \text{ kJ}$$

$$3Fe_2O_3 (s) + CO (g) \longrightarrow 2Fe_3O_4 (s) + CO_2 (g)$$
$$\Delta H = -59 \text{ kJ}$$

$$Fe_3O_4 (s) + CO (g) \longrightarrow 3FeO (s) + CO_2 (g)$$
$$\Delta H = +38 \text{ kJ}$$

calculate ΔH for the reaction,

$$FeO (s) + CO (g) \longrightarrow Fe (s) + CO_2 (g)$$

without referring to the data in Table 11.1.

11.57 Use the results of Question 11.56 and the data in Table 11.1 to compute the standard heat of formation of FeO. Express the answer in kcal/mol and kJ/mol.

11.58 Acetylene, C_2H_2, a gas used in welding torches, is produced by the action of water on calcium carbide, CaC_2. Given the following thermochemical equations, calculate ΔH_f^0 for acetylene. Express your answer in both kJ/mol and kcal/mol.

$$CaO (s) + H_2O (l) \longrightarrow Ca(OH)_2 (s) \qquad \Delta H^0 = -15.6 \text{ kcal}$$

$$CaO (s) + 3C (s) \longrightarrow CaC_2 (s) + CO (g)$$
$$\Delta H^0 = +110.5 \text{ kcal}$$

$$CaC_2 (s) + 2H_2O (l) \longrightarrow Ca(OH)_2 (s) + C_2H_2 (g)$$
$$\Delta H^0 = -30.0 \text{ kcal}$$

$$2C (s) + O_2 (g) \longrightarrow 2CO (g) \qquad \Delta H^0 = -52.8 \text{ kcal}$$

$$2H_2O (l) \longrightarrow 2H_2 (g) + O_2 (g)$$
$$\Delta H^0 = +136.6 \text{ kcal}$$

11.59 Plaster of Paris, $CaSO_4 \cdot \tfrac{1}{2}H_2O$, is mixed with water with which it combines to produce gypsum, $CaSO_4 \cdot 2H_2O$. The reaction is exothermic, which explains why a plaster cast on a broken arm becomes warm as the cast hardens. Given that for $CaSO_4 \cdot \tfrac{1}{2}H_2O$, $\Delta H_f^0 = -1573$ kJ/mol, and for $CaSO_4 \cdot 2H_2O$, $\Delta H_f^0 = -2020$ kJ/mol, calculate ΔH^0 for the reaction,

$$CaSO_4 \cdot \tfrac{1}{2}H_2O (s) + \tfrac{3}{2}H_2O (l) \longrightarrow CaSO_4 \cdot 2H_2O (s)$$

in units of kilojoules.

11.60 Important reactions in the production of ozone in polluted air are

$$2NO (g) + O_2 (g) \longrightarrow 2NO_2 (g)$$

$$NO_2 (g) \xrightarrow{h\nu} NO (g) + O (g)$$

$$O_2 (g) + O (g) \longrightarrow O_3 (g)$$

Calculate ΔH^0 (in kilojoules) for each of these processes using the data in Tables 11.1 and 11.2.

11.61 The body eliminates ethyl alcohol, C_2H_5OH, by oxidation to give water and the following series of carbon-containing products,

$$C_2H_5OH \xrightarrow{O_2} CH_3CHO \xrightarrow{O_2} CH_3COOH \xrightarrow{O_2} CO_2$$

Write balanced equations for each step in the oxidation and calculate its ΔH^0 in kilojoules. What is the overall ΔH^0 for complete oxidation to CO_2 and H_2O?

11.62 Use the data in Table 11.1 to determine how many calories are evolved in the combustion of 45.0 g of C_2H_6 (g) to produce CO_2 (g) and H_2O (g) under a constant pressure of 1.00 atm.

11.63 The average adult expends about 2000 kcal of energy per day for normal activity. If 1 g of carbohydrate provides 4 kcal of usable energy, how many grams of carbohydrates must be consumed to meet these caloric demands?

11.64 The evaporation of perspiration is one mechanism whereby the body disposes of excess thermal energy and manages to maintain a constant temperature. How many kilojoules are removed from the body by the evaporation of 10.0 g of H_2O?

*11.65 Calculate the number of kilojoules liberated during the combustion of 1.00 gal (3.79 liters) of octane (gasoline). The density of octane is 0.703 g/ml. What weight of hydrogen would have to be burned [giving H_2O (l)] to produce this same amount of heat? If this H_2 were compressed to a pressure of 2500 lb/in.^{2} (170 atm) at 25°C, what volume would it occupy? What does this suggest about the feasibility of a hydrogen fuel economy for the automobile? For octane, C_8H_{18} (l), $\Delta H_f^0 = -208.4$ kJ/mol.

*11.66 It is estimated that the body generates up to 5900 kJ of thermal energy per hour during heavy physical exercise. If the only way that this excess energy could be dissipated was through evaporation of water, how many grams of water would have to evaporate per hour to keep the body temperature constant?

*11.67 How many grams of glucose, $C_6H_{12}O_6$, would have to be metabolized per hour (to give CO_2 and H_2O) in order to generate the excess thermal energy described in the preceding problem if it is assumed that 60% of the available energy appears as excess body heat? (The remaining 40% is used by the body to do mechanical work—moving of limbs, pumping of blood, etc.). For $C_6H_{12}O_6$, $\Delta H_{combustion} = -2820$ kJ/mol.

*11.68 How many liters of natural gas (CH_4) at 25°C and 1 atm must be burned to provide sufficient energy to convert 250 ml of H_2O at 20°C (approx. 8 fluid ounces) into steam at 100°C?

*11.69 The first ionization energy of gaseous sodium atoms is 494.1 kJ/mol, and adding electrons to gaseous chlorine atoms to form gaseous chloride ions liberates 351 kJ/mol. Use this information along with the heats of formation of

gaseous Na and Cl atoms, and the heat of formation of NaCl (s), to calculate ΔH (in kJ) for the reaction

$$NaCl\ (s) \longrightarrow Na^+\ (g) + Cl^-\ (g)$$

The answer corresponds to the lattice energy of sodium chloride. (*Hint:* It will help if you write thermochemical equations for each process described in the problem.)

*11.70 An important photochemical reaction in the production of smog is

$$NO_2\ (g) + h\nu \longrightarrow NO\ (g) + O\ (g)$$

If one quantum of energy is required to cause this reaction to occur, what must the wavelength of the light be? (Use the data in Tables 11.1 and 11.2, calculate ΔE for the process in kilojoules, then calculate ν in nanometers from the Planck relationship $\Delta E = h\nu$.)

11.71 Use the data in Tables 11.2 and 11.3 to calculate ΔH_f^0 in kilojoules for acetylene, C_2H_2 (g). The structure of the molecule is $H-C\equiv C-H$.

*11.72 Benzene is often written as a resonance hybrid of two equivalent structures,

The ΔH_f^0 for gaseous benzene has been determined from its heat of combustion to be +82.8 kJ/mol.

$$6C\ (s) + 3H_2\ (g) \longrightarrow C_6H_6\ (g) \qquad \Delta H_f^0 = +82.8\ kJ/mol$$

Use the data in Tables 11.2 and 11.3 to calculate ΔH_f^0 in kJ/mol. How does your calculated value compare with the experimental value? The difference between the calculated and experimental values is called the resonance energy. What might you conclude about the stability of species that exist as a composite of two or more resonance structures?

11.73 Use data in Tables 11.2 and 11.3 to calculate ΔH_f^0 (in kJ/mol) for gaseous propylene, the substance used to make the plastic polypropylene. The structure of propylene is

11.74 Use the average bond energies in Table 11.3 to compute the standard heat of formation of C_3H_8 (g) in kJ/mol. Its structure is

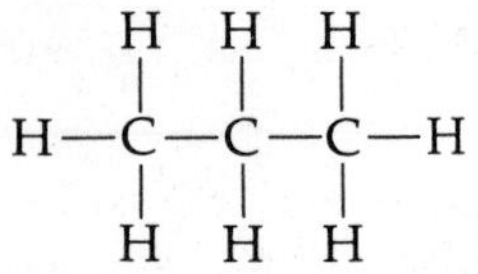

$$
\begin{array}{ccccccc}
 & H & & H & & H & \\
 & | & & | & & | & \\
H - & C & - & C & - & C & - H \\
 & | & & | & & | & \\
 & H & & H & & H &
\end{array}
$$

How well does your computed value compare with that reported in Table 11.1?

11.75 The heat of fusion of water at 0°C is 1.44 kcal/mol; its heat of vaporization is 9.72 kcal/mol at 100°C. What are ΔS for the melting and boiling of 1 mol of water? Can you explain why ΔS_{vap} is greater than $\Delta S_{melting}$?

11.76 From the data in Tables 11.1 and 11.4, calculate the boiling point of liquid bromine in °C [i.e., the temperature at which Br_2 (l) and Br_2 (g) can coexist in equilibrium with each other].

11.77 Which of the following reactions is accompanied by the greatest entropy change?
(a) SO_2 (g) + $\frac{1}{2}O_2$ (g) → SO_3 (g)
(b) CO (g) + $\frac{1}{2}O_2$ (g) → CO_2 (g)

11.78 Calculate ΔG^0 in kilojoules for the following reactions.
(a) $2Al$ (s) + Fe_2O_3 (s) → Al_2O_3 (s) + $2Fe$ (s)

(b) CaO (s) + SO_3 (g) → $CaSO_4$ (s)
(c) CuO (s) + H_2 (g) → Cu (s) + H_2O (g)

11.79 Use the results from Problems 11.52 and 11.78 to compute ΔS^0, in joules per Kelvin, for the reactions in Problem 11.78.

11.80 The standard free energy of formation of glucose is $\Delta G_f^0 = -217.54$ kcal/mol. Calculate ΔG^0 for the reaction,

$$C_6H_{12}O_6 \ (s) + 6O_2 \ (g) \longrightarrow 6CO_2 \ (g) + 6H_2O \ (l)$$

11.81 What is the maximum amount of useful work, expressed in kilojoules, that could be obtained by the oxidation of propane, C_3H_8, according to the equation,

$$C_3H_8 \ (g) + 5O_2 \ (g) \longrightarrow 3CO_2 \ (g) + 4H_2O \ (g)$$

Why is it that we always get less than this maximum amount of work in any real process that uses propane as a fuel?

11.82 Which of the following reactions could *potentially* serve as a practical method for the preparation of NO_2? (*Note:* The equations are not balanced.)
(a) N_2 (g) + O_2 (g) → NO_2 (g)
(b) HNO_3 (l) + Ag (s) → $AgNO_3$ (s) + NO_2 (g) + H_2O (l)
(c) NH_3 (g) + O_2 (g) → NO_2 (g) + NO (g) + H_2O (g)
(d) CuO (s) + NO (g) → NO_2 (g) + Cu (s)
(e) NO (g) + O_2 (g) → NO_2 (g)
(f) H_2O (g) + N_2O (g) → NH_3 (g) + NO_2 (g)

12

CHEMICAL KINETICS

The remains of a grain elevator in New Orleans, Louisiana, after it exploded in December, 1977, killing 35 people. Particle size is one of the factors that control the speeds of chemical reactions. Rapid combustion of very fine grain dust caused the explosive effects seen here. In this chapter we study the speeds of chemical reactions and the kinds of chemical information that come from such studies.

It does not take long to find a reaction that thermodynamics predicts should proceed nearly to completion but yet is not observed to occur. We know from the last chapter that hydrogen and oxygen can be kept in contact with one another almost forever without forming noticeable amounts of water, even though their reaction to produce water is accompanied by a free-energy decrease. This is an example of a chemical change where the speed of the reaction governs whether the formation of the products will or will not be observed.

Chemical kinetics, also referred to as chemical dynamics, is concerned with the speeds, or rates, of chemical reactions. In this area of chemistry we study the factors that control how rapidly chemical changes occur. These include the following:

1. *The nature of the reactants and products.* All other factors being equal, some reactions are just naturally fast and others are naturally slow, depending on the chemical makeup of the molecules or ions involved.
2. *The concentrations of the reacting species.* For two molecules to react with each other they must meet, and the probability that this will happen in a homogeneous mixture increases as their concentrations increase. For heterogeneous reactions—those in which the reactants are in separate phases—the rate also depends on the area of contact between the phases. Since many small particles have a much larger area than one large particle of the same total mass, decreasing the particle size increases the reaction rate. Sometimes this can have devastating results as we can see in the opening photograph of this chapter.
3. *The effect of temperature.* Nearly all chemical reactions take place faster when their temperatures are increased.
4. *The influence of outside agents called catalysts.* The rates of many reactions, including virtually all biochemical reactions, are affected by substances called catalysts that are not actually used up during the course of the reaction.

Studying how these factors affect the rate of a reaction serves several purposes. For example, it allows us to adjust the conditions of a reaction system to obtain the products as quickly as possible. The importance of this in the commercial manufacture of chemicals is obvious. It also allows us to adjust conditions to make a reaction occur as slowly as possible. This is helpful, for instance, in controlling the growth of fungi and other microorganisms that spoil foods.

For chemists, one of the most significant benefits that comes from studying reaction rates is knowledge about the details of how chemical changes take place. We will see later in this chapter that a chemical reaction usually does not occur in one single step that involves the simultaneous collision of all the reactant molecules described in the balanced overall equation. Instead, the net change is the result of a sequence of simple reactions. This sequence is called the **mechanism** of the reaction, and studying reaction rates gives clues to what the mechanism is. In this way we gain insight into the fundamental reasons of why substances react the way they do.

Chopping a log into small pieces of kindling with a large total surface area makes a campfire easier to start.

12.1 REACTION RATES AND THEIR MEASUREMENT

Before we examine the factors that influence rates of reaction, let's be sure that we know what is meant by "rate." In general, the rate (or speed) of any chemical reaction can be expressed as the ratio of the change in the concentration of a reactant (or product) to a change in time. This is exactly analogous to giving the speed of an automobile as the change in position (that is, the distance traveled) divided by its time of travel. Here the speed might be given in miles per hour.

With chemical reactions, the rate is usually expressed in moles per liter per second.

$$\text{speed of auto} = \text{rate of travel} = \frac{\text{distance}}{\text{time}} = \frac{\text{miles}}{\text{hour}}$$

$$\text{rate of chemical reaction} = \frac{\text{change in concentration}}{\text{time}}$$

$$= \frac{\text{moles/liter}}{\text{second}} = \frac{\text{mol/liter}}{\text{s}}$$

$$= \text{mol liter}^{-1}\text{ s}^{-1}$$

To determine the rate of a given chemical reaction, we must measure how fast the concentration of a reactant or product changes. In practice, the species whose concentration is easiest to follow is determined at various time intervals. The simplest example is a reaction in which only one reactant undergoes a change to form a single product. An example is the conversion of cyclopropane to propylene.

Cyclopropane is used as a fast-acting anesthetic.

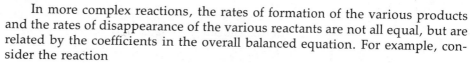

In general, the balanced equation for this type of reaction would be

$$A \longrightarrow B \qquad\qquad\text{[12.1]}$$

When the reaction is carried out, no product (B) is present initially and, as time goes on, the concentration of B increases with a corresponding decrease in the concentration of A (Figure 12.1). An inspection of Figure 12.1 reveals that the rate of this chemical reaction changes with time. For instance, near the start of the reaction the concentration of A is decreasing rapidly and the concentration of B is rising rapidly. Much later during the reaction, however, only small changes in concentration occur with time, and the rate is therefore much less. In general, this type of behavior is observed with nearly every chemical reaction—as the reactants are consumed, the rate of reaction gradually decreases.

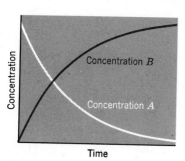

Figure 12.1

Change in the concentration of reactants and products with time for the reaction A → B.

In more complex reactions, the rates of formation of the various products and the rates of disappearance of the various reactants are not all equal, but are related by the coefficients in the overall balanced equation. For example, consider the reaction

$$N_2\ (g) + 3H_2\ (g) \longrightarrow 2NH_3\ (g)$$

We see that for every molecule of N_2 that reacts, three molecules of H_2 react. This means that the hydrogen is disappearing three times as fast as the nitrogen. The coefficients also tell us that two molecules of NH_3 are formed from each N_2, so the rate at which NH_3 is formed must be twice as fast as the rate at which N_2 disappears.

EXAMPLE 12.1   Ammonia can be made to burn according to the reaction

$$4NH_3\ (g) + 5\ O_2\ (g) \longrightarrow 4NO\ (g) + 6H_2O\ (g)$$

Suppose that at a particular moment during the reaction the ammonia is reacting at the rate of 0.24 mol liter^{-1} s^{-1}. (a) What is the rate at which oxygen is reacting? (b) What is the rate at which H_2O is being formed?

SOLUTION   The solution of this problem simply uses the coefficients of the equation to construct conversion factors relating the various numbers of moles of the reactants and products. We are given that

$$\text{rate (for } NH_3) = \frac{0.24 \text{ mol } NH_3}{\text{liter s}}$$

The coefficients in the equation allow us to construct the conversion factors

$$\frac{5 \text{ mol } O_2}{4 \text{ mol } NH_3} \quad \text{and} \quad \frac{6 \text{ mol } H_2O}{4 \text{ mol } NH_3}$$

We use these as follows:

(a) For the rate at which oxygen is consumed,

$$\text{rate (for } O_2) = \frac{0.24 \text{ mol } NH_3}{\text{liter s}} \times \left(\frac{5 \text{ mol } O_2}{4 \text{ mol } NH_3} \right)$$

$$= \frac{0.30 \text{ mol } O_2}{\text{liter s}}$$

(b) For the rate at which water is formed,

$$\text{rate (for } H_2O) = \frac{0.24 \text{ mol } NH_3}{\text{liter s}} \times \left(\frac{6 \text{ mol } H_2O}{4 \text{ mol } NH_3} \right)$$

$$= \frac{0.36 \text{ mol } H_2O}{\text{liter s}}$$

Measurement of reaction rates

$\dfrac{\text{moles}}{\text{liters}}$ = molar concentration

An accurate, quantitative estimate of the rate of reaction at any given moment during the reaction can be obtained from the slope of the tangent to the concentration-time curve at that particular instant. This is shown in Figure 12.2. Square brackets, [ ], are used here to denote concentration in moles per liter. From the tangent to the curve we can write

$$\text{rate} = \frac{\Delta[B]}{\Delta t} \tag{12.2}$$

We can also express the rate of the above reaction in terms of the concentration of the reactant A, because its concentration is also changing with time. The rate

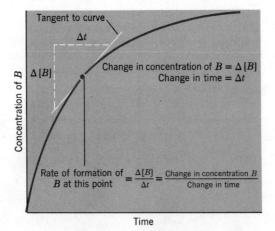

Figure 12.2

Estimation of the rate of reaction based on the change in concentration of B with time.

measured in terms of the concentration of A would be

$$\text{rate} = \frac{-\Delta[A]}{\Delta t} \qquad [12.3]$$

The minus sign indicates that the concentration of A is decreasing with time. A minus sign is always used whenever reactants are employed to express the rate.

When measuring the rate of any chemical reaction, the concentration that is monitored and the technique that is used to measure the change depends on the nature of the reaction. For example, for gaseous reactions the pressure can be followed, provided that there is a change in the number of moles of gas as the reaction proceeds. On the other hand, if a colored reactant or product is involved, the intensity of color can be monitored during the reaction. Whatever method of analysis is employed, it must be fast, accurate, and in no way interfere with the normal course of the reaction being studied.

12.2 RATE LAWS

In this section we begin to examine the factors that control the rate of reaction. Not all reactions take place at the same rate. Ionic reactions like those described in Chapter 6 are virtually instantaneous; the speed is determined by how rapidly we can mix the chemicals. Other reactions, such as the digestion of food, take place more slowly. These different rates exist primarily because of chemical differences among the reacting substances.

For any given reaction, one of the most important controlling influences is the concentrations of the reactants. Generally, if we follow a chemical reaction over a period of time, we find that its rate gradually decreases as the reactants are consumed. From this we conclude that the rate is related, in some way, to the concentrations of the reacting species. In fact, the rate is nearly always directly proportional to the concentration of the reactants raised to some power. This means that for the general reaction

$$A \longrightarrow B$$

the rate can be written as

$$\text{rate} \propto [A]^x \qquad [12.4]$$

where the exponent, x, is called the **order of the reaction.** When $x = 1$ we have a first-order reaction. An example is the decomposition of cyclopropane mentioned earlier,

$$\text{rate} \propto [\text{cyclopropane}]^1$$

In some cases an exponent can even be negative, which means that increasing the concentration of that reactant decreases the reaction rate.

Second ($x = 2$), third ($x = 3$), and higher-order reactions are also possible, as are reactions in which x is a fraction. There are also examples of zero-order reactions, where $x = 0$. For a zero-order reaction the rate is constant and does not depend on the concentration of the reactants. An example of this type of reaction is the decomposition of ammonia on a platinum or tungsten metal surface. The rate at which the ammonia decomposes is always the same, regardless of its concentration. Another example of a zero-order process is the elimination of ethyl alcohol by the body. Regardless of how much alcohol is present in the bloodstream, its rate of expulsion from the body is constant. Thus the rate is independent of concentration.

A very important fact is that there is not necessarily any direct relationship between the coefficients in the chemical equation for a reaction and the order of the reaction. *The value of x can only be determined from experiment.*

Without doing experiments, you cannot predict for sure what the order of a reaction will be.

If we consider a slightly more complex reaction, for example,

$$A + B \longrightarrow \text{products}$$

the rate usually depends on the concentrations of both A and B. Normally, increasing the concentration of either A or B will increase the reaction rate, and the rate is proportional to the product of the concentrations of A and B, each raised to some power.

$$\text{rate} \propto [A]^x[B]^y \qquad [12.5]$$

We say that the order of the reaction with respect to A is x, that the order with respect to B is y, and that the **overall order** (i.e., the sum of the individual orders) is $x + y$. Once again, x and y can have whole-number, fractional, or even zero values. When one of the exponents is zero, this simply means that the rate of the reaction is independent of the concentration of that substance. For example, for the reaction

$$NO_2\ (g) + CO\ (g) \longrightarrow CO_2\ (g) + NO\ (g)$$

at temperatures below 225°C, the relationship between concentration and rate is

$$\text{rate} \propto [NO_2]^2$$

The rate is independent of the CO concentration but depends on the *square* of the NO_2 concentration. We say that the reaction is second order with respect to NO_2 and zero order with respect to CO. Notice that there is no relationship between the coefficients and exponent. As mentioned above, the order of a reaction can only be determined experimentally.

The proportionality represented by Equation 12.5 can be converted to an equality by introducing a proportionality constant, which we call the **rate constant.** The resulting equation, termed the **rate law** for the reaction, is

$$\text{rate} = k[A]^x[B]^y$$

For example, the rate law for the reaction between ICl and H_2,

$$2ICl\ (g) + H_2\ (g) \longrightarrow I_2\ (g) + 2HCl\ (g)$$

at 230°C has been found experimentally to be

$$\text{rate} = 0.163\ \text{liter mol}^{-1}\ \text{s}^{-1}\ [ICl][H_2]$$

> The rate law lets us calculate the rate for any particular set of concentrations.

This reaction, therefore, is *first order* with respect to both ICl and H_2 (hence *second order*, overall) and has as its rate constant $k = 0.163$ liter mol^{-1} s^{-1}. It should be noted that this value of k applies only at 230°C for this particular reaction. Other reactions have other values of k, and, as we will see later, k varies with temperature.

How can a rate law such as this be determined? One way is to perform a series of experiments in which the initial concentration of each reactant is systematically varied. Once again we can use as our example the simple reaction

$$A \longrightarrow B$$

The rate law for this reaction would take the form

$$\text{rate} = k[A]^x$$

If the reaction were first order, the value of x would be 1, and the rate expression would then be

$$\text{rate} = k[A]$$

> For a first-order reaction, increasing the concentration by a factor of 2 increases the rate by a factor of 2^1.

This means that the rate of the reaction varies directly with the concentration of A raised to the first power. As a result, if we were to double the concentration of A from one experiment to another, we would also find that the rate would increase by a factor of 2. We conclude, therefore, that *when the reaction rate is doubled by doubling the concentration of a reactant, the order with respect to that reactant is 1.*

Suppose, now, that the rate law were, instead,

$$\text{rate} = k[A]^2$$

In this instance, a twofold increase in the concentration would cause a *fourfold* increase in rate. To see this, let's imagine that the initial rate was measured with the concentration of A equal to, say a moles/liter. This rate would be given by

$$\text{rate} = k(a)^2$$

Now, if the reaction were repeated with $[A] = 2a$, the rate would be

$$\text{rate} = k(2a)^2$$

or

$$\text{rate} = 4ka^2$$

For a second-order reaction, increasing the rate by a factor of 2 increases the rate by a factor of 2^2.

which is four times the previous rate. Thus, *if the rate is increased by a factor of four when the concentration of a reactant is doubled, the reaction is second order with respect to that component. Similarly, we predict that the rate of a third order reaction would undergo an eightfold increase when the concentration is doubled ($2^3 = 8$).*

The following examples illustrate how we can use these ideas to obtain the rate law for a reaction by varying the concentrations of reactants.

EXAMPLE 12.2

Below are some data collected in a series of experiments on the reaction of nitric oxide with bromine:

$$2NO\ (g) + Br_2\ (g) \longrightarrow 2NOBr\ (g)$$

at 273°C.

Experiment	Initial Concentration (mol/liter)		Initial Rate (mol liter^{-1} s^{-1})
	NO	Br$_2$	
1	0.10	0.10	12
2	0.10	0.20	24
3	0.10	0.30	36
4	0.20	0.10	48
5	0.30	0.10	108

Determine the rate law for the reaction and compute the value of the rate constant.

SOLUTION

The rate law for the reaction will have the form

$$\text{rate} = k[NO]^x[Br_2]^y$$

To determine each exponent, we will study how the rate changes when the concentration of one reactant varies while that of the other reactant stays the same. For instance, when the NO concentration is held constant, we can see how changes in the Br$_2$ concentration affect the rate and thereby determine what y must be. The value of x is determined in a similar way. With this strategy in mind, let's study the data.

In experiments 1 to 3, the concentration of NO is constant and the concentration of Br$_2$ is varied. When the concentration of Br$_2$ is doubled (experiments 1 and 2), the rate is increased by a factor of 2; when it is tripled (experiments 1 and 3), the rate is increased by a factor of 3. The only way this could happen is if the concentration of Br$_2$ appears to the first power in the rate law.

Comparing experiments 1 and 4, we see that when the Br_2 concentration is held constant, the rate increases by a factor of 4 when the NO concentration is multiplied by 2. Similarly, raising the concentration of NO by a factor of 3 causes a ninefold increase in rate (experiments 1 and 5). This means that the exponent of the NO concentration in the rate law must be 2. Therefore,

$$rate = k[NO]^2[Br_2]$$

The rate constant can be evaluated using the data from any of these experiments. Working with experiment 1, we have

$$12 \text{ mol liter}^{-1} \text{ s}^{-1} = k(0.10 \text{ mol liter}^{-1})^2(0.10 \text{ mol liter}^{-1})$$

$$12 \text{ mol liter}^{-1} \text{ s}^{-1} = k(0.0010 \text{ mol}^3 \text{ liter}^{-3})$$

Solving for k, we get

$$k = \frac{12 \text{ mol liter}^{-1} \text{ s}^{-1}}{1.0 \times 10^{-3} \text{ mol}^3 \text{ liter}^{-3}} = 1.2 \times 10^4 \text{ liter}^2 \text{ mol}^{-2} \text{ s}^{-1}$$

You might wish to verify for yourself that the same rate constant is obtained from the other data. (That's why it's called the rate *constant!*)

EXAMPLE 12.3

The following data were collected for the reaction of *t*-butyl bromide, $(CH_3)_3CBr$, with hydroxide ion at 55°C.

$$(CH_3)_3CBr + OH^- \longrightarrow (CH_3)_3COH + Br^-$$

	Initial Concentration (M)		Initial Rate of Formation of $(CH_3)_3COH$
	---	---	---
Experiment	$(CH_3)_3CBr$	OH^-	(mol liter^{-1} s^{-1})
1	0.10	0.10	0.0010
2	0.20	0.10	0.0020
3	0.30	0.10	0.0030
4	0.10	0.20	0.0010
5	0.10	0.30	0.0010

What is the rate law and rate constant for this reaction?

SOLUTION

Based on the equation, we expect a rate law of the form

$$rate = k[(CH_3)_3CBr]^x[OH^-]^y$$

To obtain x and y, we follow the same approach as in the previous example.

Let's examine experiments 1, 2, and 3 first. In each of these the OH^- concentration is the same. Doubling the $(CH_3)_3CBr$ concentration doubles the rate; tripling it triples the rate. The order with respect to $(CH_3)_3CBr$ must therefore be 1.

In experiments 1, 4, and 5, the $(CH_3)_3CBr$ concentration is the same. Changing the OH^- concentration has no effect on the rate. This means that the reaction is zero order with respect to OH^-. Therefore,

$$rate = k[(CH_3)_3CBr]^1[OH^-]^0$$

Since anything raised to the zero power is 1,

$$rate = k[(CH_3)_3CBr]^1 \cdot 1$$

When no exponent is written, it's assumed to be 1.

or

$$rate = k[(CH_3)_3CBr]$$

The final rate law contains only the concentration of $(CH_3)_3CBr$, because this is the only concentration that affects the rate. To solve for the rate constant we can use the results of any of the experiments. Using experiment 1 and substituting the rate and concentration into the rate law gives

$$0.0010 \text{ mol liter}^{-1} \text{ s}^{-1} = k(0.10 \text{ mol liter}^{-1})$$

$$k = \frac{0.0010 \text{ mol liter}^{-1} \text{ s}^{-1}}{0.10 \text{ mol liter}^{-1}}$$

$$= 0.010 \text{ s}^{-1}$$

In Example 12.1, the exponents in the rate law just happen to be the same as the coefficients in the balanced equation. This is not true in Example 12.2. Please keep in mind that the *only* way that we can find the exponents in the rate law for a chemical reaction is by experimentally measuring the way that the concentrations of the reactants affect the rate. It is also important to remember that since temperature is another factor that influences the rate, a given value of k applies only at *one* temperature (the temperature at which it was measured).

12.3 CONCENTRATION AND TIME: HALF-LIVES

The rate law for a reaction tells us how the rate of a reaction is related to the concentrations of the reactants. By applying calculus to the rate law—which we won't attempt to go through here—an expression relating the concentration to time can be derived. For example, a first-order rate law such as

$$\text{rate} = k[A]$$

gives the expression

$$\ln \frac{[A]_0}{[A]_t} = kt \qquad [12.6]$$

where $[A]_0$ is the initial concentration of A (at time t equal to zero), $[A]_t$ is the concentration at some time t after the beginning of the reaction, and the symbol "ln" tells us to take the natural logarithm of the ratio of $[A]_0$ divided by $[A]_t$. Natural logarithms occur frequently in the sciences, and if you haven't encountered them before, you should study Appendix A.3 at the back of the book.

In terms of common, or base 10 logarithms, Equation 12.6 can be written

$$2.303 \log \frac{[A]_0}{[A]_t} = kt$$

The expression ln **x** asks the question, to what power must the number **e** be raised to give **x**? The value of **e** = 2.71828. . . .

The factor 2.303 converts common logs to natural logs. If you have a scientific calculator, you will probably find it easier to work with natural logarithms.

Equation 12.6 or its counterpart in common logarithms is useful because it allows us to calculate, for a first-order reaction, the concentration of the reactant at any time during the course of the reaction, provided that we know the value of k. If we know $[A]_0$, $[A]_t$, and k, we can also compute t—the length of time that the reaction has progressed. We will see in Chapter 24 that this is useful in archaeological dating using radioactive isotopes.

You should learn the relationship ln **x** = 2.303 log **x**.

EXAMPLE 12.4   At 400°C, the first-order conversion of cyclopropane into propylene has a rate constant of $1.16 \times 10^{-6} \text{ s}^{-1}$. If the initial concentration of cyclopropane is 1.00×10^{-2} mol/liter at 400°C, what will its concentration be 24.0 hours after the reaction begins?

SOLUTION   To solve this problem we use Equation 12.6. Our data are

$$[\text{cyclopropane}]_0 = 1.00 \times 10^{-2} \text{ mol/liter}$$

$$k = 1.16 \times 10^{-6} \text{ s}^{-1}$$

$$t = 24.0 \text{ hr}$$

The first step is to solve for the ratio of concentrations.

$$\ln \frac{[\text{cyclopropane}]_0}{[\text{cyclopropane}]_t} = (1.16 \times 10^{-6} \text{ s}^{-1})(24.0 \text{ hr}) \left(\frac{3600 \text{ s}}{1 \text{ hr}} \right)$$

$$= 0.100$$

To obtain the concentration ratio we must take the antilog. If $\ln x = a$, then $x = e^a$. Therefore,

Most scientific calculators handle natural logarithms and their antilogarithms easily. See Appendix A.

$$\frac{[\text{cyclopropane}]_0}{[\text{cyclopropane}]_t} = e^{0.100}$$

$$= 1.11$$

Now we solve for $[\text{cyclopropane}]_t$.

$$[\text{cyclopropane}]_t = \frac{[\text{cyclopropane}]_0}{1.11}$$

$$= \frac{1.00 \times 10^{-2} \text{ mol/liter}}{1.11}$$

$$= 9.01 \times 10^{-3} \text{ mol/liter}$$

After 24 hours the concentration of cyclopropane will have dropped to $9.01 \times 10^{-3} \text{ } M$.

The equation relating concentration to time is different for different reaction orders. For instance, for a second-order reaction having the rate law

$$\text{rate} = k[B]^2$$

the relationship is

$$\frac{1}{[B]_t} - \frac{1}{[B]_0} = kt \qquad [12.7]$$

Even more complicated equations occur when the rate law is more complex, but we won't discuss them.

Half-lives

An important quantity, particularly for first-order reactions, is the **half-life,** $t_{1/2}$—*the length of time required for the concentration of the reactant to be decreased to half of its initial value.* At this point, $t = t_{1/2}$, and

$$[A]_t = \tfrac{1}{2}[A]_0$$

Substituting into Equation 12.6 gives

$$\ln \frac{[A]_0}{\tfrac{1}{2}[A]_0} = kt_{1/2}$$

$$\ln 2 = kt_{1/2}$$

$$0.693 = kt_{1/2}$$

Solving for $t_{1/2}$ gives

$$t_{1/2} = \frac{0.693}{k} \qquad [12.8]$$

Equation 12.8 tells us that for a first-order reaction, $t_{1/2}$ depends only on k—it is constant throughout the reaction. If the half-life for a particular first-order reaction is 1 hour, then during the first hour the concentration drops to half of its initial value. During the second hour the concentration is again cut in half, so after a total of 2 hours the concentration is $\frac{1}{4}$ of its initial value (Figure 12.3).

The half-life of a second-order reaction differs from that of a first-order process by being concentration dependent. Following the same procedure as above, we find that a second-order reaction whose rate law is

$$\text{rate} = k[B]^2$$

has a half-life given by

$$t_{1/2} = \frac{1}{k[B]_0} \qquad [12.9]$$

This means that if we double the initial concentration, the half-life is halved.

For a first-order reaction,

Number of Half-lives	Fraction of Reactant Remaining
0	1
1	1/2
2	1/4
3	1/8
4	1/16
5	1/32
⋮	⋮
n	$1/2^n$

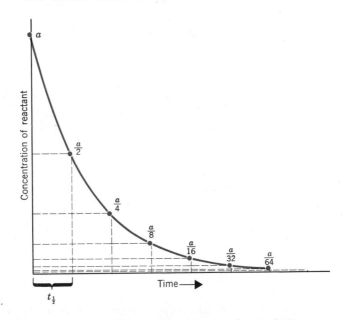

Figure 12.3

A graph of a first-order reaction illustrating the concept of half-life.

EXAMPLE 12.5

The decomposition of N_2O_5 dissolved in carbon tetrachloride is a first-order reaction. The chemical change is

$$2N_2O_5 \longrightarrow 4NO_2 + O_2$$

At 45°C the reaction was begun with an initial N_2O_5 concentration of 1.00 mol/liter. After 3.00 hours the N_2O_5 concentration had decreased to 1.21×10^{-3} mol/liter. What is the half-life of N_2O_5 expressed in minutes at 45°C?

SOLUTION

To obtain the half-life, we must know the value of k. Since the reaction is first order we can use Equation 12.6.

$$\ln\left(\frac{1.00 \text{ mol/liter}}{1.21 \times 10^{-3} \text{ mol/liter}}\right) = k(3.00 \text{ hr}) \times \left(\frac{3600 \text{ s}}{1 \text{ hr}}\right)$$

$$\ln(826) = (1.08 \times 10^4 \text{ s})k$$

$$6.72 = (1.08 \times 10^4 \text{ s})k$$

Therefore,

$$k = \frac{6.72}{1.08 \times 10^4 \text{ s}}$$

$$= 6.22 \times 10^{-4} \text{ s}^{-1}$$

Now we can calculate the half-life:

$$t_{1/2} = \frac{0.693}{6.22 \times 10^{-4} \text{ s}^{-1}}$$

$$= 1.11 \times 10^3 \text{ s}$$

This is 18.5 minutes.

12.4 COLLISION THEORY

For a chemical reaction to occur, the reacting molecules must collide with each other. This idea forms the basis of the **collision theory** of chemical kinetics. Basically, this theory states that the rate of a reaction is proportional to the number of collisions occurring each second between the reacting molecules:

$$\text{rate} \propto \frac{\text{number of collisions}}{\text{second}} \qquad [12.10]$$

As we will see shortly, this permits us to explain the dependence of reaction rate on the concentration of the reactants. In Section 12.6 we will also see that only a fraction of the total number of collisions each second are effective at producing a net chemical change, and that this fraction depends both on the nature of the reactants and on the temperature.

At this point, let's see how collision theory accounts for the way that the rate of a reaction depends on the concentrations of the reactants. Suppose that we have a reaction that occurs by the collision of two molecules, such as

$$A + B \longrightarrow \text{products}$$

In this case we are assuming that we know precisely what occurs between A and B—that is, we assume for the sake of this discussion that the products are formed in **bimolecular** (two-molecule) **collisions** between A and B.

According to our theory, the rate of the reaction is proportional to the number of collisions each second between molecules of A and B. If the concentration of A were doubled, then the number of A-B collisions would also be doubled because there would be twice as many A molecules that can collide with B. This would increase the rate by a factor of 2. Similarly, if the concentration of B were doubled, there would be a twofold increase in the number of A-B collisions and the rate would also increase by a factor of 2. From our previous discussion we conclude that the order with respect to each reactant is 1, so the rate law for this bimolecular collision process is

$$\text{rate} = k[A][B]$$

Now let's look at what would happen if we had a reaction of the type

$$2A \longrightarrow \text{products}$$

in which the reaction occurs by the collision of two A molecules. In this instance, if we double the concentration of A, we double the number of collisions that each *single* A molecule makes with its neighbors, because we have doubled the number of neighbors. However, we have also doubled the number of A

We are doubling both the number of molecules colliding and the number of collisions that each of them makes.

molecules that are colliding. The number of *A-A* collisions has therefore "doubly doubled"—that is, increased by a factor of 2 squared. Consequently, the rate law for this bimolecular reaction between identical molecules is

$$\text{rate} = k[A]^2$$

What we find, then, is that *if* we know what collision process is involved in the production of products, we can predict, on the basis of collision theory, what the rate law for that process will be. *The exponents in the rate law are equal to the coefficients in the balanced equation for that collision process.*

At this point we might ask, why is it necessary to determine the rate law for a reaction experimentally? Why can't we simply use the coefficients of the balanced overall equation to deduce the rate law? The answer is that when we begin to study a reaction, we don't know what collision processes are involved. Are *A-B* collisions important or do *A-A* collisions determine the rate? We can't answer that question until we know the mechanism of the reaction, and that's our next topic.

12.5 REACTION MECHANISMS

The overall balanced equation for a reaction represents the net chemical change that occurs as the reaction proceeds to completion. This does not mean, however, that all the reactants must come together simultaneously to undergo a change that produces the products. In fact, the net change can (and usually does) actually represent the sum of a series of simple reactions. These simple reactions are referred to as **elementary processes.** The sequence of elementary processes that ultimately leads to the formation of the products is called the **reaction mechanism.** For example, it appears that the reaction

$$2NO + 2H_2 \longrightarrow 2H_2O + N_2$$

proceeds by the three-step mechanism,

$$2NO \longrightarrow N_2O_2$$

$$N_2O_2 + H_2 \longrightarrow N_2O + H_2O$$

$$N_2O + H_2 \longrightarrow N_2 + H_2O$$

The sum of these steps in the sequence does give us the overall balanced equation.

Reaction mechanisms, such as the one just described, are usually arrived at by bringing together both theory and experiment. For instance, suppose we wished to study the following hypothetical reaction in hopes of discovering the mechanism.

$$2A + B \longrightarrow C + D$$

We begin by first determining the rate law, perhaps by studying how the rate changes as we vary the concentrations of *A* and *B* as described earlier. Let us say it turned out to be

$$\text{rate} = k[A]^2[B]$$

Next, we attempt to propose a mechanism that, by the application of the principles of collision theory, gives us a predicted rate law that is the same as the one found by experiment.

Since we are beginners at proposing mechanisms, we might be tempted to propose a one-step mechanism in which two molecules of *A* and one of *B* come together simultaneously—that is, a three-body or *termolecular collision*. This process,

$$2A + B \longrightarrow C + D$$

indeed leads to the rate law

$$\text{rate} = k[A]^2[B]$$

when we take the coefficients to be equal to the exponents, and it is the same as the rate law found from experiment. But we must now ask ourselves, is this a realistic mechanism? A simultaneous three-body (termolecular) collision is statistically a very unlikely event, and it has been generally found that reactions that must proceed by such a path are very slow. As a result, a third-order reaction such as this, if it is fairly rapid, is usually interpreted as taking place by way of a series of simple bimolecular processes. (Back to the drawing board!)

One possible sequence of reactions is

$$2A \longrightarrow A_2$$

$$A_2 + B \longrightarrow C + D$$

Here we have two steps in which we propose that some relatively unstable intermediate, A_2, is first formed by the collision of two molecules of A. In a second step a reaction between A_2 and B produces the products C and D. Again the sum of these elementary processes gives us our net overall change.

Both of these reactions are unlikely to occur at the same rate, so let's suppose that the first reaction is slow, and that once the intermediate, A_2, is formed it reacts rapidly with B in the second step to produce the products. If this were true, the rate at which the final products appear is actually determined by how fast A_2 is produced. This first step serves as a "bottleneck" in the reaction path. We refer to this slowest step as the **rate-determining step** in the reaction because it governs how rapidly the overall reaction takes place. Because the rate-determining step is an elementary process (in this instance a bimolecular collision between two A molecules), we can predict with the aid of collision theory that the rate law should be

The reaction can't proceed any faster than its slowest step.

$$\text{rate} = k[A]^2$$

If this is the rate law for the rate-determining step, it will also be the rate law for the overall reaction. However, this rate law cannot be the correct one because it is not the same as the one determined from experiment. This does not necessarily mean that our mechanism is wrong, but it does mean that the first step cannot be the rate-determining step. Let us see what we would expect to observe if the second step were slow, instead of the first. In this case the rate law would be

$$\text{rate} = k[A_2][B] \qquad [12.11]$$

However, this rate law contains the concentration of the proposed intermediate (A_2) and the experimental rate law contains only the concentration of reactants A and B. How can we express the concentration of A_2 in terms of A and B?

Once A_2 has been formed, it can react in either of two ways. Since we propose that A_2 is unstable (if it were stable we could isolate it and there would be no question at all about the path of the overall reaction) it can undergo decomposition to reform two molecules of A. The other possibility is that it undergoes a collision with B that leads to the formation of the products, C and D. Our mechanism therefore should include a reaction that allows A_2 to decompose.

$$A_2 \longrightarrow 2A$$

Our total mechanism is now

$$2A \longrightarrow A_2$$

$$A_2 \longrightarrow 2A$$

$$A_2 + B \longrightarrow C + D$$

If the rate at which the intermediate is formed from reactant A is equal to the rate at which A is formed from intermediate A_2, then these two reactions represent a state of dynamic equilibrium. We could therefore write our first two equations as an equilibrium, which would take the form

$$2A \rightleftharpoons A_2$$

We are now back to a two-step mechanism in which the first step is an equilibrium. Our mechanism is now

$$2A \rightleftharpoons A_2 \qquad \text{fast}$$

$$A_2 + B \longrightarrow C + D \qquad \text{slow}$$

Since, in an equilibrium situation, the rate of the forward reaction (rate_f) is equal to the rate of the reverse reaction (rate_r),

$$\text{rate}_f = k_f[A]^2 = \text{rate}_r = k_r[A_2]$$

or simply

$$k_f[A]^2 = k_r[A_2]$$

Solving this equation for $[A_2]$, we have

$$[A_2] = \frac{k_f[A]^2}{k_r}$$

We can now substitute this into Equation 12.11 and combine all the constants to give still another constant, let us say k'. This gives the rate expression

$$\text{rate} = k'[A]^2[B]$$

which does agree with the rate law found from experiment. Our proposed mechanism, therefore, *appears* to be a good one. However, a mechanism is, in essence, a theory. It is a sequence of steps that we dream up to explain the chemistry and to provide a rate law that agrees with experiment. It frequently happens, though, that more than one mechanism can be written to satisfy both criteria, so we can never be certain we have truly discovered the actual path of the reaction. Nevertheless, studying the kinetics of a reaction gives clues to what the mechanism may be and allows us to discard many alternatives. After that, we can only hope to gather further information that either supports (or proves wrong) our guess.

EXAMPLE 12.6   The decomposition of NO_2Cl is believed to involve the two-step mechanism,

$$NO_2Cl \longrightarrow NO_2 + Cl$$

$$NO_2Cl + Cl \longrightarrow NO_2 + Cl_2$$

What would be the observed experimental rate law if the first step were slow and the second were fast?

SOLUTION   If the first reaction is the slow step, it is also the rate-determining step. The rate law for the overall reaction should be the same as the rate law for the rate-determining step. Since only one molecule of NO_2Cl is involved, the rate law for the first reaction—as well as for the overall reaction—would be

$$\text{rate} = k[NO_2Cl]$$

12.6 EFFECTIVE COLLISIONS

If all the collisions that take place in a reaction vessel were effective in producing chemical change, all chemical reactions—including biochemical ones—would be over almost instantaneously. Since living creatures have finite life spans, it is clear that some factor (or factors) must intervene to decrease reaction rates to a reasonable level. Consider, for example, the decomposition of hydrogen iodide,

$$2HI \ (g) \longrightarrow H_2 \ (g) + I_2 \ (g)$$

At a concentration of only 10^{-3} mol/liter of HI there are approximately 3.5×10^{28} collisions per liter per second at 500°C. This is equivalent to 5.8×10^4 moles of collisions per liter per second; if each of these collisions were effective, we would expect a rate of reaction 5.8×10^4 mol liter^{-1} s^{-1}. Actually, the rate under these conditions is only about 1.2×10^{-8} mol liter^{-1} s^{-1}. This is smaller by a factor of approximately 5×10^{12} than we would observe if all collisions led to reaction, which means that only one out of every five thousand billion collisions actually leads to the formation of the products! Clearly, not all encounters between HI molecules result in the production of H_2 and I_2. In fact, only a very small fraction of the total number of collisions are effective. If we let Z be the total number of collisions that occur per second and f be the fraction of the total number of collisions that are effective, then the rate of a reaction would be

$$\text{rate} = fZ \qquad [12.12]$$

The fraction, f, is determined by the energies of the molecules that collide and, as we will see shortly, a certain minimum energy is required in order to cause a reaction to occur. In addition to this, in many instances the molecules must also collide with the proper orientation. The decomposition of a hypothetical AB molecule, whose collisions result in the formation of A_2 and B_2, can serve as an example.

$$2AB \longrightarrow A_2 + B_2$$

In order for the products A_2 and B_2 to be produced, the two atoms of A and two atoms of B must approach each other very closely so that $A—A$ and $B—B$ bonds can be formed. Suppose, now, that two $A—B$ molecules come together in a collision oriented as shown in Figure 12.4a. We certainly do not expect this collision to be effective in forming the products. However, a collision in which the AB molecules are aligned as shown in Figure 12.4b can lead to the creation of $A—A$ and $B—B$ bonds and, hence, to a net chemical change. Thus the number of effective collisions—and, therefore, the rate of the reaction—is further decreased by a factor, p, that is a measure of the importance of the molecular orientations during collision:

$$\text{rate} = pfZ \qquad [12.13]$$

Figure 12.4

Collisions between A—B molecules. (a) A collision that cannot produce a net chemical change. (b) A collision that can lead to a net reaction.

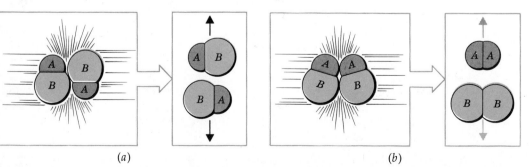

(a)   (b)

We have already seen that Z, the collision frequency, is proportional to the concentrations of the reacting molecules; therefore, in general,

$$Z = Z_0[A]^n[B]^m \ldots$$

where Z_0 is the collision frequency when all of the reactants are at unit concentration. Substituting this into Equation 12.13 gives us

$$\text{rate} = pfZ_0[A]^n[B]^m \ldots$$

or

$$\text{rate} = k[A]^n[B]^m \ldots$$

where $k = pfZ_0$. This then is the rate law derived from the principles of collision theory.

12.7 TRANSITION STATE THEORY

The collision theory, discussed in Section 12.4, focused attention primarily on the relationship between reaction rate and the *number* of collisions per second between reactant molecules. In this way the dependence of rate on concentration was explained. The **transition state theory** is concerned with what actually happens *during* a collision. It follows the energy and orientation of the reactant molecules as they collide, and in doing so seeks explanations of why so few collisions out of the many that occur are actually effective.

A collision between two molecules is quite unlike a collision between two billiard balls. The electron cloud of a molecule has no sharp boundary; its outer reaches are somewhat soft and "fuzzy." Therefore, when two molecules approach each other in a collision, the electron clouds experience a gradual increase in their mutual repulsions, and the molecules begin to slow down. As this happens, the kinetic energy of the molecules is gradually converted to potential energy—somewhat like compressing a spring. If the pair of colliding molecules had little kinetic energy to begin with—that is, if they were not moving very fast—they come to a stop before their electron clouds have penetrated each other very much, and then they simply fly apart again, chemically unchanged (Figure 12.5a).

Figure 12.5

(a) *When two slow-moving molecules collide, their electron clouds cannot interpenetrate much and they just bounce off each other, chemically unchanged. (b) When fast-moving molecules collide, atoms approach each other much more closely as their electron clouds interpenetrate. This can lead to bond making and bond breaking. The net change here is* $AB + C \rightarrow A + BC$.

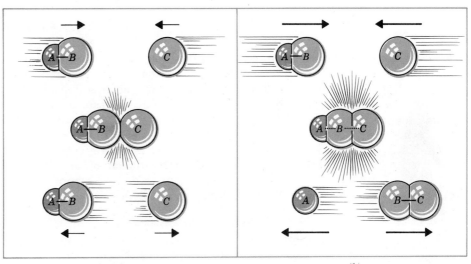

(a)   (b)

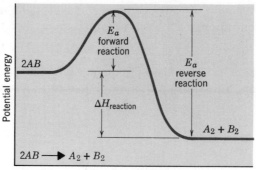

Figure 12.6

Potential energy diagram for an exothermic reaction.

When two fast-moving molecules collide, they have a lot of kinetic energy that can be converted to potential energy. This means that they are able to overcome substantial forces of repulsion between their electron clouds, and they approach each other quite closely. As illustrated in Figure 12.5b, the interpenetration of the electron clouds that occurs permits a reshuffling of electrons—old bonds are broken as new ones form and a net chemical change takes place.

From the preceding discussion we see that an effective collision—one that changes reactant molecules into product molecules—can only occur if the molecules collide with sufficient force. Expressed differently, there is a minimum kinetic energy that the molecules must possess jointly to overcome the repulsions between their electron clouds when they collide. This kinetic energy, which is changed to potential energy at the moment of impact, is called the **activation energy, E_a**.

The change in potential energy that takes place during the course of a reaction is shown in Figure 12.6. The horizontal axis is called the **reaction coordinate**—it follows the path of the reaction as molecules come together in a collision and product molecules emerge. In a sense, positions along the reaction coordinate represent the extent to which the reaction has progressed toward completion. On the left of this particular potential energy diagram we find two molecules of AB. As they approach each other, their potential energy increases to a maximum. As we continue toward the right along the reaction coordinate, the potential energy of the system decreases as the products, A_2 and B_2, move apart. When the A_2 and B_2 molecules are finally separated from one another, the total potential energy drops to essentially a constant value.

The activation energy for the decomposition of AB corresponds to the difference between the energy of the reactants and the maximum on the potential energy curve. Slow-moving molecules of AB do not possess sufficient energy to overcome this potential energy barrier, while fast-moving ones do.

In Figure 12.6 we have drawn the potential energy of the products so that it is lower than that of the reactants. The difference between them corresponds to the heat of reaction.[1] In this case, because the products are at a lower energy than the reactants, the reaction is exothermic. The energy released appears as an increase in the kinetic energy of the products; therefore, the temperature of the system rises as the reaction progresses.

In a collision, the total K.E. and P.E. are constant. If the P.E. becomes lower after the collision, the K.E. must become larger.

In the reaction mixture there are also collisions between A_2 and B_2 molecules. Such collisions, if energetic enough, can reform AB molecules. In Figure 12.6 the activation energy for the reaction

$$A_2 + B_2 \longrightarrow 2AB$$

[1] Strictly speaking, the difference in potential energy between the reactants and products corresponds to ΔE for the reaction, but we learned in Chapter 11 that ΔE and ΔH for a process are of very nearly the same magnitude.

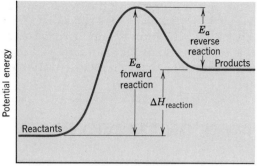

Figure 12.7

Potential energy diagram for an endothermic reaction.

An endothermic reaction is one in which K.E. is converted to P.E. In this sense a system absorbs or stores energy.

is indicated as the difference in energy between the products and the top of the potential energy hill. Since the forward reaction is exothermic, the reverse reaction is endothermic.

Figure 12.7 depicts the energy changes for a reaction that is endothermic in the forward direction. In this case the products are at a higher potential energy than the reactants. The net increase in potential energy that takes place as the products are formed occurs at the expense of kinetic energy. Consequently, there is a net overall decrease in the average kinetic energy as the reaction proceeds and the reaction mixture becomes cool.

The species that exists at the top of the potential energy barrier during an effective collision corresponds to neither the reactants nor the products but, instead, to some highly unstable combination of atoms that we speak of as the **activated complex.** This activated complex is said to exist in a **transition state** along the reaction coordinate (hence the name transition state theory).

Transition state theory views chemical kinetics in terms of the energy and geometry of the activated complex that, once it has formed, can come apart to yield the reactants again or go on to produce the products. For example, let us examine again the decomposition of hypothetical AB molecules to produce A_2 and B_2. The change that takes place along the reaction coordinate can be represented as

$$\begin{matrix} A \\ | \\ B \end{matrix} + \begin{matrix} A \\ | \\ B \end{matrix} \rightleftharpoons \begin{bmatrix} A \cdots A \\ \vdots \quad \vdots \\ B \cdots B \end{bmatrix} \longrightarrow \begin{matrix} A-A \\ + \\ B-B \end{matrix}$$

where we have used solid dashes to denote ordinary covalent bonds and dotted lines to symbolize the partially broken and partially formed bonds in the transition state, which is enclosed within brackets. Figure 12.8 illustrates this change as it occurs on the potential energy diagram for the reaction.

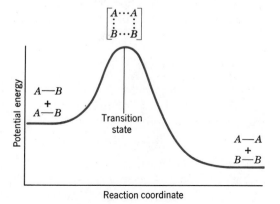

Figure 12.8

Transition state theory and the potential energy diagram for a reaction.

If the potential energy of the transition state is very high, then a great deal of energy must be available in a collision to form the activated complex. This results in a high activation energy and consequently a slow reaction. If it were possible somehow to produce an activated complex whose energy was closer to that of the reactants, the decreased activation energy would lead to a faster reaction rate.

12.8   EFFECT OF TEMPERATURE ON REACTION RATE

In nearly every instance an increase in temperature causes an increase in the rate of reaction and, as a very general rule of thumb, the rates of many reactions are about doubled by a ten-degree rise in temperature. How can this behavior be explained?

According to kinetic theory, in any system there is a distribution of kinetic energies. In the last section we interpreted the activation energy to be the minimum kinetic energy required for a collision to be effective. All molecules having kinetic energies higher than this minimum are, therefore, capable of reacting. This can be illustrated for the kinetic energy distribution in a system as shown in Figure 12.9. The total fraction of all of the molecules having energies equal to or greater than E_a corresponds to the shaded portion of the area under the curve. If we compare this area for two different temperatures, we see that the total fraction of molecules with sufficient kinetic energy to undergo effective collisions is greater at the higher temperature. As a result, the number of molecules that are capable of undergoing reaction increases with increasing temperature and, consequently, so does the reaction rate.

Measuring the activation energy

The magnitude of the rate constant, which is the rate of the reaction when all the concentrations have a value of one, depends on a number of factors. In Section 12.6 we saw that it depends in part on the collision frequency. Now we have learned that the rate of reaction also is affected by the orientations of the molecules during a collision and by the kinetic energy that they must have when they collide. These various factors are related quantitatively by the equation

$$k = Ae^{-E_a/RT}$$ [12.14]

In 1903, Arrhenius received the third Nobel Prize ever awarded in chemistry.

which is known as the **Arrhenius equation** after its discoverer, the Swedish chemist, Svante Arrhenius. In the equation, e is the base of the natural logarithms (Appendix A), R is the gas constant, T is the absolute temperature and, of course, k is the rate constant and E_a is the activation energy. The factor A is a proportionality constant whose magnitude is related to the collision frequency and also to the importance of molecular orientations during a collision.

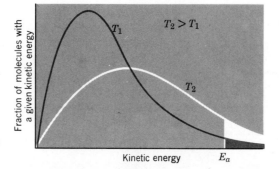

Figure 12.9

The effect of temperature on the number of molecules having kinetic energies greater than E_a.

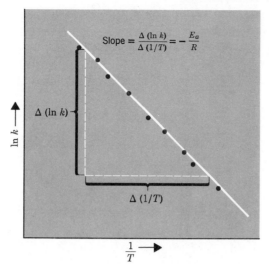

Figure 12.10

Graphical determination of the activation energy, E_a. Points on the line represent the natural logarithms of experimentally measured rate constants at various temperatures. We determine the slope of the straight line that best fits the experimental data.

The Arrhenius equation provides us with a means of determining the value of the activation energy (as well as the factor A) from measurements of the rate constant at at least two different temperatures. Taking the natural logarithm of Equation 12.14 gives

$$\ln k = \ln A - \frac{E_a}{RT} \qquad [12.15]$$

We can compare this equation to the equation for a straight line.

$$\ln k = \ln A - \frac{E_a}{R}\left(\frac{1}{T}\right)$$
$$\updownarrow \qquad \updownarrow \qquad \updownarrow \quad \updownarrow$$
$$y \; = \; b \; + \; m \quad x$$

Thus, a plot of $\ln k$ versus $1/T$ gives a straight line whose slope m is equal to $-E_a/R$ and whose intercept b with the ordinate (the vertical axis) is $\ln A$ (Figure 12.10).

We can also obtain E_a from k at two temperatures by direct computation. For any temperature, T_1, Equation 12.14 becomes

$$k_1 = Ae^{-E_a/RT_1}$$

and for any other temperature, T_2, we can write

$$k_2 = Ae^{-E_a/RT_2}$$

Dividing k_1 by k_2, we have

$$\frac{k_1}{k_2} = \frac{Ae^{-E_a/RT_1}}{Ae^{-E_a/RT_2}}$$

or

$$\frac{k_1}{k_2} = e^{(E_a/R)[(1/T_2)-(1/T_1)]} \qquad [12.16]$$

Taking the natural logarithm of both sides, we get

$$\ln\left(\frac{k_1}{k_2}\right) = \frac{E_a}{R}\left(\frac{1}{T_2} - \frac{1}{T_1}\right) \qquad [12.17]$$

This can also be expressed in terms of common logarithms (base 10 logarithms) as

$$\log\left(\frac{k_1}{k_2}\right) = \frac{E_a}{2.303R}\left(\frac{1}{T_2} - \frac{1}{T_1}\right) \qquad [12.18]$$

Equations 12.17 and 12.18 can be used to compute E_a if rate constants at two different temperatures are known. They can also be used to calculate the rate constant of some specific temperature if E_a and k at some other temperature are known. The value of R that is used in either case depends on the units of the energy. When E_a is in joules (or kilojoules), we use $R = 8.314$ J mol^{-1} K^{-1}. When E_a is in calories (or kilocalories), we use $R = 1.987$ cal mol^{-1} K^{-1}.

EXAMPLE 12.7   At 300°C the rate constant for the reaction

is 2.41×10^{-10} s^{-1}. At 400°C, k equals 1.16×10^{-6} s^{-1}. What are the values of E_a (in kilojoules per mole) and A for this reaction?

SOLUTION   We can obtain E_a by substituting values of k_1, k_2, T_1, and T_2 into either Equation 12.17 or 12.18 and then solving for E_a. To avoid confusion, let's first tabulate our data.

	k	T
1	2.41×10^{-10} s^{-1}	$300 + 273 = 573$ K
2	1.16×10^{-6} s^{-1}	$400 + 273 = 673$ K

Notice that we've converted the temperatures to kelvins. Now let's use Equation 12.18 and work in common logs. Substituting,

$$\log\left(\frac{2.41 \times 10^{-10}\ \text{s}^{-1}}{1.16 \times 10^{-6}\ \text{s}^{-1}}\right) = \frac{E_a}{2.303(8.314\ \text{J mol}^{-1}\ \text{K}^{-1})}\left(\frac{1}{673\ \text{K}} - \frac{1}{573\ \text{K}}\right)$$

$$\log(2.08 \times 10^{-4}) = \frac{E_a}{19.15\ \text{J mol}^{-1}\ \text{K}^{-1}}(0.00149\ \text{K}^{-1} - 0.00175\ \text{K}^{-1})$$

$$-3.68 = E_a(-1.36 \times 10^{-5}\ \text{J}^{-1}\ \text{mol})$$

$$E_a = \frac{-3.68}{-1.36 \times 10^{-5}\ \text{J}^{-1}\ \text{mol}}$$

$$= 2.71 \times 10^5\ \text{J/mol} = 271\ \text{kJ/mol}$$

We can now compute A from the equation[2]

$$k = Ae^{-E_a/RT}$$

[2] If your calculator has an e^x function, the simplest way to work this part of the problem is to first solve for A.

$$A = \frac{k}{e^{-E_a/RT}} = ke^{E_a/RT}$$

and then to compute A directly by multiplying k by e raised to the E_a/RT power. Thus,

$$A = (2.41 \times 10^{-10}\ \text{s}^{-1})e^{(2.71\times10^5\ \text{J mol}^{-1})/(8.314\ \text{J mol}^{-1}\ \text{K}^{-1}\times573\ \text{K})}$$

$$= (2.41 \times 10^{-10}\ \text{s}^{-1}) \times (5.07 \times 10^{24})$$

$$= 1.2 \times 10^{15}\ \text{s}^{-1}$$

The difference between this answer and the one obtained in the solution above is the result of rounding off.

Taking the natural logarithm of both sides of the equation,

$$\ln k = \ln A - \frac{E_a}{RT}$$

or, in terms of common logarithms,

$$2.303 \log k = 2.303 \log A - \frac{E_a}{RT}$$

Solving for log A,

$$\log A = \log k + \frac{E_a}{2.303RT}$$

Substituting the values for 300°C,

$$\log A = \log(2.41 \times 10^{-10}) + \frac{2.71 \times 10^5 \text{ J mol}^{-1}}{(2.303)(8.314 \text{ J mol}^{-1}\text{ K}^{-1})(573 \text{ K})}$$

$$= -9.62 + 24.7$$

$$= 15.1$$

The number of digits given after the decimal point in a logarithm is equal to the number of significant figures in the antilogarithm (see Appendix A).

Taking the antilogarithm,

$$A = 1 \times 10^{15} \text{ s}^{-1} \text{ (rounded from } 1.2 \times 10^{15})$$

Note that A must have the same units as k.

12.9 CATALYSTS

A **catalyst** is a substance that increases the rate of a reaction without being consumed; after the reaction has ceased, it can be recovered from the reaction mixture chemically unchanged. The catalyst participates in the reaction by providing a lower energy alternative mechanism for the production of the products. In Figure 12.11, note that the energy curve of the catalyzed reaction is drawn along a different reaction coordinate to emphasize that a different mechanism is involved. In addition, the energy barrier for the catalyzed path is lower than for the uncatalyzed reaction. This smaller activation energy means that in the reaction mixture there is a greater total fraction of molecules possessing sufficient kinetic energy to react (Figure 12.12). Therefore in the presence of the catalyst there are an increased number of effective collisions. Of course, an increased number of effective collisions means a greater reaction rate.

Figure 12.11

Effect of a catalyst on the potential energy diagram. The catalyst changes the reaction mechanism by providing a different, low-energy mechanism for the formation of the products. ΔH is same for each path (it must be so because ΔH is a state function).

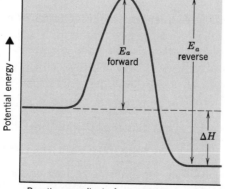

Reaction coordinate for uncatalyzed reaction

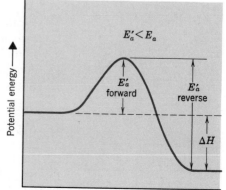

Reaction coordinate for uncatalyzed reaction

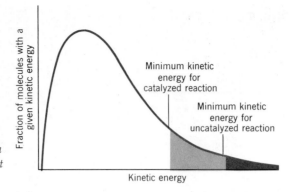

Figure 12.12

More molecules possess the minimum kinetic energy needed for an effective collision when the catalyst is present.

Since a catalyst emerges from a reaction chemically unchanged, it does not appear either as a reactant or a product in the overall balanced chemical equation. Instead, its presence is indicated by writing its name or formula over the arrow. For example, oxygen can be prepared by the thermal decomposition of potassium chlorate, $KClO_3$. In the absence of a catalyst the reaction is slow and the $KClO_3$ must be heated to a high temperature to cause it to decompose at a reasonable rate. However, if a small amount of manganese dioxide (MnO_2) is added to the $KClO_3$, the decomposition proceeds smoothly at a relatively low temperature. Analysis of the reaction mixture after the evolution of oxygen has ceased reveals that all of the MnO_2 added initially is still present, showing that MnO_2 has served as a catalyst. The equation for the catalyzed reaction is given as

$$2KClO_3 \xrightarrow{MnO_2} 2KCl + 3O_2$$

Even though a catalyst does not change the overall stoichiometry of a reaction, it does participate chemically by being consumed at one stage in the mechanism and being produced again at a later stage. This regeneration of the catalyst permits the same catalyst to be used over and over again; therefore even a small amount of catalyst can have very profound effects on reaction rate. This phenomenon is particularly significant in biological systems, where practically every reaction is catalyzed by very small quantities of highly specific biochemical catalysts called enzymes.

Catalysts may be broadly classified into two categories: homogeneous and heterogeneous catalysts. A **homogeneous catalyst** is present in the same phase as the reactants and can serve to speed up the reaction by forming a reactive intermediate with one of the reactants. For example, the decomposition of *t*-butyl alcohol, $(CH_3)_3COH$, to produce water and isobutene, $(CH_3)_2C{=}CH_2$,

$$(CH_3)_3COH \longrightarrow (CH_3)_2C{=}CH_2 + H_2O$$

is catalyzed by the presence of small amounts of HBr. In the absence of HBr the activation energy for the reaction is 65.5 kcal/mol, and below 450°C the reaction takes place at a barely perceptible rate. In the presence of HBr an activation energy of only 30.4 kcal/mol is found, and it is possible that the catalyzed reaction proceeds by an attack of HBr on the alcohol

$$(CH_3)_3COH + HBr \longrightarrow (CH_3)_3CBr + H_2O$$

followed by the rapid decomposition of the *t*-butyl bromide, $(CH_3)_3CBr$,

$$(CH_3)_3CBr \longrightarrow (CH_3)_2C{=}CH_2 + HBr$$

Thus HBr provides an alternative low-energy path for the reaction and the mechanism for the reaction when HBr is present is different from when it is absent.

Figure 12.13

Production of hydrogen atoms on a metal surface. A hydrogen molecule collides with the surface where it is adsorbed and dissociates to produce H atoms.

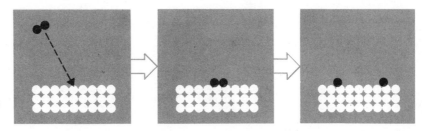

A **heterogeneous catalyst** is not in the same phase as the reactants, but provides a favorable surface on which the reaction can take place. An example of a reaction whose rate is increased by the presence of a heterogeneous catalyst is the reaction between hydrogen and oxygen to produce water. In the introduction to this chapter we pointed out that this reaction proceeds at a very slow rate when the two gases are mixed at room temperature. However, it has been found that the reaction proceeds at an appreciable rate when metals such as nickel, copper, or silver are present.

Heterogeneous catalysts appear to function through a process whereby reactant molecules are adsorbed on a surface where the reaction then takes place. The high reactivity of hydrogen, in the presence of certain metals, for example, is thought to occur by the adsorption of H_2 molecules onto the catalytic surface. On the surface of the metal the bonds between hydrogen atoms are apparently stretched or broken, as shown in Figure 12.13, so that the metal surface actually behaves as if it contained highly reactive hydrogen atoms.

Unless the reactant molecules can be adsorbed on the catalyst, no increase in reaction rate can occur. A substance whose presence during a reaction interferes with the adsorption process will therefore reduce the effectiveness of the catalysts and is thus called an **inhibitor.** These substances, by being strongly adsorbed on the catalytic surface, decrease the available space on which the reaction can occur. In some cases the catalyst eventually becomes useless and is said to be "poisoned." The destruction of catalytic activity by poisoning is very important in biological systems, as we will see in Chapter 23.

There have been many commercial and industrial applications of heterogeneous catalysts. For example, small portable flameless heaters can be purchased (for heating camping tents and the like) in which the fuel and oxygen combine on a catalytic surface. The flameless combustion evolves the same amount of heat that would be generated if the fuel were burned directly, but because no flame is involved the heaters are much safer to operate. There are drawbacks, however. Unfortunately, the catalytic oxidation of the fuel is not totally efficient, and small amounts of carbon monoxide are produced. As a result, catalytic heaters must be used with care.

Another very important application is in the control of auto exhaust emissions. Most gasoline-powered automobiles today are equipped with catalytic converters that employ a mixed metal oxide bed over which the exhaust gases pass after they are mixed with additional air (Figure 12.14). The catalyst quite effectively promotes the oxidation of CO and hydrocarbons to harmless CO_2 and H_2O. The catalysts in newer catalytic converters also remove nitrogen oxide pollutants by promoting their decomposition to nitrogen and oxygen. Catalytic mufflers suffer from the disadvantage of being poisoned by lead. As a result, lead-free fuels must be employed in autos fitted with this type of antipollution device. Another disadvantage is that they also catalyze the oxidation of SO_2 to SO_3, which then reacts with water vapor to produce a mist of sulfuric acid. Since SO_2 is produced from the combustion of high-sulfur fuels, this problem is a serious one and may ultimately lead to forced discontinuation of the use of catalytic mufflers.

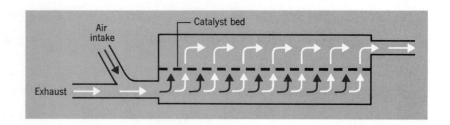

(a)

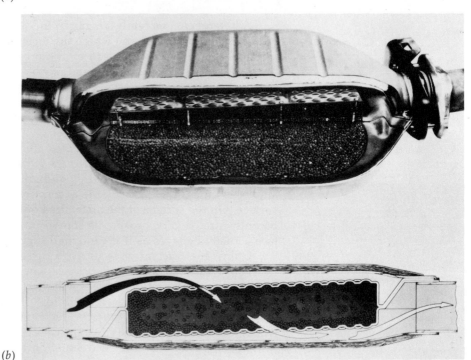

Figure 12.14

(a) In a catalytic converter, air and exhaust gases are passed over a catalyst bed where CO is oxidized to CO_2 and nitrogen oxides are decomposed to N_2 and O_2. (b) A cut-away view of a catalytic converter used by General Motors Corporation on their passenger cars and light trucks.

(b)

12.10 CHAIN REACTIONS

The reactions that we have discussed up to now have been rather simple and straightforward, with uncomplicated rate laws. There are some reactions, usually with very complex kinetics, that take place by way of an extremely reactive intermediate such as a free atom or a **free radical**—a neutral group of atoms or an ion containing one or more unpaired electrons. These reactive species can be produced either thermally (at high temperature) or by the absorption of light of an appropriate wavelength; once created they can sometimes react with other molecules to form a product *plus* yet another free atom or radical. This process, once initiated, can be repeated over and over, making the reaction *self-propagating*. The entire series of reactions that follows the production of the very reactive intermediate is called a **chain reaction.**

Free radicals produced by radiation from radioactive sources can cause serious damage to a living organism.

A chain reaction mechanism has been suggested to explain the rate equation observed for the reaction between hydrogen and bromine. The overall reaction is

$$H_2 + Br_2 \longrightarrow 2HBr$$

If this reaction proceeded simply by a bimolecular collision between H_2 and Br_2, the expected rate law would be

$$\text{rate} = k[H_2][Br_2]$$

However, the actual rate law turns out to be

$$\text{rate} = k\frac{[H_2][Br_2]^{1/2}}{1 + [HBr]/k'[Br_2]}$$

which is very complex indeed. A mechanism that has been proposed to account for this rate law is the chain reaction shown below. A dot is used to represent the unpaired electron on a highly reactive atom or free radical.

1. $Br_2 \rightarrow 2Br\cdot$	Initiation
2. $Br\cdot + H_2 \rightarrow HBr + H\cdot$	Propagation
3. $H\cdot + Br_2 \rightarrow HBr + Br\cdot$	Propagation
4. $H\cdot + HBr \rightarrow H_2 + Br\cdot$	Inhibition
5. $2Br\cdot \rightarrow Br_2$	Termination

Reaction 1 is the thermal decomposition of diatomic bromine molecules to produce bromine atoms (the reactive intermediate). The overall reaction proceeds very slowly when the two reactants are mixed at room temperature. However, at high temperature reaction 1 takes place to an appreciable extent and rapidly sets off the remaining reactions. Step 1 is therefore called the initiation step because it begins the chain. In steps 2 and 3 the product HBr is formed as well as additional free atoms that serve to keep the reaction going. These steps are then propagation steps in the chain. Step 5, which leads only to the formation of a stable species, serves to end the chain and is called the termination step. Step 4 is referred to as an inhibition step, because its occurrence removes the product and thus decreases the overall rate of production of HBr. It is included in the mechanism because the presence of HBr decreases the reaction rate (note the appearance of [HBr] in the denominator of the rate law).

In general, chain reactions are very rapid; in fact, many explosive reactions appear to occur by chain mechanisms. The production of a single reactive intermediate produces many product molecules before the chain is terminated. Consequently, the rate of production of the products is many times greater than the rate of the initiation step alone.

INDEX TO QUESTIONS AND PROBLEMS (Problem numbers in **bold type**)

General 1, 2, 5, 6, 10

Reaction rates 3, 4, **42, 43, 45**

Rate laws 7, 14, 15, **44, 46, 47, 48, 49, 50, 59**

Order of reaction 8, 11, 12

Rate constant 9, 13, **47, 53**

Concentration and time **51**

Half lives 20, 21, **52, 53**

Collision theory 16, 17

Mechanisms 18, 19, 22, 23, 24, 25, 26, 30

Effective collisions 27, 28, 29

Transition state theory 28, 29, 33, 34

Effect of temperature on rate 31, 32, 35

Arrhenius equation **54, 55, 56, 57, 58, 60, 61, 62, 63**

Catalysis 36, 37, 38, 39

Chain reactions 40, 41

REVIEW QUESTIONS

12.1 What are the four factors that control the rates of chemical reactions?

12.2 What effect does particle size have on the rate of a heterogeneous reaction?

12.3 Define *reaction rate*. What are the units?

12.4 For each of the following reactions, how would we express the reaction rate in terms of the disappearance of the reactants and the appearance of the products? Predict the role of the coefficients in the balanced overall equation in determining the relative rates of disappearance of reactants and formation of products.

(a) $2H_2 + O_2 \rightarrow 2H_2O$
(b) $2NOCl \rightarrow 2NO + Cl_2$
(c) $NO + O_3 \rightarrow NO_2 + O_2$
(d) $H_2O_2 + H_2 \rightarrow 2H_2O$

12.5 What criteria must be met by methods used to study the rate of a reaction?

12.6 Make a list of five reactions that occur in the world around you and compare their rates. Try to think of some that are fast and some that are slow.

12.7 What is a rate law? What factors affect the value of the rate constant for a given reaction?

12.8 What is meant by the *order of a reaction?*

12.9 What are the units of the rate constant for (a) a first-order reaction, (b) a second-order reaction, and (c) a third-order reaction?

12.10 Why do we say that one of the factors that influences reaction rate is the nature of the reactants?

12.11 The rate at which CO is removed from the earth's atmosphere by fungi in the soil is constant. What is the apparent order of this process?

12.12 What is the order with respect to each reactant and the overall order of the reactions described by the following rate laws:
(a) Rate = $k_1[A][B]$
(b) Rate = $k_2[E]^2$
(c) Rate = $k_3[G]^2[H]^2$

12.13 What would be the units of each of the rate constants in the preceding question if rate has the units *mol liter^{-1} s^{-1}*?

12.14 When the concentration of a reactant is doubled, by what factor would the rate of reaction be changed if the order with respect to that reactant were (a) 1 (b) 2 (c) 3 (d) 4 (e) $\frac{1}{2}$ (f) -2?

12.15 Suppose that when the concentration of a reactant was doubled the rate of reaction decreased by a factor of 2. What would be the exponent on the concentration term for that reactant in the rate law?

12.16 How does collision theory account for the dependence of rate on the concentration of the reactants?

12.17 Why can't we use the ideas of collision theory to predict in general the rate laws of chemical reactions? What must we know in order to predict the rate law?

12.18 What is meant by reaction mechanism?

12.19 Propane, C_3H_8, is a fuel used in many rural areas for cooking. It is known as LPG. It is unlikely that the combustion of propane occurs by a simple one-step mechanism,

$$C_3H_8\ (g) + 5O_2\ (g) \longrightarrow 3CO_2\ (g) + 4H_2O\ (g)$$

Explain why.

12.20 Define *half-life of a reaction.*

12.21 How is the half-life of a first-order reaction affected by the concentrations of the reactants?

12.22 A mechanism for the reaction, $2NO + Br_2 \rightarrow 2NOBr$, has been suggested to be

Step 1 $\qquad NO + Br_2 \longrightarrow NOBr_2$

Step 2 $\qquad NOBr_2 + NO \longrightarrow 2NOBr$

(a) What would be the rate law for the reaction if the first step in this mechanism were slow and the second fast?
(b) What would be the rate law if the second step were slow, with the first reaction being a rapidly established dynamic equilibrium?
(c) Experimentally, the rate law has been found to be

$$\text{rate} = k[NO]^2[Br_2]$$

What can we conclude about the relative rates of steps 1 and 2?
(d) Why do we not prefer a simple, one-step mechanism,

$$NO + NO + Br_2 \longrightarrow 2NOBr$$

(e) Can we, on the basis of the experimental rate law, definitely exclude the mechanism in part (d)?

12.23 The reaction, $NO_2(g) + CO(g) \rightarrow CO_2(g) + NO(g)$ appears to have the mechanism (at low temperature),

$$NO_2 + NO_2 \longrightarrow NO_3 + NO \qquad \text{slow}$$
$$NO_3 + CO \longrightarrow NO_2 + CO_2 \qquad \text{fast}$$

Explain why the reaction is zero-order with respect to CO.

12.24 The reaction of methyl bromide, CH_3Br, with OH^- appears to occur through a one-step mechanism involving collision of CH_3Br with OH^-,

$$CH_3Br + OH^- \longrightarrow CH_3OH + Br^-$$

The rate law for the reaction is found to be

$$\text{rate} = k[CH_3Br][OH^-]$$

In Example 12.3 we found that the rate law of the reaction of $(CH_3)_3CBr$ with OH^- has the rate law

$$\text{rate} = k[(CH_3)_3CBr]$$

Try to propose a mechanism that can account for the rate law for the reaction of $(CH_3)_3CBr$ with OH^-.

12.25 Suppose that the following sequence of reactions were proposed for a reaction.

Step 1 $\qquad 2A \longrightarrow A_2$

Step 2 $\qquad A_2 + B \longrightarrow C + 2D$

(a) What would be the overall net chemical reaction?
(b) What would the rate law be if step 1 were slow and step 2 were fast?
(c) What would the rate law be if step 2 were slow and step 1 were fast?

12.26 One of the reactions that occurs in polluted air in urban areas is $2NO_2\ (g) + O_3\ (g) \rightarrow N_2O_5\ (g) + O_2\ (g)$. It is believed that a species with the formula NO_3 is involved in the mechanism, and the observed rate law for the reaction is: rate $= k[NO_2][O_3]$. Propose a mechanism for this reaction.

12.27 How do we know that not all collisions between reactant molecules lead to chemical change? What determines whether a particular collision will be effective?

12.28 How do the orientations of molecules influence whether a collision between them can be effective at producing chemical change?

12.29 One step in the mechanism for the decomposition of NO_2Cl into NO_2 and Cl_2 appears to be $NO_2Cl + Cl \rightarrow NO_2 + Cl_2$. Given the structure of NO_2Cl,

show how molecular orientation at the moment of a collision can be important in determining whether or not the products will be formed.

12.30 Nitric acid is one of the world's most important chemicals. One of its principle uses is in making fertilizers. In 1979, over 17 billion pounds of HNO_3 were produced, mostly by oxidation of ammonia. This reaction gives nitric oxide, NO, which then reacts with oxygen to form nitrogen dioxide, NO_2. Nitrogen dioxide finally reacts with water to form nitric acid. The oxidation of NO to NO_2 follows the reaction

$$2NO\ (g) + O_2\ (g) \longrightarrow 2NO_2\ (g)$$

and has as a rate law, rate $= k[NO]^2[O_2]$. Predict a possible mechanism for this reaction.

12.31 Define the term, activation energy.

12.32 Explain qualitatively, in terms of the kinetic theory, why an increase in temperature leads to an increase in reaction rate.

12.33 Draw a potential energy diagram for an endothermic reaction. Indicate on the drawing (a) the potential energy of the reactants, (b) the potential energy of the products, (c)

the energies of activation for the forward and reverse reaction, (d) the heat of reaction.

12.34 What are meant by the terms, transition state and activated complex? Where on the potential energy diagram for a reaction will we find the transition state?

12.35 Insects, which are cold-blooded animals whose changes in body temperature tend to follow changes in the temperature of their environment, become quite sluggish in cool weather. On the basis of chemical kinetics, explain this phenomenon.

12.36 What is the difference between a homogeneous and a heterogeneous catalyst?

12.37 What is a heterogeneous catalyst? How does it function? What is an inhibitor?

12.38 How does a catalyst play a part in lowering the activation energy for a reaction?

12.39 What effect does a catalyst have on
(a) the heat of reaction?
(b) the potential energy of the reactants?
(c) the transition state?

12.40 Why are chain reactions often so fast?

12.41 The decomposition of acetaldehyde, CH_3CHO, follows the overall reaction

$$CH_3CHO \longrightarrow CH_4 + CO$$

with small amounts of H_2 and C_2H_6 also being produced. The reaction is thought to proceed by a chain reaction involving free radicals (note again that a free radical is indicated by using a dot to represent its unpaired electron). A proposed mechanism is
(1) $CH_3CHO \rightarrow CH_3\cdot + CHO\cdot$
(2) $2CH_3\cdot \rightarrow C_2H_6$
(3) $CHO\cdot \rightarrow H\cdot + CO$
(4) $H\cdot + CH_3CHO \rightarrow H_2 + CH_3CO\cdot$
(5) $CH_3\cdot + CH_3CHO \rightarrow CH_4 + CH_3CO\cdot$
(6) $CH_3CO\cdot \rightarrow CH_3\cdot + CO$
Identify (a) the initiation step, (b) the propagation step(s), and (c) the termination step(s).

REVIEW PROBLEMS (More difficult problems are marked by an asterisk)

12.42 Consider the reaction for the combustion of methane, CH_4,

$$CH_4\ (g) + 2O_2\ (g) \longrightarrow CO_2\ (g) + 2H_2O\ (g)$$

If the methane is burning at a rate of 0.16 mol liter^{-1} s^{-1}, at what rates are CO_2 and H_2O being formed?

12.43 For the reaction,

$$4NH_3\ (g) + 3O_2\ (g) \longrightarrow 2N_2\ (g) + 6H_2O\ (g)$$

it was found that at a particular instant N_2 was being formed at a rate of 0.68 mol liter^{-1} s^{-1}.

(a) At what rate was water being formed?
(b) At what rate was NH_3 reacting?
(c) At what rate was O_2 being consumed?

12.44 The rate law for a reaction was found to be

$$\text{rate} = (2.35 \times 10^{-6}\ \text{liter}^2\ \text{mol}^{-2}\ \text{s}^{-1})[A]^2[B]$$

What would the rate of reaction be if:

(a) The concentrations of A and B were 1 mol/liter?

(b) $[A] = 0.25\ M$, $[B] = 1.30\ M$?

12.45 The following data were collected for the reaction,

$$2A \rightarrow 4B + C$$

Time (min)	Concentration of A (mol/liter)	Concentration of B (mol/liter)
0	1.000	0.000
10	0.800	0.400
20	0.667	0.667
30	0.571	0.858
40	0.500	1.000
50	0.444	1.112

Make a graph of the concentrations of A and B versus time (concentrations along the vertical axis, time along the horizontal axis). Estimate the rate of disappearance of A and the rate of formation of B at $t = 25$ min and at $t = 40$ min. Compare the rates of disappearance of A and formation of B. What would you expect the rate of formation of C to be at $t = 25$ min and $t = 40$ min?

12.46 The rate constant for the reaction

$$2ICl + H_2 \longrightarrow I_2 + 2HCl$$

is 1.63×10^{-1} liter mol^{-1} s^{-1}. The rate law is given by

$$\text{rate} = k[ICl][H_2]$$

What is the rate of the reaction for each of the sets of concentrations given below?

ICl Concentration (mol/liter)	H$_2$ Concentration (mol/liter)
0.25	0.25
0.25	0.50
0.50	0.50

12.47 For the decomposition of dinitrogen pentoxide,

$$2N_2O_5 \longrightarrow 4NO_2 + O_2$$

the following data were collected:

N$_2$O$_5$ Concentration (mol/liter)	Time (s)
5.00	0
3.52	500
2.48	1000
1.75	1500
1.23	2000
0.87	2500
0.61	3000

(a) Make a graph of the concentration of N$_2$O$_5$ versus time. Draw tangents to the curve at $t = 500$, 1000, and 1500 s. Determine the rate at these different reaction times.
(b) Determine the value of the rate constant at 500, 1000, and 1500 s, given the rate law, rate $= k[N_2O_5]$.

12.48 One of the reactions that can take place in polluted air is the reaction of nitrogen dioxide, NO$_2$, with ozone, O$_3$.

$$NO_2\ (g) + O_3\ (g) \longrightarrow NO_3\ (g) + O_2\ (g)$$

The following data were collected on this reaction at 25°C:

Initial NO$_2$ Concentration (mol/liter)	Initial O$_3$ Concentration (mol/liter)	Initial Rate (mol liter^{-1} s^{-1})
5.0×10^{-5}	1.0×10^{-5}	0.022
5.0×10^{-5}	2.0×10^{-5}	0.044
2.5×10^{-5}	2.0×10^{-5}	0.022

(a) What is the rate law for the reaction?
(b) What is the value of the rate constant (give the correct units, too)?

12.49 At 27°C, the reaction, $2NOCl \rightarrow 2NO + Cl_2$, is observed to exhibit the following dependence of rate on concentration.

Initial NOCl Concentration (mol/liter)	Initial Rate (mol liter^{-1} s^{-1})
0.30	3.60×10^{-9}
0.60	1.44×10^{-8}
0.90	3.24×10^{-8}

(a) What is the rate law for the reaction?
(b) What is the rate constant?
(c) By what factor would the rate increase if the initial concentration of NOCl were increased from 0.30 to 0.45 M?

12.50 The reaction of NO with Cl$_2$ follows the equation

$$2NO + Cl_2 \longrightarrow 2NOCl$$

The following data were collected:

Initial NO Concentration (mol/liter)	Initial Cl$_2$ Concentration (mol/liter)	Initial Rate (mol liter^{-1} s^{-1})
0.10	0.10	2.53×10^{-6}
0.10	0.20	5.06×10^{-6}
0.20	0.10	10.1×10^{-6}
0.30	0.10	22.8×10^{-6}

(a) What is the rate law for the reaction?
(b) What is the value of the rate constant (be sure to give the proper units)?

12.51 The rate constant for the first-order decomposition of N_2O_5 (g) at 100°C is 1.46×10^{-1} s^{-1}.
(a) If the initial concentration of N_2O_5 in a reaction vessel was 4.50×10^{-3} mol/liter, what will the concentration be 1.00 hour after the decomposition begins?
(b) What is the half-life (in seconds) of N_2O_5 at 100°C?
(c) If the initial concentration of N_2O_5 was 4.50×10^{-3} M, what will the concentration be after three half-lives?

12.52 The decomposition of hydrogen iodide, HI, is second order. At 500°C, the half-life of HI is 2.11 min when the initial HI concentration is 0.10 M. What will be the half-life (in minutes) when the initial HI concentration is 0.010 M?

12.53 Referring to Problem 12.52, what is the rate constant for the decomposition of HI at 500°C in the units, liter mol^{-1} s^{-1}?

12.54 The rate constants for the reaction between ICl and H_2 (Question 12.46) at 230 and 240°C have been found to be 0.163 and 0.348 liter mol^{-1} s^{-1}, respectively. What are the values of E_a (in kilojoules per mole) and A for this reaction?

12.55 The rate constant for the reaction

$$CH_3I \ (g) + HI \ (g) \longrightarrow CH_4 \ (g) + I_2 \ (g)$$

at 200°C is 1.32×10^{-2} liter mol^{-1} s^{-1}. At 275°C the rate constant is 1.64 liter mol^{-1} s^{-1}. What is the activation energy (in kilojoules per mole) and the value of A?

12.56 The activation energy for the decomposition of HI

$$2HI \ (g) \longrightarrow H_2 \ (g) + I_2 \ (g)$$

is 182 kJ/mol. The rate constant for the reaction at 700°C is 1.57×10^{-3} liter mol^{-1} s^{-1}. What is the value of the rate constant at 600°C?

12.57 The activation energy for the reaction

$$HI + CH_3I \longrightarrow CH_4 + I_2$$

is 33.1 kcal/mol. At 200°C the rate constant has a value of 1.32×10^{-2} liter mol^{-1} s^{-1}. What is the rate constant at 300°C?

*12.58 A chemist was able to determine that the rate of a particular reaction at 100°C was four times faster than at 30°C. Calculate the approximate energy of activation for the reaction in kJ/mol.

*12.59 For a first order reaction, a graph of log[A] versus time (where A is a reactant) gives a straight line having a slope equal to $-k/2.303$. On the other hand, if the reaction is second-order with respect to A, a straight line is obtained when 1/[A] is plotted against time. In this case the slope of the line is equal to k. From this information, determine whether the reaction in Question 12.45 is first-order or second-order. Calculate the rate constant for the reaction.

12.60 The decomposition of C_2H_5Cl is a first-order reaction having $k = 3.2 \times 10^{-2}$ s^{-1} at 550°C and $k = 9.3 \times 10^{-2}$ s^{-1} at 575°C. What is the activation energy, in kilocalories per mole, for this reaction?

12.61 The rate constant for the reaction

$$H_2 \ (g) + I_2 \ (g) \longrightarrow 2HI \ (g)$$

was measured at a series of temperatures. The data are given below.

Temperature (°C)	k (liter mol^{-1} s^{-1})
283	1.2×10^{-4}
302	3.5×10^{-4}
355	6.8×10^{-3}
393	3.8×10^{-2}
430	1.7×10^{-1}

Graphically determine the value of E_a in kJ/mol for this reaction.

*12.62 The development of a photographic image on film is a process controlled by the kinetics of the reduction of silver halide by a developer. The time required for development at a particular temperature is inversely proportional to the rate constant for the process. Below are published data on development times for Kodak's Tri-X film using Kodak D-76 developer. From these data, estimate the activation energy for the development process in kilocalories per mole.

Temperature (°C)	Time for Development (min)
18	10
20	9
21	8
22	7
24	6

Estimate the development time at 15°C.

*12.63 The cooking of an egg involves the denaturation of a protein called albumin, and the time required to achieve a particular degree of denaturation is inversely proportional to the rate constant for the process. This reaction has a high activation energy; $E_a = 418$ kJ/mol. Calculate how long it would take to cook a traditional "three-minute egg" on top of Mt. McKinley in Alaska on a day when the atmospheric pressure there is 355 torr.

13

CHEMICAL EQUILIBRIUM

Ammonia and its salts are valuable fertilizers. Here we see gaseous ammonia being pumped directly into the soil. The manufacture of ammonia by direct combination of N_2 and H_2 constitutes one of civilization's most important chemical equilibria. In this chapter we will study the factors that affect a chemical equilibrium and how to deal with equilibria quantitatively.

When a chemical reaction takes place spontaneously, the concentrations of the reactants and products change while the free energy of the system decreases. Eventually the free energy reaches a minimum, and, as we learned in Chapter 11, the system comes to a state of equilibrium. If we follow the concentrations while this happens, we observe that they approach steady values, as shown in Figure 13.1. We find that the rate at which the reactants produce the products approaches the rate at which the products form the reactants. When equilibrium is finally reached, both forward and reverse reactions are occurring at equal rates and the concentrations no longer change. It is the continuation of the forward and reverse reactions without a change in concentrations, you recall, that identifies this as a *dynamic equilibrium*.

All chemical systems tend toward equilibrium. In this chapter we will explore the quantitative relationships that can be used to describe the equilibrium state, and we will see how the principles of kinetics and thermodynamics can be applied to a description of equilibrium.

Figure 13.1

The approach to equilibrium for the reaction $A + B \rightarrow C + D$.

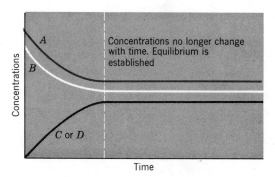

13.1 THE LAW OF MASS ACTION

a, b, e, and f are the coefficients of substances **A**, **B**, **E**, and **F**.

It was discovered experimentally some time ago that a very simple relationship governs the relative proportions of reactants and products in an equilibrium system. For the general reaction

$$aA + bB \rightleftharpoons eE + fF$$

it is observed that, at constant temperature, the condition that is fulfilled at equilibrium is

$$\frac{[E]^e[F]^f}{[A]^a[B]^b} = K_c \qquad [13.1]$$

where the quantities written within square brackets denote *equilibrium molar concentrations*. The quantity, K_c, is a constant, called the **equilibrium constant**, and the entire relationship, discovered in 1866 by the Norwegian chemists Guldberg and Waage, is known as the **law of mass action.**

The fraction appearing to the left of the equals sign in Equation 13.1 is called the **mass action expression,** and it is constructed using the coefficients in the balanced chemical equation as exponents on the appropriate concentrations. For instance, consider the reaction of nitrogen with hydrogen to form ammonia—the nitrogen fixation reaction used industrially in the production of ammonia and nitrogen fertilizers:

The mass action expression is also sometimes called the reaction quotient.

$$N_2\ (g) + 3H_2\ (g) \rightleftharpoons 2NH_3\ (g)$$

The mass action expression for this reaction is written as

$$\frac{[NH_3]^2}{[N_2][H_2]^3}$$

$$\frac{[NH_3]^2}{[N_2][H_2]^3}$$

This fraction, of course, always has some numerical value for this chemical system. For example, suppose some N_2 and H_2 were introduced into a container and permitted to react. Initially there would be no NH_3, so the value of the mass action expression would be zero. Then, as NH_3 is formed, the fraction would grow larger until, when equilibrium is finally reached, the value of the fraction would no longer change. It becomes equal to a value that we call the equilibrium constant, K_c.

$$\frac{[NH_3]^2}{[N_2][H_2]^3} = K_c \text{ (at equilibrium)} \qquad [13.2]$$

The most important point about this equation is that at a given temperature the same numerical value of the mass action expression is always obtained for *any* system containing N_2, H_2, and NH_3 in equilibrium. *There are no restrictions on the individual concentrations of any reactant or product.* The only requirement for equilibrium is that when these concentrations are substituted into the mass action expression, the fraction is numerically equal to K_c. In effect, Equation 13.2 is a condition that must be fulfilled for N_2, H_2, and NH_3 to be in equilibrium with each other, and Equation 13.2 is spoken of as the equilibrium condition or **equilibrium law** for the reaction. The data in Table 13.1 illustrate this point.

$$molarity = \frac{moles}{liters} = \frac{n}{V}$$

For an ideal gas,

$$\frac{n}{V} = \frac{P}{RT}$$

For reactions involving gases, the partial pressures of the reactants and products are proportional to their molar concentrations. The equilibrium constant expression for these reactions can therefore be written using partial pressures instead of concentrations. For example, the equilibrium condition for the reaction between N_2 (g) and H_2 (g) can also be expressed as

$$\frac{p_{NH_3}^2}{p_{N_2} p_{H_2}^3} = K_P$$

We will use the symbol, K_P, to denote equilibrium constants derived from partial pressures and K_c to indicate equilibrium constants having molar concentrations in the mass action expression. In general, K_c and K_P are not numerically equal. We will discuss this further in Section 13.4.

We have written the mass action expression with the concentrations (or partial pressures) of the products in the numerator and those of the reactants in the denominator. Since this fraction is equal to a constant at equilibrium, its

Table 13.1
Equilibrium concentrations (in moles per liter) at 500°C and the mass action expression for the reaction: N_2 (g) + $3H_2$ (g) $\rightleftharpoons$ $2NH_3$ (g)

$[H_2]$	$[N_2]$	$[NH_3]$	$\dfrac{[NH_3]^2}{[N_2][H_2]^3} = K_c$
0.150	0.750	1.23×10^{-2}	5.98×10^{-2}
0.500	1.00	8.66×10^{-2}	6.00×10^{-2}
1.35	1.15	4.12×10^{-1}	6.00×10^{-2}
2.43	1.85	1.27	6.08×10^{-2}
1.47	0.750	3.76×10^{-1}	5.93×10^{-2}
		Average	6.00×10^{-2}

reciprocal must also be a constant. Thus,

$$\frac{[NH_3]^2}{[N_2][H_2]^3} = K_C \qquad \frac{p_{NH_3}^2}{p_{N_2}\, p_{H_2}^3} = K_P$$

and

$$\frac{[N_2][H_2]^3}{[NH_3]^2} = \frac{1}{K_c} = K_c' \qquad \frac{p_{N_2}\, p_{H_2}^3}{p_{NH_3}^2} = \frac{1}{K_P} = K_P'$$

Either form is a valid description of the equilibrium state. However, chemists have chosen, somewhat arbitrarily, always to write the equilibrium expression with the concentrations or partial pressures of the products appearing in the numerator. This allows us then to tabulate equilibrium constants without always having to state explicitly the form of the mass action expression. It is only necessary to specify the chemical equation and whether we are dealing with K_c or K_p.

13.2 THE EQUILIBRIUM CONSTANT

The equilibrium constant is a quantity that must be calculated from experimental data. One method involves the use of standard free energies of reaction to calculate a *thermodynamic equilibrium constant* and is outlined in Section 13.3. Another method involves the direct measurement of equilibrium concentrations that can then be substituted into the mass action expression to obtain a numerical value for K. We will look at a sample calculation of this type in Section 13.6.

Knowing the value of an equilibrium constant is useful because it allows us to perform computations that relate the concentrations of reactants and products in an equilibrium system. But even *without* doing calculations, the magnitude of K provides us with useful qualitative information about the extent to which a reaction proceeds toward completion. For example, consider the simple reaction

$$A \rightleftharpoons B$$

for which we write

$$\frac{[B]}{[A]} = K_c$$

Suppose that $K_c = 10$. This means that

$$\frac{[B]}{[A]} = 10 = \frac{10}{1}$$

which tells us that at equilibrium the concentration of B must be ten times larger than that of A. In other words, the position of equilibrium lies in favor of the product, B. On the other hand, if $K_c = 0.1$, then

$$\frac{[B]}{[A]} = 0.1 = \frac{1}{10}$$

In this case the equilibrium concentration of A would have to be ten times larger than the concentration of B at equilibrium, and the position of equilibrium would lie in favor of the reactant, A. It is a general rule that when K is large, the position of equilibrium lies far to the right. Conversely, when K is small, only relatively small amounts of the products are present in the system at equilibrium.

The same generalization applies for both K_c and K_P.

Let's look at two examples of real chemical reactions: first, the reaction of hydrogen with chlorine,

$$H_2\ (g) + Cl_2\ (g) \rightleftharpoons 2HCl\ (g)$$

$$K_c = \frac{[HCl]^2}{[H_2][Cl_2]} = 4.4 \times 10^{32}$$

for which $K_c = 4.4 \times 10^{32}$ at 25°C. This very large value of K tells us that at equilibrium the reaction will have proceeded far toward completion. If 1 mol each of H_2 and Cl_2 are combined, very little H_2 and Cl_2 will remain unreacted at equilibrium. By way of contrast, the decomposition of water vapor at room temperature (25°C),

$$2H_2O\ (g) \rightleftharpoons 2H_2\ (g) + O_2\ (g)$$

$$K_c = \frac{[H_2]^2[O_2]}{[H_2O]^2} = 1.1 \times 10^{-81}$$

has $K_c = 1.1 \times 10^{-81}$. Examining this value of K_c, we can conclude that the decomposition takes place to only a very small degree, because in order to have such a very small value of K_c, the concentrations of the products (which appear in the numerator of the mass action expression) must be very small.

13.3 THERMODYNAMICS AND CHEMICAL EQUILIBRIUM

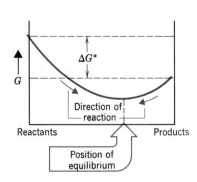

In Section 11.12 we saw, qualitatively, that there is a relationship between ΔG^0 for a reaction and the position of equilibrium. In addition, the direction in which a reaction proceeds toward equilibrium is determined by where the system lies with respect to the free-energy minimum. The reaction proceeds spontaneously only in a direction that gives rise to a decrease in free energy—that is, when ΔG is negative.

All of this is summed up quantitatively by the equation (which we will not attempt to justify)

$$\Delta G = \Delta G^0 + RT \ln Q \qquad [13.3]$$

or, in terms of common logs,

$$\Delta G = \Delta G^0 + 2.303 RT \log Q$$

The symbol Q represents the mass action expression for the reaction. For gases, Q is written with partial pressures; for reactions in solution, molar concentrations are used.[1]

Equation 13.3 tells us how ΔG varies with temperature and with the relative proportions of reactants and products. For example, for the reaction,

$$2NO_2\ (g) \rightleftharpoons N_2O_4\ (g)$$

Equation 13.3 would take the form

$$\Delta G = \Delta G^0 + RT \ln \left(\frac{p_{N_2O_4}}{p_{NO_2}^2} \right) \qquad [13.4]$$

At equilibrium the products and reactants have the same total free energy and $\Delta G = 0$ (Section 11.12), so Equation 13.4 becomes

$$0 = \Delta G^0 + RT \ln \left(\frac{p_{N_2O_4}}{p_{NO_2}^2} \right)$$

or

$$\Delta G^0 = -RT \ln \left(\frac{p_{N_2O_4}}{p_{NO_2}^2} \right)$$

[1] Actually, to make Equation 13.3 fit exactly, "effective pressures" or "effective concentrations" must be used in Q. These are called **activities**. Fortunately, at low pressures in gaseous reactions, and at low concentrations in solutions, the use of actual pressures and concentrations leads to only small errors.

At equilibrium for this reaction,

$$\frac{p_{N_2O_4}}{p_{NO_2}^2} = K_P$$

Therefore,

$$\Delta G^0 = -RT \ln K_P \qquad [13.5]$$

or

$$\Delta G^0 = -2.303 \, RT \log K_P$$

Equation 13.5, derived here for this specific example, applies to all reactions involving gases. For reactions in solution,

$$\Delta G^0 = -RT \ln K_c \qquad [13.6]$$

or

$$\Delta G^0 = -2.303 \, RT \log K_c$$

We now have a quantitative relationship between ΔG^0 and the equilibrium constant. The K computed using Equations 13.5 or 13.6 is sometimes called the **thermodynamic equilibrium constant.**

EXAMPLE 13.1

What is the thermodynamic equilibrium constant for the reaction,

$$2SO_2 \, (g) + O_2 \, (g) \rightleftharpoons 2SO_3 \, (g)$$

at 25°C?

SOLUTION

From the data in Table 11.5 we can obtain the standard free energies of formation of SO_3 and SO_2:

$$\Delta G^0_{f \, SO_3} = -370 \, \frac{kJ}{mol}$$

$$\Delta G^0_{f \, SO_2} = -300 \, \frac{kJ}{mol}$$

By definition, $\Delta G^0_{f \, O_2} = 0.0$ kJ/mol.

Using these data, we can compute ΔG^0 for the reaction:

$$\Delta G^0 = 2 \, \cancel{mol} \times \left(-370 \, \frac{kJ}{\cancel{mol}} \right) - 2 \, \cancel{mol} \times \left(-300 \, \frac{kJ}{\cancel{mol}} \right)$$

$$= -140 \, kJ$$

Next, we solve Equation 13.5 for $\ln K_P$ (we are dealing with a gaseous reaction),

$$\ln K_P = \frac{-\Delta G^0}{RT}$$

We must express ΔG^0 in joules ($\Delta G^0 = -140,000$ J), T in kelvins (298 K), and use $R = 8.314$ J/mol K). Substituting numerical values gives

$$\ln K_P = \frac{-(-140,000)}{(8.314)(298)}$$

$$= 56.5$$

Taking the antilogarithm gives

$K_P = e^{56.4} = 3 \times 10^{24}$

$$K_P = 3 \times 10^{24}$$

The magnitude of K for this reaction tells us that the position of equilibrium in the system should lie far in the direction of SO_3, and that at room temperature SO_2 should react almost completely with oxygen to form SO_3. This reaction is extremely slow at room temperature, however, but with a catalyst it becomes an important step in the industrial preparation of H_2SO_4. As mentioned in the last chapter, the same reaction takes place in the exhaust of an automobile equipped with a catalytic converter, but in this case the H_2SO_4 produced presents a health problem.

Thermodynamic data can also be used to calculate equilibrium constants at temperatures other than 25°C. This is shown in Example 13.2.

EXAMPLE 13.2

For the reaction, $2NO_2 (g) \rightleftharpoons N_2O_4 (g)$, $\Delta H^0_{298\,K} = -13.6$ kcal and $\Delta S^0_{298\,K} = -41.9$ cal/K. Calculate K_P at 100°C

SOLUTION

To calculate K_P from Equation 13.5, we must have a numerical value for the equivalent of ΔG^0, but at 100°C instead of 25°C. Let's call this quantity $\Delta G'$. In Chapter 11, we learned that

$$\Delta G^0 = \Delta H^0 - (298\ K)\ \Delta S^0$$

And for some temperature other than 298 K, we can write

$$\Delta G' = \Delta H' - T\ \Delta S'$$

It happens that ΔH and ΔS vary only slightly with temperature, so for the purposes of many calculations we can assume temperature independence and write $\Delta H^0 = \Delta H'$ and $\Delta S^0 = \Delta S'$. Thus

$$\Delta G' = \Delta H^0 - T\ \Delta S^0$$

Therefore, at 100°C (373 K),

$$\Delta G'_{373} = -13,600\ \text{cal} - (373\ K)(-41.9\ \text{cal/K})$$

$$= +2030\ \text{cal (rounded to three significant figures)}$$

Solving Equation 13.5 for $\ln K_P$ gives us

$$\ln K_P = \frac{-\Delta G'}{RT}$$

This time, we use $R = 1.987$ cal/mol K, because $\Delta G'$ is in calories. Substituting numerical values, using $T = 373$ K, gives

$$\ln K_P = \frac{-2030}{(1.987)(373)}$$

$$= -2.74$$

Taking the antilogarithm gives

$K_P = e^{-2.74} = 6.5 \times 10^{-2}$

$$K_P = 6.5 \times 10^{-2}$$

The measurement of equilibrium constants also provides a very convenient method for obtaining thermodynamic data. This is illustrated in the next example

EXAMPLE 13.3

At 25°C it was found that $K_P = 7.13$ for the reaction

$$2NO_2 (g) \rightleftharpoons N_2O_4 (g)$$

What is ΔG^0 for this reaction in kilojoules?

SOLUTION   We can calculate ΔG^0 by substituting appropriate values into Equation 13.5,

$$\Delta G^0 = -RT \ln K_P$$

Since we wish ΔG^0 in kilojoules, we must use $R = 8.314$ J/mol K. As usual, T is the absolute temperature ($T = 298$ K in this example). Substituting numerical values,

$$\Delta G^0 = -(8.314)(298) \ln (7.13)$$

$$= -4870 \text{ J (to three significant figures)}$$

The value of ΔG^0 is in joules because R is in joules. To convert to kilojoules, simply divide by 1000:

$$\Delta G^0 = -4.87 \text{ kJ}$$

13.4   THE RELATIONSHIP BETWEEN K_P AND K_c

It was stated earlier that for reactions involving gases, K_P and K_c are not necessarily equal. For the general equation,

$$aA + bB \rightleftharpoons eE + fF$$

$$K_P = \frac{p_E{}^e p_F{}^f}{p_A{}^a p_B{}^b}$$

and

$$K_c = \frac{[E]^e[F]^f}{[A]^a[B]^b}$$

Concentration, you recall, has the units *moles per liter*, or n/V. Assuming ideal gas behavior, we can use the ideal gas law,

$$PV = nRT$$

to obtain the concentration of a gas, X, in a mixture as

$$[X] = \frac{n_X}{V} = \frac{p_X}{RT}$$

where p_X is its partial pressure. From this it follows that

$$p_X = [X]RT$$

Substituting this relationship into the expression for K_P, we have

$$K_P = \frac{p_E{}^e p_F{}^f}{p_A{}^a p_B{}^b} = \frac{[E]^e(RT)^e[F]^f(RT)^f}{[A]^a(RT)^a[B]^b(RT)^b}$$

This can be rearranged to give

$$K_P = \frac{[E]^e[F]^f}{[A]^a[B]^b}(RT)^{(e+f)-(a+b)}$$

or

When $\Delta \mathbf{n}_g = 0$, $\mathbf{K}_P = \mathbf{K}_c$

$$K_P = K_c(RT)^{\Delta n_g} \qquad [13.7]$$

where Δn_g is the change in the number of moles of **gas** when going from reactants to products.

$$\Delta n_g = \text{(number of moles of gaseous products)} - \text{(number of moles of gaseous reactants)}$$

Thus, K_P and K_c are related in a very simple fashion for reactions between ideal gases, a relationship that also holds adequately for many real gases.

EXAMPLE 13.4

In Example 13.1 we determined the value of K_P for the reaction of SO_2 with O_2 to produce SO_3. What is K_c for this equilibrium at 25°C?

SOLUTION

Solving Equation 13.7 for K_c, we obtain

$$K_c = \frac{K_P}{(RT)^{\Delta n_g}} = K_P(RT)^{-\Delta n_g}$$

The chemical equation we are dealing with is

$$2SO_2 \, (g) + O_2 \, (g) \rightleftharpoons 2SO_3 \, (g)$$

To calculate Δn_g, we interpret the coefficients as moles—there are two moles of gaseous products and three moles of gaseous reactants. Therefore,

$$\Delta n_g = (2 - 3)$$

$$= -1$$

In Example 13.1, we found $K_P = 3 \times 10^{24}$. From the equilibrium expression

$$\frac{p_{SO_3}^2}{p_{SO_2}^2 \, p_{O_2}} = K_P$$

K_P has the units atm^{-1} if the partial pressures are expressed in atm. As a result, we must use $R = 0.0821$ liter atm mol^{-1} K^{-1} to obtain the proper units for K_c (why?). Thus

We don't normally include units with **K**, but in this case it helps us choose the correct value for **R**.

$$K_c = (3 \times 10^{24} \text{ atm}^{-1})[(0.0821 \text{ liter atm mol}^{-1} \text{ K}^{-1})(298 \text{ K})]^{-(-1)}$$

Therefore,

$$K_c = 7 \times 10^{25} \text{ liter mol}^{-1}$$

13.5 HETEROGENEOUS EQUILIBRIA

Up to now our discussion has focused on homogeneous reactions—reactions in which all the reactants and products are in the same phase. Heterogeneous reactions, of which there are many examples, also eventually arrive at a state of equilibrium. A typical reaction that we might consider is the decomposition of solid $NaHCO_3$ to produce solid Na_2CO_3, gaseous CO_2, and gaseous H_2O.[2]

We learned that this reaction makes $NaHCO_3$ a good fire extinguisher.

$$2NaHCO_3 \, (s) \rightleftharpoons Na_2CO_3 \, (s) + CO_2 \, (g) + H_2O \, (g)$$

Applying the law of mass action, we can write the equilibrium expression as

$$\frac{[Na_2CO_3 \, (s)][CO_2 \, (g)][H_2O \, (g)]}{[NaHCO_3 \, (s)]^2} = K_c' \qquad [13.8]$$

For reasons that will be apparent shortly, we have temporarily indicated the equilibrium constant as K_c'.

In this reaction we have an equilibrium between the two gases, CO_2 and H_2O, and the two pure solid phases, $NaHCO_3$ and Na_2CO_3. We know by now that a pure solid substance such as $NaHCO_3$ is characterized by a density that is

---

[2] Na_2CO_3 is produced commercially by this reaction. It is one of the most industrially important chemicals, ranking tenth in total production (about 17 billion pounds produced annually), and is used in the manufacture of glass and many other important products.

the same for all samples of $NaHCO_3$, regardless of their size. In addition, this density is unaffected by the nature of any chemical reaction that the substance is undergoing. This means that even during a chemical reaction the amount of $NaHCO_3$ per unit volume of the pure solid is always the same. In other words, the concentration of $NaHCO_3$ in pure solid $NaHCO_3$ is a constant. We cannot alter the number of moles per liter of $NaHCO_3$ in the pure solid, nor can we change the concentration of Na_2CO_3 in pure solid Na_2CO_3. Consequently, the concentrations of these two substances in the equilibrium expression take on constant values and can be incorporated into the equilibrium constant. Rearranging Equation 13.8 to place all the constants on the same side gives

$$[CO_2\ (g)][H_2O\ (g)] = \underbrace{K_c' \frac{[NaHCO_3\ (s)]^2}{[Na_2CO_3\ (s)]}}_{K_c}$$

or

$$[CO_2\ (g)][H_2O\ (g)] = K_c \qquad [13.9]$$

Thus we find that for heterogeneous reactions, *the equilibrium constant expression does not include the concentrations of pure solids.* Similarly, in reactions in which a reactant or product occurs as a pure liquid phase, the concentration of that substance in the pure liquid is also constant. As a result, *the concentrations of pure liquid phases also do not appear in an equilibrium constant expression.*[3] These simplifications apply *only* when we are dealing with *pure* condensed phases. When substances occur in liquid or solid solutions, their concentrations are variable and their concentration terms in the mass action expression therefore cannot be incorporated into K.

If we wish to work with K_P rather than K_c, we again need take into account only the substances present in the gas phase. For the decomposition of $NaHCO_3$, therefore, we have

$$K_P = p_{CO_2(g)}p_{H_2O(g)}$$

As noted in the last section, if we know K_c, we can evaluate K_P as

$$K_P = K_c(RT)^{\Delta n_g} \qquad [13.10]$$

where, for this reaction, $\Delta n_g = +2$.

EXAMPLE 13.5   What are the values of K_P and K_c for the "reaction"

$$H_2O\ (l) \rightleftharpoons H_2O\ (g)$$

at 25°C, given that the vapor pressure of water at 25°C equals 23.8 torr?

SOLUTION   Since liquid water is a pure liquid phase, we can write

$$K_P = p_{H_2O(g)}$$

and

$$K_c = [H_2O\ (g)]$$

(a) If we express the vapor pressure of water in atmospheres,

$$p_{H_2O} = 23.8\ torr \times \left(\frac{1\ atm}{760\ torr}\right) = 0.0313\ atm$$

---

[3] Thermodynamics handles this question in a slightly more elegant way by defining the activity of a pure solid or liquid as numerically equal to one. This simply makes terms involving pure solids or liquids disappear from the mass action expression. For instance, substituting values of 1 for $[NaHCO_3\ (s)]$ and $[Na_2CO_3\ (s)]$ in Equation 13.8 gives Equation 13.9 directly.

Therefore,

$$K_P = p_{H_2O} = 3.13 \times 10^{-2} \text{ atm}$$

Note that this equilibrium expression states that the partial pressure of water must be a constant when the liquid and vapor are in equilibrium.

(b) We can evaluate K_c as

$$K_c = K_P(RT)^{-\Delta n}$$

For this "reaction," $\Delta n_g = 1$; therefore,

$$K_c = K_P(RT)^{-1} = \frac{K_P}{RT}$$

$$= \frac{3.13 \times 10^{-2} \text{ atm}}{(0.0821 \text{ liter atm/mol K})(298 \text{ K})}$$

or

$$K_c = 1.28 \times 10^{-3} \frac{\text{mol}}{\text{liter}}$$

13.6   LE CHÂTELIER'S PRINCIPLE AND CHEMICAL EQUILIBRIUM

The equilibrium expression, in the form of either K_P or K_c, can be used to perform numerical computations of various kinds dealing with equilibrium systems. This is discussed in the next section. Often, however, it is desirable simply to be able to predict how some disturbance imposed on a system from outside will influence the position of equilibrium. For instance, we may wish to predict, in a qualitative way, the conditions that favor the greatest production of products. Should we run our reaction at high or low temperature? Should the pressure on the system be high or low? These are questions we would like to answer quickly without having to perform tedious computations. We have already seen how Le Châtelier's principle can be applied to dynamic equilibria involving such phenomena as the vapor pressure of a liquid and solubility. Changes in the position of equilibrium in chemical systems can also be understood by applying the same concepts.

Changes in the concentration of a reactant or product

In a system such as

$$H_2 (g) + I_2 (g) \rightleftharpoons 2HI (g)$$

any change in the concentration of a reactant or product will cause the system to no longer be at equilibrium. As a result, a chemical change will occur that will return the system to equilibrium. From Le Châtelier's principle we know that if a system at equilibrium is disturbed, it will attempt to undergo some change to diminish the effect of the disturbance. For example, the addition of H_2 to an equilibrium mixture of H_2, I_2, and HI upsets the equilibrium, and the system responds by using up part of the additional H_2 by reaction with I_2 to produce more HI. When equilibrium has finally been reestablished, there will be a greater concentration of HI than before, and we say that for this reaction the position of equilibrium has been shifted to the right. This is illustrated in Figure 13.2 on the next page. Notice that after equilibrium has been reestablished there continues to be more H_2 present than in the original reaction mixture. The system is never able to completely overcome the effect of a change in concentration. The final position of equilibrium differs from the original.

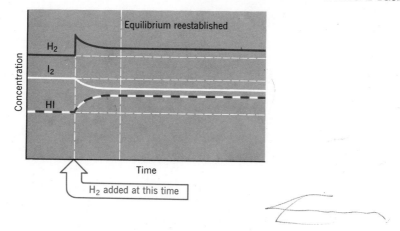

Figure 13.2

Addition of H_2 to the equilibrium, $H_2 + I_2 \rightleftharpoons 2HI$, increases the amount of HI and decreases the amount of I_2.

We can arrive at this same conclusion by considering the effect of added H_2 on the value of the mass action expression. For this reaction at equilibrium we have

$$\frac{[HI]^2}{[H_2][I_2]} = K_c$$

If H_2 is suddenly added to the system, the value of the denominator of the mass action expression becomes larger, and the entire fraction therefore becomes smaller than the equilibrium constant. The reaction that occurs to return the system to equilibrium must increase the value of the mass action expression until it once again is equal to K_c. For this to happen, the numerator must become larger and the denominator smaller. In other words, more HI will be formed at the expense of H_2 and I_2, and again we conclude that the addition of H_2 shifts the position of equilibrium in this reaction to the right.

By applying Le Châtelier's principle we can also predict the effect that removing a reactant or product will have on a system at equilibrium. For instance, if H_2 is somehow removed from the reaction vessel, the system will adjust by having some HI decompose in an effort to replenish the lost reactant. Thus, the position of equilibrium is shifted to the left when H_2 is removed.

Now we can use Le Châtelier's principle to predict what must be done to drive a reaction far toward completion—we can either add a large excess of one of the reactants or remove the products as they are formed. Recall that removing the products serves as the driving force for ionic reactions (Chapter 6) in which a product is either a precipitate, a gas, or a weak electrolyte. The creation of these products removes ions from solution and therefore forces the reaction to proceed toward completion.

Closer to home, we can see an application of Le Châtelier's principle in understanding the origin of tooth decay. Tooth enamel consists of an insoluble substance called hydroxyapatite, $Ca_5(PO_4)_3OH$. The dissolving of this substance from the teeth is called demineralization, and its formation is called remineralization. In the mouth there is an equilibrium

$$Ca_5(PO_4)_3OH\ (s) \underset{\text{remineralization}}{\overset{\text{demineralization}}{\rightleftharpoons}} 5Ca^{2+}\ (aq) + 3PO_4^{3-}\ (aq) + OH^-\ (aq)$$

which is established even with healthy teeth. However, when sugar is absorbed on teeth and ferments, H^+ is produced that upsets the equilibrium by combining with OH^- to form water and with PO_4^{3-} to form HPO_4^{2-}. Removing OH^- and PO_4^{3-} causes more of the $Ca_5(PO_4)_3OH$ to dissolve, resulting in tooth decay. Fluoride helps prevent tooth decay by replacing OH^- in hydroxyapatite. The resulting $Ca_5(PO_4)_3F$ is very resistant to acid attack.

In a chemical equation, the position of equilibrium shifts away from a substance that's been added or toward a substance that's removed.

The effect of temperature on equilibrium

Up to now we have been careful to imply that the equilibrium constant for a reaction has a fixed numerical value only as long as the temperature remains constant. This is because temperature, as well as the concentrations of reactants and products, affects the position of equilibrium. However, the temperature, unlike the concentrations of reactants and products, affects the value of the equilibrium constant itself.

Let's look at the exothermic reaction of H_2 and N_2 to form NH_3. The equation for the formation of ammonia can be written as

$$3H_2\ (g) + N_2\ (g) \rightleftharpoons 2NH_3\ (g) + 92.0\ kJ$$

where the heat of reaction is indicated as a product. If we have a system of these gases in equilibrium and wish to raise its temperature, we do so by adding heat to it from the surroundings. Le Châtelier's principle tells us that when we add this heat, the system will attempt to undergo a change that tends to use some of it up. Since the production of NH_3 is exothermic, its decomposition is endothermic. Therefore, raising the temperature will cause the position of equilibrium to shift to the left; it drives the reaction to a new position of equilibrium in which there is more N_2 and H_2 and less NH_3. *In general, an increase in temperature causes the position of equilibrium of an exothermic reaction to be shifted to the left, while that of an endothermic reaction is shifted to the right.*

We have just seen that an increase in temperature leads to a decrease in the concentration of NH_3 and an increase in the concentrations of both N_2 and H_2. This means that *at equilibrium* at the higher temperature the value of the mass action expression

$$\frac{[NH_3]^2}{[H_2]^3[N_2]}$$

will have decreased. Thus we find that for this exothermic reaction, K decreases with rising temperature. By the same token, for a reaction that is endothermic in the forward direction, K increases with increasing temperature.

Effect of pressure and volume changes on equilibrium

At constant temperature a change in the volume of a system also causes a change in pressure. We quite logically expect, therefore, that an increase in the external pressure on a system should favor any change that leads to a smaller volume (recall Boyle's law). We would not expect pressure changes to have any marked effect on the position of equilibrium in reactions where all the reactants and products are either solids or liquids, because these phases are virtually incompressible. However, pressure changes can have a very dramatic effect on equilibria that involve reactions in which gases are consumed or produced.

Let us again choose as an example the reaction for the formation of NH_3. If we have this system at equilibrium and suddenly decrease the volume of the container, we know that the pressure will go up. According to Le Châtelier's principle, we expect that a change in the system should occur that will reduce the pressure. How can this be brought about?

We know that the pressure of a gas is caused by collisions of the molecules with the walls of the container, and at a given temperature the greater the number of molecules per cubic centimeter, the greater the pressure. In the equilibrium

$$N_2\ (g) + 3H_2\ (g) \rightleftharpoons 2NH_3\ (g)$$

the number of molecules of gas decreases when the reaction proceeds from left to right—four reactant molecules produce two product molecules. This means

In a sense, we think of heat as a reactant or product in the equation.

When the external pressure on a system is increased, it causes the volume of the system to decrease.

that the pressure exerted by the gases in the system can be decreased if the position of equilibrium shifts to the right.[4] Thus *decreasing the volume of a mixture of gases that are in chemical equilibrium shifts the equilibrium in the direction of the fewest number of molecules of gas.*

Finally, note that when there are the same number of molecules of gaseous reactants and products on both sides of the equation as in the reaction between H_2 and I_2,

$$H_2 \ (g) + I_2 \ (g) \rightleftharpoons 2HI \ (g)$$

pressure changes brought about by volume changes will not influence the amounts of the various substances present in the reaction mixture at equilibrium. This is because there is no way for the system to counteract pressure changes placed on it.

Changing the external pressure on a chemical system containing only liquids and solids has virtually no effect on the position of equilibrium.

Addition of an inert gas

If an inert (nonreacting) gas is introduced into a reaction vessel containing other gases at equilibrium, it will cause an increase in the total pressure within the container. This kind of pressure increase, however, will not affect the position of equilibrium because it will not alter the partial pressures or the concentrations of any of the substances already present.

Effect of a catalyst on the position of equilibrium

In Chapter 12 we saw that a catalyst affects a chemical reaction by lowering the activation energy barrier that must be overcome in order for the reaction to proceed. A catalyst affects the rate of a chemical change. It does not, however, affect the heat of reaction, and it is the heat of reaction, ΔH^0, along with the entropy change, ΔS^0, that determine ΔG^0, which in turn fixes the position of equilibrium at any given temperature. A catalyst merely speeds the approach to the position of equilibrium that is determined by ΔG^0.

13.7 EQUILIBRIUM CALCULATIONS

This section is intended to illustrate the types of computations that one might perform either to evaluate an equilibrium constant from measured concentrations or to use the equilibrium constant to calculate the concentrations of the reactants and products in a particular equilibrium mixture. First let's see how we might evaluate K in a typical experiment.

EXAMPLE 13.6

The brown gas NO_2, an air pollutant, and the colorless gas N_2O_4 exist in equilibrium as indicated by the equation,

$$2NO_2 \rightleftharpoons N_2O_4$$

In an experiment, 0.625 mol of N_2O_4 was introduced into a 5.00-liter vessel and permitted to decompose until it reached equilibrium with NO_2. At equilibrium the concentration of N_2O_4 was 0.0750 M. What is K_c for this reaction?

---

[4] By applying Le Châtelier's principle, we find that the production of ammonia from H_2 and N_2 is favored by high pressures and low temperatures. At low temperatures, however, the reaction is very slow; therefore in the industrial preparation of NH_3, pressures of 10^2 to 10^3 atm and temperatures from 400 to 550°C are employed. Even though there is less NH_3 produced at equilibrium at these high temperatures, the speed of reaction is boosted to the point where the production of NH_3 is economically worthwhile.

SOLUTION   The equilibrium constant expression for this reaction is

$$\frac{[N_2O_4]}{[NO_2]^2} = K_c$$

Remember, the equilibrium condition is only satisfied by equilibrium concentrations.

In order to calculate K_c, we must know the equilibrium concentrations of N_2O_4 and NO_2. In working out equilibrium problems, we will generally find it useful to set up a table like that below in order to establish quantities that correspond to equilibrium concentrations. The entries in the table are obtained by reasoning from the data provided in the problem. Remember that when using K_c, molar concentrations (that is, moles/liter) must be used. In this example the concentration of N_2O_4 initially was 0.625 mol/5.00 liter = 0.125 M; the initial concentration of NO_2 was zero. These are the values in the first column of the concentration table.

In the statement of the problem we are told that the equilibrium concentration of N_2O_4 is 0.0750 M. The difference between the initial concentration (0.125 M) and the equilibrium concentration (0.0750 M) is the number of moles per liter of N_2O_4 that decomposed.

$$0.125\ M - 0.0750\ M = 0.050\ M$$

This value is entered in the change column of the table with a minus sign to indicate that the N_2O_4 concentration has decreased.

The changes in the concentrations of N_2O_4 and NO_2—the values in the "change" column of the table—are related by the stoichiometry of the reaction. For each mole of N_2O_4 that decomposes, two moles of NO_2 are formed. Therefore, when the N_2O_4 concentration decreased by 0.050 M, the NO_2 concentration increased by $2 \times 0.050\ M = 0.10\ M$. To indicate the increase, the value is written with a plus sign.

When we construct the concentration table for an equilibrium problem in this way—using a minus sign to indicate a decrease in concentration and a plus sign to indicate an increase—the equilibrium concentration is gotten by adding the value in the change column to the initial concentration.

For N_2O_4 $0.125\ M - 0.050\ M = 0.075\ M$

NO_2 $0.00\ M + 0.10\ M = 0.10\ M$

	Initial Concentrations	Change	Equilibrium Concentrations
N_2O_4	0.125 M	−0.050 M	0.075 M
NO_2	0.00 M	+0.10 M	0.10 M

Now we can substitute the equilibrium concentrations into the mass action expression to calculate K_c.

$$\frac{(0.075)}{(0.10)^2} = K_c$$

and, finally,

$$K_c = 7.5$$

Knowledge of the equilibrium constant for a reaction allows us to calculate the concentrations or partial pressures of the substances present in a reaction mixture at equilibrium. The ease with which these computations can be carried out depends on the complexity of the mass action expression, the concentrations of the various species in the reaction mixture, and the magnitude of the equilibrium constant. We will look only at some of the more simple examples of problems of this type. The following sample problems, however, illustrate the type of reasoning employed in these computations, as well as some of the concepts that have been presented up to this point.

EXAMPLE 13.7 At 25°C, $K_P = 7.13$ atm^{-1} for the reaction

$$2NO_2\ (g) \rightleftharpoons N_2O_4\ (g)$$

At equilibrium the partial pressure of NO_2 in a container is 0.15 atm. What is the partial pressure of N_2O_4 in the mixture?

SOLUTION The first step in the solution of any equilibrium problem is to write down the equilibrium expression. For K_P we have

$$K_P = \frac{p_{N_2O_4}}{p_{NO_2}^2} = 7.13 \text{ atm}^{-1}$$

We are given the equilibrium partial pressure of NO_2 ($p_{NO_2} = 0.15$ atm). There is only one unknown quantity, $p_{N_2O_4}$. Substituting,

$$\frac{p_{N_2O_4}}{(0.15 \text{ atm})^2} = 7.13 \text{ atm}^{-1}$$

$$p_{N_2O_4} = 7.13 \text{ atm}^{-1}(0.15 \text{ atm})^2$$

$$= 0.16 \text{ atm}$$

The partial pressure of N_2O_4 at equilibrium is 0.16 atm.

EXAMPLE 13.8 At a temperature of 500°C, the equilibrium constant, K_c, for the nitrogen fixation reaction for the production of ammonia,

$$3H_2\ (g) + N_2\ (g) \rightleftharpoons 2NH_3\ (g)$$

has a value of 6.0×10^{-2}. If, in a particular reaction vessel at this temperature, there are 0.250 mol/liter of H_2 and 0.0500 mol/liter of NH_3 present at equilibrium, what is the concentration of N_2?

SOLUTION Let's first write the equilibrium constant expression. For this reaction we have

$$K_c = \frac{[NH_3]^2}{[H_2]^3[N_2]} = 6.0 \times 10^{-2}$$

We wish to calculate the concentration of N_2. This can be accomplished if we know the values of the equilibrium concentrations of both NH_3 and H_2 and, in this problem, these are given to us.

In this problem we don't need the concentration table because we're given K_c and all but one equilibrium concentration.

$$\left.\begin{array}{l}[NH_3] = 0.0500\ M \\ [H_2] = 0.250\ M\end{array}\right\} \text{ at equilibrium}$$

Substituting these numerical values into the mass action expression gives us

$$\frac{(0.0500)^2}{(0.250)^3[N_2]} = 6.0 \times 10^{-2}$$

If we solve for $[N_2]$,

$$[N_2] = \frac{(0.0500)^2}{(0.250)^3(6.0 \times 10^{-2})}$$

$$= 2.7\ M$$

The equilibrium concentration of N_2 is thus 2.7 mol/liter.

EXAMPLE 13.9   At 440°C the equilibrium constant for the reaction,

$$H_2 \ (g) + I_2 \ (g) \rightleftharpoons 2HI \ (g)$$

is 49.5. If 0.200 mol of H_2 and 0.200 mol of I_2 are placed into a 10.0-liter vessel and permitted to react at this temperature, what will be the concentration of each substance at equilibrium?

SOLUTION   Our equilibrium expression is

$$\frac{[HI]^2}{[H_2][I_2]} = 49.5$$

In this example we are given the *initial* concentrations of the reactants and products. These are

$$[H_2] = \frac{0.200 \ mol}{10.0 \ liters} = 0.0200 \ M$$

$$[I_2] = \frac{0.200 \ mol}{10.0 \ liters} = 0.0200 \ M$$

$$[HI] = 0.0 \ M$$

Since no HI is present initially, we know that it will be formed from the violet-colored mixture of the H_2 and I_2 (the color is due to the I_2). Let's approach the problem by allowing x to be equal to the number of moles per liter of H_2 that react. From the stoichiometry of the reaction we realize that this x mol/liter of H_2 will react with x mol/liter of I_2 to produce $2x$ mol/liter of HI. Thus the concentrations of H_2 and I_2 both decrease by x and the concentration of HI increases by $2x$. The equilibrium concentrations are obtained by applying these changes to the initial concentrations.

The coefficients of x must be in the same ratio as the coefficients in the balanced chemical equation.

	Initial Concentrations	Change	Equilibrium Concentrations
H_2	0.0200 M	$-x$	$(0.0200 - x) \ M$
I_2	0.0200 M	$-x$	$(0.0200 - x) \ M$
HI	0.0 M	$+2x$	$0.0 + 2x = 2x \ M$

Substituting the equilibrium quantities into the mass action expression gives

$$\frac{(2x)^2}{(0.0200 - x)(0.0200 - x)} = 49.5$$

or

$$\frac{(2x)^2}{(0.0200 - x)^2} = 49.5$$

In this case, we can take the square root of both sides of the equation to obtain

$$\frac{2x}{0.0200 - x} = 7.04$$

Solving for x,

$$2x = 7.04(0.0200 - x) = 0.141 - 7.04x$$

$$2x + 7.04x = 0.141$$

$$9.04x = 0.141$$

$$x = 0.0156$$

Finally, the equilibrium concentrations are

$$[H_2] = 0.0200 - 0.0156 = 0.0044 \; M$$

$$[I_2] = 0.0200 - 0.0156 = 0.0044 \; M$$

$$[HI] = 2(0.0156) = 0.0312 \; M$$

In this last problem we employed some relatively simple algebra to help us arrive at the solution. Let us look at another example of this type.

EXAMPLE 13.10

A 10.0-liter vessel is filled with 0.40 mol of HI at 440°C. What will be the concentration of H_2, I_2, and HI at equilibrium?

SOLUTION

In this example we are concerned with the same equilibrium as in the previous problem. Initially we have no H_2 or I_2, and the reaction mixture is colorless. In order to have an equilibrium, some H_2 and I_2 must be formed from the decomposition of HI. Let's make x the number of moles per liter of HI that decomposes. The stoichiometry of the reaction tells us that 1 mol each of H_2 and I_2 are produced from every 2 mol of HI that break down, so if the HI concentration decreases by x, the H_2 and I_2 concentrations each increase by $\frac{1}{2} x$. We can use our table now to write the equilibrium concentrations. The initial concentration of HI is 0.40 mol/10.0 liter = 0.040 M.

	Initial Concentrations	Change	Equilibrium Concentrations
H_2	0.0 M	$+\frac{1}{2} x$	$0.0 + \frac{1}{2} x = \frac{1}{2} x \; M$
I_2	0.0 M	$+\frac{1}{2} x$	$0.0 + \frac{1}{2} x = \frac{1}{2} x \; M$
HI	0.040 M	$-x$	$(0.040 - x) \; M$

Substituting equilibrium quantities into the mass action expression gives us

$$\frac{(0.040 - x)^2}{(\frac{1}{2} x)(\frac{1}{2} x)} = 49.5$$

or

$$\frac{(0.040 - x)^2}{(\frac{1}{2} x)^2} = 49.5$$

Taking the square root of both sides of the equation, we have

$$\frac{0.040 - x}{\frac{1}{2} x} = 7.04$$

Solving for x, we get

$$0.040 - x = (\tfrac{1}{2} x)(7.04)$$

$$x = 0.00885$$

We now calculate the equilibrium concentrations to be

$$[H_2] = \tfrac{1}{2} x = 0.0044 \; M$$

$$[I_2] = \tfrac{1}{2} x = 0.0044 \; M$$

$$[HI] = 0.040 - x = 0.031 \; M$$

Observe that we have obtained essentially the same answers in both Examples 13.9 and 13.10. If all the H_2 and I_2 in Example 13.9 had completely reacted, it would have produced 0.40 mol of HI—the same amount of HI that we began with in Example 13.10. We find, therefore, that the same equilibrium composition can be approached from either direction.

In the last two examples the solution of the algebra was simple because we were able to take the square root of both sides of the equation. You can't always expect the algebra to work out so easily, however, and in some cases it can really prove to be quite a challenge. Fortunately, though, when the equilibrium constant is either extremely large or extremely small, it is frequently possible to greatly reduce the difficulty of even rather complex algebra by making some simple approximations. Example 13.11 illustrates the kinds of simplifying approximations that we will find useful in the Chapter 15.

EXAMPLE 13.11

The equilibrium constant, K_c, for the decomposition of gaseous water at 500°C has a value of 6.0×10^{-28}. If 2.0 mol of H_2O is placed into a 5.0-liter container, what will be the equilibrium concentrations of the three gases, H_2, O_2, and H_2O at 500°C?

SOLUTION

The equation for the reaction is

$$2H_2O \ (g) \rightleftharpoons 2H_2 \ (g) + O_2 \ (g)$$

Therefore we can write

$$\frac{[H_2]^2[O_2]}{[H_2O]^2} = 6.0 \times 10^{-28}$$

The initial H_2O concentration is 2.0 mol/5.0 liter = 0.40 M. If we let x equal the number of moles per liter of H_2O that decomposes, we will get x mol of H_2 and $0.5x$ mol of O_2. Constructing our table,

	Initial Concentrations	Change	Equilibrium Concentrations
H_2O	0.40 M	$-x$	$(0.40 - x) \ M$
H_2	0.0 M	$+x$	$x \ M$
O_2	0.0 M	$+0.5x$	$0.5x \ M$

Substituting equilibrium quantities into the mass action expression gives

$$\frac{(x)^2(0.50x)}{(0.40 - x)^2} = 6.0 \times 10^{-28}$$

Unless we can somehow simplify this equation, we have a real mess on our hands. Fortunately, in this case the problem can be made easy to solve.

Since K is very small, we know ahead of time that the reaction does not proceed very far toward completion. This means that very little H_2 and O_2 are formed, so x and $0.5x$ are going to be very small numbers. To simplify the algebra, we will make the assumption that x will be *much* smaller than 0.40, so that when x is subtracted from 0.40, the difference will still be very nearly 0.40 when rounded to the proper number of significant figures. If we neglect x when we compute the H_2O concentration, the equilibrium quantities become

We can only neglect a small x if it's <u>added</u> to our <u>subtracted from</u> some other value that's much larger.

$$[H_2O] = 0.40 - x \approx 0.40$$

$$[H_2] = x$$

$$[O_2] = 0.5x$$

Substituting these into the mass action expression gives

$$\frac{(x)^2(0.50x)}{(0.40)^2} = 6.0 \times 10^{-28}$$

from which we get

$$\frac{0.50x^3}{0.16} = 6.0 \times 10^{-28}$$

This is now simple to solve. First we solve for x^3,

$$x^3 = \frac{0.16}{0.50}(6.0 \times 10^{-28}) = 1.9 \times 10^{-28}$$

At this point x can be obtained by extracting the cube root. Many hand-held calculators can perform this operation directly or by raising 1.9×10^{-28} to the $\frac{1}{3}$ power.[5]

$$\sqrt[3]{1.9 \times 10^{-28}} = (1.9 \times 10^{-28})^{1/3} = 5.7 \times 10^{-10}$$

We see that x is, in fact, very much smaller than 0.40, thus justifying our initial assumption. The final equilibrium concentrations are

$$[H_2] = x = 5.7 \times 10^{-10} \ M$$

$$[O_2] = 0.50x = 2.8 \times 10^{-10} \ M$$

$$[H_2O] = 0.40 - (5.7 \times 10^{-10}) = 0.40 \ M$$

0.40 − 0.00000000057 = 0.39999999943. When rounded, this gives 0.40.

In working out a problem of this sort, look for any assumption that will make the algebra easier to handle. If the assumption you make is invalid, you will discover this when you check the assumption after obtaining a value for x. Sometimes no assumption of the kind we made above will be valid, and then some other method of solving the equation for x will have to be sought.

[5] If you are among that vanishing breed who still use a slide rule, you must remember to first make the exponent divisible by 3.

$$x^3 = 190 \times 10^{-30}$$

$$x = 5.7 \times 10^{-10}$$

INDEX TO QUESTIONS AND PROBLEMS (Problem numbers in **bold type**)

REVIEW QUESTIONS

13.1 What is meant by a *dynamic equilibrium?*

13.2 Write the mass action expression in terms of molar concentrations for each of the following reactions:

(a) $N_2 (g) + O_2 (g) \rightleftharpoons 2NO (g)$
(b) $2NO (g) + O_2 (g) \rightleftharpoons 2NO_2 (g)$
(c) $2H_2 (g) + S_2 (g) \rightleftharpoons 2H_2S (g)$
(d) $2N_2O_5 (g) \rightleftharpoons 4NO_2 (g) + O_2 (g)$
(e) $P_4O_{10} (g) + 6PCl_5 (g) \rightleftharpoons 10POCl_3 (g)$

13.3 Give the mass action expressions for the reactions in Question 13.2 in terms of partial pressures.

13.4 Write equilibrium constant expressions for K_P and K_c for each of the following reactions:

(a) $CO\ (g) + 2H_2\ (g) \rightleftharpoons CH_3OH\ (g)$
(b) $CO\ (g) + H_2O\ (g) \rightleftharpoons CO_2\ (g) + H_2\ (g)$
(c) $PCl_3\ (g) + Cl_2\ (g) \rightleftharpoons PCl_5\ (g)$
(d) $2NO_2\ (g) + 4H_2\ (g) \rightleftharpoons N_2\ (g) + 4H_2O\ (g)$
(e) $2H_2S\ (g) + 3O_2\ (g) \rightleftharpoons 2H_2O\ (g) + 2SO_2\ (g)$

13.5 Write equilibrium constant expressions for the reactions as written below.
(a) $H_2\ (g) + Cl_2\ (g) \rightleftharpoons 2HCl\ (g)$
(b) $\frac{1}{2}H_2\ (g) + \frac{1}{2}Cl_2\ (g) \rightleftharpoons HCl\ (g)$
How would the magnitude of the K for reaction (a) compare with that for reaction (b)?

13.6 Why do we always write the concentrations (or partial pressures) of the products in the numerator and those of the reactants in the denominator in the mass action expression?

13.7 What general information can be gathered by observing the magnitude of the equilibrium constant?

13.8 Arrange the following reactions in order of their increasing tendency to proceed toward completion:
(a) $4NH_3\ (g) + 3O_2\ (g) \rightleftharpoons 2N_2\ (g) + 6H_2O\ (g)$
$$K = 1 \times 10^{228}$$
(b) $N_2\ (g) + O_2\ (g) \rightleftharpoons 2NO\ (g) \qquad K = 5 \times 10^{-31}$
(c) $2HF\ (g) \rightleftharpoons H_2\ (g) + F_2\ (g) \qquad K = 1 \times 10^{-13}$
(d) $2NOCl\ (g) \rightleftharpoons 2NO\ (g) + Cl_2\ (g) \qquad K = 4.7 \times 10^{-4}$

13.9 What value would ΔG^0 have for a reaction if $K = 1$?

13.10 For reactions between gases, what kind of equilibrium constant is calculated from ΔG^0?

13.11 Using Equations 11.11 (p. 380) and 13.5, show that a straight line should be obtained if log K_P is plotted against $1/T$ (that is, log K_P along the vertical axis, $1/T$ along the horizontal axis). What does the slope of this line give? What does the value of log K_P at $1/T = 0$ (the y intercept of the line) give?

13.12 For which of the reactions in Questions 13.2 and 13.4 would $K_P = K_c$?

13.13 On the basis of the equilibrium

$$H_2O(l) \rightleftharpoons H_2O\ (g)$$

explain why the vapor pressure of water depends only on temperature and not on the amount of liquid water in equilibrium with its vapor.

13.14 Why is it *not* necessary to include the concentrations of pure liquid or solid phases in the equilibrium constant expression?

13.15 Write equilibrium expressions for each of the following reactions:
(a) $CaCO_3\ (s) \rightleftharpoons CaO\ (s) + CO_2\ (g)$
(b) $Ni\ (s) + 4CO\ (g) \rightleftharpoons Ni(CO)_4\ (g)$
(c) $5CO\ (g) + I_2O_5\ (s) \rightleftharpoons I_2\ (g) + 5CO_2\ (g)$

(d) $Ca(HCO_3)_2\ (aq) \rightleftharpoons CaCO_3\ (s) + H_2O\ (l) + CO_2\ (g)$
(e) $AgCl\ (s) \rightleftharpoons Ag^+\ (aq) + Cl^-\ (aq)$

13.16 Consider the equilibrium $PCl_3\ (g) + Cl_2\ (g) \rightleftharpoons PCl_5\ (g)$. How would the following affect the position of equilibrium?
(a) Addition of PCl_3
(b) Removal of Cl_2
(c) Removal of PCl_5
(d) Decrease in the volume of the container
(e) Addition of He without a change in volume

13.17 Which, if any, of the changes in Question 13.16 will change the value of the equilibrium constant for the reaction?

13.18 Indicate how each of the following changes affects the amount of H_2 in the system below, for which $\Delta H_{reaction} = +9.9$ kcal.

$$H_2\ (g) + CO_2\ (g) \rightleftharpoons H_2O\ (g) + CO\ (g)$$

(a) Addition of CO_2
(b) Addition of H_2O
(c) Addition of a catalyst
(d) Increase in temperature
(e) Decrease in the volume of the container

13.19 How will each of the changes in Question 13.18 affect the equilibrium constant?

13.20 Consider the equilibrium $2N_2O\ (g) + O_2\ (g) \rightleftharpoons 4NO\ (g)$. How will the amount of NO at equilibrium be affected by
(a) adding N_2O?
(b) removing O_2?
(c) increasing the volume of the container?
(d) adding a catalyst?

13.21 For the reaction $4NH_3\ (g) + 3O_2\ (g) \rightleftharpoons 2N_2\ (g) + 6H_2O\ (l)$, how will the amount of NH_3 at equilibrium be affected by
(a) adding O_2 to the system?
(b) adding N_2 to the system?
(c) removing H_2O from the system?
(d) decreasing the volume of the container?

13.22 Sketch a graph to show how the concentration of H_2, N_2, and NH_3 would change with time after N_2 had been added to a mixture of these gases initially at equilibrium.

13.23 In the equilibrium

$$CaCO_3\ (s) + heat \rightleftharpoons CaO\ (s) + CO_2\ (g)$$

how will the amount of $CaCO_3\ (s)$ change if
(a) CaO (s) is added?
(b) CO_2 (g) is added?
(c) the volume of the container is increased?
(d) the temperature is lowered?

REVIEW PROBLEMS (More difficult problems are marked by an asterisk)

13.24 Show that the following data, obtained for the reaction

$$PCl_5 (g) \rightleftharpoons PCl_3 (g) + Cl_2 (g)$$

demonstrate the law of mass action. What is K_c for this reaction?

Experiment	$[PCl_5]$	$[PCl_3]$	$[Cl_2]$
1	0.0023	0.23	0.055
2	0.010	0.15	0.37
3	0.085	0.99	0.47
4	1.00	3.66	1.50

13.25 Referring to Problem 13.24, calculate K_c for the reaction

$$PCl_3 (g) + Cl_2 (g) \rightleftharpoons PCl_5 (g)$$

13.26 What is the value of K_P for the reaction

$$PCl_5 (g) \rightleftharpoons PCl_3 (g) + Cl_2 (g)$$

Refer to the data in Problem 13.24. (T= 298 K)

13.27 The reaction

$$CO (g) + H_2O (g) \rightleftharpoons CO_2 (g) + H_2 (g)$$

is used industrially as a source of hydrogen. The value of K_c for this reaction at 500°C is 4.05. What is its value of K_P at this temperature?

13.28 The reaction

$$CH_4 (g) + H_2O (g) \rightleftharpoons CO (g) + 3H_2 (g)$$

is also a source of hydrogen. At 1500°C, its value of $K_c = 5.67$. What is its value of K_P at this temperature?

13.29 At 100°C, $K_P = 6.5 \times 10^{-2}$ for the reaction

$$2NO_2 (g) \rightleftharpoons N_2O_4 (g)$$

What is the value of K_c at this temperature?

13.30 At 700 K, $\Delta G'_{700 K} = -3.22$ kcal for the reaction, $CO (g) + 2H_2 (g) \rightleftharpoons CH_3OH (g)$. Calculate the value of K_P for the reaction at 700 K.

13.31 The equilibrium constant, K_P, for the reaction, $COCl_2 (g) \rightleftharpoons CO (g) + Cl_2 (g)$, has a value of 4.56×10^{-2} atm at 395°C. What is the value of $\Delta G'_{668 K}$ (in kilojoules) for this reaction?

13.32 At 527°C the reaction

$$CO (g) + H_2O (g) \rightleftharpoons CO_2 (g) + H_2 (g)$$

has $K_P = 5.10$. What is $\Delta G'_{800 K}$ for this reaction expressed in kilojoules?

13.33 Use the data in Tables 11.1 and 11.4 to compute $\Delta G'_{773 K}$ (in kilojoules) and K_P at 500°C for the reaction

$$2HCl (g) \rightleftharpoons H_2 (g) + Cl_2 (g)$$

Assume that ΔH^0 and ΔS^0 are independent of temperature.

13.34 Use the data in Table 11.5 to calculate K_P at 25°C for the reaction $2HCl (g) + F_2 (g) \rightleftharpoons 2HF (g) + Cl_2 (g)$.

13.35 Use the data in Tables 11.1 and 11.4 to compute the temperature at which $K_P = 1$ for the reaction

$$C_2H_4 (g) + H_2 (g) \rightleftharpoons C_2H_6 (g)$$

Assume that ΔH^0 and ΔS^0 are independent of temperature.

13.36 Methyl alcohol, CH_3OH, is a potential fuel that can be made from carbon monoxide (produced by burning coal) and steam. The equilibrium is

$$CO(g) + 2H_2 (g) \rightleftharpoons CH_3OH (g)$$

At 427°C (700 K) a mixture of CO, H_2, and CH_3OH having the following partial pressures was prepared: $P_{CO} = 2 \times 10^{-3}$ atm, $p_{H_2} = 1 \times 10^{-2}$ atm, $p_{CH3OH} = 3 \times 10^{-6}$ atm. For this reaction, $\Delta G'_{700 K} = -3.22$ kcal. Use Equation 13.4 to determine whether this system is at equilibrium. If not, will the reaction proceed spontaneously to the left or to the right?

13.37 At 25°C, in a mixture of N_2O_4 and NO_2 in equilibrium at a total pressure of 0.844 atm, the partial pressure of N_2O_4 is 0.563 atm. Calculate for the reaction,

$$N_2O_4 (g) \rightleftharpoons 2NO_2 (g)$$

(a) K_P, (b) K_c, (c) $\Delta G^0_{298 K}$ in kJ.

13.38 An air pollutant produced by burning high-sulfur fuels is sulfur dioxide. In smog, which contains appreciable amounts of NO_2, the sulfur dioxide can be oxidized to sulfur trioxide, which forms H_2SO_4 when it reacts with moisture. The reaction is

$$SO_2 (g) + NO_2 (g) \rightleftharpoons NO (g) + SO_3 (g)$$

Use the data in Table 11.5 to calculate K_c for this reaction at 25°C.

13.39 The following thermodynamic data apply at 25°C:

Substance	ΔG_f^0 (kcal/mol)
$NiSO_4 \cdot 6H_2O$ (s)	−531.0
$NiSO_4$ (s)	−184.9
H_2O (g)	−54.6

(a) What is ΔG^0 (in kcal) for the reaction,

$$NiSO_4 \cdot 6H_2O (s) \rightleftharpoons NiSO_4 (s) + 6H_2O (g)$$

(b) What is K_P for this reaction?
(c) What is the equilibrium vapor pressure of H_2O over solid $NiSO_4 \cdot 6H_2O$ expressed in torr?

13.40 For the reaction $PCl_5 (g) \rightleftharpoons PCl_3 (g) + Cl_2 (g)$, $K_c = 33.3$ at 760°C. In a container at equilibrium there are 1.29×10^{-3} mol/liter of PCl_5 and 1.87×10^{-1} mol/liter of Cl_2. Calculate the equilibrium concentration of PCl_3 in the vessel.

13.41 At a certain temperature the following equilibrium concentrations were found for the reactants and products in the reaction,

$$2HI\ (g) \rightleftharpoons H_2\ (g) + I_2\ (g)$$

$$[H_2] = 1.0 \times 10^{-3}\ M$$

$$[I_2] = 2.5 \times 10^{-2}\ M$$

$$[HI] = 2.2 \times 10^{-2}\ M$$

What is the value of K_c for this reaction?

13.42 In a particular experiment the following partial pressures were determined for the reaction at equilibrium

$$2NO\ (g) + Cl_2\ (g) \rightleftharpoons 2NOCl\ (g)$$

$$p_{NO} = 0.65\ atm \qquad p_{Cl_2} = 0.18\ atm \qquad p_{NOCl} = 0.15\ atm$$

What is K_P for this reaction at the temperature at which the experiment was performed?

13.43 At 25°C, 0.0560 mol of O_2 and 0.020 mol N_2O were placed in a 1.00-liter vessel and allowed to react according to the equation

$$2N_2O\ (g) + 3O_2\ (g) \rightleftharpoons 4NO_2\ (g)$$

When the system reached equilibrium, the concentration of the NO_2 was found to be 0.020 mol/liter.
(a) What were the equilibrium concentrations of N_2O and O_2?
(b) What is the value of K_c for this reaction at 25°C?

13.44 At 460°C, the reaction

$$SO_2\ (g) + NO_2\ (g) \rightleftharpoons NO\ (g) + SO_3\ (g)$$

has $K_c = 85.0$. What will be the equilibrium concentrations of the four gases if a mixture of SO_2 and NO_2 is prepared in which they each have an initial concentration of 0.0500 M?

13.45 For the reaction

$$H_2\ (g) + CO_2\ (g) \rightleftharpoons CO\ (g) + H_2O\ (g)$$

$K_c = 0.771$ at 750°C. If 1.00 mol of H_2 and 1.00 mol of CO_2 are placed into a 5.00-liter container and permitted to react, what will be the equilibrium concentrations of all four gases?

13.46 Suppose a mixture of SO_2, NO_2, NO, and SO_3 was prepared at 460°C having the following initial concentrations: $[SO_2] = 0.0100\ M$, $[NO_2] = 0.0200\ M$, $[NO] = 0.0100\ M$, and $[SO_3] = 0.0150\ M$. At this temperature the reaction

$$SO_2\ (g) + NO_2\ (g) \rightleftharpoons NO\ (g) + SO_3\ (g)$$

has $K_c = 85.0$. What will be the equilibrium concentrations of the four gases?

13.47 The reaction

$$2CO_2 \rightleftharpoons 2CO + O_2$$

has $K_c = 6.4 \times 10^{-7}$ at 2000°C. If 1.0×10^{-3} mol of CO_2 is placed into a 1.0-liter vessel at this temperature,

(a) What will be the equilibrium concentrations of CO and O_2?
(b) What fraction of the CO_2 will have decomposed?

*__13.48__ At 100°C the equilibrium constant, K_c, for the reaction

$$CO\ (g) + Cl_2\ (g) \rightleftharpoons COCl_2\ (g)$$

has a value of 4.6×10^9. If 0.20 mol of $COCl_2$ is placed into a 10.0-liter flask at 100°C, what will be the concentration of all species at equilibrium?

13.49 Sodium bicarbonate (baking soda) has many useful properties. Among them is the ability to serve as a fire extinguisher because of thermal decomposition to produce CO_2, which smothers the fire,

$$2NaHCO_3\ (s) \rightleftharpoons Na_2CO_3\ (s) + CO_2\ (g) + H_2O\ (g)$$

At 125°C the value of K_P for this reaction is 0.25 atm². What are the partial pressures of $CO_2\ (g)$ and $H_2O\ (g)$ in this system at equilibrium? Can you explain why $NaHCO_3$ is used in baking?

*__13.50__ In a 10.0-liter mixture of H_2, I_2, and HI at equilibrium at 425°C there are 0.100 mol of H_2, 0.100 mol of I_2, and 0.740 mol of HI. If 0.50 mol of HI are now added to this system, what will be the concentration of H_2, I_2, and HI once equilibrium has been reestablished?

*__13.51__ In Question 13.48 it was stated that at 100°C the value of K_c for the reaction, $CO\ (g) + Cl_2\ (g) \rightleftharpoons COCl_2\ (g)$, is 4.6×10^9. Suppose that 0.15 mol of CO and 0.30 mol of Cl_2 were placed into a 1.0-liter vessel and allowed to react. What would the concentration be of each of the gases in the system at equilibrium? (*Hint:* First assume 100% reaction; then work backwards toward equilibrium.)

*__13.52__ The production of NO by reaction of N_2 and O_2 in an automobile engine is an important source of nitrogen oxide pollution. At 1000°C the reaction, $N_2\ (g) + O_2\ (g) \rightleftharpoons 2NO\ (g)$, has $K_P = 4.8 \times 10^{-7}$. Suppose that the partial pressures of N_2 and O_2 in the cylinder of an engine after the gasoline vapor has been ignited are $p_{N_2} = 33.6$ atm and $p_{O_2} = 4.0$ atm. Assume that the temperature of the mixture is 1000°C. Calculate the partial pressure of NO in the mixture if the system has time to reach equilibrium.

*__13.53__ If it is assumed that the reactants and products in the preceding question are unable to react further when the exhaust gases are suddenly cooled as they exit the engine, calculate the partial pressure of the NO when the partial pressure of N_2 has dropped to 0.80 atm and the temperature has dropped to 150°C.

*__13.54__ At a certain temperature $K_c = 7.5$ for the reaction

$$2NO_2 \rightleftharpoons N_2O_4$$

If 2.0 mol of NO_2 are placed in a 2.0-liter container and permitted to react, what will be the concentrations of NO_2 and N_2O_4 at equilibrium? What will be the equilibrium concentrations if the size of the container is doubled? Does this conform to what you would expect from Le Châtelier's principle?

14

ACIDS AND BASES

The reaction of gaseous hydrogen chloride and gaseous ammonia, each coming from its concentrated aqueous solution, produces a white cloud of ammonium chloride. Although most acid-base reactions that we encounter take place in solution, the presence of a solvent is not always necessary, as we will learn in this chapter.

By now you've probably come to realize how important the acid-base concept is. In Chapter 6 we introduced acids and bases and discussed some of their properties as they apply to reactions in aqueous solutions. Then, in Chapter 9, we saw how useful the acid-base concept was in organizing properties in the periodic table. There are definite trends in the acidity of the oxides of the elements, as well as trends in the acidity of the binary hydrogen compounds of the nonmetals—such substances as HCl and H_2S.

In all this discussion we've restricted ourselves to considering acid-base chemistry in aqueous solutions. Because water is such a common solvent and also because it is found in all living systems, aqueous acid-base chemistry is extremely important. In fact, all of Chapter 15 deals with acid-base equilibrium in water solutions. Before we get to that subject, however, we will look at how broad the concept of acids and bases is. It can easily be extended beyond aqueous media, and even to substances that contain no hydrogen; doing this organizes and makes understandable a vast number of chemical facts.

In this chapter we will see that whether a particular substance behaves as an acid or a base depends on the way in which acids and bases are defined. There are a variety of ways of approaching this problem, with some definitions more restrictive than others. Nevertheless, several properties are characteristic of acids and bases in general. These include:

1. *Neutralization.* Acids and bases react with one another so as to cancel, or neutralize, their acidic and basic characters.
2. *Reaction with indicators.* Certain organic dyes, called indicators, give different colors depending on whether they are in an acidic or basic medium.
3. *Catalysis.* Many chemical reactions are catalyzed by the presence of acids or bases.

14.1 THE ARRHENIUS DEFINITION OF ACIDS AND BASES

In its modern version, the **Arrhenius concept** of acids and bases (sometimes referred to as the aqueous concept) defines *an acid as any substance that can increase the concentration of hydronium ion, H_3O^+, in aqueous solution.* On the other hand, *a base is a substance that increases the hydroxide ion concentration in water.* Thus, it is this concept of acids and bases that was first presented in Section 6.5. Let's review briefly some of the ideas developed there.

You will recall that HCl is an acid because it reacts with water according to the equation,

$$HCl + H_2O \longrightarrow H_3O^+ + Cl^-$$

Similarly, CO_2 is an acid because it reacts with water to form carbonic acid, H_2CO_3,

$$CO_2 + H_2O \rightleftharpoons H_2CO_3$$

which then undergoes further reaction to produce H_3O^+ and HCO_3^-.

$$H_2CO_3 + H_2O \rightleftharpoons H_3O^+ + HCO_3^-$$

In general, nonmetal oxides react with water to yield acidic solutions and are said to be **acid anhydrides** (Greek, *anydros*, waterless).

An example of an Arrhenius base is NaOH, an ionic compound containing Na^+ and OH^- ions. In water it undergoes dissociation.

$$NaOH \ (s) \xrightarrow{H_2O} Na^+ \ (aq) + OH^- \ (aq)$$

Other examples of bases include substances such as NH_3 and N_2H_4 that react with water to produce OH^-.

$$NH_3 + H_2O \rightleftharpoons NH_4^+ + OH^-$$

$$\underset{\textbf{(hydrazine)}}{N_2H_4} + H_2O \rightleftharpoons \underset{\substack{\textbf{(hydrazinium} \\ \textbf{ion)}}}{N_2H_5^+} + OH^-$$

You will also recall that metal oxides undergo reaction with water to give the corresponding hydroxides. They are called **basic anhydrides.**

$$Na_2O + H_2O \longrightarrow 2NaOH$$

$$BaO + H_2O \longrightarrow Ba(OH)_2$$

Finally, in aqueous solution the neutralization of an acid by a base takes the form of the ionic reaction,

$$H_3O^+ + OH^- \longrightarrow 2H_2O$$

14.2 BRØNSTED-LOWRY DEFINITION OF ACIDS AND BASES

The definition of acids and bases in terms of the hydronium ion and hydroxide ion in water is very restricted because it limits us to discussing acid-base phenomena in aqueous solutions only. A somewhat more general approach was that proposed independently in 1923 by the Danish chemist, J. N. Brønsted, and the British chemist, T. M. Lowry. They defined an *acid as a substance that is able to donate a proton* (i.e., a hydrogen ion, H^+) *to some other substance. A base is defined as a substance that is able to accept a proton from an acid.* Stated more simply, an acid is a proton donor and a base is a proton acceptor.

A typical example of a Brønsted-Lowry acid-base reaction occurs when HCl is added to water.

$$HCl + H_2O \longrightarrow H_3O^+ + Cl^-$$

In this reaction HCl is functioning as an acid because it is donating a proton to the water molecule. Water, on the other hand, is behaving as a base by accepting a proton from the acid.

If we have a solution of concentrated HCl and heat it, we drive off HCl gas. In other words, we can reverse the above reaction so that H_3O^+ and Cl^- react with each other to produce HCl and H_2O. This reverse reaction is also a Brønsted-Lowry reaction, with hydronium ion serving as an acid by giving up its proton, and with the chloride ion functioning as a base by accepting it. Thus, we might view the reaction of HCl with water as an equilibrium in which there are two acids and two bases, one of each on either side of the arrow.

$$\underset{\textbf{acid}}{HCl} + \underset{\textbf{base}}{H_2O} \rightleftharpoons \underset{\textbf{acid}}{H_3O^+} + \underset{\textbf{base}}{Cl^-}$$

When the acid, HCl, reacts it yields the base, Cl^-. These two substances are related to one another by the loss or gain of a single proton and constitute a **conjugate acid-base pair.** We say that Cl^- is the **conjugate base** of the acid, HCl, and similarly that HCl is the **conjugate acid** of the base, Cl^-. In this reaction we also find that H_2O and H_3O^+ form a conjugate pair. Water is the conjugate base of H_3O^+, and H_3O^+ is the conjugate acid of H_2O.

Another example of a Brønsted-Lowry acid-base reaction occurs in aqueous solutions of ammonia.

$$NH_3 + H_2O \rightleftharpoons NH_4^+ + OH^-$$

In this case water serves as an acid by giving up a proton to a molecule of NH_3, which thereby acts as a base. In the reverse reaction, on the other hand, NH_4^+ is the acid and OH^- is the base. Again we have two acid-base conjugate pairs: NH_3 and NH_4^+ is one of them, and H_2O and OH^- is the other.

In general, we can represent any Brønsted-Lowry acid-base reaction as

$$\text{acid }(X) + \text{base }(Y) \rightleftharpoons \text{base }(X) + \text{acid }(Y)$$

where acid (X) and base (X) represent one conjugate pair and acid (Y) and base (Y) the other. Notice that the members of a conjugate pair differ *only by one proton*. They are otherwise the same. Also, within a conjugate pair, the acid has one more hydrogen than the base.

In the two examples that we just examined, water functioned as a base in one instance, and as an acid in the other. Such a substance, which can serve in either capacity depending on conditions, is said to be **amphiprotic** or **amphoteric**. Water is not the only substance to behave this way. For example, water, acetic acid, and liquid ammonia all undergo **autoionization reactions** in which a proton transfer between two like molecules produces a pair of ions.[1] These reactions can be represented by the equations

> You've seen the concept of amphoterism before in Chapter 9.

$$H_2O + H_2O \rightleftharpoons H_3O^+ + OH^-$$

$$HC_2H_3O_2 + HC_2H_3O_2 \rightleftharpoons H_2C_2H_3O_2^+ + C_2H_3O_2^-$$

$$NH_3\ (l) + NH_3\ (l) \rightleftharpoons NH_4^+ + NH_2^-$$

$$(\text{acid}) + (\text{base}) \rightleftharpoons (\text{acid}) + (\text{base})$$

These reactions can also be illustrated using Lewis structures as follows:

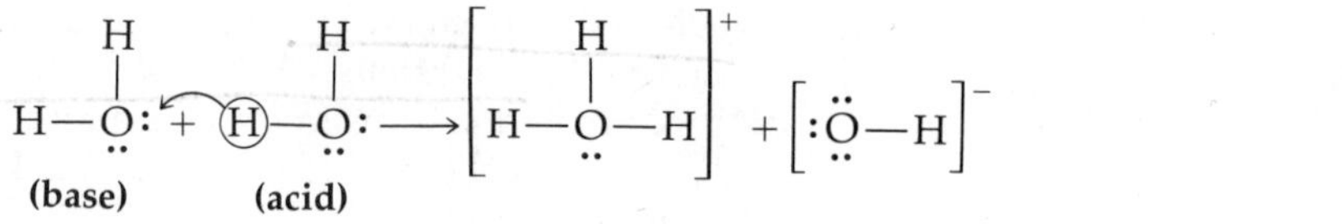

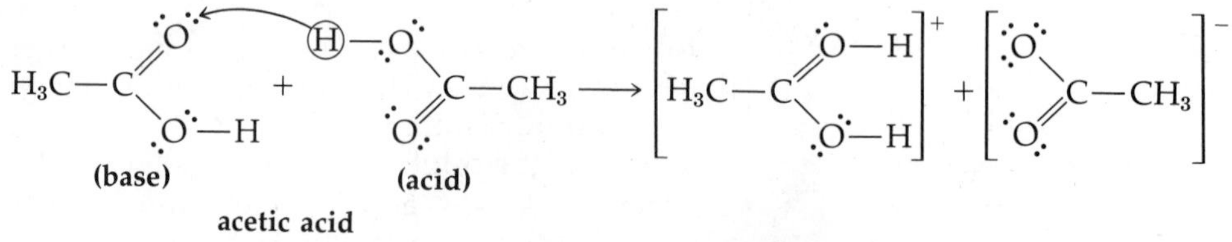

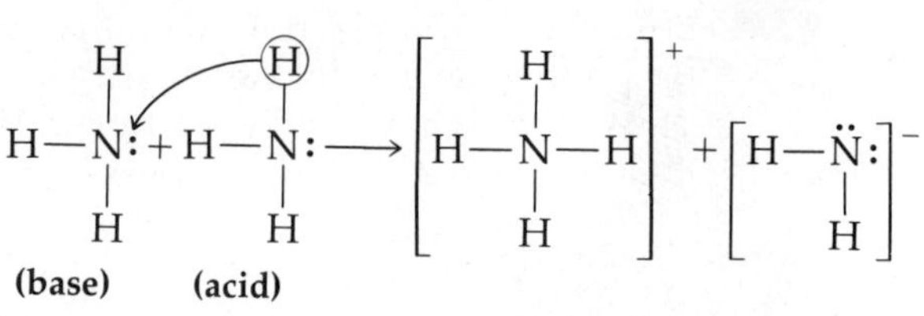

acetic acid

In each case, the substance is playing the role of both an acid and a base.

An interesting example of a Brønsted-Lowry acid-base reaction occurs in aqueous solutions that contain metal ions in high positive oxidation states. For

---

[1] In general, an autoionization reaction involves the creation of a cation-anion pair from two neutral molecules of the same substance. This occurs by the transfer of an atom and some charge from one particle to another, but the atom that is transferred doesn't have to be a proton. For example, the following is also an autoionization reaction.

$$2PCl_5 \longrightarrow [PCl_4^+][PCl_6^-]$$

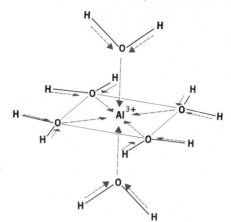

Figure 14.1

Dotted lines in color indicate how electron density is drawn toward the highly charged Al^{3+} ion and away from the O—H bonds. This increases the polarity of the O—H bonds and makes it easier to remove the hydrogens as H^+.

example, solutions of the salt $AlCl_3$ are acidic, as are solutions containing Cr^{3+} and Fe^{3+}. In these solutions the metal ions are surrounded by water molecules that have the negative ends of their dipoles directed toward the metal. The high charge on the metal ion holds these water molecules very tightly, and the metal ion with its surrounding layer of water molecules behaves like a single unit. In solutions containing Al^{3+}, for example, the aluminum ion appears to have six water molecules around it that are arranged at the vertices of an octahedron, as we saw in Chapter 10 (page 326). In fact, when salts of Al^{3+} are crystallized from aqueous solutions, their crystals usually contain the octahedral $Al(H_2O)_6{}^{3+}$ ion.

The acidity of solutions of these metal ions is explained as follows. The high charge on the metal ion draws electron density toward itself and away from the oxygen atoms of the water molecules that surround it. These oxygen atoms, in turn, draw electron density from the O—H bonds (Figure 14.1). This makes the O—H bonds even more polar than they are in an ordinary water molecule. In other words, the hydrogen atoms of the water molecules surrounding the metal ion carry an even greater partial positive charge than the hydrogen atoms in an ordinary water molecule. As a result, a hydrogen is rather easily removed as H^+ from an ion such as $Al(H_2O)_6{}^{3+}$ and the ion is acidic. Its reaction with water can be viewed as a Brønsted-Lowry acid-base reaction.

$$Al(H_2O)_6{}^{3+} + H_2O \rightleftharpoons Al(H_2O)_5OH^{2+} + H_3O^+$$

$$\text{acid} \qquad \text{base} \qquad\quad \text{base} \qquad\quad \text{acid}$$

The Brønsted-Lowry concept is more general than the Arrhenius concept because it does not limit us to water solutions. In fact, we can find acid-base reactions that occur in the absence of any solvent whatsoever. For instance, when gaseous HCl and NH_3 are mixed, they react immediately to form the white ionic solid, NH_4Cl.

$$NH_3\ (g) + HCl\ (g) \longrightarrow NH_4Cl\ (s)$$

This is the reaction shown in the photograph at the beginning of the chapter. Using Lewis structures, it can be diagrammed as

Since a proton is transferred from HCl to NH_3, this is clearly an acid-base reaction in the Brønsted-Lowry sense. However, because neither hydronium ion nor hydroxide ion ever enter the picture, the Arrhenius view of acids and bases ignores this reaction completely.

14.3 STRENGTHS OF ACIDS AND BASES

Any Brønsted-Lowry acid base reaction can be viewed as two opposing or competing reactions between acids and bases. In one sense, the two bases can be considered to be competing for a proton. When HCl reacts with water, for example, we find from conductivity measurements and freezing-point depression that essentially all the HCl has reacted with water. This means that the position of equilibrium in the reaction,

$$HCl + H_2O \rightleftharpoons H_3O^+ + Cl^-$$

lies very far to the right. In turn, this tells us that H_2O has a much stronger affinity for a proton than does a chloride ion because the water is able to capture essentially all of the available H^+. We express this relative ability to pick up a proton by saying that water is a stronger base than chloride ion.

We can also speak of the relative strengths of the two acids in this reaction, HCl and H_3O^+. Here we see that HCl is better able to donate its proton than is H_3O^+, because all the HCl has lost its protons to give H_3O^+.

A better way of judging the relative strengths of acids and bases is by comparing the positions of equilibrium in various acid-base reactions. For example, in water we find that HCl is essentially 100% ionized, but in a 1 M solution of HF, only about 3% of the HF is present as H_3O^+ and F^-.

$$HCl + H_2O \longrightarrow H_3O^+ + Cl^- \qquad (100\%)$$

$$HF + H_2O \rightleftharpoons H_3O^+ + F^- \qquad \text{(about 3\% in 1 } M \text{ HF)}$$

These results tell us that in water HCl is a much stronger acid than HF is. They also tell us that F^- is a much stronger base than Cl^-, because in solutions with the same reference acid, H_3O^+, most of the F^- is protonated and exists as HF, while none of the Cl^- is protonated at all. A useful generalization to remember from this is that *as acids become stronger their conjugate bases become weaker.* We've seen that HCl is a stronger acid than HF and that Cl^- is a weaker base than F^-.

Similar comparisons can also be made for bases. Another way of phrasing our generalization is that *as bases become stronger their conjugate acids become weaker.* For example, the amide ion, NH_2^-, is a very strong base and reacts completely with water to form its conjugate acid, NH_3.

$$NH_2^- + H_2O \longrightarrow NH_3 + OH^- \qquad (100\%)$$

Ammonia is a weaker acid than water, so we think of it as a base.

Ammonia itself is a weak base in water and, in a 1 M solution of NH_3, only about 0.4% of the NH_3 reacts to form NH_4^+ and OH^-.

$$NH_3 + H_2O \rightleftharpoons NH_4^+ + OH^- \qquad \text{(about 0.4\% in 1 } M \text{ NH}_3)$$

NH_4^+ is the conjugate acid of NH_3, and NH_3 is the conjugate acid of NH_2^-.

Comparing the reactions of NH_2^- and NH_3 with water, we find that NH_2^- is a much stronger base than NH_3. We also see from these reactions that NH_4^+ is a stronger acid than NH_3, because NH_4^+ gives up a proton to OH^- more readily than NH_3 does. The relationship between the relative strengths of various acid-base conjugate pairs is illustrated in Figure 14.2.

In general, when a Brønsted-Lowry acid-base reaction takes place, the position of equilibrium lies in the direction of the *weaker* acid and base. In other words, the stronger acid and base react to produce the corresponding weaker conjugate base and acid, respectively.

stronger acid (HA) + stronger base (B) $\longrightarrow$ weaker base (A^-) + weaker acid (HB^+)

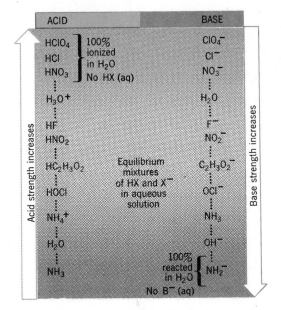

Figure 14.2

Relative strengths of acid-base pairs.

In the reaction between HCl and water, for example, we saw that the stronger acid HCl reacts 100% with the stronger base H_2O to give the weaker acid H_3O^+ and the weaker base Cl^-. Therefore, if the relative position of two acid-base conjugate pairs in Figure 14.2 is known, we can make some judgement about the position of equilibrium in an acid-base reaction involving them. For example, if a strong acid such as HNO_3 is added to a relatively strong base such as liquid NH_3, the reaction will proceed very nearly to completion.

$$HNO_3 + NH_3 \longrightarrow NH_4^+ + NO_3^-$$
(virtually 100% complete)

On the other hand, if a weak acid such as $HC_2H_3O_2$ is combined with a weak base such as Cl^- (from NaCl, for instance), essentially no reaction will be observed.

$$HC_2H_3O_2 + Cl^- \longrightarrow \text{no reaction}$$

Between these extremes, we can establish equilibria with varying amounts of both reactants and products. We saw earlier, for example, that in a 1 M solution about 3% of the HF is reacted to give F^-.

When we set about to compare the strengths of acids, such as HCl and HF, we have seen that it is best to use the same reference base. For example, we concluded that HCl is a stronger acid than HF because HCl is better able to protonate the base, H_2O, than HF is. With a given base, however, it is not possible to compare the strengths of *all* acids. For example, HCl, HNO_3, and $HClO_4$ all appear to be 100% ionized in water. With the reference base water, each of these acids appears to be of equal strength. Their differences are removed, or leveled out, and we speak of this phenomenon as the **leveling effect.**

If pure acetic acid (the substance that gives vinegar its sour taste) is used as a solvent instead of water, we find that there is an appreciable difference between the extents to which the reactions

$$HCl + HC_2H_3O_2 \rightleftharpoons H_2C_2H_3O_2^+ + Cl^-$$

$$HNO_3 + HC_2H_3O_2 \rightleftharpoons H_2C_2H_3O_2^+ + NO_3^-$$

$$HClO_4 + HC_2H_3O_2 \rightleftharpoons H_2C_2H_3O_2^+ + ClO_4^-$$

We discussed the factors that affect acid strength in Chapter 9.

The way acetic acid accepts a proton is shown on page 450.

proceed toward completion. In this instance, acetic acid is a much weaker base than water and is therefore not as easily protonated. Consequently, by using acetic acid as the solvent, it is possible to distinguish among the strengths of these three acids and, in fact, we find that their acidity increases in the order $HNO_3 < HCl < HClO_4$. For these substances water is a **leveling solvent,** while acetic acid serves as a **differentiating solvent.**

The leveling effect is not restricted to acids alone. Strong bases such as the oxide ion, O^{2-}, amide ion, NH_2^-, and hydride ion, H^-, react completely with water to give hydroxide ion:

$$O^{2-} + H_2O \longrightarrow OH^- + OH^-$$

$$NH_2^- + H_2O \longrightarrow NH_3 + OH^-$$

$$H^- + H_2O \longrightarrow H_2 + OH^-$$

The reaction of H^- with H_2O can also be viewed as a redox reaction.

These three bases are so strong that they are completely protonated by an acid as weak as water. Therefore, with water as a solvent, it is impossible to differentiate among their base strengths.

In general, any basic solvent tends to exert a leveling effect on acids, while an acidic solvent tends to level the strengths of bases. In water, for example, HF and HCl are clearly of different strengths as measured by their ability to protonate water molecules. In liquid ammonia (a basic solvent), however, the reactions

$$HF + NH_3 \longrightarrow NH_4^+ + F^-$$

$$HCl + NH_3 \longrightarrow NH_4^+ + Cl^-$$

both proceed essentially to completion. In ammonia, HF and HCl appear to be of equal strength, and both behave as strong acids.

A similar phenomenon is observed with ammonia in the solvents H_2O and $HC_2H_3O_2$. Ammonia is only slightly protonated in water, while it becomes completely protonated when it is added to pure acetic acid. In the latter solvent, NH_3 behaves as a strong base while in water it is weak.

14.4 LEWIS ACIDS AND BASES

The Brønsted-Lowry definition of acids and bases is more general than the Arrhenius definition because it removes the restriction of having to deal with reactions in aqueous solution. However, even the Brønsted-Lowry concept is restricted in scope, since it limits discussion of acid-base phenomena to proton-transfer reactions. There are many reactions that have all the earmarks of acid-base reactions but that do not fit the Brønsted-Lowry mold. The approach taken by the American chemist Gilbert N. Lewis further extends the acid-base concept to cover these cases.

This is the same G. N. Lewis who originated the Lewis symbols and structures you learned to draw in Chapter 4.

In the Lewis definition of acids and bases, primary attention is focused on the base. *A* **base** *is defined as a substance that can donate a* **pair** *of electrons to the formation of a covalent bond. An* **acid** *is a substance that can accept a pair of electrons to form the bond.*

A simple example of a Lewis acid-base reaction is the reaction of a proton with a hydroxide ion,

Lewis base — electron pair donor.
Lewis acid — electron pair acceptor.

The hydroxide ion is the Lewis base because it furnishes the pair of electrons that become shared with the hydrogen. The hydrogen ion, on the other hand, is the Lewis acid because it accepts a share of the pair of electrons when the O—H bond is created.

Another example is the reaction between BF_3 and ammonia,

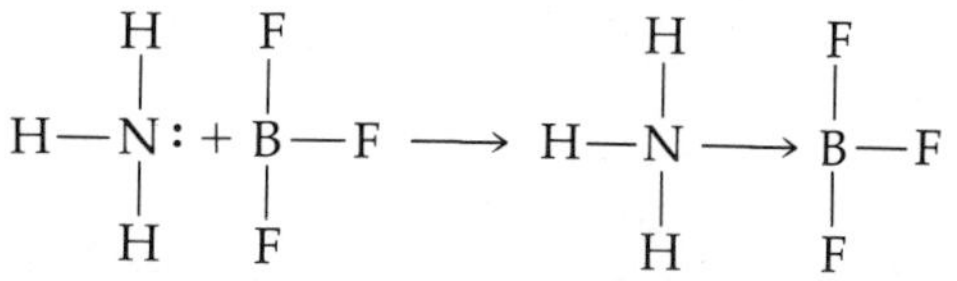

$$H-\underset{\underset{H}{|}}{\overset{\overset{H}{|}}{N}}: + \underset{\underset{F}{|}}{\overset{\overset{F}{|}}{B}}-F \longrightarrow H-\underset{\underset{H}{|}}{\overset{\overset{H}{|}}{N}} \longrightarrow \underset{\underset{F}{|}}{\overset{\overset{F}{|}}{B}}-F$$

Recall that NH_3BF_3 is an example of an addition compound.

In this case the NH_3 functions as the base and BF_3 serves as the acid. Compounds containing elements with incomplete valence shells, such as BF_3 or $AlCl_3$, tend to be Lewis acids, while compounds or ions that have lone pairs of electrons can behave as Lewis bases. When the Lewis acid-base reaction occurs, a coordinate covalent bond is formed.

Still other examples of Lewis acid-base reactions are provided by the reactions of metal oxides with nonmetal oxides. Recall that metal oxides, in water, produce hydroxides. For instance,

$$Na_2O + H_2O \longrightarrow 2NaOH$$

Nonmetal oxides react to form acids as illustrated by the reaction

$$SO_3 + H_2O \longrightarrow H_2SO_4$$

When these two solutions are mixed, neutralization occurs, with the production of the solvent plus a salt,

$$2NaOH + H_2SO_4 \longrightarrow 2H_2O + Na_2SO_4$$

The production of Na_2SO_4 from Na_2O and SO_3 can take place directly without the introduction of any water whatsoever, as shown by the equation,

$$Na_2O \ (s) + SO_3 \ (g) \longrightarrow Na_2SO_4 \ (s)$$

According to the Lewis definition, this too is a neutralization reaction between a Lewis base (oxide ion) and a Lewis acid (sulfur trioxide):

A pair of electrons in the double bond moves to the oxygen on top, allowing the sulfur to accept the share of a pair of electrons from the oxide ion.

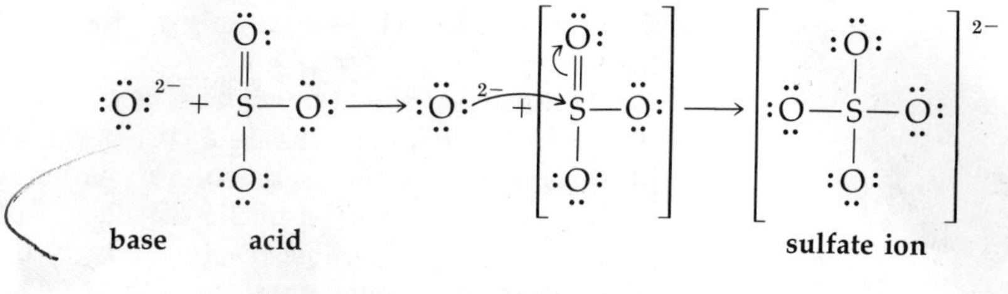

base   acid   sulfate ion

In this case we find that some electronic rearrangement must take place as the oxygen becomes attached to the sulfur. Nevertheless, the overall change can be viewed as a neutralization reaction.

Reactions of this type, between an oxide such as CaO and SO_2 or SO_3, are important for removal of sulfur oxides from the gases produced by combustion of high-sulfur fuels. For example,

$$CaO \ (s) + SO_2 \ (g) \longrightarrow CaSO_3 \ (s)$$

A reaction quite analogous to that just described has been used on spacecraft to remove carbon dioxide from the air breathed by the astronauts. In this case carbon dioxide reacts with LiOH,

$$CO_2 \ (g) + LiOH \ (s) \longrightarrow LiHCO_3 \ (s)$$
$$\textbf{(lithium}$$
$$\textbf{bicarbonate)}$$

(Lithium hydroxide is used because of the very low atomic weight of Li. This results in many moles of LiOH per pound.) This reaction too can be viewed as a Lewis acid-base reaction:

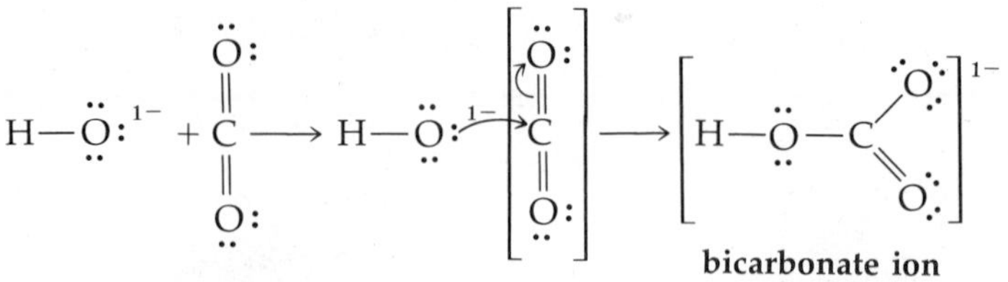

bicarbonate ion

The reaction between Na_2O and SO_3 illustrates the limitations of the Brønsted-Lowry concept. Since no protons are involved in the reaction, it would never be classified as an acid-base reaction under the Brønsted-Lowry definition.

Thus far we have looked only at simple acid-base neutralizations. The acid-base reactions discussed in Sections 14.2 and 14.3 can also be treated from the Lewis point of view. In the Brønsted-Lowry theory these reactions were regarded as competitions in which the strongest acid prevails by losing its proton. Under the Lewis definition these reactions are looked on as constituting the displacement of one base (the weaker one) by another. Referring to our reaction of HCl with H_2O, for example,

$$HCl + H_2O \longrightarrow H_3O^+ + Cl^-$$

the Lewis theory interprets the change as the result of replacing the Cl^- ion in HCl by the stronger base, H_2O.

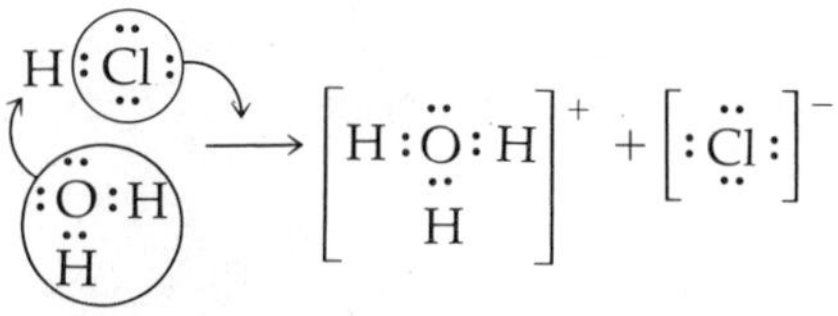

In other words, the stronger base, H_2O, pushes out the weaker one, Cl^-. Here we interpret the acid to be the H^+ ion instead of the entire HCl molecule, and in the reaction, the H^+ is changing partners as it moves from the weaker base to the stronger one.

By its definition, a Lewis base is a substance that in its reactions seeks a nucleus with which it can share a pair of electrons; it is thus said to be a **nucleophile** (nucleus-loving). A reaction in which one Lewis base displaces another is therefore called a **nucleophilic displacement.** From this point of view a very large number of chemical reactions, including *all* the proton transfer reactions of the Brønsted-Lowry type, can also be regarded simply as acid-base reactions in which a stronger Lewis base displaces a weaker one.

There are also acid-base reactions in which one Lewis acid displaces another. For example, consider the reaction,

$$AlCl_3 + COCl_2 \longrightarrow COCl^+ + AlCl_4^-$$

Using electron dot formulas this reaction can be analyzed as follows:

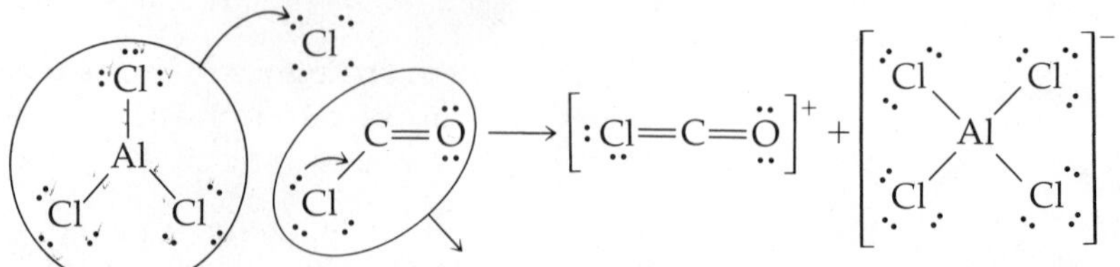

Notice that we view this reaction to take place by the displacement of the acid, $COCl^+$, by the stronger acid $AlCl_3$. In other words, we imagine the molecule, $COCl_2$, to be the "neutralization product" of the Lewis acid $COCl^+$ and the Lewis base Cl^-, and it is the base that changes partners when the displacement reaction occurs:

$$AlCl_3 + COCl_2 \longrightarrow [Cl_3Al \cdots Cl^- \cdots COCl^+] \longrightarrow AlCl_4^- + COCl^+$$

Since Lewis acids seek substances having electron pairs to which they can become bound, Lewis acids are said to be **electrophilic** (electron-loving) in character and an acid displacement reaction is said to be an **electrophilic** displacement. Although we are able to classify most acid-base reactions in water as nucleophilic displacements, electrophilic displacement reactions are important too. You will encounter a number of them if you take a course in organic chemistry.

The strengths of Lewis acids and bases can be compared by examining their tendency to form a coordinate covalent bond. When the bond is formed, electron density on the base is drawn toward the electron-poor atom of the acid. A strong base, therefore, contains an atom whose electron cloud is easily deformed, or polarized. In Section 8.3 we saw that large atoms are more easily polarized than small atoms. Thus $(CH_3)_2S$ would be expected to be a stronger Lewis base than $(CH_3)_2O$ because S is larger and more easily polarized than O. On the other hand, $(CF_3)_2S$ would be a weaker base than $(CH_3)_2S$ because the very electronegative fluorine atoms would make it more difficult to draw electron density from the S in $(CF_3)_2S$ than from the S in $(CH_3)_2S$.

Lewis acid strength is determined by the electron-attracting power of the electron-poor atom of the acid. In general, small atoms are better at attracting electrons than large atoms. The valence shell of a small atom is closer to its nucleus, and other electrons approaching this valence shell are strongly attracted. We would therefore expect BCl_3 to be a stronger acid than $AlCl_3$. Also, highly positively charged ions are better Lewis acids than positive ions of low charge. Thus Fe^{3+} would be a stronger Lewis acid than Fe^{2+}.

Lewis acids are electrophiles.

14.5 THE SOLVENT SYSTEM APPROACH TO ACIDS AND BASES

Because water is so plentiful and is such a good solvent for so many substances, much of the solution chemistry that we meet in the laboratory uses water as the solvent. Chapter 6 dealt with the various kinds of reactions that are encountered in aqueous solutions, and the definitions of acids and bases that were presented there were given in terms of the ions, H_3O^+ and OH^-.

As mentioned before, water undergoes a limited degree of autoionization that can be described by the equation,

$$H_2O + H_2O \rightleftharpoons H_3O^+ + OH^-$$

Notice that when water is the solvent, an acid is defined as a substance that produces the cation, H_3O^+, characteristic of the autoionization reaction. Similarly, a base in the solvent water gives the same anion, OH^-, that is formed in the autoionization reaction. These two observations form the basis of a more general **solvent system approach** to dealing with acid-base reactions. This approach recognizes that water is really not unique in its solvent properties, and that many reactions in other solvents can be viewed as analogs of reactions that take place in aqueous solution. The solvent system approach does this by defining a substance as an acid or a base according to the ions that are (or could be) formed when the substance is dissolved in the solvent. In any particular solvent, *an* *acid is a substance that produces the cation formed by the autoionization of the solvent, and a base is a substance that produces the anion formed in the autoionization.* Although a number of different solvent systems have been studied in terms of these definitions, we will only examine one of them here: liquid ammonia.

Liquid ammonia as a solvent

Probably the most thoroughly studied solvent besides water is liquid ammonia. At atmospheric pressure ammonia exists as a gas at room temperature; it is therefore necessary to cool the substance to its boiling point, $-33°C$, or below, in order to work with the liquid, which leads to some minor experimental difficulties.

Liquid ammonia, as a solvent, has many properties that are similar to water. Like water, it undergoes a slight degree of autoionization. When it does so, it produces ammonium ion and amide ion, NH_2^-.

$$NH_3 + NH_3 \rightleftharpoons NH_4^+ + NH_2^-$$
$$\text{amide ion}$$

By analogy with water we would predict that an acid in liquid NH_3 would be any substance capable of forming ammonium ion. Thus any ammonium salt, such as NH_4Cl or $(NH_4)_2SO_4$, should show acid properties. In fact, in liquid ammonia NH_4Cl is exactly analogous to the product of the reaction of HCl with H_2O in aqueous solution—that is, $(H_3O)Cl$.

According to the solvent system definition, a base in liquid ammonia would be any substance capable of providing amide ion—for example, KNH_2, potassium amide.

In water the neutralization of an acid and a base occurs by the reaction

$$H_3O^+ + OH^- \longrightarrow 2H_2O$$

Thus the cation and anion of the solvent combine in neutralization to produce the solvent. In fact, in *any* solvent the neutralization reaction is simply the reverse of the autoionization reaction. In liquid ammonia this corresponds to the reaction

$$NH_4^+ + NH_2^- \longrightarrow 2NH_3$$

When solutions of NH_4Cl and KNH_2 in liquid ammonia are mixed, we have the overall reaction

$$NH_4Cl + KNH_2 \longrightarrow KCl + 2NH_3$$

This is analogous to the reaction, in water, between HCl and KOH

$$HCl + KOH \longrightarrow KCl + H_2O$$

or

$$(H_3O)Cl + KOH \longrightarrow KCl + 2H_2O$$

Another way that the acid-base phenomena in water and liquid ammonia are similar is revealed by the behavior of certain organic dye molecules called indicators. For example, in water the indicator phenolphthalein is pink in basic solutions and colorless in acidic solutions. This fact is used in acid-base titrations, where there is a sharp color change when neutralization is achieved. This same indicator can also be used for performing titrations with liquid ammonia as a solvent. In basic solutions containing an excess of NH_2^- ion the phenolphthalein is pink, while in acid solutions it is colorless.

A further similarity between acid-base behavior in these two solvents is provided by the amphoteric behavior of certain metals. For example, if an aqueous solution of ZnI_2 is treated with KOH, a precipitate of $Zn(OH)_2$ is

Titrations using liquid ammonia as a solvent require special equipment.

formed that redissolves upon further addition of base to produce $K_2[Zn(OH)_4]$, as indicated below.

$$ZnI_2 + 2KOH \longrightarrow Zn(OH)_2 \text{ (s)} + 2KI$$

$$Zn(OH)_2 + 2KOH \longrightarrow K_2[Zn(OH)_4]$$

In liquid ammonia we find precisely the same behavior:

$$ZnI_2 + 2KNH_2 \longrightarrow Zn(NH_2)_2 \text{ (s)} + 2KI$$

$$Zn(NH_2)_2 + 2KNH_2 \longrightarrow K_2[Zn(NH_2)_4]$$

In each of these cases the addition of acid (H_3O^+ in water or NH_4^+ in ammonia) causes the reprecipitation of the zinc hydroxide, or amide; further addition of acid will cause these precipitates to dissolve, thereby regenerating the solvated zinc ion.

The similarities between the neutralization reactions in water and ammonia have also suggested other chemical parallels. By analogy we have established that H_3O^+ and NH_4^+ occupy corresponding positions in the water (aquo) and ammonia (ammono) systems. Similarly, OH^- and NH_2^- stand opposite one another. This leads us to postulate further that the oxide ion, O^{2-}, in the aquo system should be equivalent to the *imide ion*, NH^{2-}, or the *nitride ion*, N^{3-}, in the ammono system. In fact, extension of this reasoning leads us to a whole series of compounds that are analogous to each other in the two solvent systems. Some examples are given in Table 14.1.

In addition to acid-base neutralization there are many other parallels in the reactions that take place in water and liquid ammonia. For instance, we know

Table 14.1
Analogous compounds in the aquo and ammono systems

Aquo Compound	Ammono Compound
H_2O	NH_3
$(H_3O)Cl$ or HCl	NH_4Cl
KOH	KNH_2
Li_2O	Li_3N
P_2O_5 (actually P_4O_{10})	P_3N_5
$PO(OH)_3$ or H_3PO_4	$[PN(NH_2)_2]_3$
	$P(NH)(NH_2)_3$
CH_3OH	CH_3NH_2
C_2H_5OH	$C_2H_5NH_2$
$(CH_3)_2O$	$(CH_3)_3N$, $(CH_3)_2NH$
H_2O_2	N_2H_4
$HONO_2$ or HNO_3	HNN_2
$HOCl$	H_2NCl
$CO(OH)_2$ or H_2CO_3	$C(NH)(NH_2)_2$
$CH_3COOC_2H_5$	$CH_3C(NH)(NHC_2H_5)$
CH_3COOH	$CH_3C(NH)(NH_2)$
$Cu(H_2O)_4^{2+}$	$Cu(NH_3)_4^{2+}$
$Si(OH)_4$	$Si(NH_2)_4$
$B(OH)_3$	$B(NH_2)_3$

that in water metal oxides react with the solvent to produce hydroxides,

$$Li_2O + H_2O \longrightarrow 2LiOH$$

In liquid ammonia we find the similar reaction,

$$Li_3N + 2NH_3 \longrightarrow 3LiNH_2$$

Another example is the reaction of a metal hydride with the solvent.

$$NaH + H_2O \longrightarrow H_2 + Na^+ + OH^- \qquad \text{(Aquo system)}$$

$$NaH + NH_3 \longrightarrow H_2 + Na^+ + NH_2^- \qquad \text{(Ammono system)}$$

Still another similarity is the reaction of an active metal with an acid to yield hydrogen plus a salt.

$$Ca + 2(H_3O)Cl \longrightarrow CaCl_2 + H_2 + 2H_2O \qquad \text{(Aquo system)}$$

$$Ca + 2NH_4Cl \longrightarrow CaCl_2 + H_2 + 2NH_3 \qquad \text{(Ammono system)}$$

In Chapter 6 we saw that metathesis reactions provide a convenient route for the synthesis of certain compounds. In nonaqueous solvents, we can also have these kinds of chemical changes. Because of different solubility relationships in the aqueous and nonaqueous media, however, it is sometimes possible to prepare compounds by metathesis in a solvent such as ammonia that cannot be made in the same way in water. For example, in liquid ammonia, $Ba(NO_3)_2$ and $AgCl$ are both soluble. When their solutions are mixed, a white precipitate of $BaCl_2$ is produced.

$$Ba(NO_3)_2 + 2AgCl \xrightarrow{\text{NH}_3 \, (l)} BaCl_2 \, (s) + 2AgNO_3$$

This is, of course, exactly the reverse of the reaction that takes place in aqueous solution, where $BaCl_2$ and $AgNO_3$ are soluble and form a precipitate of $AgCl$ when their aqueous solutions are mixed.

14.6 SUMMARY

The examples in the preceding section demonstrate that the solvent system approach to acids and bases has some useful and appealing features. One of these expands our thinking to reactions in solvents other than water. However, this view of acids and bases suffers from some of the same limitations as the Arrhenius theory. For example, it does not allow us to consider, as acid-base interactions, reactions that take place in the absence of a solvent.

The Brønsted-Lowry approach frees us from the restriction of having to have a solvent present, although it limits us to dealing with systems in which there is proton transfer. Clearly, the most general approach that we have considered is the Lewis theory. All the acid-base reactions that we have examined under the various headings in this chapter can be interpreted from the Lewis point of view.

The choice of which definition of acids and bases one wishes to use in a particular instance depends largely on the sort of chemistry that is studied. For example, in a practical sense the Arrhenius and Brønsted-Lowry definitions are perfectly satisfactory for dealing with all the reactions in aqueous solution that you will encounter in the laboratory. But when you study organic chemistry you will find the Lewis concept useful on a number of occasions.

INDEX TO QUESTIONS

REVIEW QUESTIONS

14.1 Give three properties that are characteristic of acids and bases in general.

14.2 What is the Arrhenius definition of an acid and a base?

14.3 Identify each of the following as Arrhenius acids or bases. For each, write a chemical equation to show its reaction with water. If necessary, refer to Table 6.3.
(a) P_4O_{10} (d) HBr   (f) N_2O_5
(b) CaO   (e) H_2O (g) $Ba(OH)_2$
(c) NH_3OH^+

14.4 Write the net ionic equation for the neutralization of an acid and a base in aqueous solution.

14.5 Define acid anhydride, basic anhydride. What is the Brønsted-Lowry definition of an acid and a base? Why is it less restrictive than the Arrhenius concept?

14.6 Identify the two acid-base conjugate pairs in each of the following reactions:
(a) $C_2H_3O_2^- + H_2O \rightleftharpoons OH^- + HC_2H_3O_2$
(b) $HF + NH_3 \rightleftharpoons NH_4^+ + F^-$
(c) $Zn(OH)_2 + 2OH^- \rightleftharpoons ZnO_2^{2-} + 2H_2O$
(d) $Al(H_2O)_6^{3+} + OH^- \rightleftharpoons Al(H_2O)_5OH^{2+} + H_2O$
(e) $N_2H_4 + H_2O \rightleftharpoons N_2H_5^+ + OH^-$
(f) $NH_2OH + HCl \rightleftharpoons NH_3OH^+ + Cl^-$
(g) $O^{2-} + H_2O \rightleftharpoons 2OH^-$
(h) $H^- + H_2O \rightleftharpoons H_2 + OH^-$
(i) $NH_2^- + N_2H_4 \rightleftharpoons NH_3 + N_2H_3^-$
(j) $HNO_3 + H_2SO_4 \rightleftharpoons H_3SO_4^+ + NO_3^-$

14.7 Identify the acid-base conjugate pairs in each of the following reactions:
(a) $HClO_4 + N_2H_4 \rightleftharpoons N_2H_5^+ + ClO_4^-$
(b) $HSO_3^- + H_3PO_3 \rightleftharpoons H_2SO_3 + H_2PO_3^-$
(c) $C_5H_5NH^+ + (CH_3)_3N \rightleftharpoons C_5H_5N + (CH_3)_3NH^+$
(d) $CO_3^{2-} + H_2O \rightleftharpoons HCO_3^- + OH^-$
(e) $HCHO_2 + C_7H_5O_2^- \rightleftharpoons C_7H_5O_2H + CHO_2^-$
(f) $H_2C_2O_4 + CH_3NH_2 \rightleftharpoons HC_2O_4^- + CH_3NH_3^+$
(g) $H_2CO_3 + H_2O \rightleftharpoons HCO_3^- + H_3O^+$
(h) $C_2H_5OH + NH_2^- \rightarrow C_2H_5O^- + NH_3$
(i) $NO_2^- + N_2H_5^+ \rightleftharpoons HNO_2 + N_2H_4$
(j) $HCN + H_2SO_4 \rightleftharpoons H_2CN^+ + HSO_4^-$

14.8 Write autoionization reactions for the following solvents:
(a) H_2O (*l*)   (b) NH_3 (*l*)   (c) HCN (*l*)

14.9 What would be the formula of the conjugate acid of dimethylamine, $(CH_3)_2NH$? What would be the formula of its conjugate base?

14.10 From Figure 14.2, place the following reactions in order of increasing tendency to proceed toward completion:
(a) $H_2O + NH_3 \rightleftharpoons NH_4^+ + OH^-$
(b) $HClO_4 + NH_2^- \rightleftharpoons ClO_4^- + NH_3$
(c) $H_2O + NO_2^- \rightleftharpoons HNO_2 + OH^-$
(d) $NH_3 + Cl^- \rightleftharpoons NH_2^- + HCl$

14.11 Use Figure 14.2 to place the following reactions in order of increasing tendency to proceed toward completion:
(a) $OCl^- + HCl \rightleftharpoons HOCl + Cl^-$
(b) $HF + C_2H_3O_2^- \rightleftharpoons HC_2H_3O_2 + F^-$
(c) $NH_4^+ + ClO_4^- \rightleftharpoons NH_3 + HClO_4$
(d) $HNO_2 + F^- \rightleftharpoons HF + NO_2^-$

14.12 Hydrogen sulfide is a stronger acid than phosphine, PH_3. What may we conclude about the strengths of their conjugate bases, HS^- and PH_2^-?

14.13 Hydrogen cyanide, HCN, is a weaker acid in water than nitrous acid, HNO_2. What can we say about the relative strengths of CN^- and NO_2^- as bases?

14.14 Ammonia is a stronger base in water than is hydrazine, N_2H_4. If we were to use these two substances as solvents for a very weak acid, in which of them would the acid be more fully ionized?

14.15 Given the following equilibria and equilibrium constants, arrange the acids in order of increasing strength:
(a) $HOCl + H_2O \rightleftharpoons H_3O^+ + OCl^-$ $K = 3.2 \times 10^{-8}$
(b) $NH_4^+ + H_2O \rightleftharpoons H_3O^+ + NH_3$ $K = 5.6 \times 10^{-10}$
(c) $HC_2H_3O_2 + H_2O \rightleftharpoons H_3O^+ + C_2H_3O_2^-$ $K = 1.8 \times 10^{-5}$
(d) $H_2CO_3 + H_2O \rightleftharpoons H_3O^+ + HCO_3^-$ $K = 4.2 \times 10^{-7}$
(e) $HSO_4^- + H_2O \rightleftharpoons H_3O^+ + SO_4^{2-}$ $K = 1.3 \times 10^{-2}$

14.16 Based on the data in Question 14.15, where would the position of equilibrium be expected to lie in the reaction
$$HOCl + NH_3 \rightleftharpoons NH_4^+ + OCl^-$$

14.17 Arrange the conjugate bases in Question 14.15 in order of increasing base strength.

14.18 What is the leveling effect? How would it apply to the strong acids, HCl and HBr? Suggest a substance that might serve as a differentiating solvent for these two acids.

14.19 The reaction, $H_2S + H_2O \rightleftharpoons HS^- + H_3O^+$, has an equilibrium constant equal to 1.1×10^{-7}. Can you suggest a solvent where a similar reaction would have a smaller K? Can you suggest a solvent where a similar reaction would have a larger K?

14.20 Using Lewis formulas, diagram the reaction between gaseous HCN and NH_3.

14.21 It can be said that the strongest Brønsted acid that can exist in the presence of water is H_3O^+. Explain this statement.

14.22 What is the strongest base that is able to exist in the presence of water?

14.23 What is the Lewis definition of an acid and a base?

14.24 Boron trichloride, BCl_3, reacts with diethyl ether, $(C_2H_5)_2O$, to form an *addition compound,* which we can write as $Cl_3B \leftarrow O(C_2H_5)_2$. Use electron dot formulas to interpret this reaction as a Lewis acid-base neutralization.

14.25 Indicate whether the following would be expected to serve as either a Lewis acid or a Lewis base:
(a) $AlCl_3$ (e) NO^+ (i) $(CH_3)_2S$
(b) OH^- (f) CO_2 (j) SbF_5
(c) Br^- (g) NH_3
(d) H_2O (h) Fe^{3+}

14.26 Which would be expected to be a stronger Lewis base, NH_3 or NF_3? Explain.

14.27 Which would be expected to be a stronger Lewis base, $(CH_3)_3N$ or $(CH_3)_3P$? Explain.

14.28 Which would be expected to be a stronger Lewis acid
(a) BCl_3 or BBr_3?   (b) Cr^{2+} or Cr^{3+}?

14.29 Phosphorus(V) chloride exists as PCl_5 molecules in the gaseous state and as $[PCl_4^+][PCl_6^-]$ in the solid.

$$2PCl_5 \longrightarrow [PCl_4^+][PCl_6^-]$$

(a) What species is being transferred from one PCl_5 molecule to the other during this reaction?
(b) Use Lewis structures to diagram the reaction.
(c) Is the species being transferred a Lewis acid or a Lewis base?
(d) Is the reaction a nucleophilic or electrophilic displacement?

14.30 Boric acid, $B(OH)_3$, which has long been used for medicinal purposes, is a weak acid. However, it does not release protons by breaking of an O—H bond in the $B(OH)_3$ molecule. Instead, it functions as a Lewis acid by reacting with a water molecule,

$$B(OH)_3 + H_2O \rightleftharpoons B(OH)_3(H_2O)$$

$$B(OH)_3(H_2O) + H_2O \rightleftharpoons B(OH)_4^- + H_3O^+$$

Use electron dot structures to diagram this reaction. Why is $B(OH)_3$ a Lewis acid? Why is the $B(OH)_3(H_2O)$ acidic?

14.31 Explain why the reaction of CO_2 with H_2O to produce H_2CO_3, which we can also write as $CO(OH)_2$, can be viewed as a Lewis acid-base neutralization.

14.32 Define nucleophile, electrophile.

14.33 Silver ion reacts with ammonia to form a complex ion having the formula $Ag(NH_3)_2^+$ in which two ammonia molecules are bound to the Ag^+ ion. Explain how this can be viewed as a Lewis acid-base reaction.

14.34 How are an acid and a base defined in the solvent system approach?

14.35 Concentrated sulfuric acid undergoes the autoionization reaction,

$$2H_2SO_4 \rightleftharpoons H_3SO_4^+ + HSO_4^-$$

In this solvent, acetic acid behaves as a *base* and perchloric acid behaves as an *acid.* Write chemical equations to show the following:
(a) The reaction of $HC_2H_3O_2$ with the solvent, H_2SO_4
(b) The reaction of $HClO_4$ with the solvent
(c) The neutralization reaction that occurs when solutions of $HC_2H_3O_2$ and $HClO_4$ in H_2SO_4 are mixed

14.36 From the data in Table 14.1, write equations for reactions in liquid ammonia that are analogous to the following reactions that take place in aqueous solution:
(a) $P_4O_{10} + 6H_2O \rightarrow 4H_3PO_4$
(b) $Cl_2 + H_2O \rightarrow HCl + HOCl$
(c) $CH_3COOC_2H_5 + H_2O \xrightarrow{H_3O^+} CH_3COOH + C_2H_5OH$
(d) $Zn + 2HCl \rightarrow ZnCl_2 + H_2$
(e) $H_2CO_3 + 2OH^- \rightarrow CO_3^{2-} + 2H_2O$
(f) $SiCl_4 + 4H_2O \rightarrow Si(OH)_4 + 4HCl$
(g) $Zn + 2OH^- + 2H_2O \rightarrow Zn(OH)_4^{2-} + H_2$
(h) $Cu^{2+} + 4H_2O \rightarrow Cu(H_2O)_4^{2+}$
(i) $BCl_3 + 3H_2O \rightarrow 3HCl + B(OH)_3$

14.37 Liquid hydrogen cyanide can be thought to undergo the autoionization,

$$2HCN \rightleftharpoons H_2CN^+ + CN^-$$

(a) Would KCN be considered an acid or a base in this solvent?
(b) H_2SO_4 is an acid in HCN (*l*). Write the equation for the ionization of H_2SO_4 in this solvent.
(c) $(CH_3)_3N$ is a base in HCN (*l*). Write an equation for the reaction of $(CH_3)_3N$ with the solvent. Using electron dot formulas, show how this reaction takes place.
(d) What is the ionic equation for the neutralization of H_2SO_4 by $(CH_3)_3N$ in liquid HCN?
(e) What is the net ionic equation for the neutralization reaction in this solvent?

14.38 How do the Arrhenius, Brønsted-Lowry, Lewis, and solvent system concepts of acids and bases differ in terms of their general applicability?

14.39 Solutions containing the $Cr(H_2O)_6^{3+}$ ion are acidic.
(a) Explain why this is so.
(b) Write a chemical equation to show how $Cr(H_2O)_6^{3+}$ functions as a Brønsted-Lowry acid.
(c) Why would solutions containing Cr^{3+} be more acidic than solutions containing Cr^{2+}?

15

ACID-BASE EQUILIBRIA IN AQUEOUS SOLUTION

Controlling the acidity of a swimming pool with sodium bicarbonate is just one of the practical applications of the principles of acid-base equilibria discussed in this chapter.

In Chapter 6 we saw that many reactions that are of concern to us take place in aqueous solution. By now we also realize that chemical changes do not, in general, proceed entirely to completion, but instead approach a state of dynamic equilibrium. In this chapter and the next we will examine in greater detail, and on a quantitative basis, many of the ionic equilibria that can occur in aqueous solution. We begin, in this chapter, with the equilibria and reactions of acids and bases. This is very important because of the amphiprotic nature of water itself, and because many compounds of biological interest, which occur in the aqueous environment of living systems, show acid-base properties.

15.1 IONIZATION OF WATER, pH

In the last chapter we saw that some solvents can be considered to undergo autoionization. Recall, for example, that the autoionization of pure water is written as

$$H_2O + H_2O \rightleftharpoons H_3O^+ + OH^-$$

Since this is an equilibrium, we can write an equilibrium expression. Following the concepts developed in Chapter 13, this can be represented as

$$K = \frac{[H_3O^+][OH^-]}{[H_2O][H_2O]}$$

The molar concentration of water, which appears in the denominator of this expression, is very nearly constant ($\approx 55.6\ M$) in both pure water and in dilute aqueous solutions. Therefore, $[H_2O]^2$ can be included with the equilibrium constant, K, on the left side of the equation. This gives

$$K \cdot [H_2O]^2 = [H_3O^+][OH^-]$$

The left side of this expression is the product of two constants which, of course, must also equal a constant. This combined constant is written as

$$K_w = K[H_2O]^2$$

Our equilibrium condition therefore becomes

$$K_w = [H_3O^+][OH^-]$$

Since $[H_3O^+][OH^-]$ is the product of ionic concentrations, K_w is called the **ion product constant** for water, or frequently simply the **ionization constant** or **dissociation constant** of water. At 25°C, $K_w = 1.0 \times 10^{-14}$, and this is one equilibrium constant that you should be sure to memorize.

The equation for the autoionization of water is often simplified somewhat by omitting the water molecule that picks up the H^+. The dissociation reaction for water then becomes

$$H_2O \rightleftharpoons H^+ + OH^- \tag{15.1}$$

and the simplified expression for the dissociation constant of water is written simply as

$$K_w = [H^+][OH^-] \tag{15.2}$$

The equilibrium dissociation of H_2O represented by Equation 15.1 is present in any aqueous solution, and Equation 15.2 must always be fulfilled regardless of what other equilibria may also exist in solution.

Equation 15.2 can be used to calculate the molar concentrations of both the H^+ and OH^- ions in pure water. From the stoichiometry of the dissociation, we see that for each 1 mol of H^+ formed, 1 mol of OH^- is also produced. This

K_w varies with temperature. At 37°C (body temperature).

$K_w = 2.42 \times 10^{-14}$

means that at equilibrium, $[H^+] = [OH^-]$. If we let x equal the hydrogen ion concentration, then

$$x = [H^+] = [OH^-]$$

Substituting into Equation 15.2 gives

$$K_w = x \cdot x = x^2$$

or, because $K_w = 1.0 \times 10^{-14}$,

$$x^2 = 1.0 \times 10^{-14}$$

Taking the square root yields

$$x = 1.0 \times 10^{-7}$$

which means that the concentrations of hydrogen ion and hydroxide ion in pure water are

$$[H^+] = [OH^-] = 1.0 \times 10^{-7}\ M$$

Whenever the hydrogen ion concentration equals the hydroxide ion concentration, as it does in pure water, the solution is said to be *neutral*. An acid is a substance that makes the H^+ concentration greater than the OH^- concentration; conversely, a base makes the OH^- concentration greater than the H^+ concentration. However, remember that there is always *some* OH^- present in an acidic solution, just as there is always *some* H^+ present even if the solution is basic. At all times, Equation 15.2 is obeyed if the solution is at equilibrium.

In an aqueous solution of an acid we will often want to know what the H^+ concentration is. In these cases it is almost always safe to assume that essentially all the H^+ in the solution comes from the dissolved acid. In other words, it is usually safe to assume that the dissociation of water contributes a negligible amount of H^+ to the solution. This is because the presence of H^+ from an acid (for example, HCl) shifts the equilibrium

$$H_2O \rightleftharpoons H^+ + OH^-$$

to the left. Therefore, the amount of water that is dissociated in a solution of an acid is even less than in pure water, which means that the H^+ that comes from the dissociation of water is less than $10^{-7}\ M$. Similarly, the OH^- concentration in a solution of a base can be calculated just from the concentration of the solute. The OH^- contributed by the dissociation of water is negligible. Example 15.1 illustrates this point.

EXAMPLE 15.1   (a) What is the OH^- concentration in a 0.0010 M HCl solution? (b) What is the H^+ concentration derived from the dissociation of the solvent?

SOLUTION   (a) At equilibrium we must have

$$[H^+][OH^-] = 1.0 \times 10^{-14}$$

HCl is a strong acid and is essentially 100% dissociated.

$$HCl \longrightarrow H^+ + Cl^-$$

Therefore, 0.0010 mol of HCl per liter gives 0.0010 mol of H^+ per liter. The total hydrogen ion concentration, then, is 0.0010 M *plus* the amount contributed by the dissociation of water. Let's assume for the moment that this contribution is negligible, as suggested above, and that it can be ignored. This gives

$$[H^+] = 0.0010\ M + (\text{contribution from } H_2O) \approx 0.0010\ M$$

Solving for the hydroxide ion concentration,

$$[OH^-] = \frac{1.0 \times 10^{-14}}{[H^+]}$$

$$= \frac{1.0 \times 10^{-14}}{0.0010} = 1.0 \times 10^{-11} \, M$$

(b) The hydroxide ion in part (a) comes entirely from the dissociation of water. Therefore, the concentration of H^+ derived from H_2O must *also* be $1 \times 10^{-11} \, M$, as can be seen from the stoichiometry of Equation 15.1. Note that this value ($1.0 \times 10^{-11} \, M$) is indeed negligible compared to the H^+ concentration produced by the HCl ($1.0 \times 10^{-3} \, M$), so the assumption made in part (a) was valid.

The pH concept

Hydrogen ion and hydroxide ion enter into many equilibria in addition to the dissociation of water, so it is frequently necessary to specify their concentrations in aqueous solutions. These concentrations may range from relatively high values to very small ones (for example, $10 \, M$ to $10^{-14} \, M$), and a logarithmic notation[1] has been devised to simplify the expression of these quantities. In general, for some quantity X,

$$pX = \log \frac{1}{X} = -\log X \qquad [15.3]$$

For example, if we wish to specify the hydrogen ion concentration in a solution, we speak of **pH.** This is defined as

Notice that logs to the base 10 are used here, _not_ natural logs.

$$pH = \log \frac{1}{[H^+]} = -\log [H^+]$$

In a solution where the hydrogen ion concentration is $10^{-3} \, M$, we therefore have

$$pH = -\log (10^{-3}) = -(-3)$$

or

$$pH = 3$$

Similarly, if the hydrogen ion concentration is $10^{-8} \, M$, the pH of the solution is 8.

Following the same approach for the hydroxide ion concentration, we can define the **pOH** of a solution as

$$pOH = -\log[OH^-]$$

Just as the H^+ and OH^- ion concentrations in a solution are related to each other, so too are the pH and pOH. From the equilibrium expression for the dissociation of water,

$$\log K_w = \log[H^+] + \log[OH^-]$$

Multiplying through by -1 gives

$pK_w = -\log K_w$

$$(-\log K_w) = (-\log[H^+]) + (-\log[OH^-])$$

$= -\log (1.0 \times 10^{-14})$

Following our definition, $-\log K_w = pK_w$. Therefore,

$= -[\log (1.0) + \log (10^{-14})]$

$$pK_w = pH + pOH$$

$= -[0.00 + (-14)]$

Since $K_w = 1.0 \times 10^{-14}$, $pK_w = 14.00$. This gives the useful relationship

$= 14.00$

$$pH + pOH = 14.00 \qquad [15.4]$$

[1] A discussion on the use of logarithms can be found in Appendix A.

The pH of a solution is conveniently measured electronically by using a pH meter. The electrodes attached to the instrument are dipped into the solution to be tested, and the pH is read from the scale.

In a neutral solution, $[H^+] = [OH^-] = 10^{-7}$ M, and pH = pOH = 7.0, so that in a neutral solution we say the pH = 7.0. In an acidic solution the hydrogen ion concentration is greater than 10^{-7} M (for example, 10^{-3} M) and the pH is less than 7.0. By the same token, in basic solutions the $[H^+]$ is less than 10^{-7} M (for example, 10^{-10} M) and the pH is greater than 7.0. This is summarized below.

	$[H^+]$	$[OH^-]$	pH	pOH
Acidic solution	$>10^{-7}$	$<10^{-7}$	<7	>7
Neutral solution	10^{-7}	10^{-7}	7	7
Basic solution	$<10^{-7}$	$>10^{-7}$	>7	<7

Many common substances are either acidic or basic, and their degree of acidity or basicity is conveniently expressed in terms of pH (Figure 15.1). Notice that substances having a pH less than 7—that is, those that are acidic—have characteristically sour tastes. Lemon juice contains citric acid, for example, and vinegar contains acetic acid. On the other hand, basic substances such as milk of magnesia—a suspension of $Mg(OH)_2$ in water—have a bitter taste.

Figure 15.1

The pH of some common substances.

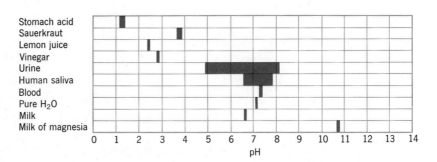

Although sour and bitter taste is the body's way of judging the acidity of foods, *never* taste chemicals in the lab; many of them are poisonous and can ruin your whole day!

Let's look now at some sample problems dealing with typical pH calculations.

EXAMPLE 15.2   What is the pH of a 0.0020 *M* HCl solution?

SOLUTION   Since HCl is a strong acid, we can write

$$[H^+] = 0.0020 \; M = 2.0 \times 10^{-3} \; M$$

We know that

$$pH = -\log[H^+]$$

and for this problem,

$$pH = -\log(2.0 \times 10^{-3})$$
$$= -(\log 2.0 + \log 10^{-3})$$
$$= -[0.30 + (-3)]$$
$$= -(-2.70)$$
$$= 2.70$$

The exponent can be considered an exact number. 10^{-3} means $10^{-3.000...}$.

EXAMPLE 15.3   What is the pH of a 5.0×10^{-4} *M* NaOH solution?

SOLUTION   Metal hydroxides are strong bases—they are completely dissociated. This means that the hydroxide ion concentration in the solution is 5.0×10^{-4} *M*. To calculate the pH, we can proceed in either of two ways: (1) knowing K_w for water and $[OH^-]$ for this solution, we can calculate $[H^+]$ using Equation 15.2 and then proceed as we did in Example 15.2; or (2) we can calculate pOH from the OH^- concentration and subtract it from pK_w to obtain pH.

METHOD 1   We know that

$$K_w = [H^+][OH^-]$$

and therefore

$$[H^+] = \frac{1.0 \times 10^{-14}}{5.0 \times 10^{-4}}$$

$$= 0.20 \times 10^{-10} \text{ or } 2.0 = 10^{-11} \; M$$

Now, proceeding as we did in Example 15.2,

$$pH = -\log (2.0 \times 10^{-11})$$
$$= -[0.30 + (-11)]$$
$$= 10.70$$

METHOD 2   By definition

$$pOH = -\log [OH^-]$$

For this problem,

$$pOH = -\log (5.0 \times 10^{-4})$$
$$= -[0.70 + (-4)]$$
$$= 3.30$$

The pH would be

$$pH = pK_w - pOH$$

$$= 14.00 - 3.30$$

$$= 10.70$$

EXAMPLE 15.4   A sample of orange juice was found to have a pH of 3.80. What were the H^+ and OH^- concentrations in the juice?

SOLUTION   In order to compute $[H^+]$ from pH we must reverse the procedure that we followed in Example 15.2. We know that

$$pH = -\log[H^+] = 3.80$$

or

$$\log[H^+] = -3.80$$

Let's write this as the sum of a decimal fraction plus a negative integer. Thus, -3.80 is the same as 0.20 plus -4.00.

$$\log[H^+] = 0.20 + (-4.00)$$

Now we take antilogs. Since 0.20 is the logarithm of 1.6, and -4 is the logarithm of 10^{-4}, we can write

$$\log[H^+] = \log 1.6 + \log 10^{-4}$$

or

$$\log[H^+] = \log(1.6 \times 10^{-4})$$

Hence

$$[H^+] = 1.6 \times 10^{-4} \ M$$

To calculate $[OH^-]$ we can divide K_w by $[H^+]$.

$$[OH^-] = \frac{1.0 \times 10^{-14}}{1.6 \times 10^{-4}}$$

$$= 6.3 \times 10^{-11}$$

Alternatively, we could first have computed the pOH of the solution,

$$pOH = 14.00 - 3.80$$

$$= 10.20$$

From pOH we can obtain $[OH^-]$,

$$pOH = -\log[OH^-] = 10.20$$

$$\log[OH^-] = -10.20 = 0.80 - 11.00$$

$$\log[OH^-] = \log(6.3) + \log(10^{-11})$$

$$[OH^-] = 6.3 \times 10^{-11} \ M$$

15.2 DISSOCIATION OF WEAK ELECTROLYTES

As a class, weak electrolytes include weak acids and bases as well as certain salts, such as $HgCl_2$ and $CdSO_4$, that are not fully dissociated in aqueous solution. In solutions of these substances there is an equilibrium between the undissociated species and its corresponding ions. For example, acetic acid ionizes according to the equation,

$$HC_2H_3O_2 + H_2O \rightleftharpoons H_3O^+ + C_2H_3O_2^-$$

The equilibrium expression for this reaction is

$$K = \frac{[H_3O^+][C_2H_3O_2^-]}{[HC_2H_3O_2][H_2O]}$$

In dilute solutions, the concentration of H_2O is not appreciably different than it is in pure water, so we may safely take it to be a constant that can be included with K on the left side of the equal sign. That is,

$$K \times [H_2O] = K_a = \frac{[H_3O^+][C_2H_3O_2^-]}{[HC_2H_3O_2]}$$

where we have used K_a to represent the **acid dissociation constant** or **ionization constant.** This same equilibrium expression can be obtained if we simplify the dissociation by omitting the solvent. Thus, for the dissociation of acetic acid we would write

$$HC_2H_3O_2 \rightleftharpoons H^+ + C_2H_3O_2^-$$

and the equilibrium expression would then be

$$K_a = \frac{[H^+][C_2H_3O_2^-]}{[HC_2H_3O_2]}$$

In general, for any weak acid, HA, the simplified dissociation reaction can be written as

$$HA \rightleftharpoons H^+ + A^-$$

and its acid dissociation constant is given by

$$K_a = \frac{[H^+][A^-]}{[HA]}$$

This same approach can also be applied to weak bases. Normally, these are substances that react with water and pick up a hydrogen ion. An example is the weak base, ammonia.

$$NH_3 + H_2O \rightleftharpoons NH_4^+ + OH^-$$

If we omit the solvent, the **base ionization constant,** K_b, for this reaction is

$$K_b = \frac{[NH_4^+][OH^-]}{[NH_3]}$$

In general, for any weak base B the ionization equilibrium can be written as

$$B + H_2O \rightleftharpoons BH^+ + OH^-$$

and the expression for K_b is

$$K_b = \frac{[BH^+][OH^-]}{[B]}$$

The extent to which a weak acid or base undergoes ionization, as well as the value of the ionization constant, must be determined experimentally. (One way of doing this is to measure the pH of a solution prepared by dissolving a known quantity of the weak acid or base in a given volume of solution, as illustrated in Examples 15.5 and 15.6.) Ionization constants of a number of weak acids and bases are listed in Table 15.1. Notice that their values are quite small—ranging from 10^{-2} to 10^{-10}. Recalling that numbers this small can be simplified by applying the logarithmic notation of Equation 15.3, we can write for any K_a,

$$pK_a = -\log K_a$$

Table 15.1
Ionization constants for some weak acids and bases

Weak Acid	Ionization	K_a	pK_a
Chloroacetic acid	$HC_2H_2O_2Cl \rightleftharpoons H^+ + C_2H_2O_2Cl^-$	1.4×10^{-3}	2.85
Hydrofluoric acid	$HF \rightleftharpoons H^+ + F^-$	6.5×10^{-4}	3.19
Nitrous acid	$HNO_2 \rightleftharpoons H^+ + NO_2^-$	4.5×10^{-4}	3.35
Formic acid	$HCHO_2 \rightleftharpoons H^+ + CHO_2^-$	1.8×10^{-4}	3.74
Lactic acid	$HC_3H_5O_3 \rightleftharpoons H^+ + C_3H_5O_3^-$	1.38×10^{-4}	3.86
Benzoic acid	$HC_7H_5O_2 \rightleftharpoons H^+ + C_7H_5O_2^-$	6.5×10^{-5}	4.19
Acetic acid	$HC_2H_3O_2 \rightleftharpoons H^+ + C_2H_3O_2^-$	1.8×10^{-5}	4.74
Butyric acid	$HC_4H_7O_2 \rightleftharpoons H^+ + C_4H_7O_2^-$	1.5×10^{-5}	4.82
Nicotinic acid	$HC_6H_4NO_2 \rightleftharpoons H^+ + C_6H_4NO_2^-$	1.4×10^{-5}	4.85
Propionic acid	$HC_3H_5O_2 \rightleftharpoons H^+ + C_3H_5O_2^-$	1.4×10^{-5}	4.85
Barbituric acid	$HC_4H_3N_2O_3 \rightleftharpoons H^+ + C_4H_3N_2O_3^-$	1.0×10^{-5}	5.00
Veronal (diethylbarbituric acid)	$HC_8H_{11}N_2O_3 \rightleftharpoons H^+ + C_8H_{11}N_2O_3^-$	3.7×10^{-8}	7.43
Hypochlorous acid	$HOCl \rightleftharpoons H^+ + OCl^-$	3.1×10^{-8}	7.51
Hydrocyanic acid	$HCN \rightleftharpoons H^+ + CN^-$	4.9×10^{-10}	9.31

Weak Base	Ionization	K_b	pK_b
Diethylamine	$(C_2H_5)_2NH + H_2O \rightleftharpoons (C_2H_5)_2NH_2^+ + OH^-$	9.6×10^{-4}	3.02
Methylamine	$CH_3NH_2 + H_2O \rightleftharpoons CH_3NH_3^+ + OH^-$	3.7×10^{-4}	3.43
Ammonia	$NH_3 + H_2O \rightleftharpoons NH_4^+ + OH^-$	1.8×10^{-5}	4.74
Hydrazine	$N_2H_4 + H_2O \rightleftharpoons N_2H_5^+ + OH^-$	1.7×10^{-6}	5.77
Hydroxylamine	$NH_2OH + H_2O \rightleftharpoons NH_3OH^+ + OH^-$	1.1×10^{-8}	7.97
Pyridine	$C_5H_5N + H_2O \rightleftharpoons C_5H_5NH^+ + OH^-$	1.7×10^{-9}	8.77
Aniline	$C_6H_5NH_2 + H_2O \rightleftharpoons C_6H_5NH_3^+ + OH^-$	3.8×10^{-10}	9.42

and for any K_b,

$$pK_b = -\log K_b$$

For example, the pK_a of acetic acid is

$$pK_a = -\log K_a = -\log(1.8 \times 10^{-5})$$
$$= 4.74$$

and for pyridine, a bad smelling liquid,

$$pK_b = -\log (1.7 \times 10^{-9})$$
$$= 8.77$$

From our discussion in Section 13.2, we know that the smaller the value of K_a or K_b, the smaller the extent of ionization and the weaker the acid or base. Relative strengths of acids and bases can also be indicated by their pK_a's and pK_b's. In this case the smaller the value of pK_a or pK_b, the *stronger* is the acid or base. Let's compare, for example, the pK_a's for acetic, chloroacetic, and dichloroacetic acids. Their pK_a's are

$$HC_2H_3O_2 \qquad pK_a = 4.74$$
$$HC_2H_2ClO_2 \qquad pK_a = 2.85$$
$$HC_2HCl_2O_2 \qquad pK_a = 1.30$$

The order of increasing acidity is, therefore,

acetic < chloroacetic < dichloroacetic acid

(A discussion of *why* the acidity of these substances increases in this fashion can be found on page 308.)

Let's look now at some examples showing how these equilibrium constants can be calculated.

EXAMPLE 15.5   A student prepared a 0.10 M acetic acid solution and experimentally measured its pH to be 2.88. Calculate K_a for acetic acid and determine its percent dissociation.

SOLUTION   The first step is to write the equation for the equilibrium.

$$HC_2H_3O_2 \rightleftharpoons H^+ + C_2H_3O_2^-$$

To evaluate K_a we must have equilibrium concentrations to substitute into the expression,

$$K_a = \frac{[H^+][C_2H_3O_2^-]}{[HC_2H_3O_2]}$$

From pH we can obtain the H^+ concentration.

$$pH = -\log[H^+] = 2.88$$

$$\log[H^+] = 0.12 - 3.00$$

$$\log[H^+] = \log 1.3 + \log 10^{-3}$$

$$[H^+] = 1.3 \times 10^{-3} \, M$$

The $[H^+]$ comes from the dissociation of $HC_2H_3O_2$ and, from the stoichiometry of the equation, we see that the concentrations of H^+ and $C_2H_3O_2^-$ must be equal because they are formed in a 1-to-1 ratio.

$$[H^+] = [C_2H_3O_2^-] = 1.3 \times 10^{-3} \, M$$

The concentration of undissociated $HC_2H_3O_2$ at equilibrium is equal to the original concentration, 0.10 M, *minus* the number of moles per liter of acetic acid that have dissociated. At equilibrium, then, we have

Equilibrium Concentrations	
H^+	$1.3 \times 10^{-3} \, M$
$C_2H_3O_2^-$	$1.3 \times 10^{-3} \, M$
$HC_2H_3O_2$	$1.0 \times 10^{-1} - 0.013 \times 10^{-1} = 1.0 \times 10^{-1} \, M$

Note that when we compute the acetic acid concentration to the *proper number of significant figures,* the amount that has dissociated is negligible compared to the amount initially present. Thus,

$$0.10 \, M - 0.0013 \, M = (0.0987 \, M) = 0.10 \, M$$

Substituting the equilibrium concentrations into the expression for K_a, we have

$$K_a = \frac{(1.3 \times 10^{-3})(1.3 \times 10^{-3})}{(1.0 \times 10^{-1})}$$

$$= 1.7 \times 10^{-5}$$

This value of **K**$_a$ differs from that given in Table 15.1 because of "round-off" error.

The **percent dissociation** of acetic acid in this solution is found by dividing the number of moles per liter of $HC_2H_3O_2$ that have dissociated by the quantity of acetic acid that was available initially, all multiplied by 100:

$$\text{percent dissociation} = \frac{(\text{moles/liter } HC_2H_3O_2 \text{ dissociated})}{(\text{moles/liter } HC_2H_3O_2 \text{ available})} \times 100$$

$$= \frac{1.3 \times 10^{-3} \, M}{1.0 \times 10^{-1} \, M} \times 100 = 1.3\%$$

EXAMPLE 15.6

A student prepared a 0.010 M NH_3 solution and, by a freezing-point-lowering experiment, determined that the NH_3 had undergone 4.2% ionization. Calculate the K_b for NH_3.

SOLUTION

Ammonia ionizes in water according to the reaction,

$$NH_3 + H_2O \rightleftharpoons NH_4^+ + OH^-$$

for which we write

$$K_b = \frac{[NH_4^+][OH^-]}{[NH_3]}$$

From the stoichiometry of the ionization we see that, at equilibrium,

$$[NH_4^+] = [OH^-]$$

The amount ionized per liter is 4.2% of 0.010 mol/liter.

Since the 0.010 M solution undergoes 4.2% ionization, the number of moles per liter of these ions present at equilibrium is

$$[NH_4^+] = [OH^-] = 0.042 \times 0.010 \, M = 4.2 \times 10^{-4} \, M$$

The number of moles per liter of NH_3 at equilibrium would be

$$[NH_3] = 1.0 \times 10^{-2} - 0.042 \times 10^{-2} = 0.958 \times 10^{-2} \, M$$

When this is rounded off to the appropriate number of significant figures, we have $[NH_3] = 1.0 \times 10^{-2} \, M$ (once again, the quantity lost by ionization is negligible). Our equilibrium concentrations are shown at the left.

When these concentrations are substituted into the equation for K_b, we have

	Equilibrium Concentrations
NH_4^+	$4.2 \times 10^{-4} \, M$
OH^-	$4.2 \times 10^{-4} \, M$
NH_3	$1.0 \times 10^{-2} \, M$

$$K_b = \frac{(4.2 \times 10^{-4})(4.2 \times 10^{-4})}{(1.0 \times 10^{-2})}$$

or

$$K_b = 1.8 \times 10^{-5}$$

In Examples 15.5 and 15.6 we computed K from a knowledge of equilibrium concentrations. We can also use our knowledge of K to calculate the concentrations in an equilibrium mixture. Let's look at some examples.

EXAMPLE 15.7

What are the concentrations of all of the species present in a 0.50 M $HC_2H_3O_2$ solution?

SOLUTION

First we write the chemical equation for the equilibrium,

$$HC_2H_3O_2 \rightleftharpoons H^+ + C_2H_3O_2^-$$

From Table 15.1 we find that $K = 1.8 \times 10^{-5}$ for $HC_2H_3O_2$. Therefore,

$$\frac{[H^+][C_2H_3O_2^-]}{[HC_2H_3O_2]} = 1.8 \times 10^{-5}$$

The quantities that we must substitute into this expression must represent equilibrium concentrations. In obtaining these we will construct a table as we did in Chapter 13. A solution labeled 0.50 M $HC_2H_3O_2$ would be prepared by dissolving 0.50 mol of $HC_2H_3O_2$ in 1.00 liter of solution, so the label gives us the concentration of acetic acid that existed before any of it dissociated. The initial concentration of $HC_2H_3O_2$ is therefore taken to be 0.50 M. Initially the solution contained no $C_2H_3O_2^-$, and we will neglect the H^+ from the dissociation of H_2O. Therefore, for the purposes of the problem there is no H^+ initially. We know that at equilibrium some of the acetic acid in the solution will have dissociated. If we therefore let x equal the number of moles per liter of $HC_2H_3O_2$ that dissociate, at equilibrium we will have produced x mol/liter of H^+, x mol/liter of $C_2H_3O_2^-$, and we will have lost x mol/liter of $HC_2H_3O_2$. At equilibrium we thus have x M H^+, x M $C_2H_3O_2^-$, and $(0.50 - x)$ M $HC_2H_3O_2$.

	Initial Molar Concentrations	Change	Equilibrium Molar Concentrations
H^+	0.0	$+x$	x
$C_2H_3O_2^-$	0.0	$+x$	x
$HC_2H_3O_2$	0.50	$-x$	$0.50 - x$

Substituting these values into the equilibrium expression gives us

$$\frac{(x)(x)}{(0.50 - x)} = 1.8 \times 10^{-5}$$

Without simplification this expression leads to a quadratic equation that can be solved using the quadratic formula. However, in Example 13.11 we saw that it is sometimes possible to make simplifying assumptions that greatly reduce the effort required to obtain solutions to problems of this type. Because K is small, very little $HC_2H_3O_2$ will have actually undergone dissociation; so x will be small. Let us assume that x will be negligible compared to 0.50; that is,

$$0.50 - x \approx 0.50$$

Our equation then becomes

$$\frac{x^2}{0.50} = 1.8 \times 10^{-5}$$

or

$$x = 3.0 \times 10^{-3}$$

If we look back on our assumption, we see that x is in fact small compared to 0.50 and that, when *rounded to the proper number of significant figures*,

$$0.50 - 0.0030 = 0.50$$

	Equilibrium Concentrations (M)
H^+	3.0×10^{-3}
$C_2H_3O_2^-$	3.0×10^{-3}
$HC_2H_3O_2$	0.50

Therefore, the equilibrium concentrations of the species involved in the dissociation of the acid are those given in the table at the left. Since the question asks for *all* concentrations, we must also calculate the concentration of OH^-, which comes from the dissociation of water. Here we use K_w.

$$[OH^-] = \frac{K_w}{[H^+]}$$

$$= \frac{1.0 \times 10^{-14}}{3.0 \times 10^{-3}}$$

$$= 3.3 \times 10^{-12} \ M$$

In the last example the only source of H^+ and $C_2H_3O_2^-$ was from the dissociation of the weak acid. Example 15.8 shows how we would handle a problem dealing with a solution where there are two sources of one of the ions.

EXAMPLE 15.8

What are the concentrations of H^+, $C_2H_3O_2^-$, and $HC_2H_3O_2$ in a solution prepared by dissolving 0.10 mol of $NaC_2H_3O_2$ and 0.20 mol of $HC_2H_3O_2$ in enough water to give a total volume of 1.00 liter?

SOLUTION

There is only one equilibrium here that we must be concerned with,

$$HC_2H_3O_2 \rightleftharpoons H^+ + C_2H_3O_2^-$$

$$\frac{[H^+][C_2H_3O_2^-]}{[HC_2H_3O_2]} = 1.8 \times 10^{-5}$$

It is very important to remember that salts are strong electrolytes.

When $NaC_2H_3O_2$ dissolves, it is completely dissociated. It is important to remember that virtually all salts are 100% dissociated in solution. Therefore 0.10 mol/liter of $NaC_2H_3O_2$ gives 0.10 mol/liter of Na^+ and 0.10 mol/liter of $C_2H_3O_2^-$. We are interested only in the $C_2H_3O_2^-$; the Na^+ is simply a *spectator ion* and we can ignore it. The initial concentrations that are of concern to us are found in the first column of our table. Since no H^+ is present, some $HC_2H_3O_2$ must ionize; so let's let x = the number of moles per liter of $HC_2H_3O_2$ that dissociates to give H^+ and $C_2H_3O_2^-$. This will increase $[H^+]$ and $[C_2H_3O_2^-]$ by x, and decrease $[HC_2H_3O_2]$ by x. The equilibrium concentrations are then found in the last column for our table.

	Initial Molar Concentrations	Change	Final Molar Concentrations
H^+	0.0	$+x$	x
$C_2H_3O_2^-$	0.10	$+x$	$0.10 + x \approx 0.10$
$HC_2H_3O_2$	0.20	$-x$	$0.20 - x \approx 0.20$

As before, we look at K_a and see that x will probably be small. We will therefore assume that $0.10 + x \approx 0.10$ and $0.20 - x \approx 0.20$. Substituting into the expression for K_a gives

$$\frac{(x)(0.10)}{(0.20)} = 1.8 \times 10^{-5}$$

$$x = 3.6 \times 10^{-5}$$

Note that x is small compared to both 0.10 and 0.20. This justifies our assumption. Finally, the equilibrium concentrations are

$$[H^+] = 3.6 \times 10^{-5} \ M$$

$$[C_2H_3O_2^-] = 0.10 \ M$$

$$[HC_2H_3O_2] = 0.20 \ M$$

EXAMPLE 15.9

What is the pH of a solution that contains 0.10 M HCl and 0.10 M $HC_2H_3O_2$? For acetic acid, $K_a = 1.8 \times 10^{-5}$.

SOLUTION

This kind of problem fools many students because they forget that HCl is a strong acid—it's completely ionized. This means that the solution contains 0.10 M H^+ just from the HCl, plus a little bit more from the weak acid, $HC_2H_3O_2$. If we let x be the number of moles per liter of $HC_2H_3O_2$ that ionizes by the reaction

$$HC_2H_3O_2 \rightleftharpoons H^+ + C_2H_3O_2^-$$

then we can construct the concentration table as follows:

	Initial Molar Concentrations	Change	Equilibrium Molar Concentrations
H^+	0.10	$+x$	$0.10 + x \approx 0.10$
$C_2H_3O_2^-$	0	$+x$	x
$HC_2H_3O_2$	0.10	$-x$	$0.10 - x \approx 0.10$

We expect x to be small, so we've assumed $0.10 \pm x \approx 0.10$. Substituting into the K_a expression,

$$1.8 \times 10^{-5} = K_a = \frac{[H^+][C_2H_3O_2^-]}{[HC_2H_3O_2]}$$

$$1.8 \times 10^{-5} = \frac{(0.10)(x)}{(0.10)}$$

$$x = 1.8 \times 10^{-5}$$

We see that x is indeed small compared to 0.10, so in the solution $[H^+] = 0.10\ M$. This gives a pH of 1.00.

15.3 DISSOCIATION OF POLYPROTIC ACIDS

Acids containing more than one atom of hydrogen that can be lost on dissociation are known as polyprotic acids. Some examples are H_2SO_4 and H_2S, both of which contain two ionizable hydrogens, and H_3PO_4, which contains three. These acids lose their hydrogens one at a time. Thus we write two steps for the dissociation of sulfuric acid, each with a corresponding equation for K_a,

$$H_2SO_4 \rightleftharpoons H^+ + HSO_4^- \qquad K_{a_1} = \frac{[H^+][HSO_4^-]}{[H_2SO_4]}$$

$$HSO_4^- \rightleftharpoons H^+ + SO_4^{2-} \qquad K_{a_2} = \frac{[H^+][SO_4^{2-}]}{[HSO_4^-]}$$

and for H_2S,

$$H_2S \rightleftharpoons H^+ + HS^- \qquad K_{a_1} = \frac{[H^+][HS^-]}{[H_2S]} \qquad [15.5]$$

$$HS^- \rightleftharpoons H^+ + S^{2-} \qquad K_{a_2} = \frac{[H^+][S^{2-}]}{[HS^-]} \qquad [15.6]$$

The three steps in the dissociation of H_3PO_4 are

$$H_3PO_4 \rightleftharpoons H^+ + H_2PO_4^- \qquad K_{a_1} = \frac{[H^+][H_2PO_4^-]}{[H_3PO_4]}$$

$$H_2PO_4^- \rightleftharpoons H^+ + HPO_4^{2-} \qquad K_{a_2} = \frac{[H^+][HPO_4^{2-}]}{[H_2PO_4^-]}$$

$$HPO_4^{2-} \rightleftharpoons H^+ + PO_4^{3-} \qquad K_{a_3} = \frac{[H^+][PO_4^{3-}]}{[HPO_4^{2-}]}$$

Table 15.2 gives some polyprotic acids and their stepwise dissociation constants. We see from this table that the first dissociation goes essentially to completion for sulfuric acid, while the second occurs only to a relatively limited de-

Table 15.2
Stepwise dissociation of some polyprotic acids at 25°C

Acid	Stepwise Dissociation	Dissociation Constant for Each Step	pK_a
Phosphoric	$H_3PO_4 \rightleftharpoons H^+ + H_2PO_4^-$	$K_{a_1} = 7.5 \times 10^{-3}$	2.13
	$H_2PO_4^- \rightleftharpoons H^+ + HPO_4^{2-}$	$K_{a_2} = 6.2 \times 10^{-8}$	7.21
	$HPO_4^{2-} \rightleftharpoons H^+ + PO_4^{3-}$	$K_{a_3} = 2.2 \times 10^{-12}$	11.66
Sulfuric	$H_2SO_4 \rightleftharpoons H^+ + HSO_4^-$	$K_{a_1} =$ very large	<0
	$HSO_4^- \rightleftharpoons H^+ + SO_4^{2-}$	$K_{a_2} = 1.2 \times 10^{-2}$	1.92
Sulfurous	$H_2SO_3 \rightleftharpoons H^+ + HSO_3^-$	$K_{a_1} = 1.5 \times 10^{-2}$	1.82
	$HSO_3^- \rightleftharpoons H^+ + SO_3^{2-}$	$K_{a_2} = 1.0 \times 10^{-7}$	7.00
Hydrosulfuric	$H_2S \rightleftharpoons H^+ + HS^-$	$K_{a_1} = 1.1 \times 10^{-7}$	6.96
	$HS^- \rightleftharpoons H^+ + S^{2-}$	$K_{a_2} = 1.0 \times 10^{-14}$	14.00
Carbonic	$H_2CO_3 \rightleftharpoons H^+ + HCO_3^-$	$K_{a_1} = 4.3 \times 10^{-7}$	6.37
	$HCO_3^- \rightleftharpoons H^+ + CO_3^{2-}$	$K_{a_2} = 5.6 \times 10^{-11}$	10.26
Ascorbic (vitamin C)	$H_2C_6H_6O_6 \rightleftharpoons H^+ + HC_6H_6O_6^-$	$K_{a_1} = 7.9 \times 10^{-5}$	4.10
	$HC_6H_6O_6^- \rightleftharpoons H^+ + C_6H_6O_6^{2-}$	$K_{a_2} = 1.6 \times 10^{-12}$	11.79

gree. Because of the virtual completion of its first dissociation, sulfuric acid is considered a strong acid. We also see that the first dissociation step of each of these acids occurs with the largest value of K_a, and that each successive step occurs with an ever-decreasing value of K_a. This trend in K_a is reasonable when we consider that it should be easiest to remove an H^+ ion from an uncharged species, and that is should become progressively more difficult to do so as the negative charge on the ion increases.

Because the equilibria involving polyprotic acids are more complex than those of monoprotic acids, equilibrium calculations are somewhat more complicated. The next example shows, however, that a number of approximations can be made that simplify the approach to these kinds of problems.

EXAMPLE 15.10

Hydrogen sulfide, H_2S, is a gas produced by anaerobic (bacterial action in the absence of air) decomposition of organic compounds. Its disagreeable odor is responsible for the terrible smell of rotten eggs. In water, H_2S is a diprotic weak acid. What are the equilibrium concentrations of H^+, HS^-, S^{2-}, and H_2S in a saturated (0.10 M) aqueous solution of H_2S?

SOLUTION

The equilibria involved are shown in Equations 15.5 and 15.6. From Table 15.2, the equilibrium constants have values: $K_{a_1} = 1.1 \times 10^{-7}$ and $K_{a_2} = 1.0 \times 10^{-14}$.

Because K_{a_1} is so *much larger* than K_{a_2}, we can safely assume that nearly all the hydrogen ion in the solution is derived from the first step in the dissociation. In addition, only very little of the HS^- formed in the first step will undergo further dissociation. On the basis of this we can calculate the H^+ and HS^- concentrations using the expression for K_{a_1} alone.

$$K_{a_1} = \frac{[H^+][HS^-]}{[H_2S]}$$

If we let x equal the number of moles per liter of H_2S that dissociate, we obtain, from the stoichiometry of the first step, x mol/liter of H^+ and x mol/liter of HS^-. At equilibrium, there will be $(0.10 - x)$ mol/liter of H_2S remaining.

	Initial Molar Concentrations	Change	Equilibrium Molar Concentrations
H^+	0.0	$+x$	x
HS^-	0.0	$+x$	x
H_2S	0.10	$-x$	$0.10 - x \approx 0.10$

Note that as before, because K_{a_1} is very small, we may assume that x will be negligible compared to 0.10 and write

$$[H_2S] = 0.10 - x \approx 0.10 \ M$$

Substituting these equilibrium quantities into the expression for K_{a_1}, we have

$$\frac{(x)(x)}{0.10} = 1.1 \times 10^{-7}$$

$$x^2 = 1.1 \times 10^{-8}$$

$$x = 1.0 \times 10^{-4}$$

Therefore, the equilibrium concentrations from this first dissociation are

$$[H^+] = 1.0 \times 10^{-4} \ M$$

$$[HS^-] = 1.0 \times 10^{-4} \ M$$

$$[H_2S] = 0.10 - 1.0 \times 10^{-4} = 0.10 \ M$$

Notice that our approximation in the H_2S concentration is valid, because 1.0×10^{-4} is in fact negligible compared to 0.10.

By employing K_{a_2}, we can now calculate the equilibrium concentration of S^{2-},

$$HS^- \rightleftharpoons H^+ + S^{2-}$$

and

$$K_{a_2} = \frac{[H^+][S^{2-}]}{[HS^-]}$$

If we let y equal the number of moles per liter of HS^- that dissociate, then, from the stoichiometry of this second dissociation step, the number of moles per liter of H^+ and S^{2-} produced would also be y. Thus the total hydrogen ion concentration from both the first and second dissociations will be $[H^+] = (1.0 \times 10^{-4} + y)$, and the concentration of HS^- that remains at equilibrium will be $(1.0 \times 10^{-4} - y)$. For this second dissociation,

	"Initial" Concentrations	Change	Equilibrium Concentrations
H^+	1.0×10^{-4}	$+y$	$1.0 \times 10^{-4} + y$
S^{2-}	0.0	$+y$	y
HS^-	1.0×10^{-4}	$-y$	$1.0 \times 10^{-4} - y$

These quantities may be simplified by recognizing that because K_{a_2} is so very small, the amount of HS^- that will dissociate will also be very small. We can therefore make the assumption that y will be negligible compared to 1.0×10^{-4}. Our equilibrium concentrations then become

$$[H^+] = 1.0 \times 10^{-4} + y \approx 1.0 \times 10^{-4} \ M$$

$$[S^{2-}] = y$$

$$[HS^-] = 1.0 \times 10^{-4} - y \approx 1.0 \times 10^{-4} \ M$$

Note that the value of **y** is very much smaller than the value of **x** obtained for the first step in the dissociation.

Substituting, we obtain

$$K_{a_2} = \frac{(1.0 \times 10^{-4})(y)}{(1.0 \times 10^{-4})} = 1.0 \times 10^{-14}$$

$$y = 1.0 \times 10^{-14}$$

Therefore,

$$[S^{2-}] = 1.0 \times 10^{-14} \ M$$

In summary, the concentrations of all solute species present at equilibrium in a 0.10 M H$_2$S solution are

$$[H^+] = 1.0 \times 10^{-4} \ M$$

$$[HS^-] = 1.0 \times 10^{-4} \ M$$

$$[S^{2-}] = 1.0 \times 10^{-14} \ M$$

$$[H_2S] = 0.10 \ M$$

Reviewing this last example, we see that in any solution containing H$_2$S as the only solute, the concentration of the sulfide ion will be equal to K_{a_2}. In fact, for any polyprotic acid where $K_{a_2} \ll K_{a_1}$, the concentration of the anion formed in the second dissociation will always equal K_{a_2}, provided, of course, that the acid is the only solute. For example, K_{a_2} for H$_3$PO$_4$ has a value of 6.2×10^{-8}, and in a solution containing only H$_3$PO$_4$ and H$_2$O, the concentration of HPO$_4^{2-}$ is $6.2 \times 10^{-8} \ M$.

Saturated solutions of H$_2$S are sometimes used in chemical analysis to detect the presence of certain cations by the formation of an insoluble sulfide precipitate. In these analyses the concentration of the S^{2-} is critical and, therefore, must be controlled. By applying Le Châtelier's principle to the dissociation of H$_2$S, we see that any increase in the H$^+$ concentration (perhaps by the addition of a strong acid) will cause a shift in the equilibrium to the left, favoring the formation of more H$_2$S and decreasing the concentrations of both the S^{2-} and HS$^-$ species. Conversely, lowering the H$^+$ concentration will increase [HS$^-$] and [S^{2-}]. Thus, we can control the concentration of the S^{2-} in a saturated H$_2$S solution by varying the H$^+$ concentration. A useful equation expressing the relationship that exists between H$^+$ and S^{2-} in an H$_2$S solution can be derived by multiplying K_{a_1} by K_{a_2} for H$_2$S. Thus

$$K_a = K_{a_1} \times K_{a_2} = \frac{[H^+][\cancel{HS^-}]}{[H_2S]} \times \frac{[H^+][S^{2-}]}{[\cancel{HS^-}]}$$

$$= \frac{[H^+]^2[S^{2-}]}{[H_2S]} = 1.1 \times 10^{-21} \qquad [15.7]$$

A word of caution about the use of Equation 15.7 is in order. *This equation can be used only when two of the three equilibrium concentrations are given and we wish to calculate the third.* It cannot be used to determine, for example, both the H$^+$ and S^{2-} concentrations in solutions of known H$_2$S concentrations. You can verify this for yourself by using it to calculate the concentrations of H$^+$ and S^{2-} that are present in a 0.10 M H$_2$S solution and then comparing your answers with those that we calculated using the two dissociation constants. The proper use of Equation 15.7 is illustrated in Example 15.11.

Be sure you know when you can and cannot use Equation 15.7.

EXAMPLE 15.11   Calculate the S^{2-} concentration in a saturated solution (0.10 M) of H_2S whose pH was adjusted to 2.00 by the addition of HCl.

SOLUTION   Since this is a saturated H_2S solution with a known H^+ concentration, we can use Equation 15.7.

$$K_a = \frac{[H^+]^2[S^{2-}]}{[H_2S]}$$

Rearranging the equation to solve for $[S^{2-}]$, we have

$$[S^{2-}] = \frac{K_a[H_2S]}{[H^+]^2}$$

From the pH of this acidic solution we calculate that $[H^+] = 1.0 \times 10^{-2}$ M; since the solution is saturated with H_2S, we know that $[H_2S] = 0.10$ M. Substituting these values and the value of K_a into our equation, we have

$$[S^{2-}] = \frac{(1.1 \times 10^{-21})(1.0 \times 10^{-1})}{(1.0 \times 10^{-2})^2}$$

or

$$[S^{2-}] = 1.1 \times 10^{-18} \ M$$

15.4 BUFFERS

Any solution that contains both a weak acid and a weak base has the ability to absorb small amounts of either a strong acid or strong base with very little change in pH. When small quantities of a strong acid are added, they are neutralized by the weak base, while small quantities of a strong base are neutralized by the weak acid. Such solutions are said to be **buffers** because they resist significant changes in the pH.

A buffer whose pH is less than 7 generally can be prepared by mixing a weak acid with the salt of that weak acid—for example, acetic acid and sodium acetate. A buffer whose pH is greater than 7 generally can be prepared by mixing a weak base with the salt of that weak base—for example, ammonia and ammonium chloride. When H^+ or OH^- are added to an acetic acid-acetate buffer, the following neutralization reactions take place:

Acetic acid is a weak acid and acetate ion is its weak conjugate base.

$$H^+ + C_2H_3O_2^- \longrightarrow HC_2H_3O_2$$

$$OH^- + HC_2H_3O_2 \longrightarrow H_2O + C_2H_3O_2^-$$

Similarly, for the NH_3, NH_4Cl, alkaline buffer we have

$$H^+ + NH_3 \longrightarrow NH_4^+$$

and

Ammonia is a weak base and ammonium ion is its weak conjugate acid.

$$OH^- + NH_4^+ \longrightarrow H_2O + NH_3$$

In an acid buffer the H^+ concentration (and pH) is determined by the relative concentrations of the weak acid and its conjugate base—that is, the anion. For example, for acetic acid we have

$$K_a = \frac{[H^+][C_2H_3O_2^-]}{[HC_2H_3O_2]}$$

Solving for $[H^+]$ gives

$$[H^+] = K_a\left(\frac{[HC_2H_3O_2]}{[C_2H_3O_2^-]}\right) \qquad [15.8]$$

To calculate $[H^+]$, and then pH, we must know K_a for the weak acid (from Table 15.1) as well as the ratio of the concentrations of the weak acid and its anion.

Since salts completely dissociate in aqueous solution, the number of moles of the anion in the solution from this source is determined by the formula of the salt and by the number of moles of salt dissolved. Thus a 1.0 M $NaC_2H_3O_2$ solution contains 1.0 mol/liter of $C_2H_3O_2^-$, while a 1.0 M $Ca(C_2H_3O_2)_2$ solution contains 2.0 mol/liter of $C_2H_3O_2^-$. In the buffer there is also an additional amount of acetate ion that comes from the dissociation of $HC_2H_3O_2$. The amount of H^+ and $C_2H_3O_2^-$ stemming from this source is very small even in solutions containing only acetic acid, and this small amount is reduced even further in the buffer because of the presence of the large concentration of $C_2H_3O_2^-$ from the salt.[2] The total anion concentration in the buffer is essentially determined by the salt concentration alone, since the contribution from the dissociation of the weak acid is negligible. As in our previous calculations on weak acids, the concentration of $HC_2H_3O_2$ will not be reduced appreciably by its dissociation and, in a mixture of 1.0 M $NaC_2H_3O_2$ and 1.0 M $HC_2H_3O_2$, for example, the concentrations of both the molecular acid and the anion are 1.0 M. The H^+ concentration in such a buffer could be found from Equation 15.8 using $K_a = 1.8 \times 10^{-5}$ (Table 15.1).

$$[H^+] = (1.8 \times 10^{-5}) \frac{1.0}{1.0}$$

$$= 1.8 \times 10^{-5} \, M$$

The pH of this solution is 4.74.

If you are taking a biology or biochemistry course, you will probably encounter an equation derived (essentially) from Equation 15.8.

$$pH = pK_a + \log \frac{[anion]}{[acid]} \qquad [15.9]$$

This is called the **Henderson-Hasselbalch equation.**[3] For acetic acid, $pK_a = 4.74$, and for a buffer having $[HC_2H_3O_2] = [C_2H_3O_2^-] = 1.0 \, M$

$$pH = 4.74 + \log \left(\frac{1.0}{1.0} \right)$$

$$= 4.74 + \log 1$$

$$= 4.74$$

log 1 = 0.

As before, we find the pH equal to 4.74.

Notice that when the concentrations of the acid and anion are the same in a buffer, the H^+ concentration in that solution is equal to the K_a of the weak acid, and the pH = pK_a. Thus, if a buffer was prepared by mixing 0.1 mol of formic acid ($K_a = 1.8 \times 10^{-4}$ from Table 15.1) and 0.1 mol of sodium formate into a liter of solution, the resulting H^+ concentration would be

$$[H^+] = 1.8 \times 10^{-4} \, M$$

and

$$pH = 3.74$$

---

[2] This can easily be seen by applying Le Châtelier's principle to the dissociation of $HC_2H_3O_2$. The presence of acetate ion from a salt causes the dissociation equilibrium of the acid to be shifted to the left and actually suppresses the dissociation.

[3] A similar equation that would apply to a basic buffer (e.g., NH_3, NH_4Cl) is

$$pOH = pK_b + \log \frac{[cation, BH^+]}{[base, B]}$$

We can also use Equation 15.8 or 15.9 to calculate the concentrations of acid and salt that would be needed to achieve a certain pH buffer. This is illustrated in Example 15.12.

EXAMPLE 15.12	What ratio of acetic acid to sodium acetate concentration is needed to form a buffer whose pH is 5.70?
SOLUTION	To solve this problem we need to rearrange Equation 15.8 to solve for the ratio of concentrations. Thus

$$\frac{[HC_2H_3O_2]}{[C_2H_3O_2^-]} = \frac{[H^+]}{K_a}$$

The H^+ concentration when the pH is 5.70 is

$$[H^+] = 2.0 \times 10^{-6} \ M$$

Therefore,

$$\frac{[HC_2H_3O_2]}{[C_2H_3O_2^-]} = \frac{2.0 \times 10^{-6}}{1.8 \times 10^{-5}} = \frac{2.0 \times 10^{-6}}{18 \times 10^{-6}}$$

or

$$\frac{[HC_2H_3O_2]}{[C_2H_3O_2^-]} = \frac{1}{9}$$

As long as this ratio is maintained, the pH of an acetic acid-sodium acetate buffer is 5.70. For example, if 0.2 mol of $HC_2H_3O_2$ and 1.8 mol of $NaC_2H_3O_2$ are dissolved in 1 liter of solution, the pH is 5.70. This same pH will result if 0.1 mol of $HC_2H_3O_2$ and 0.9 mol of $NaC_2H_3O_2$ are dissolved.

We have seen that by adjusting the ratio of concentrations of the weak acid to that of the salt, a buffer of almost any desired pH can be achieved. For example, a buffer composed of 0.01 mol of $HC_2H_3O_2$ and 1.00 mol of $NaC_2H_3O_2$ would have a pH of 6.74. However, when as little as 0.01 mol of base is added to this buffer, all of the acetic acid is neutralized and a large change in the pH of the buffer results. Therefore, *the most effective pH range for any buffer is at or near the pH where the acid and salt concentrations are equal* (that is, pK_a). Also, in order to be most effective, the amounts of weak acid and base used to prepare the buffer must be considerably greater than the amounts of acid or base that may later be added to the buffer.

Let us now see how effective a buffer is at holding the pH nearly constant. Suppose we have an acetic acid-acetate buffer in which the concentrations of $HC_2H_3O_2$ and $C_2H_3O_2^-$ are each 1.00 M. We saw earlier that its pH will be 4.74 and $[H^+] = 1.8 \times 10^{-5} \ M$. What happens to the pH if we add, say 0.20 mol of HCl to a liter of this buffer?

When a strong acid such as HCl is added, its H^+ reacts with acetate ion.

$$H^+ + C_2H_3O_2^- \longrightarrow HC_2H_3O_2$$

This decreases the $C_2H_3O_2^-$ concentration and increases the $HC_2H_3O_2$ concentration. Below are the number of moles per liter of all species before and after the addition.

The larger the concentrations of the components of the buffer, the more effective it is at resisting changes in pH.

Initial	Final
$[H^+] = 1.8 \times 10^{-5} \ M$	$[H^+] = x$
$[C_2H_3O_2^-] = 1.00 \ M$	$[C_2H_3O_2^-] = 1.00 - 0.20 \ M = 0.80 \ M$
$[HC_2H_3O_2] = 1.00 \ M$	$[HC_2H_3O_2] = 1.00 + 0.20 \ M = 1.20 \ M$

Substituting these final concentrations into Equation 15.8, we have

$$[H^+] = (1.8 \times 10^{-5}) \times \left(\frac{1.20}{0.80}\right)$$

$$= 2.7 \times 10^{-5} \ M$$

The pH of the buffer after the 0.20 mol of H^+ is added is 4.57—a change of only 0.17 pH units from its initial pH of 4.74.

Suppose now that we were to add 0.20 mol of H^+ to 1 liter of a solution of HCl whose pH = 4.74 (that is, a $1.8 \times 10^{-5} \ M$ HCl solution). Since Cl^- has virtually no tendency to react with H^+, the final H^+ concentration will be 0.20 M ($0.20 + 1.8 \times 10^{-5} = 0.20$), and the pH of the solution will be 0.70. The change in pH in this case is 4.04 pH units, as opposed to a change of only 0.17 pH units when the same quantity of H^+ is added to the buffer.

Additions of strong base are also absorbed by the buffer. When 0.20 mol of OH^- is added to the 1 liter of our original buffer, it is neutralized according to the reaction

$$OH^- + HC_2H_3O_2 \longrightarrow H_2O + C_2H_3O_2^-$$

The number of moles per liter before the addition are the same as above, but the number of moles per liter after 0.20 mol of OH^- is added would be

Concentration After Addition of 0.20 mol of OH^-
$[H^+] = x$
$[C_2H_3O_2^-] = 1.00 + 0.20 = 1.20 \ M$
$[HC_2H_3O_2] = 1.00 - 0.20 = 0.80 \ M$

Substituting these values into Equation 15.8, we find that the H^+ concentration after the addition is $1.2 \times 10^{-5} \ M$ and the resulting pH is 4.92—once again, a change of 0.17 pH units. Finally, note that the addition of an acid to the buffer lowers the pH, while addition of a base raises the pH. Although the change in pH is small, the *direction* of the change is as expected.

EXAMPLE 15.13 A buffer was prepared by mixing exactly 200 ml of a 0.60 M NH_3 solution and 300 ml of a 0.30 M NH_4Cl solution. (a) What is the pH of this buffer, assuming a final volume of 500 ml? (b) What will be the pH after 0.020 mol of H^+ is added?

SOLUTION The number of moles of NH_3 added to this solution is

$$0.60 \ \frac{mol}{liter} \times 0.200 \ liter = 0.12 \ mol$$

and the number of moles of NH_4^+ added is

$$0.30 \ \frac{mol}{liter} \times 0.300 \ liter = 0.090 \ mol$$

Therefore, the concentrations of these ions in the 500 ml is

$$[NH_3] = \frac{0.12 \ mol}{0.500 \ liter} = 0.24 \ M$$

$$[NH_4^+] = \frac{0.090 \ mol}{0.500 \ liter} = 0.18 \ M$$

(a) The OH^- concentration for this alkaline buffer is found by using the K_b for NH_3:

$$NH_3 + H_2O \rightleftharpoons NH_4^+ + OH^-$$

$$K_b = \frac{[NH_4^+][OH^-]}{[NH_3]}$$

Rearranging and solving for $[OH^-]$, we have

$$[OH^-] = K_b \frac{[NH_3]}{[NH_4^+]}$$

Substituting K_b for NH_3 from Table 15.1 and the concentrations of NH_3 and NH_4^+ for this buffer into this equation gives

$$[OH^-] = (1.8 \times 10^{-5}) \times \left(\frac{0.24}{0.18}\right)$$

$$= 2.4 \times 10^{-5}\ M$$

$$pOH = 4.62$$

$$pH = 9.38$$

(b) The neutralization reaction for H^+ in this buffer is

$$H^+ + NH_3 \longrightarrow NH_4^+$$

We are adding 0.020 mol of H^+ to 500 ml, or 0.040 mol of H^+ per liter. Therefore, the concentrations before and after the addition of the acid are

Initial	Final
$[OH^-] = 2.4 \times 10^{-5}\ M$	$[OH^-] = x$
$[NH_3] = 0.24\ M$	$[NH_3] = 0.24 - 0.040\ M = 0.20\ M$
$[NH_4^+] = 0.18\ M$	$[NH_4^+] = 0.18 + 0.040\ M = 0.22\ M$

and the OH^- concentration is

$$[OH^-] = (1.8 \times 10^{-5}) \times \left(\frac{0.20}{0.22}\right)$$

$$= 1.6 \times 10^{-5}\ M$$

$$pOH = 4.80$$

The pH is therefore 9.20, a decrease of 0.18 pH units.

HCO_3^- is itself a buffer. In the photo at the beginning of the chapter, $NaHCO_3$ is being added to a swimming pool to control the pool's pH.

Buffers find many important applications. Living systems employ buffers to maintain nearly constant pH so that biochemical reactions can follow their correct paths. For example, blood contains, among other things, a H_2CO_3/HCO_3^- buffer system that helps maintain the pH at 7.4.

In the laboratory many inorganic and organic chemical reactions are performed in buffered solutions to minimize any adverse effects caused by acids or bases that might be consumed or produced during reaction.

15.5 HYDROLYSIS

It was pointed out in Section 6.6 that salts are produced during an acid-base neutralization reaction. For example, NaCl is considered the salt of the strong acid HCl and the strong base NaOH. In a similar fashion we say that $NaC_2H_3O_2$ is the salt of a weak acid ($HC_2H_3O_2$) and a strong base (NaOH), and NH_4Cl is the salt of a strong acid (HCl) and a weak base (NH_3). A salt such as $NH_4C_2H_3O_2$ is the salt of a weak acid and a weak base. When these salts are added to water, the pH of the resulting solution is found experimentally to depend on the type of salt dissolved. For example, the pH of an aqueous solution of a salt of a strong acid and a strong base is always very close to 7, while the pH for a salt of a weak acid and a strong base is greater than 7. The pH that results when each type of salt is dissolved in water is summarized as follows:

Type of Salt	pH of Aqueous Solution
Strong acid—strong base	7
Weak acid—strong base	>7
Strong acid—weak base	<7
Weak acid—weak base	Depends on salt

How might we account for these differences in pH?

When a salt dissolves in water, it dissociates fully to produce cations and anions that may subsequently react chemically with the solvent in a process called **hydrolysis.** For example, the cation of a salt may undergo the reaction

$$M^+ + H_2O \rightleftharpoons MOH + H^+$$

while an anion may react according to the equation

$$X^- + H_2O \rightleftharpoons HX + OH^-$$

Since the H^+ and OH^- ions produced in these reactions influence the pH of the salt solution, the extent to which the hydrolysis reactions take place determines whether the pH will be greater than, less than, or equal to 7.

Consider, for example, NaCl, the salt of a strong acid and a strong base. If this salt were to hydrolyze, the products would be NaOH and HCl. Both of these are strong electrolytes, however, and are completely dissociated. The result is that there will be the same amount of H^+ and OH^- in the solution—a condition that is fulfilled only when their concentrations are each 10^{-7} M. Consequently, the pH of the NaCl solution is 7. Since this is the same as pure water, the net effect is that no hydrolysis actually takes place at all. In general, we may conclude that *anions and cations of strong acids and bases, respectively, do not undergo hydrolysis, and the salts derived from strong acids and bases yield neutral solutions.* This, however, is not the case for other types of salts.

Salts of weak acids and strong bases: anion hydrolysis

For this type of salt we are concerned only with anion hydrolysis because, in light of the previous discussion, cations of strong bases do not undergo hydrolysis. Taking $NaC_2H_3O_2$ as an example, we would write for the hydrolysis reaction of the anion

$$C_2H_3O_2^- + H_2O \longrightarrow HC_2H_3O_2 + OH^-$$

The acetate ion, which is the conjugate base of acetic acid, is sufficiently basic to pull some protons away from water molecules, so that when equilibrium is established an excess of OH^- is present. As a result the solution is basic. We can write an equilibrium constant expression in the usual fashion:

$$K_h' = \frac{[HC_2H_3O_2][OH^-]}{[C_2H_3O_2^-][H_2O]}$$

As usual, the concentration of H_2O may be included with the equilibrium constant K_h', and we obtain

$$K_h' \times [H_2O] = K_h = \frac{[HC_2H_3O_2][OH^-]}{[C_2H_3O_2^-]}$$

where K_h is called the **hydrolysis constant.**

This same equation can be derived simply by dividing the equation for K_w by the K_a of the weak acid (acetic acid in this case). Thus

$$\frac{K_w}{K_a} = \frac{[\cancel{H^+}][OH^-]}{[\cancel{H^+}][C_2H_3O_2^-]/[HC_2H_3O_2]}$$

or

$$\frac{K_w}{K_a} = \frac{[HC_2H_3O_2][OH^-]}{[C_2H_3O_2^-]}$$

We see, therefore, that the hydrolysis constant for the anion is equal to

$$K_h = \frac{K_w}{K_a}$$

and can be calculated simply from a knowledge of K_w and K_a of the weak acid from which the anion is formed.

From K_h and the concentration of the anion in the salt solution we may then calculate the concentration of OH^- and eventually determine the pH of the salt solution, as shown in Example 15.14.

EXAMPLE 15.14

Calculate the pH of a 0.10 M $NaC_2H_3O_2$ solution.

SOLUTION

The first step is to inspect the salt to determine what equilibrium has to be considered. When $NaC_2H_3O_2$ is dissolved in water, we know that it dissociates completely to give Na^+ and $C_2H_3O_2^-$. (Don't make the mistake of writing an equilibrium involving the salt!) Next we ask ourselves what acid and base would be combined to form the salt. The answer here is $HC_2H_3O_2$ and $NaOH$. Since $HC_2H_3O_2$ is a weak acid, we must consider the hydrolysis of its anion, but we don't have to worry about the Na^+ because $NaOH$ is a strong base and Na^+ therefore doesn't hydrolyze. The Na^+ is just a spectator ion and we ignore it. Now we can write the equation for the equilibrium—the hydrolysis of $C_2H_3O_2^-$.

Only ions of <u>weak</u> acids or bases hydrolyze.

$$C_2H_3O_2^- + H_2O \rightleftharpoons HC_2H_3O_2 + OH^-$$

Next we write the equilibrium expression

$$K_h = \frac{[HC_2H_3O_2][OH^-]}{[C_2H_3O_2^-]}$$

Before this equation becomes useful, we must first calculate K_h. We can do this with the equation

$$K_h = \frac{K_w}{K_a}$$

Substituting the values of K_w and K_a (from Table 15.1) into this equation, we obtain

$$K_h = \frac{1.0 \times 10^{-14}}{1.8 \times 10^{-5}} = 5.6 \times 10^{-10}$$

Because K_h is so small, we know that the position of equilibrium lies far to the left in favor of $C_2H_3O_2^-$. Therefore, if we let x equal the number of moles per liter of $C_2H_3O_2^-$ that undergo hydrolysis, we can set up our table of concentrations and expect that x will be small compared to 0.10.

	Initial Molar Concentrations	Change	Equilibrium Molar Concentrations
$HC_2H_3O_2$	0.0	$+x$	x
OH^-	0.0	$+x$	x
$C_2H_3O_2^-$	0.10	$-x$	$0.10 - x \approx 0.10$

Substituting these values along with K_h into the hydrolysis equation, we have

$$K_h = \frac{(x)(x)}{0.10} = 5.6 \times 10^{-10}$$

$$x^2 = 5.6 \times 10^{-11}$$

$$x = 7.5 \times 10^{-6} \; M$$

Thus

$$[OH^-] = 7.5 \times 10^{-6} \; M$$

$$pOH = 5.12$$

and therefore

$$pH = 8.88$$

Thus the pH of this solution indicates that it is basic.

Salts of strong acids and weak bases: cation hydrolysis

From our previous discussion we know that only the cation in this type of salt undergoes hydrolysis. For example, the hydrolysis reaction that takes place in an aqueous solution of NH_4Cl is

$$NH_4^+ + H_2O \rightleftharpoons H_3O^+ + NH_3$$

As a result of the hydrolysis of the cation, some H_2O molecules are converted into H_3O^+ which, of course, make the solution acidic. The equilibrium condition for this hydrolysis is written as

$$K_h = \frac{[H_3O^+][NH_3]}{[NH_4^+]}$$

which can also be derived by dividing K_w by K_b of the weak base, NH_3.

$$K_h = \frac{K_w}{K_b} = \frac{[H_3O^+][\cancel{OH^-}]}{[NH_4^+][\cancel{OH^-}]/[NH_3]} = \frac{[H_3O^+][NH_3]}{[NH_4^+]}$$

Therefore in order to calculate the pH for this type of sale we must know K_w, K_b, and the concentration of the salt, as illustrated in Example 15.15.

EXAMPLE 15.15   What is the pH of a 0.10 M N_2H_5Cl solution?

SOLUTION   We first must recognize that this is the salt of a strong acid (HCl) and a weak base (N_2H_4); therefore, only the cation undergoes hydrolysis. The hydrolysis reaction is

$$N_2H_5^+ + H_2O \rightleftharpoons H_3O^+ + N_2H_4$$

for which we can write

$$K_h = \frac{[H_3O^+][N_2H_4]}{[N_2H_5^+]}$$

For this salt we know that

$$K_h = \frac{K_w}{K_b}$$

where K_b is for the weak base, N_2H_4. Using the value of K_b from Table 15.1 gives

$$K_h = \frac{1.0 \times 10^{-14}}{1.7 \times 10^{-6}} = 5.9 \times 10^{-9}$$

Once again, because of the size of K_h, the hydrolysis equilibrium lies mainly to the left. If x equals the number of moles per liter of $N_2H_5^+$ that undergo hydrolysis, then

	Initial Molar Concentrations	Change	Equilibrium Molar Concentrations
N_2H_4	0.0	$+x$	x
H_3O^+	0.0	$+x$	x
$N_2H_5^+$	0.10	$-x$	$0.10 - x \approx 0.10$

Substituting equilibrium concentrations into the expression for K_h gives

$$\frac{(x)(x)}{0.10} = 5.9 \times 10^{-9}$$

$$x^2 = 5.9 \times 10^{-10}$$

$$x = 2.4 \times 10^{-5}$$

Therefore,

$$[H_3O^+] = 2.4 \times 10^{-5} \ M$$

and

$$pH = 4.62$$

Salts of weak acids and weak bases: cation and anion hydrolysis

Solutions of this type of salt can either be acidic, neutral, or basic, because both the cation and the anion of the salt undergo hydrolysis. The pH of such a salt solution is determined by the relative extent of the hydrolysis reactions of each ion. By applying what we have learned from the last two types of salts to these, we should be able to predict, at least qualitatively, the pH of their aqueous solutions. If the K_a of the weak acid and the K_b of the weak base of the salt are identical, then the extent of cation and anion hydrolysis is exactly the same (the K_h for the cation is exactly equal to the K_h for the anion) and the solution will be neutral. For example, in the case of $NH_4C_2H_3O_2$, where the K_b of NH_3 is 1.8×10^{-5} and the K_a of $HC_2H_3O_2$ is 1.8×10^{-5}, the value of K_h for both ions is 5.6×10^{-10}, and an aqueous solution of this salt, regardless of the concentration, is neutral. On the other hand, we would predict that an aqueous solution of NH_4CN would be basic because of the relative values of the K_h's of the cation and anion. For NH_4^+ we have

$$K_h = \frac{K_w}{K_b} = \frac{1.0 \times 10^{-14}}{1.8 \times 10^{-5}} = 5.6 \times 10^{-10}$$

and for CN^- we have

$$K_h = \frac{K_w}{K_a} = \frac{1.0 \times 10^{-14}}{4.9 \times 10^{-10}} = 2.0 \times 10^{-5}$$

Since the CN^- undergoes more extensive hydrolysis than NH_4^+, the reaction

$$CN^- + H_2O \rightleftharpoons HCN + OH^-$$

goes further to completion than does the reaction

$$NH_4^+ + H_2O \rightleftharpoons H_3O^+ + NH_3$$

More OH^- ions are produced than H_3O^+ and therefore the solution is basic.

In a similar fashion, we predict that an aqueous solution of ammonium formate, NH_4CHO_2, is acidic because of the relative values of the K_h's for the cation and anion. The K_h for NH_4^+ from above is 5.6×10^{-10}, while the K_h for the formate ion, CHO_2^-, is

$$K_h = \frac{K_w}{K_a} = \frac{1.0 \times 10^{-14}}{1.8 \times 10^{-4}}$$

$$= 5.6 \times 10^{-11}$$

Thus the hydrolysis reaction,

$$NH_4^+ + H_2O \rightleftharpoons H_3O^+ + NH_3$$

occurs to a slightly greater extent than does the reaction,

$$CHO_2^- + H_2O \rightleftharpoons HCHO_2 + OH^-$$

Therefore, there is a small excess of H_3O^+ ions and solutions of this salt are acidic.

Hydrolysis of salts of polyprotic acids

An example of this type of salt is Na_2S, the salt of a weak acid, H_2S, and a strong base, $NaOH$. Since only the anion undergoes hydrolysis, the equilibrium reaction is

$$S^{2-} + H_2O \rightleftharpoons HS^- + OH^-$$

We see that this reaction produces another anion, HS^-, that can also undergo hydrolysis. Its equilibrium reaction is

$$HS^- + H_2O \rightleftharpoons H_2S + OH^-$$

Notice that to compute K_{h_1} we use K_{a_2}.

The hydrolysis constant, K_{h_1}, for the first reaction is

$$K_{h_1} = \frac{K_w}{K_{a_2}} = \frac{[HS^-][OH^-]}{[S^{2-}]}$$

where K_{a_2} is the acid dissociation constant for the weak acid, HS^-. The equilibrium constant for the second step in the hydrolysis is

$$K_{h_2} = \frac{K_w}{K_{a_1}} = \frac{[H_2S][OH^-]}{[HS^-]}$$

where K_{a_1} in this case is the dissociation constant for the weak acid, H_2S. Substituting the values of the K_a's from Table 15.2 into these equations, we obtain

$$K_{h_1} = \frac{1.0 \times 10^{-14}}{1.0 \times 10^{-14}} = 1.0$$

and

$$K_{h_2} = \frac{1.0 \times 10^{-14}}{1.1 \times 10^{-7}} = 9.1 \times 10^{-8}$$

The relative magnitudes of these two equilibrium constants indicate to us that the second hydrolysis reaction occurs to a negligible extent compared to the first; therefore, only K_{h_1} need be used to determine the pH of an aqueous solution of this salt, as shown in our next example.

EXAMPLE 15.16 Sodium sulfide, Na_2S, is a chemical used in dehairing animal hides. What is the pH of a 0.20 M solution of Na_2S?

SOLUTION From our previous discussion we know that only the first hydrolysis reaction

$$S^{2-} + H_2O \rightleftharpoons HS^- + OH^-$$

is important in determining the pH of this solution.

The equation for the hydrolysis constant is

$$K_{h_1} = \frac{[HS^-][OH^-]}{[S^{2-}]} = 1.0$$

As usual, we let x equal the number of moles per liter of S^{2-} that hydrolyze. Then,

	Initial Molar Concentrations	Change	Equilibrium Molar Concentrations
HS^-	0.0	$+x$	x
OH^-	0.0	$+x$	x
S^{2-}	0.20	$-x$	$0.20 - x$

Our first reaction here is to make the simplifying assumption that $0.20 - x \approx 0.20$. However, if we do so and then solve for x, we obtain $x = 0.45$. This is clearly impossible because the equilibrium sulfide concentration becomes $-0.25\ M$. Negative concentrations are absurd—we can't have less than nothing! Therefore, we cannot simplify the S^{2-} concentration term in our usual way, and when we substitute the quantities corresponding to the equilibrium concentrations into the mass action expression we obtain

$$\frac{(x)(x)}{(0.20 - x)} = 1.0$$

Multiplying both sides by $(0.20 - x)$ gives

$$x^2 = (0.20 - x)1.0$$

which can be rearranged as

$$x^2 + x - 0.20 = 0$$

The roots of the quadratic equation, $ax^2 + bx + c = 0$ are given by the quadratic formula

$$x = \frac{-b \pm \sqrt{b^2 - 4ac}}{2a}$$

This is a quadratic equation whose roots can be obtained using the quadratic formula. This gives

$$x = \frac{-1 \pm \sqrt{(1)^2 - 4(1)(-0.20)}}{(2)(1)}$$

$$x = \frac{-1 \pm \sqrt{1.8}}{2} = \frac{-1 \pm 1.34}{2}$$

Notice that two values of x are obtained.

$$x = \frac{-2.34}{2} = -1.17$$

$$x = \frac{0.34}{2} = 0.17$$

The first value of x makes no sense. It has no physical meaning because it tells us that the concentrations of HS^- and OH^- are negative. As we said before, we cannot have less than nothing. The second value of x is meaningful, and we conclude that

$$x = 0.17\ M$$

and therefore,

$$[OH^-] = 0.17\ M$$

from which we obtain

Sulfide ion is a pretty strong base.

$$pOH = 0.77$$

Thus, the pH of the solution is 13.23.

15.6 ACID-BASE TITRATIONS: THE EQUIVALENCE POINT

In Chapter 6 we saw that a titration is a useful and accurate way of determining the concentrations of acids and bases, provided that the equivalence point can be detected. The equivalence point, you should remember, occurs when equal numbers of equivalents of acid and base have been combined. In this section we will see how the pH of a solution changes during the course of typical acid-base titrations and what the pH is at the equivalence point.

Titration of a strong acid with a strong base

A typical example of a titration of a strong acid with a strong base occurs when 25.00 ml of 0.10 M HCl is titrated with 0.10 M NaOH. We can mathematically determine the pH throughout the titration by calculating the H^+ concentration present in the flask each time a quantity of NaOH is added to the HCl. For example, the number of moles of H^+ present in the 25 ml of a 0.10 M HCl solution is

$$\left(\frac{0.10\ \text{mol}}{1000\ \text{ml}}\right) \times 25\ \text{ml} = 2.5 \times 10^{-3}\ \text{mol of } H^+$$

When 10 ml of the 0.10 M NaOH are added, we in fact have added

$$\left(\frac{0.10\ \text{mol}}{1000\ \text{ml}}\right) \times 10\ \text{ml} = 1.0 \times 10^{-3}\ \text{mol of } OH^-$$

The neutralization reaction,

$$H^+ + OH^- \longrightarrow H_2O$$

occurs, and the amount of H^+ remaining is

$$(2.5 \times 10^{-3}) - (1.0 \times 10^{-3}) = 1.5 \times 10^{-3}\ \text{mol of } H^+$$

The molar concentration of H^+ is now

$$[H^+] = \frac{1.5 \times 10^{-3}\ \text{mol}}{0.035\ \text{liter}} = 4.3 \times 10^{-2}\ M$$

The total volume

and the pH is calculated to be 1.37. The concentrations of H^+ after further additions of NaOH have occurred are summarized in Table 15.3.

Our calculations show that the pH increases slowly at first, then rises rapidly near the equivalence point, and finally levels off gradually after the equivalence point is reached.

Table 15.3
Titration of 25 ml of 0.10 M HCl with a 0.10 M NaOH solution

Volume of HCl	Volume of NaOH	Volume Total	Moles of H^+	Moles of OH^-	Molarity of Ion in Excess	pH
25.00	0.00	25.00	2.5×10^{-3}	0	0.10 (H^+)	1.00
25.00	10.00	35.00	2.5×10^{-3}	1.0×10^{-3}	4.3×10^{-2} (H^+)	1.37
25.00	24.99	49.99	2.5×10^{-3}	2.499×10^{-3}	2.0×10^{-5} (H^+)	4.70
25.00	25.00	50.00	2.5×10^{-3}	2.50×10^{-3}	0	7.00
25.00	25.01	50.01	2.5×10^{-3}	2.501×10^{-3}	2.0×10^{-5} (OH^-)	9.30
25.00	26.00	51.00	2.5×10^{-3}	2.60×10^{-3}	2.0×10^{-3} (OH^-)	11.30
25.00	50.00	75.00	2.5×10^{-3}	5.0×10^{-3}	3.3×10^{-2} (OH^-)	12.52

If a graph is drawn of pH versus the volume of base added, we obtain the plot shown in Figure 15.2. The equivalence point occurs, in this case, at a pH of 7. At the equivalence point the solution is neutral because neither of the ions of the salt that's left in solution (NaCl) undergoes hydrolysis.

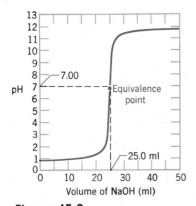

Figure 15.2

Titration of 0.10 M HCl with 0.10 M NaOH.

Titration using a weak acid and a strong base

In an acid-base titration in which one substance is strong and the other is weak, the solution is not neutral at the equivalence point because of hydrolysis of the salt. For example, consider the titration of 25.0 ml of 0.10 M $HC_2H_3O_2$ with 0.10 M NaOH. Before any base is added the only solute is acetic acid. The pH of the solution is calculated as shown in Section 15.2—for 0.10 M $HC_2H_3O_2$, the pH = 2.89.

When we begin to add NaOH, acetic acid molecules are converted to acetate ions.

$$HC_2H_3O_2 + OH^- \longrightarrow H_2O + C_2H_3O_2^-$$

Since the solution then contains both $HC_2H_3O_2$ and $C_2H_3O_2^-$, it is a buffer, and we've also learned how to calculate the pH for this kind of mixture. For instance, when 10.0 ml of 0.10 M NaOH have been added, 1.0×10^{-3} mol of OH^- has been supplied. This "neutralizes" 1.0×10^{-3} mol of $HC_2H_3O_2$ and converts it to 1.0×10^{-3} mol of $C_2H_3O_2^-$. The original 25.0 ml of 0.10 M $HC_2H_3O_2$ contained 2.5×10^{-3} mol $HC_2H_3O_2$, so the amount that is left is $(2.5 \times 10^{-3}$ mol) $-$ $(1.0 \times 10^{-3}$ mol) $= 1.5 \times 10^{-3}$ mol $HC_2H_3O_2$. The concentrations of acetic acid and acetate ion in the total volume of 35.0 ml are therefore

$$[HC_2H_3O_2] = \frac{1.5 \times 10^{-3} \text{ mol}}{0.0350 \text{ liter}} = 4.3 \times 10^{-2} \ M$$

$$[C_2H_3O_2^-] = \frac{1.0 \times 10^{-3} \text{ mol}}{0.0350 \text{ liter}} = 2.9 \times 10^{-2} \ M$$

If we solve the K_a expression for acetic acid for the H^+ concentration and substitute these values for $[HC_2H_3O_2]$ and $[C_2H_3O_2^-]$, we obtain

$$[H^+] = K_a \times \frac{[HC_2H_3O_2]}{[C_2H_3O_2^-]}$$

$$= 1.8 \times 10^{-5} \left(\frac{4.3 \times 10^{-2}}{2.9 \times 10^{-2}} \right)$$

$$= 2.7 \times 10^{-5} \ M$$

Therefore,

$$pH = 4.57$$

From the time of the first addition of base until the equivalence point is reached, the solution contains both acetic acid and acetate ion, and the pH may be computed in this fashion.

When a total of 25.0 ml of NaOH are added, all the acetic acid is "neutralized," and we have produced 2.5×10^{-3} mol of $NaC_2H_3O_2$ in 50.0 ml of solution. The resulting $0.050\ M\ NaC_2H_3O_2$ solution undergoes hydrolysis because it contains the anion of a weak acid. We have seen that for this solute the equilibrium is

$$C_2H_3O_2^- + H_2O \rightleftharpoons HC_2H_3O_2 + OH^-$$

From the last section we know that

$$K_h = \frac{[HC_2H_3O_2][OH^-]}{[C_2H_3O_2^-]} = \frac{K_w}{K_a} = 5.6 \times 10^{-10}$$

We can calculate the OH^- concentration in this solution as we did previously by letting x equal the number of moles per liter of $C_2H_3O_2^-$ that hydrolyze. This allows us to construct our table.

For the $NaC_2H_3O_2$,

$$\frac{2.5 \times 10^{-3}\ mol}{0.050\ liter} = \underline{0.050\ M.}$$

	Initial Molar Concentrations	Change	Equilibrium Molar Concentrations
$HC_2H_3O_2$	0.0	$+x$	x
OH^-	0.0	$+x$	x
$C_2H_3O_2^-$	0.050	$-x$	$0.050 - x \approx 0.050$

Substituting into the K_h expression gives

$$K_h = \frac{(x)(x)}{0.050} = 5.6 \times 10^{-10}$$

$$x^2 = 2.8 \times 10^{-11}$$

$$x = 5.3 \times 10^{-6}$$

This means that $[OH^-] = 5.3 \times 10^{-6}\ M$, from which we obtain

$$pOH = 5.28$$

and finally,

$$pH = 8.72$$

Thus the pH at which the equivalence point occurs is greater than 7. We find that this is true for any weak acid-strong base titration.

Thus far we have discussed only the first half of the titration (see Table 15.4). What takes place beyond the equivalence point? As soon as all the weak acid has been neutralized, any further addition of NaOH suppresses the hydrolysis of the anion and the pH is then solely dependent on the concentration of OH^- coming from the added NaOH. Thus we generate the last half of Table 15.4 in the same manner as we did Table 15.3 in the HCl/NaOH titration.

A graph of these data is shown in Figure 15.3, where we have plotted pH versus volume of base added. From both Table 15.4 and Figure 15.3, we can see that the change in pH near the equivalence point is not as drastic as in the case of the HCl/NaOH titration. This less rapid change near the equivalence point becomes even more pronounced for weaker acids such as HCN.

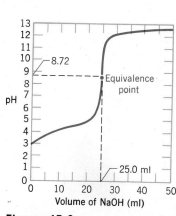

Figure 15.3

Titration of 25 ml of 0.10 M acetic acid with 0.10 M sodium hydroxide.

Table 15.4
Titration of 25.0 ml
of 0.10 M $HC_2H_3O_2$ with 0.10 M NaOH

Milliliters of Base Added	Molar Concentration of Species in Parentheses	pH
0.0	1.3×10^{-3} (H^+)	2.89
10.0	2.7×10^{-5} (H^+)	4.57
24.99	7.2×10^{-9} (H^+)	8.14
25.0	5.3×10^{-6} (OH^-)	8.72
25.01	2.0×10^{-5} (OH^-)	9.30
26.0	2.0×10^{-3} (OH^-)	11.30

Titration of a weak base with a strong acid

When a weak base is titrated with a strong acid, the titration curve that is generated is very similar in shape to that obtained by reaction of a weak acid with a strong base. During the initial addition of acid the solution contains unreacted weak base and its salt; it therefore constitutes a buffer. At the equivalence point the solution contains the salt of the weak base, and the pH of the mixture is determined by the hydrolysis of the cation. Finally, beyond the equivalence point the pH of the solution is controlled by the excess hydrogen ion from the strong acid. The shape of the titration curve for such a titration is shown in Figure 15.4 for the titration of 25.0 ml of 0.10 M NH_3 with 0.10 M HCl. We can show that the pH at the equivalence point is less than 7 by considering the hydrolysis of the NH_4Cl produced during the reaction.

From the last section we recall that the K_h for NH_4^+ is written as

$$K_h = \frac{[H_3O^+][NH_3]}{[NH_4^+]} = \frac{K_w}{K_b} = 5.6 \times 10^{-10}$$

All the NH_3 is "neutralized" in this titration when exactly 25.0 ml of the 0.10 M HCl (2.5×10^{-3} mol HCl) have been added. At this point, the concentration of NH_4^+ is

$$\frac{2.5 \times 10^{-3} \text{ mol}}{0.0500 \text{ liter}} = 5.0 \times 10^{-2} M$$

If we let x equal the number of moles per liter of NH_4^+ that undergo hydrolysis

$$[H_3O^+] = x$$

$$[NH_3] = x$$

$$[NH_4^+] = 5.0 \times 10^{-2} - x = 5.0 \times 10^{-2} M$$

Substituting these concentrations into the equation for K_h gives

$$K_h = \frac{(x)(x)}{5.0 \times 10^{-2}} = 5.6 \times 10^{-10}$$

$$x^2 = 28.0 \times 10^{-12}$$

$$x = 5.3 \times 10^{-6}$$

$$[H_3O^+] = 5.3 \times 10^{-6} M$$

and

$$pH = 5.28$$

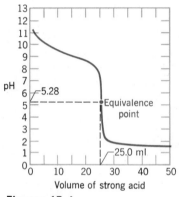

Figure 15.4

Titration of 25 ml of 0.10 M
NH_3 with 0.10 M HCl.

The pH at the equivalence point of this titration is less than 7, which is typical for all weak base-strong acid titrations.

15.7 ACID-BASE INDICATORS

Some indicators, like thymol blue, have two color changes over separate pH ranges.

Indicators are often used, in very small amounts, to detect the equivalence point in an acid-base titration. They are usually weak organic acids or bases that change color on going from an acidic medium to a basic medium. Not all indicators change color at the same pH, however. The choice of indicator for a particular titration depends on the pH at which the equivalence point is expected to occur. A list of some common indicators, with their color changes and the pH ranges over which the color changes are observed, is found in Table 15.5. Let us examine briefly how these indicators work.

If we denote an indicator by the general formula HIn, we have the dissociation reaction

$$HIn \rightleftharpoons H^+ + In^-$$

Applying Le Châtelier's principle to this equilibrium, we see that in an acid solution (excess H^+) the predominant species is HIn. On the other hand, in basic solutions the equilibrium is shifted to the right and the predominant species is In^-. Therefore, HIn is said to be the "acid form" and In^- the "basic form" of the indicator. The ability of HIn to function as an indicator is based on the difference in color between acid and basic forms. For example, with litmus the acid form (HIn) is pink while the basic form (In^-) is blue.

The dissociation constant, K_a, for an indicator is

$$K_a = \frac{[H^+][In^-]}{[HIn]}$$

Let's solve this for the ratio $[In^-]/[HIn]$.

$$\frac{[In^-]}{[HIn]} = \frac{K_a}{[H^+]}$$

Table 15.5
Some common indicators

Indicator	Color Change	pH Range in Which Color Change Occurs
Thymol blue	Red to yellow	1.2–2.8
Bromophenol blue	Yellow to blue	3.0–4.6
Congo red	Blue to red	3.0–5.0
Methyl orange	Red to yellow	3.2–4.4
Bromocresol green	Yellow to blue	3.8–5.4
Methyl red	Red to yellow	4.8–6.0
Bromocresol purple	Yellow to purple	5.2–6.8
Bromothymol blue	Yellow to blue	6.0–7.6
Cresol red	Yellow to red	7.0–8.8
Thymol blue	Yellow to blue	8.0–9.6
Phenolphthalein	Colorless to pink	8.2–10.0
Alizarin yellow	Yellow to red	10.1–12.0

We have seen that as we pass through the equivalence point, the pH changes very rapidly. For example, in the NaOH/HCl titration described earlier—Table 15.3 and Figure 15.2—the pH changed from 4.7 to 9.3 with the addition of only 0.02 ml of base, which is only about one-half drop of solution! This pH change corresponds to a change in $[H^+]$ from 2×10^{-5} M to 5×10^{-10} M. How does this affect the $[In^-]/[HIn]$ ratio?

Suppose that we were using an indicator whose $K_a = 1 \times 10^{-7}$. Then, before the equivalence point,

$$\frac{[In^-]}{[HIn]} = \frac{1 \times 10^{-7}}{2 \times 10^{-5}} = \frac{1}{200}$$

Visually, we observe the color of the species that is present in largest amount.

This tells us that there is 200 times as much HIn as In^-, and the color observed is that resulting from HIn.

After the equivalence point,

$$\frac{[In^-]}{[HIn]} = \frac{1 \times 10^{-7}}{5 \times 10^{-10}} = \frac{200}{1}$$

Now there is 200 times as much In^- as HIn, and the color that we see is due to In^-. Thus, as we pass through the equivalence point, there is a sudden change in the relative amounts of the acid and basic forms of the indicator, which we notice as a change in color.

If the indicator changes color at the equivalence point, the end point in the titration—that point at which we observe the color change—occurs at the same pH as the equivalence point. Often, however, we find ourselves using an indicator whose color change takes place at a pH slightly different from that of the equivalence point. This is shown in Figure 15.5 for phenolphthalein. When the color change occurs we have actually gone slightly past the equivalence point.

In choosing an indicator, we wish to have it change color very close to the equivalence point. Phenolphthalein, for example, would be a poor choice of indicator for the titration depicted in Figure 15.4, because its color change would occur long before the equivalence point. We would find that we had stopped adding acid before the equivalence point had been reached, thereby defeating the purpose of using an indicator. A better choice would be an indicator such as methyl red, where the center of the color change range occurs very near the pH at the equivalence point.

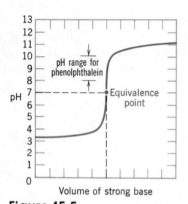

Figure 15.5

Titration curve for the titration of a strong acid with a strong base.

INDEX TO QUESTIONS AND PROBLEMS (Problem numbers are in **bold type**)

REVIEW QUESTIONS

15.1 Why can we almost always ignore the H^+ contributed by the dissociation of water when we calculate the H^+ concentration in solutions containing an acid? Under what conditions would we have to consider the H^+ from the dissociation of H_2O, even though the solute was an acid?

15.2 How is pH defined? How is pOH defined? Why does pH + pOH = 14?

15.3 Identify the following as representing acidic, basic, or neutral solutions:

(a) pH = 3.54   (b) pH = 8.25

(c) pOH = 7.00 (d) pOH = 10.43
(e) pOH = 2.25

15.4 Arrange the solutions in Question 15.3 in order of increasing acidity.

15.5 Refer to Table 15.1 and write the appropriate equilibrium constant expressions for the ionization of:

(a) benzoic acid (d) Veronal
(b) hydrazine (e) pyridine
(c) formic acid

15.6 Write appropriate mass action expressions for K_{a_1} and K_{a_2} for ascorbic acid (vitamin C).

15.7 Citric acid, which is present in many fruits and vegetables, has the formula, $H_3C_6H_5O_7$. It is a triprotic acid. Write the three equilibria for the dissociation of the acid and write the appropriate equilibrium constant expression for each step.

15.8 What is a buffer? Explain how the following solutes function as buffers:

(a) $NaCHO_2$ and $HCHO_2$
(b) C_5H_5N and C_5H_5NHCl
(c) $NH_4C_2H_3O_2$
(d) $NaHCO_3$

15.9 Would a solution containing a mixture of NaCl and HCl be an effective buffer? Explain.

15.10 Blood contains, among others, the buffer system $HPO_4^{2-}/H_2PO_4^-$. Explain how this pair of ions can serve as a buffer.

15.11 What is hydrolysis? Without performing any computations, predict whether the following solutions will be acidic, basic, or neutral:

(a) KCl (c) $NaC_4H_7O_2$
(b) NH_4NO_3 (d) $C_6H_5NH_3NO_3$

15.12 If the concentration of each solute in Question 15.11 were 0.10 M, which solution would be most basic? Which would be most acidic?

15.13 Why is it necessary to consider only the first step in the hydrolysis of a salt such as Na_2SO_3?

15.14 Is it possible to have a pH other than 7 at the equivalence point in an acid-base titration?

15.15 In a titration, the equivalence point and the end point are often not exactly the same. Justify this statement.

15.16 Explain how an indicator works. Why do we want to use as little of the indicator as possible when we perform a titration?

15.17 What indicators might be acceptable for the titration depicted in Figure 15.3? Why would we not wish to use congo red as an indicator?

15.18 Would congo red be an acceptable indicator for the titration depicted in Figure 15.5? Explain your answer.

REVIEW PROBLEMS (More difficult problems are marked by an asterisk)

15.19 Calculate the H^+ and OH^- concentrations and the pH of the following solutions of strong acids and bases:
(a) 0.0010 M HCl
(b) 0.125 M HNO_3
(c) 0.0031 M NaOH
(d) 0.012 M $Ba(OH)_2$
(e) 2.1×10^{-4} M $HClO_4$
(f) 1.3×10^{-5} M HCl
(g) 8.4×10^{-3} M NaOH
(h) 4.8×10^{-2} M KOH

15.20 Calculate the H^+ and OH^- concentrations in a solution having a pH equal to:
(a) 1.30 (d) 7.80
(b) 5.73 (e) 10.94
(c) 4.00 (f) 12.61

15.21 What is the pOH of each solution in Problem 15.20?

*__15.22__ A dilute solution of hydrochloric acid was labeled 1.0×10^{-8} M HCl. Is this solution acidic or basic? What is its pH?

15.23 A weak acid has an equilibrium constant $K_a = 3.8 \times 10^{-9}$. What is the pK_a of the acid?

15.24 A base has $pK_b = 3.84$. What is K_b of the base?

15.25 What is the H^+ concentration in each of the following solutions:
(a) 0.30 M HNO_2

(b) 1.00 M HF
(c) 0.025 M HCN
(d) 0.10 M butyric acid
(e) 0.050 M barbituric acid

15.26 Calculate the OH^- concentration in the following solutions:
(a) 0.15 M NH_3
(b) 0.20 M N_2H_4
(c) 0.80 M CH_3NH_2
(d) 0.35 M hydroxylamine
(e) 0.010 M pyridine

15.27 What is the OH^- concentration in each solution in Problem 15.25?

15.28 What is the pH of each solution in Problem 15.26?

15.29 A 0.25 M solution of a monoprotic weak acid was observed to have a pH = 1.35. What is K_a for this acid?

15.30 A 0.10 M solution of a weak monoprotic acid was found to have a pH = 5.37. What is K_a for the acid?

15.31 A weak base was found to give a solution with pH = 8.75 when its concentration was 0.10 M. What is K_b for the base?

15.32 What is the percent ionization of the acid in each of the following six solutions:
(a) 1.0 M formic acid
(b) 0.010 M propionic acid

(c) 0.025 M HCN
(d) 0.35 M nicotinic acid
(e) 0.50 M HOCl
(f) 0.25 M HNO$_3$

15.33 Calculate the percent ionization of each of the following acetic acid solutions. What conclusions can you draw? Can you explain on a molecular level why you obtain these results? Can you explain this using Le Châtelier's principle?

(a) 1.00 M HC$_2$H$_3$O$_2$
(b) 0.10 M HC$_2$H$_3$O$_2$
(c) 0.010 M HC$_2$H$_3$O$_2$

15.34 Calculate the hydrogen ion concentration in mol/liter for each of the following solutions of an acid or base and its salt.

(a) 0.25 M HC$_2$H$_3$O$_2$, 0.15 M NaC$_2$H$_3$O$_2$
(b) 0.50 M HCHO$_2$, 0.50 M NaCHO$_2$
(c) 0.30 M HNO$_2$, 0.40 M NaNO$_2$
(d) 0.25 M NH$_3$, 0.15 M NH$_4$Cl
(e) 0.30 M N$_2$H$_4$, 0.50 M N$_2$H$_5$NO$_3$

15.35 The pH of a 0.012 M solution of a weak base, BOH, was experimentally determined to be 11.40. Calculate K_b for the base.

15.36 How many grams of HCl gas would have to be dissolved in 500 ml of 1.0 M NaC$_2$H$_3$O$_2$ to give a solution having a pH = 4.74?

15.37 Calculate the pH obtained by dissolving a 500-mg tablet of vitamin C in 250 ml (approx. 8 oz) of H$_2$O.

15.38 In the stomach the fluids have a pH $\approx$ 1.0 due to the strong acid, HCl. What fraction of the vitamin C in a 500-mg tablet will be dissociated if the volume of fluid in the stomach is 200 ml?

15.39 If 10 mg of sodium barbituate is swallowed, what fraction is converted to barbituric acid if the pH of the stomach is 1.0 and there are 250 ml of fluid in the stomach?

15.40 Nicotinic acid is another name for the important vitamin, niacin. What is the pH of a 0.010 M solution of nicotinic acid?

15.41 What is the molarity of a solution of acetic acid whose pH is 2.5?

15.42 What molar concentration of hydrazine, N$_2$H$_4$, yields a solution whose pH = 10.64?

15.43 A 0.010 M solution of a weak acid, HA, is found to have a pH of 4.55. What is the value of K_a for this acid?

15.44 Calculate the molar concentrations of all of the species in a 0.050 M solution of the diprotic acid, vitamin C.

15.45 Calculate the molar concentrations of all species present in a 1.0 M H$_3$PO$_4$ solution.

15.46 Selenious acid, H$_2$SeO$_3$, has $K_{a_1} = 3 \times 10^{-3}$ and $K_{a_2} = 5 \times 10^{-8}$. What is the pH of a 0.50 M solution of H$_2$SeO$_3$? What are the equilibrium molar concentrations of H$_2$SeO$_3$, HSeO$_3^-$, and SeO$_3^{2-}$?

15.47 What is the HCO$_3^-$ concentration (in mol/liter) in a 0.10 M solution of H$_2$CO$_3$ whose pH = 3.00? What is the CO$_3^{2-}$ concentration in this solution?

15.48 Suppose that we wish the sulfide ion concentration to be 8.4×10^{-15} M in a saturated (0.10 M) solution of H$_2$S. What hydrogen ion concentration must be maintained by a buffer to give this S^{2-} concentration?

15.49 What is the sulfide ion concentration in a saturated H$_2$S solution (0.10 M H$_2$S) whose pH has been adjusted to a value of 4.60 by the addition of a buffer?

15.50 What ratio of lactic acid to sodium lactate is required to give a solution having a pH = 4.25?

15.51 Calculate the pH of each of the following buffers prepared by placing, in 1.0 liter of solution,
(a) 0.10 mol of NH$_3$ and 0.10 mol of NH$_4$Cl
(b) 0.20 mol of HC$_2$H$_3$O$_2$ and 0.40 mol of NaC$_2$H$_3$O$_2$
(c) 0.15 mol of N$_2$H$_4$ and 0.10 mol of N$_2$H$_5$Cl
(d) 0.20 mol of HCl and 0.30 mol of NaCl

15.52 How many grams of NaC$_2$H$_3$O$_2$ must be added to 1.00 mol of HC$_2$H$_3$O$_2$ in order to prepare 1.00 liter of a buffer whose pH equals 5.15?

15.53 What must the ratio of NH$_3$ to NH$_4^+$ be to have a buffer with a pH of 10.0?

*15.54 How many moles of HCl must be added to 1.0 liter of a mixture containing 0.010 M HC$_2$H$_3$O$_2$ and 0.010 M NaC$_2$H$_3$O$_2$ in order to give a solution whose pH = 3.0?

15.55 Calculate the pH change produced by adding 0.10 mol of solid NaOH to each of the following buffers.
(a) 500 ml of 1.00 M HC$_2$H$_3$O$_2$ and 1.00 M NaC$_2$H$_3$O$_2$
(b) 500 ml of 0.50 M HC$_2$H$_3$O$_2$ and 0.50 M NaC$_2$H$_3$O$_2$
(c) 500 ml of 0.30 M HC$_2$H$_3$O$_2$ and 0.70 M NaC$_2$H$_3$O$_2$
(d) 500 ml of 0.20 M HC$_2$H$_3$O$_2$ and 0.80 M NaC$_2$H$_3$O$_2$
(e) 500 ml of 0.10 M HC$_2$H$_3$O$_2$ and 0.90 M NaC$_2$H$_3$O$_2$

15.56 How much would the pH change if 0.10 mol of HCl were added to 1.0 liter of a formic acid-sodium formate buffer containing 0.45 mol of HCHO$_2$ and 0.55 mol of NaCHO$_2$?

15.57 How much would the pH change if 0.20 mol of NaOH were added to the original buffer in Question 15.56?

15.58 Determine the pH of each of the following salt solutions:
(a) 1.0×10^{-3} M NaC$_2$H$_3$O$_2$
(b) 0.125 M NH$_4$Cl
(c) 0.10 M Na$_2$CO$_3$
(d) 0.10 M NaCN
(e) 0.20 M NH$_3$OHCl

15.59 What is the percent hydrolysis of a 0.10 M solution of pyridinium chloride, C$_5$H$_5$NHCl? For pyridine, C$_5$H$_5$N, the value of $K_b = 1.7 \times 10^{-9}$.

15.60 A 0.10 M solution of the sodium salt of a weak monoprotic acid has a pH of 9.35. What is the K_a of the weak acid?

15.61 Liquid chlorine bleach is really nothing more than a dilute solution of NaOCl, usually about 5% NaOCl by weight. A particular sample of bleach was found to contain 0.67 mol/liter NaOCl. Calculate the pH of the solution.

15.62 Veronal, a barbiturate drug, is generally administered as its sodium salt. What is the pH of a solution of $NaC_8H_{11}N_2O_3$ that contains 10 mg of the drug in 250 ml of solution? For veronal, $HC_8H_{11}N_2O_3$, the value of $K_a = 3.7 \times 10^{-8}$.

15.63 What would be the concentration of barbituric acid in a 0.0010 M solution of sodium barbiturate?

15.64 What is the pH of a 0.20 M solution of sodium ascorbate?

15.65 What would be the pH of a 0.50 M solution of Na_3PO_4?

15.66 What would be the pH of a 0.0010 M solution of potassium cyanide (a deadly poison)?

15.67 A 15.0-ml portion of a solution of 0.0200 M HNO_3 is titrated with 0.0100 M KOH.
(a) What will be the pH at the equivalence point?
(b) How many milliliters of the base will be required to reach the equivalence point?
(c) What will be the pH when 10.0 ml of the KOH solution has been added?
(d) What will be the pH when 35.0 ml of the KOH solution has been added?

15.68 What would be the pH at the equivalence point if 25.0 ml of 0.010 M barbituric acid is titrated with 0.020 M NaOH?

15.69 When 50.0 ml of 0.200 M HF is titrated with 0.100 M NaOH, what is the pH
(a) After 5.0 ml of base has been added?
(b) When half of the HF has been neutralized?
(c) At the equivalence point?

15.70 Sodium benzoate is often used as a preservative in packaged food products. What would be the pH of a 0.020 M solution of sodium benzoate?

15.71 Plot a curve showing the pH of a solution of 100 ml of 0.10 M butyric acid that is gradually neutralized by the addition of solid NaOH. Do this by calculating the pH after addition of 0.0, 0.0010, 0.0050, 0.0090, 0.010, and 0.011 mol of NaOH. Assume no change in volume. What is the pH at the equivalence point? What indicator in Table 15.5 could be used for this titration?

15.72 Determine the shape of the titration curve for the titration of 50.0 ml of 0.10 M HCl with 0.10 M NaOH.

15.73 Using the data in Tables 15.1 and 15.5, choose an indicator that is suitable for the titration of:

(a) Hydrocyanic acid with sodium hydroxide
(b) Aniline with hydrochloric acid

15.74 An indicator, HIn, has an ionization constant, K_a, equal to 1×10^{-5}. If the molecular form of the indicator is yellow and the In^- ion is green, what is the color of a solution containing this indicator when its pH is 7.0?

15.75 Calculate the H^+ concentration in 0.0010 M $HC_2H_3O_2$.

15.76 Calculate the molar concentrations of all species in 0.010 M formic acid solution.

15.77 Calculate the pH of 0.50 M $NaHCO_3$. How much will the pH change if 0.05 mol/liter of HCl is added?

15.78 How many milliliters of 6.0 M HCl are required to be added to 100 ml of 0.10 M $NaC_2H_3O_2$ to give a solution having a pH = 4.25?

15.79 A sample of arterial blood was found to contain 2.6×10^{-2} mol of dissolved CO_2 per liter. The pH of the sample was 7.43. If it is assumed that in solution the CO_2 forms H_2CO_3, what is the HCO_3^- concentration in this blood sample?

15.80 Calculate the pH of 0.10 M NH_4NO_2.

15.81 Determine the shape of the titration curve when 100 ml of 0.20 M H_2CO_3 is titrated with 0.10 M NaOH. Determine the pH at each equivalence point.

16

SOLUBILITY AND COMPLEX ION EQUILIBRIA

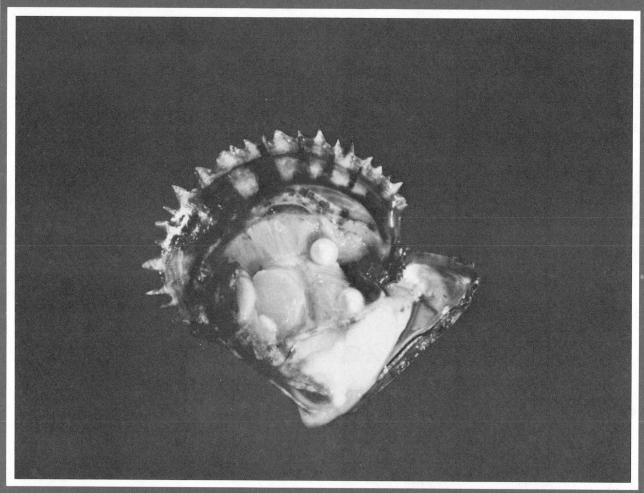

Oysters extract Ca^{2+} and CO_3^{2-} from sea water and deposit them on irritating grains of sand to form beautiful pearls that are composed mostly of insoluble $CaCO_3$. In this chapter we will study, quantitatively, the solubility of salts such as calcium carbonate and learn how it is possible to predict the conditions that are necessary for a precipitate to form from a solution.

In Chapter 15 we studied ionic equilibria involving acids and bases. These are not, however, the only dynamic equilibria that can take place in aqueous solutions. In this chapter we will turn our attention to the equilibria involved with salts that have very low solubilities—those we considered to be insoluble in water in our discussions in Chapter 6. We will also look at equilibria involving species that we call "complex ions"—ions composed of a metal atom surrounded by a number of anions, or neutral molecules, to which the metal is bound.

16.1 SOLUBILITY PRODUCT

In Chapter 6 you were presented with a list of solubility rules that described certain salts as soluble and others as quite insoluble. Even the most insoluble salts, however, dissolve in water to at least some degree, and their saturated solutions constitute dynamic equilibria that can be studied by the same principles that we applied to acid-base equilibria in the last chapter.

Nearly all salts are completely dissociated in water. Exceptions such as $HgCl_2$ and $CdSO_4$ are rare. For simplicity, therefore, our discussions will not include them, and we will assume that in a saturated solution an equilibrium exists between the solid salt and its dissolved ions. For example, in a saturated solution of silver chloride we have the equilibrium

$$AgCl\ (s) \rightleftharpoons Ag^+\ (aq) + Cl^-\ (aq)$$

It is safe to assume that there is no molecular AgCl in the solution.

for which we can write

$$K = \frac{[Ag^+][Cl^-]}{[AgCl\ (s)]}$$

In Section 13.5 we saw that the *concentration* of a pure solid is independent of the amount of solid present. In other words, the concentration of the solid is a constant and can therefore be included with the constant K, so that

$$K[AgCl\ (s)] = K_{sp} = [Ag^+][Cl^-]$$

The equilibrium constant K multiplied by the concentration of solid AgCl is still another constant called the **solubility product constant,** given the label K_{sp}. For example, we can obtain the expression for the K_{sp} of silver acetate from its solubility equilibrium,

$$AgC_2H_3O_2\ (s) \rightleftharpoons Ag^+\ (aq) + C_2H_3O_2{}^-\ (aq)$$

The equilibrium condition is therefore,

$$K_{sp} = [Ag^+][C_2H_3O_2{}^-]$$

In the case of an insoluble solid such as $Mg(OH)_2$, the coefficients in the dissociation equilibrium are not all equal to one:

$$Mg(OH)_2\ (s) \rightleftharpoons Mg^{2+}\ (aq) + 2OH^-\ (aq)$$

The K_{sp} for $Mg(OH)_2$ is then given by

$$K_{sp} = [Mg^{2+}][OH^-]^2$$

Thus the solubility product constant is equal to the product of the concentrations of the ions produced in a saturated solution, each raised to a power equal to its coefficient in the balanced equation. A list of some ionic solids and their K_{sp}'s at temperatures ranging between 18 and 25°C is given in Table 16.1.

A suspension of magnesium hydroxide in water.

Table 16.1
Solubility product constants

Compound	K_{sp}	Compound	K_{sp}
$Al(OH)_3$	2×10^{-33}	PbS	7×10^{-27}
$BaCO_3$	8.1×10^{-9}	$Mg(OH)_2$	1.2×10^{-11}
$BaCrO_4$	2.4×10^{-10}	MgC_2O_4	8.6×10^{-5}
BaF_2	1.7×10^{-6}	$Mn(OH)_2$	4.5×10^{-14}
$BaSO_4$	1.5×10^{-9}	MnS	7×10^{-16}
CdS	3.6×10^{-29}	Hg_2Cl_2	2×10^{-18}
$CaCO_3$	9×10^{-9}	HgS	1.6×10^{-54}
CaF_2	1.7×10^{-10}	NiS	2×10^{-21}
$CaSO_4$	2×10^{-4}	$AgC_2H_3O_2$	2.3×10^{-3}
CoS	7×10^{-23}	Ag_2CO_3	8.2×10^{-12}
CuS	8.5×10^{-36}	AgCl	1.7×10^{-10}
Cu_2S	2×10^{-47}	AgBr	5×10^{-13}
$Fe(OH)_2$	2×10^{-15}	AgI	8.5×10^{-17}
$Fe(OH)_3$	1.1×10^{-36}	Ag_2CrO_4	1.9×10^{-12}
FeC_2O_4	2.1×10^{-7}	AgCN	1.6×10^{-14}
FeS	3.7×10^{-19}	Ag_2S	2×10^{-49}
$PbCl_2$	1.6×10^{-5}	$Sn(OH)_2$	5×10^{-26}
$PbCrO_4$	1.8×10^{-14}	SnS	1×10^{-26}
PbC_2O_4	2.7×10^{-11}	$Zn(OH)_2$	4.5×10^{-17}
$PbSO_4$	2×10^{-8}	ZnS	1.2×10^{-23}

Calculations involving K_{sp} can be divided into three categories:

1. Calculating K_{sp} from solubility data
2. Calculating solubility from K_{sp}
3. Problems dealing with precipitation

We will begin (as you might expect) with the first type.

EXAMPLE 16.1

It was experimentally determined that at 25°C the solubility of $BaSO_4$ in water is 0.0091 g/liter. What is the value of K_{sp} for barium sulfate?

SOLUTION

From the solubility we can calculate the number of moles of $BaSO_4$ that are dissolved in 1 liter of solution.

By solubility we mean the amount of salt needed to give a saturated solution.

$$\left(0.0091 \, \frac{g}{liter}\right) \times \left(\frac{1 \, mol}{233 \, g}\right) = 3.9 \times 10^{-5} \, \frac{mol}{liter}$$

The solubility equilibrium for $BaSO_4$ is

$$BaSO_4 \, (s) \rightleftharpoons Ba^{2+} \, (aq) + SO_4^{2-} \, (aq)$$

so that for every mole of $BaSO_4$ that dissolves, 1 mol of Ba^{2+} and 1 mol of SO_4^{2-} are produced. Therefore the molar concentrations of Ba^{2+} and SO_4^{2-} in this saturated solution at 25°C are

The solubility in moles/liter is called the **molar solubility.**

$$[Ba^{2+}] = 3.9 \times 10^{-5} \, M$$

$$[SO_4^{2-}] = 3.9 \times 10^{-5} \, M$$

and the K_{sp} would be

$$K_{sp} = [Ba^{2+}][SO_4^{2-}]$$
$$= (3.9 \times 10^{-5})(3.9 \times 10^{-5})$$
$$= 1.5 \times 10^{-9}$$

EXAMPLE 16.2   The solubility of lead iodate, $Pb(IO_3)_2$, is 4.0×10^{-5} mol per liter at 25°C. What is the value of K_{sp} for this salt?

SOLUTION   First we write the chemical equation and the K_{sp} expression.

$$Pb(IO_3)_2 \ (s) \rightleftharpoons Pb^{2+} \ (aq) + 2IO_3^- \ (aq)$$

$$K_{sp} = [Pb^{2+}][IO_3^-]^2$$

When the $Pb(IO_3)_2$ dissolves, we get 1 mol of Pb^{2+} and 2 mol of IO_3^- for each mole of $Pb(IO_3)_2$. Therefore, when 4.0×10^{-5} mol of $Pb(IO_3)_2$ is dissolved in 1 liter, we obtain

$$[Pb^{2+}] = 4.0 \times 10^{-5} \ M$$

$$[IO_3^-] = 2(4.0 \times 10^{-5}) = 8.0 \times 10^{-5} \ M$$

These quantities are now substituted into the K_{sp} expression.

$$K_{sp} = (4.0 \times 10^{-5})(8.0 \times 10^{-5})^2$$
$$= 2.6 \times 10^{-13}$$

Let us now look at how we can determine solubility from a known value of K_{sp}—the second type of problem on our list.

EXAMPLE 16.3   What is the molar solubility of AgCl in water at 25°C?

SOLUTION   We are concerned with the equilibrium

$$AgCl \ (s) \rightleftharpoons Ag^+ \ (aq) + Cl^- \ (aq)$$

for which

$$K_{sp} = [Ag^+][Cl^-] = 1.7 \times 10^{-10}$$

In working problems of this type we will again set up a concentration table similar to those we've used in previous equilibrium calculations. In this particular example, the AgCl is dissolving into water that contains neither Ag^+ nor Cl^-, so the entries in the first column are both zero. Next, let's allow x to be the molar solubility—the number of moles of AgCl that dissolves per liter. Since one Ag^+ and one Cl^- are produced for each AgCl that dissolves, the concentrations of Ag^+ and Cl^- will each increase by x; both entries in the change column are therefore $+x$. Finally, the values in the "equilibrium concentrations" column are obtained by adding the change to the initial concentration—a rather trivial operation in this particular instance.

	Initial Molar Concentrations	Change	Equilibrium Molar Concentrations
Ag^+	0.0	$+x$	x
Cl^-	0.0	$+x$	x

Substituting the equilibrium quantities and the value of K_{sp} from Table 16.1 into the

K_{sp} expression,

$$K_{sp} = (x)(x) = 1.7 \times 10^{-10}$$

$$x^2 = 1.7 \times 10^{-10}$$

$$x = 1.3 \times 10^{-5}$$

Therefore, the molar solubility of AgCl in water is 1.3×10^{-5} M.

EXAMPLE 16.4

What are the concentrations of Ag^+ and CrO_4^{2-} in a saturated solution of Ag_2CrO_4 at 25°C?

SOLUTION

Ag_2CrO_4 dissolves in water according to the equilibrium

$$Ag_2CrO_4 \text{ (s)} \rightleftharpoons 2Ag^+ \text{ (aq)} + CrO_4^{2-} \text{ (aq)}$$

and the K_{sn} expression is

$$K_{sp} = [Ag^+]^2[CrO_4^{2-}]$$

Once again, neither of the ions involved in the equilibrium is present in the solution before the salt is added, so the initial concentrations are zero. Next, we let x be the molar solubility. Since two Ag^+ and one CrO_4^{2-} are produced for each Ag_2CrO_4 that dissolves, then when x mol of Ag_2CrO_4 dissolves per liter, the Ag^+ concentration increases by $2x$ and the CrO_4^{2-} concentration increases by x. Finally, we add the change to the initial concentration to obtain the equilibrium concentration in the last column.

> Taking **x** to be the molar solubility, the coefficients of **x** in the change column are the coefficients of the ions in the chemical equation for the equilibrium.

	Initial Molar Concentrations	Change	Equilibrium Molar Concentrations
Ag^+	0.0	$+2x$	$2x$
CrO_4^{2-}	0.0	$+x$	x

Substituting the value of K_{sp} from Table 16.1 and solving for x, we have

$$K_{sp} = (2x)^2(x) = 1.9 \times 10^{-12}$$

> $(2x)^2 = 4x^2$

$$(4x^2)x = 4x^3 = 1.9 \times 10^{-12}$$

$$x^3 = 0.48 \times 10^{-12}$$

and

$$x = 7.8 \times 10^{-5}$$

Therefore,

$$[Ag^+] = 2(7.8 \times 10^{-5}) = 1.6 \times 10^{-4} \text{ M}$$

$$[CrO_4^{2-}] = 7.8 \times 10^{-5} \text{ M}$$

We now turn our attention to determining when a precipitate can form in a solution of two salts. You should recall from earlier discussions that a saturated solution is one in which the undissolved solute is in dynamic equilibrium with the solution. This is precisely the situation to which we apply K_{sp}. In other words, a saturated solution exists *only* when the **ion product**—*the product of the concentrations of the dissolved ions each raised to its proper power*—is exactly equal to K_{sp}. When the ion product is less than K_{sp}, the solution is unsaturated, because more salt would have to dissolve in order to raise the concentrations to the point where the ion product equals K_{sp}. On the other hand, when the ion product exceeds K_{sp}, a supersaturated solution exists because some of the salt would have to precipitate to lower the concentrations until the ion product is equal to K_{sp} once again.

In a solution, a precipitate will be formed only when the mixture is supersaturated. Therefore, we can use the value of the ion product in a solution to tell us whether or not precipitation will occur. In summary, we find that

$$\left.\begin{array}{l} \text{Ion product} < K_{sp} \\ \text{Ion product} = K_{sp} \end{array}\right\} \quad \text{No precipitate will form}$$
$$\text{Ion product} > K_{sp} \quad \text{Precipitation will occur}$$

EXAMPLE 16.5

Will a precipitate of $PbCl_2$ form in a solution having a $Pb(NO_3)_2$ concentration of 0.010 M and an HCl concentration of 0.010 M? For $PbCl_2$, $K_{sp} = 1.6 \times 10^{-5}$.

SOLUTION

For lead chloride, we would write the following equilibrium

$$PbCl_2 \ (s) \rightleftharpoons Pb^{2+} \ (aq) + 2Cl^- \ (aq)$$

so the ion product that we will use in our test for precipitation is $[Pb^{2+}][Cl^-]^2$. In a 0.010 M $Pb(NO_3)_2$ solution, $[Pb^{2+}] = 0.010$ M and in 0.010 M HCl, $[Cl^-] = 0.010$ M. Using these values, we now compute the ion product

We can ignore the H^+ and NO_3^- because they are simply spectator ions here.

$$[Pb^{2+}][Cl^-]^2 = (0.010)(0.010)^2 = 1.0 \times 10^{-6}$$

Since this is less than the value of K_{sp} (1.6 $\times 10^{-5}$), we conclude that no precipitate will form.

EXAMPLE 16.6

Will a precipitate of $PbSO_4$ form when exactly 100 ml of a 0.0030 M $Pb(NO_3)_2$ solution is mixed with exactly 400 ml of 0.040 M Na_2SO_4?

SOLUTION

For lead sulfate, the ion product that we must examine is

$$[Pb^{2+}][SO_4^{2-}]$$

but this time we have to take into account the fact that one solution dilutes the other when they are mixed. We approach this by imagining that the solutions can be combined before any reaction can take place, and then we look at the value of the ion product in the final mixture. The first step is to calculate the concentrations of Pb^{2+} and SO_4^{2-} in our total volume of 500 ml. The original 100 ml of the 0.0030 M $Pb(NO_3)_2$ solution contains

$$(0.100 \text{ liter}) \times \left(0.0030 \frac{\text{mol}}{\text{liter}}\right) = 0.00030 \text{ mol of } Pb^{2+}$$

(This solution also contains 0.0006 mol of NO_3^-, but this species is unimportant in this calculation—it is a spectator ion.)

The 400 ml of Na_2SO_4 contains

$$(0.400 \text{ liter}) \times \left(0.040 \frac{\text{mol}}{\text{liter}}\right) = 0.016 \text{ mol of } SO_4^{2-}$$

(This solution also contains 0.032 mol of Na^+, but it too is unimportant in this problem.)

The concentration of the Pb^{2+} in the 500 ml is then

$$\frac{0.00030 \text{ mol}}{0.500 \text{ liter}} = 0.00060 \ M = 6.0 \times 10^{-4} \ M$$

and the concentration of SO_4^{2-} in the 500 ml is

$$\frac{0.016 \text{ mol}}{0.500 \text{ liter}} = 0.032 \ M = 3.2 \times 10^{-2} \ M$$

The ion product in the final solution is therefore

$$[Pb^{2+}][SO_4^{2-}] = (6.0 \times 10^{-4})(3.2 \times 10^{-2}) = 1.9 \times 10^{-5}$$

When we compare the value of the ion product to the value of K_{sp} for PbSO$_4$ (from Table 16.1, $K_{sp} = 2 \times 10^{-8}$), we find that the ion product is greater than K_{sp} and therefore a precipitate will form.

Separation of ions by precipitation

From the solubility rules presented in Chapter 6 we know that it is possible to separate certain ions from each other when they are present together in solution. For instance, the addition of chloride ion to a solution containing both Na$^+$ and Ag$^+$ yields a precipitate of AgCl, thereby removing most of the Ag$^+$ from the mixture. In this case one possible product, NaCl, is soluble while the other, AgCl, is quite insoluble.

Even when both products are "insoluble," it is still frequently possible to achieve some degree of separation. Consider, for example, the salts CaSO$_4$ and BaSO$_4$. Although both have very low solubilities, as evidenced by their respective K_{sp}'s, we can compute that CaSO$_4$ is about 1000 times more soluble, on a mole basis, than BaSO$_4$. As a result, if we had a solution containing equal concentrations of Ca^{2+} and Ba^{2+}, we would find that as the SO$_4^{2-}$ concentration was increased in the solution, BaSO$_4$ would precipitate first. Conceivably, one could separate Ca^{2+} and Ba^{2+} by appropriately adjusting the SO$_4^{2-}$ concentration so that the Ca^{2+} would remain in solution while nearly all the Ba^{2+} would be removed as BaSO$_4$. This general concept is used often in the separation of ions in qualitative analysis.

When the anion employed in a separation is derived from a weak acid, it is possible to control its concentration by appropriately adjusting the hydrogen ion concentration. This is illustrated for the selective precipitation of metal sulfides in Example 16.7.

EXAMPLE 16.7   A solution containing 0.10 M Sn^{2+} and 0.10 M Zn^{2+} is saturated with H$_2$S ([H$_2$S] = 0.10 M). What range of hydrogen ion concentrations will permit the selective precipitation of one of these ions as its sulfide?

SOLUTION   From Table 16.1 we have

$$SnS \qquad K_{sp} = 1 \times 10^{-26}$$

$$ZnS \qquad K_{sp} = 1.2 \times 10^{-23}$$

In this problem the sulfide ion concentration must be controlled so that one ion will precipitate while the other remains in solution. Let us, therefore, calculate for each salt the value of [S^{2-}] that will make the ion product equal to K_{sp}. For tin we have

$$K_{sp} = [Sn^{2+}][S^{2-}] = 1 \times 10^{-26}$$

Substituting the Sn^{2+} concentration into the expression gives

$$(0.10)[S^{2-}] = 1 \times 10^{-26}$$

$$[S^{2-}] = 1 \times 10^{-25} \ M$$

In a similar fashion for zinc, we obtain

$$[S^{2-}] = 1.2 \times 10^{-22} \ M$$

These numbers tell us that if the sulfide ion concentration is *greater* than $1 \times 10^{-25} \ M$ but *less than or equal to* $1.2 \times 10^{-22} \ M$, only SnS will precipitate.

We saw in Equation 15.7 that the sulfide ion concentration is directly related to the hydrogen ion concentration; that is,

$$\frac{[H^+]^2[S^{2-}]}{[H_2S]} = K_{a_1}K_{a_2} = 1.1 \times 10^{-21}$$

We can use this expression to calculate the $[H^+]$ that gives us our desired $[S^{2-}]$.

For the lower limit, $[S^{2-}] = 1 \times 10^{-25}$ M. Since the solution is saturated with H_2S, we have $[H_2S] = 0.10$ M. Substituting gives

$$\frac{[H^+]^2(1 \times 10^{-25})}{(0.10)} = 1.1 \times 10^{-21}$$

$$[H^+]^2 = \frac{(1.1 \times 10^{-21})(0.10)}{1 \times 10^{-25}}$$

$$= 1.1 \times 10^3$$

$$[H^+] = 3.3 \times 10^1 = 33\ M$$

This calculation implies that in order to prevent SnS from precipitating, the H^+ concentration must be 33 M! This concentration is impossible to achieve; therefore, SnS *must* precipitate, no matter how acidic the solution is.

To prevent ZnS from forming, the S^{2-} concentration cannot be larger than 1.2×10^{-22} M. Using this value for $[S^{2-}]$ in Equation 15.7, we have

$$\frac{[H^+]^2(1.2 \times 10^{-22})}{0.10} = 1.1 \times 10^{-21}$$

$$[H^+]^2 = 0.92$$

$$[H^+] = 0.96\ M$$

Thus, when $[H^+] = 0.96$ M, the S^{2-} concentration will be 1.2×10^{-22} M, the highest value it can have without causing the Zn^{2+} to precipitate. A hydrogen ion concentration *greater* than 0.96 M will produce a sulfide ion concentration *less* than 1.2×10^{-22} M. In summary, to achieve a separation,

$$[H^+] < 33\ M$$

and

$$[H^+] \geq 0.96\ M$$

The preceding example illustrates how two metal ions whose sulfides differ rather widely in solubility can be separated from one another by precipitating only one of them. This separation is really quite complete. For example, at a hydrogen ion concentration of 1 M, which will prevent zinc sulfide from forming, the solubility of SnS in the saturated H_2S solution is only 9×10^{-5} M, which means that 99.91% of the Sn^{2+} originally in solution would be precipitated as SnS! Table 16.2 lists some metal ions that can be separated from one another in this manner. They are divided into groups according to the solubilities of their sulfides in an acidic solution. Those in the first column of the table are often referred to as metals having acid-insoluble sulfides. They can be separated from the metal ions in the second column, which precipitate as insoluble sulfides at higher pH and are referred to as metals with basic-insoluble sulfides.

Table 16.2
Metal ions that can be separated
according to the solubilities of their sulfides

Metal Ions with "Acid-Insoluble Sulfides"			Metal Ions with "Basic-Insoluble Sulfides"		
Metal Ion	Metal Sulfide	K_{sp}	Metal Ion	Metal Sulfide	K_{sp}
Cu^{2+}	CuS	8.5×10^{-36}	Zn^{2+}	ZnS	1.2×10^{-23}
Bi^{3+}	Bi_2S_3	2.9×10^{-70}	Co^{2+}	CoS	7.0×10^{-23}
Pb^{2+}	PbS	7×10^{-27}	Ni^{2+}	NiS	2×10^{-21}
Hg^{2+}	HgS	1.6×10^{-54}	Fe^{2+}	FeS	3.7×10^{-19}
Sn^{2+}	SnS	1×10^{-26}	Mn^{2+}	MnS	7×10^{-16}

16.2 COMMON ION EFFECT AND SOLUBILITY

When a salt is dissolved in a solution that already contains one of its ions, its solubility is less than in pure water. Silver chloride, for example, is less soluble in a solution of NaCl than it is in pure water. In this case both solutes have an ion in common: chloride ion. The reduction in the solubility in the presence of a **common ion** is called the **common ion effect.**

The effect of a common ion on solubility can easily be understood on the basis of Le Châtelier's principle. Suppose that solid silver chloride is placed into pure water and allowed to come to equilibrium with its ions in solution.

$$AgCl\ (s) \rightleftharpoons Ag^+\ (aq) + Cl^-\ (aq)$$

If a soluble chloride salt such as NaCl is now added to this solution, the chloride ion concentration will increase and drive this equilibrium to the left, thereby causing some AgCl to precipitate. In other words, AgCl is less soluble when NaCl is in the solution than when it is placed in pure water.[1] Let's look at a few examples.

EXAMPLE 16.8

SOLUTION

NaCl is soluble and completely dissociated. Don't write an equilibrium for it!

What is the molar solubility of AgCl in a 0.010 M solution of NaCl?

For this salt we have

$$K_{sp} = [Ag^+][Cl^-] = 1.7 \times 10^{-10}$$

Before any AgCl dissolves we have an initial Cl^- concentration of 0.010 M. We can ignore the Na^+ because it is not involved in the equilibrium. We now let x equal the number of moles per liter of AgCl that dissolve. This increases both $[Cl^-]$ and $[Ag^+]$ by x. Thus,

	Initial Molar Concentrations	Change	Equilibrium Molar Concentrations
Ag^+	0.0	$+x$	x
Cl^-	0.010	$+x$	$0.010 + x \approx 0.010$

---

[1] A similar effect was seen in the section on buffers in Chapter 15 when a common ion was added to the equilibrium dissociation of a weak acid.

Note that we have assumed that we can neglect x in computing the equilibrium Cl^- concentration. We make this assumption because the value of K_{sp} is very small. In doing so we greatly simplify the algebra. Substituting the equilibrium concentrations into the expression for K_{sp} gives

$$(x)(0.010) = 1.7 \times 10^{-10}$$

or

$$x = 1.7 \times 10^{-8} \ M$$

Thus, the molar solubility of AgCl is $1.7 \times 10^{-8} \ M$. We might compare this to the molar solubility of AgCl in pure water, which we found to be $1.3 \times 10^{-5} \ M$ in Example 16.3. The solubility of AgCl is indeed much less in a solution containing a common ion.

EXAMPLE 16.9

What is the molar solubility of $Mg(OH)_2$ in 0.10 M NaOH?

SOLUTION

The K_{sp} for $Mg(OH)_2$ is

$$K_{sp} = [Mg^{2+}][OH^-]^2$$

From this point on, such problems seem to mystify some students, but if you approach your construction of the concentration table systematically, you should have no difficulty. First, ask yourself what the initial concentrations are before any $Mg(OH)_2$ is added. Since the solution initially contains 0.10 M NaOH, $[Na^+] = 0.10 \ M$ and $[OH^-] = 0.10 \ M$. There is no magnesium ion, so $[Mg^{2+}] = 0.0 \ M$. We are only interested in the concentrations of OH^- and Mg^{2+}, so we enter their values in the first column.

Next, how do the concentrations change? If x mol/liter of $Mg(OH)_2$ dissolves, then $[Mg^{2+}]$ increases by x and $[OH^-]$ increases by $2x$. These are entered in the center column, and the equilibrium concentrations are obtained by adding the quantities in the first two columns (as usual).

	Initial Molar Concentrations	Change	Equilibrium Molar Concentrations
Mg^{2+}	0.0	$+x$	x
OH^-	0.10	$+2x$	$0.10 + 2x \approx 0.10$

Again, we simplify the algebra by assuming that $2x$ is negligible compared to 0.10. Substituting the equilibrium concentrations and the K_{sp} for $Mg(OH)_2$ from Table 16.1 into the solubility product expression gives

$$(x)(0.10)^2 = 1.2 \times 10^{-11}$$

or

$$x = \frac{1.2 \times 10^{-11}}{(0.10)^2}$$

$$= 1.2 \times 10^{-9} \ M$$

Thus 1.2×10^{-9} mol per liter of $Mg(OH)_2$ dissolves in a 0.10 M solution of NaOH.

EXAMPLE 16.10

What is the molar solubility of PbI_2 in 0.10 M $Pb(NO_3)_2$ solution? For lead iodide, $K_{sp} = 1.4 \times 10^{-8}$.

SOLUTION

The K_{sp} expression is

$$K_{sp} = [Pb^{2+}][I^-]^2$$

To construct the solubility table, we begin by asking, "What are the initial concentrations of Pb^{2+} and I^-?" The solution initially contains 0.10 M $Pb(NO_3)_2$, so the initial

Pb^{2+} concentration is 0.10 M. No iodide is present initially, so the initial I^- concentration is 0.0 M. These values go in the first column.

Next, we let x be the molar solubility of PbI_2. When x mol/liter of PbI_2 dissolves, $[Pb^{2+}]$ increases by x and $[I^-]$ increases by $2x$. These quantities are entered in the change column. Then the first and second columns are added to give the equilibrium concentrations.

	Initial Molar Concentrations	Change	Equilibrium Molar Concentrations
Pb^{2+}	0.10	$+x$	$0.10 + x \approx 0.10$
I^-	0.0	$+2x$	$2x$

After making our usual simplification, the equilibrium quantities are substituted into the K_{sp} expression.

$$K_{sp} = (0.10)(2x)^2 = 1.4 \times 10^{-8}$$

$$4x^2 = 1.4 \times 10^{-7}$$

$$x^2 = 3.5 \times 10^{-8}$$

$$x = 1.9 \times 10^{-4}$$

The molar solubility is 1.9×10^{-4} M.

16.3 COMPLEX IONS

Many metal ions, particularly those of the transition elements, are able to combine with one or more other molecules or ions to produce more complex species that are called **complex ions,** or simply **complexes.** The substances that combine with the metal ion are called **ligands** and are usually Lewis bases. They can be either (a) neutral molecules such as H_2O and NH_3, (b) monatomic anions such as Cl^- and Br^-, or (c) polyatomic anions such as CN^- and $C_2O_4^{2-}$. One example of a complex ion that we saw earlier is $Al(H_2O)_6^{3+}$. Another example, containing fewer ligands, is formed when NH_3 is added to a solution containing Ag^+. Its formula is $Ag(NH_3)_2^+$. The charge on a complex ion such as this is the algebraic sum of the charges of the metal ion and the ligands. Thus Ag^+ also forms a complex ion with CN^- having the formula $Ag(CN)_2^-$.

We will discuss the details of structure and bonding of complex ions in Chapter 21. For now we will focus our attention on their dissociation equilibria and the effect that their formation has on the solubility of salts.

There are two ways of dealing with the equilibria involving complex ions. One is to consider their dissociation equilibria. For example, the overall reaction for the equilibrium dissociation of $Ag(NH_3)_2^+$ can be written as

$$Ag(NH_3)_2^+ \ (aq) \rightleftharpoons Ag^+ \ (aq) + 2NH_3 \ (aq)$$

The equilibrium constant for this reaction is called an **instability constant.** This is because the larger the value of K_{inst}, the less stable the complex is as reflected by its tendency to dissociate. For the $Ag(NH_3)_2^+$ ion the equilibrium expression is

$$K_{inst} = \frac{[Ag^+][NH_3]^2}{[Ag(NH_3)_2^+]}$$

The value of the instability constant for this complex has been found to be 6.0×10^{-8}. We can see by the size of this constant that this particular complex is quite

Table 16.3
Instability constants
and formation constants at 25°C

Complex Ion	K_{inst}	K_{form}
AlF_6^{3-}	1.5×10^{-20}	6.7×10^{19}
$Cd(CN)_4^{2-}$	1.3×10^{-17}	7.7×10^{16}
$Co(NH_3)_6^{2+}$	1.3×10^{-5}	7.7×10^4
$Co(NH_3)_6^{3+}$	2.0×10^{-34}	5.0×10^{33}
$Cu(NH_3)_4^{2+}$	2.1×10^{-13}	4.8×10^{12}
$Cu(CN)_2^{-}$	1.0×10^{-16}	1.0×10^{16}
$Fe(CN)_6^{4-}$	1.0×10^{-35}	1.0×10^{35}
$Fe(CN)_6^{3-}$	1.1×10^{-42}	9.1×10^{41}
$Ni(NH_3)_4^{2+}$	1.1×10^{-8}	9.1×10^7
$Ni(NH_3)_6^{2+}$	2.0×10^{-9}	5.0×10^8
$Ag(NH_3)_2^{+}$	6.0×10^{-8}	1.7×10^7
$Ag(CN)_2^{-}$	1.9×10^{-19}	5.3×10^{18}
$Zn(OH)_4^{2-}$	3.6×10^{-16}	2.8×10^{15}

stable and will readily form whenever Ag^+ and NH_3 are added to the same solution. Other examples of complex ions and their instability constants can be seen in Table 16.3.

An alternative way of writing the equilibrium for a complex ion is as an equation representing its formation. For example,

$$Ag^+ \, (aq) + 2NH_3 \, (aq) \rightleftharpoons Ag(NH_3)_2^+ \, (aq)$$

The equilibrium expression, of course, is simply the reciprocal of the K_{inst} expression. In this case the equilibrium constant (which equals the reciprocal of K_{inst}) is called a **formation constant** or **stability constant**. These are also given in Table 16.3.

$$K_{form} = \frac{[Ag(NH_3)_2^+]}{[Ag^+][NH_3]^2}$$

$$K_{form} = \frac{1}{K_{inst}}$$

In the chemical literature the equilibrium constants for complex ions are sometimes tabulated as instability constants and at other times as formation constants or stability constants. You should know the difference between them.

16.4 COMPLEX IONS AND SOLUBILITY

When a complex ion is formed in a solution of an insoluble salt, it reduces the concentration of free metal ion. As a result, more solid must dissolve in order to replenish the amount of metal ion lost, until that concentration required by the K_{sp} of the salt is achieved. Thus the solubility of an insoluble salt generally increases when complex ions are formed. To see this more clearly, let us see what effect adding NH_3 has on a saturated solution of AgCl.

Before any NH_3 is added, we have the equilibrium

$$AgCl \, (s) \rightleftharpoons Ag^+ + Cl^-$$ [16.1]

Because NH_3 forms such a stable complex with the free silver ion, when NH_3 is added to this system a second equilibrium is established, namely,

$$Ag^+ + 2NH_3 \rightleftharpoons Ag(NH_3)_2{}^+ \qquad [16.2]$$

The creation of this new equilibrium upsets the first by removing some of the Ag^+, thereby causing the first equilibrium to shift to the right. As a result, some of the solid AgCl dissolves.

The two equilibria represented by Equations 16.1 and 16.2 can be combined into one overall equilibrium by adding them together.

$$\begin{array}{r} AgCl\ (s) \rightleftharpoons Ag^+ + Cl^- \\ Ag^+ + 2NH_3 \rightleftharpoons Ag(NH_3)_3{}^+ \\ \hline AgCl\ (s) + 2NH_3 \rightleftharpoons Ag(NH_3)_2{}^+ + Cl^- \end{array}$$

The equilibrium constant for this overall reaction is

$$K_c = \frac{[Ag(NH_3)_2{}^+][Cl^-]}{[NH_3]^2}$$

We can obtain this same expression by multiplying the K_{sp} of AgCl by the K_{form} of the complex ion.[2] Thus,

$$K_{sp} \times K_{form} = [\cancel{Ag^+}][Cl^-] \times \frac{[Ag(NH_3)_2{}^+]}{[\cancel{Ag^+}][NH_3]^2} = K_c$$

Therefore, with a knowledge of K_{sp} of the salt, K_{form} of the complex ion (or K_{inst}), and the concentration of NH_3, it is possible to calculate the concentrations of Ag^+ and Cl^- present at equilibrium and thus determine the solubility of AgCl in NH_3, as shown by the next example.

EXAMPLE 16.11 What is the molar solubility of AgCl in 1 liter of 1.0 M NH_3 at 25°C?

SOLUTION As we have seen, the overall equilibrium reaction for this problem is

$$AgCl\ (s) + 2NH_3\ (aq) \rightleftharpoons Ag(NH_3)_2{}^+ (aq) + Cl^- (aq)$$

for which we write

$$K_c = \frac{[Ag(NH_3)_2{}^+][Cl^-]}{[NH_3]^2}$$

where

$$K_c = K_{sp} \times K_{form} = (1.7 \times 10^{-10}) \times (1.7 \times 10^7) = 2.9 \times 10^{-3}$$

If we let x equal the number of moles per liter of AgCl that dissolves, then we have the following initial and equilibrium concentrations:

	Initial Molar Concentrations	Change	Equilibrium Molar Concentrations
NH_3	1.0	$-2x$	$(1.0 - 2x)$
$Ag(NH_3)_2{}^+$	0.0	$+x$	x
Cl^-	0.0	$+x$	x

---

[2] In general, if some equilibrium equation is obtained as the sum of two or more equations, the K_c for the final equation is equal to the *product* of the K's of the equations that were added together.

Substituting the concentrations at equilibrium into the K_c equation, we have

$$K_c = \frac{(x)(x)}{(1.0 - 2x)^2} = \frac{x^2}{(1.0 - 2x)^2} = 2.9 \times 10^{-3}$$

Taking the square root of both sides, we have

$$\frac{x}{1.0 - 2x} = 5.4 \times 10^{-2}$$

from which we obtain

$$x = 0.049$$

Therefore, we find that 0.049 mol of AgCl will dissolve in 1 liter of 1.0 M NH_3.

In Example 16.11 we assumed that when the AgCl dissolves in the ammonia solution, essentially all the Ag^+ becomes complexed by NH_3. In other words, we said that the chloride ion concentration was equal to the concentration of $Ag(NH_3)_2^+$. Note that this assumption is valid only if K_{form} is very large, indicating that the complex is very stable.

EXAMPLE 16.12

Zinc hydroxide is amphoteric.

How many moles of solid NaOH must be added to 1.0 liter of H_2O in order to dissolve 0.10 mol of $Zn(OH)_2$ according to the reaction,

$$Zn(OH)_2 + 2OH^- \rightleftharpoons Zn(OH)_4^{2-}$$

SOLUTION

The applicable equilibrium constants are

$$Zn(OH)_2 \qquad K_{sp} = 4.5 \times 10^{-17}$$

$$Zn(OH)_4^{2-} \qquad K_{inst} = 3.6 \times 10^{-16}$$

The two equilibria involved in this system are

$$Zn(OH)_2 \, (s) \rightleftharpoons Zn^{2+} + 2OH^- \qquad K_{sp} = 4.5 \times 10^{-17}$$

$$Zn^{2+} + 4OH^- \rightleftharpoons Zn(OH)_4^{2-} \qquad K_{form} = \frac{1}{3.6 \times 10^{-16}} = 2.8 \times 10^{15}$$

As before, the overall reaction can be written as the sum of these two equilibria,

$$Zn(OH)_2 \, (s) + 2OH^- \rightleftharpoons Zn(OH)_4^{2-}$$

for which

$$K_c = K_{sp} \times K_{form} = (4.5 \times 10^{-17})(2.8 \times 10^{15})$$

$$= 1.3 \times 10^{-1}$$

Therefore,

$$\frac{[Zn(OH)_4^{2-}]}{[OH^-]^2} = 1.3 \times 10^{-1}$$

In this problem we know that 0.10 mol of zinc goes into solution where it is present as either free Zn^{2+} or $Zn(OH)_4^{2-}$. Because K_{form} is so very large, essentially all of the zinc will be present as the complex ion; therefore we can write

$$[Zn(OH)_4^{2-}] = 0.10 \, M$$

Substituting this into the equilibrium expression gives

$$1.3 \times 10^{-1} = \frac{0.10}{[OH^-]^2}$$

Therefore,

$$[OH^-]^2 = \frac{0.10}{1.3 \times 10^{-1}} = 7.7 \times 10^{-1}$$

and

$$[OH^-] = 0.88 \ M$$

This corresponds to the equilibrium concentration of free OH^-. In this solution, however, we also have 0.10 mol of $Zn(OH)_4^{2-}$, which contains an additional 0.40 mol of OH^-, 0.20 mol of which was contained in the original 0.10 mol of $Zn(OH)_2$ that dissolved. Therefore, the total number of moles of NaOH that must be *added* to the water is $0.88 + 0.20 = 1.08$ mol.

INDEX TO QUESTIONS AND PROBLEMS (Problem numbers are in **bold type**)

REVIEW QUESTIONS

16.1 Why can the concentration of the solid be left out of the equilibrium expression for the solubility of a salt?

16.2 Write the K_{sp} expression for each of the following substances.
(a) Ag_2S
(b) CaF_2
(c) $Fe(OH)_3$
(d) MgC_2O_4
(e) Bi_2S_3
(f) $BaCO_3$

16.3 Write the K_{sp} expression for these salts:
(a) PbF_2
(b) Cu_2S
(c) $Fe_3(PO_4)_2$
(d) Li_2CO_3
(e) $Ca(IO_3)_2$
(f) $Ag_2Cr_2O_7$

16.4 What condition must be met to have a precipitate form in a solution?

16.5 What is a complex ion? What is a ligand? What kinds of substances are found as ligands?

16.6 On the basis of Le Châtelier's principle, explain why the addition of solid NH_4Cl to a beaker containing solid $Mg(OH)_2$ in contact with water causes the $Mg(OH)_2$ to dissolve.

16.7 Silver forms a relatively stable complex ion, AgI_2^-. When a solution containing this ion is diluted with water, AgI precipitates. Explain why this happens in terms of the equilibria that are involved.

16.8 What is the common ion effect?

16.9 Write equilibrium expressions corresponding to K_{form} for the complex ions:
(a) $AgCl_2^-$
(b) $Ag(S_2O_3)_2^{3-}$
(c) $Zn(NH_3)_4^{2+}$

16.10 Write equilibrium expressions corresponding to K_{inst} for the complex ions:
(a) $Fe(CN)_6^{4-}$
(b) $CuCl_4^{2-}$
(c) $Ni(NH_3)_6^{2+}$

REVIEW PROBLEMS (More difficult problems are marked by an asterisk)

16.11 The solubility of CuCl in water is 1.0×10^{-3} mol/liter. What is its value of K_{sp}?

16.12 The solubility of $PbCO_3$ is 1.8×10^{-7} mol/liter. What is K_{sp} for $PbCO_3$?

16.13 The solubility of barium oxalate, BaC_2O_4, is 0.0781 g/liter. Calculate K_{sp} for BaC_2O_4.

16.14 The solubility of $CaCrO_4$ is 1.0×10^{-2} mol/liter. What is K_{sp} for $CaCrO_4$?

16.15 The solubility of lead iodide, PbI_2, in water is 1.5×10^{-3} mol/liter. Calculate its K_{sp}.

16.16 A student determined that 0.0981 g of PbF_2 was dissolved in 200 ml of saturated PbF_2 solution. What is K_{sp} for PbF_2?

16.17 The solubility of MgF_2 is 7.6×10^{-2} g/liter. Calculate K_{sp} for this salt.

16.18 The solubility of Bi_2S_3 is 2.5×10^{-12} g/liter. What is K_{sp} for Bi_2S_3?

16.19 The pH of a saturated solution of $Ni(OH)_2$ is 8.83. Calculate K_{sp} for $Ni(OH)_2$.

*16.20 A 500-ml portion of $0.0020 M$ $Na_2C_2O_4$ (sodium oxalate) solution is able to dissolve 0.47 g of MgC_2O_4. What is K_{sp} for MgC_2O_4?

16.21 Using the data in Table 16.1, calculate the molar solubility in water of each of the following:

(a) PbS
(b) $Fe(OH)_2$
(c) $BaSO_4$
(d) Hg_2Cl_2 (which yields Hg_2^{2+} and $2Cl^-$)
(e) $Al(OH)_3$
(f) MgC_2O_4

16.22 Milk of magnesia is a suspension of solid $Mg(OH)_2$ in water. Calculate the pH of the aqueous phase, assuming that it is saturated with $Mg(OH)_2$.

16.23 How many grams of $CaSO_4$ will dissolve in 600 ml of water?

16.24 What volume of saturated HgS solution contains a single Hg^{2+} ion?

16.25 What is the molar solubility of $CaCO_3$ in $0.50 M$ Na_2CO_3?

16.26 What is the molar solubility of AgCl in $0.020 M$ $AlCl_3$? Assume that $AlCl_3$ gives Al^{3+} and Cl^- in solution.

*16.27 What is the molar solubility of $PbCl_2$ in $0.020 M$ $AlCl_3$? Assume that $AlCl_3$ gives Al^{3+} and Cl^- in solution.

16.28 How many moles of Ag_2CrO_4 will dissolve in 1.0 liter of $0.10 M$ $AgNO_3$?

16.29 How many moles of Ag_2CrO_4 will dissolve in 1.0 liter of $0.10 M$ Na_2CrO_4?

16.30 What is the molar solubility of CaF_2 in $0.010 M$ NaF?

16.31 How many grams of NaF must be added to 1.00 liter of solution to reduce the molar solubility of BaF_2 to 6.8×10^{-4} mol/liter?

16.32 Would a precipitate form in the following solutions?
(a) 5.0×10^{-2} mol of $AgNO_3$ and 1.0×10^{-3} mol of $NaC_2H_3O_2$ dissolved in 1.0 liter of solution
(b) 1.0×10^{-2} mol of $Ba(NO_3)_2$ and 2.0×10^{-2} mol of NaF dissolved in 1.0 liter of solution
(c) 500 ml of $1.4 \times 10^{-2} M$ $CaCl_2$ and 250 ml of $0.25 M$ Na_2SO_4 mixed to give a final volume of 750 ml

16.33 What is the minimum pH necessary to cause a precipitate of $Fe(OH)_2$ to form in a $0.010 M$ $FeCl_2$ solution?

16.34 A solution is prepared by mixing 100 ml of $0.20 M$ $AgNO_3$ with 100 ml of $0.10 M$ HCl. What are the molar concentrations of all species present in the solution when equilibrium is reached?

16.35 Will a precipitate form in a solution containing:
(a) $0.025 M$ $CaCl_2$ and $0.0050 M$ Na_2CO_3
(b) $0.010 M$ $Pb(NO_3)_2$ and $0.030 M$ $CaCl_2$
(c) $1.5 \times 10^{-3} M$ $FeCl_2$ and $2.2 \times 10^{-3} M$ $Na_2C_2O_4$

16.36 Which will precipitate first when Na_2CrO_4 (s) is gradually added to a solution containing $0.010 M$ Pb^{2+} and $0.010 M$ Ba^{2+}? What will be the molar concentration of the ion precipitated first when the other ion just begins to form a precipitate?

16.37 A solution is known to contain $0.010 M$ Pb^{2+} and $0.010 M$ Ni^{2+}. How must the pH be adjusted to achieve the maximum separation when the solution is saturated with H_2S ($[H_2S] = 0.10 M$)?

16.38 A solution containing $0.10 M$ Zn^{2+} and $0.10 M$ Fe^{2+} is saturated with H_2S. What must the H^+ concentration be to separate these ions by selectively precipitating ZnS? What is the smallest Zn^{2+} concentration that can be achieved without precipitating any of the Fe^{2+} as FeS?

16.39 What would the H^+ concentration have to be in order to prevent the precipitation of HgS when a $0.0010 M$ $Hg(NO_3)_2$ solution is saturated with H_2S? Can you explain why HgS is insoluble in concentrated (12 M) HCl?

16.40 Show that ZnS is soluble in concentrated (12 M) HCl.

16.41 Magnesium oxalate, MgC_2O_4, has $K_{sp} = 8.6 \times 10^{-5}$ and calcium oxalate, CaC_2O_4, has $K_{sp} = 2.3 \times 10^{-9}$. What pH must be maintained to achieve *maximum* separation of Ca^{2+} from Mg^{2+} if both have a concentration of $0.10 M$ and the concentration of oxalic acid, $H_2C_2O_4$, is maintained at $0.10 M$? For $H_2C_2O_4$, $K_{a_1} = 6.5 \times 10^{-2}$ and $K_{a_2} = 6.1 \times 10^{-5}$.

16.42 Use the data in Tables 16.1 and 16.3 to determine the molar solubility of AgI in $0.010 M$ KCN solution.

16.43 The solubility of $Zn(OH)_2$ in $1.0 M$ NH_3 is 5.7×10^{-3} mol/liter. Determine the value of the instability constant of the complex ion, $Zn(NH_3)_4^{2+}$. Ignore the reaction, $NH_3 + H_2O \rightleftharpoons NH_4^+ + OH^-$.

*16.44 What is the molar solubility of $Mg(OH)_2$ in $0.10 M$ NH_3 solution? Remember that NH_3 is a weak base.

*16.45 Will a precipitate form in a solution formed by dissolving 1.0 mol of $AgNO_3$ and 1.0 mol $HC_2H_3O_2$ in 1.0 liter of solution?

*16.46 How much solid sodium acetate would have to be added to 200 ml of a solution containing $0.200 M$ $AgNO_3$ and $0.10 M$ nitric acid to cause silver acetate to begin to precipitate. For $HC_2H_3O_2$, $K_a = 1.8 \times 10^{-5}$ and for $AgC_2H_3O_2$, $K_{sp} = 2.3 \times 10^{-3}$.

*16.47 How many grams of solid potassium fluoride must be added to 200 ml of a solution that contains $0.20 M$ $AgNO_3$ and $0.10 M$ acetic acid to cause silver acetate to begin to precipitate? For HF, $K_a = 6.5 \times 10^{-4}$; for $HC_2H_3O_2$, $K_a = 1.8 \times 10^{-5}$; for $AgC_2H_3O_2$, $K_{sp} = 2.3 \times 10^{-3}$.

*16.48 How many moles of HCl must be added to 1.0 liter of water to dissolve completely 0.20 mol of FeS? Remember that a saturated H_2S solution is 0.10 M.

*16.49 How many moles of solid NH_4Cl must be added to 1.0 liter of water in order to dissolve 0.10 mol of solid $Mg(OH)_2$? *Hint:* Consider the simultaneous equilibria:

$$Mg(OH)_2 \rightleftharpoons Mg^{2+} + 2OH^-$$

$$NH_3 + H_2O \rightleftharpoons NH_4^+ + OH^-$$

*16.50 Plaster is composed of $CaSO_4$. Suppose that there was a leak above a ceiling through which water was seeping at the rate of 2.0 liters/day. If the plaster in the ceiling is 1.50 cm thick, how long would it take to dissolve a circular hole 1 cm in diameter? Assume that the density of the plaster is 0.97 g/ml.

*16.51 25.0 ml of 0.10 M HCl is added to 1.000 liter of saturated $Mg(OH)_2$ in contact with more than enough $Mg(OH)_2$ (s) to react with all the HCl. After reaction has ceased, what will be the molar concentration of Mg^{2+}? What will be the pH of the solution?

*16.52 2.20 g of NaOH (s) are added to 250 ml of 0.10 M $FeCl_2$ solution. What weight of $Fe(OH)_2$ will be formed? What will be the molar concentration of Fe^{2+} in the final solution?

*16.53 1.75 g of NaOH (s) are added to 250 ml of 0.10 M $NiCl_2$ solution. What mass, in grams, of $Ni(OH)_2$ will be formed? What will be the pH of the final solution? For $Ni(OH)_2$, $K_{sp} = 1.6 \times 10^{-14}$.

*16.54 Solid $Mn(OH)_2$ is added to a solution of 0.100 M $FeCl_2$. After reaction, what will be the molar concentrations of Mn^{2+} and Fe^{2+} in the solution? What will be the pH of the solution?

17

ELECTROCHEMISTRY

We routinely rely on pocket calculators to do arithmetic on exams, and we take for granted the chemical reactions occurring inside the batteries that provide the electricity to power them. In this chapter we will study how electrical energy can cause nonspontaneous chemical changes to occur, and how spontaneous reactions can serve as a source of electricity.

Electrochemistry is concerned with the conversion of electrical energy into chemical energy in **electrolytic cells,** as well as with the conversion of chemical energy into electrical energy in **galvanic** or **voltaic cells.** In an electrolytic cell a process called electrolysis takes place in which the passage of electricity through a solution provides sufficient energy to cause an otherwise nonspontaneous oxidation-reduction reaction to take place. A galvanic cell, on the other hand, provides a source of electricity that results from a spontaneous oxidation-reduction reaction taking place in solution.

Electrochemical processes have a practical importance in chemistry and in everyday life. Electrolytic cells can provide us with information about the chemical environment as well as the energy that is required for many important oxidation-reduction reactions to occur. In addition, electrolysis is used to make many important chemicals that find their way into our lives. Examples are lye, NaOH, which is used to make soap, paper, and many other chemicals, and liquid bleach, NaOCl. For years now, galvanic cells such as the dry cell and "nicad" battery have powered our flashlights, radios, electronic calculators, wristwatches, cameras, and children's toys. The familiar lead storage battery has achieved widespread applications, especially in the automotive industry. More recently, fuel cells, in which the energy available from the combustion of fuels is converted directly into electricity, are finding many uses, especially in space vehicles. Electrochemical know-how has aided scientists in producing modern equipment for pollution analysis and biomedical research. With the aid of tiny electrochemical probes, scientists are beginning to study the chemical reactions taking place in living cells.

All these processes will be discussed in this chapter. However, before we begin let us first understand, qualitatively, how electrolytic solutions conduct electricity.

17.1 METALLIC AND ELECTROLYTIC CONDUCTION

For a substance to be classified as a conductor of electricity, it must be able to allow electrical charges within it to be moved from one point to another for the purpose of completing an electrical circuit. From our earlier discussion of solids, we know that most metals are conductors of electricity because of the relatively free movement of their *electrons* throughout the metallic lattice. This conduction is simply called **metallic conduction.** We also know from Chapter 6 that solutions containing electrolytes have the ability to conduct electricity. In this case, however, there are no "free" electrons in the solution to carry the current. How, then, do they conduct?

We can determine whether or not a solution is a conductor of electricity by using an apparatus similar to that shown in Figure 6.3 (p. 173). When the two electrodes are connected to a source of electricity and are dipped into a solution, we can observe whether or not the bulb of the apparatus lights. When this is done, we find that the bulb burns brightly when the solution contains a salt such as NaCl, but not at all when the solution contains a molecular compound such as sugar. Only when there are mobile ions present is electrical conduction possible, and this condition is fulfilled only by solutions of electrolytes and by molten salts.

When the source of electricity to the electrodes of a conductivity apparatus is a battery or other direct current (D.C.) source, each ion in the liquid tends to move toward the electrode of opposite charge, as shown in Figure 17.1. Thus, when the voltage is applied, the positive ions migrate toward the negative electrode and the negative ions move toward the positive electrode. This movement

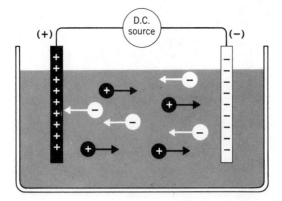

Figure 17.1

Ion flow in an electrolytic cell.

of ionic charges through the liquid, brought about by the application of electricity, is called **electrolytic conduction.**

When electrolytic conduction occurs chemical reactions take place as the ions in the liquid come in contact with the electrodes. At the positive electrode—where a deficiency of electrons exists—the negative ions are forced to give up electrons and are therefore oxidized. At the negative electrode—which has an excess of electrons—the positive ions pick up electrons and are reduced. Thus, during electrolytic conduction, oxidation is occurring at the positive electrode and reduction is taking place at the negative electrode. The liquid will continue to conduct electricity only as long as the oxidation-reduction reactions occurring at the electrodes continue.

The electrons that are deposited during the oxidation reaction are pumped out of the electrode by the voltage source and transferred to the negative electrode. During electrolytic conduction we have electrons flowing through the exterior wire and ions flowing through the solution. This situation is illustrated in Figure 17.2*a*.

The ionic movement, as well as the reactions at the electrodes, must take place so that electrical neutrality is maintained. This means that even in the most minute part of the liquid, whenever a negative ion moves away, a positive ion must also leave, or another negative ion must immediately take its place (Figure 17.2*b*). In this way every portion of the liquid is electrically neutral at all times. During the reactions at the electrodes, electrical neutrality is assured by having equal numbers of electrons deposited and picked up. For example,

If redox didn't occur, the electrodes would become neutralized by the layer of oppositely charged ions and no further ion migration would tend to occur.

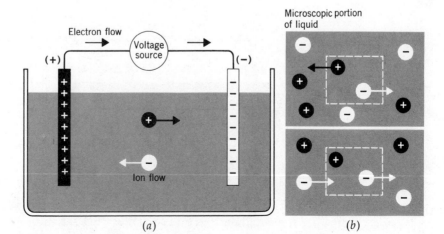

Figure 17.2

Electrolytic conduction (a) Electrolytic cell. (b) Maintaining electrical neutrality on a microscopic scale.

whenever one electron is deposited at the positive electrode, one electron must simultaneously be taken from the negative electrode. We focus our attention next on the chemical consequences of these last two processes.

17.2 ELECTROLYSIS

The chemical reactions that occur at the electrodes during electrolytic conduction constitute **electrolysis.** When liquid (molten) sodium chloride, for example, is *electrolyzed,* we find that the Na^+ ions move toward the negative electrode and the Cl^- ions move toward the positive electrode (Figure 17.3). The reactions that take place at the electrodes are

Positive electrode	$2Cl^- \longrightarrow Cl_2 + 2e^-$	oxidation
Negative electrode	$Na^+ + e^- \longrightarrow Na$	reduction

In electrochemistry we always assign the terms **cathode** and **anode** according to the chemical reaction that is taking place at the electrode. *Reduction always takes place at the cathode and oxidation always takes place at the anode.* Thus, in our electrolysis reactions above, we label the negative electrode the cathode and the positive electrode the anode.

The net chemical change that takes place in the electrolytic cell is called the **cell reaction.** It is obtained by adding together the anode and cathode reactions in such a way that the same number of electrons are gained and lost. This is the same procedure we used in the ion electron method of balancing oxidation-reduction reactions in Chapter 6. Thus, in this case we must multiply the reduction half-reaction by 2 to get

As in any redox reaction, total electron gain must equal total electron loss.

$$2Cl^- \,(l) \longrightarrow Cl_2 \,(g) + 2e^-$$
$$\underline{2Na^+ \,(l) + 2e^- \longrightarrow 2Na \,(l)}$$
$$2Na^+ \,(l) + 2Cl^- \,(l) \longrightarrow Cl_2 \,(g) + 2Na \,(l)$$

Therefore, in this electrolytic cell sodium is formed at the cathode and chlorine gas is produced at the anode. This is one of the major sources of pure sodium metal and chlorine gas in the United States (see Section 17.3).

The electrolysis of aqueous solutions of electrolytes is somewhat more complex because of the ability of water to be oxidized as well as reduced. The oxidation reaction for water is

$$2H_2O \,(l) \longrightarrow O_2 \,(g) + 4H^+ \,(aq) + 4e^- \qquad [17.1]$$

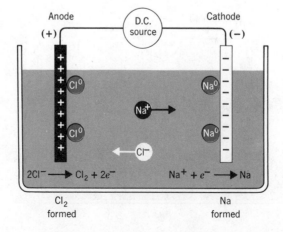

Figure 17.3
Electrolysis of molten NaCl.

and the reduction reaction takes the form

$$2H_2O\ (l) + 2e^- \longrightarrow H_2\ (g) + 2OH^-(aq) \qquad [17.2]$$

In acidic solutions another reaction that may take place is the reduction of H^+, which is

$$2H^+\ (aq) + 2e^- \longrightarrow H_2\ (g) \qquad [17.3]$$

Reaction 17.3 is not, however, a major reaction in most dilute aqueous solutions that we will consider.

In aqueous solution, we have the possible oxidation and reduction of the solvent in addition to the possible oxidation and reduction of the ions of the solute. Whether the solute anion or water is going to be oxidized, or whether the solute cation or water is going to be reduced, depends on the relative ease of the two competing reactions, as we will see in the next few examples.

Electrolysis of aqueous NaCl

In the electrolysis of aqueous NaCl, the following two anode (oxidation) reactions are possible:

$$(1)\ 2Cl^-\ (aq) \longrightarrow Cl_2\ (g) + 2e^-$$

$$(2)\ 2H_2O\ (l) \longrightarrow O_2\ (g) + 4H^+\ (aq) + 4e^-$$

and the following two cathode (reduction) reactions are possible:

$$(3)\ Na^+\ (aq) + e^- \longrightarrow Na\ (s)$$

$$(4)\ 2H_2O\ (l) + 2e^- \longrightarrow H_2\ (g) + 2OH^-\ (aq)$$

We can, of course, experimentally determine the outcome of this electrolysis by simply examining the products that are formed at the electrodes. Here we find that, in concentrated NaCl solutions (brine), chlorine gas is produced at the anode and hydrogen gas at the cathode (Figure 17.4). Therefore, during the electrolysis of such an aqueous solution of NaCl, the two half-reactions and the cell reaction are

$$\begin{array}{r} 2Cl^-\ (aq) \longrightarrow Cl_2\ (g) + 2e^- \\ 2H_2O\ (l) + 2e^- \longrightarrow H_2\ (g) + 2OH^-\ (aq) \\ \hline 2H_2O\ (l) + 2Cl^-\ (aq) \longrightarrow Cl_2\ (g) + H_2\ (g) + 2OH^-\ (aq) \end{array}$$

This tells us that Na^+ is more difficult to reduce than water, and under these conditions, Cl^- is more easily oxidized than water—that's why water is reduced and chloride ion is oxidized.

It is not necessarily an easy matter to predict what reactions will actually occur at the electrodes.

As the NaCl becomes more dilute, reaction (2) begins to compete and some O_2 is formed as well.

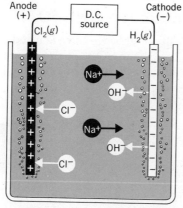

Figure 17.4

Electrolysis of aqueous sodium chloride.

Cathode $2H_2O + 2e^- \longrightarrow H_2(g) + 2OH^-(aq)$

Anode $2Cl^-(aq) \longrightarrow Cl_2(g) + 2e^-$

Electrolysis of aqueous $CuSO_4$

As in our last example, there are two possible oxidation and two possible reduction reactions for the electrolysis of $CuSO_4$. These are

Oxidation (1) $2SO_4^{2-}$ (aq) $\longrightarrow$ $S_2O_8^{2-}$ (aq) + $2e^-$

(2) $2H_2O$ (l) $\longrightarrow$ O_2 (g) + $4H^+$ (aq) + $4e^-$

and

Reduction (3) Cu^{2+} (aq) + $2e^-$ $\longrightarrow$ Cu (s)

(4) $2H_2O$ (l) + $2e^-$ $\longrightarrow$ H_2 (g) + $2OH^-$ (aq)

During this electrolysis we find experimentally that oxygen gas bubbles from the anode and a reddish coating of copper metal is deposited on the cathode. Therefore, we would write for the electrolysis of aqueous $CuSO_4$:

$$2H_2O\ (l) \longrightarrow O_2\ (g) + 4H^+\ (aq) + 4e^- \qquad \text{(Anode)}$$
$$\underline{2Cu^{2+}\ (aq) + 4e^- \longrightarrow 2\ Cu\ (s) \qquad\qquad \text{(Cathode)}}$$
$$2H_2O\ (l) + 2Cu^{2+}\ (aq) \longrightarrow O_2\ (g) + 4H^+\ (aq) + 2Cu\ (s) \quad \text{(Cell reaction)}$$

Notice that we multiplied the equation for the reduction of Cu^{2+} by 2 so that we have equal numbers of electrons lost and gained.

Because of the products formed, we can conclude that in the electrolysis of aqueous $CuSO_4$, the H_2O is more easily oxidized than the SO_4^{2-}, and the Cu^{2+} is more easily reduced than the H_2O.

Electrolysis of aqueous $CuCl_2$

At this point, we should be able to apply what we have learned about the electrolysis of aqueous solutions of NaCl and $CuSO_4$ to the electrolysis of $CuCl_2$. We would expect that the two species that could be oxidized are water and chloride ion, and that the two species that could be reduced are water and Cu^{2+}. Since we already know that Cl^- is more easily oxidized than H_2O and that Cu^{2+} is more easily reduced than H_2O, we expect the following reactions.

$$2Cl^-\ (aq) \longrightarrow Cl_2\ (g) + 2e^- \qquad \text{(Anode)}$$
$$\underline{Cu^{2+}\ (aq) + 2e^- \longrightarrow Cu\ (s) \qquad \text{(Cathode)}}$$
$$Cu^{2+}\ (aq) + 2Cl^-\ (aq) \longrightarrow Cl_2\ (g) + Cu\ (s) \quad \text{(Cell reaction)}$$

This is exactly what is found experimentally.

Electrolysis of aqueous Na_2SO_4

Once again we call on what we have learned previously. We know that water is more easily oxidized than SO_4^{2-} at the anode and that water is more easily reduced than Na^+ at the cathode. Therefore, in this solution H_2O is both oxidized and reduced, giving us

The electrolysis of aqueous Na_2SO_4 is shown in Color Plate 12.

$$2H_2O\ (l) \longrightarrow O_2\ (g) + 4H^+\ (aq) + 4e^- \qquad\qquad \text{(Anode)}$$
$$\underline{4H_2O\ (l) + 4e^- \longrightarrow 2H_2\ (g) + 4OH^-\ (aq) \qquad\qquad\quad \text{(Cathode)}}$$
$$6H_2O\ (l) \longrightarrow O_2\ (g) + 2H_2\ (g) + 4OH^-\ (aq) + 4H^+\ (aq) \quad \text{(Cell reaction)}$$

Notice that we had to multiply the reduction reaction (Equation 17.2) by 2 to have the same number of electrons as in the oxidation reaction.

The overall reaction, as written above, can be simplified further by recalling that H^+ and OH^- will react to give H_2O.

$$4OH^- + 4H^+ \longrightarrow 4H_2O$$

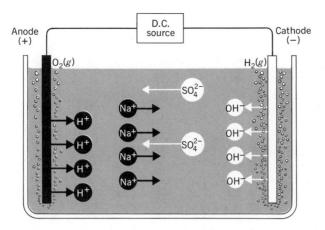

Figure 17.5

Electrolysis of aqueous Na_2SO_4. The sodium ions and sulfate ions are needed to counter the charges of the ions formed at the electrodes and thereby maintain electrical neutrality.

Thus the net overall reaction for the electrolysis of a stirred aqueous Na_2SO_4 solution is

$$2H_2O\ (l) \longrightarrow O_2\ (g) + 2H_2\ (g)$$

which is simply the reaction for the electrolysis of H_2O. Sodium sulfate does not participate in this electrolysis in the sense that it is not consumed at the electrodes. Yet we would find experimentally that it or some other similar salt is needed if electrolysis of water is to occur. What, then, is the role of the Na_2SO_4? The Na_2SO_4 is needed to maintain electrical neutrality (Figure 17.5). During the oxidation of H_2O, H^+ ions are produced in the immediate vicinity of the anode. A negative ion must also be present in that vicinity to neutralize the positive charges. This is fulfilled by SO_4^{2-} ions. Likewise, at the cathode, where OH^- ions are produced, there must be positive ions present to neutralize the charges on the OH^- ions, and thereby keep the solution electrically neutral.

17.3 PRACTICAL APPLICATIONS OF ELECTROLYSIS

Each day our lives are touched either directly or indirectly by the products of electrolysis reactions. For example, the drinking water in most places is treated with chlorine to kill bacteria, and chlorine is used to manufacture many chemicals, from pesticides that protect crops to plastics such as polyvinyl chloride—usually just called vinyl. Yet elemental chlorine does not occur free in nature—it must be extracted from its compounds, and this is done most economically by electrolysis. In this section we will study how chlorine and some other commercially important substances are made.

Electrolysis of molten sodium chloride

The "chemistry" of this process was described in our introduction to the discussion of electrolysis reactions on page 520. The products—sodium and chlorine—are both commercially important. Sodium is used as a heat transfer medium for cooling nuclear reactors and in sodium vapor lights (see Color Plate 4). Chlorine, as mentioned above, is used in many chemical processes.

Sodium and chlorine are both very reactive chemicals. Therefore, when they are produced from NaCl they must be kept apart, otherwise they will react and reform sodium chloride. The Down's cell, illustrated in Figure 17.6, accomplishes this. The chlorine, of course, comes off as a gas. Because of the temperature at which the cell operates, the sodium is formed as a liquid and is also easily removed. This allows the cell to operate continuously as fresh sodium chloride is added and the products are taken away.

Cathode:
  $Na^+ + e^- \rightarrow Na\ (\ell)$
Anode:
  $2Cl^- \rightarrow Cl_2\ (g) + 2e^-$

Melting points:
  NaCl   801°C
  Na   98°C

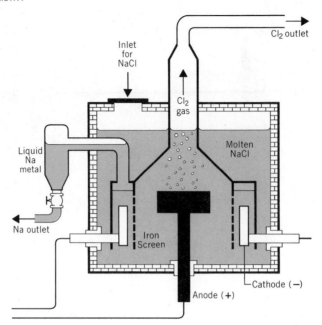

Figure 17.6

The Downs cell used for the electrolysis of molten sodium chloride. The cell is constructed to keep the metallic sodium and gaseous chlorine from reacting with each other after they are formed in the electrolysis reaction.

In 1980, production of NaOH by this reaction amounted to about 13 million tons!

Electrolysis of brine (NaCl solution)

Earlier we discussed the net cell reaction for the electrolysis of aqueous sodium chloride. If we look at the overall reaction, however, we can better appreciate its commercial importance.

$$2Na^+ + 2Cl^- + 2H_2O \longrightarrow Cl_2\ (g) + H_2\ (g) + 2Na^+ + 2OH^-$$

The products are chlorine, hydrogen, and sodium hydroxide (NaOH)—all important industrial chemicals. Some uses of chlorine have already been mentioned. Hydrogen is used to make ammonia, and sodium hydroxide is used in huge quantities to neutralize acids in various chemical processes, to process pulp and paper, to purify aluminum ores, and in the manufacture of textiles and the refining of petroleum.

Although the electrolysis of aqueous NaCl in an apparatus like that in Figure 17.4 produces NaOH in solution, it is always contaminated by unreacted NaCl. A purer, more concentrated NaOH solution can be prepared using a mercury cell, illustrated in Figure 17.7. In this cell, sodium is actually the substance that is reduced, and it dissolves in the liquid mercury as it is formed. The solution of sodium in mercury—it's called sodium amalgam—is pumped to a separate vessel where the metallic sodium at the surface of the mercury can react with water. This reaction liberates hydrogen and leaves pure NaOH in solution.

$$2Na + H_2O \longrightarrow 2Na^+ + 2OH^- + H_2\ (g)$$

The mercury cell has a disadvantage of posing a serious threat of mercury water pollution and so must be carefully monitored.

If the electrolysis of brine is carried out in a vigorously stirred solution, the OH⁻ produced at the cathode reacts with the Cl₂ formed at the anode. The reaction is

$$Cl_2 + 2OH^- \longrightarrow Cl^- + OCl^- + H_2O$$

Continued electrolysis therefore gradually converts nearly all the chloride ion to hypochlorite ion, OCl⁻, and the sodium chloride solution is changed to a solution of sodium hypochlorite. When diluted to about 5 to 6 percent by weight, this is sold as liquid laundry bleach (e.g., Clorox®).

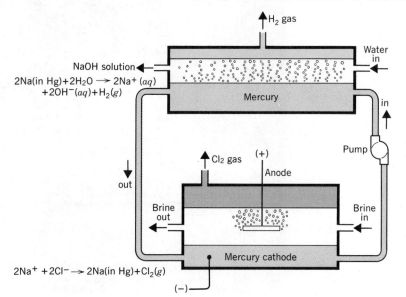

Figure 17.7

Electrolysis of aqueous sodium chloride using a mercury cell. Chlorine gas is evolved at the anode where chloride ions are oxidized. At the cathode, sodium ions are reduced to sodium atoms, which dissolve in the mercury. The mercury is pumped to a tank where it comes in contact with water. Sodium atoms react there with water to liberate hydrogen and produce sodium hydroxide.

Aluminum

As you are undoubtedly aware, aluminum finds many important uses as a structural metal because of its strength and low density ("light weight"). Its commercial availability has been made possible through the application of electrochemical reduction.

If we were to electrolyze an aqueous solution of an aluminum salt, such as $AlCl_3$, we would find that H_2O is more easily reduced than the Al^{3+}, so an aqueous solution of an aluminum salt cannot be used to produce the metal. A 22-year-old graduate of Oberlin College, Charles Hall, invented a process, in 1886, whereby molten Al_2O_3 is used. He prepared a mixture of Al_2O_3 with *cryolite*, Na_3AlF_6, and electrolyzed it in the molten state. The cryolite, he found, reduced the melting temperature from 2000°C for Al_2O_3 to 1000°C for the mixture. A diagram of the electrolysis cell is shown in Figure 17.8. The vessel holding the molten mixture is made of iron lined with carbon and serves as the cathode. Carbon rods that serve as the anode are inserted into the melt. As the oxidation-reduction reactions proceed, pure aluminum is produced at the cathode and sinks to the bottom of the vessel. The reactions at the electrodes are

At the high temperature of the cell, the O_2 attacks the carbon electrodes and they must be replaced periodically.

$$\begin{array}{rll} 3O^{2-}\,(l) &\longrightarrow \tfrac{3}{2}O_2\,(g) + 6e^- & \text{(Anode)} \\ 2Al^{3+}\,(l) + 6e^- &\longrightarrow 2Al\,(l) & \text{(Cathode)} \\ \hline 2Al^{3+}\,(l) + 3O^{2-}\,(l) &\longrightarrow \tfrac{3}{2}O_2\,(g) + 2Al\,(l) & \end{array}$$

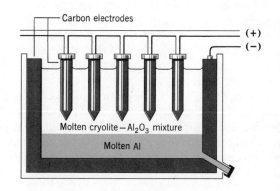

Figure 17.8

Production of aluminum by the Hall process.

Graphite electrodes can be seen projecting from the tops of the electrolysis cells in this aluminum pot line at Alcan's Arvida Works in Quebec.

Today other materials are used in place of the cryolite. These materials permit operation at still lower temperatures and are less dense than the cryolite used by Hall. This lower density of the electrolyte mix permits easier separation of the molten aluminum.

Magnesium

Mg^{2+} is the third most abundant ion in sea water.

Magnesium, another structural metal that is important because of its light weight, occurs to an appreciable extent in sea water. Magnesium ions are precipitated from sea water as the hydroxide and the $Mg(OH)_2$ is then converted to the chloride by treatment with hydrochloric acid. After evaporation of the water, the $MgCl_2$ is melted and electrolyzed. Magnesium is produced at the cathode and chlorine is evolved at the anode. The overall net reaction is simply

$$MgCl_2\ (l) \longrightarrow Mg\ (l)\ +\ Cl_2\ (g)$$

Copper

An interesting application of electrolysis is the refining, or purification, of copper metal. When first separated from its ore, copper metal is about 99% pure, with iron, zinc, silver, gold, and platinum as major impurities. In the refining process the impure copper is used as the anode in an electrolytic cell containing aqueous copper sulfate as the electrolyte. The cathode of the cell is constructed of high-purity copper (Figure 17.9).

When electrolysis is carried out, the voltage across the cell is adjusted so that only copper and other more active metals, such as iron or zinc, are able to dissolve at the anode. The silver, gold, and platinum do not dissolve, and simply fall off and settle to the bottom of the electrolysis cell. At the cathode only the most easily reduced species, Cu^{2+}, is caused to pick up electrons; hence, only copper is deposited.

The net result of the operation of this cell is that copper is transferred from the anode to the cathode while the Fe and Zn impurities remain in solution as Fe^{2+} and Zn^{2+}. Afterwards the silver, gold, and platinum "sludge" is removed from the apparatus and sold for enough money to nearly pay for the cost of the electricity required in the electrolysis. As a result, the purification of copper

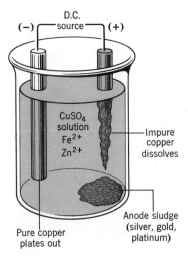

Figure 17.9

Purification of copper by electrolysis.

Pure copper starter plates (cathodes) are lowered into an electrolysis cell between anodes of impure copper. It takes about 28 days for the anodes to dissolve and for the copper to be deposited on the cathodes.

(about 99.95% pure) costs nearly nothing! Nevertheless, the total production cost of copper is still considerable, because it includes the mining of the crude ore and its initial purification.

Electroplating

We have just seen how copper can be "plated out" on an electrode in an electrolysis cell. The plating out of a metal in this fashion is called **electroplating.** If we replaced the cathode in the cell in Figure 17.9 with another metal, the surface of that metal will also become covered with a layer of pure copper when the current is applied. Other metals can be electroplated as well as copper, which makes this process of great commercial importance. In the manufacturing of automobiles, for example, various parts, such as steel bumpers, are electroplated with chromium for beauty as well as for protection against corrosion.

A silver plated server.

17.4 QUANTITATIVE ASPECTS OF ELECTROLYSIS

Michael Faraday was the first to describe, in a quantitative fashion, the relationship that exists between the amount of current used and the extent of the chemical change that takes place at the electrodes during electrolysis. He found that the amount of chemical change that occurs in an electrolysis reaction is related to the amount of electricity that passes through the cell. In modern terms, we say that the amount of change is related to the number of moles of electrons lost or gained in the oxidation-reduction reactions. For example, in the reduction of silver ion to silver metal,

$$Ag^+ \ (aq) + e^- \longrightarrow Ag \ (s)$$

1 mol of electrons "reacts" with 1 mol of silver ions to give 1 mol, or 107.87 g, of solid silver. Thus, in this case, when 107.87 g of silver are deposited on the cathode, we know that 1 mol of electrons must have passed through the cell.

The amount of electricity that must be supplied to a cell in order to deliver 1 mol *of electrons is called a* **faraday** ($\mathscr{F}$). In the above example, 1 $\mathscr{F}$ must be supplied to produce the 107.87 g of silver, and it would take 2 $\mathscr{F}$ to produce 215.74 g of silver, and so on. In other words, "faraday" is just another way of saying "a mole of electrons."

$$1 \ \mathscr{F} \equiv 1 \text{ mol of electrons}$$

Another important unit is the coulomb (C). This is the SI unit of electric charge. One coulomb is the amount of charge that moves past any given point in a circuit when a current of 1 ampere (1 A) is supplied for 1 second (1 s). Thus

$$1 \text{ coulomb} = 1 \text{ ampere} \times 1 \text{ second}$$

$$1 \text{ C} = 1 \text{ A} \cdot \text{s}$$

Experimentally, it is found that 1 $\mathscr{F}$ is equivalent to 96,487 coulombs, or 96,500 when rounded off to three significant figures. Thus

$$1 \ \mathscr{F} \sim 96,500 \text{ C} \sim 1 \text{ mol of electrons}$$

Let's look at some examples of how these concepts can be applied.

EXAMPLE 17.1

In an electrolytic cell like that in Figure 17.2, how many grams of Cu will be deposited from a solution of $CuSO_4$ by a current of 1.5 A flowing for 2.0 hr?

SOLUTION

First, we must have an equation representing the reaction that takes place. Since metallic copper is being deposited from a solution containing Cu^{2+}, the copper ions must gain electrons and thereby undergo reduction. This lets us write the following half-reaction.

$$Cu^{2+}(aq) + 2e^- \longrightarrow Cu \ (s)$$

This equation is necessary because it gives the important relationship

$$1 \text{ mol Cu} \sim 2 \text{ mol of electrons}$$

or simply

$$1 \text{ mol Cu} \sim 2 \ \mathscr{F}$$

Our procedure now is to use the current and time to calculate the number of coulombs that are delivered. Then we calculate the number of faradays supplied. Finally we compute the number of moles of Cu, followed by the number of grams of Cu.

$$1.5 \text{ A} \times 2.0 \text{ hr} \times \left(\frac{3600 \text{ s}}{1 \text{ hr}}\right) = 11,000 \text{ A} \cdot \text{s (rounded)}$$

$$11,000 \text{ A} \cdot \text{s} \times \left(\frac{1 \text{ C}}{1 \text{ A} \cdot \text{s}}\right) = 11,000 \text{ C}$$

$$11,000 \text{ C} \times \left(\frac{1 \text{ ℱ}}{96,500 \text{ C}}\right) = 0.11 \text{ ℱ}$$

$$0.11 \text{ ℱ} \times \left(\frac{1 \text{ mol Cu}}{2 \text{ ℱ}}\right) \times \left(\frac{63.55 \text{ g Cu}}{1 \text{ mol Cu}}\right) \sim 3.5 \text{ g Cu}$$

EXAMPLE 17.2 How long would it take to produce 25.0 g of Cr from a solution of $CrCl_3$ by a current of 2.75 A?

SOLUTION Again, we begin by writing the equation for the reaction. It too is a reduction;

$$Cr^{3+} (aq) + 3e^- \longrightarrow Cr (s)$$

from which we can say that

$$1 \text{ mol Cr} \sim 3 \text{ ℱ}$$

First we calculate the number of faradays required.

$$25.0 \text{ g Cr} \times \left(\frac{1 \text{ mol Cr}}{52.0 \text{ g Cr}}\right) \times \left(\frac{3 \text{ ℱ}}{1 \text{ mol Cr}}\right) \sim 1.44 \text{ ℱ}$$

Next we calculate the number of coulombs required.

$$1.44 \text{ ℱ} \times \left(\frac{96,500 \text{ C}}{1 \text{ ℱ}}\right) \sim 139,000 \text{ C}$$

Since 1 C is equal to 1 A · s,

$$139,000 \text{ C} \times \left(\frac{1 \text{ A} \cdot \text{s}}{1 \text{ C}}\right) \times \left(\frac{1}{2.75 \text{ A}}\right) \sim 50,500 \text{ s}$$

If we convert this to hours, we get

$$50,500 \text{ s} \times \left(\frac{1 \text{ hr}}{3600 \text{ s}}\right) = 14.0 \text{ hr}$$

We can experimentally determine the weight of a substance that has been deposited on an electrode during electrolysis by weighing the electrode before and after the current has been supplied. The apparatus used in experiments of this kind is called a **coulometer.** In Figure 17.10 we see two such coulometers connected in series so that the same current, and thus the same number of faradays, passes through both cells. With the aid of this apparatus it is possible to use a known oxidation-reduction reaction in one cell to provide an experimental measure of the equivalent weight of an unknown in the other cell. This type of analysis is illustrated in the next example.

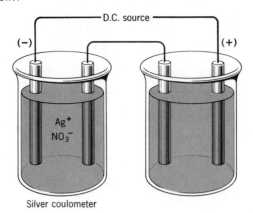

Figure 17.10

Two coulometers connected in series. The same current must pass through both cells. When 107.87 g of silver are deposited on the cathode in the cell at the left we know that 1 ℱ has passed through both cells.

Silver coulometer

$$\left[\begin{array}{c}1\ \mathcal{F}\ \text{deposits}\ 107.87\ \text{g Ag}\\ \text{on a cathode}\end{array}\right]$$

1 ℱ passes through this cell when 107.87 g of Ag (1 mole Ag) is deposited in the other cell

---

EXAMPLE 17.3   In the left cell of Figure 17.10 we place a solution containing Ag^+ ions, and in the right cell a solution whose metal ion is unknown (X). The same current is passed through both cells for the same amount of time. When the current is turned off and the electrodes are rinsed, dried, and weighed, it is found that 3.50 g of silver are deposited during the same period of time that 2.50 g of element X are deposited. What is the equivalent weight of element X?

SOLUTION   This rather long-winded problem has a relatively simple solution. Since the current and time are the same for both cells in this series circuit, the same number of faradays are passed through both. We can calculate the number of faradays supplied by using the information from the Ag^+ cell.

From our earlier discussion in this section, we know that the reduction reaction for Ag^+ is

$$Ag^+\ (aq) + e^- \longrightarrow Ag\ (s)$$

Therefore,

$$1\ \text{mol Ag} \sim 1\ \mathcal{F}$$

or

$$107.87\ \text{g Ag} \sim 1\ \mathcal{F}$$

The faraday equivalent of 3.50 g of silver is then

$$3.50\ \cancel{g} \times \left(\frac{1\ \mathcal{F}}{107.87\ \cancel{g}}\right) = 0.0324\ \mathcal{F}$$

This means that 0.0324 ℱ of electricity was passed through both cells. In the cell containing the unknown, then, 0.0324 ℱ deposited 2.50 g of the substance. Thus, for the unknown,

$$0.0324\ \mathcal{F} \sim 2.50\ \text{g}\ X$$

and 1 ℱ is equivalent to

$$1\ \cancel{\mathcal{F}} \times \left(\frac{2.50\ \text{g}\ X}{0.0324\ \cancel{\mathcal{F}}}\right) = 77.2\ \text{g}\ X$$

Recall that the equivalent weight of a substance undergoing a redox reaction is equal to that weight of the substance that gains or loses 1 mol of electrons—that is, 1 ℱ. The equivalent weight of the unknown metal is therefore 77.2 g.

17.5 GALVANIC CELLS

Luigi Galvani (1737–1798) was an Italian anatomist who pioneered in the field of electrophysiology — the study of the relationship between electricity and living organisms.

In our previous discussion of electrolytic cells, chemical changes occurred because a voltage placed across electrodes forced an otherwise nonspontaneous chemical reaction to take place. We now turn to the opposite situation, in which electron flow is produced as a result of spontaneous oxidation-reduction reactions in galvanic cells.

An example of a spontaneous oxidation-reduction reaction taking place in a solution can be seen simply by placing a piece of metallic zinc into a solution of $CuSO_4$ (see Color Plate 11). A dark brown, spongelike layer begins to form on the piece of zinc and, at the same time, the blue color of the $CuSO_4$ begins to disappear. The brownish substance forming on the zinc is metallic copper, and if we were to analyze the solution, we would find that it now contains Zn^{2+}. The reaction that is taking place, therefore, is

$$Cu^{2+} (aq) + Zn (s) \longrightarrow Cu (s) + Zn^{2+} (aq)$$

This can be divided into a pair of redox half-reactions:

$$Cu^{2+} (aq) + 2e^- \longrightarrow Cu (s)$$
$$Zn (s) \longrightarrow Zn^{2+} (aq) + 2e^-$$

We see from these reactions that the Cu^{2+} ions are spontaneously removed from the solution and are replaced by the colorless Zn^{2+} ions. Thus the blue color of the solution gradually disappears as more and more Zn^{2+} ions are formed.

As long as these reactions take place at the surface of the zinc, no useful flow of electrons can be obtained. The reaction simply generates heat. We can, however, take advantage of the spontaneous electron transfer by making the oxidation and reduction half-reactions occur in separate compartments of a cell, as shown in Figure 17.11. Each compartment is called a **half-cell,** and when they are properly connected the electrons produced by the oxidation of the zinc must travel through the wire and into the electrode in the $CuSO_4$ solution. The electrons are then picked up by the Cu^{2+} ions and reduction takes place. The electrons flowing through the external wire constitute an electric current, and the galvanic cell therefore serves as a source of electricity.

Although the zinc and copper have to be separated to obtain a useful flow of electrons, complete isolation of the two species would lead to an electrical imbalance at the electrodes, and the electron flow would soon cease. We can see

Figure 17.11

A galvanic cell that employs the reaction

Zn (s) + Cu^{2+} (aq) →
    Zn^{2+} (aq) + Cu (s)

The salt bridge in (a) serves the same purpose as the porous partition in (b) —they each allow electrical neutrality to be maintained as the reactions occur at the electrodes.

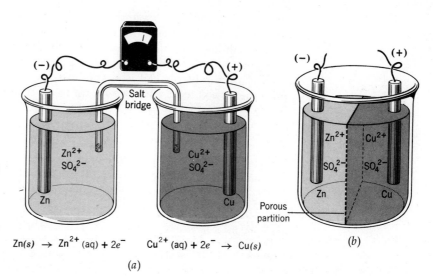

Zn(s) → Zn^{2+} (aq) + 2e$^-$ Cu^{2+} (aq) + 2e$^-$ → Cu(s)

(a)

(b)

how electrical imbalance would occur if we imagine that the two half-cells were completely isolated from each other and the oxidation-reduction reactions still continued to take place. On the left side of this hypothetical setup, Zn^{2+} ions entering the solution would give the solution an overall positive charge. This would prevent additional Zn^{2+} from entering. On the right we would find that when Cu^{2+} leaves the solution the SO_4^{2-} ions left behind would give the solution a negative charge and the electrode would become positively charged. This would cause the electrode to repel Cu^{2+} ions and prevent their further removal from the solution.

From the preceding discussion we see that a continuous flow of current, accompanied by a continuous chemical activity, can only occur if the solution around each electrode is kept electrically neutral. For this to happen, however, ions must flow into or out of the cell compartments. In the zinc half-cell, for example, Zn^{2+} must leave the electrode compartment or anions must enter it. Similarly, in the copper half-cell, cations must enter to balance the charge of the SO_4^{2-}, or SO_4^{2-} ions must leave.

Although ions must be able to diffuse from one cell compartment to another, the two solutions cannot be allowed to mix freely, otherwise Cu^{2+} would react directly at the Zn electrode and no flow of electrons through the external circuit would have to occur.

The salt bridge in Figure 17.11a and the porous partition in Figure 17.11b allow for the slow mixing of the ions in the two solutions. A salt bridge is usually a tube filled with an electrolyte such as KNO_3 or KCl in gelatin. Cations from the salt bridge can move into one compartment to compensate for the excess negative charge, while the anions from the salt bridge diffuse into the other compartment to neutralize the excess positive charge. The porous partition in Figure 17.11b serves the same purpose as the salt bridge. With either the salt bridge or the porous partition in place, there is a continuous electron flow through the external wire and an ion flow through the solution as a result of the spontaneous oxidation-reduction reactions taking place at the electrodes.

The signs of the electrodes in galvanic cells

Earlier we defined the anode in electrochemistry as the electrode at which oxidation takes place and the cathode as the one at which reduction occurs. This applies regardless of whether the cell is an electrolytic cell or a galvanic cell. In the galvanic cell just described, oxidation takes place in the zinc compartment, so the zinc bar would be the anode and the copper electrode would be the cathode. As zinc ions leave the solid zinc anode and enter the solution, electrons are left behind and the zinc electrode acquires a negative charge. At the copper cathode, Cu^{2+} ions become attached to the electrode and seek electrons to become reduced. This gives the copper electrode a positive charge. Thus we see that in a galvanic cell, the anode is negative and the cathode is positive, quite the opposite of what we found to be true in electrolytic cells.[1]

17.6 CELL POTENTIALS

The electric current obtained from a galvanic cell is a result of electrons being pushed or forced to flow from the negative electrode, through an external wire,

---

[1] This labeling of electrodes in galvanic cells is, however, consistent with electrolytic cells when we consider the movement of the ions within the solution. The Zn^{2+} ions produced at the anode and the SO_4^{2-} ions freed at the cathode must mingle with each other if electrical neutrality is to prevail in the solution. To accomplish this, some of the Zn^{2+} ions must move toward the cathode and some of the SO_4^{2-} ions must move toward the anode. Thus we have cations moving toward the cathode and anions moving toward the anode, which is precisely the same situation as in electrolytic cells.

to the positive electrode. The "force" with which these electrons move through the wire is called the **electromotive force,** or **emf,** and is measured in **volts** (V). Actually, the volt is a measure of the energy that is capable of being extracted from the flowing electric charge. If the emf is 1 V, the passage of 1 coulomb is able to accomplish 1 joule of work.

$$1 \text{ volt} = \frac{1 \text{ joule}}{\text{coulomb}}$$

$$1 \text{ V} = 1 \text{ J/C} \qquad\qquad [17.4]$$

The emf produced by a galvanic cell is called the **cell potential,** $\mathscr{E}_{cell}$. This emf depends on the concentrations of the ions in the cell, the temperature, and the partial pressures of any gases that might be involved in the cell reactions. When all ion concentrations are 1 M, all partial pressures of gases are 1 atm, and the temperature of the cell is 25°C, the emf is called the **standard cell potential,**[2] designated as $\mathscr{E}^0_{cell}$.

Notice that the standard conditions here are the same as those used in defining standard thermodynamic quantities.

To measure the cell potential accurately, care must be taken to avoid drawing current from the cell. This is because some of the cell's voltage is required to overcome the cell's own internal resistance when current is drawn. The remaining voltage that can be measured under these conditions is less than the maximum.

One device that can usually be used to measure the emf of a cell is called a **potentiometer.** In this instrument the potential generated by the cell is balanced by an opposing potential from within the potentiometer. When the two opposing potentials are equal, no current flows and the cell potential is equal to the opposing emf, which can be read directly from the instrument. The voltage that is measured in this way is the maximum emf of the cell. Today modern advances in electronics have led to a variety of other devices that are able to measure the emf of a cell quickly and simply, without drawing significant amounts of current.

17.7 REDUCTION POTENTIALS

A very important and useful concept can be developed if we attempt to answer the question, "What is the origin of the cell potential?" To answer this question we will use the Zn/Cu cell described earlier. In this cell we have a solution containing Zn^{2+} ions around one electrode and a solution containing Cu^{2+} ions around the other. Each of these ions has a certain tendency to acquire electrons from its respective electrode and become reduced. In other words, each reduction half-reaction,

$$Zn^{2+} \, (aq) + 2e^- \longrightarrow Zn \, (s)$$

and

$$Cu^{2+} \, (aq) + 2e^- \longrightarrow Cu \, (s)$$

has a certain intrinsic tendency to proceed from left to right that we can describe by its **reduction potential.** The larger the reduction potential for any half-reaction, the greater its tendency to undergo reduction.

When the cell reaction takes place, what we are actually observing is a kind of "tug-of-war." Each of the species in solution attempts to pull electrons from its electrode so as to become reduced. The species with the greater tendency to

---

[2] In footnote 1 in Chapter 13, it was mentioned that activities ("effective" concentrations and pressures) should be used in the mass action expression when computing ΔG. This applies also to the effect of concentration on $\mathscr{E}$. The standard potential is obtained when all species are at unit activity. Only a small error is introduced, however, by using actual concentrations when solutions are relatively dilute.

acquire electrons—that is, the one with the larger reduction potential—wins the tug-of-war and does undergo reduction. The loser, on the other hand, must supply the electrons to the winner, and that substance is oxidized. In the zinc-copper cell, therefore, copper must have a larger reduction potential than zinc, because copper is reduced.

The potential that we measure for a cell corresponds to the *difference* in the tendencies of the two ions to become reduced, and is equal to the reduction potential for the substance that actually undergoes reduction minus the reduction potential for the substance that is forced to undergo oxidation. In terms of standard reduction potentials,

$$\mathscr{E}^0_{cell} = \mathscr{E}^0_{substance\ reduced} - \mathscr{E}^0_{substance\ oxidized} \qquad [17.5]$$

In the zinc-copper cell, therefore,

$$\mathscr{E}^0_{cell} = \mathscr{E}^0_{Cu} - \mathscr{E}^0_{Zn}$$

where $\mathscr{E}^0_{Cu}$ is the standard reduction potential for copper and $\mathscr{E}^0_{Zn}$ is the standard reduction potential for zinc. Since $\mathscr{E}^0_{Cu}$ is larger than $\mathscr{E}^0_{Zn}$, $\mathscr{E}^0_{cell}$ is positive.

Experimentally, it is only possible to measure overall cell potentials. This means that we are only capable of obtaining differences between the reduction potentials for any two half-reactions. Then how can we obtain the reduction potential for any specific half-reaction? Clearly, if the cell potential and the $\mathscr{E}^0$ for one of the half-reactions are known, the $\mathscr{E}^0$ for the other half-reaction can be calculated. What has been done, therefore, is to choose a half-reaction arbitrarily and assign to it a standard reduction potential of zero volts. All other half-reactions can then be compared to this standard and a set of relative values of $\mathscr{E}^0$ obtained.

The electrode chosen to be the standard is called the **hydrogen electrode,** shown in Figure 17.12a. It consists of a platinum wire encased in a glass sleeve with hydrogen gas passing through it at a pressure of 1 atm. The platinum wire is attached to a platinum foil that is coated with a black velvet-looking layer of finely divided platinum, which serves as a catalyst for the reaction,

$$2H^+\ (aq) + 2e^- \rightleftharpoons H_2\ (g)$$

This assembly is then immersed in an acid solution whose hydrogen ion concentration is 1 M.

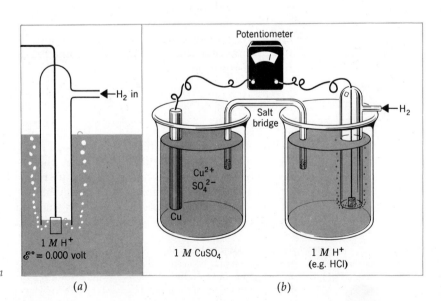

Figure 17.12

(a) *The hydrogen electrode.*
(b) *The hydrogen electrode used in a galvanic cell.*

When the hydrogen electrode is paired with another half-cell in a galvanic cell, it can undergo either oxidation or reduction, depending on the reduction potential of the other half-cell. For instance, if the reduction potential of the species in the other half-cell is greater than that for the hydrogen electrode—that is, if its $\mathscr{E}^0$ is positive—the hydrogen electrode is forced to undergo oxidation. The corresponding half-reaction for the oxidation of the hydrogen electrode is

$$H_2\,(g) \longrightarrow 2H^+\,(aq) + 2e^-$$

On the other hand, if the reduction potential of the other half-reaction is less than 0.000 V, this species has a negative reduction potential and the hydrogen electrode undergoes reduction.

$$2H^+\,(aq) + 2e^- \longrightarrow H_2\,(g)$$

This causes the other species to become oxidized.

To illustrate how the hydrogen electrode is used, let's examine the galvanic cell in Figure 17.12b. When we connect a potentiometer to this cell to measure its potential, proper readings can only be obtained if we connect the terminal labeled (+) to the positive electrode and the terminal labeled (−) to the negative electrode. That's the way these instruments are designed. In this case, we find that to obtain a proper reading the (+) terminal must be connected to the copper electrode and the (−) terminal to the hydrogen electrode. From what we learned earlier, this tells us that the copper electrode is the cathode and the hydrogen electrode is the anode. The spontaneous half-reactions in this cell are therefore

> The cathode carries the positive charge in a galvanic cell.

$$Cu^{2+}\,(aq) + 2e^- \longrightarrow Cu\,(s) \qquad \text{(Cathode)}$$

$$H_2\,(g) \longrightarrow 2H^+\,(aq) + 2e^- \qquad \text{(Anode)}$$

Since copper is reduced, when we apply Equation 17.5 we must write

$$\mathscr{E}^0_{cell} = \mathscr{E}^0_{Cu} - \mathscr{E}^0_{H_2}$$

> A _measured_ cell potential is always a positive number.

The measured cell potential (as read from the potentiometer) is 0.34 V. Therefore

$$0.34\ V = \mathscr{E}^0_{Cu} - 0.000\ V$$

or

$$\mathscr{E}^0_{Cu} = +0.34\ V$$

Now let's look at the cell in Figure 17.13. Here we have a zinc half-cell paired with the hydrogen electrode. To obtain a reading with the potentiometer we find, by experiment, that the positive terminal has to be connected to the

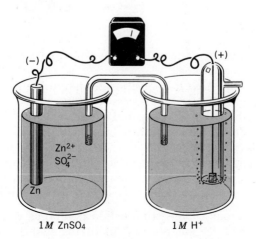

Figure 17.13

A galvanic cell that can be used to determine the standard reduction potential of Zn^{2+}.

$1M$ ZnSO$_4$ $1M$ H$^+$

hydrogen electrode, so this time it is the cathode. The spontaneous half-reactions in this cell are therefore

$$2H^+ \ (aq) + 2e^- \longrightarrow H_2 \ (g) \qquad \text{(Cathode)}$$

$$Zn \ (s) \longrightarrow Zn^{2+} \ (aq) + 2e^- \qquad \text{(Anode)}$$

In other words, zinc ion is more difficult to reduce than H^+.

The measured value of the cell potential, as read from the potentiometer, is 0.76 V. Using Equation 17.5 again,

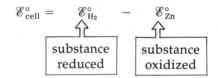

Substituting,

$$0.76 \text{ V} = 0.000 \text{ V} - \mathscr{E}^0_{Zn}$$

Solving for $\mathscr{E}^0_{Zn}$ gives

$$\mathscr{E}^0_{Zn} = -0.76 \text{ V}$$

The negative sign for $\mathscr{E}^0_{Zn}$ reflects the fact that Zn^{2+} is more difficult to reduce than H^+.

With a knowledge of the reduction potentials for the zinc and copper electrodes, we can now *predict* the cell potential and the spontaneous cell reaction for the Zn/Cu cell. This can be done even if we had no previous knowledge of the species undergoing oxidation or reduction.

First, simply by examining the reduction potentials,

$$Cu^{2+} \ (aq) + 2e^- \rightleftharpoons Cu \ (s) \qquad \mathscr{E}^0_{Cu} = +0.34 \text{ V}$$

$$Zn^{2+} \ (aq) + 2e^- \rightleftharpoons Zn \ (s) \qquad \mathscr{E}^0_{Zn} = -0.76 \text{ V}$$

we know immediately that Cu^{2+} is more easily reduced than Zn^{2+}. This is because Cu^{2+} has the higher (more positive) reduction potential. The cell reaction must therefore be

$$Zn \ (s) + Cu^{2+} \ (aq) \longrightarrow Cu \ (s) + Zn^{2+} \ (aq)$$

Next, the only way we can obtain a positive $\mathscr{E}^0_{cell}$ is to subtract $\mathscr{E}^0_{Zn}$ (the $\mathscr{E}^0$ of the substance oxidized) from $\mathscr{E}^0_{Cu}$ (the $\mathscr{E}^0$ of the substance reduced).

$$\mathscr{E}^0_{cell} = \mathscr{E}^0_{Cu} - \mathscr{E}^0_{Zn}$$

$$\mathscr{E}^0_{cell} = +0.34 \text{ V} - (-0.76 \text{ V})$$

$$\mathscr{E}^0_{cell} = +1.10 \text{ V}$$

This is precisely the value that we observe experimentally when we measure the potential of the cell.

We are now in a position to determine the reduction potentials for many different half-reactions, because all we have to do is construct galvanic cells in which the reduction potential of one half-cell is known, relative to the hydrogen electrode. Some standard reduction potentials, $\mathscr{E}^0$, determined in this manner are given in Table 17.1. In this table the reduction potential of the hydrogen electrode is placed in the middle, with the species more difficult to reduce than hydrogen listed below it and those more easily reduced placed above it. Such a table of reduction potentials serves many useful purposes.

1. From a table of reduction potentials we can, at a glance, pick out substances that are good oxidizing agents and those that are good reducing

Table 17.1
Standard reduction potentials at 25°C

Half-reaction	$\mathscr{E}°$ (volts)
$F_2 + 2e^- \rightleftharpoons 2F^-$	2.87
$S_2O_8^{2-} + 2e^- \rightleftharpoons 2SO_4^{2-}$	2.00
$H_2O_2 + 2H^+ + 2e^- \rightleftharpoons 2H_2O$	1.78
$PbO_2 + SO_4^{2-} + 4H^+ + 2e^- \rightleftharpoons PbSO_4 + 2H_2O$	1.69
$8H^+ + MnO_4^- + 5e^- \rightleftharpoons Mn^{2+} + 4H_2O$	1.49
$2ClO_3^- + 12H^+ + 10e^- \rightleftharpoons Cl_2 + 6H_2O$	1.47
$Cl_2 (g) + 2e^- \rightleftharpoons 2Cl^-$	1.36
$Cr_2O_7^{2-} + 14H^+ + 6e^- \rightleftharpoons 2Cr^{3+} + 7H_2O$	1.33
$MnO_2 + 4H^+ + 2e^- \rightleftharpoons Mn^{2+} + 2H_2O$	1.28
$O_2 + 4H^+ + 4e^- \rightleftharpoons 2H_2O$	1.23
$Br_2 (aq) + 2e^- \rightleftharpoons 2Br^-$	1.09
$Ag^+ + e^- \rightleftharpoons Ag$	0.80
$Fe^{3+} + e^- \rightleftharpoons Fe^{2+}$	0.77
$I_2 (aq) + 2e^- \rightleftharpoons 2I^-$	0.54
$Cu^+ + e^- \rightleftharpoons Cu$	0.52
$Cu^{2+} + 2e^- \rightleftharpoons Cu$	0.34
$Hg_2Cl_2 + 2e^- \rightleftharpoons 2Hg + 2Cl^-$	0.27
$AgCl + e^- \rightleftharpoons Ag + Cl^-$	0.22
$2H^+ + 2e^- \rightleftharpoons H_2$	0.00
$Fe^{3+} + 3e^- \rightleftharpoons Fe$	−0.04
$Pb^{2+} + 2e^- \rightleftharpoons Pb$	−0.13
$Sn^{2+} + 2e^- \rightleftharpoons Sn$	−0.14
$Ni^{2+} + 2e^- \rightleftharpoons Ni$	−0.25
$PbSO_4 + 2e^- \rightleftharpoons Pb + SO_4^{2-}$	−0.36
$Fe^{2+} + 2e^- \rightleftharpoons Fe$	−0.44
$Cr^{3+} + 3e^- \rightleftharpoons Cr$	−0.74
$Zn^{2+} + 2e^- \rightleftharpoons Zn$	−0.76
$2H_2O + 2e^- \rightleftharpoons H_2 + 2OH^-$	−0.83
$Mn^{2+} + 2e^- \rightleftharpoons Mn$	−1.03
$Al^{3+} + 3e^- \rightleftharpoons Al$	−1.67
$Mg^{2+} + 2e^- \rightleftharpoons Mg$	−2.38
$Na^+ + e^- \rightleftharpoons Na$	−2.71
$Ca^{2+} + 2e^- \rightleftharpoons Ca$	−2.76
$Ba^{2+} + 2e^- \rightleftharpoons Ba$	−2.90
$K^+ + e^- \rightleftharpoons K$	−2.92
$Li^+ + e^- \rightleftharpoons Li$	−3.05

agents. Any species that appears on the left of the double arrow serves as an oxidizing agent if it undergoes reduction during the course of a chemical reaction. Since the substances at the top left side of the table are more easily reduced than those at the bottom, their ability to serve as oxidizing agents decreases as we proceed down the table. Thus we could conclude, from their positions in this table, that H^+ is a better oxidizing agent than Zn^{2+}, and that F_2 is a better oxidizing agent than Cl_2. In brief, *good oxidizing agents are those species on the left of the double arrow at the top of the table.*

Each of the half-reactions listed in Table 17.1 is reversible. We saw, for example, that H_2 is oxidized to H^+ when placed in a cell with copper, and that H^+ is reduced to H_2 when placed against zinc. When the reactions in Table 17.1 are forced to proceed from right to left—that is, when they are caused to be the oxidation step in the overall reaction—then the species appearing at the right in Table 17.1 are functioning as reducing agents by being oxidized. *All the substances appearing on the right side of the reactions in Table 17.1 could behave as reducing agents; those species at the bottom right of the table, such as Li, are the best and those at the top right, such as F^-, are the poorest.*

2. Using Table 17.1 we can find rather quickly which combinations of reactants lead to spontaneous oxidation-reduction reactions (when the concentrations of the reactants and products are 1 M and the partial pressures of any gases involved are 1 atm). We can also determine whether or not a given reaction, as written, will proceed spontaneously in the forward direction.

Consider, for example, a mixture consisting of pieces of solid zinc and solid chromium in contact with a solution containing 1 M Zn^{2+} and 1 M Cr^{3+}. What reaction will occur in this mixture? To answer this question we look at the following half-reactions found in Table 17.1.

$$Cr^{3+} (aq) + 3e^- \rightleftharpoons Cr (s) \qquad \mathscr{E}^0_{Cr} = -0.74 \text{ V}$$

$$Zn^{2+} (aq) + 2e^- \rightleftharpoons Zn (s) \qquad \mathscr{E}^0_{Zn} = -0.76 \text{ V}$$

The reduction potentials tell us that Cr^{3+} is more easily reduced than Zn^{2+}, so in this mixture reduction of Cr^{3+} will occur and the zinc half-reaction will be forced to occur as oxidation. To obtain the cell reaction we combine the half-reactions in a way that makes the total number of electrons gained equal the total number lost.

$2 \times [Cr^{3+} (aq) + 3e^- \longrightarrow Cr (s)]$	(Reduction)
$3 \times [Zn (s) \longrightarrow Zn^{2+} (aq) + 2e^-]$	(Oxidation)

$$2Cr^{3+} (aq) + 3Zn (s) \longrightarrow 2Cr(s) + 3Zn^{2+} (aq) \qquad \text{Cell reaction}$$

To calculate the $\mathscr{E}^0_{cell}$ for this reaction, we use Equation 17.5.

$$\mathscr{E}^0_{cell} = \mathscr{E}^0_{\text{substance reduced}} - \mathscr{E}^0_{\text{substance oxidized}}$$

$$= \mathscr{E}^0_{Cr} - \mathscr{E}^0_{Zn}$$

$$= (-0.74 \text{ V}) - (-0.76 \text{ V})$$

$$= +0.02 \text{ V}$$

Notice that $\mathscr{E}^0_{cell}$ is positive, as it must be for a spontaneous change. Also notice that even though the half-reactions are multiplied by factors before they are combined, the reduction potentials are not. They are simply subtracted one from the other. This is because reduction potentials are intensive quantities and are therefore independent of the number of moles of reactants and products involved.

We have also said that we can determine whether a reaction, as written, will occur spontaneously. Let's consider the possible reaction

$$Fe^{2+} (aq) + Ni (s) \longrightarrow Fe (s) + Ni^{2+} (aq)$$

−0.74 V is more positive than −0.76 V.

If the reaction is spontaneous, the calculated $\mathscr{E}^0_{cell}$ will be positive, as it was above. On the other hand, if the reaction is not spontaneous in the direction written, the resulting calculated $\mathscr{E}^0_{cell}$ will be negative.

For the reaction we are considering, the first step is to divide the overall equation into two half-reactions.

$$Fe^{2+} (aq) + 2e^- \longrightarrow Fe (s)$$

$$Ni (s) \longrightarrow Ni^{2+} (aq) + 2e^-$$

We see that the first equation is a reduction and we can find its reduction potential, $\mathscr{E}^0_{Fe} = -0.44$ V, from Table 17.1. The second equation is an oxidation. If we were to rewrite it as a reduction, we would also be able to find its reduction potential, $\mathscr{E}^0_{Ni} = -0.25$ V. Since in our overall equation iron(II) is reduced and nickel is oxidized, when we substitute into Equation 17.5, we get

$$\mathscr{E}^0_{cell} = \mathscr{E}^0_{Fe} - \mathscr{E}^0_{Ni}$$

$$= -0.44 - (-0.25)$$

$$= -0.19 \text{ V}$$

Because $\mathscr{E}^0_{cell}$ is computed to be negative, the reaction of $Fe^{2+} (aq)$ with $Ni (s)$ is *not* spontaneous. In fact, it is the reverse reaction that is spontaneous.

EXAMPLE 17.4

What will be the spontaneous reaction between the following set of half-reactions? What is the value of $\mathscr{E}^0_{cell}$?

(1) $Cr^{3+} (aq) + 3e^- \rightleftharpoons Cr (s)$

(2) $MnO_2 (s) + 4H^+ (aq) + 2e^- \rightleftharpoons Mn^{2+} (aq) + 2H_2O (l)$

SOLUTION

From Table 17.1, reaction (1) has $\mathscr{E}^0 = -0.74$ V and reaction (2) has $\mathscr{E}^0 = +1.28$ V. Since the reduction potential of reaction (2) is larger (more positive) than that of reaction (1), reaction (2) will occur as reduction. Reaction (1) must be reversed and written as oxidation. To obtain the net overall reaction we multiply by appropriate coefficients so that electrons cancel.

$$3[MnO_2 (s) + 4H^+ (aq) + 2e^- \longrightarrow Mn^{2+} (aq) + 2H_2O (l)]$$

$$\underline{2[Cr (s) \longrightarrow Cr^{3+} (aq) + 3e^-]}$$

$$2Cr (s) + 3MnO_2 (s) + 12H^+ (aq) \longrightarrow 2Cr^{3+} (aq) + 3Mn^{2+} (aq) + 6H_2O (l)$$

To calculate $\mathscr{E}^0_{cell}$ we can subtract reduction potentials to obtain a positive value.

$$\mathscr{E}^0_{cell} = +1.28 - (-0.74)$$

$$= +2.02 \text{ V}$$

---

3. In the process of combining half-reactions from Table 17.1, we see that some of the reactants in the spontaneous reaction appear on the left side of one half-reaction while the rest of the reactants are found on the right side of another half-reaction. Among the reactants in Example 17.4, for instance, we have $MnO_2 + 4H^+$ from the left side of one half-reaction and Cr (s) from the right side of the other. The order of these reactions in

Table 17.1 is

$$MnO_2 \; (s) + 4H^+ \; (aq) + 2e^- \rightleftharpoons Mn^{2+} \; (aq) + 2H_2O \; (l) \qquad \mathscr{E}^0 = 1.28 \; V$$

$$Cr^{3+} \; (aq) + 3e^- \rightleftharpoons Cr \; (s) \qquad \mathscr{E}^0 = -0.74 \; V$$

and we see that the reactants in the overall spontaneous reaction are those substances related by the diagonal line (colored arrow) running from upper left to lower right. As a general statement, *we can say that when comparing reactants and products having unit concentrations, any species on the left of a given half-reaction will react spontaneously with a substance that is found on the right of a half-reaction located below it in Table 17.1.* We could use this rule of thumb, for example, to tell us that Br_2 will react spontaneously with I^- to produce Br^- and I_2, while Br_2 will *not* react spontaneously with Cl^-. Our rule, therefore, permits us to determine the course of a reaction without having to worry about subtracting electrode potentials in the proper sequence.

4. A point worth noting is that a collection of half-reactions, such as that found in Table 17.1, enables us to predict the outcome of many chemical reactions when we know only a relatively few half-reactions and their corresponding reduction potentials. From the 36 half-reactions listed in Table 17.1, for example, we can predict the results of 630 different chemical reactions! A table of this type, therefore, provides us with a very compact way of storing chemical information.

> The reduction potentials of hundreds of half-reactions have been measured.

5. With a knowledge of the standard reduction potentials listed in Table 17.1, we account for the course of electrolysis reactions. For example, we know from experiment that we can produce copper by electrolyzing an aqueous solution containing Cu^{2+}, but that we cannot obtain aluminum in this same fashion. From Table 17.1 we see that the reduction potential of copper is $+0.34 \; V$ and that for H_2O it is $-0.83 \; V$. Thus copper ion is more readily reduced than H_2O and will plate out on the electrode according to the half-reaction,

> There are complicating factors we haven't discussed that make it difficult to make accurate predictions about what will or will not be formed at an electrode during electrolysis.

$$Cu^{2+} \; (aq) + 2e^- \longrightarrow Cu \; (s)$$

In the case of aluminum, however, we find that the reduction potential for Al^{3+} is $-1.66 \; V$, which makes it more difficult to reduce than water. This means that when an aqueous solution containing Al^{3+} ions is electrolyzed, the H_2O will preferentially be reduced.

17.8 SPONTANEITY OF OXIDATION-REDUCTION REACTIONS

It was pointed out in Section 11.9 that the thermodynamic criterion for spontaneity of a chemical reaction is that the change in free energy, ΔG, has to be a negative quantity. In Section 11.10 we saw that ΔG also represents the maximum amount of useful work obtainable from a chemical reaction. The relationship between ΔG and maximum useful work (W_{max}) for any system takes the form

$$\Delta G = -W_{max}$$

But what is W_{max} for an electrochemical cell?

The work derived from an electrochemical cell might be compared to that obtained from a waterwheel, shown in Figure 17.14. The amount of work that can be obtained from this waterwheel depends on two things: (1) the volume of water flowing over the blades of the wheel, and (2) the energy given to the wheel per unit volume of water as it drops to the lower level of the stream:

$$\text{work} = (\text{volume of water}) \times \left(\frac{\text{energy released}}{\text{unit volume}}\right)$$

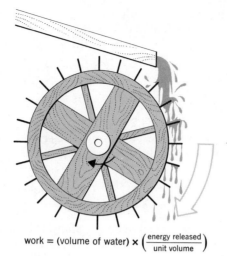

Figure 17.14

Work obtained from a waterwheel.

$$\text{work} = (\text{volume of water}) \times \left(\frac{\text{energy released}}{\text{unit volume}}\right)$$

Similarly, the work done by an electrochemical cell is dependent on (1) the number of coulombs that flow and (2) the energy available per coulomb:

$$\text{work} = (\text{number of coulombs}) \times \left(\frac{\text{energy available}}{\text{coulomb}}\right)$$

The number of coulombs that flow is equal to the number of moles of electrons that are involved in the redox reaction, n, multiplied by the faraday (which is the number of coulombs per mole of electrons):

$$\text{number of coulombs} = n\mathscr{F}$$

The value of n depends on the nature of the half-reactions taking place in the cell and can be derived once the specific reactions are known. For example, in the Zn/Cu cell there are two electrons involved in both of the half-reactions; therefore, n for this cell is 2.

The energy available per coulomb is simply the emf of the cell, because the volt is equal to the energy per coulomb (Equation 17.4):

$$\frac{\text{energy available}}{\text{coulomb}} = \text{emf}$$

When the emf is a maximum, the work derived from the cell is also a maximum. In Section 17.6 we saw that the maximum emf is the cell potential, $\mathscr{E}_{\text{cell}}$.

Thus the equation for maximum work for an electrochemical cell is

$$
\begin{array}{ccccccc}
W_{\text{max}} = & n & \times & \mathscr{F} & \times & \mathscr{E} \\
& \updownarrow & & \updownarrow & & \updownarrow \\
\text{joules} = & (\text{moles of electrons}) & \times & \left(\dfrac{\text{coulombs}}{\text{mole}}\right) & \times & \left(\dfrac{\text{joules}}{\text{coulomb}}\right)
\end{array}
$$

Since $\Delta G = -W_{\text{max}}$, then

$$\Delta G = -n\mathscr{F}\mathscr{E}_{\text{cell}} \qquad\qquad [17.6]$$

When all species are at unit concentration as identified by the superscript zero in $\mathscr{E}^0_{\text{cell}}$, then ΔG becomes the standard free-energy change for the reaction, ΔG^0. Thus, Equation 17.6 becomes

$$\Delta G^0 = -n\mathscr{F}\mathscr{E}^0_{\text{cell}} \qquad\qquad [17.7]$$

With the aid of Equation 17.7 we can calculate the standard free-energy change for an oxidation-reduction reaction from a knowledge of its standard cell potential. Consider, for example, the Zn/Cu cell.

$$\text{Zn } (s) \longrightarrow \text{Zn}^{2+} (aq) + 2e^-$$
$$\underline{\text{Cu}^{2+} (aq) + 2e^- \longrightarrow \text{Cu } (s)}$$
$$\text{Zn } (s) + \text{Cu}^{2+} (aq) \longrightarrow \text{Zn}^{2+} (aq) + \text{Cu } (s)$$

The n for this reaction is 2 (because two electrons are transferred), $\mathscr{F} = 96{,}500$ C/mol of electrons, and $\mathscr{E}^0_{cell}$, either derived from Table 17.1 or determined experimentally, is $+1.10$ V; therefore,

$$\Delta G^0 = -2 \text{ mol } e^- \times \left(\frac{96{,}500 \text{ C}}{\text{mol } e^-}\right) \times \left(\frac{+1.10 \text{ J}}{\text{C}}\right)$$
$$= -212{,}000 \text{ J}$$

This relationship between the standard cell potential and ΔG^0 is extremely important, because it ties together two different aspects of spontaneity while at the same time it gives us a readily accessible pathway for calculating standard free-energy changes. Experimentally, we have only to measure the standard cell emf, from which we can then compute the value of ΔG^0. But still more important, perhaps, is that through this equation we can derive even more useful thermodynamic quantities.

17.9 THERMODYNAMIC EQUILIBRIUM CONSTANTS

In Section 13.3 we saw that for reactions in solution,

$$\Delta G^0 = -2.303RT \log K_c \qquad [17.8]$$

Combining Equations 17.7 and 17.8 gives us

$$\Delta G^0 = -n\mathscr{F}\mathscr{E}^0 = -2.303RT \log K_c$$

or simply

$$n\mathscr{F}\mathscr{E}^0 = 2.303RT \log K_c$$

Solving for $\mathscr{E}^0$, we have

$$\mathscr{E}^0 = \frac{2.303RT}{n\mathscr{F}} \log K_c \qquad [17.9]$$

If we choose to restrict ourselves to discussing reactions that take place at 25°C (298 K), the quantity $2.303RT/\mathscr{F}$ becomes a constant,

$$\frac{2.303RT}{\mathscr{F}} = \frac{2.303(8.314 \text{ J/mol K})(298 \text{ K})}{96{,}500 \text{ C/mol}} = 0.0592 \text{ J/C}$$

Since 1 V = 1 J/C,

$$\frac{2.303RT}{\mathscr{F}} = 0.0592 \text{ V}$$

Thus, at 25°C, Equation 17.9 becomes

$$\mathscr{E}^0 = \frac{0.0592}{n} \log K_c$$

Solving for K_c gives

$$\log K_c = \frac{n\mathscr{E}^0}{0.0592} \qquad [17.10]$$

We see, therefore, that from a knowledge of the standard cell potential, the equilibrium constant for the cell reaction can be calculated. For the Zn/Cu cell, we have

$$\log K_c = \frac{n \mathscr{E}^0}{0.0592} = \frac{2(+1.10)}{0.0592} = 37.2$$

Hence,

$$K_c \approx 1 \times 10^{37}$$

From the magnitude of this equilibrium constant, we could certainly say that the spontaneous Zn/Cu cell reaction will go very nearly to completion.

EXAMPLE 17.5   Using Table 17.1, determine whether the oxidation-reduction reaction

$$Sn \ (s) + Ni^{2+} \longrightarrow Sn^{2+} + Ni \ (s)$$

is spontaneous, and calculate its equilibrium constant at 25°C.

SOLUTION   The two half-reactions for this overall reaction are

$$Sn \ (s) \longrightarrow Sn^{2+} \ (aq) + 2e^- \quad \text{(Oxidation)}$$
$$Ni^{2+} \ (aq) + 2e^- \longrightarrow Ni \ (s) \quad \text{(Reduction)}$$

From Table 17.1 we find the reduction potential for the half-reaction involving Sn and Sn^{2+} to be $\mathscr{E}^0_{Sn} = -0.14$ V; for the nickel half-reaction $\mathscr{E}^0_{Ni} = -0.25$ V. Applying Equation 17.5, we get

$$\mathscr{E}^0_{cell} = -0.25 - (-0.14)$$
$$= -0.11 \text{ V}$$

This means that under *standard conditions* (unit concentrations), the reaction in this direction is nonspontaneous. We can still calculate the equilibrium constant in the same fashion as outlined above. Since two electrons are transferred during the reaction,

$$\log K_c = \frac{2(-0.11)}{0.0592} = -3.7$$

Taking the antilogarithm gives

$$K_c = 2 \times 10^{-4}$$

From the size of this equilibrium constant, we can say that this reaction will not occur to an appreciable extent in the forward direction.

17.10 CONCENTRATION EFFECT ON CELL POTENTIAL

Thus far we have limited our discussion to those cells containing reactants at unit concentration. In the laboratory, however, we usually do not restrict ourselves to only this one set of conditions, and it is found that the cell emf, and in fact the direction of the cell reaction, can be controlled by the concentrations of the species taking part in the reaction. Let us examine this now from a quantitative point of view.

An equation that summarizes how the free energy of the reactants and products of a given reaction varies with temperature and concentration was given in Section 13.3. For the generalized reaction,

$$aA + bB \longrightarrow eE + fF$$

this equation takes the form

$$\Delta G = \Delta G^0 + 2.303RT \log \left(\frac{[E]^e[F]^f}{[A]^a[B]^b} \right)$$

The quantity, $\frac{[E]^e[F]^f}{[A]^a[B]^b}$, is the mass action expression for the reaction.

Equations 17.6 and 17.7 (in Section 17.8) show the relationship between ΔG and $\mathscr{E}$, and ΔG^0 and $\mathscr{E}^0$, respectively, for an oxidation-reduction reaction. Substituting these expressions for ΔG and ΔG^0 into the above equation, we have

$$-n\mathscr{F}\mathscr{E} = -n\mathscr{F}\mathscr{E}^0 + 2.303RT \log \left(\frac{[E]^e[F]^f}{[A]^a[B]^b} \right)$$

which can be rearranged to give

$$\mathscr{E} = \mathscr{E}^0 - \frac{2.303RT}{n\mathscr{F}} \log \left(\frac{[E]^e[F]^f}{[A]^a[B]^b} \right) \qquad [17.11]$$

It was Walter Nernst who discovered the third law of thermodynamics.

This equation, first developed by Walter Nernst in 1889, now bears his name and is called the **Nernst equation.**

At 25°C we have seen that the numerical value of $2.303RT/\mathscr{F}$ is 0.0592. Therefore, at 25°C the Nernst equation becomes

$$\mathscr{E} = \mathscr{E}^0 - \frac{0.0592}{n} \log \left(\frac{[E]^e[F]^f}{[A]^a[B]^b} \right) \qquad [17.12]$$

$\log 1 = 0$

We can see from Equation 17.12 that when all ionic species are present at unit concentration, the log term becomes zero and the emf of the cell becomes $\mathscr{E}^0$—that is, at unit concentration $\mathscr{E} = \mathscr{E}^0$. This, of course, must be true in light of our basic definition of $\mathscr{E}^0$. When the species in a cell are not present at unit concentration, $\mathscr{E}$ is generally not equal to $\mathscr{E}^0$ and the Nernst equation must be employed to calculate $\mathscr{E}$. For example, in the case of the Zn/Cu cell, whose cell reaction is

$$Zn\ (s)\ +\ Cu^{2+}\ (aq) \longrightarrow Cu\ (s)\ +\ Zn^{2+}\ (aq)$$

the Nernst equation takes the form

$$\mathscr{E} = \mathscr{E}^0 - \frac{0.0592}{n} \log \frac{[Zn^{2+}]}{[Cu^{2+}]}$$

Note that as usual we omit the concentrations of pure solids from the mass action expression.[3] Since two electrons are transferred in the reaction, n = 2 and

$$\mathscr{E} = \mathscr{E}^0 - 0.0296 \log \frac{[Zn^{2+}]}{[Cu^{2+}]}$$

Thus, $\mathscr{E}$ can be calculated for any particular set of concentrations of Zn^{2+} and Cu^{2+}. This use of the Nernst equation is illustrated in the next example.

EXAMPLE 17.6   Calculate the emf of the Zn/Cu cell under the following conditions:

$$Zn\ (s)\ +\ Cu^{2+}\ (0.020\ M) \longrightarrow Cu\ (s)\ +\ Zn^{2+}\ (0.40\ M)$$

SOLUTION   We have just seen that for this system the Nernst equation is

$$\mathscr{E} = \mathscr{E}^0 - 0.0296 \log \frac{[Zn^{2+}]}{[Cu^{2+}]}$$

[3] The Nernst equation applies exactly only if we use activities. The activity of any pure solid or liquid is equal to 1. Errors introduced by using the concentrations of the ions instead of their activities are small, as mentioned before, provided that the solutions are relatively dilute.

We can also calculate $\mathscr{E}^0_{cell} = +1.10$ V for the equation as written. Substituting this $\mathscr{E}^0$ and the concentrations of the Zn^{2+} and Cu^{2+} into the Nernst equation, we have

$$\mathscr{E} = 1.10 - 0.0296 \log \frac{(0.40)}{(0.020)}$$

$$= 1.10 - 0.0296 \log 20$$

$$= 1.10 - 0.0296(1.30)$$

$$= 1.10 - 0.0385$$

$$= 1.06 \text{ V}$$

Thus we see that under these conditions of concentration the voltage obtained from this cell is slightly less than that obtained at unit concentration.

17.11 APPLICATIONS OF THE NERNST EQUATION

Just as the cell emf is dependent on the concentrations of the ions involved in the half-reactions, so we find that the reduction potential of the individual half-reactions is also determined by the concentrations of the ions involved. This effect of the concentration on the reduction potential can also be given by the Nernst equation. For example, if we consider the half-reaction,

$$Zn^{2+} (aq) + 2e^- \rightleftharpoons Zn (s)$$

the Nernst equation at 25°C takes the form

$$\mathscr{E}_{Zn} = \mathscr{E}^0_{Zn} - \frac{0.0592}{2} \log \frac{1}{[Zn^{2+}]}$$

As usual, we have omitted the concentration of the solid from the mass action expression.

Because the reduction potential of an electrode depends on the concentrations of the ions in solution, it is possible to construct a cell in which the cathode and anode compartments contain the same electrode materials but different concentrations of the ions. Such a cell is called a **concentration cell** and is illustrated in Figure 17.15.

In Figure 17.15 we have a cell composed of two zinc electrodes dipping into separate solutions of $ZnSO_4$ whose Zn^{2+} concentrations are different. The con-

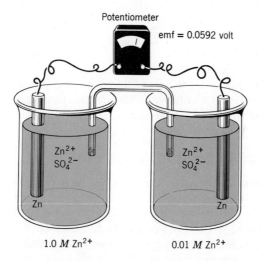

Potentiometer

emf = 0.0592 volt

Zn^{2+}
SO_4^{2-}

Zn^{2+}
SO_4^{2-}

Zn

Zn

Figure 17.15

A concentration cell.

1.0 M Zn^{2+} 0.01 M Zn^{2+}

centration of the Zn^{2+} on the left (1.0 M) is 100 times greater than the Zn^{2+} concentration in the right compartment, and when the circuit is completed a spontaneous reaction takes place in a direction that tends to make the two Zn^{2+} concentrations become equal. Thus, in the more concentrated side Zn^{2+} ions disappear forming Zn (s), in order to decrease the Zn^{2+} concentration, and in the more dilute side, more Zn^{2+} will be produced. Thus, in the more concentrated compartment, we have

$$Zn^{2+} \text{ (1 } M) + 2e^- \longrightarrow Zn \text{ (s)} \quad \text{(Reduction)}$$

and on the more dilute side,

$$Zn \text{ (s)} \longrightarrow Zn^{2+} \text{ (0.01 } M) + 2e^- \quad \text{(Oxidation)}$$

From our earlier discussions we know that the potential of a cell is found by subtracting the reduction potential of the half-cell in which oxidation occurs from the reduction potential of the half-cell that undergoes reduction. For this concentration cell

$$\mathscr{E}_{cell} = \mathscr{E}_{conc} - \mathscr{E}_{dil}$$

where $\mathscr{E}_{conc}$ and $\mathscr{E}_{dil}$ are the electrode potentials of the concentrated and dilute half-cells, respectively. These are given as

$$\mathscr{E}_{conc} = \mathscr{E}^0_{Zn} - \frac{0.0592}{2} \log \frac{1}{[Zn^{2+}]_{conc}}$$

and

$$\mathscr{E}_{dil} = \mathscr{E}^0_{Zn} - \frac{0.0592}{2} \log \frac{1}{[Zn^{2+}]_{dil}}$$

Therefore,

$$\mathscr{E}_{cell} = (\mathscr{E}^0_{Zn} - \mathscr{E}^0_{Zn}) - \frac{0.0592}{2} \left(\log \frac{1}{[Zn^{2+}]_{conc}} - \log \frac{1}{[Zn^{2+}]_{dil}} \right)$$

or

$$\mathscr{E}_{cell} = - \frac{0.0592}{2} \log \frac{[Zn^{2+}]_{dil}}{[Zn^{2+}]_{conc}}$$

Substituting the concentrations of Zn^{2+} into this expression allows us to compute the cell potential.

$$\mathscr{E}_{cell} = - \frac{0.0592}{2} \log \frac{(0.01)}{(1)}$$

$$= 0.0592 \text{ V}$$

In general, for any concentration cell we could write

$$\mathscr{E}_{cell} = - \frac{0.0592}{n} \log \frac{[M^{n+}]_{dil}}{[M^{n+}]_{conc}}$$

The voltage obtained from this type of cell is usually small and will continually decrease as the concentrations in the two compartments approach each other. The voltage becomes zero when the concentrations of the ions in each compartment of the cell are the same.

Solubility product constant

The Nernst equation can also be useful in determining the solubility product constant of an insoluble salt. To find the K_{sp} of $PbSO_4$, for example, an experiment might be designed in the following fashion: a galvanic cell is prepared consisting of Pb/Pb^{2+} versus an Sn/Sn^{2+} electrode with a salt bridge connecting them. In the tin compartment the Sn^{2+} concentration is held constant at 1 M. In the lead compartment SO_4^{2-} is added to precipitate $PbSO_4$ and thereby establish

the equilibrium

$$PbSO_4 \; (s) \rightleftharpoons Pb^{2+} \; (aq) + SO_4{}^{2-} \; (aq)$$

The $SO_4{}^{2-}$ concentration in the lead compartment is then adjusted until it is 1 M, and the emf of the cell is found to be $+0.22$ V. It is also observed that the Pb electrode is negative with respect to the Sn electrode, thereby indicating that the Pb is undergoing oxidation while the Sn^{2+} is reduced. The cell reaction must therefore be

$$Pb \; (s) + Sn^{2+} \; (1 \; M) \longrightarrow Pb^{2+} \; (?) + Sn \; (s)$$

and the calculated $\mathscr{E}^0$ is

$$\mathscr{E}^0 = \mathscr{E}_{Sn} - \mathscr{E}_{Pb}$$
$$= (-0.14 \; V) - (-0.13 \; V)$$
$$= -0.01 \; V$$

(Note that *if* the Sn^{2+} and Pb^{2+} concentrations were both 1 M, the reaction would be spontaneous from right to left, rather than from left to right.)

We can calculate the concentration of Pb^{2+} by using the Nernst equation for this cell reaction which takes the form

$$\mathscr{E} = \mathscr{E}^0 - \frac{0.0592}{n} \log \frac{[Pb^{2+}]}{[Sn^{2+}]}$$

We know $\mathscr{E}$, $\mathscr{E}^0$, and $[Sn^{2+}]$ and, because there are two electrons transferred in this reaction, the above equation becomes, after substitution,

$$0.22 \; V = -0.01 \; V - \frac{0.0592}{2} \log \frac{[Pb^{2+}]}{(1)}$$

Next we solve for $\log[Pb^{2+}]$.

$$-0.22 \; V = 0.01 \; V + 0.0296 \log[Pb^{2+}]$$

or

$$\log[Pb^{2+}] = \frac{-0.22 \; V - 0.01 \; V}{0.0296 \; V}$$

and

$$\log[Pb^{2+}] = -7.8$$

Taking the antilogarithm, we find that the concentration of Pb^{2+} in this cell is

$$[Pb^{2+}] = 2 \times 10^{-8} \; M$$

The expression for K_{sp} of $PbSO_4$ is

$$K_{sp} = [Pb^{2+}][SO_4{}^{2-}]$$

Since

$$[Pb^{2+}] = 2 \times 10^{-8}$$
$$[SO_4{}^{2-}] = 1 \; M$$

then

$$K_{sp} = (2 \times 10^{-8})(1) = 2 \times 10^{-8}$$

which is the value that was given in Table 16.1.

Determination of pH

An extremely important application of the Nernst equation is its use in calculating the concentration of a single ionic species from the experimentally measured potential of a carefully designed cell. We have already seen one example of this in the determination of K_{sp}. If we were to use the Cu/H_2 cell, discussed in Section 17.7, we could determine the H^+ concentration of a solution and then calculate its pH. The cell reaction for the Cu/H_2 cell is

$$Cu^{2+} (aq) + H_2 (g) \longrightarrow Cu (s) + 2H^+ (aq)$$

and the corresponding form of the Nernst equation at 25°C is

p_{H_2} is the partial pressure of H_2.

$$\mathscr{E} = \mathscr{E}^0 - \frac{0.0592}{n} \log \frac{[H^+]^2}{[Cu^{2+}]p_{H_2}}$$

If the concentration of Cu^{2+} is 1 M and the pressure of H_2 is 1 atm, this equation reduces to

$$\mathscr{E} = \mathscr{E}^0 - \frac{0.0592}{n} \log[H^+]^2$$

which is the same as

$\log x^2 = 2 \log x$

$$\mathscr{E} = \mathscr{E}^0 - \frac{(0.0592)(2)}{n} \log[H^+]$$

Let's rewrite this equation as

$$\mathscr{E} = \mathscr{E}^0 + \frac{(0.0592)(2)}{n} (-\log[H^+])$$

We see that because $\mathscr{E}^0$ and the quantity $0.0592(2)/n$ are both constant for a specific reaction, then

$$\mathscr{E} \propto -\log[H^+]$$

By definition, pH $= -\log[H^+]$ and, therefore, we have

$$\mathscr{E} \propto pH$$

Thus, by measuring the emf of a galvanic cell containing a reference electrode (such as the Cu, Cu^{2+} electrode here) and the hydrogen electrode, the pH of a solution can be calculated. One application is shown in the next example.

EXAMPLE 17.7   A galvanic cell consisting of a Cu versus a hydrogen electrode was used to determine the pH of an unknown solution. The unknown was placed in the hydrogen electrode compartment and the pressure of the hydrogen gas was controlled at 1 atm. The concentration of Cu^{2+} was 1 M and the emf of the cell at 25°C was determined to be +0.48 V. Calculate the pH of this unknown solution.

SOLUTION   The cell reaction for the Cu/H_2 cell is

$$Cu^{2+} (1\ M) + H_2 (g)(1\ atm) \longrightarrow Cu (s) + 2H^+ (?M)$$

for which we write the Nernst equation as

$$\mathscr{E} = \mathscr{E}^0 - \frac{0.0592}{n} \log \frac{[H^+]^2}{[Cu^{2+}]p_{H_2}}$$

Because $[Cu^{2+}] = 1\ M$ and $p_{H_2} = 1$ atm,

$$\mathscr{E} = \mathscr{E}^0 - \frac{(0.0592)(2)}{n} \log[H^+]$$

The $\mathscr{E}^0$ for this cell is +0.34 V; the value of n is 2. Substituting these values as well as the measured value of $\mathscr{E}$ into the equation, we have

$$+0.48 = +0.34 - 0.0592 \log[H^+]$$

and hence

$$-\log[H^+] = \frac{0.48 - 0.34}{0.0592} = 2.4$$

Therefore the pH of this solution is 2.4.

17.12 ION-SELECTIVE ELECTRODES

The last example represents only a single case where, with the proper choice of electrodes, the concentration of a single ionic species can be selectively measured. Through many years of research in this particular area of electrochemistry, scientists have developed many practical electrodes whose emf depends on the concentration of only one species. Such electrodes are called **ion-selective electrodes** and are used in conjunction with a reference electrode, whose potential always remains constant and is of a known value. Thus, when an ion-selective electrode is placed in a solution with a reference electrode, only the ion-selective electrode changes in emf, and the measured voltage can immediately be used to calculate the concentration of the species being determined. Such electrodes have been found to be of great importance in such areas as chemical analysis, pollution analysis, clinical measurements, oceanography, and geology.

One type of ion-selective electrode, shown in Figure 17.16, consists of a very thin-walled membrane that is sealed onto one end of a hollow tube. Inside

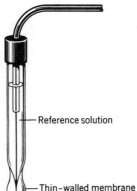

— Reference solution

Figure 17.16

Construction of an ion-selective electrode.

— Thin-walled membrane

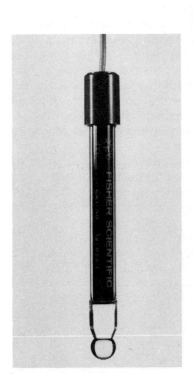

A glass electrode.

the tube in contact with the membrane is a reference solution and immersed into this solution is a wire. The wire extends from the reference solution through the tube and out the top, to make electrical contact with the outside circuitry. The material used to make the membrane as well as the composition of the reference solution depends on the species that is to be measured. Some of the cations and anions whose concentration can be determined by these electrodes are listed in Table 17.2.

The **glass electrode** is one ion-selective electrode. The membrane in this electrode is made of an extremely thin piece of glass, and the reference solution inside is a dilute HCl solution whose H^+ concentration is known and which remains constant. The wire electrode is a silver wire coated with silver chloride. The emf of the glass electrode is sensitive to the relative concentrations of H^+ in-

- Glass electrode sensitive to ammonium ion

- Enzyme suspended in gel that coats the thin-walled membrane

Figure 17.17

An enzyme-substrate electrode.

Table 17.2
Ions whose concentration can be determined by an ion-selective electrode

Cations	Anions
Cadmium (Cd^{2+})	Bromide (Br$^-$)
Calcium (Ca^{2+})	Chloride (Cl$^-$)
Copper (Cu^{2+})	Cyanide (CN$^-$)
Hydrogen (H$_3$O$^+$)	Fluoride (F$^-$)
Lead (Pb^{2+})	Nitrate (NO$_3^-$)
Mercury (Hg^{2+})	Perchlorate (ClO$_4^-$)
Potassium (K$^+$)	Sulfide (S^{2-})
Silver (Ag$^+$)	Thiocyanate (SCN$^-$)
Sodium (Na$^+$)	

side and outside across the thin glass membrane. Since the H$^+$ concentration inside is constant, the emf of the electrode, in effect, is determined by the concentration of H$^+$ in the solution in contact with the membrane on the outside.

The glass electrode can be made selective to various other monovalent cations such as Na$^+$, K$^+$, and NH$_4^+$ by suitable changes in the composition of the glass. Still other ions can be detected by electrodes if the glass membrane is replaced by a solid crystal. For example, when a solid crystal such as LaF$_3$ is used, fluoride ion concentrations can be measured, and when Ag$_2$S is used, silver and sulfide ion concentrations can be determined. Other ions such as chloride, cyanide, and lead can be determined using a membrane made of silver mixed with silver sulfide.

Ion-selective electrodes have also become very important and useful in the study of biological processes. One of these, called an **enzyme-substrate electrode,** employs a glass electrode sensitive to ammonium ion coated with a thin layer of a gel containing an enzyme, as shown in Figure 17.17. Enzymes are large organic molecules that catalyze very specific chemical reactions in biochemical systems and if, for example, the enzyme in the gel is urease, the decomposition of urea to produce ammonium ion will occur when the solution around the electrode contains urea. This ammonium ion is detected by the electrode and, in effect, the electrode becomes sensitive to the presence of urea in the solution being tested.

The application of electrochemistry to biochemical research is still in its infancy. An illustration of how this field of research is progressing is the relatively recent development of miniature ion-selective electrodes that can actually be placed into a living cell to monitor changes in the concentrations of ions during the life process.

The field of medicine is also putting ion-selective electrodes to use. Miniaturized glass electrodes have been developed that fit within a small hypodermic syringe and that can monitor pH in capillary blood vessels. It is anticipated that they will be useful in monitoring respiratory distress (which shows up as a blood pH change) in the human fetus.

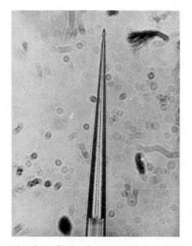

A microelectrode so small that it is able to monitor activity within a cell.

17.13 SOME PRACTICAL GALVANIC CELLS

As the final section in this chapter, let's discuss some galvanic cells that play an important part in our lives by providing us with electrical power.

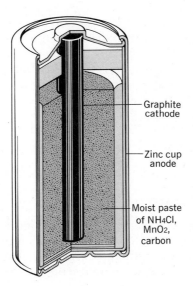

Figure 17.18
The dry cell.

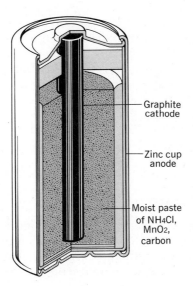

Graphite cathode

Zinc cup anode

Moist paste of NH4Cl, MnO2, carbon

The common dry cell is technically known as a Leclanché cell.

The physical size of a cell doesn't affect its voltage, but it does affect the amount of current that can be delivered. A small AA battery doesn't produce as much current as a larger D-size cell.

We say the cell has become polarized.

Zinc-carbon dry cell

This type of cell is used in flashlights, portable radios, toys, and the like. A cutaway diagram of a typical zinc-carbon dry cell is shown in Figure 17.18. It has an exterior layer of either cardboard or metal which serves only as a seal against the atmosphere. Inside this outer shell is a zinc cup that serves as the anode. The zinc cup is filled with a moist paste consisting of ammonium chloride, manganese dioxide, and finely divided carbon. Immersed in this paste is a graphite rod, which serves as the cathode. The chemical reactions that take place when the circuit is completed are actually quite complex and, in fact, are not completely understood. The following, however, is perhaps a reasonable estimate of what occurs.

At the anode zinc is oxidized,

$$Zn\ (s) \longrightarrow Zn^{2+}\ (aq) + 2e^- \qquad \text{(Anode)}$$

while at the carbon cathode the MnO_2/NH_4Cl mixture undergoes reduction to give a complex mixture of products. One of these reactions appears to be

$$2MnO_2\ (s) + 2NH_4^+\ (aq) + 2e^- \longrightarrow Mn_2O_3\ (s) + 2NH_3\ (aq) + H_2O\ (l)\ \text{(Cathode)}$$

The ammonia produced at the cathode reacts with part of the Zn^{2+} formed at the anode to give the complex ion, $Zn(NH_3)_4^{2+}$. Because of the complex nature of the dry cell, no simple overall cell reaction can be written.

The common zinc-carbon dry cell suffers from the disadvantage that under heavy use it rather quickly appears to become "dead." Yet, if allowed to rest for a while, it appears to come back to life and can deliver additional current. When the cell delivers current the reaction products can't diffuse away from electrodes very quickly. As they accumulate it becomes more difficult for the electrode reactions to occur, and the voltage of the cell drops. After sitting idle for a while, however, these products diffuse away from the electrodes and the cell regains the ability to function.

Dry cells cannot be effectively recharged and, therefore, have a relatively short lifetime (as compared to the rechargeable lead storage and nickel-cadmium batteries, for example).

Alkaline battery

Another type of dry cell that uses zinc and manganese dioxide as reactants is the alkaline dry cell. Once again, zinc serves as the anode and manganese dioxide functions as the cathode. The electrolyte, however, contains potassium hydroxide and is therefore basic (alkaline). The zinc anode is also slightly porous, giving it a larger effective area. This allows the cell to deliver more current than the common zinc cell. As you've probably learned from TV commercials, these batteries are able to stand up better under heavy use and have a longer shelf life. The reactions in the alkaline battery are

$$Zn\ (s) + 2OH^-\ (aq) \longrightarrow Zn(OH)_2\ (s) + 2e^- \qquad \text{(Anode)}$$

$$2MnO_2\ (s) + 2H_2O + 2e^- \longrightarrow 2MnO(OH)\ (s) + 2OH^-\ (aq) \qquad \text{(Cathode)}$$

The cell produces an emf of about 1.5 V.

Silver oxide battery

These tiny and rather expensive batteries (Figure 17.19) have become popular as power sources in electronic wristwatches, auto exposure cameras, and electronic calculators. The cathode reactant is silver oxide, Ag_2O, and the anode once again is zinc. The electrode reactions occur in a basic electrolyte.

$$Zn\ (s) + 2OH^-\ (aq) \longrightarrow Zn(OH)_2\ (s) + 2e^- \qquad \text{(Anode)}$$

$$Ag_2O\ (s) + H_2O + 2e^- \longrightarrow 2Ag\ (s) + 2OH^-\ (aq) \qquad \text{(Cathode)}$$

The emf of this battery is about 1.5 V.

Lead storage battery

The common automobile battery is a lead storage battery that usually delivers either 6 or 12 V, depending on the number of cells used in its construction. The inside of the battery consists of a number of galvanic cells connected to each other in series (Figure 17.20).

To increase the current output, each of the individual cells contains a number of lead anodes connected together, plus a number of cathodes, composed of PbO_2, also joined together. These electrodes are immersed in an electrolyte composed of dilute sulfuric acid (actually about 30% by weight in a fully charged cell). A single lead storage cell delivers 2 V, so that a 12-V battery contains six such cells connected in series.

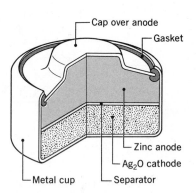

Figure 17.19

A silver oxide battery.

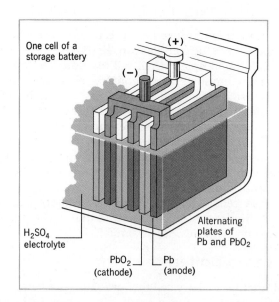

Figure 17.20

Lead storage battery.

When the external circuit is complete and the battery is in operation, the following oxidation-reduction reactions take place:

(Anode) $Pb\ (s) + SO_4^{2-}\ (aq) \longrightarrow PbSO_4\ (s) + 2e^-$

(Cathode) $PbO_2\ (s) + 4H^+\ (aq) + SO_4^{2-}\ (aq) + 2e^- \longrightarrow PbSO_4\ (s) + 2H_2O$

and the overall reaction is

$$Pb\ (s) + PbO_2\ (s) + 4H^+\ (aq) + 2SO_4^{2-}\ (aq) \longrightarrow 2PbSO_4\ (s) + 2H_2O$$

A hydrometer being used to check the state of charge of a lead storage battery.

These batteries have the advantage that the electrode reactions can be reversed by placing a voltage across the electrodes that is slightly larger than that which the battery can deliver. The recharging operation is performed in such a way that the negative external voltage is applied to the negative pole and the positive voltage to the positive pole. In doing this, the H_2SO_4 that is used up while the battery is in operation is restored. Recharging is accomplished by the generator or alternator of the car, or, if the battery is really run down, with the aid of a battery charger.

A convenient method of estimating the degree to which the battery has been discharged is by checking the density (or specific gravity) of the electrolyte. If the battery is in a weakened state, the electrolyte will be mostly water—a product of the overall reaction—and will have a density somewhere near 1 g/ml. If, however, the battery is in good operating order, with a full charge, the density of the electrolyte will be somewhat higher than 1 g/ml (the density of concentrated sulfuric acid is 1.8 g/ml). The mechanic in a garage can perform this test with the aid of a **hydrometer,** a device having a float that sinks to a depth that is a function of the density of the liquid in which it is immersed.

Nickel-cadmium cell

A storage cell that has acquired widespread use in recent years is the "nicad," or nickel-cadmium battery. The anode in the cell is composed of cadmium, which undergoes oxidation in an alkaline (basic) electrolyte.

(Anode) $Cd\ (s) + 2OH^-\ (aq) \longrightarrow Cd(OH)_2\ (s) + 2e^-$

The cathode is composed of NiO_2, which undergoes reduction.

(Cathode) $NiO_2\ (s) + 2H_2O + 2e^- \longrightarrow Ni(OH)_2\ (s) + 2OH^-\ (aq)$

The net cell reaction during discharge is therefore

$$Cd\ (s) + NiO_2\ (s) + 2H_2O \longrightarrow Cd(OH)_2\ (s) + Ni(OH)_2\ (s)$$

The voltage of the cell is about 1.4 V, somewhat less than the dry cell.

The nicad battery has some appealing features. First, it has a longer life than a lead storage battery. Second, it can be packaged in a sealed unit, much like the common dry cell. These advantages have made the nicad the choice among manufacturers of such devices as rechargeable calculators and electronic flash units in photography.

Fuel cells

Fuel cells are another means by which chemical energy may be converted into electrical energy. When gaseous fuels, such as H_2 and O_2, are allowed to undergo reaction in a carefully designed environment, electrical energy can be obtained. This type of cell finds great importance in space vehicles, where the fuels used in such cells can be the same as those used to power the rockets.

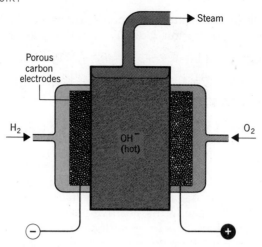

Figure 17.21

Hydrogen-oxygen fuel cell.

A diagram of a H_2/O_2 fuel cell is shown in Figure 17.21. In this cell there are three compartments separated from one another by porous electrodes. The hydrogen gas is fed into one compartment and the oxygen gas is fed into another. These gases then diffuse (not bubble) slowly through the electrodes and react with an electrolyte that is in the center compartment. The electrodes are made of a conducting material, such as carbon, with a sprinkling of platinum to act as a catalyst, and the electrolyte is an aqueous solution of a base.

At the cathode the oxygen undergoes reduction, producing OH^- ions. This can be expressed as

$$O_2\,(g) + 2H_2O\,(l) + 4e^- \longrightarrow 4OH^-\,(aq)$$

These OH^- ions travel to the anode where they undergo reaction with H_2:

$$H_2\,(g) + 2OH^-\,(aq) \longrightarrow 2H_2O\,(l) + 2e^-$$

The net reaction in the cell is

$$2H_2\,(g) + O_2\,(g) \longrightarrow 2H_2O\,(l)$$

The fuel cell is operated at a high temperature so that the water that is formed as a product of the cell reaction evaporates and may be condensed and used as drinking water for an astronaut. A number of these cells are usually connected together so that several kilowatts of power can be produced.

Fuel cells offer several advantages over other sources of energy. Unlike the dry cell or storage battery, the cathode and anode reactants may be continually supplied so that, in principle, energy can be withdrawn indefinitely from a fuel cell as long as the outside supply of fuel is maintained. Another advantage of the fuel cell is that the energy is extracted from the reactants under more nearly reversible conditions. Therefore, the thermodynamic efficiency of the reaction, in terms of producing useful work, is higher than when the reactants such as H_2 and O_2 are burned to produce heat that must be subsequently harnessed to produce work. These two advantages suggest that the development of fuel cells will probably continue at an accelerated pace in the future, particularly in light of recent energy shortages caused by greater demands for energy.

Fuel cells have efficiencies that can approach 75%, whereas power plants that burn fuels have efficiencies of only about 40%.

INDEX TO QUESTIONS AND PROBLEMS (Problem numbers are in **bold type**)

REVIEW QUESTIONS

17.1 Distinguish between: (a) electrolytic and galvanic cells, (b) metallic and electrolytic conduction, (c) oxidation and reduction.

17.2 Why must oxidation-reduction occur to maintain a steady flow of electricity during electrolytic conduction?

17.3 How do we define anode and cathode?

17.4 Write equations for the half-reactions for the oxidation and reduction of water.

17.5 From the reactions discussed in Section 17.2, predict the products that you would obtain in the electrolysis of an aqueous solution of H_2SO_4.

17.6 What is the function of an electrolyte such as Na_2SO_4 or H_2SO_4 during the electrolysis of water? Why can't we carry out electrolysis on pure H_2O?

17.7 Write equations for the electrode reactions and the net reaction for (a) electrolysis of an unstirred brine solution; (b) electrolysis of a stirred brine solution.

17.8 What are the advantages and disadvantages of the mercury electrolysis cell used to produce NaOH?

17.9 What function does the design of the Downs cell serve?

17.10 Why is cryolite mixed with the Al_2O_3 prior to its electrolysis to produce Al?

17.11 Why can't Al be produced by electrolysis of an aqueous solution containing a salt such as $Al_2(SO_4)_3$?

17.12 Write a series of chemical equations representing the reactions involved with the recovery of Mg from sea water.

17.13 Describe the electrolytic purification of metallic copper. Why is the process economically feasible?

17.14 What is a faraday?

17.15 How does a coulometer work? What advantages are there to the use of a coulometer?

17.16 Compare the signs of the anode and cathode in galvanic and electrolytic cells.

17.17 In Section 17.5 we saw that electrical energy can be extracted from a Zn/Cu cell. If Cu^{2+} is brought into contact with metallic Zn, it is reduced to Cu while the Zn is oxidized, without the generation of electricity. In this case, what happens to the energy that is *not* being extracted as electrical energy?

17.18 What is the function of a salt bridge in a galvanic cell?

17.19 Sketch a galvanic cell in which the following net reaction occurs: $Ni^{2+} (aq) + Fe (s) \longrightarrow Ni (s) + Fe^{2+} (aq)$
(a) Label the cathode and anode.
(b) Indicate the charges on the electrodes.
(c) Indicate the direction of electron flow.
(d) Indicate the direction of the flow of cations and anions.
(e) If the concentrations of the ions are each 1 M, what is the potential of the cell?

17.20 What is a volt? What is an ampere?

17.21 Without computing $\mathscr{E}^0$, determine which reactions will occur spontaneously among the following sets of reactants in aqueous solution.
(a) Al (s), Ni (s), $NiSO_4$ (aq), $Al_2(SO_4)_3$ (aq)
(b) PbO_2 (s), $K_2Cr_2O_7$ (aq), H_2SO_4 (aq), $PbSO_4$ (s), $Cr_2(SO_4)_3$ (aq)
(c) Ag (s), $AgNO_3$ (aq), Pb (s), $Pb(NO_3)_2$ (aq)
(d) MnO_2 (s), HCl (aq), Cl_2 (g), $MnCl_2$ (aq)
(e) Mn (s), HCl (aq), $MnCl_2$ (aq), H_2 (g)

17.22 Without computing $\mathscr{E}^0$, determine whether the following reactions will occur spontaneously.
(a) $2Fe^{3+} + Sn \rightarrow 2Fe^{2+} + Sn^{2+}$
(b) $Cu + 2H^+ \rightarrow Cu^{2+} + H_2$
(c) $3Mg^{2+} + 2Al \rightarrow 3Mg + 2Al^{3+}$
(d) $Mn + Zn^{2+} \rightarrow Mn^{2+} + Zn$
(e) $PbO_2 + SO_4^{2-} + 4H^+ + 2Hg + 2Cl^- \rightarrow Hg_2Cl_2 + PbSO_4 + 2H_2O$

17.23 Without computing $\mathscr{E}^0$, determine whether the following reactions will occur spontaneously.
(a) $Ca^{2+} + Mg \rightarrow Ca + Mg^{2+}$
(b) $Pb^{2+} + 2Cl^- \rightarrow Pb + Cl_2$
(c) $2Cl^- + S_2O_8^{2-} \rightarrow Cl_2 + 2SO_4^{2-}$
(d) $6Mn^{2+} + 5Cr_2O_7^{2-} + 22H^+ \rightarrow$
$$6MnO_4^- + 10Cr^{3+} + 11H_2O$$
(e) $O_2 + 4Cl^- + 4H^+ \rightarrow 2H_2O + 2Cl_2$

17.24 How can one identify *experimentally* which electrode in a galvanic cell is the anode and which is the cathode?

17.25 Which is the better oxidizing agent?
(a) Li^+ or Ca^{2+}
(b) Cl_2 or F_2
(c) H_2O or Al^{3+}
(d) $S_2O_8^{2-}$ or Cl_2
(e) Br_2 or H_2O

17.26 Which is the better oxidizing agent?
(a) Cl_2 or ClO_3^-
(b) O_2 or $Cr_2O_7^{2-}$
(c) MnO_4^- or $Cr_2O_7^{2-}$
(d) PbO_2 or Hg_2Cl_2

17.27 Which is the better reducing agent?
(a) Ni or Fe
(b) H_2 or Mg
(c) Br^- or I^-
(d) SO_4^{2-} or F^-
(e) Sn or Mn

17.28 Which is the better reducing agent?
(a) Na or Cr
(b) $PbSO_4$ or Cl_2
(c) Ag or Cu
(d) I^- or Sn
(e) H_2 or H_2O

17.29 What is a concentration cell?

17.30 Describe the anode and cathode reactions in the ordinary dry cell (the Leclanché cell).

17.31 Why does a "dead" dry cell seem to come back to life if left idle for a while?

17.32 What are the anode, cathode, and net cell reactions in the alkaline zinc-manganese dioxide battery? What is the electrolyte?

17.33 What are the cathode, anode, and net cell reactions in the silver oxide battery?

17.34 What are the reactions that take place during the discharge of the lead storage battery? What reactions occur when this battery is being charged?

17.35 Write the anode, cathode, and overall cell reaction for the discharge of the nickel-cadmium battery.

17.36 What is a fuel cell? What advantages does a fuel cell offer over the lead storage battery?

17.37 What possible advantages do fuel cells offer over current electrical power plants?

REVIEW PROBLEMS (More difficult problems are marked by an asterisk)

17.38 How many faradays would be required to reduce 1 mol of each of the following to the indicated product?
(a) Cu^{2+} to Cu^0
(b) Fe^{3+} to Fe^{2+}
(c) MnO_4^- to Mn^{2+}
(d) F_2 to $2F^-$
(e) NO_3^- to NH_3

17.39 Calculate the number of electrons that corresponds to 1 coulomb of charge.

17.40 How many faradays would be required to oxidize 1 mol of each of the following to give the indicated product?
(a) Cu^+ to Cu^{2+}
(b) Pb to PbO_2
(c) Cl_2 to $2ClO_3^-$
(d) O_2 to H_2O_2 (hydrogen peroxide)
(e) NH_3 to NO_3^-

17.41 How many faradays are given by
(a) 8950 C
(b) a current of 1.5 A for 30 s
(c) a current of 14.7 A for 10 min

17.42 State how many minutes it would take to
(a) deliver 10,500 C using a current of 25 A
(b) deliver 0.65 $\mathscr{F}$ using a current of 15 A
(c) reduce 0.20 mol of Cu^{2+} to Cu using a current of 12 A

17.43 State how many minutes it would take to
(a) deliver 84,200 C using a current of 6.30 A
(b) deliver 1.25 $\mathscr{F}$ using a current of 8.40 A
(c) produce 0.50 mol of Al from molten $AlCl_3$ using a current of 18.3 A

17.44 How many faradays are required to produce the following?
(a) 10.0 ml of O_2 gas (at STP) from aqueous Na_2SO_4
(b) 10.0 g of Al from molten Al_2O_3 (in cryolite)
(c) 5.00 g of Na from molten NaCl
(d) 5.00 g of Mg from molten $MgCl_2$

17.45 How many grams of Na and Cl_2 would be produced if a current of 25 A was applied for 8.0 hr to the cell shown in Figure 17.3?

17.46 How many grams of O_2 and H_2 are produced in 1.0 hr when water is electrolyzed at a current of 0.50 A? What would be the volume in liters, at STP, of O_2 and H_2?

17.47 How many grams of copper could be purified by a current of 115 A for 8.00 hr? Refer to Figure 17.9.

17.48 How many grams of silver could be plated out on a serving tray by electrolysis of a solution containing Ag in the 1+ oxidation state for a period of 8.00 hr at a current of 8.46 A? What area would this cover, assuming that the density of Ag is 10.5 g/cm³ and the thickness of the silver plate is 0.00100 in. (0.00254 cm)?

17.49 How many seconds would it take to deposit 21.4 g of Ag from a solution of $AgNO_3$ by a current of 10.0 A?

17.50 How many hours would it take to deposit 35.3 g of Cr from a solution of $CrCl_3$ at a current of 6.00 A?

17.51 How many minutes would it take to plate out 5.00 g of copper from a solution of $CuSO_4$ at a current of 5.00 A?

17.52 What current is required to deposit 0.225 g of Ni from a solution of $NiSO_4$ in 10.0 min?

17.53 What current is required to produce 1.33 g of Cl_2 from a solution of NaCl in 45.0 min?

17.54 In an experiment two coulometers were connected in series, one containing $CuSO_4$, the other an unknown salt. It was found that 1.25 g of copper was plated out during the same period of time that 3.42 g of the unknown metal was plated out.

(a) How many faradays passed through this coulometer?
(b) If the oxidation state of the unknown metal ion in the solution was 2+, what is the atomic weight of the unknown?

17.55 Two coulometers were connected in series so that the same current passes through each of them. In an experiment, 0.125 mol of Cu was deposited from a solution of $CuSO_4$ in one of the coulometers. How many moles of Cr were deposited at the same time from a $Cr_2(SO_4)_3$ solution in the other?

17.56 What current is required to produce 50.0 ml of O_2 gas, measured at STP, by the electrolysis of H_2O for a period of 3.00 hr?

***17.57** A current of 0.250 A is passed through 400 ml of a 0.250 M solution of NaCl for 35.0 min. What will be the pH of the solution after the current is turned off?

***17.58** An unstirred solution of NaCl was electrolyzed for a period of 25.0 minutes and then titrated with 0.250 M HCl. The titration required 15.5 ml of the acid. What was the current during the electrolysis?

17.59 Given the following sets of half-reactions, write the net cell reaction and calculate $\mathscr{E}^0$ for the spontaneous changes that will occur.

(a) $Hg_2Cl_2 + 2e^- \rightleftharpoons 2Hg + 2Cl^-$
$\quad\ PbSO_4 + 2e^- \rightleftharpoons Pb + SO_4^{2-}$
(b) $AgCl + e^- \rightleftharpoons Ag + Cl^-$
$\quad\ Cu^{2+} + 2e^- \rightleftharpoons Cu$
(c) $Mn^{2+} + 2e^- \rightleftharpoons Mn$
$\quad\ Cl_2\ (g) + 2e^- \rightleftharpoons 2Cl^-$
(d) $Al^{3+} + 3e^- \rightleftharpoons Al$
$\quad\ Br_2\ (aq) + 2e^- \rightleftharpoons 2Br^-$

17.60 Determine the value of $\mathscr{E}^0$ for each of the spontaneous reactions in Question 17.21.

17.61 Determine the value of $\mathscr{E}^0$ for each of the reactions as written from left to right in Question 17.23.

17.62 Calculate the equilibrium constants for the following cell reactions:

(a) $Ni\ (s) + Sn^{2+}\ (aq) \rightleftharpoons Ni^{2+}\ (aq) + Sn\ (s)$
(b) $Cl_2\ (g) + 2Br^-\ (aq) \rightleftharpoons Br_2\ (aq) + 2Cl^-\ (aq)$
(c) $Fe^{2+}\ (aq) + Ag^+\ (aq) \rightleftharpoons Ag\ (s) + Fe^{3+}\ (aq)$

17.63 Calculate the equilibrium constants for the reactions in Question 17.22.

17.64 Calculate the equilibrium constants for the reactions in Question 17.23.

17.65 Calculate ΔG_{298}^0 in kilojoules for each reaction in Question 17.22.

17.66 Calculate ΔG_{298}^0 in kilocalories for each reaction in Question 17.23.

17.67 Write the Nernst equation, calculate $\mathscr{E}^0$ and $\mathscr{E}$ for the following reactions:

(a) $Cu^{2+}\ (0.1\ M) + Zn\ (s) \rightarrow Cu\ (s) + Zn^{2+}\ (1.0\ M)$
(b) $Sn^{2+}\ (0.5\ M) + Ni\ (s) \rightarrow Sn\ (s) + Ni^{2+}\ (0.01\ M)$
(c) $F_2\ (g,\ 1\ atm) + 2Li\ (s) \rightarrow 2Li^+\ (1\ M) + 2F^-\ (0.5\ M)$
(d) $Zn\ (s) + 2H^+\ (0.01\ M) \rightarrow Zn^{2+}\ (1\ M) + H_2\ (1\ atm)$
(e) $2H^+\ (1.0\ M) + Fe\ (s) \rightarrow H_2\ (1\ atm) + Fe^{2+}\ (0.2\ M)$

17.68 Calculate $\mathscr{E}^0$, $\mathscr{E}$, and ΔG (in kilojoules) for the following cell reactions (not balanced):

(a) $Al\ (s) + Ni^{2+}\ (0.80\ M) \rightarrow Al^{3+}\ (0.020\ M) + Ni\ (s)$
(b) $Ni\ (s) + Sn^{2+}\ (1.10\ M) \rightarrow Sn\ (s) + Ni^{2+}\ (0.010\ M)$
(c) $Cu^+\ (0.050\ M) + Zn\ (s) \rightarrow Cu\ (s) + Zn^{2+}\ (0.010\ M)$

17.69 Calculate the cell potential for the following:

(a) $Sn\ (s) + Pb^{2+}\ (0.050\ M) \rightarrow Sn^{2+}\ (1.50\ M) + Pb\ (s)$
(b) $3Zn\ (s) + 2Cr^{3+}\ (0.010\ M) \rightarrow 3Zn^{2+}\ (0.020\ M) + Cr\ (s)$
(c) $PbO_2(s) + SO_4^{2-}\ (0.010\ M) + 4H^+\ (0.10\ M) + Cu(s) \rightarrow$
$\qquad\qquad PbSO_4(s) + 2H_2O + Cu^{2+}(0.0010\ M)$

17.70 Calculate the potential generated by a concentration cell consisting of a pair of iron electrodes dipping into two solutions, one containing 0.10 M Fe^{2+} and the other containing 0.0010 M Fe^{2+}.

17.71 Calculate the potential of a concentration cell containing 0.0020 M Cr^{3+} in one compartment and 0.10 M Cr^{3+} in the other compartment with Cr (s) electrodes dipping into each solution.

17.72 The solubility product constant of AgBr is 5×10^{-13}. What will be the potential of a cell constructed using the H_2 electrode ($[H^+] = 1.0\ M$, $p_{H_2} = 1$ atm) versus a half-cell containing a silver wire coated with AgBr immersed in 0.010 M HBr?

17.73 What is the reduction potential of a half-cell composed of a copper wire dipping into 2×10^{-4} M $CuSO_4$?

17.74 A cell was constructed using the standard hydrogen electrode ($[H^+] = 1.0\ M$, $p_{H_2} = 1$ atm) in one compartment and a lead electrode in a 0.10 M K_2CrO_4 solution in contact with undissolved $PbCrO_4$. The potential of the cell was measured to be 0.51 V with the Pb electrode as the anode. Determine the K_{sp} of $PbCrO_4$ from these data.

17.75 A galvanic cell was set up using silver as one electrode dipping into 200 ml of 0.100 M $AgNO_3$ and magnesium as the other electrode dipping into 250 ml of 0.100 M $Mg(NO_3)_2$ solution.

(a) What is the potential of the cell?
(b) Suppose that current was drawn from the cell for a period of time until 1.00 g of silver plated out on the silver electrode. What is the potential of the cell now?
(c) Suppose that the original magnesium electrode had only weighed 0.080 g (it consisted of 0.080 g of magnesium deposited on an inert platinum electrode). What would the potential of the cell be just before the last tiny bit of magnesium dissolved?

*17.76 The standard reduction potential for Ag^+ is 0.80 V. Compute the standard reduction potential for the half-reaction,

$$Ag_2S \; (s) + 2e^- \rightleftharpoons 2Ag \; (s) + S^{2-} \; (aq)$$

in a solution buffered to a pH of 3.00.

*17.77 A student set up an electrolysis apparatus and passed a current of 1.22 A through a 3 M H_2SO_4 solution for 30.0 min. He collected the H_2 evolved and found that it occupied a volume, over water at 27°C, of 288 ml at a total pressure of 767 torr. Use these data to calculate the charge on the electron, expressed in the units, coulombs.

*17.78 How many hours will a 25-watt light bulb burn if it is powered by a lead storage battery that has available 25.0 g of Pb that can react as an anode. Assume a constant voltage of 1.5 V. (1 watt = 1 J/s)

*17.79 What current would be required to deposit 1 m² of chrome plate having a thickness of 0.050 mm in 25 min from a solution containing H_2CrO_4? The density of Cr is 7.19 g/ml.

*17.80 What weights of H_2 and O_2 in grams would have to react each second in a fuel cell at 110°C to provide 1.0 kilowatt (kW) of power, assuming a thermodynamic efficiency of 70%. (*Hint:* Use the data in Chapter 10 to compute ΔG^0 for the reaction, $H_2 \; (g) + \frac{1}{2}O_2 \; (g) \rightarrow H_2O \; (g)$ at 110°C. 1 watt = 1 J/s.)

*17.81 A hydrogen electrode is immersed in a 0.10 M solution of acetic acid. This electrode is connected to another consisting of an iron nail dipping into 0.10 M $FeCl_2$. What will be the measured emf of this cell? Assume $p_{H_2} = 1$ atm.

*17.82 How much work, expressed in kilojoules, is able to be accomplished by a 5.00-min flow of electricity having a voltage of 110 V and a current of 1.00 A?

*17.83 Assuming that the typical electric generating plant has an efficiency of only about 30%, what volume (in liters) of fuel having an average formula of $C_{12}H_{26}$ must be burned, giving $H_2O \; (g)$ and $CO_2 \; (g)$, to produce 1.0 kilowatt hour (kWh) of electricity? Assume ΔH_f^0 of $C_{12}H_{26} \; (l) = 291$ kJ/mol and a density of 0.74 g/ml. 1 watt = J/s.

*17.84 How many minutes would it take to remove all the Cr from 500 ml of 0.270 M $Cr_2(SO_4)_3$ by a current of 3.00 A?

*17.85 Calculate the value of ΔG (in kilojoules) for a system containing the following species: Mn^{2+} (0.10 M), $Cr_2O_7^{2-}$ (0.010 M), MnO_4^- (0.0010 M), Cr^{3+} (0.0010 M). The pH of the solution is 6.00. The reaction that you should consider is

$$6Mn^{2+} \; (aq) + 5Cr_2O_7^{2-} \; (aq) + 22H^+ \; (aq) \rightleftharpoons$$
$$6MnO_4^- \; (aq) + 10Cr^{3+} \; (aq) + 11H_2O \; (l)$$

Which direction will this reaction proceed to get to equilibrium from the starting conditions given above?

*17.86 A Ag/AgCl electrode dipping into 1 M HCl has a standard reduction potential of +0.22 V [$AgCl \; (s) + e^- \rightleftharpoons Ag \; (s) + Cl^- \; (aq)$]. A second Ag/AgCl electrode is dipped into a solution containing Cl^- at an unknown concentration. The cell generates a potential of 0.0435 V, with the electrode in the unknown solution serving as the anode. What is the molar concentration of Cl^- in the unknown?

*17.87 A student set up a galvanic cell to measure the K_{sp} of CuS. On one side of the cell she had a copper electrode dipping into a 0.10 M Cu^{2+} solution and on the other side a zinc electrode in a Zn^{2+} solution. The Zn^{2+} concentration was held constant at 1.0 M and the [Cu^{2+}] brought to a minimum by saturating the Cu^{2+} solution with H_2S. The emf of the cell was read as +0.67V, with the Cu electrode serving as the cathode. Calculate the Cu^{2+} concentration and the K_{sp} of CuS. Compare your answer to the K_{sp} reported in Table 16.1. In a saturated solution the concentration of H_2S is 0.10 M. The solution in which the CuS was formed was not buffered.

18

CHEMICAL PROPERTIES OF THE REPRESENTATIVE METALS

A novel use for a metal — as a rocket propellant! White clouds of aluminum oxide produced by burning aluminum powder billow from the solid booster rockets that lift the space shuttle <u>Columbia</u> from its launch pad at the Kennedy Space Center in Florida. The representative metals and their compounds find many practical uses, as we will discover in this chapter.

In Chapter 9 we discussed some general properties of the elements and the way these properties vary within the periodic table. There the goal was to study trends, rather than the specific behavior of individual elements. Now that you have acquired a broader understanding of basic chemical principles— thermodynamics, kinetics, and equilibrium, for example—we again turn our attention toward the properties of the elements. But this time we will examine more closely their individual chemical characteristics. Our goal is not only to learn the chemistries of the elements, but also to see how intimately our daily lives are tied to chemistry.

In this chapter we examine the representative metals—the metals found in the A-groups of the periodic table. These include the elements in Group IA and IIA (excluding hydrogen) as well as the heavier members of Groups IIIA, IVA, and VA.

18.1 PREPARATION OF METALS

Most metals, including those of both the representative and transition elements, are always found in nature in the combined state. The oceans, for example, provide a huge storehouse of minerals in which the metals occur primarily as soluble sulfates and halides. The major source of magnesium, for instance, is the oceans, and in the future greater attention will no doubt be focused on the sea as a source of raw materials as supplies of ore deposits on land are depleted. There is already a great deal of interest in mining the "manganese nodules" that seem to line the ocean floor. The oceans also provide a source of important nonmetals such as chlorine, bromine, and iodine.

On land, some metals occur as deposits of their carbonates. Limestone, for example, is primarily $CaCO_3$. A mixed $CaCO_3$, $MgCO_3$ limestone, called dolomite, is often ground up and used on farms and lawns to decrease the acidity of the soil (recall that acids react with carbonates) and to provide a source of Mg, which is needed in the production of chlorophyll.

Oxides are also important sources of metals. Two important examples are aluminum (Al_2O_3) and iron (Fe_2O_3). Sulfides are the primary sources of lead (as PbS) and copper (as Cu_2S). Regardless of the type of ore, however, metals almost always exist in positive oxidation states, and in order to produce the free element a chemical reduction must be brought about. The nature of this reduction process depends on the ease with which the metal can be reduced.

Some metals are so easily reduced that many of their compounds can be decomposed just by heating them at relatively low temperatures. Priestley, for ex-

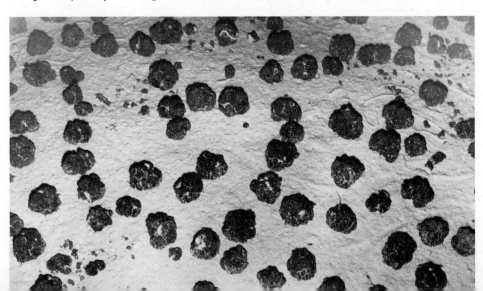

Manganese nodules lying on the floor of the north Pacific Ocean. The depth here is about 16,400 feet.

ample, in his experiments on oxygen, produced metallic mercury and oxygen from mercuric oxide by simply heating it with sunlight focused on the HgO by means of a magnifying glass. In this case, HgO decomposes quite spontaneously at elevated temperatures according to the equation

$$2HgO\ (s) \longrightarrow 2Hg\ (g) + O_2\ (g)$$

The practicality of using a thermal decomposition reaction of this type to produce a free metal depends on the extent to which the reaction proceeds to completion at a given temperature. In Chapter 11 we saw that at 25°C the position of equilibrium in a reaction is governed by the sign and magnitude of ΔG^0. If we take $\Delta G'$ to be the equivalent of ΔG^0, but at some other temperature, we have the relationship

$$\Delta G' = \Delta H' - T\,\Delta S'$$

where $\Delta H'$ and $\Delta S'$ are the heat and entropy changes that accompany the reaction. As discussed in Chapter 13, for most systems ΔH and ΔS do not change much with temperature, so $\Delta H'$ and $\Delta S'$ can reasonably be approximated by ΔH^0 and ΔS^0. We have seen that when ΔG^0 for a reaction is negative, the reaction is feasible from a practical standpoint because significant amounts of products will be formed. Extending this idea to temperatures other than 25°C, we can say that a thermal decomposition reaction will be feasible when $\Delta G'$ is negative, since under these conditions an appreciable amount of product will be formed. We must now look at the magnitudes of $\Delta H'$ and $\Delta S'$, because they control the sign and magnitude of $\Delta G'$.

Since a gas (O_2) and sometimes the metal vapor is produced in the decomposition of an oxide, the process occurs with a sizable increase in entropy, so $\Delta S'$ will be positive. The enthalpy change for the decomposition, $\Delta H'$, is simply the negative of the heat of formation of the oxide. Since $\Delta H_f'$ is generally negative for metal oxides, $\Delta H'$ for the decomposition reaction will be positive. As a result, the sign of $\Delta G'$ is determined by the difference between two positive quantities, $\Delta H'$ and $T\,\Delta S'$.

Remember, T is always positive.

If the metal oxide has a large negative heat of formation—that is, if a great deal of energy is evolved when the oxide is formed—then $\Delta H'$ for the decomposition will have a large positive value. As a result, the value of $\Delta G'$, which is given by the difference $\Delta H' - T\,\Delta S'$, will be negative *only* at very high temperatures, where $T\,\Delta S'$ is larger than $\Delta H'$. We express this by saying that the metal oxide is very stable with respect to thermal decomposition. On the other hand, if the $\Delta H_f'$ of the metal oxide is relatively small, as with HgO and certain other oxides (for example, Ag_2O, CuO, and Au_2O_3), then $\Delta H'$ for the decomposition reaction is a small positive quantity, and $\Delta G'$ for the reaction becomes negative at relatively low temperatures. These oxides, therefore, are said to have relatively low thermal stabilities.

| EXAMPLE 18.1 | Above what temperature would the decomposition of Ag_2O be expected to proceed to an appreciable extent toward completion? At 25°C, ΔH_f^0 for Ag_2O is -30.5 kJ/mol, $\Delta S_f^0 = -66.1$ J/mol K. |

SOLUTION   Since the decomposition of Ag_2O,

$$Ag_2O\ (s) \longrightarrow 2Ag\ (s) + \tfrac{1}{2}O_2\ (g)$$

is the reverse of formation, we have for this reaction

$$\Delta H^0 = -\Delta H_f^0 = +30.5\ \text{kJ/mol}$$

$$\Delta S^0 = -\Delta S_f^0 = +66.1\ \text{J/mol K}$$

Now let's calculate the temperature at which $\Delta G' = 0$, because above that temperature $\Delta G'$ will be negative. We will assume, as stated in the text, that $\Delta H'$ and $\Delta S'$ are approximately independent of temperature so that we can use ΔH^0 and ΔS^0 in the equation for $\Delta G'$.

$$\Delta G' = \Delta H^0 - T\Delta S^0$$

When $\Delta G' = 0$

$$0 = \Delta H^0 - T\,\Delta S^0$$

Solving for T,

$$T = \frac{\Delta H^0}{\Delta S^0}$$

$$= \frac{30,500 \text{ J/mol}}{66.1 \text{ J/mol K}}$$

$$= 461 \text{ K}$$

Notice that we converted kJ to J so that the units cancel properly.

Because ΔH^0 and ΔS^0 are both positive, $\Delta G'$ will become negative at temperatures above 461 K (188°C). This means that above 461 K the reaction should become feasible, with much of the Ag_2O undergoing decomposition.

EXAMPLE 18.2

Above what temperature would $\Delta G'$ be negative for the reaction

$$Au_2O_3 \,(s) \longrightarrow 2Au \,(s) + \tfrac{3}{2}O_2 \,(g)$$

At 25°C, $\Delta H_f^0 = +80.8$ kJ/mol for Au_2O_3. Also for Au_2O_3, $S^0 = 125$ J/mol K; for Au, $S^0 = 47.7$ J/mol K; for O_2, $S^0 = 205$ J/mol K.

SOLUTION

First, let's calculate ΔH^0 and ΔS^0 for the decomposition reaction. ΔH^0 is simply the negative of ΔH_f^0—that is, $\Delta H^0 = -80.8$ kJ for the decomposition of 1 mol of Au_2O_3. The value of ΔS^0 is

$$\Delta S^0 = (2S_{Au}^0 + \tfrac{3}{2}S_{O_2}^0) - (S_{Au_2O_3}^0)$$

$$= (2 \text{ mol}) \times \left(\frac{47.7 \text{ J}}{\text{mol K}}\right) + \tfrac{3}{2}\text{mol} \times \left(\frac{205 \text{ J}}{\text{mol K}}\right) - 1 \text{ mol} \left(\frac{125 \text{ J}}{\text{mol K}}\right)$$

$$= 278 \text{ J/K} = 0.278 \text{ kJ/K}$$

Again, we obtain $\Delta G'$ by assuming that $\Delta H'$ and $\Delta S'$ are the same as ΔH^0 and ΔS^0. Therefore,

$$\Delta G' = \Delta H^0 - T\,\Delta S^0$$

Substituting,

$$\Delta G' = -80.8 \text{ kJ} - T(0.278 \text{ kJ/K})$$

Note that regardless of the temperature $\Delta G'$ will be negative because the absolute temperature is always a positive quantity. What this tells us is that Au_2O_3 is unstable with respect to decomposition at any temperature. It exists only because at low temperatures its rate of decomposition is very slow.

Except in a few cases, thermal decomposition is not a practical way of producing the free metals. Instead, their compounds are normally reacted with some substance that is a better reducing agent than the metal being sought. One of the most common agents used for the reduction of metal oxides is carbon. Tin and lead, for example, can be produced by heating their oxides with carbon.

$$2SnO + C \xrightarrow{\text{heat}} 2Sn + CO_2$$

$$2PbO + C \xrightarrow{\text{heat}} 2Pb + CO_2$$

Carbon is used in large quantities in commercial metallurgy because of its abundance and low cost. Its importance in the reduction of iron ore and in steel making will be examined in Chapter 21.

Hydrogen is another reducing agent that can be used to liberate metals of moderate chemical activity from their compounds. For instance, tin and lead oxides will also be reduced when heated under a stream of H_2.

$$SnO + H_2 \xrightarrow{\text{heat}} Sn + H_2O$$

$$PbO + H_2 \xrightarrow{\text{heat}} Pb + H_2O$$

The use of a more active metal to carry out the reduction is also possible. In Chapter 16 we saw that a galvanic cell could be established between two different metals, for example, Zn and Cu. In that cell the more active reducing agent, Zn, causes the less active one, Cu^{2+}, to be reduced. Aluminum was first prepared in 1825 by the reaction of aluminum chloride with the more active metal, potassium.

$$AlCl_3 + 3K \longrightarrow 3KCl + Al$$

As a practical source of metals, the reduction of compounds with other elements that are better reducing agents suffers from a serious limitation, specifically, the availability (and cost) of the reducing agent. By this method, each possible reducing agent must be generated by reacting one of its compounds with a still better reducing agent. Ultimately, of course, there must be some "best" reducing agent. How could this substance be prepared if there were no better reducing agent available that could be used to reduce its compounds?

The solution to this dilemma is electrolysis. By applying a suitable potential, virtually any oxidation-reduction process can be brought about. Consequently, metals that themselves are very powerful reducing agents are nearly always prepared by electrolysis. Among the representative elements these include the very active metals in Groups IA and IIA.

The preparation of sodium by electrolysis of molten NaCl was described in Chapter 17.

Although less active than the alkali and alkaline earth metals, the metallic elements in Group IIIA are generally produced by electrolysis too. When a molten salt is used for the electrolysis, a halide is generally employed because of its relatively low melting point. The production of aluminum by the Hall process described in Chapter 17 is an exception. Here, you recall, molten Na_3AlF_6 is used as a solvent for Al_2O_3. It serves to lower the melting point of the Al_2O_3, which is the substance that is actually reduced electrolytically.

18.2 GROUP IA: THE ALKALI METALS

The elements of Group IA consist of hydrogen, lithium, sodium, potassium, rubidium, cesium, and francium. Except for hydrogen, all of them are very reactive metals. They are called alkali metals because their oxides are quite soluble in water and produce very basic (alkaline) solutions. For example, sodium oxide—the basic anhydride of sodium hydroxide—reacts as follows.

$$Na_2O \ (s) + H_2O \ (l) \longrightarrow 2Na^+ \ (aq) + 2OH^- \ (aq)$$

The oxides of the other alkali metals react in a similar fashion.

Each of the alkali metals has only a single, rather loosely held electron in its outer shell. Loss of this electron gives an ion with a charge of 1+. This is the only oxidation state (other than zero, of course) that these elements exhibit, which leads to some rather simple chemistry. In fact, the chemical and physical similarities among the metals of Group IA illustrate in a very striking way the empirical basis of the periodic table—elements having similar properties are placed in the same vertical column.

Occurrence and preparation

The most abundant of the alkali metals are sodium and potassium. It is not surprising, therefore, that they are also the most biologically important. In animals, for instance, both sodium and potassium ions are needed, and the proper balance between them must be maintained in order for the organism to function properly. In plants, potassium is significantly more important than sodium.

Lithium, rubidium, and cesium are present in the earth's crust in much smaller amounts than sodium and potassium. They and their compounds are therefore more difficult to come by and, as a result, they are more expensive. Because of this, they have few practical applications.

Francium is radioactive, and even its longest-lived isotope, ^{223}Fr, has a half-life of only 21 minutes. (It is produced in the radioactive decay of another radioactive isotope, ^{227}Ac.) Because of francium's short half-life it is estimated that at any given time there is less than 30 g of it in the entire Earth's crust!

The alkali metals never occur free in nature. They always exist in compounds, which are found in both the Earth's crust as well as in the ocean. For example, gigantic salt deposits (Figure 18.1*a*) are located in many places below the surface of the Earth—in Louisiana, New York, Michigan, Oklahoma, California, and Texas, to name just some. Where the climate is arid, salt deposits even occur on the surface (Figure 18.1*b*). Nearly all the compounds of the alkali metals are soluble in water, and where rainfall is plentiful they have been largely washed away into the oceans and salt lakes, or leached into subterranean waters. The recovery of the alkali metals therefore occurs from both land and water—from salt mines like that shown in Figure 18.1, by the evaporation of sea water in large ponds (Figure 18.2), and from brine wells.

The recovery of the alkali metals from their compounds can be accomplished by electrolytic reduction of a molten compound. Normally, halides are

Sodium is not required at all by plants, except for some salt marsh species.

Molar Concentrations in Sea Water	
Li$^+$	6×10^{-5}
Na$^+$	0.47
K$^+$	0.010
Rb$^+$	$\approx 10^{-6}$
Cs$^+$	$\approx 10^{-8}$

Clays, mica, and silicate ores also contain alkali metal ions.

Figure 18.1

Sodium chloride occurs in large deposits both below and above ground. (a) *The interior of a salt mine located in Texas.* (b) *The Bonneville Salt Flats in Utah.*

(a)

(b)

Figure 18.2

Salt is harvested from the sea in many parts of the world, including the United States. Here we see brine being agitated by hand in seawater evaporation ponds in the Canary Islands. Most of the salt collected by these people is used in fish factories.

chosen because they have relatively low melting points (compared to oxides, for instance). An example is the electrolysis of molten sodium chloride using the Downs cell, which was described in Chapter 17.

The Downs cell does not work well for the electrolysis of molten KCl, RbCl, and CsCl because at the temperatures required to melt these salts, the metals are very volatile. This makes their collection somewhat difficult. Instead, the molten chlorides are exposed to sodium vapor. The equilibrium

$$\text{Na } (g) + M\text{Cl } (l) \rightleftharpoons \text{NaCl } (l) + M(g)$$

is shifted to the right because potassium, rubidium, and cesium are considerably more volatile than sodium—sodium condenses and the other metal evaporates. What makes this reaction especially interesting is that it is driven to the right even though sodium is not as strong a reducing agent as potassium, rubidium, or cesium when they are all compared under identical conditions.

Physical properties

The alkali metals exhibit many of the typical properties that we've come to expect of metals: high luster and good thermal and electrical conductivity. Nevertheless, applications of the free metals rarely exploit these properties because the metals are so reactive. An exception has been the use of sodium in cooling nuclear reactors, which takes advantage of its low melting point, its relatively high boiling point, and its good thermal conductivity. These characteristics

Table 18.1
Some physical properties of the alkali metals

Element	Ionization Energy (kJ/mol)	$\mathscr{E}°$ (V)	M^+ Radius (Å)	Melting Point (°C)	Boiling Point (°C)
Lithium	520.1	−3.05	0.60	180.5	1326
Sodium	495.8	−2.71	0.95	97.8	883
Potassium	418.8	−2.92	1.33	63.7	756
Rubidium	402.9	−2.99	1.48	38.98	688
Cesium	375.6	−3.02	1.69	28.59	690

The operation of a nuclear reactor is described further in Chapter 24.

allow the metal to be easily melted and pumped through pipes that pass through the hot core of the reactor where the sodium quickly absorbs heat. The sodium is then pumped through the pipes of a heat exchanger outside the reactor where the heat is transferred to water, which can be made into steam to generate electricity.

As we mentioned in Chapter 9, the softness and low melting points of the alkali metals (Table 18.1) reflect the existence in the metallic lattice of singly charged cations that attract the surrounding "electron sea" weakly. These cations are formed by the loss of the single s electrons from the outer shells of the atoms. Because these outer-shell electrons are only weakly held, the alkali metals also have low ionization energies, electron affinities, and electronegativities.

An important physical property of the alkali metals is their emission spectra, which can be produced by passing an electric discharge through their vapors or by introducing one of their salts into a bunsen burner flame. Lithium salts, for example, impart a beautiful red color to a flame, sodium salts give a brilliant yellow color, whereas potassium salts produce a violet colored flame. These colors are intense enough to serve as useful qualitative tests called **flame tests** that can be used in analyzing mixtures of unknown composition. For example, if a drop of a solution of an unknown is placed into a flame and the yellow color is observed, sodium ions are in the unknown. If no yellow color is seen, then sodium is absent. The violet color of the potassium flame test is not as bright as the yellow sodium flame and is easily masked, even by traces of sodium in the unknown. Viewing the flame through blue glass—called *cobalt glass*—filters out the yellow and allows the violet potassium flame to be seen.

The brilliant yellow emission by sodium accounts for one of this element's growing commercial uses—in sodium vapor lamps (Color Plate 4). These are becoming more and more widely used throughout the country for street lighting because they are much more economical to operate than incandescent lamps. An incandescent lamp, such as an ordinary light bulb, gives off much of its energy in the form of invisible infrared radiation, so much of the electrical energy used to operate it is wasted. However, in a sodium vapor lamp, which is really nothing more than a gas discharge tube containing sodium vapor as the gas, most of the energy of the electrical discharge appears as visible yellow light (λ = 589 nm). This corresponds to a pair of closely spaced lines in the emission spectrum of sodium. See Color Plate 3.

Chemical properties and important compounds

The alkali metals are the most reactive metals. They are very powerful reducing agents and, as we learned in Chapter 9, they are able to reduce water with the evolution of hydrogen. The general reaction is

M = alkali metal.

$$2M\ (s) + 2H_2O\ (l) \longrightarrow 2M^+\ (aq) + 2OH^-\ (aq) + H_2\ (g)$$

The effectiveness of the alkali metals as reducing agents is reflected in the extremely negative reduction potentials of their ions (Table 18.1). These suggest that the reactions

$$M^+\ (aq) + e^- \longrightarrow M\ (s)$$

occur with great difficulty, and therefore that the oxidation reactions

$$M(s) \longrightarrow M^+\ (aq) + e^-$$

take place easily.

A close examination of the trends in the ionization energy and in $\mathscr{E}^0$ reveals an apparent contradiction, however. Note that the ionization energy (IE) decreases as we proceed down within the group, suggesting that it becomes progressively easier to strip an electron from the atom as we go from Li to Cs. In Chapter 3 we saw that this is, in fact, expected. We would also anticipate that the reduction potentials should become more negative as the IE becomes smaller since the elements should become more easily oxidized. This trend is indeed followed from Na downward; however, Li has an $\mathscr{E}^0$ that is more negative than Na (or any of the other alkali metals for that matter). Why is this so?

The ionization energy, remember, is a measure of the ease with which a *gaseous* atom loses electrons to produce a *gaseous* cation. The reduction potential, on the other hand, is concerned with the transfer of electrons between the *solid* metal and the corresponding cation in *aqueous solution*, where it is hydrated by the water molecules surrounding it. This latter process is more complex than simply removing an electron from the isolated metal atom. To understand the trends in the reduction potentials, we must break down the overall reaction into several steps. If we concentrate on the enthalpy changes involved in the reaction, we can construct the diagram in Figure 18.3. We see that the net enthalpy change is the sum of three energy terms. Two of these are endothermic—the sublimation energy, ΔH_{subl}, which is the energy needed to convert the solid into gaseous atoms, and the ionization energy, which we have already examined. The third quantity, called the hydration energy, is strongly exothermic. It corresponds to the energy *released* when the cation is placed into the solvent cage where it is surrounded by the water dipoles oriented in such a way that their negative ends are directed at the positive ion (Figure 18.4).

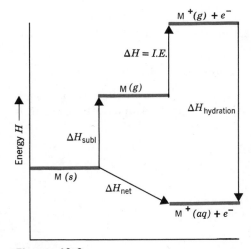

Figure 18.3

Enthalpy diagram for the reaction:
M (s) → M$^+$(aq) + e$^-$.

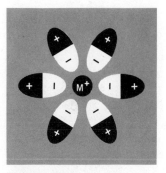

Figure 18.4

Solvation of a cation by water dipoles.

Among the alkali metals the sublimation energy remains approximately constant as we descend the group while the ionization energy decreases. To reach the peak on the energy diagram, we require the greatest amount of energy for Li and the least amount for Cs. However, because of its small size and high ionic potential, upon hydration Li^+ interacts much more strongly with the water dipoles than any of the other Group IA ions do. As a result, the hydration energy of Li^+ is unusually large, and much greater than for the other M^+ ions in the group. Therefore, the net overall enthalpy change is most exothermic for Li. This in turn causes Li to be more easily oxidized than any other alkali metal. In other words, the extraordinarily high hydration energy of the small Li^+ ion more than compensates for its relatively high ionization energy, and causes Li to have an unexpectedly large negative $\mathscr{E}^0$.

> Recall that the ionic potential is the ratio of an ion's charge to its radius.

Associated with the easy loss of electrons from the alkali metals is their interesting behavior in liquid ammonia. We have already seen that these metals are capable of reducing water to liberate H_2. Ammonia is not as easily reduced as water and, when placed into this solvent, alkali metals dissolve without reaction to form deep blue solutions. It is generally agreed that this color, which is identical for liquid ammonia solutions of all the alkali metals (as well as for Ca, Sr, and Ba from Group IIA), is a result of the presence of free electrons that have become solvated by ammonia molecules. Apparently, when the metal dissolves in NH_3, it loses its valence electron to become a cation. This electron becomes surrounded by NH_3 molecules arranged so that the positive ends of their dipoles are directed at the negatively charged electron, as shown in Figure 18.5, thereby stabilizing it through solvation. Solutions containing alkali metals in liquid ammonia are, as we would expect from the presence of readily available electrons, excellent reducing agents.

The strong tendency of the alkali metals to undergo oxidation permits them to react readily with most of the elemental nonmetals. The design of the Downs cell, you recall, is based on the need to keep Cl_2 and Na apart after they are formed by electrolysis. All the halogens (F_2, Cl_2, Br_2, and I_2) react with all the alkali metals to form the corresponding salts.

> For example,
> $2Na (s) + Cl_2 (g) \rightarrow 2NaCl (s)$.

Among the most interesting reactions of the alkali metals is their behavior toward elemental oxygen. Only lithium burns in air to form the "normal oxide," Li_2O.

$$4Li\ (s) + O_2\ (g) \longrightarrow 2Li_2O\ (s)$$

> Na_2O has a strong affinity for moisture and is an effective drying agent.

Sodium will undergo a similar reaction, but only if the supply of oxygen is limited. In the presence of excess oxygen sodium forms the pale yellow peroxide.

$$2Na\ (s) + O_2\ (g) \longrightarrow Na_2O_2\ (s)$$
$$\textbf{sodium peroxide}$$

This solid contains the peroxide ion, O_2^{2-}, which itself is an effective oxidizing agent. Sodium peroxide is therefore used commercially as an oxidizing agent. It

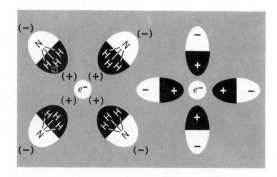

Figure 18.5
Solvated electron in liquid ammonia.

is also used as a bleaching agent, because the oxidation of intensely colored molecules often gives colorless reaction products. When dissolved in water, the peroxide ion hydrolyzes extensively:

$$Na_2O_2 \; (s) + 2H_2O \; (l) \longrightarrow 2Na^+ \; (aq) + 2OH^- \; (aq) + H_2O_2 \; (aq)$$

This causes solutions of Na_2O_2 to be very alkaline.

Potassium, rubidium, and cesium react with oxygen to form *superoxides*. For example,

$$K \; (s) + O_2 \; (g) \longrightarrow KO_2 \; (s)$$
potassium superoxide

These compounds contain the paramagnetic superoxide ion, O_2^-. When the superoxides are dissolved in water, oxygen is evolved.

$$2MO_2 \; (s) + 2H_2O \; (l) \longrightarrow 2M^+ \; (aq) + 2OH^- \; (aq) + O_2 \; (g) + H_2O_2 \; (aq)$$

Because of this reaction, potassium superoxide has found applications in breathing equipment designed to recirculate the air of the user. When air is exhaled, it contains both moisture and carbon dioxide. This is circulated through a canister containing KO_2. Moisture reacts as above, and carbon dioxide is removed from the air and replaced with oxygen by the overall reaction

$$4KO_2 \; (s) + 2CO_2 \; (g) \longrightarrow 2K_2CO_3 \; (s) + 3O_2 \; (g)$$

This allows the user to continue to breathe without having to draw in possibly contaminated air from outside the apparatus.

Because of their vigorous reactions with both moisture and oxygen, the alkali metals must be stored under an inert (nonreactive) liquid—often an oil or kerosene. Use is made of their affinity for H_2O in drying solvents. Sodium, for instance, is often employed to effectively remove traces of moisture from laboratory solvents. Commercially, the alkali metals are used as *getters* in the production of electronic vacuum tubes. Getters combine with the last traces of H_2O and O_2 that would otherwise interfere with the operation of the tube.

Besides combining with the halogens and oxygen, the alkali metals react directly with virtually all the other nonmetals as well. Lithium, however, is the only one that combines directly with gaseous nitrogen to form a *nitride*.

$$6Li \; (s) + N_2 \; (g) \longrightarrow 2Li_3N \; (s)$$
lithium nitride

Most of the important compounds of the alkali metals are those of sodium and potassium. This is simply because among the Group IA metals these two elements have the largest abundance. Sodium chloride is by far the most readily available and easily recoverable of all the alkali metal compounds, which makes it an inexpensive raw material for preparing other sodium compounds. For this reason, sodium compounds are considerably cheaper than those of potassium or the other alkali metals. Since there are such close similarities among the chemical properties of alkali metal compounds of a given type, when given a choice industry will almost always select a sodium compound simply because of economics.

The primary raw material for the preparation of potassium compounds is potassium chloride. It is obtained from *sylvite,* a mineral form of KCl, and from *carnallite,* $KCl \cdot MgCl_2 \cdot 6H_2O$. These are important as fertilizers because of the need of plants for both potassium and magnesium.

Sodium hydroxide—made from salt by the electrolysis of brine—is industry's most important strong base. Its common name is *lye* or *caustic soda.* Around the home it is found in oven and drain cleaners. It is used industrially to make soap and detergents, pulp and paper, textiles, in removing sulfur from

A recirculating breathing apparatus like this allows the user to avoid breathing toxic fumes.

Over 24 billion pounds of Cl_2 and NaOH are made annually from NaCl by electrolysis!

petroleum, and in the manufacture of myriad other chemicals. In the majority of its applications, NaOH is used to neutralize acids.

Potassium hydroxide, which is produced by electrolysis of KCl solution, is more expensive than NaOH, so its uses have been limited. It serves as the electrolyte in alkaline Zn/MnO_2 batteries, as we learned in Chapter 17.

The carbonates of the alkali metals constitute another class of important compounds. Once again, the sodium salts are used most widely because of their low cost. The principal source of sodium carbonate is from *Trona ore*, a mixture of sodium carbonate and sodium bicarbonate with the composition, $Na_2CO_3 \cdot NaHCO_3 \cdot 2H_2O$. Large amounts of this ore are mined from deposits in Wyoming.

Sodium carbonate is also made chemically from salt and limestone ($CaCO_3$) by the **Solvay process,** which takes advantage of the fact that $NaHCO_3$ is less soluble in cold water than is NaCl. The process begins by thermally decomposing limestone to give CaO and CO_2.

$$CaCO_3 \xrightarrow{\text{heat}} CaO \ (s) + CO_2 \ (g)$$

The carbon dioxide, along with ammonia, is bubbled into a concentrated solution of salt. The ammonia partially neutralizes carbonic acid that is formed from the carbon dioxide.

$$CO_2 \ (g) + H_2O \ (l) \longrightarrow H_2CO_3 \ (aq)$$

$$H_2CO_3 \ (aq) + NH_3 \ (aq) \longrightarrow NH_4^+ \ (aq) + HCO_3^- \ (aq)$$

As the concentration of HCO_3^- builds up, the less soluble $NaHCO_3$ begins to precipitate. The overall reaction can be written

$$Na^+ \ (aq) + Cl^- \ (aq) + NH_4^+ \ (aq) + HCO_3^- \ (aq) \longrightarrow NaHCO_3 \ (s) + NH_4^+ \ (aq) + Cl^- \ (aq)$$

After the solid $NaHCO_3$ is separated by filtration it is heated to convert it to sodium carbonate.

$$2NaHCO_3 \ (s) \xrightarrow{\text{heat}} Na_2CO_3 \ (s) + H_2O \ (g) + CO_2 \ (g)$$

Meanwhile, the solution containing NH_4Cl is treated with the calcium oxide left over from the decomposition of the limestone. This regenerates ammonia, which is recycled and used again. The reactions are

$$CaO + H_2O \longrightarrow Ca(OH)_2$$

$$Ca(OH)_2 + 2NH_4Cl \longrightarrow CaCl_2 + 2H_2O + 2NH_3 \ (g)$$

If care is taken not to lose any ammonia, the process consumes NaCl and $CaCO_3$ and produces Na_2CO_3 and $CaCl_2$. The net overall change is

$$2NaCl + CaCO_3 \longrightarrow Na_2CO_3 + CaCl_2$$

The hydrate, $Na_2CO_3 \cdot 10H_2O$, is called washing soda. Its solutions are basic because of hydrolysis of the CO_3^{2-} ion.

Sodium carbonate is used in huge amounts by industry (about 17 billion pounds of it are produced each year). About half is used to manufacture glass; the rest is used to manufacture other chemicals, to process pulp and paper, and to make soap and detergents.

Sodium bicarbonate, the intermediate product in the Solvay process is also a useful chemical. A solution of it is mildly basic and serves as a buffer because of the ability of the HCO_3^- ion to react with both acids and bases.

The bicarbonate ion is simultaneously a weak Brønsted acid and a weak Brønsted base.

$$HCO_3^- + H^+ \longrightarrow H_2CO_3$$

$$HCO_3^- + OH^- \longrightarrow H_2O + CO_3^{2-}$$

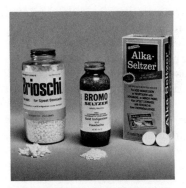

Sodium bicarbonate is an ingredient in these products which also contain a weak acid. When they dissolve in water CO₂ is produced, which is why they fizz.

For this reason, sodium bicarbonate is recommended as an additive to swimming pools because it helps control the pH of the water when other chemicals are added to destroy bacteria.

The common name for sodium bicarbonate is *baking soda*. When added to dough it decomposes during baking to release the gas carbon dioxide $[2NaHCO_3 \; (s) \xrightarrow{heat} Na_2CO_3 \; (s) + CO_2 \; (g) + H_2O \; (g)]$. The CO₂ produces tiny bubbles throughout the dough and makes the baked product rise and become ''light'' and appealing.

The carbonates and bicarbonates of the other alkali metals have properties similar to the sodium compounds. Potassium carbonate, K_2CO_3, is called *potash* and is a major constituent of wood ashes. When needed in quantity it is prepared by first reacting potassium hydroxide with carbon dioxide to give potassium bicarbonate.

$$KOH + CO_2 \longrightarrow KHCO_3$$

Thermal decomposition gives K_2CO_3.

Lithium carbonate, Li_2CO_3, has been found to be useful as a drug in the treatment of manic depression.

18.3 GROUP IIA: THE ALKALINE EARTH METALS

The elements of Group IIA consist of beryllium, magnesium, calcium, strontium, barium, and radium. All are very reactive metals, although not as reactive as the metals of Group IA. They are called *alkaline earth* metals because their oxides are basic, and because many of their compounds are of low solubility in water and are therefore found in mineral deposits in the Earth's crust.

Each of the alkaline earth metals has two electrons in an *s* subshell that lies outside a noble gas core. Each exhibits only a 2+ oxidation state in its compounds and, with the exception of the compounds of beryllium and some of those of magnesium, their compounds are largely ionic. As in Group IA, there are strong similarities among all members of this group, but among the heavier ones—calcium through radium—the similarities are particularly close.

Occurrence and preparation

Calcium is the third most abundant metal in the Earth's crust.

Calcium and magnesium are among the most abundant of the elements in the Earth's crust, ranking fifth and eighth in terms of percentage by weight. They are found in many locations in large mineral deposits of various compositions. Examples are gypsum ($CaSO_4 \cdot 2H_2O$), limestone ($CaCO_3$), dolomite ($CaCO_3 \cdot MgCO_3$), and carnallite ($MgCl_2 \cdot KCl \cdot H_2O$). In sea water, Ca^{2+} and Mg^{2+} are both major constituents among the dissolved ions. On a mole basis, Mg^{2+} is the third most abundant ion in the sea, and Ca^{2+} places sixth. Calcium and magnesium are also the most biologically important alkaline earth metals. Calcium is found in the bones of animals, and shellfish extract Ca^{2+} from sea water to form their $CaCO_3$ shells. Magnesium is vital to plants where it is found at the center of the chlorophyll molecule—the substance that captures solar energy and begins the biological food chain.

In sea water, the concentration of Mg^{2+} is 0.056 M and that of Ca^{2+} is 0.011 M.

The major source of beryllium is the mineral *beryl*, $Be_3Al_2(SiO_3)_6$. Sometimes beautiful large pure crystals of this substance are found and, when polished, become the gems emerald and aquamarine (Color Plate 13).

Radium was discovered by Pierre and Marie Curie.

Strontium and barium are recovered from deposits of their insoluble sulfates and carbonates. Radium, however, occurs principally as an impurity in *pitchblend*—a mineral from which uranium is extracted. Radium, which itself is radioactive, is formed as a product of the radioactive decay of heavier elements.

For example, ^{226}Ra is radium's longest-lived isotope, with a half-life of 1600 years. It is one of a chain of isotopes produced from ^{238}U when it decays (see Chapter 24).

Of the metals in Group IIA, only beryllium and magnesium are produced and used in significant quantities. This is because they are the only alkaline earth metals that do not react rapidly with air and moisture at room temperature.

Beryllium is obtained by the electrolysis of molten beryllium chloride. However, sodium chloride must be added to the melt as an electrolyte because $BeCl_2$ is primarily covalent and therefore is a very poor electrical conductor. During the electrolysis, the less active metal, Be, is produced at the cathode and Cl_2 is evolved at the anode.

$$BeCl_2\ (l) \xrightarrow[\text{(NaCl)}]{\text{electrolysis}} Be\ (l) + Cl_2\ (g)$$

Beryllium-copper alloy tools can be used in situations like this because they are nonsparking if accidentally struck against other metal objects.

In recent times beryllium has found a variety of practical uses. It is a very lightweight, strong metal that has structural applications in missiles and spacecraft. It absorbs X rays less than any other metal that's stable toward air, and is used as windows for X-ray tubes. Beryllium-copper alloy is fashioned into springs, electrical contacts, and welding rods. It is also made into tools—hammers, for example—that will not create sparks when used. This is particularly important when workers must labor in an explosive atmosphere. A major drawback of beryllium, however, is that its compounds are highly toxic, and stringent safety requirements must be maintained when it or its alloys are machined.

Magnesium is extracted both from its land-based ores and from the sea. On land its principle sources are its chloride, found in carnallite, $MgCl_2 \cdot KCl \cdot 6H_2O$, and its carbonate in dolomite, $CaCO_3 \cdot MgCO_3$. Extraction from dolomite involves first heating the ore strongly—a process called *calcining*—which decomposes the carbonates and forms the oxides.

$$CaCO_3 \cdot MgCO_3\ (s) \xrightarrow{\text{heat}} CaO \cdot MgO\ (s) + 2CO_2\ (g)$$

The mixed oxides are then treated with an excess of sea water, which contains appreciable amounts of dissolved Mg^{2+}. The water converts the oxides to the hydroxides.

$$CaO\ (s) + H_2O \longrightarrow Ca^{2+}\ (aq) + 2OH^-\ (aq)$$

$$MgO\ (s) + H_2O \longrightarrow Mg(OH)_2\ (s)$$

Calcium hydroxide, although not extremely soluble, is appreciably more soluble than $Mg(OH)_2$. As it dissolves, it makes the water basic, causing the Mg^{2+} that's in the sea water to precipitate as $Mg(OH)_2$. The combined precipitate of $Mg(OH)_2$ is then filtered and dissolved in hydrochloric acid to convert it to the chloride. This solution is evaporated and the solid $MgCl_2$ is melted and electrolyzed, giving magnesium and chlorine. The chlorine produced is made into HCl again and recycled.

In the absence of dolomite, the magnesium in sea water can be recovered by making the water basic with calcium oxide. This is obtained by calcining limestone, $CaCO_3$, or even sea shells, which are also composed of $CaCO_3$.

$$CaCO_3\ (s) \xrightarrow{\text{heat}} CaO\ (s) + CO_2\ (g)$$

$$CaO\ (s) + H_2O + Mg^{2+}\ (aq) \longrightarrow Ca^{2+}\ (aq) + Mg(OH)_2\ (s)$$

A modern plant that produces magnesium from sea water. In the foreground we see the magnesium hydroxide settling ponds.

The precipitate of $Mg(OH)_2$ is treated as described above.

Magnesium metal has a number of practical uses. Its low density (light weight) and moderate strength when alloyed with aluminum make it a useful structural metal. (Perhaps you or your family own a magnesium alloy steplad-

Table 18.2
Some properties of the alkaline earth metals

| Element | Ionization Energy (kJ/mol) | | $\mathscr{E}°$ (V)^a | M^{2+} Radius (Å) | Melting Point (°C) |
	First	Second			
Beryllium	900	1757	−1.70	0.31^b	1278
Magnesium	737.6	1450	−2.38	0.65	651
Calcium	589.5	1146	−2.76	0.99	843
Strontium	549	1064	−2.89	1.13	769
Barium	503	965	−2.90	1.35	725
Radium	509	979	−2.92	1.40	700

^{a} For M^{2+} (aq) + $2e^-$ → M (s).
^{b} Estimated.

Figure 18.6

Combustion of the fine magnesium wire inside the flashbulb produces a flash of light and leaves the interior of the bulb coated with magnesium oxide.

Barium is used as a getter to remove traces of oxygen in the manufacture of vacuum tubes.

der.) Presently magnesium is more expensive than aluminum, but its virtually inexhaustible supply in the sea is likely to make it comparatively less expensive as land-based supplies of aluminum ore are depleted.

Another application of magnesium—one you've surely seen—is in flashbulbs and signal flares. The combustion of magnesium

$$Mg\ (s)\ +\ O_2\ (g)\ \longrightarrow\ MgO\ (s)$$

produces not only a great deal of heat, but also intense light. A flashbulb (Figure 18.6) contains a fine magnesium wire in an atmosphere of pure oxygen. The flashbulb is fired by passing a small electrical current through the wire, which heats it and sets off the combustion reaction. Afterwards, the interior of the bulb is coated with a thin deposit of the white powder MgO.

Calcium, strontium, and barium have very few commercial applications as free metals because of their reactivity toward oxygen and moisture. As a result, they are prepared only in small quantities, usually by electrolysis of their molten chlorides.

Physical properties

Table 18.2 lists some physical properties of the alkaline earth metals. We see that in general they have higher melting points and higher densities than their neighbors in Group IA. This is not difficult to explain. The greater effective nuclear charges experienced by their outer electrons makes the atoms of the alkaline earth metals smaller than those of the alkali metals alongside them in the periodic table, so more mass is packed into a smaller volume, which leads to higher densities. In Chapter 9 we learned that the 2+ charge on the cations in the metallic lattice of an alkaline earth metal makes it more difficult to pull them apart than the 1+ cations in the metallic lattice of an alkali metal. The Group IIA metals therefore have the higher melting points. The Group IIA metals are also harder than the Group IA metals for the same reason, and they have larger ionization energies.

Salts of the heavier alkaline earth metals, like those of the metals of Group IA, produce striking colors when introduced into a bunsen burner flame. These colors serve as flame tests for them. Calcium salts, for example, give a brick-red color; strontium salts produce a crimson flame; and barium salts give a yellowish-green flame. Salts of these metals are often used to give spectacular colors to fireworks displays (Color Plate 14).

Chemical properties and important compounds

As we've already noted, the alkaline earth metals are all very reactive elements. They are easily oxidized and therefore serve as excellent reducing agents, as evidenced by their very negative reduction potentials. The heavier of them, in fact, have reduction potentials comparable to those of the alkali metals. At first glance, this may appear somewhat surprising, because the sum of the first two ionization energies of an alkaline earth metal is considerably greater than just the first ionization energy of an alkali metal. In other words, removing two electrons from an isolated alkaline earth metal atom is much more difficult than removing a single electron from an isolated metal atom of Group IA. However, as we learned earlier in our discussion of lithium's unusually negative $\mathscr{E}^0$, the size of the reduction potential is controlled by more than simply the ionization energy. The hydration energy of the ion also plays a very important role.

The magnitude of the hydration energy of an ion depends both on its size and on its charge. A small ion can get closer to the water dipoles than a large ion can, so the smaller ion interacts more strongly with the solvent and its hydration energy is larger. A highly charged ion attracts the water dipoles more strongly than one of low charge, so the higher the ion's charge, the larger its hydration energy. Therefore, because the ions of the alkaline earth metals are both smaller *and* more highly charged than those of their neighbors in Group IA, their hydration energies are much larger. This serves to offset their much larger ionization energies. As a result, the alkaline earth metals have reduction potentials almost as negative as those of the alkali metals.

As reducing agents, the Group IIA metals are all powerful enough to reduce water, at least in principle. However, beryllium and magnesium both form insoluble oxide coatings that protect them from attack by water. Beryllium, in particular, is quite resistant to oxidation, even by acids, because of its BeO coating. Magnesium is more reactive than beryllium. Even though it is not attacked by cold water, magnesium reacts slowly with boiling water and quite rapidly with steam to liberate hydrogen.

The remaining alkaline earth metals—calcium through radium—form oxides that are at least moderately soluble in water, so their oxides are unable to protect them. As a result, they reduce water and liberate hydrogen. The general reaction is

$$M \text{ (s)} + 2H_2O \longrightarrow M(OH)_2 \text{ (aq)} + H_2 \text{ (g)}$$

The vigor with which this reaction occurs increases from calcium to strontium to barium.

Like the alkali metals, the Group IIA elements react directly with most elemental nonmetals. For example, magnesium, we learned, reacts directly with oxygen to form MgO. It also is able to react with nitrogen to give magnesium nitride.

$$3Mg \text{ (s)} + N_2 \text{ (g)} \longrightarrow Mg_3N_2 \text{ (s)}$$

With sulfur it gives MgS, and with the halogens it yields MgX_2 (X = halide ion). The other Group IIA metals react similarly.

An interesting phenomenon among the Group IIA elements is the variation in their metallic character. Although physically they all exhibit metallic characteristics, chemically we find that beryllium, and, to some slight extent, magnesium exhibit a degree of nonmetallic character as well.

Beryllium and its oxide are amphoteric—they dissolve in both acids and bases. For example, in base they react as follows.

$$Be + 2H_2O + 2OH^- \longrightarrow Be(OH)_4^{2-} + H_2 \text{ (g)}$$

$$BeO + H_2O + 2OH^- \longrightarrow Be(OH)_4^{2-}$$

When $\mathscr{E}^0$ has a large negative value, the metal ion is difficult to reduce and the metal itself is easily oxidized.

Ionic Radii (Å)			
Li$^+$	0.60	Be^{2+}	0.31*
Na$^+$	0.95	Mg^{2+}	0.65
K$^+$	1.33	Ca^{2+}	0.99
Rb$^+$	1.48	Sr^{2+}	1.13
Cs$^+$	1.69	Ba^{2+}	1.35

* Estimated.

Their protective oxide coatings permit beryllium and magnesium to serve as useful structural metals.

The ion Be(OH)$_4^{2-}$ is called the beryllate ion.

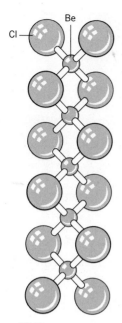

Figure 18.7

The structure of $(BeCl_2)_x$.

The remainder of the alkaline earth metals and their oxides are not amphoteric, so beryllium is chemically less metallic than the other members of its group.

Another way that beryllium is less metallic than the other Group IIA elements is in the degree of covalence of its compounds. There is no evidence that beryllium exists as Be^{2+} in any of its compounds—they all show a significant degree of covalent character. Beryllium chloride, for example, is a poor conductor when melted, and we saw that an electrolyte—sodium chloride—has to be added to molten $BeCl_2$ so that it can be electrolyzed. In the solid state, $BeCl_2$ exists as a covalently linked chain of $BeCl_2$ units in which Be completes its octet by forming coordinate covalent bonds with chlorine atoms on adjacent $BeCl_2$ molecules.

$$\overset{\cdot\cdot}{\underset{\cdot\cdot}{Cl}} \quad \overset{\cdot\cdot}{\underset{\cdot\cdot}{Cl}} \quad \overset{\cdot\cdot}{\underset{\cdot\cdot}{Cl}} \quad \overset{\cdot\cdot}{\underset{\cdot\cdot}{Cl}}$$
$$Be \qquad Be \qquad Be \qquad Be$$
$$\underset{\cdot\cdot}{Cl} \quad \underset{\cdot\cdot}{Cl} \quad \underset{\cdot\cdot}{Cl} \quad \underset{\cdot\cdot}{Cl} \quad \underset{\cdot\cdot}{Cl}$$

(Arrows indicate coordinate covalent bonds)

Since each Be atom has four separate electron pairs around it, the arrangement of the chlorines around the Be is tetrahedral, as shown in Figure 18.7.

The covalent character of bonds to beryllium can be related to the small size and high charge that a true Be^{2+} ion would have. We saw in Chapter 9 that this would give the Be^{2+} a very large ionic potential—sufficiently large to polarize other ions next to it and cause the bonds to become mostly covalent.

Although most magnesium compounds are ionic, magnesium does form a variety of covalently bonded compounds in which portions of organic molecules—molecules derived from methane (CH_4), ethane (C_2H_6), and other hydrocarbons—are bonded to magnesium. Examples are C_2H_5MgBr and $Mg(C_2H_5)_2$. These are called *organomagnesium compounds* and are important reagents in organic chemistry. The compounds of the rest of the alkaline earth metals are nearly all predominantly ionic.

The most important compounds of the alkaline earth metals are their carbonates, oxides, hydroxides, and sulfates. The carbonates are all quite insoluble in water, but there are trends in the solubilities of the oxides, hydroxides, and sulfates that influence their practical applications.

The solubilities of the oxides increase going down the group. BeO and MgO are insoluble, but CaO, SrO, and BaO react with water to form the corresponding hydroxides.

$$O^{2-} + H_2O \rightarrow 2\,OH^-$$

$$MO + H_2O \longrightarrow M(OH)_2 \qquad (M = Ca, Sr, Ba)$$

The solubilities of the hydroxides also increase going down the group. Magnesium hydroxide is insoluble, calcium hydroxide is slightly soluble, and the hydroxides of strontium and barium are even more soluble. In contrast, the solubilities of the sulfates vary in the opposite direction—$BaSO_4$, $SrSO_4$ and $CaSO_4$ are insoluble, but their K_{sp}'s increase from Ba to Ca. Magnesium sulfate, on the other hand, is quite soluble.

	K_{sp}
$CaSO_4$	2×10^{-4}
$SrSO_4$	2.9×10^{-7}
$BaSO_4$	1.5×10^{-9}

We learned earlier that calcium and magnesium occur in large deposits of their carbonates—$CaCO_3$ (limestone and marble) and $MgCO_3 \cdot CaCO_3$ (dolomite). Calcium carbonate is a very common chemical. Seashells are composed almost entirely of $CaCO_3$, as is the chalk your teacher uses to write on the blackboard. Calcium carbonate is also used as a mild abrasive in toothpaste and household cleansers, and as an antacid. Limestone is an extremely important raw material in many industrial reactions. As we saw earlier, its thermal decomposition produces calcium oxide and carbon dioxide.

Crushed limestone or dolomite is often spread on lawns and gardens to make the soil less acidic.

$$CaCO_3 \xrightarrow[900°C]{heat} CaO + CO_2$$

Some things that are composed of or contain calcium carbonate.

About 19 million tons of CaO are produced each year.

Calcium oxide is called *lime*, or sometimes *quicklime*, and its annual production ranks second among industrial chemicals because it is an inexpensive, relatively strong base. When treated with water—a process called *slaking*—calcium hydroxide (*slaked lime*) is formed with considerable evolution of heat.

$$CaO + H_2O \longrightarrow Ca(OH)_2$$

Calcium oxide is an ingredient in portland cement, and this reaction is among the first to occur when water is added to the cement.

Magnesium oxide can also be formed by decomposing its carbonate. However, unlike CaO, magnesium oxide is unreactive toward water. It is used as a component in refractory bricks—those used to line the interiors of high temperature furnaces. It is also used in the manufacture of paper, and medicinally as an antacid.

MgO melts at about 2800°C.

Magnesium hydroxide, $Mg(OH)_2$, is much less soluble in water than $Ca(OH)_2$. It too is used as an antacid and as a laxative—it is the creamy white substance in milk of magnesia.

Important sulfates of the alkaline earth metals are those of magnesium, calcium, and barium. Magnesium sulfate in the form of its hydrate, $MgSO_4 \cdot 7H_2O$, is called *epsom salts*, and is used to treat fabrics so that they readily accept dyes, to fireproof fabrics, as a fertilizer, and medicinally.

The mineral gypsum is $CaSO_4 \cdot 2H_2O$ and is used to make plaster and plasterboard—a building material commonly called sheet rock. When gypsum is partially dehydrated by heating, *plaster of paris* is formed.

In plaster of paris there is 1 mol of H_2O for each 2 mol of $CaSO_4$.

$$2CaSO_4 \cdot 2H_2O \xrightarrow{\text{heat}} (CaSO_4)_2 \cdot H_2O$$
$$\textbf{gypsum} \qquad\qquad \textbf{plaster of paris}$$

The formula for plaster of paris is normally written $CaSO_4 \cdot \frac{1}{2}H_2O$. When water is added to $CaSO_4 \cdot \frac{1}{2}H_2O$ the crystals absorb it and reform the dihydrate. This is an exothermic reaction, and anyone who has had a broken limb set in a plaster cast has probably noticed how warm the cast became as it hardened.

Barium sulfate has a number of uses that are based on its whiteness and low solubility in water. It is used as a whitener in photographic papers and as a filler in papers and polymeric fibers. Medicinally it is used for X-ray diagnosis of intestinal tract disorders because $BaSO_4$ is quite opaque to X rays. Even though barium salts are normally quite poisonous, $BaSO_4$ is so insoluble that a suspension of it can be swallowed without harm. This is because so little Ba^{2+} is in solution. As the suspension of $BaSO_4$ passes through a patient's intestines, its path can be followed by X-ray photographs like that in Figure 18.8.

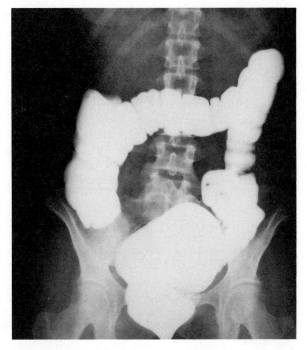

Figure 18.8
An X-ray photograph of a patient's large intestine that's been filled with a suspension of barium sulfate. The $BaSO_4$ is opaque to X-rays and allows the shape of the intestine to be seen.

18.4 METALS OF GROUPS IIIA, IVA, AND VA

In Chapter 9 we learned that the metallic character of the elements decreases from left to right in a period and increases from top to bottom in a group. The impact of these trends is especially apparent in Groups IIIA, IVA, and VA, where fewer metallic elements are found as the group number increases (Figure 18.9). In Group IIIA, for example, the metals are aluminum, gallium, indium, and thallium. In Group IVA, only tin and lead are metals, and in Group VA the only metal is bismuth. In general, the metals in these three groups tend to be less reactive and less metallic in their chemical behavior than the metals of Groups IA and IIA. For instance, many are amphoteric and many form covalent compounds.

There are no genuine metals in Groups VIA or VIIA.

Aluminum has three electrons in its valence shell outside a neon core ($[Ne]3s^23p^1$), and in its compounds aluminum always occurs in the 3+ oxidation state. The metals in periods 4 (Ga, In, Tl), 5 (Sn, Pb), and 6 (Bi) occur after a row of transition elements and are called *post-transition metals*. They have pseudonoble gas cores beneath their valence shells, and their chemistry is characterized by the occurrence of two oxidation states. In each case, the lower oxidation state corresponds to the loss of the outer p electron(s) and the higher one corre-

The pseudonoble gas core is
$ns^2np^6nd^{10}$.

III A	IV A	V A	VI A	VII A	0
					He
B	C	N	O	F	Ne
Al	Si	P	S	Cl	Ar
Ga	Ge	As	Se	Br	Kr
In	Sn	Sb	Te	I	Xe
Tl	Pb	Bi	Po	At	Rn

Figure 18.9
The metals of Groups IIIA, IVA, and VA

sponds to the further loss of the pair of *s* electrons. Thus, in Group IIIA the two oxidation states are 1+ and 3+; in Group IVA they are 2+ and 4+; and in Group VA they are 3+ and 5+.

Among the post-transition metals the relative stability of the lower oxidation state increases with increasing atomic number within a group. For instance, Ga^{3+} is more stable than Ga^+, while Tl^+ is more stable than Tl^{3+}. This trend in the relative stabilities of the high and low oxidation states persists in Groups IVA and VA, and results, it appears, from a decreasing stability of $M-X$ bonds with increasing size of the metal atom. Since atoms become larger going down a group, it becomes increasingly more difficult to recover, by bond formation, the extra energy that must be expended to remove the pair of *s* electrons. This causes the lower oxidation state to become much more stable toward the bottom of a group.

Occurrence and preparation

Aluminum Of the metals in Group IIIA, aluminum is the only one of much practical importance and the only one whose chemistry we will examine in much detail. On a weight basis, aluminum is the third most abundant element in the Earth's crust. For example, it is found combined with silicon and oxygen in *aluminosilicates*, which occur in rock such as granite and in various clays. Unfortunately, no practical method has yet been found to extract aluminum from these sources.

The Earth's crust is composed of about 8.8% aluminum by weight, which makes aluminum the most abundant metal in the crust.

The major ore of aluminum is *bauxite*, which contains its oxide Al_2O_3. However, before the metal can be obtained electrolytically by the Hall process, which was described in Chapter 17, it must first be purified. The method of purification takes advantage of the amphoteric nature of aluminum and its compounds. The ore is first treated with a concentrated NaOH solution, which dissolves the Al_2O_3 and leaves most of the impurities behind.

Purification of bauxite consumes about 1.2 billion pounds of NaOH each year!

$$Al_2O_3 \ (s) + 2OH^- \longrightarrow \underset{\text{aluminate ion}}{2AlO_2^-} + H_2O$$

Then the basic aluminum-containing solution is acidified, which precipitates the insoluble hydroxide.

$$AlO_2^- + H_3O^+ \longrightarrow Al(OH)_3 \ (s)$$

After filtration the aluminum hydroxide is heated. This drives off water and gives the oxide again, which is now pure.

$$2Al(OH)_3 \ (s) \xrightarrow{\text{heat}} Al_2O_3(s) + 3H_2O \ (g)$$

The Al_2O_3 is then dissolved in molten cryolite and electrolyzed as described in Chapter 17.

Aluminum finds many uses in modern society. Large amounts of it serve as a structural metal in kitchen utensils, automobiles, aircraft, beverage cans, aluminum foil, and other consumer products. Aluminum is a good electrical conductor and is used in electrical wiring. Alloyed with magnesium, it is employed in structural applications, and an alloy called *alnico*—50% Fe, 20% Ni, 20% Al, 10% Co—forms powerful magnets.

Tin and Lead Tin occurs as SnO_2 in an ore called *cassiterite*. To recover the metal, the ore is first heated strongly in air to drive off volatile oxides of some of its impurities—for example, arsenic and sulfur. Afterwards, the SnO_2 is reduced with carbon.

$$SnO_2 + C \longrightarrow Sn + CO_2$$

The metal can be further purified in a manner similar to the electrolytic purification of copper. The impure tin is made the anode in an electrolytic cell and pure tin is made the cathode. Operation of the cell causes the impure tin to dissolve and pure tin to plate out on the cathode.

In Chapter 9 we saw that elemental tin occurs in three different physical forms. This phenomenon, you recall, is termed allotropism, and the various forms of the element are called allotropes. The most common allotrope of tin is called *white tin* or *malleable tin*—you've seen it as the shiny coating on "tin" cans. Below 13.2°C, the white form very gradually changes to a powdery, nonmetallic form called *gray tin.* White tin and gray tin were the two forms of this element mentioned in Chapter 9. A third form, called *brittle tin,* is obtained when white tin is heated. Its properties are obvious from its name.

The allotropes of tin differ in their crystal structures.

One of the principal uses of tin is as a protective coating over steel—the familiar tin can, for example. The coating is thin and is usually applied electrolytically. It protects the steel simply by excluding air and moisture. However, once the coating is scratched and the steel below is exposed, corrosion occurs very rapidly. Iron is more easily oxidized than tin, so when both metals are exposed to moisture a galvanic cell is established in which iron is the anode and is therefore oxidized in preference to the tin. Other applications of tin are in alloys such as bronze (copper and tin) and solder (tin and lead).

Lead is found in nature as its sulfate, $PbSO_4$, its carbonate, $PbCO_3$, and its sulfide, PbS. The principal ore of lead is called *galena* and contains PbS. The metal is obtained from the ore by first heating it strongly in air, which converts it to the oxide.

$$2PbS + 3O_2 \longrightarrow 2PbO + 2SO_2$$

The oxide is then reduced with carbon.

$$2PbO + C \longrightarrow 2Pb + CO_2$$

The metallic lead that comes from this process contains impurities of silver, gold, and other metals. It too can be purified by electrolysis in the same manner as copper and tin. The silver and gold, of course, are recovered and help offset the cost of the electricity used in the purification process.

Metallic lead is used to make lead storage batteries and as a raw material in the manufacture of tetraethyl lead, $Pb(C_2H_5)_4$—the additive in "leaded" gasoline. It is also used to make compounds found in "lead based" paints. Lead's ability to absorb high energy radiation makes it a useful shielding material for X rays and high energy radiation produced in nuclear reactors.

Bismuth Bismuth is found as its oxide, Bi_2O_3, and sulfide, Bi_2S_3. The sulfide must first be heated in air to convert it to the oxide before reduction.

$$2Bi_2S_3 + 9O_2 \longrightarrow 2Bi_2O_3 + 6SO_2$$

$$2Bi_2O_3 + 3C \longrightarrow 4Bi + 3CO_2$$

Bismuth is also obtained as a byproduct in the production of lead.

Bismuth is a hard, brittle, slightly reddish-colored metal. It is fairly resistant to corrosion and is one of only a few known substances that expands when the molten metal freezes. It is used to make alloys whose volumes stay very nearly constant when they solidify.

In general, alloys of bismuth have low melting points. One of them, called *Wood's metal,* is composed of 50% Bi, 25% Pb, and $12\frac{1}{2}$% each of Sn and Cd. It has a melting point of about 70°C, well below the boiling point of water, and is used in the triggering mechanism of automatic sprinkler systems. Similar alloys are used in fuses designed to protect electrical circuits. When too much current is drawn through the alloy wire in the fuse, it becomes hot and quickly melts. This breaks the electrical circuit and prevents damage to the device that it's meant to protect.

Chemical properties and compounds

Aluminum Aluminum is a very reactive element, a fact that prevented the simple recovery of the metal from its compounds until the invention of the Hall process. It is a good reducing agent, and its reduction potential ($\mathscr{E}^0 = -1.67$ V) is sufficiently negative that aluminum should react with water and liberate hydrogen. However, it forms a tough oxide coating that adheres to the metal and protects it from attack both by moisture and by oxygen.[1] This allows aluminum to be a useful structural metal. In fact, this oxide coating is sometimes deliberately made especially thick by having aluminum be the anode in an electrolytic cell—the product is called *anodized aluminum*.

In Chapter 9 you learned that aluminum metal is amphoteric and is able to dissolve in both dilute acids and in base. Both reactions liberate hydrogen.

$$2Al\ (s) + 6H^+\ (aq) \longrightarrow 2Al^{3+}\ (aq) + 3H_2\ (g)$$

$$2Al\ (s) + 2OH^-\ (aq) + 2H_2O \longrightarrow 2AlO_2^-\ (aq) + 3H_2\ (g)$$

However, toward concentrated HNO_3 aluminum appears passive—that is, it doesn't react. The strong oxidizing power of HNO_3 creates an oxide coating that protects the metal from further attack.

Besides protecting the metal from oxidation, aluminum oxide is an important compound in its own right. It occurs in two principal forms, which are designated as γ-Al_2O_3 and α-Al_2O_3. The γ-Al_2O_3 is formed by dehydrating aluminum hydroxide at a relatively low temperature.

$$2Al(OH)_3\ (s) \xrightarrow{<450°C} \gamma\text{-}Al_2O_3\ (s) + 3H_2O\ (g)$$

This form of the oxide is quite reactive, dissolving readily in both acids and bases.

If γ-Al_2O_3 is heated strongly, its crystal structure changes to that of the α-Al_2O_3 form, which is called *corundum*. This substance has a very high melting point (about 2050°C), is very hard and is quite inert, especially toward acids. Corundum from deposits that are found in nature is used as an abrasive in sandpaper and in making refractory bricks that line furnace interiors.

A number of familiar gemstones are also composed of almost pure α-Al_2O_3. For example, ruby consists of α-Al_2O_3 with small amounts of dissolved Cr^{3+}. Other impurities in the α-Al_2O_3 produce gems with other colors. Sapphire, for instance, contains traces of Fe^{2+} and Ti^{4+}, oriental topaz contains Fe^{3+}, and oriental amethyst contains Mn^{3+} (see Color Plate 15).

Aluminum oxide has a very large exothermic heat of formation. ($\Delta H_f^0 = -1676$ kJ/mol). This allows aluminum to extract oxygen from other metal oxides with the simultaneous release of large amounts of heat—enough to melt the products of the reaction. For example, the reaction of Fe_2O_3 with aluminum

$$2Al\ (s) + Fe_2O_3\ (s) \longrightarrow Al_2O_3\ (l) + 2Fe\ (l)$$

produces temperatures approaching 3000°C! This is called the **thermite reaction** and is used to weld large masses of iron and steel. Thermite bombs have also

The thermite reaction being used to weld reinforcing rods for concrete at a construction site.

[1] A freshly exposed aluminum surface can be amalgamated—coated with a film of mercury in which the metallic aluminum dissolves. The Al_2O_3 that's formed on exposure to air does not adhere to the amalgamated surface and the aluminum reacts easily with oxygen, corroding very rapidly. This very exothermic reaction can make the aluminum very hot.

been used by the military as incendiary devices because of the intense heat of the reaction.

The large exothermic heat of formation of Al_2O_3 has had another interesting application in recent times—it provides the thrust for the booster rockets that enable the space shuttle Columbia to take off. The solid propellant in these booster rockets is a mixture of powdered aluminum (the fuel) and ammonium perchlorate, NH_4ClO_4 (the oxidizer). They are mixed with a small amount of iron oxide catalyst, and the entire mixture is held together in a solid mass by an epoxy plastic. When the rocket is ignited, the aluminum is oxidized, and the formation of Al_2O_3 liberates large amounts of heat that cause the gases that are also formed to expand with great force. That is what lifts the rocket.

The anhydrous aluminum halides are interesting compounds because of their tendency to form *dimeric species*—molecules formed by the pairing of two AlX_3 units. This is especially so in the vapor and in solutions of these compounds in nonpolar solvents, such as benzene and carbon tetrachloride. The pairing occurs in a way similar to the linking together of $BeCl_2$ molecules in solid $BeCl_2$—by the formation of a coordinate covalent bond from the halogen on one AlX_3 unit to the aluminum atom of another. For example, with $AlCl_3$ the species Al_2Cl_6 is formed.

The white cloud seen billowing beneath the rising space shuttle in the photograph at the beginning of this chapter is composed of fine particles of Al_2O_3.

The chemistries of beryllium and aluminum are similar in many ways. This has been attributed to their similar ionic potentials.

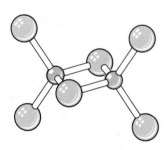

Structure of Al_2X_6 (small spheres, Al; large spheres, X).

When aluminum chloride or other aluminum salts are dissolved in water, their solutions are acidic. In Chapter 14 we saw that this could be accounted for by considering how the Al^{3+} ion polarizes the water molecules that surround it in solution, making it easier for the hydrogens to be removed as H^+.

$$H_2O + Al(H_2O)_6^{3+} \longrightarrow Al(H_2O)_5(OH)^{2+} + H_3O^+$$

As base is added to a solution containing Al^{3+}, it neutralizes H_3O^+ and gradually strips protons from the water molecules that surround the Al^{3+} until the insoluble hydroxide is produced.

$$Al(H_2O)_6^{3+} + 3OH^- \longrightarrow Al(H_2O)_3(OH)_3\ (s) + 3HOH$$

Because of all the water contained within it, the aluminum hydroxide formed in this way is gelatinous (gelatinlike), rather than crystalline.

When base is added to the aluminum hydroxide precipitate, it dissolves, presumably because the OH^- extracts a proton from yet another water molecule that is attached to the Al^{3+}.

$$Al(H_2O)_3(OH)_3 + OH^- \longrightarrow Al(H_2O)_2(OH)_4^- + HOH$$

The species $Al(H_2O)_2(OH)_4^-$ is called the aluminate ion, although its precise composition is uncertain. Notice, for example, the equivalence

$$Al(H_2O)_2(OH)_4^- \text{ and } AlO_2^- + 4H_2O$$

The reactions described above can be reversed by the addition of acid. For instance, when a basic solution containing aluminate ion is gradually neutralized the hydroxide precipitates and then redissolves as more acid is added,

$$Al(H_2O)_2(OH)_4^- + H_3O^+ \longrightarrow Al(H_2O)_3(OH)_3\ (s) + H_2O$$

$$Al(H_2O)_3(OH)_3\ (s) + H_3O^+ \longrightarrow Al(H_2O)_4(OH)_2^+ + H_2O$$

and ultimately, in solutions that are sufficiently acidic,

$$Al(H_2O)_5OH^{2+} + H_3O^+ \longrightarrow Al(H_2O)_6^{3+} + H_2O$$

$Al_2(SO_4)_3$ ranks about 37 in total production among industrial chemicals.

A compound of aluminum produced in large quantities, much of which is destined to be changed to the hydroxide, is aluminum sulfate. It is usually made from bauxite and sulfuric acid.

$$Al_2O_3 \ (s) + 3H_2SO_4 \ (aq) \longrightarrow Al_2(SO_4)_3 \ (aq) + 3H_2O$$

When crystallized it forms hydrates with as many as 18 water molecules ($Al_2(SO_4)_3 \cdot 18H_2O$).

The principal use of $Al_2(SO_4)_3$ is in the paper industry, where it is used to adjust acidity and to treat paper to make it water resistant. Aluminum sulfate is also employed in municipal water treatment. It is added to the water, which is then made basic by the addition of lime (CaO). This causes a gelatinous aluminum hydroxide precipitate to be formed that settles slowly to the bottom, taking fine sediment and bacteria with it.

When an aqueous mixture of aluminum sulfate and sodium sulfate is evaporated, crystals having the composition $NaAl(SO_4)_2 \cdot 12H_2O$ are formed. Similar crystals are also formed if $(NH_4)_2SO_4$ or K_2SO_4 are substituted for the Na_2SO_4 and if sulfates of Cr^{3+} or Fe^{3+} are substituted for $Al_2(SO_4)_3$. These crystals are called **alums.** They are characterized by the general formula $M^+M^{3+}(SO_4)_2 \cdot 12H_2O$, and are examples of **double salts.** If conditions are right, large well-formed octahedrally shaped crystals can be produced, as shown in Figure 18.10.

$M^+ = Na^+, K^+, NH_4^+$
$M^{3+} = Al^{3+}, Cr^{3+}, Fe^{3+}$

One use of sodium alum is in baking powder, where it is combined with sodium bicarbonate. When added to moist dough, the acidity of the aluminum ion in water causes carbon dioxide to be released by reaction of hydronium ion with the bicarbonate ion,

$$HCO_3^- + H_3O^+ \longrightarrow 2H_2O + CO_2 \ (g)$$

As mentioned earlier, when the dough is baked the small bubbles of CO_2 give the finished product a light and porous texture.

Figure 18.10
An octahedrally shaped crystal of potassium alum, $KAl(SO_4)_2 \cdot 12H_2O$, shown in actual size.

Tin, Lead, and Bismuth The chemistries of these elements are characterized by the existence of two oxidation states. Tin and lead, for example, form compounds in both the 2+ and 4+ states, while bismuth forms compounds in the 3+ and 5+ states. The relative stabilities of their higher and lower oxidation states are reflected in the way the metals react with various substances. For example, tin reacts with nonoxidizing acids such as HCl to form Sn^{2+}, but with oxidizing acids the 4+ state is produced.

$$Sn + 2HCl \longrightarrow SnCl_2 + H_2$$

$$Sn + 4HNO_3 \longrightarrow SnO_2 + 4NO_2 + 2H_2O$$

Lead, however, forms only lead(II) compounds, even when strong oxidizing acids such as concentrated HNO_3 are used.

$$3Pb + 8HNO_3 \longrightarrow 3Pb(NO_3)_2 + 2NO + 4H_2O$$

This tells us that when tin and lead are compared, the higher oxidation state is relatively more stable for tin than for lead. This is also revealed in their reaction products with chlorine. Tin combines with chlorine to form $SnCl_4$, but lead reacts to produce $PbCl_2$.

$$Sn + 2Cl_2 \longrightarrow SnCl_4$$

$$Pb + Cl_2 \longrightarrow PbCl_2$$

In the case of bismuth, the 3+ state is much more stable than the 5+ one. Direct combination of bismuth with oxygen or chlorine, for example, produces Bi_2O_3 and $BiCl_3$, respectively. The oxidation of Bi_2O_3 to Bi_2O_5 requires very severe oxidizing conditions, and the compound $BiCl_5$ does not exist at all because the 5+ oxidation state of bismuth is such a powerful oxidizing agent—that is, it has such a strong tendency to be reduced—that it would oxidize Cl^- to Cl_2.

Besides dissolving in acids, tin and lead also dissolve in base with the evolution of hydrogen.

Tin and lead are both amphoteric.

$$Sn + 2OH^- + 2H_2O \longrightarrow Sn(OH)_4^{2-} + H_2 \,(g)$$

$$Pb + 2OH^- + 2H_2O \longrightarrow Pb(OH)_4^{2-} + H_2 \,(g)$$

The *stannite ion*, $Sn(OH)_4^{2-}$, is a very powerful reducing agent and is easily oxidized. On the other hand, the *plumbite ion*, $Pb(OH)_4^{2-}$, is less easily oxidized, which again reveals that the 2+ state is more stable for lead than it is for tin.

Compounds of tin, lead, and bismuth have a variety of practical uses. For example, tin(II) fluoride, SnF_2, also called stannous fluoride, is the decay-inhibiting ingredient in some "fluoride" toothpastes. (The way fluoride ion helps prevent tooth decay was discussed in Chapter 13, page 435.) Some tin compounds are also useful fungicides. An example is $(C_4H_9)_3SnO$, whose chemical name is tributyltin oxide. It is used in some antifouling paints that are applied to boat hulls to prevent marine growth, and in preparations that are applied to wood to prevent rotting.

Among the most useful compounds of lead are its oxides. Lead(II) oxide, PbO, a yellow powder that is also called litharge, is used in pottery glazes and in making fine lead crystal. If PbO is heated carefully in air, it can be oxidized to Pb_3O_4. This is a mixed oxide containing both lead(II) and lead(IV). Its common name is *red lead*, and it is used in corrosion-inhibiting paints that are applied to structural steel.

If solutions of plumbite ion are subjected to strong oxidizing conditions, PbO_2 can be prepared. For example,

OCl$^-$ is a strong oxidizing agent.

$$Pb(OH)_4^{2-} + OCl^- \longrightarrow PbO_2 + H_2O + 2OH^- + Cl^-$$

Unlike SnO_2, PbO_2 is a strong oxidizing agent. Its most common use is as the cathode material in lead storage batteries. Recall that during the discharge of this battery the net reaction is

$$Pb + PbO_2 + 2H_2SO_4 \longrightarrow 2PbSO_4 + 2H_2O$$

As you are probably aware, in relatively recent times there has been a great deal of concern over the past use of lead-based paints, particularly on surfaces in living areas children may be exposed to. This is because lead compounds are very toxic. The pigment most used in lead-based paints is a white basic carbonate, $Pb_3(OH)_2(CO_3)_2$, and its use in interior paints is now restricted. Although it provides excellent covering power as a pigment, it suffers from the disadvantage of being darkened on contact with H_2S produced in the environment by decaying vegetation. The darkening is caused by the production of black PbS. Another lead-based pigment that is used in artists' oil colors is lead chromate, $PbCrO_4$. This compound has a bright yellow color.

Bismuth compounds are utilized frequently in the cosmetic and pharmaceutical industry. In fact, these industries account for about 30 percent of the bismuth produced each year. For example, a substance known as bismuth subnitrate, produced by partial hydrolysis of $Bi(NO_3)_3$, is used medicinally as an antacid in the treatment of gastric ulcers. Its exact composition varies according to how it is prepared, but it can be approximately formulated as $BiO(NO_3)$.

The hydrolysis of bismuth(III) compounds is not unusual. When $BiCl_3$ is dissolved in water it hydrolyzes, producing the bismuthyl ion, BiO^+. As the solution is diluted an insoluble precipitate of $BiOCl$ is formed that redissolves if the solution is made more acidic by the addition of hydrochloric acid.

The oxide of bismuth, Bi_2O_3, is formed when bismuth is heated in air. As mentioned earlier, under extreme oxidizing conditions this can be oxidized to give bismuth in the 5+ oxidation state. Compounds containing bismuth(V), such as $NaBiO_3$ (*sodium bismuthate*), are extremely powerful oxidizing agents as a result of bismuth's strong tendency to revert to the 3+ oxidation state by acquiring electrons from other substances.

INDEX TO QUESTIONS AND PROBLEMS (Problem numbers are in **bold type**)

REVIEW QUESTIONS

18.1 How are most metals found in nature? What are some important sources of metals?

18.2 What property must a metal compound possess to be easily decomposed thermally?

18.3 Why must metal compounds always be reduced to extract the metals from them?

18.4 What is implied thermodynamically when we say that a particular compound is thermally stable?

18.5 Why is carbon a preferred reducing agent in commercial metallurgy?

18.6 Write equations showing the use of carbon and hydrogen as reducing agents in the extraction of a metal from one of its compounds.

18.7 Why isn't sodium produced commercially by reduction of its compounds with a reducing agent that is stronger than sodium?

18.8 Why are halide salts often used when preparing metals by electrolysis?

18.9 Write equations for the commercial electrolytic production of sodium and aluminum.

18.10 Why are the Group IA elements called *alkali metals?* What oxidation states are observed for the Group IA metals?

18.11 Which alkali metals are most abundant? Which one is least abundant? Why?

18.12 Why do compounds of lithium, rubidium, and cesium have little commercial importance?

18.13 Where do the alkali metals occur in nature?

18.14 Write a chemical equation to show how elemental potassium, rubidium, and cesium are usually made.

18.15 Describe two applications of metallic sodium related to its physical properties.

18.16 What colors are given to a bunsen burner flame by compounds of (a) sodium, (b) lithium, and (c) potassium?

18.17 How can potassium ion in a mixture be detected if the mixture also contains sodium ion?

18.18 Write a chemical equation for the reaction of rubidium with water.

18.19 Why is the reduction potential of lithium more negative than the reduction potential of sodium?

18.20 What happens when an alkali metal is added to liquid ammonia? Why are these solutions such good reducing agents?

18.21 Write chemical equations to show the reaction of lithium and sodium with (a) bromine, (b) sulfur, and (c) nitrogen.

18.22 Write equations that illustrate the reactions of each of the alkali metals with oxygen (present in excess).

18.23 What compound of the alkali metals is used in a recirculating breathing apparatus? Write chemical equations to show how it functions.

18.24 Why are solutions of Na_2O_2 basic?

18.25 What is a *getter*?

18.26 Why are sodium compounds more important, commercially, than compounds of the other alkali metals?

18.27 What are other common names for sodium hydroxide? What are some of its uses?

18.28 What is trona ore? Give the chemical reactions involved in the Solvay process. What is the net reaction in the Solvay process?

18.29 How does sodium bicarbonate serve as a buffer? What is the common name for sodium bicarbonate? How can $NaHCO_3$ serve as a fire extinguisher?

18.30 What is potash? How is it made?

18.31 What alkali metal compound is used to treat manic depression?

18.32 Why are the Group IIA elements called alkaline earth metals? How do their densities, hardness, melting points, and ionization energies compare to their neighbors in Group IA?

18.33 Why do the alkaline earth metals have reduction potentials that are nearly as negative as the metals in Group IA?

18.34 Where are calcium and magnesium found? What is the source of radium?

18.35 How is magnesium recovered from dolomite? How is magnesium recovered from sea water?

18.36 Define *calcining*. What is lime? What happens when water is added to lime? Why is lime such an important industrial chemical?

18.37 What reaction takes place inside a flashbulb when it is fired?

18.38 What color is given to a bunsen burner flame by compounds of (a) calcium, (b) strontium, and (c) barium?

18.39 What chemical fact is responsible for the structural uses of metallic beryllium and magnesium? Why are metallic calcium, strontium, and barium not used as structural metals?

18.40 Write chemical equations for the reaction of magnesium with elemental oxygen, sulfur, and nitrogen.

18.41 Write chemical equations that illustrate the amphoteric behavior of metallic beryllium.

18.42 What is the structure of solid $BeCl_2$? How does this support the statements in earlier chapters that molecules of $BeCl_2$ have less than an octet in the valence shell of beryllium?

18.43 Why are beryllium compounds covalent? What are organomagnesium compounds?

18.44 How do the solubilities of the alkaline earth hydroxides vary from top to bottom in the group? How do the solubilities of the sulfates vary?

18.45 What are some uses of calcium carbonate?

18.46 What is milk of magnesia composed of?

18.47 What is gypsum? Write chemical equations showing how plaster of paris is made and what happens when it combines with water during hardening.

18.48 What property of barium sulfate allows it to be used for X-ray diagnosis of intestinal tract disorders?

18.49 What is *epsom salts*? What are some uses of magnesium oxide?

18.50 In what way are the metals in Groups IIIA, IVA, and VA less metallic than those in Groups IA and IIA?

18.51 What oxidation states are observed for the metals in Groups IIIA, IVA, and VA?

18.52 Which are the post-transition metals? Why do their lower oxidation states become more stable relative to the higher ones going down a group?

18.53 What is the ore of aluminum? Write chemical equations to show how it is purified.

18.54 What are some commercial uses of metallic aluminum?

18.55 Write chemical equations showing how tin, lead, and bismuth are recovered from their ores.

18.56 Why does a tin can rust so rapidly if the tin coating is scratched, exposing the steel beneath it?

18.57 What are allotropes? Which of the representative metals exhibits allotropism?

18.58 What unusual property is possessed by metallic bismuth? What is Wood's metal? What are its properties and uses?

18.59 Why doesn't aluminum corrode rapidly in the presence of air and moisture? How can it be made to corrode very quickly?

18.60 Write chemical equations showing the amphoteric behavior of metallic aluminum.

18.61 How do the properties of γ-Al_2O_3 and α-Al_2O_3 differ. What gems are composed primarily of aluminum oxide?

18.62 What is the thermite reaction?

18.63 What is the structure of dimeric aluminum chloride?

18.64 Why are solutions of aluminum salts acidic?

18.65 Write chemical equations for the gradual neutralization of $Al(H_2O)_6^{3+}$, and the dissolving in base of the gelatinous aluminum hydroxide precipitate.

18.66 What are two ways of writing the formula of the aluminate ion?

18.67 How is aluminum sulfate used in water treatment plants?

18.68 What is an alum? Give an example. Which alum is used in baking powders and how does it function?

18.69 How do the elements tin and lead differ in their behavior toward nitric acid and chlorine? Illustrate using chemical equations.

18.70 Why doesn't $BiCl_5$ exist?

18.71 Write chemical equations for the reactions of metallic tin and lead with base.

18.72 What are the oxides of lead and what are some of their uses?

18.73 Why do lead-based paints gradually darken over a period of time?

18.74 What are some uses of bismuth compounds? What is the formula of the bismuthyl ion?

REVIEW PROBLEMS (More difficult problems are marked by an asterisk.)

18.75 Given the following thermodynamic data, calculate the hydration energy for the Na^+ ion in kcal/mol.

$$\Delta H_f^0 \text{ of } Na^+ (aq) = -57.28 \text{ kcal/mol}$$

$$\Delta H_{atom}^0 \text{ of } Na = 25.98 \text{ kcal/mol}$$

$$IE \text{ of } Na = 118.0 \text{ kcal/mol}$$

18.76 Using the data for the atomization energy of Na (Problem 18.75) and the first and second ionization energies for Na in Table 3.7, compute the value of the hydration energy (in kcal/mol) required to produce a negative ΔH_f for Na^{2+} (aq).

18.77 From the data in Tables 11.1 and 11.4, calculate the temperature (in °C) above which the thermal decomposition of ZnO should become feasible.

18.78 Given the data below, determine the temperature in °C at which $K_P = 1$ for the reaction,

$$CuO (s) \longrightarrow Cu (s) + \tfrac{1}{2}O_2 (g)$$

For CuO (s), $\Delta H_f^0 = -155$ kJ/mol. Absolute entropies; CuO (s), 43.5 J/mol K; Cu (s), 33.3 J/mol K; O_2 (g), 205.0 J/mol K.

*18.79** Calculate K_P at 100, 500, and at 2000°C for the reaction,

$$MoO_3 (s) \longrightarrow Mo (s) + \tfrac{3}{2}O_2 (g)$$

given the following data:

	ΔH_f^0 (kcal/mol)	S^0 (cal/mol K)
MoO_3 (s)	−180.3	18.68
Mo (s)	0.0	6.83
O_2 (g)	0.0	49.00

*18.80** The standard reduction potential of potassium is −2.92 V. ΔH_{hyd}^0 of K^+ in 1 M aqueous solution is −759 kJ/mol. The atomization energy of K is +90.0 kJ/mol and the ionization energy of K (g) is +418 kJ/mol. Calculate ΔS^0 (in J/mol K) for the process,

$$K (s) \longrightarrow K^+ (1 M) + e^-$$

THE CHEMISTRY OF SELECTED NONMETALS, PART I: HYDROGEN, CARBON, OXYGEN, AND NITROGEN

Carbon, hydrogen, oxygen, and nitrogen touch our lives closely. Not only are they found in nearly all the molecules in our bodies, they are also involved, one way or another, in the production of photochemical smog, seen here hanging over a western city. How photochemical smog is formed is one of the topics discussed in this chapter.

In Chapter 18 we discussed the chemical and physical properties of the representative metals. We now turn our attention, in this chapter and the next, to the remainder of the representative elements: the nonmetals and metalloids. In many ways, the chemistry of these nonmetallic elements is more interesting than that of the A-group metals because of the large variety of compounds that they form. Not only do they combine with metals to form substances that are often predominantly ionic, they also combine with each other by way of covalent bonds to form molecules and polyatomic ions that range from simple clusters of atoms to gigantic molecules, such as the DNA that controls heredity and guides the chemical functions of our cells.

In this chapter we begin by studying the properties of hydrogen, carbon, oxygen, and nitrogen. Although our discussions will not be biased toward biology, these four nonmetals are uniformly important to all life as we know it, and much of our interest in their chemical properties is tied to how they and their compounds influence our lives and affect the quality of our environment.

19.1   HYDROGEN

How stars produce energy is discussed further in Chapter 24.

Hydrogen, first recognized as an element by the English chemist, Henry Cavendish (1731–1810), is the most abundant of all the elements in the universe—approximately 93% if counted by atoms. It is the principal element in the solar atmosphere, and it is the nuclear fuel that stars consume in their generation of energy. Here on Earth, hydrogen is much less abundant—about 3% by atoms, or about 0.14% by mass—presumably because the Earth's gravity was insufficient to hold onto most of the hydrogen that was present when the planet was formed.

Virtually all the Earth's hydrogen exists in the combined state. Its reactivity, especially toward oxygen, is too great to allow it to be present as the free element in the atmosphere except in trace amounts. Water, of course, is two-thirds hydrogen on an atom basis (about 11% by weight) and the oceans represent a huge storehouse of this element. Hydrogen is also a principal element in all organic material. This includes living things, both animal and vegetable, as well as their fossils—petroleum and natural gas.

The official name for H_2 is dihydrogen.

Only helium has a lower boiling point than hydrogen.

In its elemental state hydrogen exists as diatomic molecules of H_2. Its boiling point is $-253°C$ and its freezing point (melting point) is $-259°C$. At room temperature, of course, hydrogen is a gas, and its very small molecular weight makes it the least dense gas of all (it is only half as dense as helium). Hydrogen therefore has great "lifting power" in balloons, but as we learned in Chapter 1 (Figure 1.11, page 25), its extreme flammability poses a threat of disaster!

H $1s^1$

In the periodic table, hydrogen is placed in Group IA because its valence shell has the same electron configuration as the other members of this group. Here any similarities cease, however. Hydrogen has no electrons below its valence shell, and for this reason its chemistry does not resemble that of the Group IA metals at all. In fact, hydrogen really does not fit well into any group within the periodic table.

Isotopes of hydrogen

Many elements occur in nature as mixtures of isotopes, but hydrogen is the only one whose isotopes have their own names. The nucleus of ordinary hydrogen, $_1^1H$, consists of a single proton. It is the most abundant of hydrogen's three isotopes, and on rare occasions is called **protium.**

Deuterium, $_1^2H$, has a nucleus consisting of one proton and one neutron, and is often given the symbol D. For example, the formula for deuterium oxide—also

Deuterium is called heavy hydrogen.

known as **heavy water**—is usually written D_2O. Only about 1 of every 5000 atoms of naturally occurring hydrogen is deuterium.

The third isotope of hydrogen, ^3_1H, is called **tritium** (symbol T). It is radioactive, and because of its relatively short half-life of 12.3 years, it is found in only very minute amounts in naturally occurring hydrogen. However, it can be made in nuclear reactions—for example, by bombarding lithium or boron with neutrons. In fact, tritium is a by-product of the operation of nuclear power plants, where it is produced by a variety of nuclear reactions including ones involving lithium and boron additives in the reactor cooling system.

Only 1 of every 10 million atoms of naturally occurring hydrogen is ^3_1H.

Chemically, the isotopes of hydrogen are identical except for small differences in the rates at which they react. This is both a blessing and a curse. One beneficial use of hydrogen's isotopes is in the study of reaction mechanisms. For example, a compound can be "labeled" by replacing one or more of the hydrogens in its structure with atoms of deuterium. After this compound is allowed to react, the locations of the labels in the product molecules can be determined using a mass spectrometer, and the information gathered in this way can help a chemist deduce the mechanism of the reaction.

Replacing H with D doesn't alter the chemical properties of a compound.

A danger also exists in the chemical similarities of isotopes. A large-scale use of nuclear power plants will almost surely increase the concentration of tritium in the environment. This isotope can easily be incorporated into biological molecules because it behaves chemically just like ordinary hydrogen, and the radiation that it would give off within an organism could cause many problems, including cancer and other radiation-related maladies.

Preparation and uses

Hydrogen is an important industrial chemical. Its principal source is natural gas, from which it is extracted by reactions with steam at high temperature in the presence of a catalyst.

$$CH_4\ (g) + H_2O\ (g) \xrightarrow[\text{catalyst}]{\text{heat}} CO\ (g) + 3H_2\ (g)$$
methane
(natural gas)

Similar reactions can occur with other hydrocarbons.

$$CO\ (g) + H_2O\ (g) \xrightarrow[\text{catalyst}]{\text{heat}} CO_2\ (g) + H_2\ (g)$$

The hydrogen and carbon dioxide can be easily separated from each other by bubbling the gas mixture through water, in which CO_2 is fairly soluble and H_2 is virtually insoluble.

Hydrogen can also be extracted from water by allowing steam to react with carbon (from coal, for instance) at temperatures of about 1000°C.

$$C\ (s) + H_2O\ (g) \xrightarrow{1000°C} CO\ (g) + H_2\ (g)$$

The mixture of CO and H_2 is called *water gas* and the reaction is referred to as the *water gas reaction*. It is used by industry because it changes a solid fuel (coal), which is awkward to handle in large quantities, into a gaseous combustible mixture that is easily piped to where it is needed.

Another way of obtaining hydrogen from water is by electrolysis. In Chapter 17 we saw that the electrolysis of brine is used to produce huge quantities of caustic soda (sodium hydroxide). A second product of this electrolysis is hydrogen, which also becomes part of the industrial supply of this element. The overall cell reaction is

Combustion of water gas:

$$2CO + O_2 \longrightarrow 2CO_2$$

$$2H_2 + O_2 \longrightarrow 2H_2O$$

$$2NaCl\ (aq) + 2H_2O \longrightarrow 2NaOH\ (aq) + Cl_2\ (g) + H_2\ (g)$$

If hydrogen is needed in small quantities in the laboratory, it can be conveniently made by reacting an active metal with an acid. Any metal that is below

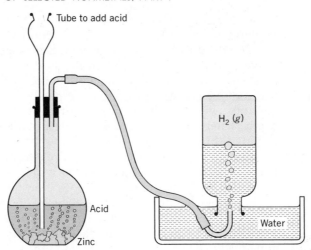

Figure 19.1

Laboratory preparation of hydrogen by the reaction of a metal such as zinc with an acid.

Zinc metal readily dissolves in hydrochloric acid with the evolution of bubbles of hydrogen.

hydrogen in Table 17.1 is a better reducing agent than hydrogen, and, in principle, should cause H^+ to be reduced. However, those metals having reduction potentials near zero (lead and tin, for example) react with acids very slowly. At the other extreme, those metals near the bottom of Table 17.1 are such active reducing agents that they react very vigorously with water (as we saw in the last chapter) and almost explosively with an acid. Metals having intermediate reduction potentials react smoothly with acids and serve as practical sources of hydrogen. A metal frequently used for this purpose is zinc. With dilute sulfuric acid it reacts as follows.

$$H_2SO_4 \ (aq) + Zn \ (s) \longrightarrow ZnSO_4 \ (aq) + H_2 \ (g)$$

Figure 19.1 shows the type of apparatus that can be used to prepare H_2 in the laboratory. The gas is collected by allowing it to displace water from the inverted bottle, which has its mouth below the surface of the water in the tray. This method of collection works because H_2 has a very low solubility in water.

The greatest single use of hydrogen is in the production of ammonia from nitrogen.

$$3H_2 \ (g) + N_2 \ (g) \longrightarrow 2NH_3 \ (g)$$

This reaction, which consumes approximately two-thirds of the annual world production of hydrogen, is discussed further in Section 19.4.

Hydrogen is also used in large quantities in the manufacture of methanol (also called methyl alcohol or wood alcohol[1]). The reaction combines carbon monoxide and hydrogen at high pressure and temperature over a catalyst.

Using appropriate catalysts, CO and H_2 can be made to form other alcohols and hydrocarbons. Industrially, mixtures of CO and H_2 are also known as **synthesis gas.**

$$CO + 2H_2 \xrightarrow[\substack{200-300 \text{ atm} \\ \text{catalyst}}]{300-400°C} CH_3OH$$

This reaction is important because it provides a simple route by which coal can be converted to a liquid fuel. First the coal can be converted to a mixture of CO and H_2 by the water gas reaction. Then CO and H_2 can be combined to produce methanol, which is itself a potentially useful fuel. What makes matters even more interesting is a catalytic process discovered by Mobil Oil Corporation that is able to change CH_3OH into high octane gasoline.[2] At the present time, how-

---

[1] At one time methanol was obtained as one of the products formed by heating wood in the absence of air—a process called *destructive distillation*. Although methanol is still called wood alcohol in many places, very little of it is currently produced by this older method.

[2] Direct conversion of coal to liquid hydrocarbon fuels by catalytic reactions of carbon in the form of coal dust with hydrogen has also been under study, and some success has been achieved.

ever, gasoline produced from methanol is more expensive than that obtained directly from petroleum.

Still another commercial use of hydrogen is the **hydrogenation** of vegetable oils, in which hydrogen is added chemically to carbon-carbon double bonds. For example, ethylene can be converted to ethane by hydrogenation.

Because organic molecules having double bonds possess the ability to take on additional hydrogen, they are said to be *unsaturated*. On the other hand, molecules with only carbon-carbon single bonds cannot react with more hydrogen, so they are termed *saturated*. Hydrogenation therefore converts unsaturated vegetable oils into saturated fats. We will say more about this in Chapter 22.

Compounds of hydrogen

Hydrogen is found in more compounds than any other element. Virtually every organic compound contains hydrogen—they are either hydrocarbons or are formed from hydrocarbons by replacing some of their hydrogen atoms with other elements. Hydrogen is found in both binary acids and oxoacids, and hydrogen even combines directly with active metals.

Binary compounds with hydrogen are called **hydrides,** and can be divided into two main types: ionic or saltlike hydrides and covalent hydrides. The ionic hydrides are formed from hydrogen and an active metal. Because of hydrogen's rather high electronegativity, it combines directly with the active metals in Groups IA and IIA by acquiring an electron and forming the **hydride ion,** H^-, thereby completing its valence shell.

$$2Li~(s) + H_2~(g) \longrightarrow 2LiH~(s)$$

$$Ca~(s) + H_2~(g) \longrightarrow CaH_2~(s)$$

These compounds are very sensitive toward moisture because of the basicity of the hydride ion. In water their overall reactions are as follows.

$$LiH~(s) + H_2O \longrightarrow LiOH~(aq) + H_2~(g)$$

$$CaH_2~(s) + 2H_2O \longrightarrow Ca(OH)_2~(aq) + 2H_2~(g)$$

but the reaction is really one between H^- and H_2O.

$$H^- + H_2O \longrightarrow H_2 + OH^-$$

This can be viewed as either an acid-base reaction or a redox reaction.

Hydrogen, of course, can also complete its $1s$ subshell by electron sharing. You no doubt recall drawing many Lewis structures showing hydrogen covalently bonded to another atom. Except for the very special case of boron and the hydrogen difluoride ion HF_2^-, in which a H atom lies equidistant between two F atoms (that is, $[F—H—F]^-$), hydrogen is capable of binding to only one other atom by way of electron sharing, because it seeks only one additional electron. As a result, the structural chemistry of the simple hydrides is rather straightforward and, perhaps, even somewhat mundane. With the halogens it forms compounds with the general formula HX (e.g., HF, HCl). Similarly, we find, quite expectedly, that the Group VIA elements form molecules of general formula H_2X (e.g., H_2O, H_2S), the elements of Group VA form H_3X (usually

Table 19.1
Catenation among nonmetal hydrides

Group IVA				
CH_4	SiH_4	GeH_4	SnH_4	
C_2H_6	Si_2H_6	Ge_2H_6	Sn_2H_6	
C_3H_8	Si_3H_8	Ge_3H_8		
.	.			
.	.			
.	.			
C_nH_{2n+2}	Si_6H_{14}			
$+$				
many others				
Group VA				
NH_3	PH_3	AsH_3	SbH_3	BiH_3
N_2H_4	P_2H_4			
Group VIA				
H_2O	H_2S	H_2Se	H_2Te	H_2Po
H_2O_2	H_2S_2			
	H_2S_n $(n = 1 - 6)$			

written XH_3, for example, NH_3, PH_3), and those in Group IVA form H_4X (or XH_4, for example, CH_4, SiH_4). The geometries of these molecules are all readily predicted by the VSEPR theory discussed in Section 5.2.

In addition to these simple hydrides, which contain a single atom of the nonmetal, there are others that possess two or more nonmetal atoms. Some examples are given in Table 19.1. All these compounds are characterized by nonmetal atoms of the same element linked directly to one another, a phenomenon called **catenation.** Thus hydrogen peroxide, H_2O_2, has the Lewis structure

$$
\begin{array}{c}
\text{H} \\
| \\
\ddot{\text{O}}\text{---}\ddot{\text{O}}\text{:} \\
| \\
\text{H}
\end{array}
$$

Similarly, there are others such as

$$
\begin{array}{ccc}
\text{H} \quad \text{H} & \text{H} \quad \text{H} & \text{H} \quad \text{H} \quad \text{H} \\
| \quad | & | \quad | & | \quad | \quad | \\
\text{H---N---N---H} & \text{H---Si---Si---H} & \text{H---C---C---C---H} \\
.. \quad .. & | \quad | & | \quad | \quad | \\
& \text{H} \quad \text{H} & \text{H} \quad \text{H} \quad \text{H} \\
\textbf{hydrazine} & \textbf{disilane} & \textbf{propane}
\end{array}
$$

The ability of nonmetals to form compounds in which they bond to other like atoms varies greatly. You will notice, for example, that in Group VIA only oxygen and sulfur form such compounds. In Group VA we find that both nitrogen and phosphorus catenate, but the chain length seems to be limited to two atoms. When we proceed to Group IVA, all the elements, down to and including tin, exhibit this property and here we find chains containing three, four, and even more atoms. We also see that the tendency toward catenation generally decreases downward in a group, as evidenced by the trend toward shorter chains demonstrated by the heavier elements in Group IVA, Ge and Sn.

Of all the elements, carbon has the greatest capacity to form bonds to itself. In fact, the broad area of organic chemistry is concerned entirely with hydrocarbons and compounds that are derived from them by substituting other elements for hydrogen. Organic compounds are compounds in which the molecular framework consists primarily of carbon-carbon chains. The unique ability of

We will see later that catenation is not restricted to nonmetal hydrides.

carbon to form such diverse compounds containing these long, stable carbon chains is undoubtedly the reason why life has evolved around the element carbon instead of around another element such as silicon.

Preparation of nonmetal hydrides

Nonmetal hydrides are formed as products of many different chemical reactions; however, we will consider only two general methods of preparation here. One of these is the direct combination of the elements, as illustrated, for example, by the reaction of hydrogen with either chlorine,

$$H_2 + Cl_2 \longrightarrow 2HCl$$

or with oxygen,

$$2H_2 + O_2 \longrightarrow 2H_2O$$

However, this method is not applicable to all the hydrides, as we can see by examining some of their thermodynamic properties shown in Table 19.2. Here we see that only the hydrides of the more active nonmetals possess negative free energies of formation. Those lying below the heavy colored line in the table have positive free energies of formation and from a practical standpoint cannot be prepared directly from the free elements. Instead an indirect procedure must be employed.

The rates of reaction toward hydrogen vary substantially among the nonmetals. In period 2, for instance, fluorine reacts immediately with hydrogen when they are placed in contact. On the other hand, H_2 and O_2 mixtures are stable virtually indefinitely, unless the reaction is initiated in some way—for example, by applying heat or introducing a catalyst.

Nitrogen is even less reactive than oxygen, not only toward hydrogen, but toward nearly all other chemical reagents as well. Presumably this is because of the high stability of the N_2 molecule that arises as a consequence of its strong triple bond (the bond energy of N_2 is 946 kJ/mol, compared to 502 and 159 kJ/mol for O_2 and F_2, respectively).

The second method of preparation of nonmetal hydrides involves the addition of protons, from a Brønsted acid, to the conjugate base of a nonmetal hy-

Table 19.2
Standard enthalpies and
free energies of formation of nonmetal hydrides

	XH_n ΔG_f^0(kJ/mol) ΔH_f^0(kJ/mol)			
BH_3 Not stable, simplest hydride is B_2H_6	CH_4 −74.9 −50.6	NH_3 −46.0 −16	H_2O −242 −228	HF −271 −273
	SiH_4 +34 +56.9	PH_3 +5.4 +13	H_2S −21 −33	HCl −92.5 −95.4
	GeH_4 (positive)	AsH_3 +66.5 +69.0	H_2Se +30 +16	HBr −36 −53.6
		SbH_3 (?)	H_2Te +154 +138	HI +26 +2

dride, a reaction that we might depict as

$$X^{n-} + nHA \longrightarrow H_n X + nA^-$$

where X^{n-} is the conjugate base of the hydride $H_n X$, and HA is the Brønsted acid. Let's look at some examples.

The hydrogen halides are commonly prepared in the laboratory by treating a halide salt with a nonvolatile acid such as sulfuric or phosphoric acid.

$$NaCl\ (s) + H_2SO_4\ (l) \longrightarrow HCl\ (g) + NaHSO_4\ (s)$$

$$NaCl\ (s) + H_3PO_4\ (l) \longrightarrow HCl\ (g) + NaH_2PO_4\ (s)$$

In these examples HCl is removed as a gas, which causes the reaction to proceed to completion.

With the heavier halogens—bromine and iodine—sulfuric acid cannot be used because it is a sufficiently strong oxidizing agent to oxidize the halide ion to the free halogen. For example, when treated with concentrated H_2SO_4, I^- reacts as follows:

$$2I^- + HSO_4^- + 3H^+ \longrightarrow I_2 + SO_2 + 2H_2O$$

Phosphoric acid, being a much weaker oxidizing agent than H_2SO_4, simply supplies protons to I^-, and HI can therefore be produced in a reaction analogous to the production of HCl above, that is,

$$NaI\ (s) + H_3PO_4\ (l) \longrightarrow HI\ (g) + NaH_2PO_4\ (s)$$

As we proceed from right to left across a period—for example, from fluorine toward carbon—we have seen that the acid strengths of the $H_n X$ compounds decrease. Thus HF is a stronger acid than H_2O which, in turn, is stronger than NH_3, and so forth. This means that the strengths of their corresponding conjugate bases *increase* from right to left ($C^{4-} > N^{3-} > O^{2-} > F^-$). As a result, the strength of the Brønsted acid required to react with the anion of the nonmetal to produce the hydride decreases. For example, the production of HF, whose conjugate base, F^-, is relatively weak, requires a strong acid such as H_2SO_4. Oxide ion, on the other hand, is a much stronger base than F^- and when treated with even a relatively weak source of protons—for example, acetic acid—the oxide ion gobbles them up to produce water.

$$O^{2-} + 2HC_2H_3O_2 \longrightarrow H_2O + 2C_2H_3O_2^-$$

This is a reaction we have seen before in Chapters 6 and 14.

Nitride ion, N^{3-}, is expected to be even a stronger base than O^{2-}. Therefore it is not surprising to find that Mg_3N_2 reacts with an acid as weak as H_2O to produce NH_3 in a reaction that we can interpret as a hydrolysis of the N^{3-} ion.

$$Mg_3N_2 + 6H_2O \longrightarrow 3Mg(OH)_2 + 2NH_3$$

Metal carbides, which can be prepared by heating an active metal with carbon, also react with water in the same fashion. Aluminum carbide, for instance, which contains C^{4-} ions, hydrolyzes according to the reaction

$$Al_4C_3 + 12H_2O \longrightarrow 4Al(OH)_3 + 3CH_4$$

This general method of preparation also extends to the third, fourth, and fifth periods, too, with the same trends in the strength of the Brønsted acid required to liberate the hydride. In period 3 we have these anions:

Group	IV	V	VI	VII
Anion	Si^{4-}	P^{3-}	S^{2-}	Cl^-

We again expect the anions to become increasingly basic as we move from right to left (from Cl^- to Si^{4-}); therefore the strength of the Brønsted acid needed to protonate the anion decreases. To form HCl from NaCl, a strong acid is required. Sulfide ion, on the other hand, is sufficiently basic to be extensively hydrolyzed in aqueous solution (as we found in Chapter 15), and solutions containing a soluble sulfide such as Na_2S always are very basic and have a strong odor of H_2S because of the reaction[3]

$$S^{2-} + 2H_2O \longrightarrow H_2S + 2OH^-$$

The ability of S^{2-} to pick up protons also explains why many insoluble metal sulfides dissolve in acids with the evolution of H_2S.

Phosphides, like sulfides, also hydrolyze on contact with water. However, because the P^{3-} ion is more basic than the S^{2-} ion, the hydrolysis proceeds essentially to completion. Thus aluminum phosphide, AlP, reacts with water to produce phosphine, PH_3.

$$AlP + 3H_2O \longrightarrow Al(OH)_3 + PH_3$$

Moving left to Group IVA, we again find that a hydrolysis reaction serves to prepare silicon hydrides. A metal silicide such as Mg_2Si (which can be formed by simply heating Mg and Si together) reacts with water to generate a mixture of silanes; SiH_4, Si_2H_6, Si_3H_8, and so on, up to Si_6H_{14}.

The heavier nonmetals behave in much the same fashion as those above them. Thus H_2Se and H_2Te, like H_2S, can be prepared by adding an acid to the metal selenide or telluride. Arsine, AsH_3, like phosphine, PH_3, is made by the hydrolysis of a metal arsenide such as Na_3As or AlAs, and the germanes, GeH_4, Ge_2H_6, and Ge_3H_8 are produced by the action of dilute HCl on Mg_2Ge.

A hydrogen economy

As the world's supplies of petroleum dwindle and become increasingly more expensive, the search for alternative fuels grows more urgent. Coal, and its conversion to synthetic petroleumlike products, has been mentioned as a means of providing substitutes, as we've noted earlier. Another interesting alternative is suggested by proponents of the large-scale use of hydrogen as a fuel, around which virtually our entire energy economy could be built.

Hydrogen has some very attractive features as a fuel. Its reaction with oxygen

$$2H_2\ (g) + O_2\ (g) \longrightarrow 2H_2O\ (l) \qquad \Delta H^0 = -572\ kJ$$

is highly exothermic and produces no pollutants. The gas could be either burned according to this reaction to produce heat or it could be used as a fuel in hydrogen-oxygen fuel cells to generate electricity directly. Since hydrogen is a gas, it could be pumped through the already existing network of pipelines that crisscross the nation carrying natural gas. Hydrogen is also available in virtually infinite amounts from water.

Despite all these advantages, however, there are problems. One of them is that hydrogen is not a very convenient portable fuel for use in an automobile. High pressure tanks of gaseous hydrogen are heavy and can carry only relatively small amounts of this fuel. Other approaches are being tried, including combining hydrogen with certain metals to form metal hydrides that can be decomposed when the hydrogen is needed, but many technical difficulties still exist.

The main problem with hydrogen as a fuel is that it is not a primary energy source like petroleum and natural gas, which already exist in a state ready for

It makes no sense, of course, to burn petroleum to make electricity to produce hydrogen by electrolysis!

---

[3] Actually, there is a two-step equilibrium involving both HS^- and H_2S. See Chapter 15.

use. Hydrogen can only be obtained by first investing energy to extract it from water, so hydrogen will only become a viable fuel if some inexpensive method can be found to produce it. A number of approaches to this problem have been proposed that make use of solar or nuclear energy.

Solar energy can be harnessed in several ways. One is to employ large arrays of solar cells to generate electricity that can be used in the electrolysis of water. Another is to focus solar energy by mirrors into solar furnaces that would heat water vapor to very high temperatures where it would decompose into H_2 and O_2. The mixture of gases would then be rapidly cooled before they could recombine to form H_2O. Nuclear reactors could also be used either to generate the electricity needed to decompose water by electrolysis, or to produce the high temperatures necessary to "crack" water into H_2 and O_2.

Recently, advances have been made in harnessing solar energy directly to split water into H_2 and O_2. Using special catalysts, scientists in France have been able to achieve efficient decomposition of water under visible and ultraviolet illumination. If this process can be made industrially practicable, a convenient means of converting solar energy directly to a useful form of stored chemical energy will be available.

19.2   CARBON

The element carbon is found in every living thing; all life as we know it is based on carbon-containing compounds. It is this fact that is responsible for the term *organic* that is used to describe hydrocarbons, and those compounds that come from hydrocarbons by substituting other atoms for some of the hydrogen atoms in their molecules. At one time it was believed that such substances could only be made within *living* organisms. This has since been shown to be false, but the name organic is still used in discussing these kinds of compounds, and the name *inorganic* is used to describe all the rest. This section deals with the inorganic compounds of carbon and the properties of the element itself. Organic chemistry is discussed in Chapter 22.

Carbon is found in period 2 at the head of Group IVA. Its atoms each have four valence electrons that they tend to share with other atoms in the formation of four covalent bonds. Although in most of its compounds carbon forms four bonds, we will see some exceptions in this section.

Carbon occurs in nature as both the free element and in compounds. Coal contains elemental carbon, and when it is heated strongly in the absence of air, volatile substances are driven off and the material that remains—called **coke**—is almost all carbon. Diamonds like those in Color Plate 10 are, for all practical purposes, pure carbon. In the combined state, we find carbon in all living things and in fossil fuels such as methane, CH_4, and petroleum. Carbon also occurs in large amounts in carbonates such as limestone.

The free element

At room temperature and pressure, graphite is the most thermodynamically stable form of carbon.

In Chapter 9 we saw that carbon can exist in two different allotropic forms. One of them is graphite—a soft black slippery solid that is a reasonably good conductor of electricity. The other is diamond—an extremely hard, nonconducting substance well known for its gem quality crystals. Their different properties, you recall, can be traced to differences in the way the carbon atoms are bonded to each other.

The loosely stacked layer structure of graphite, with extensive delocalized π bonding within each layer, accounts for both its slippery feel and its electrical conductivity. Not surprisingly, graphite is used commercially as a lubricant and to construct electrodes—for example, for use in ordinary dry cells.

In diamond, the interlocking network of covalent bonds produces large

The thermodynamics of the conversion of graphite to diamond were discussed in Chapter 11.

A slurry of granular activated carbon (activated charcoal) is pumped into a water treatment facility where it will be used to remove offensive odors and toxic impurities from municipal drinking water.

Carbon monoxide is particularly dangerous because it is both colorless and odorless.

crystals that are actually single molecules! As we learned earlier, this makes diamond extremely hard. Aside from its decorative use in jewelry, diamond is one of industry's most important abrasives, and is used to make cutting and grinding tools.

Slippery graphite and brilliant diamonds represent quite pure forms of elemental carbon. Less pure, predominantly graphitic forms of carbon are also known. One of these—charcoal—is formed by heating wood strongly in the absence of air. Charcoal has a particularly open structure with an enormous amount of surface area for a given mass of carbon. *Activated charcoal*—a finely pulverized form having a surface area of about 1000 m² per gram—has many commercial uses. Its large surface area permits small amounts of it to adsorb large numbers of molecules, a property that allows it to remove molecules from the air that have offensive odors and to remove toxic impurities from water. In fact, several municipalities located in areas polluted by chemical spills have had to install activated carbon filtration systems to make contaminated well water pure enough to drink.

When a hydrocarbon is burned in a very limited supply of oxygen the hydrogen combines with the oxygen to form water, and finely divided carbon—called *carbon black*—remains.

$$CH_4\ (g) + O_2\ (g) \longrightarrow C\ (s) + 2H_2O\ (g)$$

Carbon black is used as a pigment in black inks, and large amounts of it are used in making automobile tires.

Oxides and oxoacids of carbon

Carbon forms two principal oxides: carbon monoxide and carbon dioxide. Carbon monoxide has the Lewis structure

$$:C\equiv O:$$

and is one of a rather small number of species in which carbon has only three bonds. Carbon monoxide is a nearly nonpolar substance with a low melting point and boiling point (mp = −199°C, bp = −192°C). It is unreactive toward water and has a low solubility. It is also flammable and burns with a hot flame, which is why the mixture of CO and H_2 from the water gas reaction (Section 19.1) is a useful industrial fuel.

One way that carbon monoxide can be formed is by burning carbon or a hydrocarbon in a limited supply of oxygen. These are exactly the conditions that exist in the fire in a charcoal barbecue or in a gasoline engine, and both produce some carbon monoxide. As you are probably aware, carbon monoxide is quite toxic—it binds to hemoglobin in the blood, thereby preventing it from carrying oxygen—and that is why it is dangerous to be in an enclosed space with either a charcoal fire or a running automobile engine.

In the laboratory, small amounts of carbon monoxide can be made by treating formic acid, HCO_2H, with concentrated sulfuric acid. Sulfuric acid has a strong affinity for water and actually extracts the components of water (two hydrogens and an oxygen) from a formic acid molecule.

$$HCO_2H\ (l) \xrightarrow{\text{H}_2\text{SO}_4} H_2O\ (l) + CO\ (g)$$

Industrially, carbon monoxide is proving to be an extremely important chemical. More and more interest is being shown in reactions like those discussed in the last section in which CO and H_2 are combined catalytically to form hydrocarbons—a replacement for petroleum. We saw in the last section that

CO can be made by reacting steam with white-hot carbon in the form of coke.

$$C\ (s)\ +\ H_2O\ (g)\ \xrightarrow{1000°C}\ CO\ (g)\ +\ H_2\ (g)$$

At high temperatures, carbon monoxide is an effective reducing agent. For example,

$$Fe_2O_3\ (s)\ +\ 3CO\ (g)\ \xrightarrow{heat}\ 2Fe\ (s)\ +\ 3CO_2\ (g)$$

This overall reaction is used to extract iron from its ore and is discussed in greater detail in Chapter 21.

An interesting chemical property of carbon monoxide is its ability to form covalent compounds with transition metals in which the metal is in a low (or zero) oxidation state. These are called **metal carbonyl compounds.** An example is $Ni(CO)_4$, nickel carbonyl, which is formed by simply warming metallic nickel in the presence of CO. The compound is quite volatile and very toxic.

$$Ni\ (s)\ +\ 4CO\ (g)\ \longrightarrow\ Ni(CO)_4\ (g)$$

Some other examples of metal carbonyls are $Cr(CO)_6$, $Mn_2(CO)_{10}$, $Fe(CO)_5$, and $Co_2(CO)_8$.

The second major oxide of carbon is carbon dioxide.

$$\ddot{O}=C=\ddot{O}$$

Dry ice—solid carbon dioxide at a temperature of −78°C—sublimes at atmospheric pressure.

As predicted by VSEPR theory, CO_2 is a linear molecule. It is also nonpolar, because of its symmetrical structure. The triple point of CO_2 is above 1 atm, and when CO_2 is cooled at atmosphereic pressure it condenses to a solid rather than to a liquid. Solid carbon dioxide is called **Dry Ice** and its temperature is −78°C—the temperature at which CO_2 sublimes at 1 atm. Because CO_2 is more dense ("heavier") than air and because it doesn't support combustion, CO_2 is used to smother fires, especially those that are difficult to extinguish using water.

Carbon dioxide can be formed in a number of ways. Complete combustion of carbon, a hydrocarbon, or carbon monoxide gives CO_2.

$$C\ +\ O_2\ \longrightarrow\ CO_2$$

$$CH_4\ +\ 2O_2\ \longrightarrow\ CO_2\ +\ 2H_2O$$

$$2CO\ +\ O_2\ \longrightarrow\ 2CO_2$$

Industrially it is often made by the thermal decomposition of limestone.

$$\underset{\textbf{limestone}}{CaCO_3\ (s)}\ \xrightarrow{heat}\ CaO\ (s)\ +\ CO_2\ (g)$$

Carbon dioxide is also formed naturally by decomposing organic matter.

In the laboratory, CO_2 can be conveniently prepared by reaction of an acid with a carbonate—for example, calcium carbonate.

$$CaCO_3\ (s)\ +\ 2HCl\ (aq)\ \longrightarrow\ CaCl_2\ (aq)\ +\ H_2O\ (l)\ +\ CO_2\ (g)$$

Similar reactions with bicarbonates (especially $NaHCO_3$) were discussed in Chapter 18 (page 571).

Carbon dioxide is consumed in large quantities by industry. We learned in Chapter 18 that it is used to make sodium carbonate by the Solvay process. Major uses also include refrigeration (including Dry Ice) and beverage carbonation. The process of carbonation consumes about 35% of a total annual production of about 2.5 million tons of CO_2.

CO_2 has replaced chlorofluorocarbons as a propellant in many aerosol products.

In nature, green plants consume carbon dioxide in photosynthesis and produce glucose, $C_6H_{12}O_6$, a sugar from which they manufacture starch, cellulose, and other chemicals. In the process, oxygen is released to the atmosphere.

$$6CO_2 \ (g) + 6H_2O \ (l) \xrightarrow[\text{chlorophyll}]{\text{light}} C_6H_{12}O_6 \ (aq) + 6O_2 \ (g)$$

The oceans represent a huge reservoir of dissolved CO_2.

Carbon dioxide is moderately soluble in water, and its solutions are slightly acidic due to the formation of carbonic acid.

$$CO_2 \ (aq) + H_2O \ (l) \rightleftharpoons H_2CO_3 \ (aq)$$
$$\text{carbonic acid}$$

Carbonic acid, as we've seen earlier, is a weak diprotic acid that ionizes in two steps. Neutralization with a base is able to give two types of salts: carbonates, formed by complete neutralization, and bicarbonates (hydrogen carbonates), formed by partial neutralization.

$$NaOH + H_2CO_3 \longrightarrow NaHCO_3 + H_2O$$

$$2NaOH + H_2CO_3 \longrightarrow Na_2CO_3 + 2H_2O$$

Bicarbonates such as $NaHCO_3$, you recall, are easily decomposed by heating to give carbonates ($2NaHCO_3 \rightarrow Na_2CO_3 + H_2O + CO_2$).

In Chapter 6 it was pointed out that when groundwater containing dissolved CO_2 trickles slowly through large deposits of limestone it gradually dissolves the $CaCO_3$ and forms huge limestone caverns. The equilibrium equations are

$$H_2O + CO_2 \ (g) \rightleftharpoons H_2CO_3 \ (aq)$$

$$H_2CO_3 \ (aq) + CaCO_3 \ (s) \rightleftharpoons Ca(HCO_3)_2 \ (aq)$$

When a drop of water containing a dilute solution of $Ca(HCO_3)_2$ collects on the ceiling of a limestone cave, it gradually evaporates. This causes the first of these equilibria to be shifted to the left as H_2CO_3 decomposes and gaseous CO_2 is released. The loss of H_2CO_3 also affects the second reaction. That equilibrium is also shifted to the left, causing $CaCO_3$ to precipitate. Over a period of many years, drop after drop evaporates and a stalactite composed of calcium carbonate is slowly formed. Drops that fall to the floor evaporate, too, and give rise to stalagmites. Limestone caves are not the only places to observe the growth of stalactites. Calcium carbonate is one of the substances formed during the curing of concrete, and when water seeps through the concrete of a highway overpass, stalactites looking a bit like whiskers are formed that hang from the overpass above the roadway, as shown in Figure 19.2.

Figure 19.2

Photograph of stalagtites formed on the ceilings of highway overpass.

Figure 19.3

Photograph of boiler scale.

Because of reactions like those just described, the groundwater in many parts of the country contains dissolved calcium as Ca^{2+}, as well as other ions such as Mg^{2+} and Fe^{3+}. If their concentrations are sufficiently large, these ions can interfere with the action of ordinary soap by forming a precipitate with it. The water is then described as **hard water,** and the ions causing the problem are called **hardness ions.** Hardness ions can be removed from water in a number of ways. If the solution also contains bicarbonate ion—as it would if the ions in it came from the dissolving of limestone—heating the water will drive off CO_2 and cause a precipitate to form. For example,

$$Ca(HCO_3)_2 \ (aq) \xrightarrow{\text{heat}} CaCO_3 \ (s) + H_2O \ (l) + CO_2 \ (g)$$

This type of hard water causes serious problems because the precipitate (called *boiler scale*) can clog hot water pipes and make it more difficult to heat water in a boiler (see Figure 19.3).

If the hard water doesn't contain HCO_3^-, the ions can still be precipitated by adding *washing soda*—a hydrate of sodium carbonate, $Na_2CO_3 \cdot 10H_2O$. The carbonate ion precipitates the hardness ions,

$$Ca^{2+} \ (aq) + CO_3^{2-} \ (aq) \longrightarrow CaCO_3 \ (s)$$

and, by removing them, allows the soap to do its job.

Other inorganic compounds of carbon

Carbides Binary compounds formed between carbon and a metal or metalloid are generally referred to as carbides. They are of three types. In *covalent carbides*, carbon is bonded covalently to the other element. An example is silicon carbide, SiC, known commercially as **carborundum.** It is made by heating silicon dioxide (sand) and carbon to very high temperatures.

$$SiO_2 \ (s) + 3C \ (s) \xrightarrow{\text{heat}} SiC \ (s) + 2CO \ (g)$$

This solid has the same structure as diamond, except that silicon atoms alternate with carbon atoms (Figure 19.4). Like diamond, silicon carbide is very hard and is used as an abrasive—for example, in sandpaper.

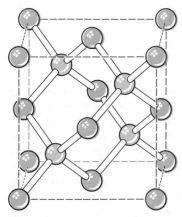

Figure 19.4

The structure of carborundum. Gray spheres represent silicon and the colored spheres represent carbon. In diamond all the spheres would be dark.

Ionic carbides (also called *saltlike carbides*) are formed from carbon and active metals, and contain more or less discrete ions of carbon in their solids. An example mentioned earlier (page 594) is Al_4C_3, which contains the anion C^{4-}. When placed into water it hydrolyzes and methane is formed. A particularly important saltlike carbide is calcium carbide, CaC_2, which contains the ion, $[:C \equiv C:]^{2-}$, called acetylide ion. It is made by reacting calcium oxide (lime) with carbon at high temperatures.

$$CaO \ (s) + 3C \ (s) \longrightarrow CaC_2 \ (s) + CO \ (g)$$

Calcium carbide reacts with water to form acetylene, the gas used in welding torches.

$$CaC_2 \ (s) + 2H_2O \ (l) \longrightarrow \underset{\textbf{acetylene}}{C_2H_2 \ (g)} + Ca(OH)_2 \ (s)$$

The third type of carbide is called an *interstitial carbide*. In these substances, carbon atoms fit into voids (empty spaces called *interstices*) between the metal atoms in a crystal. An example is tungsten carbide, WC, which is extremely hard and brittle and is used to make cutting tools used to machine other metals.

Cyanides Hydrogen cyanide, HCN, can be made by a number of reactions, but one that is especially important commercially is the reaction of ammonia with

H—C≡N:
hydrogen cyanide

[:C≡N:]⁻
cyanide ion

CN⁻ forms many stable complex ions.

methane. It is carried out at high temperatures over a platinum catalyst.

$$NH_3\ (g) + CH_4\ (g) \xrightarrow[\text{Pt}]{1200°C} HCN\ (g) + 3H_2\ (g)$$

Hydrogen cyanide is an intermediate in the synthesis of a number of well-known plastics, including nylon. It is also a very potent, fast-acting poison that both deactivates critical enzymes in the body and binds irreversibly to iron in hemoglobin in the blood. In water, HCN is a weak acid and salts such as NaCN can be formed by neutralization. Like HCN, NaCN is a deadly poison. It is used in certain electroplating baths and to extract gold from its ores.

Carbon Disulfide Carbon reacts directly with sulfur to form carbon disulfide, CS_2.

$$C + 2S \xrightarrow[\text{temp}]{\text{high}} CS_2$$

:S̈=C=S̈:
carbon disulfide

It is a useful solvent, but it is extremely flammable—boiling water is hot enough to ignite it! When it burns, CO_2 and SO_2 are formed. Besides being a solvent, carbon disulfide is a useful chemical reagent. It is used commercially to manufacture carbon tetrachloride.

$$CS_2 + 3Cl_2 \longrightarrow \underset{\substack{\text{carbon} \\ \text{tetrachloride}}}{CCl_4} + S_2Cl_2$$

19.3 OXYGEN

The atmosphere is 20.9% oxygen by volume.

There are very few people who are unaware of the importance of oxygen to our existence. Breathing oxygen keeps us alive, and fuels couldn't be burned without it. Oxygen is literally everywhere—not only as a free element in the atmosphere, but also in all living creatures as well as in most of the substances that surround them. Water, which covers much of the Earth, is one-third oxygen on an atom basis, but about 89% oxygen by mass. In the Earth's crust, which is composed mostly of silicate minerals, oxygen is the *most* abundant element—46.6% by mass, 62.6% if counted by atoms, and an amazing 93.8% if measured by volume!

In its ground state, the element oxygen has the electron configuration $1s^2 2s^2 2p^4$, and therefore needs only two electrons to complete its octet. As you know, it is able to accomplish this either by acquiring electrons or by electron sharing. In the free state, oxygen exists as diatomic molecules, O_2 (called dioxygen). The molecular orbital energy diagram for O_2 was discussed in Chapter 5 and accounts for both the double bond between the oxygen atoms as well as the fact that O_2 is paramagnetic, with two unpaired electrons. When cooled, oxygen forms a pale blue liquid that boils at $-183°C$. Oxygen freezes at $-219°C$, giving a pale blue solid.

The bond length and bond energy both suggest a double bond in O_2.

Preparation and uses

Because it is so freely available, O_2 is normally produced at the plant site where it will be used.

The most obvious source of oxygen when it is needed in large quantities is the atmosphere, and the commercial preparation of oxygen involves separating it from liquefied air. Nitrogen, which has a lower boiling point than O_2, is removed from the liquid air by allowing it to boil off, thereby leaving behind liquid oxygen contaminated with small amounts of N_2 and argon (another component of air). Warming the liquid, of course, converts the oxygen to a gas. Most of the oxygen prepared this way (about 85%) is used in the steel industry and in metal fabrication. Some is also used to make chemical intermediates in the

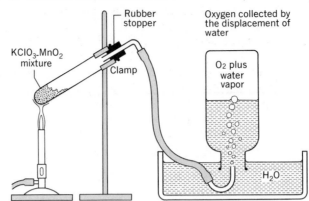

Figure 19.5

The laboratory preparation of oxygen by the thermal decomposition of potassium chlorate, $KClO_3$, using manganese dioxide, MnO_2, as a catalyst. Care must be taken not to heat the rubber stopper or allow the hot $KClO_3$ to come in contact with it.

Liquid oxygen was used to oxidize the fuel in this modified Saturn V rocket, which put the Skylab space station into orbit in 1973. More recently, it has also been used along with liquid hydrogen to power the space shuttle Columbia.

manufacture of plastics, in waste water treatment, for life support in hospitals, and in bleaching pulp and paper.

In the laboratory, small amounts of pure oxygen can be prepared in a number of ways. Joseph Priestley (1733–1804), credited with the discovery of oxygen, obtained the gas by thermally decomposing mercury(II) oxide. Usually, however, oxygen is made in the laboratory either by electrolysis of water or by thermally decomposing potassium chlorate, $KClO_3$, using manganese dioxide as a catalyst.

$$2KClO_3\,(s) \xrightarrow[\text{heat}]{MnO_2} 2KCl\,(s) + 3O_2\,(g)$$

Like hydrogen, oxygen can be collected by displacement of water because of its relatively low solubility (Figure 19.5).

In nature, oxygen is generated by green plants during photosynthesis. The chemical equation for the conversion of CO_2 and H_2O into glucose and O_2 was described in the last section. Large forests and the plankton in the sea are responsible for maintaining the balance of oxygen in the Earth's atmosphere.

Ozone

When an electric discharge is passed through molecular oxygen a second allotrope of the element is formed that is called **ozone, O_3**.

$$3O_2\,(g) \longrightarrow 2O_3\,(g) \qquad \Delta H = +284 \text{ kJ}$$

Its pungent odor can sometimes be detected after a severe thunderstorm or near electrical machinery. In high concentrations it is poisonous. In Chapter 9 we saw that the structure of ozone can be represented by the resonance formulas

VSEPR theory predicts that O_3 will be nonlinear.

As we would expect from the number of groups of electrons around the central oxygen atom, the molecule is nonlinear.

When ozone is formed, a large amount of energy must be supplied, and an equally large amount of energy, of course, is released when the ozone decomposes. Therefore, since highly exothermic reactions tend to be quite spontaneous, ozone decomposes very easily. Ozone is also a very powerful oxidizing agent—considerably more powerful than ordinary O_2. This has both advantages and disadvantages. On the positive side, the oxidizing power of ozone shows promise as an alternative to chlorine in the treatment of municipal drinking water supplies. It has been found that Cl_2 in drinking water is able to form chlorine compounds with some of the organic substances that are also present in small amounts. Some of these products, such as chloroform, $CHCl_3$, have been shown to be carcinogenic, and the long-range toxic effects of others are open to question. All these problems are avoided if bacteria in the water are killed by O_3 instead of Cl_2. Another problem arises, however, because little residual O_3 remains after treatment, so any bacteria that get into the water after the treatment process are not destroyed.

On the negative side, the powerful oxidizing ability of ozone can cause extensive damage to plants and to articles made of natural rubber. As we will see in the next section, ozone is one of the major constituents of smog, so in locations that experience severe smog episodes, damage caused by ozone can be particularly troublesome.

In the Earth's upper atmosphere, at altitudes ranging from about 9 to 15 miles, ozone is formed in appreciable amounts from O_2 by absorption of ultraviolet radiation from the sun. The light energy first splits oxygen molecules into oxygen atoms.

$$O_2 \xrightarrow{h\nu} 2O$$

Reaction of oxygen atoms with oxygen molecules produces O_3.

$$O_2 + O \longrightarrow O_3$$

Ozone also absorbs ultraviolet light, especially at wavelengths that prove harmful to living organisms. This causes the O_3 to decompose and form O_2 again. The absorption of UV radiation by ozone converts the energy of the UV light into heat and also protects the inhabitants of our planet from the radiation's harmful effects.

You are probably aware of the controversy that has arisen in recent years concerning the effects of certain human activity on this ozone shield. For a time there was worry that high-flying suspersonic airliners such as the Concorde would be emitting nitrogen oxide pollutants in quantities that would have a significant effect on the ozone concentration. Even small amounts of NO could have damaging effects because of such reactions as

$$NO + O_3 \longrightarrow O_2 + NO_2$$
$$NO_2 + O \longrightarrow O_2 + NO$$

Thus, molecules of NO remove not only O_3, but also oxygen atoms needed to reform the O_3. Since the product of the second reaction is the reactant in the first, the cycle can be repeated many times by each NO molecule. Despite these early concerns it appears that the Concorde, as presently operated, has not caused any noticeable change in the stratospheric ozone concentration.

Another danger to the ozone shield has been the release into the atmosphere of Freons. These substances are composed of carbon, fluorine, and chlorine, and are called chlorofluorocarbons. They have been widely used as propellants in aerosol cans and as refrigerants. An example is Freon-11, $CFCl_3$, which is colorless, virtually odorless, and unreactive under ordinary conditions. However, as these chlorofluorocarbons diffuse into the upper atmosphere they can absorb ultraviolet radiation, which ruptures carbon-chlorine bonds to give chlorine atoms.

$$CFCl_3 \xrightarrow{h\nu} CFCl_2 + Cl$$

Both products are reactive, but the chlorine atoms are believed to be involved in the destruction of ozone. Possible reactions are

$$Cl + O_3 \longrightarrow ClO + O_2$$
$$ClO + O \longrightarrow Cl + O_2$$

Recall that carcinogenic means cancer-causing.

The ozone concentration at these altitudes approaches 27% by mass.

Destruction of stratospheric ozone could lead to increased incidents of skin cancer and a rise in the Earth's average temperature.

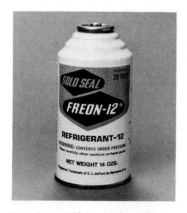

Freon-12, CCl_2F_2, is the refrigerant used in the air conditioners in automobiles. Containers of it can be purchased in any automotive store.

Once again we have a cycle by which a single species—a chlorine atom—destroys not only O_3, but oxygen atoms needed to replace O_3. Concern over the threat posed by Freons has led to a ban on their use as aerosol propellants in the United States.

Compounds of oxygen

Oxygen forms compounds with every element except helium, neon, and argon. These compounds are called **oxides** and are generally of two types: ionic or covalent. *Ionic oxides* are formed with many metals and can usually be made by direct combination of the elements, for example,

$$4Li\ (s) + O_2\ (g) \longrightarrow 2Li_2O\ (s)$$

$$2Ca\ (s) + O_2\ (g) \longrightarrow 2CaO\ (s)$$

Metal oxides are basic and, if soluble, give solutions containing hydroxide ion caused by the virtual complete hydrolysis of O^{2-},

$$Li_2O\ (s) + H_2O\ (l) \longrightarrow 2LiOH\ (aq) \qquad \text{(Molecular equation)}$$

$$O^{2-}\ (aq) + H_2O\ (l) \xrightarrow{100\%} 2OH^-\ (aq) \qquad \text{(Net ionic equation)}$$

Remember, metal oxides are basic anhydrides.

Even insoluble metal oxides are basic because they neutralize acids. Iron(III) oxide, for example, dissolves in acids such as HCl or H_2SO_4.

$$Fe_2O_3\ (s) + 6H^+\ (aq) \longrightarrow 2Fe^{3+}\ (aq) + 3H_2O$$

This reaction, in fact, is often used to remove rust (Fe_2O_3) from iron or steel prior to being given a protective coating of zinc or tin. The acid treatment is called **pickling.**

Some oxides of metals show acidic properties as well as basic ones, and are said to be amphoteric. Examples that we have discussed earlier are the oxides of beryllium and aluminum, which dissolve in both acids and bases. For instance,

$$Al_2O_3 + 6H^+ \longrightarrow 2Al^{3+} + 3H_2O$$

$$Al_2O_3 + 2OH^- \longrightarrow 2AlO_2^- + H_2O$$

Covalent oxides are generally formed by the nonmetals. In them, oxygen completes its octet by electron sharing—normally by forming either two single bonds (as in H_2O) or one double bond (as in CO_2). Exceptions are CO and NO, which contain triple bonds. Table 19.3 contains a list of many oxides of the nonmetallic elements.

Table 19.3
Typical oxides of the nonmetallic elements

Group III	B_2O_3			
Group IV	CO	SiO_2	GeO_2	
	CO_2			
Group V	N_2O	P_4O_6	As_4O_6	Sb_4O_6
	NO	P_4O_{10}	$As_2O_5{}^a$	$Sb_2O_5{}^a$
	N_2O_3			
	NO_2; (N_2O_4)			
	N_2O_5			
Group VI	O_2	SO_2	SeO_2	TeO_2
	O_3	SO_3	SeO_3	TeO_3
Group VII	OF_2	Cl_2O	Br_2O	I_2O_5
	O_2F_2	ClO_2	BrO_2	I_2O_7
		Cl_2O_7		

^{a} Molecular structure unknown.

Table 19.4

**Thermodynamic properties
of some nitrogen oxides**

Oxide	ΔH_f^0(kJ/mol)	ΔG_f^0(kJ/mol)
N_2O (g)	+81.5	+104
NO (g)	90.4	86.8
NO_2 (g)	38	51.9
N_2O_4 (g)	9.7	98.3
N_2O_5 (g)	11	115

There are several ways that oxides of the nonmetallic elements can be made. Except for the halogens and the noble gases most oxides can be prepared simply by the direct union of the elements, as typified by the reactions

$$S + O_2 \longrightarrow SO_2$$

$$C + O_2 \longrightarrow CO_2 \qquad \text{(Excess oxygen)}$$

$$2C + O_2 \longrightarrow 2CO \qquad \text{(Limited supply of oxygen)}$$

$$2H_2 + O_2 \longrightarrow 2H_2O$$

Not all oxides can be prepared effectively in this manner, however. For example, in Table 19.4 we see that many oxides of nitrogen have positive free energies of formation and, therefore, from what we know of thermodynamics, they cannot be synthesized in significant amounts directly from the elements.[4] In these instances, and in others too, indirect methods of preparation are employed.

The indirect procedures are, expectedly, many in number. However, a few generalizations can be made. In some cases an oxide can be prepared from a lower oxide by further reaction with oxygen. The synthesis of SO_3, for example, consists of catalytic oxidation of SO_2,

$$2SO_2 + O_2 \longrightarrow 2SO_3$$

Earlier (page 417) it was mentioned that this reaction is promoted in catalytic mufflers originally designed to speed up the conversion of CO to CO_2.

$$2CO + O_2 \longrightarrow 2CO_2$$

The combustion of CO, you recall, is an important industrial reaction because CO is often used as a fuel.

Another technique that serves to produce oxides is the combustion of nonmetal hydrides. Methane, the chief constituent of natural gas, and other hydrocarbons burn to produce CO_2 and H_2O when an excess of O_2 is present.

$$\underset{\textbf{methane}}{CH_4} + 2O_2 \longrightarrow CO_2 + 2H_2O$$

$$\underset{\substack{\textbf{octane} \\ \text{(gasoline)}}}{2C_8H_{18}} + 25O_2 \longrightarrow 16CO_2 + 18H_2O$$

When insufficient O_2 is available, as in an automobile engine, CO may be produced instead of CO_2.

---

[4] Since they also have positive ΔS_f^0, it should be possible to prepare them at very high temperatures where $\Delta H - T \Delta S$, and hence ΔG, is negative. This is the case with NO, which is formed in small amounts near 3000°C. The high-temperature production of NO in motor vehicle engines is a major source of urban air pollution.

Table 19.5

Simple oxoacids and oxoanions of the nonmetallic elements

Group IIIA	H_3BO_3 (no simple borates)			
Group IVA	H_2CO_3 (CO_3^{2-})	H_4SiO_4[a] (SiO_4^{4-})	H_4GeO_4[a] (GeO_4^{4-})	
Group VA	HNO_2 (NO_2^-) HNO_3 (NO_3^-)	H_3PO_3 HPO_3^{2-} H_3PO_4 (PO_4^{3-})	H_3AsO_4 (AsO_4^{3-})	
Group VIA		H_2SO_3 (SO_3^{2-}) H_2SO_4 (SO_4^{2-})	H_2SeO_3 (SeO_3^{2-}) H_2SeO_4 (SeO_4^{2-})	H_2TeO_3[a] (TeO_3^{2-}) $Te(OH)_6$ $(TeO(OH)_5^-)$
Group VIIA	HOF	HOCl (OCl^-) $HClO_2$ (ClO_2^-) $HClO_3$ (ClO_3^-) $HClO_4$ (ClO_4^-)	HOBr (OBr^-) $HBrO_2$ (BrO_2^-) $HBrO_3$ (BrO_3^-) $HBrO_4$ (BrO_4^-)	HOI (OI^-) HIO_3 (IO_3^-) H_5IO_6 $(H_2IO_6^{3-})$ HIO_4 (IO_4^-)

[a] Not observed.

A reaction of this general type that is of great commercial importance is the oxidation of ammonia. In this case a platinum catalyst is used and the reaction is

$$4NH_3 + 5O_2 \xrightarrow{\text{Pt}} 4NO + 6H_2O$$

The NO formed in this reaction is readily oxidized further to produce NO_2,

$$2NO + O_2 \longrightarrow 2NO_2$$

These reactions are essential to the production of nitric acid and nitrates, which are used in manufacturing many chemicals from explosives to fertilizers. We will discuss them in the next section.

Finally, another indirect method of obtaining nonmetal oxides makes use of oxidation-reduction reactions. For instance, when nitric acid serves as an oxidizing agent, the nitrate ion is reduced and, depending on conditions, nitrogen in various oxidation states may be produced. When concentrated nitric acid is used, the reduction product is frequently NO_2. Dilute solutions of HNO_3 often yield NO as the reduction product.

$$4HNO_3 + Cu \longrightarrow Cu(NO_3)_2 + 2NO_2 + 2H_2O \qquad \text{(Concentrated)}$$

$$8HNO_3 + 3Cu \longrightarrow 3Cu(NO_3)_2 + 2NO + 4H_2O \qquad \text{(Dilute)}$$

Similarly, hot concentrated sulfuric acid is a fairly potent oxidizing agent, and the reduction product is usually SO_2; for example,

$$Cu + 2H_2SO_4 \longrightarrow CuSO_4 + SO_2 + 2H_2O$$

In Chapter 14 we saw that nonmetal oxides are acid anhydrides; when they react with water they produce oxoacids. For example,

$$SO_2 + H_2O \longrightarrow H_2SO_3$$

Neutralization gives the corresponding oxoanion. Thus, sulfurous acid (H_2SO_3) gives sulfite ion (SO_3^{2-}). Table 19.5 provides a summary of the simple oxoacids and oxoanions of the nonmetallic elements. As indicated in the table, some of

In the absence of a catalyst, combustion of ammonia in air gives N_2 and H_2O.

$$4NH_3\,(g) + 3O_2(g) \longrightarrow 2N_2\,(g) + 6H_2O\,(g)$$

the oxoacids cannot actually be isolated, although their corresponding anions can be.

Other compounds of oxygen

Peroxides and Superoxides In addition to forming "normal" oxides with metals (i.e., compounds that contain the O^{2-} ion), oxygen combines with the more active alkali metals to form peroxides and superoxides, which contain the O_2^{2-} and O_2^{-} ions, respectively. Sodium forms the yellow peroxide.

Peroxides and superoxides of the alkali metals were discussed on page 569.

$$2Na\ (s) + O_2\ (g) \longrightarrow Na_2O_2\ (s)$$
sodium peroxide

Potassium, rubidium, and cesium form yellow to orange superoxides.

$$K\ (s) + O_2 \longrightarrow KO_2\ (s)$$
potassium superoxide

Hydrogen Peroxide In Chapter 18 we saw that when Na_2O_2 is placed in water the peroxide ion hydrolyzes to form hydrogen peroxide, H_2O_2.

$$Na_2O_2\ (s) + 2H_2O\ (l) \longrightarrow 2NaOH\ (aq) + H_2O_2\ (aq)$$

Hydrogen peroxide is a molecular substance having an oxygen-oxygen bond.

$$\overset{\displaystyle H}{\underset{\displaystyle H}{:\!\overset{..}{O}\!-\!\overset{..}{O}\!:}}$$

The structure of the molecule is shown in Figure 19.6. In water, H_2O_2 is a weak acid,

$$H_2O_2 \rightleftharpoons H^+ + HO_2^- \qquad K_a = 1.5 \times 10^{-12}$$

Pure H_2O_2 is a colorless liquid with a boiling point of 150°C, but in its pure state it is an extremely hazardous substance to work with. Its decomposition to water and oxygen can occur explosively.

$$2H_2O_2\ (l) \longrightarrow 2H_2O\ (l) + O_2\ (g)$$

This decomposition is promoted by heat or by traces of many heavy metal ions.

Usually, hydrogen peroxide is purchased as a solution in water. Solutions of 3% H_2O_2 by weight can be purchased in drug stores for use as an antiseptic. Its oxidizing power destroys bacteria, while blood catalyzes its decomposition. The fizzing action caused by escaping O_2 dislodges dirt and other foreign

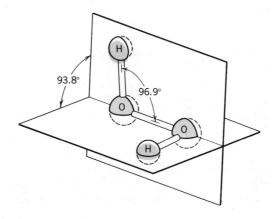

Figure 19.6

The structure of hydrogen. peroxide.

matter from a wound. More concentrated solutions of H_2O_2 are effective oxidizing agents and are used as bleaches for hair and industrially in the bleaching of cotton fabrics.

19.4 NITROGEN

Nitrogen is another of the principal elements found in all living creatures. It is an essential constituent of amino acids—the primary building blocks of proteins—and is the key element in a precise "lock-and-key" fit between molecules that cells use to decipher the genetic code incorporated in the DNA residing in their nuclei. Because of the importance of nitrogen to the growth of plants and animals, it is not surprising that nitrogen compounds—particularly fertilizers—rank very high in total annual commercial production.

The element nitrogen has as its ground state electron configuration, $1s^2 2s^2 2p^3$. It achieves a completed octet either by gaining electrons to form ionic nitrides that contain the N^{3-} ion or by covalent bonding. In many of its compounds, nitrogen forms three bonds.

The nitrogen in organic matter represents a small fraction of the Earth's total.

Very little nitrogen is present in a combined state in the Earth's crust. Instead, most occurs as diatomic molecules of N_2 (dinitrogen) in the atmosphere. This is a colorless, odorless, quite unreactive gas that boils at $-196°C$ and freezes at $-210°C$. Its lack of reactivity is attributed to the strength of the nitrogen-nitrogen triple bond

$$:N \equiv N:$$

The bond energy of N_2 (946 kJ/mol) is very high, so the molecule is very stable. For example, at 25°C (room temperature), N_2 reacts directly only with lithium to form Li_3N, although the nitrogen-fixing bacteria in the root nodules of clover, beans, peas, and certain other plants are also able to convert N_2 into usable nitrogen compounds. The high bond energy of N_2 also causes many nitrogen compounds to have positive enthalpies and free energies of formation.

Preparation and uses of elemental nitrogen

The Earth's atmosphere is composed of about 78% N_2 by volume and, as with oxygen, the commercial source of N_2 is from the liquefaction of air. As liquid air boils, the lower boiling nitrogen escapes and is collected. Generally, commercial nitrogen from this source consists of about 99% N_2, with a small amount of argon and traces of oxygen.

Solid NH_4NO_2 explodes if warmed, and produces the same products.

In the laboratory, nitrogen can be made chemically by warming a solution that contains an ammonium salt (such as NH_4Cl) and a nitrite salt (such as $NaNO_2$). The net ionic reaction corresponds to the decomposition of NH_4NO_2.

$$NH_4^+ \ (aq) + NO_2^- \ (aq) \xrightarrow{\text{warm}} N_2 \ (g) + 2H_2O \ (l)$$

Gaseous N_2 is often used in chemical apparatus in the laboratory to provide an inert atmosphere.

Commercially, the leading use of nitrogen is in the manufacture of ammonia, which is described later in this section. Because of nitrogen's low reactivity, it is also used in large quantities as an inert gaseous blanket to exclude oxygen during the manufacture of chemicals, the fabrication of metals, and in the production of electronic devices. Large amounts of liquid nitrogen are employed in the food industry where its low temperature ($-196°C$) is used to rapidly freeze foods.[5]

---

[5] Businesses have been established that, for a fee, offer to freeze your body in liquid nitrogen after you've died, and maintain it in a frozen state until (hopefully) a cure is found for whatever killed you. No one has yet been returned to life, however, and most scientists question whether the cell damage caused by rapid freezing can ever be reversed in humans.

Ionic compounds of nitrogen

Nitrides Elemental nitrogen combines directly with some of the very reactive metals on the left side of the periodic table. Earlier it was mentioned that lithium reacts with N_2 at room temperature to form an ionic nitride.

$$6Li\ (s) + N_2\ (g) \longrightarrow 2Li_3N$$

Similar nitrides are formed at higher temperatures by magnesium, calcium, strontium, and barium, and have the general formula M_3N_2. When placed into water the nitrides immediately hydrolyze, liberating ammonia.

The N^{3-} ion is a very strong Brønsted-Lowry base.

$$Li_3N\ (s) + 3H_2O\ (l) \longrightarrow 3LiOH\ (aq) + NH_3\ (g)$$

$$Mg_3N_2\ (s) + 6H_2O\ (l) \longrightarrow 3Mg(OH)_2\ (s) + 2NH_3\ (g)$$

Covalent compounds of nitrogen

Nitrogen forms covalent compounds with many nonmetals. The most important are those with hydrogen and oxygen, and among them we find nitrogen in every oxidation state from $3-$ to $5+$, as shown in Table 19.6. Actually, the oxidation numbers have no real physical significance for these substances—for example, the nitrogen in NH_3 certainly doesn't carry a charge of $3-$. Nevertheless, the oxidation numbers are useful in balancing redox equations and in organizing the discussion of the chemistry of nitrogen.

Ammonia (3− Oxidation State) Ammonia (NH_3) is by far the most important compound of nitrogen. It serves as a route to virtually all other nitrogen compounds and is itself a useful fertilizer. Ammonia is prepared industrially by the **Haber process,** which combines hydrogen and nitrogen on a catalytic iron surface.

About 18 million tons of ammonia are produced each year.

$$N_2\ (g) + 3H_2\ (g) \xrightarrow[\text{catalyst}]{\text{iron}} 2NH_3\ (g) \qquad \Delta H^0 = -92\ kJ$$

The reaction is run at pressures of several hundred atmospheres, which favors the production of NH_3. It is also run at temperatures between 400°C and 500°C in order to cause the reaction to proceed at a reasonable rate, even though in

Table 19.6
Oxidation states of nitrogen

Oxidation State	Example	Preparative Reaction
3−	NH_3 (ammonia)	$N_2 + 3H_2 \rightarrow \mathbf{2NH_3}$
2−	N_2H_4 (hydrazine)	$2NH_3 + NaOCl \rightarrow \mathbf{N_2H_4} + NaCl + H_2O$
1−	NH_2OH (hydroxylamine)	$NaNO_2 + NaHSO_3 + SO_2 + 2H_2O \rightarrow 2NaHSO_4 + \mathbf{NH_2OH}$
0	N_2 (dinitrogen)	$NH_4NO_2 \xrightarrow{\text{heat}} \mathbf{N_2} + 2H_2O$
1+	N_2O (nitrous oxide)	$NH_4NO_3 \xrightarrow{\text{heat}} \mathbf{N_2O} + 2H_2O$
2+	NO (nitric oxide)	$4NH_3 + 5O_2 \rightarrow \mathbf{4NO} + 6H_2O$
3+	N_2O_3 (dinitrogen trioxide)	$NO + NO_2 \xrightarrow{-30°C} \mathbf{N_2O_3}$
4+	NO_2 (nitrogen dioxide) N_2O_4 (dinitrogen tetroxide)	$2NO + O_2 \rightarrow \mathbf{2NO_2} \rightleftharpoons \mathbf{N_2O_4}$
5+	HNO_3 (nitric acid)	$3NO_2 + H_2O \rightarrow \mathbf{2HNO_3} + NO$

principle the position of equilibrium is less favorable than at lower temperatures.

Ammonia is a colorless gas with a powerful irritating odor that most people would recognize immediately. It has a boiling point of $-33.35°C$ and freezes at $-77.7°C$. As a liquid it has many solvent properties similar to those of water, and the parallels between reactions in liquid ammonia and in water were discussed in Section 14.5.

The Lewis structure of ammonia is

$$H—\overset{..}{N}—H$$
$$|$$
$$H$$

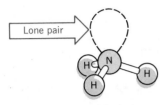

Lone pair

The pyramidal ammonia molecule.

and, as we would expect from VSEPR theory, NH_3 is a pyramidal molecule. It is quite polar and is extremely soluble in water. A saturated solution at room temperature contains about 28% w/w of NH_3 and is about 15 molar. Its high solubility in water is the result of its ability to form hydrogen bonds with water molecules, both of the type $O—H\cdots N$ as well as $N—H\cdots O$.

Aqueous solutions of ammonia are basic, as you recall from Chapter 15. The reaction between NH_3 and H_2O is

$$NH_3 + H_2O \rightleftharpoons NH_4^+ + OH^- \qquad K_b = 1.8 \times 10^{-5}$$

Bottles of concentrated ammonia purchased from chemical supply companies are almost always labeled "ammonium hydroxide"; however, there is no evidence that the species NH_4OH actually exists. These solutions simply consist of molecules of NH_3 dissolved in water (along with small amounts of NH_4^+ and OH^- produced by the ionization of NH_3).

Ammonia can serve as either a proton donor or a proton acceptor, depending on conditions, and therefore forms two types of salts. When acting as a proton donor, NH_3 forms *amide salts* such as $NaNH_2$ which contains the NH_2^- ion. These can only be made in nonaqueous media, because NH_2^- is a very strong base and is completely hydrolyzed in water to give NH_3.

$$\underset{\textbf{amide ion}}{NH_2^-} + H_2O \longrightarrow NH_3 + OH^-$$

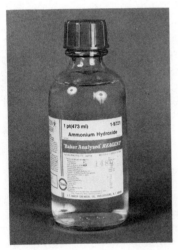

Reagent bottles of concentrated aqueous ammonia are usually labeled "ammonium hydroxide."

When acting as a proton acceptor, ammonia forms ammonium salts—for example, NH_4Cl.

$$NH_3 + H_3O^+ \longrightarrow NH_4^+ + H_2O \qquad K = 1.8 \times 10^9$$

As we learned in Chapter 15, NH_4^+ also hydrolyzes slightly, so solutions of NH_4^+ tend to be slightly acidic (providing the anion present is that of a strong acid).

$$NH_4^+ + H_2O \rightleftharpoons NH_3 + H_3O^+ \qquad K = 5.5 \times 10^{-10}$$

In the laboratory, ammonia can be prepared by reacting an ammonium salt with a strong base such as oxide ion or hydroxide ion.

$$CaO + 2NH_4Cl \longrightarrow 2NH_3 + CaCl_2 + H_2O$$

$$NaOH + NH_4Cl \longrightarrow NH_3 + NaCl + H_2O$$

These strong bases displace the weaker base, NH_3, from the NH_4^+ ion. The reaction of ammonium ion with a base serves as part of the qualitative test for ammonium ion in a mixture. If a sample of the mixture is warmed with base, an odor of ammonia serves to confirm the presence of NH_4^+ in the mixture. Alternatively, the NH_3 can be detected by the basic reaction of its vapors with moist litmus paper (pink $\rightarrow$ blue).

One of the principal reactions of ammonia is its catalytic oxidation to nitric oxide in the **Ostwald process** (see Color Plate 16).

This is a very exothermic reaction.

$$4NH_3 \ (g) + 5O_2 \ (g) \xrightarrow[\substack{\text{catalyst} \\ 750-900°C}]{\text{Pt}} 4NO \ (g) + 6H_2O \ (g)$$

The NO is quickly oxidized to NO_2 in the presence of excess oxygen,

$$2NO \ (g) + O_2 \ (g) \longrightarrow 2NO_2 \ (g)$$

and when the NO_2 is dissolved in water it **disproportionates**—that is, it enters into a redox reaction with itself so that some of it is oxidized while the rest is reduced.

N is oxidized from 4+ to 5+

$$3NO_2 \ (g) + H_2O \ (l) \longrightarrow \underbrace{2H^+ \ (aq) + 2NO_3^- \ (aq)}_{2HNO_3} + NO \ (g)$$

N is reduced from 4+ to 2+

The commercial application of this sequence of reactions accounts for the major source of nitric acid and nitrates used in the manufacture of explosives, fertilizers, plastics, and many other useful substances. In fact, the development of this process in Germany by Wilhelm Ostwald, accompanied by the successful preparation of NH_3 from N_2 and H_2 by Haber, is said to have prolonged World War I, since the Allied blockade of Germany was unable to halt the German manufacture of munitions that had depended, prior to these processes, on the importation of nitrates from other countries.

Hydrazine (2 – Oxidation State) Hydrazine has the formula N_2H_4 and can be considered the nitrogen analog of hydrogen peroxide (although hydrazine's reactions are vastly different than those of H_2O_2). Its Lewis structure is

$$\begin{array}{cc} H & H \\ | & | \\ H-\underset{..}{N}-\underset{..}{N}-H \end{array}$$

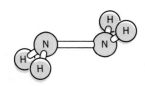

Figure 19.7

The structure of hydrazine, N_2H_4.

and its molecular geometry is shown in Figure 19.7. The staggered arrangement of the hydrogens probably arises because of the tendency of the lone pairs of the nitrogens to be as far apart as possible so that the repulsions between them are a minimum.

Hydrazine is prepared by the reaction of ammonia with hypochlorite ion in aqueous solutions.

$$2NH_3 + NaOCl \longrightarrow N_2H_4 + NaCl + H_2O$$

Hydrazine is a violent poison, and this reaction is one of the reasons that warnings are given about the dangers of mixing household cleaning agents—specifically household ammonia and liquid bleach, which contains NaOCl. The danger of such a mixture is somewhat lessened, fortunately, because an intermediate in the reaction (NH_2Cl) has a tendency to react with N_2H_4 yielding harmless NH_4Cl, especially if traces of Cu^{2+} are present.

Traces of Cu^{2+} are common in water that passes through the copper water pipes found in many homes.

Pure hydrazine is a liquid that freezes at 2°C and boils at 114°C. It has a strong affinity for water and its solutions are weakly basic because of the reaction,

$$N_2H_4 + H_2O \rightleftharpoons N_2H_5^+ + OH^- \qquad K_b = 1.7 \times 10^{-6}$$

hydrazinium ion

A major industrial use of hydrazine is as an oxygen scavenger in high temperature boilers.

monomethylhydrazine — one of the fuels used by the space shuttle Columbia.

In basic solutions, hydrazine is a powerful reducing agent. For example, it reacts with iodine to liberate nitrogen.

$$N_2H_4 + 2I_2 + 4OH^- \longrightarrow N_2 + 4H_2O + 4I^-$$

The combustion of hydrazine is very exothermic.

$$N_2H_4\ (l) + O_2\ (g) \longrightarrow N_2\ (g) + 2H_2O\ (g) \qquad \Delta H^0 = -534\ \text{kJ}$$

Because of this, hydrazine as well as some compounds derived from it have been used as rocket fuels.

Hydroxylamine (1– Oxidation State) Hydroxylamine can be thought of as an ammonia molecule in which one hydrogen has been replaced by an —O—H group.

$$\text{H}-\overset{\cdot\cdot}{\text{N}}-\text{OH}$$
$$|$$
$$\text{H}$$

It can be made by reduction of nitrites with SO_2.

$$NaNO_2 + NaHSO_3 + SO_2 + 2H_2O \longrightarrow 2NaHSO_4 + NH_2OH$$

Pure hydroxylamine is a white solid that melts at 33°C, but it decomposes very easily. In water it is a weak base

$$NH_2OH + H_2O \rightleftharpoons NH_3OH^+ + OH^- \qquad K_b = 1.1 \times 10^{-8}$$

NH_3OH^+ is called the hydroxylammonium ion.

and forms salts such as $[NH_3OH]Cl$ and $[NH_3OH]_2SO_4$. The salts are stable and are used as mild reducing agents in photography. The sulfate salt is used for removing hair from animal hides.

Nitrous Oxide (1+ Oxidation State) Nitrous oxide (N_2O is made by decomposing molten ammonium nitrate, NH_4NO_3,

$$NH_4NO_3\ (l) \xrightarrow{\text{heat}} N_2O\ (g) + 2H_2O\ (g)$$

NH_4NO_3 is used as a high explosive.

Although this reaction proceeds smoothly under most circumstances, NH_4NO_3 can be made to explode if detonated by another explosive.

The structure of N_2O can be described by the resonance formulas

$$:\overset{\cdot\cdot}{\text{N}}\!=\!\text{N}\!=\!\overset{\cdot\cdot}{\text{O}}: \longleftrightarrow :\text{N}\!\equiv\!\text{N}\!-\!\overset{\cdot\cdot}{\underset{\cdot\cdot}{\text{O}}}:$$

As the VSEPR theory would predict, it is a linear molecule.

Thermodynamically, N_2O is unstable with respect to decomposition to the elements. This is because its heat of formation and free energy of formation are both positive ($\Delta H_f^0 = +81.5$ kJ/mol, $\Delta G_f^0 = +104$ kJ/mol). Nitrous oxide is stable at room temperature only because its *rate* of decomposition is extremely slow, but at elevated temperatures it decomposes easily to N_2 and O_2 with the *release* of heat. Oxygen, of course, supports combustion, so burning a fuel with N_2O releases more heat than with just O_2 because of the added heat released by the decomposition of the N_2O. This is the reason that race car drivers often use N_2O as an oxidizer for their fuel—it gives their engines extra power.

Nitrous oxide is also used as an anesthetic. It produces a mild intoxicating effect and is commonly called "laughing gas." Another application is as a propellant in aerosol cans.

Nitric Oxide (2+ Oxidation State) and Nitrogen Dioxide (4+ Oxidation State) Nitric oxide, NO, and nitrogen dioxide, NO_2, are the two most important oxides

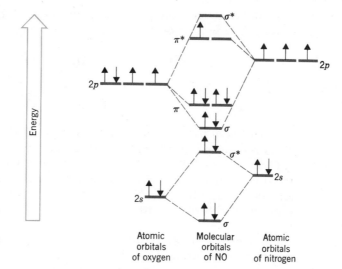

Figure 19.8

Molecular orbital energy level diagram for NO. Note that the energies of the atomic orbitals on the separate nitrogen and oxygen atoms are not the same.

of nitrogen. They are intermediates in the conversion of ammonia to nitric acid, and both play a major role in the formation of a type of air pollution called photochemical smog. In this context, they are generally referred to together as NO_x.

The path from ammonia to nitric acid was discussed earlier in this section. Oxidation of ammonia on a platinum catalyst gives NO, which is rapidly oxidized to NO_2 by excess oxygen. Dissolving the NO_2 in water gives nitric acid plus NO (which is oxidized again to NO_2). This can be represented schematically as

$$NH_3\ (g) \xrightarrow[\text{catalyst}]{O_2} NO\ (g) \xrightarrow{O_2} NO_2\ (g) \xrightarrow{H_2O} HNO_3\ (aq) + NO\ (g)$$

Nitric oxide is a fairly reactive, colorless gas. It has an odd number of electrons, so they can't all be paired. NO is therefore paramagnetic. The bonding in nitric oxide, like that in O_2, is perhaps best described by molecular orbital theory. The molecular orbital energy level diagram for NO is given in Figure 19.8. The net bond order in the molecule is 2.5 because the electron in the π^* antibonding MO cancels the effect of one of the electrons in the π bonding orbitals. It is interesting that NO loses an electron rather easily to form the NO^+ (nitrosonium) ion. This is because the electron that is lost is from a high energy π^* orbital. Removal of the antibonding electron raises the bond order to 3.0, and the NO^+ ion actually has a shorter, stronger bond than NO.

Nitrogen dioxide is a toxic, reddish-brown gas. Its structure is given by the resonance formulas

Compounds containing the NO^+ ion can be isolated.

$$\ddot{N} \qquad \longleftrightarrow \qquad \ddot{N}$$
$$:\!\ddot{O} \qquad\quad \ddot{O}\!: \qquad\qquad :\!\ddot{O}\!: \qquad\quad \ddot{O}\!\cdot$$

and in the gas phase there is an equilibrium between NO_2 and its dimer, dinitrogen tetroxide, N_2O_4. (A *dimer* is a molecule formed by joining two simpler ones.)

$$\underset{\textbf{reddish-brown}}{2NO_2\ (g)} \rightleftharpoons \underset{\textbf{colorless}}{N_2O_4\ (g)} + 57\ kJ$$

The formation of N_2O_4 from NO_2 can be explained by the tendency of the unpaired electron in NO_2, which spends most of its time on the nitrogen, to become paired with another electron from a neighboring NO_2 molecule.

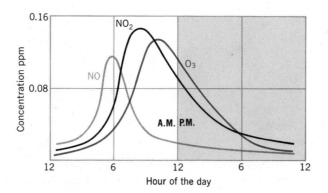

Dinitrogen tetroxide is a low-boiling liquid (bp = 21°C) that is deep brown at its boiling point caused by some NO_2 produced by its dissociation (pure N_2O_4 is colorless). It is a good oxidizing agent and has been used as an oxidizer in liquid rockets. In fact, the space shuttle Columbia burns monomethylhydrazine (see page 612) with N_2O_4 in the rocket engine that it uses to drop out of orbit on its return to Earth.

Photochemical Smog Within the last several decades, the growing number of automobiles in urban areas has produced a kind of air pollution never before experienced by civilization. It is characterized by a reddish-brown haze that contains substances irritating to the eyes, nose, and lungs, and that cause extensive damage to vegetation and to rubber products not containing antioxidants. This haze has come to be known as **photochemical smog** (or often just smog) and has been traced to abnormally high levels of ozone and oxides of nitrogen in the atmosphere.

A typical smog episode begins in the early morning when urban rush-hour traffic spews out the primary pollutant, nitric oxide. This is formed in the gasoline engine during combustion by the direct combination of N_2 and O_2, which are present in the air that is drawn into the engine to burn the fuel. Even though the reaction

$$N_2\ (g) + O_2\ (g) \rightleftharpoons 2NO\ (g)$$

has an extremely small equilibrium constant at ordinary temperatures, the K increases with increasing temperature, so at the high temperature inside the engine small amounts of NO are produced. When the exhaust gases leave the engine, they cool so rapidly that the NO doesn't have an opportunity to decompose back to N_2 and O_2. As a result, it is released into the atmosphere. In Figure 19.9 we see that during the early hours the NO concentration rises.

The next step in the sequence of reactions is the oxidation of NO to NO_2. This occurs slowly as the morning wears on, and in Figure 19.9 we see that the NO concentration starts to drop as the NO_2 concentration rises. The presence of the NO_2 in smog is what gives this form of air pollution its characteristic reddish-brown color (see Color Plate 17).

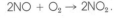

NO_2 is paramagnetic, but N_2O_4 is not.

$2NO + O_2 \rightarrow 2NO_2$.

Figure 19.9

The concentrations of pollutants during a typical photochemical smog episode.

As the sun climbs higher in the sky, the NO_2 begins to undergo a photochemical reaction—a reaction brought on by the absorption of light. In this case, ultraviolet light from the sun's rays causes NO_2 to decompose.

$$NO_2 \xrightarrow{h\nu} NO + O$$

hν is the energy absorbed from a photon of frequency ν.

The oxygen atoms produced in this reaction combine with oxygen molecules and ozone is formed.

$$O_2 + O \longrightarrow O_3$$

Over 99% of the oxygen atoms react with O_2 to form O_3.

In Figure 19.9 we see that the ozone concentration peaks about noon as the NO_2 concentration begins to decline.

As we learned in Section 19.3, ozone is an extremely reactive substance. It reacts with NO to form NO_2, and this reaction helps moderate the rate of buildup of O_3.

$$O_3 + NO \longrightarrow NO_2 + O_2$$

Ozone also attacks other substances and is particularly hard on vegetation because it reacts with chlorophyll. It is especially reactive toward molecules that contain carbon-carbon double bonds. Until antipollution devices were installed on automobiles, considerable amounts of hydrocarbons (including those with double bonds) were released into the atmosphere by the evaporation of gasoline and as unburned hydrocarbons in auto exhausts. These substances react with ozone in a series of complex reactions to give final products with the formula

$$\begin{array}{c} :\!O\!: \\ \| \\ R\!-\!\overset{..}{C}\!-\!\overset{..}{O}\!-\!\overset{..}{O}\!-\!NO_2 \end{array}$$

$$\begin{array}{c} :\!O\!: \\ \| \\ CH_3\!-\!\overset{..}{C}\!-\!\overset{..}{O}\!-\!\overset{..}{O}\!-\!NO_2 \\ \text{peroxyacetylnitrate} \end{array}$$

where R is a portion of a hydrocarbon molecule—for example, $-CH_3$, $-C_2H_5$, and so on. They are collectively called *peroxyacylnitrates*, or PAN for short. Molecules of PAN are oxidizing agents—a property that is typical of compounds having an oxygen-oxygen bond (e.g., H_2O_2)—and contribute to the oxidizing nature of photochemical smog. Molecules of PAN are also eye irritants, even at concentration levels of parts per billion. This is one of the reasons that smog is so unpleasant.

Finally, as late afternoon approaches, the O_3 concentration begins to fall off as it reacts with airborn hydrocarbons and other substances, and the smog attack gradually subsides—until daybreak the next day.

Dinitrogen Trioxide, Nitrous Acid, and Nitrites (the 3+ Oxidation State) Dinitrogen trioxide, N_2O_3, is produced by condensing an equimolar mixture of NO and NO_2 at very low temperatures (below $-30°C$).

$$NO + NO_2 \longrightarrow N_2O_3$$

It is a blue liquid in which there are molecules having the structure

$$\begin{array}{ccc} :\!\overset{..}{O} & & :\!\overset{..}{O}\!\cdot \\ \| & & \| \\ \overset{..}{N}\!-\!N & & \\ & & \overset{..}{O}\!\cdot \end{array}$$

$N_2O_3 + H_2O \rightarrow 2HNO_2$.

At least in a formal sense, N_2O_3 is the acid anhydride of nitrous acid, HNO_2. In fact, if an equimolar mixture of NO and NO_2 is bubbled into water, nitrous acid is produced.

Nitrous acid is a weak acid ($K_a = 4.5 \times 10^{-4}$) and is only stable in solution and in the gas phase. It cannot be isolated in pure liquid form because it decomposes by a disproportionation reaction.

$$3HNO_2 \longrightarrow HNO_3 + H_2O + 2NO$$

When HNO_2 is needed for a chemical reaction it is usually generated in aqueous solution by adding a strong acid, such as HCl, to a salt of HNO_2 such as $NaNO_2$ (sodium nitrite). Hydrogen ion and nitrite ion combine in the net reaction

$$H^+ + NO_2^- \longrightarrow HNO_2$$

The formation of the weak electrolyte HNO_2 drives this reaction to the right.

Nitrous acid is able to function as either an oxidizing agent or a reducing agent, depending on the ease of oxidation or reduction of the other reactant. For example, it is oxidized to HNO_3 by permanganate ion,

$$H^+ + 5HNO_2 + 2MnO_4^- \longrightarrow 5NO_3^- + 2Mn^{2+} + 3H_2O$$

Here HNO_2 is oxidized and is a reducing agent.

but, in the presence of iodide ion, HNO_2 is reduced.

$$2H^+ + 2HNO_2 + 2I^- \longrightarrow I_2 + 2NO + 2H_2O$$

Here HNO_2 is reduced and is an oxidizing agent.

Neutralization of nitrous acid gives nitrite ion, although nitrite salts are normally made by reducing a metal nitrate. For example,

$$NaNO_3 + C \xrightarrow{\text{heat}} NaNO_2 + CO$$

The nitrite ion has an angular structure that is predicted by VSEPR theory from either of its resonance formulas

A major but controversial use of nitrites is in preserving meats such as ham, frankfurters, bologna, and bacon. The nitrite serves two functions. The most important one is that it inhibits the growth of bacteria, especially *clostridium botulinum*, which produces the very poisonous botulinus toxin that causes fatal food poisoning. The second is that it preserves the red color of the meat and thereby maintains the food's appetizing appearance.

The controversy over the use of nitrites in cured meat products stems from the effect that nitrous acid has on organic compounds called amines. The reaction of HNO_2 with amines produces other compounds called nitrosoamines, which have been shown to cause cancer, mutations, and birth defects in experimental animals. Meat and body fluids contain many amines, and it is feared by some scientists that when the NO_2^- contacts the acidic condition in the stomach some of the HNO_2 that is formed might react with amines that are there and thereby give nitrosoamines. Although no direct evidence has been found linking nitrites in meat to cancer in either animals or humans, the FDA has set limits on the maximum allowable NO_2^- concentrations in these food products.

Some scientists feel that the risk of food poisoning from meats untreated with nitrites is greater than the risk of cancer from meats that are treated.

Dinitrogen Pentoxide, Nitric Acid, and Nitrates (the 5+ Oxidation State) Dinitrogen pentoxide, N_2O_5, exists in molecular form in the vapor. Its structure is

In the solid state, N_2O_5 dissociates into ions and exists as $NO_2^+NO_3^-$. When allowed to react with water, N_2O_5 gives nitric acid, HNO_3,

N_2O_5 is the acid anhydride of HNO_3.

$$N_2O_5 + H_2O \longrightarrow 2HNO_3$$

This reaction can be reversed, and N_2O_5 is produced by removing the components of water from a pair of HNO_3 molecules. This requires a very powerful dehydrating agent such as P_4O_{10}. (The reaction of P_4O_{10} with water is discussed in the next chapter.)

About 8.5 million tons of HNO_3 are made annually.

Nitric acid is one of the world's most vital chemicals, and huge amounts of it are produced each year. Much of it is made into ammonium nitrate to be used as a nitrogen fertilizer, because plants can utilize both NH_4^+ and NO_3^- ion. The manufacture of explosives such as TNT and nitroglycerine also require nitric acid, and nitrates are used in curing meats along with nitrites.

The commercial production of nitric acid by dissolving NO_2 in water was described earlier in our discussion of the Ostwald process. In the laboratory, nitric acid can be made by heating a mixture of KNO_3 and concentrated sulfuric acid.

$$KNO_3\ (s) + H_2SO_4\ (l) \xrightarrow{\text{heat}} KHSO_4\ (s) + HNO_3\ (g)$$

The nitric acid vapors condense to a liquid when they are cooled.

Pure nitric acid is a colorless liquid that decomposes easily above 0°C into NO_2, H_2O, and O_2. The concentrated laboratory reagent (Color Plate 18) is about 70% HNO_3 and is often slightly yellow in color from the presence of small amounts of NO_2 formed by a photochemical decomposition.

$$\underset{\textbf{colorless}}{4HNO_3} \xrightarrow{h\nu} \underset{\substack{\textbf{red-brown}\\ \text{(appears yellow}\\ \text{when dilute)}}}{4NO_2} + O_2 + 2H_2O$$

Nitric acid is a strong acid, and the presence of nitrate ion in its solutions makes it an especially powerful oxidizing agent. It is therefore able to dissolve many metals that fail to dissolve in an acid such as HCl, which contains H^+ as its strongest (and only) oxidizing agent. For example, copper dissolves in both concentrated and dilute HNO_3, but the reduction products differ. These reactions were described previously.

$$Cu + 2NO_3^- + 4H^+ \longrightarrow Cu^{2+} + 2NO_2 + 2H_2O \qquad \text{(Concentrated)}$$

$$3Cu + 2NO_3^- + 8H^+ \longrightarrow 3Cu^{2+} + 2NO + 4H_2O \qquad \text{(Dilute)}$$

With stronger reducing agents, the nitrogen can be reduced all the way to the $3-$ oxidation state. For example,

$$4Zn + NO_3^- + 10H^+ \longrightarrow 4Zn^{2+} + NH_4^+ + 3H_2O$$

A mixture of one part by volume of concentrated HNO_3 and three parts by volume of concentrated HCl is called aqua regia. As discussed earlier, this mixture is able to dissolve the noble metals such as gold and platinum that fail to dissolve in concentrated HNO_3 alone. The chloride ion in aqua regia forms complex ions with the ions of these metals and that helps "draw" them into solution. For instance, gold dissolves as follows.

$$4H^+\ (aq) + 4Cl^-\ (aq) + NO_3^-\ (aq) + Au\ (s) \longrightarrow AuCl_4^-\ (aq) + NO\ (g) + 2H_2O$$

INDEX TO QUESTIONS AND PROBLEMS (Problem numbers are in **bold type**)

REVIEW QUESTIONS

19.1 Which is the most abundant element in the universe?

19.2 What is the probable reason why the abundance of hydrogen on Earth is less than it is in the rest of the universe?

19.3 Why is there little free hydrogen in the Earth's atmosphere?

19.4 What advantage does hydrogen have over helium for use in lighter-than-air balloons? What is a disadvantage?

19.5 Why are the properties of hydrogen so different from the properties of the other elements in Group IA? Why is hydrogen placed in Group IA?

19.6 Give the names and symbols of the three isotopes of hydrogen. Which one is radioactive? Why does it potentially pose an environmental problem?

19.7 Give chemical reactions for the preparation of hydrogen from water and (a) methane (natural gas); (b) coal.

19.8 Write chemical equations for the combustion of water gas.

19.9 In what way is the commercial production of caustic soda related to the commercial production of hydrogen?

19.10 What is the greatest industrial use for hydrogen? Give the appropriate chemical equation.

19.11 Write a chemical equation for the laboratory preparation of hydrogen. Sketch the apparatus you would use.

19.12 How is hydrogen used in the synthesis of methanol? What importance is this commercially?

19.13 Use Lewis structures to illustrate the hydrogenation of acetylene, C_2H_2. What effect does hydrogenation have on vegetable oils?

19.14 What general name is given to binary compounds of hydrogen? What would be the name for the compound NaH? How could it be formed? What is its reaction with water?

19.15 How many covalent bonds does hydrogen normally form? Why?

19.16 Predict the molecular structures of (a) H_2S; (b) PH_3; (c) SiH_4.

19.17 Explain how the reaction $H^- + H_2O \rightarrow H_2 + OH^-$ can be interpreted as both an acid-base reaction and a redox reaction.

19.18 Define *catenation*. Which element has the greatest tendency to catenate? How does the tendency to catenate vary within a group in the periodic table?

19.19 Why can't all the nonmetal hydrides be made by direct combination of the elements? Which nonmetal hydrides can be made by this method?

19.20 How do the basicities of the conjugate bases of the nonmetal hydrides vary within periods and groups in the periodic table? Give two examples.

19.21 What are the advantages of hydrogen as a fuel? Give two disadvantages (see Problem 11.65 on page 391).

19.22 What is an *organic* compound?

19.23 How is coke made?

19.24 How do the structures of diamond and graphite differ?

19.25 What are two uses of graphite?

19.26 What is activated charcoal? What property does it possess that makes it commercially useful?

19.27 How is carbon black made? What are two of its uses?

19.28 Give the structures of carbon monoxide and carbon dioxide.

19.29 How can carbon monoxide be prepared in the laboratory?

19.30 Write a chemical equation showing the reducing properties of carbon monoxide toward Fe_2O_3.

19.31 What is a metal carbonyl compound? Give an example.

19.32 How is CO_2 usually made industrially? How can it be made conveniently in the laboratory?

19.33 Give three major industrial uses of CO_2.

19.34 How do green plants use CO_2?

19.35 Give the equations for the dissolving of limestone by dissolved carbon dioxide.

19.36 How are stalagmites and stalactites formed in limestone caves?

19.37 What is hard water? How can hardness ions be removed from hard water?

19.38 Can you suggest a chemical method for removing boiler scale from the insides of pipes and water boilers?

19.39 What is carborundum? How is its structure related to that of diamond? How is carborundum made?

19.40 Write an equation for the reaction of aluminum carbide with water.

19.41 How is acetylene made?

19.42 What is an interstitial carbide? Give an example.

19.43 How is hydrogen cyanide synthesized? How does it function as a poison?

19.44 Give the structure of carbon disulfide. What makes CS_2 a hazardous solvent with which to work?

19.45 What are the commercial sources of oxygen and nitrogen?

19.46 What are the correct chemical names for H_2, O_2, and N_2?

19.47 How is oxygen conveniently prepared in the laboratory?

19.48 What is the largest use for O_2 industrially?

19.49 Give the structure for ozone. How can it be made in the laboratory? What advantage does ozone have over Cl_2 for the purification of drinking water?

19.50 How is ozone formed in the Earth's upper atmosphere? How does it protect the Earth's inhabitants from ultraviolet radiation from the sun? How could nitrogen oxide pollutants affect the Earth's ozone layer (give equations)? How could Freons affect the ozone layer (give equations)?

19.51 Write a chemical equation for the hydrolysis of oxide ion.

19.52 What does pickling mean with respect to the treatment of iron and steel objects?

19.53 Write equations that show the amphoteric behavior of a metal oxide.

19.54 Give three ways that the oxide, CO_2, can be made. Illustrate each with a chemical equation.

19.55 Why can't N_2O be synthesized from its elements?

19.56 Write an equation for the oxidation of ammonia with oxygen.

19.57 Write chemical equations for the reaction of copper with (a) concentrated HNO_3; (b) dilute HNO_3.

19.58 Give equations showing reactions of oxygen to form (a) a metal peroxide; (b) a metal superoxide.

19.59 What is the structure of hydrogen peroxide?

19.60 Write an equation for the decomposition of hydrogen peroxide.

19.61 Why is dinitrogen so unreactive?

19.62 Give three commercial uses of N_2.

19.63 What are *nitrogen-fixing bacteria*?

19.64 How can small amounts of N_2 be made in the laboratory?

19.65 Which is the only element N_2 reacts with at room temperature?

19.66 What happens when an ionic nitride such as Mg_3N_2 is placed into water?

19.67 Give an example of a nitrogen compound in which the oxidation number of nitrogen is (a) 3−, (b) 1−, (c) 1+, (d) 3+, and (e) 5+.

19.68 Give the chemical equation and reaction conditions in the Haber process. Why are those particular conditions chosen?

19.69 Why is ammonia so soluble in water?

19.70 Give the formula for the principal nitrogen-containing species in a solution of *ammonium hydroxide*.

19.71 Give the formula for potassium amide. Why can't potassium amide be made in an aqueous solution?

19.72 How can ammonia be made in the laboratory? Could it be collected by displacement of water in the same manner as hydrogen?

19.73 How could you determine whether a particular unknown solid was an ammonium salt?

19.74 Give the chemical reactions in the Ostwald process.

19.75 What is a disproportionation reaction?

19.76 What influence did the Haber and Ostwald processes have on the waging of World War I?

19.77 Give the structure of hydrazine.

19.78 How is hydrazine prepared? Why has hydrazine been used as a rocket fuel?

19.79 Give the structure of hydroxylamine. Compare the basicities of ammonia, hydrazine, and hydroxylamine.

19.80 Why is it potentially dangerous to mix household ammonia with liquid laundry bleach?

19.81 Give a chemical equation for the preparation of nitrous oxide. What are the resonance structures for nitrous oxide?

19.82 Look up the standard free energies of formation of N_2O, NO, and NO_2. What accounts for the apparent stability of these oxides of nitrogen?

19.83 Construct molecular orbital energy level diagrams for NO and for O_2. (If necessary, refer to Figure 19.8 and Figure 5.22.) Compare expected bond orders, bond lengths, and bond energies for (a) the series NO^-, NO, NO^+, (b) the series O_2^+, O_2, O_2^-, O_2^{2-}.

19.84 Give the Lewis structures for NO_2 and N_2O_4. Write the equilibrium that exists between these two molecules.

19.85 Describe the series of chemical reactions responsible for photochemical smog.

19.86 What substance causes the reddish-brown haze of photochemical smog? What is PAN?

19.87 What is the structure of N_2O_3? Under what conditions can it be formed? Of which acid is N_2O_3 the acid anhydride?

19.88 Write a chemical equation for the decomposition of nitrous acid.

19.89 Write Lewis formulas for NO_2 and NO_2^-. On the basis of the principles of VSEPR theory, which one should have the larger bond angle?

19.90 Write equations that show nitrous acid functioning as (a) a reducing agent, (b) an oxidizing agent.

19.91 How is $NaNO_2$ normally made? What are the functions of $NaNO_2$ in meat products such as bologna? What are the pros and cons concerning its use in such products?

19.92 Give the Lewis structure for N_2O_5 as it exists in the vapor. How does N_2O_5 exist in the solid state?

19.93 How can N_2O_5 be made?

19.94 Write a chemical equation for the reaction of N_2O_5 with water.

19.95 Why does concentrated HNO_3 usually have a pale yellow color?

19.96 How could pure HNO_3 be prepared in the laboratory?

19.97 Give three uses of nitric acid and nitrates.

19.98 What nitrogen-containing product is formed when zinc reacts with HNO_3?

19.99 Why does nitric acid dissolve metals such as silver and copper which are unaffected by hydrochloric acid?

19.100 What is aqua regia? How does it function in dissolving noble metals such as gold?

REVIEW PROBLEMS

19.101 For octane, $\Delta H_f^0 = -255.1$ kJ/mol. Use this and the data in Table 11.1 to calculate the standard heats of reaction for the following:

$$C_8H_{18}\ (l) + 12\tfrac{1}{2}O_2\ (g) \longrightarrow 8CO_2\ (g) + 9H_2O\ (g)$$

$$C_8H_{18}\ (l) + 25N_2O\ (g) \longrightarrow 8CO_2\ (g) + 9H_2O\ (g) + 25N_2\ (g)$$

THE CHEMISTRY OF SELECTED NONMETALS, PART II: PHOSPHORUS, SULFUR, THE HALOGENS, THE NOBLE GASES, AND SILICON

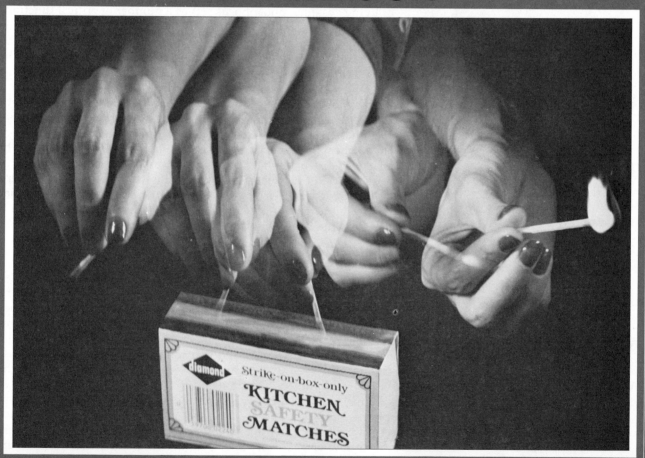

The simple act of striking a match involves many of the elements discussed in this chapter. The head of the match contains $KClO_3$ as an oxidizing agent, and glue and sulfur compounds as fuel. The match is ignited by rubbing it across a striking surface composed of powdered glass (made from SiO_2) and red phosphorus. The friction ignites the phosphorus, which ignites the match head.

The nonmetallic elements discussed in this chapter cover a broad range of chemical reactivity and commercial importance. They range from fluorine—the most reactive element—to helium—the least reactive. Here we find sulfuric and phosphoric acids—chemicals that consistently rank among the top 10 in total tonnage manufactured each year. At the other extreme we have compounds of the heavier noble gases—substances that have so far found no practical uses. (In fact, the ability of any of the noble gases to react at all was a bombshell in the chemical world only two decades ago.)

We also see in this chapter a range of biological importance. Compounds of phosphorus and sulfur occur in all living systems, but many of those containing the halogens are lethal to living organisms. On the other hand, living things are almost indifferent toward the noble gases.

20.1   PHOSPHORUS

The Earth's crust contains about 0.1% phosphorus.

Phosphorus is a relatively abundant element, ranking 12th in the Earth as a whole. It is always found in the combined state, normally in *phosphate rock*, which consists primarily of minerals that contain calcium phosphate, $Ca_3(PO_4)_2$. In living systems, phosphorus serves a number of critical functions: Phosphate units play a role in the structure of DNA, which directs the chemistry of our cells and transmits genetic information from one generation to another. Phosphate units are also present in phospholipids—substances that make up cell membranes—and phosphorus-oxygen bonds store the energy that we derive from the metabolism of foods.

About 85% of the phosphate rock mined in the United States now comes from Florida and North Carolina.

Because phosphorus is so important to biological systems of all kinds, it is not surprising that most industrial applications of phosphorus compounds—especially phosphates—are ultimately related in one way or another to providing sufficient phosphorus for the growth of agricultural products. For this reason also, deposits of phosphate rock are an important national resource. Until recently, some rich U.S. deposits of phosphate rock appeared on the verge of depletion during the 1990s. However, discoveries of huge phosphate deposits covering hundreds of square miles below the ocean floor within 60 miles of North Carolina have relieved fears of imminent shortages.

The free element

Phosphorus atoms in their ground state have the electron configuration, [Ne] $3s^2 3p^3$, and therefore require three electrons to complete an octet. In the free element this is accomplished by linking each phosphorus atom to three others by single covalent bonds. As we learned in Chapter 9, this gives rise to three principal allotropes.

White phosphorus consists of individual tetrahedral P_4 molecules. It is a waxy solid that melts at 44.1°C and boils at 280°C. Because of the small, highly strained 60° P—P—P bond angle in P_4, this allotrope is extremely reactive, especially toward oxygen, and must be stored under water. When exposed to air, white phosphorus ignites spontaneously, and contact with the skin produces painful slow-healing burns. White phosphorus is also very toxic, and long-term exposure to even low levels of P_4 vapor produces a gradual deterioration and softening of the bones of the jaw.

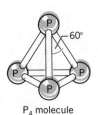

P_4 molecule

About 0.05 g of white phosphorus is a fatal dose.

White phosphorus was first prepared (unintentionally) in 1669 by an alchemist named Hennig Brand, who heated a mixture of dried urine and sand and condensed the vapors that were given off by passing them through water. The new element was named phosphorus, from the Greek *phosphoros* = light bringer, because the dry solid in a sealed bottle glows in the dark. The glow is actually caused by a slow oxidation of the phosphorus surface by residual oxygen in the container.

Brand's mixture also contained phosphates, carbon, and silica.

Modern methods of producing white phosphorus differ only slightly from Brand's. A mixture of phosphate rock, $Ca_3(PO_4)_2$, silica, SiO_2, and coke is heated to about 1300°C in an **electric furnace**—a furnace in which the contents are heated to very high temperatures by passing an electric current through them. The reaction is

$$2Ca_3(PO_4)_2\ (s) + 6SiO_2\ (s) + 10C\ (s) \longrightarrow 6CaSiO_3\ (l) + 10CO\ (g) + P_4\ (g)$$

The CO and P_4 vapor is passed through water, which causes the P_4 to condense. Most of the phosphorus produced in this way is ultimately used to make phosphoric acid.

Red phosphorus is formed when white phosphorus is heated or exposed to ultraviolet light. It is amorphous and as we learned earlier, probably consists of P_4 tetrahedra joined at their corners. It is also relatively nonpoisonous and is considerably less reactive than white phosphorus. Red phosphorus is used to make incendiary devices (bombs, fireworks, flares, etc.) and, when mixed with fine sand, is used on the striking surfaces of matchbooks.

Black phosphorus is the least reactive form of phosphorus. It consists of layers of phosphorus atoms. Within each layer the phosphorus atoms are covalently bonded to each other, but the attractions between layers are the result of much weaker London forces. This gives it a flaky, graphitelike appearance.

Compounds of phosphorus

Phosphorus combines with most of the nonmetals, as well as with active metals. With the metals of Groups IA and IIA, for example, binary *phosphides* are formed that are predominantly ionic and contain the *phosphide ion*, P^{3-}. In water, this ion hydrolyzes to give *phosphine*, PH_3, a very bad-smelling poisonous gas.

$$Na_3P\ (s) + 3H_2O\ (l) \longrightarrow 3NaOH\ (aq) + PH_3\ (g)$$

The most important compounds of phosphorus are formed with nonmetals, particularly oxygen and the halogens.

Oxides of Phosphorus When any of the allotropes of phosphorus are burned in an excess supply of oxygen, white, powdery phosphorus(V) oxide, P_4O_{10}, is formed. For example,

$$P_4\ (s) + 5O_2\ (g) \longrightarrow P_4O_{10}\ (s)$$

This oxide is often called *phosphorus pentoxide* because its empirical formula, P_2O_5, was recognized long before its molecular formula and structure were discovered. The structure of P_4O_{10} is shown in Figure 20.1. Notice that it is related to the basic P_4 tetrahedron by insertion of an oxygen bridge between phosphorus atoms along each edge of the tetrahedron, plus an additional oxygen at each vertex.

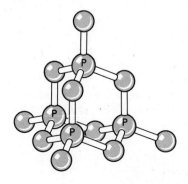

Figure 20.1

The structure of phosphorus(V) oxide, P_4O_{10}.

The most striking property of P_4O_{10} is its strong affinity for water. When placed into water it reacts vigorously and exothermically to form phosphoric acid, H_3PO_4.

$$P_4O_{10} \ (s) + 6H_2O \ (l) \longrightarrow 4H_3PO_4 \ (l)$$

The systematic name of P_4O_{10} is tetraphosphorus decaoxide.

This oxide is such a strong dehydrating agent that it is even able to extract the components of water—two hydrogens and an oxygen—from other molecules. For example, in Chapter 19 we saw that P_4O_{10} is able to remove hydrogen and oxygen from HNO_3 molecules to form N_2O_5.

$$P_4O_{10} + 12HNO_3 \longrightarrow 4H_3PO_4 + 6N_2O_5$$

In the laboratory, P_4O_{10} is often used as a **desiccant**—an agent that removes moisture from air or other gases. When the moist air or gas mixture comes in contact with the P_4O_{10}, the water vapor reacts to form H_3PO_4, thereby leaving the gas virtually moisture-free.

If white phosphorus is burned in a limited supply of oxygen, not all of it is oxidized to P_4O_{10}. About half is oxidized to phosphorus(III) oxide, P_4O_6 (often called *phosphorus trioxide* because its empirical formula is P_2O_3). The structure of P_4O_6, shown in Figure 20.2, is similar to that of P_4O_{10}, except that the oxygen at each apex of the tetrahedron is missing in P_4O_6.

The systematic name of P_4O_6 is tetraphosphorus hexaoxide.

Phosphorus(III) oxide is a crystalline solid that melts at 23.8°C and boils at 175°C. Its vapors are quite poisonous. When stirred with cold water, P_4O_6 reacts to form phosphorus acid, H_3PO_3.

Note the spelling: phosphor<u>us</u> is the name of the element; phosphor<u>ous</u> implies the lower oxidation state of phosphorus in its oxoacids.

$$P_4O_6 \ (s) + 6H_2O \ (l) \longrightarrow 4H_3PO_3 \ (aq)$$

Phosphoric Acid and Phosphates The most important oxoacid of phosphorus, by far, is **phosphoric acid,** H_3PO_4 (also called **orthophosphoric acid**). It is produced in huge amounts—about 10 million tons annually—for use in the manufacture of fertilizers such as ammonium phosphate, as a food additive, and for the production of detergents.

Most phosphoric acid is made directly from phosphate rock by reaction with sulfuric acid.

$$Ca_3(PO_4)_2 \ (s) + 3H_2SO_4 \ (aq) + 6H_2O \longrightarrow 3CaSO_4 \cdot 2H_2O \ (s) + 2H_3PO_4 \ (aq)$$

After reaction, the insoluble calcium sulfate is separated from the mixture by filtration and the phosphoric acid is then concentrated by evaporating water from the solution. This gives a concentrated solution that is about 85% H_3PO_4 by weight. It is known as *syrupy phosphoric acid* because of its viscous nature.

85% of the elemental phosphorus made each year is used to manufacture H_3PO_4.

A purer form of phosphoric acid is obtained from elemental phosphorus by burning the element to give P_4O_{10} and then dissolving the oxide in water. As we saw earlier, P_4O_{10} reacts with water to give H_3PO_4. The phosphoric acid produced in this way is used to make detergents and in food preparations. For example, H_3PO_4 is added to many carbonated beverages to give them tartness.

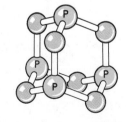

Figure 20.2

The structure of phosphorus(III) oxide, P_4O_6.

The acid is also used to make the acid salt $Ca(H_2PO_4)_2$, which is used in baking powders.

Pure phosphoric acid is a solid at room temperature; its clear, colorless crystals melt at 42.4°C. In water, H_3PO_4 is a weak triprotic acid and is a very poor oxidizing agent. Its structure is

$$H-\overset{\cdot\cdot}{\underset{\cdot\cdot}{O}}-\overset{\overset{\displaystyle :\overset{\cdot\cdot}{O}:}{|}}{\underset{\underset{\displaystyle H}{|}}{\underset{\displaystyle :O:}{|}}}{P}-\overset{\cdot\cdot}{\underset{\cdot\cdot}{O}}-H$$

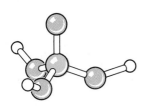

Phosphoric acid

○ = Hydrogen

◐ = Oxygen

◑ = Phosphorus

Neutralization of H_3PO_4 with a base produces three different ions, depending on the amount of base that is added.

$$H_3PO_4 \xrightarrow{OH^-} \underset{\begin{pmatrix}\text{dihydrogen}\\\text{phosphate ion}\end{pmatrix}}{H_2PO_4^{2-}} \xrightarrow{OH^-} \underset{\begin{pmatrix}\text{hydrogen}\\\text{phosphate ion}\end{pmatrix}}{HPO_4^{2-}} \xrightarrow{OH^-} \underset{\begin{pmatrix}\text{phosphate ion}\\\text{[orthophosphate ion]}\end{pmatrix}}{PO_4^{3-}}$$
$$+ \qquad\qquad + \qquad\qquad +$$
$$H_2O \qquad\qquad H_2O \qquad\qquad H_2O$$

It is possible to isolate salts of each of the three anions.

Salts of phosphoric acid have a variety of applications, and one of the most common is in fertilizers. Phosphate rock itself can be pulverized and used as a phosphate fertilizer, but because $Ca_3(PO_4)_2$ has a very low solubility it is able to deliver phosphate only slowly and in small amounts. However, treatment of $Ca_3(PO_4)_2$ with dilute sulfuric acid gives a fertilizer known as *superphosphate*—a mixture of $CaSO_4$ and $Ca(H_2PO_4)_2$.

$$Ca_3(PO_4)_2 + 2H_2SO_4 + 4H_2O \longrightarrow \underbrace{2CaSO_4 \cdot 2H_2O + Ca(H_2PO_4)_2}_{\textbf{superphosphate}}$$

Because $Ca(H_2PO_4)_2$ is water soluble, this mixture is a more effective fertilizer than phosphate rock—hence the name superphosphate.

The anions of phosphoric acid are also found in the blood where they serve as one of the buffer systems. This system consists of the ions $H_2PO_4^-$ and HPO_4^{2-}. If base is added to the blood it is able to react with $H_2PO_4^-$.

$$H_2PO_4^- + OH^- \longrightarrow HPO_4^{2-}$$

and if acid is added, it reacts with HPO_4^{2-}.

$$HPO_4^{2-} + H^+ \longrightarrow H_2PO_4^-$$

These reactions help prevent large changes in the pH of the blood.

Trisodium phosphate, Na_3PO_4, often called TSP, is an effective water softener and cleansing agent. In hard water it reacts with hardness ions such as Ca^{2+}, and Mg^{2+} and Fe^{3+} to form precipitates or complex ions, which removes the ions from solution so they can't interfere with the action of the soap. Solutions of Na_3PO_4 are basic because of the hydrolysis of the PO_4^{3-} ion.

$$PO_4^{3-} + H_2O \rightleftharpoons HPO_4^{2-} + OH^-$$

A 1.0 M Na_3PO_4 solution has a pH of about 12.8. This is very basic, so you should always wear rubber gloves when working with concentrated TSP solutions.

Polymeric Phosphoric Acids and Their Anions Orthophosphoric acid is the parent of a host of more complex acids and anions that contain more than one

$H_4P_2O_7$ is also called
diphosphoric acid.

phosphorus atom. They are formed by eliminating the components of water from −OH groups on neighboring acid molecules and linking together adjacent PO_4 tetrahedra by the mutual sharing of an oxygen atom. For example, when H_3PO_4 is heated to 250°C, **pyrophosphoric acid**, $H_4P_2O_7$, is formed.

$$2H_3PO_4 \xrightarrow{250°C} H_2O + H_4P_2O_7$$

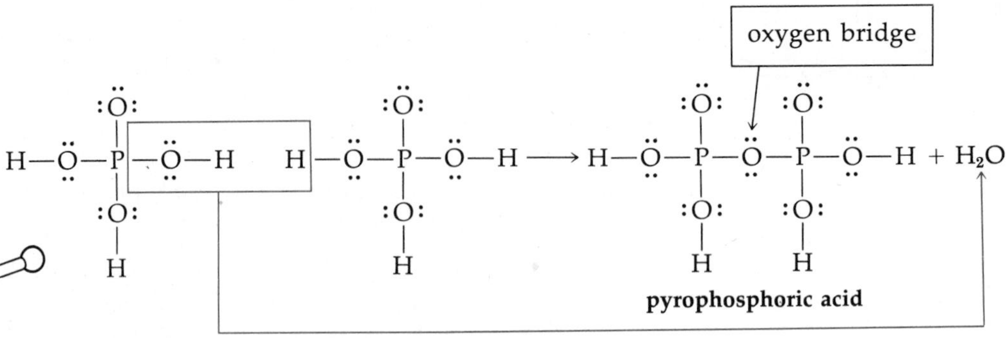

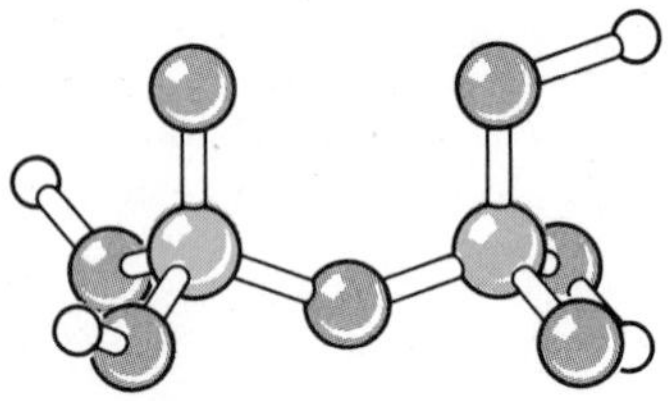

oxygen bridge

pyrophosphoric acid

Pyrophosphoric acid

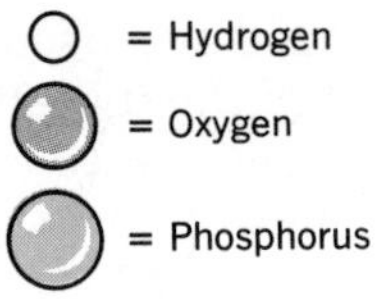

○ = Hydrogen

◑ = Oxygen

◓ = Phosphorus

Pyrophosphoric acid is a colorless, glassy solid that is very soluble in water. It forms salts such as $Na_4P_2O_7$ and $Na_2H_2P_2O_7$. Both the acid and its salts are very slowly hydrolyzed to phosphoric acid or its anions. The reaction is slow enough, however, so that $Na_4P_2O_7$ is used as the phosphate ingredient in many liquid detergents.

If orthophosphoric acid is heated above 400°C, extensive polymerization occurs by elimination of water, as shown below. The product is called **metaphosphoric acid**. Its formula is $(HPO_3)_n$ where n is a large number.

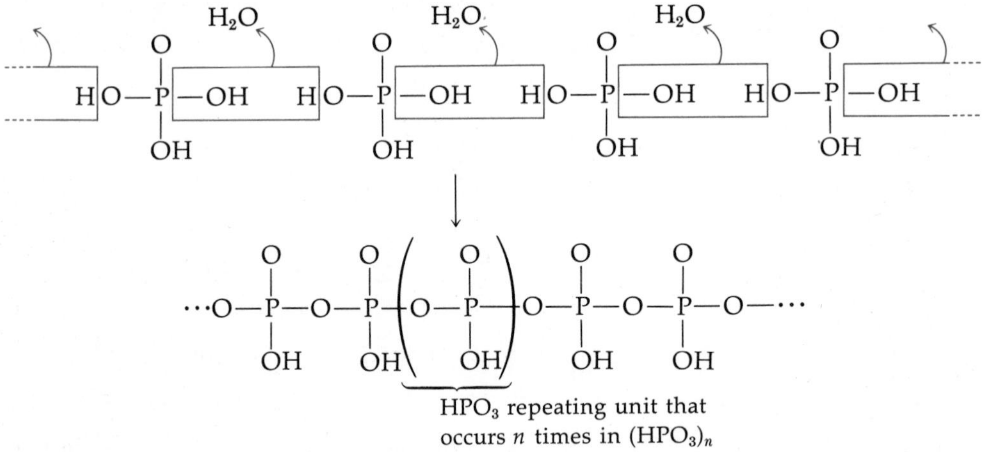

HPO_3 repeating unit that
occurs n times in $(HPO_3)_n$

Hydrolysis adds water to the
P—O—P bridge to regenerate
the pair of P—OH units.

The sodium salt of metaphosphoric acid, which contains the $(PO_3^-)_n$ ion, is normally made by heating sodium dihydrogen phosphate.

$$n \; NaH_2PO_4 \xrightarrow{heat} (NaPO_3)_n + n \; H_2O$$

Polyphosphates of intermediate chain length can also be made. For example,

$$NaH_2PO_4 + 2Na_2HPO_4 \xrightarrow{heat} Na_5P_3O_{10} + 2H_2O$$
**sodium
tripolyphosphate**

Sodium tripolyphosphate is used in many solid detergent mixtures and contains the ion

$$\overset{\overset{\displaystyle :\ddot{O}:}{|}}{\underset{(-) \ :\ddot{O}:}{(-) \ :\ddot{O}—P—\ddot{O}—P—\ddot{O}—P—\ddot{O}: \, (-)}} \quad \overset{\overset{\displaystyle :\ddot{O}:}{|}}{\underset{(-) \ :\ddot{O}:}{}} \quad \overset{\overset{\displaystyle :\ddot{O}:}{|}}{\underset{(-) \ :\ddot{O}:}{}}$$

In many parts of the United States the use of phosphate-based detergents has been banned or severely restricted because of effects that high phosphate concentrations have on lakes. In general, the rate of algae growth in lakes is determined by the nutrient present in the most limited amount. Normally, this limiting nutrient is phosphate, and when soluble phosphates from fertilizers and detergents become washed into lakes, episodes of rapid growth of algae—*algae blooms*—can occur. When the algae die and settle to the bottom they begin to decompose. This causes the waters to be depleted of oxygen, so other marine organisms—fish, for example—begin to die. These processes speed the natural aging or *eutrophication* of the lake and gradually reduce the oxygen level to the point where no aquatic life can survive in the waters.

Phosphorous Acid We learned earlier that **phosphorous acid,** H_3PO_3, is formed when phosphorus(III) oxide is dissolved in water.

$$P_4O_6 \ (s) + 6H_2O \ (l) \longrightarrow 4H_3PO_3 \ (aq)$$

Despite the way its formula is written, H_3PO_3 is only a diprotic acid. Its structure is

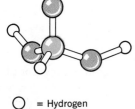

= Hydrogen

= Oxygen

= Phosphorus

$$\begin{array}{c} H \\ | \\ H—\ddot{O}—P—\ddot{O}—H \\ | \\ :\ddot{O}: \end{array}$$

Only the O—H bonds break to give H^+; the hydrogen attached directly to the phosphorus is not acidic. This means that only two kinds of salts can be prepared—for example, NaH_2PO_3 and Na_2HPO_3. Phosphorous acid and phosphites are reasonably good reducing agents. For example, they reduce silver compounds to metallic silver.

Halogen Compounds Phosphorus forms two kinds of binary compounds with the halogens: the trihalides, PX_3 (X = F, Cl, Br, and I) and the pentahalides, PX_5 (X = F, Cl, and Br). The compound PI_5 is not known, presumably because iodine atoms are so large that five of them simply cannot be packed around a phosphorus atom.

The most important halogen compounds of phosphorus are the chlorides, PCl_3 and PCl_5. Their structures are shown in Figure 20.3. Phosphorus tri-

Figure 20.3

The structures of phosphorus trichloride and phosphorus pentachloride.

PCl_3

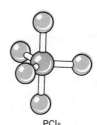

PCl_5

chloride is made by reacting molten phosphorus with chlorine. If an excess of chlorine is present, the pentachloride can also be formed.

$$P_4 \ (l) + 6Cl_2 \ (g) \longrightarrow 4PCl_3 \ (g)$$

$$PCl_3 \ (g) + Cl_2 \ (g) \rightleftharpoons PCl_5 \ (g)$$

The second reaction is an equilibrium—PCl_5 decomposes if heated.

Phosphorus trichloride is a volatile liquid that boils at 76°C. It is used as a starting material for the preparation of many other phosphorus compounds. When exposed to water, PCl_3 hydrolyzes to give phosphorous acid.

Many phosphorus-containing pesticides are made using PCl_3.

$$PCl_3 + 3H_2O \longrightarrow H_3PO_3 + 3HCl$$

In fact, this is the method usually used to make phosphorous acid.

Phosphorus trichloride reacts with oxygen to give **phosphoryl chloride,** $POCl_3$ (also called **phosphorus oxychloride**). About half of the PCl_3 manufactured each year is oxidized to $POCl_3$ and much of that is used to make compounds that are employed as flame retardants.

Phosphorus pentachloride in the vapor or liquid has the trigonal bipyramidal structure shown in Figure 20.3. In the solid, however, PCl_5 appears to exist as an ionic compound composed of tetrahedral PCl_4^+ ions and octahedral PCl_6^- ions (i.e., in the solid, PCl_5 is really $PCl_4^+PCl_6^-$). One of the principal uses of PCl_5 is in the manufacture of $POCl_3$ by the reaction

$$P_4O_{10} + 6PCl_5 \longrightarrow 10POCl_3$$

20.2   SULFUR

The element sulfur is widely distributed in nature, although it is only about half as abundant as phosphorus. It occurs as the free element, and long before sulfur was recognized as an element it was known as *brimstone,* meaning "stone that burns." In the combined state it is found in mineral sulfides (those of lead and copper, for example), in sulfates such as gypsum and epsom salts, and as sulfate ion in the ocean. Sulfur compounds also occur as impurities in natural gas, petroleum, and coal.

Brimstone is mentioned in the Bible.

Somewhat more than half of the sulfur used by industry each year is mined from large underground deposits of the free element, chiefly in Texas and Louisiana. A rather clever method to accomplish this was invented in 1890 by an American engineer, Herman Frasch, and has come to bear his name. It involves pumping superheated water under pressure into the sulfur deposit, which causes the sulfur to melt (Figure 20.4). The hot sulfur-water mixture is then foamed to the surface with compressed air and sprayed into huge piles to dry (Color Plate 19).

Sulfur is also found in Spain, Mexico, Japan, and Italy.

The rest of the industrial supply of sulfur is recovered during the purification of natural gas and petroleum, which contain sulfur compounds such as H_2S that would create pollution problems if burned along with these fuels.

The free element

Sulfur atoms in their ground state have the electron configuration $[Ne] \ 3s^2 3p^4$. As we saw in Chapter 9, these combine with each other by forming two covalent bonds to separate sulfur atoms, and eight member rings are formed. The most stable allotrope of sulfur at room temperature is known as rhombic sulfur in which the S_8 rings are stacked in a way that gives a rhombic crystal structure.

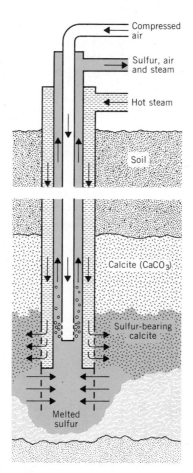

Figure 20.4

Sulfur is extracted from deposits of the free element deep below the surface by the Frasch process.

Crystals of yellow rhombic sulfur are shown in Figure 20.5 and Color Plate 10. A second allotrope of sulfur is known as monoclinic sulfur in which S_8 rings are stacked in a monoclinic crystal structure. Rhombic sulfur melts about 112°C and

Figure 20.5

(a) Crystals of rhombic sulfur, the most stable crystalline form of sulfur at room temperature.
(b) Needlelike crystals of monoclinic sulfur.

(a)   *(b)*

when heated above 120° and then allowed to cool slowly, needlelike crystals of the monoclinic allotrope (mp 119°C) are formed.

The behavior of elemental sulfur as it is heated from its melting point to its boiling point is quite interesting. When it first melts, a yellow, relatively non-viscous liquid is formed. As this is heated it gradually darkens and becomes molasseslike. At still higher temperatures it thins out again and the dark red liquid finally boils at 445°C. These changes are shown in Color Plate 20.

The changes in the sulfur as it is heated are explained as follows. When the sulfur first melts, the liquid is composed of S_8 rings in the usual jumbled arrangement characteristic of liquids in general. As its temperature is raised, thermal energy increases the vibrational motion of the sulfur atoms in the rings, and sulfur-sulfur bonds begin to break. This gives chains of sulfur atoms that have an unpaired electron at each end.

When an end sulfur atom of one chain encounters another from a different chain a covalent bond is formed and an S_{16} chain is produced. Coupling can continue and long chains of S_{24}, S_{32}, S_{40}, and so on are formed that intertwine and cause the liquid to become very viscous. At still higher temperatures the more violent motions of the sulfur atoms cause the long chains to start to break into smaller fragments, and the liquid becomes relatively nonviscous again.

It is interesting to note that if the thickened liquid sulfur is cooled rapidly—for example, by being poured into cold water—the sulfur atoms do not have a chance to rearrange into S_8 rings. As a result, a supercooled liquid called amorphous sulfur is produced that has many of the elastic properties of rubber. When allowed to stand, the S_x chains in the amorphous sulfur gradually revert to the thermodynamically more stable S_8 rings of the rhombic form.

Compounds of Sulfur

Sulfur Dioxide and Sulfurous Acid Elemental sulfur is easily ignited and burns with a blue flame to give **sulfur dioxide,** SO_2, a colorless gas with a choking, irritating odor. If you've ever gotten a whiff of the fumes from an igniting match, you've smelled SO_2—sulfur is one of the substances used in matches. Sulfur dioxide is a nonlinear molecule and therefore is polar. This gives it a relatively high boiling point ($-10°C$) and allows it to be easily liquefied under pressure. For this reason, SO_2 has been used as a gas in refrigeration systems.

Sulfur dioxide is also produced when sulfur compounds are burned and when metal sulfides are heated in air—an important step in the extraction of metals such as lead and copper from their ores. The SO_2 formed in metallurgical process is now recovered, but the SO_2 released into the atmosphere by combustion of high sulfur fuels—both petroleum based and coal—presents a major pollution problem in some areas.

When sulfur dioxide is dissolved in water, in which it is quite soluble, it forms sulfurous acid, a weak diprotic acid.

$$SO_2 \ (aq) + H_2O \rightleftharpoons H_2SO_3 \ (aq)$$

Pure sulfurous acid cannot be isolated—it decomposes into SO_2 and H_2O as its solutions become more concentrated by evaporation of the water. Neutralization of H_2SO_3 by base produces two kinds of salts—for example, Na_2SO_3

Another name for amorphous sulfur is plastic sulfur.

$S \ (s) + O_2 \ (g) \rightarrow SO_2 \ (g).$

resonance structures of SO_2

Heating metal sulfides in air to convert them to their oxides is called smelting and is a common metallurgical process, but the SO_2 that is released must not be allowed to escape into the atmosphere. Here we see the effects of a longtime smelting operation at Copper Hill, Tennessee.

NaHSO₃ is used in the manufacture of paper and pulp.

(sodium sulfite) and $NaHSO_3$ (sodium hydrogen sulfite or sodium bisulfite). These neutralization reactions are easily reversed, and the most convenient method of preparation of SO_2 in the laboratory is by reacting a sulfite or bisulfite with a concentrated strong acid such as H_2SO_4. For example,

$$Na_2SO_3 \ (s) + H_2SO_4 \ (aq) \longrightarrow Na_2SO_4 \ (aq) + H_2O + SO_2 \ (g)$$

Sulfurous acid and sulfites are both rather good reducing agents. On exposure to air they are slowly oxidized to sulfuric acid and sulfates.

Acid Rain The high solubility of SO_2 in water, along with the acidity of its solutions, produces serious environmental problems in locations where significant quantities of high sulfur fuels are burned. As mentioned, combustion of sulfur compounds gives sulfur dioxide as the principle sulfur-containing product. When it rains, the SO_2 is washed from the atmosphere and falls to earth as a dilute solution of sulfurous acid. In cities this can cause structural damage to vehicles, buildings, and statues (Figure 20.6), and in the countryside it can damage trees and cause lakes to become too acidic to support fish. This is a

Figure 20.6

How polluted air and acid rain cause decay is seen in this statue made of Baumberg sandstone at the Herten Castle in Westphalia, West Germany. On the left is its appearance in 1908 after 206 years of exposure to the atmosphere. On the right is its appearance 60 years later after exposure to air pollution produced by European heavy industry.

Rain as acidic as vinegar once fell during a rainstorm in Pitlochry, Scotland, in April, 1974.

particularly serious problem in the northern United States, Canada, and Scandinavia.

One method of removing SO_2 from the exhaust gases of industrial furnaces is to allow it to pass over moist limestone. It is absorbed by the moisture and reacts as follows.

$$CaCO_3 \text{ (s)} + SO_2 \text{ (g)} \longrightarrow CaSO_3 \text{ (s)} + CO_2 \text{ (g)}$$

This is essentially the same reaction that occurs when acid rain falls on limestone and marble building materials.

Sulfur Trioxide and Sulfuric Acid At room temperature sulfur trioxide is a solid consisting of SO_3 units linked together in various complex chain structures. The solid is easily vaporized and in the gas phase SO_3 exists as discrete planar triangular molecules.

Sulfur trioxide is made by the oxidation of sulfur dioxide with oxygen. As noted earlier, this is a highly favorable reaction thermodynamically, as can be seen by its large negative ΔG^0.

$$2SO_2 \text{ (g)} + O_2 \text{ (g)} \longrightarrow 2SO_3 \text{ (g)} \qquad \Delta G^0 = -140 \text{ kJ/mol}$$

The reaction is slow in the absence of a catalyst, which explains the apparent stability of SO_2 in air. However, in the presence of an appropriate catalyst the oxidation of SO_2 occurs quickly. As mentioned in Chapter 12, catalysts used in automobile catalytic converters accelerate this reaction and the SO_3 produced combines with moisture in the exhaust gases to form a mist of H_2SO_4.

$$SO_3 \text{ (g)} + H_2O \text{ (g)} \longrightarrow H_2SO_4 \text{ (l)}$$

The oxidation of SO_2 to SO_3 and its subsequent conversion to sulfuric acid are, by far, industry's most important chemical reactions. The amount of sulfuric acid produced annually—approximately 42 million tons—is more than *twice* that of any other single chemical! About 60% of it is used to treat phosphate rock in the production of phosphate fertilizers and in making ammonium sulfate (also a fertilizer). Sulfuric acid is also used in the refining of petroleum, in the steel industry, in lead storage batteries, and in chemical reactions involved in the manufacture of paints, explosives, plastics, drugs, and many other chemicals.

Most sulfuric acid is made by the **contact process.** The raw material is sulfur, which is first burned in air to form sulfur dioxide. The SO_2 and additional oxygen (in air) are then passed over a vanadium pentoxide (V_2O_5) catalyst, which oxidizes the SO_2 to SO_3. Next, the SO_3 is absorbed into concentrated H_2SO_4, with which it reacts to form **pyrosulfuric acid,** $H_2S_2O_7$.

$$H_2SO_4 \text{ (l)} + SO_3 \text{ (g)} \longrightarrow H_2S_2O_7 \text{ (l)}$$

The H_2SO_4 is used to trap the SO_3 because it is more effective than water. Finally, the $H_2S_2O_7$ is diluted with water, which converts it to sulfuric acid.

$$H_2S_2O_7 \text{ (l)} + H_2O \text{ (l)} \longrightarrow 2H_2SO_4 \text{ (l)}$$

Pure H_2SO_4 is a dense, colorless, viscous liquid that tends to decompose into SO_3 and H_2O when heated. The concentrated H_2SO_4 found in bottles on laboratory shelves is 96% H_2SO_4 by weight (18 M). Dilution of the concentrated

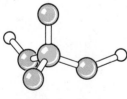

Sulfuric acid

○ = Hydrogen

◔ = Oxygen

◉ = Sulfur

Figure 20.7
Effects of H_2SO_4 on sugar.   (a)   (b)

acid with water is *very* exothermic because of the reaction

$$H_2SO_4 \text{ (conc.)} + H_2O \text{ (l)} \longrightarrow H_3O^+ \text{ (aq)} + HSO_4^- \text{ (aq)}$$

Care should always be taken to add concentrated H_2SO_4 *to the water*—never the other way around. If water is added to the concentrated acid, the heat generated causes the water to boil and the expanding steam spatters the neighborhood with H_2SO_4!

The strong affinity that H_2SO_4 has for water makes the concentrated acid an effective dehydrating agent. For example, if concentrated H_2SO_4 is poured onto sugar, $C_{12}H_{22}O_{11}$, it extracts the components of water and leaves a blackened, charred mass behind (Figure 20.7).

$$C_{12}H_{22}O_{11} \xrightarrow{\text{conc. } H_2SO_4} 12C + 11H_2O$$

Sulfuric acid is a strong diprotic acid. The first step in its dissociation is complete in aqueous solutions, but the second step only occurs to about 10%. This gives a 1.0 M solution of H_2SO_4 an H_3O^+ concentration of approximately 1.1 M.

By controlling the amount of base during the neutralization of H_2SO_4, two series of salts can be formed: sulfates such as Na_2SO_4 and hydrogen sulfates or bisulfates such as $NaHSO_4$. Solutions of bisulfates are quite acidic because the HSO_4^- ion is a relatively strong acid.

Concentrated sulfuric acid is a mild oxidizing agent when cold, but a fairly strong oxidizing agent when hot. For example, cold concentrated H_2SO_4 oxidizes iodide ion to iodine and, to some extent, bromide ion to bromine.

$$2X^- + 3H_2SO_4 \longrightarrow X_2 + SO_2 + 2H_2O + 2HSO_4^- \qquad (X = I \text{ or } Br)$$

Cold H_2SO_4 doesn't attack metallic copper, but the hot concentrated acid does. The reaction, which causes the sulfur to be reduced from the 6+ state to the 4+ state, is

$$Cu + 2H_2SO_4 \xrightarrow{\text{heat}} CuSO_4 + SO_2 + 2H_2O$$

A stronger reducing agent such as zinc reduces the sulfuric acid even further to free sulfur (oxidation number = zero) or even hydrogen sulfide (oxidation number of sulfur = 2−).

Other Compounds of Sulfur Sulfur combines with most metals and nonmetals to

$HSO_4^- \rightleftharpoons H^+ + SO_4^{2-}$

$K_a = 1.2 \times 10^{-2}$

A 0.10 $\underline{M}$ solution of $NaHSO_4$ has a pH of 1.4.

form a large variety of different compounds. We will look only briefly at some of the more important types.

Binary compounds of metals with sulfur are sulfides and contain the sulfide ion, S^{2-}. They can often be prepared by direct reaction of the metal with sulfur. For example, zinc and sulfur react vigorously to produce zinc sulfide.

Metal sulfides tend to be less ionic than their oxides.

$$Zn\ (s) + S\ (s) \longrightarrow ZnS\ (s)$$

Since most metal sulfides are insoluble in water, they can also be formed by precipitation reactions using hydrogen sulfide, H_2S. The use of hydrogen sulfide in separating metal ions by selective precipitation was discussed in Chapter 16.

Hydrogen sulfide itself is a poisonous gas with an odor of rotten eggs. It is more poisonous than carbon monoxide, but its bad odor allows it to be detected at low concentrations. In nature, hydrogen sulfide is released into the air during volcanic eruptions and by the decomposition of organic matter in the absence of air—that's why rotten eggs smell of H_2S. Airborn H_2S is responsible for the gradual tarnishing of silver and the darkening of lead-based paints. In each case, a black metal sulfide (Ag_2S and PbS) is formed.

We learned in Chapter 15 that H_2S is a weak diprotic acid.

In the laboratory, H_2S can be prepared by reacting a metal sulfide with a strong nonoxidizing acid—for example, HCl.

$$FeS\ (s) + 2H^+\ (aq) \longrightarrow Fe^{2+}\ (aq) + H_2S\ (g)$$

Another method, commonly used when one wants to generate the H_2S in an aqueous solution, is the hydrolysis of an organic compound called thioacetamide.

$$\underset{\textbf{thioacetamide}}{CH_3-\overset{\overset{\textstyle S}{\|}}{C}-NH_2\ (aq)} + 2H_2O \longrightarrow H_2S\ (aq) + NH_4^+\ (aq) + \underset{\textbf{acetate ion}}{\left(CH_3\overset{\overset{\textstyle O}{\|}}{C}-O\right)^-(aq)}$$

The advantage of this reaction is that it avoids the release of significant amounts of toxic H_2S into the atmosphere.

The prefix *thio* in a chemical name implies that sulfur replaces oxygen in a compound. For example,

$$\underset{\textbf{acetamide}}{CH_3-\overset{\overset{\textstyle O}{\|}}{C}-NH_2} \qquad \underset{\textbf{thioacetamide}}{CH_3-\overset{\overset{\textstyle S}{\|}}{C}-NH_2}$$

An important oxoanion that fits this nomenclature pattern is the thiosulfate ion, $S_2O_3^{2-}$. It is formed by boiling sulfur with a solution containing sulfite ion.

$$S\ (s) + SO_3^{2-}\ (aq) \longrightarrow S_2O_3^{2-}\ (aq)$$

The Lewis structure for the $S_2O_3^{2-}$ ion is

$$\left[\begin{array}{c} :\overset{..}{\underset{}{O}}: \\ | \\ :\overset{..}{\underset{..}{S}}-\overset{}{\underset{}{S}}-\overset{..}{\underset{..}{O}}: \\ | \\ :\overset{}{\underset{..}{O}}: \end{array}\right]^{2-}$$

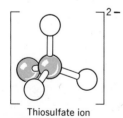

Thiosulfate ion

○ Oxygen

◉ Sulfur

Notice that it can be viewed as a sulfate ion in which one oxygen is replaced by sulfur—hence, *thio*sulfate.

Thiosulfate ion forms quite stable complex ions with metal ions. The one with silver is particularly important in photography. Silver bromide is the usual light-sensitive substance found in photographic films and paper. After an image is developed, the photosensitive emulsion still contains unexposed silver bromide that must be removed to prevent the picture from darkening gradually when viewed in the light. Solutions of sodium thiosulfate—known as *hypo* to photographers—are used to remove the unexposed silver bromide. The equilibria involved are

$$AgBr \ (s) \rightleftharpoons Ag^+ \ (aq) + Br^- \ (aq)$$

$$Ag^+ \ (aq) + 2S_2O_3^{2-} \ (aq) \rightleftharpoons Ag(S_2O_3)_2^{3-} \ (aq)$$

Le Châtelier's principle explains how the thiosulfate works. As $S_2O_3^{2-}$ is added, the second equilibrium shifts to the right and the Ag^+ concentration drops. This causes the AgBr to dissolve in an attempt to restore the Ag^+ concentration to its former value. If sufficient thiosulfate is present, equilibrium is never reestablished and all the AgBr dissolves. Afterwards, the $Ag(S_2O_3)_2^{3-}$ complex ion and residual thiosulfate are washed from the film or paper, leaving only the desired image.

Thiosulfate ion is also a good reducing agent. It reacts with strong oxidizing agents such as chlorine to give sulfate ion.

$$4Cl_2 \ (g) + S_2O_3^{2-} + 5H_2O \longrightarrow 8Cl^- + 2SO_4^{2-} + 10H^+$$

Thiosulfate ion is often used in titrations of the weaker oxidizing agent iodine. Its reaction with iodine is

$$2S_2O_3^{2-} + I_2 \longrightarrow S_4O_6^{2-} + 2I^-$$

thiosulfate   tetrathionate
ion   ion

tetrathionate ion

20.3 THE HALOGENS

Until now we've discussed the nonmetallic elements one at a time because the variations in their properties are rather substantial. For example, although oxygen and sulfur are in the same group and are similar in some respects, they also have many differences, and the same applies to nitrogen and phosphorus. Among the halogens—the Group VIIA elements—many close similarities and trends in properties exist that are easy to follow, so we will discuss them as a group.

The halogens—fluorine, chlorine, bromine, and iodine—are never found free in nature because of their high reactivity. Until humans began their study of chemistry and began to produce the free halogens, these elements occurred almost exclusively in inorganic salts. In fact, the name halogen comes from the Greek *halos*, meaning salt. Today, halogens have been incorporated into many useful organic compounds, ranging from Teflon and polyvinyl chloride plastics to Freon refrigerants and aerosol propellants to insecticides such as DDT. In recent times it has been found that some of these halogenated organic compounds—DDT, chloroform, and polychlorinated biphenyls (PCB's), for example—can have very harmful effects on our environment and its inhabitants—us!

The principal sources of the halogens are their salts. Fluorine is found in the earth's crust principally in deposits of fluorspar, CaF_2, cryolite, Na_3AlF_6,

Astatine, At, is also a halogen, but it is rare and radioactive. We will not include it in our discussions.

Cryolite, you recall, was used by Hall in his process to recover aluminum from Al_2O_3 by electrolysis.

and fluoroapatite, $Ca_5(PO_4)_3F$. The main source of chlorine is NaCl which, as we learned earlier, is recovered from the sea and from large underground deposits formed, presumably, by evaporation of ancient seas. Bromine and iodine are also found in the sea, but their concentrations as Br^- and I^- are *much* less than that of Cl^-. (Sea water contains these concentrations of the halide ions: 0.53 M Cl^-, 8.1×10^{-4} M Br^-, and 5×10^{-7} M I^-). Bromine and iodine are also found in the waters of brine (salt) wells, and iodine is obtained from sodium iodate, $NaIO_3$, that is recovered from deposits of saltpeter, $NaNO_3$, imported from Chile.

Properties of the free elements

Halogen atoms each have seven valence electrons and therefore need only one more to achieve a noble gas configuration. As a result, in the elemental state they form singly bonded diatomic molecules, X_2.

X = F, Cl, Br, or I.

$$:\ddot{X}:\ddot{X}:$$

Some physical properties of the halogens are given in Table 20.1. We see that as their molecules become larger, the melting points and boiling points increase, corresponding to an increase in the strengths of the London forces.

At room temperature fluorine is a pale yellow gas. It is the most reactive of all the elements because of the low F—F bond energy—that is, little energy is needed to split the molecule into very reactive fluorine atoms. This low bond energy is believed to be caused by electron-electron repulsions between the small, compact, electron-rich valence shells of the bonded fluorine atoms.

Chlorine is a pale green gas at room temperature (Color Plate 10). Chlorine molecules have a slightly larger bond energy than F_2 molecules, and chlorine atoms are less electronegative than fluorine. As a result, Cl_2 is somewhat less reactive than F_2.

At room temperature, bromine is a volatile, nonviscous red liquid, and iodine forms dark metallic-looking crystals that easily sublime to give a purple vapor (Color Plate 10). Chemical reactivity decreases from chlorine to bromine to iodine, following the decrease in their electronegativities.

Because of the high electronegativities of the halogens compared to other elements, they tend to gain electrons from other substances and thereby serve as oxidizing agents. The ability of the halogens to serve as oxidizing agents decreases going down the group. As a result we find that a given halogen is able to oxidize the anions of the halogens below it in Group VIIA. Thus F_2 will oxidize Cl^-, Br^-, and I^-, while Cl_2 will oxidize only Br^- and I^- but not F^-. This is illustrated by these typical reactions:

Table 20.1
Some physical properties of the halogens

Halogen	Melting Point (°C)	Boiling Point (°C)	Bond Energy (kJ/mol)	Electronegativity
Fluorine (F_2)	−233	−188	157	4.0
Chlorine (Cl_2)	−103	−34.6	242	3.0
Bromine (Br_2)	−7.2	58.8	193	2.8
Iodine (I_2)	113.5	184.4	150	2.5

Fluorine

$$F_2 + \begin{Bmatrix} 2NaCl \\ 2NaBr \\ 2NaI \end{Bmatrix} \longrightarrow 2NaF + \begin{Bmatrix} Cl_2 \\ Br_2 \\ I_2 \end{Bmatrix}$$

Chlorine

$$Cl_2 + NaF \longrightarrow \text{No reaction}$$

$$Cl_2 + \begin{Bmatrix} 2NaBr \\ 2NaI \end{Bmatrix} \longrightarrow 2NaCl + \begin{Bmatrix} Br_2 \\ I_2 \end{Bmatrix}$$

Bromine

$$Br_2 + \begin{Bmatrix} NaF \\ NaCl \end{Bmatrix} \longrightarrow \text{No reaction}$$

$$Br_2 + 2NaI \longrightarrow 2NaBr + I_2$$

Iodine

$$I_2 + \begin{Bmatrix} NaF \\ NaCl \\ NaBr \end{Bmatrix} \longrightarrow \text{No reaction}$$

Preparation of the free elements

Fluorine Elemental fluorine is such an active oxidizing agent that it can only be made by electrolysis. The raw material is hydrogen fluoride, which is dissolved in molten KF. Electrolysis produces fluorine gas at the anode and hydrogen gas at the cathode. The KF in the mixture serves as an electrolyte because pure hydrogen fluoride is molecular and therefore is nonconducting.

$$2HF \xrightarrow[\text{KF}]{\text{electrolysis}} H_2\,(g) + F_2\,(g)$$

Fluorine is used to make a variety of fluorine-containing organic compounds. Examples are Teflon and the Freons. As we learned in Chapter 19, Freons are chlorofluorocarbons that are used as refrigerants and aerosol propellants, and they may have harmful effects on the Earth's ozone layer as they diffuse into the stratosphere.

Chlorine The production of chlorine by electrolysis both of molten NaCl and of aqueous NaCl (brine) has been described previously. The reactions are

$$2NaCl\,(l) \xrightarrow{\text{electrolysis}} 2Na\,(l) + Cl_2\,(g)$$

$$2Na^+\,(aq) + 2Cl^-\,(aq) + 2H_2O \xrightarrow{\text{electrolysis}} 2Na^+\,(aq) + 2OH^-\,(aq) + Cl_2\,(g) + H_2\,(g)$$

Industrially, chlorine ranks about eighth in total annual tons produced.

Over 12 million tons of chlorine are produced each year. Its major use has been in the manufacture of chemical intermediates—chemicals that are used to make other chemicals. Chlorine is also used to treat municipal drinking water, to make solvents and plastics such as polyvinyl chloride ("vinyl" plastics), and to manufacture pesticides.

In the laboratory chlorine can be made by oxidation of chloride ion in an acidic solution by a strong oxidizing agent such as manganese dioxide, MnO_2, or potassium permanganate, $KMnO_4$. The simplest method is to react concentrated hydrochloric acid with MnO_2.

$$MnO_2\,(s) + 2Cl^-\,(aq) + 4H^+\,(aq) \longrightarrow Mn^{2+}\,(aq) + Cl_2\,(g) + 2H_2O$$

Bromine Although bromide ion occurs in a low concentration in sea water, bromine can be recovered from it by taking advantage of the ease of oxidation of Br^- by chlorine and of the volatility of Br_2. First, chlorine is dissolved in the water and oxidizes bromide ion to bromine.

$$2Br^- \ (aq) + Cl_2 \ (aq) \longrightarrow Br_2 \ (aq) + 2Cl^- \ (aq)$$

Air is then blown through the water and the volatile bromine and residual unreacted chlorine are flushed out. Cooling the air causes the Br_2 to condense to a liquid. Today, most bromine is extracted from brines obtained from wells in Arkansas and Michigan. These contain bromide ion in concentrations that are 50 to 60 times greater than in sea water.

About half the bromine produced each year is used to make ethylene dibromide, $C_2H_4Br_2$, which is an additive in "leaded" gasoline. Its purpose is to prevent deposits of lead compounds from forming inside the engine. During combustion lead bromide, $PbBr_2$, is produced, which is volatile at temperatures that exist in the cylinders and therefore leaves the engine as a gas in the exhaust. Leaded gasoline cannot be used in cars equipped with catalytic converters because the lead compounds would be adsorbed on the catalyst surface and destroy its catalytic activity. Bromine is also used to make silver bromide, the principal ingredient in the light sensitive emulsions on photographic film and paper.

In the laboratory, Br_2 can be made by oxidation of a bromide salt by MnO_2 in an acidic solution (e.g., a solution containing H_2SO_4).

$$MnO_2 \ (s) + 2Br^- + 4H^+ \longrightarrow Mn^{2+} + Br_2 + 2H_2O$$

Iodine Iodide ion is present in very low concentrations in sea water, but seaweed extracts it from the water and concentrates it. Commercial quantities of iodine are recovered from the ashes of burned seaweed, in which the iodide concentrations approach 1%. The I^- is oxidized to I_2 using chlorine or other oxidizing agents.

Another commercial source of iodine is Chilean saltpeter, which contains sodium iodate, $NaIO_3$. The iodate is reduced to free iodine using sodium bisulfite, $NaHSO_3$, as the reducing agent.

$$2IO_3^- + 5HSO_3^- \longrightarrow I_2 + 5SO_4^{2-} + H_2O + 3H^+$$

The recovery of iodine is expensive and its applications are limited. It is used to make various medicinal products (tincture of iodine, for example) and silver iodide, which is also employed in photographic film.

Compounds of the halogens

Binary Halides of Metals Halogen atoms easily gain an electron to form the halide ions—fluoride, chloride, bromide, and iodide—and their compounds with metals are quite common. Most metal halides are ionic, providing the metal is in a low oxidation state. When the metal is in a high oxidation state, polarization of the anion often produces covalently bonded species. For example, in the vapor, aluminum chloride exists as Al_2Cl_6 molecules. Tin(IV) chloride ($SnCl_4$) and titanium(IV) chloride ($TiCl_4$) are both covalent liquids at room temperature.

Hydrogen Halides The binary hydrogen compounds of the halogens have the general formula, HX. They can be prepared by direct combination of the elements

$$H_2 + X_2 \longrightarrow 2HX$$

but the vigor of the reaction varies substantially from fluorine to iodine.

HCl is sometimes made commercially by the reaction

$$H_2 + Cl_2 \longrightarrow 2HCl$$

Fluorine reacts violently with hydrogen as soon as the two gases are mixed. Chlorine and hydrogen react at a much slower rate, however, provided their mixtures are not heated or exposed to ultraviolet light. Light or heat splits Cl_2 molecules into Cl atoms and initiates a chain reaction that is explosively fast.

The chain mechanism for the reaction of H_2 with Br_2 was discussed in Chapter 12.

Chain mechanisms are also involved in the reaction of H_2 with Br_2 and I_2, but the reactions are less vigorous than with chlorine.

The hydrogen halides can also be made from their binary salts by reaction with a nonvolatile acid. For example, hydrogen fluoride is prepared by reacting CaF_2 with sulfuric acid.

$$CaF_2\ (s)\ +\ H_2SO_4\ (l) \longrightarrow CaSO_4\ (s)\ +\ 2HF\ (g)$$

Similarly, HCl is evolved if concentrated H_2SO_4 is added to NaCl.

$$NaCl\ (s)\ +\ H_2SO_4\ (l) \longrightarrow HCl\ (g)\ +\ NaHSO_4\ (s)$$

Additional HCl can be produced by adding more salt to the $NaHSO_4$ and heating the mixture.

$$NaCl\ (s)\ +\ NaHSO_4\ (s) \xrightarrow{\text{heat}} Na_2SO_4\ (s)\ +\ HCl\ (g)$$

Concentrated sulfuric acid is too powerful an oxidizing agent, even when cold, to be used to generate HBr and HI from their salts. Oxidation of the halide ion to free Br_2 or I_2 occurs. Phosphoric acid—a very poor oxidizing agent—can be used in place of sulfuric acid, but the mixtures must be warmed to expel the hydrogen bromide or hydrogen iodide.

Hydrogen fluoride has a substantially higher boiling point than the other hydrogen halides because of extensive hydrogen bonding that produces long staggered chains of HF molecules in the liquid.

	Boiling Points (°C)
HF	+19.7
HCl	−85.1
HBr	−66.8
HI	−35.4

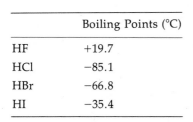 (Dots indicate hydrogen bonds)

Water solutions of the hydrogen halides are the *hydrohalic acids*—hydrofluoric acid, hydrochloric acid, hydrobromic acid, and hydriodic acid. In water, HF is a weak acid, whereas the others are all 100% ionized. Despite being a weak acid, hydrofluoric acid attacks glass and sand. In this case the reaction produces silicon tetrafluoride, SiF_4, a volatile substance that can escape as a gas.

$$SiO_2\ (s)\ +\ 4HF\ (aq) \longrightarrow SiF_4\ (g)\ +\ 2H_2O\ (l)$$
$$\text{(in sand}$$
$$\text{and glass)}$$

Concentrated solutions of HF contain the HF_2^- ion in which a hydrogen ion is shared between two fluoride ions.

$$[F\cdots H\cdots F]^-$$

Solutions of hydrochloric acid can be purchased in hardware stores under the name muriatic acid.

Hydrochloric acid is the most important of the hydrohalic acids because it is relatively inexpensive. It is used to remove rust from iron and steel before the metal is coated with zinc (galvanizing) and in the manufacture of many other chemicals. Hydrochloric acid is also produced in the stomach and helps us digest our foods.

Oxoacids and Oxoanions The halogens form four kinds of oxoacids and anions (Table 20.2). Those of chlorine are the most familiar and the most important.

Table 20.2
Oxoacids and oxoanions of the halogens

Fluorine	Chlorine	Bromine	Iodine
HOF	HOCl	HOBr	HOI
	(OCl^-)	(OBr^-)	(OI^-)
	$HClO_2$	$HBrO_2$	HIO_3
	(ClO_2^-)	(BrO_2^-)	(IO_3^-)
	$HClO_3$	$HBrO_3$	H_5IO_6
	(ClO_3^-)	(BrO_3^-)	$(H_2IO_6^{3-})$
	$HClO_4$	$HBrO_4$	HIO_4
	(ClO_4^-)	(BrO_4^-)	(IO_4^-)

HOCl OCl^-	hypochlorous acid hypochlorite ion	$H\!-\!\ddot{O}\!-\!\ddot{Cl}\!:$
$HClO_2$ (or HOClO) ClO_2^-	chlorous acid chlorite ion	$H\!-\!\ddot{O}\!-\!\ddot{Cl}\!-\!\ddot{O}\!:$
$HClO_3$ (or $HOClO_2$) ClO_3^-	chloric acid chlorate ion	$H\!-\!\ddot{O}\!-\!\underset{}{\overset{:\ddot{O}:}{Cl}}\!-\!\ddot{O}\!:$
$HClO_4$ (or $HOClO_3$) ClO_4^-	perchloric acid perchlorate ion	$H\!-\!\ddot{O}\!-\!\underset{:\ddot{O}:}{\overset{:\ddot{O}:}{Cl}}\!-\!\ddot{O}\!:$

As we learned earlier, the strengths of these acids increase from $HOCl$, which is a weak acid, to $HClO_4$, which is an extremely powerful acid. Recall that the explanation given is that the electron withdrawing effect of the lone oxygens leads to an increased polarization of the O—H bond as the number of lone oxygens becomes larger.

Hypochlorous acid is formed by a disproportionation reaction—a reaction in which the same chemical is both oxidized and reduced—when chlorine is dissolved in cold water.

$$Cl_2\ (aq) + H_2O \rightleftharpoons HOCl\ (aq) + H^+\ (aq) + Cl^-\ (aq)$$

Similar reactions of Br_2 and I_2 occur to much lesser extents.

Approximately 30% of the dissolved chlorine exists in the form of $HOCl$ and Cl^-. When chlorine is dissolved in base, the equilibrium is shifted to the right because the $HOCl$ is neutralized. The net reaction is

$$Cl_2 + 2OH^- \longrightarrow OCl^- + Cl^- + H_2O$$

Electrolysis of a stirred NaCl solution, you recall, produces Cl_2, which reacts with the OH^- formed at the cathode. In this way, the aqueous NaCl is gradually changed to aqueous NaOCl, which is diluted and sold as liquid laundry bleach (e.g., Clorox®).

Reaction of Cl_2 with lime, CaO, produces a calcium salt that is sold as a solid laundry bleach, an algicide for backyard swimming pools, and in antimildew preparations.

$$CaO\ (s) + Cl_2\ (g) \longrightarrow CaCl(OCl)\ (s)$$

HOCl is too unstable to be isolated in pure form.

Hypochlorous acid and its salts are powerful oxidizing agents. This is why hypochlorites are used as bleaches—the oxidation of colored compounds often produces colorless oxidation products. This same strong oxidizing ability also kills bacteria and fungi.

The hypohalites (OCl^-, OBr^-, OI^-) all tend to disproportionate to form the corresponding halides (X^-) and halates (ClO_3^-, BrO_3^-, IO_3^-).

$$3OX^- \longrightarrow XO_3^- + 2X^- \qquad (X = Cl, Br, I)$$

The equilibrium constants for this reaction are large for all three of these halogens; however, the *rates* of disproportionation differ greatly. The reaction of OI^- is very rapid at all temperatures, while OBr^- reacts moderately fast at room temperature. (Solutions of OBr^- ion can be prepared only if they are kept cold.) At room temperature the disproportionation of OCl^- is very slow, so its solutions can be stored for reasonable periods, which explains why they can be sold as liquid bleaches. This is an interesting example of "stability" being determined by the slow rate of reaction rather than by thermodynamics.

Chlorous acid and the chlorite ion are among the lesser important compounds of chlorine. Reaction of chlorine dioxide with base produces ClO_2^- ion.

$$2ClO_2\ (g) + 2OH^-\ (aq) \longrightarrow ClO_2^-\ (aq) + ClO_3^-\ (g) + H_2O$$

Chlorous acid is made from its barium salt by a metathesis reaction using H_2SO_4.

$$2H^+ + SO_4^{2-} + Ba(ClO_2)_2\ (s) \longrightarrow BaSO_4\ (s) + 2HClO_2\ (aq)$$

The reaction proceeds because of the very low solubility of $BaSO_4$. Like HOCl, $HClO_2$ is unstable and decomposes when attempts are made to isolate it in pure form.

The halate ions, XO_3^-, are obtained when Cl_2, Br_2, and I_2 are dissolved in hot basic solutions.

$$3X_2 + 6OH^-\ (aq) \longrightarrow 5X^-\ (aq) + XO_3^-\ (aq) + 3H_2O$$

Iodic acid, HIO_3, is the only halic acid that can be isolated in the pure state.

Chloric acid, like chlorous acid, can be made in solution from its barium salt.

$$H_2SO_4\ (aq) + Ba(ClO_3)_2\ (aq) \longrightarrow BaSO_4\ (s) + 2HClO_3\ (aq)$$

or

$$2H^+\ (aq) + SO_4^{2-}\ (aq) + Ba^{2+}\ (aq) + 2ClO_3^-\ (aq) \longrightarrow BaSO_4\ (s) + 2H^+\ (aq) + 2ClO_3^-\ (aq)$$

Chloric acid is a very powerful oxidizing agent. It is a strong acid, and it too cannot be isolated in pure form.

Perchlorates are commonly prepared by electrolytic oxidation of chlorates. Disproportionation of $KClO_3$ at moderate temperatures in the absence of catalysts also produces $KClO_4$.

$$4KClO_3 \longrightarrow 3KClO_4 + KCl$$

In the presence of a catalyst (MnO_2) or at high temperatures, $KClO_3$ decomposes to KCl and O_2.

Perchloric acid itself can be obtained pure, but it is unstable and tends to explode. In concentrated solutions it is a very strong oxidizing agent, although its dilute solutions, for unknown reasons, have very little oxidizing power.

Other halogen compounds of the nonmetals

In Table 20.3 you will find a list (although not an exhaustive one) of many of the compounds that are formed between the nonmetals and the halogens. The structures of the substances found in this table can, without exception, be predicted on the basis of the valence shell electron-pair repulsion theory.

Table 20.3
Halogen compounds of the nonmetals

Group IIIA	BX_3 (X = F, Cl, Br, I) BF_4^-			
Group IVA	CX_4 (X = F, Cl, Br, I)	SiF_4 SiF_6^{2-} $SiCl_4$	GeF_4 GeF_6^{2-} $GeCl_4$	
Group VA	NX_3 (X = F, Cl, Br, I) N_2F_4	PX_3 (X = F, Cl, Br, I) PF_5 PCl_5 PBr_5	AsF_3 AsF_5	SbF_3 SbF_5
Group VIA	OF_2 (O_2F_2) OCl_2 OBr_2	SF_2 SCl_2 S_2F_2 S_2Cl_2 SF_4 SCl_4 SF_6	SeF_4 SeF_6 $SeCl_2$ $SeCl_4$ $SeBr_4$	TeF_4 TeF_6 $TeCl_4$ $TeBr_4$ TeI_4
Group VIIA	ICl IBr BrF BrCl ClF	ClF_3 BrF_3 ICl_3 IF_3	ClF_5 BrF_5 IF_5	IF_7

In Table 20.3 we see that most of the nonmetals form more than one compound with a given halogen. The number of halogen atoms that can become bound to any particular nonmetal can be related to two factors. One is the electronic structures of the elements that are combined together and the other has to do with the sizes of the atoms.

Each halogen atom contains seven electrons in its valence shell and requires only one more to achieve the stable noble gas configuration. As a result, there is little tendency for them to form multiple bonds with other nonmetals. Furthermore, the halogens ordinarily do not accept electrons in the formation of coordinate covalent bonds because this would mean the addition of two electrons to a valence shell that already contains seven, thereby exceeding the stable octet by one electron.[1]

On this basis, we can divide the halogen compounds into two groups: those that obey the octet rule and those that do not. The compositions of the compounds in the first category are determined by the number of electrons that a given nonmetal requires to reach an octet, since each halogen atom furnishes one electron. For example, in Group VIIA—the halogens themselves—only one electron is needed and only one bond is formed. Thus the halogens are diatomic, and substitution of one halogen for another is possible, as we see for substances such as ClF, BrF, BrCl, BrI, and ICl.

$$:\ddot{Cl}\!:\!\overset{\times\times}{\underset{\times\times}{\ddot{F}}}\!\times$$

In Group VIA each element requires two electrons to reach an octet, so compounds such as OF_2 and SCl_2 are formed, and in Group VA, three electrons are given to the central atom by three halogens in molecules such as NF_3, PF_3,

[1] The halide ions (for example, Cl^-) do *furnish* electron pairs toward the formation of coordinate covalent bonds and are common ligands in complex ions.

AsF_3, and so on. Similarly, the Group IVA elements pick up four electrons from four halogen atoms in compounds such as CCl_4 and $SiCl_4$.

Boron, in Group IIIA, is a special case. Since the boron atom has only three valence electrons, it forms only three ordinary covalent bonds to the halogens. With fluorine, however, BF_3 can add on an additional F^- ion to form the BF_4^- (tetrafluoroborate) anion.

We see that the fourth B—F bond can be considered a coordinate covalent bond (although by now you know that we really cannot distinguish the source of the electrons once the bond has been formed). Since the boron halides have less than an octet of electrons in the valence shell of boron, they are all powerful Lewis acids and the formation of the BF_4^- ion is a typical Lewis acid-base reaction.

In the second category of halogen compounds we have substances in which more than four pairs of electrons surround the central atom. These are limited to those nonmetals beyond the second period, because the second-period elements have a valence shell that can contain a maximum of only eight electrons corresponding to the completion of the $2s$ and $2p$ subshells. The elements below the second period, however, also have in their valence shell a low-energy set of vacant d orbitals as well as the s and p subshells. These d orbitals may be used, through hybridization, to make additional electrons available for bonding.

The second factor influencing the number of halogen atoms that can become bound to a nonmetal is the relative sizes of the different atoms. In Table 20.3 we see that the compounds containing a large number of halogen atoms are formed from nonmetals found toward the bottom of the periodic table. This makes sense because these nonmetals are large and are therefore able to accommodate a relatively large number of bonded atoms with a minimum of crowding. On the other hand, an element near the top of a group is much smaller, so only a relatively few halogen atoms could be expected to be packed about it.

An interesting facet of the chemistry of the nonmetal halides is their reactivity toward compounds containing an —OH group, the most familiar of which is water. Here once again we find that both thermodynamics and kinetics are involved in determining the course of reactions.

The kind of reaction we will focus our attention on is the hydrolysis of the nonmetal halide to produce either the oxoacid, or an oxide, plus the corresponding hydrogen halide. Some examples are the reactions of PCl_5, $SiCl_4$, and SF_4 with water.

$$PCl_5 + 4H_2O \longrightarrow H_3PO_4 + 5HCl$$

$$SiCl_4 + 2H_2O \longrightarrow SiO_2 + 4HCl$$

$$SF_4 + 2H_2O \longrightarrow SO_2 + 4HF$$

These reactions occur very rapidly and proceed to completion with the evolution of considerable amounts of heat. In fact, it is quite common for many halogen compounds of the elements below period 2 to react very rapidly in this same way. For example, the tin(IV) and lead(IV) halides, which are covalent, also hydrolyze in this manner with the formation of a mixture of species including complexes of Sn^{4+} with the halide ion.

There are also some nonmetal halogen compounds that are quite *unreactive* toward water, for example, CCl_4, SF_6, and NF_3. In these cases it is unfavorable kinetics, instead of thermodynamics, that prevents the hydrolysis from taking place.

Consider, for example, the potential hydrolysis reactions of CCl_4 and $SiCl_4$. Calculations based on thermodynamics imply that both should proceed very nearly to completion.

$$SiCl_4 \ (l) + 2H_2O \ (l) \longrightarrow SiO_2 \ (s) + 4HCl \ (aq) \qquad \Delta G^0 = -282 \text{ kJ}$$

$$CCl_4 \ (l) + 2H_2O \ (l) \longrightarrow CO_2 \ (g) + 4HCl \ (aq) \qquad \Delta G^0 = -377 \text{ kJ}$$

In fact, we see from the values of ΔG^0 that the hydrolysis of CCl_4 is even more "spontaneous" than $SiCl_4$. Kinetically, however, the hydrolysis of CCl_4 is essentially prohibited. This is attributed to the absence of a low-energy path for the hydrolysis of CCl_4. Attack by water on the carbon atom of CCl_4 is prevented by the crowding of the Cl atoms. In $SiCl_4$, on the other hand, the larger Si atom provides a greater opportunity for attack and, in addition, the presence of low-energy $3d$ orbitals in the valence shell of the Si atom permits a temporary bonding of the water molecule to the Si atom prior to the expulsion of a molecule of HCl. The mechanism of this hydrolysis is believed to be

Repetition of this process eventually yields $Si(OH)_4$ (orthosilicic acid), which loses water spontaneously to give a hydrated SiO_2.

$$Si(OH)_4 \longrightarrow SiO_2 + 2H_2O$$

SiO_2 is a complex network solid.

The stability of SF_6 and NF_3 toward hydrolysis can also be attributed to the absence of a low-energy reaction mechanism. Like CCl_4, SF_6 should also undergo hydrolysis quite spontaneously, with the value of ΔG^0 for the reaction being tremendous.

$$SF_6 \ (g) + 4H_2O \ (l) \longrightarrow H_2SO_4 \ (aq) + 6HF \ (g) \qquad \Delta G^0 = -423 \text{ kJ}$$

However, the crowding of the fluorine atoms around the sulfur atom apparently prevents attack by water (even up to 500°C), as well as by most other reagents. This crowding is absent with SF_4, and hydrolysis by water is instantaneous.

The resistance of NF_3 toward attack by water cannot be attributed to interference by the fluorine atoms as in SF_6, since the NF_3 molecule is pyramidal, with the nitrogen atom quite openly exposed to an attacking water molecule. We might compare NF_3 with NCl_3, which *does* hydrolyze (if it doesn't explode first—NCl_3 is extremely unstable). The mechanism for this reaction appears to involve the initial formation of a hydrogen bond from water to the lone pair on the nitrogen atom, followed by expulsion of hypochlorous acid.

The ultimate products of the hydrolysis are ammonia and HOCl. This mechanism is not favorable for NF_3 because of its very low basicity. In this case the highly electronegative fluorine atoms draw electron density from the nitrogen atom. As a result, it has been suggested that the lone pair of electrons on nitrogen, in NF_3, may not be available to serve as a point of attachment for the H_2O molecule which, as we see above, is a necessary step in the mechanism proposed for the hydrolysis of NCl_3.

20.4 NOBLE GAS COMPOUNDS

In our discussion of the nonmetals we have not mentioned compounds of the noble gases. These are rather unusual substances because, on the basis of the electronic structure of the noble gases, we would perhaps not have predicted their existence. In fact, until 1962 most chemists firmly believed that these elements were totally incapable of forming compounds (other than several **clathrates,** in which the noble gas atoms are trapped in cagelike sites within a crystalline lattice). For this reason, chemists had referred to them as the *inert gases*. Today they are spoken of as the noble gases in recognition of the fact that although some do react, they nevertheless possess a very low degree of reactivity.

The first real chemistry of the noble gases was discovered in 1962 by Neil Bartlett at the University of British Columbia. He had found that molecular oxygen, O_2, reacts with PtF_6 to form an orange-red compound, O_2PtF_6, containing the ion, O_2^+. Since the ionization energies of O_2 and Xe are nearly the same (1210 and 1170 kJ/mol, respectively), he reasoned that Xe should react in the same way O_2 does. When he reacted Xe with PtF_6 he isolated a yellow compound, containing Xe, that was formulated as $XePtF_6$.

After the initial report by Bartlett, it was not long before chemists at Argonne National Laboratory found that Xe also reacts directly with fluorine at elevated temperatures. This reaction yields a series of fluorides: XeF_2, XeF_4, and XeF_6. Other reactions and compounds were soon discovered and a partial list of the known Xe compounds is given in Table 20.4. The oxides and oxofluorides result from the hydrolysis of the fluorides

$$XeF_6 + 3H_2O \longrightarrow XeO_3 + 6HF$$

$$XeF_6 + H_2O \longrightarrow XeOF_4 + 2HF$$

Table 20.4
Some compounds of xenon

	Melting Point (°C)	Physical Form
Fluorides		
XeF_2	140	Colorless crystals
XeF_4	114	Colorless crystals
XeF_6	47.7	Colorless crystals
Oxides		
XeO_2	Explodes	Colorless crystals
XeO_4	Explodes	Colorless gas
Salts		
$XePtF_6{}^a$	—	Red-orange crystals
$CsXeF_7$	Decomp $> 50°C$	Colorless solid
Cs_2XeF_8	Decomp $> 400°C$	Yellow solid

^{a} Since shown to be more complex; $Xe(PtF_6)_x$, where x lies between 1 and 2.

Some of these compounds are quite unstable and tend to decompose. This is particularly true for the oxides XeO_3 and XeO_4, which explode (XeO_3 has $\Delta H_f^0 = +400$ kJ/mol). Others, on the other hand, appear quite stable. For example, Cs_2XeF_8 does not decompose even when heated to 400°C, and the fluorides have moderately high melting points, suggesting a modest degree of thermal stability.

The structure and bonding in these compounds is quite interesting. Since Xe has four pairs of electrons in its valence shell, corresponding to a completed $5s$ and $5p$ subshell, unpairing of electrons and expansion of the octet must occur to provide unpaired electrons for bonding. Let us consider XeF_2 and XeF_4.

The electronic structure of Xe can be represented as

$$\text{Xe} \quad \underset{5s}{\underline{\uparrow\downarrow}} \qquad \underset{5p}{\underline{\uparrow\downarrow}\ \underline{\uparrow\downarrow}\ \underline{\uparrow\downarrow}} \qquad \underset{5d}{\underline{\ }\ \underline{\ }\ \underline{\ }\ \underline{\ }\ \underline{\ }}$$

In order to form XeF_2, one electron must be promoted to the $5d$ subshell, followed by hybrid orbital formation. The smallest hybrid set that will accommodate all the electrons is sp^3d.

$$\text{Xe} \quad \underset{5s}{\underline{\uparrow\downarrow}} \qquad \underset{5p}{\underline{\uparrow\downarrow}\ \underline{\uparrow\downarrow}\ \underline{\uparrow}} \qquad \underset{5d}{\underline{\uparrow}\ \underline{\ }\ \underline{\ }\ \underline{\ }\ \underline{\ }}$$

gives

$$\text{Xe} \quad \underset{sp^3d}{\underline{\uparrow\downarrow}\ \underline{\uparrow\downarrow}\ \underline{\uparrow\downarrow}\ \underline{\uparrow}\ \underline{\uparrow}} \qquad \underset{\text{unhybridized } 5d}{\underline{\ }\ \underline{\ }\ \underline{\ }\ \underline{\ }}$$

The two unpaired electrons can now be used in bonding to fluorine to give

$$\text{Xe (in XeF}_2\text{)} \quad \underset{sp^3d}{\underline{\uparrow\downarrow}\ \underline{\uparrow\downarrow}\ \underline{\uparrow\downarrow}\ \underline{\uparrow\downarrow}\ \underline{\uparrow\downarrow}} \qquad \underset{\text{unhybridized } 5d}{\underline{\ }\ \underline{\ }\ \underline{\ }\ \underline{\ }}$$

(Colored arrows represent fluorine electrons)

In Chapter 5 we saw that the sp^3d hybrids point to the vertices of a trigonal bipyramid. In terms of the valence shell electron-pair repulsion theory, these five electron pairs will also be situated in this fashion, and from our rules (presented on p. 146), we expect that the three lone pairs will locate themselves in the triangular plane with the fluorine atoms above and below (Figure 20.8). The XeF_2 molecule is therefore linear.

In the case of XeF_4 we must provide four unpaired electrons for bonding to fluorine. This requires promotion of two electrons to the $5d$ subshell and the formation of sp^3d^2 hybrid orbitals.

$$\text{Xe} \quad \underset{5s}{\underline{\uparrow\downarrow}} \qquad \underset{5p}{\underline{\uparrow\downarrow}\ \underline{\uparrow}\ \underline{\uparrow}} \qquad \underset{5d}{\underline{\uparrow}\ \underline{\uparrow}\ \underline{\ }\ \underline{\ }\ \underline{\ }}$$

gives

$$\text{Xe} \quad \underset{sp^3d^2}{\underline{\uparrow\downarrow}\ \underline{\uparrow\downarrow}\ \underline{\uparrow}\ \underline{\uparrow}\ \underline{\uparrow}\ \underline{\uparrow}} \qquad \underset{\text{unhybridized } 5d}{\underline{\ }\ \underline{\ }\ \underline{\ }}$$

Finally, bonding with fluorine gives

$$\text{Xe (in XeF}_4\text{)} \quad \underset{sp^3d^2}{\underline{\uparrow\downarrow}\ \underline{\uparrow\downarrow}\ \underline{\uparrow\downarrow}\ \underline{\uparrow\downarrow}\ \underline{\uparrow\downarrow}\ \underline{\uparrow\downarrow}} \qquad \underset{\text{unhybridized } 5d}{\underline{\ }\ \underline{\ }\ \underline{\ }}$$

In XeF_4 there are six electron pairs about the Xe. Both valence bond theory, with its sp^3d^2 hybrids, as well as the VSEPR theory predict that these are directed toward the corners of an octahedron. As we have seen before, the two lone pairs occupy positions on opposite sides of a square plane containing the four ligand atoms, so that XeF_4 has a square planar structure (Figure 20.9).

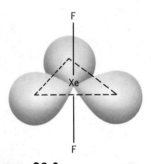

Figure 20.8

Molecular structure of XeF_2.

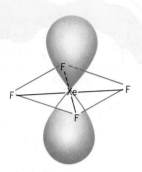

Figure 20.9
Molecular structure of XeF₄.

Since the initial discovery of noble gas compounds by Bartlett, three noble gases have been demonstrated to form compounds, Rn, Xe, and Kr. The lighter ones—helium, neon, and argon—do not appear able to form chemical compounds because of their much higher ionization energies.

Bartlett's work on Xe has taught chemists an important lesson. So firmly convinced were they that the noble gases were totally unreactive that after some initial experiments attempting to react Xe with fluorine had failed in the 1930s, no further efforts were made to explore the possibility that they were not inert. It is interesting that the noble gas compounds obtained do not present any particular problem in bonding. In fact, many of the interhalogen compounds that had already been found (for example, BrF_5, ICl_4^-, IF_7) have the same number of electrons in the valence shell of the central atom as do the noble gas compounds. The same concepts that we applied to other compounds in which the octet was exceeded, therefore, work with the noble gas compounds too. The failure by chemists to recognize the possibility that the noble gases might react reflects a "blind spot" in their thinking that was probably founded in an overzealous acceptance of the stability and inertness of the ns^2np^6 octet of electrons.

20.5 SILICON

Silica rock (top). Silicon (bottom). The silicon is reduced from the mineral silica through reaction with carbon in an electric arc furnace.

The second most abundant element in the Earth's crust is silicon. It constitutes approximately 27.7% of the crust, by mass, where it occurs in rocks of various kinds in the form of silicates—silicon-oxygen compounds. The free element is normally obtained by reduction of silicon dioxide (sand) using carbon in an electric furnace. The reaction is

$$SiO_2 \ (s) + 2C \ (s) \xrightarrow{\text{heat}} Si \ (s) + 2CO \ (g)$$

Silicon is a dark, metallic looking solid that melts at 1410°C. In Chapter 9 we learned that it is a semiconductor, a fact that has made possible the fantastic world of miniature electronic devices such as pocket calculators and microcomputers. The silicon used in these devices must be extremely pure. It is usually made from a tetrahalide such as silicon tetrachloride ($SiCl_4$) by high temperature reduction with hydrogen.

$$SiCl_4 \ (g) + 2H_2 \ (g) \longrightarrow Si \ (s) + 4HCl \ (g)$$

The silicon is then further purified by an interesting process called **zone refining** (Figure 20.10). This method is based on the fact that when a solution freezes, the crystal lattice of the solid formed does not accommodate the impu-

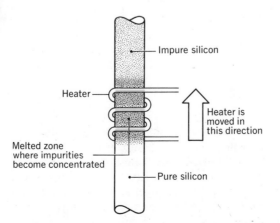

Impure silicon

Heater

Heater is moved in this direction

Melted zone where impurities become concentrated

Pure silicon

Figure 20.10
Zone refining.

rities very well, so the impurities tend to become concentrated in the remaining solution. In zone refining, a bar of silicon is placed in a device that melts a thin cross-sectional wafer of the solid near one end. The heat source is then gradually moved toward the other end of the bar, and as it moves the melted zone follows along. Behind this molten section, the silicon solidifies in a very pure crystalline state, while the impurities collect in the molten band. In this way the impurities are brought to one end of the bar. After the procedure is repeated several times the impure end is cut off and discarded. The rest of the bar consists of very pure silicon with impurity levels ranging from a few parts per million to as little as a few parts per billion.

Chemical properties

Silicon is found in Group IVA of the periodic table, beneath carbon. Its atoms in their ground state have the electron configuration [Ne] $3s^2 3p^2$. Unlike carbon, silicon has virtually no tendency to form pi bonds by overlap of p atomic orbitals ($p\pi$—$p\pi$ bonds). As a result, silicon only forms single bonds. In the free element, each silicon atom completes its octet by bonding to four separate silicon atoms located at the corners of a tetrahedron, and, as we saw in Chapter 9, elemental silicon crystallizes in a diamond-type lattice.

Silicon is not a very reactive element at room temperature. It is unaffected by acids, but it does dissolve in hot basic solutions (NaOH, for example) to give hydrogen plus a mixture of silicates.

$$Si\ (s) + 4OH^-\ (aq) \longrightarrow SiO_4^{4-}\ (aq) + 2H_2\ (g)$$
(plus other silicates)

At high temperatures silicon also combines with halogens to produce the tetrahalides (for example, $SiCl_4$) and with hydrogen to form hydrides called silanes (SiH_4, Si_2H_6, etc.). The hydrides are not very stable, and the ability of silicon atoms to link to each other in these compounds is limited. The longest chain to be observed is in Si_6H_{14}, which tends to decompose to SiH_4.

Compounds with silicon oxygen bonds

Silicon has a strong affinity for oxygen because it forms very stable silicon-oxygen bonds. In fact, all the naturally occurring silicon compounds are silicates in which the basic structural unit is the SiO_4 tetrahedron. Although some are rather complex, the structures of these naturally occurring silicates can be understood by considering them as polymers of more basic silicate units.

The simplest of the silicates contain the *orthosilicate ion*, SiO_4^{4-}, which is the anion of orthosilicic acid.

Basic SiO_4 tetrahedron

orthosilicic acid
(cannot be isolated)

orthosilicate ion

An example of a substance containing the SiO_4^{4-} ion is the gem *zircon*, which consists of crystals of $ZrSiO_4$ (Color Plate 21).

In Section 20.1 we saw that phosphoric acid units are able to be polymerized by the removal of the components of water from a pair of —OH groups on neighboring molecules and the formation of an oxygen bridge between the

Notice that in the various silicates discussed in this section each nonbridging oxygen carries a negative charge.

phosphorus atoms. Among the silicates this kind of polymerization also occurs, and as the pH of a solution of sodium silicate, Na_4SiO_4, is lowered the SiO_4 units become joined by similar oxygen bridges. For example, the pyrosilicate ion, $Si_2O_7^{6-}$, can be considered to be formed as follows:

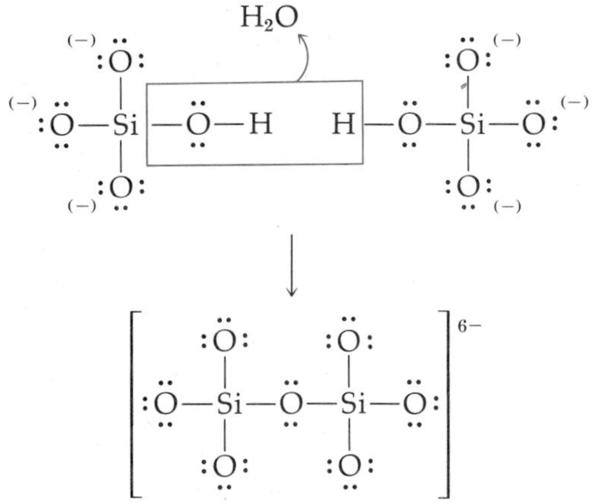

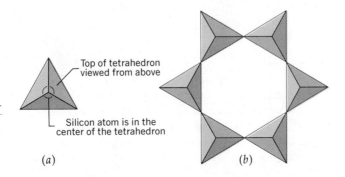

In this anion a pair of SiO_4 tetrahedra share a corner in common (Figure 20.11). In nature, the $Si_2O_7^{6-}$ ion is found in the mineral $Sc_2Si_2O_7$, called thortveitite.

When one SiO_4 tetrahedron shares *two* of its corners with other SiO_4 tetrahedra, rings and long chain structures are able to be formed. For example, the polymeric anion $Si_6O_{18}^{12-}$, shown in Figure 20.12, is present in beryl, $Be_3Al_2(Si_6O_{18})$, which forms crystals of gem quality that are known as emeralds (Color Plate 13).

Linking the SiO_4 tetrahedra in long chains gives huge polymeric anions having the empirical formula SiO_3^{2-}.

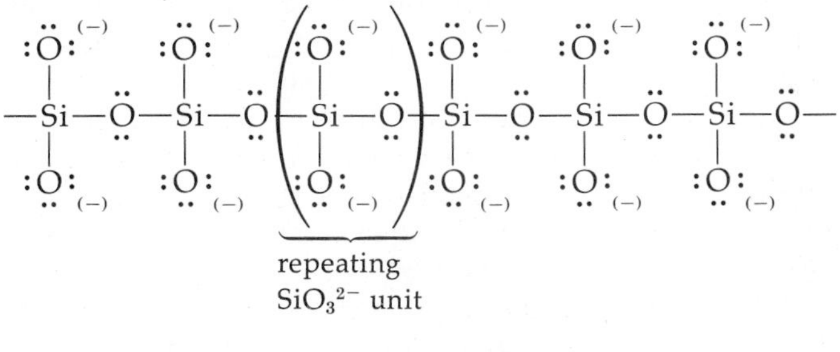

Figure 20.11

The structure of the pyrosilicate anion, $Si_2O_7^{6-}$. Colored spheres are silicon atoms. Each nonbridging oxygen atom carries a negative charge.

The staggered structure of an $(SiO_3)_x^{2x-}$ chain is shown in Figure 20.13. As you might expect, minerals such as $MgSiO_3$ and $LiAl(SiO_3)_2$, which contain these "linear" SiO_3^{2-} chains, have a fibrous appearance. In addition to the simple

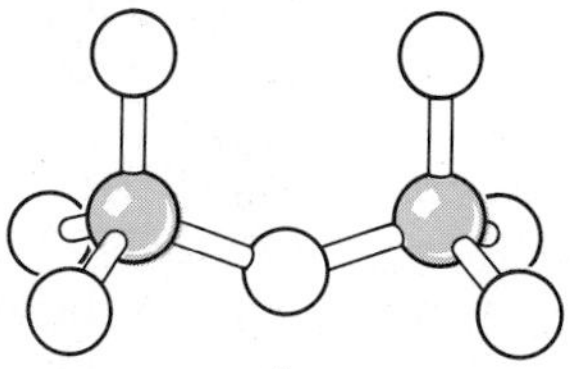

Figure 20.12

(a) A representation of the SiO_4 tetrahedron as viewed from above. (b) The structure of the $Si_6O_{18}^{12-}$ ion showing the SiO_4 tetrahedra linked by oxygen bridges at their corners.

Top of tetrahedron viewed from above

Silicon atom is in the center of the tetrahedron

(a) (b)

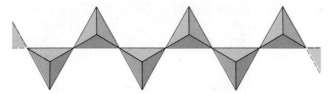

Figure 20.13

The staggered arrangement of SiO_4 tetrahedra in a "linear" $(SiO_3^{2-})_x$ metasilicate chain.

strands of SiO_4 tetrahedra, double strands, formed by the sharing of three corners by *every other* SiO_4 unit, are also found. This time, an infinite *double* chain results, a small segment of which is illustrated in Figure 20.14. Once again, each unshared oxygen atom carries a negative charge, and the repeating unit along the chain is $Si_4O_{11}^{6-}$. This anion is found in asbestos, $[Ca_2Mg_5(Si_4O_{11})_2(OH)_2]_x$, and, as we might expect, asbestos has a fiberlike nature because of the presence of the long $(Si_4O_{11}^{6-})_x$ chains that line up more or less parallel to one another.

Asbestos has many useful properties. Its fibrous nature makes it excellent as a reinforcing filler in brake and clutch linings. Because it is fireproof it is spun and woven into fabrics used in fireproof curtains and ironing board covers. It is used in insulation and sealers in cars, trucks, and planes. Once again, however, nature takes as well as gives. Asbestos fibers breathed into the lungs have been implicated in many cases of lung cancer. There is also evidence that if consumed in food it can produce stomach cancer.

Still another type of silicate is formed if *each* SiO_4 tetrahedron shares three of its corners so that each Si is attached to three others by oxygen bridges. When this occurs a planar sheet of SiO_4 units results. A portion of one of these is shown in Figure 20.15 with the repeating unit, $Si_2O_5^{2-}$, outlined by the rectangle.

A number of different minerals are known to contain these $(Si_2O_5^{2-})_x$ sheets. They differ in the way the silicate layers are stacked, and the nature of the cations and other anions that are also present in the structure: however, all have certain similarities to each other. Some examples are talc (used in bath powder) and soapstone, in which the $(Si_2O_5^{2-})_x$ sheets are packed together with cations in such a way that there is a minimum of attractive forces between successive layers. These layers therefore slide over each other easily and both these minerals feel slippery.

In other related minerals there is substitution of another element for Si. In mica, for instance every fourth Si atom is replaced by an Al^{3+} ion. The properties of mica are therefore different from talc; however, the layer structure is still apparent. A mica-type material you may be familiar with is *vermiculite*, used in

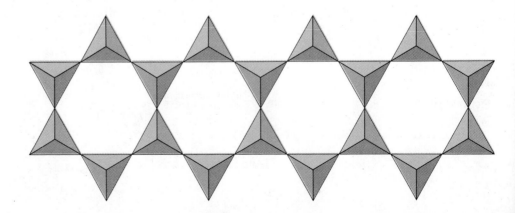

Figure 20.14

Linear double chain of the $(Si_4O_{11}^{6-})_x$ anion found in asbestos.

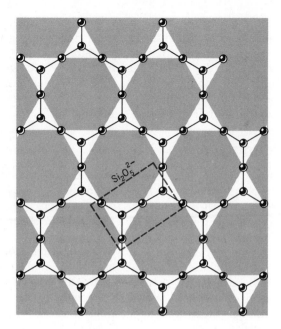

Figure 20.15
Planar sheet silicate formed by sharing three corners on every SiO_4 tetrahedron.

Figure 20.16
Quartz crystals.

place of soil in propagating house plants and as a cushioning filler in packaging items for shipment. This solid flakes into thin flat layers characteristic of mica.

When Si finally shares all four of its oxygen atoms with other Si atoms, a three-dimensional framework is produced that has the empirical formula SiO_2. Silicon dioxide, also called silica, sometimes occurs in large beautiful crystals of quartz (Figure 20.16), although most people are more familiar with quartz in a finer state of subdivision—sand.

Quartz is composed of spiral chains of SiO_4 tetrahedra that are linked to each other. Since the spiral can twist either clockwise or counterclockwise (Figure 20.17), two types of crystals are produced that are exactly alike in all but one

Figure 20.17
Left- and right-handed helixes in quartz. The spiral chains of SiO_2 tetrahedra can twist in either of two directions. This gives two kinds of quartz crystals known as d-quartz and l-quartz.

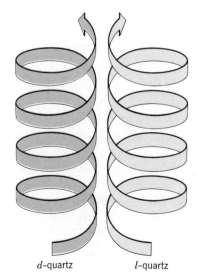

d-quartz   *l*-quartz

respect. They differ in the same way that your left and right hands differ. If you examine your two hands with their palms facing you, the thumbs point in opposite directions—otherwise they are essentially the same. That's why a left-handed glove doesn't fit your right hand. The two kinds of quartz crystals have a similar "handedness," and we will see later that this phenomenon extends to certain individual molecules as well. The topic of molecular "handedness" is discussed further in the next chapter.

Silicones

Organic polymers, such as polyethylene, nylon, and the polyesters, have many desirable physical properties that make them ideally suited for packaging materials and the production of synthetic fibers used in clothing. There are drawbacks, however. They are generally flammable and they do not stand up well to high temperatures.

Inorganic polymers—polymers in which atoms other than carbon make up the primary chain or "backbone" of the molecules—often do not suffer from these disadvantages. A type of inorganic polymer that has achieved widespread use are the silicones. These are polymers in which the backbone is made up of alternating silicon and oxygen atoms. Their thermal stability arises from the strength and inertness of Si—O bonds. They are formed by hydrolysis of compounds such as $(CH_3)_2SiCl_2$. The formation of silicone polymers can be thought of as proceeding by formation of a hydroxy intermediate that then eliminates water.

Silicones have many medical applications. For example silicone oils are used to lubricate the skin of burn victims.

$$\underset{\underset{CH_3}{|}}{\overset{\overset{CH_3}{|}}{Cl-Si-Cl}} + 2H_2O \longrightarrow \underset{\underset{CH_3}{|}}{\overset{\overset{CH_3}{|}}{HO-Si-OH}} + 2HCl$$

Depending on the chain length, and the degree to which chains may be cross-linked to one another, the silicones may be oils, greases, or rubbery solids. They are useful in waterproofing garments. In low-temperature applications the oils remain fluid (hydrocarbon oils become very viscous), and the rubbers retain their elastic properties (ordinary rubber becomes brittle). Silicones are also unaffected by hydrocarbon solvents and greases that soften ordinary rubber and they are not attacked by ozone in the air, which causes ordinary rubber to crack.

Water drops on fabric

INDEX TO QUESTIONS

REVIEW QUESTIONS

20.1 What is the formula for the phosphorus-containing compound in phosphate rock?

20.2 What kinds of roles does phosphorus play in living systems?

20.3 Give the electron configuration of (a) phosphorus, (b) sulfur, (c) fluorine, (d) chlorine, and (e) silicon.

20.4 What is the structure of the molecules found in white phosphorus? Why is white phosphorus so reactive?

20.5 How is white phosphorus made? Give a chemical equation.

20.6 What is an *electric furnace?*

20.7 Describe red phosphorus and black phosphorus. How do their reactivities compare to white phosphorus?

20.8 Write a chemical equation for the reaction of sodium with phosphorus? What happens when the product of this reaction is placed in water?

20.9 Write chemical equations for the reaction of white phosphorus with oxygen (a) when the O_2 is present in excess and (b) when there is a limited amount of O_2.

20.10 Sketch the structures of the molecules found in (a) phosphorus(V) oxide and (b) phosphorus(III) oxide.

20.11 Write equations for the reaction of water with (a) phosphorus(V) oxide and (b) phosphorus(III) oxide.

20.12 What is a *desiccant?* Why is P_4O_{10} a good desiccant?

20.13 Give a chemical equation showing how phosphoric acid is made directly from phosphate rock.

20.14 What is *syrupy phosphoric acid?*

20.15 How is the phosphoric acid used in food products manufactured?

20.16 Give three commercial uses of phosphoric acid.

20.17 Give the formulas and names of three salts that can be formed from $Mg(OH)_2$ and phosphoric acid.

20.18 What is orthophosphoric acid?

20.19 What ions make up the phosphate buffer in the blood? How do they serve to help maintain a nearly constant blood pH?

20.20 What are some uses of TSP? Give a chemical equation to explain why its solutions are basic.

20.21 Give Lewis structures for H_3PO_4 and H_3PO_3.

20.22 What is superphosphate fertilizer? Give a chemical equation showing how it is made.

20.23 Why is phosphate rock itself a poor phosphate fertilizer?

20.24 Use Lewis structures to illustrate how pyrophosphoric acid is formed from phosphoric acid.

20.25 What is the basic structural unit in the polymeric phosphoric acids and polyphosphates?

20.26 What is the empirical formula for the metaphosphate ion? Give its Lewis structure and show how it can be considered to be formed from phosphoric acid.

20.27 Give the structure of the tripolyphosphate ion. What uses does it have?

20.28 What mole ratio of NaH_2PO_4 and Na_2HPO_4 should be chosen to obtain a linear polyphosphate containing five phosphorus atoms?

20.29 Explain why phosphate pollution can be very harmful to lakes.

20.30 What is the name of the acid, H_3PO_3? What salts can be formed by reacting it with $Mg(OH)_2$?

20.31 How do H_3PO_4 and H_3PO_3 compare as oxidizing and/or reducing agents?

20.32 Sketch the structures of PCl_3 and PCl_5. How does PCl_5 exist in the solid state? What kinds of hybrid orbitals does phosphorus use in PCl_3 and PCl_5?

20.33 What is *brimstone?* What does this name mean?

20.34 How does the element sulfur occur in nature?

20.35 What are the two allotropic forms of sulfur? How do they differ? Outline the physical changes that take place when sulfur is gradually heated to its boiling point and relate them to the structural changes that occur.

20.36 What is plastic sulfur?

20.37 Describe the Frasch process for mining sulfur.

20.38 Which air pollutant is produced by the combustion of high-sulfur fuels?

20.39 Write chemical equations for the production of H_2SO_4 using O_2, S, and H_2O as starting materials.

20.40 Why is SO_2 stable toward oxidation to SO_3 in the air?

20.41 How can SO_2 be conveniently prepared in the laboratory?

20.42 Write chemical equations for the reaction of water with (a) sulfur dioxide and (b) sulfur trioxide.

20.43 Describe *acid rain*—how it is formed and what damage it does.

20.44 How do the acidities of H_2SO_3 and H_2SO_4 compare? How can the differences be explained?

20.45 List four uses for sulfuric acid.

20.46 Draw the resonance structures for and describe the shapes of SO_2 and SO_3.

20.47 What is the structure of pyrosulfuric acid? (*Hint:* What is the structure of pyrophosphoric acid?)

20.48 What is the proper way to prepare dilute H_2SO_4 from concentrated H_2SO_4?

20.49 Even though H_2SO_4 isn't fully dissociated into H^+ and SO_4^{2-}, it is considered a strong acid. Why?

20.50 Write an equation showing the dehydrating action of concentrated H_2SO_4 on the sugar, glucose, $C_6H_{12}O_6$.

20.51 The cyanate ion has the Lewis structure $\left[:N\equiv C-\ddot{O}:\right]^-$. What would be the structure of the thiocyanate ion?

20.52 Write an equation for the generation of H_2S by the hydrolysis of thioacetamide.

20.53 Why are hydrogen sulfide fumes to be avoided in the laboratory?

20.54 What are the names of these salts?
(a) Na_2SO_3 (c) Na_2HPO_4
(b) $NaHSO_4$ (d) NaH_2PO_4

20.55 What is the structure of the thiosulfate ion? Write a chemical equation showing how thiosulfate ion is made.

20.56 Explain the function of sodium thiosulfate, $Na_2S_2O_3$, in photography.

20.57 Write balanced chemical equations for the oxidation of $S_2O_3^{2-}$ by
(a) Cl_2
(b) I_2

20.58 What products, if any, are formed in the following reactions?
(a) cold concentrated H_2SO_4 and NaI
(b) hot concentrated H_2SO_4 and copper
(c) hot concentrated H_2SO_4 and zinc
(d) cold dilute H_2SO_4 and zinc
(e) cold dilute H_2SO_4 and copper

20.59 How are the halogens normally found in nature?

20.60 What are the principal sources of each of the halogens?

20.61 How is fluorine prepared?

20.62 Describe how you could prepare chlorine in the laboratory. How is chlorine made industrially? What are three commercial uses for chlorine?

20.63 Describe the physical characteristics of the halogens.

20.64 Complete and balance the following equations. If no reaction occurs, write N.R.
(a) $Cl_2 + KI \rightarrow$
(b) $F_2 + KBr \rightarrow$
(c) $I_2 + NaCl \rightarrow$
(d) $Br_2 + NaI \rightarrow$

20.65 Why is fluorine so reactive?

20.66 How is bromine obtained industrially? Give two principal uses of Br_2. How can Br_2 be made in the laboratory? (Give a balanced equation.)

20.67 What are the commercial sources of iodine?

20.68 How do the reactivities of the halogens toward hydrogen compare?

20.69 How is HF made?

20.70 How would you prepare HCl in the laboratory? What are two commercial uses of hydrochloric acid?

20.71 How would you prepare HBr and HI in the laboratory?

20.72 Why is glass attacked by hydrofluoric acid?

20.73 Compare the boiling points of the hydrogen halides. How do HF molecules exist in liquid HF?

20.74 Name these compounds.
(a) HOBr   (e) HIO_4
(b) NaOCl   (f) $HBrO_3$
(c) $KBrO_3$ (g) $NaIO_3$
(d) $Mg(ClO_4)_2$ (h) $KClO_2$

20.75 What is a disproportionation reaction?

20.76 Write a chemical equation for the reaction that takes place when Cl_2 is dissolved in
(a) cold water
(b) cold NaOH solution

20.77 CaCl(OCl) is a solid water-soluble bleach. What reaction would occur if acid were added to this solid?

20.78 Compare the stabilities of the ions OCl^-, OBr^-, and OI^- in cold aqueous solutions. What accounts for their relative stabilities?

20.79 What reaction occurs if $KClO_3$ is heated at moderate temperatures? What happens if MnO_2 is present while the $KClO_3$ is heated?

20.80 Use the VSEPR theory to predict the molecular shapes of

(a) SF_2

(b) PBr_5

(c) SiF_6^{2-}

(d) AsF_3

(e) BrF_3

(f) $TeBr_4$

(g) BrF_5

(h) $GeCl_4$

20.81 What is the likely reason that IF_7 can be made but ClF_7 cannot?

20.82 Suggest two reasons why oxygen doesn't form OF_4 even though sulfur forms SF_4.

20.83 Predict what products would be formed in the reaction between $GeCl_4$ and H_2O.

20.84 Why does $SiCl_4$ hydrolyze but CCl_4 does not?

20.85 Compare the effectiveness of NH_3 and NF_3 as Lewis bases.

20.86 What is a *clathrate?*

20.87 Predict the molecular structures of XeF_4 and XeF_2 using the VSEPR theory.

20.88 Why were chemists so surprised when it was found that the noble gases were not inert?

20.89 How is elemental silicon prepared? Give an equation.

20.90 Describe *zone refining.*

20.91 Why doesn't silicon form graphitelike crystals?

20.92 What is the structure of the orthosilicate ion?

20.93 Use Lewis structures to describe the linking together of SiO_4 tetrahedra in the formation of the pyrosilicate ion.

20.94 Sketch the structure of the cyclic $Si_6O_{18}^{12-}$ anion. Which mineral contains this anion?

20.95 Sketch a portion of the anion found in asbestos. What is the repeating unit in the structure?

20.96 Sketch a portion of the polymeric metasilicate anion, $(SiO_3^{2-})_x$.

20.97 What kind of structure is formed when each SiO_4 tetrahedron shares three of its corners with neighboring SiO_4 tetrahedra? What are some common minerals containing this structure?

20.98 What is the empirical formula for quartz? Why do quartz crystals occur in left- and right-"handed" forms?

20.99 What structure is found along the "backbone" of the silicone polymers?

20.100 What compound would be formed by hydrolysis of

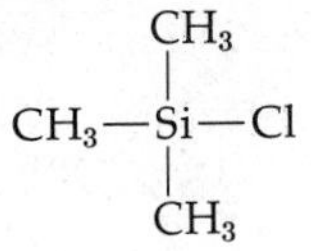

$$CH_3-\underset{\underset{\displaystyle CH_3}{|}}{\overset{\overset{\displaystyle CH_3}{|}}{Si}}-Cl$$

21

THE TRANSITION ELEMENTS

Many of our most familiar metals are transition metals. The most common of all is iron. Here we see molten iron from a blast furnace being poured into a "basic oxygen furnace" where, mixed with scrap iron and selected additives, it will be refined into steel. The making of steel is one of the topics discussed in this chapter.

In this chapter we consider the collection of elements, generally called the transition elements, that fit into the periodic table between Groups IIA and IIIA. We saw in Chapter 3 that these elements arise as a consequence of the gradual filling of d and f subshells and, as a rule, a transition element is usually considered to be one that possesses an incompletely filled d or f subshell in either the free state or in one of its compounds. We will also include in this discussion the elements in Group IIB, zinc, cadmium, and mercury, which are found at the extreme right of the transition elements and which complete a transition series (a horizontal row) by having their outer d subshells filled.

21.1 GENERAL PROPERTIES

In discussing the transition elements it is convenient to divide them into two categories: the **d-block elements** (or **main transition elements**), which in our condensed version of the periodic table are located in the main body of the table between Groups IIA and IIIA, and the **inner transition elements,** which correspond to the two long rows of 14 elements each that are placed just below the table (Figure 21.1). The d-block elements themselves consist of three rows frequently referred to as the first, second, and third transition series.

Like the representative elements, most of the d-block transition elements possess certain vertical similarities in chemical and physical properties and are therefore divided into groups, designated as B groups. They begin with Group IIIB on the left and proceed through Group VIIB. Next is a set of nine elements collectively termed Group VIII, and finally, on the right, we find Groups IB and IIB. The group numbers are chosen to correspond to the highest positive oxidation state that their elements normally exhibit.

The division of the periodic table into A and B groups—for example, IIIA and IIIB—suggests that there may be certain parallels between the two, and to a limited degree this is true. The similarities, however, are restricted primarily to likenesses of composition, structure, and maximum positive oxidation state, instead of chemical reactivity. Some examples of these similarities are found in Table 21.1.

The Group VIII elements, which lie between Groups VIIB and IB, are classed differently from the other d-block elements because they have no counterparts among the representative elements. Within this group there are greater *horizontal similarities* than vertical ones, and the description of the behavior of these elements is usually organized on the basis of horizontal groups of three elements each, called **triads.** Each triad is named after the best-known element

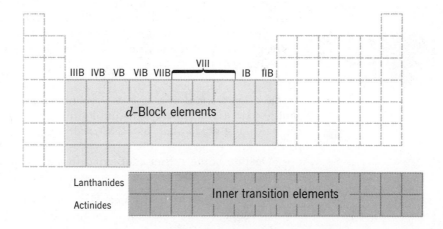

Figure 21.1

The transition elements.

Table 21.1
Some similarities of chemical composition between A and B group compounds

Group	Compounds	Group	Compounds
IVA	CCl_4, $SnCl_4$, CO_2	IVB	$TiCl_4$, TiO_2
VA	PO_4^{3-}, $POCl_3$	VB	VO_4^{3-}, $VOCl_3$
VIA	SO_4^{2-}, $S_2O_7^{2-}$	VIB	CrO_4^{2-}, $Cr_2O_7^{2-}$
VIIA	ClO_4^-, Cl_2O_7	VIIB	MnO_4^-, Mn_2O_7
IA	NaCl	IB	CuCl, AgCl
IIA	$CaCl_2$	IIB	$ZnCl_2$

within it. Thus we have the iron triad, the palladium triad, and the platinum triad.

As a class, the transition elements are all typical metals; they possess a characteristic metallic luster and are good conductors of heat and electricity. Silver has the highest electrical and thermal conductivity of any metal, followed closely by copper. Copper, of course, is used in vast quantities in electrical wiring. Silver is the preferred coating for mirrors because of its high reflectivity.

About one-quarter of the silver consumed by industry each year is used to make electronic equipment.

The chemical and physical properties of the transition elements cover a wide range and account for the large variety of uses to which they are applied. Some of these metals are very hard and strong and are used as structural metals, either in the pure state or as alloys. Iron is the prime example; steels with a variety of different properties are formed by incorporating iron with other transition elements such as chromium, cobalt, and nickel. Even copper, which is very soft when pure, can be made very strong by forming an alloy with beryllium. Beryllium-copper alloys are used in place of steel in nonsparking tools for use in explosive atmospheres and as high-quality springs in cameras and other precision instruments.

Tungsten has the highest melting point of any metal.

The melting points of the transition elements also vary over a wide range, as we saw in Chapter 9. Most are high-melting. Tungsten, with a melting point of approximately 3400°C, is used for filaments in light bulbs. At the other extreme is mercury, which is a liquid at room temperature and is used in thermometers.

The chemical reactivity of the free elements varies greatly too. Most react directly with nonmetals such as oxygen and the halogens to produce the corresponding oxides and halides. In fact, some transition elements are so easily oxidized that they react with water to liberate hydrogen. This is true for scandium (Sc), yttrium (Y), lanthanum (La), and the lanthanide elements (atomic numbers 58 to 71). They have very negative reduction potentials and react according to the equation

$$M + 3H_2O \longrightarrow \tfrac{3}{2} H_2 + M(OH)_3$$

Some other transition elements—such as platinum and gold—are very resistant to oxidation. They are insoluble in both protonic acids such as HCl as well as oxidizing acids such as HNO_3, although they do dissolve slowly in *aqua regia*.

Despite some rather marked differences in behavior, the transition elements have several characteristics in common with each other:

1. **Multiple oxidation states.** With only a few exceptions, the transition elements tend to exhibit more than one oxidation state.

2. **Many of their compounds are paramagnetic.** Because the transition elements tend to have partially completed d or f subshells in both the free state and in their compounds, the metal atoms often possess unpaired electrons. These impart the property of paramagnetism.
3. **Many (if not most) of their compounds are colored.** The origin of the colors of complex ions of the transition elements will be discussed in Section 21.12.
4. **They have a strong tendency to form complex ions.** As a group, these elements form a huge number of complex ions of varying degrees of complexity. The last six sections of this chapter are devoted to a discussion of their structures and bonding.

21.2 ELECTRONIC STRUCTURE AND OXIDATION STATES

Table 21.2

Electronic structures of the elements in the first transition series

Sc	$[Ar]3d^14s^2$	Fe	$[Ar]3d^64s^2$
Ti	$[Ar]3d^24s^2$	Co	$[Ar]3d^74s^2$
V	$[Ar]3d^34s^2$	Ni	$[Ar]3d^84s^2$
Cr	$[Ar]3d^54s^1$	Cu	$[Ar]3d^{10}4s^1$
Mn	$[Ar]3d^54s^2$	Zn	$[Ar]3d^{10}4s^2$

In Chapter 3 we saw that as we proceed from left to right across a period through the main transition elements, there is a gradual filling of the d subshell that lies just below the outer shell. In period 4, for example, this corresponds to the $3d$ subshell, as seen for the electronic structures of the first row elements given in Table 21.2. Each of these elements possesses a completed argon core with additional electrons in the $3d$ and the $4s$ subshells. Notice once again that chromium and copper are anomalous because of the extra stability associated with half-filled and filled subshells. Similar irregularities are found in the second and third transition series (Table 21.3), although other factors in addition to those having to do with half-filled and filled subshells are apparently involved, so no simple correlations can be made.

Among the inner transition elements, the lanthanides and actinides—so named because lanthanum and actinium have properties more or less typical of their respective series—there is a gradual filling of an f subshell that lies *two* shells below the outer shell. Thus, in Table 21.4 we see that as we pass through

Table 21.4

Electronic structures of the lanthanide and actinide elements

Lanthanides		Actinides	
La	$[Xe]5d^16s^2$	Ac	$[Rn]6d^17s^2$
Ce	$[Xe]4f^26s^2$	Th	$[Rn]6d^27s^2$
Pr	$[Xe]4f^36s^2$	Pa	$[Rn]5f^26d^17s^2$
Nd	$[Xe]4f^46s^2$	U	$[Rn]5f^36d^17s^2$
Pm	$[Xe]4f^56s^2$	Np	$[Rn]5f^57s^2$
Sm	$[Xe]4f^66s^2$	Pu	$[Rn]5f^67s^2$
Eu	$[Xe]4f^76s^2$	Am	$[Rn]5f^77s^2$
Gd	$[Xe]4f^75d^16s^2$	Cm	$[Rn]5f^76d^17s^2$
Tb	$[Xe]4f^96s^2$	Bk	$[Rn]5f^86d^17s^2$
Dy	$[Xe]4f^{10}6s^2$	Cf	$[Rn]5f^{10}7s^2$
Ho	$[Xe]4f^{11}6s^2$	Es	$[Rn]5f^{11}7s^2$
Er	$[Xe]4f^{12}6s^2$	Fm	$[Rn]5f^{12}7s^2$
Tm	$[Xe]4f^{13}6s^2$	Md	$[Rn]5f^{13}7s^2$
Yb	$[Xe]4f^{14}6s^2$	No	$[Rn]5f^{14}7s^2$
Lu	$[Xe]4f^{14}5d^16s^2$	Lr	$[Rn]5f^{14}6d^17s^2$

Table 21.3

Electronic structures of elements of the second and third transition series

Period 5 Second Transition Series		Period 6 Third Transition Series	
Y	$[Kr]4d^15s^2$	La	$[Xe]5d^16s^2$
Zr	$[Kr]4d^25s^2$	Hf	$[Xe,4f^{14}]5d^26s^2$
Nb	$[Kr]4d^45s^1$	Ta	$[Xe,4f^{14}]5d^36s^2$
Mo	$[Kr]4d^55s^1$	W	$[Xe,4f^{14}]5d^46s^2$
Tc	$[Kr]4d^65s^1$	Re	$[Xe,4f^{14}]5d^56s^2$
Ru	$[Kr]4d^75s^1$	Os	$[Xe,4f^{14}]5d^66s^2$
Rh	$[Kr]4d^85s^1$	Ir	$[Xe,4f^{14}]5d^76s^2$
Pd	$[Kr]4d^{10}5s^0$	Pt	$[Xe,4f^{14}]5d^96s^1$
Ag	$[Kr]4d^{10}5s^1$	Au	$[Xe,4f^{14}]5d^{10}6s^1$
Cd	$[Kr]4d^{10}5s^2$	Hg	$[Xe,4f^{14}]5d^{10}6s^2$

IIIB	IVB	VB	VIB	VIIB	VIII			IB	IIB
Sc	Ti	V	Cr	Mn	Fe	Co	Ni	Cu	Zn
3+	2+	1+	2+	2+	2+	2+	2+	1+	2+
	3+	2+	3+	3+	3+	3+	3+	2+	
	4+	3+	6+	4+	4+				
		4+		6+	6+				
		5+		7+					
Y	Zr	Nb	Mo	Tc	Ru	Rh	Pd	Ag	Cd
3+	2+	2+	2+	2+	2+	1+	2+	1+	2+
	3+	3+	3+	3+	3+	2+	3+	2+	
	4+	4+	4+	4+	4+	3+	4+	3+	
		5+	5+	5+	5+	4+			
			6+	6+	6+	5+			
			8+	7+	7+	6+			
					8+				
La	Hf	Ta	W	Re	Os	Ir	Pt	Au	Hg
3+	3+	2+	2+	3+	2+	1+	2+	1+	1+
	4+	3+	3+	4+	3+	2+	3+	3+	2+
		4+	4+	5+	4+	3+	4+		
		5+	5+	6+	5+	4+	5+		
			6+	7+	6+	5+	6+		
					8+	6+			

Figure 21.2

Oxidation states of the transition metals. The most stable oxidation states are shown in color.

The 2+ oxidation state is common because many transition elements have a pair of **s** electrons that are the first to be lost.

the lanthanides in period 6, the 4f subshell is completed. In the following period, as we pass through the actinides, the 5f subshell becomes populated.

The chemical and physical properties of the transition elements are controlled, of course, by their electronic structures. With the d-block elements the outer s and underlying d subshells are of nearly equal energy. Therefore, when these elements react, the d electrons are able to participate in bonding. The importance of the d electrons in determining the chemistry of these elements accounts for their varied chemical properties, including the multiplicity of oxidation states.

The wide variety of different oxidation states found for the transition elements is illustrated in Figure 21.2. Don't be overwhelmed—we are only interested here in looking at trends in the relative stabilities of oxidation states. When we compare these stabilities, we can see two important trends.

1. Elements at the left of a row of transition elements prefer the highest oxidation states. As we proceed to the right, the lower oxidation states become increasingly more stable relative to the higher ones. (Remember, the *highest* oxidation state is usually equal to the group number.)
2. Going down a group of transition elements, the higher oxidation states become increasingly more stable than the lower ones.

Knowing these trends is useful because it helps us make comparisons of the strengths of oxidizing agents. For example, in the first row of transition elements (the first *transition series*), we see an increasing tendency toward a stable 2+ oxidation state and fewer and fewer high oxidation states. Suppose we use this to compare the relative stabilities of Fe^{2+} and Fe^{3+} with Ni^{2+} and Ni^{3+}. The horizontal trend suggests that Ni^{3+} has a greater tendency to become Ni^{2+} than Fe^{3+} has to become Fe^{2+}. In other words, Ni^{3+} has a greater tendency to gain an electron than Fe^{3+}. Since these species are functioning as oxidizing agents when they gain electrons, we conclude that Ni^{3+} is a stronger oxidizing agent than Fe^{3+}. In fact, this is what is found experimentally.

Remember, an oxidizing agent becomes reduced in a redox reaction.

We can also make vertical comparisons. For example, suppose we consider the ions MnO_4^- and ReO_4^-. Since high oxidation states become more stable going down, ReO_4^- should be a weaker oxidizing agent than MnO_4^-. This is also found to be true experimentally.

MnO_4^- is permanganate ion.
ReO_4^- is perrhenate ion.

EXAMPLE 21.1 Which would you expect to be a more powerful oxidizing agent, TiO_2 or MnO_2?

SOLUTION First we locate Ti and Mn in the periodic table. Since the higher oxidation states become less stable relative to the lower ones going from left to right, Mn(IV) should be relatively less stable than Ti(IV) toward reduction—Mn(IV) should have a greater tendency to be reduced than Ti(IV). Therefore, MnO_2 should be the stronger oxidizing agent. (Actually, TiO_2 is a stable white pigment used in paint, and MnO_2 is the oxidizing agent in the common dry cell.)

In contrast to the wide range of chemical properties of the *d*-block elements, the lanthanides exhibit a remarkable sameness of properties. The 4*f* subshell, which is only partially filled for most of these elements, is buried beneath the outer 5*d* and 6*s* subshells and does not interact to an appreciable extent with the surrounding chemical environment. Consequently, the chemistry of the lanthanides, like that of lanthanum itself, is predominantly that of the 3+ ion, and differences in behavior depend primarily on differences in ionic size.

The actinide elements exhibit a greater variation in oxidation numbers than the lanthanides do (e.g., uranium forms compounds in the 3+, 4+, 5+, and 6+ oxidation states). An explanation sometimes given for this is that the 5*f* orbitals of the actinides project further toward the outer parts of the atoms than the 4*f* orbitals of the lanthanides. As a result, the 5*f* orbitals are able to become involved to a greater degree in chemical bonding, so more complex chemistry is observed.

21.3 ATOMIC AND IONIC RADII

We have seen before that many trends in properties can be correlated with variations that occur in atomic and ionic radii. This is true among the transition elements as well as the representative elements.

In Chapter 3 the horizontal and vertical trends in atomic size were discussed. Let's briefly review them here. As we move across a given transition series there is only a gradual decrease in atomic radius. This is because the 3*d* electrons that are added to the atom shield the outer 4*s* electrons quite well from the increasing nuclear charge and, as a result, the effective nuclear charge experienced by the outer electrons rises only slowly. Therefore, only a small size decrease occurs (Table 21.5).

Vertically, we find a rather large increase in size among the *d*-block elements going from period 4 to period 5. However, between periods 5 and 6 there is only a very small size increase and, in some cases, none at all. As we learned

Table 21.5
Atomic Radii (Å)

Sc	1.62	Y	1.80	La	1.87
Ti	1.47	Zr	1.60	Hf	1.58
V	1.34	Nb	1.46	Ta	1.46
Cr	1.27	Mo	1.39	W	1.39
Mn	1.26	Tc	1.36	Re	1.37
Fe	1.26	Ru	1.34	Os	1.35
Co	1.25	Rh	1.34	Ir	1.36
Ni	1.24	Pd	1.37	Pt	1.38
Cu	1.28	Ag	1.44	Au	1.44
Zn	1.38	Cd	1.54	Hg	1.57

earlier, this is a consequence of the lanthanide contraction—the gradual decrease in size that occurs across the lanthanide series. Apparently this just cancels the size increase that would be otherwise expected as we go from period 5 to period 6, so the period 6 elements that follow the lanthanides are essentially the same size as those above them in period 5.

These variations in size have some very pronounced chemical and physical consequences. They can be correlated, for example, with variations in ionization energies (IE), shown in Table 21.6. Here we see that the gradual decrease in size along a period that is associated with an increase in effective nuclear charge is also accompanied by an increase in IE. As we might expect, this increasing difficulty encountered in removing an outer electron from the isolated atoms is reflected in a gradual, although not altogether uniform, rise in their standard reduction potentials. In other words, as we proceed across the table from left to right, it generally becomes more difficult to oxidize the elements.

Trends in the ease of oxidation of the metals were summarized in Figure 9.8 on page 292.

The effect of the lanthanide contraction is demonstrated in these properties as well. Among the representative elements the IE generally decreases as we proceed down a group, paralleling the increase in size. This phenomenon is also observed among the transition elements on going from period 4 to period 5. However, from periods 5 to 6 there is an increase in nuclear charge without an accompanying increase in size, so the IE increases. This in turn manifests itself in reduction potentials that tend to be quite high for the third transition series elements, thereby accounting for their virtually inert behavior toward many oxidizing agents.

Table 21.6
Ionization energy (kJ/mol)

Sc	632	Y	616	La	540
Ti	660	Zr	672	Hf	675
V	651	Nb	665	Ta	763
Cr	653	Mo	694	W	771
Mn	718	Tc	720	Re	761
Fe	763	Ru	711	Os	842
Co	760	Rh	720	Ir	868
Ni	737	Pd	805	Pt	866
Cu	746	Ag	732	Au	891
Zn	907	Cd	869	Hg	1008

21.4 METALLURGY

Metallurgy is the process whereby a metal is extracted from its ore and brought to the point where it can be put to practical use. The desirable physical properties of many transition metals, such as high strength, hardness, and high melting points, make them extremely important in modern technology, and some of the methods and procedures used to obtain them from their ores are worth examining.

An **ore** is a substance that contains a particular desirable constituent in a high enough concentration that its extraction from the ore is economically worthwhile. For example, many minerals may contain small amounts of iron, but only those that are rich in iron would be considered iron ores and would serve as economical sources of this metal. As the earth's reserves of rich ores are consumed, it will become the job of the chemist to devise new ways to obtain metals such as iron from less rich ores.

Generally, we can divide metallurgical processes into three categories:

1. **Concentration.** Ores that contain substantial amounts of impurities, such as rock, must often be treated to concentrate the metal-bearing constituent. Pretreatment of an ore is also carried out to convert some metal compounds into substances that can be more easily reduced.
2. **Reduction.** In their compounds, metals nearly always exist in positive oxidation states. Therefore, to obtain a metal from its ore it must be reduced. The particular procedure employed for a given metal depends on its ease of reduction to the free state.
3. **Refining.** Often, during reduction, substantial amounts of impurities become introduced into the metal. Refining is the process whereby these impurities are removed and the composition of the metal adjusted (alloys formed) to meet specific applications.

Let us now take a brief look at each of these steps as they apply to some important metals.

Concentration

Not all ores have to be subjected to a pretreatment step prior to reduction, although most of them must. These pretreatment procedures involve the separation of the metal-bearing component of the ore from unwanted or interfering impurities. This is particularly important for low-grade ores in which the desired metal is present only in small amounts.

As expected, different methods are applied to different ores, depending on the specific properties of the impurities and the metal compounds. We can divide these procedures into two classes: *physical separations*, in which the chemical compositions of the constituents are not altered, and *chemical separations*, which use the chemical properties of the different substances in the ore.

Some metals, such as silver and gold, are found in deposits as the free element, and their recovery simply involves removing them from the rock and sand with which they are mixed. One of the earliest forms of physical separation was used by the "forty-niners" in panning for gold. A mixture of sand containing (hopefully!) particles of metallic gold was placed in a shallow pan with water. The mixture was swirled about and the sand was washed over the rim, leaving the gold dust in the bottom of the pan. The success of this procedure is based on the fact that gold is about nine times as dense as the sand and gravel impurities. As a result, the lighter impurities are more easily washed away than the more dense metal.

Another way of removing metallic gold and silver from their ores is to treat the mixture with metallic mercury—a liquid in which silver and gold dissolve

A rare photograph of a nineteenth century miner panning for gold. The technique is still used occasionally, especially by amateur prospectors.

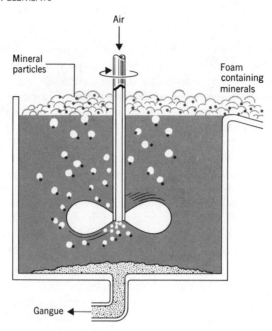

Figure 21.3
The flotation process.

to form an alloy called an **amalgam.** The silver and gold are later recovered by distilling away the mercury, which is reclaimed and used again. You are probably familiar with silver and gold amalgams as the material used by dentists to fill teeth.

A physical separation technique that can be applied to the sulfide ores of zinc, copper, and lead is called **flotation.** In this process, illustrated in Figure 21.3, the ore is pulverized and added to a mixture of water and oil containing suitable additives. The metal-bearing component of the ore becomes coated by the oil while the unwanted material, called the **gangue,** is wetted by the water. A stream of air is then blown through the mixture and the oil-covered mineral is carried to the surface by bubbles where it is trapped in a froth that can be removed to recover the metal compound. The gangue, on the other hand, simply settles to the bottom of the apparatus and is later discarded.

Chemical methods of concentrating the metal-bearing component of an ore vary considerably because of the variety of chemical properties exhibited by the metals and their compounds. For example, aluminum, whose electrolytic reduction was described in Chapter 17, occurs in deposits of bauxite, a form of Al_2O_3. In this case the ore is concentrated by taking advantage of the amphoteric behavior of aluminum. The bauxite is treated with concentrated base, which dissolves the Al_2O_3 to produce aluminate ion, AlO_2^-.

$$Al_2O_3 + 2OH^- \longrightarrow 2AlO_2^- + H_2O$$

The purification of Al_2O_3 was also discussed in Chapter 18.

After it is removed from the gangue, the solution is acidified. This precipitates $Al(OH)_3$, which yields pure Al_2O_3 when heated.

$$2Al(OH)_3 \xrightarrow{\text{heat}} Al_2O_3 + 3H_2O$$

This purified aluminum oxide serves as the charge in the Hall process discussed previously.

Another chemical pretreatment, often given to a sulfide ore, is called **roasting.** Here the ore is heated in air, converting the metal sulfide to an oxide that is more conveniently reduced.

$$2PbS + 3O_2 \longrightarrow 2PbO + 2SO_2$$
$$2ZnS + 3O_2 \longrightarrow 2ZnO + 2SO_2$$

As we saw in the last chapter, this industrial process is a severe source of air pollution unless the SO_2 is recovered.

Reduction

The reduction of metal compounds is accomplished commercially either by electrolysis or by the use of a substance that is a better reducing agent than the metal being sought. Commercial electrolysis reactions like those described in Chapter 17 are generally reserved for metals that are particularly difficult to reduce. Most metals are extracted from their compounds by a chemical reducing agent. Titanium, for example, is prepared by reacting $TiCl_4$ with the more active metal, magnesium. The $TiCl_4$ is produced from rutile, a fairly pure source of TiO_2, by reaction with chlorine gas and carbon.

$$TiO_2 + 2C + 2Cl_2 \longrightarrow TiCl_4 + 2CO$$

The titanium tetrachloride is a volatile liquid (boiling point = 136°C) and can be separated from impurities by distillation. The final reduction to the metal follows the equation

$$TiCl_4 + 2Mg \longrightarrow Ti + 2MgCl_2$$

Titanium is a very useful metal because it is considerably lighter than steel (d_{Fe} = 7.86 g/ml, d_{Ti} = 4.51 g/ml), yet does not lose its strength at high temperature as aluminum does. It is used in large quantities in jet engines, and replaces aluminum and steel in other aircraft applications. Its oxide, TiO_2, is also used in large quantities as a white pigment in paint. It is better than white lead, $Pb_3(OH)_2(CO_3)_2$, because it appears to be of low toxicity and because it doesn't darken in the presence of H_2S as the lead-based pigments do.

Active metals such as magnesium are very expensive reducing agents because they themselves are difficult and costly to prepare. As a result, less expensive reducing agents are employed whenever possible. One of the cheapest of all is carbon, in the form of coke, which is produced from coal by heating it at high temperatures in the absence of air. This treatment drives off the volatile components of the coal (from which other important chemicals are derived), leaving nearly pure carbon behind.

Some typical reductions of metal oxides with carbon were illustrated in Section 18.4, and many of the transition metals (for example, Zn, Cd, Fe, Co, Ni, Mo) can be prepared in this way. Undoubtedly the most important chemical reduction brought about with carbon is that of iron oxide, Fe_2O_3, to iron. This is accomplished in the **blast furnace,** developed in about 1300 A.D. and which in modern times takes the form shown in Figure 21.4.

The mixture of ingredients added to the furnace is called the **charge,** and consists of limestone, coke, and iron ore. The ore is normally composed primarily of Fe_2O_3 with impurities of SiO_2 (sand, about 10%) and smaller amounts of compounds containing sulfur, phosphorus, aluminum, and manganese. Heated air is forced in at the bottom of the furnace, where it reacts with carbon in a very exothermic reaction to produce carbon dioxide.

$$C + O_2 \longrightarrow CO_2 \qquad \Delta H = -394 \text{ kJ}$$

The large amount of heat generated in this region of the furnace raises the temperature to nearly 1900°C. As the hot gases rise, the CO_2 reacts with additional carbon in an endothermic reaction to form carbon monoxide, the active reducing agent in the furnace.

$$CO_2 + C \longrightarrow 2CO \qquad \Delta H = +173 \text{ kJ}$$

The reduction of the iron oxide takes place in a series of steps. Near the top

TiCl$_4$ was used by the U.S. Navy during WWII to make smoke screens because its vapor hydrolyzes readily in moist air to give a dense white cloud of solid TiO$_2$ particles.

This hot blast of air gives the blast furnace its name.

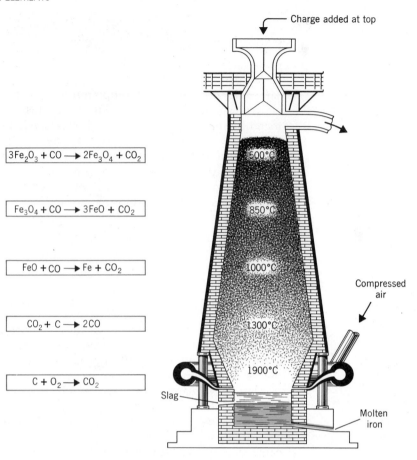

$$3Fe_2O_3 + CO \longrightarrow 2Fe_3O_4 + CO_2$$

$$Fe_3O_4 + CO \longrightarrow 3FeO + CO_2$$

$$FeO + CO \longrightarrow Fe + CO_2$$

$$CO_2 + C \longrightarrow 2CO$$

$$C + O_2 \longrightarrow CO_2$$

Figure 21.4

The blast furnace.

of the furnace, Fe_2O_3 is reduced to Fe_3O_4.

$$3Fe_2O_3 + CO \longrightarrow 2Fe_3O_4 + CO_2$$

Farther down, in a hotter region of the furnace, this is reduced to FeO.

$$Fe_3O_4 + CO \longrightarrow 3FeO + CO_2$$

Finally, still farther down the FeO is reduced to the metal that, at these high temperatures, is a liquid and trickles down to form a pool of molten metal at the base of the tower.

$$FeO + CO \longrightarrow Fe + CO_2$$

The function of the limestone in the furnace is to provide a basic medium with which acidic oxides, such as SiO_2 and P_4O_{10}, or amphoteric oxides such as Al_2O_3 can react. At elevated temperatures limestone, $CaCO_3$, decomposes to form lime (CaO) and CO_2 according to the equation

$$CaCO_3 \longrightarrow CaO + CO_2$$

The lime then reacts as follows,

$$CaO + SiO_2 \longrightarrow CaSiO_3$$

$$6CaO + P_4O_{10} \longrightarrow 2Ca_3(PO_4)_2$$

$$CaO + Al_2O_3 \longrightarrow Ca(AlO_2)_2$$

The products of these reactions have relatively low melting points and are liquids when they are formed. The mixture, called **slag,** also runs to the base of the furnace, where it floats atop the molten iron. As these two layers are formed,

Some blast furnaces are as high as a 15-story building and produce up to 2400 tons of iron each day.

the charge in the furnace settles and additional limestone-coke-ore mixture is added at the top. In this way the blast furnace operates continuously, with fresh charge being added at the top and molten iron and slag being tapped off at the bottom. These furnaces are often run for months at a time before they are shut down for routine maintenance.

The liquid iron, when it is withdrawn from the blast furnace, is called **pig iron** and consists of about 95% Fe and approximately 4% carbon, with small amounts of silicon, manganese, phosphorus, and sulfur. This somewhat impure iron is very hard and can be poured into molds as **cast iron.** The slag that comes from the furnace can be used in making cement.

Refining

In the process of separating a metal from its ore, impurities are often introduced that impart undesirable properties to the final product. Therefore, it is generally necessary to purify the metal before it can be put to practical use. This purification process is called **refining.**

The specific procedure employed for refining a given metal depends on the chemical and physical properties of the metal as well as the properties of the impurities. As a result, there is no single method applicable to a very large number of different metals. We saw in Chapter 17 that copper can be economically refined electrolytically. This occurs, however, primarily because the silver and other precious metals recovered from the electrolytic cell offset the generally high cost of electricity.

Carbonyl compounds were discussed in Chapter 19.

An interesting process for refining nickel, called the Mond process, makes use of the relative ease of formation of nickel carbonyl—a compound formed between nickel and carbon monoxide.

$$Ni + 4CO \longrightarrow Ni(CO)_4$$

Besides being easily formed, nickel carbonyl is also very volatile (and very poisonous). The impure nickel is therefore treated with CO at a moderately low temperature of 60°C, where the $Ni(CO)_4$ that is formed exists as a gas. This is circulated to another portion of the apparatus, where it is heated to about 200°C and decomposes to give pure nickel plus CO, which can be recycled through the process.[1]

The most important commercial refining processes involve the conversion of pig iron into steel. This requires the removal of impurities such as silicon, sulfur, and phosphorus and lowering the carbon content significantly from the approximately 4% introduced into the pig iron in the blast furnace.

Modern steel making began with the introduction of the **Bessemer converter** in England in 1856. A batch of molten pig iron from the blast furnace, weighing about 25 tons, is transferred to a tapered cylindrical vessel containing a refractory lining (Figure 21.5). The composition of the lining is determined in part by the nature of the impurities in the iron. Since these impurities are usually silicon, phosphorus, and sulfur, whose oxides are acidic, a basic lining of dolomite (a $MgCO_3$, $CaCO_3$ mineral) is generally used. A blast of air (or oxygen) is blown through the metal from a set of small holes at the bottom of the vessel. The oxygen passing through the molten metal converts the silicon, phosphorus, and sulfur to oxides that then react with the lining to form a slag. The carbon in the pig iron is also oxidized to CO, so its concentration is reduced, too. The conversion of the pig iron to steel by this process is rapid, requiring about 15 minutes, and gives rise to a spectacular display of fire and showers of sparks. The reaction is difficult to control, however, and the quality of the steel produced in the Bessemer converter can be quite variable.

[1] For a time it was believed that extremely toxic $Ni(CO)_4$ was responsible for the so-called Legionnaires' disease that killed a group of people attending an American Legion Convention in Philadelphia in 1976. Later, however, this idea was abandoned.

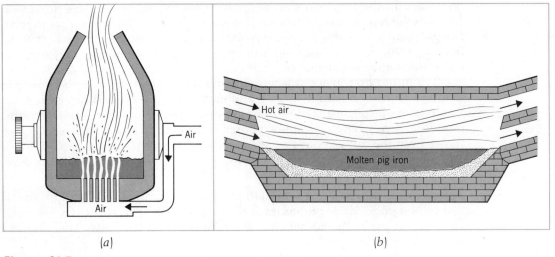

Figure 21.5

(a) Bessemer converter. (b) Open hearth furnace.

A somewhat newer method that virtually replaced the Bessemer process employs an **open hearth furnace,** a large, shallow hearth usually lined with a basic oxide refractory (for example, MgO, CaO). The furnace is charged with a mixture of pig iron, Fe_2O_3, scrap iron, and limestone. A mixture of burning gases and hot air is played over the surface of the charge to maintain it in a molten state while a series of chemical reactions take place. Impurities in the steel are oxidized by the Fe_2O_3 and air. Carbon dioxide, formed by oxidation of the carbon in the pig iron, bubbles out of the mixture, keeping it stirred, while the SiO_2 and other acidic oxides combine with CaO (from the limestone) and the refractory lining to form a slag. This entire process takes much longer than the Bessemer process, requiring 8 to 10 hours to complete. However, the quality of the steel is much more easily controlled because chemical analyses can be constantly carried out on samples of the mixture. The increased length of time required to process a batch of steel is also offset by the fact that much larger quantities (about 200 tons) can be handled at one time. In addition, prior to being poured from the furnace, other metals (e.g., cobalt, chromium, nickel, vanadium, and tungsten) can be added to the steel to form alloys with special properties. A typical stainless steel, for instance is composed of approximately 72% iron, 19% chromium, and 9% nickel.

Modern methods of chemical analysis, making use of high-speed computers, have enabled a return to a modified form of the Bessemer process called the **basic oxygen process.** This newer procedure, which has largely replaced the open hearth furnace because of its speed, involves forcing a mixture of powdered $CaCO_3$ and oxygen gas into the molten pig iron. This rapidly burns away the impurities, which form a slag. The characteristic emission spectra of the elements in the steel permit rapid chemical analysis, and additives can be incorporated into the steel in the proper proportions to give a product with the desired properties. This process takes only about 20 to 25 minutes to complete, thereby yielding very substantial savings in time (and, of course, money) over the open hearth process.

21.5 MAGNETISM

In Chapter 3 we saw that the presence of unpaired electrons in an atom or molecule imparts the property called paramagnetism to the substance. The tiny electron magnets cause the atom or molecule as a whole to behave as a small magnet.

When these are placed into a magnetic field, the microscopic magnets tend to align themselves with and be attracted toward the field. However, thermal motion operates to randomize the orientations of the little magnets. The net result is that only a relatively small fraction of the number of tiny magnets are aligned with the field at any particular instant, so an ordinary paramagnetic substance is drawn only weakly into an external magnetic field.

Characteristically, the transition elements and their compounds possess a partially filled d or f subshell, so many of them exhibit the phenomenon of paramagnetism. The prediction of magnetic properties of transition metal compounds is therefore one requisite of a theory of bonding that is applicable to these substances. We will explore this a little further when we discuss complex ions later in this chapter.

Related to the property of paramagnetism is the phenomenon called **ferromagnetism,** observed for the three pure elements, iron, cobalt, and nickel. Ferromagnetic materials, like paramagnetic ones, are also attracted to a magnetic field; however, the magnitude of the interaction for a ferromagnetic substance is approximately a million times stronger than it is with paramagnetic materials. How does this occur?

The origin of ferromagnetism is the same as paramagnetism—that is, the existence of unpaired electrons in the ferromagnetic material. In these substances it is believed that regions exist, called **domains,** that contain very large numbers of paramagnetic atoms with their atomic magnets all lined up in the same direction, as illustrated in Figure 21.6. Ordinarily, these domains are randomly oriented in a ferromagnetic solid, so even though each domain behaves as a relatively large magnet, their combined effects cancel. When the ferromagnetic material is placed in a magnetic field, the domains tend to become aligned, much the same as the atomic magnets of a paramagnetic substance. In this case, however, each time one domain becomes aligned with the field, millions of tiny atomic magnets become aligned all at once. As a result, the interaction between the ferromagnetic solid and the magnetic field is very much larger than that experienced by paramagnetic substances.

When the magnetic field is removed from a paramagnetic substance, its atomic magnets very quickly become randomly oriented and no permanent magnetism is induced. For a ferromagnetic substance, however, the domains tend to remain in the orientation in which they found themselves when the external magnetic field was present. This alignment of domains in the absence of

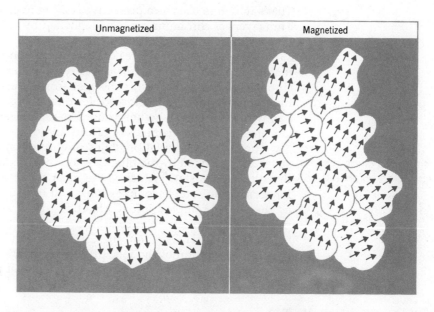

Figure 21.6

Domains in a ferromagnetic solid.

an external field causes the substance to possess a residual magnetism, and we say that it has become *permanently magnetized*. Any piece of iron (e.g., a pin) can be magnetized simply by stroking it with another permanent magnet.

A permanent magnet is not really permanent, because the magnetism may be destroyed either by heating the solid or by pounding it. In the first case the increased thermal motion causes the domains to become randomly oriented, while in the second instance violent vibrational motions cause the domains to twist and turn and become disoriented.

The phenomenon of ferromagnetism is associated only with the solid state. Iron, for example, is no longer ferromagnetic when it is melted. Instead it exhibits only paramagnetism. Melting of the solid thus appears to destroy the domains, and each individual atom in the liquid behaves more or less independently of the others nearby.

Even in the solid state not all elements containing unpaired electrons are ferromagnetic. Manganese, for example, possesses five unpaired electrons compared to only four for iron; yet pure iron is ferromagnetic but pure manganese is not. Apparently a requirement for ferromagnetism is for the spacings between paramagnetic ions to be just right so that they may lock onto each other to form a domain. Nonferromagnetic metals, in which the ions are too close together, can sometimes be made ferromagnetic by forming an alloy. This is the case with manganese, where the addition of the proper amount of copper permits the Mn^{2+} ions in the metallic lattice to interact strongly and form domains, thereby producing a ferromagnetic alloy.

21.6 PROPERTIES OF SOME TRANSITION METALS

Many transition metals have useful applications that depend on both their chemical and physical properties. We find them almost anywhere we look— iron in the many steel products that are all around us, chromium on automobile bumpers, the zinc coating on galvanized steel, and the titanium dioxide pigment that is in nearly all paints. In this section we will examine the chemical and physical properties of some of the more important transition elements.

Chromium

Chromium is a hard, brittle, lustrous metal that is very resistant to corrosion. That's why it is used as a protective coating over steel for automobile bumpers. Thin layers of chromium are also deposited by electroplating on brass or bronze objects for decorative purposes.

One of the principal uses of chromium is in stainless steel—a type of steel that is very resistant to corrosion. A typical stainless steel contains about 19% chromium, 9% nickel, with the rest iron. Unlike ordinary iron and steel, a high quality stainless steel is not ferromagnetic. A small magnet is therefore a handy tool to check whether or not a metal object claimed to be made of stainless steel is, in fact, composed of this alloy.

Cr^{2+} is also called chromous ion; Cr^{3+} is called chromic ion.

The principal oxidation states of chromium are 2+, 3+, and 6+. The 2+ state, characterized by the blue Cr^{2+} ion in aqueous solutions, is very easily oxidized to the 3+ state, which is the most stable oxidation state. Chromium(III) ion forms many stable complex ions, and in aqueous solutions it actually exists as the violet complex ion, $Cr(H_2O)_6^{3+}$. This ion is what gives many chromium(III) salts their violet color.

Like the $Al(H_2O)_6^{3+}$ ion, the $Cr(H_2O)_6^{3+}$ ion is slightly acidic. When base is added to a solution of the ion, protons are extracted from the water molecules attached to the chromium and a pale blue-violet precipitate is formed. The net

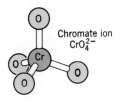

The $Cr(H_2O)_6^{3+}$ ion

reaction is

$$Cr(H_2O)_6^{3+} (aq) + 3OH^- (aq) \longrightarrow Cr(H_2O)_3(OH)_3 (s) + 3H_2O$$

This precipitate redissolves in additional base *or* in acid, so chromium(III) is amphoteric.

In base

$$Cr(H_2O)_3(OH)_3 (s) + OH^- (aq) \longrightarrow Cr(H_2O)_2(OH)_4^- (aq) + H_2O$$

In acid

$$Cr(H_2O)_3(OH)_3 (s) + H^+ (aq) \longrightarrow Cr(H_2O)_4(OH)_2^+ (aq) + H_2O$$

$$Cr(H_2O)_4(OH)_2^+ (aq) + H^+ (aq) \longrightarrow Cr(H_2O)_5OH^{2+} (aq) + H_2O$$

$$Cr(H_2O)_5OH^{2+} (aq) + H^+ (aq) \longrightarrow Cr(H_2O)_6^{3+} (aq) + H_2O$$

In general, oxides of metals in high oxidation states tend to be acidic instead of basic.

In the 6+ oxidation state, chromium forms the red oxide, CrO_3. Chromium(VI) oxide is a strong oxidizing agent and the acid anhydride of *chromic acid*, H_2CrO_4. In very acidic solutions, H_2CrO_4 is the principal species. As the pH is raised, however, two other species are formed. One is the yellow *chromate ion*, CrO_4^{2-}, and the other is the red-orange *dichromate ion*, $Cr_2O_7^{2-}$. Both are strong oxidizing agents, and exist in equilibrium with each other.

$$2CrO_4^{2-} + 2H^+ \rightleftharpoons Cr_2O_7^{2-} + H_2O$$

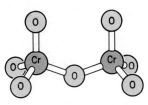

Chromate ion
CrO_4^{2-}

Dichromate ion
$Cr_2O_7^{2-}$

By Le Châtelier's principle, we see that $Cr_2O_7^{2-}$ predominates at low pH and CrO_4^{2-} is the major species at high pH.

Manganese

Manganese is much less corrosion resistant than its neighbor chromium. It corrodes in moist air, and dilute acids dissolve it, much like iron. Manganese is used mostly in making alloys such as the ferromagnetic manganese-copper alloy mentioned in the last section.

The most stable oxidation state of manganese is 2+. In solution it exists as the very pale pink $Mn(H_2O)_6^{2+}$ ion. An important compound of manganese in the 4+ oxidation state is MnO_2, commonly called manganese dioxide. This is the substance, you recall, that undergoes reduction at the cathode in the ordinary dry cell (see Section 17.13).

When MnO_2 is added to molten KOH and oxidized with O_2 or KNO_3, the green manganate ion, MnO_4^{2-}, is formed. It is stable only in very alkaline (basic) solutions, and when they are acidified the MnO_4^{2-} disproportionates—some of it is oxidized to MnO_4^- (permanganate ion) while the rest is reduced to MnO_2.

$$3MnO_4^{2-} + 4H^+ (aq) \longrightarrow 2MnO_4^- (aq) + MnO_2 (s) + 2H_2O$$

Solutions of permanganate ion have a deep violet color (Color Plate 5), and are strong oxidizing agents. When the MnO_4^- is reduced in an acidic solution, the pale pink Mn^{2+} ion is formed.

$$\underset{\text{deep violet}}{MnO_4^- (aq)} + 8H^+ (aq) + 5e^- \longrightarrow \underset{\text{very pale pink}}{Mn^{2+} (aq)} + 4H_2O$$

This makes MnO_4^- a useful analytical reagent for performing redox titrations. As the aqueous MnO_4^- is added from a buret to a solution of a reducing agent, reduction of the MnO_4^- takes place. This produces Mn^{2+} ion whose pale pink color is invisible at the concentrations used in the titration. Therefore, when the MnO_4^- is reduced, its violet color disappears and the solution appears colorless.

When all the reducing agent has finally been consumed, the next drop of MnO_4^- solution gives an excess of this ion and that makes the solution appear pink, signaling the endpoint.

Reduction of MnO_4^- in neutral or somewhat basic solutions produces a dark brown precipitate of MnO_2.

$$MnO_4^- \ (aq) + 4H^+ \ (aq) + 3e^- \longrightarrow MnO_2 \ (s) + 2H_2O$$

Because a precipitate is formed, MnO_4^- is not used for redox titrations in neutral or basic solutions. The dark brown color of the precipitate obscures the endpoint.

Iron

Iron is the second most abundant metal and the fourth most abundant element in the Earth's crust.

Iron is the most used of the transition metals because it is relatively abundant and easily extracted from its ores. In the pure state, iron is not very hard, but when small amounts of carbon and other metals are added to it, strong steel alloys are formed.

Iron is a fairly reactive metal. It forms compounds principally in two oxidation states: 2+ and 3+. Generally, iron(II) compounds are easily oxidized to give the corresponding iron(III) compounds. Three oxides of iron are formed: FeO, Fe_2O_3, and Fe_3O_4. Iron(II) oxide is difficult to prepare and disproportionates into Fe and Fe_2O_3 when heated.

$$3FeO \ (s) \longrightarrow Fe \ (s) + Fe_2O_3 \ (s)$$

Iron(III) oxide is the principal component of most iron ores, and its hydrated form is produced when iron rusts (see below). The oxide Fe_3O_4, known as *magnetite*, contains iron in two different oxidation states, and can be formulated as $Fe^{II}Fe_2^{III}O_4$ (Roman numerals are used here to express the oxidation states of the iron atoms). As its name suggests, magnetite is magnetic. It is an important iron ore because a strong magnet can easily separate it from useless rock and because it is rich in iron.

Metallic iron dissolves in nonoxidizing acids such as HCl or H_2SO_4 with the release of hydrogen (see photo on page 292).

$$Fe \ (s) + 2H^+ \ (aq) \longrightarrow Fe^{2+} \ (aq) + H_2 \ (g)$$

An open-pit iron mine in the Mesabi Range in Minnesota.

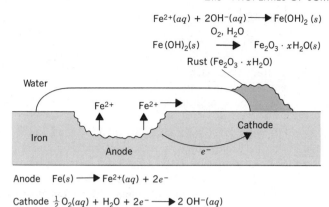

Figure 21.7

Corrosion of iron. Iron dissolves at anode sites and diffuses through the water as Fe^{2+} to cathodic sites where it precipitates as $Fe(OH)_2$. Iron(II) hydroxide is easily oxidized by O_2 in the presence of water to give the hydrated iron(III) oxide, $Fe_2O_3 \cdot xH_2O$, called rust.

Anode $Fe(s) \longrightarrow Fe^{2+}(aq) + 2e^-$

Cathode $\frac{1}{2}O_2(aq) + H_2O + 2e^- \longrightarrow 2\,OH^-(aq)$

One of the most important reactions of iron, at least from an economic point of view, is its corrosion in the presence of air and moisture. The mechanism for the reaction appears to be electrochemical, as shown in Figure 21.7. The iron—in contact with water—is oxidized to the 2+ state.

$$Fe\ (s) \longrightarrow Fe^{2+}\ (aq) + 2e^-$$

Iron is acting as the anode in a galvanic cell. That's why it is oxidized.

The electrons released by the iron during this oxidation are transmitted to sites on the iron that are in contact with oxygen and moisture, where reduction of O_2 to hydroxide ions occurs.

$$\tfrac{1}{2}O_2\ (aq) + H_2O + 2e^- \longrightarrow 2\ OH^-\ (aq)$$

The iron(II) ions diffuse through the water and when they contact the OH^- a precipitate of $Fe(OH)_2$ is formed, which is very easily oxidized to $Fe(OH)_3$ by oxygen.

$$4Fe(OH)_2\ (s) + O_2\ (aq) + 2H_2O \longrightarrow 4Fe(OH)_3\ (s)$$

Notice that complete dehydration of $Fe(OH)_3$ would give Fe_2O_3:

$2Fe(OH)_3 \rightarrow Fe_2O_3 + 3H_2O$

Iron(III) hydroxide is easily dehydrated and is better represented as a hydrated oxide, $Fe_2O_3 \cdot xH_2O$, in which the proportion of water is somewhat variable. This hydrated oxide is **rust.**

This mechanism for the rusting of iron explains some interesting observations. First, rusting only occurs if *both* oxygen and water are present. Iron won't rust in dry air or in oxygen free water. Second, the formation of rust often occurs at a site somewhat removed from the location where the iron is pitting. Have you noticed, for example, that a car "rusts out" *under* the paint around places where the paint has been scratched. The iron that dissolves migrates under the paint to the place of the scratch where it can be oxidized by O_2 to give the hydrated oxide.

Rusting has taken place beneath the paint on this very badly corroded car.

Cobalt

Cobalt is an important metal because it is used in many alloys having special properties. *Stellite,* for example, is an alloy containing cobalt, chromium, and tungsten that retains its hardness even when very hot. This property allows it to be used to make cutting tools such as drill bits for high-speed machining of steel parts. Cobalt is also used in catalysts.

The principle oxidation states of cobalt are 2+ and 3+. In the absence of complex ion forming substances other than water, the 2+ state is most stable. In most complex ions, the 3+ state is very easily formed and is the most stable.

Nickel

Nickel is very useful because it resists corrosion, and because its alloys have desirable properties. Electroplating of steel parts with nickel gives them a thin protective coating. Nickel, along with chromium, is added to iron to produce stainless steel, and iron-nickel alloys are used for armor plating because they are impact resistant. Even the familiar "nickel" five-cent piece is a copper-nickel alloy.

Nickel salts are added to glass to give it a green color.

The most stable oxidation state of nickel is 2+. Like the other transition elements, nickel forms many complex ions and many nickel salts are green because they contain the $Ni(H_2O)_6^{2+}$ ion. An important compound of nickel in the 4+ oxidation state is NiO_2—the cathode material in the nickel-cadmium battery. When the battery discharges, the nickel is reduced to the 2+ state.

$$NiO_2 \ (s) + Cd \ (s) + 2H_2O \longrightarrow Ni(OH)_2 \ (s) + Cd(OH)_2 \ (s)$$

The coinage metals – copper, silver, and gold

The name "coinage metals" given to the Group IB elements is not surprising since these metals have been used for many centuries to make coins as well as jewelry. They are relatively unreactive and their reactivity decreases going down the group. As a result, gold is found free in nature, as are silver and copper, but silver and especially copper also are found in deposits of their compounds.

All the coinage metals have practical uses. As mentioned earlier, copper has the second highest electrical conductivity of any metal and is used extensively for electrical wiring. Silver is an even better conductor than copper, but its lower abundance and its value as a decorative metal makes it too expensive to use generally in place of copper. Nevertheless, U.S. industries use over 175 million ounces of silver each year—about 24 percent is consumed by the electronics industry. Large amounts of silver are also used to make photographic film and paper. Even gold has practical applications. Low voltage electrical contacts are often gold plated because of gold's resistance to corrosion. Even a thin film of corrosion on contacts made of other metals would seriously impede the flow of electricity in low voltage circuits.

Photofinishers recover and recycle large amounts of silver from the films and prints they process.

Copper, silver, and gold have more positive reduction potentials than hydrogen, which means that they cannot displace hydrogen from acids such as HCl or H_2SO_4 in which the strongest oxidizing agent is H^+. Copper and silver do dissolve in HNO_3, however, because it contains NO_3^-, which serves as the oxidizing agent.

$$3Cu \ (s) + 8H^+ \ (aq) + 2NO_3^- \ (aq) \longrightarrow 3Cu^{2+} \ (aq) + 2NO \ (g) + 4H_2O$$

$$3Ag \ (s) + 4H^+ \ (aq) + NO_3^- \ (aq) \longrightarrow 3Ag^+ \ (aq) + NO \ (g) + 2H_2O$$

Gold is much more difficult to oxidize than copper or silver, and even concentrated HNO_3 is not a strong enough oxidizing agent to do the job. However, aqua regia—a 3-to-1 mixture of concentrated HCl and NHO_3—does dissolve gold slowly because the chloride ion stabilizes the Au^{3+} by forming a complex ion with it.

$$Au \ (s) + 6H^+ \ (aq) + 3NO_3^- \ (aq) + 4Cl^- \ (aq) \longrightarrow AuCl_4^- \ (aq) + 3H_2O + 3NO_2 \ (g)$$

Each of the Group IB metals forms compounds in the 1+ oxidation state. Examples are CuCl, AgCl, and AuCl, all of which are insoluble in water. For copper and gold, the 1+ oxidation states are not especially stable and tend to disproportionate. In aqueous solution, for example, copper(I) ion spontaneously gives metallic copper and copper(II).

$$2Cu^+ (aq) \longrightarrow Cu (s) + Cu^{2+} (aq)$$

The most stable oxidation state of gold is 3+, so gold(I) compounds tend to disproportionate to give the free metal and gold(III).

By far, the most stable oxidation state for silver is the 1+ state. The most important compound of silver is silver nitrate, $AgNO_3$—made by dissolving silver in nitric acid. It serves as the starting material for nearly all other silver compounds—for instance, the halides AgCl, AgBr, and AgI that are used in photographic films and papers.

The ions of the Group IB metals form many complex ions. Copper salts are pale blue in aqueous solutions because of the pale blue ion $Cu(H_2O)_4^{2+}$. Addition of ammonia produces the deep blue $Cu(NH_3)_4^{2+}$ ion (see Color Plate 22). The qualitative test for silver ion in a solution uses a similar complex between Ag^+ and NH_3. First the solution to be tested is made acidic with HCl, which precipitates white AgCl, as well as the chlorides of lead ($PbCl_2$) and mercury(I), (Hg_2Cl_2) if any of these metal ions are in the solution. To determine whether the precipitate contains any AgCl, it is treated with aqueous NH_3. This causes AgCl to dissolve, but not $PbCl_2$ or Hg_2Cl_2 because lead and mercury do not form soluble complex ions with ammonia.

The precipitate might be just $PbCl_2$ and/or Hg_2Cl_2.

$$AgCl (s) + 2NH_3 (aq) \rightleftharpoons Ag(NH_3)_2^+ (aq) + Cl^- (aq)$$

The solution, which would contain $Ag(NH_3)_2^+$ and Cl^-, is then acidified. If any AgCl had dissolved when the aqueous NH_3 was added, it is now reprecipitated because the equilibrium is shifted to the left as NH_3 is converted to NH_4^+ by the acid.

$NH_3 + H^+ \rightarrow NH_4^+$

Zinc, cadmium, and mercury

These metals are often not considered true transition elements because their d subshells are completed. Each has a valence shell consisting of only two electrons in an s orbital, so their highest oxidation state is 2+. In fact, zinc and cadmium show only a 2+ oxidation state (other than zero, of course), while mercury has oxidation states of 2+ and 1+.

Zinc and cadmium are both silvery, rather reactive metals. They dissolve readily in nonoxidizing acids such as HCl and H_2SO_4 with the evolution of hydrogen. In fact, the reaction of zinc with dilute H_2SO_4 is a common way of preparing hydrogen in the laboratory.

$$Zn (s) + H_2SO_4 (aq) \longrightarrow ZnSO_4 (aq) + H_2 (g)$$

The reactivities of zinc and cadmium account for one of their principal uses as free metals—providing corrosion protection for iron and steel. Coating an iron or steel object with zinc is called **galvanizing.** The zinc protects the metal in two ways. First, it reacts with moisture and CO_2 to form a film of $Zn_2(OH)_2CO_3$ that prevents oxygen and moisture from coming in contact with and reacting with the zinc below or with the iron. But even if this zinc coating is scratched through to the iron, the zinc still protects the iron by electrolytic action. Zinc is more easily oxidized than iron, so when the metals are in contact the zinc becomes the anode of a galvanic cell and iron becomes the cathode. Since oxidation always occurs at the anode, zinc is oxidized, but iron is protected by being the cathode. The phenomenon is called **cathodic protection.** Cadmium plating on steel functions in a similar way, but cadmium is used less often than zinc for a number of reasons. One is that cadmium is less abundant than zinc, and is therefore more expensive. Another is that cadmium compounds are very toxic—they cause high blood pressure, heart disease, and can even lead to painful death.

The shiny zinc coating on a galvanized steel object such as a garbage pail becomes dull as the zinc reacts with air and moisture to form $Zn_2(OH)_2CO_3$.

Cadmium generally occurs in nature as an impurity in zinc ores.

Zinc wire being placed beside the Alaskan pipeline. When connected electrically to the pipeline, it prevents corrosion by providing cathodic protection.

Zinc and cadmium have uses other than as protective coatings over other metals. Zinc is the metal used as the anode in the common dry cell, and it is alloyed with copper to produce *brass* and with copper and tin to produce *bronze.* Cadmium is used to make rechargeable nickel-cadmium batteries.

Mercury is the only metal that is a liquid at ordinary room temperatures. It freezes at $-38.9°C$ and boils at $357°C$. This liquid range has led to its widespread use as the fluid in thermometers. As we learned earlier, mercury has the ability to dissolve many metals to form solutions called amalgams.

Not <u>all</u> metals dissolve in mercury. Iron is an example of one that doesn't.

Cadmium and zinc have many similar chemical properties, which differ considerably from those of mercury. For instance, we have seen that both Zn and Cd dissolve in dilute acids with the evolution of hydrogen. Mercury, however, is considerably less reactive. It doesn't dissolve in acids such as HCl or H_2SO_4, although it does react with nitric acid.

$$3Hg\ (l) + 8H^+\ (aq) + 2NO_3^-\ (aq) \longrightarrow 3Hg^{2+}\ (aq) + 2NO\ (g) + 4H_2O$$

One important difference between zinc and cadmium is that zinc is amphoteric but cadmium is not. For example, zinc dissolves in base.

$$Zn\ (s) + 2OH^-\ (aq) + 2H_2O \longrightarrow Zn(OH)_4^{2-}\ (aq) + H_2\ (g)$$

Its hydroxide is also amphoteric. When base is added to Zn^{2+} in aqueous solution, a precipitate of $Zn(OH)_2$ is formed.

$$Zn^{2+}\ (aq) + 2OH^-\ (aq) \longrightarrow Zn(OH)_2\ (s)$$

The zinc hydroxide dissolves either in acid or additional base.

$$Zn(OH)_2\ (s) + 2H^+\ (aq) \longrightarrow Zn^{2+}\ (aq) + 2H_2O$$

$$Zn(OH)_2\ (s) + 2OH^-\ (aq) \longrightarrow Zn(OH)_4^{2-}\ (aq)$$

This chemical difference between zinc and cadmium explains why metals are sometimes cadmium plated rather than zinc plated. Cadmium is used if the metal is to be exposed to an alkaline environment, because the base will destroy a zinc coating but not a cadmium one.

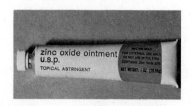

Tubes of zinc oxide ointment can be purchased for use as sun screens.

Many zinc compounds have important commercial uses. Zinc oxide, which forms when zinc is heated in air, is used as a paint pigment, in creams that are spread on the skin as sun screens, and in fast-setting dental cements. Zinc chloride is used in many different ways, including deodorants, embalming, and fireproofing lumber.

As mentioned earlier, mercury forms compounds in both the 1+ and 2+ oxidation states. In the 1+ state, two mercury(I) ions are joined by a covalent bond to give Hg_2^{2+}. The existence of this ion is supported by equilibrium studies (see Review Problem 21.93), by X-ray determination of crystal structures, and by the fact that mercury(I) compounds are diamagnetic. If there were a simple Hg^+ ion, it would have an odd number of electrons, so all the electrons couldn't be paired and Hg^+ would be paramagnetic. Pairing of the odd electron with an electron from another Hg^+ gives a covalent bond and a diamagnetic Hg_2^{2+} species.

Mercury compounds have quite different properties from those of zinc and cadmium. They are considerably less ionic, for example, and $HgCl_2$ in aqueous solution exists primarily (99%) as $HgCl_2$ molecules.

HgCl₂ is a weak electrolyte.

$$HgCl_2 \ (aq) + H_2O \rightleftharpoons Hg(OH)Cl + H^+ + Cl^-$$

As you probably know, mercury and its compounds are very toxic. Mercury spills should be avoided in the laboratory because the vapor pressure of mercury (about 10^{-3} torr at room temperature) is sufficient to cause mercury poisoning if the vapors are breathed for long periods of time. Soluble mercury compounds, such as mercury(II) chloride, are especially poisonous because they can quickly provide sufficient mercury to the body to cause death. By contrast, mercury(I) chloride, Hg_2Cl_2, is very insoluble in water and at one time—before the discovery of penicillin—it was used medicinally as a treatment for syphilis. Its low solubility prevents the body from absorbing lethal doses of mercury. Mercury is a cumulative poison, however, so even small amounts absorbed over extended periods can lead to serious medical problems.

One or two grams of HgCl₂ is a fatal dose.

Because Hg_2Cl_2 is insoluble in water, the presence of Hg_2^{2+} ion in a solution can be tested by treating the solution with HCl, which causes the Hg_2Cl_2 to precipitate. However, since AgCl and $PbCl_2$ are also white and insoluble, and would also be formed if Ag^+ and Pb^{2+} were present, the precipitate must be tested further. Addition of aqueous ammonia—which is part of the test for silver—causes the Hg_2Cl_2 to disproportionate.

Hg(NH₂)Cl is called mercury(II) amido chloride.

$$Hg_2Cl_2 + 2NH_3 \longrightarrow Hg(NH_2)Cl + Hg + NH_4^+ + Cl^-$$

The mixture of Hg and $Hg(NH_2)Cl$ appears black or gray because the black color of the finely divided mercury masks the white of the $Hg(NH_2)Cl$.

21.7 COORDINATION COMPOUNDS

As we've noted before, the transition elements are known for their ability to form many complex ions—substances in which a metal cation is surrounded by two or more ions or molecules that are referred to in general as ligands. Actually, the name complex *ion* is too restrictive because a number of these complex species are actually neutral. Therefore, to avoid problems of description they are often simply called *complexes*. They are also frequently called **coordination compounds,** because from the point of view of the valence bond theory—the first bonding theory to be applied to them—they are considered to be held together by *coordinate* covalent bonds between the ligands and the metal. Because of the extremely large number of these coordination compounds, as well as their unusual colors, magnetic properties, structures, and chemical reactions, their study has become one of the major areas of inorganic chemical research.

Coordination compounds have a number of important uses. You've learned that unexposed silver salts in photographic film and paper are removed by dissolving these salts in a solution containing thiosulfate ion, with which Ag^+ forms a complex ion. Complex ions are used in water softening (phosphates binding to iron and manganese ions) and as catalysts in a variety of industrial processes. The formation of complex ions has also been used to alleviate poisoning produced by beryllium and lead and to retard the spoilage of foods. As the study of biochemistry has progressed, it has also become evident that many biologically important molecules owe their biological activity to a metal ion held in a "complex ion" within the molecule. Hemoglobin, containing iron(II) atoms, is a well-known example. The importance of metal ions in biosystems (not to mention the increased availability of research funding in biochemistry), has recently turned many inorganic chemists into bioinorganic chemists. We will take a closer look at mental-containing biomolecules in Chapter 23.

The father of modern coordination chemistry was Alfred Werner, who received the Nobel Prize in chemistry in 1913 for his work on these compounds. Werner was the first to recognize that metal ions could combine with other molecules or ions through more than one type of "valence" to produce relatively stable complex species. He was also the first to propose structures of complex ions that were consistent with their properties.

Metal complexes are formed with many kinds of ligands, including ions such as Cl^-, CN^-, and NO_2^-, as well as neutral molecules such as H_2O or NH_3. Nearly all ligands, however, have one thing in common: they possess a lone pair of electrons that may be shared with the metal cation in coordinate covalent bonds. In this sense the formation of a complex can be viewed as a Lewis acid-base reaction; in general, we can expect that the ligands in coordination compounds will all be Lewis bases with few exceptions.

In a complex, the ligands attached to the metal are considered in a **first coordination sphere,** and in solution they are held tightly by the metal ion, compared to other ions and molecules that might also be present nearby in the mixture. When the formula of a metal complex is written, we usually indicate the species that are bonded to the metal in the first coordination sphere by enclosing them and the metal ion within square brackets. An example is the ion

$$[CoCl_6]^{3-}$$

One can usually tell from the context of a discussion whether or not square brackets mean molar concentration.

These brackets are *not* to be confused with those that we used earlier when we wished to denote molar concentration.[2] Note that the charge on the complex is indicated *outside* the brackets, showing that the entire complex ion carries, in this example, a charge of -3.

Ligands such as Cl^- or NH_3, which have one atom that can bond to a metal cation, are said to be **monodentate** (one "tooth") ligands. There are also many molecules and ions that are able to attach themselves to a metal ion through more than one donor atom to produce a cyclic *ring* type of arrangement. Two very common examples that have been much studied are oxalate ion, $C_2O_4^{2-}$,

and ethylenediamine, $H_2N—CH_2—CH_2—NH_2$,

[2] In our earlier discussions of complex ion equilibria, we avoided this notation specifically to prevent such confusion.

Figure 21.8

Structures of $[Co(C_2O_4)_3]^{3-}$ and $[Co(en)_3]^{3+}$. (a) $[Co(C_2O_4)_3]^{3-}$. The colored atom is cobalt, solid gray atoms are carbon, and white atoms are oxygen. (b) $[Co(en)_3]^{3+}$. The colored atom is cobalt, solid gray atoms are nitrogen, the large white atoms are carbon, and the small white atoms are hydrogen.

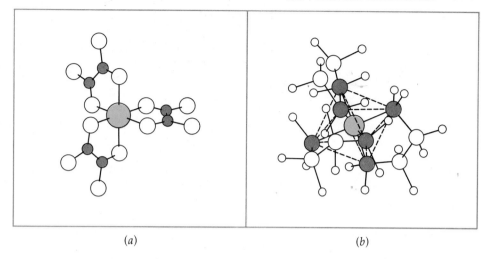

(a)   (b)

They bond to a metal ion as shown below.[3]

With cobalt(III), for example, these two ligands form complexes such as $[Co(C_2O_4)_3]^{3-}$ and $[Co(H_2NCH_2CH_2NH_2)_3]^{3+}$, whose structures are illustrated in Figure 21.8. Ethylenediamine is such a common ligand in coordination chemistry that in writing formulas containing it the abbreviation, **en,** is almost always used. The latter ion is therefore normally written as simply $[Co(en)_3]^{3+}$.

Molecules or ions such as ethylenediamine or oxalate ion, which have two atoms that may coordinate to a metal ion, are said to be **bidentate** ligands. There are also more complex **polydentate** ligands containing three, four, or even more donor atoms. Table 21.7 contains a list of some common mono- and bidentate ligands, as well as a few examples of polydentate ligands.

A particularly important polydentate ligand is ethylenediaminetetraacetic acid (EDTA). Its six coordinating atoms firmly attach themselves to free metal ions, and EDTA has been used as an antidote in lead poisoning because it binds Pb^{2+}, thereby preventing it from inhibiting certain important enzyme functions. EDTA also binds to iron and calcium and is used as water softeners in products such as shampoos. EDTA is added to foods to tie up metal ions that catalyze oxidation (and hence deterioration) of the food product. It has also been found that EDTA increases the storage life of whole blood by removing free Ca^{2+}, which promotes clotting.

To prevent EDTA from extracting calcium from the body, $CaNa_2EDTA$ is the salt that's added to food products such as salad dressings.

21.8 COORDINATION NUMBER

The term **coordination number (C.N.)** refers to the total number of ligand atoms that are bound to a given metal ion in a complex. These atoms may be supplied by either monodentate or polydentate ligands, or both. Thus, in the three complexes, $[CoCl_6]^{3-}$, $[Co(en)_2Cl_2]^+$, and $[Co(en)_3]^{3+}$, the C.N. of cobalt is the same, since in each case there are *six* donor atoms about the Co^{3+} ion.

[3] The origin of chemical terminology is sometimes rather colorful. Complexes of this general type are often called **chelates,** from the Greek *chele,* meaning claw. The ligand in this case bites the metal with two claws (donor atoms) much like a crab.

Table 21.7
Common ligands found in complex ions

Monodentate	H_2O	Water	Br^-	Bromide
	NH_3	Ammonia	I^-	Iodide
	CN^-	Cyanide	NO_2^-	Nitrite
	OH^-	Hydroxide	SCN^-	Thiocyanate
	F^-	Fluoride	$S_2O_3^{2-}$	Thiosulfate
	Cl^-	Chloride		

Bidentate

oxalate

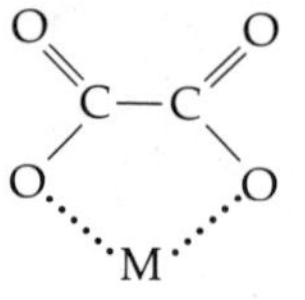

ethylenediamine

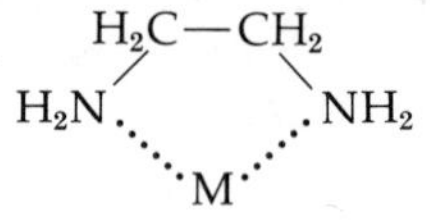

o-phenanthroline

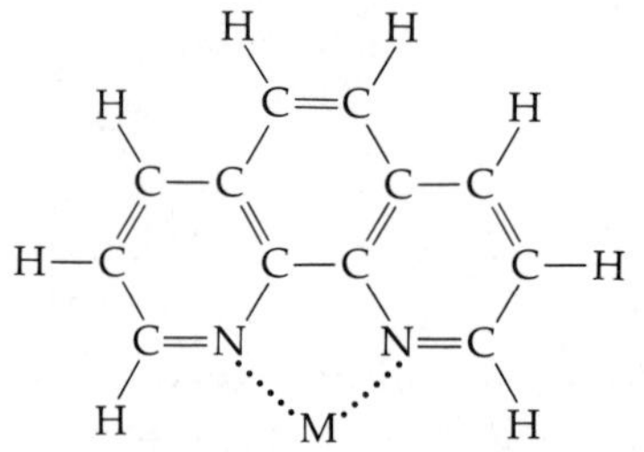

dipyridyl

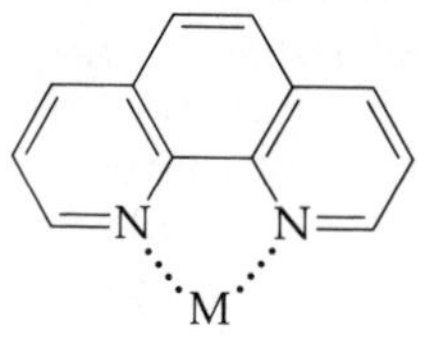

or

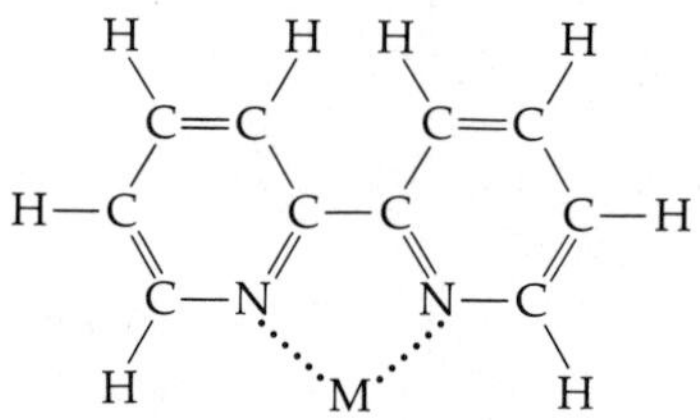

or

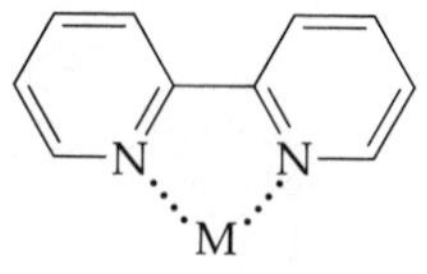

Polydentate

(coordinating atoms indicated with asterisks)

Diethylenetriamine (three coordinating atoms)

$$H_2\overset{*}{N}-CH_2-CH_2-\overset{*}{N}H-CH_2-CH_2-\overset{*}{N}H_2$$

Ethylenediaminetetraacetate (six coordinating atoms)

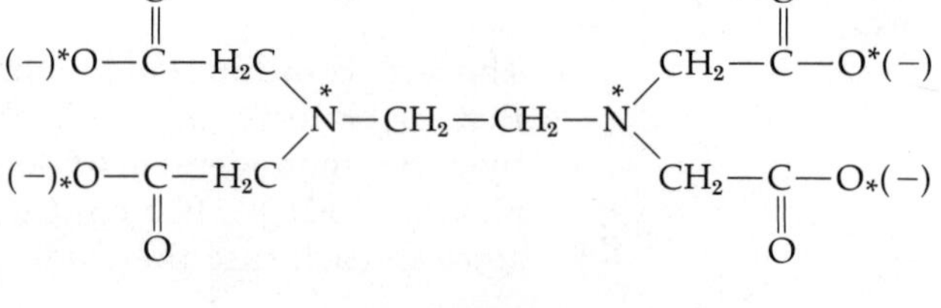

also called EDTA

C.N. = 2	Linear	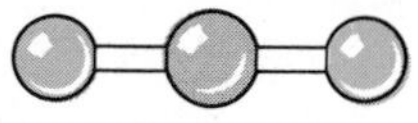
C.N. = 4	Tetrahedral	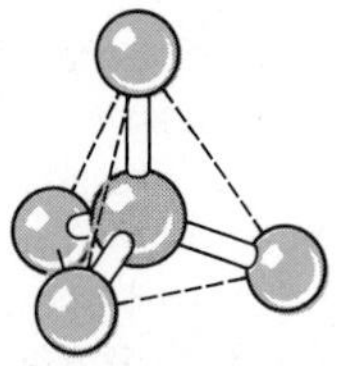
	Square planar	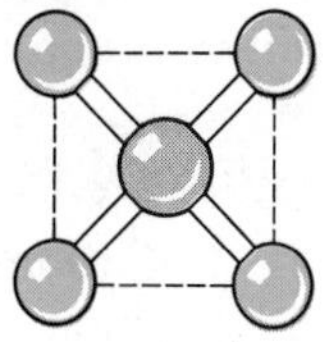
C.N. = 6	Octahedral	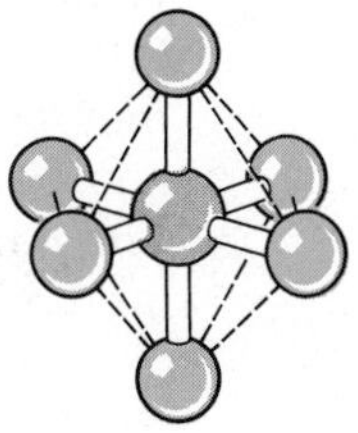

Figure 21.9
Structure types for coordination 2, 4, and 6.

Coordination numbers ranging from 2 to more than 8 are observed in various coordination compounds, with the C.N. in any given instance determined by the nature of the metal ion, its oxidation state, and to some extent, the ligands and the environment surrounding the complex. The most common coordination numbers are observed to be 2, 4, and 6. The basic structural types found for them are shown in Figure 21.9.

By far the most frequently occurring coordination number in transition metal complexes is 6, and the geometry that is observed in nearly all instances is octahedral. A simple two-dimensional way of representing the octahedral geometry was described in Chapter 5 on page 142 and is shown in Figure 21.10. The dashed rectangle represents the square plane that joins the upper and lower pyramids in the octahedron. The six solid lines connect the center of the metal cation to the coordinated ligand atoms. This arrangement is illustrated in Figure 21.10 for the complex ion, $[CoCl_6]^{3-}$.

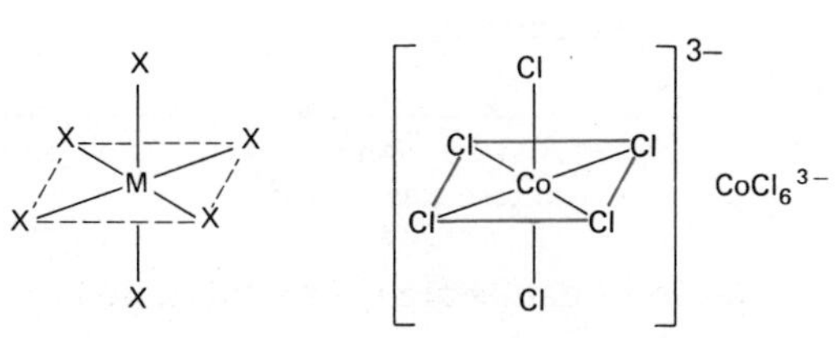

Figure 21.10
Two-dimensional representation of octahedral coordination.

21.9 NOMENCLATURE

The IUPAC is composed of a group of chemists drawn from all over the world. One of their tasks is to meet periodically to discuss current problems in nomenclature.

The naming of chemical compounds was introduced in Chapter 4 where we discussed the nomenclature system for simple inorganic compounds. This system, developed and kept up to date by the International Union of Pure and Applied Chemistry (IUPAC) has been extended to cover names for complexes, too. Below are some rules that have been developed by the IUPAC to name coordination complexes. A few of the names assigned following these rules may sound odd, and even funny. Remember, however, that we are primarily interested in formulating a name that is able to transmit the maximum amount of information in the shortest possible name. The end result is therefore sometimes difficult to pronounce.

Rules of nomenclature of coordination compounds

1. **Cationic species are named before anionic species.** This is just as in other cases of ionic compounds, such as NaCl, which is named as sodium chloride (cation, anion).

2. **Within a complex ion, the ligands are named first in alphabetical order followed by the metal ion.** This is opposite to the sequence in which they appear in the formula. For example, the complex $[Co(NH_3)_6]^{3+}$ is named by specifying the ammonia first, then the cobalt.

3. **The names of anionic ligands end in the suffix -o.**
   (a) Ligands whose names end in *ide* have this suffix replaced by *-o.*

Anion		Ligand
Chloride	Cl^-	Chloro
Bromide	Br^-	Bromo
Cyanide	CN^-	Cyano
Oxide	O^{2-}	Oxo

   (b) Ligands whose named end in *-ite* or *-ate* become *-ito* and *-ato*, respectively.

Anion		Ligand
Carbonate	CO_3^{2-}	Carbonato
Thiosulfate	$S_2O_3^{2-}$	Thiosulfato
Thiocyanate	SCN^-	Thiocyanato (bonded through sulfur) Isothiocyanato (bonded through nitrogen)
Oxalate	$C_2O_4^{2-}$	Oxalato
Nitrite (bonded through oxygen, ONO^-)[a]	NO_2^-	Nitrito

   [a] An exception to this is NO_2^- when bonded through nitrogen, in which case it is named as *nitro.*

4. **Neutral ligands are given the same names as the neutral molecule.** Thus ethylenediamine as a ligand is called ethylenediamine in the name of

the complex. Two very important exceptions to this, however, are

In all but the most recent literature, water is named as <u>aquo</u>.

| H_2O | Aqua | NH_3 | Ammine (note double *m*) |

5. **When there are more than one of a particular ligand, their number is specified by di = 2, tri = 3, tetra = 4, penta = 5, hexa = 6, and so forth. When confusion might result, the prefixes bis = 2, tris = 3, tetrakis = 4, and so forth, are employed.** Thus the presence of two chloride ions is specified as *dichloro*. However, because ethylenediamine already contains the term *di*, two of these molecules are indicated by placing the name of the ligand in parentheses preceded by the term *bis*; that is, *bis(ethylenediamine)*.

6. **Negative (anionic) complex ions always end in the suffix -ate.** This suffix is appended to the English name of the metal atom in most cases.

Element	Metal as Named in Anionic Complex
Aluminum	Aluminate
Chromium	Chromate
Manganese	Manganate
Nickel	Nickelate
Cobalt	Cobaltate
Zinc	Zincate
Molybdenum	Molybdate
Tungsten	Tungstate

For some metals, the *ate* is appended to the Latin stem.

Except for mercury, metals whose symbols are derived from their Latin names are specified in anionic complexes by using the Latin stem.

Element	Stem	Metal as Named in Anionic Complex
Iron	Ferr-	Ferrate
Copper	Cupr-	Cuprate
Lead	Plumb-	Plumbate
Silver	Argent-	Argentate
Gold	Aur-	Aurate
Tin	Stann-	Stannate

In neutral or positively charged complexes the metal *always* appears with the common English name for the element.

7. **The oxidation number of the metal in the complex is written in Roman numerals within parentheses following the name of the metal.** For example,

$[Co(H_2O)_6]^{3+}$ is the hexaaquacobalt(III) ion.

$[CoCl_6]^{3-}$ is the hexachlorocobaltate(III) ion.

Note that the charge on the complex is obtained as the *algebraic sum* of the oxidation number of the metal and the charges on the ligands.

Some additional examples illustrate these rules.

Notice that the alphabetical order of the ligands is determined by the ligand name, not the number-prefix.

$[Ni(CN)_4]^{2-}$	Tetracyanonickelate(II) ion
$[Co(NH_3)_4Cl_2]^+$	Tetraamminedichlorocobalt(III) ion
$Na_3[Cr(NO_2)_6]$	Sodium hexanitrochromate(III)
$[Ag(NH_3)_2]^+$	Diamminesilver(I) ion
$[Ag(CN)_2]^-$	Dicyanoargentate(I) ion
$[Co(en)_3]Cl_3$	Tris(ethylenediamine)cobalt(III) chloride
$[Cr(NH_3)_3Cl_3]$	Triamminetrichlorochromium(III)

21.10 ISOMERISM AND COORDINATION COMPOUNDS

When two different compounds have the same molecular formula, but differ in the way that their atoms are arranged, they are said to be **isomers** of one another. For example, there are two compounds with the general formula

$$Cr(NH_3)_5SO_4Br$$

One of these we should formulate as

$$[Cr(NH_3)_5SO_4]Br$$

because it yields a precipitate of AgBr when treated in aqueous solution with $AgNO_3$ but does not give a precipitate of $BaSO_4$ when treated with $Ba(NO_3)_2$. This last observation means that the $SO_4{}^{2-}$ is not free in the solution and, hence, must be bound to the chromium.

The second compound is written as

$$[Cr(NH_3)_5Br]SO_4$$

and produces $BaSO_4$ when treated with $Ba(NO_3)_2$. On the other hand, addition of $AgNO_3$ to a solution of the compound does not yield AgBr.

The two compounds just described have different chemical properties and are clearly different chemical substances, even though they are composed of the same number of the same kinds of atoms. This particular type of isomerism is not uncommon among coordination compounds and is called **ionization isomerism.**

Another type of isomerism that is very important is called **stereoisomerism,** and results when a given molecule or ion can exist in more than one structural form in which the same atoms are bonded to one another but find themselves oriented differently in space. To illustrate this, we will focus our attention on octahedral complexes because they represent the most common structural type.

The simplest form of stereoisomerism results when a complex has the general formula Ma_4b_2 in which a and b represent monodentate ligands. An example would be the ion, $[Co(NH_3)_4Cl_2]^+$. (How would you name it?) This complex can exist in two different isomeric forms, called **geometrical isomers,** as shown in Figure 21.11. As you can see, in one of these isomers the two b ligands are located across from one another on opposite sides of the metal ion. Such an

Figure 21.11

Cis-trans isomers for complexes Ma_4b_2.

cis–Ma_4b_2

trans–Ma_4b_2

cis   trans

N ⌒ N represents the bidentate ethylenediamine ligand

Figure 21.12

Cis-trans isomerism in
$[Cr(en)_2Cl_2]^+$.

isomer is given the designation **trans** (Latin *trans* means "across"). The other isomer has the two *b* ligands adjacent to one another and is referred to as the **cis** isomer (L. *cis* = on the same side). Thus the two isomers would be specified as

$$trans\text{-}[Co(NH_3)_4Cl_2]^+$$

and

$$cis\text{-}[Co(NH_3)_4Cl_2]^+$$

Because *cis-* and *trans-*isomers possess different structures, they are different chemical species, each with its own set of chemical and physical properties. While these properties may often be similar, the fact that they are not identical clearly tells us that the two structures represent truly different compounds.

Geometrical isomers also occur when there are bidentate ligands in a complex, as illustrated by the *cis* and *trans* forms of the ion $[Cr(en)_2Cl_2]^+$ shown in Figure 21.12. Once again, in the *trans* form we see that the chloride ligands are on opposite sides of the metal while in the *cis* form they are alongside each other.

A second form of stereoisomerism is called **optical isomerism.** As we will see, optical isomers affect polarized light differently and bear the same structural relationship to each other as do your left and right hands—that is, they are *nonsuperimposable mirror images* of one another. To see what this means, try this simple experiment. Place your right hand in front of a mirror, with the palm toward the mirror, and hold your left hand alongside with the palm facing you (Figure 21.13). Notice that the image of your *right* hand in the mirror looks the same as your left hand. That is why we say that your left and right hands are mirror images of each other. The nonsuperimposable aspect arises because your

Bidentate ligands always span adjacent or <u>cis</u> positions in a complex.

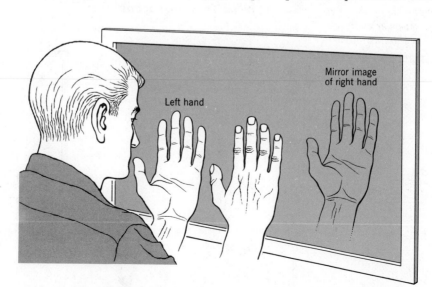

Left hand

Mirror image of right hand

Figure 21.13

Illustration of nonsuperimposable mirror images.

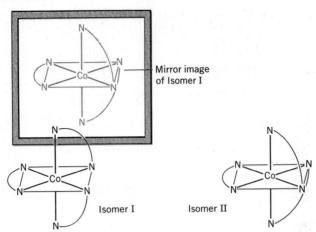

Mirror image of Isomer I

Isomer I   Isomer II

Figure 21.14

The mirror image relationship between the isomers of the chiral $[Co(en)_3]^{3+}$ ion. Isomer II is the same as the mirror image of Isomer I, but I and II are not identical. They are nonsuperimposable isomers.

If you make a model of a molecule and use its reflection as a guide in constructing its mirror image, you can test for superimposability.

Figure 21.15

Isomers of $[Co(en)_2Cl_2]^+$. The mirror image of the trans isomer is identical to the original, so the trans isomer is not chiral. The mirror image of the cis isomer is not superimposable on the original, so the cis isomer exists as two nonidentical optical isomers.

left and right hands, while similar in appearance, do not match exactly when one is placed over the other, both with palms down; the thumbs point in different directions. This difference is perhaps seen even more clearly if you attempt to place your right hand into a left-hand glove; it doesn't fit properly. Thus your left and right hand, and optical isomers too, cannot be superimposed on each other.

Molecules or ions that can have two structures related to each other in the same way that your left and right hands are related—that is, as nonsuperimposable mirror images—are said to be **chiral** (from the Greek *cheir*, meaning "hand"). An example of a chiral complex is the $[Co(en)_3]^{3+}$ ion. The mirror image relationship between the two isomers is shown in Figure 21.14 in which the hydrogen atoms have been omitted for simplicity. A pair of isomers related in this manner are called **enantiomers.**

In the complex $[Co(en)_3]^{3+}$ it is the arrangement of the chelate rings that gives rise to the chirality, or optical isomerism. In fact, any octahedral complex containing three bidentate ligands is chiral and can exist as two optical isomers.

Optical isomerism is also important for the *cis* form of complex ions containing two bidentate ligands and two monodentate ligands—for example, *cis*-$[Co(en)_2Cl_2]^+$ (Figure 21.15). The *trans* form of this complex is not chiral and does not exhibit optical isomerism, however, because it and its mirror image are identical. They are superimposable.

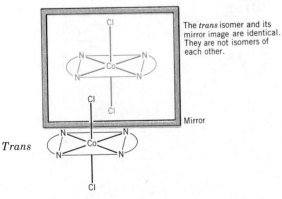

Trans

The *trans* isomer and its mirror image are identical. They are not isomers of each other.

Mirror

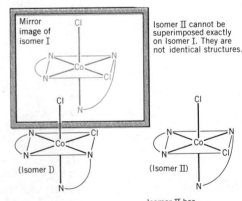

Cis

Mirror image of isomer I

Isomer II cannot be superimposed exactly on Isomer I. They are not identical structures.

(Isomer I)   (Isomer II)

Isomer II has the same structure as this mirror image of isomer I

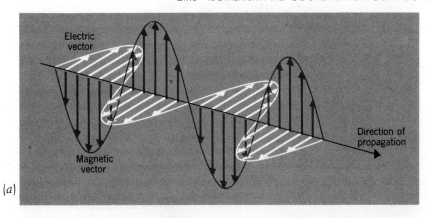

(a)

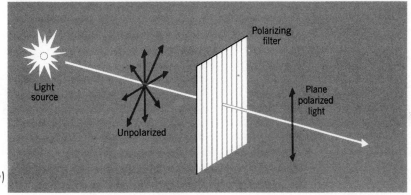

(b)

Figure 21.16

Polarized light. (a) Electromagnetic radiation composed of electric and magnetic vectors. (b) Orientation of electric vectors in unpolarized and polarized light.

Quartz crystals occur in two nonsuperimposable forms and they are optically active.

In general, the properties of optical isomers are identical except for the way in which they interact with outside influences that are able to distinguish between left and right handedness. The situation here is analogous to having a group of baseball players, some of whom are left-handed and some right-handed. Since they are all able to toss a baseball with equal ease, a baseball will not differentiate between left- and right-handed players. The same applies to a bat, since there are no left- or right-handed baseball bats. A fielder's glove, however, will fit only one hand. A glove designed to be worn on the left hand cannot be used by a player who catches the ball in the right hand. In this case, the glove differentiates between these two kinds of players because it too has a left or right handedness to it. In this same fashion, optical isomers interact in an identical way with most chemical reagents and physical probes. They do differ, however, in the way they react toward other optically active chemicals and toward polarized light.

Light, in general, is composed of electromagnetic radiation that possesses both electric and magnetic components that behave like vectors. These vectors oscillate in a sinusoidal fashion perpendicular to the direction in which the light wave is traveling (Figure 21.16). If we examine the electric vectors, all different orientations are observed in an unpolarized beam. However, when such a beam is passed through a polarizing medium, only the vibrations in one plane remain. The result is called **plane polarized light.** A unique feature of optical isomers is that when plane polarized light is passed through them (or their solutions) the plane of polarization is rotated through some angle, θ, as shown in Figure 21.17. Substances that affect polarized light this way are said to be optically active. One enantiomer (optical isomer) causes the light to be rotated to the right—that is, clockwise when viewed down the axis of the oncoming light

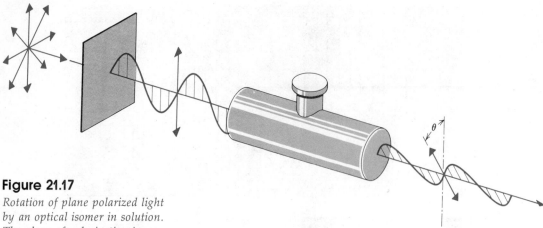

Figure 21.17

Rotation of plane polarized light by an optical isomer in solution. The plane of polarization is rotated by an angle θ as it passes through the solution containing the optically active compound. In this example the light is rotated to the left; the substance in the solution is said to be levorotatory.

beam—and is said to be **dextrorotatory.** The other enantiomer causes the polarized beam to be rotated to the left and is described as **levorotatory.** The two isomers are therefore designated as *d* or *l* depending on the direction of rotation of the polarized light.

An equal mixture of two enantiomers tends to rotate polarized light to both the left and right simultaneously. These effects therefore cancel each other, and such a mixture shows no optical activity; it is said to be **racemic.** In almost all cases, when enantiomers are produced in a chemical reaction, they are formed in equal numbers so that a racemic mixture results. One of the arts in chemistry is the separation of optical isomers from one another.

21.11 BONDING IN COORDINATION COMPOUNDS: VALENCE BOND THEORY

There are three important properties of transition metal complexes that must be explained by a bonding theory: (1) structure, (2) magnetic properties, and (3) color.

In the earliest theories of coordination complexes the metal was considered to be attached to the ligands by way of coordinate covalent bonds, and the first serious attempt to explain the structures and magnetic properties of complexes was made by applying the concepts of valence bond theory. Much of what follows is simply an extension of some of the ideas that were developed in Chapter 5.

In valence bond theory, you recall, a bond is formed by the overlap of two orbitals and the subsequent sharing of a pair of electrons between the two atoms in the region of overlap. Ligands, as a rule, do not possess unpaired electrons, so the bonding in the complex must result from the overlap of ligand orbitals containing lone pairs of electrons with vacant orbitals on the metal ion, thereby giving rise to a coordinate covalent bond as illustrated in Figure 21.18. To see how this occurs, let's consider an octahedral ion such as the blue-violet $[Cr(H_2O)_6]^{3+}$, characteristic of simple Cr(III) salts in aqueous solution. The electronic structure of a free chromium atom is

Cr [Ar] ↑ ↑ ↑ ↑ ↑   ↑   __ __ __   __ __ __ __ __
                      3*d*   4*s*   4*p*   4*d*

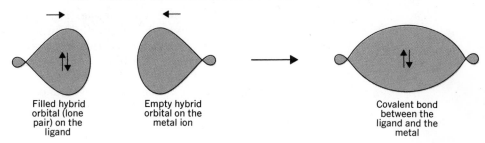

Figure 21.18

Formation of a coordinate covalent bond by the overlap of a filled ligand orbital and an empty hybrid orbital of the metal ion.

Filled hybrid orbital (lone pair) on the ligand

Empty hybrid orbital on the metal ion

Covalent bond between the ligand and the metal

The **4s** electrons are lost before any of the **3d** electrons.

where we have shown the empty $4p$ and $4d$ subshells as well as those that are occupied by electrons. The central Cr^{3+} ion in our complex results from the loss of three electrons to give

$$Cr \quad [Ar] \quad \uparrow \ \uparrow \ \uparrow \ \_\ \_ \quad \_ \quad \_\ \_\ \_ \quad \_\ \_\ \_\ \_\ \_$$
$$ 3d 4s 4p 4d$$

In Chapter 5 we saw that an octahedral structure is formed when the central atom uses hybrid orbitals derived from two atomic d orbitals, an s orbital, and three p orbitals. We called them sp^3d^2 hybrids. We have the additional requirement here that the orbitals used by the metal to create the hybrid must be empty so that the electron pairs from the ligands can be placed into them when the bonds are formed. In this case, suitable hybrids can be constructed using the two vacant $3d$ orbitals. The $3d$ orbitals are preferred over the $4d$ because they are lower in energy; stronger bonds are formed when the $3d$ orbitals are used instead of the $4d$. In this case we will call the hybrids d^2sp^3, recognizing that the d orbitals come from the shell beneath the one that supplies the s and p orbitals. If we use dots to represent ligand electrons, we can now write the electronic structure for the complex as

$$\overset{\textstyle d^2sp^3}{}$$
$$[Cr(H_2O)_6]^{3+} \quad \uparrow \ \uparrow \ \uparrow \ \overset{..}{}\ \overset{..}{} \quad \overset{..}{} \quad \overset{..}{}\ \overset{..}{}\ \overset{..}{} \quad \_\ \_\ \_\ \_\ \_$$
$$\phantom{[Cr(H_2O)_6]^{3+} \quad} 3d 4s \phantom{\overset{..}{x}} 4p \phantom{\overset{..}{x}\ \overset{..}{x}} 4d$$

Experimental measurements demonstrate the presence of three unpaired electrons in this ion, as suggested by our bonding picture.

Let's consider next the emerald-green complex ion, $[Ni(H_2O)_6]^{2+}$. The electron configuration of the Ni^{2+} ion, obtained in the same manner as Cr^{3+} above, is

$$Ni^{2+} \quad [Ar] \quad \uparrow\downarrow \ \uparrow\downarrow \ \uparrow\downarrow \ \uparrow \ \uparrow \quad \_ \quad \_\ \_\ \_ \quad \_\ \_\ \_\ \_\ \_$$
$$\phantom{Ni^{2+} \quad [Ar] \quad} 3d 4s 4p 4d$$

Once again we must have a set of six vacant hybrids in order to obtain the octahedral geometry. However, this time we cannot use a pair of $3d$ orbitals to form the hybrid set. At best, we could obtain only one empty $3d$ orbital if we paired all the electrons in the $3d$ subshell. When the hybrids are created, both d orbitals must come from the *same* subshell. This means that two $4d$ orbitals must be used. The electronic structure of the complex then becomes

$$\overset{\textstyle sp^3d^2}{}$$
$$[Ni(H_2O)_6]^{2+} \quad \uparrow\downarrow \ \uparrow\downarrow \ \uparrow\downarrow \ \uparrow \ \uparrow \quad \overset{..}{} \quad \overset{..}{}\ \overset{..}{}\ \overset{..}{} \quad \overset{..}{}\ \overset{..}{}\ \_\ \_\ \_$$
$$\phantom{[Ni(H_2O)_6]^{2+} \quad} 3d 4s \phantom{\overset{..}{x}} 4p \phantom{\overset{..}{x}\ \overset{..}{x}} 4d$$

Note that the complex contains two unpaired electrons. This is also in agreement with experiment.

We have now seen two complex ions that can be considered to employ an octahedral set of hybrids for bonding. In valence bond language, when $3d$ orbitals are used to form the hybrids, an **inner orbital complex** results. On the other hand, when the $4d$ orbitals are used an **outer orbital complex** is formed.

In these last two examples there really was no choice as to which type of bonding (inner or outer orbital) would occur. Let's look at a situation now where we do have a choice. An example is Co(III). The electronic structure of the Co^{3+} ion is

$$Co^{3+} \quad [Ar] \quad \underset{3d}{\uparrow\downarrow \; \uparrow \; \uparrow \; \uparrow \; \uparrow} \quad \underset{4s}{\_\_} \quad \underset{4p}{\_\_ \; \_\_ \; \_\_} \quad \underset{4d}{\_\_ \; \_\_ \; \_\_ \; \_\_ \; \_\_}$$

In this case the hybrids can be formed in either of two ways. One is to make use of two $4d$ orbitals, thereby giving an outer orbital complex. The second is to pair the electrons together to produce two vacant $3d$ orbitals that can be used in the hybrids. This gives rise to an inner orbital complex.

Outer orbital

The Roman numeral superscript indicates the oxidation state of the cobalt.

$$Co^{III}X_6 \quad \underset{3d}{\uparrow\downarrow \; \uparrow \; \uparrow \; \uparrow \; \uparrow} \quad \overbrace{\underset{4s}{\cdot\cdot} \quad \underset{4p}{\cdot\cdot \; \cdot\cdot \; \cdot\cdot} \quad \underset{4d}{\cdot\cdot \; \cdot\cdot} \; \_\_ \; \_\_ \; \_\_}^{sp^3d^2}$$

Inner orbital

$$Co^{III}X_6 \quad \underset{3d}{\uparrow\downarrow \; \uparrow\downarrow \; \uparrow\downarrow} \overbrace{\cdot\cdot \; \cdot\cdot \quad \underset{4s}{\cdot\cdot} \quad \underset{4p}{\cdot\cdot \; \cdot\cdot \; \cdot\cdot}}^{d^2sp^3} \quad \underset{4d}{\_\_ \; \_\_ \; \_\_ \; \_\_ \; \_\_}$$

Notice that we can distinguish experimentally between these two possibilities by examining the number of unpaired electrons in the complex. The outer orbital complex has four unpaired electrons and is paramagnetic; the inner orbital complex has none and is diamagnetic.

Both inner and outer orbital complexes are possible whenever the metal ion contains either four, five, or six d electrons. In each case, two empty $3d$ orbitals can be created by the pairing of electrons. Failure to pair them, however, leads to the formation of outer orbital complexes.

When there is a choice, what determines whether inner or outer orbital bonding will occur? To answer this question we must consider two opposing factors:

1. As stated earlier, when $3d$ orbitals are used to form the hybrid orbitals, stronger metal-ligand bonds result than when $4d$ orbitals are used. This favors the pairing of electrons and the production of inner orbital complexes.

2. The pairing of electrons required to produce the necessary vacant $3d$ orbitals for inner orbital bonding also requires an input of energy. This is because electrons, which repel each other, are being forced to occupy the same orbital. Since this pairing energy doesn't have to be invested if the $4d$ orbitals are used to form the hybrids, this factor favors the production of outer orbital complexes.

The way in which these two factors come into play is illustrated in Figure 21.19. In the first drawing we see that the energy released when the bonds are formed using hybrids composed of $3d$ orbitals is so great that it more than compensates for the pairing energy, and the resulting inner orbital complex is of lower energy than the outer orbital complex. In this case the preferred complex

Lower energy means greater stability.

Figure 21.19

Energy changes in the production of inner orbital and outer orbital complexes. (a) inner orbital favored; $E_{outer} > E_{inner}$. (b) Outer orbital favored; $E_{outer} < E_{inner}$. (E_b = Energy released upon bond formation. P = pairing energy.)

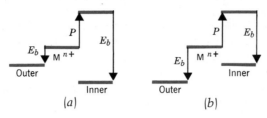

(a)   (b)

would be the inner orbital one. On the right in Figure 21.19 we find the other situation—the energy released on inner orbital bond formation does not lead to an overall lower energy than that achieved with the formation of an outer orbital complex. In this case, outer orbital bonding would occur in preference to inner orbital bonding.

As a general rule most ligands tend to give inner orbital complexes with first-row transition elements having either d^4 or d^6 configurations. Exceptions are the ligands H_2O and F^-, which usually produce outer orbital complexes.

With a $3d^5$ configuration the subshell is half-filled, and we have noted earlier that a half-filled subshell in which all electrons have the same spin possesses a certain extra stability. As a result, these electron configurations are difficult to disturb and electron pairing is difficult to accomplish. Consequently, metal ions with a d^5 structure tend to keep their electrons spread out with parallel spins and thus tend to form outer orbital complexes with most ligands. An exception to this occurs with cyanide ion. This particular ligand forms very stable metal-ligand bonds,[4] sufficiently stable that electron pairing of the d^5 configuration can occur. This is illustrated by the hexacyanoferrate(III) ion, $[Fe(CN)_6]^{3-}$. For iron(III) we have

The $[Fe(CN)_6]^{3-}$ ion is also called the <u>ferricyanide</u> ion.

$$Fe^{3+}\quad [Ar] \quad \underset{3d}{\uparrow\ \uparrow\ \uparrow\ \uparrow\ \uparrow} \quad \underset{4s}{—} \quad \underset{4p}{—\ —\ —} \quad \underset{4d}{—\ —\ —\ —\ —}$$

and for the $[Fe(CN)_6]^{3-}$ inner orbital complex,

$$[Fe(CN)_6]^{3-}\quad \underset{3d}{\underset{}{\uparrow\downarrow\ \uparrow\downarrow\ \uparrow}}\ \overbrace{\underset{..\ ..}{}\quad \underset{4s}{..}\quad \underset{4p}{..\ ..\ ..}}^{d^2sp^3}\quad \underset{4d}{—\ —\ —\ —\ —}$$

Other geometries

Valence bond theory can also be applied to geometries other than octahedral. For example, the complex ion, $[Ni(CN)_4]^{2-}$, is found to have a square planar shape and is diamagnetic. To have a square planar geometry, a dsp^2 set of hybrid orbitals must be used by the nickel ion. Once again we have nickel(II) and hence a d^8 electron configuration.

$$Ni^{2+}\quad [Ar] \quad \underset{3d}{\uparrow\downarrow\ \uparrow\downarrow\ \uparrow\downarrow\ \uparrow\ \uparrow} \quad \underset{4s}{—} \quad \underset{4p}{—\ —\ —}$$

We can create the one empty d orbital needed for the hybrids by pairing the electrons in the $3d$ subshell, so that for the complex we have

$$[Ni(CN)_4]^{2-}\quad \underset{3d}{\uparrow\downarrow\ \uparrow\downarrow\ \uparrow\downarrow\ \uparrow\downarrow}\ \overbrace{\underset{..}{}\quad \underset{4s}{..}\quad \underset{4p}{..\ ..}}^{dsp^2}\ —$$

[4] Cyanides, such as KCN or HCN, are very poisonous because of the ability of CN^- to irreversibly bind to iron atoms in hemoglobin and because CN^- is able to inactivate certain enzymes.

Since we have paired all the electrons, the complex should be diamagnetic, which it is.

Tetrahedral complexes such as $[CoCl_4]^{2-}$ can also be accounted for. This ion, containing cobalt(II) with a d^7 configuration, makes use of tetrahedral sp^3 hybrids and has the electronic structure

$$[CoCl_4]^{2-} \quad \underset{3d}{\uparrow\downarrow \; \uparrow\downarrow \; \uparrow \; \uparrow \; \uparrow} \quad \overbrace{\underset{4s}{\cdot\cdot} \quad \underset{4p}{\cdot\cdot \; \cdot\cdot \; \cdot\cdot}}^{sp^3}$$

From our discussion, valence bond theory appears to be quite effective at accounting for the structures and magnetic properties of complex ions. However, there are some problems with the theory. One very serious drawback is that it does not allow us to explain why complex ions exist in such a wide profusion of colors, even where they contain the same metal ion in the same oxidation state (Color Plate 23). There are also certain complex ions that are quite difficult to account for in a reasonable and satisfying way. The ion $[Co(NO_2)_6]^{4-}$ is one. This ion contains Co(II), a d^7 ion,

$$Co^{2+} \quad [Ar] \quad \underset{3d}{\uparrow\downarrow \; \uparrow\downarrow \; \uparrow \; \uparrow \; \uparrow} \quad \underset{4s}{—} \quad \underset{4p}{—\;—\;—} \quad \underset{4d}{—\;—\;—\;—\;—}$$

With water we saw that cobalt(II) forms an outer orbital complex containing three unpaired electrons. The $[Co(NO_2)_6]^{4-}$ ion, however, contains only one unpaired electron; therefore, it must be postulated that two of the three unpaired electrons in Co^{2+} become paired and that the third is promoted to the $4d$ subshell to make two $3d$ orbitals available for inner orbital complex formation. The final result looks like this:

$$[Co(NO_2)_6]^{4-} \quad \underset{3d}{\uparrow\downarrow \; \uparrow\downarrow \; \uparrow\downarrow} \quad \overbrace{\underset{4s}{\cdot\cdot \; \cdot\cdot} \quad \underset{}{\cdot\cdot} \quad \underset{4p}{\cdot\cdot \; \cdot\cdot \; \cdot\cdot}}^{d^2sp^3} \quad \underset{4d}{\uparrow \;—\;—\;—\;—}$$

Thus with valence bond theory we can account for the magnetic properties of this ion, but they certainly are not what we would have predicted. Let us now look at another bonding theory that manages to avoid some of the pitfalls of valence bond theory as applied to these transition metal complexes.

21.12 CRYSTAL FIELD THEORY

A second theory of bonding in transition metal complexes, that has been extensively applied over the past 25 years, is called **crystal field theory** (CFT). It was developed by physicists in the early 1930s to deal with metal ions trapped in crystalline lattices, but it was not until the early 1950s that chemists realized that it could be applied to coordination complexes in general. It differs from valence bond theory in that it views the complex as held together by purely electrostatic attractions, that is, *in its simplest form, CFT ignores covalent bonding*. The most significant aspect of the theory, however, is its concern with the effect that the ligands have on the energies of the d orbitals of the metal.

Generally, the ligands in a transition metal complex are either anions, or they are polar molecules. When they are polar molecules the negative ends of the ligand dipoles point in the direction of the metal cation. Let's examine how these ligands affect the d orbitals. One of the simplest complex ions that we can consider for this purpose is the $[Ti(H_2O)_6]^{3+}$ cation, consisting of a Ti^{3+} ion surrounded octahedrally by six water molecules. Titanium(III) has a single $3d$ elec-

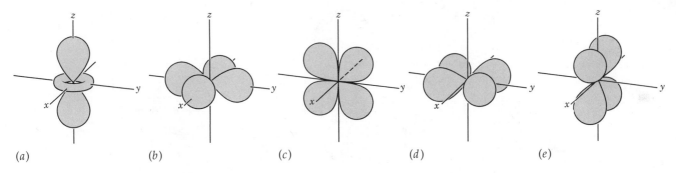

(a)   (b)   (c)   (d)   (e)

Figure 21.20

Directional properties of the d orbitals. (a) d_{z^2}. (b) $d_{x^2-y^2}$. (c) d_{yz}. (d) d_{xy}. (e) d_{xz}.

The labels for the **d** orbitals have their origins in the mathematics of quantum mechanics.

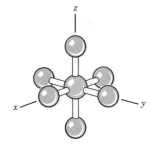

Figure 21.21

An octahedral arrangement of ligands about a central metal ion produced by placing the ligands along the x, y, and z axes.

tron,

$$Ti^{3+} \quad \uparrow \; \_\_ \; \_\_ \; \_\_ \qquad \_\_ \qquad \_\_ \; \_\_ \; \_\_$$
$$\qquad\qquad 3d \qquad\qquad 4s \qquad 4p$$

Which of the five $3d$ orbitals will this electron prefer to occupy? Before we can answer this question, we must examine the shapes and directional properties of d orbitals.

In Chapter 3 we discussed the shapes of the p orbitals; each one consists of a pair of lobes directed along a coordinate axis. The d orbitals are somewhat more complex, as we can see in Figure 21.20. Four of them, labeled d_{xy}, d_{xz}, d_{yz}, and $d_{x^2-y^2}$, have the same shape, and are composed of four lobes each. The fifth, the d_{z^2}, consists of two large lobes directed along the positive and negative z axis plus a donut of charge in the xy plane. For our purposes here, it is important to notice that two of these d orbitals have lobes that are pointed along the coordinate axes—the $d_{x^2-y^2}$ and d_{z^2} orbitals. The other three—the d_{xy}, d_{xz}, and d_{yz}—have lobes that point between the axes at 45° angles to them.

Now let's consider constructing an octahedral complex by placing the six ligands along the xyz axes as shown in Figure 21.21. Since the ligands are negatively charged, or have the negative ends of their dipoles pointing at the metal ion, when the complex is formed, electrons in the d orbitals will feel an electrostatic repulsion and their energies will be raised. What is especially important, however, is that an electron in a $d_{x^2-y^2}$ or d_{z^2} orbital will be repelled *more* than an electron in one of the d_{xy}, d_{xz}, or d_{yz} orbitals because the $d_{x^2-y^2}$ and d_{z^2} orbitals point directly at the ligands (Figure 21.22). As a result, the energies of the $d_{x^2-y^2}$ and d_{z^2} are raised more than the energies of the d_{xy}, d_{xz}, and d_{yz} orbitals, as shown in Figure 21.23. This splits the d subshell into two energy levels. The lower one consists of the d_{xy}, d_{xz}, and d_{yz} orbitals and the higher one has the $d_{x^2-y^2}$ and d_{z^2} orbitals. For reasons beyond the scope of this book, in an octahedral complex the lower level is labeled t_{2g} and the upper level is labeled e_g. The energy difference between the t_{2g} and e_g levels is called the **crystal field splitting** and is usually indicated by the symbol Δ.

Returning to the $[Ti(H_2O)_6]^{3+}$ ion, we see that its single d electron will have the lowest energy if it occupies one of the orbitals of the t_{2g} level. We can also understand what happens when the complex absorbs light (Figure 21.24). If the energy of the light that strikes the complex is equal to Δ, the light can be absorbed and the electron can be raised from the t_{2g} level to the e_g level. The energy of this absorbed light, of course, depends on the magnitude of Δ and, as you learned in Chapter 3, the energy of a light wave, E, is also related to its frequency, ν.

$$E = h\nu \qquad (h \text{ is Planck's constant})$$

For most complexes, the magnitude of Δ is such that the frequencies of light that are absorbed lie in the visible portion of the spectrum. Since the color of light is related to its frequency, the color of the complex depends on the fre-

Figure 21.22
Interaction of the ligands with the d orbitals of the metal.
(a) d_{z^2}. *(b)* $d_{x^2-y^2}$. *(c)* d_{yz}.
(d) d_{xz}. *(e)* d_{xy}.

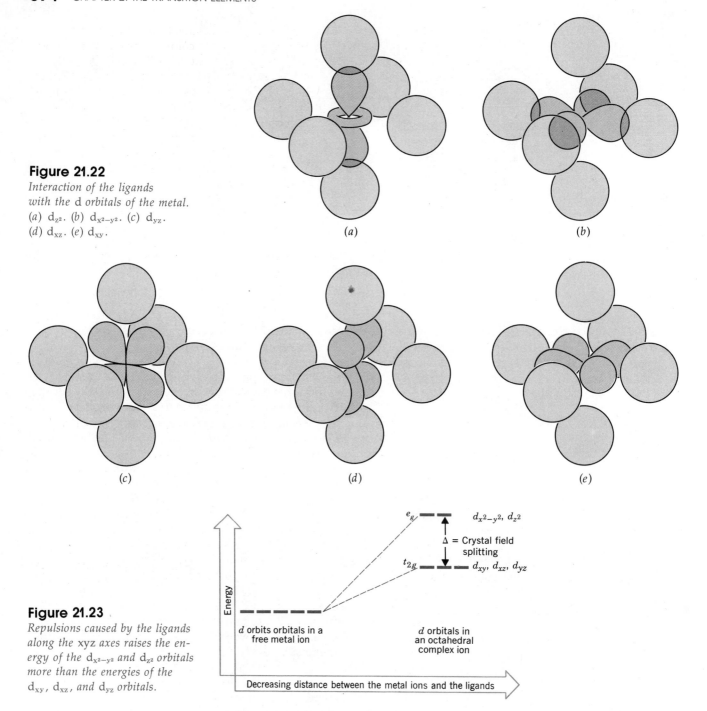

(a)

(b)

(c)

(d)

(e)

Figure 21.23
Repulsions caused by the ligands along the xyz axes raises the energy of the $d_{x^2-y^2}$ and d_{z^2} orbitals more than the energies of the d_{xy}, d_{xz}, and d_{yz} orbitals.

Energy

e_g ———— $d_{x^2-y^2}$, d_{z^2}

Δ = Crystal field splitting

t_{2g} ———— d_{xy}, d_{xz}, d_{yz}

d orbits orbitals in a free metal ion

d orbitals in an octahedral complex ion

Decreasing distance between the metal ions and the ligands

Figure 21.24
Absorption of light (hν) by the $[Ti(H_2O)_6]^{3+}$ ion promotes the electron from the t_{2g} to the e_g level.

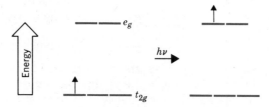

Energy

e_g

$hν$

t_{2g}

The perception of the color of compounds was discussed in Section 9.6.

quencies that are absorbed when white light is reflected from it or passes through it. In other words, we see the complement of the color of the light that is absorbed. For example, the magnitude of Δ for the $[Ti(H_2O)_6]^{3+}$ ion corresponds to the energy of light in the yellow-green portion of the spectrum. Therefore, when white light passes through a solution of this complex, yellow-green light is absorbed and the light that emerges appears violet.

For a given metal ion, different ligands have different effects on the magnitude of the splitting of the d orbitals—that is, on Δ. By examining the absorption spectra of various complexes, we can arrange the ligands in the order of their ability to produce a large Δ. This series is called the **spectrochemical series** and can be given in abbreviated form as

$$I^- < Br^- < Cl^- < F^- < OH^- < H_2O < NH_3 < en < NO_2^- < CN^-$$

Thus I^- is poorest at splitting the energies of the t_{2g} and e_g levels, and CN^- is best. What is particularly interesting is that this same series applies for essentially any metal in any oxidation state. However, although the order is usually the same, the actual magnitude of Δ for a given complex in a given geometry depends on the ligand, the metal, and its oxidation state.

As with nearly any generalization we might make in attempting to describe chemical properties, there are exceptions. This is true here with the order of ligands in the spectrochemical series, because in some instances the relative positions of neighboring ligands in the series is reversed. With cobalt(III), for example, Cl^- appears to produce a greater crystal field splitting than F^-. Nevertheless, the spectrochemical series often serves as a useful guide in understanding, and sometimes even predicting, the properties of complexes. For instance, we have just seen how CFT accounts for the colors of complexes. We can explain their magnetic properties as well. Consider, for example, the cobalt(III) complexes of F^- and Cl^-. The metal ion here contains six d electrons and, in a weak crystal field—one that produces a small Δ—they will be unpaired as much as possible, as shown in Figure 21.25a, to give a complex with four unpaired electrons. This is what occurs with F^- in the $[CoF_6]^{3-}$ ion.

When the ligand produces a large crystal field splitting we have the possibility of pairing all of the d electrons in the t_{2g} level (Figure 21.25b) to produce a diamagnetic complex. This will occur if the magnitude of Δ is greater than the energy needed to pair the electrons in a given orbital. In other words, when the pairing energy (let's call it P) is less than Δ, more energy is required to place the electron in the e_g orbital than is required to pair them and place them in the t_{2g} level. This happens when the ligand is Cl^-.

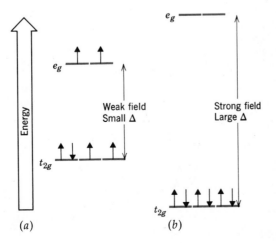

Figure 21.25

Electron configuration of Co(III) in weak and strong crystal fields.

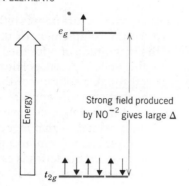

Figure 21.26

Pairing of electrons in t_{2g} level in $[Co(NO_2)_6]^{4-}$.

For Co(III) complexes (or for that matter, any d^6 system), a paramagnetic complex with four unpaired electrons will occur whenever $\Delta < P$; diamagnetic complexes will be formed when $\Delta > P$. In general, we speak of these two possibilities as **high-spin** complexes (minimum pairing of electrons) and **low-spin** complexes (maximum pairing of electrons from the e_g into the t_{2g}). Comparing them to the valence bond treatment, we find that low-spin complexes correspond to inner orbital complexes whereas high-spin complexes correspond to outer orbital complexes.

The possibility of both low-spin and high-spin complexes exists when the central metal ion contains four, five, six, or seven d electrons. The electron configurations of the t_{2g} and e_g levels in these species are left to you as an exercise

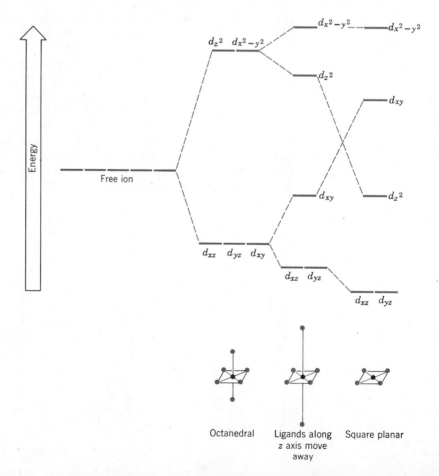

Figure 21.27

The splitting pattern of the d *orbitals changes as the geometry of the complex changes.*

(Question 21.92). For d^1, d^2, and d^3 systems the electrons will naturally prefer the three low-energy t_{2g} orbitals because no pairing is required; therefore, only one type of electron configuration will be found for them. Likewise, with a d^8 or d^9 configuration, six electrons will be forced to occupy the t_{2g} (thereby filling it) and the e_g level will contain either two or three electrons, respectively. Once again we see that d^8 and d^9 ions will each have only one type of electron configuration in an octahedral complex.

With this as background, we note that the magnetic properties of the $[Co(NO_2)_6]^{4-}$ ion, which presented such a problem with the valence bond theory, are easily explained in terms of the CFT. Recall that this complex contains the Co^{2+} ion, a d^7 system. From the spectrochemical series we also note that NO_2^- produces a very strong crystal field; therefore, we would expect that Δ is probably quite large, larger in fact than the pairing energy. Under these circumstances there will be pairing of electrons in the t_{2g} level, as shown in Figure 21.26. Since the t_{2g} level can accommodate only six electrons (three pairs), the seventh electron is forced to occupy the e_g level. The complex is therefore low-spin (analogous to inner orbital of the valence bond theory) and will contain a single unpaired electron, in agreement with experiment. Thus we see that there is really nothing unusual about the $[Co(NO_2)_6]^{4-}$ ion, quite opposite to what we would have concluded based on the valence bond theory.

Crystal field theory can be extended to other geometries besides octahedral; the difference is that other splitting patterns are observed. For example, a square planar complex can be thought of as being derived from an octahedral complex by removing the ligands that lie along the z axis. As shown in Figure 21.27, when this occurs the energies of the d_{z^2}, d_{xz} and d_{yz} orbitals decrease because an electron placed into them experiences less repulsion than in an octahedral complex. Also, by removing the ligands along the z axis, those along the x and y axes can move in slightly and therefore the energies of the $d_{x^2-y^2}$ and d_{xy} orbitals rise somewhat. In the $[Ni(CN)_4]^{2-}$ ion, the energy separation between the d_{xy} and $d_{x^2-y^2}$ is large enough so that the eight d electrons of the Ni^{2+} ion can exist as four pairs (Figure 21.28).

Finally, in tetrahedral complexes the splitting pattern of the d orbitals is that shown in Figure 21.29. Notice that the order of the energy levels is exactly opposite to that found in octahedral complexes. The magnitude of Δ is also considerably smaller (actually, $\Delta_{\text{tet}} \approx \frac{4}{9}\Delta_{\text{oct}}$ for the same ligands and metal ion). The small Δ observed for tetrahedral complexes is always less than the pairing energy, and tetrahedral complexes are always of the high-spin variety. Note that this agrees with the valence bond theory in which sp^3 hybrids were used by the metal thereby making all the d orbitals available for spreading out the d electrons.

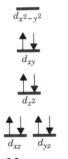

Figure 21.28

Electron distribution among the d *orbitals of the diamagnetic* $[Ni(CN)_4]^{2-}$ *complex.*

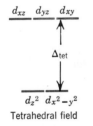

Figure 21.29

Splitting pattern for a tetrahedral field. $\Delta_{\text{tet}} \approx \frac{4}{9}\Delta$ *oct.*

INDEX TO QUESTIONS AND PROBLEMS (Problem numbers are in **bold type**)

REVIEW QUESTIONS

21.1 In general, what distinguishes a transition element from a representative element?

21.2 Which elements are called *inner transition elements?*

21.3 What similarities exist between elements in the A and B groups?

21.4 Why do the Group VIII elements consist of *three* columns?

21.5 Which elements are in the iron triad?

21.6 Make a list of the *d*-block elements you are familiar with and give as many applications of each as you can.

21.7 What are the four general properties we normally associate with the transition metals?

21.8 Write the electron configurations of the members of the first transition series.

21.9 Why do many transition metals exhibit a 2+ oxidation state in their compounds?

21.10 Which *d*-block elements react with water with the evolution of hydrogen? Give a chemical equation for the reaction.

21.11 Why do many of the transition metals exhibit multiple oxidation states?

21.12 How do the relative stabilities of high and low oxidation states vary within the *d*-block elements?

21.13 Which is probably the better oxidizing agent, CrO_4^{2-} or WO_4^{2-}?

21.14 Which would you expect to be the stronger oxidizing agent, Cr^{3+} or Ni^{3+}?

21.15 Which would you expect to be more easily oxidized, Cr^{2+} or Fe^{2+}?

21.16 Which would you expect to be a better oxidizing agent, Cu^{3+} or Au^{3+}?

21.17 Why are the chemical properties of the lanthanide elements so similar?

21.18 Construct a graph showing how atomic radius varies among the *d*-block elements. Plot atomic radius vertically and atomic number, from 21 to 30, horizontally. On the graph, plot the radii of the elements for each of the three transition series. Do you see evidence of the lanthanide contraction? What general relationship is there between atomic radius and the number of unpaired *d* electrons?

21.19 The chemistries of Zr and Hf are very similar; the elements are always found together in nature and are difficult to separate from one another. What explanation can be offered to account for the similar properties of Zr and Hf compounds?

21.20 What is an *ore?*

21.21 What are the three steps involved in extracting a metal from its ore and making it ready for practical use?

21.22 Why is "panning" able to separate sand and mud from tiny particles of gold?

21.23 What is an amalgam? How is it used in the recovery of gold from gold ores?

21.24 Describe the process called flotation. What is meant by *roasting* as applied to metallurgy?

21.25 Write chemical equations for the purification of bauxite, Al_2O_3.

21.26 Why is TiO_2 a better paint pigment than white lead? Why is titanium metal useful in the manufacture of aircraft?

21.27 Write chemical equations for (a) the reduction of Fe_2O_3 in the blast furnace and (b) the production of slag from SiO_2 and $CaCO_3$.

21.28 Why must pig iron be refined to be useful as a strong structural metal? Describe the Bessemer converter; the open hearth furnace; the basic oxygen process. Which of these is the principal method used today to make steel?

21.29 Describe the Mond process.

21.30 Compare paramagnetism and ferromagnetism. Why can a ferromagnetic material become permanently magnetized?

21.31 Why is chromium used to coat other metals such as steel? What are the most important oxidation states of chromium?

21.32 What is stainless steel? How does it differ from ordinary steel?

21.33 Why is chromium(III) ion in aqueous solution acidic?

21.34 Write chemical equations showing the amphoteric behavior of chromium(III) hydroxide.

21.35 CrO_3 is the anhydride of $CrO_2(OH)_2$, which was written as H_2CrO_4 in the text because it is acidic. On the other hand, Cr_2O_3 is the anhydride (at least in a formal sense) of $Cr(OH)_3$, which exhibits both basic and acidic properties. On the basis of what you learned in Chapter 9 about the factors that affect the acidity of X—O—H bonds, explain why $CrO_2(OH)_2$ is more acidic than $Cr(OH)_3$.

21.36 Based on your answer to Question 21.35, explain why the oxides of metals in high oxidation states tend to be acid anhydrides instead of basic anhydrides.

21.37 When solutions containing chromate ion (CrO_4^{2-}) are acidified, dichromate ion ($Cr_2O_7^{2-}$) is produced. Use structural formulas to indicate how this polymerization occurs. (*Hint:* How does polymerization of H_3PO_4 occur? See Section 20.1 if necessary.)

21.38 What oxoanions of sulfur are analogous to CrO_4^{2-} and $Cr_2O_7^{2-}$?

21.39 What are the principal oxidation states of manganese? Which one is most stable with respect to redox in aqueous solution?

21.40 Why is $KMnO_4$ a useful titrant for redox reactions in acidic solutions? Why is it not used for reactions in neutral or basic solutions?

21.41 Write an equation for the disproportionation of manganate ion in an acidic solution.

21.42 Why does iron have so many practical uses?

21.43 What are the oxides of iron? Which one is magnetic?

21.44 What is the apparent mechanism for the rusting of iron in moist air?

21.45 Write chemical equations for the reaction of hydrochloric acid with (a) iron (b) manganese.

21.46 Why is cobalt an important metal? What are the principal oxidation states of cobalt?

21.47 Why is nickel such a useful metal?

21.48 Why are aqueous solutions of many nickel salts green? What practical application is there for the compound, NiO_2?

21.49 How does the ease of oxidation of the coinage metals vary? How is this related to how they are found in nature?

21.50 Give a practical application for each of the metals: copper, silver, and gold.

21.51 What oxidation states are observed for each of the coinage metals?

21.52 Why can't hydrochloric acid be used to dissolve the coinage metals? Which of them react with nitric acid? Give chemical equations.

21.53 Write a chemical equation for the reaction of gold with aqua regia.

21.54 Which compounds of silver are used in photography?

21.55 Describe how a solution can be tested for the presence of Ag^+. Give all the important chemical equations.

21.56 What happens when concentrated aqueous ammonia is added to a solution containing copper(II) ion?

21.57 What reactions occur, if any, when dilute sulfuric acid is added to (a) zinc (b) cadmium (c) mercury?

21.58 What is *cathodic protection*?

21.59 What is galvanizing? How does it protect steel?

21.60 Why is cadmium sometimes used in place of zinc as a protective coating over steel? Why is cadmium not used more often as a protective coating over other metals?

21.61 What are two common alloys that contain zinc?

21.62 What are some common uses for zinc oxide?

21.63 Aqueous solutions of $HgCl_2$ are poor conductors of electricity. Why? What equilibrium accounts for the small degree of conductivity that is observed?

21.64 Describe how a solution can be tested for the presence of Hg_2^{2+}.

21.65 Define ligand, first coordination sphere, coordination compound, monodentate ligand, polydentate ligand, chelate, and coordination number.

21.66 What are some applications of coordination compounds?

21.67 What are some uses of EDTA?

21.68 Sketch the common structures found for complex ions with coordination number 4 and coordination number 6.

21.69 Sketch the structure of an octahedral EDTA complex.

21.70 Nitrilotriacetic acid, NTA (structure shown below), was used by detergent manufacturers for a while in place of phosphates because it is biodegradable and does not promote the growth of algae. However, it was found to increase the solubility of some heavy metals that are poisonous to a variety of life forms, and its use has been discontinued. Can you suggest, using appropriate chemical equations and structural formulas, how NTA dissolves metal compounds?

nitrilotriacetate ion

21.71 Give IUPAC names for each of the following:
(a) $[Ni(NH_3)_6]^{2+}$
(b) $[CrCl_3(NH_3)_3]^0$
(c) $[Co(NO_2)_6]^{3-}$
(d) $[Mn(C_2O_4)_3]^{3-}$
(e) MnO_4^-

21.72 What are the IUPAC names for the following:
(a) $[AgI_2]^-$
(b) $[Cr(NH_3)_5Cl]^{2+}$
(c) $[Co(H_2O)_4(NH_3)_2]Cl_2$
(d) $[Co(en)_2(H_2O)_2]_2(SO_4)_3$
(e) $[Cr(NH_3)_4Cl_2]Cl$

21.73 Write chemical formulas for the following:
(a) Tetraaquadicyanoiron(III) ion
(b) Tetraammineoxalatonickel(II)
(c) Potassium hexacyanomanganate(III)
(d) Tetrachlorocuprate(II) ion
(e) Tetraoxochromate(VI) ion

21.74 Write chemical formulas for the following:
(a) Tetrachloroaurate(III) ion
(b) Bis(ethylenediamine)dinitroiron(III) sulfate
(c) Tetraamminecarbonatocobalt(III) nitrate
(d) Ethylenediaminetetraacetatoferrate(II) ion
(e) Dithiosulfatoargentate(I) ion

21.75 What is meant by *isomer*? What are stereoisomers?

21.76 Sketch the isomers of $[Co(NH_3)_2Cl_4]^-$. Identify *cis* and *trans* isomers. How many isomers are there for the complex, $[Co(NH_3)_3Cl_3]$? Sketch them.

21.77 What condition must be met for a molecule or ion to be *chiral*?

21.78 Why are chiral substances said to be optically active?

21.79 Sketch the isomers of $[Cr(en)_2Cl_2]^+$. Identify *cis* and *trans* isomers and indicate any isomers that exhibit optical isomerism.

21.80 Draw the two optical isomers of $[Co(EDTA)]^-$.

21.81 What are enantiomers? What is meant by *racemic*?

21.82 What is the difference between an inner orbital complex and an outer orbital complex?

21.83 Use valence bond theory to predict the electron configuration, the type of bonding (inner orbital or outer orbital), and the number of unpaired electrons for each of the following:

(a) $[VCl_6]^{3-}$
(b) $[Ni(NH_3)_6]^{2+}$
(c) $[Fe(NH_3)_6]^{3+}$
(d) $[Co(CN)_6]^{3-}$
(e) $[CrCl_6]^{3-}$

21.84 Predict the number of unpaired electrons in (a) $[Cr(H_2O)_6]^{2+}$ (b) $[Cr(CN)_6]^{4-}$

21.85 What magnetic properties would you predict for the square planar complex $[Cu(NH_3)_4]^{2+}$?

21.86 Sketch on appropriate coordinate axes the shapes of the five *d* orbitals.

21.87 Diagram the crystal field splitting of the *d* orbitals, and indicate the electron population of each energy level, in the paramagnetic complex $[Mn(H_2O)_6]^{3+}$. Label the energy levels.

21.88 How does crystal field theory account for the colors of complex ions?

21.89 What relationship exists between Δ (the crystal field splitting) and the pairing energy in determining whether a given complex will be paramagnetic or diamagnetic?

21.90 What are meant by high-spin and low-spin complexes? How do these compare with inner orbital and outer orbital complexes in the valence bond theory?

21.91 Sketch the CFT splitting patterns of the *d* orbitals for
(a) square planar and
(b) tetrahedral complexes.

21.92 Using the CFT splitting pattern for an octahedral complex, indicate the high-spin and low-spin distribution of electrons among the t_{2g} and e_g levels for the configurations: (a) d^4, (b) d^5, (c) d^6, (d) d^7.

REVIEW PROBLEM

21.93 Saturated solutions of mercurous chloride were prepared by adding the solid to solutions having various chloride ion concentrations. The total concentration of mercury in each solution was then determined. Use the concepts that you learned having to do with equilibrium and solubility product to show that the data at the right are only consistent with mercury(I) having the formula Hg_2^{2+}, [and hence mercury(I) chloride being Hg_2Cl_2], and not with Hg^+ (and therefore HgCl).

Chloride Ion Concentration	Moles of Mercury per Liter
1.0 *M*	2.2×10^{-18}
0.5 *M*	8.8×10^{-18}
0.2 *M*	5.5×10^{-17}
0.1 *M*	2.2×10^{-16}

22

ORGANIC CHEMISTRY

Petroleum is a complex mixture of hydrocarbons that is separated into its various components in oil refineries such as this. These components serve not only as fuels but also as important raw materials for many consumer products, from synthetic fibers and plastics to medicinals. In this chapter we will learn about some of the kinds of organic compounds and their uses.

As far back as the eighteenth century chemists were able to distinguish between two types of compounds: those derived from plants and animals and those that come from the mineral constituents of the earth. The second kind, called inorganic substances, have received much of our attention in this book. This chapter and the next are devoted to a discussion of organic compounds, many of which are found in nature but most of which have been synthesized in the laboratory.

One does not have to delve very far into the chemistry of life before observing that the element carbon is present in all the molecules in life's makeup. Of the 100 odd elements, carbon is the only one found universally in these substances. Thus, organic chemistry has become known as the study of carbon and its compounds. Besides carbon, organic molecules contain relatively few other elements; among the most prevalent are hydrogen, oxygen, nitrogen, and, to a lesser extent, phosphorus and sulfur.

Until 1828 chemists believed that the only source of these organic compounds was from nature itself. It was thought to be impossible to synthesize them in the laboratory because nature's "vital force" was missing. In 1828 Friedrich Wöhler first synthesized urea—a compound found in urine—from inorganic materials. During one of his experiments, Wöhler evaporated a solution of the inorganic salt ammonium cyanate, NH_4CNO. When he heated the solid he discovered that another substance was formed, which he analyzed and found to be urea.

$$NH_4^+CNO^- \xrightarrow{\text{heat}} H_2N-\overset{\overset{\textstyle O}{\|}}{C}-NH_2$$
ammonium cyanate   urea

Once the results of Wöhler's synthesis were known, many scientists began to attempt to prepare other organic materials with a good deal of success. Gradually, the vital force theory was abandoned, and during the next several years a great profusion of organic compounds were made. Today there are over two million known organic compounds—compared to about 100,000 inorganic substances—and new ones are being discovered every day.

The great abundance of organic compounds is a result of carbon's ability to bond to itself to form long chains, rings, and complex combinations of both. By comparison, silicon (which is below carbon in the periodic table) is able to form only relatively short chains when bonded to itself. Silicon is not able to form chains containing thousands of atoms bonded together as carbon does as, for example, in rubber, plastics, and synthetic fibers.

In organic chemistry we are fortunate to be able to systematically categorize a gigantic number of compounds into a relatively small number of groups quite successfully. In this chapter we will survey a number of these groups and look at some sample organic reactions. The goal here will be simply to provide an overview of some topics that are of importance in this vast field of organic chemistry.

Friedrich Wöhler—the father of organic chemistry.

22.1 HYDROCARBONS

We begin our classification with the **hydrocarbons,** which are compounds containing only carbon and hydrogen. All the remaining types of organic compounds can then be looked on as being derived from the hydrocarbons. Hydrocarbons can be divided into two main categories: **aliphatic hydrocarbons,** which include straight-chain, branched-chain, and cyclic compounds, and **aromatic hydrocarbons,** which contain highly stable rings of carbon atoms. The ali-

In Section 19.1 we saw that unsaturated hydrocarbons can combine with additional hydrogen until they have formed saturated hydrocarbons.

phatic hydrocarbons can be further subdivided into two groups based on the multiplicity of the carbon–carbon bond: **saturated hydrocarbons** that contain only carbon–carbon single bonds; and **unsaturated hydrocarbons** that possess at least one carbon–carbon double bond or triple bond.

Saturated hydrocarbons

The compounds that constitute the saturated hydrocarbons are collectively called the **alkanes** or **paraffins.** Except for methane, they contain chains of carbon atoms linked by single bonds. The simplest ones have all the carbons in one continuous chain and are known as *normal* or *straight chain* alkanes. More complex alkanes have chains that branch. As you might expect, they are called *branched chain* alkanes. The first ten members of the straight-chain alkanes are listed in Table 22.1 in order of an increasing number of carbon atoms in the chain.

Ordinary candle wax — paraffin — is composed of saturated alkanes.

The alkanes are important as fuels and as raw materials in the synthesis of other organic compounds. They occur in abundance in petroleum from which they may be separated to a large degree by fractional distillation. Methane, a gas at room temperature, is the "natural gas" used for cooking and as a heating fuel. Propane, which liquefies under high pressure, is used as a fuel in many rural areas where tanks of LPG (liquefied petroleum gas) are delivered to homes. Butane liquefies more easily than propane and is used in cigarette lighters. Octane, of course, has a boiling point that places it in the range of gasoline fuel. Heavier alkanes are found in kerosene, diesel fuel, lubricating oils, and in the paraffins used to make candles.

The names of each of the members of the alkanes are composed of two parts. The first part, *meth-, eth-, prop-,* and so on in Table 22.1, reflects the number of carbon atoms in the chain. The second part, which is the same for all the members, is *ane* after the parent name alk*ane*. Thus we have methane, an alkane with one carbon, ethane having two carbons, propane consisting of three carbons, and so on. You should become familiar with the names of all ten of these simple alkanes, because they serve as the basis for naming many of the remaining organic compounds.

Remember: meth = 1, eth = 2, prop = 3, but = 4. When there are more than four carbon atoms, the stems of the Greek number prefixes are used — for example, pent = 5, hex = 6, hept = 7, oct = 8, non = 9, dec = 10.

In listing the alkanes by increasing number of carbon atoms, two things become apparent. First, the molecular formula for each member of this series can be represented by a single general formula, $C_n H_{2n+2}$, where n is the number of

Table 22.1
The first ten members of the straight-chain alkanes

Formula	Name	Boiling Point (°C) at 1 atm
CH_4	Methane	−161
C_2H_6	Ethane	−89
C_3H_8	Propane	−44
C_4H_{10}	Butane	−0.5
C_5H_{12}	Pentane	36
C_6H_{14}	Hexane	68
C_7H_{16}	Heptane	98
C_8H_{18}	Octane	125
C_9H_{20}	Nonane	151
$C_{10}H_{22}$	Decane	174

carbons in the molecular chain. For example, the formula for the alkane with four carbon atoms would be C_4H_{10} and, according to Table 22.1, would be called butane. Second, we see that any two successive members of the series differ from each other by a single CH_2 group (this becomes quite apparent after ethane). Such a series, in which one member differs from the next by the same repeating cluster of atoms, is called a **homologous series.** The alkanes form such a series.

In organic chemistry it is important for us to be able to write molecular formulas, which indicate the number of each of the various atoms present in a molecule, as well as structural formulas, which show the relative positions of each of the atoms in a molecule. Since carbon forms the backbone of all organic compounds, it is mainly the shape of the carbon skeleton that is responsible for the overall shape of the various molecules. Carbon normally forms four covalent bonds, and in the alkanes these are single bonds. Each carbon atom is sp^3 hybridized, with the hybrid orbitals pointing to the corners of a tetrahedron.

$$\underset{1s}{\underline{\uparrow\downarrow}}\ \ \underset{2s}{\underline{\uparrow\downarrow}}\ \ \underset{2p}{\underline{\uparrow\ \uparrow\ \_}} \qquad \underset{1s}{\underline{\uparrow\downarrow}}\ \ \underset{sp^3\ \text{hybrid}}{\underline{\uparrow\ \uparrow\ \uparrow\ \uparrow}}$$

**unhybridized
carbon**

Thus in the straight- and branched-chain alkanes, carbon is tetrahedrally surrounded by bonded atoms. There are several ways of representing tetrahedral carbon in two dimensions, as illustrated in Figure 22.1 for methane. The most common, and the simplest, two-dimensional representation used for most organic compounds is the structural formula shown as Figure 22.1a. The re-

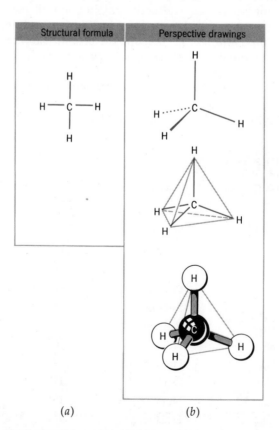

Figure 22.1

Two-dimensional representations of tetrahedral carbon. Methane is chosen as an example. (a) A structural formula. (b) Perspective drawings.

(a)   (b)

Figure 22.2

Two- and three-dimensional illustrations of the structure of butane. (a) A two-dimensional structural formula. (b) A three-dimensional drawing. Note that carbon atoms are not actually in a straight line.

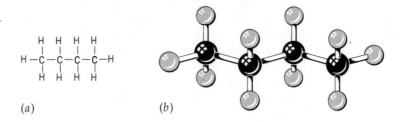

(a)　　　　　　　　(b)

maining members of the straight-chain alkanes are drawn simply by connecting several tetrahedral carbons along with their respective hydrogens.

In Figure 22.2 we see two representations of the four-carbon alkane, butane. Although the two-dimensional representation drawn in Figure 22.2*a* suggests a flat molecule with a straight carbon chain, you should remember that such molecules are not planar. The tetrahedral geometry about the carbon atoms prevents them from being in a straight line, as shown in Figure 22.2*b*.

"Straight chain" doesn't mean the carbons are in a straight line — only that they are in one continuous sequence.

Very often we find it convenient to condense structural formulas somewhat by not drawing all the C—H bonds. For instance, the butane formula can be condensed to either

$$CH_3{-}CH_2{-}CH_2{-}CH_3$$

or

$$CH_3CH_2CH_2CH_3$$

We will often use condensed formulas in this chapter. Their advantage is that they save time (and space), yet they still convey structural information.

Unsaturated hydrocarbons

The unsaturated hydrocarbons can be divided into two groups: **alkenes** or **olefins,** which contain at least one carbon-carbon double bond and **alkynes,** which contain at least a carbon-carbon triple bond.

Following a procedure similar to that used for the alkanes, we can introduce the alkenes by considering a series of compounds with one carbon-carbon double bond and an increasing number of carbon atoms in the chain. The simplest alkene contains two carbons:

$CH_2{=}CH_2$

$$\begin{array}{c} \text{H} \qquad\qquad \text{H} \\ \diagdown\quad\diagup \\ \text{C}{=}\text{C} \qquad (C_2H_4) \\ \diagup\quad\diagdown \\ \text{H} \qquad\qquad \text{H} \end{array}$$

ethene
(ethylene)

which is then followed by the three-carbon alkene,

$CH_3{-}CH{=}CH_2$

$$\begin{array}{c} \text{H} \qquad \text{H} \qquad \text{H} \\ \diagdown\quad |\quad\diagup \\ \text{H}{-}\text{C}{-}\text{C}{=}\text{C} \qquad (C_3H_6) \\ |\qquad\quad\diagdown \\ \text{H} \qquad\qquad \text{H} \end{array}$$

propene
(propylene)

Next come the alkenes with the formulas C_4H_8, C_5H_{10}, C_6H_{12}, and so on. Thus we see that the alkenes, like the alkanes, form a homologous series, with a CH_2 group as the difference between any two successive members. We also see that, like the alkanes, the alkenes can collectively be represented by one general formula, C_nH_{2n}, where n is the number of carbons in the molecule.

The names of the straight-chain alkenes also consist of two parts. The first part, which is the same as that used in the naming of the alkanes, indicates the number of carbons in the chain: *eth-,* two; *prop-,* three; *but-,* four; and so on. To this stem we add *ene,* which tells us that there is a carbon–carbon double bond present. The first two members of the alkenes are thus ethene and propene. These are also called by their older names, ethylene and propylene, and are probably the best known of the alkenes because of their use in making the plastics, polyethylene and polypropylene. We will discuss the nonmenclature of these and other hydrocarbons in greater detail in Section 22.3.

The bonding in ethene was discussed earlier in Section 5.5, so it will be reviewed only briefly here. The carbon atoms participating in the double bond are each sp^2 hybridized. The overlap of the sp^2 hybrid orbitals between the two carbon atoms produces a σ bond along the C—C axis. Each carbon also possesses an unhybridized pure p orbital. These overlap sideways to produce a π bond with electron density concentrated above and below the bond axis. Because of the geometry of the sp^2 hybrids and the restriction that the unhybridized p orbitals must overlap in the π bond, a planar configuration of atoms with approximately 120° bond angles is formed.

$$
\begin{array}{c}
\text{H} \sim 120° \; \text{H} \\
\sim 120° \;\; \text{C} = \text{C} \\
\text{H} \sim 120° \;\; \text{H}
\end{array}
$$

Besides the straight-chain alkenes there also exist alkenes with more than one double bond, alkenes that have other groups attached to the straight chain, alkenes with branched chains, and alkenes that are cyclic in structure.

The second type of unsaturated hydrocarbon that still remains to be discussed here is the **alkynes**—a series of compounds containing a carbon-carbon triple bond. The simplest, but truly one of the most important alkynes is HC≡CH, which is called ethyne or, more commonly, acetylene. This is the fuel used in welding torches. The next two members of this group are CH_3—C≡CH, propyne, and CH_3—CH_2—C≡CH, butyne. In naming these compounds we once again use the stems *eth-, prop-,* and *but-* to mean two, three, and four carbons, respectively. To this stem is added *yne* to denote the existence of the carbon-carbon triple bond.

The bonding in the carbon-carbon triple bond was also discussed earlier (Section 5.5). The hybridization used by the two carbons in the triple bond is sp. Thus one of the bonds—a σ bond—is formed by sp-sp overlap. The remaining two bonds are both π bonds that are formed by p_x-p_x and p_y-p_y overlap.

22.2 ISOMERS IN ORGANIC CHEMISTRY

Beginning with hydrocarbons containing four carbon atoms, we find that besides the normal straight-chain structures, there also exist branched-chain structures bearing the same molecular formula. For example, we find that there are two compounds with the formula C_4H_{10}.

Recall that isomers are compounds with the same formula but with distinctly different properties.

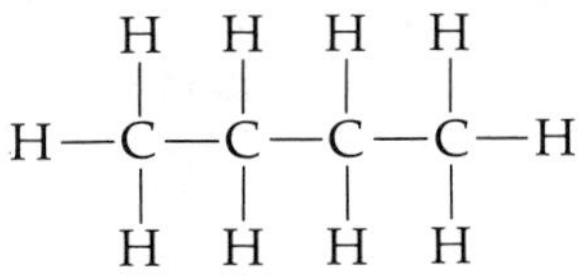

(1)
straight chain
m.p. = −138.3°C
b.p. = −0.5°C

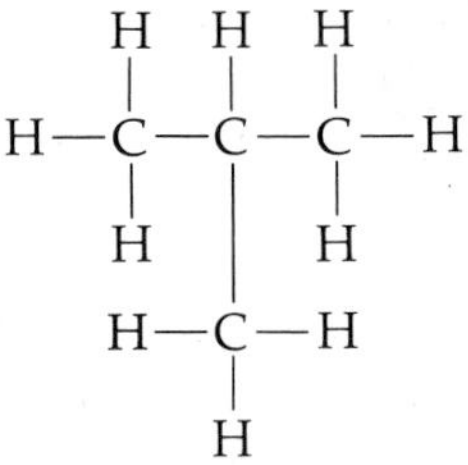

(2)
branched chain
m.p. = − 159°C
b.p. = −12°C

Two or more compounds that have the same molecular formula but that differ in the sequence in which the atoms are joined together are said to be **structural isomers.** Thus there are two structural isomers of butane, each with its own chemical and physical properties. This is an important point to keep in mind. Each structural isomer is a unique chemical compound.

Each of the remaining members of the alkane series show an even greater number of isomers. With C_5H_{12}, for example, we can write three structures:

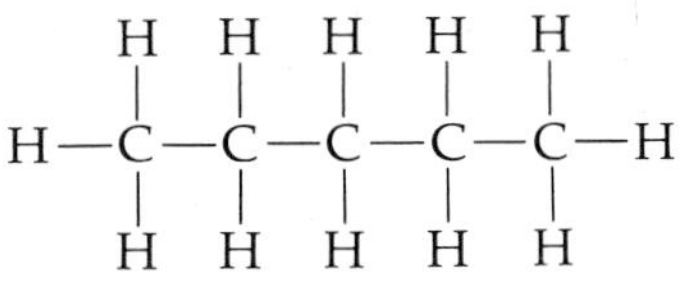

(1)

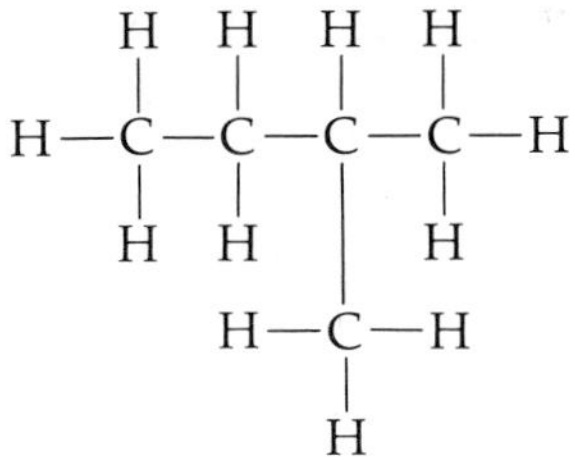

(2)

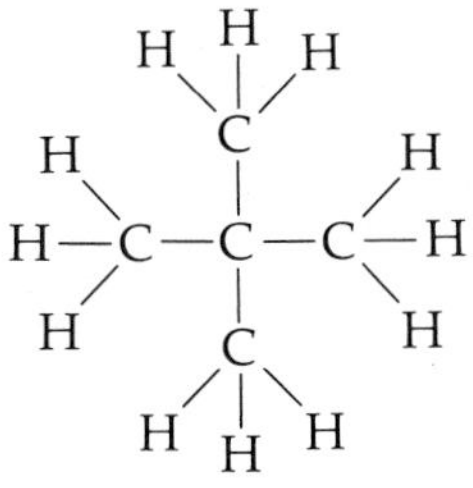

(3)

In the case of C_6H_{14} we find five isomers, which are listed in Table 22.2. In this table we have left room for you to add the names of each of the isomers following our discussion of nomenclature in the next section.

Structural isomerism also exists in the alkenes. Butene (C_4H_8), for example, can be written as a straight chain with the double bond between the first and second carbon atoms.

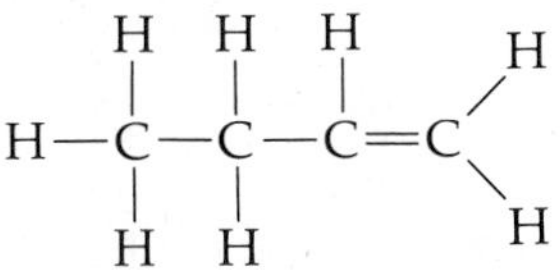

or $CH_3—CH_2—CH=CH_2$

isomer 1

Table 22.2
The five isomers of hexane

Isomer	Name (Fill in this column after reading Section 22.3)
H—C—C—C—C—C—C—H (straight chain, all H)	hexane
H—C—C—C—C—C—H with H—C—H branch	2, metal pentane
H—C—C—C—C—C—H with H—C—H branch	3, methyl pentane
H—C—C—C—C—H with two H—C—H branches	2,3 methyl butan
H—C—C—C—C—H with two H—C—H branches (one top, one bottom)	2,2 methat butane

There is also an isomer with the double bond between the middle two carbons,

$$H—\underset{\underset{H}{|}}{\overset{\overset{H}{|}}{C}}—\overset{H}{C}=\overset{H}{C}—\underset{\underset{H}{|}}{\overset{\overset{H}{|}}{C}}—H \quad \text{or} \quad CH_3—CH{=}CH—CH_3$$

isomer 2

and another is a branched isomer in which two CH_3 groups are attached to one of the carbon atoms participating in the carbon-carbon double bond.

$$H-\underset{\underset{H}{|}}{\overset{\overset{H}{|}}{C}}-\underset{\underset{\underset{\underset{H}{|}}{\underset{H}{C}-H}}{|}}{\overset{\overset{H}{|}}{C}}=\overset{\overset{H}{|}}{C}-H \qquad \text{or} \qquad CH_3-\underset{\underset{CH_3}{|}}{C}=CH_2$$

isomer 3

Notice that the arrangement of atoms in the molecules is different in all three cases even though the molecular formula, C_4H_8, is the same. Thus in alkenes we have isomers that result from branching of the carbon chains as well as isomers that result from a difference in the relative position of the double bond.

Another type of isomerism exists in organic compounds in which the sequence of atoms in the molecule is the same but in which the relative positions of the atoms or groups of atoms are different. In general, this type of isomerism is called **stereoisomerism,** and is found in two forms: geometrical isomerism and optical isomerism (see Section 21.10).

Geometrical isomerism

In organic compounds *cis* and *trans* geometrical isomers can occur in molecules that possess one or more carbon–carbon double bonds. This is illustrated by one of the isomers of butene,

Rotation about a C—C single bond is relatively unrestricted, so geometrical isomers of

$$\underset{H}{\overset{H_3C}{\diagdown}}\underset{\overset{|}{H}}{C}-\underset{\overset{|}{H}}{C}\overset{CH_3}{\diagup}$$

can't be isolated.

$$\underset{H}{\overset{H_3C}{\diagdown}}C=C\underset{H}{\overset{CH_3}{\diagup}} \qquad \underset{H}{\overset{H_3C}{\diagdown}}C=C\underset{CH_3}{\overset{H}{\diagup}}$$

cis **isomer**   *trans* **isomer**

These are not identical molecules because rotation of the atoms about the axis of the C=C double bond is restricted. When one of the $CH(CH_3)$ groups begins to rotate, the p orbitals that overlap sideways to form the π bond become misaligned—the overlap is lost. This destroys the bond, so in effect, conversion of *cis* to *trans* in this manner would require the breaking of a bond. The energy available through molecular collisions is generally not sufficient to accomplish this, so the rate of interconversion between *cis* and *trans* isomers is usually so slow that the individual isomers can be isolated.

Optical isomerism

Optical isomers, as we saw in Section 21.10, are molecules that have the same formula but possess structures that are nonsuperimposable mirror images of each other. As a rule, in organic compounds the presence of an **asymmetric carbon atom** is responsible for optical isomerism. An asymmetric carbon atom occurs when the carbon is bonded to four *different* atoms or groups of atoms; for example,

$$A-\underset{\underset{D}{|}}{\overset{\overset{B}{|}}{C}}-E$$

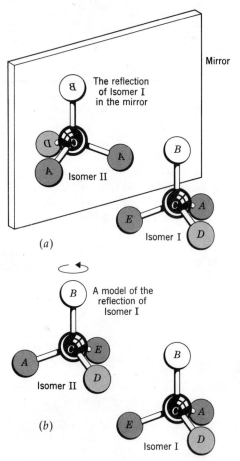

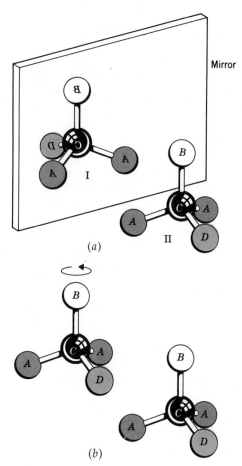

Figure 22.3

Optical isomers. (a) Isomer II is a reflection of Isomer I in the mirror; that is, it is the mirror image of Isomer I. (b) When Isomer II is rotated about the B—C bond so that atoms D in both I and II are in same relative position, Isomers I and II are not superimposable. Atoms B, C, and D match, but A and E do not.

Figure 22.4

(a) Lack of optical isomers when two groups attached to carbon are the same. (b) "Isomer" I rotated about B—C bond so that atoms D are matched. These two are identical and are not isomers of each other.

where *A*, *B*, *D*, and *E* represent different groups. For instance, the isomer of heptane with the structural formula

$$H-\overset{\overset{\displaystyle H}{|}}{\underset{\underset{\displaystyle H}{|}}{C}}-\overset{\overset{\displaystyle H}{|}}{\underset{\underset{\displaystyle H}{|}}{C}}-\overset{\overset{\displaystyle H}{|}}{\underset{\underset{\displaystyle |}{C^*}}{}}-\overset{\overset{\displaystyle H}{|}}{\underset{\underset{\displaystyle H}{|}}{C}}-\overset{\overset{\displaystyle H}{|}}{\underset{\underset{\displaystyle H}{|}}{C}}-\overset{\overset{\displaystyle H}{|}}{\underset{\underset{\displaystyle H}{|}}{C}}-H$$

$$H-\overset{\overset{\displaystyle }{}}{\underset{\underset{\displaystyle H}{|}}{C}}-H$$

has one asymmetric carbon atom—the one marked with an asterisk. This can be seen more readily if we rewrite the structural formula for this isomer in a slightly more condensed fashion.

$$H_5C_2 - \overset{\overset{\displaystyle H}{|}}{\underset{\underset{\displaystyle CH_3}{|}}{C^*}} - C_3H_7$$

Let's see how compounds with an asymmetric carbon atom give rise to optical isomers and why molecules without one cannot. To show this, we compare the structures of two different "isomers" of a compound to see if one is the nonsuperimposable mirror image of the other. If this condition is fulfilled, then they are optical isomers. If not, then they are identical and the compound is not optically active.

In the case of our general molecule, the two structures shown in Figure 22.3 are mirror images of each other with nonsuperimposable structures. Note that when one is placed over the other, the D's and B's line up, but the E's and A's do not. Furthermore, we would find that, regardless of the amount of manipulation of the two structures, we could never bring about the situation where one structure would be exactly superimposable on the other.

If there are two groups that are alike in each structure, the asymmetry of the central carbon is lost and the two structures no longer represent optical isomers; they are, in fact, identical. For example, by replacing the E in our general formula with another A, we would then have the two structures shown in Figure 22.4, which can be superimposed on each other. Thus we see that optical isomerism will occur only if an asymmetric carbon atom is present.

22.3 NOMENCLATURE

We have already seen some of the basic elements of naming organic compounds in our discussion of the hydrocarbons. The complexities produced by isomerism, and the introduction of atoms other than carbon and hydrogen, create a need for a systematic procedure for naming compounds. The guidelines presented below are currently followed in the modern chemical literature. Unfortunately, many of the older, nonsystematic *common names* are still often used in less formal situations. For instance, ethyne, C_2H_2, is also called acetylene; ethene and propene are also known as ethylene and propylene. A student of organic chemistry must be aware of this dual nomenclature.

The systematic nomenclature of organic compounds, like that of coordination compounds, has been established by the IUPAC (page 682). The application of the IUPAC system to the saturated and unsaturated hydrocarbons is defined by the following rules.

Rules of nomenclature of organic compounds

1. **The longest unbroken chain of carbon atoms in a molecule serves as the parent name for any hydrocarbon or its derivative.** With the alkenes this longest chain must contain the double bond; in the alkynes the triple bond must be included. For example, each of the following three compounds would be named as a derivative of pentane, because in each case the longest carbon chain consists of five carbon atoms.

$$CH_3 - \overset{\overset{\displaystyle CH_3}{|}}{\underset{\underset{\displaystyle CH_3}{|}}{C}} - CH_2 - CH_2 - CH_3$$

$$CH_3 - CH_2 - \overset{\overset{\displaystyle CH_3}{|}}{CH} - \overset{\overset{\displaystyle CH_3}{|}}{CH} - CH_3$$

$$CH_3 - \overset{\overset{\displaystyle CH_3}{|}}{\underset{\underset{\displaystyle CH_3}{|}}{C}} - CH_2 - \overset{\overset{\displaystyle CH_3}{|}}{\underset{\underset{\displaystyle CH_3}{|}}{C}} - CH_3$$

Sometimes, when a structural formula is drawn, the longest carbon chain might not be written in a straight line. For instance, the compound below has an eight-carbon chain (can you find it?) and would be named as a derivative of octene (note the double bond).

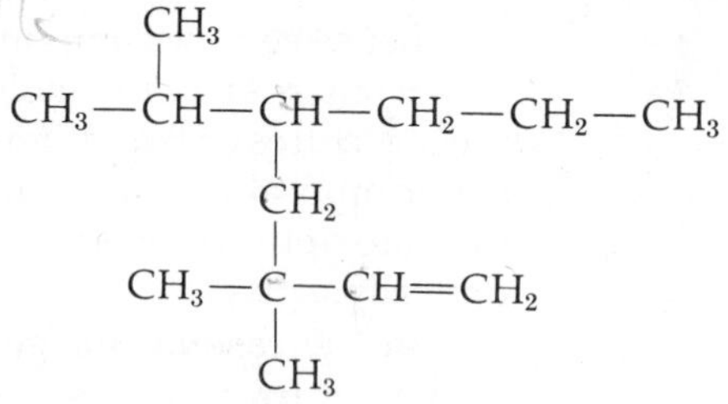

1-dien, 3,3methyl 5,propyl, octene

$$CH_3-CH-CH-CH_2-CH_2-CH_3$$

with CH_3 above the second carbon, CH_2 below the third carbon leading to $CH_3-C-CH=CH_2$ with CH_3 above and CH_3 below the central C.

2. **The number of carbon atoms in the chain is indicated by the first part of the parent name.** We've already seen that *meth* implies one carbon atom, *eth* implies two, *prop* implies three, and so forth. (See page 703.)

3. **Saturation and unsaturation is indicated by the second part of the parent name.** The ending *-ane* means that only single bonds between carbon atoms occur in the chain, *-ene* means there is a carbon-carbon double bond, and *-yne* means there is a carbon-carbon triple bond. When two double bonds are present the ending is **-diene,** and when three are present the ending is **-triene.** Thus we would have

$$CH_3-CH_2-CH_2-CH_3 \quad \text{butane}$$

$$CH_3-CH_2-CH=CH_2 \quad \text{butene}$$

$$CH_3-CH_2-C\equiv CH \quad \text{~~butyne~~} \;\; propane$$

$$CH_2=CH-CH=CH_2 \quad \text{butadiene}$$

4. **Branched isomers are named as derivatives of straight-chain hydrocarbons in which one or more hydrogen atoms are replaced by hydrocarbon fragments.** For example, isomer 2 of butane (on page 707) can be considered to be derived from propane by replacing one of the middle hydrogen atoms with a CH_3 group.

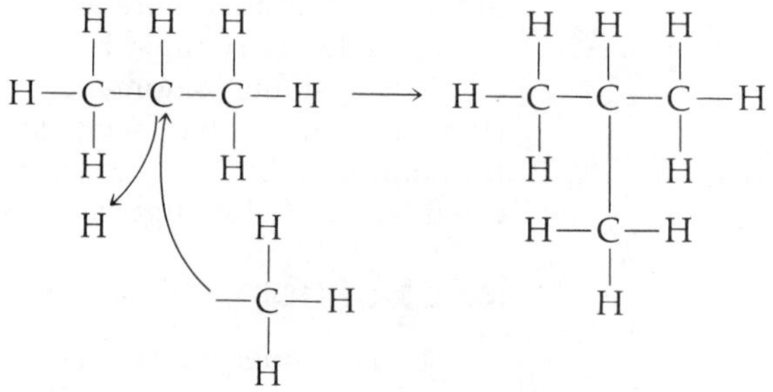

Hydrocarbon groups that are derived from members of the alkane series are called **alkyl groups.** The CH_3 group is derived from methane and is called a *methyl group.*

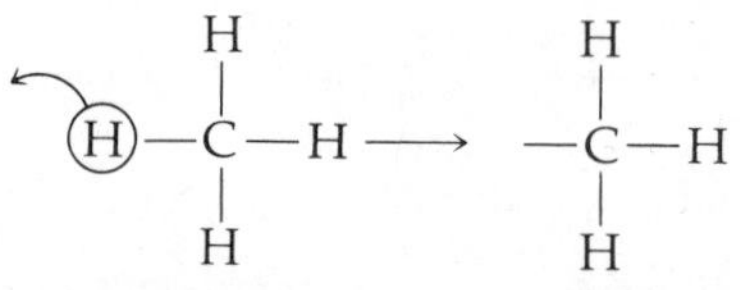

methane   methyl group

Similarly, the C_2H_5 group is derived from ethane and is called an *ethyl group*.

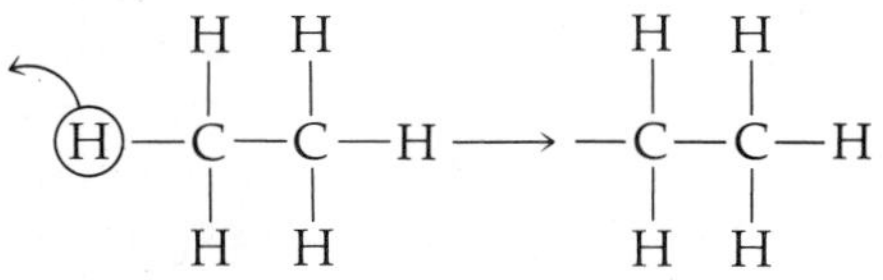

ethane   ethyl group

Thus, in naming an alkyl group the ending **-yl** is added to the alkane stem that indicates the number of carbon atoms in the group. Some additional alkyl groups are listed in Table 22.3.

Referring back to the branched isomer of butane, we see that it has a three-carbon chain as its longest and a CH_3 group attached to the middle carbon; therefore it would be called

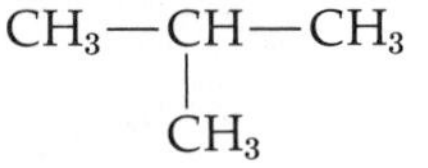

$$CH_3-CH-CH_3 \qquad \text{methylpropane}$$
$$| $$
$$CH_3$$

Likewise, we have

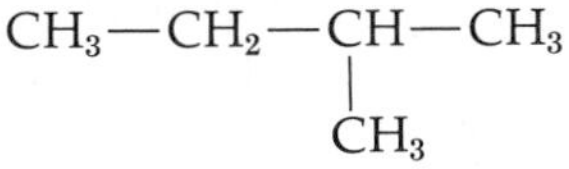

$$CH_3-CH_2-CH-CH_3 \qquad \text{methylbutane}$$
$$|$$
$$CH_3$$

and

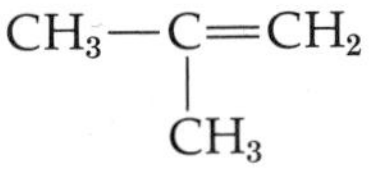

$$CH_3-C=CH_2 \qquad \text{methylpropene}$$
$$|$$
$$CH_3$$

Table 22.3
The names of some alkyl groups

Alkyl Group	Condensed Formula	Name
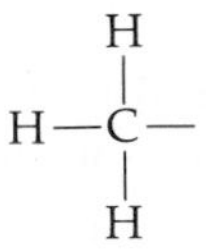 H—C— (with H above and H below the C)	$CH_3—$	Methyl
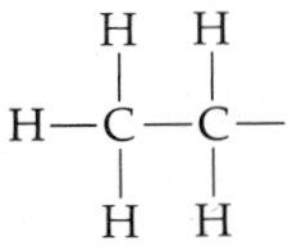 H—C—C— (with H's above and below)	$CH_3—CH_2—$	Ethyl
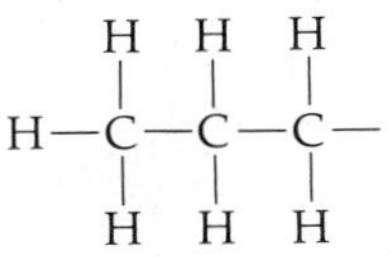 H—C—C—C— (with H's above and below)	$CH_3—CH_2—CH_2—$	Propyl
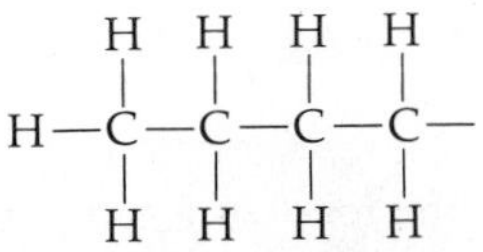 H—C—C—C—C— (with H's above and below)	$CH_3—CH_2—CH_2—CH_2—$	Butyl

5. **To denote the positions of alkyl groups attached to the parent chain, as well as the positions of double and triple bonds, a numbering system is used.** In alkanes the numbering starts from the end of the molecule that gives the lowest numbers to the alkyl groups. When multiple bonds are present the numbering begins from the end of the molecule that gives the lowest numbers to the multiple bonds. In the case of an alkyl group, the number identifying its position immediately precedes its name, while the numbers identifying a double or triple bond precede the name of the parent chain. Examples of this include

$$CH_3-CH_2-CH_2-CH-CH_2-CH_3 \qquad \text{3-methylhexane}$$
$$|$$
$$CH_3$$

$$CH_3-CH{=}CH-CH_3 \qquad \text{2-butene}$$

$$CH_3-CH{=}CH-CH_2-CH{=}CH_2 \qquad \text{1,4-hexadiene}$$

$$CH_3-CH-CH{=}CH-CH_3 \qquad \text{4-methyl-2-pentene}$$
$$|$$
$$CH_3$$

6. **When two or more identical alkyl groups are attached to a carbon chain, their number is specified by the Greek prefixes di- (two), tri- (three), tetra- (four), and so on. The position of each is also specified.** For example,

$$\qquad CH_3 \quad CH_3$$
$$\qquad | \qquad |$$
$$CH_3-CH-CH-CH_3 \qquad \text{2,3-dimethylbutane}$$

$$\qquad CH_3 \quad CH_3$$
$$\qquad | \qquad |$$
$$CH_3-CH-C-CH_3 \qquad \text{2,2,3-trimethylbutane}$$
$$\qquad\qquad |$$
$$\qquad\qquad CH_3$$

7. **When different alkyl groups are present along the parent chain, they are given in alphabetical order.** For example,

$$CH_3-CH_2-CH-CH_2-CH_2-CH_2-CH-CH_3 \qquad \text{6-ethyl-2-methyloctane}$$
$$\qquad\qquad | \qquad\qquad\qquad\qquad\qquad |$$
$$\qquad\qquad CH_2 \qquad\qquad\qquad\qquad\qquad CH_3$$
$$\qquad\qquad |$$
$$\qquad\qquad CH_3$$

$$\qquad CH_3$$
$$\qquad |$$
$$CH_3-C--CH-CH_2-CH_3 \qquad\qquad \text{3-ethyl-2,2-dimethylpentane}$$
$$\qquad | \qquad |$$
$$\qquad CH_3 \quad CH_2-CH_3$$

EXAMPLE 22.1   (a) Give the IUPAC name for the following compounds:

(1) 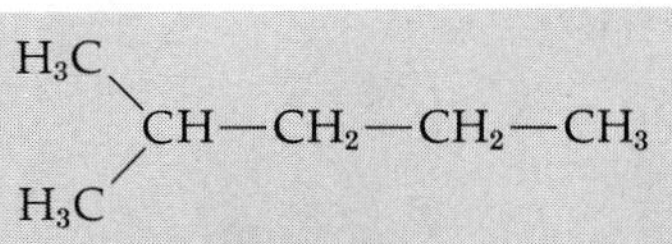   H₃C
         \
          CH—CH₂—CH₂—CH₃
         /
      H₃C

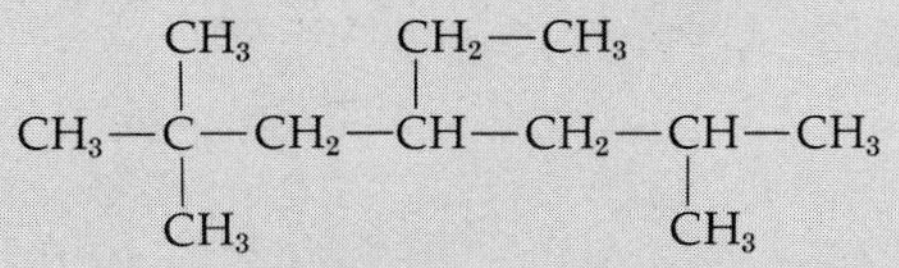

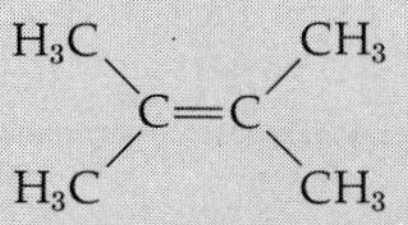

(2)
           CH₃   CH₂—CH₃
            | |
   CH₃—C—CH₂—CH—CH₂—CH—CH₃
            | |
           CH₃   CH₃

(3)   H₃C   CH₃
         \   /
          C=C
         /   \
      H₃C   CH₃

(4)   CH₃—CH—CH=CH—CH=CH₂
              |
             CH₃

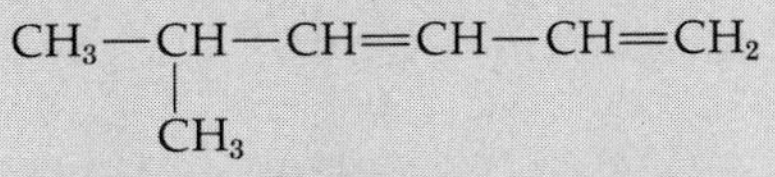

(5)   CH₃—CH₂—C≡C—CH₂—CH₃

(b) Write structural formulas for the following:
   (1) 2,3-dimethylbutane
   (2) 2-pentyne
   (3) 2-ethyl-1-butene
   (4) 1,5-octadiene
   (5) 2-ethyl-3-methyl-1-pentene

SOLUTION (a) (1) 2-methylpentane
              (2) 4-ethyl-2,2,6-trimethylheptane
              (3) 2,3-dimethyl-2-butene
              (4) 5-methyl-1,3-hexadiene
              (5) 3-hexyne

(b) (1) 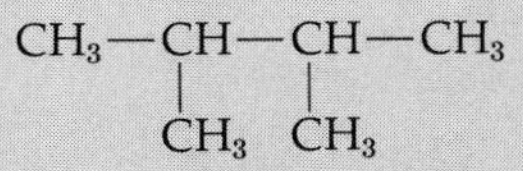   CH₃—CH—CH—CH₃
               | |
              CH₃ CH₃

    (2)   CH₃—CH₂—C≡C—CH₃

    (3)   CH₂—CH₃
                        |
           CH₃—CH₂—C=CH₂

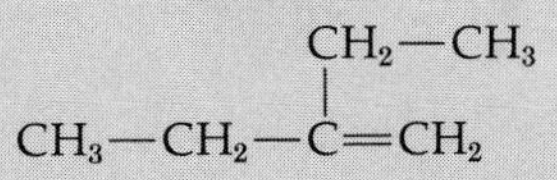

    (4)   CH₃—CH₂—CH=CH—CH₂—CH₂—CH=CH₂

    (5) 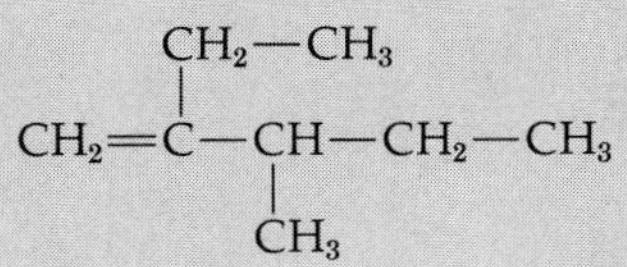   CH₂—CH₃
                  |
        CH₂=C—CH—CH₂—CH₃
                  |
                 CH₃

As an exercise, you should now go back and fill in the names of all the isomers of hexane in Table 22.2.

Common names

As mentioned earlier, quite a large number of organic compounds were known prior to the introduction of the IUPAC rules. As a result, many had already been named using other systems of nomenclature, and some of these names are carried over into current usage. For example, methylpropane,

$$CH_3-\underset{\underset{\textstyle CH_3}{|}}{CH}-CH_3$$

It is not necessary to specify that the methyl group is on carbon number 2, because if it were on the end, the compound would be named butane.

is more commonly referred to as isobutane. This name indicates that there are four carbons present (*but*ane), but that they are not in straight chain and that the molecule is an isomer of butane, hence *iso*. In general, alkanes that possess the arrangement

$$\underset{H_3C}{\overset{H_3C}{\diagdown}} CH - \underset{\underset{\textstyle H}{|}}{\overset{\overset{\textstyle H}{|}}{C}} - \ldots$$

are called *iso* compounds. Thus we have

$$\underset{H_3C}{\overset{H_3C}{\diagdown}} CH-CH_3 \qquad \underset{H_3C}{\overset{H_3C}{\diagdown}} CH-CH_2-CH_3 \qquad \underset{H_3C}{\overset{H_3C}{\diagdown}} CH-CH_2-CH_2-CH_3$$

$$\textbf{isobutane} \qquad\qquad \textbf{isopentane} \qquad\qquad\qquad \textbf{isohexane}$$

In the case of the alkynes, the first member, acetylene, is so important that its name is used in the common nomenclature of all the remaining members. Thus we have

Alkyne	Common	IUPAC
$CH_3-C\equiv CH$	Methylacetylene	Propyne
$CH_3-CH_2-C\equiv CH$	Ethylacetylene	1-Butyne
$CH_3-CH_2-CH_2-C\equiv CH$	Propylacetylene	1-Pentyne

We will see that the use of common names carries over to organic compounds that also contain elements in addition to carbon and hydrogen.

22.4   CYCLIC HYDROCARBONS

Cyclic alkanes

With alkanes containing three or more carbon atoms it is possible for the atoms to be arranged in a ring. An example is

$$\underset{\underset{\textstyle H \quad\quad H}{\diagup\;\;\diagdown}}{\overset{\overset{\textstyle H \quad\quad H}{\diagdown\;\;\diagup}}{\underset{H}{\overset{}{\diagup}}\underset{}{\overset{C}{}}\underset{H}{\overset{}{\diagdown}}}}$$

Table 22.4
Some cycloalkanes

Structural Formula	Molecular Formula	Name
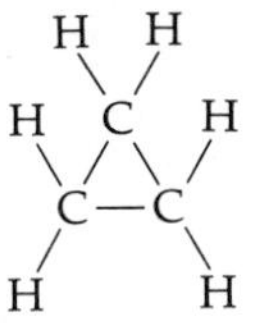	C_3H_6	Cyclopropane
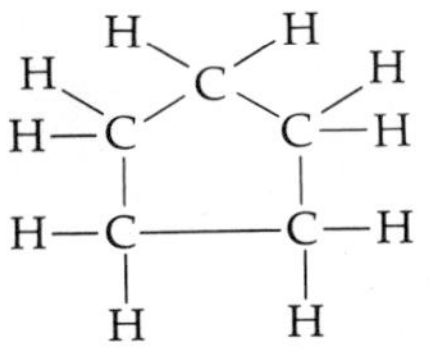	C_5H_{10}	Cyclopentane
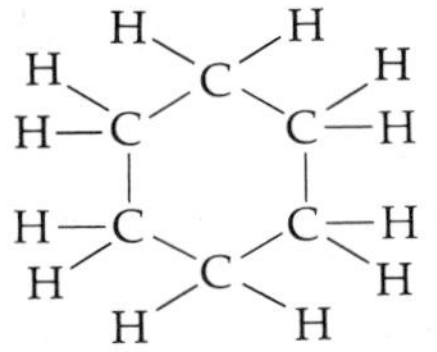	C_6H_{12}	Cyclohexane
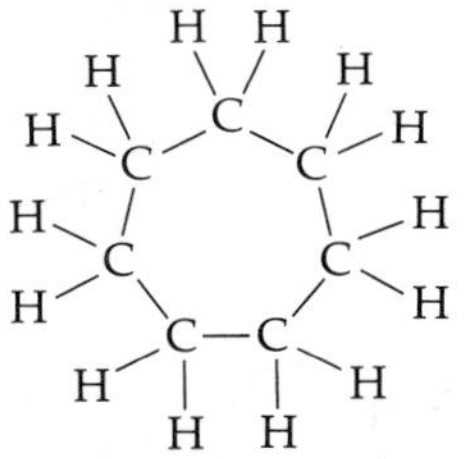	C_7H_{14}	Cycloheptane

This compound, as well as the other members of this group, contains only carbon-carbon single bonds, and is therefore saturated. These hydrocarbons are named in the same manner as the straight-chain alkanes, but with the prefix *cyclo-* added. The compound above would therefore be named cyclopropane (*cyclo* meaning ringed, *prop* for three carbon atoms, and *ane* because it is saturated). Some other cycloalkanes, with their molecular and structural formulas, are listed in Table 22.4. Notice that the general formula for the cycloalkanes is C_nH_{2n}. This is the same as the general formula for the noncyclic alkenes and demonstrates that caution must be exercised when writing structures from molecular formulas alone.

Cyclic hydrocarbons are often represented by a polygon, where it is assumed that at each vertex there is a carbon plus a sufficient number of hydrogens to give a total of four bonds to the carbon.

 means 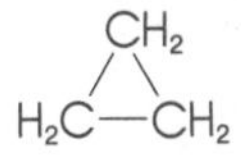

 means 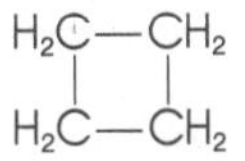

In cyclopropane, shown above, the C—C—C bond angle is 60°, while in cyclobutane,

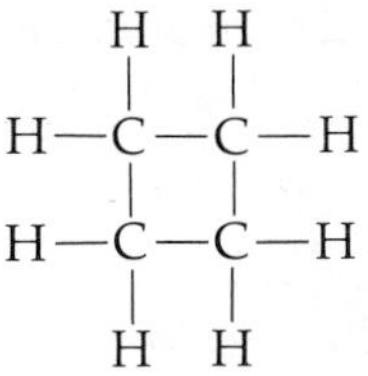

the C—C—C bond angle is about 90°. In both cases the bond angle is much less than the stable tetrahedral bond angle of 109° that is exhibited by carbon in the noncyclic alkanes. These small bond angles produce a relatively poor overlap of

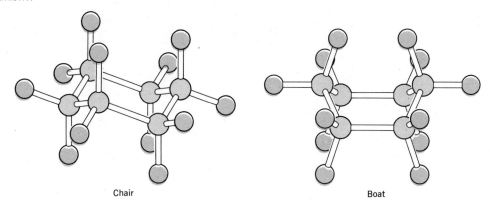

Figure 22.5

Two structures of cyclohexane that are free of "angle strain."

Chair   Boat

the orbitals used to form the bonds. This is because the orbitals do not point directly at one another and, as a result, the bonds are weak. In the language of organic chemistry, the bonds are said to be *strained,* and as a rule such molecules possess a high degree of reactivity. Cyclopropane, for example, which is used as a very fast-acting anesthetic, forms extremely flammable mixtures with air. The straight-chain propane, on the other hand, is much less reactive.

The next two members that follow cyclobutane (i.e., cyclopentane and cyclohexane) are quite stable. In cyclopentane, which has a nonplanar pentagonal structure, the C—C—C bond angle is very nearly 109° and consequently we expect it to be stable. In cyclohexane we find that in order to achieve more closely the tetrahedral angle of 109°, the ring is warped or puckered. In a planar or flat hexagon,

cyclohexane

the C—C—C bond angle would have to be 120° and, as a result, the structure would be strained. Two structures of cyclohexane that are free of this "angle strain" are shown in Figure 22.5.

Cyclic alkenes

Cyclic alkenes are formed containing three or more carbon atoms and are named in a fashion similar to that for the cycloalkanes. For example, cyclopentene would be

cyclopentene

Other examples include

1,3-cyclopentadiene

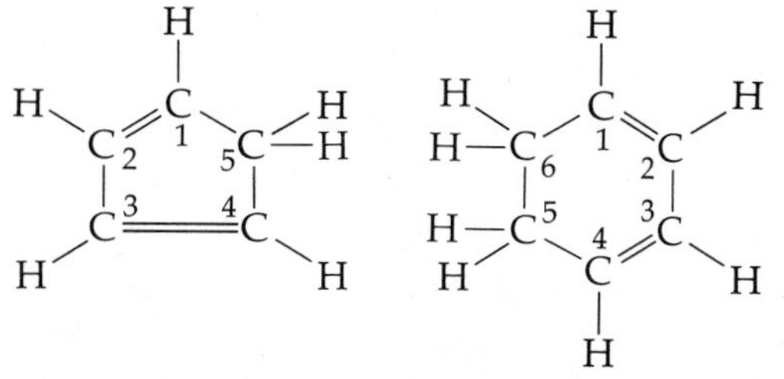

1,3-cyclopentadiene **1,3-cyclohexadiene**

In these compounds the numbering begins at the carbon with the first double bond and continues in the direction that leads to the smallest numbers for the remaining double bonds. Following this idea, the compound

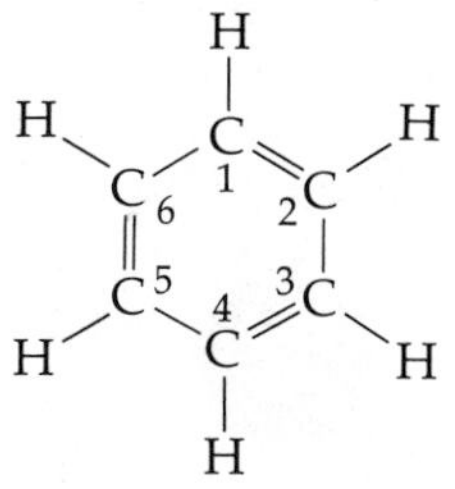

would be called 1,3,5-cyclohexatriene. However, the orientation of the orbitals that would be involved in the bonding in this molecule gives rise to some special properties. This compound is called benzene and forms the basis of another entire series of organic compounds.

22.5 AROMATIC HYDROCARBONS

Benzene has properties that are unlike those of the other cycloalkenes and is placed in a separate class. Benzene, and the host of benzenelike compounds (those containing a "benzene ring"), are collectively called the *aromatic compounds* because many of them have pleasing "aromatic" odors. In spite of this, however, benzene and many other aromatic compounds are quite toxic.

The main structural feature in the aromatic compounds, which is responsible for their distinctive chemical properties, is the benzene ring. We saw in the last section that the stoichiometry of benzene corresponds to that of the cyclotriene,

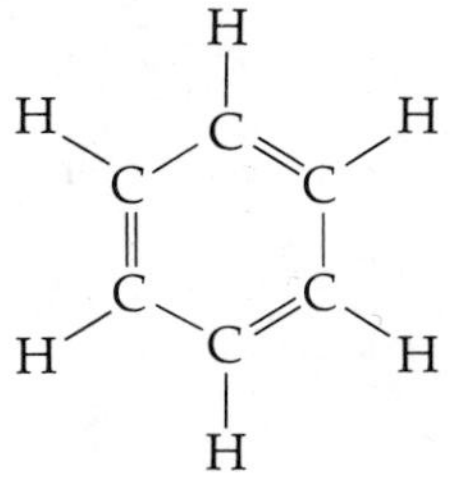

Physical and chemical evidence, however, reveals that this does not give an accurate representation of the molecule. For instance, it is found experimentally that all the C—C bond distances in benzene are the same. If there were, in fact,

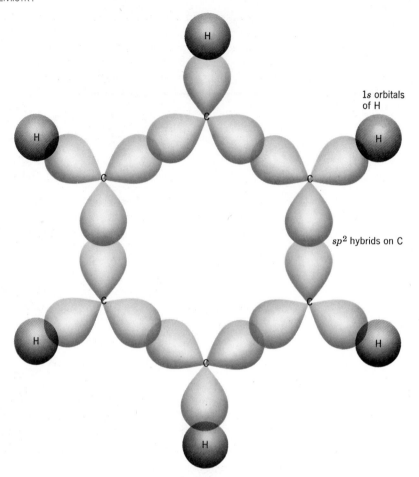

1s orbitals of H

sp^2 hybrids on C

Figure 22.6

Sigma-bond framework in benzene. Each carbon uses sp² hybrid orbitals.

alternating double and single bonds, some bond distances (C=C double bonds) would be shorter than others. In addition, the benzene molecule does not readily undergo chemical reactions typical of molecules containing double bonds.

The uniform bond distances in benzene can be accounted for by resonance. The two resonance structures (also called Kekulé structures[1]) that are usually written are

These are usually represented simply as

and

---

[1] These structures are named after the German chemist, August Kekulé, who first proposed them in 1865.

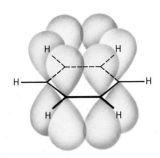

Figure 22.7

Electrons in unhybridized p or-bitals on C atoms in benzene.

It is understood that at each vertex of the hexagon there is a carbon atom bound to a hydrogen atom.

According to *valence bond theory*, to achieve a C—C—C bond angle of 120°, the carbon atoms must use sp^2 hybrid orbitals to form the σ bond framework of the ring. Thus on each of the carbons in the benzene ring we have three of the valence electrons in sp^2 hybrids and a fourth in a pure p orbital. The carbon skeleton showing only the hybridization can be seen in Figure 22.6. This same skeleton showing the usual dash for the σ bonds as well as the electrons in the pure p orbitals is illustrated in Figure 22.7. The two resonance structures of ben-zene would be formed by the overlap of pairs of adjacent p orbitals, as shown in Figure 22.8.

In Section 5.7 it was pointed out that the *molecular orbital theory* can quite successfully explain the bonding in polyatomic molecules without resorting to resonance. According to this theory, the six unhybridized p atomic orbitals in benzene overlap to form one continuous molecular orbital that extends over the entire molecule. This is illustrated in Figure 22.9, where we see the σ bonds as the solid lines between the carbons and the π bonds as a donut-shaped, cloud with electron density above and below the carbon ring. You may recall that electrons belonging to a molecular orbital that extends over several nuclei (such as the six π electrons in benzene) are said to be delocalized. The benzene ring is frequently drawn as

to emphasize the delocalized nature of the π electrons.

The delocalization of the π-electron cloud in benzene leads to a very stable ring structure. In fact, thermochemical calculations show that benzene is more stable than the hypothetical 1,3,5-cyclohexatriene by approximately 150 kJ/mol.

Aromatic compounds are important in many ways. We often find them in solvents of various kinds as well as in many plastics. The benzene ring is also in

This extra stability is called the resonance energy or delocali-zation energy.

Figure 22.8

Valence bond resonance struc-tures of benzene.

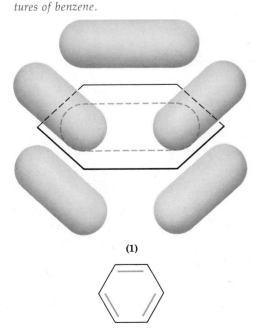

(1)

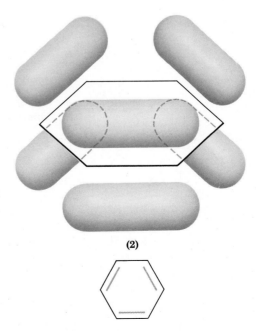

(2)

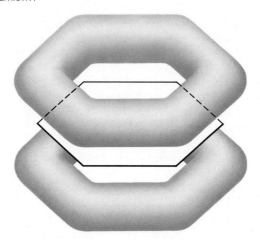

Figure 22.9

Molecular orbital delocalized electron cloud for benzene.

many biologically important compounds such as vitamins, proteins, and hormones. Unfortunately, the human body does not possess the ability to synthesize the benzene ring and we must therefore obtain them from an outside source. Plants have the ability to synthesize some of the needed aromatics and they are therefore essential to our diets.

Not all aromatics are beneficial. In fact, some of them are extremely harmful. One example is the compound 1,2-benzopyrene, composed of five benzene rings fused together.

1,2-benzopyrene

This compound has been found in cigarette smoke, in the exhaust from gasoline engines, and even on charcoal-broiled steaks. It is one of the most potent carcinogens (cancer-producing agents) known. For example, even small amounts applied to a shaved area of a mouse produces skin cancer very nearly 100% of the time.

Nomenclature of some benzene derivatives

The nomenclature of the aromatic hydrocarbons follows much the same pattern as we outlined for the alkanes and alkenes. Here, *benzene is generally taken to be the parent and the attached groups are named as before.* For example, the compound

is called methylbenzene. Its common name is *toluene.* Another example is ethylbenzene,

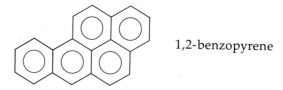

In a compound having more than one substituent attached to the ring, two systems may be used to identify their positions. In one of them the carbon atoms in the ring are numbered 1 to 6, beginning with the carbon that is bonded to the first group and continuing in such a direction as to lead to the lowest numbers for the remaining ones. For example, the compound,

would be called 1,2-dimethylbenzene, whereas the compound

would be named 1,3-dimethylbenzene.

The second system uses the names *ortho, meta,* and *para* to identify positions that are adjacent to the first group. For example, the *ortho, meta,* and *para* positions on toluene are

Thus the compounds

would be called *ortho*-methyltoluene and *meta*-methyltoluene, respectively. These compounds also have common names, which are *ortho*-xylene and *meta*-xylene (or simply *o*-xylene and *m*-xylene).

In some compounds it is sometimes easier to consider the benzene ring as the attached group and the longest carbon chain as the parent. In this case the benzene ring (C_6H_5) is called **phenyl** (rhymes with kennel), and the same rules developed earlier for the alkanes and alkenes are employed. For example, we have

which we would call phenylethene. Its more common name is *styrene*. This compound is used to manufacture the plastic *polystyrene*, which is easily molded into various forms (e.g., model airplanes) and foamed to give the thermal insulating product, *styrofoam*.

Some other examples are

1,2-diphenylethene   **biphenyl**

The second compound, biphenyl, has become the villain in a very serious pollution problem. When the hydrogen atoms in biphenyl are replaced by chlorine atoms, polychlorinated biphenyls (PCBs) are produced. These have many desirable physical and chemical properties that have led to their wide-spread use in many consumer products. For example, they are excellent flame retardants and, as oils, have been used in many types of electric motors; for example, in refrigerators, air conditioners, clothes washers and dryers, and furnace blowers. They are also used in electrical transformers such as those seen on utility poles that feed electricity into homes. Over the past five decades large quantities of these very stable PCBs have found their way into the environment where they are just now beginning to be seen as a potential hazard of immense proportions. It has been discovered that they produce severe acne and loss of hair in humans, while in test animals, PCBs have caused birth detects, cancer, and death. Their presence in high concentrations in sediments in the Hudson River has already led to a ban on commercial fishing in this river. At the same time, PCBs are beginning to show up in fish caught in other areas too.

PCB's are nonpolar and become concentrated in the fatty tissue of animals.

22.6   HYDROCARBON DERIVATIVES

Only a small fraction of all organic compounds contain just carbon and hydrogen; most contain other elements as well. In attempting to organize these compounds it is most convenient to view them as being derived from a hydrocarbon by replacing one or more of the hydrogens of the parent molecule by other atoms or groups of atoms. That is why we call them *hydrocarbon derivatives.* Usually the attached atoms or groups bestow some characteristic property to the molecule so that any molecule with the same grouping will react chemically in a similar fashion. Groups bestowing such properties on organic compounds are called **functional groups.** The study of organic chemistry is greatly simplified by examining the properties and reactions of various functional groups.

We have already seen two functional groups within the hydrocarbons—the carbon-carbon double bond and the carbon-carbon triple bond. Some other important functional groups, such as the halogens (X), hydroxyl group (—OH, in alcohols) and so on, are listed in Table 22.5.

For convenience, organic compounds can be viewed as composed of two parts: a hydrocarbon fragment that is generally denoted as R (aliphatic) or Ar (aromatic) plus one or more functional groups, such as those listed in Table 22.5. Thus we have R—X or Ar—X, an alkyl halide or aryl halide, respectively, R—OH, an alcohol, and so forth. Let's now take a brief look at some of these functional groups and some of the properties they impart to organic compounds.

Table 22.5
Some functional groups in organic compounds

Functional Group	Compound Class	Example	Name
$-\overset{\displaystyle\mid}{C}=\overset{\displaystyle\mid}{C}-$	Alkenes	$CH_2{=}CH_2$	Ethene
$-C{\equiv}C-$	Alkynes	$CH{\equiv}CH$	Ethyne
F, Cl, Br, I	Halides	CH_3Cl	Chloromethane
$-OH$	Alcohols	CH_3OH	Methanol
$-\overset{\displaystyle O}{\underset{\displaystyle H}{\overset{\|}{C}}}$	Aldehydes	$CH_3-\overset{\displaystyle O}{\overset{\|}{C}}-H$ or CH_3CHO	Ethanal (acetaldehyde)
$-\overset{\displaystyle O}{\overset{\|}{C}}\diagdown$	Ketones	$CH_3-\overset{\displaystyle O}{\overset{\|}{C}}-CH_3$	Propanone (acetone)
$-\overset{\displaystyle O}{\underset{\displaystyle OH}{\overset{\|}{C}}}$	Carboxylic acids	$CH_3-\overset{\displaystyle O}{\overset{\|}{C}}-OH$	Ethanoic acid (acetic acid)
$-\overset{\displaystyle\mid}{\underset{\displaystyle\mid}{C}}-N\overset{\displaystyle H}{\underset{\displaystyle H}{\diagup}}$	Amines	CH_3NH_2	Methylamine
$-\overset{\displaystyle\mid}{\underset{\displaystyle O}{C}}-N\overset{\displaystyle H}{\underset{\displaystyle H}{\diagup}}$	Amides	$CH_3-\overset{\displaystyle O}{\overset{\|}{C}}-NH_2$	Ethanamide (acetamide)
$-\overset{\displaystyle\mid}{\underset{\displaystyle\mid}{C}}-O-\overset{\displaystyle\mid}{\underset{\displaystyle\mid}{C}}-$	Ethers	CH_3-O-CH_3	Dimethyl ether
$-\overset{\displaystyle\mid}{\underset{\displaystyle\mid}{C}}-\overset{\displaystyle O}{\overset{\|}{C}}-O-\overset{\displaystyle\mid}{\underset{\displaystyle\mid}{C}}-$	Esters	$CH_3-\overset{\displaystyle O}{\overset{\|}{C}}-O-CH_3$	Methyl acetate

Alkenes and alkynes

The double and triple bonds in alkenes and alkynes show a characteristic tendency to undergo addition reactions, that is, reactions in which the components of a reactant molecule become incorporated into the alkene or alkyne. For example, we have already been that molecules having carbon-carbon double or triple bonds can be made to react with hydrogen.

The most common catalysts for hydrogenation reactions are finely divided nickel, palladium, platinum, ruthenium, and rhodium.

$$\overset{H}{\underset{H}{}}C{=}C\overset{H}{\underset{H}{}} + H_2 \xrightarrow[\text{pressure}]{\text{catalyst}} H-\overset{H}{\underset{H}{\overset{\|}{C}}}-\overset{H}{\underset{H}{\overset{\|}{C}}}-H$$

$$H-C{\equiv}C-H + 2H_2 \xrightarrow[\text{pressure}]{\text{catalyst}} H-\overset{H}{\underset{H}{\overset{\|}{C}}}-\overset{H}{\underset{H}{\overset{\|}{C}}}-H$$

Hydrogen is not the only substance that can be added to a double bond. Some other examples are

$$CH_3—CH\!\!=\!\!CH—CH_3 + HBr \longrightarrow CH_3—\underset{\underset{H}{|}}{C}H—\underset{\underset{Br}{|}}{C}H—CH_3$$

$$H—C\!\!\equiv\!\!C—H + Br_2 \longrightarrow H—\underset{\underset{Br}{|}}{C}\!\!=\!\!\underset{\underset{Br}{|}}{C}—H$$

$$H—C\!\!\equiv\!\!C—H + 2Br_2 \longrightarrow H—\overset{\overset{Br}{|}}{\underset{\underset{Br}{|}}{C}}—\overset{\overset{Br}{|}}{\underset{\underset{Br}{|}}{C}}—H$$

$$CH_2\!\!=\!\!CH_2 + H_2O \xrightarrow{\text{H}^+} CH_3CH_2OH$$

In this last case H$^+$ serves as a catalyst and is written over the arrow.

When a substance such as HBr is added to an unsymmetrical alkene there are two possibilities for reaction.

$$CH_3—CH\!\!=\!\!CH_2 + HBr \longrightarrow CH_3—CHBr—CH_3$$
2-bromopropane

$$CH_3—CH\!\!=\!\!CH_2 + HBr \longrightarrow CH_3—CH_2—CH_2Br$$
1-bromopropane

In this situation it happens that very little of the 1-bromo product is formed. Nearly 90% of the product is 2-bromopropane. The products of such reactions can be predicted by *Markovnikov's*[2] *rule,* which states that during an addition to ordinary alkenes the hydrogen of the reactant becomes attached to that carbon of the alkene that is already bonded to the most hydrogen atoms. In propene, the end carbon has two hydrogen atoms while the middle carbon has only one. The H of the HBr therefore becomes attached to the end carbon, while the Br becomes bonded to the middle carbon. Other unsymmetrical reagents that follow this rule are H$_2$O, HCN, and the other hydrogen halides. For example, we find that the addition of water to 2-methylpropene gives the following:

$$CH_3—\overset{\overset{\text{CH}_3}{|}}{C}\!\!=\!\!CH_2 + H_2O \xrightarrow[\text{acid}]{} CH_3—\overset{\overset{\text{CH}_3}{|}}{\underset{\underset{\text{OH}}{|}}{C}}—CH_3$$

The tendency of the unsaturated hydrocarbons to undergo addition reactions can be contrasted with the tendency of the alkanes to react through substitution. For example, methane will react with Cl$_2$ to give a variety of products in which Cl replaces H.

$$CH_4 + Cl_2 \longrightarrow \begin{Bmatrix} CH_3Cl \\ CH_2Cl_2 \\ CHCl_3 \\ CCl_4 \end{Bmatrix} + HCl$$

[2] The publication stating Vladimir W. Markovnikov's rule was published in 1905, one year after his death.

Under similar conditions the alkenes and alkynes have little tendency to undergo substitution.

22.7 HALOGEN DERIVATIVES

A *halogen derivative* is a molecule in which a halogen atom has replaced a hydrogen in the parent hydrocarbon. Some examples of these were seen in the last section as products of reactions of hydrocarbons with halogens or hydrogen halides. In naming these compounds, the halogen is specified as *fluoro, chloro, bromo,* or *iodo.*

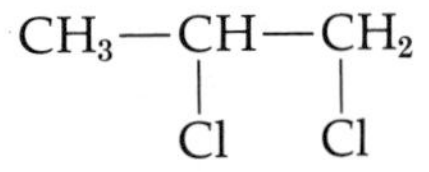

$$CH_3-CH-CH_2 \qquad Cl-\overset{\displaystyle Cl}{\underset{\displaystyle Cl}{C}}-H$$
$$\underset{\displaystyle Cl}{|} \quad \underset{\displaystyle Cl}{|}$$

1,2-dichloropropane 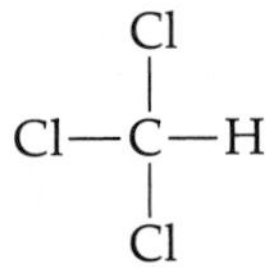   **trichloromethane**
(chloroform)

$$Cl-\overset{\displaystyle Cl}{\underset{\displaystyle Cl}{C}}-Cl \qquad \overset{\displaystyle H}{\underset{\displaystyle Cl}{}}C=C\overset{\displaystyle Cl}{\underset{\displaystyle Cl}{}} \qquad F-\overset{\displaystyle F}{\underset{\displaystyle Cl}{C}}-Cl$$

tetrachloromethane   **trichloroethene**   **dichlorodifluoromethane**
(carbon tetrachloride)   **(trichloroethylene)**   *(Freon-12)*

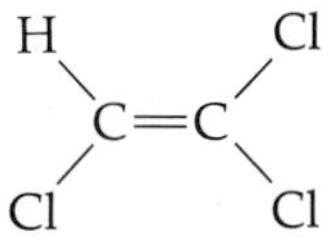

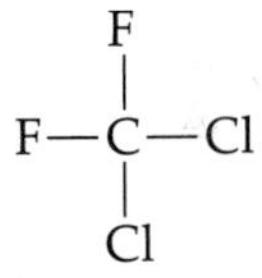

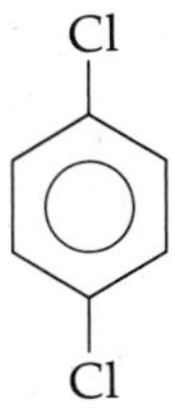

p-**dichlorobenzene** 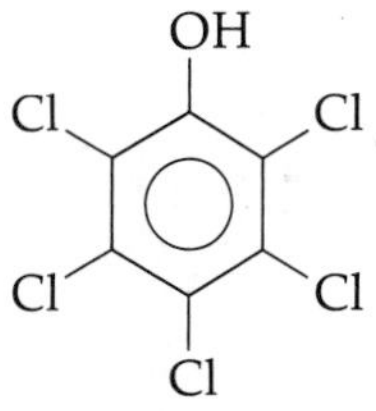   **pentachlorophenol**

Many halogenated hydrocarbons have important commercial applications. Trichloroethene, for example, is a common dry cleaning solvent. The *Freons* are used as refrigerants and propellants in aerosol products. As we discussed earlier, there is considerable controversy over whether these materials are seriously depleting the ozone shield in the earth's upper atmosphere (see page 603).

Halogenated compounds are also toxic. Carbon tetrachloride, for example, is no longer used as a dry cleaning solvent because it is a cumulative poison. Many insecticides contain halogenated compounds. For example, *p*-dichlorobenzene (above) has been used in moth balls. Pentachlorophenol, also shown above, is used as a wood preservative because it is toxic to creatures that attack wood.

The alkyl halides can be prepared by addition of the halogen or hydrogen halides to alkenes, as we have seen, as well as by substitution of a halogen for a hydrogen in an alkane.

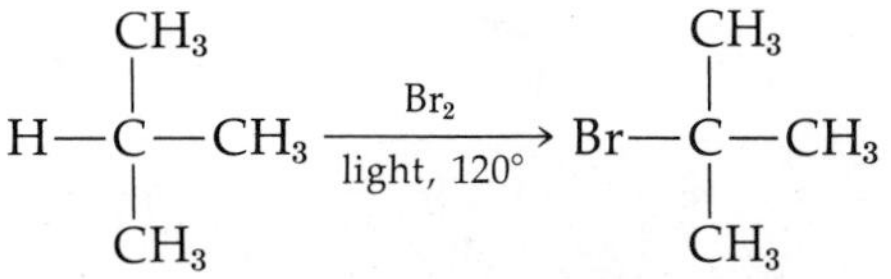

$$H-\overset{\displaystyle CH_3}{\underset{\displaystyle CH_3}{C}}-CH_3 \xrightarrow[\text{light, 120°}]{Br_2} Br-\overset{\displaystyle CH_3}{\underset{\displaystyle CH_3}{C}}-CH_3$$

Here the Br replaces the hydrogen that is bonded least strongly to the carbon in the molecule.

Perhaps the most important method of preparing alkyl halides is from alcohols. During this reaction the OH of the alcohol is displaced by a halide ion. Examples of this are

$$CH_3—CH_2—CH_2—CH_2—OH + HBr \longrightarrow CH_3—CH_2—CH_2—CH_2—Br + H_2O$$

$$\underset{\displaystyle \overset{|}{CH_3}}{\overset{\displaystyle \overset{CH_3}{|}}{CH_3—C—OH}} + HCl \longrightarrow \underset{\displaystyle \overset{|}{CH_3}}{\overset{\displaystyle \overset{CH_3}{|}}{CH_3—C—Cl}} + H_2O$$

22.8 IMPORTANT OXYGEN- CONTAINING DERIVATIVES

Many common organic compounds have functional groups that contain oxygen—for example, alcohols, aldehydes and ketones, carboxylic acids and esters, and ethers. Many are related to each other and therefore we discuss them together in this section.

Alcohols

Alcohols are characterized by the functional group, —OH. Some typical alcohols with their IUPAC and common names are

Formula	IUPAC Name	Common Name
$CH_3—OH$	Methanol	Methyl alcohol
$CH_3—CH_2—OH$	Ethanol	Ethyl alcohol
$CH_3—CH_2—CH_2—OH$	1-Propanol	n-Propyl alcohol

As you can see, the name of an alcohol is obtained from the name of the parent alkane by replacing the e by ol. Also, when necessary, the position of the OH group is identified by number. Thus we have

$$\underset{\displaystyle }{\overset{\displaystyle \overset{OH}{|}}{CH_3—CH—CH_3}} \qquad \text{2-propanol}$$

$$\underset{\displaystyle }{\overset{\displaystyle \overset{OH}{|}}{CH_3—CH—CH_2—CH_3}} \qquad \text{2-butanol}$$

$$\underset{\displaystyle }{\overset{\displaystyle \overset{OH}{|}}{CH_3—CH_2—CH—CH_2—CH_3}} \qquad \text{3-pentanol}$$

$$\underset{\displaystyle H_3C}{\overset{\displaystyle H_3C}{{}^{\diagdown}{}_{\diagup}}}CH—CH_2—OH \qquad \text{2-methyl-1-propanol}$$

There are many important alcohols. Ethanol, of course, is found in alcoholic beverages. Compared with most other alcohols, ethanol is relatively nontoxic.

When it is consumed even in small quantities, however, it causes the blood vessels to dilate, resulting in a lowering of the blood pressure followed by a general feeling of relaxation. In larger amounts ethanol causes intoxication, and excessively prolonged use can permanently damage the liver and eventually lead to death.

Methanol, also known as wood alcohol, is a deadly poison, while the remaining alcohols, other than ethanol, are somewhat "milder poisons." Methanol can cause blindness and eventually total loss of motor control and death. Perhaps one familiar use of methanol is in "dry gas," which is added to gasoline to cause suspended water droplets in the fuel to dissolve. In this way water that may have condensed in a fuel tank can pass harmlessly through the engine.

Methanol is prepared in large quantities from the reaction between carbon monoxide and hydrogen in the presence of a metal oxide catalyst at high temperature and pressure. The balanced equation for this preparation is

$$CO + 2H_2 \xrightarrow[\substack{3000 \text{ psi} \\ 350-400°C}]{ZnO/Cr_2O_3} CH_3OH$$

This reaction was discussed in Section 19.1.

The common method of preparing ethanol is through the fermentation of carbohydrates (sugars or starch). Sugars are converted into ethanol and carbon dioxide by the action of yeast in the absence of oxygen. For example,

$$C_6H_{12}O_6 \xrightarrow{yeast} 2C_2H_5OH + 2CO_2$$

glucose   ethanol

(a carbohydrate)

Another important alcohol is 2-propanol, commonly known as isopropyl alcohol, which is used widely as rubbing alcohol.

The number of groups attached to the carbon to which the OH is bonded aids us in classifying the alcohols. The C—OH grouping is called the **carbinol group,** and the carbon of this group is referred to as the **carbinol carbon.** Compounds in which there is one hydrocarbon group (R) attached to the carbinol carbon are known as **primary alcohols.** Alcohols that contain two such R groups are known as **secondary alcohols. Tertiary alcohols,** on the other hand, have three R groups bonded to the carbinol carbon.

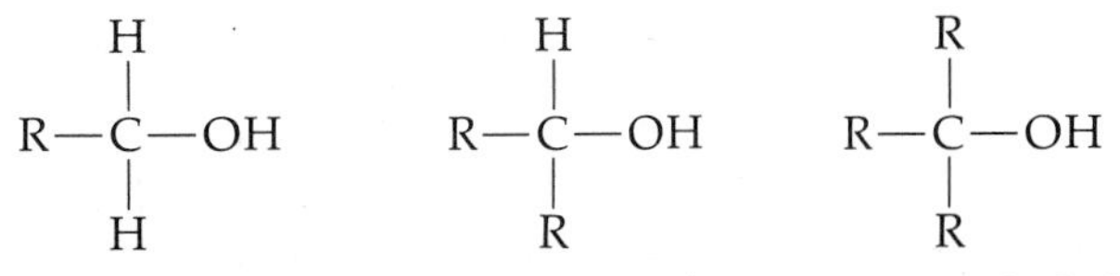

primary alcohol   secondary alcohol   tertiary alcohol

Some specific examples are

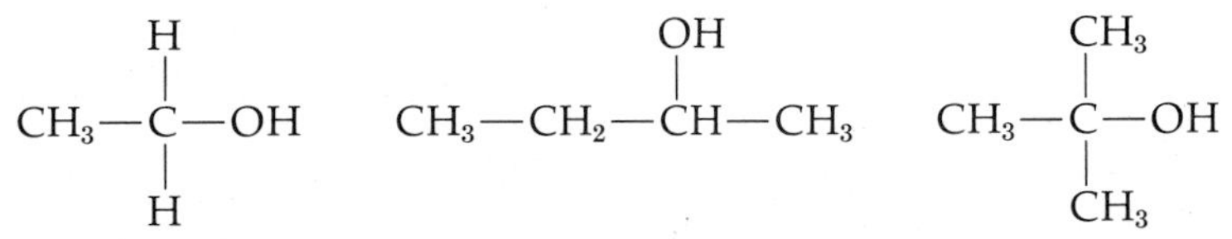

a primary alcohol   a secondary alcohol   a tertiary alcohol

Following the IUPAC nomenclature, these alcohols are called ethanol, 2-butanol, and 2-methyl-2-propanol. Their common names are ethyl alcohol, *sec*-butyl alcohol, and *tert*-butyl alcohol.

Alcohols containing more than one hydroxyl group are also possible and are called polyhydroxy alcohols. Important examples are ethylene glycol (1,2-ethanediol),

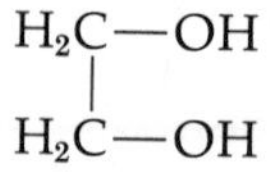

$$H_2C\text{---}OH$$
$$|$$
$$H_2C\text{---}OH \qquad \text{ethylene glycol}$$

propylene glycol (1,2-propanediol)

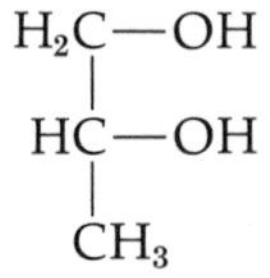

$$H_2C\text{---}OH$$
$$|$$
$$HC\text{---}OH \qquad \text{propylene glycol}$$
$$|$$
$$CH_3$$

and glycerol (1,2,3-propanetriol)

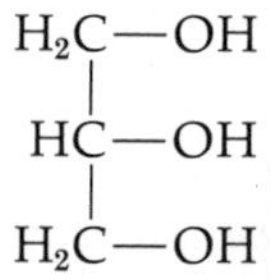

$$H_2C\text{---}OH$$
$$|$$
$$HC\text{---}OH \qquad \text{glycerol (also, glycerin)}$$
$$|$$
$$H_2C\text{---}OH$$

Ethylene glycol is toxic, but propylene glycol and glycerol are not.

These compounds are both soluble in water in all proportions. Ethylene glycol is used widely as an automotive antifreeze and in the manufacture of polyesters such as *Dacron* and *Mylar*. Propylene glycol is used as a nontoxic antifreeze and as a food additive to prevent the growth of mold. Glycerol occurs in fats, as we will see in the next chapter. It too is used in antifreeze applications where a nontoxic antifreeze is necessary, and glycerol is also sometimes added to alcoholic beverages to promote "smoothness."

Polyhydroxy alcohols can be *nitrated* by their cautious addition to a mixture of nitric and sulfuric acids.

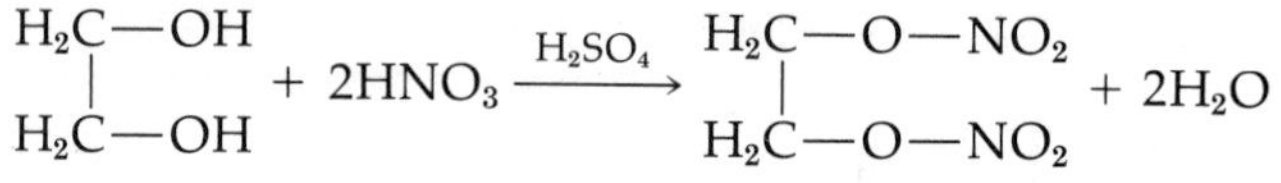

$$
H_2C\text{---}OH \atop H_2C\text{---}OH
\ + 2HNO_3 \ \xrightarrow{H_2SO_4} \
H_2C\text{---}O\text{---}NO_2 \atop H_2C\text{---}O\text{---}NO_2
\ + 2H_2O
$$

glycol dinitrate

$$
\begin{matrix} H_2C\text{---}OH \\ HC\text{---}OH \\ H_2C\text{---}OH \end{matrix}
\ + 3HNO_3 \ \xrightarrow{H_2SO_4} \
\begin{matrix} H_2C\text{---}O\text{---}NO_2 \\ HC\text{---}O\text{---}NO_2 \\ H_2C\text{---}O\text{---}NO_2 \end{matrix}
\ + 3H_2O
$$

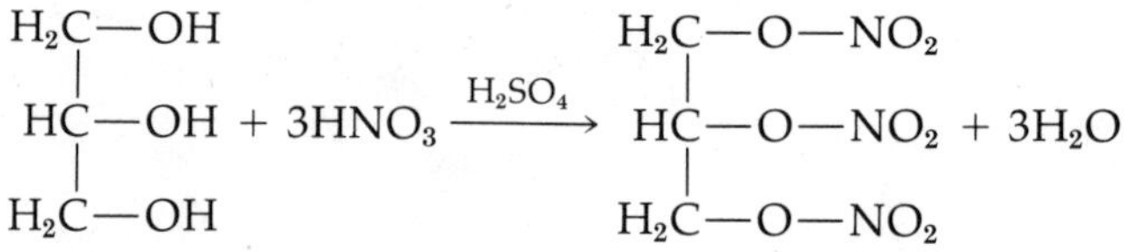

glyceryl trinitrate

The products of these reactions must be handled with extreme caution. Glyceryl trinitrate is also called nitroglycerin and is used with glycol dinitrate in the production of dynamite.

Aldehydes and ketones

Aldehydes and ketones are characterized by the presence of a **carbonyl group,** $>C=O$. In aldehydes this occurs on an end carbon,

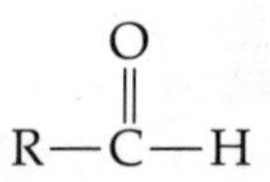

$$
\begin{matrix} O \\ \| \\ R\text{---}C\text{---}H \end{matrix} \qquad \text{aldehyde}
$$

while in ketones it occurs on one of the middle carbon atoms,

$$R-\overset{\overset{\displaystyle O}{\|}}{C}-R' \qquad \text{ketone}$$

(R and R' indicate the possibility of having different alkyl groups attached to the carbonyl group.)

Many aldehydes have pleasant odors, particulary those in which the R group is aromatic.

benzaldehyde   **vanillin**   **cinnamaldehyde**
(bitter almonds)   **(vanilla bean)**   **(cinnamon)**

Ketones often have very desirable solvent properties. Acetone, for example, is found in nail polish remover; methyl ethyl ketone is a solvent in airplane glue.

$$CH_3-\overset{\overset{\displaystyle O}{\|}}{C}-CH_3 \qquad CH_3-\overset{\overset{\displaystyle O}{\|}}{C}-CH_2-CH_3$$

acetone   **methyl ethyl ketone**

In naming aldehydes and ketones, the $-e$ of the corresponding hydrocarbon parent is dropped and replaced by **-al** for aldehydes or **-one** for ketones. Thus we have

$$CH_3-\overset{\overset{\displaystyle O}{\|}}{C}-H \qquad \text{ethanal (common: acetaldehyde)}$$

The IUPAC names of the ketones, acetone and methyl ethyl ketone, would be propanone and butanone, respectively. When there are possible alternative locations of the carbonyl group, its position is indicated by number.

$$CH_3-\overset{\overset{\displaystyle O}{\|}}{C}-CH_2-CH_2-CH_3 \qquad CH_3-CH_2-\overset{\overset{\displaystyle O}{\|}}{C}-CH_2-CH_3$$

2-pentanone   **3-pentanone**

Aldehydes and ketones are related to alcohols through oxidation and reduction. Oxidation of a primary alcohol can yield an aldehyde (although these are usually difficult to isolate because they are generally easily oxidized further).

$$CH_3-\overset{\overset{\displaystyle H}{|}}{\underset{\underset{\displaystyle H}{|}}{C}}-OH \xrightarrow{(O)} CH_3-\overset{\overset{\displaystyle H}{|}}{C}=O$$

ethanol   **ethanal**

Secondary alcohols are oxidized to ketones.

$$CH_3-\underset{\underset{\text{2-propanol}}{}}{\overset{\overset{OH}{|}}{CH}}-CH_3 \xrightarrow{(O)} CH_3-\underset{\underset{\substack{\text{propanone}\\\text{(acetone)}}}{}}{\overset{\overset{O}{\parallel}}{C}}-CH_3$$

The symbol (O) is used here to indicate oxidation without specifying the oxidizing agent. Oxidation of tertiary alcohols, which is considerably more difficult, breaks down the carbon chain.

The reason for the various products seems to be the number of hydrogens that are attached to the carbinol carbon. During the oxidation reaction one hydrogen is eliminated from the carbinol carbon as well as the one on the OH. With a primary alcohol, therefore, one hydrogen remains on the carbon after oxidation, whereas with a secondary alcohol none remain. Since no hydrogens are attached to the carbinol carbon in a tartiary alcohol, no reaction occurs when H_2CrO_4 or other similar oxidizing materials are added.

The oxidation of alcohols to aldehydes and ketones can be reversed through reduction. Often a metal hydride is used (for example, $NaBH_4$, sodium borohydride) because of the availability, at least in principle, of electron-rich H^- ions.

$$CH_3-\overset{\overset{O}{\parallel}}{C}-H \xrightarrow{NaBH_4} CH_3-\underset{\underset{H}{|}}{\overset{\overset{H}{|}}{C}}-OH$$

$$CH_3-\overset{\overset{O}{\parallel}}{C}-CH_3 \xrightarrow{NaBH_4} CH_3-\overset{\overset{OH}{|}}{CH}-CH_3$$

Organic acids and esters

Organic acids are characterized by the presence of the **carboxyl group,** —COOH. Structurally this is

$$R-\overset{\overset{O}{\parallel}}{C}-O-H$$

The formula for a carboxylic acid is sometimes written RCO_2H.

The presence of the lone oxygen bonded to the carbon that is attached to the —OH group polarizes the O—H bond and permits the H to be lost as a proton.

$$R-\overset{\overset{:O:}{\parallel}}{C}-\overset{..}{\underset{..}{O}}-H + H_2O \longrightarrow \left(R-\overset{\overset{:O:}{\parallel}}{C}-\overset{..}{\underset{..}{O}}:\right)^- + H_3O^+$$

Many important organic compounds are acids or their salts. Some examples are.

citric acid
(citrus fruits)

acetylsalicylic acid
(aspirin)

sodium benzoate
(a food preservative)

monosodium glutamate, MSG
(*Áccent*, a flavor enhancer)

Organic acids derived from hydrocarbons are named by dropping the $-e$ from the end of the name of the parent and adding **-oic acid.**

$$CH_2\overset{O}{\overset{\|}{-}C}-OH \quad \text{ethanoic acid (acetic acid)}$$

$$CH_3-CH_2-CH_2\overset{O}{\overset{\|}{-}C}-OH \quad \text{butanoic acid (butyric acid, from rancid butter)}$$

One method of preparing an acid is by oxidation of an aldehyde (or by thorough oxidation of a primary alcohol).

$$CH_3CH_2OH \xrightarrow{(O)} CH_3CHO \xrightarrow{(O)} CH_3COOH$$

ethanol   **ethanal**   **ethanoic acid**

Esters are characterized by the presence of the functional group,

$$R\overset{O}{\overset{\|}{-}C}-O-R'$$

They are products of the acid-catalyzed elimination of water from between a carboxylic acid and an alcohol, a process called **esterification.**

$$CH_3\overset{O}{\overset{\|}{-}C}+O-H + H+O-CH_2CH_3 \underset{}{\overset{H^+}{\rightleftharpoons}} CH_3\overset{O}{\overset{\|}{-}C}-O-CH_2CH_3 + H_2O$$

acetic acid   **ethyl alcohol**   **ethyl acetate**

This reaction leads to an equilibrium. The position of equilibrium can be shifted to the right by employing a dehydrating agent to remove H_2O from the reaction mixture as it is formed. This type of reaction, where molecules are

The polymerization of H_3PO_4 discussed in Chapter 20 is also an example of a condensation reaction.

joined with the simultaneous elimination of another smaller molecule, is also called a **condensation reaction.**

When an alcohol is treated with an inorganic acid an inorganic ester is produced. For example, when nitric acid is *cautiously* added to ethyl alcohol,

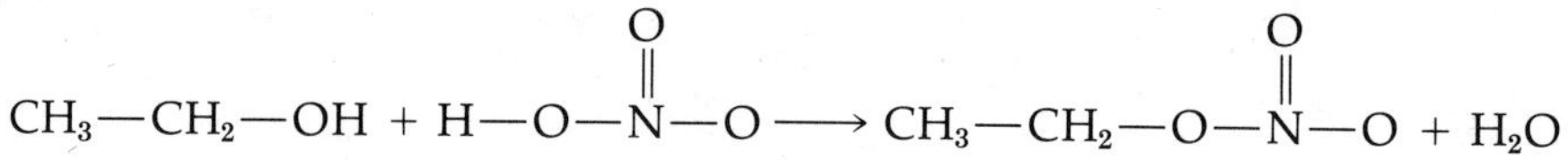

$$CH_3-CH_2-OH + H-O-\overset{\overset{O}{\|}}{N}-O \longrightarrow CH_3-CH_2-O-\overset{\overset{O}{\|}}{N}-O + H_2O$$

the product is ethyl nitrate (quite explosive). Organic phosphate esters of the type

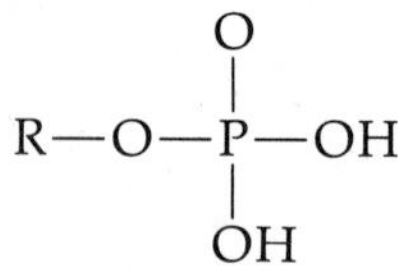

$$R-O-\overset{\overset{O}{|}}{\underset{\underset{OH}{|}}{P}}-OH$$

are very important in biological systems as are esters of the trihydroxy alcohol, glycerol.

The reaction to produce an ester is reversible and the insertion of an H_2O molecule into the ester to give the acid and alcohol is termed hydrolysis. An example is the hydrolysis of methyl acetate.

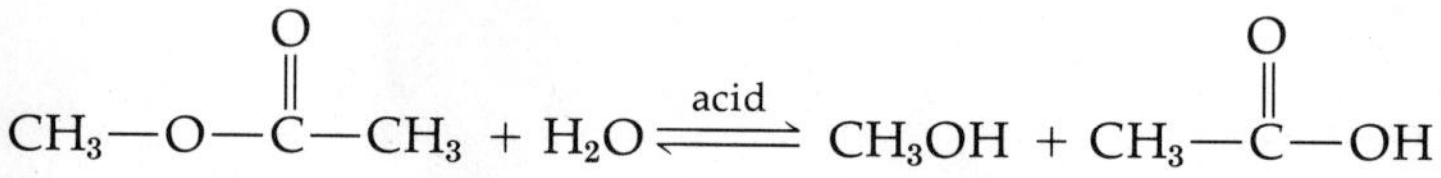

$$CH_3-O-\overset{\overset{O}{\|}}{C}-CH_3 + H_2O \underset{}{\overset{acid}{\rightleftharpoons}} CH_3OH + CH_3-\overset{\overset{O}{\|}}{C}-OH$$

methyl acetate   **methanol**   **acetic acid**

This equilibrium is established rapidly, and in order to drive the reaction to the right, the alcohol can be removed by distillation.

In the base-catalyzed hydrolysis, also called **saponification,** the acid that is produced in the forward reaction is neutralized, thereby shifting the position of equilibrium to the right. For example, using molecular formulas,

Sodium acetate is really $Na^+CH_3CO_2^-$.

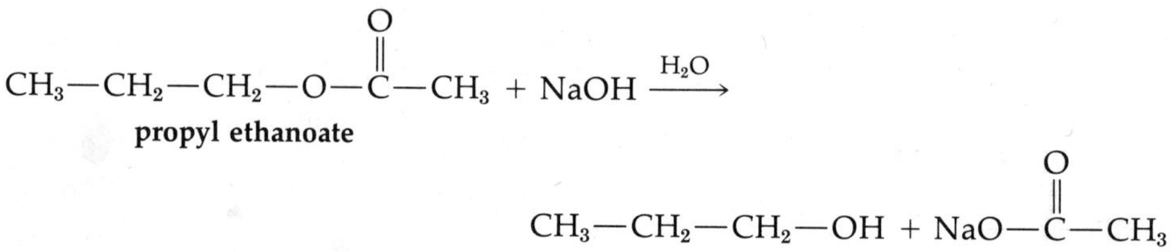

$$CH_3-CH_2-CH_2-O-\overset{\overset{O}{\|}}{C}-CH_3 + NaOH \overset{H_2O}{\longrightarrow}$$

propyl ethanoate

$$CH_3-CH_2-CH_2-OH + NaO-\overset{\overset{O}{\|}}{C}-CH_3$$

sodium acetate

Esters, particularly those of low molecular weight, generally have pleasant, rather agreeable odors, as seen in Table 22.6. Some of them have very desirable solvent properties and are used in paints and varnishes.

Table 22.6
Odors of some common esters

Name	Formula	Odor
n-Amyl acetate	$CH_3COOCH_2(CH_2)_3CH_3$	Banana
n-Octyl acetate	$CH_3COOCH_2(CH_2)_6CH_3$	Orange
iso-Amyl butyrate	$CH_3(CH_2)_2COOCH(CH_3)CH_2CH_2CH_3$	Pear

Ethers

We saw that esters are produced by a condensation reaction between an alcohol and an acid. Alcohols can also undergo self-condensation to give **ethers** in which two hydrocarbon units are joined by an oxygen bridge.

$$R-OH + HO-R \longrightarrow R-O-R + H_2O$$
$$\text{ether}$$

A specific example is

$$2CH_3-CH_2-OH \xrightarrow[\text{H}_2\text{SO}_4]{\text{conc}} CH_3-CH_2-O-CH_2-CH_3 + H_2O$$
$$\textbf{diethyl ether}$$

The concentrated sulfuric acid is used as a dehydrating agent to help in the removal of the H_2O as it is formed. Diethyl ether is used as an anesthetic. It must be used with care because it is extremely flammable. Most ethers are used primarily as solvents.

22.9 AMINES AND AMIDES

The **amines** are a group of compounds that can be viewed as derivatives of ammonia. Below are some typical amines.

Formula	IUPAC Name	Common Name
CH_3-NH_2	Aminomethane	Methylamine
$CH_3-CH_2-NH_2$	Aminoethane	Ethylamine
$CH_3-CH_2-NH-CH_2-CH_3$	Ethylaminoethane	Diethylamine
$(CH_3)_3C-NH_2$	2-Amino-2-methylpropane	t-Butylamine
$H_2N-CH_2-CH_2-NH_2$	1,2-Diaminoethane	Ethylenediamine
⟨◯⟩$-NH_2$	Aminobenzene	Aniline

Amines can be classified as being primary, secondary, or tertiary, depending on the number of R groups attached to the nitrogen.

primary amine   secondary amine   tertiary amine

All the amines in our list above are primary amines except diethylamine, which is a secondary amine.

Amines also exist in which the nitrogen is a member of a ring. Examples of this type are

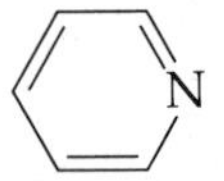

pyridine

piperidine

These compounds are referred to as **heterocycles** because not all the atoms in the ring are identical.

Amines, like ammonia, are weak bases. Most of them also have very unpleasant odors. The stench of decaying protein, for instance, can be traced to compounds like those below.

$$H_2N—CH_2—CH_2—CH_2—CH_2—NH_2 \qquad \text{putrescine}$$

$$H_2N—CH_2—CH_2—CH_2—CH_2—CH_2—NH_2 \qquad \text{cadaverine}$$

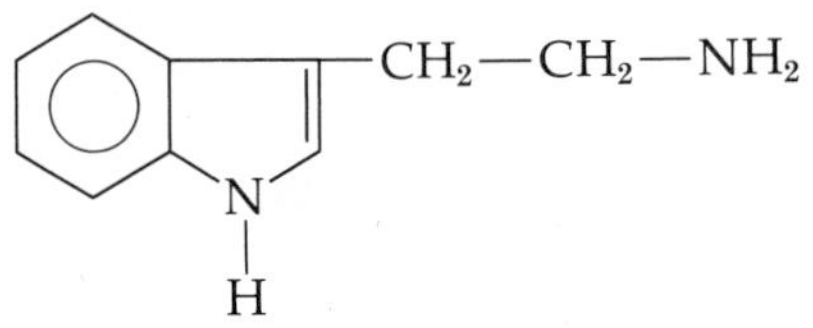

skatole (in feces)

Amides are identified by the functional group $—\overset{\overset{\displaystyle O}{\|}}{C}—NH_2$. Some examples of this type of compound are

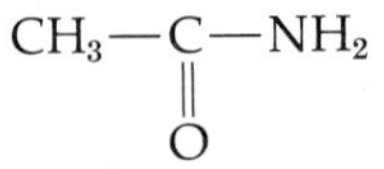

acetamide 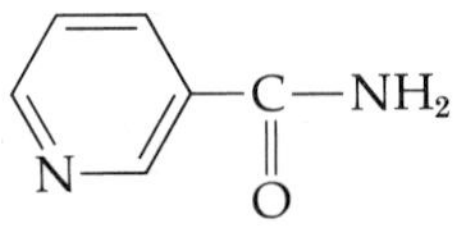   nicotinamide

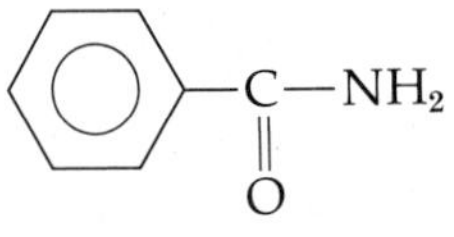

benzamide 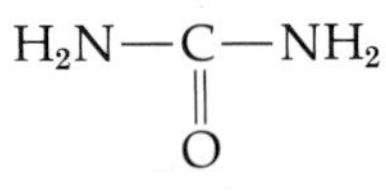   urea

The functional groups of the amines and amides are found in many important biological compounds that will be discussed in the next chapter. These compounds include the nucleic acids, the amino acids, thiamin, riboflavin, and biotin. The heterocyclic amines are also found as a basic unit of a group of compounds known as alkaloids. Alkaloids are rather complex compounds containing nitrogen that are found in plants. Compounds such as nicotine, codeine, morphine, and lysergic acid diethylamide (LSD) are all alkaloids.

22.10   POLYMERS

Polymers are very large molecules that are made by bonding together many smaller molecules called **monomers.** For example, polyethylene (plastic food wrap) is prepared by linking together a large number of $CH_2=CH_2$ monomer units to give a hydrocarbon having the general formula, $(-CH_2-CH_2-)_n$. Many naturally occurring substances are polymers, for example, rubber, starch, proteins, and the nucleic acids. Artificial polymers include such familiar materials as Bakelite, Melmac, Nylon, Dacron, Plexiglass, Teflon, and polyvinyl chloride (PVC).

Polymers can be classified as addition polymers or as condensation polymers. **Addition polymers** are formed by the direct linking together of the monomer units. An example is polyvinyl chloride (PVC), which is formed by polymerizing vinyl chloride.

$$CH_2=CH$$
$$|$$
$$Cl$$
vinyl chloride

In the presence of a suitable initiator (generally, a peroxide) the double bond opens and the monomer units link end-to-end.

$$n\ CH_2=CH \xrightarrow{\text{peroxide}} \left(CH_2-CH \right)_n$$
$$\qquad\quad | \qquad\qquad\qquad\quad |$$
$$\qquad\quad Cl \qquad\qquad\qquad\quad Cl$$

This material finds many uses, for example, in phonograph records and plastic pipe. It can also be mixed with esters that soften the polymer. The softened material finds uses in such products as plastic garden hoses, tablecloths, raincoats and "vinyl leather" products.

When a **condensation polymer** is formed, two monomers are joined with the simultaneous elimination of a small molecule such as water or methanol. Usually two different monomers are joined and the polymer is said to be a **copolymer.** An example is Nylon, which is formed when a dicarboxylic acid (i.e., a carboxylic acid that contains two—COOH groups) reacts with a diamine (an amine with two —NH_2 groups). The overall reaction for the production of Nylon is

$$n\text{HOOC}(CH_2)_4\text{COOH} + \quad n\text{H}_2\text{N}(CH_2)_6\text{NH}_2 \quad \xrightarrow[280°C]{-H_2O}$$
adipic acid \qquad\qquad **hexamethylenediamine**

$$\left(\begin{matrix} & & & H & & H \\ & & & | & & | \\ C(CH_2)_4C & - & N(CH_2)_6N \\ \| & \| \\ O & O \end{matrix} \right)_n + n\text{H}_2\text{O}$$

Nylon

Dacron, a **polyester,** is prepared by the reaction of methyl terephthalate (a diester) with ethylene glycol in the presence of an acid or base.

$$n\text{CH}_3\text{OOC} \bigcirc \text{COOCH}_3 + n\text{HOCH}_2\text{CH}_2\text{OH} \xrightarrow[\text{base}]{\text{acid or}} \left(\begin{matrix} C \bigcirc C - \text{OCH}_2\text{CH}_2\text{O} \\ \| \qquad\quad \| \\ O \qquad\quad O \end{matrix} \right)_n + n\text{CH}_3\text{OH}$$

methyl terephthalate \qquad **ethylene glycol** \qquad\qquad\qquad\qquad **Dacron** \qquad **methanol**

Table 22.7
Compositions of some common polymers

Monomer	Polymer	Type
$CH_2{=}CH_2$ ethylene	Polyethylene	Addition
$CH_2{=}CHCl$ vinyl chloride	Polyvinyl chloride (PVC)	Addition
$F_2C{=}CF_2$ tetrafluoroethylene	Teflon	Addition
⬡—$CH{=}CH_2$ styrene	Polystyrene	Addition
$CH_2{=}C(CH_3)COOCH_3$ methyl methacrylate	Plexiglass	Addition
$HOOC{-}(CH_2)_4{-}COOH$ + $NH_2{-}(CH_2)_6{-}NH_2$ adipic acid · · · hexamethylenediamine	Nylon	Condensation
$HOOC{-}C_6H_4{-}COOH$ + $HO{-}CH_2CH_2{-}OH$ terephthalic acid · · · ethylene glycol	Dacron	Condensation
C_6H_5OH + HCHO phenol · · · formaldehyde	Bakelite	Condensation
C_6H_5OH + furfural phenol	Durite	Condensation
melamine + HCHO formaldehyde	Melmac	Condensation

In the production of Nylon, H_2O is eliminated; and with Dacron, methanol is eliminated. Table 22.7 contains a number of important polymers, the reactants needed for their production, and whether they are addition or condensation polymers.

Using appropriate monomers—ones that are able to link to more than two other species—bonds can be formed between adjacent polymer strands. This binds the strands in place, so the greater the degree of this **cross-linking** between parallel rows of polymer molecules, the stronger the material. Bakelite,

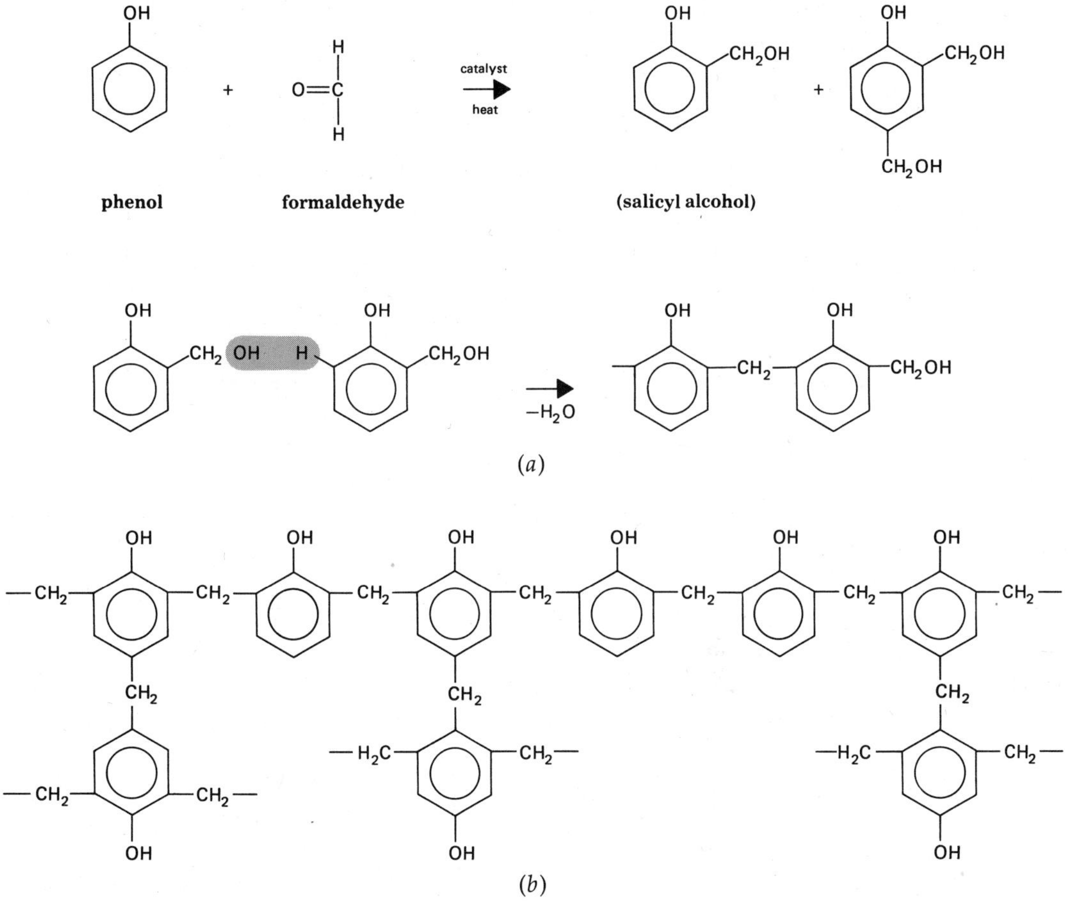

Figure 22.10
Bakelite. (a) Polymerization of salicyl alcohol gives a linear polymer. (b) Continued polymerization, with cross-linking, gives a rigid three-dimensional structure called Bakelite.

for example, owes its strength and hardness to a three-dimensional network of covalent bonds throughout the entire polymer, as illustrated in Figure 22.10.

Natural rubber, too, can be made harder and stronger by a process known as **vulcanization.** In this reaction sulfur bridges between different chains create cross-links that lead to a tougher material.

INDEX TO QUESTIONS

Hydrocarbons 1, 2, 3, 4, 12
Isomerism 5, 6, 10, 11, 16, 26
Structure and bonding 7, 8, 9, 20, 21, 32
Nomenclature 10, 14, 15, 17, 18, 19, 25, 30
Familiar properties and applications 13, 22, 28, 29
Functional groups 24, 35, 36, 41
Reactions 27, 31, 33, 34, 37, 38, 39, 40, 41
Polymers 43, 44, 45, 46, 47, 48

REVIEW QUESTIONS

22.1 What is the difference between a saturated and an unsaturated hydrocarbon?

22.2 The straight-chain alkanes are nonpolar molecules. How do we explain the fact that their boiling points increase from CH_4 to $C_{10}H_{22}$?

22.3 What would be the molecular formulas for: (a) an alkane having 30 carbon atoms, (b) an alkene having 27 carbon atoms, (c) an alkyne having 33 carbon atoms?

22.4 What would be the molecular formula for a straight-chain hydrocarbon having 17 carbon atoms and (a) all

C—C single bonds, (b) one C—C double bond, (c) one C—C triple bond, (d) three C—C double bonds, (e) two C—C triple bonds?

22.5 How does optical isomerism arise in organic compounds? Draw the optical isomers of

$$
\begin{array}{c}
\text{H} \\
| \\
\text{Br}-\text{C}-\text{I} \\
| \\
\text{Cl}
\end{array}
$$

22.6 Sketch the *cis* and *trans* isomers of 2-pentene.

22.7 What geometry do we expect for the molecules described in Question 22.5?

22.8 The molecule C_2Cl_4 is planar. Why?

22.9 The carbon atoms in 2-butyne lie in a straight line while those in butane do not. Explain why this is so.

22.10 Draw the structural formulas for and name the nine isomers of heptane. Which of these isomers would give rise to optical isomerism?

22.11 Draw all the possible isomers of hexene and show geometric isomers wherever possible.

22.12 What is a homologous series?

22.13 Name some uses of the alkanes.

22.14 Name the following compounds:

(a)
$$
\begin{array}{c}
\text{H}_3\text{C} \\
\diagdown \\
\text{CH}-\text{CH}_2-\text{CH}-\text{CH}_2-\text{CH}_3 \\
\diagup | \\
\text{H}_3\text{C} \text{CH}_3
\end{array}
$$

(b)
$$
\begin{array}{c}
\text{CH}_3-\text{CH}-\text{CH}_2-\text{CH}-\text{CH}_2-\text{CH}_3 \\
| | \\
\text{CH}_2 \text{CH}_3 \\
| \\
\text{CH}_3
\end{array}
$$

(c)
$$
\begin{array}{c}
\text{CH}_3 \\
 | \\
\text{CH}_3-\text{CH}_2-\text{CH}-\text{CH}_2-\text{CH}-\text{CH}_2-\text{CH}_3 \\
 | \\
\text{CH}_2-\text{CH}_2-\text{CH}_3
\end{array}
$$

(d)
$$
\begin{array}{c}
\text{CH}_3-\text{CH}_2-\text{CH}=\text{CH}-\text{CH}-\text{CH}_3 \\
 | \\
\text{CH}_2-\text{CH}_3
\end{array}
$$

(e)
$$
\begin{array}{c}
\text{CH}_3 \\
 | \\
\text{CH}_3-\text{CH}_2-\text{CH}-\text{CH}_2-\text{CH}-\text{CH}_3 \\
 | \\
\text{CH}_3
\end{array}
$$

22.15 Name the following compounds:

(a)
$$
\begin{array}{c}
\text{CH}_2-\text{CH}_3 \\
 | \\
\text{CH}_3-\text{C}=\text{CH}-\text{CH}_2-\text{CH}_3 \\
\\
\text{CH}_3-\text{C}=\text{CH}-\text{CH}_3
\end{array}
$$

(b)
$$
\begin{array}{c}
\text{CH}_3-\text{CH}_2-\text{C}\equiv\text{C}-\text{CH}-\text{CH}_3 \\
 | \\
\text{CH}_2-\text{CH}_3
\end{array}
$$

(c)
$$
\begin{array}{c}
\text{CH}_3 \text{CH}_3 \text{CH}_3 \\
 | | | \\
\text{CH}_3-\text{CH}-\text{C}-\text{C}-\text{CH}_2-\text{CH}_3 \\
 | | \\
\text{CH}_3 \text{CH}_3
\end{array}
$$

(d)
$$
\begin{array}{c}
\text{CH}_3 \\
 | \\
\text{CH}_3-\text{C}\equiv\text{C}-\text{CH} \\
 | \\
\text{CH}_3
\end{array}
$$

(e)
$$
\begin{array}{c}
\text{H}_3\text{C} \text{CH}=\text{CH} \\
\diagdown \diagup \diagdown \\
\text{C}=\text{C} \text{CH}_2-\text{CH}_3 \\
\diagup \diagdown \\
\text{CH}_3-\text{CH}_2 \text{CH}_3
\end{array}
$$

22.16 Draw the remaining isomers of the compound

$$
\begin{array}{c}
\text{CH}_3 \text{CH}_3 \\
 | | \\
\text{CH}_3-\text{C}=\text{C}-\text{CH}_3
\end{array}
$$

22.17 Write structural formulas for the following:
(a) 2-methylpentane
(b) 2,3-dimethylbutane
(c) 2,3-dimethyl-2-butene
(d) 1,3,5-octatriene
(e) 3,3,4-trimethyl-1-pentyne

22.18 What is the proper IUPAC name for 2,4-diethyl-3,3,4-trimethylpentane?

22.19 Write the structural formula for (a) *cis*-1,2-dichloropropene, (b) isohexane, (c) isopropyl alcohol, (d) *m*-dichlorobenzene.

22.20 What would be the C—C—C bond angles in *planar* cyclopropane, cyclobutane, cyclopentane, cyclohexane?

22.21 The chair form of cyclohexane is more stable (of lower energy) than the boat form by several kilocalories per mole. By examining these two structures, can you suggest why the boat form has a higher energy?

22.22 What is a carcinogen? In what way is the structure of 1,2-benzopyrene similar to graphite? Can you suggest why incomplete combustion of many hydrocarbons produces products similar to 1,2-benzopyrene?

22.23 Describe the bonding in benzene. How does it compare to the bonding in graphite?

22.24 What is a functional group? Give the structural formula of
(a) An aldehyde

(b) A ketone
(c) A carboxylic acid
(d) An amine
(e) An alcohol
(f) An ester
(g) An ether

22.25 Write structural formulas for the following compounds:
(a) 2-methyl-1-butene
(b) 2,3-dimethyl-2-butanol
(c) 2-bromo-1-phenylpropane
(d) 3-methyl-2-pentanone
(e) 1,3,5-tribromo-2,4,6-trichlorobenzene

22.26 Draw *all* the structural isomers, including cyclic structures, of $C_3H_4Cl_2$.

22.27 What type of reaction is characteristic of the alkenes and alkynes? What type of reaction is characteristic of the alkanes?

22.28 What are some uses of halogenated hydrocarbons?

22.29 Examine the labels of a number of common household insecticides and make a list of their active ingredients. Identify the kinds of functional groups present in these ingredients.

22.30 Name the following compounds using IUPAC rules.

(a)

(b) CH₃ ... Cl

(c)

$$CH_3-CH_2-\overset{\displaystyle O}{\overset{\displaystyle \|}{C}}-CH_3$$

(d)

$$CH_3-\overset{\displaystyle CH_3}{\overset{\displaystyle |}{CH}}-CH_2-NH_2$$

22.31 Using Markovnikov's rule, complete the following addition reactions. Name the reactants and products.

(a) $CH_3-CH_2-CH=CH_2 + HI \longrightarrow$

(b) $CH_3-CH=CH_2 + H_2O \xrightarrow{H_2SO_4}$

(c)

$$CH_3-CH_2-CH=C\overset{\displaystyle CH_3}{\underset{\displaystyle CH_3}{}} + H_2O \xrightarrow{H_2SO_4}$$

22.32 Among the alcohols,

$$CH_3-(CH_2)_x-CH_2OH$$

as x increases the molar solubility in water decreases. Why does this occur?

22.33 What products (if any) are obtained by the *mild* oxidation, using $K_2Cr_2O_7$ in acid solution, of each of the following:

(a) CH_3-CH_2-OH

(b) $CH_3-\overset{\displaystyle CH_3}{\overset{\displaystyle |}{CH}}-CH_2-OH$

(c) $CH_3-\overset{\displaystyle OH}{\overset{\displaystyle |}{CH}}-CH_3$

(d) CH_3-CH_2-CHO

(e) $CH_3-\overset{\displaystyle CH_3}{\underset{\displaystyle OH}{\overset{\displaystyle |}{\underset{\displaystyle |}{C}}}}-CH_3$

(f) CH_3-COOH

(g) $CH_3-\overset{\displaystyle O}{\overset{\displaystyle \|}{C}}-CH_3$

22.34 What chemical reactions discussed in this chapter could be used to distinguish between the following:

(a) 2-propanol and 2-methyl-2-propanol
(b) 1-butanol and 2-butanol
(c) *n*-butane and 1-butene
(d) ethanal and 2-propanone

22.35 Vanillin (page 731) contains three types of functional groups. What are they?

22.36 Draw the structures of the esters formed from
(a) acetic acid and 2-propanol
(b) acetic acid and 1-pentanol
(c) benzoic acid and methanol
(d) formic acid (methanoic acid) and methanol

22.37 Write equations for the saponification of:

(a)

$$CH_3-CH_2-\overset{\displaystyle O}{\overset{\displaystyle \|}{C}}-O-CH_3$$

(b)

$$CH_3-CH_2-O-\overset{\displaystyle O}{\overset{\displaystyle \|}{C}}-CH_2-CH_2-\overset{\displaystyle O}{\overset{\displaystyle \|}{C}}-O-CH_2-CH_3$$

22.38 The synthesis of organic compounds from simple starting materials is an important aspect of organic chemistry. From the reactions described in this chapter, describe how you could prepare
(a) dichloroethane from ethene
(b) propanoic acid from 1-propanol
(c) 2-propanol from 1-chloropropane
(d) ethyl acetate from ethanal
(e) methyl ethyl ketone from 1-bromobutane

22.39 Predict the results of the following reactions:

(a)

$$CH_3-CH_2OH + CH_3-\overset{\displaystyle O}{\overset{\displaystyle \|}{C}}-OH \xrightarrow{H^+}$$

(b) $CH_3—CH_2—CH_2—OH \xrightarrow[\Delta]{KMnO_4}$

(c)

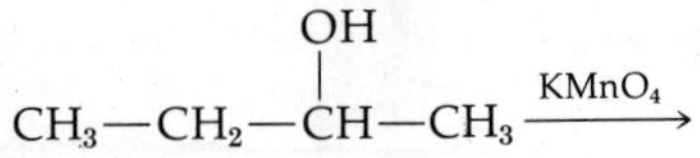

$$CH_3—CH_2—\overset{\overset{\displaystyle OH}{|}}{CH}—CH_3 \xrightarrow{KMnO_4}$$

(d) $CH_3—CH_2—CH_2—CH_2—OH \xrightarrow[conc]{H_2SO_4}$

22.40 Write a chemical equation showing the saponification of propyl acetate. What drives this reaction toward completion?

22.41 Compare the products obtained by reduction of an aldehyde and a ketone with hydrogen.

22.42 Diethylamine gives a basic solution in water. Why?

22.43 Styrene (phenylethene) forms an addition polymer. Sketch the structure of the repeating unit in polystyrene.

22.44 What is the difference between addition polymerization and condensation polymerization?

22.45 *Saran* is an addition copolymer of vinyl chloride and vinylidene chloride, $CH_2{=\!=}CCl_2$. Sketch a portion of the polymer chain showing several monomer units.

22.46 If ethylene glycol and dimethylmalonate were to form a condensation polymer by the elimination of methanol, what would be the structure of the repating unit in the polymer chain?

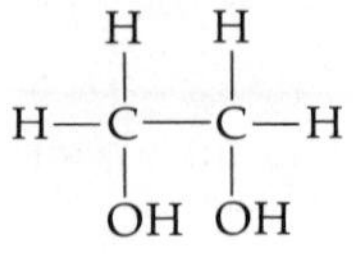

$$H—\overset{\overset{\displaystyle H}{|}}{\underset{\underset{\displaystyle OH}{|}}{C}}—\overset{\overset{\displaystyle H}{|}}{\underset{\underset{\displaystyle OH}{|}}{C}}—H$$

ethylene glycol

$$CH_3—O—\overset{\overset{\displaystyle O}{\|}}{C}—CH_2—\overset{\overset{\displaystyle O}{\|}}{C}—O—CH_3$$

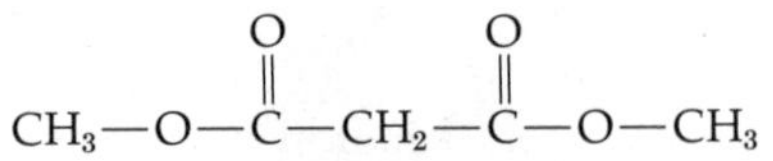

dimethylmalonate

22.47 Nylon stockings appear practically to disintegrate when hydrochloric or sulfuric acid is accidentally spilled on them. Actually, they dissolve very rapidly. Can you suggest what occurs chemically when nylon is dissolved by acid?

22.48 What is cross-linking? What effect does it have on the physical properties of a polymer?

23

BIOCHEMISTRY

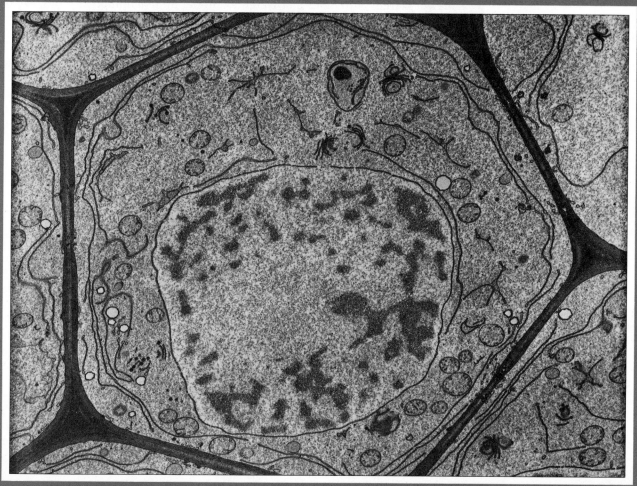

A living cell, like the one shown here, is a complex chemical factory. In this chapter we will examine some of the kinds of molecules that are found in cells and the roles they play in the life process.

Without question, one of the most active areas of chemical research today is the field of biochemistry which, as its name implies, is concerned with the chemistry that takes place in living systems. The modern biochemist views a living organism as a collection of organic molecules that interact with each other and with their environment in a very unique and special way. When isolated from a living system these biomolecules are themselves lifeless. They obey all of the laws of chemistry and thermodynamics that we have examined up to now, and it is the goal of the biochemist to understand the functions and intricate interactions of these molecules that give rise to the phenomenon we call life.

The field of biochemistry is very large and quite complex, and we certainly cannot hope to explore it fully in a single chapter. Instead, we will be content to examine some of the types of biomolecules and their apparent functions in the operations of a living cell.

Nearly all the compounds found in a living system are composed primarily of carbon, hydrogen, oxygen, nitrogen, and some sulfur and, for the most part, they depend on carbon for their molecular backbone. In general these molecules are very large, with molecular weights ranging up to a million or more. We will see, however, that in many cases these **macromolecules** are constructed using a relatively small number of different simple molecules. For the purpose of discussing the various kinds of biomolecules we can place nearly all of them into one or another of four classes: **proteins, carbohydrates, lipids,** and **nucleic acids.**

Before we proceed, remember that you are not expected to memorize the names and formulas of all the different compounds that we will discuss in this chapter. Instead, concentrate on the types of molecules that are involved, the way that they combine with each other, and the general features of the resulting structures.

23.1 PROTEINS

Proteins are very large molecules having molecular weights ranging from about 6000 to approximately 1,000,000. They constitute nearly 50% of the dry weight of cells and, depending on the individual protein, serve a variety of different functions within a living organism. Some of them are hormones, like insulin, which serve as chemical messengers that coordinate certain biochemical activities. Insulin, for example, controls the level of sugar in the bloodstream by promoting its absorption into cells. Others are enzymes that act as catalysts for biochemical reactions. We will discuss these further in the next section. Some proteins serve to transport substances through the organism. Hemoglobin, for instance, carries oxygen in the bloodstream and delivers it to different parts of the body. There are long fibrous proteins, such as *actin* and *myosin,* that are found in muscle. Another fibrous protein, *α-keratin,* serves as the major constituent of hair, nails, and skin, while *collagen* is the prime constituent of tendons. Proteins are also found in toxins (poisonous materials) such as botulinus toxin as well as in antibodies.

Despite their wide range of functions, all proteins have something in common with one another. They are polymers made up by linking together, in various combinations, a number of different simple monomeric units called **α-amino acids.**

An amino acid is a bifunctional organic molecule that contains both a carboxyl group, —COOH, as well as an amine group, —NH$_2$. In an α-amino acid the amine group is located on the carbon atom adjacent to the carboxyl group (the α-carbon atom). This gives a structure that we can generalize as

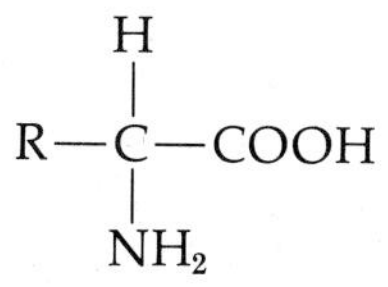

$$R-\overset{\overset{\displaystyle H}{|}}{\underset{\underset{\displaystyle NH_2}{|}}{C}}-COOH$$

Since the —NH$_2$ group (like ammonia) is basic and the —COOH group is acidic, in neutral solution the amino acid exists in an internal ionic form called a *zwitterion* where the proton of the —COOH group is transferred to the NH$_2$ to give

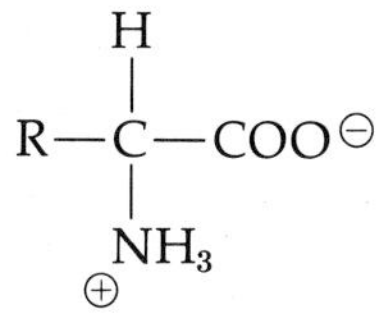

$$R-\overset{\overset{\displaystyle H}{|}}{\underset{\underset{\displaystyle \overset{NH_3}{\oplus}}{|}}{C}}-COO^{\ominus}$$

A very interesting and important fact is that in all organisms nearly all proteins are constructed using as building blocks a set of only 20 α-amino acids. These are shown in Figure 23.1. Note that except for proline, all fit the general

Figure 23.1

The 20 amino acids found in most proteins.

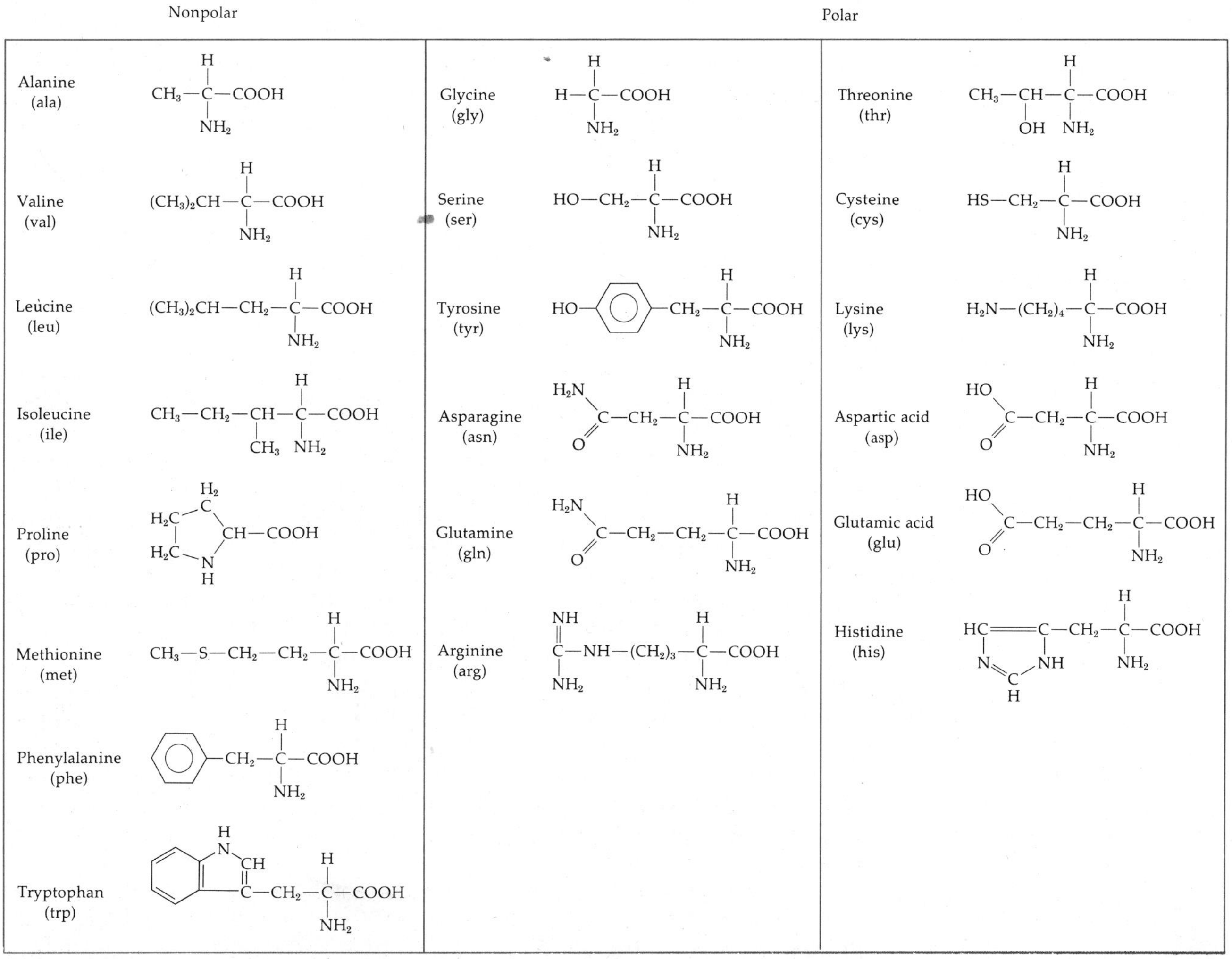

Nonpolar		Polar	
Alanine (ala)	$CH_3-\underset{NH_2}{\overset{H}{C}}-COOH$	Glycine (gly)	$H-\underset{NH_2}{\overset{H}{C}}-COOH$	Threonine (thr)	$CH_3-\underset{OH}{CH}-\underset{NH_2}{\overset{H}{C}}-COOH$
Valine (val)	$(CH_3)_2CH-\underset{NH_2}{\overset{H}{C}}-COOH$	Serine (ser)	$HO-CH_2-\underset{NH_2}{\overset{H}{C}}-COOH$	Cysteine (cys)	$HS-CH_2-\underset{NH_2}{\overset{H}{C}}-COOH$
Leucine (leu)	$(CH_3)_2CH-CH_2-\underset{NH_2}{\overset{H}{C}}-COOH$	Tyrosine (tyr)	$HO-\bigcirc-CH_2-\underset{NH_2}{\overset{H}{C}}-COOH$	Lysine (lys)	$H_2N-(CH_2)_4-\underset{NH_2}{\overset{H}{C}}-COOH$
Isoleucine (ile)	$CH_3-CH_2-\underset{CH_3}{CH}-\underset{NH_2}{\overset{H}{C}}-COOH$	Asparagine (asn)	$\underset{O}{\overset{H_2N}{C}}-CH_2-\underset{NH_2}{\overset{H}{C}}-COOH$	Aspartic acid (asp)	$\underset{O}{\overset{HO}{C}}-CH_2-\underset{NH_2}{\overset{H}{C}}-COOH$
Proline (pro)	structure	Glutamine (gln)	$\underset{O}{\overset{H_2N}{C}}-CH_2-CH_2-\underset{NH_2}{\overset{H}{C}}-COOH$	Glutamic acid (glu)	$\underset{O}{\overset{HO}{C}}-CH_2-CH_2-\underset{NH_2}{\overset{H}{C}}-COOH$
Methionine (met)	$CH_3-S-CH_2-CH_2-\underset{NH_2}{\overset{H}{C}}-COOH$	Arginine (arg)	$\underset{NH_2}{\overset{NH}{C}}-NH-(CH_2)_3-\underset{NH_2}{\overset{H}{C}}-COOH$	Histidine (his)	structure
Phenylalanine (phe)	$\bigcirc-CH_2-\underset{NH_2}{\overset{H}{C}}-COOH$				
Tryptophan (trp)	structure				

formula above. Another point of interest is that except for glycine, all these amino acids have four different groups attached to the α-carbon atom, which is therefore an asymmetric carbon atom. Each of these amino acids can therefore exist in two different isomeric forms (optical isomers). In proteins, however, only one isomer of each is commonly observed to occur. Apparently, substitution of one isomer for another destroys the biological activity of the protein molecule.

All the α-amino acids except glycine are chiral.

In a protein molecule, amino acids are linked together to form a long chain. This can be viewed as the result of the elimination of a water molecule from between the —NH₂ group of one amino acid molecule and the —COOH group of another,

$$H_2N-\underset{\underset{R_1}{|}}{\overset{\overset{H}{|}}{C}}-\overset{\overset{O}{\|}}{C}-\boxed{O-H \quad H}-\underset{\underset{H}{|}}{N}-\underset{\underset{R_2}{|}}{\overset{\overset{H}{|}}{C}}-COOH \xrightarrow{-H_2O} H_2N-\underset{\underset{R_1}{|}}{\overset{\overset{H}{|}}{C}}-\overset{\overset{O}{\|}}{C}-\underset{\underset{H}{|}}{N}-\underset{\underset{R_2}{|}}{\overset{\overset{H}{|}}{C}}-COOH$$

peptide bond or peptide linkage

The molecule that results is called a **peptide** and the group of atoms within the dotted line constitutes a **peptide bond** (amide bond), or **peptide linkage.** In the particular example above, the peptide is composed of two amino acids and is said to be a **dipeptide.** Since one end of this molecule contains a carboxyl group and the other a free —NH₂ group, additional amino acids may be joined to give ultimately a **polypeptide,** a long chain composed of many amino acid molecules linked by peptide bonds. A segment of such a chain could be indicated as

Notice that the "amide bond" that connects monomers in polypeptides is the same as the linkage that connects monomers in the polymer Nylon (page 737)!

$$\cdots-N-\underset{\underset{H}{|}}{\overset{\overset{H}{|}}{C}}-\overset{\overset{O}{\|}}{C}-\underset{}{N}-\underset{\underset{H}{|}}{\overset{\overset{H}{|}}{C}}-\overset{\overset{O}{\|}}{C}-N-\underset{\underset{H}{|}}{\overset{\overset{H}{|}}{C}}-\overset{\overset{O}{\|}}{C}-N-\underset{\underset{H}{|}}{\overset{\overset{H}{|}}{C}}-\overset{\overset{O}{\|}}{C}-\cdots$$

with R₁, R₂, R₃, R₄ groups.

The backbone of the chain is thus the same series of atoms repeated over and over again, with only the R groups changing as we move along the chain.

Before we continue, we should note that in the formation of a protein the linking together of the different amino acids is not a random process. Each molecule of a given protein has the same sequence of amino acids along its polypeptide chain. In fact, it is this very sequence that imparts to a particular protein its own specific properties.

The amino acid sequence that exists in a polypeptide is called its **primary structure.** In addition to this, the polypeptide chain twists and turns and assumes a **secondary structure** that is determined by hydrogen bonding that occurs between different groups along the chain. An example of this is found in the fibrous protein α-keratin (the major component of hair), in which the polypeptide chains coil themselves into the **α-helix,** shown in Figure 23.2. The hydrogen bonding, in this case, takes place between the oxygen atom in a carbonyl group (⊃C=O) and a hydrogen atom attached to a nitrogen atom that lies in an adjacent loop of the helix (Figure 23.3). This therefore serves to hold the chain in its coiled shape.

Altering the temperature or changing the properties of the solvent (e.g., by changing the pH) can alter the shape of the protein and destroy its biological activity. We say the protein is denatured.

In globular proteins, so named because of their overall shape, coiled polypeptide chains are also folded to give a complex three-dimensional structure, referred to as its **tertiary structure.** This is shown in Figure 23.4 for the protein myoglobin, a substance that stores oxygen in muscle tissue until it is needed in

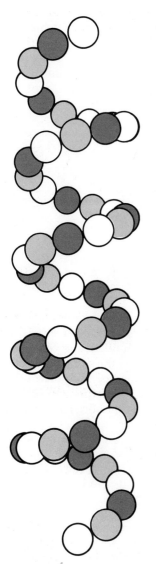

Figure 23.2

The α-helix, composed of a repetition of amino acid units.

$$-\overset{|}{\underset{R}{C}}-\overset{O}{\overset{\|}{C}}-\overset{H}{\underset{}{N}}-\left(\overset{|}{\underset{\underset{R\ \ C\ \ N}{\underbrace{\quad\quad}}}{C}}-\overset{O}{\overset{\|}{C}}-\overset{H}{\underset{}{N}}\right)_n\overset{|}{\underset{R}{C}}-\overset{O}{\overset{\|}{C}}-\overset{H}{\underset{}{N}}-$$

In the illustration N is white, C is pale orange, and R is dark orange.

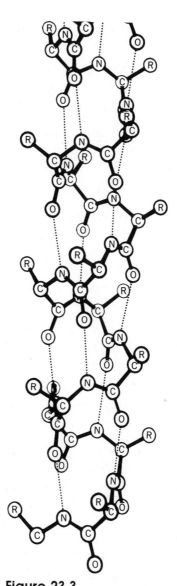

Figure 23.3

Hydrogen bonding (dotted lines) in the α-helix. Courtesy of Carroll K. Johnson, Oak Ridge National Laboratory, Oak Ridge, Tennessee.

metabolic oxidation. It is the presence of large amounts of myoglobin in the leg and thigh muscles of birds, for instance, that gives this meat a darker color than the breast meat. Myoglobin is also responsible for the red color of beef steak.

The tertiary structure of a protein is controlled by several different kinds of interactions that serve to hold the folded segments of the chain in place. For example, besides hydrogen bonding there are also ionic attractions that occur

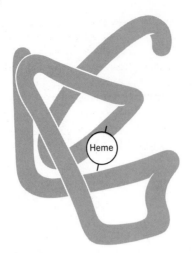

Figure 23.4

The tertiary structure of myoglobin. The polypeptide helix folds and turns to produce a globular protein.

Nonpolar substances that repel water are said to be <u>hydrophobic</u> (water hating). Biochemists call interactions involving nonpolar groups <u>hydrophobic</u> interactions.

between a negatively charged deprotonated carboxyl group (like that found in the R group of glutamic acid) and a positively charged protonated amine group (like that found in lysine). This is shown in Figure 23.5.

The solvent is also important in determining the shape of the protein molecule. In the presence of the polar solvent water, nonpolar R groups such as the phenyl ring in phenylalanine (see Figure 23.1) are forced toward the center of the folded polypeptide chain, away from the solvent. This is the same phenomenon, you may remember, that leads to the low solubility of nonpolar substances in polar solvents. It too helps determine the tertiary structure of proteins because the polypeptide chain tends to fold in such a way that nonpolar groups do not contact the solvent.

Still another type of interaction that maintains the folded conformation of the protein molecule is the formation of covalent bonds between cysteine molecules located at different points along the chain. This occurs by partial oxidation of the —SH (thiol) group,

$$\text{R—SH} + \text{HS—R} \xrightarrow{\text{oxidation}} \text{H}_2\text{O} + \text{R—S—S—R}$$

The resulting linkage is called a **disulfide bridge.**

Disulfide bridges not only help to keep the polypeptide chain folded but can also bind two such chains together. For example, beef insulin (Figure 23.6) consists of two polypeptide chains that are cross-linked at two points by these disulfide bridges.[1]

There is an interesting sidelight to the subject of the disulfide bridge in protein chemistry. The curl (or lack of curl) in hair is determined by protein

---

[1] The elucidation of the primary structure of this protein by Frederick Sanger won him the Nobel Prize in 1958.

Figure 23.5

Ionic interactions hold portions of the polypeptide chains together in the tertiary structure of proteins.

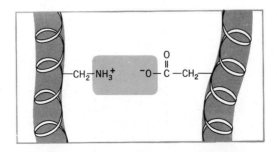

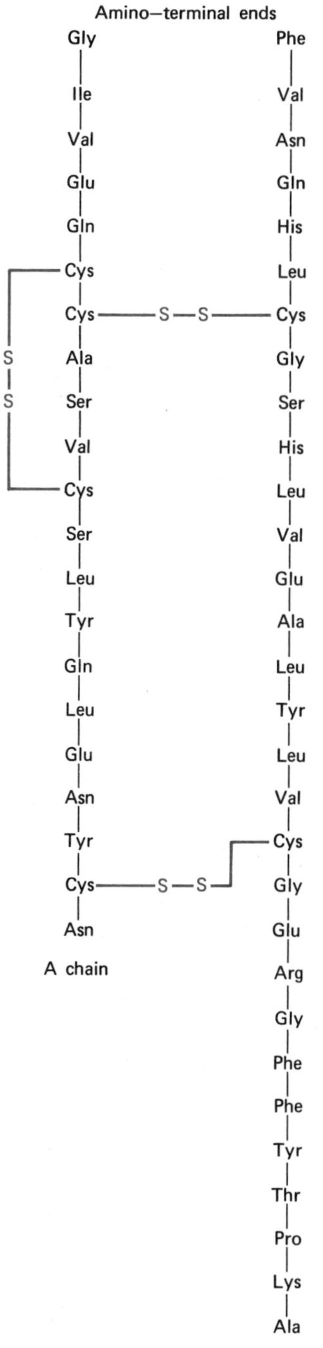

conformation locked in place by disulfide bridges. The "permanent wave" treatment that women (and, recently, men) use to produce curly hair involves two chemical reactions. The hair is first treated with a chemical able to break the disulfide bridges so that the protein chains in the hair are free to twist into any desired shape, as determined by the curlers. A second solution is then applied that produces a mild oxidation, thereby causing the disulfide bridges to be reestablished. These newly formed disulfide bridges "set" the hair in the newly curled shape. The only reason a permanent wave isn't really "permanent" is because new hair grows that hasn't received the setting treatment.

Proteins that contain more than one independent polypeptide chain exhibit still another degree of structural sophistication, called **quaternary structure.** This is determined by the way in which the folded chains orient themselves with respect to one another. A good example of this occurs in hemoglobin, Figure 23.7. This protein consists of four polypeptide chains; two α-chains each containing 141 amino acids and two β-chains each with 146 amino acids.

In addition to the polypeptide chains, this protein also contains groupings of atoms, called **heme groups,** that serve to bind oxygen so that it can be transported through the bloodstream and be deposited at oxygen-poor cells. As depicted in Figure 23.7, the four folded polypeptide chains with their heme groups are packed together in a roughly tetrahedral fashion.

The heme group in hemoglobin is also found in myoglobin and accounts for the red color of blood and muscle tissue. The structure of heme is

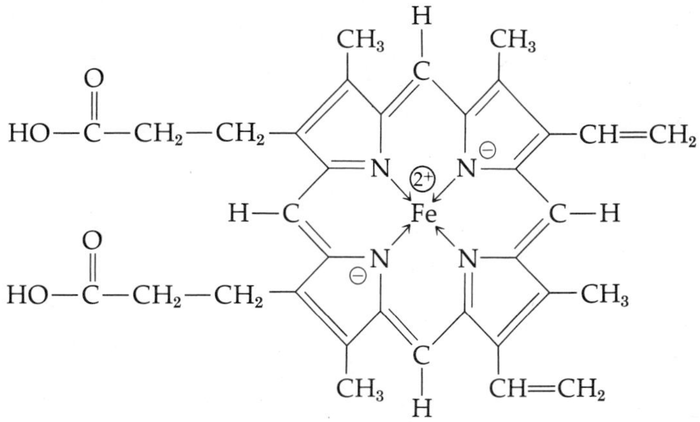

Amino—terminal ends

Gly — Ile — Val — Glu — Gln — Cys — Cys — Ala — Ser — Val — Cys — Ser — Leu — Tyr — Gln — Leu — Glu — Asn — Tyr — Cys — Asn

A chain

Phe — Val — Asn — Gln — His — Leu — Cys — Gly — Ser — His — Leu — Val — Glu — Ala — Leu — Tyr — Leu — Val — Cys — Gly — Glu — Arg — Gly — Phe — Phe — Tyr — Thr — Pro — Lys — Ala

B chain

Cys — S—S — Cys (interchain)
Cys — S—S — Cys (interchain)
Cys — S—S — Cys (intrachain A)

Figure 23.6

Amino acid sequence in beef insulin.

Figure 23.7

Quaternary structure of hemoglobin. Four globular protein molecules containing heme groups are packed into the hemoglobin structure. Adapted from R. E. Dickerson and I. Geis, The Structure and Action of Proteins, *W. A. Benjamin, Inc., Menlo Park, Calif., 1969. Original illustration copyright 1969 by R. E. Dickerson and I. Geis.*

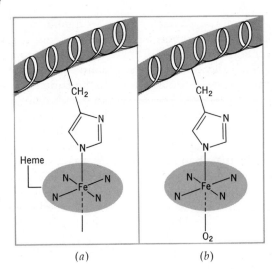

Figure 23.8

Attachment of a heme group to the hemoglobin protein by coordination of histidine to the iron atom of heme. (a) Sixth coordination site is vacant. (b) When hemoglobin carries oxygen, the O₂ is bound to the Fe of heme at the sixth coordination site.

Normally, metals needed by the body in trace amounts are present in our cells in the form of complex ions.

It is, in fact, a complex ion containing iron(II) enclosed within a square planar grouping of nitrogen atoms. In hemoglobin each heme group is attached to its polypeptide by additional coordination to the nitrogen atom of a histidine, as shown in Figure 23.8. The sixth coordination site about the iron(II) is empty and is used to bind an oxygen molecule.

The basic square planar ligand structure in heme is called a **porphyrin** and forms very stable complexes with several different metal ions. For example, a structure very similar to heme containing Mg^{2+} instead of Fe^{2+} is found in chlorophyll, the green pigment in plants that is used in photosynthesis. Still another porphyrin structure, containing Co^{2+} in the center, exists in a substance called vitamin B_{12} coenzyme (Figure 23.9). Thus, even though most of the structures of biomolecules are made up of carbon, hydrogen, nitrogen, and oxygen, some metals are also critically important to the well-being of a living organism.

23.2 ENZYMES

Enzymes are globular proteins that serve to catalyze specific biochemical reactions with what can only be judged as amazing effectiveness. In some cases

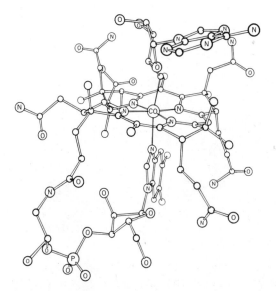

Figure 23.9

Vitamin B-12 coenzyme. Courtesy of Carroll K. Johnson, Oak Ridge National Laboratory, Oak Ridge, Tennessee.

reactions are speeded up, with respect to their uncatalyzed paths, by factors ranging from 10^9 to 10^{20}! Competing side reactions are not affected and are very slow by comparison. The result is that essentially 100% of the reactants are funneled through the same reaction path. In this way a buildup of by-products that would otherwise cause a waste-disposal problem for the organism is avoided. Enzymes also provide the organism with a way of controlling the rates of the reactions that take place, because biochemical reactions do not occur at appreciable rates in the absence of the catalyst. Removal, or at least temporary blockage, of a critically important enzyme "turns off" the chemistry which that particular enzyme catalyzes. Thus, in a very real sense, enzymes direct the chemical reactions that take place in a living cell.

Some enzymes require an additional substance called a **coenzyme** in order to function. Many vitamins that we must ingest to maintain good health are precursors of coenzymes. An example is vitamin B_{12}, whose absence from the diet leads to a deficiency disease known as pernicious anemia. In the body vitamin B_{12} is converted to its coenzyme, with the structure we saw in Figure 23.9. Many metals, like cobalt, are needed in small amounts to promote enzyme activity.

The mechanism by which an enzyme acts has long been the subject of intense research. It appears to depend on the ability of the enzyme to bind very selectively to a reactant molecule (called the **substrate** of the enzyme). There thus seems to be a "lock-and-key" relationship between an enzyme and its substrate, where the substrate molecule just precisely fits into (or onto) the folded globular protein. There is evidence that when this occurs, there is a slight alteration in the shape of the enzyme that strains certain key bonds in the substrate, thereby making them more susceptible to chemical attack.

Some enzymes are very specific in their activity, affecting the rate of reaction of a single compound. Others are less choosy and simply promote a certain kind of chemical reaction on a whole class of compounds having similar structures. This behavior, too, is the direct result of the lock-and-key relationship between the enzyme and its substrate. An example is the enzyme *chymotrypsin*, which accelerates the hydrolysis of the dotted bond in the compounds,

$$\langle\!\!\bigcirc\!\!\rangle\text{---}CH_2\text{---}\underset{\underset{H}{|}}{\overset{\overset{R}{|}}{C}}\text{---}\underset{\underset{O}{\|}}{C}\cdots X\text{---}R' \qquad \text{where } X = N \text{ or } O$$

An example is

$$\langle\!\!\bigcirc\!\!\rangle\text{---}CH_2\text{---}\underset{\underset{H}{|}}{\overset{\overset{NH_3^+}{|}}{C}}\text{---}\underset{\underset{O}{\|}}{C}\cdots NH\text{---}\underset{\underset{H}{|}}{\overset{\overset{CH_3}{|}}{C}}\text{---}COO^-$$

In this case it is thought that the hydrophobic benzene ring serves to position the substrate molecule on the enzyme (Figure 23.10), which then interacts with the $>C=O$ group in a way that makes the dotted bond more susceptible to hydrolysis. Since the function of the enzyme depends on the hydrophobic tail and the proper location of the $>C=O$ group, a family of similar compounds are affected by the enzyme.

Enzyme inhibition

When a substance other than the enzyme substrate becomes bound to the active site of an enzyme, the catalytic activity is lost and the enzyme is said to be inhib-

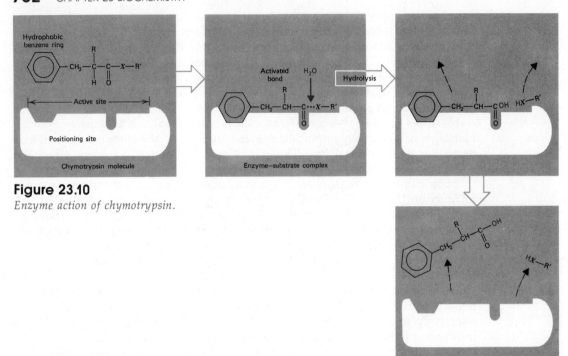

Figure 23.10

Enzyme action of chymotrypsin.

ited. In some cases this inhibition is irreversible. This occurs when the inhibitor becomes permanently bound to the enzyme, by covalent bond formation, and cannot be displaced. An example is diisopropylfluorophosphate,

$$H-\underset{\underset{CH_3}{|}}{\overset{\overset{CH_3}{|}}{C}}-O-\underset{\underset{O}{|}}{\overset{\overset{F}{|}}{P}}-O-\underset{\underset{CH_3}{|}}{\overset{\overset{CH_3}{|}}{C}}-H$$

Botulinus toxin — the most powerful poison yet discovered — blocks the production of acetylcholine, so it too is a nerve poison.

a highly toxic nerve poison and an ingredient in some nerve gases. This molecule reacts with and poisons (inhibits) an enzyme called *acetylcholine esterase,* which is required for the transport of impulses along nerve tissue.

$$enzyme-O-(H + F-\underset{\underset{\underset{CH(CH_3)_2}{|}}{\overset{\overset{O}{|}}{P}}}{\overset{\overset{CH(CH_3)_2}{|}}{\underset{|}{O}}}-O \longrightarrow HF + enzyme-O-\underset{\underset{\underset{CH(CH_3)_2}{|}}{\overset{\overset{O}{|}}{P}}}{\overset{\overset{CH(CH_3)_2}{|}}{\underset{|}{O}}}-O$$

A second type of enzyme inhibition is called **competitive inhibition,** in which there is a competition between the inhibitor and the substrate for the enzyme active site. This is a system in which there are two simultaneous equilibria, one between the enzyme (E) and the substrate (S),

$$E + S \rightleftharpoons ES$$

and one between the enzyme and the inhibitor (I),

$$E + I \rightleftharpoons EI$$

As we would predict from Le Châtelier's principle, increasing the substrate concentration displaces the inhibitor.

An example of competitive inhibition is the action of the sulfa drug sulfanilamide. This molecule, shown below, bears a very close similarity to *p*-aminobenzoic acid, which is acted on by an enzyme to produce an important coenzyme required by bacteria.

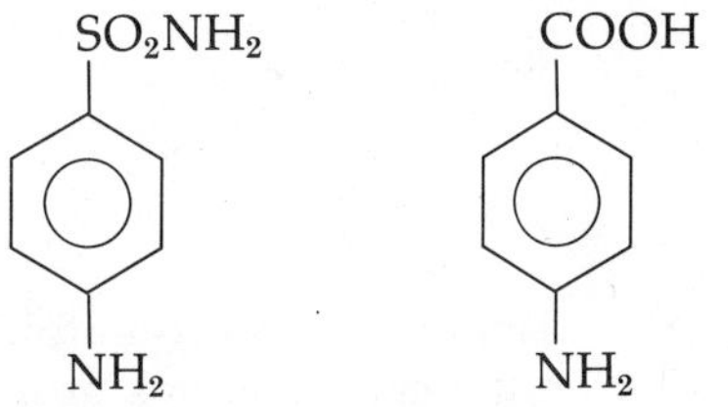

sulfanilamide   *p*-aminobenzoic acid

The sulfanilamide, by occupying the active site of the enzyme that works on the *p*-aminobenzoic acid, prevents the production of the required coenzyme and hence leads to the demise of the bacterium.

As a final note, enzyme activity is also affected by temperature and pH. These factors alter the conformation and shape of the globular protein structure; therefore, changes in temperature or pH can destroy the precise fit that must exist between the enzyme and its substrate in order to obtain the desired catalytic activity.

23.3 CARBOHYDRATES

The carbohydrates form an important class of compounds that are used by living organisms in a variety of ways—as a source of energy, as a source of carbon to be used in the synthesis of other biomolecules, and as a structural element in cells and tissues. Historically the name carbohydrate arose as a consequence of the empirical formula exhibited by many of them, $C_n(H_2O)_m$, which suggested that they were *hydrates of carbon*. Examples are glucose, $C_6H_{12}O_6$, having the empirical formula CH_2O, and sucrose (ordinary cane sugar), $C_{12}H_{22}O_{11}$, with the empirical formula $C_{12}(H_2O)_{11}$. The name carbohydrate has remained with these substances even though it is now known that they do not contain intact water molecules.

Most carbohydrates, such as starch and cellulose, are very large molecules having enormous molecular weights. However, like the proteins, they are composed of many relatively simple units arranged in long polymeric chains. The simplest of these units are called **monosaccharides,** and constitute the **simple sugars.** As a class, the monosaccharides are polyhydroxy aldehydes or ketones, the simplest of which is glyceraldehyde,

The Latin <u>saccharum</u> means sugar.

$$
\begin{array}{c}
\text{CHO} \\
| \\
\text{H}-\text{C}-\text{OH} \\
| \\
\text{CH}_2\text{OH}
\end{array}
$$

Glyceraldehyde is a **triose,** *tri* denoting three carbon atoms and *ose*, the characteristic ending used in naming the sugars (for example, gluc*ose*, sucr*ose*, and fruct*ose*).

Glyceraldehyde, like the other saccharides, contains an asymmetric carbon atom and exhibits optical isomerism. In the two-dimensional structural formulas written for the saccharides, the H and OH units that are attached to the

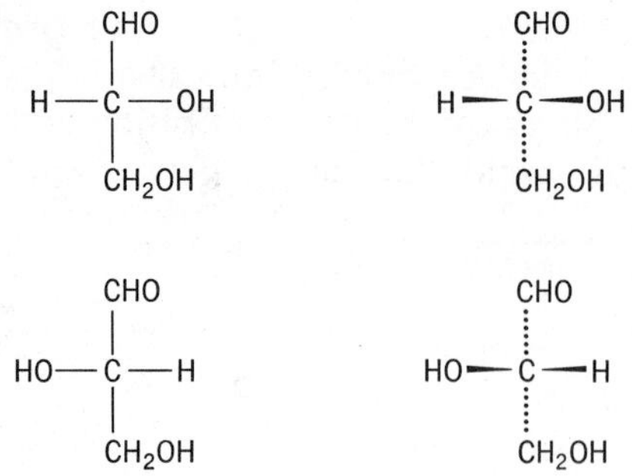

$$\begin{array}{c} CHO \\ | \\ H-C-OH \\ | \\ CH_2OH \end{array} \qquad \begin{array}{c} CHO \\ | \\ H-C-OH \\ | \\ CH_2OH \end{array}$$

$$\begin{array}{c} CHO \\ | \\ HO-C-H \\ | \\ CH_2OH \end{array} \qquad \begin{array}{c} CHO \\ | \\ HO-C-H \\ | \\ CH_2OH \end{array}$$

Figure 23.11

Optical isomers of glyceralde-hyde.

asymmetric carbon atoms project upward from the paper while the bonds to other carbon atoms project downward. For example, the two optical isomers of glyceraldehyde are shown in Figure 23.11.

Among the saccharides it is generally observed that one optical isomer is significantly more important than the others. Glucose, for example, contains four asymmetric carbon atoms (indicated by asterisks).

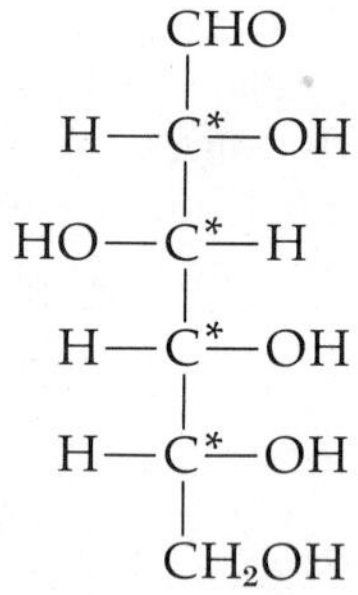

$$\begin{array}{c} CHO \\ | \\ H-\overset{*}{C}-OH \\ | \\ HO-\overset{*}{C}-H \\ | \\ H-\overset{*}{C}-OH \\ | \\ H-\overset{*}{C}-OH \\ | \\ CH_2OH \end{array}$$

D-**glucose**

There are 16 possible optical isomers of glucose; and the most important one is D-glucose, shown above.

By far the most important monosaccharides are those containing five and six carbon atoms, the pentoses and hexoses, respectively. Some of the more prominent ones are found in Table 23.1. The most common hexose is glucose, whose structure is shown above. Glucose, however, like most of the other pentoses and hexoses, exists predominantly in a cyclic structure in which the molecule turns on itself as shown in Figure 23.12. When the ring is closed, the —OH group that is created from the aldehyde functional group can point either up or down (this is the —OH group on the rightmost carbon atom in the structures drawn in Figure 23.12). Two isomers are thus created, α-D-glucose and β-D-glucose. As we will see, the orientation of this —OH group is quite significant in the polysaccharides, starch and cellulose.

Another important six-carbon sugar is fructose. In its open-chain structure the molecule is a ketone.

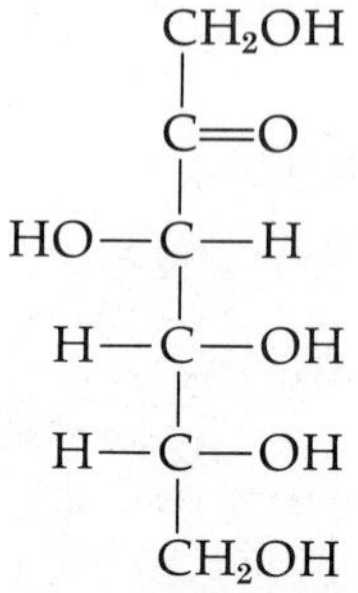

Fructose is found along with glucose and sucrose in honey and fruit juices.

$$\begin{array}{c} CH_2OH \\ | \\ C=O \\ | \\ HO-C-H \\ | \\ H-C-OH \\ | \\ H-C-OH \\ | \\ CH_2OH \end{array}$$

Table 23.1
Some important monosaccharides

Pentoses

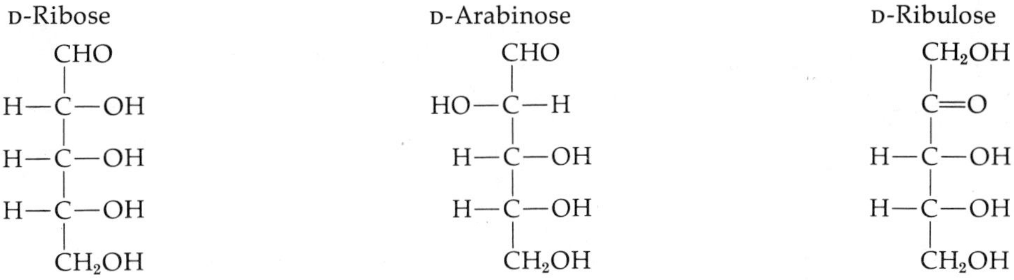

D-Ribose	D-Arabinose	D-Ribulose

```
      CHO              CHO            CH₂OH
       |                |               |
  H—C—OH          HO—C—H           C=O
       |                |               |
  H—C—OH           H—C—OH          H—C—OH
       |                |               |
  H—C—OH           H—C—OH          H—C—OH
       |                |               |
     CH₂OH            CH₂OH           CH₂OH
```

Hexoses

D-Glucose	D-Mannose	D-Galactose	D-Fructose

```
      CHO            CHO            CHO           CH₂OH
       |              |              |              |
  H—C—OH         HO—C—H         H—C—OH          C=O
       |              |              |              |
 HO—C—H         HO—C—H        HO—C—H         HO—C—H
       |              |              |              |
  H—C—OH          H—C—OH        HO—C—H          H—C—OH
       |              |              |              |
  H—C—OH          H—C—OH         H—C—OH          H—C—OH
       |              |              |              |
     CH₂OH          CH₂OH          CH₂OH          CH₂OH
```

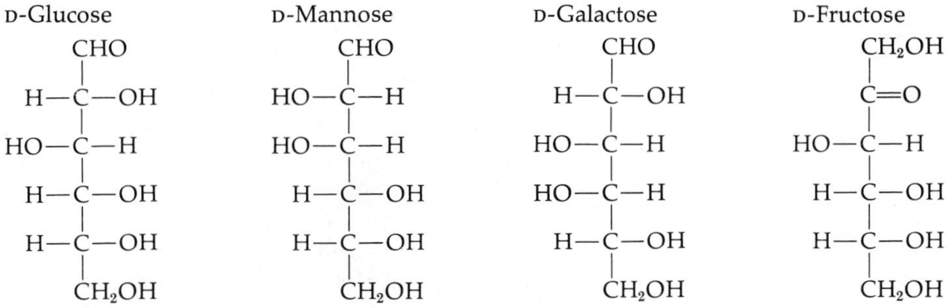

Like glucose, however, it too prefers a cyclic structure, as shown in Figure 23.13. In this case a five-membered ring is formed.

The five-membered ring also occurs in two very important pentoses, **ribose**

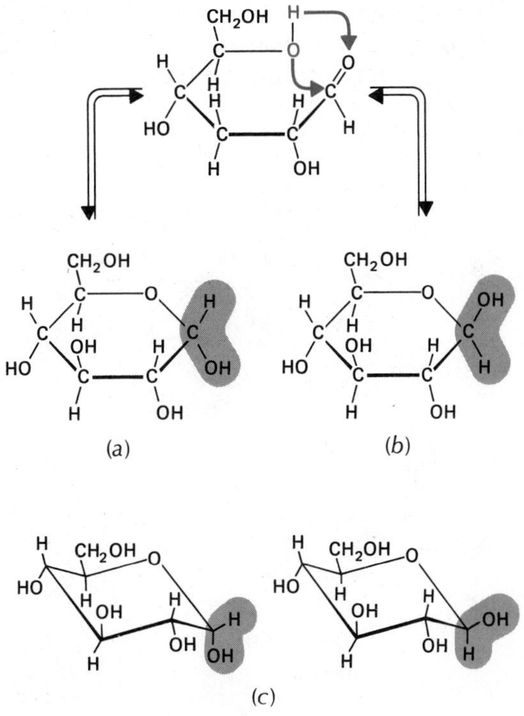

(a)   (b)

Figure 23.12

Cyclic structures for glucose. (a) α-D-glucose. (b) β-D-glucose. (c) Puckered ring. Note orientation of H and OH on the rightmost carbon.

(c)

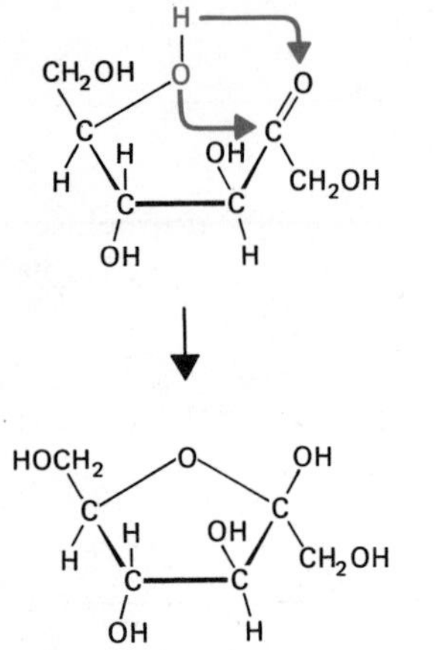

Figure 23.13

Cyclic structure for fructose.

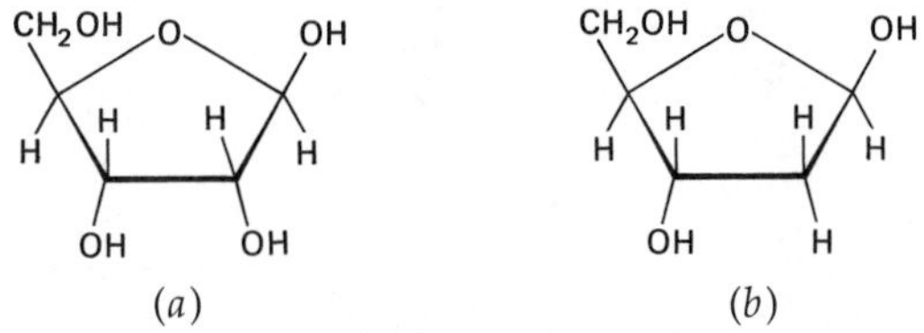

(a) *(b)*

Figure 23.14

(a) Ribose and (b) deoxyribose.

and **deoxyribose** (Figure 23.14), sugars that are part of the backbone of RNA and DNA, respectively. We'll examine the structures of these in Section 23.5.

In the more complex sugars and the polysaccharides, monosaccharide units are condensed together by way of C—O—C bridges called **glycoside linkages.** Sucrose, for example, is a disaccharide consisting of a glucose and a fructose unit joined by eliminating H_2O from an —OH group on the glucose and an —OH group on the fructose. This is illustrated in Figure 23.15. As indicated, addition of H_2O to the glycoside linkage (hydrolysis) splits the sucrose molecule into the simple monosaccharides from which it is formed. This hydrolysis reaction is accelerated by the presence of dilute acid and, in general, polysaccharides can be broken down into their simple sugars by this reaction. Special enzymes in saliva start "digesting" carbohydrates in the mouth by this hydrolysis reaction.

The formation of two glycoside linkages by a single monosaccharide unit permits the formation of long polymeric chains called polysaccharides, the two most important of which are starch and cellulose. Starch (amylose) is composed of α-D-glucose units strung together, while cellulose is composed of β-D-glucose units, as shown in Figure 23.16.

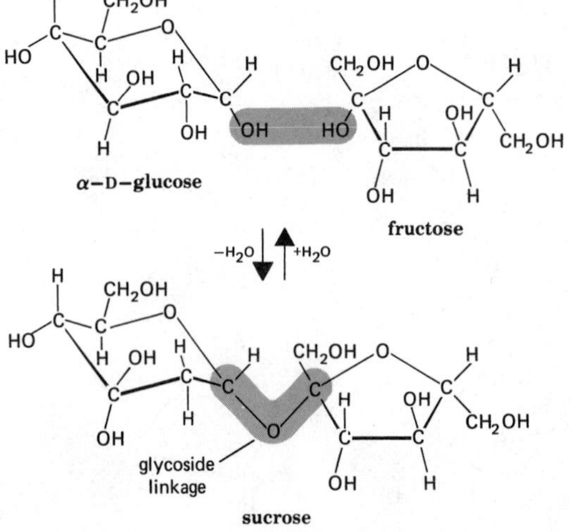

Figure 23.15

Sucrose.

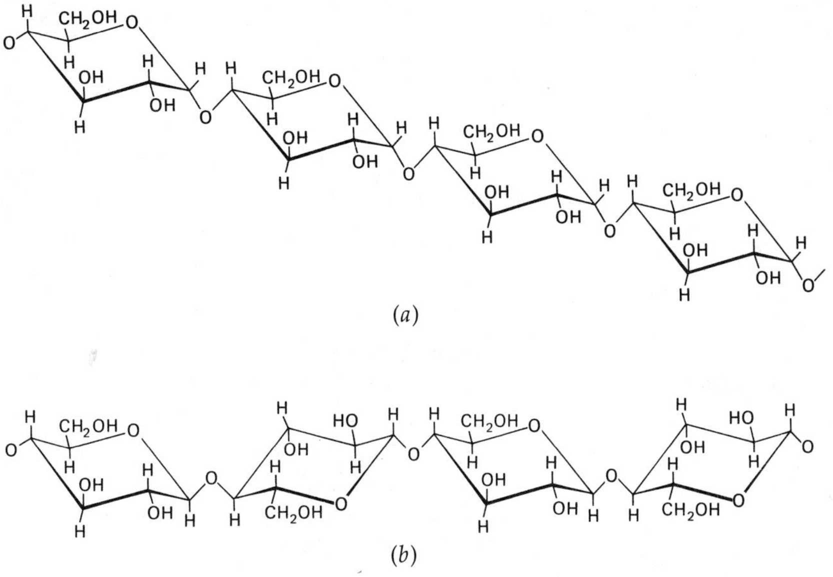

(a)

Figure 23.16

Structures of the polysaccharides, starch and cellulose. (a) Amylose (starch). (b) Cellulose.

(b)

The difference between these two structures is rather subtle, but nevertheless has very profound effects. In starch the polysaccharide chains tend to coil in a helical structure with the polar —OH groups pointing outward. When placed into water these —OH groups on the starch molecule interact strongly with the polar solvent and cause the starch to be slightly water-soluble. Cellulose, on the other hand, forms linear chains that interact with each other via hydrogen bonding. This phenomenon gives wood—which is composed of approximately 50% cellulose—its structural strength.

The relatively minor structural differences between starch and cellulose also account for the fact that starch can be digested by humans but cellulose cannot. In the digestive tract the starch molecule is hydrolyzed enzymatically, which requires a certain fit between the carbohydrate molecule and the enzyme. With cellulose this necessary fit is not achieved and, hence, cellulose is unaffected. In termites, cows, and many other animals, however, cellulose is hydrolyzed and digested with the aid of bacteria in their digestive tract.

23.4 LIPIDS

A third class of biomolecules is made up of the **lipids**—water-insoluble substances that can be extracted from other cell components by nonpolar organic solvents (hydrocarbon solvents, carbon tetrachloride, etc.). Lipids serve mainly as storage of energy-rich fuel for use in metabolism (for example, in fats) and as a major structural element in cell membranes.

As was true with the proteins and carbohydrates, most lipids are composed of simpler substances. The primary building blocks of the lipids are called **fatty acids,** long unbranched hydrocarbon chains, from 12 to 28 carbon atoms long, terminated at one end with the carboxyl group characteristic of organic acids. Nearly all the naturally occurring fatty acids have an even number of carbon atoms and occur with both saturated and unsaturated chains. Some typical examples are shown in Table 23.2.

Most lipids can be classed as either **neutral lipids** or **polar lipids.** Fats, for example, are neutral lipids and are esters of the fatty acids with the alcohol, glycerol.

Table 23.2
Some naturally occurring fatty acids

Fatty Acid		Melting Point (°C)
Saturated		
Lauric acid (coconut or palm) (kernal oil)	$CH_3(CH_2)_{10}COOH$	44
Myristic acid (nutmeg fat)	$CH_3(CH_2)_{12}COOH$	54
Palmitic acid (palm oil, animal fats)	$CH_3(CH_2)_{14}COOH$	63
Stearic acid (animal fats)	$CH_3(CH_2)_{16}COOH$	70
Unsaturated		
Palmitoleic acid (butter fat)	$CH_3(CH_2)_5CH{=}CH(CH_2)_7COOH$	−1
Oleic acid (olive oil, animal fats)	$CH_3(CH_2)_7CH{=}CH(CH_2)_7COOH$	13.4
Linoleic acid (linseed oil)	$CH_3(CH_2)_4CH{=}CHCH_2CH{=}CH(CH_2)_7COOH$	−5
Linolenic acid (linseed oil)	$CH_3(CH_2)CH{=}CHCH_2CH{=}CHCH_2CH{=}CH(CH_2)_7COOH$	−11

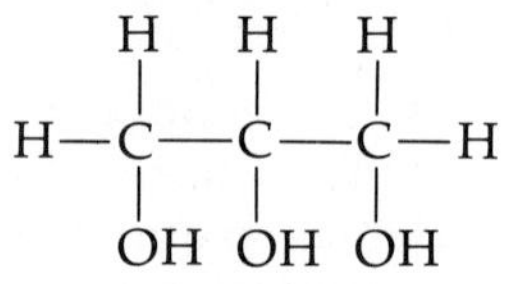

$$
\begin{array}{ccccc}
 & H & H & H & \\
 & | & | & | & \\
H-&C&-C&-C&-H \\
 & | & | & | & \\
 & OH & OH & OH &
\end{array}
$$

glycerol

The resulting triester is called a **triglyceride.**

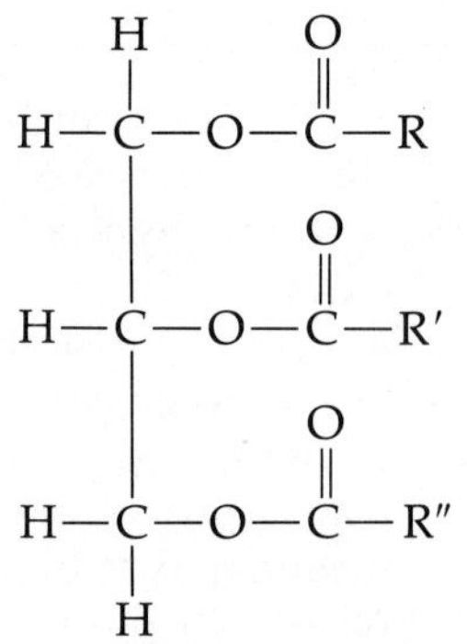

$$
\begin{array}{l}
\quad\; H \qquad\quad O \\
\quad\; | \qquad\qquad \| \\
H-C-O-C-R \\
\quad\; | \qquad\qquad O \\
\quad\; | \qquad\qquad \| \\
H-C-O-C-R' \\
\quad\; | \qquad\qquad O \\
\quad\; | \qquad\qquad \| \\
H-C-O-C-R'' \\
\quad\; | \\
\quad\; H
\end{array}
$$

As you might expect, many different triglycerides are found to occur, as determined by the nature and location of the fatty acids attached to the glycerol molecule. Lipids containing saturated fatty acids, such as tristearin (glycerol esterified with three stearic acid molecules),

$$H-\underset{\underset{H}{|}}{\overset{\overset{H}{|}}{C}}-O-\overset{\overset{O}{\|}}{C}-C_{17}H_{35}$$

are solids, while those containing three unsaturated fatty acids are liquids at room temperature. An example of this liquid type is *triolein,* in which glycerol is esterified with three oleic acid molecules. This substance is the major constituent of olive oil. The liquid unsaturated triglycerides that are found in vegetable oils, such as olive oil and corn oil, serve as the basis of oleomargarine. Addition of hydrogen to the double bonds of unsaturated vegetable oils produces saturated chains and hence solid fats.

Recall that one of the principal commercial uses of hydrogen is the hydrogenation of vegetable oils.

In recent years the growth of the processed food industry has led to widespread use, as food additives, of monoglycerides and diglycerides, in which only one or two of the —OH groups of glycerol are esterified. These are added to foods as emulsifiers to improve texture and to keep oils suspended.

Fats, like other esters, can be **saponified** on treatment with aqueous base. The products of this reaction are glycerol plus the anions of the fatty acids that were bound to the glycerol in the fat.

$$\begin{array}{l} H_2COOCC_{17}H_{35} \\ | \\ HCOOCC_{17}H_{35} \\ | \\ H_2COOCC_{17}H_{35} \end{array} \xrightarrow[H_2O]{OH^-} \begin{array}{l} H_2COH \\ | \\ HCOH \\ | \\ H_2COH \end{array} + 3C_{17}H_{35}COO^-$$

Hydrophilic means "water loving."

These anions constitute a soap and have rather peculiar properties that result from having a polar hydrophilic "head" and a nonpolar hydrophobic "tail."

$$CH_3-CH_2-\cdots-CH_2-CH_2-CH_2-\overset{\overset{O}{\|}}{C}\diagdown_{O^-}$$

$$\underbrace{\hspace{5cm}}_{\text{tail}} \quad \underbrace{\hspace{2cm}}_{\text{head}}$$

The polar end of the anion tends to be water-soluble while the other end, the nonpolar hydrocarbon tail, tends to be insoluble in water but soluble in nonpolar solvents. As a result, in water these anions group themselves into small globules called **micelles** in which the nonpolar tails "dissolve" in each other, leaving the polar heads facing outward toward the aqueous surroundings. This is illustrated in Figure 23.17. The same properties that lead to micelle formation are also responsible for the ability of soap to dissolve grease. In this case the nonpolar tails dissolve in the grease particle and the polar heads dissolve in water (Figure 23.18). This keeps the grease particle suspended in water so that it can be rinsed away.

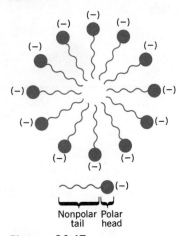

Nonpolar Polar
tail head

Figure 23.17
Micelle formation with soap.

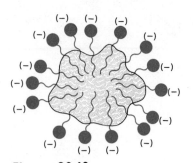

Figure 23.18
The dissolving of a grease globule by soap.

Another very important class of lipids are the **phospholipids.** These are polar lipids and, like the fats, are esters of glycerol. In this case, however, only two fatty acid molecules are esterified to glycerol, at the first and second carbon atom. The remaining end position of the glycerol is esterified to a molecule of phosphoric acid, which in turn is also esterified to another alcohol. This gives a general structure,

$$
\begin{array}{c}
\text{O} \\
\| \\
\text{O}-\text{P}-\text{O}-\text{R}'' \\
| \\
\text{O} \\
| \\
\text{H}_2\text{C}-\text{CH}-\text{CH}_2 \\
| \quad\quad | \\
\text{O} \quad\quad \text{O} \\
| \quad\quad | \\
\text{O}=\text{C} \quad \text{C}=\text{O} \\
| \quad\quad | \\
\text{R} \quad\quad \text{R}'
\end{array}
$$

An example is the phospholipid *phosphatidyl ethanolamine,*

$$
\begin{aligned}
&\text{CH}_3-\text{CH}_2-\text{CH}_2-\text{CH}_2-\text{CH}_2-\text{CH}_2-\text{CH}_2-(\text{CH}_2)_8\overset{\displaystyle\overset{\text{O}}{\|}}{\text{C}}-\text{O}-\overset{\displaystyle\overset{\text{H}}{|}}{\underset{\displaystyle|}{\text{C}}}-\text{H} \\
&\text{CH}_3-\text{CH}_2-\text{CH}_2-\text{CH}_2-\text{CH}_2-\text{CH}_2-\text{CH}=\text{CH}-(\text{CH}_2)_7\overset{\displaystyle\overset{\text{O}}{\|}}{\text{C}}-\text{O}-\overset{|}{\underset{|}{\text{C}}}-\text{H} \quad \text{O} \\
&\quad\quad\quad\quad\quad\quad\quad\quad\quad\quad\quad\quad\quad\quad\quad\quad\quad\quad\quad\quad\quad\quad\quad\quad \text{H}-\overset{|}{\underset{|}{\text{C}}}-\text{O}-\overset{\displaystyle|}{\underset{\displaystyle|}{\text{P}}}-\text{O}^{(-)} \\
&\quad\quad\quad\quad\quad\quad\quad\quad\quad\quad\quad\quad\quad\quad\quad\quad\quad\quad\quad\quad\quad\quad\quad\quad\quad\text{H} \quad\quad \text{O} \\
&\quad\quad\quad\quad\quad\quad\quad\quad\quad\quad\quad\quad\quad\quad\quad\quad\quad\quad\quad\quad\quad\quad\quad\quad\quad\quad\quad\quad \text{CH}_2 \\
&\quad\quad\quad\quad\quad\quad\quad\quad\quad\quad\quad\quad\quad\quad\quad\quad\quad\quad\quad\quad\quad\quad\quad\quad\quad\quad\quad\quad \text{CH}_2 \\
&\quad\quad\quad\quad\quad\quad\quad\quad\quad\quad\quad\quad\quad\quad\quad\quad\quad\quad\quad\quad\quad\quad\quad\quad\quad\quad\quad \text{NH}_2{}^{(+)}
\end{aligned}
$$

nonpolar tail   **polar head**

As indicated, this type of lipid contains a polar head and nonpolar tail, much the same as the anions of the fatty acids.

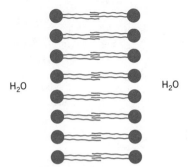

Figure 23.19

Formation of bilayer by phospholipids. Nonpolar tails dissolve in each other. Polar heads are exposed to aqueous environment.

Cell membranes are composed of phospholipids and proteins in about equal proportion. The phospholipids in the membrane appear to be arranged in a double layer, or *bilayer* (Figure 23.19) in which the nonpolar tails face each other, thereby exposing the polar heads to the aqueous environment on either side of the membrane. As shown in Figure 23.20, the proteins found in the membrane are embedded in the mosaic formed by the lipids. Much research today is centered on the mechanism of transport of matter and energy across such membranes.

Finally, there is a third class of lipids that do not contain fatty acids and glycerol, and that do not undergo saponification when treated with a base. Included in this group are the **steroids,** complex substances that possess unusually high biological activity.

The steroids have in common a fused ring structure.

Examples of some important steroids, whose names you may have come across before, are the following:

estrone

estradiol

(female sex hormones: note similarity in structures)

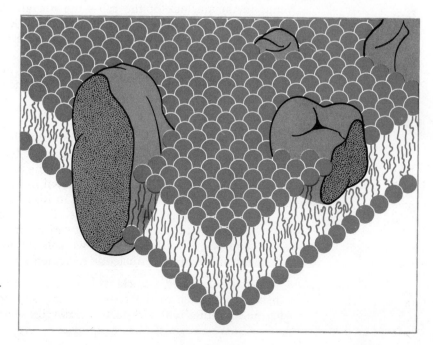

Figure 23.20

Bilayer structure of a cell membrane. Lipid mosaic model: irregularly shaped proteins float randomly in a lipid sea. From S. J. Singer, Annals of the New York Academy of Sciences, *Vol. 195, p. 21, 1962. Used by permission of the publisher and author.*

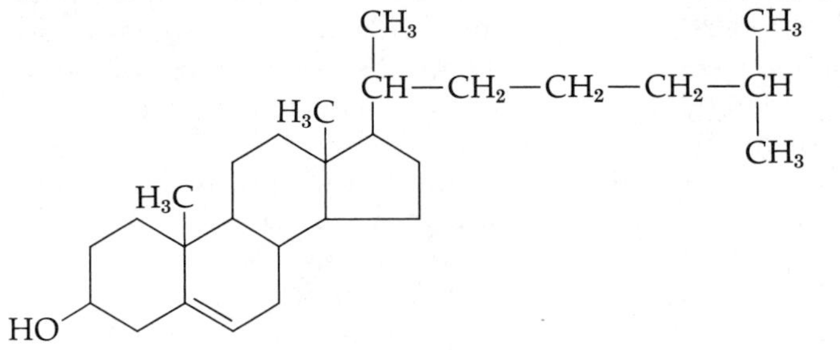

cholesterol

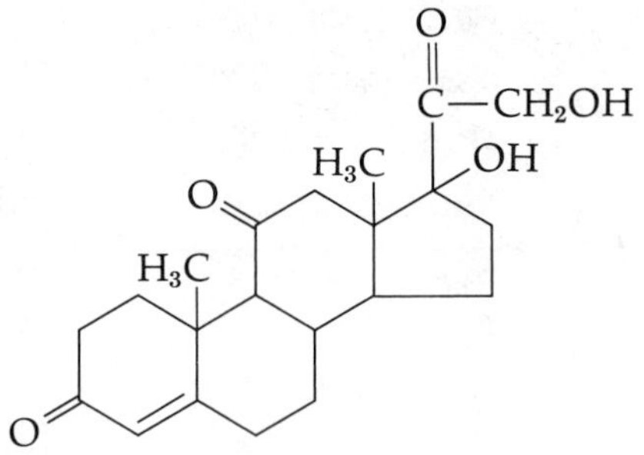

cortisone
(affects protein metabolism)

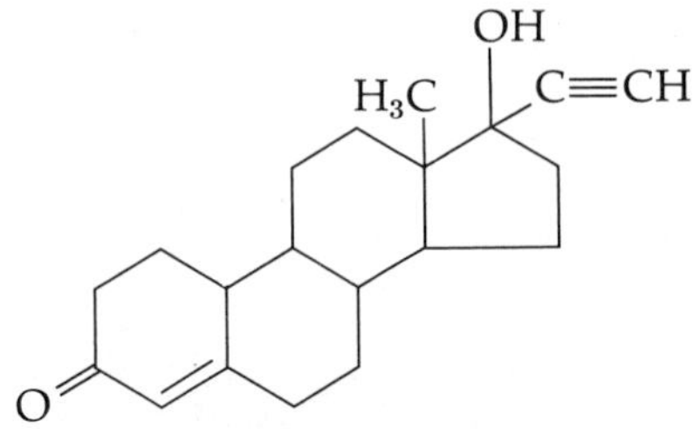

norethindrone: an
oral contraceptive
("the pill")

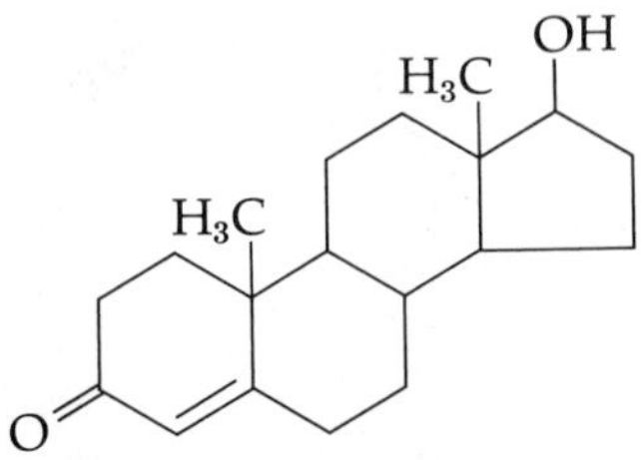

testosterone
(male sex hormone)

23.5 NUCLEIC ACIDS

One of the most intriguing aspects of biochemistry has been, from its very beginnings, the mechanism whereby an organism transmits its genetic information from one generation to the next during cell division. It is now believed, with a good deal of confidence, that this process is controlled by a substance found in the nucleus of the cell—the **nucleic acid, DNA** (deoxyribonucleic acid). Furthermore, DNA in conjunction with **RNA** (ribonucleic acid, another type of nucleic acid) is responsible for the synthesis of the proteins that are characteristic of a given organism.

Nucleic acids, like the proteins and carbohydrates that we looked at earlier, are polymers. The simpler units that make up the nucleic acid are called **nucleotides,** and are themselves composed of three even simpler molecules. They include the following:

1. **A nitrogenous base.** These are heterocyclic organic compounds with two or more nitrogen atoms in the ring skeleton. They are called bases because the lone pairs of electrons on the nitrogen atoms make them Lewis bases. These substances are shown in Figure 23.21.
2. **A five-carbon sugar (pentose).** In RNA this sugar is ribose, whereas in DNA the sugar is deoxyribose. These two are shown in Figure 23.22. Notice that they differ only at carbon atom number 2 in the ring.
3. **Phosphoric acid.** H_3PO_4, as we will see, forms esters to —OH groups of the sugar to bind nucleotide segments together.

A molecule called a **nucleoside** is formed from these components by condensing a molecule of the base with the appropriate pentose. For example, adenine combines with ribose and deoxyribose at carbon number 1 to give the

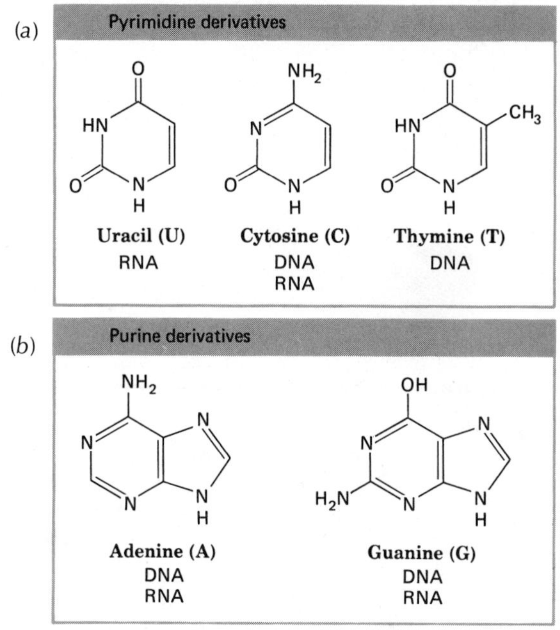

Figure 23.21

Nitrogenous bases found in DNA and RNA. (a) Pyrimidine derivatives. (b) Purine derivatives.

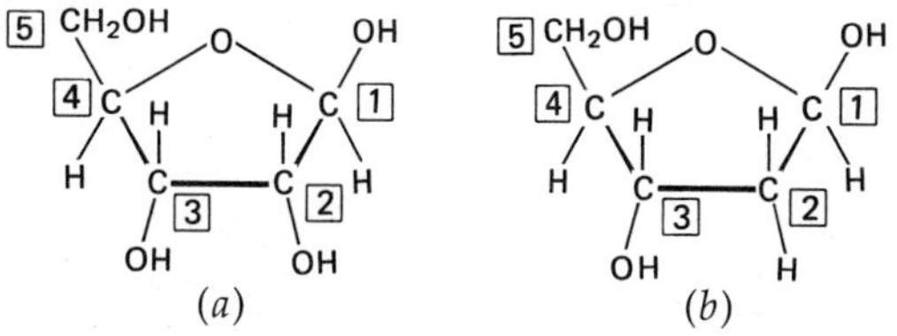

Figure 23.22

(a) Ribose and (b) deoxyribose.

compounds shown in Figure 23.23a. Adenosine is an important constituent of ATP (adenosine triphosphate) and ADP (adenosine diphosphate), which are both involved in energy-transfer processes in a cell. Finally, linkage of phos-

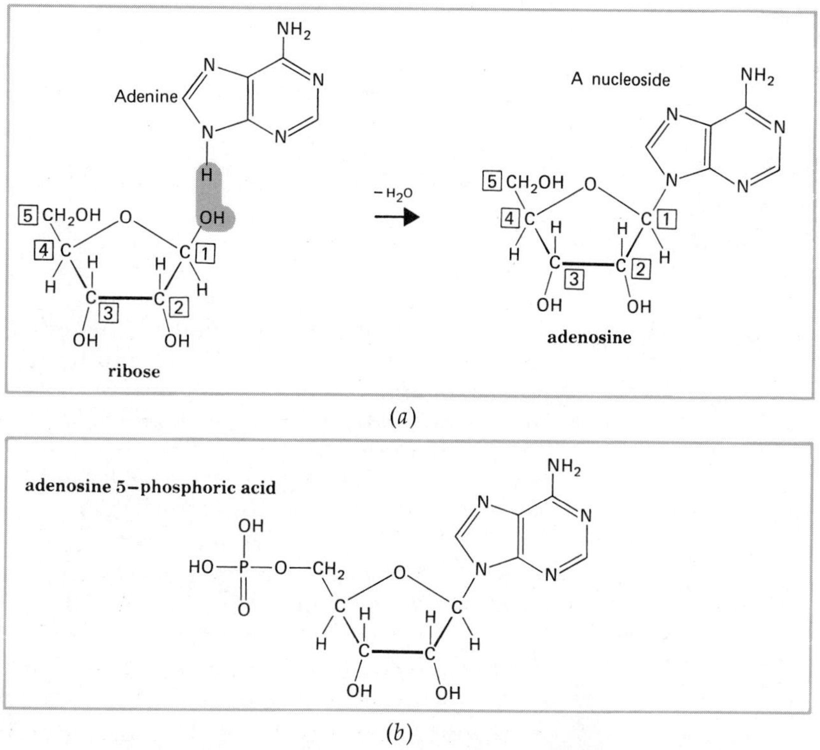

Figure 23.23

Nucleosides and nucleotides. (a) Adenine combines with ribose to form a nucleoside. (b) Linkage of phosphoric acid to carbon 5 gives a nucleotide.

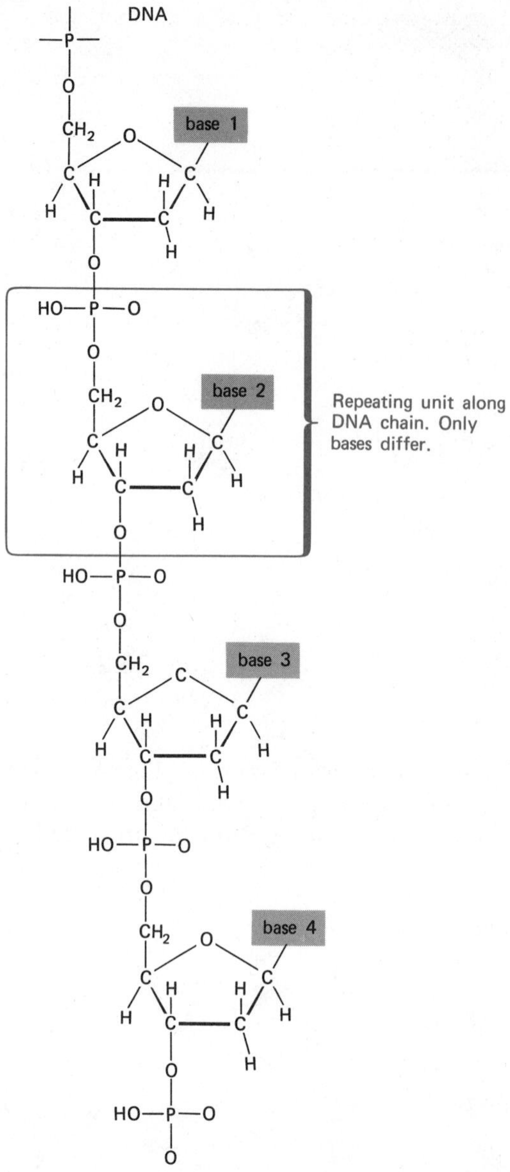

Figure 23.24

Polymerization of nucleotides gives nucleic acids.

phoric acid to carbon atom number 5 (Figure 23.23*b*) produces a **nucleotide,** the basic building block of both DNA and RNA.

The nucleic acids are condensation polymers of the nucleotide monomers and are formed by the creation of an ester linkage from the phosphoric acid residue on one nucleotide to the hydroxy group on carbon number 3 in the pentose of the second nucleotide, as illustrated in Figure 23.24. The result is a very long polymeric chain, possessing up to a billion or so nucleotide units in DNA!

In Figure 23.21 it was indicated that the base compositions of DNA and RNA are not the same. In DNA the organic bases adenine (A), guanine (G), cytosine (C), and thymine (T) occur bound to the deoxyribose ring, whereas in RNA the bases adenine, guanine, cytosine, and uracil (U) are found. As in the proteins, the sequence of bases along the DNA or RNA chain establishes its primary structure, which controls the specific properties of the nucleic acid. For instance, the base sequence in DNA contains coded information that the cell utilizes in the synthesis of its own characteristic proteins. We will look at this more closely in the next section.

The truly remarkable properties of DNA that account for its ability to reproduce itself exactly are dictated by its secondary structure, in which two strands of DNA intertwine into the now much celebrated **double helix** proposed by Watson and Crick, for which they received the Nobel Prize in 1962.[2] The key to the formation of the double helix, as well as the function of DNA and RNA in protein synthesis, lies in the interaction of the nitrogenous bases by way of hydrogen bonding.

Consider, for example, the bases cytosine and guanine, situated across from one another on separate DNA strands. The structure of these bases, as shown below, allows them to be ideally suited to interact with each other through the formation of hydrogen bonds.

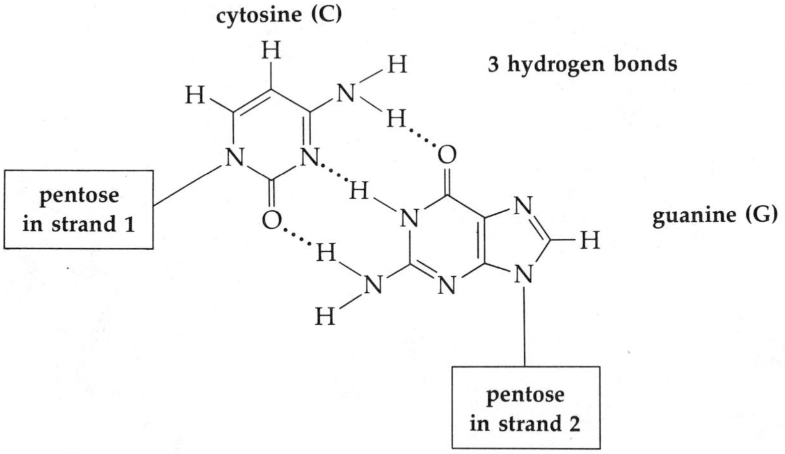

A similar relationship holds for the bases thymine and adenine.

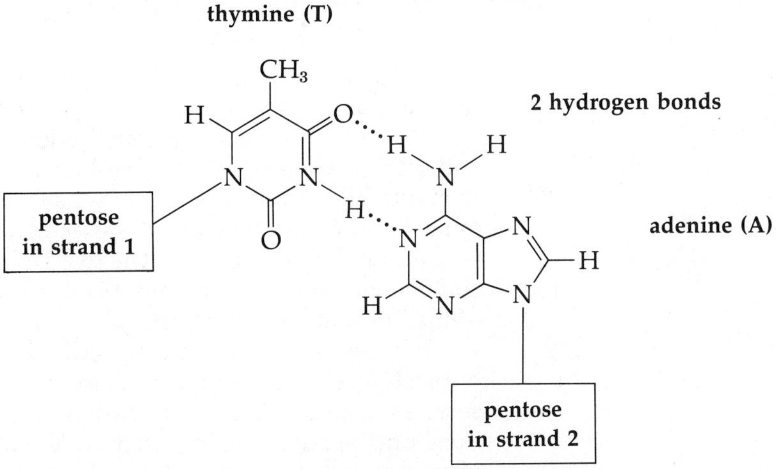

Thus cytosine and guanine fit together like a hand in a glove, as do thymine and adenine. Now, in DNA there is always the same amount of C as G. In addition, the quantities of T and A are the same. Furthermore, the total amount of C and T together is always equal to the combined total of G and A. If you think

[2] Watson and Crick shared the 1962 Nobel Prize with Wilkins. They used Rosalind Franklin's X-ray data in the deduction of the double-helix structure of DNA.

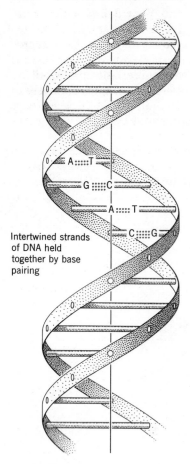

Figure 23.25

The double helix. Interwined strands of DNA held together by base pairing.

Intertwined strands of DNA held together by base pairing

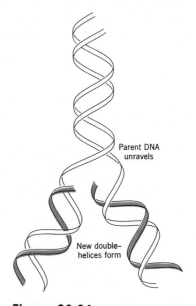

Parent DNA unravels

New double-helices form

Figure 23.26

Replication of DNA.

about this for a while, you will see that this suggests that C and G are paired together, as are T and A.

These observations, in conjunction with X-ray diffraction data, led Watson and Crick to propose the double-helical structure of DNA, shown schematically in Figure 23.25. In this structure we see that for each G on one strand there is a C opposite it, across the axis of the helix, on the other strand. A similar relationship also holds for T and A. The two DNA strands are not identical, but instead complement one another, and it is this property that accounts for the replication of the DNA on cell division.

It is believed that during cell division the two DNA strands begin to unravel, as shown in Figure 23.26, giving the two complementary chains that serve as templates for the construction of two new daughter chains. The restrictions on the base pairing (that is, T with A and C with G) causes the newly formed strands to be identical to the departing complementary parent chain and, as a result, a pair of DNA double helices are produced that are exact copies of the original.

23.6 PROTEIN SYNTHESIS

The DNA found in the nucleus of a cell serves indirectly to determine the makeup of the proteins that are synthesized at *ribosomes* located outside the nucleus. The genetic information that determines the amino acid sequence in each of the enzymes is stored in the DNA in a genetic code that is made up of the sequence of bases along a DNA strand. The transcription of this code to the site of protein synthesis, as well as the decoding and construction of the polypeptide chain of the protein, is accomplished by the other nucleic acids—the RNA.

Unlike DNA, RNA occurs only in single strands. In addition, there are several types of RNA. One of these is called **messenger RNA, mRNA,** and serves to carry the genetic code from the DNA template within the nucleus to the ribosomes outside. Another type of RNA molecule, called **transfer RNA, tRNA,** is much smaller than mRNA. It acts as an amino acid carrier and through a decoding mechanism that we will examine momentarily, adds the amino acid to the growing polypeptide chain at just the right place at just the right time.

The mechanism by which this process is believed to occur is not really too different from that involved in the duplication of DNA itself. It is known, for instance, that the production of mRNA takes place within the nucleus on an untwisted segment of a DNA chain. This segment corresponds to the gene that is characteristic of the particular protein to be synthesized. The mRNA strand produced contains a sequence of bases that is determined, through base pairing, by the sequence of bases in the DNA; however, the pairing scheme in this case is

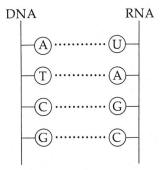

Thus uracil occurs in RNA instead of thymine. Once formed, the mRNA is transported from the nucleus to the active site of protein synthesis.

It is now known that the genetic code that directs the insertion of amino acids in the proper sequence in the growing polypeptide chain consists of sets of three bases (called **codons**). The amino acids are brought to the proper place by tRNA, which is able to decipher the code.

A molecule of tRNA contains from 75 to 90 nucleotide units containing the bases U, A, G, C, and, in addition, many minor bases. What is interesting is that although the tRNAs corresponding to different amino acids have different minor bases and base sequences, they can all be brought into a characteristic cloverleaf shape if maximum base pairing is assumed. For example, the yeast tRNA for the amino acid alanine has the structure shown in Figure 23.27. The molecule is folded in such a way as to give maximum base pairing (colored lines). On the stem of this RNA molecule, alanine becomes attached by the action of the proper enzyme. The two arms to either side of the site of attachment of the amino acid appear to be important in the interaction of the tRNA with this enzyme. The bottom loop contains the three-unit anticodon that attaches itself to the complementary codon on the mRNA chain.

The sequence of operations that leads to the synthesis of a polypeptide is summarized in Figure 23.28. In the bacterium, *E. coli*, for example, it is known that the amino acid sequence is initiated by a derivative of methionine. The tRNA containing this derivative becomes attached at the head of the mRNA through its codon-anticodon pairing. Once this is in place the next amino acid in the sequence is brought into place by its tRNA, which becomes bound to the

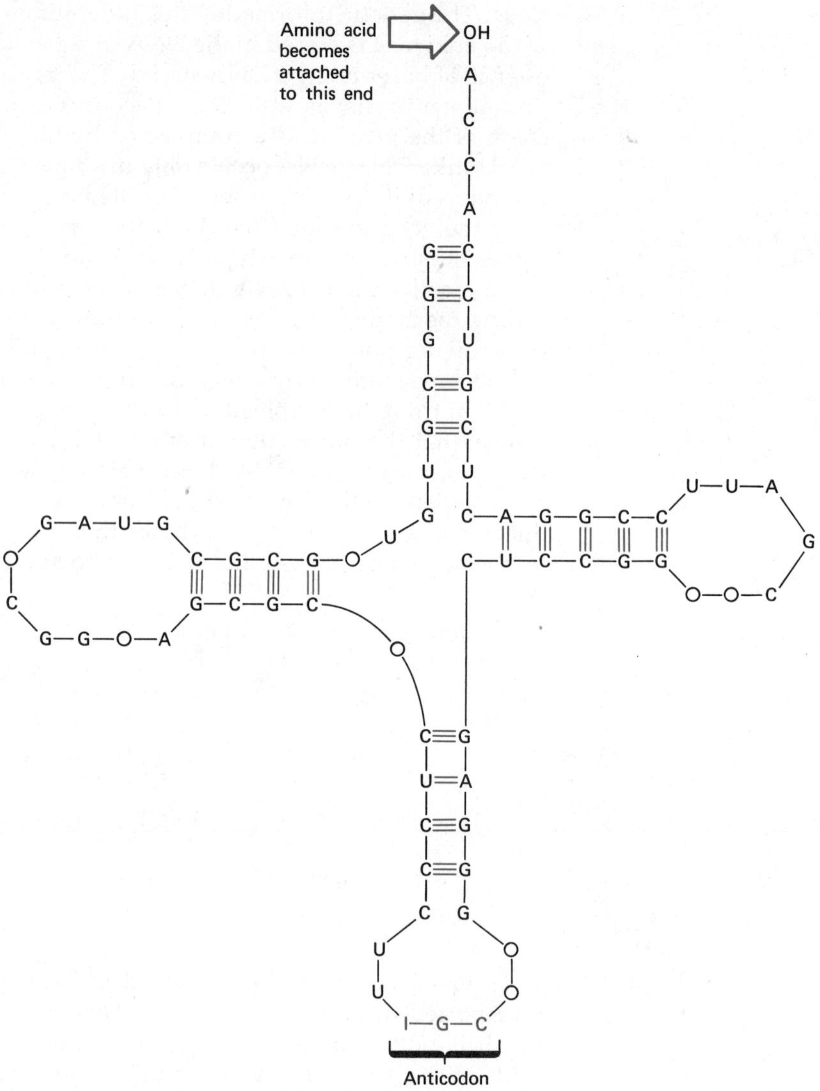

Figure 23.27

The base sequence in a yeast alanine tRNA.

mRNA at the second codon site. This is followed by an enzyme-induced formation of the peptide linkage and the subsequent departure of the tRNA from the first codon. Next, the third amino acid is delivered by its tRNA, which attaches itself to the third codon. Again a peptide linkage is formed and the second tRNA leaves. This process is repeated over and over along the mRNA as the polypeptide chain grows in size. The chain is finally terminated when a "nonsense" codon (one that cannot be recognized by any tRNA molecule) is encountered.

The genetic code

Through a series of very clever experiments, which earned Nirenberg, Holley, and Khorana a Nobel Prize in 1968, the base sequences in the genetic code triplets were determined. These are given in Table 23.3. Notice that most amino acids are specified by more than one code "word." For these, apparently, the attachment of the tRNA to the mRNA is determined primarily by the first two bases in the sequence (which are usually the same for a given amino acid). The

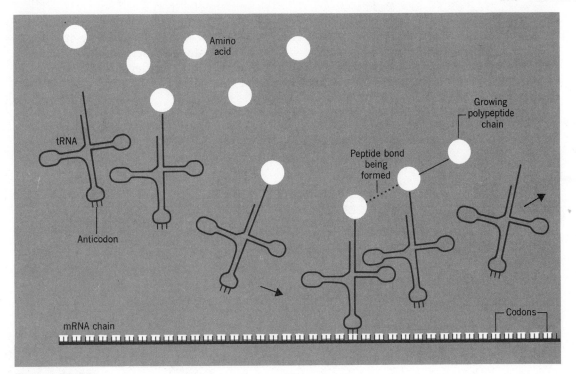

Figure 23.28
Synthesis of polypeptide.

Table 23.3
The genetic code

		Second Base in Codon			
		U	C	A	G
First Base in Codon	U	UUU $\big\}$ Phe UUC UUA $\big\}$ Leu UUG	UCU $\Big\}$ Ser UCC UCA UCG	UAU $\big\}$ Tyr UAC UAA[b] UAG[b]	UGU $\big\}$ Cys UGC UGA[b] UGG Trp
	C	CUU $\Big\}$ Leu CUC CUA CUG	CCU $\Big\}$ Pro CCC CCA CCG	CAU $\big\}$ His CAC CAA $\big\}$ Gln CAG	CGU $\Big\}$ Arg CGC CGA CGG
	A	AUU $\big\}$ Ile AUC AUA AUG Met[a]	ACU $\Big\}$ Thr ACC ACA ACG	AAU $\big\}$ Asn AAC AAA $\big\}$ Lys AAG	AGU $\big\}$ Ser AGC AGA $\big\}$ Arg AGG
	G	GUU $\Big\}$ Val GUC GUA GUG	GCU $\Big\}$ Ala GCC GCA GCG	GAU $\big\}$ Asp GAC GAA $\big\}$ Glu GAG	GGU $\Big\}$ Gly GGC GGA GGG

[a] AUG appears to initiate an amino acid sequence with methionine.
[b] Nonsense codons: these do not code for any amino acid. They serve to terminate peptide chains.

interaction between the third base in the mRNA codon and the tRNA anticodon thus does not appear to be as critical as the first two in determining specificity.

With this code we can now see how the amino acid sequence in a polypeptide is fixed by the primary structure of the DNA. Consider, for example, a segment of DNA having the base sequence

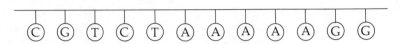

The mRNA formed from it will have the base sequence

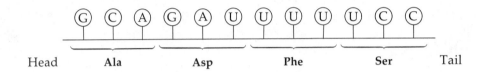

which, as you can see, is composed of the code words for the polypeptide

Ala—Asp—Phe—Ser

The triplet code presented in Table 23.3 has been shown to apply to such diverse species as the bacterium, *E. coli,* the tobacco plant, the guinea pig, and even humans. It is widely believed that this code is universal for all species.

Mutations

Any factor that will have the effect of altering the base sequence in a DNA molecule will, in effect, alter the mRNA transcribed from it and thereby cause a change in the amino acid sequence produced using that mRNA as a template. Provided that this does not prove fatal to the cell, a mutation wll have occurred that will be transmitted from one generation to another as the DNA reproduces itself on cell division.

Often these mutations prove harmful, although not immediately lethal to the organism in which they occur, and give rise to symptoms that cause them to be referred to as diseases. Since the origin of the diseases is in the genetic material of the cell, they are said to be genetic diseases. Many such diseases are recognized. Some of the more well known are cystic fibrosis, hemophelia, and sickle-cell anemia. Some evidence even suggests that schizophrenia may be of genetic origin.

In sickle-cell anemia, for example, the red blood cells assume a crescent shape instead of the flat disklike shape of normal cells. This is caused by a mutated DNA in the gene that is responsible for the synthesis of hemoglobin. It has been found that in sickle-cell hemoglobin a *glutamic acid* unit in one of the hemoglobin chains is replaced by *valine.* This alters the secondary and tertiary structure of the protein and reduces its ability to carry oxygen.

Sickle-cell anemia is thus a disease with its origin in the DNA of the cell nucleus and is therefore passed from one generation to the next because of its genetic nature.

INDEX TO QUESTIONS

REVIEW QUESTIONS

23.1 What is an α-amino acid? Indicate how amino acids are linked together to give a dipeptide; a polypeptide.

23.2 Lye, NaOH, is able to dissolve proteins such as hair that are lodged in a sink drain. What chemical reaction is involved?

23.3 Describe what is meant by the primary structure of a protein. What is meant by secondary structure, tertiary structure, and quaternary structure?

23.4 Indicate three functions served by proteins in a living organism.

23.5 What is a zwitterion? Why do amino acids form zwitterions?

23.6 Describe the α-helix structure found in many polypeptides. What holds the polypeptide chain in this helical conformation?

23.7 How do the properties of the solvent affect the tertiary structures of proteins? What interactions in addition to hydrogen bonding determine tertiary structure?

23.8 What is meant by a quaternary structure? What proteins exhibit this kind of structural sophistication?

23.9 How do the roles played by hemoglobin and myoglobin differ? In what way are they similar?

23.10 What is a porphyrin? Name two biologically important porphyrin structures.

23.11 In what sense do enzymes guide the chemistry that takes place in living organisms? What problems would a living cell encounter without the existence of enzymes?

23.12 Describe in qualitative terms how an enzyme operates. What is meant by *enzyme substrate?* What is enzyme inhibition? How do the sulfa drugs function?

23.13 Monosodium glutamate (MSG), which is the monosodium salt of glutamic acid, is a popular flavor enhancer. However, only the L-isomer is effective. Is this particularly surprising? Explain your answer.

23.14 How does diisopropylflurophosphate function as a nerve gas? Many insecticides contain organic phosphates. Can you guess how they function?

23.15 Only one of the basic set of 20 amino acids does not exhibit optical isomerism. Which one is it? Why is it not optically active?

23.16 What is a monosaccharide? Give an example of (a) a pentose and (b) a hexose.

23.17 How many optical isomers of fructose are there?

23.18 Compare the structures of starch and cellulose. What is the major difference between them?

23.19 Using the method described on page 754, draw formulas for the remaining optical isomers of

$$
\begin{array}{c}
\text{CHO} \\
\mid \\
\text{HO} - \text{C} - \text{H} \\
\mid \\
\text{H} - \text{C} - \text{OH} \\
\mid \\
\text{HO} - \text{C} - \text{H} \\
\mid \\
\text{CH}_2\text{OH}
\end{array}
$$

23.20 Sucrose, cane sugar, is used in large amounts in food products of all kinds. What monosaccharides are produced by hydrolysis of a sucrose molecule?

23.21 What functions do the carbohydrates serve in living organisms?

23.22 What is a lipid? What functions do they serve in living systems?

23.23 What is a fatty acid? Give an example of a triglyceride. What difference in physical properties exists between the saturated and unsaturated triglycerides?

23.24 Write a chemical equation for the saponification of tristearin.

23.25 Give the structure of triolein. What product is formed upon addition of H_2 to the double bonds in triolein?

23.26 What is a soap? How does it form micelles? What is responsible for the ability of soap to dissolve grease?

23.27 How does a phospholipid differ from a triglyceride?

23.28 How are phospholipids believed to contribute to the structure of cellular membranes?

23.29 What characteristic structural feature is found among the steroids?

23.30 Describe the structure of a *nucleotide*. What is the difference between the nucleotide units in DNA and RNA?

23.31 Imagine that a single DNA strand contained a segment with the composition

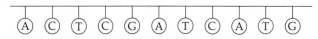

What would be the base sequence in the complementary strand?

23.32 What holds the two DNA strands together in the double helix? Illustrate the base pairing between C and G; between T and A.

23.33 What occurs with DNA during cell division? How is the DNA replicated?

23.34 What is meant by the *genetic code?* Describe the functions of mRNA and tRNA.

23.35 Give a base sequence that would have to exist on mRNA to give a polypeptide with the amino acid sequence,

Arg—Leu—Lys—Gly—Cys

23.36 If a DNA strand contains the base sequence

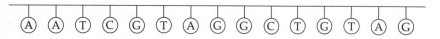

what will be the base sequence transcribed onto the mRNA?

23.37 What amino acid sequence is specified by the base sequence (from left to right) in the following mRNA strand?

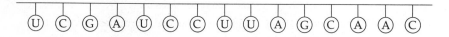

23.38 What base sequence on a DNA strand would give a mRNA that would produce a polypeptide having the amino acid sequence,

Ala—Pro—Asp—Tyr—Ile—Gly

23.39 Suppose that through some mutational change the first base was removed from the sequence in the DNA strand depicted in Question 23.36. What amino acid sequence would result in the polypeptide synthesized using this modified (mutant) DNA template?

23.40 What amino acid sequence is specified by the base sequence (starting at the left) in the DNA strand in Question 23.36?

23.41 What is a genetic disease? Must it always be fatal to the organism?

24

NUCLEAR CHEMISTRY

The ages of once-living things found at archaeological sites, such as this one in Peru, are often determined by measuring the ratio of radioactive carbon-14 to ordinary carbon-12. This technique, described further in this chapter, is just one of the many practical uses of radioactivity.

In our discussion of chemistry up to now we have paid very little attention to the atomic nuclei, other than to note that their charge determines the number of electrons in the neutral atom. There are, however, nuclear phenomena that have applications to chemistry and we will explore them in this chapter. At the same time, we will also explore the nuclear changes that occur in radioactive elements as well as those that can cause an otherwise stable element to become radioactive.

24.1 SPONTANEOUS RADIOACTIVE DECAY

Natural radioactivity was discovered, quite by accident, by a French physicist named Antoine Henri Becquerel (1852–1908). Becquerel found that when uranium salts were left in contact with photographic plates, the plates were darkened in the same manner as when they were exposed to X rays. He reasoned correctly that uranium spontaneously emits a penetrating form of radiation that was responsible for blackening the photographic plates. Two colleagues of Becquerel, Pierre and Marie Curie, were successful in isolating two other radioactive elements from uranium ore—polonium (Po) and radium (Ra)—and found that they were even more intensely radioactive. For their discoveries, Becquerel and the two Curies were awarded the 1903 Nobel Prize for physics.

We now know that the radiation emitted from radioactive substances is of three main types: **alpha particles** (α), **beta particles** (β), and **gamma rays** (γ). Experimentally it is found that the α-particle carries a positive charge, the β-particle is negatively charged, and the γ-rays carry no charge. The actual charges and masses of these and other particles we will be concerned with are listed in Table 24.1.

When a substance spontaneously emits either an α-particle or a β-particle there is a change in the charge on the nucleus, and therefore a change in atomic number. For example, nuclei of the most abundant isotope of uranium, $^{238}_{92}\text{U}$, spontaneously emit α-particles. This removes two units of charge and four units of mass from the nucleus, and gives the isotope $^{234}_{90}\text{Th}$. Thus the emission of an alpha particle changes a uranium atom into a thorium atom—we say that $^{238}_{92}\text{U}$ *decays* to $^{234}_{90}\text{Th}$.

The changes that occur during a nuclear reaction such as the decay of $^{238}_{92}\text{U}$ can be represented by a nuclear equation. For example,

$$^{238}_{92}\text{U} \longrightarrow {}^{4}_{2}\text{He} + {}^{234}_{90}\text{Th}$$
$$\textbf{\textalpha-particle}$$

In balancing nuclear equations, notice that both the total charge and the total mass must balance.

Mass number
(sum of protons + neutrons)

$$^{238}_{92}\text{U}$$

Atomic number
(number of protons)

The algebraic sum of the subscripts must be the same on both sides, and the algebraic sum of the superscripts must match.

Table 24.1
Principal types of radiation emitted by radioactive nuclei

Radiation	Approximate Mass (amu)	Charge	Symbol	Type
Alpha	4	2+	${}^{4}_{2}\text{He}$	Particle
Beta	0	1−	${}^{0}_{-1}e$	Particle
Gamma	0	0	γ	Electromagnetic radiation
Neutron	1	0	${}^{1}_{0}n$	Particle
Proton	1	1+	${}^{1}_{1}p$ (${}^{1}_{1}\text{H}$)	Particle
Positron	0	1+	${}^{0}_{1}e$	Particle

The thorium produced in the decay of $^{238}_{92}U$ is itself radioactive and decays by β-emission. The nuclear equation for the change is

$$^{234}_{90}Th \longrightarrow \, ^{0}_{-1}e + \, ^{234}_{91}Pa$$
β-particle

Thus β-emission increases the atomic number by one unit, but has (essentially) no effect on the mass.

γ-rays are emitted by nearly all radioactive substances.

Gamma radiation, as we learned in Chapter 3, is really nothing more than a very energetic form of electromagnetic radiation. Its emission from a nucleus doesn't change the charge or mass number, so gamma radiation is often omitted from nuclear equations.

In discussions of nuclear reactions and radioactive decay, certain terms are used frequently, so you should become familiar with them. The most common is **nuclide**—a general term used when referring to the nucleus of a particular isotope. Radioactive nuclei are called **radionuclides** and the atoms having these nuclei are called **radioisotopes.** In a radioactive decay, the isotope that decays is often called the **parent isotope** and the isotope that is formed is referred to as the **daughter isotope.** Thus in the decay of uranium-238, the nuclide $^{238}_{92}U$ is the parent and $^{234}_{90}Th$ is the daughter.

EXAMPLE 24.1

(a) ^{234}U decays by α-emissions. What is its daughter isotope?

(b) ^{214}Pb decays to ^{214}Bi. By what type of radiation is this accomplished?

SOLUTION

(a) We begin by writing a balanced nuclear equation, keeping in mind that the total mass and charge must be the same on both sides. Using X to represent the symbol for the daughter isotope,

$$^{234}_{92}U \longrightarrow \, ^{4}_{2}He + \, ^{230}_{90}X$$

The element with $Z = 90$ is thorium. The daughter isotope is therefore $^{230}_{90}Th$.

(b) Again, we write a balanced nuclear equation. This time we will let X be the symbol for the type of radiation.

$$^{214}_{82}Pb \longrightarrow \, ^{214}_{83}Bi + \, ^{0}_{-1}X$$

From Table 24.1 we see that $^{0}_{-1}X$ corresponds to $^{0}_{-1}e$; therefore the decay occurs by β-emission.

Radioactive decay series

We have seen that ^{238}U decays to ^{234}Th, which is also radioactive and decays to ^{234}Pa. This isotope is also unstable and decays to ^{234}U (how?), which is also radioactive, and so on. This decay process continues until a stable (nonradioactive) isotope of an element is formed. This entire scheme, where one isotope decays to another and that to another, and so on, is called a **radioactive series** or **decay series.** ^{238}U, for example, decays by some 14 steps to stable ^{206}Pb.

Kinetics of radioactive decay

In Chapter 12 we saw that the rate of a chemical reaction is given by the change in the concentration of a reactant per unit time. For a radioactive substance, its concentration is directly proportional to the number of particles or rays that it emits per minute. These emissions can be detected and counted by a suitable device such as the Geiger-Müller counter, shown schematically in Figure 24.1.

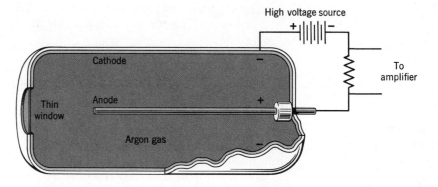

Figure 24.1

The Geiger-Müller counter. An α- or β-particle enters the Geiger tube through the thin window shown at the left of the apparatus. As the particle passes through the gas inside the tube, it ionizes argon atoms along its path. These ions cause *an electrical breakdown (discharge) between the wire and the wall of the tube, thereby producing a current pulse. This current pulse is readily amplified and counted electronically.*

Experimentally, the concentration of the radioactive isotope is proportional to the number of counts per minute (cpm) recorded by the counter.

If, for a particular radioactive isotope, we construct a graph in which the number of counts per minute is plotted against time, we obtain a curve similar to that shown in Figure 24.2a. If the log of the cpm is graphed versus time, a straight line is obtained as shown in Figure 24.2b. Both these plots are indicative of a first-order process and, as it turns out, all radioactive decay reactions obey first-order kinetics.

In Chapter 12, we saw that for a first-order process the concentration of a reactant (for example, a radioactive isotope) at any time t is given by the equation

$$\log \frac{[A]_0}{[A]} = \frac{kt}{2.303}$$

[24.1]

In this equation $[A]_0$ is the initial concentration of the isotope (concentration at time zero) and $[A]$ is its concentration at any time t. Equation 24.1 can be employed to calculate rate constants as shown in the following example.

EXAMPLE 24.2   A chemist determined that after exactly 1 week an initial 10.0 μg (1 μg = 10^{-6} g) of ^{222}Rn had decayed, and there was now 2.82 μg of the radon. What is the rate constant for the α-decay of ^{222}Rn?

SOLUTION   The rate constant can easily be obtained by using Equation 24.1, where $[A]_0$ = 10.0 g, $[A]$ = 2.82 g, and t = 7.00 days. Solving Equation 24.1 for k, we have

$$k = \frac{2.303}{t} \log \frac{[A]_0}{[A]}$$

and substituting the values given above into this equation, we have

$$k = \left(\frac{2.303}{7.00 \text{ days}}\right) \log \left(\frac{10.0}{2.82}\right)$$

or

$$k = 0.181 \text{ days}^{-1}$$

Figure 24.2
*Kinetics of radioactive decay.
(a) A graph of concentration expressed in counts per minute versus time. (b) A logarithmic plot of concentration versus time. The straight line indicates a first-order process.*

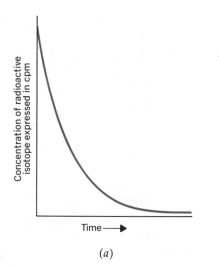

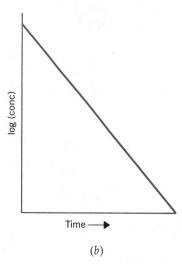

(a)

(b)

In Chapter 12 we also learned that the *half-life*, $t_{1/2}$, of a reactant in a first-order process is independent of the initial concentration of the reactant. In other words, the time required for half of a radioactive isotope to decay is the same regardless of the amount of the isotope that is present. Thus, if we were to begin with 10.0 g of a particular isotope, after one half-life only 5.0 g would remain. In a time equal to another half-life this would decay to 2.5 g, and so on. This is represented graphically in Figure 24.3. In this figure we see, in general, how an isotope decays through many half-lives. We have seen that the half-life of any substance is inversely proportional to the rate constant for its decay. The relationship is

$$t_{1/2} = \frac{0.693}{k}$$

[24.2]

From this equation we see that if the rate constant for decay is large, its half-life will be small and vice versa. Also, we see that if either $t_{1/2}$ or k is known, the other can be calculated.

Figure 24.3
A graph of a first-order radioactive decay illustrating the concept of half-life. The initial concentration is a, which drops to $\frac{a}{2}$ after one half-life, to $\frac{a}{4}$ after the second half-life, and so forth.

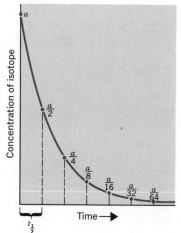

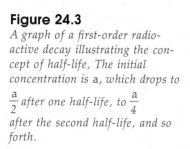

EXAMPLE 24.3 (a) What is the half-life of ^{222}Rn and (b) What is the fraction of a sample of Rn that decays in 1 week?

SOLUTION (a) We can calculate the half-life of Rn directly from Equation 24.2 by substituting in the value of k calculated in Example 24.2,

$$t_{1/2} = \frac{0.693}{k}$$

and $k = 0.181$ days^{-1}, so

$$t_{1/2} = \frac{0.693}{0.181 \text{ days}^{-1}}$$

or

$$t_{1/2} = 3.83 \text{ days}$$

(b) We first have to calculate the fraction left after decay. By this we mean the ratio $[A]/[A]_0$. Therefore we employ Equation 24.1 for this purpose and solve it for $[A]_0/[A]$.

$$\log \frac{[A]_0}{[A]} = \frac{kt}{2.303}$$

Substituting in $k = 0.181$ days^{-1} and $t = 7.00$ days, we have

$$\log \frac{[A]_0}{[A]} = \frac{(0.181 \text{ days}^{-1})(7.00 \text{ days})}{2.303}$$

or

$$\log \frac{[A]_0}{[A]} = 0.550$$

Taking the antilog gives

$10^{0.550} = 3.55.$

$$\frac{[A]_0}{[A]} = 3.55$$

The reciprocal, therefore, is

$$\frac{[A]}{[A]_0} = 0.282$$

This is the fraction left after 1 week, so the fraction that decayed is simply $1 - 0.282$, or 0.718.

Half-lives are important for two reasons. First, they indicate the stability of a particular isotope; the longer the half-life the more stable the nuclide is. Second, the half-life can be effectively used to measure the age of such things as rocks, bones, ancient pieces of art, and the like.

Uses of radioactive decay

The age of rocks containing uranium can be dated by determining the ratio of ^{238}U to ^{206}Pb. Remember that ^{206}Pb is the stable isotope to which ^{238}U eventually decays. Using a ^{238}U/^{206}Pb ratio of 1 to 0 as corresponding to the ratio at zero time and a ^{238}U/^{206}Pb ratio of 1 to 1 as corresponding to the ratio after one half-life—that is, 4.5×10^9 years—the age of these rocks can be closely approximated. The oldest rocks that have been found on earth have an age of 4.55×10^9 years using this method.

Rocks that do not contain uranium are presently being dated using a potassium-argon method. This makes use of the reaction

$$^{40}_{19}\text{K} + ^{0}_{-1}e \longrightarrow ^{40}_{18}\text{Ar} \qquad t_{1/2} = 1.3 \times 10^9 \text{ years}$$

The electron in this case comes from the $1s$ orbital of the potassium and is captured by the unstable $^{40}_{19}K$ nucleus. In the dating procedure the ratio of ^{40}K to argon is measured, and the age of the rock is determined in the same manner as the uranium dating described above.

The age of materials that were once living, such as bones and wood, can be estimated quite accurately by measuring their ratio of ^{14}C to ^{12}C. Carbon-14 is radioactive and is constantly being produced in the upper atmosphere by the bombardment of cosmic neutrons upon $^{14}_{7}N$, which is present there in large amounts. The equation for this reaction is

$$^{14}_{7}N + ^{1}_{0}n \longrightarrow ^{14}_{6}C + ^{1}_{1}p$$

The carbon-14 thus produced immediately begins to decay.

$$^{14}_{6}C \longrightarrow ^{14}_{7}N + ^{0}_{-1}e \qquad t_{1/2} = 5770 \text{ years}$$

The steady-state concentration of ^{14}C is 15 cpm per gram of carbon.

Because ^{14}C is being both formed in the atmosphere and removed by its decay, a constant concentration is maintained. We say that a *steady-state* concentration is achieved. This ^{14}C becomes incorporated in carbon dioxide in the atmosphere where it can be taken in by plants through the process of photosynthesis. The intake of ^{14}C into animals is by the consumption of such plants or by the consumption of plant-eating animals. While they are alive, plants and animals consume and excrete carbon so that they also maintain a steady-state concentration of ^{14}C and are thus in equilibrium with their surroundings. Once they die, however, the ^{14}C that they possess is not replaced as the organisms decay, so the ^{14}C concentration begins to decrease. The half-life of the ^{14}C is 5770 years; therefore, if we find that the carbon-14 concentration in an object that had one been living has dropped to half its initial value, we could conclude that the object is 5770 years old.

Atmospheric testing of nuclear weapons has made it impossible for our current history to be dated this way by future archaeologists.

EXAMPLE 24.4

A piece of charcoal from the ruins of a settlement in Japan was found to have a $^{14}C/^{12}C$ ratio that was 0.617 times that found in living organisms. How old is this piece of charcoal?

SOLUTION

The answer to this problem can be obtained once again by employing Equations 24.2 and 24.1. From Equation 24.2 we can obtain the k for the decay of ^{14}C.

$$k = \frac{0.693}{5770 \text{ years}} = 1.20 \times 10^{-4} \text{ years}^{-1}$$

Next, we need values to substitute for $[A]_0$ and $[A]$. Fortunately we don't need actual values for them. If we *arbitrarily* take the $^{14}C/^{12}C$ ratio in a living organism to be equal to 1.000, then according to the data in the problem, the charcoal has a $^{14}C/^{12}C$ ratio of $0.617 \times 1.000 = 0.617$. Taking the ratio of these numbers is the *same* as taking the ratio of the concentrations. Therefore, substituting into Equation 24.1, we have

$$\log \frac{[A]_0}{[A]} = \frac{kt}{2.303}$$

$$\log \left(\frac{1.000}{0.617} \right) = \log 1.62 = 0.210 = \frac{(1.20 \times 10^{-4} \text{ years}^{-1})t}{2.303}$$

or

$$t = 4030 \text{ years}$$

24.2   NUCLEAR TRANSFORMATIONS

Lord Rutherford, in 1919, performed the first known **nuclear transformation** of one element into another. Rutherford found that by bombarding $^{14}_{7}N$ with α-particles he produced the isotope of oxygen, $^{17}_{8}O$. The equation for this process is

$$^{14}_{7}N + {}^{4}_{2}He \longrightarrow [{}^{18}_{9}F] \longrightarrow {}^{17}_{8}O + {}^{1}_{1}H$$

The square brackets indicate that the isotope $^{18}_{9}F$ is a very unstable intermediate that rapidly decays to the products above.

In 1933 Irene Curie and her husband Frederick Joliot showed that other light elements could similarly be transformed by bombardment with α-particles. For example, $^{27}_{13}Al$ can be transformed into $^{30}_{15}P$ by this process.

> Irene Joliot-Curie was the daughter of Marie and Pierre Curie — the discoverers of polonium and radium.

$$^{27}_{13}Al + {}^{4}_{2}He \longrightarrow {}^{30}_{15}P + {}^{1}_{0}n$$

Reactions of this type, where an α-particle is used for bombardment and a neutron is one of the products, is known as an *alpha, neutron reaction*— symbolized by (α,n). A shorthand notation for the reaction is $^{27}_{13}Al(\alpha,n)^{30}_{15}P$.

Since 1933 many isotopes have been produced by bombardment reactions, some in which particles other than α-particles have been used. One of the main problems with such experiments is that a positively charged nucleus is being bombarded with positively charged particles. Heavy elements, with their very highly positive nuclei, will repel particles with positive charges like the α-particle. One way to circumvent this problem is to use neutrons as the bombarding particles.

Neutrons, which have no charge, are not repelled by the nucleus and, therefore, are excellent materials for bombardment reactions. The supply of neutrons for these reactions can be obtained from either of two sources—from a transformation reaction in which neutrons are produced or from a nuclear reactor in which fission reactions (which will be examined in Section 24.6) occur at a controlled rate. Examples of neutron-producing reactions are the $^{27}_{13}Al(\alpha,n)^{30}_{15}P$ reaction mentioned above and

$$^{9}_{4}Be + {}^{4}_{2}He \longrightarrow {}^{12}_{6}C + {}^{1}_{0}n$$

in which the beryllium isotope undergoes an α,n reaction.

A second way that nuclear transformations can be brought about is through the use of particle accelerators. Particle accelerators, such as the cyclotron illustrated in Figure 24.4, speed up particles to extremely high velocities and then direct them at target nuclei. At these speeds the positive particles are able to

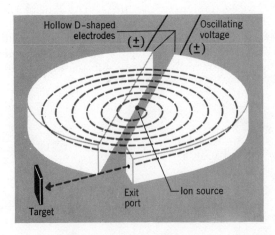

Figure 24.4
Diagram of a cyclotron.

Table 24.2
Elements produced by particle accelerators

$$^{238}_{92}U + ^{4}_{2}He \longrightarrow ^{239}_{94}Pu + 3\,^{1}_{0}n$$

$$^{239}_{94}Pu + ^{4}_{2}He \longrightarrow ^{240}_{95}Am + ^{1}_{1}H + 2\,^{1}_{0}n$$

$$^{239}_{94}Pu + ^{4}_{2}He \longrightarrow ^{242}_{96}Cm + ^{1}_{0}n$$

$$^{244}_{96}Cm + ^{4}_{2}He \longrightarrow ^{245}_{97}Bk + ^{1}_{1}H + 2\,^{1}_{0}n$$

$$^{238}_{92}U + ^{12}_{6}C \longrightarrow ^{246}_{98}Cf + 4\,^{1}_{0}n$$

$$^{238}_{92}U + ^{14}_{7}N \longrightarrow ^{247}_{99}Es + 5\,^{1}_{0}n$$

$$^{238}_{92}U + ^{16}_{8}O \longrightarrow ^{249}_{100}Fm + 5\,^{1}_{0}n$$

$$^{253}_{99}Es + ^{4}_{2}He \longrightarrow ^{256}_{101}Md + ^{1}_{0}n$$

$$^{246}_{96}Cm + ^{13}_{6}C \longrightarrow ^{254}_{102}No + 5\,^{1}_{0}n$$

$$^{252}_{98}Cf + ^{10}_{5}B \longrightarrow ^{257}_{103}Lw + 5\,^{1}_{0}n$$

$$^{249}_{98}Cf + ^{12}_{6}C \longrightarrow ^{257}_{104}Unq + 4\,^{1}_{0}n$$

overcome the coulombic repulsion of a nucleus and collide with it. Accelerators have been used extensively by Dr. Glenn Seaborg and his colleagues at the University of California in producing many of the *transuranium elements*—that is, elements 93 to 105. In these accelerators positive ions of such isotopes as $^{2}_{1}H$ (deuterium), $^{12}_{6}C$, $^{13}_{6}C$, $^{16}_{8}O$, $^{14}_{7}N$, $^{10}_{6}B$, as well as $^{4}_{2}He$ have been used in producing new, artificial, elements. A list of these bombardment reactions and the elements they produce is shown in Table 24.2. Many of these elements are extremely short-lived and, as a result, only a few atoms, especially of the high-atomic-numbered isotopes, have ever been formed.

24.3 NUCLEAR STABILITY

Experimentally it is observed that all the elements with atomic numbers greater than 83 (bismuth) are radioactive and possess no known stable isotopes. On the other hand, all the lighter elements, with the exception of technetium ($Z = 43$) and promethium ($Z = 61$), have one or more stable, nonradioactive isotopes. In addition, radioactive isotopes undergo nuclear transformations that lead ultimately to stable nuclei. Sometimes this is accomplished by a simple one-step process, while in other cases a series of nuclear reactions occur before a stable isotope is reached. A question that naturally arises from these observations is; what factors give rise to stable or unstable nuclei?

Little is known about the nature of the forces that hold a nucleus together. Some interesting facts concerning nuclear stability emerge, however, if we examine the numbers of protons and neutrons found in stable nuclei. For example, if we make a graph of the number of neutrons versus the number of protons in different nuclei, we find that all the stable isotopes fall in a narrow band, which we might call a **band of stability,** as shown in Figure 24.5. In this illustration we see that at low atomic numbers stable nuclei possess approximately equal numbers of protons and neutrons. Above about $Z = 20$, however, the number of neutrons always exceeds the number of protons, and the neutron-to-proton ratio gradually increases to about 1.5 at the upper end of the band of stability. Apparently, as the number of protons in the nucleus increases, there must be more and more neutrons present to help overcome the strong repulsive forces between the protons. It also seems that there is an upper limit to the number of protons that can exist in a stable nucleus, with that number being reached at bismuth.

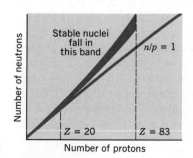

Figure 24.5
Band of stability.

Nuclei that lie outside the band of stability are unstable and decay in a manner that tends to give them a stable neutron-to-proton (n/p) ratio. On this basis, we can understand why certain nuclei undergo the type of radioactive decay that they do. For instance, a nucleus that lies above the band of stability must either lose neutrons or gain protons to achieve stability. Thus we can understand why elements such as ^{14}C (which lies above the band) decay by β-emission, because this process converts a neutron into a proton (^1_1p).

$$^1_0n \longrightarrow {}^1_1p + {}^{\ 0}_{-1}e$$

For ^{14}C we have

$$^{14}_6C \longrightarrow {}^{14}_7N + {}^{\ 0}_{-1}e$$

Another way that an element located above the band can achieve a stable n/p ratio is by emitting a neutron, although this particular mode of decay is rare. An example is the decay of ^{137}I.

$$^{137}_{53}I \longrightarrow {}^{136}_{53}I + {}^1_0n$$

Elements located *below* the band of stability must increase their n/p ratio to achieve stability. This is accomplished generally in either of two ways. One involves the emission of a positron—a particle having the same mass as the electron but with a unit positive charge. The positron is symbolized as ^0_1e. The ejection of a positron by an unstable nucleus converts a proton into a neutron,

$$^1_1p \longrightarrow {}^1_0n + {}^0_1e$$

An example is the decay of ^{11}C.

$$^{11}_6C \longrightarrow {}^{11}_5B + {}^0_1e$$

The second mode of decay that results in an increased n/p ratio is called **electron capture.** In this case the unstable nucleus captures an electron, usually from its own 1s orbital. Since the captured electron most often originates in the K shell, the process is also called **K-capture.** The addition of this electron to the nucleus transforms a proton into a neutron.

$$^1_1p + {}^{\ 0}_{-1}e \longrightarrow {}^1_0n$$

Two examples of decay by K-capture are

$$^7_4Be + {}^{\ 0}_{-1}e \xrightarrow{\ K\text{-capture}\ } {}^7_3Li$$

and

$$^{40}_{19}K + {}^{\ 0}_{-1}e \xrightarrow{\ K\text{-capture}\ } {}^{40}_{18}Ar$$

The vacancy created in the 1s subshell as a result of K-capture is only temporary, and electrons from higher energy levels quickly drop to fill the 1s orbital. Since electrons are falling from higher energy levels to lower ones, energy is emitted in the form of electromagnetic radiation (light)—in this instance in the X-ray region of the spectrum.

Elements having atomic numbers higher than 83—that is, those beyond the end of the band of stability—cannot find their way to a stable n/p ratio by any of the decay modes we just discussed. In these cases the unstable nuclei must lose both protons *and* neutrons. As a result, their decay usually involves emission of α-particles, since each α-emission removes two protons and two neutrons simultaneously. Earlier, for example, we saw this type of decay process for uranium,

$$^{238}_{92}U \longrightarrow {}^4_2He + {}^{234}_{90}Th$$

Another type of nuclear transformation that is available to the heavy elements is **fission,** in which a heavy nucleus splits into several much lighter fragments, many of which may also lie outside the band of stability and hence may be radioactive. The smaller nuclei that are produced by fission, if they are unstable, are able to undergo the simpler types of decay in order to produce a stable nucleus. We will take a closer look at nuclear fission in Section 24.6.

We may also observe that nuclei with even numbers of protons and neutrons are apparently more stable than those containing an odd number of these particles. For example, there are 157 stable isotopes in which there are even numbers of both protons and neutrons, 52 isotopes having an even number of protons and an odd number of neutrons, and 50 with an even number of neutrons but an odd number of protons. By contrast, there are only five stable nuclides in which there are odd numbers of both protons and neutrons.

| Protons | Even | Even | Odd | Odd |
Neutrons	Even	Odd	Even	Odd
Stable nuclei	157	52	50	5

This phenomenon suggests that in stable nuclei, protons and neutrons each tend to be paired, in much the same way that electrons become paired in the outer region of the atom. Apparently, extra stability, as evidenced by the number of stable nuclides, results when pairing takes place with both protons and neutrons. On the other hand, when pairing cannot occur, as must be true when the numbers of protons and neutrons are both odd, very few stable isotopes occur (most isotopes having odd numbers of both protons and neutrons are radioactive).

A final observation on nuclear stability is that nuclei that contain certain specific numbers of protons and neutrons possess a degree of extra stability. These so-called *magic numbers* for protons and neutrons are 2, 8, 20, 28, 50, and 82, with an additional magic number of 126 for neutrons. When nuclei contain a magic number of both protons and neutrons, they are said to be *doubly magic* and are extremely stable. Examples are ^4_2He, $^{16}_8O$, $^{40}_{20}Ca$, and $^{208}_{82}Pb$.

The occurrence of these magic numbers suggests a shell structure for the nucleus somewhat akin to the shell structure exhibited by electrons. For example, we have seen that very stable (unreactive) electron configurations occur when an atom contains magic numbers of 2, 8, 18, 36, or 54 electrons, corresponding to the noble gases, He through Kr. In the nucleus, it seems that nuclear shells of either protons or neutrons become completed when the nuclear magic numbers are reached and that a particularly stable nucleus occurs whenever there is a completed shell of either neutrons or protons. Exceptionally stable nuclei result when the nucleus contains filled shells of protons and neutrons simultaneously.

24.4 EXTENSION OF THE PERIODIC TABLE

The heaviest naturally occurring element is uranium and, as we saw earlier in this chapter, elements beyond $Z = 92$ are all prepared artificially by bombarding lighter nuclei with protons, α-particles, and the positive ions of some of the second-period elements. The discovery of these new elements quite expectedly prompted chemists to begin to think about a whole host of new elements with new and interesting properties to be studied. However, it soon become apparent that as the atomic number of the artificial element became higher, its half-life became shorter, so the prospects for stable elements of very high atomic numbers became dim.

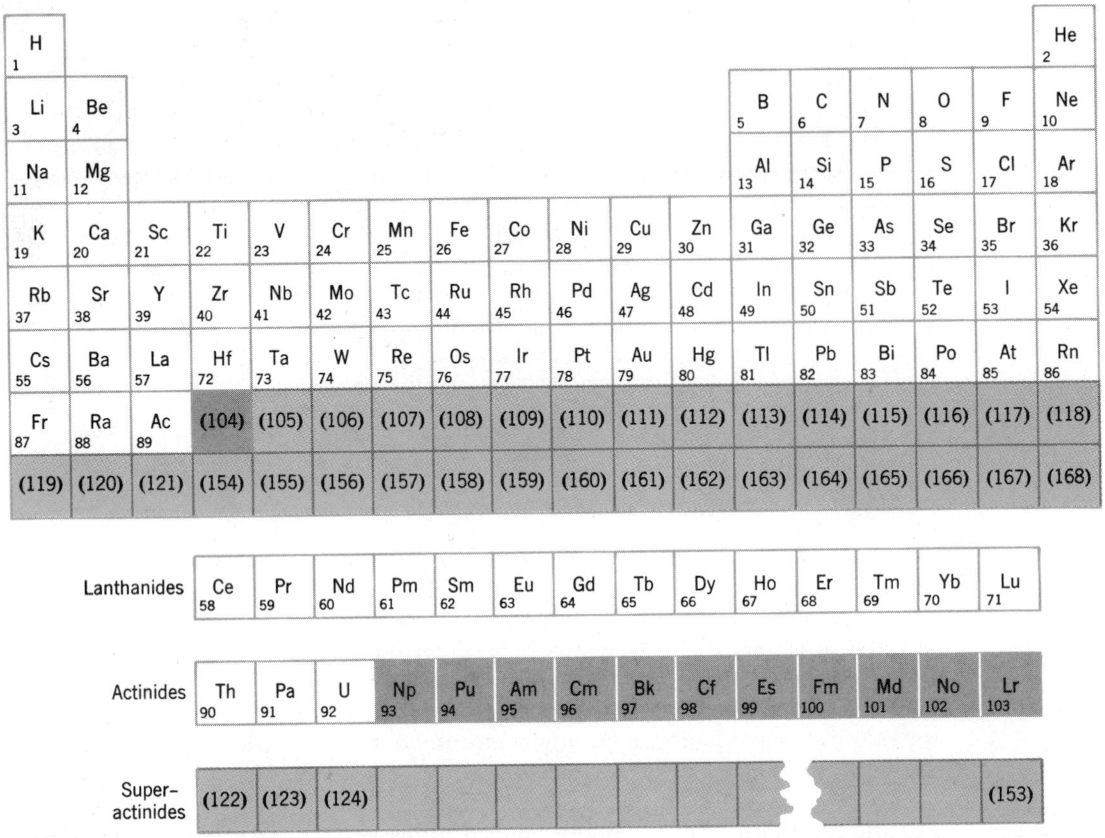

H 1																	He 2
Li 3	Be 4											B 5	C 6	N 7	O 8	F 9	Ne 10
Na 11	Mg 12											Al 13	Si 14	P 15	S 16	Cl 17	Ar 18
K 19	Ca 20	Sc 21	Ti 22	V 23	Cr 24	Mn 25	Fe 26	Co 27	Ni 28	Cu 29	Zn 30	Ga 31	Ge 32	As 33	Se 34	Br 35	Kr 36
Rb 37	Sr 38	Y 39	Zr 40	Nb 41	Mo 42	Tc 43	Ru 44	Rh 45	Pd 46	Ag 47	Cd 48	In 49	Sn 50	Sb 51	Te 52	I 53	Xe 54
Cs 55	Ba 56	La 57	Hf 72	Ta 73	W 74	Re 75	Os 76	Ir 77	Pt 78	Au 79	Hg 80	Tl 81	Pb 82	Bi 83	Po 84	At 85	Rn 86
Fr 87	Ra 88	Ac 89	(104)	(105)	(106)	(107)	(108)	(109)	(110)	(111)	(112)	(113)	(114)	(115)	(116)	(117)	(118)
(119)	(120)	(121)	(154)	(155)	(156)	(157)	(158)	(159)	(160)	(161)	(162)	(163)	(164)	(165)	(166)	(167)	(168)

Lanthanides	Ce 58	Pr 59	Nd 60	Pm 61	Sm 62	Eu 63	Gd 64	Tb 65	Dy 66	Ho 67	Er 68	Tm 69	Yb 70	Lu 71
Actinides	Th 90	Pa 91	U 92	Np 93	Pu 94	Am 95	Cm 96	Bk 97	Cf 98	Es 99	Fm 100	Md 101	No 102	Lr 103

Super-actinides	(122)	(123)	(124)							(153)

Figure 24.6

Extended periodic table. From G. Seaborg, Journal of Chemical Education, *Vol. 46, p. 626, October 1969. Used by permission.*

Calculations by many nuclear physicists, based on the nuclear shell model, now suggest that a closed nuclear shell for protons exists at $Z = 114$ and that one for neutrons occurs at 184. As a result, chemists have once again begun to speculate about the possibilities of new stable elements.

One proposed extension of the periodic table to include these heavy and superheavy elements is shown in Figure 24.6. Recall that the actinide series, which occurs as the result of the filling of the $5f$ subshell, ends at lawrencium, $Z = 103$. The next element, 104, therefore lies under hafnium if we follow the scheme for the filling of subshells developed in Chapter 3. Elements 104 to 112, therefore, would correspond to the filling of the $6d$ subshell. Next we have six elements in the p-block (113 to 118), which would have their $7p$ subshell gradually filled. Elements 119 and 120, in period 8, correspond to the completion of the $8s$ subshell, and following element 121 a sequence of 32 inner transition elements occurs, from $Z = 122$ to 153. These *superactinides* would be accounted for by the completion of first the $6f$ subshell (14 elements) followed by the filling of a $5g$ subshell (a g subshell would contain nine orbitals that can accommodate 18 electrons; therefore we would have 18 more elements, 136 to 153). After the superactinides we would fill the $7d$ subshell (elements 154 to 162) and then the $8p$ subshell (163 to 168).

In the search for new elements it is expected that nuclides that differ much from $Z = 114$ would be extremely unstable and would decompose by fission with very short half-lives. However, in the vicinity of $Z = 114$ it has been suggested that fission should not occur and that the half-lives of these elements with respect to α- and β-decay should be long enough so that it should be possible to detect them and perhaps even investigate their chemical properties.

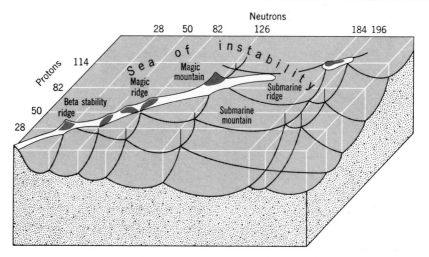

Figure 24.7

Known and predicted regions of nuclear stability, surrounded by a sea of instability. From G. Seaborg, Journal of Chemical Education, *Vol. 46, p. 626, October 1969. Used by permission.*

There is even the possibility that these superheavy elements will not be radioactive at all.

The relative stabilities of nuclides containing differing numbers of protons and neutrons have been dramatized in a drawing (Figure 24.7) published by Dr. Glenn T. Seaborg, formerly head of the U. S. Atomic Energy Commission. Here the stable nuclei of our band of stability are shown as a long peninsula extending out into a sea of instability. Stable nuclei correspond to points above sea level, whereas submerged regions constitute unstable nuclei. Notice that nuclei with magic numbers of either protons or neutrons are shown as higher, more stable ridges, while doubly magic nuclei are shown as mountains of stability.

The superheavy elements with approximately a magic number of 114 protons and either 184 or 196 neutrons are depicted as an island of stability separated from the peninsula by a region of high nuclear instability. As a result, in order to reach this island we cannot bombard stable (or relatively stable) nuclei with light particles such as $_{2}^{4}$He, because this simply places the product nuclei into the sea of instability where they decompose rapidly before any additional mass and charge can be added. Consequently, the jump to the island must be made in one step. At the present time this presents substantial problems, because bombarding nuclei must contain a n/p ratio of at least 1.6 (184/114 = 1.61). However, light nuclei such as $_{18}^{40}$Ar, while possessing sufficient protons to give the desired atomic number by a reaction such as

$$_{96}^{248}\text{Cm} + {}_{18}^{40}\text{Ar} \longrightarrow {}_{114}^{284}X + 4_{0}^{1}n$$

do not contain enough neutrons to place the product isotope within the island of stability. For example, $_{114}^{284}X$ contains only $284 - 114 = 170$ neutrons, 14 less than the 184 that we would want to achieve a doubly magic nucleus. Research today is directed at obtaining suitable target and projectile nuclei that will give not only $Z = 114$ but also a sufficient number of neutrons to place the product nucleus within the bounds of the predicted island of stability.

While physicists continue their search for ways to synthesize these superheavy elements, other scientists are looking for evidence of their current or past existence in the universe. In 1975 evidence was discovered by Dr. Edward Anders at the University of Chicago's Fermi Institute that suggested the one-time existence of element 114 (or perhaps 115 or 113) in a meteorite that fell in Mexico in 1969. Other scientists at Oak Ridge National Laboratories have found crystals that may once have contained elements 116 and 126.

24.5   CHEMICAL APPLICATIONS

Ordinarily, nuclear changes such as those involved in radioactive decay have very little direct effect on chemical reactions, although in some cases the high-energy radiation emitted by radioactive nuclei can influence the products of reaction. This radiation is generally capable of disrupting chemical bonds. If the cleavage of a chemical bond occurs in DNA, for example, mutations of the DNA strand can be brought about. Mutational changes can also take place by reactions between DNA and the products of other cleavage reactions. For instance, free radicals are produced by splitting an H—O bond in water. This takes place through a series of steps, the first of which is the absorption of radiation by the water molecule and the subsequent ejection of an electron

$$H_2O + h\nu \longrightarrow H_2O^+ + e^-$$
$$\underbrace{\qquad\qquad}_{\text{energy from radiation}}$$

The final products of the reaction are a hydrogen atom, H· (the dot indicates the unpaired electron in the radical) and a *hydroxyl radical*, ·OH. The overall change is

$$H_2O + h\nu \longrightarrow H\cdot + \cdot OH$$

Free radicals are extremely reactive because of the presence of their unpaired electrons, and their interactions with DNA can cause mutations or otherwise disrupt the replication of the DNA strands.

It is only in rare cases that the chemist makes direct use of the energy emitted in nuclear transformations. Most chemical applications of radioactive nuclides stem from their ease of identification and detection, even when they are present in very small amounts. Hence radioactive isotopes are usually employed in **tracer studies,** where they may be added in very small amounts and used to follow, or trace, the course of a chemical reaction. The range of applications of these tracer techniques is limited only by the imagination and ingenuity of the experimenter. Let's take a brief look at some examples that demonstrate the scope of these applications.

Analytical chemistry

There are many examples of analytical uses for radioactive isotopes. One of these techniques, called **isotope dilution,** can be used when it is impossible to completely separate a desired substance from a mixture. In this case, a small measured amount of the substance containing a known quantity of a radioactive isotope is *added* to the mixture. After making sure that complete mixing has occurred, a small amount of the *pure* desired substance is separated from the mixture. This sample will contain some of the added radioactive isotope, and from the proportion of the labeled isotope present in the sample the total quantity of the substance in the original mixture can be computed.

Consider, for instance, a mixture of salts of similar solubilities, such as a mixture of KNO_3 and NaCl. By fractional crystallization only a portion of the KNO_3 can be separated from the mixture. As a result, we cannot determine, in a simple fashion, how much of this salt is in the mixture.

Suppose, now, that 1.0 g of KNO_3 containing a small amount of radioactive ^{40}K is added to the salt mixture and then some KNO_3 (now containing K from the original mixture as well as from the added tagged KNO_3) is separated by fractional crystallization. If the **specific activity** of this KNO_3—that is, the number of counts per minute per gram (cpm/g)—has dropped to 1% of the specific activity of the added KNO_3, then we know that only 1% of the added solid has been recovered in our KNO_3 sample and that the other 99% of the KNO_3 must have been present in the original mixture. In other words, after we had added the 1 g of labeled KNO_3 there was a 99-to-1 ratio of unlabeled to labeled salt. Therefore the original mixture must have contained 99 g of KNO_3.

Isotope dilution methods are also used when the volume of a liquid in an irregular container must be measured. A small known volume of radioactive material is added and after mixing is complete, the extent of dilution allows one to calculate backwards to find the initial liquid volume. This method has been used to measure blood volumes in living animals and the volumes of underground reservoirs of water.

Another technique applicable to analytical chemistry is called **neutron activation analysis.** When nonradioactive isotopes are bombarded by neutrons, heavy isotopes of these elements can be produced. The product of this reaction may lie outside the band of stability and hence be radioactive. Even if another nonradioactive nucleus is produced, however, the absorption of these neutrons generally gives nuclei that are excited and that emit γ-radiation in much the same way that an excited atom emits light when it returns to the ground state.

$$^A_Z X + {}^1_0 n \longrightarrow {}^{A+1}_Z X^* \qquad \text{(the asterisk indicates an excited nucleus)}$$

$$^{A+1}_Z X^* \longrightarrow {}^{A+1}_Z X + h\nu \qquad (h\nu = \gamma \text{ photon})$$

Since each element has its own characteristic γ-emission spectrum, an analysis of the energies of the γ-emissions from the activated sample allows its composition to be determined. In addition, from the intensity of the emitted γ-radiation, the concentration of each element can be computed.

This technique has some very useful advantages. First, it is nondestructive. Since the number of nuclei that must be activated to perform the analysis is small, most of the sample is unaffected. Second, as implied in the preceding sentence, the method is very sensitive and is, therefore, well suited to the analysis of trace amounts of impurities. In some cases, sensitivities of the order of 10^{-12} g can be achieved.

Descriptive chemistry

Many elements having atomic numbers greater than $Z = 83$ (bismuth) have short half-lives and, therefore, are not observed to occur naturally. Instead, they must be synthesized in particle accelerators. As a result, only extremely small quantities of these elements have ever been prepared. A question then arises: How can we study their chemistry if we cannot even obtain enough to be able to see them?

To arrive at the solution to this problem, let us consider the element astatine. Astatine was first produced in the cyclotron by the reaction

$$^{209}_{83} Bi + {}^4_2 He \longrightarrow {}^{211}_{85} At + 2 {}^1_0 n$$

in which the ^{211}At produced has a half-life of only about 7.5 hr. The most stable isotope, ^{210}At, has a half-life of only 8.3 hr, so large quantities of the element cannot be accumulated.

Since astatine occurs in Group VIIA, we expect the element to be similar in some of its properties to iodine. To verify this the astatine is added as a tracer in reactions involving iodine, and the fate of the At is followed as the iodine undergoes reactions. If in a given reaction the At occurs in the products along with the iodine, we conclude that, in this reaction, At behaves just as I does. We have discovered something about the chemical behavior of an element that we cannot even see! For instance, it is observed that, like iodine, elemental astatine is rather volatile, since it is carried with the iodine when I_2 is sublimed. In solu-

How would you study the chemical properties of francium, Fr $(Z = 87)$?

tion At⁻ is carried from solution along with I⁻ upon the addition of Ag⁺. Thus we conclude that AgAt is insoluble just as is AgI.

Reaction mechanisms

In Chapter 12 we saw that a study of the effect of the concentrations of the reactants on the rate of a chemical reaction can often give some insight into the mechanism of the reaction. Such studies, however, seldom answer all the questions that we might ask about the reaction mechanism. Consider, for example, the reaction of an alcohol and an organic acid to produce an ester and water,

$$R\text{—OH} + HO\text{—}\overset{\displaystyle O}{\overset{\displaystyle \|}{C}}\text{—R}' \longrightarrow R\text{—O—}\overset{\displaystyle O}{\overset{\displaystyle \|}{C}}\text{—R}' + H_2O$$

Upon the formation of the ester molecule, two hydrogen atoms and one oxygen atom are removed from the alcohol and acid to become a molecule of water. There seems little doubt about the origin of the two hydrogen atoms; however, there is a question about which one of the —OH oxygen atoms is removed and finds its way into the H_2O molecule.

This question can be resolved by carrying out the reaction with a labeled oxygen (for example, ¹⁸O) incorporated into the OH group of either the alcohol or the acid. For instance, if the alcohol is labeled with ¹⁸O, it is found that all the labeled oxygen becomes incorporated into the ester. On the other hand, if the acid contains ¹⁸O in the OH group, all the labeled oxygen ends up in the water with none in the ester. It is clear, therefore, that the reaction involves the removal of the OH from the acid and the H from the alcohol.

Reaction using labeled alcohol

$$R\text{—O*—H} + H\text{—O—}\overset{\displaystyle O}{\overset{\displaystyle \|}{C}}\text{—R}' \longrightarrow R\text{—O*—}\overset{\displaystyle O}{\overset{\displaystyle \|}{C}}\text{—R}' + H_2O$$

Reaction using labeled acid

$$R\text{—O—H} + H\text{—O*—}\overset{\displaystyle O}{\overset{\displaystyle \|}{C}}\text{—R}' \longrightarrow R\text{—O—}\overset{\displaystyle O}{\overset{\displaystyle \|}{C}}\text{—R}' + H_2O*$$

Many similar experiments using tagged atoms have been employed to aid in the elucidation of a large number of reaction mechanisms, including biological and biochemical processes. For instance, labeled water can be added to the root system of a plant, and its progression into the stem—and ultimately into the leaves—can be traced. Experiments using ¹⁴C-labeled CO_2 have been used to follow the course of carbon in photosynthesis in plants. In this case plants are exposed to CO_2 and, at various intervals, are killed and their cellular components separated to determine which compounds have had ¹⁴C built into them. In this way the sequence of reactions in photosynthesis can be unraveled.

24.6 NUCLEAR FISSION AND FUSION

In the late 1930s in Germany, two chemists, Otto Hahn and Fritz Strassmann, and a physicist, Lise Meitner, found that when ²³⁵U was bombarded with neutrons the unexpected products were the isotopes ¹³⁹Ba and ⁹⁴Kr, as well as three neutrons:

$$^{235}_{92}U + ^{1}_{0}n \longrightarrow ^{139}_{56}Ba + ^{94}_{36}Kr + 3\,^{1}_{0}n$$

The significance of this accidental discovery was explained by Meitner and her nephew, Otto Frisch, who pointed out that fragmentation (splitting) of the ^{235}U was taking place, and that a very large amount of energy was emitted during the process. The splitting of an atom into two approximately equal parts is known as **fission.** The results of the fussion of ^{235}U were tested and substantiated by Enrico Fermi at Columbia University in New York City and by physicists at Berkeley in California. Unfortunately, scientists saw military applications of the fission process, and it was Albert Einstein who alerted President Roosevelt to its possibilities. Roosevelt responded by establishing the Manhattan Project, whose research efforts led to the two bombs that were dropped on Hiroshima and Nagasaki, and ultimately to the production of energy by controlled nuclear fission.

Since three neutrons are produced during each fission of ^{235}U, and since neutrons are required as a reactant for each fission process, then, potentially at least, the initial reaction is capable of triggering several additional reactions, as we can see in Figure 24.8. These reactions can in turn trigger many more and so on, permitting a nuclear chain reaction to take place. This is indeed what occurs if enough pure ^{235}U is present. Each fission reaction causes several others to take place, with the evolution of a tremendous amount of energy.

We mentioned earlier that the most abundant isotope of uranium found in its naturally occurring ores is ^{238}U. This isotope is nonfissionable and can, in fact, prohibit the fission chain reaction of the ^{235}U from occurring. The ^{238}U absorbs the neutrons emitted during the fission reaction, thus preventing the chain from continuing. There is, then, a minimum quantity of ^{235}U that must be present in order for the fission reaction to sustain itself. This minimum quantity needed for fission is called the isotope's **critical mass.**

Nuclear reactors

In a nuclear reactor, fission reactions take place, but at a controlled rate: slow enough to avoid a chain explosion but fast enough to produce usable heat. The fission reactions are controlled by the use of control rods made of such materials as cadmium, which absorb neutrons and thus prohibit the chain from occurring too rapidly. When these rods are extended all the way into the reactor (or pile), fission occurs very slowly. The rate can be increased by withdrawing the rods. When they are pulled out, they absorb fewer and fewer neutrons and the reac-

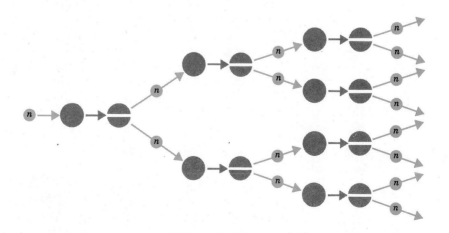

Figure 24.8

Fission chain reaction.

The Big Rock Nuclear Power Plant on the shore of Lake Michigan at Charlevoix, a town on the northwestern tip of Michigan's lower peninsula. The reactor itself is located in the spherically shaped containment building at the right.

tion occurs faster and faster. The ideal position of the rods is at that point where the fission reaction is just able to sustain itself at a desired level.

We are all familiar with the problems created by a breakdown in the cooling system as occurred at the nuclear reactor at Three Mile Island in Pennsylvania.

The large amount of energy generated in the controlled nuclear fission reaction appears primarily as heat, and hence nuclear reactors must be cooled. There are, of course, many current applications of this released thermal energy in the production of electrical power. The heat removed from the reactor is used to convert water to steam, which is then used to drive turbines that generate electricity (Figure 24.9).

^{235}U is the only naturally occurring fissionable isotope.

As with our supply of fossil fuels, there is only a limited supply of the fissionable ^{235}U, and nuclear reactors would face a somewhat uncertain future were it not possible to generate other fissionable isotopes. In a **breeder reactor** some of the control rods are replaced with rods containing ^{238}U. Some of the neutrons produced in the fission reaction are absorbed by the ^{238}U and give the reaction,

$$^{238}_{92}U + ^{1}_{0}n \longrightarrow ^{239}_{92}U$$

The $^{239}_{92}U$ decays rapidly to yield ultimately $^{239}_{94}Pu$ which, like $^{235}_{92}U$, is fissionable.

Plutonium is extremely poisonous. Even minute amounts produce cancer with near certainty.

$$^{239}_{92}U \longrightarrow ^{239}_{93}Np + ^{0}_{-1}e$$

$$^{239}_{93}Np \longrightarrow ^{239}_{94}Pu + ^{0}_{-1}e$$

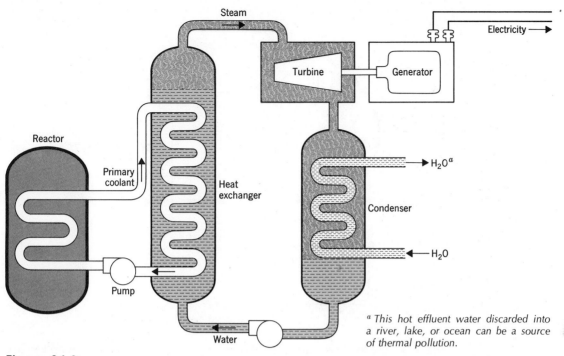

Figure 24.9

Application of nuclear fission to the production of electricity.

^a *This hot effluent water discarded into a river, lake, or ocean can be a source of thermal pollution.*

Breeder reactors thus have the useful property that they produce as much or *more* fissionable isotopes as they consume.

Nuclear reactors, in general, have the undesirable effect of producing highly radioactive by-products, some with extremely long half-lives, that are nearly impossible to dispose of safely. Nevertheless, the energy crisis in recent years has placed increasing pressure on the need to develop additional sources of electrical power, including the exploitation of nuclear energy.

Nuclear fusion

Quite the opposite of nuclear fission (fragmentation) is nuclear fusion. In this process two isotopes, usually very light ones, are brought together to form a heavier one. In so doing, a very large amount of energy is released. The fusion reaction known as the hydrogen bomb is the reaction

$$\text{}^2_1\text{H} + \text{}^3_1\text{H} \longrightarrow \text{}^4_2\text{He} + \text{}^1_0n + \text{energy}$$

This is also one of the reactions taking place on the sun and accounts for the production of a good deal of its energy.

Fusion reactions—not surprisingly—possess a high energy of activation mainly because of the electrostatic repulsion between the two nuclei that are being joined. As a result, they occur only at extremely high temperatures where the kinetic energies of the nuclei that are being joined are sufficient to overcome this repulsion. In fact, it is estimated that temperatures of approximately 200 million degrees Celsius are needed for fusion to occur (by comparison, the average temperature of the sun is $4 \times 10^6 \,°\text{C}$). The temperatures required to initiate such fusion reactions can be supplied by using an atomic (fission) bomb as a sort of "nuclear match." The energy obtained from one fusion reaction is sufficient to cause other reactions to occur; thus a chain reaction is set up, resulting in a thermonuclear explosion. In controlled fusion applications, such as the generation of electrical power, the use of an atomic bomb to initiate the fusion

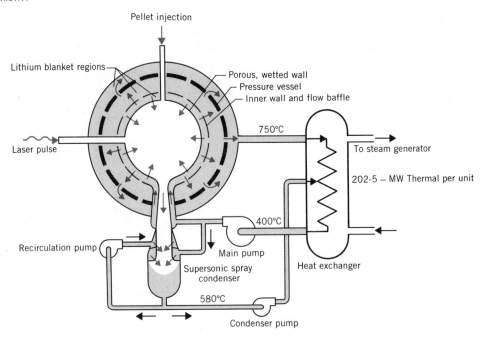

Figure 24.10

Fusion by laser implosion.

process is, to say the least, unacceptable. Currently work is centering on the use of multiple high-energy lasers to provide the high temperatures required to get the fusion process started (Figure 24.10)

As a potential source of commercial electrical power, the fusion process has several advantages over the fission reaction. First, the quantity of energy liberated in nuclear fusion is much greater than in fission. Another important advantage is that fusion reactions are relatively "clean" in the sense that the products of the fusion reaction are generally not radioactive. In fission reactions, on the other hand, many products and by-products are unstable radioactive nuclei. Fission reactors, therefore, pose a waste-disposal problem not anticipated for potential fusion reactors. As a result, scientists are attempting to establish controlled fusion reactors. One of the main obstacles to this, however, is the lack of a container that is able to hold a reaction mass that is at a temperature of $2 \times 10^{8}°C$! Current research is aimed at maintaining the reacting mass of ions (called a **plasma**) suspended and enclosed within a powerful magnetic field (See Figure 24.11)

Figure 24.11

Fusion by magnetic confinement of a plasma. This is an artists conception of a full scale Ormak reactor nearing the end of its assembly. In the foreground are two partially assembled sectors of the donut-shaped magnetic confinement apparatus. During operation the plasma will circulate through the circular tunnel under the influence of powerful magnetic forces.

24.7 NUCLEAR BINDING ENERGY

Nuclei are composed of protons and neutrons. Naturally we would expect that if we added up the mass of all the protons that go into forming a nucleus, and then added in the mass of all the neutrons, we would obtain the mass of the nucleus. Actually, however, the mass of the nucleus is always somewhat *less* than the total mass of the individual protons and neutrons. How can this be?

Within the nucleus the nuclear particles are bound together by very strong forces, the nature of which is not understood very well. Nevertheless, enormous amounts of energy have to be supplied to separate the nucleus into its component protons and neutrons. It follows that if we were to form a nucleus, this same large amount of energy would be released.

Einstein showed that mass and energy are related by his now famous equation, $E = mc^2$, where c is the speed of light. The energy liberated when the nucleus is formed comes at the expense of some of the mass of the nucleons. In other words, some mass is converted to the energy that is liberated as the nucleus is formed. Consequently, the final mass of the nucleus is less than we might have expected it to be.

Nucleons are the individual particles found in the nucleus.

The energy needed to decompose the nucleus (or the energy released when it is formed) is called the **binding energy.** The difference between the actual mass of a nucleus and the sum of the masses of its individual protons and neutrons is termed the **mass defect.** Let's look at an example.

A $^4_2$He atom is composed of two protons, two neutrons, and two electrons. These individual particles have the following masses:

$$p \qquad 1.007277 \text{ amu}$$
$$n \qquad 1.008665 \text{ amu}$$
$$e^- \qquad 0.0005486 \text{ amu}$$

The calculated weight of a $^4_2$He atom is therefore

$$(2 \times 1.007277 \text{ amu}) + (2 \times 1.008665 \text{ amu})$$
$$+ (2 \times 0.0005486 \text{ amu}) = 4.032981 \text{ amu}$$

$$\text{calculated mass } ^4_2\text{He} = 4.032981 \text{ amu}$$

The actual mass of $^4_2$He, as measured with a mass spectrometer, is 4.002603 amu. The mass defect is the difference between the computed and measured mass; that is,

$$4.032981 \text{ amu} - 4.002603 \text{ amu} = 0.030378 \text{ amu}$$

Thus when a helium nucleus is formed from two protons and two neutrons, 0.030378 amu of mass is converted to energy and is released. How much energy does this represent?

Suppose that we were to form 1 mol of He atoms. The total mass lost would then be 0.030378 g or 3.0378×10^{-5} kg. We can use Einstein's equation to calculate the energy equivalent. Using $c = 2.9979 \times 10^8$ m/s, we have

$$E = (3.0378 \times 10^{-5} \text{ kg}) \times (2.9979 \times 10^8 \text{ m/s})^2$$
$$= 2.7302 \times 10^{12} \text{ kg m}^2/\text{s}^2$$

In the SI, 1 joule = 1 kilogram meter^2/second^2. Therefore,

The binding energy of a mole of helium is about three million times larger than the energy liberated by burning a mole of methane!

$$E = 2.73 \times 10^{12} \text{ J/mol}$$
$$= 2.73 \times 10^9 \text{ kJ/mol}$$

For comparison, combustion of 1 mol of CH_4 liberates only 8.9×10^2 kJ. The binding energy, therefore, represents a huge amount of energy.

Let us now look at the average binding energy per nucleon that occurs for various atoms. This is usually expressed in energy units of *MeV per nucleon*. Nuclear physicists generally deal in energy units of MeV. One MeV is 1 million **electron volts,** where the electron volt is the kinetic energy that an electron would acquire if it were accelerated from one electrode to another across a potential difference of 1 volt. Then,

1 MeV per molecule is equivalent to 9.65×10^7 kJ/mol!

$$1 \text{ MeV} = 10^6 \text{ eV}$$

From Einstein's equation, the energy equivalence of 1 amu can be calculated. Expressed in MeV, this is

$$1 \text{ amu} \sim 931 \text{ MeV}$$

For ^4_2He the binding energy is therefore

$$0.030378 \text{ amu} \times \left(\frac{931 \text{ MeV}}{1 \text{ amu}}\right) = 28.3 \text{ MeV}$$

Since ^4_2He is composed of four nucleons ($2p$, $2n$),

$$\text{average binding energy per nucleon} = \frac{28.3 \text{ MeV}}{4 \text{ nucleons}} = 7.07 \text{ MeV/nucleon}$$

The average binding energy per nucleon varies, of course, for different atoms. Figure 24.12 is a plot of this average binding energy versus mass number.

Examination of Figure 24.12 leads to several observations. We see that atoms of intermediate mass have larger binding energies than either very light atoms or very heavy ones. This means that when light nuclei are combined, as during fusion, there is a net increase in binding energy. As the heavier nucleus is formed, an amount of energy equal to this extra binding energy is released. This is the origin of the energy given off during a fusion reaction.

We can also see in Figure 24.12 that when a very heavy nucleus is split to give lighter fragments, the fission products have a greater binding energy per nucleon. Energy equal to the increased binding energy is therefore released during the fission process.

In summary, the energy changes in both fission and fusion are the result of changes in the binding energy experienced by the protons and neutrons as the products of these nuclear reactions are formed.

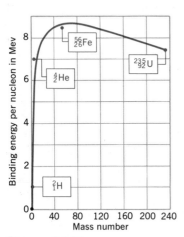

Figure 24.12

Average nuclear binding energy per nucleon for atoms of different mass number.

INDEX TO QUESTIONS AND PROBLEMS (problem numbers in **bold type**)

REVIEW QUESTIONS

24.1 What are the three main types of radiation emitted by radioactive nuclei? What are their properties?

24.2 What differences and similarities exist between β-particles and positrons?

24.3 Complete and balance the following nuclear equations:

(a) $^{81}_{36}\text{Kr} + ^{0}_{-1}e \rightarrow$?

(b) $^{104}_{47}\text{Ag} \rightarrow ^{0}_{1}e +$?

(c) $^{73}_{31}\text{Ga} \rightarrow ^{0}_{-1}e +$?

(d) $^{104}_{48}\text{Cd} \rightarrow ^{104}_{47}\text{Ag} +$?

(e) ? $+ ^{0}_{-1}e \rightarrow ^{54}_{24}\text{Cr}$

24.4 Complete and balance the following nuclear equations:

(a) $^{47}_{20}\text{Ca} \rightarrow ^{47}_{21}\text{Sc} + ?$

(b) $^{55}_{27}\text{Co} \rightarrow ^{55}_{26}\text{Fe} + ?$

(c) $^{220}_{86}\text{Rn} \rightarrow ^{116}_{84}\text{Po} + ?$

(d) $^{54}_{26}\text{Fe} + ^{1}_{0}n \rightarrow ^{1}_{1}\text{H} + ?$

(e) $^{46}_{20}\text{Ca} + ^{1}_{0}n \rightarrow ?$

24.5 Complete and balance the following nuclear equations:

(a) $^{125}_{53}\text{I} \rightarrow ^{135}_{54}\text{Xe} + ?$

(b) $^{245}_{97}\text{Bk} \rightarrow ^{4}_{2}\text{He} + ?$

(c) $^{238}_{92}\text{U} + ^{12}_{6}\text{C} \rightarrow ^{246}_{98}\text{Cf} + ?$

(d) $^{96}_{42}\text{Mo} + ^{2}_{1}\text{H} \rightarrow ^{1}_{0}n + ?$

(e) $^{20}_{8}\text{O} \rightarrow ^{20}_{9}\text{F} + ?$

24.6 Complete and balance the following nuclear equations:

(a) $^{35}_{17}\text{Cl} + ^{1}_{0}n \rightarrow ^{35}_{16}\text{S} + ?$

(b) $^{40}_{19}\text{K} \rightarrow ^{0}_{-1}e + ?$

(c) $^{98}_{42}\text{Mo} + ^{1}_{0}n \rightarrow ^{0}_{-1}e + ?$

(d) $^{229}_{90}\text{Th} \rightarrow ^{4}_{2}\text{He} + ?$

(e) $^{184}_{80}\text{Hg} \rightarrow ^{184}_{79}\text{Au} + ?$

24.7 Write balanced equations for the nuclear decay reactions below.

(a) Alpha emission by $^{11}_{5}\text{B}$

(b) Beta emission by $^{98}_{38}\text{Sr}$

(c) Neutron absorption by $^{107}_{47}\text{Ag}$

(d) Neutron emission by $^{88}_{35}\text{Br}$

(e) Electron absorption by $^{116}_{51}\text{Sb}$

(f) Positron emission by $^{70}_{33}\text{As}$

(g) Proton emission by $^{41}_{19}\text{K}$

24.8 Write nuclear equations for the following processes:

(a) $^{27}_{13}\text{Al}(\alpha,n)^{30}_{15}\text{P}$

(b) $^{209}_{83}\text{Bi}(d,n)^{210}_{84}\text{Po}$; ($d = $ deuteron, $^{2}_{1}\text{H}$)

(c) $^{15}_{7}\text{N}(p,\alpha)^{12}_{6}\text{C}$

(d) $^{12}_{6}\text{C}(p,\gamma)^{13}_{7}\text{N}$

(e) $^{14}_{7}\text{N}(\alpha,p)^{17}_{8}\text{O}$

24.9 Write nuclear equations for the following:

(a) $^{242}_{96}\text{Cm}(\alpha,n)^{245}_{98}\text{Cf}$

(b) $^{108}_{48}\text{Cd}(n,\gamma)^{109}_{48}\text{Cd}$

(c) $^{14}_{7}\text{N}(n,p)^{14}_{6}\text{C}$

(d) $^{27}_{13}\text{Al}(d,\alpha)^{25}_{12}\text{Mg}$

(e) $^{249}_{98}\text{Cf}(^{18}_{8}\text{O},4n)^{263}_{106}\text{Xe}$

24.10 Describe how the Geiger-Müller counter works.

24.11 Show that Equation 24.1 reduces to Equation 24.2 if we take $[A] = \frac{1}{2}[A]_0$.

24.12 What is the significance of the *band of stability?* What decay processes are likely to occur for nuclides that have n/p ratios that place them above the band of stability?

24.13 Elements with atomic numbers greater than 83 generally decay by either α-emission or fission. Why are the other forms of decay less likely for these nuclides?

24.14 What is a *magic number?* What magic numbers occur for protons? For neutrons? What magic numbers do we observe for orbital electrons?

24.15 In the absence of any specific information about their actual stability, rank the following nuclides in their expected order of decreasing stability.

$^{4}_{2}\text{He}$ $\quad$ $^{39}_{20}\text{Ca}$ $\quad$ $^{10}_{5}\text{B}$ $\quad$ $^{71}_{32}\text{Ge}$ $\quad$ $^{58}_{28}\text{Ni}$

24.16 What would you anticipate for the order of increasing nuclear stability for the following nuclides?

$^{3}_{2}\text{He}$ $\quad$ $^{40}_{20}\text{Ca}$ $\quad$ $^{116}_{50}\text{Sn}$ $\quad$ $^{13}_{6}\text{C}$ $\quad$ $^{192}_{77}\text{Ir}$

24.17 What chemical and physical properties would you predict for element number 114? If any of this element were formed at the time the universe came into being, where among earthly minerals would be a likely place to search for it?

24.18 If element 116 is found, what would be the expected formula of (a) its sodium salt, (b) its simple hydride, (c) its oxide? Would 116 be a metal or a nonmetal?

24.19 What would be the probable mass numbers of the most stable isotopes of element 114?

24.20 The element $^{34}_{17}\text{Cl}$ emits γ-radiation with energies of 0.14, 1.15, 2.27, 3.22, and 4.80 MeV. How does this observation support the nuclear shell theory?

24.21 Explain the difficulties that must be overcome if a stable element with $Z = 114$ is to be made by nuclear bombardment.

24.22 Technetium and promethium do not possess any stable isotopes. In light of the discussion in Section 24.3, can you comment on this observation?

24.23 Dinitrogen trioxide, N_2O_3, is largely dissociated into NO and NO_2 in the gas phase where there exists the equilibrium, $N_2O_3 \rightleftharpoons NO + NO_2$. In an effort to determine the structure of N_2O_3, a mixture of NO and *NO_2 was prepared containing isotopically labeled N in the NO_2. After a period of time the mixture was analyzed and found to contain substantial amounts of both *NO and *NO_2. Explain how this is consistent with the structure for N_2O_3 being ONONO.

24.24 The reaction $(CH_3)_2Hg + HgI_2 \rightarrow 2CH_3HgI$ is believed to occur through a transition state with the structure

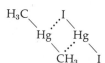

If this is so, what should be observed if CH_3HgI and *HgI_2 are mixed? Explain your answer.

24.25 *Racemization* is a chemical reaction in which one optical isomer of a compound is converted into its mirror image. One possible mechanism for the racemization of octahedral complex ions containing three bidentate ligands involves the temporary loss of one of the ligands,

$$d\text{-}[M(AA)_3] \longrightarrow \begin{pmatrix} M(AA)_2 \\ + \\ AA \end{pmatrix} \longrightarrow l\text{-}[M(AA)_3]$$

Figure 24.13

A possible mechanism for the racemization of an octahedral [M(AA)₃] complex (AA = bidentate ligand).

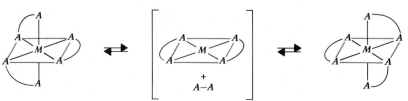

This can be pictured as shown in Figure 24.13. Can you suggest a simple experiment, making use of radioisotopes, that would be able to confirm whether or not this mechanism is operative in the racemization of the $[Co(C_2O_4)_3]^{3-}$ ion?

REVIEW PROBLEMS (More difficult problems are marked by an asterisk)

24.26 Cobalt-60 has a half-life of 5.26 years. If 1.00 g of ^{60}Co was allowed to decay, how many grams would be present after (a) one half-life, (b) three half-lives, (c) five half-lives?

24.27 Selenium-75 has a half-life of 120.0 days. If we began with 8.00 g of ^{75}Se, how many grams would remain after (a) 240 days, (b) 480 days, (c) 960 days?

24.28 The rate constant for the decay of ^{45}Ca is 4.23×10^{-3} days^{-1}. What is the half-life of ^{45}Ca expressed in days.

24.29 The rate constant for the decay of ^{36}Cl is 2.30×10^{-6} year^{-1}; what is the half-life of ^{36}Cl (expressed in years)?

24.30 The half-life of ^{51}Cr is 27.72 days. What is the rate constant for decay of ^{51}Cr in units of seconds^{-1}?

24.31 The half-life for the decay of ^{109}Cd is 470 days. What is the value for the rate constant for this decay in units of days^{-1}?

24.32 A sample of rock was found to contain 2.07×10^{-5} mol of ^{40}K and 1.15×10^{-5} mol of ^{40}Ar. If we assume that all of the ^{40}Ar came from the decay of ^{40}K, what is the age of the rock in years ($t_{1/2} = 1.3 \times 10^9$ years for ^{40}K)?

24.33 The ^{14}C content of an ancient piece of wood was found to be one-eighth of that in living trees. How many years old is this piece of wood ($t_{1/2} = 5770$ years for ^{14}C)?

24.34 When an electron and a positron (a positively charged electron) encounter each other, they destroy each other with the production of energy. How much energy, in joules, results from such an encounter? The rest mass of the electron is 9.1096×10^{-28} g.

24.35 Calculate the binding energy in kJ/mol and in MeV for the following isotopes, given the following masses: $p = 1.007277$ amu, $n = 1.008665$ amu, $e = 5.4859 \times 10^{-4}$ amu.

Isotope	Actual Atomic Mass (amu)
$^{7}_{3}$Li	7.01600
$^{14}_{7}$N	14.003074
$^{19}_{9}$F	18.99840

24.36 What is the binding energy of $^{56}_{26}$Fe expressed in MeV (atomic mass, 55.9349 amu)? Where is this on the binding energy versus mass number curve? Why don't we have to worry about an enemy that claims they have developed the iron bomb?

24.37 Calculate the energy (in kilojoules) liberated in the fusion reaction to produce 1 mol of helium from deuterium.

$$^{2}_{1}H + {}^{2}_{1}H \longrightarrow {}^{4}_{2}He$$

The accurate atomic masses are

$$^{2}_{1}H = 2.014102 \text{ amu}$$

$$^{4}_{2}He = 4.002603 \text{ amu}$$

*24.38 The following data were obtained for the decay of ^{47}Ca.

Time (hr)	cpm
0.0	4720
8.0	4485
12.0	4372
24.0	4050
48.0	3475
72.0	2983
96.0	2560

Determine (a) the rate constant for the decay, (b) the half-life of ^{47}Ca.

*24.39 A large, complex piece of apparatus has built into it a cooling system containing an unknown volume of cooling liquid. It is desired to measure the volume of the coolant without draining the lines. To the coolant was added 10 ml of methanol labeled with ^{14}C and having a specific activity of 580 cpm per gram. The coolant was permitted to circulate to assure complete mixing before a sample was withdrawn that was found to have a specific activity of 29 cpm per gram. Calculate the volume of coolant in the system in milliliters. The density of methanol is 0.792 g/ml and the density of the coolant is 0.884 g/ml.

*24.40 A complex ion of chromium(III) with oxalate ion was prepared from ^{51}Cr-labeled $K_2Cr_2O_7$, having a specific activity of 843 cpm/gram, and ^{14}C-labeled oxalic acid, $H_2C_2O_4$, having a specific activity of 345 cpm/gram. Chromium-51 decays by electron capture with the emission of a γ-ray, whereas ^{14}C is a pure β-emitter. Because of the characteristics of the β- and γ-detectors, each of these isotopes may be counted independently. A sample of the complex ion was observed to give a γ-count of 165 cpm and an β-count of 83 cpm. From these data, determine the number of oxalate ions bound to each Cr(III) in the complex ion. (*Hint:* For the starting materials calculate the cpm per mole of Cr and oxalate, respectively.)

*24.41 Compare the energies liberated in the following fusion reactions:

(a) $2{}_1^2H \rightarrow {}_2^4He$

(b) $2{}_6^{12}C \rightarrow {}_{12}^{24}Mg$

Which reaction produces the most energy per mole of product? If each were equally feasible from an engineering standpoint for producing energy by controlled fusion, which would be preferable? Actual atomic masses: ${}_1^2H = 2.014102$, ${}_2^4He = 4.002603$, ${}_{12}^{24}Mg = 23.98504$.

*24.42 How many gallons of gasoline, C_8H_{18}, having a density of 0.703 g/ml, would have to be burned (to give H_2O (l) + CO_2 (g)) to produce the same amount of energy as in the production of 1 mol of ${}_2^4He$ by the fusion reaction described in Problem 24.37? 1 gallon = 3.79 liters. ΔH_f^0 of C_8H_{18} (l) is -208.4 kJ/mol.

APPENDIX A

MATHEMATICS FOR GENERAL CHEMISTRY

For many students, solving numerical problems is often the most difficult part of any chemistry course. In this appendix we review some of the mathematical concepts that you will find useful in your study of chemistry.

A.1 THE FACTOR-LABEL METHOD OF PROBLEM SOLVING

Even after learning the principles of chemistry, students sometimes have difficulty in correctly setting up the arithmetic to give the proper numerical answer to a problem. The "factor-label" method uses the units associated with numbers as a guide in working out the arithmetic. The method is based on the idea that *units cancel from numerator and denominator in a fraction, just as numbers do*. For example, if the units *feet* appear in both numerator and denominator, they may be cancelled

$$\frac{3 \; ft}{2 \; ft} = \frac{3}{2}$$

Numerical problem solving uses this idea by employing valid relationships between units to create **conversion factors.** For instance, to convert 4 *yards* into *feet* the relationship,

$$3 \text{ ft} = 1 \text{ yard}$$

is used to create a conversion factor (a fraction) by which the 4 yards is multiplied. We do this by dividing both sides of this equation by "1 yard."

$$\frac{3 \text{ ft}}{1 \text{ yard}} = \frac{1 \; yard}{1 \; yard} = 1$$

Notice that this fraction is equal numerically to 1. If we multiply a quantity by 1, we do not change its magnitude. Thus we can multiply 4 yards by this fraction without altering the length. The only effect is to change units.

$$4 \; yards \left(\frac{3 \text{ ft}}{1 \; yard} \right) = 12 \text{ ft}$$

Note also that the factor was constructed deliberately so that *yards* cancelled and only the desired units, *feet*, remained. Had we inverted this conversion factor, we would obtain the wrong numerical answer *and* the wrong units.

$$4 \text{ yard} \left(\frac{1 \text{ yard}}{3 \text{ ft}} \right) = \frac{4}{3} \frac{\text{yard}^2}{\text{ft}}$$

Creation of a conversion factor can be accomplished from any *valid* relationship between a set of units. This can be an equality, as in the relationship between feet and yards (that is 3 ft *equal* 1 yard). It can also be an equivalency. For instance, for a student who earns 5 dollars per hour, there is an equivalence between dollars and time.

5 dollars are equivalent to 1 hour

We will use the symbol $\sim$ to stand for "are equivalent to." Thus, in the example just cited,

5 dollars $\sim$ 1 hr

If this student works 12 hr, we can use the dollar-hour relationship to construct a conversion factor that allows us to calculate the student's pay (before taxes!)

$$12 \; \cancel{\text{hrs}} \left(\frac{5 \text{ dollars}}{1 \; \cancel{\text{hr}}} \right) = 60 \text{ dollars}$$

Note that hours cancel. You will see many examples in the text in which numerical problems are solved using this "factor-label" technique.

A.2 EXPONENTIAL NOTATION (SCIENTIFIC NOTATION)

Quite often in science it is necessary to deal with numbers that are very large, such as Avogadro's number,

602,300,000,000,000,000,000,000

or numbers that are very small, such as the mass of a single molecule of water,

0.000 000 000 000 000 000 000 03 g

These numbers are very cumbersome and difficult to work with without making mistakes in arithmetic computations. To aid us in handling these large and small numbers, a system called either **exponential notation** or **scientific notation** is employed. In this system, a number is expressed as a decimal part multiplied by 10 raised to an appropriate power. Thus

$$200 = 2 \times 10 \times 10 = 2 \times 10^2$$

$$205{,}000 = 2.05 \times 100{,}000 = 2.05 \times (10 \times 10 \times 10 \times 10 \times 10)$$

$$= 2.05 \times 10^5$$

To determine the exponent on the 10, we can simply count the number of places the decimal must be moved to produce the number that precedes the 10 when the number is expressed in the scientific notation

$$205 \; 000 \; = 2.05 \times 10^5$$
5 places

Note that the exponent on the 10 is positive when the decimal is moved to the left. When it is moved to the right, the exponent is negative.

$$0 \; 000000315 = 3.15 \times 10^{-7}$$
7 places

Most students today perform their calculations using a calculator, and virtually every "scientific calculator" is designed to handle calculations involving numbers expressed in scientific notation. Nevertheless, someday the batteries in your calculator may go dead, so it is a good idea to know the rules for arithmetic using scientific notation.

Multiplication

In multiplication, the decimal portions of the number are multiplied and the exponents on the 10 are *added* algebraically.

$$(2.0 \times 10^4) \times (3.0 \times 10^3) = (2.0 \times 3.0) \times 10^{(4+3)} = 6.0 \times 10^7$$

$$(4.0 \times 10^8) \times (-2.0 \times 10^{-5}) = (4.0 \times (-2.0)) \times 10^{(8+(-5))} = -8.0 \times 10^3$$

Division

The decimal portions are divided, and the exponent on 10 in the denominator is *subtracted algebraically* from the exponent on 10 in the numerator.

$$\frac{8.0 \times 10^7}{4.0 \times 10^3} = \left(\frac{8.0}{4.0}\right) \times 10^{(7-3)} = 2.0 \times 10^4$$

$$\frac{6.0 \times 10^5}{2.0 \times 10^{-3}} = \left(\frac{6.0}{2.0}\right) \times 10^{(5-(-3))} = 3.0 \times 10^8$$

$$\frac{9.0 \times 10^{-4}}{3.0 \times 10^{-6}} = \left(\frac{9.0}{3.0}\right) \times 10^{(-4-(-6))} = 3.0 \times 10^2$$

You have probably noticed that the usual practice is to express a number with the decimal point located between the first and second digit. There are, of course, other ways that these numbers can be written that are all equivalent, and you will undoubtedly find occasions where it is convenient to use a number in other than its standard form. An example of a few equivalent expressions of the same number are

$$3.15 \times 10^{-7} = 315 \times 10^{-9} = 0.0315 \times 10^{-5}$$

Notice that in converting from one to another, one part of the number is increased while the other is decreased. For instance, to change 8.25×10^6 to 825×10^4, multiply *and* divide by 100 (or 10^2)

$$8.25 \times 10^6 \left(\frac{100}{100}\right) = (8.25 \times 100) \times \left(\frac{10^6}{10^2}\right) = 825 \times 10^4$$

Addition and subtraction

When carrying out addition and subtraction, each quantity must first be written with the same power of 10. Then addition or subtraction is performed on the decimal parts; the power of 10 remains the same. For example,

$$(2.17 \times 10^5) + (3.0 \times 10^4) = ?$$

If we express both numbers with the same power of 10, we have

$$
\begin{array}{c}
2.17 \times 10^5 \\
+\,0.30 \times 10^5 \\
\hline
2.47 \times 10^5
\end{array}
\quad \text{or} \quad
\begin{array}{c}
21.7 \times 10^4 \\
+\,3.0 \times 10^4 \\
\hline
24.7 \times 10^4
\end{array}
$$

Taking a root

To extract a root (e.g., the square root), the exponent on the 10 is made to be divisible by the desired root. For instance, to take the square root of 3.7×10^7, we first change the number so that the power of 10 is divisible by 2. Then we take the square root of the decimal part and divide the exponent by 2.

$$\sqrt{3.7 \times 10^7} = \sqrt{37 \times 10^6} = \sqrt{37} \times 10^3 = 6.1 \times 10^3$$

A.3 LOGARITHMS

A logarithm is an exponent! **Common logarithms** are exponents to which 10 must be raised to give a specified number. For instance, the log (100) = 2 because $10^2 = 100$. Similarly, log (1000) = log (10^3) = 3.

Since logarithms are exponents, when we perform mathematical operations the same rules that apply to exponents also apply to logarithms. Thus we have

Multiplication $\begin{cases} \text{add exponents} \\ \text{add logarithms} \end{cases}$

Division $\begin{cases} \text{subtract exponents} \\ \text{subtract logarithms} \end{cases}$

For example,

$$10^3 \times 10^4 = 10^{3+4} = 10^7$$
$$\log(10^3 \times 10^4) = \log(10^3) + \log(10^4) = 3 + 4 = 7 = \log(10^7)$$

Similarly, for division

$$\frac{10^8}{10^6} = 10^{8-6} = 10^2$$
$$\log\left(\frac{10^8}{10^6}\right) = \log(10^8) - \log(10^6) = 8 - 6 = 2 = \log(10^2)$$

For decimal numbers between 1 and 10 their logarithms lie between 0 and 1, since

$$\log(1) = 0 \qquad (1 = 10^0)$$
$$\log(10) = 1 \qquad (10 = 10^1)$$

For example, log 2 = 0.3010 or

$$10^{0.3010} = 2$$

The common logarithm of 2 and other numbers between 1 and 10 can be obtained from the table of logarithms in Appendix B.

To use this table to find the logarithm of a number, we use the extreme left column to locate the first two digits of the number, and the top horizontal row to locate the third digit. The value in the table corresponding to these is the logarithm of our number. For example, if we want log (4.61), we would locate 46 in the left column and proceed to the right until we were in the column headed by 1. The answer is

$$\log(4.61) = 0.6637$$

This table is extremely easy to use as long as our numbers are expressed in this fashion, that is, as a decimal number between 1 and 10. If the number whose logarithm we seek does not appear this way, we can first express the number in exponential notation and then take its logarithm. For example, what is log(728)?

$$\log(728) = \log(7.28 \times 10^2)$$

$$\log(7.28 \times 10^2) = \log(7.28) + \log(10^2)$$

$$\log(7.28) = 0.8621 \quad \text{(from table)}$$
$$+\log(10^2) = \underline{2.0000}$$

therefore, $\quad\quad\quad\quad \log(728) = 2.8621$

What would be the value of $\log(0.00583)$? Once again we first express the number in exponential notation:

$$\log(0.00583) = \log(5.83 \times 10^{-3})$$

$$\log(5.83 \times 10^{-3}) = \log(5.83) + \log(10^{-3})$$

$$\log(5.83) = +0.7657$$
$$+\log(10^{-3}) = \underline{-3.0000}$$

Adding these algebraically, we get

$$\log(0.00583) = -2.2343$$

Sometimes it is necessary to obtain the number whose logarithm is known. This is called taking the **antilogarithm.** The procedure is simply the reverse of that given above. For example, suppose that we wish to find the number whose logarithm is 3.253.

$$\log x = 3.253$$

First, we divide the number into two parts, a positive integer and a decimal.

$$3.253 = 3 + 0.253 = 0.253 + 3$$

$$\log x = (0.253 + 3)$$

We locate 0.253 in the body of the log table and find that it is the log of 1.79; we also know that 3 is the log of 10^3. Therefore,

$$\log x = \log(1.79) + \log(10^3) = \log(1.79 \times 10^3)$$

$$x = 1.79 \times 10^3$$

If the logarithm of the number is negative, the procedure for taking the antilogarithm is just slightly different. For example, suppose we have the problem,

$$\log x = -8.475$$

Once again, we divide the logarithm into two parts—a *positive decimal* and a *negative integer*.

$$-8.475 = (+0.525) + (-9)$$

Therefore,

$$\log x = (+0.525) + (-9)$$

Next, we locate 0.525 in the body of the log table and find that it is the log of 3.35. The -9 becomes the exponent on the 10, so our answer is

$$x = 3.35 \times 10^{-9}$$

Natural logarithms

A system of logarithms encountered frequently in the sciences, known as natural logarithms, has as its base $e = 2.71828. \ldots$. In other words, natural logarithms are exponents to which e must be raised to give a number. The relationship between common logs and natural logs is seen below

$$\log_{10}(10) = 1 \quad \text{or} \quad 10^1 = 10$$

$$\log_e(10) = 2.303 \quad \text{or} \quad e^{2.303} = 10$$

With common logarithms we usually omit the base and write log 10 = 1. With natural logarithms the base e is omitted, and they are written

$$\ln 10 = 2.303$$

The conversion from base e to base 10 logarithm is accomplished by the equation

$$\ln x = 2.303 \log x$$

A.4 THE QUADRATIC EQUATION

When an equation can be written in the form

$$ax^2 + bx + c = 0$$

in which the coefficients a, b, and c are known, two values (called roots) of the variable x can be obtained by substituting the values of a, b, and c into the expression

$$x = \frac{-b \pm \sqrt{b^2 - 4ac}}{2a}$$

For example, given the equation

$$x^2 - 5x + 4 = 0$$

what is the value of x? In this equation $a = 1$, $b = -5$, and $c = 4$. Thus

$$x = \frac{-(-5) \pm \sqrt{(-5)^2 - 4\,(1)(4)}}{2\,(1)} = \frac{5 \pm \sqrt{25 - 16}}{2}$$

$$= \frac{5 \pm \sqrt{9}}{2} = \frac{5 \pm 3}{2}$$

Therefore,

$$x = \frac{2}{2} = 1 \quad \text{and} \quad x = \frac{8}{2} = 4$$

Both values of x are mathematically correct. Usually when a quadratic equation is encountered in a chemical problem, only one of the roots has any real significance. Generally, the other root will be clearly meaningless—for instance, a negative concentration, which is impossible (you can't have a smaller amount of matter than no matter at all!).

A.5 ELECTRONIC CALCULATORS

Today much of the tiresome work of arithmetic is relieved by the use of small, hand-held electronic calculators. Scientific calculators possessing remarkable computational power are available for less than $20 (the price of a reasonably good slide rule not too many years ago), but to get the most out of them you should be aware of some simple mathematical relationships. Those that are most useful to you in chemistry are mentioned below. Since operational procedures differ on various calculators, you will have to refer to the direction booklet that accompanies your calculator for specific instructions about how to apply these relationships.

Logarithms and antilogarithms

We've seen that a logarithm is an exponent and that there are two systems of logarithms generally encountered, base e and base 10. If your calculator possesses logarithm capabilities, you will probably find a key labeled LN for base e (natural) logarithms and a key labeled LOG for base 10 (common) logarithms. Generally, if a number is entered and the LN key depressed, the display will show the natural log of that number. The common log would have appeared had you depressed the LOG key.

Useful relationships among logarithms are

$$10^{\log X} = X$$

$$e^{\ln X} = X$$

For example, log 2 = 0.3010, ln 2 = 0.6931.

$$10^{\log 2} = 10^{0.3010} = 2$$

$$e^{\ln 2} = e^{0.6931} = 2$$

These provide a means to obtain the antilogarithms. If you have the natural log of a number, enter it and depress the "e^x" key. If you have the common log, enter it and depress the "10^x" key. In each case you will obtain the antilogarithm.

If your calculator does not have a "10^x" key, but does have an "x^y" (or "y^x") key, enter 10 and raise it to an exponent that corresponds to the common log. The result will be the antilogarithm.

Exponents and roots

Most calculators have X^2 and $\sqrt{X}$ keys, and these operations are simple. For higher powers and roots you can use either of two methods.

1. **Using the x^y key.** To compute, $X = a^b$, enter a and raise it to the power b. To compute $X = \sqrt[b]{a}$, enter a and raise it to the power, $1/b$. For example,

   $$X = 2^3 = 8$$

   $$X = \sqrt[3]{2} = 2^{1/3} = 2^{0.3333...3} = 1.25992$$

2. **Using logarithms.** To compute $X = a^b$ with natural logarithms, we use the relationship that

   $$\ln X = \ln a^b = b \ln a$$

   Therefore,

   $$X = e^{\ln X}$$

   $$X = e^{b \ln a}$$

Let's suppose that we wished to compute 3^5. On a typical calculator we would perform this computation in the following sequence:

1. Take ln 3	ln 3 = 1.098612
2. Multiply ln 3 by 5	5 ln 3 = 5.493061
3. Raise e to this exponent	$e^{5 \ln 3} = 243$

$$3^5 = 243$$

These calculations, of course, can also be done using common logarithms, in which case

$$X = 10^{b \log a}$$

To compute a root, $X = \sqrt[b]{a}$, we find that

$$X = e^{(\ln a)/b}$$

For example, suppose that we wished to calculate $\sqrt[5]{12}$. The sequence of operations is

1. Take ln 12 $\ln 12 = 2.484907$

2. Divide ln 12 by 5 $\dfrac{(\ln 12)}{5} = 0.496981$

3. Raise e to this exponent $e^{(\ln 12)/5} = 1.643752$

$$\sqrt[5]{12} = 1.643752$$

APPENDIX B

COMMON LOGARITHMS

	0	1	2	3	4	5	6	7	8	9
10	0000	0043	0086	0128	0170	0212	0253	0294	0334	0374
11	0414	0453	0492	0531	0569	0607	0645	0682	0719	0755
12	0792	0828	0864	0899	0934	0969	1004	1038	1072	1106
13	1139	1173	1206	1239	1271	1303	1335	1367	1399	1430
14	1461	1492	1523	1553	1584	1614	1644	1673	1703	1732
15	1761	1790	1818	1847	1875	1903	1931	1959	1987	2014
16	2041	2068	2095	2122	2148	2175	2201	2227	2253	2279
17	2304	2330	2355	2380	2405	2430	2455	2480	2504	2529
18	2553	2577	2601	2625	2648	2672	2695	2718	2742	2765
19	2788	2810	2833	2856	2878	2900	2923	2945	2967	2989
20	3010	3032	3054	3075	3096	3118	3139	3160	3181	3201
21	3222	3243	3263	3284	3304	3324	3345	3365	3385	3404
22	3424	3444	3464	3483	3502	3522	3541	3560	3579	3598
23	3617	3636	3655	3674	3692	3711	3729	3747	3766	3784
24	3802	3820	3838	3856	3874	3892	3909	3927	3945	3962
25	3979	3997	4014	4031	4048	4065	4082	4099	4116	4133
26	4150	4166	4183	4200	4216	4232	4249	4265	4281	4298
27	4314	4330	4346	4362	4378	4393	4409	4425	4440	4456
28	4472	4487	4502	4518	4533	4548	4564	4579	4594	4609
29	4624	4639	4654	4669	4683	4698	4713	4728	4742	4757
30	4771	4786	4800	4814	4829	4843	4857	4871	4886	4900
31	4914	4928	4942	4955	4969	4983	4997	5011	5024	5038
32	5051	5065	5079	5092	5105	5119	5132	5145	5159	5172
33	5185	5198	5211	5224	5237	5250	5263	5276	5289	5302
34	5315	5328	5340	5353	5366	5378	5391	5403	5416	5428
35	5441	5453	5465	5478	5490	5502	5514	5527	5539	5551
36	5563	5575	5587	5599	5611	5623	5635	5647	5658	5670
37	5682	5694	5705	5717	5729	5740	5752	5763	5775	5786
38	5798	5809	5821	5832	5843	5855	5866	5877	5888	5899
39	5911	5922	5933	5944	5955	5966	5977	5988	5999	6010
40	6021	6031	6042	6053	6064	6075	6085	6096	6107	6117
41	6128	6138	6149	6160	6170	6180	6191	6201	6212	6222
42	6232	6243	6253	6263	6274	6284	6294	6304	6314	6325
43	6335	6345	6355	6365	6375	6385	6395	6405	6415	6425
44	6435	6444	6454	6464	6474	6484	6493	6503	6513	6522
45	6532	6542	6551	6561	6571	6580	6590	6599	6609	6618
46	6628	6637	6646	6656	6665	6675	6684	6693	6702	6712
47	6721	6730	6739	6749	6758	6767	6776	6785	6794	6803
48	6812	6821	6830	6839	6848	6857	6866	6875	6884	6893
49	6902	6911	6920	6928	6937	6946	6955	6964	6972	6981
50	6990	6998	7007	7016	7024	7033	7042	7050	7059	7067
51	7076	7084	7093	7101	7110	7118	7126	7135	7143	7152
52	7160	7168	7177	7185	7193	7202	7210	7218	7226	7235
53	7243	7251	7259	7267	7275	7284	7292	7300	7308	7316
54	7324	7332	7340	7348	7356	7364	7372	7380	7388	7396

	0	1	2	3	4	5	6	7	8	9
55	7404	7412	7419	7427	7435	7443	7451	7459	7466	7474
56	7482	7490	7497	7505	7513	7520	7528	7536	7543	7551
57	7559	7566	7574	7582	7589	7597	7604	7612	7619	7627
58	7634	7642	7649	7657	7664	7672	7679	7686	7694	7701
59	7709	7716	7723	7731	7738	7745	7752	7760	7767	7774
60	7782	7789	7796	7803	7810	7818	7825	7832	7839	7846
61	7853	7860	7868	7875	7882	7889	7896	7903	7910	7917
62	7924	7931	7938	7945	7952	7959	7966	7973	7980	7987
63	7993	8000	8007	8014	8021	8028	8035	8041	8048	8055
64	8062	8069	8075	8082	8089	8096	8102	8109	8116	8122
65	8129	8136	8142	8149	8156	8162	8169	8176	8182	8189
66	8195	8202	8209	8215	8222	8228	8235	8241	8248	8254
67	8261	8267	8274	8280	8287	8293	8299	8306	8312	8319
68	8325	8331	8338	8344	8351	8357	8363	8370	8376	8382
69	8388	8395	8401	8407	8414	8420	8426	8432	8439	8445
70	8451	8457	8463	8470	8476	8482	8488	8494	8500	8506
71	8513	8519	8525	8531	8537	8543	8549	8555	8561	8567
72	8573	8579	8585	8591	8597	8603	8609	8615	8621	8627
73	8633	8639	8645	8651	8657	8663	8669	8675	8681	8686
74	8692	8698	8704	8710	8716	8722	8727	8733	8739	8745
75	8751	8756	8762	8768	8774	8779	8785	8791	8797	8802
76	8808	8814	8820	8825	8831	8837	8842	8848	8854	8859
77	8865	8871	8876	8882	8887	8893	8899	8904	8910	8915
78	8921	8927	8932	8938	8943	8949	8954	8960	8965	8971
79	8976	8982	8987	8993	8998	9004	9009	9015	9020	9025
80	9031	9036	9042	9047	9053	9058	9063	9069	9074	9079
81	9085	9090	9096	9101	9106	9112	9117	9122	9128	9133
82	9138	9143	9149	9154	9159	9165	9170	9175	9180	9186
83	9191	9196	9201	9206	9212	9217	9222	9227	9232	9238
84	9243	9248	9253	9258	9263	9269	9274	9279	9284	9289
85	9294	9299	9304	9309	9315	9320	9325	9330	9335	9340
86	9345	9350	9355	9360	9365	9370	9375	9380	9385	9390
87	9395	9400	9405	9410	9415	9420	9425	9430	9435	9440
88	9445	9450	9455	9460	9465	9469	9474	9479	9484	9489
89	9494	9499	9504	9509	9513	9518	9523	9528	9533	9538
90	9542	9547	9552	9557	9562	9566	9571	9576	9581	9586
91	9590	9595	9600	9605	9609	9614	9619	9624	9628	9633
92	9638	9643	9647	9652	9657	9661	9666	9671	9675	9680
93	9685	9689	9694	9699	9703	9708	9713	9717	9722	9727
94	9731	9736	9741	9745	9750	9754	9759	9763	9768	9773
95	9777	9782	9786	9791	9795	9800	9805	9809	9814	9818
96	9823	9827	9832	9836	9841	9845	9850	9854	9859	9863
97	9868	9872	9877	9881	9886	9890	9894	9899	9903	9908
98	9912	9917	9921	9926	9930	9934	9939	9943	9948	9952
99	9956	9961	9965	9969	9974	9978	9983	9987	9991	9996

APPENDIX C

ANSWERS TO EVEN-NUMBERED NUMERICAL PROBLEMS

CHAPTER 1

1.26 (a) 7.7 (b) 73.3 (c) 0.785 (d) 3.478 (e) 81.4

1.28 (a) 4.0×10^{-4} (b) 3×10^{-10} (c) 2.146×10^{-3}
(d) 3.28×10^{-5} (e) 9.1×10^{-13}

1.30 (a) 3.0×10^3 m (b) 3.00×10^3 m (c) 3.0×10^5 cm

1.32 (a) 5.56×10^3 (b) 2.9×10^4 (c) 1.49×10^{10}
(d) 3.8×10^{-6} (e) 9.0×10^{-31}

1.34 (a) 2140 (b) 12100 (c) 41 (d) 5.9 (e) 261

1.36 (a) 140 cm (b) 2.800 m (c) 0.185 liter (d) 1.8×10^{-2} kg
(e) 8.4 m^2 (f) 6.34×10^6 in. (g) 14 mi/hr (h) 4×10^9 m^3
(i) 1.8×10^3 cm/s (j) 25 dm^3

1.38 $N = kg\ m\ s^{-2}$

1.40 56 km

1.42 260 lb

1.44 4.54 g/ml

1.46 (a) 136 g (b) 13.7 cm^3

1.48 (a) 9.647 ml (b) 0.8489 g/ml

1.50 (a) 0.787 (b) 0.787 g/ml

1.52 140 g carbon

1.54 carbon, 0.6322 amu; hydrogen, 0.05305 amu

1.56 1.54×10^3 J

1.58 $\dfrac{gC}{gF} = 0.158$ $\qquad \dfrac{gC}{gCl} = 0.0857$ $\qquad \dfrac{gF}{gCl} = 0.542$

1.60 NO_2

1.62 35.5 amu

1.64 $-78°C = -108°F = 195$ K

1.66 1.2×10^4 cal, 5.0×10^4 J

1.68 6.30×10^4 J, 63.0 kJ

1.70 6.5×10^5 cal

1.72 $°N = (°C - 80) \times \frac{100}{138}$; F.p. $= -58°N$, b.p. $= 14°N$

CHAPTER 2

2.16 (a) 2.00 mol S (b) 0.720 mol Fe (c) 6.00 mol S
(d) 3.00 mol FeS_2

2.18 3.00 mol S

2.20 1.00 mol CO_2

2.22 (a) 24.3 g Mg (b) 12.0 g C (c) 55.8 g Fe (d) 35.5 g Cl
(e) 32.1 g S (f) 87.8 g Sr

2.24 (a) 40.31 (b) 110.98 (c) 208.22 (d) 135.02 (e) 163.94

2.26 262 g

2.28 1.91×10^3 g

2.30 2.88 mol

2.32 0.870 mol

2.34 1.68 mol

2.36 7.80 mol FeS_2

2.38 1.64×10^{15} atom C

2.40 0.844 g Cu

2.42 (a) 92.27 % C, 7.73 % H
(b) 52.14 % C, 13.13 % H, 34.73 % 0
(c) 26.58 % K, 35.35 % Cr, 38.07 % 0
(d) 63.34 % Xe, 36.66 % F
(e) 40.04 % Ca, 12.00 % C, 47.96 % 0

2.44 2.13 g H

2.46 SO_3

2.48 CCl_2

2.50 SCl_2

2.52 $C_8H_8O_3$

2.54 CH_3NO

2.56 (a) $Na_2S_4O_6$ (b) $C_6H_4Cl_2$ (c) $C_6H_3Cl_3$ (d) $Na_{12}Si_6O_{18}$
(e) $Na_3P_3O_9$

2.58 (a) 2.50 mol C_2H_2
(b) 13.0 g C_2H_2
(c) 6.40 mol H_2O
(d) 79.7 g $Ca(OH)_2$

2.60 (a) $P_4 + 5\ O_2 \rightarrow P_4O_{10}$
(b) 0.100 mol P_4O_{10}
(c) 21.8 g P_4
(d) 19.4 g P_4

2.62 (a) 52.5 mol CO
(b) 1.50 mol Fe_2O_3
(c) 45.5 g Fe_2O_3
(d) 0.911 mol CO
(e) 24.7 g Fe

2.64 1575 kg DDT

2.66 (a) $(CH_3)_2NNH_2 + 2N_2O_4 \rightarrow 4H_2O + 2CO_2 + 3N_2$
(b) 153 kg N_2O_4

2.68 (a) 1.35 mol C_2H_2
(b) 84.1 g C_3H_3Cl
(c) 1.9 g HCl

2.70 (a) 30.1 g C_6H_5Br
   (b) 0.828 g C_6H_6
   (c) 28.5 g C_6H_5Br
   (d) 94.7 %
2.72 20 g $C_2H_2Br_4$, 24.9 g $C_2H_2Br_2$
2.74 (a) 12.5 g (b) 12.8 g CO_2
2.76 (a) 63 g CCl_4 (b) 200 g CCl_4 (c) 32 % (d) 269 g Cl_2
2.78 3660 lb NH_3
2.80 (a) 0.551 M NH_4Cl (b) 0.569 M $AgNO_3$
   (c) 0.300 M KCl   (d) 0.0229 M $NaHCO_3$
2.82 (a) 0.0625 mol Li_2CO_3
   (b) 11.6 g Li_2CO_3
   (c) 40.0 ml soln
   (d) 4.33 ml
2.84 79.1 g $Ca(C_2H_3O_2)_2$
2.86 18.5 g $MgSO_4 \cdot 7H_2O$

CHAPTER 3

3.74 151.9 amu
3.76 207 amu
3.78 51.82 % ^{107}Ag, 48.18 % ^{109}Ag
3.80 2.965 m, 341 m
3.82 9.09×10^{-28} g
3.84 7459.85 nm, 4653.77 nm
3.86 1.94×10^{-18} J
3.88 2×10^{-18} J, 1×10^{-5} m
3.90 0.095 mi
3.92 2.37×10^5 tons
3.94 3.0×10^{19} seconds

CHAPTER 4

4.52 101.4 kcal are evolved
4.54 -2265 kJ/mol
4.56 Δ(bond energy) increases from LiH $\rightarrow$ RbH, therefore Δ(electronegativity) increases from LiH $\rightarrow$ RbH. This means that electronegativity must decrease from Li $\rightarrow$ Rb.

CHAPTER 6

6.50 (a) 0.750 M (b) 0.992 M (c) 0.556 M (d) 1.36 M
   (e) 0.962 M
6.52 4.77 g Na_2CO_3
6.54 24.01 M
6.56 0.0825 %, 0.0106 M
6.58 108 ml
6.60 (a) 850 ml (b) 567 ml (c) 283 ml
6.62 150 ml
6.64 (a) $AgNO_3 + NaCl \rightarrow AgCl(s) + NaNO_3$
   (b) 4.00×10^{-3} mol AgCl
   (c) 0.573 g AgCl
   (d) Na^+, 0.120 M; Cl^-, 0.0400 M; NO_3^-, 0.0800M
6.66 (a) 2.80 g $BaSO_4$
   (b) SO_4^{2-}, 0.129 M; Cl^-, 0.253 M; Fe^{3+}, 0.171 M

6.68 (a) 116 g/mol (b) $C_6H_{12}O_2$
6.70 $TiCl_3$
6.72 (a) 6.66×10^{-4} mol HCl
   (b) 0.00933 g N
   (c) 18.7 % N; Same % N as glycine
6.74 34.2 % $CuBr_2$
6.76 0.400 eq
6.78 0.280 eq
6.80 21.6 g Na_2CrO_4
6.82 1.96 g
6.84 (a) 0.452 N (b) 0.50 N (c) 0.300 N (d) 1.42 N
   (e) 0.538 N (f) 0.338 N
6.86 37.5 ml
6.88 26.8 %
6.90 36.0 % $MgCO_3$, 64.0 % $CaCO_3$
6.92 48.0 %
6.94 12.5 ml
6.96 67 ml
6.98 204 ml
6.100 1.56 liter

CHAPTER 7

7.30 158 torr
7.32 239 mm
7.34 882 torr
7.36 288 ml
7.38 802 torr
7.40 61.4 cm
7.42 1.6 liter
7.44 374 torr
7.46 35 psig
7.48 375°C
7.50 659 torr
7.52 1.54 g/liter
7.54 850 torr
7.56 680 torr
7.58 81.5 ml
7.60 $X_{N_2} = 0.749$, $X_{O_2} = 0.153$, $X_{CO_2} = 0.037$, $X_{H_2O} = 0.062$
7.62 (a) 4.48 liter (b) 3.92 liter (c) 3.36 liter
7.64 2.59 g/liter
7.66 8.312 Pa m^3/mol K
7.68 4.1×10^4 torr, 54 atm
7.70 403 ml
7.72 (a) 65.8 % C, 15.2 % H, 19.2 % N
   (b) $C_4H_{11}N$
7.74 200 ml N_2 and 600 ml H_2
7.76 (a) 0.144 g (b) 0.368 g
7.78 He effuses 2.25 times faster
7.80 1.99 g/mol
7.82 0.9925 atm (ideal gas, P = 1.0000 atm)
7.84 19.6 ml
7.86 1560 torr (to 3 significant figures)
7.88 255 ml
7.90 530 ml
7.92 53.1 liter
7.94 22.38 liter

CHAPTER 8

8.80 11.0 kcal

8.82 59.3 kJ/mol, 14.2 kcal/mol

8.84 600 cal

8.86 40.3°C

8.88 206 pm, 153 pm, 121 pm

8.90 124.9 pm

8.92 144.20 pm

8.94 176 pm

8.96 (a) 7.50 g/cm^3 (b) 9.70 g/cm^3 (c) 10.6 g/cm^3; face centered cubic

8.98 primitive, 0.48 Å^3; body centered, 0.50 Å^3; face centered 0.75 Å^3

8.100 for primitive cubic, calculated density = 3.99 g/cm^3 which matches experimentally measured density.

8.102 4 formula units of CaF_2

CHAPTER 10

10.30 (a) 36.0 % C_6H_6, 64.0 % C_7H_8
(b) $X_{C_6H_6}$ = 0.399, $X_{C_7H_8}$ = 0.601
(c) 7.20 m C_6H_6

10.32 X_{CuCl_2} = 7.4 × 10^{-3}, 0.42 m $CuCl_2$, 5.3 % $CuCl_2$

10.34 52.6 %, 18.5 m

10.36 X_{NaCl} = 0.101, w_{NaCl} = 0.268

10.38 (a) 6.17 × 10^{-3} M (b) 6.17 × 10^{-3} m

10.40 15.3 %, $X_{NH_4NO_3}$ = 0.0390, X_{H_2O} = 0.9610

10.42 17.9 m, 9.10 M

10.44 24.1 kJ evolved

10.46 88g

10.48 572 torr

10.50 0.216 g/liter

10.52 0.188

10.54 549 torr

10.56 128 g/mol, $C_8H_4N_2$

10.58 10.9 g, 100.21°C

10.60 −0.200°C

10.62 2.0 × 10^3 g/mol

10.64 1.03 (i factors for weak electrolytes are slightly larger than 1.00.)

10.66 28.4 atm

10.68 13.5 g

10.70 69 torr

10.72 (a) 0.0285 (b) 0.0294 (c) 280 g/mol

10.74 7.6 %

CHAPTER 11

11.38 ΔE = 0, q_1 = w_1 = 75.0 liter atm, q_2 = w_2 = 130 liter atm

11.40 w = −0.0182 kcal, q = 3.00 kcal, ΔE_{system} = 3.02 kcal, ΔE_{surr} = −3.02 kcal

11.42 5.29 × 10^4 cal, ΔE = −529 kcal/mol

11.44 ΔH = −77.2 kcal/mol, ΔE = −77.8 kcal/mol

11.46 −39.4 kJ or −3940 kJ/mol

11.48 ΔH = −106 kJ/mol

11.50 −213.1 kcal

11.52 (a) −854 kJ (b) −1429 kJ (c) −402 kJ (d) −87 kJ (e) −136 kJ

11.54 −94 kcal

11.56 −17 kJ

11.58 ΔH_f° = 54.2 kcal/mol = 227 kJ/mol

11.60 (a) −113 kJ (b) 306 kJ (c) −107 kJ

11.62 5.12 × 10^5 cal evolved

11.64 24 kJ

11.66 2600 g H_2O

11.68 19.8 liters

11.70 395 nm

11.72 calculated ΔH_f° = 242 kJ/mol. Resonance energy = 160 kJ/mol. Such species are much more stable than predicted.

11.74 calculated ΔH_f° = −127 kJ/mol; from table, ΔH_f° = −104 kJ/mol

11.76 59°C

11.78 (a) −836 kJ (b) −346 kJ (c) −101 kJ

11.80 −688 kcal/mol

11.82 (a) ΔG° = +12.4 kcal (No) (b) ΔG° = −13.8 kcal (Yes) (c) ΔG° = −122.7 kcal (Yes) (d) ΔG° = +22.1 kcal (No) (e) ΔG° = −8.3 kcal (Yes) (f) ΔG° = +393 kcal (No)

CHAPTER 12

12.42 Rate (CO_2) = 0.16 mol liter^{-1}s^{-1}, Rate (H_2O) = 0.32 mol liter^{-1}s^{-1}

12.44 (a) 2.35 × 10^{-6} mol liter^{-1}s^{-1} (b) 1.91 × 10^{-7} mol liter^{-1}s^{-1}

12.46 (a) 1.0 × 10^{-2} mol liter^{-1}s^{-1} (b) 2.0 × 10^{-2} mol liter^{-1}s^{-1} (c) 4.1 × 10^{-2} mol liter^{-1}s^{-1}

12.48 Rate = $k[NO_2][O_3]$, k = 4.4 × 10^7 liter mol^{-1}s^{-1}

12.50 (a) Rate = $k[NO]^2[Cl_2]$ (b) k = 2.5 × 10^{-3} liter^2 mol^{-2}s^{-1}

12.52 21 min

12.54 163 kJ/mol, 1.28 × 10^{16} liter^2 mol^{-2}s^{-1}

12.56 1.19 × 10^{-4} liter mol^{-1}s^{-1}

12.58 19 kJ/mol

12.60 59 kcal/mol

12.62 15 kcal/mol; 14 min at 15°C

CHAPTER 13

13.24 All give K_c = 5.5

13.26 1.3 × 10^2

13.28 1.20 × 10^5

13.30 10.1

13.32 −10.8 kJ

13.34 2.9 × 10^{61}

13.36 ΔG = +552 cal, therefore not at equilibrium; spontaneous from right to left.

13.38 1.4 × 10^6

13.40 0.230 mol/liter

13.42 K_P = 0.30

13.44 [NO] = [SO_3] = 0.0451 M, [SO_2] = [NO_2] = 0.0049 M

13.46 $[SO_2] = 0.0005\ M$, $[NO_2] = 0.0105\ M$, $[NO] = 0.0195\ M$, $[SO_3] = 0.0245\ M$

13.48 $[COCl_2] = 0.020\ M$, $[CO] = [Cl_2] = 2.1 \times 10^{-6}\ M$

13.50 $[H_2] = [I_2] = 0.0153\ M$, $[HI] = 0.113\ M$

13.52 8.0×10^{-3} atm

13.54 $[NO_2] = 0.23\ M$, $[N_2O_4] = 0.39\ M$ at V = 2.0 liter
$[NO_2] = 0.15\ M$, $[N_2O_4] = 0.17\ M$ at V = 4.0 liter
For V = 2.0 liters, total moles of gas = 0.46 + 0.78 = 1.24 mol
V = 4.0 liters, total moles of gas = 0.60 + 0.68 = 1.28 mol
Increasing volume increases the total number of moles of gas.

CHAPTER 15

15.20
	$[H^+]$	$[OH^-]$
(a)	0.050 M	$2.0 \times 10^{-13}\ M$
(b)	$1.9 \times 10^{-6}\ M$	$5.4 \times 10^{-9}\ M$
(c)	$1.0 \times 10^{-4}\ M$	$1.0 \times 10^{-10}\ M$
(d)	$1.6 \times 10^{-8}\ M$	$6.3 \times 10^{-7}\ M$
(e)	$1.1 \times 10^{-11}\ M$	$8.7 \times 10^{-4}\ M$
(f)	$2.5 \times 10^{-13}\ M$	$4.1 \times 10^{-2}\ M$

15.22 pH = 6.98

15.24 1.4×10^{-4}

15.26 (a) 1.6×10^{-3} (b) 5.8×10^{-4} (c) 1.7×10^{-2} (d) 6.2×10^{-5} (e) 4.1×10^{-6}

15.28 (a) 11.20 (b) 10.76 (c) 12.23 (d) 9.79 (e) 8.62

15.30 1.8×10^{-10}

15.32 (a) 1.3 % (b) 3.7 % (c) 0.014 % (d) 0.63 % (e) 0.025 % (f) 100 %

15.34 (a) 3.0×10^{-5} (b) 1.8×10^{-4} (c) 3.4×10^{-4} (d) 3.3×10^{-10} (e) 9.8×10^{-9}

15.36 9.1 g HCl

15.38 fraction = 7.9×10^{-4} (0.079 %)

15.40 3.43

15.42 0.11 M

15.44 $[H^+] = [HC_6H_6O_6^-] = 2.0 \times 10^{-3}\ M$, $[H_2C_6H_6O_6] = 0.050\ M$, $[C_6H_6O_6^{2-}] = 1.6 \times 10^{-12}\ M$

15.46 pH = 1.4, $[H^+] = [HSeO_3^-] = 0.04\ M$, $[H_2SeO_3] \approx 0.46\ M$, $[SeO_3^{2-}] = 5 \times 10^{-8}\ M$

15.48 $1.1 \times 10^{-4}\ M$

15.50 0.41

15.52 210 g $NaC_2H_3O_2$

15.54 9.6×10^{-3} mol HCl

15.56 ΔpH = 0.17

15.58 (a) 7.87 (b) 5.08 (c) 11.63 (d) 11.15 (e) 3.37

15.60 2.0×10^{-6}

15.62 8.86

15.64 12.51

15.66 10.12

15.68 8.41

15.70 8.23

15.72 See Figure 15.2.

15.74 Green; $[In^-] = 100 \times [HIn]$

15.76 $[H^+] = [HCO_2^-] = 1.3 \times 10^{-3}\ M$, $[HCO_2H] = 8.7 \times 10^{-3}\ M$, $[OH^-] = 7.7 \times 10^{-12}\ M$

15.78 1.3 ml

15.80 6.21

CHAPTER 16

16.12 3.2×10^{-14}

16.14 1.0×10^{-4}

16.16 3.20×10^{-8}

16.18 3.1×10^{-70}

16.20 8.7×10^{-5}

16.22 10.46

16.24 1.3×10^3 liters

16.26 $2.8 \times 10^{-9}\ M$

16.28 1.9×10^{-10} mol

16.30 1.7×10^{-6} mol/liter

16.32 (a) ion product = $5 \times 10^{-5} < K_{sp}$; no ppt.
(b) ion product = $4.0 \times 10^{-6} > K_{sp}$; ppt. forms
(c) ion product = $7.8 \times 10^{-4} > K_{sp}$; ppt. forms

16.34 $[Ag^+] = [H^+] = 0.050\ M$, $[NO_3^-] = 0.10\ M$, $[Cl^-] = 3.4 \times 10^{-9}\ M$

16.36 $PbCrO_4$ first, $[Pb^{2+}] = 7.5 \times 10^{-7}\ M$

16.38 $[H^+] \geq 5.5 \times 10^{-3}\ M$, $[Zn^{2+}] = 3.2 \times 10^{-6}\ M$

16.40 ZnS will dissolve until $[Zn^{2+}]$ is 16 M

16.42 $4.9 \times 10^{-3}\ M$ ($\approx 5 \times 10^{-3}\ M$)

16.44 7.1×10^{-6} mol/liter

16.46 1.82 g

16.48 0.41 mol

16.50 0.3 days

16.52 2.2 g $Fe(OH)_2$, $[Fe^{2+}] = 5 \times 10^{-12}\ M$

10.54 $[Fe^{2+}] = 4 \times 10^{-3}\ M$, $[Mn^{2+}] = 0.1\ M$, pH = 7.8

CHAPTER 17

17.38 (a) 2 (b) 1 (c) 5 (d) 2 (e) 8

17.40 (a) 1 (b) 4 (c) 10 (d) 2 (e) 8

17.42 (a) 7.0 min (b) 70 min (c) 54 min

17.44 (a) $1.79 \times 10^{-3}\ \mathscr{F}$ (b) $1.11\ \mathscr{F}$ (c) $0.217\ \mathscr{F}$ (d) $0.411\ \mathscr{F}$

17.46 0.15 g O_2, 0.019 g H_2; 0.11 liters O_2, 0.21 liters H_2 (rounded)

17.48 272 g Ag, 1.02×10^4 cm²

17.50 9.10 hr

17.52 1.23 A

17.54 (a) $0.0393\ \mathscr{F}$ (b) 174 g/mol

17.56 0.0798 A

17.58 0.249 A

17.60 (a) $\mathscr{E}° = 1.42$ V (b) $\mathscr{E}° = 0.36$ V (c) $\mathscr{E}° = 0.93$ V (d) $\mathscr{E}° = 0.08$ V (e) $\mathscr{E}° = 1.03$ V

17.62 (a) $K_c = 5.2 \times 10^3$ (b) $K_c = 1.3 \times 10^9$ (c) $K_c = 3$

17.64 (a) $K_c = 1.5 \times 10^{-13}$ (b) $K_c = 4.6 \times 10^{-51}$ (c) $K_c = 4.2 \times 10^{21}$ (d) $K_c = 8.3 \times 10^{-82}$ (e) $K_c = 1.6 \times 10^{-9}$

17.66 (a) 17.5 kcal (b) 68.7 kcal (c) −29.6 kcal (d) 111 kcal (e) 12.0 kcal

17.68
	$\mathscr{E}°$	$\mathscr{E}$	ΔG
(a)	1.42 V	1.45 V	−840 kJ
(b)	0.11 V	0.17 V	−32.9 kJ
(c)	1.28 V	1.26 V	−244 kJ

17.70 0.06 V

17.72 0.19 V

17.74 1.5×10^{-14}

17.76 -0.17 V

17.78 0.39 hr

17.80 0.013 g H_2 and 0.10 g O_2

17.82 33.0 kJ

17.84 434 min

17.86 5.43 M

CHAPTER 18

18.76 For Na^{2+}, ΔH_{hyd} would have to be larger than -1236 kcal/mol

18.78 1.41×10^3 °C

18.80 103 J mol^{-1}K^{-1}

CHAPTER 24

24.26 (a) 0.500 g (b) 0.125 g (c) 0.0313 g

24.28 164 days

24.30 2.894×10^{-7} s^{-1}

24.32 8.3×10^8 yr

24.34 1.6375×10^{-13} J

24.36 492.2 MeV (8.79 MeV/nucleon). Largest binding energy per nucleon, so fission or fusion can't produce energy.

24.38 $k = 6.5 \times 10^{-3}$ hr^{-1}, $t_{1/2} = 1.1 \times 10^2$ hr

24.40 two $C_2O_4^{2-}$ per Cr(III)

24.42 17,800 gallons of gasoline

PHOTO CREDITS

CHAPTER 18

Opener: NASA.

Page 560: National Science Foundation.

Figure 18.1: (b) © 1974, A. Satterwhite/The Image Bank.

Figure 18.2: © Robert Davis/Photo Researchers.

Page 569: Scott Aviation, A Division of A-T-O, Inc.

Page 572: (top) Berylco Company. (bottom) The Dow Chemical Company.

Figure 18.16: © 1976, Fundamental Photographs.

Page 576: (from left to right) Jean-Antoine Houdon, *Comtesse de Cayla*, © 1977, The Frick Collection, New York. Courtesy of The American Museum of Natural History. Peter Lerman.

Figure 18.8: Lester V. Bergman & Associates, Inc.

Page 580: U.S. Thermite, Inc.

Figure 18.10 Kathy Bendo

CHAPTER 19

Opener: Courtesy of Environmental Protection Agency.

Page. 590: Peter Lerman.

Page 597: Dick Breho/Associated Photographers. Courtesy of Calgon Corporation.

Page 598: Peter Lerman.

Figure 19.2: James Brady.

Figure 19.3: Courtesy of The Permutit Company, Division of Sybron Company.

Page 602: NASA.

Pages 603 and 610: Peter Lerman.

CHAPTER 20

Opener: Richard Megna/Fundamental Photographs.

Figure 20.5: (a) Ruck/Schoenberger/Grant Heilman. (b) Mario Fantain/Photo Researchers.

Page 631: Kenneth Murray/Nancy Palmer.

Figure 20.6: Schmidt-Thomsen, Landesdenkmalamt, Westfalen-Lippe, Muenster, Germany.

Figure 20.7: Peter Lerman.

Page 647: The Dow Chemical Company.

Figure 20.16: Ward's Natural Science Establishment, Inc.

Page 653: Kathy Bendo.

CHAPTER 21

Opener: American Iron and Steel Institute.

Page 663: Courtesy of History Division, Los Angeles County Museum of Natural History.

Page 672: United States Steel.

Page 673: Daniel S. Brody/Stock, Boston.

Page 676: (top) Alyeska Pipeline Service Company. (bottom) Kathy Bendo.

CHAPTER 22

Opener: Tenneco Oil Company.

Page 702: The Bettmann Archive.

CHAPTER 23

Opener: Courtesy of Gordon Whaley, University of Texas, Austin.

CHAPTER 24

Opener: M. E. Mosley/Anthro Photo.

Page 790: Department of Energy.

Figure 24.11: U.S. Department of Energy.

INDEX

Italicized references are to tables.

$$H_2O \rightleftharpoons H^+ + OH^-$$

~~$H_2O + NaOH$~~

NH_4OH

$$NaOH \xrightarrow{H_2O} Na^+ + OH^-$$
$$NH_4OH \xrightarrow{H_2O} NH_4^+ + OH^-$$

5.25% in 1.5 gallons

$NaSlO_4$

$1 L = 1.057$ quart $\quad 1.057\overline{)15000}$ ... 1.4.2

1057
4430
4228
2020

14.2

5.6 L = 1.5 g

.0525
.05.25 × 5.6
3150
2625
26400

.26 L

0.375
1.5
12
30
3220

1.5
4
0.0

6 × 1L
1.057

5.68 L

1.057)6.000
5285
7150
6342
8080
7399

.26 L   Na 22.9

.10 L   cl 35.4

          0   16.0 ×4
              64.0
              74.3

$NaClO_4$ 74.3 g/mol

H_2O 18.0 /mol